NUTRIENT REQUIREMENTS OF BEEF CATTLE

Eighth Revised Edition

Committee on Nutrient Requirements of Beef Cattle

Board on Agriculture and Natural Resources

Division on Earth and Life Studies

The National Academies of
SCIENCES · ENGINEERING · MEDICINE

THE NATIONAL ACADEMIES PRESS
Washington, DC
www.nap.edu

THE NATIONAL ACADEMIES PRESS 500 Fifth Street, NW Washington, DC 20001

This activity was supported by Contract No. HHSP233201400020B/HHSP23337015 with the DHHS/FDA Center for Veterinary Medicine and grants from the American Society of Animal Science, Canadian Beef Cattle Research Council, Institute for Feed Education and Research, Illinois Beef Association, Illinois Corn Marketing Board, Iowa Corn Promotion Board and Iowa Corn Marketing Association, National Cattlemen's Beef Association, Nebraska Corn Board and Nebraska Corn Growers Association, and the Minnesota Corn Research & Promotion Council and Minnesota Corn Growers Association with additional support from the National Animal Nutrition Program (NRSP-9). Any opinions, findings, conclusions, or recommendations expressed in this publication do not necessarily reflect the views of any organization or agency that provided support for the project.

International Standard Book Number-13: 978-0-309-31702-3
International Standard Book Number-10: 0-309-31702-9
International Standard Book Number-13: 978-0-309-27335-8 (Paperback)
Digital Object Identifier: 10.17226/19014

Library of Congress Cataloging-in-Publication Data

Names: National Academies of Sciences, Engineering, and Medicine (U.S.). Committee on Nutrient Requirements of Beef Cattle, issuing body.
Title: Nutrient requirements of beef cattle / Committee on Nutrient Requirements of Beef Cattle, Board on Agriculture and Natural Resources, Division on Earth and Life Studies, National Academies of Sciences, Engineering, and Medicine.
Description: Eighth revised edition. | Washington, DC : National Academies Press, 2016. | Includes bibliographical references and index.
Identifiers: LCCN 2015043170| ISBN 9780309317023 (hard cover : alk. paper) | ISBN 0309317029 (hard cover : alk. paper)
Subjects: LCSH: Beef cattle—Feeding and feeds.
Classification: LCC SF203 .N87 2016 | DDC 636.2/13—dc23 LC record available at http://lccn.locgov/2015043170

Additional copies of this report are available for sale from the National Academies Press, 500 Fifth Street, NW, Keck 360, Washington, DC 20001; (800) 624-6242 or (202) 334-3313; http://www.nap.edu.

Printed in the United States of America

Suggested citation: National Academies of Sciences, Engineering, and Medicine. 2016. *Nutrient Requirements of Beef Cattle, Eighth Revised Edition.* Washington, DC: The National Academies Press. doi: 10.17226/19014.

The National Academies of

SCIENCES • ENGINEERING • MEDICINE

The **National Academy of Sciences** was established in 1863 by an Act of Congress, signed by President Lincoln, as a private, nongovernmental institution to advise the nation on issues related to science and technology. Members are elected by their peers for outstanding contributions to research. Dr. Ralph J. Cicerone is president.

The **National Academy of Engineering** was established in 1964 under the charter of the National Academy of Sciences to bring the practices of engineering to advising the nation. Members are elected by their peers for extraordinary contributions to engineering. Dr. C. D. Mote, Jr., is president.

The **National Academy of Medicine** (formerly the Institute of Medicine) was established in 1970 under the charter of the National Academy of Sciences to advise the nation on medical and health issues. Members are elected by their peers for distinguished contributions to medicine and health. Dr. Victor J. Dzau is president.

The three Academies work together as the **National Academies of Sciences, Engineering, and Medicine** to provide independent, objective analysis and advice to the nation and conduct other activities to solve complex problems and inform public policy decisions. The Academies also encourage education and research, recognize outstanding contributions to knowledge, and increase public understanding in matters of science, engineering, and medicine.

Learn more about the National Academies of Sciences, Engineering, and Medicine at **www.national-academies.org.**

COMMITTEE ON NUTRIENT REQUIREMENTS OF BEEF CATTLE

MICHAEL L. GALYEAN, *Chair,* Texas Tech University, Lubbock
KAREN A. BEAUCHEMIN, Agriculture and Agri-Food Canada, Lethbridge, Alberta
JOEL CATON, North Dakota State University, Fargo
NOEL ANDY COLE, Agricultural Research Service, U.S. Department of Agriculture (Retired)
JOAN H. EISEMANN, North Carolina State University, Raleigh
TERRY ENGLE, Colorado State University, Fort Collins
GALEN E. ERICKSON, University of Nebraska, Lincoln
CLINTON R. KREHBIEL, Oklahoma State University, Stillwater
RONALD P. LEMENAGER, Purdue University, West Lafayette, Indiana
LUIS O. TEDESCHI, Texas A&M University, College Station

Staff

CAMILLA YANDOC ABLES, Study Director
KATI REIMER, Senior Program Assistant (until January 2014)

External Support

AUSTIN J. LEWIS, Consultant to the Board on Agriculture and Natural Resources

Contents

APPENDIXES

FIGURES

Preface

This eighth revised edition of the *Nutrient Requirements of Beef Cattle* reflects an effort by the Committee on Nutrient Requirements of Beef Cattle, commissioned by the National Academies of Sciences, Engineering, and Medicine, to substantially update and expand the previous National Research Council beef cattle publications.[1] Although many approaches used to calculate nutrient requirements in the seventh revised edition and the *Update 2000* remain unchanged (e.g., energy requirements for maintenance and growth), extensive analyses of published data were used to update several other areas (e.g., estimation of microbial growth, nitrogen recycling, changes in body energy and protein reserves in beef cows relative to body condition score, and effects of ionophores on energy requirements). A major effort was made to update and expand feed composition data using extensive analysis of samples analyzed at commercial laboratories to provide information on average composition and associated variability of nutrients in common feedstuffs used in beef cattle production. All chapters were updated (with some rearrangement of original material into separate chapters—e.g., maintenance and growth) to reflect relevant literature published since the release of the seventh revised edition and the *Update 2000*, and a large amount of new information has been added as additional chapters that were not part of the previous edition and the *Update 2000*. New areas include in-depth reviews of beef cattle production systems, beef quality, and safety; ruminant physiology, digestion, and metabolism; carbohydrates and lipids; compounds that modify ruminant digestion and metabolism; nutrition and the environment (including new prediction equations for methane); and nutritional value of byproduct feeds. In addition, the computer software has been updated in a more intuitive and user-friendly format. Consistent with the Statement of Task, the committee also has identified research areas needed to fill the significant gaps in knowledge that remain as a challenge for subsequent revisions.

Similar to other NRC reports in the animal nutrient requirements series, the committee did not change requirements established by the previous edition and the *Update 2000* unless sufficient evidence in the published literature or analyses of published and unpublished data suggested a change was justified. Whenever possible, recommendations were evaluated with independent data.

Establishing nutrient requirements for beef cattle poses significant challenges. Beef cattle production systems vary considerably across North America and around the globe, and the wide variety of feedstuffs used in beef production, along with diversity in breeds, environmental conditions, and management strategies make it virtually impossible to define all possible scenarios that might affect nutrient requirements. Despite these challenges, the eighth revised edition reflects a significant step forward in our understanding of the nutrient requirements of beef cattle, and it should serve as an important resource to scientists and beef cattle producers for many years to come.

[1]NRC (National Research Council). 1996. *Nutrient Requirements of Beef Cattle*, 7th Rev. Ed. Washington, DC: National Academy Press; NRC. 2000. *Nutrient Requirements of Beef Cattle, Update 2000.* Washington, DC: National Academy Press.

Acknowledgments

This report has been reviewed in draft form by individuals chosen for their diverse perspectives and technical expertise, in accordance with procedures approved by the National Academies of Science, Engineering, and Medicine's Report Review Committee. The purpose of this independent review is to provide candid and critical comments that will assist the institution in making its published report as sound as possible and to ensure that the report meets institutional standards for objectivity, evidence, and responsiveness to the study charge. The review comments and draft manuscript remain confidential to protect the integrity of the deliberative process. We wish to thank the following individuals for their review of this report:

Steve Armbruster, Steve Armbruster Consulting, Stillwater, OK
John Arthington, University of Florida, Ona
Antonello Cannas, University of Sassari, Sardinia, Italy
Ed Charmley, Commonwealth Scientific and Industrial Research Organisation, Townsville, Queensland, Australia
Tony Bryant, JBS Five Rivers Cattle Feeding, Greeley, CO
George Fahey, Jr., University of Illinois, Urbana
Mary Beth Hall, U.S. Dairy Forage Research Center, Madison, WI
David Harmon, University of Kentucky, Lexington
Kristen Johnson, Washington State University, Pullman
John McKinnon, University of Saskatchewan, Saskatoon, Canada
James Oltjen, University of California, Davis
Eric Scholljegerdes, New Mexico State University, Las Cruces

Although the reviewers listed above have provided many constructive comments and suggestions, they were not asked to endorse the conclusions or recommendations nor did they see the final draft of the report before its release. The review of this report was overseen by **Dale E. Bauman**, Cornell University. Appointed by the National Academies, he was responsible for making certain that an independent examination of this report was carried out in accordance with institutional procedures and that all review comments were carefully considered. Responsibility for the final content of this report rests entirely with the authoring committee and the institution.

The committee wishes to thank the Illinois Corn Marketing Board, Illinois Beef Association, Iowa Corn Promotion Board and Iowa Corn Marketing Association, Minnesota Corn Growers/Minnesota Corn Research and Promotion Council, Nebraska Corn Board and Nebraska Corn Growers Association, Institute for Feed Education and Research, Canadian Beef Cattle Research Council, Food and Drug Administration Center for Veterinary Medicine, American Society of Animal Science Foundation, and the National Cattlemen's Beef Association. Support from the Plains Nutrition Council, Amarillo, TX, in providing meeting facilities and a meal for the committee at a meeting in San Antonio, TX, is greatly appreciated. The assistance of the National Animal Nutrition Program (NANP), a National Research Support Project (NRSP-9) administered by the U.S. Department of Agriculture National Institute of Food and Agriculture was vital to the completion of this project. In particular, the effort by Phil Miller of the NANP Feed Composition Committee and Huyen Tran to summarize feed composition data was essential to the committee's mission.

The committee is indebted to Camilla Yandoc Ables, Program Officer for this project. Her commitment to producing a report of the highest quality in a timely manner was the "glue" that held this project together, and her pleasant personality and helpful approach to the job were inspirational throughout the process. Likewise, the committee appreciates the guidance and support of Austin Lewis, Consultant to the Board on Agriculture and Natural Resources (BANR), whose wisdom, sound advice, and encouragement were vital to producing the report. The committee also thanks Robin Schoen, Director of BANR, for the support and guidance she provided during the preparation of the report.

Summary

For several decades, the National Research Council's series on the nutrient requirements of beef cattle has provided vital information to nutritionists and academicians that has been essential for improving the economic and environmental sustainability of the beef industry. Each new edition in the series has proved its worth despite extensive changes in production and feeding practices, as well as biological types and physiological potential of cattle over time. The committee responsible for producing the eighth revised edition of the *Nutrient Requirements of Beef Cattle* has invested significant mental and physical capital into the revision process in an effort to ensure that the current edition continues to meet the high standards set by previous publications in the series.

The Statement of Task for the committee is in Appendix B. The basic charge was to update the seventh revised edition of the *Nutrient Requirements of Beef Cattle*, which was published in 1996 and the *Update 2000* by reviewing the scientific literature on the nutrition of beef cattle for all life phases and types of production. Specific nutrients identified were amino acids, lipids, minerals, vitamins, and water, and an update on energy systems used in beef cattle nutrition was considered an essential component. A summary of the composition of feed ingredients, mineral supplements, and feed additives routinely fed to beef cattle, as well as information regarding byproduct feeds, particularly those from the biofuels industry, were additional areas identified as needing new or revised information. New data on nutrient metabolism (ruminal and postruminal) and consideration of feeding strategies to minimize nutrient losses in manure and decrease greenhouse gas production were to be included. Discussion of the nutritional quality and food safety of beef and future areas of needed research also were noted as essential components of the report, as was updating the computer model to calculate nutrient requirements.

As a result of the committee's work, the text of the current edition has been expanded significantly, adding new topics and increasing discussion of components in the previous edition. New sections on beef cattle production systems, food quality, and safety; ruminant anatomy and digestion; carbohydrates; lipids; compounds that modify digestion and metabolism; nutrition and the environment; and byproduct feed ingredients are significant additions to previous reports in the series. Moreover, chapters that had been included in the previous edition were updated and expanded, with substantial effort to provide improved prediction equations for modeling various aspects of nutrient supply and metabolism and to evaluate new and existing equations. Specific changes in various chapters are described briefly in the following paragraphs.

Information on beef cattle production systems in North America is included in Chapter 1. This new section provides an important backdrop for understanding the settings from which data are derived to establish nutrient requirements and the typical conditions in which recommendations are applied. Additional new information on beef safety in terms of foodborne pathogens and antimicrobial resistance has been included, as well as a brief review of the nutritional profile of beef products. This chapter also points out the significance of animal welfare issues to the beef industry.

Chapter 2 is a comprehensive review of ruminant anatomy, digestion, and nutrient utilization. This new chapter provides the reader with a conceptual basis for establishing nutrient requirements of beef cattle and is particularly important for understanding the application of mechanistic models to beef cattle production.

Chapter 3 provides an update on energy terminology and concepts and includes updated information where available. The review of current data suggests that research efforts need to be focused on estimating the ratio of metabolizable energy to digestible energy across a range of dietary energy values.

Chapter 4 is a new chapter that provides a wide-ranging overview of the various types of carbohydrates used in ruminant production. As carbohydrates are the major source of energy in beef cattle diets, understanding their physiochemical properties and metabolism by ruminants is essential to understanding nutrient supply and development of models to predict requirements and performance of beef cattle. The role of processing carbohydrate sources to alter nutrient supply and the importance of forage fiber in maintaining healthy rumen function also are reviewed in this chapter.

Information on the role of lipids in beef cattle nutrition is included in Chapter 5, which is also new to this edition. Processes of lipid digestion and absorption are described, and practical information on the energy value of lipid sources and supplementation of lipids in beef cattle diets is reviewed. Moreover, the importance of biohydrogenation of fatty acids by the ruminal microbial population is also considered, with the opportunity for implementation of fatty acid metabolism in future editions of the computer model.

The metabolizable protein (MP) system was adopted in the previous edition in the series as the basis for establishing the protein requirements of beef cattle. The MP system is reviewed and updated in Chapter 6. Important changes include new equations for predicting microbial protein synthesis, as well as an equation for predicting urea nitrogen used for anabolism (i.e., recycled nitrogen that is incorporated into microbial protein or other microbial products), which is a new and significant addition to the publication.

Chapter 7 provides an update of macro- and micromineral requirements for beef cattle. Factors that affect mineral requirements are discussed, as well as mineral-specific diseases that can influence beef cattle production. A significant amount of new information has been added relative to the role of sulfur in beef cattle production, which is particularly relevant with increased use of high-sulfur byproduct feeds.

Chapter 8 provides an update of beef cattle vitamin nutrition, with new information regarding the fat- and water-soluble vitamins. The review articulates issues associated with specific deficiencies and excesses and suggests areas for additional research. Of special note is the greater clarity that has been provided with respect to recommendations for provision of vitamin E in various production settings.

Of the six essential nutrient classes, water is the single most important nutrient for beef cattle. Chapter 9 provides an update of equations to predict water intake by beef cattle and examines factors that influence water intake, including the role of water quality in beef cattle production.

Accurate prediction of feed intake is essential to defining nutrient requirements and predicting performance. Factors affecting feed intake are reviewed in Chapter 10, and a larger database than used in the previous edition was assembled to reevaluate existing equations based on dietary net energy for maintenance concentrations and develop new ones for growing-finishing beef cattle. Equations for beef cows remain the same; however, additional guidance has been provided for predicting intake by beef cows, particularly those grazing forages.

Considerations related to energy and protein requirements for maintenance are discussed in Chapter 11. Factors affecting maintenance requirements are included in the discussion, equations to estimate requirements are provided, and gaps in our understanding are indicated. Greater discussion and cautionary statements about applying adjustments for cold stress and physical activity are important additions to this chapter.

Energy and protein requirements for growing and finishing cattle are reviewed in Chapter 12. No changes were made to the equations used in the previous edition; however, new data were used to provide additional evaluation of equations in predicting retained energy and protein.

The role of nutrition in reproduction of beef cows is vital to beef production, as reproductive efficiency is a major factor limiting productivity of beef herds. Chapter 13 provides a comprehensive review of the use of body condition scoring (BCS) as the basis for assessing the protein and energy requirements of beef cows. A significant change to the BCS-based system was the adoption of a fixed percentage of shrunk body weight (SBW) change per unit of BCS. New information on the role of nutrition in developmental programming also was added.

Information on compounds that modify digestion and metabolism of ruminants is the focus of Chapter 14. The chapter consolidates this information into one location and includes a more complete review of this topic than the previous edition. Updated guidelines for adjustments to dietary energy values associated with the use of ionophores are provided, as well as recommendations for adjustments to final SBW estimates associated with the use of implants and β-agonists in beef production.

Nutrient requirements as affected by stress, particularly those associated with weaning, marketing, transportation, new environments, and disease, are reviewed in Chapter 15. Adjustments to nutrient requirements associated with these stressors are suggested, with particular emphasis on accounting for effects of decreased feed intake in stressed beef cattle.

Chapter 16 is an important new addition to the publication that highlights the potential effects of livestock operations on the environment. Comprehensive reviews of factors affecting nutrient losses in feces and urine and emissions of greenhouse gases, ammonia, and other volatile compounds are provided. Significant effort was devoted to developing and evaluating prediction equations to estimate enteric methane production from low- and high-forage diets.

The increasing use of byproduct feeds in beef cattle production, especially those derived from production of grain-based biofuels, necessitated a thorough review of the utilization of these important feed ingredients in Chapter 17. Application in cow-calf, stocker, and growing-finishing settings is considered, and potential interactions with other ingredients and management practices, as well as the role of excess concentrations of some nutrients (e.g., sulfur) are considered.

An extensive, statistically based evaluation of a large amount of data from commercial laboratories was undertaken to provide the new feed composition tables that are included in Chapter 18. Information on the effects of grain processing on nutrient availability has been updated, and new data on composition of grazed forages were added.

The new computer model (Beef Cattle Nutrient Requirements Model) is described in detail in Chapter 19. In addition, sensitivity analyses are included to illustrate components of the model that have the greatest effect on predictions. The model software, which is spreadsheet-based, is substantially more intuitive and user-friendly than the software used in the previous edition. Users have a greater ability to turn specific calculations "on or off" and to change some coefficients to meet their needs. The software includes an optimizer to assist with diet formulation and balancing, an ability to perform stochastic modeling, and a table generator that allows the user to create tables of nutrient requirements through an optimization procedure that is described in Chapter 20. Finally, areas that need additional research to clarify mechanisms related to developing estimates of nutrient requirements and to more fully refine recommendations are highlighted in Chapter 21.

1

Beef Production: Systems, Quality, and Safety

INTRODUCTION

The beef industry in North America is segmented and can be categorized into three major components: cow-calf (includes both purebred and commercial), stocker/backgrounding, and finishing phases. These three segments span a multitude of environments varying in soil type, topography, temperature, precipitation, length of growing season, and elevation. The beef animal is adaptable to an assortment of environments that are not suitable for many of the domesticated nonruminant species. Both the cow-calf and stocker/backgrounding segments utilize high-fiber forages largely produced on lands that are not suitable for row-crop production, in addition to crop residues following grain harvest, and cover-crop forages used on lands used for grain production to improve soil characteristics and decrease soil erosion. The finishing segment in the United States and Canada is largely accomplished in feedlots using nutrient-dense diets that contain grains and smaller amounts of forages, byproduct feeds, vitamins, and minerals. These production segments have evolved into a total production system that encompasses social responsibility, environmental stewardship, and economic sustainability.

The term sustainability has been used by many around the world to argue for, and against, the use of technologies, human and animal activities, industries, as well as environmental and financial well-being. The Brundtland Report (UN, 1987) provided the most widely used definition of sustainable development (CAST, 2013), which is that it "meets the needs of the present without compromising the ability of future generations to meet their own needs." The United Nations (UN, 2005) further defined the concept of sustainability by partitioning it into three components: environmental stewardship, economic viability, and social responsibility. If one or more of these components are misaligned, ignored, or made the sole focal point, then the system cannot achieve long-term sustainability (CAST, 2013).

The UN (2011) predicted that the world will need to support more than 9 billion people by the year 2050, and the population will exceed 10 billion people by the year 2100. As global population increases, it has been further predicted that the demand for food, fuel, and fiber will increase by 60% before the year 2050. This challenge is enhanced by the assertions that animal agriculture competes directly with the production of other forms of human food for renewable and nonrenewable resources, and that animal agriculture is an inherently inefficient method of food production (CAST, 2013).

The Food and Agriculture Organization of the United Nations (FAO, 2012) estimated that 26% of the world land area and 70% of agricultural land are covered by grasslands. Of the total land area of the United States (0.785 billion hectares), 23.3% is water or federal land, 21.0% is classified as rangeland, 21.0% is forestland, 18.6% is cropland, 6.2% is pastureland, 5.8% is developed land, 1.4% is in the conservation reserve program, and 2.6% is rural land used for other purposes (USDA, 2013e). Nonfederal lands are predominantly rangeland (34.7%), forestland (34.7%), and cropland (30.6%). Claassen et al. (2010) reported that most of the highly productive land is used for cultivated crop production (80%), while most of the low-productivity land is rangeland, used primarily for grazing (73%). Land classified as medium-productivity land is used in cultivated crop production (52%), forage production (hay, pasture, and range, 42.5%), and conservation reserve program (5.5%). Conversion of the more marginal lands, used primarily for pasture and range, to cultivation of grain crops would alter the ecosystem, eliminate a major feed resource for grazing ungulates (including livestock), destroy the habitat for wildlife and other species, increase the risk of soil and wind erosion, increase nutrient runoff, and decrease soil carbon storage (Claassen et al., 2010). In addition to forages produced on pastureland, there are more than 4.54 million metric tons of crop residues available each year in the United States following grain harvest (USDA, 2006).

The ruminants' unique symbiotic relationship with its ruminal microflora allows them to utilize high-fiber feeds (cellulose and hemicellulose) compared to nonruminant, hindgut fermenters, such as horses, which utilize forages less efficiently. These forages are digested and converted in the rumen to volatile fatty acids that are used to meet a significant portion, if not all, of the animal's energy requirements. Ruminants are also capable of utilizing plant protein where up to 70% of the nitrogen is bound to fiber, which varies in availability, or in nonprotein nitrogen forms (Beever, 1993). Therefore, two of the major nutrient constituents (energy and protein) required for growth, reproduction, and milk production in ruminants can be partially or wholly met in an efficient and economical manner by beef cattle consuming forages from grasslands and crop residues (CAST, 2013). In addition, there is ample opportunity to use current technologies and practices, as well as develop new ones, that will improve low-quality forage digestibility. Chemical treatments such as anhydrous ammonia, calcium hydroxide, and more recently, calcium oxide (Sewell et al., 2009; Russell et al., 2011; Shreck et al., 2013) have been used to improve fiber digestion and animal performance. The beef animal is also uniquely positioned to utilize byproducts resulting from human food, fiber, and fuel production that are not suitable for direct human consumption because of safety, quality, cultural, or digestibility considerations (CAST, 2013). Gill (1999) estimated that 37 kg of byproducts suitable as a livestock feed are produced from every 100 kg of plants grown for human food. In summary, high-fiber feeds, which are largely grown on lands not suitable for direct human food production, or feeds that are produced as a byproduct of human food production, can be utilized by beef cattle to produce safe, wholesome, nutrient-dense food for a growing global population.

BEEF BREEDS

Cattle are considered one of the first animals to be domesticated by humans for meat, milk, hides, and draft purposes. It is thought that the first domestication took place in Europe and Asia about 8,500 years ago. Domesticated cattle belong to the family *Bovidae*, genus *Bos*, subgenera *Taurine,* and one of two species—*taurus* and *indicus*. All breeds originating from Britain and Europe belong to the *taurus* species. In the United States and Canada, the most popular British breeds include Angus, Hereford, Shorthorn, Red Poll, and Devon, whereas the most popular European beef breeds include Simmental, Charolais, Gelbvieh, Limousin, Maine Anjou, Chianina, Braunvieh, Tarentaise, Pinzgauer, and Salers. Breeds originating in the tropical countries belong to the *indicus* (zebu) species, which are considered to be more heat tolerant and parasite resistant than *Bos taurus* cattle. The most popular breeds in the southern United States and Mexico that fall into this category include the U.S. Brahman (Guzerat, Nelore, Gyr), and Tuli. In addition, there are a number of contemporary breeds that have been developed by crossing one or more of the *taurus* and *indicus* breeds to capitalize on breed complementarity and heterosis.

The most extensive comparison of breeds in the United States was conducted by the U.S. Meat Animal Research Center (see Cundiff et al., 1993, 1998; Cundiff and Gregory 1999; and reviews by Marshall, 1994, and Arthur et al., 1999). According to the U.S. Department of Agriculture (USDA, 2008a) report, the U.S. cowherd can be classified as 17.5% purebred or straightbred, 13.3% composite breeds, 44.9% two-breed crossbreds, and 24.3% three- or more breed crossbreds. In general, breeds, cow size (frame and weight), and level of milk production are matched to available feed resources and the environment, resulting in variable cow weights. An estimated 45% of the cows in the U.S. herd weigh less than 500 kg, 43.8% are between 500 and 590 kg, and 11.2% weigh more than 590 kg, with a combined estimated average cow weight at weaning of 520 kg (USDA, 2010a). These data indicate that cow weights tend to be lighter in operations that utilize native grass range and pasture than operations that have ready access to improved pastures (grasses and legumes), as well as cereal grains, silages, and byproducts. Therefore, cow size tends to be larger in the Corn Belt and Canada compared with the U.S. High Plains, U.S. Southwest, and Mexico.

PRODUCTION SYSTEMS

Conventional Beef Production

The most widely recognized and utilized production system in North America is referred to as the conventional beef production system. This system is classically divided into cow-calf, stocker/background, and feedlot segments. The cow-calf segment encompasses both a purebred and commercial (crossbred) component. The stocker/background and feedlot segments of the industry also include dairy beef. The conventional production system utilizes technologies such as implants, feed additives, and antimicrobials, which have been extensively tested and found to be safe and efficacious when used according to label directions, to produce a safe, wholesome beef product in an efficient manner that minimizes the impact on environmental resources.

Cow-Calf Segment

In the conventional production system, cow herds consume a diet consisting of forages (pasture, hay, ensiled forages, and crop residues) with supplemental energy, protein, vitamins, and minerals as needed to meet their nutrient requirements. The land base utilized in this system is largely not suitable for grain and row crop (soybean, corn, wheat, sorghum, barley, oats, cotton, sunflower, canola, etc.) production and is dispersed across the North American continent. As a result, the cow-calf segment of the industry will likely not see the consolidation that has occurred in the pork and poultry industries because of the capital required for land.

Small farms and ranches tend to predominate in the eastern and southern regions of the United States and Canada where they utilize introduced forages (annuals, perennials, and crop residues). Large ranches tend to predominate in the western regions of the United States and Canada, as well as the arid and semi-arid regions of northern and western Mexico, where they utilize native grasses and forbs. The USDA (2014a) reported that 34.5% of the 2.11 million farms in the United States (nearly 728,000 farms) had beef cows, with cattle and calf sales accounting for over 19% of the total value of agricultural products sold, and ranking first in sales among all commodities. The average herd size in the United States was nearly 40 cows and 74 cows plus calves (calculated, 85% calf crop) with 35.8% of the farms having a cow inventory of fewer than 10 cows, 84.5% fewer than 50 cows, and 96.3% with fewer than 100 cows (USDA, 2014a). In Canada, the national cow herd averages 129 cows plus calves per operation (Statistics Canada, 2011).

The ideal time of year for calving in a beef cow-calf operation depends on available labor, environment, feed supply, and the intended target market. According to the USDA (2009a) report, approximately 60% (range was 58.8 to 63.9% across the reporting years of 1992, 1997, 2007) of the calves in the United States are born during February, March, and April, often referred to as the spring calving season. The primary reasons for the timing of this calving season are: (1) to utilize forages during the growing season to support lactation; and (2) to minimize the negative effect of higher temperatures experienced when cows are bred during late July, August, and early September on reproductive efficiency. A majority of the remaining calves (10 to 15%) are born in what is often referred to as a fall calving season (September to November). The remaining calves (25 to 30%) are born to operations that claim year-round calving with a portion of these calves also falling into a spring or fall calving season.

The percentage of females that calved as a percentage of cows exposed to the bull, or artificially inseminated, averaged 91.5% in 2007 (USDA, 2009a); the range across 3 survey years was 91.5 to 92.6%), with 96.5% of all calves born live and 96.8% of live calves surviving to weaning. This results in an estimated calf crop weaned per cow exposed of nearly 85.5%, which closely mirrors the trend across years of 80 to 85% calf crop weaned per cow exposed reported by Bevers (2012).

The national average culling rate of cows in a herd averages between 15 and 20% per year. When reviewing the factors that affect culling decisions (USDA, 2010a), approximately 72% of the cows are culled from the herd because of reproductive failure, age, teeth concerns, or overall lack of productivity. The remaining approximately 28% of the cows are culled primarily as a result of producing light-weight calves (typically because of late calving or poor milk production); temperament, udder, or eye concerns; or herd reduction caused by economic reasons (drought, market conditions, input costs). Cow longevity is an economically important trait in beef cow herds, and the USDA (2010a) report suggests that of the cows sold for purposes other than breeding (culls), 15.6% are culled when they are less than 5 years of age, 31.8% are culled when they are between the ages of 5 and 9 years, and 52.6% are culled when they are 10 or more years of age.

Beef calves typically graze alongside their dam until weaning at 5 to 9 months of age when they weigh from 175 to 300 kg. The USDA (2008a) data reported an average weaning age of 207 d and weaning weight of 240 kg of all calves born in 2007, which is similar to the Standardized Performance Analysis data reported by Bevers (2012) for New Mexico, Oklahoma, and Texas. At weaning, approximately 50% of all beef operations market their calves either privately or through an auction market (USDA, 2010a), whereas approximately 50% of all beef operations retain calves on the farm/ranch to complete a preconditioning or backgrounding program, with more than 80% of all calves sold within 60 d of weaning (USDA, 2010a). Preconditioned calves, which represented an average of 18.4% of all calves over the period from 1993 to 2007 (USDA, 2009a), are typically vaccinated twice, dewormed, castrated, dehorned, and trained to eat from a feed bunk and drink from a water trough.

Heifer and bull calves identified as herd replacements are typically placed on a high-forage diet to promote skeletal growth and lean tissue gain, with minimal fattening from weaning to initiation of their first breeding season. As a result of differences between forage species and variety, stage of maturity, and environmental conditions, forages can be extremely variable in available nutrient profile. Therefore, forage-based diets are often supplemented with feedstuffs to provide additional energy, protein, vitamins, and minerals to meet the nutrient requirements for targeted gains of 0.45 to 0.90 kg/d from weaning to initiation of the first breeding season, in order to achieve a target weight of 55 to 65% of mature weight by 13 to 15 months of age. This target weight at breeding is critical to attaining puberty and optimizing reproductive efficiency. Replacement heifers continue on high-forage diets that are supplemented to reach approximately 80% of their mature weight at first calving (22 to 24 mo) and mature weight by year 4 to 5 of age. See Chapter 13 (Reproduction) for a more complete discussion of nutritional management for reproduction.

When there is a drought, early weaning of calves has often been recommended to extend forage supplies for the cow herd, eliminate the nutrient requirements for lactation in the dam, and improve cow body condition during a time when her nutritional requirements for pregnancy are low. It is estimated that dry matter intake (DMI) decreases 25% in cows when calves are early weaned (R. P. Lemenager,

Animal Sciences, Purdue University, West Lafayette, IN, personal communication, May 16, 2014). When one adds the decrease in trampling losses as a result of calf hoof traffic, and the elimination of grazed forage consumption by the calf, a decrease of more than 30% in forage disappearance could be expected; or conversely, more than a 30% increase in available forage supply when calves are early weaned. This was confirmed by work done in Florida where early weaned primiparous females consumed 36.6% less dry matter (DM; 5.7 vs. 9.0 kg) than normal weaned cow-calf pairs (Arthington and Minton, 2004), and a second study (J. Arthington, Range Cattle Research and Education Center, University of Florida, Ona, FL, personal communication, July 22, 2014) that suggested a 25% greater stocking density when cows had their calves weaned early. Early weaned calves are highly efficient at converting feed to live weight gain and, in some instances, improved carcass quality grades have been observed (Myers et al., 1999a,b; Wertz et al., 2001; Meyer et al., 2005). It has been reported that intramuscular fat is deposited at a faster rate relative to subcutaneous fat in calves compared with yearlings (Wertz et al., 2001), and that the rate of intramuscular fat accretion is greater in cattle when projected hot carcass weight is less than 300 kg (Bruns et al., 2004). This would not only suggest that intramuscular fat accretion is nonlinear, but also that management and nutrition early in life could be as important as management during the finishing phase in determining carcass quality.

Stocker/Background Segment

The stocker/backgrounding phase, which includes steers and nonreplacement heifers, serves as a bridge between the cow-calf and finishing segments of the industry. In the United States and Canada, calves are often placed into the forage-based stocker/backgrounding phase of the beef industry (Sip and Pritchard, 1991) to more evenly distribute the supply of cattle into the finishing phase. Forage-based systems allow cattle to graze native and introduced forage species during the growing season, as well as consume harvested forages during the winter feeding period. In the Southeastern and Southern Plains regions of the United States, the warmer climate and lower occurrence of snowfall allow extended grazing into the winter. In the Northern Plains and Midwestern regions of the United States, as well as in Canada, forages are often harvested as hay or ensiled to provide a winter feed supply. This phase is particularly important for the smaller-framed breeds to increase lean and skeletal growth before entry into the finishing phase. During this growing phase, cattle typically gain between 0.35 and 1.15 kg/d, depending on the quality of forage used in the diet, and normally weigh 300 to 400 kg when they enter the finishing phase at 12 to 15 months of age.

Although high-roughage systems are often used to achieve adequate lean gain during the growing phase, high-concentrate diets fed in drylot often provide a cheaper source of energy on an equal energy unit basis (Sip and Pritchard, 1991). Therefore, when grain prices are low relative to forages, providing nutrient requirements by limit-feeding nutrient-dense, high-energy diets for growing cattle can be more economical. In a review by Galyean (1999), restricted feeding was defined as any method of feed intake management where intake is restricted relative to actual or anticipated ad libitum intake. Program feeding, however, was defined as a method in which net energy (NE) equations are used to calculate quantities of feed required to meet the needs for maintenance and a predetermined rate of gain. High-concentrate diets fed at programmed levels can be fed to target specific rates of gain using the NE (net energy required for maintenance [NEm] and net energy required for gain [NEg]) system similar to that of high-roughage diets fed ad libitum (Loerch, 1998). Some of the advantages for the use of program feeding of high-concentrate diets versus grazing and forage-based systems include cheaper costs at times when pasture costs are high or in short supply, easier adaptation to a finishing diet, improvements in feed efficiency (Galyean, 1999), as well as improved carcass characteristics and palatability of retail beef cuts (Coleman et al., 1995).

Feedlot Segment

Before World War II, beef finishing was primarily accomplished on grazed forages (pasture or range), but as the need for food increased, land became more intensively managed to produce more grain and animal products. This evolution led to grain finishing of beef in feedlots. Beef production systems continued to evolve with an increased understanding of nutrient requirements and nutrient interactions, animal and forage production and management, inheritance of economically important traits, metabolic pathways, gene expression and regulation, factors affecting product safety and quality, and the interaction of all these processes with the environment. When expressed on a per animal basis, the combination of research and innovation has resulted in an 80% increase in beef production over a 50-year time frame (Elam and Preston, 2004). In addition, advancements in beef production have been significantly affected by the use of technologies such as genetic predictors (expected progeny difference [EPD]), DNA markers, estrous regulators, implants, ionophores, antibiotics, repartitioning agents, parasiticides, and vaccines, which have resulted in improved reproductive performance in the cow herd; younger, faster-gaining cattle with improved feed conversion; heavier carcasses; beef products with improved taste, tenderness, and juiciness; as well as a decrease in carbon footprint and greenhouse gas emission per unit of edible product compared with the earlier pasture-finished system. The USDA (2014b) estimated that 24% of the world's beef supply is produced in the United States, Canada, and Mexico and that the United States is the world's largest producer of beef, especially high-quality grain-fed beef, with the largest fed-cattle industry in the world. At the same time, however,

4.63 kg × 158 d ≈ 732 kg of grain DM, 2.95 kg/d × 158 d ≈ 466 kg of nonforage, nongrain ingredient DM, and 0.841 kg × 158 d ≈ 133 kg of forage DM.

- Total forage DMI to produce one grain-fed feedlot animal can be calculated as 4,995 kg + 357 kg + 926 kg + 133 kg ≈ 6,411 kg.
- If one assumes that cows during the winter and calves between weaning and entry into the feedlot (stocker/background phase) are supplemented with byproduct feeds at an average of 1 kg/d during a 150-d period (=150 kg for cows) and 147 d (=147 kg for stocker/background calves), the total production system feed needed to produce one finished feedlot steer is 6,411 kg/(6,411 kg total forage + 732 kg feedlot grain + 466 kg nonforage, nongrain feedlot ingredients + 150 kg/0.86 cow byproduct feed + 147 kg stock/background byproduct feed) ≈ 80.8% forage, 732 kg/(6,411 kg + 732 kg + 466 kg + 150 kg/0.86 + 147 kg feed) ≈ 9.3% grain, and (466 kg + 150 kg/0.86 + 147 kg)/(6,408 kg + 732 kg + 466 kg + 150 kg/0.86 + 147 kg) ≈ 9.9% of nonforage, nongrain ingredients on a DM basis.

Alternative Production Systems

Agriculture will continue to evolve in response to new challenges and opportunities as consumers demand products with various attributes that are perceived to be beneficial. The industry and its producers will adapt to meet realistic consumer demands for product attributes, and policymakers will seek solutions that meet the food and resource challenges of an increasing global population. Evidence suggests that some consumers are willing to pay premium prices above conventionally produced beef to obtain product attributes that are consistent with their preferences (Sparling et al., 2002; Brewer and Calkins, 2003; McCluskey et al., 2005; Dutton et al., 2007; Springer et al., 2009; Umberger et al., 2009; Abidoye et al., 2011). These preferences are generally related to an increased interest in where their food comes from, how it is produced, and how they perceive the impact of alternative production systems on food safety, nutrient profile, animal welfare, the environment, and cost. As a result, niche and alternative production systems, largely associated with the finishing phase of the beef industry, have evolved to address various combinations of beef attributes desired by consumers. Consumer demand for beef finished without antibiotics, growth-promoting technologies, and grains (conventionally selected or genetically modified) has motivated the development of a small, but growing niche market. Consumer preference for differentiated product attributes might boil down to trust and how they understand and perceive the factors that affect food safety and wholesomeness, environmental stewardship, and animal well-being.

Lister et al. (2014) conducted a consumer survey to evaluate the concept of ranking food by employing a specific set of food values. In their survey, they considered four meat and dairy products: ground beef, beef steak, chicken breast, and milk. Although not a terminal human value, which has been defined by Rokeach (1973) as an end state of being, food values can be used to determine consumption choices that are derived from, and lead to, basic human values (Lister et al., 2014). This should, in turn, offer a more stable understanding of consumer demand. The consumer survey included the food value categories of freshness, taste, price, safety, convenience, nutrition, health, origin/traceability, hormone/antibiotic-free, animal welfare, and environmental impact. Results of the survey indicated that the food values, ranked as most and least important, were similar across the four food products evaluated. Safety and freshness were the dominant consumer food values in this survey. Consumers differed, however, in their food value ratings based on price sensitivity. Those rating price as relatively more important also ranked health, hormone/antibiotic-free, animal welfare, and taste as less important, and vice versa. Environmental impact, animal welfare, origin/traceability, and convenience were considered least important by consumers in this survey.

The most recognized alternative production systems in the marketplace are grass-fed, natural, and certified organic. Although numbers are hard to document, the USDA (2013a) reported that beef from alternative production systems—natural, organic (grain-fed or otherwise), and grass/forage-fed (including cattle finished on grasses/forages to a specific quality standard)—accounts for about 3% of the U.S. beef market and has grown approximately 20% per year in recent years, according to the referenced Irish Food Board/Bord Bia (FeedInfo News Service, 2010). Other terms used alone, or in combination, include no antibiotics, no hormones, locally raised, family farm, and humanely raised. The availability of unique beef attributes can be positive for both the beef industry and consumers because some consumers would not eat beef if alternatives to conventional beef production were unavailable. If consumers are willing to pay a premium for beef products from alternative production systems, this demand can create a profit opportunity for both producers and processors. A key issue facing consumers is how these product attributes are promoted and marketed. Scientifically unsubstantiated claims that imply that a product is healthier or safer because it is chemical-free, hormone-free, grass-fed, or organic are often misleading because all plant and animal derived foods naturally contain chemicals and hormones, and all beef is grass-fed, at least in part, during various stages of the production chain.

Grass-Fed

Because most cattle consume forages nearly all their lives, a distinction should be noted between grass-fed animals and grass-finished animals (Mathews and Johnson, 2013). Grass-fed has been defined at least two different ways. The USDA (2008b) defines "grass-fed" as ruminant animals, and the meat products derived from these animals, that have solely consumed forages throughout their life, with the exception of milk (or milk replacer) consumed before weaning. For this claim, animals cannot be fed grain or grain products and must have continuous access to pasture during the growing season. The American Grassfed Association (AGA, 2013) further defines grass-fed beef, and their beef products, into a three-tier system that specifies how and when approved non-forage supplements can be used. Most grass-fed programs allow vaccinations, but most do not allow the use of antibiotics (therapeutic or subtherapeutic), growth-promoting implants, or ionophores. To qualify for a grass-fed branded program, source and management verification is required (USDA, 2008b). Compliance with the requirements of a branded grass-fed beef program is often monitored with audits performed by the affiliated marketing alliance or a certifying agency. For animals in a grass-fed production system to be acceptable, they must obtain adequate levels of carcass finish (equivalent to USDA Select or Choice) within an economically feasible time frame. To accomplish this, careful attention must be given to production and storage of a high-quality, year-round supply of forages, as well as selection of cattle breeds that have the ability to grow rapidly and marble easily on forages.

Natural Beef

The definition of "natural" within the context of beef production is more ambiguous than the definition of "organic." According to the USDA Food Safety Inspection Service (USDA, 2013c,d), all fresh meat qualifies as "natural" if it is minimally processed and does not contain any artificial flavors or flavorings, coloring ingredients, chemical preservatives, or other artificial or synthetic ingredients. This USDA definition of natural beef refers only to the beef product, and does not specify any animal production practices. By this definition, most fresh beef products qualify as natural. Unless otherwise stated on a label, beef in the retail market can be assumed to be natural. Some marketing organizations, however, promote their product as "natural" if the animals have not received antibiotics, ionophores, or growth-enhancing hormones, or if they were finished on pastures rather than in feedlots. The terms "no hormones administered," "no antibiotics," or "never-ever" can be approved for use on beef product labels if sufficient documentation is provided to the USDA Agricultural Marketing Service Process Verified Program (USDA, 2013c,d).

Certified Organic Beef

A small, but growing subset of consumers (<3%) believes that organic foods are safer and healthier than conventionally grown foods. A systematic review of the available literature submitted to the Food Safety Standards Agency (FSSA, 2009) concluded that while minor differences in nutrient content and other substances were noted between organic and conventionally produced crops and livestock, they are most likely related to crop or animal management and soil characteristics. The report went on to state that the differences would not likely be relevant to individuals consuming a normal, varied diet with regard to nutrient intake or human health. These observations are supported by a review conducted by Dangour et al. (2009), who concluded that organically and conventionally produced foodstuffs are broadly comparable in their nutrient content, and Van Loo et al. (2012), who concluded that there is no evidence that organic food is safer, healthier, or more nutritious.

The certified organic production system is much more regimented and requires significantly more time, effort, and documentation than conventional, grass-fed, or natural production systems. The USDA National Organic Program (USDA, 2014c) outlines the livestock production and handling requirements for animals to qualify for marketing under the organic label. In brief, the program requires that animals are to be raised under organic standards from the last third of gestation without the use of antibiotics or growth hormones; land used for grain and forage fed to animals must be 100% organic for at least 3 years before harvest; at least 30% of the animal's forage needs must be met through pasture during the grazing season (grain-finished beef cattle are excluded from this requirement during the last 20% or 120 d of their lives, whichever is shorter); and processors of organic meat must be organic certified.

Locally Grown

Many of the alternative production systems are smaller operations that use local slaughter and processing plants. Locally grown beef products are often defined by region, company, marketing channel, and by consumer definitions that can vary by scale of production, supply chain, and marketing outlet (Mathews and Johnson, 2013). "Local" often implies a conventional production system, but a different marketing system. Marketing of beef directly from a producer to a local consumer (e.g., freezer beef, farmer's market) is the primary method, but marketing can also represent a more complex arrangement, such as where a group of producers raises animals in a designated production system for a local meat brand, or market fresh to restaurants, retailers, or other food service establishments. Limitations in local slaughter and processing are often cited by producers as one of the key barriers to the marketing and expansion of alternatively produced beef (Mathews and Johnson, 2013).

FOOD SAFETY

Food safety is a multifaceted attribute that is a primary concern to all consumers, and it involves not only the production system, but also the method of transport and sale, processing, and food preparation. Much attention has been given to both preharvest production and postharvest processing. All animals naturally harbor some quantity of foodborne pathogens in the digestive tract, but many of these pathogens do not cause symptomatic responses in the host animal (CAST, 2013). The United States is generally recognized as having one of the world's safest food supplies, thanks to the combined efforts of producers working in cooperation with the USDA, the Food and Drug Administration (FDA), and the U.S. food industry to minimize adulteration by bacteria and antibiotic residues. Similarly, the Canadian Food Inspection Agency ensures food safety in Canada. Beef production in the United States embraces beef quality assurance practices that address not only the judicious use of pharmaceutical products (proper product administration and withdrawal times), but also animal handling and well-being, as well as environmental stewardship principles.

There is not only a strong economic incentive, but also a moral incentive to optimize animal health and well-being in the beef industry. Antimicrobial products, commonly referred to as antibiotics, have been used in beef production to treat disease, for disease prevention, and for growth promotion. Although an integral part of some alternative beef production systems, the use of pharmaceutical products is most often specific to conventional beef production (Mathews and Johnson, 2013). In general, pasture cattle (cow-calf and stocker/backgrounding segments) use less pharmaceutical products than cattle in feedlots, with the exception of vaccines. The use of some pharmaceutical compounds is virtually universal among large commercial feedlots, but the beef industry (producers, nutritionists, veterinarians, and allied industries) has focused on decreasing both the incidence of disease and the need for antibiotic use in later stages of production by implementing a holistic approach to production. This includes diet formulations that meet or exceed animal nutrient requirements, production practices that minimize animal stress, development and utilization of vaccines in a more comprehensive health program, and the judicious use of antibiotics.

Foodborne Pathogens

Batz et al. (2012) studied the 14 most costly U.S. foodborne pathogens and concluded that *Escherichia coli* O157:H7 (shiga-toxin-producing *E. coli* [STEC]) and *Salmonella enterica* were the most significant beef pathogens. There are some data (Diez-Gonzalez et al., 1998) that suggest acid-resistant *E. coli* is more prevalent in cattle fed grain-based diets than in those fed forage-based diets, but specific pathogens were not evaluated. In a review conducted by Jacob et al. (2009), it was concluded that studies chronicling variations in pathogen shedding in cattle fed grain- vs. forage-based diets are not repeatable, and that the complexity of the hindgut ecosystem is affected by many factors. The Council for Agricultural Science and Technology (CAST, 2013) report further concluded that there is little evidence to suggest that pathogen presence in cattle is significantly altered by diets containing either primarily grains or primarily forages.

Cattle are considered a major reservoir of *E. coli* O157:H7 and manure is a source of contamination for food and water (Reinstein et al., 2009). Beef muscle and fat are not contaminated before slaughter, but beef products are sometimes contaminated with foodborne pathogens through exposure and cross contamination from hides, digestive system contents, or manure during processing. Humans consuming contaminated beef can become infected with these pathogens if the beef products, especially ground beef products, are not handled and cooked properly at the consumer level. Antibiotics are used to treat human illness, and resistance of these pathogens to various antibiotics has become a relevant food-safety issue. In general, resistance seems to be greater in the beef systems that use antimicrobial drugs compared with those that do not, but the differences are not always large or statistically significant, and results vary by drug (Reinstein et al., 2009; Rao et al., 2010; Morley et al., 2011).

Commercial North American meat processors have dramatically decreased the levels of foodborne contamination and animal welfare concerns by following a manufacturing process called the Hazard Analysis and Critical Control Point (HACCP) system. Intervention strategies employed by processors include, but are not limited to, eliminating the processing of "downer animals" and the implementation of microbial reduction safeguards such as thermal pasteurization, hot water and acid rinses, and steam vacuums. Irradiation, sometimes referred to as cold pasteurization, is an efficacious technology that was approved by the Food Safety and Inspection Service (64 Fed. Reg. 72149 [1999]) for treating refrigerated or frozen uncooked meat, meat byproducts, and certain other meat food products to reduce levels of foodborne pathogens and to extend shelf life; however, its use has been limited because of negative public perception.

Preslaughter interventions and strategies have the potential to decrease foodborne pathogens entering the processing plant, and this is an area that is receiving a significant amount of research attention. Several of these interventions show substantial promise, such as specific strains of direct-fed microbials (DFM), vaccine technology, sodium chlorate, and neomycin sulfate, whereas others such as brown seaweed or chlorination of water have little or no detectable benefit (Loneragan and Brashears, 2005). Many of the strategies and technologies that have been developed rely on the natural competitive nature of bacteria to eliminate pathogens that negatively affect either animal production or food safety (Callaway et al., 2008). Feed additives that have been evaluated with varying degrees of success include prebiotics, probiotics, and competitive exclusion cultures.

Prebiotics are often referred to as neutraceuticals, or functional foods. In general, these are organic compounds that are unavailable to the animal, but that can be utilized by a portion of the gastrointestinal microbial population of the animal. The use of prebiotics in cattle has been limited because of cost and by the ability of the rumen microflora to degrade most of these products (Callaway et al., 2008), but future development of rumen-protected products might allow these compounds to be used. Probiotics are a general category of dietary products that have been defined by Schrezenmeir and De Vrese (2001) as a preparation, or a product-containing viable, defined microflora (by implantation or colonization) in a compartment of the host that exerts a beneficial health effect in the host.

Probiotics fall into three general classifications: live cultures of yeast or bacteria, heat-treated (or otherwise inactivated) cultures of yeast or bacteria, or fermentation end products from incubation of yeast or bacteria (Callaway et al., 2008). Direct-fed microbials are similar, but do not contain fermentation end products. The key to successful pathogenic microbial reduction with DFM seems to be related to dose and selection of bacterial strains (Loneragan and Brashears, 2005). The *Lactobacillus*-based DFM have been shown to have some efficacy in controlling *E. coli* O157:H7, and their cost is at least partially offset by a small improvement in animal performance (Krehbiel et al., 2003; Younts-Dahl et al., 2004; Peterson et al., 2007; Vasconcelos et al., 2008). The *Lactobacillus acidophilus* (NP51)-based DFM products are currently being used by some feedlots across the United States and Canada (Callaway et al., 2008) to decrease *E. coli* O157:H7 shedding and hide contamination. Brashears et al. (2003a,b) reported a decrease of approximately 50% in *E. coli* O157:H7 shedding in feedlot cattle when fed a *Lactobacillus*-based DFM. These results are supported by Ransom et al. (2003), who reported a decrease in the percentage of animals (46% vs. 13%) with fecal shedding of *E. coli* O157:H7. Younts-Dahl et al. (2005) also reported a decrease of *E. coli* O157:H7 shedding by 57% and isolation from hides by up to 75%. Similarly, Stephens et al. (2007) reported a decrease in *Salmonella* shedding, while Tabe et al. (2008) reported a decrease in new cases of *Salmonella* infections when cattle were provided a DFM.

Until recently, anti-foodborne pathogen vaccines had not been widely used by the beef industry because of the absence of an economic incentive, as well as increased animal stress and lower performance associated with additional animal processing to administer vaccines. Nonetheless, the concept of vaccine development to decrease pathogen shedding has some potential (Potter et al., 2004; Oliver et al., 2009; Khare et al., 2010). The introduction of an oral vaccine seems to have the potential to make animal immunization economically viable for many diseases, including foodborne pathogens (Oliver et al., 2009).

Several studies suggest that diets containing wet or dry distillers grains result in higher pathogen (STEC) prevalence in cattle (Jacob et al., 2008a,b, 2010; Varel et al., 2008; Wells et al., 2009; Chaney et al., 2011). While the evidence suggests the distillers grains can affect pathogen prevalence, some studies suggest that the prevalence is related to the quantity of distillers grains in the diet and the normal seasonal variation associated with fecal shedding of *E. coli* O157:H7. More research is needed to elucidate the mechanisms involved in fecal shedding of *E. coli* O157:H7 when distillers grains are included in the diet to allow development of mitigation strategies.

Antimicrobial Use

Some consumers perceive that food safety and animal well-being are compromised by the use of pharmaceutical products in some production systems. It should be noted, however, that all pharmaceutical products undergo a rigorous evaluation for safety, efficacy, and environmental impact by the FDA Center for Veterinary Medicine before approval and adoption by the industry. Although the threat of antibiotic resistance is real, the judicious use of antibiotics, safeguarding of important classes of antibiotics for human use only, and the judicious use of antibiotics when treating humans to minimize resistance can help mitigate some of these issues.

Wileman et al. (2009) conducted a meta-analysis using 51 studies of conventional, organic, and natural beef production with untreated control groups and reported a significant increase in efficiency and a decrease in cost when technologies (mainly antibiotics and implants) were used in beef production. Their analysis indicated an improvement of 17% in ADG and 9% in gain efficiency from a single implant in steers. Results also indicated a 53% decrease in morbidity and a 27% decrease in mortality from metaphylaxis (whole-group treatment with an antimicrobial) on arrival of cattle at the feedlot. Low-level antibiotic feeding is often used to prevent liver abscesses and subsequent liver condemnation associated with rapid digestion of high-starch (grain) diets, production of lactic acid, and a sudden drop in ruminal pH. In their analysis, feeding an antibiotic (tylosin) to feedlot cattle decreased the risk of liver abscesses by 8%, without a consistent advantage in ADG over control groups.

The Institute of Food Technologists (IFT) published an expert report (Doyle et al., 2006) and an updated report (Doyle et al., 2013) that summarized current scientific knowledge surrounding the effect of antimicrobial use in the food system on antimicrobial resistance and public health. In addition, concerns about the public-health implications of microbial resistance to antibiotics used in both human medicine and food-animal agriculture have led to the publication of the World Health Organization's (WHO, 2011) List of Critically Important Antimicrobials for Human Medicine, and the World Organisation for Animal Health's (2015) List of Antimicrobials of Veterinary Importance. In 2003, the FDA created tables that categorized various classes of antimicrobials as important, highly important, and critically important for use in humans. Since that time, FDA has updated the lists and issued rules that prohibit most extra-label uses of the critically important human antimicrobials (e.g., fluoroquinolones and cephalosporins) in food animal species.

in the food industry as food consumers represent the primary demand for agricultural markets (Lister et al., 2014). McNeill et al. (2012) reviewed the evolution of lean beef in today's U.S. marketplace and observed that its nutritional role in the American diet is often underappreciated and misunderstood. Inaccurate and outdated estimates of total fat, saturated fat, cholesterol content, and nutrient profile of beef available in the marketplace foster misunderstanding and confusion among dieticians, the medical community, and consumers. Changes in cattle breeding and management, coupled with extensive trimming of visible fat from retail cuts have resulted in beef cuts that have decreased total fat content and a favorable fatty acid profile that provides a lean beef product with unique nutrient profile that can easily fit into a healthy diet (Wolmarans et al., 1999; Cassady et al., 2007).

The USDA determines and regulates which beef cuts can be classified as "lean" or "extra-lean" based on their total fat, saturated fat, and cholesterol content. The lean designation must contain less than 10 g of total fat, 4.5 g of saturated fat, and 95 mg of cholesterol, whereas the extra-lean designation must contain less than 5 g of total fat, 2 g of saturated fat, and 95 mg of cholesterol per 100 g (3.5 oz) serving. The USDA Nutrient Data Set for Retail Beef Cuts (USDA, 2013f) is updated regularly and serves as a Standard Reference that focuses on beef cuts identified by the USDA Food Safety and Inspection Service for nutrition labeling, which also includes data for the newer Beef Value Cuts. Based on this database, there are 29 lean cuts of beef in the U.S. marketplace that can be used by consumers to meet their cardiovascular health goals.

Lean beef is a nutrient-dense food that is an excellent source of essential amino acids, vitamin B_{12}, selenium, zinc, niacin, vitamin B_6; and a good source of phosphorus, choline, iron, and riboflavin. Compositional information on various beef products can be found at the USDA National Nutrient Database for Standard Reference, Release 27 (USDA, 2015). BeefNutrition.org (2014) reported that an 85-g (3-oz) serving of beef provides more than 20% (considered an excellent source) of the daily value for protein (48%, with all of the essential amino acids represented), vitamin B_{12} (44%), selenium (40%), zinc (36%), niacin (26%), vitamin B_6 (22%), and more than 10% (considered a good source) of phosphorus (19%), choline (16%), iron (12%), and riboflavin (10%).

Fatty Acids

Taste is likely the most common reason that Americans consume beef, but total and saturated fat content of beef can also be among the reasons Americans choose to eat less beef in their diet (McNeill et al., 2012). Fat is an essential dietary nutrient that not only provides energy, but also facilitates absorption of fat-soluble vitamins and formation of hormones necessary for normal body function. The 2010 Dietary Guidelines Advisory Committee Report (USDA, 2010b) and the American Heart Association (Gidding et al., 2009) recommended that all healthy Americans limit their intake of saturated fatty acids (SFA) to less than 10% of energy intake and cholesterol to less than 300 mg/d. Consumers have often interpreted this recommendation to mean that they should decrease or eliminate beef and other red meats from their diet (Guenther et al., 2005). Only one-third of the fatty acids in beef are cholesterol-raising fatty acids, and one-third of the SFA in beef consists of stearic acid, which neither raises nor lowers serum cholesterol levels (Kris-Etherton et al., 2005; Daley et al., 2010). Beef is the single largest source of monounsaturated fatty acids (MUFA) in the U.S. diet (Cotton et al., 2004).

Roussell et al. (2012) conducted a well-controlled consumption study to evaluate the low-density lipoprotein (LDL)-lowering effects of four low-SFA dietary regimens: (a) Dietary Approaches to Stop Hypertension (DASH; considered the gold standard; 28 g of lean beef/d); (b) Beef in an Optimal Lean Diet (BOLD; 113 g lean beef/d); (c) BOLD plus additional protein (BOLD+; 153 g lean beef/d); and (d) control diet considered to be a Healthy American Diet (HAD; 20 g beef/d) to serve as a control. They reported that low-SFA, heart-healthy dietary patterns that contain lean beef (DASH, BOLD, BOLD+) were almost identical in their ability to lower total cholesterol and LDL-cholesterol concentrations compared to the HAD diet. These results, in conjunction with the beneficial effects they observed on apolipoprotein cardiovascular disease risk factors (decrease in A-1, C-III, and C-III bound to apolipoprotein particles) after consumption of the BOLD and BOLD+ diets, which were greater with the BOLD+ diet, provide support for including lean beef in a heart-healthy dietary regimen (Roussell et al., 2012).

Grass-fed beef is often regarded as being healthier because of its fatty acid content. When expressed on a milligram per gram of fat basis, grass-fed beef has consistently lower SFA, higher concentrations of polyunsaturated fatty acids (PUFA), and higher concentrations of omega-3 fatty acids (Shantha et al., 1997; Razminowicz et al., 2006; Alfaia et al., 2009; Daley et al., 2010) than conventionally raised beef. In the review conducted by Daley et al. (2010), the authors reported that grass-fed beef on a milligram per gram of fat basis had more conjugated linoleic acid (CLA) isomers (18:2), *trans*-vaccenic acid (C18:1 t11, a precursor to CLA), omega-3 (n-3) fatty acids, slightly higher cholesterol-neutral SFA (stearic, C18:0), and less of the cholesterol-elevating SFA (myristic, C14:0 and palmitic, C16:0) than grain-fed beef.

When Elswyk and McNeill (2014) scrutinized the literature to compare the nutrient and sensory quality of grass-fed versus grain-feed beef, they realized that the previous literature reviews comparing nutrient profile of grass- versus grain-fed have either combined the results from studies conducted throughout the world (Daley et al., 2010), throughout Europe (Scollan et al., 2006), or those focused on one of the several countries practicing primarily fresh pasture feeding

(Ponnampalam et al., 2006). Because there is variation in forage species, forage availability, forage maturity, growing season, cattle breed types, as well as stage of growth and fat content of beef at slaughter among countries, there are possible confounding factors affecting the nutrient composition of the diet consumed by the grazing animal, and therefore affecting meat nutritional and sensory quality (DeSmet et al., 2004; Mir et al., 2006; Luciano et al., 2011). In addition, because grass/forage feeding typically results in a leaner product (Daley et al., 2010), the review by Elswyk and McNeill (2014) only summarized the literature from the United States and then reported the nutritional characteristics on a milligram per 100-g portion size rather than a milligram per gram of fat basis. In their review, only one out of four studies (Rule et al., 2002; vs. Leheska et al., 2008; Duckett et al., 2009, 2013) reported a significantly lower cholesterol content in beef from grass/forage-fed beef (20 mg decrease per 8-oz steak). When lean cuts were compared across these same four studies, there was a 2- to 4-g decrease in total fat per 100 g (as-consumed) in grass-fed versus grain-finished beef. Elswyk and McNeill (2014) concluded that the effect of grass feeding on fatty acid profile of beef cannot be easily generalized because it seems that the fatty acid profile of meat is influenced by breed, type of grass (Itoh et al., 1999), and variation among muscles (Lorenzen et al., 2007), as well as fat type and location within the carcass (Jiang et al., 2010).

Approximately one-third of the SFA content of beef, regardless of feeding regimen, is stearic acid (Elswyk and McNeill, 2014), which has been recognized as neutral with regard to plasma LDL cholesterol (IOM, 2005; FAO, 2010; USDA, 2010b). In their review, Elswyk and McNeill (2014) concluded that, from the four U.S. studies reporting SFA content, grass-fed beef has a higher percent SFA content, but has a 1.4 g/100 g serving lower total SFA content than grain-finished beef. It also seems from these studies that grass-fed beef might contain a greater proportion of SFA neutral stearic acid compared with grain-finished beef in some steaks.

Beef is a significant source of MUFA in the U.S. diet, with the most common fatty acid being oleic acid (18:1 n-9; USDA, 2010b). Elswyk and McNeill (2014) summarized the results of two recent randomized controlled studies by Adams et al. (2010) and Gilmore et al. (2011), which suggested that the higher MUFA content in grain-finished beef could be important for increasing plasma high-density lipoprotein (HDL) cholesterol among consumers, which is a desirable result. In contrast, as a result of an altered MUFA:SFA ratio, beef produced exclusively from a grass-fed program could result in lowered HDL, increased triglycerides, and increased LDL among consumers. Collectively, these studies strongly suggest that the nutrient profile of grass-fed beef is not superior to grain-fed beef as some would advocate.

The PUFA content is low and averages less than 5% of total fatty acids in beef (Scollan et al., 2006). Elswyk and McNeill (2014) concluded from their review that the omega-6 fatty acid, linoleic acid (18:2 n-6), is the primary PUFA in both U.S. grass/forage-fed and grain-finished beef, providing up to 60 to 85% of total PUFA. From three U.S. studies reporting PUFA concentrations (Leheska et al., 2008; Duckett et al., 2009, 2013), Elswyk and McNeill (2014) concluded that grass/forage-fed beef compared with grain-fed beef has more (16 to 26 vs. 4 to 13 mg/100 g) short-chain omega-3 fatty acid (α-linolenic acid [ALA]; 18:3 n-3) than grain-fed beef, but only slight increases in the long-chain omega-3 fatty acids (n-3 LCPUFA $\geq$ C20). The contribution of ALA to cardiovascular health is debatable (Calder et al., 2010), but the role of long-chain omega-3 fatty acids in the prevention of heart disease is more convincing (FAO, 2010). According to the Institute of Medicine (IOM, 2005), ALA intake by Americans is adequate (1.6 g/d), but intake of n-3 LCPUFA is less than 130 mg/d (USDA, 2010b) vs. the recommended 250-mg/d intake (EFS, 2010; FAO, 2010; USDA, 2010b). Scollan et al. (2006) reported that beef cattle have a limited ability to accumulate significant amounts of n-3 LCPUFA because of dietary unsaturated fatty acid biohydrogenation in the rumen, and the limited ability of most animals to convert eicosapentaenoic acid (C20:5 n-3) to docosahexaenoic acid (C22:6 n-3). Elswyk and McNeil (2014) concluded that lean cuts from either grass/forage-fed beef or grain-fed beef can make only a modest contribution to circulating plasma concentration of n-3 LCPUFA in healthy adults (Welch et al., 2010; McAfee et al., 2011).

Conjugated linoleic acid accumulates in the fat and muscle of ruminant animals as a result of both ruminal biohydrogenation of linoleic acid and the delta-9 desaturation of *trans*-vaccenic acid to *cis*-9,*trans*-11 CLA, the predominant CLA isomer in both grass/forage-fed and grain-finished beef (Elswyk and McNeill, 2014). Daley et al. (2010), Duckett et al. (2009), and Lorenzen et al. (2007) reported a significant increase (up to twofold) in the percentage of CLA (mainly, *cis*-9,*trans*-11) in grain-finished beef. Because of the lower fat content of most grass-fed beef, however, the total amount of CLA from either grass/forage-fed or grain-fed U.S. beef is nearly identical and only makes a minor contribution to CLA intake levels that were observed to be beneficial in clinical trials (Elswyk and McNeill, 2014).

Protein

Although many Americans consume more protein than the Recommended Daily Allowance, certain population subgroups including adolescent females and older women have inadequate protein intake (Fulgoni, 2008). In addition, there is a growing body of scientific evidence that suggests that higher protein intake can benefit weight management, muscle mass maintenance, cholesterol and triglyceride levels, and satiety (Westerterp-Plantenga et al., 2009), as well as minimize sarcopenia (Paddon-Jones et al., 2008) and other physiologic functions (Wolfe et al., 2008).

Vitamins and Minerals

There are several reviews that detail the importance of vitamins and minerals in the human diet as they relate to fitness and well-being (Fairfield and Fletcher, 2002; Huskisson et al., 2007; Dennehy and Tsourounis, 2010), cognitive ability and mental function (Calvaresi and Bryan, 2001; Dangour et al., 2010), and cancer (Mamede et al., 2011). Iron (Fe) deficiency is a common condition worldwide among young children and women of child-bearing age where monotonous, plant-based diets provide low amounts of available Fe (Tapiero et al., 2001; Zimmermann and Hurrell, 2007). Iron plays a vital role in many biological functions, but most importantly, it facilitates oxygen transport to the cells, brain development, and immune system support. Beef is a good source of Fe that is more readily available and more easily absorbed than Fe from plant sources (Zimmermann and Hurrell, 2007). Moreover, when beef is added to the diet (heme-Fe), it increases the absorption of Fe (nonheme-Fe) from plant sources (Beard et al., 1996; Tapiero et al., 2001). Zinc (Zn) is an essential nutrient that fuels metabolic processes that include building muscles, healing wounds, maintaining the immune system, and contributing to cognitive function. Deficiencies of Fe and Zn can affect cognition and may be associated with neuropsychological impairments in infants, preschool- and school-age children, adolescents and adults (Tapiero et al., 2001).

The B-complex vitamins (folic acid, B_{12}, and B_6 in particular) are involved with optimum function of the central nervous system and play specific roles in the methylation and decarboxylation processes that govern the integrity of DNA; and the synthesis of DNA, proteins, phospholipids as well as monoamine and catecholamine neurotransmitters (Mattson and Shea, 2003). These three B-complex vitamins are also instrumental in the remethylation and metabolism of the potentially toxic amino acid homocysteine, which can contribute to a range of neurodegenerative and psychological disorders (Mattson and Shea, 2003). In addition, Selhub et al. (2000) reported that B-complex vitamins are involved in metabolic processes associated with normal function of cells, the central nervous system, the body's ability to fight infection, and lowering blood levels of homocysteine that can cause an increased risk of heart disease and dementia.

TASTE AND TENDERNESS

The U.S. and Canadian marketing systems are predicated on the use of grain-fed cattle and use quality grade (U.S. Select, Choice, Prime; Canadian A, AA, AAA, Prime) in addition to yield grade (U.S. 1 to 5; Canadian 1 to 3) to determine carcass value (USDA, 1997; Canadian Beef, 2009). Quality grade is largely based on external fat cover and marbling (intramuscular fat) and maturity (physiological age of the carcass) as a way to predict taste, tenderness, juiciness, and flavor. A distinguishing feature of U.S. and Canadian beef in the global marketplace has been grain-fed cattle. Grass-finished cattle can grade USDA Choice when provided quality forages, but grass-finishing typically results in smaller carcasses, yellow vs. white fat, less fat cover, lower marbling score, and advanced physiological maturity compared with grain-fed beef. Berthiaume et al. (2006) reported that feeding grain to cattle decreased the length of the feeding period by 21% compared with grass-fed cattle. The longer feeding period required for grass-fed cattle typically results in advanced carcass maturity and lower fat cover, which can result in cold shortening of muscles during the postexsanguination chill, thereby potentially decreasing tenderness. Nonetheless, Faucitano et al. (2008) observed no statistically significant difference in tenderness scores between beef from cattle fed grass and silage compared with those fed grain when cattle were fed to the same (8-mm) external fat cover.

ANIMAL WELL-BEING

Beef producers have a vested interest in animal well-being and care of their animals. Carroll and Forsberg (2007) and Duff and Galyean (2007), in their reviews, discussed the influence of stress and nutrition on cattle immunity and animal performance. It is easily concluded from these reviews of the literature that animals under stress are more vulnerable to disease-causing organisms, perform at a lower level, and produce a less-valuable carcass. Although profit is often cited as a reason to criticize management practices and production systems, it could easily be argued as "the" reason to make sure animals are cared for in the best possible manner. Consumers have expressed concerns about animals in confinement systems and the animal's ability to move, exercise, eat, drink, and interact socially. Grass-finishing systems often emphasize their lack of confinement and imply that the open space enhances the well-being of cattle and the value of their products (Mathews and Johnson, 2013). Crowding of cattle is a natural behavior and is not unique to cattle in confinement. Cattle are not as gregarious as sheep, but there is a herding instinct present. For example, cattle will crowd under a tree for shade or to avoid flies, along a windbreak to decrease wind chill, follow one another to feed and water, and crowd together when threatened by predators. Cattle on pasture are more vulnerable to extremes in temperature, humidity, and precipitation; internal and external parasites; and predators. In addition, forage quality and quantity are affected by season of year, environmental conditions, and stage of maturity. Individually and collectively, these factors can sometimes affect the animal's ability to consume nutrients and remain healthy. In contrast, animals in pasture systems have some advantage in the area of disease transmission between animals and animal injury because they can individually be more isolated. Cattle in confinement systems are, by default, more crowded and have the potential for more animal-to-animal transmission of disease, increased likelihood of injury, and have a higher concentration of manure. The advantage of a confinement system, however, is closer and more frequent animal observation and care, and daily delivery of diets that are consistently formulated to keep animals healthy, productive, and efficient.

FEED ADDITIVES AND GROWTH MODIFIERS

The reader is referred to Chapter 14 (Compounds That Modify Digestion and Metabolism) for a detailed discussion of how feed additives and growth-modifying products used by the beef industry work to alter metabolism, promote growth and feed efficiency, increase carcass value, and provide a safe, wholesome beef product to consumers at the lowest possible price.

ENVIRONMENTAL IMPACT OF PRODUCTION SYSTEMS

The reader is referred to Chapter 16 (Environment) for a detailed discussion of how feeding, production, and management practices used within the North American beef industry interact with the environmental conditions where cattle reside to affect nutrient losses, greenhouse gas emissions, and the carbon footprint.

REFERENCES

AAAS (American Association for the Advancement of Science). 2012. Legally Mandating GM Food Labels Could "Mislead and Falsely Alarm Consumers." Available online at http://www.aaas.org/news/aaasboard-directors-legally-mandating-gm-foodlabels-could-%E2%80%9Cmislead-and-falselyalarm. Accessed on May 5, 2014.

Abidoye, B. O., H. Bulut, J. D. Lawrence, B. Mennecke, and A. M. Townsend. 2011. U.S. consumers' valuation of quality attributes in beef products. *Journal of Agricultural and Applied Economics* 43:1-12.

Adams, T. H., R. L. Walzem, D. R. Smith, S. Tseng, and S. M. Smith. 2010. Hamburger high in total, saturated and trans-fatty acids decreases HDL cholesterol and LDL particle diameter, and increases TAG, in mildly hypercholesterolaemic men. *British Journal of Nutrition* 103:91-98.

AGA (American Grassfed Association). 2013. Grassfed and Grass Pastured Ruminant Standards. Available online at http://www.americangrassfed.org/wp-content/uploads/2014/03/AGA-Grassfed-Standards-October-2013.pdf. Accessed on May 8, 2014.

Alfaia, C. P., S. P. Alves, S. I. Martins, A. S. Costa, C. M. Fontes, J. P. Lemos, R. J. Bessa, and J. A. Prates. 2009. Effects of feeding system on intramuscular fatty acids, and conjugated linoleic acid isomers of beef cattle, with an emphasis on their nutritional value and discriminatory ability. *Food Chemistry* 114:939-946.

Ali, A., and A. Abdulai. 2010. The adoption of genetically modified cotton and poverty reduction in Pakistan. *Journal of Agricultural Economics* 61:175-192.

Arthington, J. D., and J. E. Minton. 2004. The effect of early calf weaning on feed intake, growth, and postpartum interval in thin, Brahman-crossbred primiparous cows. *The Professional Animal Scientist* 20:34-38.

Arthur, P. F., J. A. Archer, E. C. Richardson, and R. M. Herd. 1999. Potential for selection to improve efficiency of feed use in beef cattle: A review. *Australian Journal of Agricultural Research* 50:147-163.

Batz, M. B., S. Hoffmann, and J. Glenn Morris, Jr. 2012. Ranking the disease burden of 14 pathogens in food sources in the United States using attribution data from outbreak investigations and expert elicitation. *Journal of Food Protection* 7:1278-1291.

Beard, J. L., H. Dawson, and D. J. Pinero. 1996. Iron metabolism: A comprehensive review. *Nutrition Reviews* 54:295-317.

BeefNutrition.org. 2014. Beef: Big Nutrient Power in a Small Package. Ten Essential Nutrients and What They Do for You. Available online at http://www.beefnutrition.org/CMDocs/BeefNutrition/Updated%20Materials/Beef%20Nutrients/101613Nutrient%20Power%20Fact%20Sheet.pdf. Accessed on May 30, 2014.

Beever, D. E. 1993. Ruminal animal production from pastures. Current opportunities and future perspectives. Pp.158-164 in *Forages for Our World: Proceedings of the XVII International Grasslands Congress*, M. J. Baker, ed. Wellington, New Zealand: SIR Publishing,

Berthiaume, R., I. Mandell, L. Faucitano, and C. Lafreniere. 2006. Comparison of alternative beef production systems based on forage finishing or grain-forage diets with or without growth promotants: 1. Feedlot performance, carcass quality, and production costs. *Journal of Animal Science* 84:2168-2177.

Bevers, S. J. 2012. Standardized Performance Analysis (SPA) for Decision Making. 2012 Beef Cattle Short Course, August 8, 2012. Available online at http://agrisk.tamu.edu/files/2012/05/SPA-Informing-Decision-Makers.pdf. Accessed on May 20, 2014.

Brashears, M. M., M. L. Galyean, G. H. Loneragan, J. E. Mann, and K. Killinger-Mann. 2003a. Prevalence of *Escherichia coli* O157:H7 and performance by beef feedlot cattle given *Lactobacillus* direct-fed microbials. *Journal of Food Protection* 66:748-754.

Brashears, M. M., D. Jaroni, and J. Trimble. 2003b. Isolation, selection, and characterization of lactic acid bacteria for a competitive exclusion product to reduce shedding of *Escherichia coli* O157:H7 in cattle. *Journal of Food Protection* 66:355-363.

Brewer, P., and C. Calkins. 2003. Quality traits of grain- and grass-fed beef: A review. Pp. 74-77 in *Nebraska Beef Cattle Report MP80-A*. University of Nebraska-Lincoln.

Brookes, G., and P. Barfoot. 2013. The global income and production effects of genetically modified (GM) crops 1996-2011. *GM Crops and Food: Biotechnology in Agriculture and the Food Chain* 4:74-83.

Bruns, K. W., R. H. Pritchard, and D. L. Boggs. 2004. The relationship among body weight, body composition, and intramuscular fat content in steers. *Journal of Animal Science* 82:1315-1322.

Burachik, M. 2010. Experience from use of GMOs in Argentinian agriculture, economy and environment. *New Biotechnology* 27:588-592.

Calder, P. C., A. D. Dangour, C. Diekman, A. Eilander, B. Koletzko, G. W. Meijer, D. Mozaffarian, H. Niinikoski, S. M. J. Osendarp, P. Pietinen, J. Schuit, and R. Uauy. 2010. Essential fats for future health. Proceedings of the 9th Unilever Nutrition Symposium, May 26-27. *European Journal of Clinical Nutrition* 64:S1-S13.

Callaway, T. R., T. S. Edrington, R. C. Anderson, R. B. Harvey, K. J. Genovese, C. N. Kennedy, D.W. Venn, and D. J. Nisbet. 2008. Probiotics, prebiotics and competitive exclusion for prophylaxis against bacterial disease. *Animal Health Research Reviews* 9:217-225.

Calvaresi, E., and J. Bryan. 2001. B vitamins, cognition, and aging. *Journal of Gerontology: Psychological Sciences and Social Sciences* 56:327-339.

Canadian Beef. 2009. The Canadian Beef Grading System. Available online at http://www.canadabeef.ca/OrderCentre_files/public/151748-en.pdf. Accessed on May 30, 2014.

CanFax. 2014. Alberta and Saskatchewan Feedlot Demographics. Available online at http://www.canfax.ca/CattleOnFeed/BunkCapacity.aspx. Accessed on April 28, 2014.

Carpenter, J. E. 2013. The socio-economic impacts of currently commercialised genetically engineered crops. *International Journal of Biotechnology* 12:249-268.

Carroll, J. A., and N. E. Forsberg. 2007. Influence of stress and nutrition on cattle immunity. *Veterinary Clinics of North America: Food Animal Practice* 23:105-149.

Cassady, B. A., N. L. Charboneau, E. E. Brys, K. A. Crouse, D. C. Beitz, and T. Wilson. 2007. Effects of low carbohydrate diets high in red meats or poultry, fish and shellfish on plasma lipids and weight loss. *Nutrition and Metabolism* 4:23.

CAST (Council for Agricultural Science and Technology). 2001. Evaluation of the U.S. Regulatory Process for Crops Developed through Biotechnology. Council for Agricultural Science and Technology Issue Paper 19.

CAST. 2013. Animal Feed vs. Human Food: Challenges and Opportunities in Sustaining Animal Agriculture Toward 2050. Council for Agricultural Science and Technology Issue Paper No. 53.

CAST. 2014. The Potential Impacts of Mandatory Labeling for Genetically Engineered Food in the United States. Council for Agricultural Science and Technology Issue Paper No. 54.

Chaney, E. W., G. H. Loneragan, R. McCarthy, M. F. Miller, B. J. Johnson, J. C. Brooks, and M. M. Brashears. 2011. Effects of corn-based distillers grain (DG) inclusion into feeding rations on the burden of *Escherichia coli* O157:H7 in commercial feedlot settings. 98th Annual Meeting of the International Association for Food Protection (IAFP), July 31-August 3, Milwaukee, WI.

Chassy, B. 2010. Food safety risks and consumer health. *New Biotechnology* 27:534-544.

Chassy, B. M., J. J. Hlywka, G. A. Kleter, E. J. Kok, H. A. Kuiper, M. McGoughlin, I. C. Munro, R. H. Phipps, and J. E. Reid. 2004. Nutritional and safety assessments of foods and feeds nutritionally improved through biotechnology. *Comprehensive Reviews in Food Science and Food Safety* 3:35-104.

Claassen, R., F. Carriazo, and K. Ueda. 2010. *Grassland Conversion for Crop Production in the United States: Defining Indicators for Policy Analysis.* Prepared for OECD Agri-environmental Indicators: Lessons Learned and Future Directions, March 23-26, 2010, Leysin Switzerland Available online at http://www.oecd.org/tad/sustainable-agriculture/44807867.pdf. Accessed on June 13, 2014.

CODEX Alimentarius Commission. 2014. International Food Standards. Available online at http://www.codexalimentarius.org/standards/en/. Accessed on June 24, 2014.

Coleman, S. W., B. C. Evans, and J. J. Evans. 1993. Body and carcass composition of Angus and Charolais steers as affected by age and nutrition. *Journal of Animal Science* 71:86-95.

Coleman, S. W., R. H. Gallavan, W. A. Phillips, J. D. Volesky, and S. Rodriguez. 1995. Silage or limit-fed grain growing diets for steers: II. Empty body and carcass composition. *Journal of Animal Science* 73:2621-2630.

Cotton, P. A., A. F. Subar, J. E. Friday, and A. Cook. 2004. Dietary sources of nutrients among US adults, 1994 to 1996. *Journal of the American Dietetic Association* 104:921-930.

Cundiff, L.V., and K. E. Gregory. 1999. What Is Systematic Crossbreeding? Presented at Cattlemen's College: Cattle Industry Annual Meeting, National Cattlemen's Beef Association, February 11, 1999. Charlotte, NC.

Cundiff, L. V., F. Szabo, K. E. Gregory, R. M. Koch, M. E. Dikeman, and J. D. Crouse. 1993. Breed comparisons in the germplasm evaluation program at MARC. Pp. 124-136 in *Proceedings of Beef Improvement Federation 25th Anniversary Conference*, May 26-29, 1993, Asheville, NC.

Cundiff, L.V., K. E. Gregory, and R. M. Koch. 1998. Germplasm evaluation in beef cattle—cycle IV: Birth and weaning traits. *Journal of Animal Science* 76:2525-2535.

Daley, C. A., A. Abbott, P. S. Doyle, G. A. Nader, and S. Larson. 2010. A review of fatty acid profiles and antioxidant content in grass-fed and grain-fed beef. *Nutrition Journal* 9:10. Available online at http://www.nutritionj.com/content/9/1/10. Accessed on June 13, 2014.

Dangour, A. D., S. K. Dodhia, A. Hayter, E. Allen, K. Lock, and R. Uauy. 2009. Nutritional quality of organic foods: A systematic review. *American Journal of Clinical Nutrition* 90:680-685.

Dangour, A. D., P. J. Whitehouse, K. Rafferty, S. A. Mitchell, L. Smith, S. Hawkesworth, and B. Vellas. 2010. B vitamins and fatty acids in the prevention and treatment of Alzheimer's disease and dementia: A systematic review. *Journal of Alzheimer's Disease* 22:205-224.

Dennehy, C., and C. Tsourounis. 2010. A review of selected vitamins and minerals used by postmenopausal women. *Maturitas* 66:370-380.

DeSmet, S., K. Raes, and D. Demeyer. 2004. Meat fatty acid composition as affected by fatness and genetic factors: A review. *Animal Research* 53:81-98.

Diez-Gonzalez, F., T. R. Callaway, M. G. Kizoulis, and J. B. Russell. 1998. Grain feeding and the dissemination of acid-resistant *Escherichia coli* from cattle. *Science* 281:1666-1668.

Doyle, M. P., F. Busta, B. R. Cords, P. M. Davidson, J. Hawke, H. S. Hurd, R. E. Isaacson, K. Matthews, J. Maurer, J. Meng, T. J. Montville, T. R. Shryock, J. N. Sofos, A. K. Vidaver, and L. Vogel. 2006. Antimicrobial resistance: Implications for the food system—An expert report. *Comprehensive Reviews in Food Science and Food Safety* 5:71-137.

Doyle, M. P., G. H. Loneragan, H. M. Scott, and R. S. Singer. 2013. Antimicrobial resistance: Challenges and perspectives—An expert report. *Comprehensive Reviews in Food Science and Food Safety* 12:234-248.

Duckett, S. K., J. P. S. Neel, J. P. Fontenot, and W. M. Clapham. 2009. Effects of winter stocker growth rate and finishing system on: III. Tissue proximate, fatty acid, vitamin and cholesterol content. *Journal of Animal Science* 87:2961-2970.

Duckett, S. K., J. P. S. Neel, R. M. Lewis, J. P. Fontenot, and W. M. Clapham. 2013. Effects of forage species or concentrate finishing on animal performance, carcass and meat quality. *Journal of Animal Science* 91:1454-1467.

Duff, G. C., and M. L. Galyean. 2007. Board-invited review: Recent advances in management of highly stressed, newly received feedlot cattle. *Journal of Animal Science* 85:823-840.

Dutton, J. M., C. E. Ward, and J. L. Lusk. 2007. Implicit value of retail beef brands and retail meat product attributes. *Proceedings of the NCCC-134 Conference on Applied Commodity Price Analysis, Forecasting, and Market Risk Management*, Chicago, IL. Available online at http://ageconsearch.umn.edu/bitstream/37571/2/confp15-07.pdf. Accessed on December 2, 2014.

EFS (European Food Safety). 2010. Scientific opinion on dietary reference values for fats, including saturated fatty acids, polyunsaturated fatty acids, monounsaturated fatty acids, trans fatty acids and cholesterol. European Food Safety Panel on Dietetic Products, Nutrition, and Allergies (NDA). *EFSA Journal* 1461:1-107.

Elam, T. E., and R. L. Preston. 2004. Fifty Years of Pharmaceutical Technology and Its Impact on the Beef We Provide to Consumers. Growth Enhancement Technology Information Team. Available online at http://www.sustainablebeef.org/_assets/SBRC-Fifty-Years-Technology-Impact.pdf. Accessed on June 13, 2014.

Elswyk, M. E., and S. H. McNeill. 2014. Impact of grass/forage feeding versus grain finishing on beef nutrients and sensory quality: The U.S. experience. *Meat Science* 96:535-540.

Erickson, G. E., V. R. Bremer, T. J. Klopfenstein, A. Stalker, and R. Rasby. 2010. Feeding Corn Milling Co-products to Feedlot Cattle. Available online at http://www.nebraskacorn.org/internally-linked-pages/corn-co-product-manuals/. Accessed on May 14, 2014.

Fairfield, K. M., and R. H. Fletcher. 2002. Vitamins for chronic disease prevention in adults: Scientific review. *Journal of the American Medical Association* 287:3116-3126.

FAO (Food and Agriculture Organization of the United Nations). 2010. Fats and Fatty Acids in Human Nutrition: Report of an Expert Consultation. Food and Nutrition Paper 91. Food and Agriculture Organization. Available online at http://www.fao.org/docrep/013/i1953e/i1953e00.pdf. Accessed on April 28, 2014.

FAO. 2012. *Plant Production and Protection Division: Grasslands, Rangelands and Forage Crops.* Rome, Italy: Food and Agriculture Organization. Available online at http://www.fao.org/agriculture/crops/

thematic-sitemap/theme/spi/grasslands-rangelands-and-forage-crops/en/. Accessed on April 28, 2014.
Faucitano, L., P. Y. Chouinard, J. Fortin, I. B. Mandell, C. Lafreniere, C. L. Girard, and R. Berthiaume. 2008. Comparison of alternative beef production systems based on forage finishing or grain-forage diets with or without growth promotants: 2. Meat quality, fatty acid composition, and overall palatability. *Journal of Animal Science* 86:1678-1689.
FDA (U.S. Food and Drug Administration). 2013. *Guidance for Industry (GFI) #213.* U.S. Department of Health and Human Services, Food and Drug Administration, Center for Veterinary Medicine. December 2013. Available online at http://www.fda.gov/downloads/animalveterinary/guidancecomplianceenforcement/guidanceforindustry/ucm299624.pdf. Accessed on July 16, 2014.
FeedInfo News Service. 2010. U.S. Grass Fed Beef Demand Strengthening. Available online at http://www.thefreelibrary.com/US%3A+Grass+Fed+Beef+Demand+Strengthening.-a0225390375. Accessed on July 16, 2014.
Fernandez-Cornejo, J., S. J. Wechsler, M. Livingston, and L. Mitchell. 2014. Genetically Engineered Crops in the United States. Economic Research Report No. 162. U.S. Department of Agriculture. Available online at http://www.ers.usda.gov/publications/err-economic-research-report/err162.aspx. Accessed on May 5, 2014.
Ferrell, C. L., and T. G. Jenkins. 1998. Body composition and energy utilization by steers of diverse genetypes fed a high-concentrate diet during the finishing period: II. Angus, Boran, Brahman, Hereford, and Tuli sires. *Journal of Animal Science* 76:647-657.
Flachowsky, G., H. Schafft, and U. Meyer. 2012. Animal feeding studies for nutritional and safety assessments of feeds from genetically modified plants: A review. *Journal für Verbraucherschutz und Lebensmittelsicherheit* 7:179-194.
FSSA (Food Safety Standards Agency). 2009. Comparison of Composition (Nutrients and Other Substances) of Organically and Conventionally Produced Foodstuffs: A Systematic Review of the Available Literature. Prepared for the Food Safety Standards Agency by the Nutrition and Public Health Intervention Research Unit, School of Hygiene and Tropical Medicine, London. Available online at http://www.nutriwatch.org/04Foods/fsa/nutrient.pdf. Accessed on December 2, 2014.
Fulgoni, V. L., III. 2008. Current protein intake in America: Analysis of the National Health and Nutrition Examination Survey, 2003-2004. *American Journal of Clinical Nutrition* 87:1554S-1557S.
Galyean, M. L. 1999. Review: Restricted and programmed feeding of beef cattle—definitions, application, and research results. *The Professional Animal Scientist* 15:1-6.
Galyean, M. L., C. Ponce, and J. Schutz. 2011. The future of beef production in North America. *Animal Frontiers* 1:29-36.
Gidding, S. S., A. H. Lichtenstein, M. S. Faith, A. Karpyn, J. A. Mennella, B. Popkin, J. Rowe, L. Van Horn, and L. Whitsel. 2009. Implementing American Heart Association pediatric and adult nutrition guidelines: A scientific statement from the American Heart Association Nutrition Committee of the Council on Nutrition, Physical Activity and Metabolism, Council on Cardiovascular Disease in the Young, Council on Arteriosclerosis, Thrombosis and Vascular Biology, Council on Cardiovascular Nursing, Council on Epidemiology and Prevention, and Council for High Blood Pressure Research. *Circulation* 119:1161-1175.
Gill, M. 1999. Meat production in developing countries. *Proceedings of the Nutrition Society* 58:371-376.
Gilmore, L. A., R. L. Walzem, S. F. Crouse, D. R. Smith, T. H. Adams, V. Vaidyanathan, X. Cao, and S. B. Smith. 2011. Consumption of high-oleic acid ground beef increases HDL-cholesterol concentration but both high- and low-oleic acid ground beef decrease HDL particle diameter in normocholesterolemic men. *Journal of Nutrition* 141:1188-1194.
Guenther, P. M., H. H. Jensen, S. P. Batres-Marquez, and C. F. Chen. 2005. Sociodemographic, knowledge, and attitudinal factors related to meat consumption in the United States. *Journal of the American Dietetic Association* 105:1266-1274.
Gunter, S. A., M. L. Galyean, and K. J. Malcolm-Callis. 1996. Factors influencing the performance of feedlot steers limit-fed high-concentrate diets. *The Professional Animal Scientist* 12:167.
Herman, R. A., and W. D. Price. 2013. Unintended compositional changes in genetically modified (GM) crops: 20 years of research. *Journal of Agricultural and Food Chemistry* 61:11695-11701.
Huang, J., J. Mi, H. Lin, Z. Wang, R. Chen, R. Hu, S. Rozelle, and C. Pray. 2010. A decade of Bt cotton in Chinese fields: Assessing the direct effects and indirect externalities of Bt cotton adoption in China. *Science China Life Sciences* 53:981-991.
Huskisson, E., S. Maggini, and M. Ruf. 2007. The role of vitamins and minerals in energy metabolism and well-being. *Journal of International Medical Research* 35:277-289.
IOM (Institute of Medicine). 2005. *Dietary Reference Intake for Energy, Carbohydrate, Fiber, Fat, Fatty Acids, Cholesterol, Protein, and Amino Acids (Macronutrients).* Washington, DC: The National Academies Press.
IOM/NRC (Institute of Medicine and National Research Council). 2004. *Safety of Genetically Engineered Foods: Approaches to Assessing Unintended Health Effects.* Washington, DC: The National Academies Press.
ISAAA (International Service for the Acquisition of Agribiotech Applications). 2009. Global Status of Commercialized Biotech/GM Crops. Brief 41. International Service for the Acquisition of Agri-Biotech Applications. Available online at http://www.isaaa.org/resources/publications/briefs/41/download/isaaa-brief-41-2009.pdf. Accessed on July 15, 2014.
ISAAA. 2014. GM Crop Events Approved in United States of America. International Service for the Acquisition of Agribiotech Applications. Available online at http://www.isaaa.org/gmapprovaldatabase/approvedeventsin/default.asp?CountryID=US&Country=United%20States%20of%20America. Accessed on May 5, 2014.
Itoh, M., C. B. Johnson, G. P. Cosgrove, P. D. Muir, and R. W. Purchas. 1999. Intramuscular fatty acid composition of neutral and polar lipids for heavy-weight Angus and Simmental steers finished on pasture or grain. *Journal of the Science of Food and Agriculture* 79:821-827.
Jacob, M. E., J. T. Fox, J. S. Drouillard, D. G. Renter, and T. G. Nagaraja. 2008a. Effects of dried distillers' grains on fecal prevalence and growth of *Escherichia coli* O157 in batch culture fermentations from cattle. *Applied and Environmental Microbiology* 74:38-43.
Jacob, M. E., J. T. Fox, S. K. Narayanan, J. S. Drouillard, D. G. Renter, and T. G. Nagaraja. 2008b. Effects of feeding wet corn distillers grains with solubles, with or without monensin and tylosin, on the prevalence and antimicrobial susceptibilities of fecal foodborne pathogenic and commensal bacteria in feedlot cattle. *Journal of Animal Science* 86:1182-1190.
Jacob, M. E., T. R. Callaway, and T. G. Nagaraja. 2009. Dietary interactions and interventions affecting *Escherichia coli* O157 colonization and shedding in cattle. *Foodborne Pathogens and Disease* 6:785-792.
Jacob, M. E., Z. D. Paddock, D. G. Renter, K. F. Lechtenberg, and T. G. Nagaraja. 2010. Inclusion of dried or wet distillers' grains at different levels in diets of feedlot cattle affects fecal shedding of *Escherichia coli* O157:H7. *Applied and Environmental Microbiology* 76:7238-7242.
Jiang, T., J. R. Busboom, M. L. Nelson, J. O'Fallon, T. P. Ringkob, D. Joss, and K. Piper. 2010. Effect of sampling fat location and cooking on fatty acid composition of beef steaks. *Meat Science* 84:86-92.
Kathage, J., and M. Qaim. 2012. Economic impacts and impact dynamics of Bt (*Bacillus thuringiensis*) cotton in India. *Proceedings of the National Academy of Sciences of the United States of America* 109:11652-11656.
Khare, S., W. Alali, S. Zhang, D. Hunter, R. Pugh, F. C. Fang, S. J. Libby, and L. G. Adams. 2010. Vaccination with attenuated *Salmonella enterica* Dublin expressing *E coli* O157:H7 outer membrane protein Intimin induces transient reduction of fecal shedding of *E coli* O157:H7 in cattle. *BMC Veterinary Research* 6:35-43.
Krehbiel, C. R., S. R. Rust, G. Zhang, and S. E. Gilliland. 2003. Bacterial direct-fed microbials in ruminant diets: Performance response and mode of action. *Journal of Animal Science* 81:120-132.

Kris-Etherton, P. M., A. E. Griel, T. L. Psota, S. K. Gebauer, J. Zhang, and T. D. Etherton. 2005. Dietary stearic acid and risk of cardiovascular disease: Intake, sources, digestion, and absorption. *Lipids* 40:1193-1200.

Leheska, J. M., L. D. Thompson, J. C. Howe, E. Hentges, J. Boyce, J. C. Brooks, B. Shriver, L. Hoover, and M. F. Miller. 2008. Effects of conventional and grass-feeding systems on the nutrient composition of beef. *Journal of Animal Science* 86:3575-3585.

Lister, G., G. Tonsor, M. Brix, T. Schroeder, and C. Yang. 2014. Food Values Applied to Livestock Products. Working Paper. Available online at http://www.agmanager.info/livestock/marketing/WorkingPapers/WP1_FoodValues-LivestockProducts.pdf. Accessed on May 8, 2014.

Loerch, S. C. 1998. Effects of programming intake on performance and carcass characteristics of feedlot cattle. *Journal of Animal Science* 76:371-377.

Loneragan, G. H., and M. M. Brashears. 2005. Pre-harvest interventions to reduce carriage of *E. coli* O157 by harvest-ready feedlot cattle. *Meat Science* 71:72-78.

Lorenzen, C. L., J. W. Golden, F. A. Martz, I. U. Grun, M. R. Ellersieck, J. R. Gerrish, and K. C. Moore. 2007. Conjugated linoleic acid content of beef differs by feeding regime and muscle. *Meat Science* 75:159-167.

Luciano, G., A. P. Moloney, A. Priolo, F. T. Rohrle, V. Vasta, L. Biondi, P. Lopez-Andres, S. Grasso, and F. J. Monahan. 2011. Vitamin E and polyunsaturated fatty acids in bovine muscle and the oxidative stability of beef from cattle receiving grass or concentrated-based rations. *Journal of Animal Science* 89:3759-3768.

MacDonald, J. M., and W. D. McBride. 2009. The Transformation of U.S. Livestock Agriculture: Scale, Efficiency, and Risks. Economic Information Bulletin No. 43. U.S. Department of Agriculture-Economic Research Service, Washington, DC. Available online at http://www.ers.usda.gov/publications/eib43/. Accessed on April 28, 2014.

Mamede, A. C., S. D. Tavares, A. M. Abrantes, J. Trindade, J. M. Maia, and M. F. Botelho. 2011. The role of vitamins in cancer: A review. *Nutrition and Cancer* 63:479-494.

Marshall, D. M. 1994. Breed differences and genetic parameters for body composition traits in beef cattle. *Journal of Animal Science* 72:2745-2755.

Mathews, K. H., Jr., and R. J. Johnson. 2013. *Alternative Beef Production Systems: Issues and Implications.* LDPM-218-01. Washington, DC: U.S. Department of Agriculture, Economic Research Service. Available online at http://www.ers.usda.gov/media/1071057/ldpm-218-01.pdf. Accessed on April 28, 2014.

Mattson, M. P., and T. B. Shea. 2003. Folate and homocysteine metabolism in neural plasticity and neurodegenerative disorders. *Trends in Neurosciences* 26:137-146.

McAfee, A. J., E. M. McSorely, G. J. Cuskelly, A. M. Fearon, B. W. Moss, J. A. M. Beattie, J. M. W. Wallace, M. P. Bonham, and J. J. Strain. 2011. Red meat from animals offered a grass diet increases plasma and platelet n-3 PUFA in healthy consumers. *British Journal of Nutrition* 105:80-89.

McCluskey, J. J., T. I. Wahl, Q. Li, and P. R. Wandschneider. 2005. U.S. grass-fed beef: Marketing health benefits. *Journal of Food Distribution Research* 36:1-8.

McNeill, S. H., K. B. Harris, T. G. Field, and M. E. Van Elswyk. 2012. The evolution of lean beef: Identifying lean beef in today's U.S. marketplace. *Meat Science* 90:1-8.

Meissner, H. H., M. Smuts, and R. J. Coertze. 1995. Characteristics and efficiency of fast-growing feedlot steers fed different dietary energy concentrations. *Journal of Animal Science* 73:931-936.

Meyer, D. L., M. S. Kerley, E. L. Walker, D. H. Keisler, V. L. Pierce, T. B. Schmidt, C. A. Stahl, M. L. Linville, and E. P. Berg. 2005. Growth rate, body composition, and meat tenderness in early vs. traditionally weaned beef calves. *Journal of Animal Science* 83:2752-2761.

Mir, P. S., S. Bittman, D. Hunt, T. Entz, and B. Yip. 2006. Lipid content and fatty acid composition of grasses sampled on different dates through the early part of the growing season. *Canadian Journal of Animal Science* 86:279-290.

Morley, P. S., D. A. Dargatz, D. R. Hyatt, G. A. Dewell, J. G. Patterson, B. A. Burgess, and T. E. Wittum. 2011. Effects of restricted antimicrobial exposure on antimicrobial resistance in fecal *Escherichia coli* from feedlot cattle. *Foodborne Pathogens and Disease* 8:87-98.

Myers, S. E., D. B. Faulkner, F. A. Ireland, L. J. Berger, and D. F. Parrett. 1999a. Production systems comparing early weaning to normal weaning with or without creep feeding for beef steers. *Journal of Animal Science* 77:300-310.

Myers, S. E., D. B. Faulkner, F. A. Ireland, and D. F. Parrett. 1999b. Comparison of three weaning ages on cow-calf performance and steer carcass traits. *Journal of Animal Science* 77:323-329.

Nicolia, A., A. Manzo, F. Veronesi, and D. Rosellini. 2014. An overview of the last 10 years of genetically engineered crop safety research. *Critical Reviews in Biotechnology* 34:77-88.

Oliver, S. P., D. A. Patel, T. R. Callaway, and M. E. Torrence. 2009. ASAS Centennial Paper: Developments and future outlook for pre-harvest food safety. *Journal of Animal Science* 87:419-437.

Paddon-Jones, K., R. Short, W. W. Campbell, E. Volpi, and R. R. Wolfe. 2008. Role of dietary protein in the sarcopenia of aging. *American Journal of Clinical Nutrition* 87:1562S-1566S.

Peel, D. S. 2005. The Mexican Cattle and Beef Industry: Demand, Production, and Trade. Western Economic Forum, Spring 2005. Milwaukee, WI: Western Agricultural Economics Association. Available online at http://ageconsearch.umn.edu/bitstream/27996/1/04010014.pdf. Accessed on April 28, 2014.

Peel, D. S., R. J. Johnson, and K. H. Matthews, Jr. 2010. *Cow-Calf Beef Production in Mexico.* LPD-M-196-01. Washington, DC: U.S. Department of Agriculture, Economic Research Service. Available online at http://www.ers.usda.gov/media/134973/ldpm19601.pdf. Accessed on April 28, 2014.

Peterson, R. E., T. J. Klopfenstein, G. E. Erickson, J. Folmer, S. Hinkley, R. A. Moxley, and D. R. Smith. 2007. Effect of *Lactobacillus acidophilus* strain NP51 on *Escherichia coli* O157:H7 fecal shedding and finishing performance in beef feedlot cattle. *Journal of Food Protection* 70:287-291.

Ponnampalam, E. N., N. J. Mann, and A. J. Sinclair. 2006. Effect of feeding systems on omega-3 fatty acids, conjugated linoleic acid and trans fatty acids in Australian beef cuts, potential impact on human health. *Asia Pacific Journal of Clinical Nutrition* 15:21-29.

Potter, A. A., S. Klashinsky, Y. Li, E. Frey, H. Townsend, D. Rogan, G. Erickson, S. Hinkley, T. Klopfenstein, R. A. Moxley, D. R. Smith, and B. B. Finley. 2004. Decreased shedding of *Escherichia coli* O157:H7 by cattle following vaccination with type III secreted proteins. *Vaccine* 22:362-369.

Qaim, M., and S. Kouser. 2013. Genetically modified crops and food security. *PLoS ONE* 8(6):e64879, doi:10.1371/journal.pone.0064879.

Rao, S., J. VanDonkersgoed, V. Bohaychuk, T. Besser, X. M. Song, B. Wagner, D. Hancock, D. Renter, D. Dargatz, and P. S. Morley. 2010. Antimicrobial drug use and antimicrobial resistance in enteric bacteria among cattle from Alberta feedlots. *Foodborne Pathogens and Disease* 7:449-457.

Ransom, J. R., K. E. Belk, J. N. Sofos, J. A. Scanga, M. L. Rossman, G. C. Smith, and J. D. Tatum. 2003. Investigation of On-Farm Management Practices as Pre-harvest Beef Microbiological Interventions. Research Fact Sheet. Centennial, CO: National Cattlemen's Beef Association. Available online at https://www.beefusa.org/uDocs/ACF3A9B.pdf. Accessed on December 2, 2014.

Razminowicz, R. H., M. Kreuzer, and M. R. L. Scheeder. 2006. Quality of retail beef from two grass-based production systems in comparison with conventional beef. *Meat Science* 73:351-361.

Reinhart, C., and J. Waggoner. 2012. Focus on Feedlots. Kansas State University. Available online at http://www.asi.k-state.edu/about/newsletters/focus-on-feedlots/. Accessed on May 30, 2014.

Reinstein, S., J. T. Fox, X. Shi, M. J. Alam, D. G. Renter, and T. G. Nagaraja. 2009. Prevalence of *Escherichia coli* O157:H7 in organically and naturally raised beef cattle. *Applied and Environmental Microbiology* 75:5421-5423.

Rokeach, M. 1973. *The Nature of Human Values*. New York: The Free Press.

Roussell, M. A., A. M. Hill, T. L. Gaugler, S. G. West, J. P. Vanden Heuvel, P. Alaupovic, P. J. Gillies, and P. M. Kris-Etherton. 2012. Beef in an optimal lean diet study: Effects on lipids, lipoproteins, and apolipoproteins. *American Journal of Clinical Nutrition* 95:9-16.

Rule, D. C., K. S. Broughton, S. M. Shellito, and G. Maiorano. 2002. Comparison of muscle fatty acid profiles and cholesterol concentrations of bison, beef cattle, elk, and chicken. *Journal of Animal Science* 80:1202-1211.

Russell, J., D. Loy, J. Anderson, and M. Cecava. 2011. *Potential of Chemically Treated Corn Stover and Modified Distillers' Grains as a Partial Replacement for Corn Grain in Feedlot Diets*. Animal Industry Report ASL R2586. Iowa State University. Available online at http://lib.dr.iastate.edu/ans_air/vol657/iss1/10/. Accessed on March 3, 2014.

Schoonmaker, J. P., M. J. Cecava, F. L. Fluharty, H. N. Zerby, and S. C. Loerch. 2004. Effect of source and amount of energy and rate of growth in the growing phase on performance and carcass characteristics of early- and normal-weaned steers. *Journal of Animal Science* 82:273-282.

Schrezenmeir, J., and M. De Vrese. 2001. Probiotics, prebiotics, and synbiotics—approaching a definition. *American Journal of Clinical Nutrition* 73(Suppl.):354s-361s.

Scollan, N., J. P. Hocquette, K. Nuernberg, D. Dannenberger, I. Richardson, and A. Moloney. 2006. Innovations in beef production systems that enhance the nutritional and health value of beef lipids and their relationship with meat quality. *Meat Science* 74:17-33.

Selhub, J., L. C. Bagley, J. Miller, and I. H. Rosenberg. 2000. B vitamins, homocysteine, and neurocognitive function in the elderly. *American Journal of Clinical Nutrition* 71:614-620.

Sewell, J. R., L. L. Berger, T. G. Nash, M. J. Cecava, P. H. Doane, J. L. Dunn, M. K. Dyer, and N. A. Pyatt. 2009. Nutrient digestion and performance by lambs and steers fed thermochemically treated crop residues. *Journal of Animal Science* 87:1024-1033.

Shantha, N. C., W. G. Moody, and Z. Tabeidi. 1997. A research note: Conjugated linoleic acid concentration in semimembranosus muscle of grass- and grain-fed and zeranol-implanted beef cattle. *Journal of Muscle Foods* 8:105-110.

Shreck, A. L., C. J. Schneider, B. L. Nuttelman, D. B. Burken, G. E. Erickson, T. J. Klopfenstein, and M. J. Cecava. 2013. Varying proportions and amounts of distillers' grains and alkaline-treated forage as substitutes for corn grain in finishing cattle diets. Pp. 56-57 in *Nebraska Beef Cattle Report*. University of Nebraska-Lincoln. Available online at https://beef.unl.edu/c/document_library/get_file?uuid=c014882f-87f8-46fc-8c68-9fb5eee4355b&groupId=4178167&.pdf. Accessed on April 2, 2015.

Sip, M. L., and R. H. Pritchard. 1991. Nitrogen utilization by ruminants during restricted intake of high concentrate diets. *Journal of Animal Science* 69:2655-2662.

Snell, C., A. Bernheim, J. B. Bergé, M. Kuntz, G. Pascal, A. Paris, and A. E. Ricroch. 2012. Assessment of the health impact of GM plant diets in long-term and multigenerational animal feeding trials: A literature review. *Food and Chemical Toxicology* 50:1134-1148.

Sparling, E., J. L. Grannis, and D. D. Thilmany. 2002. Regional Demand for Natural Beef Products: Urban vs. Rural Willingness to Pay and Target Consumers. Paper presented at the Annual Meeting of the Western Agricultural Economics Association, July 28-31, 2002, Long Beach, CA. Available online at http://ageconsearch.umn.edu/bitstream/36544/1/sp02sp01.pdf. Accessed on December 2, 2014.

Springer, J., J. Biermacher, D. Alkire, and D. Childs. 2009. Premiums Being Offered for Natural Beef Cattle. The Samuel Roberts Noble Foundation. Available online at http://www.noble.org/ag/economics/naturalbeef/index.html. Accessed on April 28, 2014.

Statistics Canada. 2011. Cattle and calves, average number per farm reporting. Table 10. Available online at http://www.statcan.gc.ca/pub/23-012-x/2010002/t091-eng.htm. Accessed on April 28, 2014.

Stephens, T. P., G. H. Loneragan, E. Karunasena, and M. M. Brashears. 2007. Reduction of *Escherichia coli* O157 and *Salmonella* in feces and on hides of feedlot cattle using various doses of a direct-fed microbial. *Journal of Food Protection* 70:2386-2391.

Tabe, E. S., J. Oloya, D. K. Doetkott, M. L. Bauer, P. S. Gibbs, and M. L. Khaitsa. 2008. Comparative effect of direct-fed microbials on fecal shedding of *Escherichia coli* O157:H7 and *Salmonella* in naturally infected feedlot cattle. *Journal of Food Protection* 71:539-544.

Tapiero, H., L. Gate, and K. D. Tew. 2001. Iron: Deficiencies and requirements. *Biomedicine and Pharmacotherapy* 55:324-332.

Umberger, W. J., P. C. Boxall, and R. C. Lacy. 2009. Role of credence and health information in determining US consumers' willingness-to-pay for grass-finished beef. *Australian Journal of Agricultural and Resource Economics* 53:603-623.

UN (United Nations). 1987. *Our Common Future*. United Nations World Commission on Environment and Development. Oxford University Press. Available online at http://www.un-documents.net/our-common-future.pdf. Accessed on May 5, 2015.

UN. 2005. *2005 World Summit Outcome*. New York: United Nations Publications.

UN. 2011. *World Population Prospects: The 2010 Revision, Volume I: Comprehensive Tables*. U.N. Department of Economic and Social Affairs, Population Division. Available online at http://esa.un.org/unpd/wpp/Documentation/pdf/WPP2010_Volume-I_Comprehensive-Tables.pdf. Accessed on November 26, 2014.

USDA (U.S. Department of Agriculture). 1997. U.S. Standard for Grades of Carcass Beef. USDA Agricultural Marketing Service, Livestock and Seed Division. Available online at http://www.ams.usda.gov/AMSv1.0/getfile?dDocName=STELDEV3002979. Accessed on May 30, 2014.

USDA. 2006. Crop Residue Removal for Biomass Energy Production: Effects on Soils and Recommendations. Soil Quality—Agronomy Technical Note No. 19. USDA Natural Resources Conservation Service. Available online at http://www.nrcs.usda.gov/Internet/FSE_DOCUMENTS/nrcs142p2_053253.pdf. Accessed on May 30, 2014

USDA. 2008a. *Beef 2007-08 Part I: Reference of Beef Cow-Calf Management Practices in the United States, 2007-08*. USDA Animal and Plant Health Inspection Service. Available online at http://www.aphis.usda.gov/animal_health/nahms/beefcowcalf/downloads/beef0708/Beef0708_dr_PartI_rev.pdf. Accessed on May 30, 2014.

USDA. 2008b. Grass-Fed Marketing Claim Standard. USDA Agricultural Marketing Service. Available online at http://www.ams.usda.gov/AMSv1.0/ams.fetchTemplateData.do?template=TemplateN&navID=GrassFedMarketingClaimStandards&rightNav1=GrassFedMarketingClaimStandards&topNav=&leftNav=GradingCertificationandVerfication&page=GrassFedMarketingClaims&resultType=&acct=lss. Accessed on May 30, 2014.

USDA. 2009a. *Beef 2007-08 Part II: Reference of Beef Cow-Calf Management Practices in the United States, 2007-08*. USDA Animal and Plant Health Inspection Service. Available online at http://www.aphis.usda.gov/animal_health/nahms/beefcowcalf/downloads/beef0708/Beef0708_dr_PartII.pdf. Accessed on May 30, 2014.

USDA. 2009b. *Beef 2007-08 Part III: Changes in the U.S. Beef Cow-Calf Industry, 1993-2008*. USDA Animal and Plant Health Inspection Service. Available online at http://www.aphis.usda.gov/animal_health/nahms/beefcowcalf/downloads/beef0708/Beef0708_dr_PartIII.pdf. Accessed on May 30, 2014.

USDA. 2010a. *Beef 2007-08 Part IV: Reference of Beef Cow-Calf Management Practices in the United States, 2007-08*. USDA Animal and Plant Health Inspection Service. Available online at http://www.aphis.usda.gov/animal_health/nahms/beefcowcalf/downloads/beef0708/Beef0708_dr_PartIV.pdf. Accessed on May 30, 2014.

USDA. 2010b. *Dietary Guidelines for Americans, 2010*. USDA Center for Nutrition Policy and Promotion. Available online at http://www.cnpp.usda.gov/DGAs2010-PolicyDocument.htm. Accessed on May 30, 2014.

USDA. 2013a. *Alternative Beef Production Systems: Issues and Implications*. USDA Economic Research Service. Available online at http://

www.ers.usda.gov/media/1071057/ldpm-218-01.pdf. Accessed May 20, 2014.

USDA. 2013b. *Feedlot 2011: Part III: Trends in Health and Management Practices on U.S. Feedlots, 1994-2011*. USDA Animal and Plant Health Inspection Service. Available online at http://www.aphis.usda.gov/animal_health/nahms/feedlot/downloads/feedlot2011/Feed11_dr_Part%20III.pdf. Accessed on May 16, 2014.

USDA. 2013c. Grading, Certification and Verification; LPS Process Verified Program. USDA Agricultural Marketing Service. Available online at http://www.ams.usda.gov/AMSv1.0/ams.fetchTemplateData.do?template=TemplateN&navID=GradingCertificationandVerfication&leftNav=GradingCertificationandVerfication&page=ProcessVerified.usda.govHomePage. Accessed on May 30, 2014.

USDA. 2013d. Meat and Poultry Labeling Terms. USDA Food Safety and Inspection Service. Available online at http://www.fsis.usda.gov/wps/portal/fsis/topics/food-safety-education/get-answers/food-safety-fact-sheets/food-labeling/meat-and-poultry-labeling-terms/meat-and-poultry-labeling-terms. Accessed on May 30, 2014.

USDA. 2013e. *Summary Report: 2010 National Resources Inventory*. Natural Resources Conservation Service, Washington, DC, and Center for Survey Statistics and Methodology, Iowa State University, Ames, IA. Available online at http://www.nrcs.usda.gov/Internet/FSE_DOCUMENTS/stelprdb1167354.pdf. Accessed on December 3, 2014.

USDA. 2013f. USDA Nutrient Data Set for Retail Beef Cuts from SR, Release 3.0. Nutrient Data Laboratory. Agricultural Research Service, USDA Agricultural Marketing Service. Available online at http://www.ars.usda.gov/services/docs.htm?docid=18961. Accessed on May 5, 2014.

USDA. 2014a. *2012 Census of Agriculture, Volume 1. Geographic Area Series: Part 51 (AC-12-A-51)*. U.S. Department of Agriculture, National Agricultural Statistics Service (NASS). Available online at http://www.agcensus.usda.gov/Publications/2012/Full_Report/Volume_1,_Chapter_1_US/usv1.pdf. Accessed on June 6, 2014.

USDA. 2014b. Livestock and Poultry: World Markets and Trade. USDA Foreign Agricultural Service. Available online at http://apps.fas.usda.gov/psdonline/circulars/livestock_poultry.pdf. Accessed on April 28, 2014.

USDA. 2014c. National Organic Program Standards. USDA Agricultural Marketing Service. Available online at http://www.ams.usda.gov/AMSv1.0/nop. Accessed on May 16, 2014.

USDA. 2015. USDA National Nutrient Database for Standard Reference Release 27. USDA Agricultural Research Service. Available online at http://www.ars.usda.gov/Services/docs.htm?docid=8964. Accessed on June 16, 2015.

Van Loo, E. J., W. Alali, and S. C. Ricke. 2012. Food safety and organic meats. *Annual Review of Food Science and Technology* 3:203-225.

Varel, V. H., J. E. Wells, E. D. Berry, M. J. Spiehs, D. N. Miller, C. L. Ferrell, S. D. Shackelford, and M. Koohmaraie. 2008. Odorant production and persistence of *Escherichia coli* in manure slurries from cattle fed zero, twenty, forty, or sixty percent wet distillers' grains with solubles. *Journal of Animal Science* 86:3617-3627.

Vasconcelos, J. T., and M. L. Galyean. 2007. Nutritional recommendations of feedlot nutritionists: The 2007 Texas Tech University survey. *Journal of Animal Science* 85:2772-2781.

Vasconcelos, J. T., N. A. Elam, M. M. Brashears, and M. L. Galyean. 2008. Effects of increasing dose of live cultures of *Lactobacillus acidophilus* (Strain NP 51) combined with a single dose of *Propionibacterium freudenreichii* (Strain NP 24) on performance and carcass characteristics of finishing beef steers. *Journal of Animal Science* 86:756-762.

Welch, A. A., S. Shakya-Shrestha, A. H. Lentjes, J. N. Wareham, and K. T. Khaw. 2010. Dietary intake and status on n-3 polyunsaturated fatty acids in a population of fish-eating and non-fish-eating meat-eaters, vegetarians, and vegans and the precursor-product ratio of α-linolenic acid to long-chain n-3 polyunsaturated fatty acids: Results from the EPIC-Norfolk cohort. *American Journal of Clinical Nutrition* 92:1040-1051.

Wells, J. E., S. D. Schackelford, E. D. Berry, N. Kalchayanand, M. N. Guerini, V. H. Varel, T. M. Arthur, J. M. Bosilevac, H. C. Freetly, T. L. Wheeler, C. L. Ferrell, and M. Koohmaraie. 2009. Prevalence and level of *Escherichia coli* O157:H7 in feces and on hides of feedlot steers fed diets with or without wet distillers grains with solubles. *Journal of Food Protection* 72:1624-1633.

Wertz, E., L. L. Berger, P. M. Walker, D. B. Faulkner, F. K. McKeith, and S. Rodriguez-Zas. 2001. Early weaning and postweaning nutritional management affect feedlot performance of Angus × Simmental heifers and the relationship of 12th rib fat and marbling score to feed efficiency. *Journal of Animal Science* 79:1660-1669.

Westerterp-Plantenga, M. S., A. Nieuwenhuizen, D. Tome, S. Soenen, and K. R. Westerterp. 2009. Dietary protein, weight loss, and weight maintenance. *Annual Review of Nutrition* 29:21-41.

WHO (World Health Organization). 2011. *Critically Important Antimicrobials for Human Medicine,* 3rd Rev. Geneva, Switzerland: World Health Organization. Available online at http://apps.who.int/iris/bitstream/10665/77376/1/9789241504485_eng.pdf. Accessed on November 28, 2014.

Wileman, B. W., D. U. Thomson, C. D. Reinhardt, and D. G. Renter. 2009. Analysis of modern technologies commonly used in beef cattle production: Conventional beef production versus nonconventional production using meta-analysis. *Journal of Animal Science* 87:3418-3426.

Wolfe, R. R., S. L. Miller, and K. B. Miller. 2008. Optimal protein intake in the elderly. *Clinical Nutrition* 27:675-684.

Wolmarans, P., J. A. Laubscher, S. van der Merwe, J. A. Kriek, C. J. Lombard, M. Marais, H. H. Vorster, H. Y. Tichelaar, M. A. Dhansay, and A. J. Spinnler Benade. 1999. Effects of a prudent diet containing either lean beef and mutton or fish and skinless chicken on the plasma lipoproteins and fatty acid composition of triacylglycerol and cholesteryl ester of hypercholesterolemic subjects. *Journal of Nutritional Biochemistry* 10:598-608.

World Organisation for Animal Health. 2015. OIE List of Antimicrobial Agents of Veterinary Importance. Paris: OiE. Available online at http://www.oie.int/fileadmin/Home/eng/Our_scientific_expertise/docs/pdf/Eng_OIE_List_antimicrobials_May2015.pdf. Accessed on November 10, 2015.

Younts-Dahl, S. M., M. L. Galyean, G. H. Loneragan, N. A. Elam, and M. M. Brashears. 2004. Dietary supplementation with lactobacillus- and propionibacterium-based direct-fed microbials and prevalence of *Escherichia coli* O157 in beef feedlot cattle and on hides at harvest. *Journal of Food Protection* 67:889-893.

Younts-Dahl, S. M., G. D. Osborn, M. L. Galyean, J. D. Rivera, G. H. Loneragan, M. M. Brashears. 2005. Reduction of *Escherichia coli* O157 in finishing beef cattle by various doses of *Lactobacillus acidophilus* in direct-fed microbials. *Journal of Food Protection* 1:6-10.

Zimmermann, M. B., and R. F. Hurrell. 2007. Nutritional deficiency. *The Lancet* 370:511-520.

Zurek, L., and A. Ghosh. 2014. Insects represent a link between food animal farms and the urban environment for antibiotic resistance traits. *Applied and Environmental Microbiology.* DOI:10.1128/AEM.00600-14. Available online at http://aem.asm.org/content/early/2014/04/01/AEM.00600-14.full.pdf+html. Accessed on May 5, 2014.

2

Anatomy, Digestion, and Nutrient Utilization

INTRODUCTION

The ruminant has a digestive tract that allows digestion of plant cell wall biomass, such as forages and fibrous byproduct feeds. Long-term health and productivity of cattle depends on the consumption of fiber to stimulate normal digestive function, although feedlot cattle can be very productive when fed low-fiber, high-grain diets. Ruminants themselves do not produce the enzymes required for the degradation of complex plant cell wall polysaccharides. Rather, they have developed a symbiotic relationship with anaerobic (i.e., oxygen intolerant) microorganisms that establish residence in the ruminant digestive tract soon after birth. Fiber is selectively retained in the reticulorumen (often referred to simply as the rumen), the main pregastric site of digestion, where it is anaerobically fermented by microorganisms. Rumination is another important distinguishing characteristic of ruminants. Rumination allows the animal to consume feed quickly when available, and later regurgitate and rechew the feed, thereby decreasing particle size and increasing surface area, which is necessary for microbial digestion and passage from the rumen.

The symbiotic relationship between the animal, its microbial population, and diet are critical to effective utilization of feedstuffs. Microorganisms in the reticulorumen ferment plant carbohydrates ingested by the animal, yielding end products of fermentation that are important precursors and regulators for metabolism and synthesis. Through this process, beef cattle have the ability to transform cellulosic materials and relatively low-quality dietary protein into high-quality meat for human consumption. As such, ruminants utilize marginal grasslands, pastures, conserved forages, fibrous byproduct feeds, and nonprotein nitrogen (NPN) sources that cannot be used by other major food-producing species.

Fermentation of feed in the reticulorumen offers unique advantages to the host animal. It provides energy from fibrous materials and supports microbial growth, which in turn provides the animal with amino acids (AA) and B-complex vitamins. However, the anaerobic environment in the digestive tract results in the production of end products that are not completely oxidized. Anaerobic microorganisms in the reticulorumen produce carbon dioxide (CO_2), methane (CH_4), volatile fatty acids (VFA), and microbial cells as end products. The VFA are mainly acetate, propionate, and butyrate, with lesser amounts of formate, isobutyrate, 2-methylbutyrate, isovalerate, valerate, and caproate. Other products are also produced (e.g., ethanol, lactate, succinate), but concentrations are usually very low because they are used by other microorganisms as substrates (a process known as cross-feeding) (Russell, 2002).

Fortuitously, ruminants have evolved mechanisms that allow the digestion, absorption, and metabolism of the VFA. There are also disadvantages associated with pregastric fermentation. Anaerobic fermentation only produces 3 to 4 moles of adenosine triphosphate (ATP) per mole of glucose fermented for microbial use, as opposed to 30 moles of ATP from the aerobic metabolism of glucose (McDonald et al., 2010, Table 9.1). The ruminal microbes extract sufficient ATP for growth, thereby providing the animal with energy and protein for physiological processes. In addition, significant losses of energy in the form of CH_4, heat of fermentation, and production of indigestible microbial cell wall occur during fermentative digestion; substrates such as starch, sugar, and proteins are used less efficiently in the reticulorumen than in the small intestine (SI); biological value of high-quality proteins is decreased; and carbohydrate fermentation is not always well controlled, and as such, can result in metabolic dysfunction (e.g., ruminal acidosis).

Ruminants in nature can be divided into three main groups based on their feeding preferences: concentrate selectors, intermediate feeders, and forage eaters (Hofmann, 1988). Although all three feeding groups possess some ability to digest cellulosic materials, forage eaters (e.g., cattle and sheep) have highly developed digestive systems that allow them to consume and digest predominantly cellulosic diets. In contrast, concentrate selectors (e.g., white-tailed deer, moose) consume highly digestible diets like new-growth leaves and fruits from browse plants that are higher in starch, sugar, and lipid content, whereas intermediate feeders (e.g., goats) consume a mix of more digestible plant components and lower-quality forages, with these animals often being very selective. In comparison, cattle are considered to be relatively nonselective, and although they possess the ability to consume all-roughage diets, they also can digest concentrates and mixed diets.

This chapter provides a brief review of the anatomy and digestive physiology of the gastrointestinal tract of cattle including a general discussion of animal, diet, and microbial factors involved in ruminal function and nutrient digestion kinetics. Understanding the form and function of the reticulorumen is very important because these dynamics are ultimately responsible for providing nutrients to the host animal.

MECHANICS AND ROLE OF EATING

Having evolved as forage eaters, cattle have short, relatively immobile lips and a relatively small mouth opening compared with ruminants classified as concentrate selectors (Hofmann, 1988). The mouth, with its lips, tongue, lower incisor teeth, and dental pad in front of the hard palate, allow cattle to prehend and masticate feed during eating (Hofmann, 1988). Although incisors are absent from the upper jaw in cattle, the dental pad aids the tongue in grasping feed during ingestion, and the morphology of the tongue and hard pallet allow processing of feed in the mouth.

Once in the mouth, the feed is chewed by a sideways swinging of the mandible, causing a lateral chewing action that shears, rather than cuts, the plant tissues. Cattle chew only with their molar teeth and only on one side of their mouth at a given time (Hofmann, 1988). The shearing of feed during eating releases cell solubles and exposes the internal plant cell walls to microbial digestion once the feed particles enter the reticulorumen (Beauchemin, 1992). During the chewing process, a large quantity of saliva is secreted, which enables bolus formation and swallowing. Thus, chewing during eating plays a key role in the digestion and passage of feed through the gastrointestinal tract of beef cattle because it helps decrease feed particle size and increases salivation rate.

ANATOMY AND PHYSIOLOGY OF THE GASTROINTESTINAL TRACT

The main distinguishing feature of the ruminant digestive system is a four-compartment stomach, but all portions of the gastrointestinal tract show some degree of specialization and adaptation (Hofmann, 1988). A detailed description of the anatomy of the gastrointestinal tract in cattle is given by Hofmann (1988) and Budras et al. (2003). As shown in Figure 2-1, the stomach of cattle is composed of the reticulum, rumen, omasum, and abomasum. The digestive tract accounts for a larger proportion of total body weight in ruminants than it does in most other mammals (Van Soest, 1994). In beef cattle, the volume of contents in the rumen is <1 L at birth, increasing to approximately 7 L at 3 months of age, and approximately 60 L at maturity (Lyford, 1988), and accounts for approximately 70% of the total digestive tract volume.

Components of the Ruminant Stomach

The first three compartments of the ruminant stomach (reticulum, rumen, and omasum) are considered the forestomach, as they are the site of anaerobic fermentation. The reticulum is a small but distinct pouch connected to the craniad portion of the rumen (Figure 2-1). Feed boluses enter the reticulum after swallowing and passage through the esophagus. The reticulum is easily distinguished by its unique honeycomb epithelial lining. The rumen is the largest compartment of the forestomach and, together with the reticulum, forms one large, functionally integrated sac (i.e., reticulorumen) containing microbes, digesta, and gas. The reticulorumen occupies most (greater than three-fourths) of the left side of the abdominal cavity, except a small area occupied by the spleen and SI. The left side of the reticulorumen rests on the diaphragm, abdominal wall, and spleen, whereas the right side is more irregular and rests on the liver, left kidney, omasum, abomasum, pancreas, aorta, and posterior vena cava. The peritoneum and connective tissue firmly attach this organ to the diaphragm and sublumbar muscles. The reticulum rests against the diaphragm on the left side of the median plane and is connected to the omasum by a short narrow neck called the reticulo-omasal orifice. The omasum is spherical and lies on the right side. The abomasum is an elongated sac, which lies primarily on the abdominal floor.

Grooves, or internal pillars, separate the reticulorumen into sacs (Figure 2-2). The most conspicuous is the rumino-reticular fold or groove, which separates the reticulum from the rumen. Although the rumino-reticular groove is prominent, dorsally there is no separation between the two compartments, and they join at the entrance of the cardia, which is the terminal end of the esophagus and entrance into the reticulorumen. Thus, the opening between the reticulum and rumen is large, allowing digesta to move easily between the two compartments. The rumen has five main compartments: cranial sac (or atrium), dorsal sac, caudodorsal blind sac, ventral sac (or ruminal recess) and caudoventral blind sac. Cranial, longitudinal, and caudal pillars or grooves form an almost complete circle surrounding the interior rumen externally, dividing the rumen into dorsal and ventral sacs. Dorsal and ventral coronary grooves separate the caudodorsal blind sac and the caudoventral blind sac from the dorsal and ventral regions of the rumen, respectively. The pillars are muscles that aid in the movement and contraction of the reticulorumen. When contracted, the pillars are well defined, but when relaxed, they are not prominent. The ventricular (or reticular) groove begins at the cardia and extends 18 to 20 cm, ending at the reticulo-omasal orifice, which

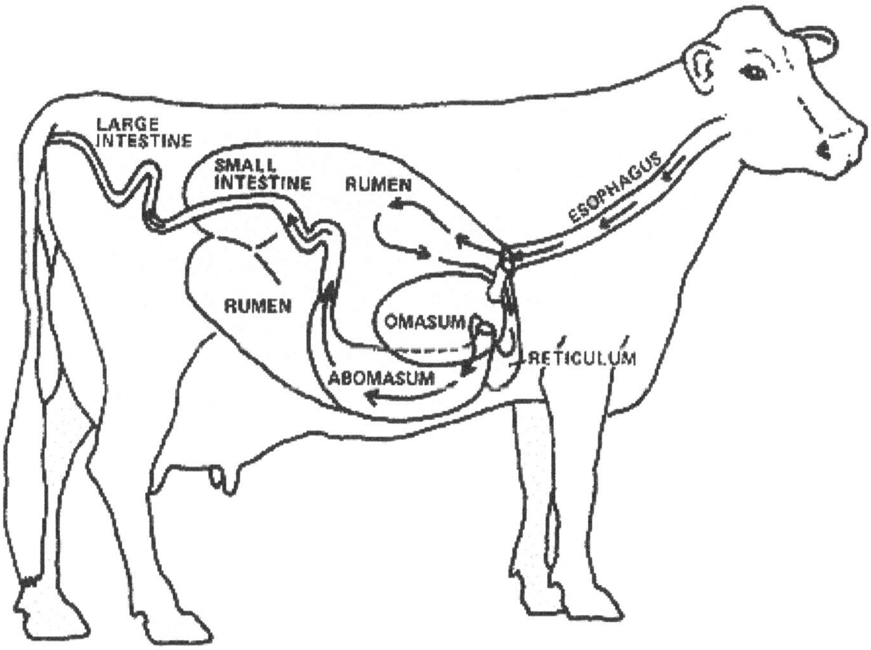

FIGURE 2-1 The digestive system of the beef cow. Courtesy of the University of Minnesota Extension.

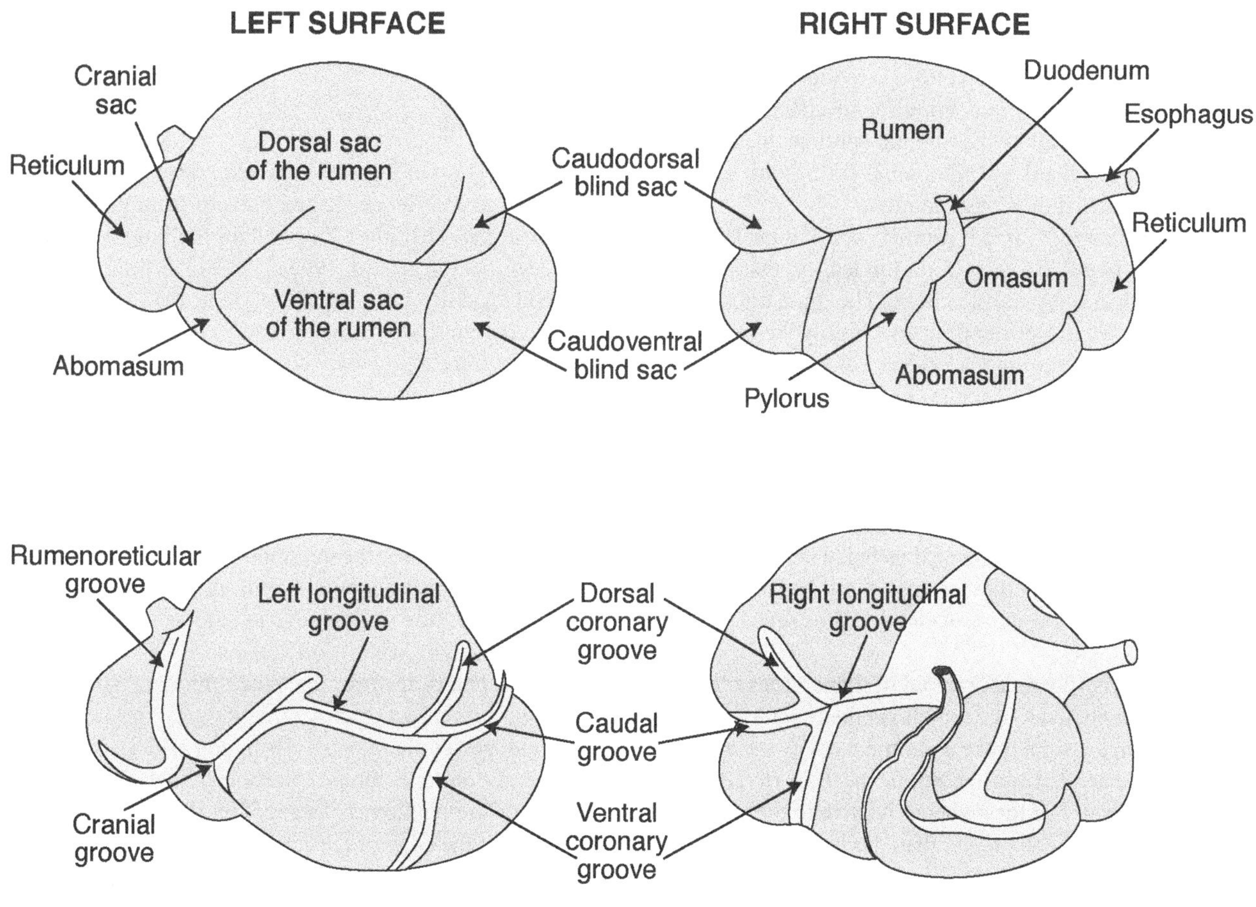

FIGURE 2-2 The reticulum, rumen, omasum, and abomasum of the bovine. Adapted from the University of Bristol (1998).

connects the rumen to the omasum. Normally, the ventricular groove is open, but closure of the lips of the groove results in a tube that connects the cardia with the reticulo-omasal opening, allowing liquids to pass to the omasum and bypass the reticulorumen.

The reticulorumen and omasum are lined with a stratified, squamous epithelium (Hofmann, 1988). The cells are not glandular, and it is thought that they do not produce mucus. The epithelium has important physiological functions, including absorption, transport, metabolic activity, and protection by creating a major protective barrier separating the external from the internal environment. The cellular structure of the rumen epithelium is complex, consisting of four distinct cell layers. From the luminal surface these are the stratum corneum, stratum granulosum, stratum spinosum, and stratum basale (Steven and Marshall, 1970; Graham and Simmons, 2005). When examined using light microscopy, these layers seem to merge into each other without distinct boundaries, such that only the stratum corneum and stratum basale are clearly distinct (Graham and Simmons, 2005). In the bovine rumen, the sodium-potassium (Na^+/K^+) pumps are concentrated in the cells of the stratum basale, and decrease in density as the strata approach the lumen (Graham and Simmons, 2005). Mitochondria are abundant in epithelial cells, especially in the stratum basale, which is consistent with the energy requirements of primary active ion pumps (Fell and Weekes, 1975; Graham and Simons, 2005). A permeability barrier of tight cell junctions is found in the stratum granulosum, whereas gap junctions found concentrated in the stratum spinosum allow the ruminal epithelia to function as a syncytium (Graham and Simmons, 2005) with a basal membrane positioned on a basal lamina.

The luminal surface of the ruminal epithelium is lined with numerous papillae (10 to 15 mm in length) that greatly increase the absorptive surface area. The squamous cells on the surface of the epithelium are highly keratinized, and extensive cell sloughing is evidence that the outermost keratinized cell layer (the stratum corneum) is unlikely to act as an epithelial permeability barrier. The papillae have a rich blood and lymphatic supply in intimate contact with the basal layer of the stratified squamous epithelium. The size, number, and distribution of the papillae vary with location within the rumen and the type of diet fed. The papillae are particularly well developed in the ventral sac and in the floor of the caudodorsal blind sac, as well as on the cranial groove and adjacent surfaces.

The omasum is spherical or ovoid in shape (Becker et al., 1963) and is connected to the rumen via the reticulo-omasal orifice. It lies on the right in the abdominal cavity against the diaphram and extends from the 8th to the 10th rib (Becker et al., 1963). The size of the omasum relative to the reticulorumen in cattle is distinctly larger than it is in sheep and goats (Engelhardt and Hauffe, 1975). The function of the omasum is not well understood, but it is thought to act as a bottleneck for the outflow of particles from the reticulorumen. The inner structure of the omasum contains numerous parallel sheets of tissue, or laminae, of various sizes that extend into the omasal canal (Bost, 1970). The laminae bear small papillae that further increase the absorptive surface area of the omasum. It is estimated that omasal absorption of H_2O, Na^+, K^+, and NH_3 is about 20 to 30% of that absorbed from the reticulorumen, whereas absorption of VFA from the omasum accounts for 10 to 15% of total VFA absorption (Engelhardt and Hauffe, 1975). The omasum also functions to decrease particle size of feeds before passing to the abomasum (Becker et al., 1963).

The abomasum is pear-shaped and consists of fundic, body, and pyloric regions (Leek, 2004). It is connected to the omasum via the omasal-abomasal orifice, which is 3 to 4 times larger than the reticulo-omasal orifice. A constriction called the pylorus separates the abomasum from the duodenum. The internal surface of the abomasum is characterized by flaps of tissue or folds that run spirally over the fundic and body parts, decreasing in height toward the pyloric region. This organ is different from the rumen, reticulum, and omasum in that it contains secretory tissue. The glandular gastric mucosa is supplied with specialized secretory cells that produce mucus, pepsinogen, and hydrochloric acid (HCl). The fundic region contains gastric glands (HCl-producing parietal cells) and the pyloric region contains thick glandular mucosa and the pepsin-secreting chief cells. The histology and physiology of the abomasum is similar to the nonruminant stomach.

Small Intestine

The SI is an elongated tube extending from the pyloric sphincter to the ileocecal junction. In ruminants, the SI is approximately 20 times longer than the length of the body of the animal (Argenzio, 1993). The SI has three regions that proceed caudally from the abomasum: the duodenum, jejunum, and ileum. Stomach contents pass from the abomasum and through the SI by way of peristaltic muscle contractions that start at the pyloric sphincter. The ruminant intestine is contained in the supraomental recess to the right of the rumen (Hofmann, 1988).

The duodenum makes a loop as it turns to cross from the right to the left side, forming an S-shaped curve (sigmoid flexure) that contains the common bile and pancreatic ducts. The descending duodenum begins at the craniad portion of the sigmoid flexure and moves caudodorsally. The duodenum passes from right to left around the caudal border of the root of the mesentery to form the caudal flexure. The ascending duodenum runs in a cranial direction to join the jejunum at the duodeno-jejunal flexure (Hofmann, 1988).

The middle and longest portion of the SI is called the jejunum. The jejunum is arranged in the form of numerous loops along the edge of the common mesenteric plate, which carries the spiral colon. The most craniad jejunal coils lie adjacent to the rumen, abomasum, omasum, and pancreas. The caudal jejunal coils are attached to a long extension of the mesenteric plate, which permits them to move outside the supraomental recess. The ileum is the most distal and final section of the SI that enters the cecum at the ileocecal junction. Peyer's patches are organized lymphoid nodules found in the ileum, and can be used to differentiate the ileum from the duodenum and jejunum.

The muscular tunic of the SI is composed of a thicker inner circular and a weaker outer longitudinal layer (Hofmann, 1988). The inside layer of the SI is composed of an epithelial cell layer known as the mucosa. The submucosa is a connective tissue layer that provides space for blood vessels, lymph vessels, and nerve fibers. The mucosa can shift on the muscular tube through the loose connective tissue of the submucosal layer, which permits the formation of temporary folds. These folds, along with their covering of intestinal villi on the muscularis mucosa, increase the surface area for nutrient absorption. In addition, individual epithelial cells that cover the villi have their own microvilli on the luminal surface. The microvilli greatly amplify the surface area of the SI and constitute the brush border. The villi are fused at their base portions forming villous crests. Between adjacent villi, the mucosa embeds simple tubular crypts of Lieberkühn, which extend down to the lamina muscularis mucosa. New cells for the villi are provided by cell migration from the crypts of Lieberkühn toward the tips of the villi.

Large Intestine

The large intestine consists of the cecum and colon. The cecum and colon are distinguished from other segments of the gastrointestinal tract by the absence of villi, by semicircular mucosal folds, by an increasing proportion of goblet cells, and by long tubular crypts of Lieberkühn (Hofmann, 1988). Although individual lymph nodules are found in all portions of the intestinal mucosa, there is a concentration of Peyer's patches (aggregate nodules) in the ileum, which continue into the cecal mucosa. The cecum is a blind sac that extends caudodorsally from the ileocecal junction along the right flank. Its rounded blind end lies at the right side of the pelvic inlet. The cecum is attached dorsally by a short cecocolic fold to the proximal loop of the colon. The colon continues from the cecum to its termination at the anus. The colon consists of ascending, transverse, and descending parts. The initial portion of the colon is wide and forms the first two limbs of the ansaproximalis coli, an S-shaped tube, which at first is the same size as that of the cecum, but decreases in size before it continues as the spiral colon. Ventral to the right kidney, the proximal loop of the colon doubles back on itself dorsolaterally between the cecum and the descending duodenum. Then it doubles on itself once more mediodorsally, and passes in a cranial direction on the left side of the mesentery to become continuous with the spiral loop. Its short continuation is the transverse colon, which is related dorsally to the pancreas. The descending colon stays on the left side of the cranial mesenteric root in a high dorsal position. It is attached to the ascending duodenum. On its caudal course its mesentery gradually gets longer, resulting in a ventral curve called the colon sigmoideum. The longer mesentery permits rectal palpation in larger ruminants.

The descending colon is followed by the anus. The descending colon located within the pelvis is known as the rectum. The muscular tunic and the muscularis mucosae reach their maximum thickness in the retroperitoneal portion of the rectum. Subsequently, this musculature condenses to form the internal anal sphincter, which is added to by the voluntary external anal sphincter. The anus is the junction of the terminal digestive tract with the skin. A muscular sphincter composed of both smooth and striated muscle allows the anus to close (Hofmann, 1988).

Reticuloruminal Motility

Coordinated cyclical contractions of the reticulorumen play an important role in the digestion process. Contractions mix newly ingested food with the ruminal contents, thereby helping to ensure microbial access to feed particles. Movement of the ruminal contents helps position water and end products of digestion near the epithelium for absorption and, likewise, moves secretions of bicarbonate from the epithelium into the ruminal contents. Contractions of the reticulorumen are an essential component of eructation and rumination. Finally, contractions propel fluid, soluble compounds, microbial cells, and undigested feed residues into the omasum. If motility is suppressed for a significant length of time, ruminal impaction (i.e., the rumen becomes abnormally packed with digesta, leading to cessation of eating, bloating, and eventually death) can result.

The muscular activity of the reticulorumen is a combination of primary contractions, referred to as the mixing cycle, and secondary contractions, called eructative contractions. The primary cycle of contractions is initiated by a sharp contraction of the reticulum and the reticuloruminal fold, causing the reticulum to contract to about half its normal size. A second, more powerful contraction of the reticulum follows, leading to a series of contractions of the various pillars and sacs of the rumen. This wave of contractions is followed by a wave of relaxation, so as parts of the rumen are contracting, other sacs are dilating. The complete sequence of events has been documented using radiography and is reported by Wyburn (1980). The contractions cause the ingesta to flow from the reticulum into the cranial sac, from the cranial sac into the dorsal sac, and from the dorsal sac back to the cranial sac, and then to the reticulum, or in the case of less buoyant material, into the ventral sac (Ruckebusch, 1988). The entire cycle of contractions is completed in about 30 to 50 s when the animal is resting (Ruckebusch, 1988). Muscular activity of the reticulorumen is influenced by eating, ruminating, bloat, and deprivation of food or water, and in some cases, only partial cycles occur.

Secondary contractions occur in parts of the rumen and are associated with eructation, the process whereby gases (mainly CH_4 and CO_2) produced in the rumen during microbial digestion of feed are expelled (Ruckebusch, 1988). Unlike primary contractions, secondary contractions do not involve contraction of the reticulum. Eructation is stimulated by gas pressure in the rumen and occurs with contraction of the caudoventral blind sac followed by a contraction running in a cranial direction across the rumen displacing gas for eructation. Eructation of gases occurs as the dorsal sac contracts and the cardia relaxes, moving the gases from the dorsal rumen into the esophagus. About one-third of the gas in the rumen is exhaled directly, while the remainder is shunted into the lungs where it is respired through the nose and mouth (Dougherty et al., 1962; Ruckebusch, 1988). Some of the eructated CH_4 and CO_2 reaching the lungs are absorbed into the blood (Dougherty et al., 1964). Thus, the exhalation pattern of ruminants comprises a low but regular flow of respired CO_2 and CH_4, with less frequent but strong spikes of eructated CO_2 and CH_4 (Ulyatt et al., 1999). Unlike belching in humans, eructation is an almost silent process.

Mechanics and Role of Rumination

Rumination or cud chewing is an important distinguishing characteristic of ruminants. It is the process of rechewing ruminal contents that were swallowed in previous meals and is characterized by the cyclical process of regurgitation, remastication, and reswallowing (Ruckebusch, 1988). Digesta is regurgitated into the mouth as a result of an extra contraction of the reticulum that precedes the usual biphasic contraction of the mixing cycle (i.e., primary contractions), combined with negative pressure within the esophagus caused by inspiration with the glottis closed (Ruckebusch, 1988). At the height of this initial contraction, digesta passes through the cardia into the esophagus and up to the mouth (Wyburn, 1980). Some of the regurgitated liquid and small particles are swallowed during the rechewing process. The remaining material is remasticated for about 30 to 70 s, and the bolus is reswallowed. One to three seconds later, the next bolus is regurgitated, and the cycle is repeated. Bailey and Balch (1961a) reported that during a period of rumination, mastication occurred on only one side of the mouth, but that mastication on one side was invariably followed by mastication on the other side during the next period of rumination. Chewing during rumination is generally slower, and more deliberate, than during eating.

Rumination periods (or bouts) range from 30 s to more than 2 h, with 10 to 20 periods each day, to a maximum of about 8 to 9 h/d for forage-fed cattle (Welch, 1982). Cattle can ruminate 20 to 40 g/d of plant cell wall constituents per kilogram of metabolic weight depending on the age, with older cattle ruminating more fiber relative to body weight than younger ones (Welch, 1982).

The pattern of rumination is repetitive from day to day. There is a circadian rhythm of rumination, but most rumination occurs at night and during periods of drowsiness (Beauchemin, 1991). The number of rumination periods, the duration of each period, the number of ruminated boli, and the total rumination time each day are proportional to particle size and cell wall content of the diet. Feeding concentrates and finely ground forages decreases rumination time; however, the urge to ruminate in cattle is very strong, and rumination occurs even in feedlot cattle fed grain-based diets. Beauchemin et al. (1994) reported that mature cattle fed all-concentrate diets ruminated 1.25 h/d when fed whole corn grain, and 2 to 2.5 h/d when fed whole wheat or whole barley grain, whereas Hironaka et al. (1992) reported that smaller cattle ruminated 6 h/d when fed whole barley grain.

Rumination is triggered or inhibited when the gastric centers in the brain integrate neural activity originating from tension receptors in the muscle layers and mucosal (epithelial) receptors in the luminal surface of the reticulorumen (Ash and Kay, 1959; Leek and Harding, 1975; Leek, 1986). The tension receptors activate reticuloruminal motility and are excited by mild tactile stimulation and low to moderate distension of the rumen by the ruminal contents. The epithelial receptors slow down ruminal contractions and respond to extreme distension of the rumen, such as that which occurs during bloat, and various chemical stimuli, such as elevated concentrations of VFA and urea, and high osmotic pressure (Ash, 1959; Gregory, 1987). Thus, feeding highly fermentable diets can be inhibitory to rumination and ruminal motility, and causes the period of latency after feeding before rumination begins, despite a large quantity of feed in the rumen. Likewise, excessive fermentation rates leading to high concentrations of VFA cause ruminoreticular stasis in the early stages of ruminal acidosis (Leek, 1986).

The function of mastication during eating and rumination differ. Eating is primarily responsible for preparing food for swallowing, releasing soluble constituents, and damaging plant tissues for microbial digestion, whereas rumination has the primary function of decreasing particle size and dry weight of individual particles so that refractory material can pass from the reticulorumen (Ulyatt et al., 1986). Thus, rumination, and not eating, is usually the principal means by which feed particles are decreased in size by cattle (Lee and Pearce, 1984; McLeod and Minson, 1988). Although microbial digestion in the rumen contributes to decreasing the dry weight of individual feed particles, its direct contribution to decreasing the length of particles is small (<20%; McLeod and Minson, 1988). It mainly serves to weaken the cell wall structure so that particle breakdown during rumination is facilitated.

The efficiency of comminution (measured as the reduction of large particles to a size that would enable passage from the rumen) during eating is lower for cattle than sheep. Cattle have a lower frequency of chewing and are less effective in decreasing particle size than sheep (Ulyatt et al.,

1986). It is estimated that in cattle, 15 to 55% of particle size reduction occurs during eating (as summarized by Beauchemin, 1991). Thus, more than half the swallowed feed particles require further particle size reduction during rumination. Because cattle do not decrease the particle size of feed to a threshold size before swallowing during eating, the reticulorumen is presented with a range of particle sizes in newly ingested feed. The particle size of the swallowed masticate during eating depends on the rate of ingestion and the initial particle size, moisture content, and the chemical composition of the feed (Murphy and Kennedy, 1993), and to some extent, the physical and chemical differences between forages predetermine their response to comminution during chewing. In general, larger feed particles are decreased in size to a greater extent during eating than smaller ones, fresh forage is more resistant than hay to breakdown during eating, and less digestible forages are decreased in size to a greater extent than highly digestible forages (Dong, 1990, as cited by Beauchemin, 1991), although some exceptions have been noted (e.g., barley straw; Lee and Pearce, 1984). For ground and finely chopped forages, very little particle size reduction occurs during eating. There is also considerable variability among animals in terms of the extent to which they reduce particle size during eating (Lee and Pearce, 1984), which could be related to anatomical differences such as mouth size, jaw movement, teeth action, and differences in eating rate.

Salivation

Saliva is added to feed during chewing and plays an important role in the digestive function in cattle. It lubricates feed, which allows cattle to swallow particles, provides one means of recycling nitrogen (N) from plasma into the rumen, buffers the VFA produced during microbial digestion of feeds, adds fluid to the ruminal environment for fermentation, provides nutrients for the ruminal microorganisms, inhibits foam formation and prevents bloat, and facilitates passage of digesta through the gastrointestinal tract (see reviews by Church, 1988, and Beauchemin, 1991). Like most animals, cattle have three major pairs of salivary glands that differ in the type of saliva produced: parotid, submaxillary, and sublingual glands. The parotid glands, which extend from the base of the ear to the posterior end of the mandible, produce more than half the total daily saliva (Cook, 1995). Although parotid saliva is very thin and watery, it has a pH of about 8.2, and is strongly buffered between pH 6 and 7 because it contains a high bicarbonate and moderate phosphate content (McDougall, 1948). As such, saliva plays an important role in buffering pH of the rumen contents and preventing ruminal acidosis (refer to Ruminal pH and Acidosis section). Mixed saliva has a dry matter (DM) content of 1.0 to 1.4 g/100 mL (McDougall, 1948), pH of 8.3 to 8.5 (Bailey and Balch, 1961b), and contains Na (161 to 166 meq/L), K (6 to 8 meq/L), bicarbonate (113 to 125 meq/L), phosphate (6 to 180 meq/L), and other inorganic compounds (Bailey and Balch, 1961b; Cook, 1995). Anion concentration shows a slight flow rate dependency, whereas phosphate concentration is more highly variable (Cook, 1995). The N content of saliva is variable, but it is usually on the order of 0.1 to 0.2%, of which 60 to 85% is urea N (Bailey and Balch, 1961a; Church, 1988). Amylase is not present in the saliva of cattle.

Parotid saliva production increases dramatically at the start of eating, but decreases soon afterward, and then increases again during rumination. In beef cattle, the rate of saliva secretion during eating and rumination is 2 to 4 times greater than during resting (resting rate: 40 to 70 mL/min, Bailey and Balch, 1961b; 30 mL/min, Yarns et al., 1965a; 20 to 35 mL/min, Rumsey et al., 1969; 31 mL/min, Rumsey et al., 1972). In contrast, salivation during eating in dairy cows is only 1.3 to 2 times greater than during resting, but resting salivation rates are considerably higher in mature dairy cows (151 mL/min, Cassida and Stokes, 1986; 70 mL/min, Maekawa et al., 2002a; 107 mL/min, Maekawa et al., 2002b; 138 mL/min, Bowman et al., 2003) than in beef cattle. Total saliva production increases with increasing level of feed intake, but at a decreasing rate, such that the amount of saliva secreted per kilogram of intake decreases (Putnam et al., 1966). Thus, the relative role of saliva secretion in regulating ruminal pH may be less in animals fed at high levels of intake (e.g., dairy cows) compared with those fed at maintenance intake (e.g., beef cow), because production of VFA would be high, but saliva secretion relative to feed intake (and VFA production) would be low. In addition, there is considerable variability among animals in terms of salivation rate (Yarns et al., 1965b).

The amount of saliva produced by cattle each day depends on the physical and chemical characteristics of the feeds, such as moisture content, neutral detergent fiber (NDF) content, and particle size. The amount of saliva added to concentrates during eating is much less (0.76 to 1.12 mL saliva/g of DM) than for forages (3.40 to 7.23 mL saliva/g of DM; Bailey, 1961; Beauchemin et al., 2008). Among the forages, there is about a threefold difference between silage and hay (Bailey, 1961; Beauchemin et al., 2008), but when compared on a DM basis the difference is only minor. Similarly, there is a twofold difference between hay and straw but when equalized for NDF content, straw does not seem to be more effective than other forages in promoting salivation (Beauchemin et al., 2008). Feed characteristics influence total daily salivary secretion mainly by affecting the time spent eating and ruminating, rather than the salivation rate (mL/min) directly. Greater insalivation of feed occurs as eating time increases (Beauchemin et al., 2008) and the swallowed long particles entering the rumen promote rumination, which increases total salivary secretion. For example, long-particle forages (e.g., long-stemmed hay) and forages with high NDF content are consumed more slowly in longer meals compared with finely chopped forages or concentrates. Considerably less saliva is produced when cattle are fed grain diets compared with forage-based diets. Beauchemin et al. (2008) reported that cows consume concentrates about 3 to 12 times faster than forage (on a DM basis) depending on the source of forage. When particle size of feed does not slow eating rate (e.g., ground feeds), salivation, not particle size reduction, becomes the main limiting factor for swallowing. Feed needs a certain level of lubrication before swallowing (Carter and Grovum, 1990). The moisture content of swallowed masticated concentrates is typically 50 to 65% compared with 80 to 90% for forages that require mastication for particle size reduction (Balch, 1958; Beauchemin et al., 2008). The slower eating rate of forages compared with concentrates seems to be a result of the need to decrease particle size rather than lubricate feed because the lower DM content of the concentrate masticates indicates that cows are able to swallow less-well-ensalivated feed.

Increasing the time spent eating and ruminating in beef cattle is expected to increase the volume of saliva secreted each day. Although it is difficult to measure total daily saliva secretion, it can be crudely estimated from the salivation rate and total daily chewing time. The time spent eating by beef cattle is highly variable and is affected by numerous factors such as how eating time is defined and measured, level of intake (i.e., restricted vs. ad libitum feeding), feeding frequency, proportion of forage in the diet, the particle size of the feed, method of providing feed to the animal (e.g., trough, grazing), and so forth. Feedlot cattle fed high-grain diets for ad libitum intake spent from 0.5 to 5 h/d eating in most studies (Gibb et al., 1998; Krause et al., 1998; Cooper et al., 1999; Beauchemin et al., 2001; Schwartzkopf-Genswein et al., 2004), and up to 9 h/d in one study (Erickson et al., 2003). Cattle fed a mixed concentrate and forage diet spent 1 to 2 h/d eating in the study by Gonyou and Stricklin (1981) and almost 4 h/d in the study by Chase et al. (1976), whereas cattle fed a silage diet spent about 5 h/d eating (Wilson and Flynn, 1975), while cattle grazing low-quality forage spent up to 11 h/d eating (Chacon and Stobbs, 1976). Total rumination time is also highly variable, ranging from 2 to 6 h/d for feedlot cattle fed high-grain diets (Sudweeks et al., 1975; Hironaka et al., 1992; Krause et al., 1998; Beauchemin et al., 2001) and from 4 to 10 h/d for cattle fed a roughage diet, depending upon the quality of the feed (Welch et al., 1982). Total chewing time (sum of eating and ruminating) rarely exceeds 18 h/d (Kristensen and Nørgaard, 1987). Assuming a resting salivation rate of 30 mL/min and threefold higher salivation rate during chewing, total saliva production in most beef cattle would be expected to rarely exceed 100 L/d.

THE RUMINAL ENVIRONMENT AND MICROBIOME

The reticulorumen is a complex dynamic anaerobic ecosystem. Trace amounts of oxygen that enter the rumen with ingested feed, or through diffusion from the bloodstream across the rumen wall, are quickly consumed by facultative anaerobic bacteria residing in the rumen (Baldwin and Emery, 1960). Temperature in the rumen is closely regulated at approximately 39°C, and pH is usually between 5.8 and 6.8 in cattle fed predominantly forage diets. In feedlot cattle fed concentrate-based diets, pH typically drops below this range for a portion of the day, especially in cases of ruminal acidosis (see Ruminal pH and Acidosis section). Because the ruminal environment is anaerobic and oxygen is not available as an electron acceptor, other means of oxidation that are closely associated with reduction reactions are used. Ruminal fluid is highly reduced, with an oxidation-reduction potential (Eh) of –250 to –450 mv (Van Soest, 1994; Eh is expressed as the potential difference between a platinum electrode and a reference electrode of calomel or silver chloride). The Eh values reported in the literature can be confusing because other reference electrodes are sometimes used. For example, using a hydrogen (H_2) electrode, Eh values of ruminal fluid ranged from –145 to –260 mv (Marden et al., 2005). The osmolality of ruminal contents is usually maintained within a range of 280 to 300 mOsm similar to that of blood (Kahn, 2010), but it can increase up to 350 mOsm after feeding.

The feed consumed by ruminants is digested by a diverse group of anaerobic microorganisms, and in turn, some of the end products of fermentation and the microbial biomass are used by the host. The unique ability of ruminants to utilize a wide variety of feeds is a result of the highly diverse ruminal ecosystem, consisting of bacteria, methanogenic Archaea, and eukaryotes (i.e., protozoa and fungi; Qi et al., 2009). The ruminal microbiota utilize specialized enzymes and enzyme complexes to convert feed components to end products of digestion and microbial cells. As noted previously, the main products of fermentation are VFA with the principal ones being acetate, propionate, and butyrate. The VFA are usually referred to as their dissociated ions, the undissociated forms being acetic, propionic, and butyric acids. In addition, ammonia (NH_3), CO_2, CH_4, and H_2 gas are produced. Long-chain fatty acids are released from lipids, and many other end products result from microbial degradation of minor components of the feed. Hydrogen produced during microbial fermentation of feed is used by methanogens in a syntrophic process known as interspecies H_2 transfer. The methanogens reduce CO_2 with 2H (electrons are represented as reduced protons and referred to here as 2H) to produce CH_4, which is removed from the rumen by eructation and expiration. Formate can also be used by methanogens, but its contribution as a precursor for CH_4 is much less than that of hydrogen (Janssen, 2010). The VFA are largely absorbed across the ruminal wall, thereby supplying an important part of the ruminant's energy requirements. Amino acids, oligopeptides, and NH_3 released by microbial degradation of proteins and NPN compounds are used by microorganisms and converted to microbial protein. The microbial protein and ingested plant proteins supply the animal with dietary AA. A portion of the microbes, undigested feed particles, and compounds dissolved in the ruminal liquid, can be digested in the abomasum and lower digestive tract.

Details of ruminal fermentation and the microbes involved have been published many times, and in much greater detail elsewhere (e.g., Stewart and Bryant, 1988; Russell, 2002; Firkins and Yu, 2006). Early studies were focused on culturing and isolating ruminal microorganisms with the aim of enhancing the understanding of this complex microbial ecosystem (e.g., Bryant and Burkey, 1953; Hungate, 1966; Dehority and Orpin, 1988; Stewart and Bryant, 1988). Recent advances in culture-independent molecular methods (whole-genome sequencing, metagenomic sequencing, proteomics, and transcriptomics) have facilitated the study of the ruminal microbiome and its functional genes (Krause et al., 2013). Continued development of molecular procedures is helping to characterize the ruminal microbiome as affected by the host, diet, and environment and is creating a new understanding of the complexity of this unique ecosystem (Krause et al., 2013).

Bacteria account for the largest proportion of the microbiota with 10^9 to 10^{11} cells/mL, representing more than 50 genera (Stewart et al., 1997; Russell, 2002). Bacterial mass is in the range of 14 to 18 mg bacterial dry weight per milliliter (7 to 9 mg bacterial protein/mL; Russell, 2002). The bacteria contain about 100 g N/kg organic matter (OM) or 62.5% crude protein (CP; Clark et al., 1992), with about 80% of CP being true protein and 20% being nucleic acids (Russell, 2002).

Ruminal bacteria vary widely in their substrate specificity. Most bacteria use monomers and oligomers released from the hydrolysis of plant polymers by polymer-fermenting bacteria (Stewart et al., 1997). Glycogen, poly-β-hydroxybutyrate, and polyphosphate are accumulated in most bacteria as storage materials (Gottschalk, 1986). Of the culturable rumen bacteria, *Fibrobacter succinogenes*, *Ruminococcus albus*, and *R. flavefaciens* are considered the predominant cellulolytic bacteria, although this understanding may change as nonculturable cellulolytic bacteria are increasingly characterized. The cellulolytic bacteria degrade plant fiber by adhering to the substrate, which concentrates cellulolytic enzymes at the digestion site, decreases loss of hydrolytic products by diffusion, and helps protect the bacteria from protozoal predation (Qi et al., 2009).

Ciliated and flagellated protozoa are found in the rumen (Bohatier, 1991), but the majority of the protozoa are ciliates (10^4 to 10^6 cells/mL) with over 100 species from about 25 genera represented (Williams, 1986; Qi et al., 2009). Because ruminal protozoa are substantially larger (20 to 200 μm) than bacteria, they can account for up to 50% of the microbial mass in the rumen (Nolan, 1993; Russell, 2002). Because they can sequester in the rumen, they represent a smaller proportion of the total microbial flow from the rumen to the duodenum relative to their mass, and thus make a smaller contribution to the animal's protein supply than bacteria (Nolan, 1993). Protozoa are predators of ruminal bacteria and engulf large numbers of bacteria as their main N source (Firkins and Yu, 2006). This N recycling process decreases the efficiency of N use in the rumen. Protozoa have the ability to engulf and degrade sugars and starch, and larger entodiniomorphids and holotrichs degrade fiber (Williams and Coleman, 1988). Similar to bacteria, protozoa accumulate glycogen (a branched homoglucan, similar to amylopectin) when provided with sugar or fructan substrates (Masson and Oxford, 1951; Williams and Coleman, 1988). Protozoa also participate in a symbiotic relationship with ruminal methanogens in that they possess hydrogenosomes that generate H_2 for use by methanogens.

Six genera of anaerobic fungi (10^3 to 10^5 zoospores/mL) have also been identified, accounting for 8 to 20% of the total microbial biomass in the rumen (Stewart et al., 1995; Qi et al., 2009; Krause et al., 2013). Anaerobic fungi contribute substantially to fiber degradation in the rumen, particularly when animals are fed low-quality, fibrous diets (Krause et al., 2013). Zoosporic fungi colonize and germinate to produce hyphae that penetrate and disrupt plant tissues, facilitating bacterial colonization. They also possess a range of hydrolytic polysaccharidases and complementary glycosidase activities (Stewart et al., 1995).

Members of the domain Archaea make up only a small part of the microbial biomass in the rumen (Janssen, 2010), contributing up to 3.3% of the microbial small subunit (16S and 18S) rRNA (Janssen and Kirs, 2008). Most Archaea within the rumen are methanogens that have the central role of removing H_2 from the rumen. Methanogens possess hydrogenases that split H_2 into H^+ and electrons at very low concentrations (Stewart, 1991) using the electrons derived from H_2 and sometimes formate to reduce CO_2 to CH_4. This process helps maintain a low concentration of H_2 (<0.1% vol/vol; Weimer, 1998) in the rumen, thereby eliminating the inhibitory effect of H_2 on microbial fermentation (Janssen and Kirs, 2008). The majority (>90%) of Archaea detected in ruminal contents can be placed in three genus-level groups: *Methanobrevibacter*, *Methanomicrobium*, and a large group of uncultured Archaea referred to as cluster C (Janssen and Kirs, 2008).

No single organism is responsible for the complete degradation of complex feed substrates in the rumen. Rather, feed digestion is accomplished by a myriad of complementary organisms that form a complex microbial community or biofilm on the surface of plant tissues, throughout the fluid phase, and on the ruminal epithelium (McAllister et al., 1994). These biofilms comprise a multispecies community that produces the array of enzymes needed for digestion of complex feeds. Complex feedstuffs are colonized and broken into oligomers and eventually tri-, di-, or monomers of each component, which are then used by other members of the microbial population to produce VFA, vitamins, or cofactors needed by other microorganisms (Krause et al., 2013). This cross-feeding activity among community members prevents the accumulation of digestive end products that would inhibit digestion and thereby accelerates the digestive process.

The ruminal microbiome is compartmentalized in relation to nutrient acquisition (Cheng and McAllister, 1997). Ruminal fluid contains soluble compounds and large amounts of particulate material that certain bacteria adhere to and colonize in the process of digestion. Within the population of particle-associated bacteria there is differentiation between bacteria that are firmly and loosely adhered to feed particles. Adherent microbial populations are numerically predominant and account for up to 70% of the total microbial population (Russell, 2002). There are also many nonadherent bacteria in ruminal fluid, some of which are likely transitioning from one recently digested feed particle to another. Sloughing of biofilms is a common mechanism whereby bacteria enhance their ability to come in contact with fresh substrate in aquatic environments. Yet another community within the rumen is the epimural bacteria that colonize the luminal surface of the stratified squamous epithelium. This bacterial community performs a variety of functions necessary for host health, including hydrolysis of urea, scavenging of oxygen, and recycling of epithelial tissue (Petri et al., 2013a).

As part of the particulate-associated bacterial fraction, cellulolytic bacteria attach to and degrade plant cell wall polymers; however, substrate inaccessibility represents a major constraint to this process (Russell, 2002). Cellulolytic bacteria attach directly to fiber particles in a highly ordered fashion and form biofilms (McAllister et al., 1994; Krause et al., 2003). The bacteria within close proximity of substrate can adhere nonspecifically to the substrate by wedging into feed cavities and exposed inner surfaces. The bacterial glycocalyx on the outer membrane of gram-negative cells and the peptidoglycan of gram-positive cells seem to be involved in the initial binding process (Miron et al., 2001). Bacteria adhere specifically to digestible tissue via the production of more extensive linkages and adhesins (Miron et al., 2001). In the rumen, specific adhesion of cellulolytic bacteria to cellulose occurs in various ways, including via large multicomponent complexes called cellulosomes, fimbriae or pili adhesins, carbohydrate epitopes of bacterial glycocalyx layer, and enzyme-binding domains (Miron et al., 2001). The predominant culturable cellulolytic bacteria *F. succinogenes*, *R. flavefaciens*, and *R. albus* in the rumen each uses a specific combination of these mechanisms to specifically adhere to cellulose as detailed by Miron et al. (2001) and Flint and Bayer (2008). Many anaerobic cellulolytic bacteria possess unique extracellular multienzyme complexes within cellulosomes (Lynd et al., 2002). Cellulosomes have two functional domains integrated into a cohesive complex: a catalytic domain that integrates the various cellulases and xylanases and a large noncatalytic scaffolding protein that anchors the bacterium to the substrate (Russell, 2002). Within the scaffolding subunit, a cellulose-binding module facilitates the interaction of substrate with the active site (Lynd et al., 2002). Adhesion of bacteria to feed particles is critical, in particular, for digestion of fiber in the rumen, and as such, the cellulosome mediates the proximity of the bacterium with its enzyme complex to the substrate. Specific adhesion to cellulosic substrates imparts strategic advantages: cellulolytic enzymes are concentrated on the substrate, other microbes are excluded from the site of hydrolysis, and attached bacteria may have a measure of protection from predatory ruminal protozoa. Enzymes within the biofilm are likely protected from ruminal proteases, and feed-attached bacteria have a longer retention time in the rumen than do free bacteria in the fluid phase (Miron et al., 2001).

The development of molecular techniques to investigate ecological microbial communities has provided a new understanding of the diversity and complexity of the rumen microbial ecosystem, and this information is expected to evolve rapidly in the coming years. For example, it is now known that as cattle age, the ruminal microbial community becomes more diverse, but there is increased similarity of the microbiome among animals within a population (Jami et al., 2013). Mature ruminants are considered to have a core ruminal microbiome, although taxa abundance can vary greatly across animals (Jami and Mizrahi, 2012; Petri et al., 2013a). The dominant phyla of bacteria found in cattle of all age groups are Firmicutes, Bacteroidetes, and Proteobacteria (Jami et al., 2013).

DIGESTION KINETICS

The digestive process in cattle is complex, with numerous animal, dietary, and management factors interacting to affect the rate and extent of feed digestion in the rumen and total digestive tract. Diet digestibility directly affects nutrient supply to the animal; thus, the ability to estimate the site and extent of nutrient digestibility in the digestive tract is important. Models that integrate an understanding of the mechanisms involved in digestion are ultimately the best way of integrating the information (Baldwin, 1995), but these models are inherently complex and their practical use for feed formulation is somewhat limited for beef cattle production (Tedeschi et al., 2005). Thus, dynamic empirical models are often used to represent the kinetic aspects of feed digestion in the rumen, as discussed in detail by others (Owens and Goetsch, 1986; Mertens, 1993b; Firkins et al., 1998; Huhtanen et al., 2006).

Feed components vary in the rate and extent they are digested in the rumen and digestive tract. Sugars (mono- and oligosaccharides) and soluble cell-wall carbohydrates (e.g., pectic substances, β-glucans, galactans, fructans) are almost completely degraded in the rumen (Van Soest, 1994). Digestion of starch in the rumen is variable depending on feed source and processing method; however, in most cases extensive postruminal digestion of starch compensates for lower ruminal digestion, such that digestion of starch in the total tract is usually high for processed grains (>80%; Firkins et al., 2001). In contrast, digestion of fiber in the rumen depends on the physical and chemical attributes of the fiber and the ruminal microbial population. Fiber escaping ruminal digestion is only partially degraded postruminally

(Huhtanen et al., 2006). Digestion of NDF primarily occurs in the reticulorumen, with typically less than 10% of NDF digestion occurring in the hindgut of cattle (Huhtanen et al., 2006). Microbial digestion of fiber in ruminants is a relatively slow process, and a long retention time in the rumen increases the extent of fiber digestion but lowers the DM intake (DMI) as a result of limits imposed by ruminal fill. As a result of differential rates and extents of digestion of feed components in the rumen, total tract DM digestibility coefficients can range from 0.35 to 0.50 for high-fiber, low-quality forages such as straw (Waiss et al., 1972; Horton, 1978) to 0.75 to 0.90 for high-starch, grain-based diets fed to feedlot cattle (Zinn, 1990; Beauchemin et al., 2001; Corona et al., 2006).

Digestion of feed in the reticulorumen is determined by the rate of microbial digestion of feed and the rate at which feed passes from the rumen. The rate of digestion (kd, h^{-1}) and the rate of passage (kp, h^{-1}) for feed are competitive processes; thus, a faster passage rate or a slower digestion rate will decrease ruminal digestion of feed, whereas a slower passage rate or a faster digestion rate will increase ruminal digestion. Most non-cell wall components in feeds are rapidly digested with rates that are 3 to 10 times faster than rates of passage (Mertens, 1993b). Conversely, the rate of cell wall digestion is generally of the same magnitude as the rate of passage, although the rate of plant cell wall digestion is highly variable depending on plant species (e.g., grass vs. legumes) and physiological stage of maturity of the forage (Mertens, 1993b).

The digestion process is often assumed to follow first-order kinetics with the assumption that the pool of potentially digestible feed is homogeneous and is digested as a linear function of time in the rumen (Waldo et al., 1972). Feeds are considered to have a potentially digestible fraction (PD) and an indigestible (1 − PD) fraction with ruminal digestibility estimated as (Waldo et al., 1972):

$$\text{Ruminal digestibility} = \text{kd}/(\text{kd} + \text{kp}),$$

and the amount of potentially digestible matter that is digested in the rumen is then

$$\text{PD} \times \text{kd}/(\text{kd} + \text{kp}).$$

Disappearance of the indigestible fraction can only occur by passage. This single-compartment, first-order representation is a simplification of the system, but it is useful in practical diet formulation. For feed components that contain a soluble fraction (e.g., DM, CP), the model can be expanded to include a fraction that is considered instantaneously available (not subjected to digestion and passage rate). For fiber, a discrete lag time can be incorporated into the first-order digestion model to accommodate any delay in digestion. These expanded models are described by Mertens (1993a,b).

Digestion Rate

The digestion rate (kd) can be measured for diets or single feeds using in vitro and in situ methods that characterize digestion curves (NRC, 1985; López, 2005). With the in vitro methods, feed is incubated in buffered ruminal fluid to obtain feed digestion data over a time course (Grant and Weidner, 1992; Cherney et al., 1993), whereas with the in situ technique, feed is incubated in small porous bags in the rumen of ruminally cannulated cattle, and feed disappearance from the bags (i.e., degradability) is measured over time (Vanzant et al., 1998; Ørskov, 2000). The time course usually spans from a zero time measurement to determine initial solubility to a final measurement at 48 or 72 h for rapidly digested components (Nocek, 1988), and 96 to 240 h for slowly digested components (Van Soest et al., 2005). Frequent measurement in the early hours of incubation is required to determine the lag time, defined as the initial period during which digestion is slow or nonexistent, usually considered the time required for bacterial attachment to substrate (Owens and Goetsch, 1986; Mertens, 1993a). Consideration of lag time is particularly important when measuring kinetics of cell wall digestion. The data from the in vitro or in situ incubations are fit to models to estimate the digestion kinetics of the feed (Nocek, 1988; Mertens, 1993a,b; Dhanoa et al., 1995). Full methodological details of these in vitro (Grant and Weidner, 1992; Cherney et al., 1993) and in situ (Lindberg and Varvikko, 1982; Weakley et al., 1983; Lindberg, 1985; Nocek, 1985, 1988; Vanzant et al., 1998; Ørskov, 2000) techniques have been published.

As discussed by Mertens (1993b), the primary advantage of the in vitro technique is that the environment of the system (e.g., temperature and pH) is easy to control and the fermentation vessel prevents infiltration or loss of materials. The disadvantage is that the technique may not adequately represent the conditions of the ruminal environment or the variability among animals. In addition, the microbial inoculum requires care in preparing and handling to ensure it remains reduced and anaerobic so that digestion in early and late stages of the fermentation is not limited. The advantage of the in situ technique is that it measures the combined effects of the intrinsic characteristics of the feed and the unique ruminal fermentation characteristics of the animal (i.e., animal variability) on digestion kinetics. Nonetheless, the rumen is rarely under steady-state conditions, and thus disappearance of feed from the bags can be affected by the time samples are placed in the rumen. In addition, suboptimal ruminal environments (e.g., low pH, deficiency in N supply) can slow the rate of digestion of substrates, and the bags suspended in the rumen are porous and allow escape and infiltration of fine particles that hamper calculations. The reader is referred to Nocek and English (1986), Mertens (1993a,b), Dhanoa et al. (1995), and Firkins et al. (1998) for a more comprehensive discussion of the limitations and strengths of the various techniques used to estimate ruminal digestion kinetics.

Passage Rate

The ruminal contents of forage-fed ruminants occur as distinct phases: a liquid phase, a floating mat (raft) formed by particles whose buoyancy promotes their selective retention in the dorsal rumen, and a pool of small particles dispersed within the fluid phase ventral to the floating mat (Vieira et al., 2008). The passage of feed from the rumen is a complicated process involving several compartments such that the various feed components pass through the digestive tract at differential rates. Soluble components dissolve and pass with the fluid phase. Ground concentrates and forages pass from the rumen faster than long forage particles that are selectively retained in the ruminal raft (Mertens, 1993b). Selective retention of undigested fiber allows ruminants to increase ruminal fiber digestion, but extended ruminal retention times of the retained fiber can decrease feed intake because of ruminal distention (i.e., filling effect; Allen, 1996). The rate of passage of digesta through the digestive tract increases with increasing feed intake (Balch and Campling, 1965), which allows the animal to increase its energy intake, even though ruminal and total tract digestibility can be decreased (Firkins et al., 1998). Several factors contribute to the probability of escape of digesta from the rumen. Particles sequestered within the fibrous mat have a lesser likelihood of leaving the rumen. As particles are decreased in size through digestion and rumination, fermentation gases are released, functional specific gravity is increased, and particles sink to the ventral rumen and are moved by ruminal contractions toward the reticuloruminal orifice where they are more likely to pass from the rumen (Faichney, 1986, 1993; Sutherland, 1987). Factors that control the passage rate of particles include composition of the diet, particle size of feed, feed additives, feeding method, breed of animal, ambient temperature, level of feed intake, and animal variability (Owens and Goetsch, 1986).

The rate of passage (kp) is determined in vivo by calculating the recovery of an external or internal marker. For example, a feed (or a portion thereof) can be labeled with a nonabsorbable marker (e.g., rare earth soaking, chromium mordanting), dosed into the rumen, and ruminal digesta sampled over time. The passage rate is calculated as the outflow of the marker from the rumen (g/h, mL/h) divided by pool size dosed (total milliliters or grams in the rumen). The decrease in marker concentration in ruminal digesta is fit to models to estimate fractional passage rate. The marker can also be fed as a pulse dose, with fecal samples taken to eliminate the need for ruminally cannulated cattle; however, ruminal sampling is often considered more appropriate because postruminal flow and mixing can introduce variability (Firkins et al., 1998). Because indigestible feed can only leave the rumen via passage, an internal feed marker such as indigestible acid detergent fiber (ADF) or NDF (Cochran et al., 1986) can also be used to estimate passage rate from the rumen. In this case, passage rate of the indigestible feed fraction is determined from the duodenal flow of the marker and the ruminal pool size of the marker estimated by evacuation of ruminal contents (Voelker and Allen, 2003). Estimates of passage rate can also be highly variable because of errors associated with using markers, as discussed by others (Beauchemin and Buchanan-Smith, 1989; Mertens, 1993b; Firkins et al., 1998). Despite the technical difficulties associated with measuring the passage rate, for most beef cattle diets, kp of feed particles varies from 2 to 6%/h.

A number of models have been proposed to describe the passage of digesta through the gastrointestinal tract of ruminants, as discussed in detail by others (Ellis et al., 1979; Pond et al., 1988; Mertens, 1993a). A single-compartment model may be satisfactory when marker is dosed into the rumen, the rumen is the site of sampling, and near instantaneous mixing within the rumen is assumed. Such is often the case when estimating free fluid (i.e., liquid not imbibed in particles) kinetics; however, most kinetic models for particulate matter assume that the passage of plant cell walls is a multicompartmental process, with at least two distinct compartments. Grovum and Williams (1973) proposed a model containing two mixing compartments in sequence with a discrete time delay (i.e., transit time). The model assumes an age-independent process, meaning that simple dilution of intake by the compartmental mass determines the competition for escape; thus, all feed particles have equal opportunity for escape regardless of their age in the compartment. Grovum and Williams (1973) assigned the compartment with the slower turnover rate to the rumen and designated its turnover rate kp1, with turnover rate kp2 being assigned to the faster turnover compartment (cecum and proximal colon). Matis (1972) and others (Pond et al., 1988; Matis et al., 1989; Ellis et al., 1994) suggested that passage through the digestive tract is an age-dependent process in which the probability for escape of a particle increases with its residence time in the reticulorumen. When using an age-dependent model, it is generally assumed that the two largest compartments represent distinct pools within the rumen: a mixing compartment whereby newly ingested particles are mixed with the accumulated larger mass of digesta and a pool of small particles eligible for escape from the rumen (Pond et al., 1988; Ellis et al., 1994). Other models have been proposed that consider multiple compartments (e.g., Dhanoa et al., 1985; France et al., 1985; Vieira et al., 2008).

The *Nutrient Requirements of Dairy Cattle, 7th Revised Edition* (NRC, 2001) used three empirical equations to predict the rates of passage for dry forages, wet forages, and concentrates. The current Committee on Nutrient Requirements of Beef Cattle has adopted the revised equations from Seo et al. (2006), which are based on an extensive database of studies that used external markers, and 553, 195, and 766 forages, concentrates, and fluid, respectively, as treatments. Although the predictability of these equations needs to be improved, these equations are useful in diet formulation programs when measured values of kp are not available.

FERMENTATIVE PROCESSES

Approximately 60 to 75% of digestible energy is derived from fermentation in the rumen (Sutton, 1979). Carbohydrates, including cellulose, hemicellulose, pectin, starch, and sugars, are the main substrates utilized by ruminal microbes, with the population generally deriving less energy from lipids and proteins. As much as 90% of carbohydrate digestion occurs in the rumen, although this value can be much lower for diets that contain highly processed forages or if grains such as corn and sorghum, which have slow rates of digestion, are not adequately processed before consumption (Sutton, 1979). Carbohydrates entering the rumen are subjected to hydrolytic digestion by the resident population of microorganisms. The main end products of ruminal fermentation are VFA, CH_4, CO_2, NH_3, and microbial cells (Wolin et al., 1997; Figure 2-3). The principal VFA in descending order of concentration are acetate, propionate, and butyrate. Lactic acid is produced as well from pyruvate, but its concentration in ruminal fluid is usually very low (except during acute acidosis; refer to Ruminal pH and Acidosis section), as lactate is rapidly converted to acetate and propionate (Piveteau, 1999). The fate of ruminal lactate varies depending on ruminal conditions resulting in different proportions of acetate and propionate. Other compounds (e.g., succinate, formate, ethanol) are produced by some ruminal bacteria but these are present in very low concentrations in the rumen (Van Soest, 1994).

Before being fermented to VFA, carbohydrates are degraded to hexoses and, in some cases, pentoses to the intermediate pyruvate. The carbohydrate moiety of lipids (glycerol and galactose) is also fermented to VFA, but the contribution to the energy supply of the animal is moderate because lipids are usually restricted to <6% of dietary DM. Proteins are hydrolyzed to peptides and AA, which are incorporated into bacterial cells or deaminated to form NH_3, CO_2, and VFA. The branched-chain VFA (isobutyrate, isovalerate, and 2-methylbutyrate) that arise from branched-chain AA are important growth factors for some ruminal bacteria (Allison, 1969).

A simplified balance for fermentation of carbohydrates in the rumen can be calculated as follows assuming that all substrates are composed of hexose units and the molar proportion of VFA are 65 acetate: 20 propionate: 15 butyrate (adapted from Wolin, 1960):

$$57.5 \text{ hexose } (C_6H_{12}O_6) + NH_3 \rightarrow \text{Microbes} + 65 \text{ Acetate} + 20 \text{ Propionate} + 15 \text{ Butyrate} + 60\ CO_2 + 35\ CH_4 + 25\ H_2O.$$

In practice, the relative rate of conversion of hexose to end products depends on the substrate and the microbial population. Thus, the actual yield of end products per mole of hexose fermented will vary from that shown in the equation above.

During microbial fermentation in the rumen, approximately 75 to 85% of feed energy is converted to VFA with the remainder lost as heat and CH_4 (Sutton, 1979). Ruminal concentrations of VFA represent the balance between production and removal (through neutralization, absorption,

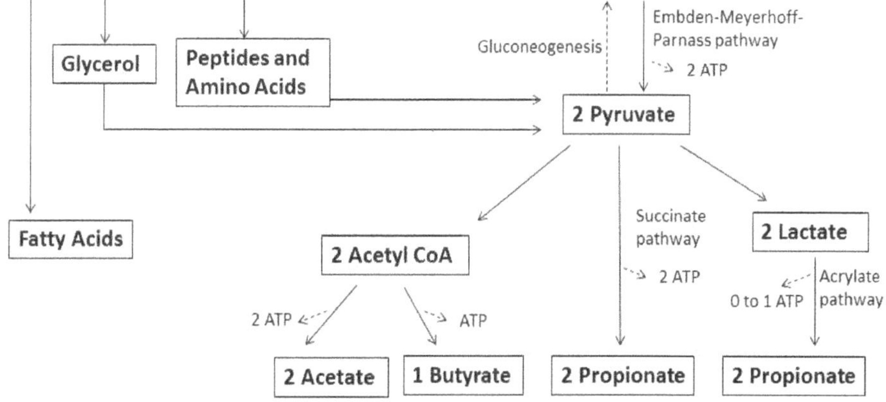

FIGURE 2-3 Schematic diagram of the major pathways of fermentation in the rumen. Adapted from Van Soest (1994) and Ungerfeld and Kohn (2006).

and passage), as well as their interconversion. As a result, ruminal concentrations of VFA do not reflect ruminal VFA production.

The total VFA concentration in the rumen of beef cattle is normally between 70 and 130 m*M*, but can range from 30 to 200 m*M* (France and Siddons, 1993). The relative proportions of VFA produced as a result of ruminal fermentation depends on the chemical composition of the diet, use of feed additives (e.g., ionophores), and factors affecting the ruminal environment (e.g., pH, dilution rate; Sutton, 1979). High-forage diets promote the formation of acetate. Thus, acetate:propionate:butyrate proportions in ruminal fluid are typically 70:20:10 to 65:25:10 in cattle fed high-forage diets, with an acetate:propionate ratio above 3:1 (Owens and Goetsch, 1988; France and Siddons, 1993; Wolin et al., 1997). In contrast, high-concentrate diets promote propionate production at the expense of acetate. Thus, acetate:propionate:butyrate proportions in feedlot cattle are approximately 50:40:10, such that acetate:propionate ratio is <2:1 (e.g., Bevans et al., 2005; Wierenga et al., 2010; Li et al., 2011). High propionate proportion is typically observed during periods of dietary transition, when protozoa are few, and when ruminal pH is low (Schwartz and Gilchrist, 1975). Higher butyrate concentrations are typically observed with diets containing high-sugar ingredients (e.g., molasses; Demeyer, 1991) and when there is a large protozoal population composed of Entodinia whose principal end products of metabolism are butyrate and acetate (Schwartz and Gilchrist, 1975).

Most ruminal microorganisms use the Embden-Meyerhof-Parnas pathway to oxidize sugars to pyruvate (i.e., glycolysis) yielding two molecules of pyruvate for each molecule of glucose metabolized (Russell and Wallace, 1988; Miller, 1995), as shown:

$$\text{Glucose} + 2\ \text{NAD}^+ + 2\ \text{ADP} + 2\ \text{P}_i \rightarrow 2\ \text{Pyruvate} + 2\ \text{NADH} + 2\ \text{H}^+ + 2\ \text{ATP} + 2\ \text{H}_2\text{O},$$

where NAD^+ is nicotinamide adenine dinucleotide, ADP is adenosine diphosphate, NADH is reduced nicotinamide adenine dinucleotide, and ATP is adenosine triphosphate. The free energy released in this process is used to form the high-energy compounds ATP and NADH. The pyruvate generated as an end product of glycolysis is a substrate for the Krebs cycle, which results in an additional ATP molecule and five redox equivalents including NADH and $FADH_2$ (reduced flavin adenine dinucleotide). Under aerobic conditions, NAD^+ is regenerated by the transfer of the electrons of NADH to the electron acceptors of the respiratory chain, resulting in the reduction of molecular oxygen to water and the generation of proton gradient across the mitochondrial membrane. This gradient is subsequently used to drive ATP synthesis (Russell and Wallace, 1988). In total, one mole of glucose gives rise to 30 ATP during aerobic respiration (McDonald et al., 2010). In contrast, anaerobic organisms lack a respiratory chain, and to allow glycolysis to continue, they must be able to oxidize cofactors, such as NADH, back to NAD^+. If this did not occur, all the oxidized cofactors would eventually be reduced, and digestion would cease. Thus, pyruvate generated as an end product of glycolysis acts as an electron sink, being further reduced to provide for the regeneration of NAD^+ and other cofactors, and the general removal of excess electrons, thereby yielding ATP. During anaerobic fermentation, oxidation of one mole of glucose results in net gain of two ATP in most cases. For lactate, three molecules of lactate on average will produce two molecules of propionate and one molecule of acetate with generation of three ATP (Piveteau, 1999).

A significant amount of H^+ (and electrons) is produced from the oxidation of NADH to NAD^+. Therefore, anaerobic microbes work with a surplus of reducing equivalents (e.g., NADH) and utilize a variety of reactions to dispose of surplus protons. Hydrogen, taken up from the ruminal environment, acts as a reducing agent that drives electron transport. A number of ruminal bacteria, fungi, and protozoa possess hydrogenases producing H_2 as a sink for reducing equivalents. In H_2 formation, the electron transfer between H_2-producing microorganisms and H_2-utilizing species of methanogens occurs by the reduction of H^+ to H_2, diffusion of H_2 through the biofilm to the methanogen, and then oxidation of H_2 by the methanogen (Ellis et al., 2008). The methanogens use H_2 for growth, leading to CH_4 production and removal of H_2. The NADH-mediated production of H_2 is thermodynamically regulated by the partial pressure of the gaseous production of H_2 (Ungerfeld and Kohn, 2006). Even trace amounts of H_2 in the rumen inhibit hydrogenase activity and limit the oxidation of sugar when pathways for the disposal of H_2 are absent. Methanogens are very efficient at utilizing H_2 gas; thus the concentration of dissolved H_2 gas in the rumen is low (about 0.1 to 50 µM; Janssen, 2010) with ruminal H_2 partial pressure ranging from 1 to 10 Pa (Ellis et al., 2008). Propionate formation is an alternative to H_2 formation because it also serves as an electron acceptor (Table 2-1). Therefore, increases in propionate formation are strongly associated with decreases in CH_4 production. Formation of acetate and butyrate results in the net production of H_2 and are associated with increases in CH_4 production, although formation of butyrate from acetate is an electron sink. Therefore, the relationship between butyrate concentration and CH_4 production can be variable. There are other highly competitive processes in the rumen for reducing equivalents; sulfates and nitrates are reduced to sulfides and NH_3 and unsaturated fatty acids are biohydrogenated. Synthesis of microbial cells also provides a sink for H^+. Thus, changes in diet composition and microbial populations in the rumen will alter the end products of fermentation and the production of CH_4.

The process whereby some microbes use H_2 produced by other microbes is known as interspecies H_2 transfer. One such example of interspecies H_2 transfer is the facilitated electron transfer from protozoa to methanogens owing to their close

TABLE 2-1 VFA Production and Reductive Processes in the Rumen[a]

Substrate	Products	Reaction
VFA production		
$C_6H_{12}O_6 + 2\ H_2O \rightarrow$	$2\ C_2H_4O_2 + 2\ CO_2 + 4\ H_2$	Acetate production
$C_6H_{12}O_6 + 2\ H_2 \rightarrow$	$2\ C_3H_6O_2 + 2\ H_2O$	Propionate production
$C_6H_{12}O_6 \rightarrow$	$C_4H_8O_2 + 2\ CO_2 + 2\ H_2$	Butyrate production
Reductive processes		
$CO_2 + 4\ H_2 \rightarrow$	$CH_4 + 2\ H_2O$	Methane production
$2\ CO_2 + 4\ H_2 \rightarrow$	$C_2H_4O_2 + 2\ H_2O$	Reductive acetogenesis
$SO_4^{2-} + 4\ H_2 + 2\ H^+ \rightarrow$	$H_2S^- + 4\ H_2O$	Sulfate reduction
$NO_3^- + H_2 \rightarrow$	$NO_2^- + H_2O$	Nitrate reduction
$NO_2^- + 3\ H_2 + 2\ H^+ \rightarrow$	$NH_4^+ + 2\ H_2O$	Nitrite reduction

[a]Adapted from Ungerfeld and Kohn (2006) and Weimer (1998).

association. This relationship is advantageous to both organisms: the protozoa can rapidly dispose of H_2 produced in hydrogenosomes and the methanogens can rapidly utilize the H_2 as an energy source. Furthermore, protozoa are selectively retained in the rumen, which is favorable for methanogens because they grow slowly (Ellis et al., 2008).

The fermentation of dietary carbohydrates to VFA has a major effect on the glucose supply to the ruminant and accentuates the importance of gluconeogenesis (the formation of glucose from substances other than sugars). Propionate is the only VFA that makes a net contribution to glucose synthesis; about 27 to 54% of the glucose in the ruminant is formed from propionate (Lindsay, 1970). Of the propionate formed in the rumen, 80 to 95% is absorbed unchanged through the ruminal epithelium and is converted in the liver to phosphoenolpyruvate via methylmalonyl-CoA, succinate, and oxaloacetate (Fahey and Berger, 1988; Kristensen and Harmon, 2004). Acetate and butyrate do not contribute to glucose supply. Only a small amount of acetate absorbed from the rumen is converted to ketone bodies; most is carried by the portal circulation to the liver unchanged. Butyrate is metabolized extensively by the ruminal epithelium, with the production of the ketone bodies acetoacetate and β-hydroxybutyrate, which are excreted into the circulation and oxidized in many tissues for energy production. In addition to propionate, lactate, glycerol from lipids, and some AA from extrahepatic tissues are used to produce glucose (Brockman, 2005). Studies have reported that about 5 to 23% of the glucose is derived from glycerol, and protein contributes up to 20% of glucose (Lindsay, 1970).

RUMINAL pH AND ACIDOSIS

Acidosis, characterized by low ruminal pH (Owens et al., 1998; Nagaraja and Titgemeyer, 2007), is thought to be a prevalent digestive disorder in feedlot cattle fed high-grain diets. It is commonly classified as either subacute (subclinical) or acute (clinical or lactic) acidosis based on the extent of the depression in ruminal pH, the length of time that ruminal pH remains depressed, and whether lactate accumulates in ruminal fluid (Kleen et al., 2003; Plaizier et al., 2008; Aschenbach et al., 2011). For feedlot cattle, a drop in ruminal pH below 5.8 (Beauchemin et al., 2001; Moya et al., 2011), 5.6 (Owens et al., 1998; Cooper et al., 1999; Brown et al., 2000; Bevans et al., 2005; Nagaraja and Titgemeyer, 2007) or 5.5 (Hibbard et al., 1995; Goad et al., 1998; Coe et al., 1999; Wierenga et al., 2010; Zhang et al., 2013a,b) has been used to signify subacute ruminal acidosis. Most researchers use a drop in pH below 5.0 (Nocek, 1997; Nagaraja and Titgemeyer, 2007; Aschenbach et al., 2011) or 5.2 (Owens et al., 1998; Penner et al., 2007; Wierenga et al., 2010) as an indicator of acute acidosis. Others have suggested that acute acidosis can be identified by an increase in ruminal lactate concentration (>5 m*M* used by Aschenbach et al., 2011; >50 m*M* used by Goad et al., 1998 and Nagaraja and Titgemeyer, 2007). During subclinical acidosis, ruminal pH usually returns to prefeeding values slowly after feeding, whereas pH usually remains very low during acute acidosis and the animal ceases to eat unless intervention occurs. Some authors have characterized an episode of subacute ruminal acidosis as pH remaining below the threshold for more than 3 (Castillo-Lopez et al., 2014) or 4 h/d (Paton et al., 2006).

Low ruminal pH alters the ruminal microbiome (Nagaraja and Titgemeyer, 2007; Petri et al., 2013b) in a manner that decreases the rate and extent of fiber digestion (Russell and Wilson, 1996). Furthermore, low pH (<5.6) can damage the ruminal epithelia, disrupting barrier function that supports nutrient and water transport while preventing microbial contamination of the interstitial tissues. Consequently, absorption of VFA decreases (Aschenbach and Gäbel, 2000; Penner et al., 2010; Wilson et al., 2012). Low ruminal pH is associated with numerous animal health issues, including lameness (Brent, 1976; Nocek, 1997), rumenitis (Jensen et al., 1954; Kleen et al., 2003; Plaizier et al., 2008), liver abscess (Robinson et al., 1951; Nocek, 1997; Nagaraja and Chengappa, 1998), and an acute-phase protein response (Plaizier et al., 2008).

Ruminal pH is regulated through behavioral (González et al., 2012) and physiological (Aschenbach et al., 2011) mechanisms. For example, cattle decrease DMI (Fulton et al., 1979a,b) and selectively consume long particles when ruminal pH is low (DeVries et al., 2008). The risk for ruminal acidosis in beef cattle is increased by feeding diets rich in rapidly fermentable nonstructural carbohydrates, with the risk of acidosis high in feedlot cattle fed diets containing >90% DM from nonforage ingredients. Implementing a gradual dietary transition (Burrin et al., 1988; Bartle and Preston, 1992; Bevans et al., 2005; Holtshausen et al., 2013) can encourage both microbial (Allison et al., 1964; Towne et al., 1990; Goad et al., 1998; Brown et al., 2006) and epithelial (Dirksen et al., 1985; Bannink et al., 2008) adaptation. Extending the duration of time that cattle are fed a high-grain diet helps stabilize ruminal pH, but does not decrease the susceptibility of cattle to a bout of ruminal acidosis should a disruption in feeding management occur (Schwaiger et al., 2013a). Feeding management practices such as avoiding abrupt changes in diet formulation, maintaining consistent timing of feed delivery, delivering feed at multiple times per day, including ionophore antibiotics in the diet, processing grain consistently and avoiding fine particles, and increasing the dietary proportion of forage help decrease ruminal acidosis in feedlot cattle (Schwartzkopf-Genswein et al., 2003).

The primary non-NDF carbohydrate fed to cattle is starch, which is much more rapidly fermented in the rumen than structural carbohydrates such as cellulose and hemicellulose (Sniffen et al., 1992). Rapid fermentation of carbohydrates (mainly starch) leads to rapid production of VFA, and under normal feeding conditions, VFA readily dissociate (i.e., release a proton), causing a decrease in ruminal pH (Aschenbach et al., 2011). When the dissociation of VFA is greater than the removal of protons (i.e., acids) from the rumen (through neutralization, absorption, and passage) ruminal pH decreases (Penner et al., 2007, 2009b) and osmolality increases (Argyle and Baldwin, 1988; Brown et al., 2000; Bevans et al., 2005).

The physiological mechanisms regulating ruminal pH primarily involve the neutralization of hydrogen ions with bicarbonate supplied in saliva (Bailey and Balch, 1961a; Erdman, 1988) and from ruminal bicarbonate secretion by the ruminal epithelium (Penner et al., 2009a). Protons are primarily removed from the rumen in a condensation reaction with ruminal bicarbonate (producing water and CO_2), the majority of which is supplied through salivary secretion (Bailey and Balch, 1961a,b; Erdman, 1988; Dijkstra et al., 2012) and in exchange for VFA across the ruminal epithelium (Gäbel et al., 1991; Allen, 1997; Dijkstra et al., 2012). The relative supply of bicarbonate from saliva and epithelial secretion is likely dependent on a number of factors, including ruminal pH (Schwaiger et al., 2013b).

Saliva is an important physiological mechanism for buffering the pH of the rumen (Turner and Hodgetts, 1955; Hibbard et al., 1995; Allen, 1997), as discussed (refer to Salivation section). It has been estimated that saliva production is responsible for the neutralization of up to 40% of the acid produced in the rumen (Allen, 1997; Gäbel et al., 2002; Aschenbach et al., 2011). Mixed ruminant saliva contains bicarbonate and hydrogen phosphate (Bailey and Balch, 1961a,b), and has a pH of 8.4 (Bailey and Balch, 1961a,b) and a pK_a of 6.1 (Aschenbach et al., 2011). The bicarbonate equilibrium in saliva is in its ionic form and likewise hydrogen phosphate is also primarily unbound to hydrogen ions (pK_a = 7.2; Kohn and Dunlap, 1998). When saliva is mixed with ruminal contents, however, both equilibriums shift so that bicarbonate and hydrogen phosphate bind with hydrogen ions. Because bicarbonate has an effective pK_a that mirrors physiological pH (approximately 5.5 to 7.0), it is an ideal ruminal buffer. Bicarbonate reacts with hydrogen ions producing water and CO_2 to resist a decrease in ruminal pH (Allen, 1997). For this reason, bicarbonate is thought to be the most important buffering component of saliva (Turner and Hodgetts, 1955; Allen, 1997; Aschenbach et al., 2011). Because hydrogen phosphate has a pK_a well above physiological pH, it is not an appropriate ruminal buffer but rather a weak base (Counotte et al., 1979; Erdman, 1988); however, as a weak base, it can help neutralize protons at virtually any physiological pH. By doing so, hydrogen phosphate is thought to combine with hydrogen ions and remove them from the rumen indirectly, through passage to the omasum (Counotte et al., 1979; Allen, 1997).

It is estimated that VFA absorption through the ruminal epithelium removes about 50 to 55% of protons from the rumen at a pH of 6.0 (Gäbel et al., 1991; Allen, 1997). Although the transport mechanisms governing ruminal absorption of VFA have not yet been fully elucidated, they include both passive diffusion and facilitated transport. Undissociated VFA (HVFA) are lipid-soluble and able to passively diffuse across biological membranes. Lipophilic permeability decreases in the order of butyrate > propionate > acetate (Walter and Gutknecht, 1986), as does intraepithelial metabolism (Britton and Krehbiel, 1993; Kristensen and Harmon, 2004). It was once thought that there was little or no absorption of VFA in their dissociated (ionized) form (Danielli et al., 1945; Masson and Phillipson, 1951; Ash and Dobson, 1963), but this theory has since been disproved. It is now known that, in fact, under physiological conditions the majority of VFA are absorbed in the ionized form (Aschenbach et al., 2011). Using the Henderson-Hasselbalch equation ($pH = pK_a + \log_{10}([VFA^-]/[HVFA])$) and a pK_a of 4.8 (Aschenbach et al., 2011), the proportion of ionized VFA^- can be calculated for any ruminal pH. At a pH of 7.0, 99% of VFA are ionized (VFA^-), and only 1% are lipophilic (HVFA). As pH decreases to 5.5, 83% of VFA are ionized, and 17% are lipophilic.

Cellular bicarbonate is secreted into the rumen by ruminal epithelial cells (Gäbel et al., 1991; Bilk et al., 2005). Absorption of a portion of the butyrate (Sehested et al., 1999; Penner et al., 2009a), propionate (Kramer et al., 1996), and acetate (Aschenbach et al., 2009; Penner et al., 2009a) occurs in exchange for bicarbonate. It is estimated that for every mole of VFA absorbed, 0.53 mole of bicarbonate is secreted (Gäbel et al., 1991). As ruminal pH declines, VFA (Dijkstra et al., 1993) and lactate (Williams and Mackenzie, 1965; Harmon et al., 1985) absorption initially increase, decreasing the risk for ruminal acidosis (Penner et al., 2009a). In addition, increased expression of enzymes involved in VFA metabolism (Penner et al., 2009b) might also be associated with decreased severity of ruminal acidosis. Nonetheless, there is a growing body of evidence suggesting that in response to a severe acidotic challenge, VFA absorption rates might actually decrease (Wilson et al., 2012; Schwaiger et al., 2013b) possibly as a strategy to protect cells in the ruminal epithelium from acidification. Although a decrease in VFA absorption rate will not help to prevent ruminal acidosis or buffer the rumen, it could represent an important homeostatic mechanism. Other physiological buffering mechanisms that influence ruminal pH to a lesser degree include ruminal NH_3 and phosphates, liquid dilution, and passage of VFA from the reticulorumen (Allen, 1997; Aschenbach et al., 2011).

RUMINAL NITROGEN METABOLISM

The extensive pregastric fermentation that occurs in the ruminant forestomach transforms dietary protein and NPN into microbial protein and NH_3, altering the profile of protein reaching the SI. Microbial protein is digested postruminally, providing the animal with a source of good-quality protein, as all 10 essential AA are synthesized by ruminal microorganisms (Nolan, 1993). Thus, the ruminant animal has a unique evolutionary advantage of being able to subsist on diets that contain no true protein. Furthermore, there is extensive recycling of N within the body (refer to Chapter 6, Protein and Amino Acids), which in theory can extend the period of survival of the animal when fed a low-protein diet, or when dietary carbohydrate and protein are asynchronous. Nonetheless, the effectiveness of this N conservation mechanism to overcome a deficit in dietary N, especially over a longer period of time (several weeks), should not be overestimated or relied upon in practical beef production. When ruminants are fed very-high-protein diets, increased N recycling and excretion are associated with higher energetic costs (Reynolds and Kristensen, 2008).

Although ruminants have the unique ability to transform relatively low-quality dietary protein and NPN into high-quality meat protein for human consumption, the efficiency of use of feed N for growth is relatively low (ranging from 10 to 20%). As a result, 80 to 90% of dietary N is excreted in urine and feces. Hence, there has been considerable research aimed at improving N efficiency of ruminants (see reviews by Satter et al., 2002; Schwab et al., 2005; and refer to Chapter 16, Environment). In general, decreasing N excretion by beef cattle requires closely matching the supply of feed N to the N requirements of the ruminal microorganisms and the AA requirements of the animal.

Feed Nitrogen Dynamics

A substantial portion of dietary protein and virtually all NPN (urea, nucleic acids, amides, amines, AA, and nitrate) are degraded in the rumen to AA, peptides, and NH_3 by the ruminal microorganisms. This fraction of feed N is referred to as ruminally degradable protein (RDP); the remaining feed protein that escapes ruminal degradation and flows to the intestine is referred to as ruminally undegradable protein (RUP). The physical and chemical characteristics of the feed affect the solubility characteristics of the CP and its susceptibility to ruminal degradation, as reviewed elsewhere (Broderick et al., 1991; NRC, 2001). The process of feed CP degradation in the rumen is complex and depends on a number of factors as reviewed previously (NRC, 1985); hence, the proportion of dietary CP that is degraded in the rumen is extremely variable.

The kinetics of predicting ruminal CP degradability using a three-pool, in situ ABC technique is described in detail by NRC (1985, 2001), and a similar approach was adopted in Chapter 6. Briefly, ruminal CP degradation of feedstuffs is described using a first-order disappearance model (Ørskov and McDonald, 1979). Fraction A is the percentage of CP that is instantaneously or completely degraded in the rumen (NPN and true protein that escapes from the in situ bag because it is highly soluble or has a very fine particle size); the B fraction is assumed to be potentially degraded CP, with the amount degraded at time *t* determined from its rate of digestion (kd). The C fraction is the percentage of CP that is completely undegraded and considered to pass in its entirety to the SI. Ruminally degradable protein consists of the entire A fraction and the portion of fraction B that is actually degraded in the rumen, whereas RUP consists of the portion of B fraction that passes from the rumen before it is digested and the entire C fraction.

The in situ ABC technique is subject to numerous inherent sources of error, which must be controlled if reproducible results are to be obtained. Sources of variation that need to be considered are sample size, bag size, porosity of the bag material, treatment of the bags following removal from the rumen, diet and associated ruminal environment of the test animals, and so forth. Standardized procedures have been proposed to help reduce the variability in these measurements (Lindberg and Varvikko, 1982; Weakley et al., 1983; Lindberg, 1985; Nocek, 1985, 1988; AFRC, 1992; Vanzant et al., 1998; Ørskov, 2000).

In contrast to the simplified system adopted by the current committee, the NRC (1996, 2000) adopted the protein fractionation used by the Cornell Net Carbohydrate and Protein System (CNCPS) model (Sniffen et al., 1992). The original CNCPS model considered five protein fractions (A, B1, B2, B3, and C) based on in vitro solubility in buffer and detergent solutions (Sniffen et al., 1992). The A fraction is NPN; the C fraction is unavailable protein and determined chemically as the percentage of CP bound to the ADF (ADIP); B1 is the protein soluble in borate-phosphate buffer; and PB2 and PB3 are the potentially degradable protein with varying rates of fermentation.

To increase the escape of limiting AA to the intestine, various methods for treating feedstuffs to decrease their RDP contents have been evaluated including heating to invoke the Maillard reaction between amino groups of lysine and carbonyl compounds, chemical treatment, use of tannins and other inhibitory compounds, physical encapsulation, and other treatments (Broderick et al., 1991). Use of ionophore antibiotics are known to decrease proteolysis of feeds by decreasing deamination. Further discussion on the manipulation of ruminal fermentation to affect protein degradation through the use of feed additives is in Chapter 14 (Compounds That Modify Digestion and Metabolism).

Microbial Protein Synthesis

Numerous ruminal microorganisms (bacteria, protozoa, and fungi) with a diverse array of proteolytic enzymes act synergistically to degrade feed CP in the rumen (Walker et al., 2000). The process of CP degradation by ruminal microorganisms begins with extracellular proteases that hydrolyze proteins, releasing oligopeptides. The released oligopeptides are degraded to smaller peptides and free AA before cellular uptake occurs. Once inside the cell, peptides are hydrolyzed to free AA. Intracellular free AA are either used for protein synthesis or catabolized to NH_3 and carbon skeletons (i.e., deamination), with the resultant carbon skeletons fermented to VFA. The NH_3 is used to resynthesize AA, or it is diffused out of the bacterial cell (NRC, 2001).

About 30 to 50% of isolated bacterial species exhibit proteolytic activity (Nolan, 1993; Schwab et al., 2005), but the bacteria most extensively studied for their role in protein degradation in the rumen are *Ruminobacter amylophilus*, *Butyrivibrio fibrisolvens*, *Streptococcus bovis*, and *Prevotella* spp. (Walker et al., 2000). It is generally accepted that deamination is primarily carried out by a small group of bacteria with high deamination activity (e.g., *Acidaminococcus fermentans*, *Clostridium sticklandii*, *C. aminophilim*, *Eubacterium* spp., *Peptostreptococcus anaerobius*), referred to as the hyperammonia-producing bacterial populations (Chen and Russell, 1989; Walker et al., 2000). Although protozoa also possess deaminase activity (Walker et al., 2000), their most important role in protein degradation is their ability to engulf and degrade insoluble particulate proteins, as well as bacterial and fungal cells (Wallace, 1991; Walker et al., 2000). Engulfment and subsequent digestion of bacteria by ciliate protozoa cause substantial recycling of bacterial protein in the rumen. As much as 50% of the microbial protein formed in the rumen is recycled via protozoa, autolysis, and lysis of the bacteria by bacteriophages and mycoplasmas (Walker at al., 2000). The fungi possess proteolytic activity, although their role in protein degradation is thought to be minor, except in the degradation of very resistant protein-carbohydrate matrices (Walker et al., 2000).

Ruminal bacteria derive between 38 and 80% of their N from NH_3 (Hristov and Jouany, 2005). Many ruminal bacterial species have an obligate NH_3 requirement, including the three major cellulolytic species *F. succinogenes*, *R. flavefaciens* and *R. albus*. Furthermore, most species of bacteria are capable of growth using only NH_3 as a source, as long as carbon skeletons such as branched-chained VFA are available (Wallace, 1991; Nolan, 1993). Despite their ability to synthesize the AA they require, given the opportunity, most ruminal microorganisms will preferentially use short peptides and AA to meet their N requirements. Thus, although NH_3 can provide all bacterial N requirements, peptides and AA are beneficial to ruminal fermentation because they stimulate the growth of ruminal microorganisms, particularly the fibrolytic bacteria (Carro and Miller, 1999; Ranilla et al., 2001; Griswold et al., 2003). The microbial requirement for peptides and AA does not seem to be related to a requirement for specific limiting AA (Argyle and Baldwin, 1988). In contrast to bacteria, protozoa are not able to use NH_3 (Bach et al., 2005).

The ruminal concentration of NH_3 is highly variable depending on the DMI of the animal, protein and RDP contents of the diet, assimilation of NH_3 by the ruminal microbes, and absorption and passage of NH_3 from the rumen. Ruminal concentration of NH_3 in beef cattle typical ranges from 2 to 10 m*M* (3 to 16 mg N/100 mL); however, cattle consuming diets rich in RDP can have a ruminal concentration of NH_3 greater than 30 m*M* (50 mg N/100 mL; Chung et al., 2013; Hünerberg et al., 2013). The animal must expend energy to convert excess NH_3 to urea in the liver. Conversely, when ruminal concentration of NH_3 is not adequate, uncoupled fermentation can occur, resulting in fermentation without useful ATP production (Stern and Hoover, 1979). The optimum ruminal concentration of NH_3 needed for microbial growth has been studied and reviewed extensively (Stern and Hoover, 1979; NRC, 1985). A ruminal concentration of 5 m*M* NH_3-N is generally considered the minimal concentration needed to promote microbial growth, although Schwab et al. (2005) proposed, based on a summary of the literature, that ruminal NH_3-N concentrations of 5 to 11 m*M* (8 to 18 mg N/100 mL) are needed to maximize flows of microbial N from the rumen particularly when diets containing highly fermentable carbohydrates are fed. Nonetheless, these higher ruminal NH_3 concentrations will result in increased N losses from the rumen. The optimal ruminal concentration of peptides and AA for efficient microbial protein synthesis has not been defined (Firkins et al., 2007).

Diets that provide high amounts of ruminally fermentable carbohydrates enhance NH_3 capture and promote greater microbial protein synthesis in the rumen. Rapidly fermentable carbohydrates, such as starch or sugars, are more effective than other carbohydrate sources, such as cellulose, in promoting microbial growth (Stern and Hoover, 1979). Yet, an excess of rapidly fermentable carbohydrate can decrease efficiency of microbial growth by decreasing ruminal pH and, if RDP is limiting, by increasing the nongrowth energy expenditure, a process referred to as energy spilling (Russell, 1998). Many bacteria spill energy when it is in excess, as excess carbohydrate can be toxic.

Microbial crude protein (MCP) synthesis represents the daily total flow of MCP from the rumen, whereas efficiency of MCP synthesis is the amount of MCP produced per unit of available energy. Efficiency of MCP synthesis is most commonly expressed as grams of N per kilogram of OM fermented in the rumen. Fermentable OM is usually expressed as truly fermented OM, which represents the dietary OM fermented in the rumen, and is calculated by correcting for the OM in the rumen contributed by microbial sources. Sometimes microbial efficiency is expressed on the basis of apparent OM fermented in the rumen, total carbohydrates digested in the rumen (Hoover and Stokes, 1991), or total digestible nutrients (TDN) such as in the dairy NRC (2001), the previous beef NRC (1996, 2000), and the current edition of *Nutrient Requirements of Beef Cattle*.

Efficiency of MCP synthesis is a function of the energy used by the microbes for maintenance and growth (Stouthamer, 1973; Hespell and Bryant, 1979). The energy required to maintain bacterial cells in a healthy state of growth includes functions such as motility, intracellular turnover, production of extracellular enzymes, active transport of nutrients into the cell, oxidation of substrates, storage of carbohydrates as microbial glycogen, and so forth. In the complex ruminal ecosystem, additional maintenance energy is expended in the synthesis of extracellular polysaccharides for attachment and development of microbial consortia communities. Additional energy is also expended as a result of recycling of cells within the rumen (protozoal predation, lytic bacteriophages), inefficient phosphorylation, and energetic uncoupling (Harmeyer, 1986). Uncoupled fermentation refers to a situation in which energy is released faster than it can be used by the ruminal bacteria. For maximum efficiency of microbial growth to occur, N and energy availability in the rumen must be balanced. Other nutritional factors, such as sulfur supply, and nonnutritional factors, such as ruminal pH and dilution rate, also play an important role in microbial protein synthesis, as reviewed in detail by others (Bach et al., 2005).

Estimates of MCP synthesis in the literature are highly variable, ranging from 12 to 54 g N/kg OM truly fermented in the rumen (NRC, 2001; Bach et al., 2005). In a summary of the literature for beef cattle, Galyean and Tedeschi (2014) reported that MCP synthesis ranged from about 19.3 to 206 g N/d (mean of 81.1 g N/d), corresponding to an estimated mean efficiency of approximately 19 g N/kg OM truly fermented in the rumen (refer to Chapter 6 for additional information). Bach et al. (2005) concluded that optimum bacterial growth in the rumen occurs when efficiency of MCP synthesis is 29 g of bacterial N/kg fermented OM, which would be rarely observed in beef cattle. Differences in total flow of MCP and efficiency of MCP synthesis observed among studies are partly a result of the measurement techniques used. The main technical limitation of the method used to measure efficiency of MCP synthesis is that the technique is indirect, and is based on measuring flow of OM at the duodenum using digestibility markers that are known to be variable (Titgemeyer, 1997) and subject to error.

Total flow of MCP decreases with increased proportion of forage in the diet, whereas efficiency of MCP synthesis is usually optimized between 30 and 70% forage in the diet (Clark et al., 1992). Although efficiency of MCP synthesis can be less for high- and low-forage diets, compared with intermediate diets, Galyean and Tedeschi (2014) recently showed that the published data for beef cattle do not support a quadratic fit. In that study, MCP efficiency decreased slightly as TDN increased.

For high-forage diets, microbial efficiency tends to be lower because of slower particulate passage rate. In vivo, flow of MCP is highly correlated to particulate passage rate (Sniffen and Robinson, 1987), because of the close association of bacteria to feed particles. Slower particulate passage results in higher bacterial recycling in the rumen and slower bacterial growth, resulting in higher maintenance energy requirements for bacteria. Theoretically, efficiency of microbial synthesis would also be less for poorer-quality forage than for higher-quality forage for the same reasons; however, in vivo studies indicate that the relationship between total tract OM digestibility and efficiency of MCP synthesis is poor (NRC, 2001). Therefore, in practice, differences in efficiency of MCP as a result of forage quality are likely not important. Decreased efficiency of MCP synthesis for high-concentrate diets has been attributed to low ruminal pH (Pitt et al., 1996), slower microbial turnover (Owens and Goetsch, 1988), and uncoupled fermentation (Clark et al., 1992). Owens and Goetsch (1986) indicated that the passage rate of concentrate decreased from 6.9 to 3.1%/h, and the passage rate of forage decreased from 3.7 to 2.9%/h as concentrate level in the diet increased from 50% to greater than 80%. Effects of ruminal pH on microbial synthesis are discussed in more detail in relation to physically effective fiber.

Metabolizable Protein

A detailed discussion of metabolizable protein (MP) is given in Chapter 6. Briefly, the flow of nitrogenous compounds from the rumen consists of NH_3-N, RUP, MCP synthesized in the rumen, and undegraded endogenous protein (mucoproteins in saliva, epithelial cells from the respiratory tract, sloughed cells, and enzymatic secretions). Metabolizable protein is considered to be the true protein (i.e., nonammonia N) absorbed by the intestine and it is composed of MCP and RUP. Because the contribution of endogenous protein to MP in beef cattle is small (4.4 to 6.0 g N/kg DMI) it was not considered part of MP by the current committee. The supply of MP is affected by dietary factors that affect MCP synthesis and the supply of RUP. In many feeding scenarios, as much as two-thirds to three-quarters of the AA absorbed by beef cattle are from MCP.

In MP systems, protein requirements are partitioned into the needs of the ruminal microorganisms and the needs of the host animal. Protein supplementation strategies require that NH_3 and peptide requirements of the microorganisms be satisfied relative to the supply of fermentable carbohydrate. In addition to feed protein, recycled urea, intraruminally recycled microbial protein, and to a lesser extent, endogenous salivary proteins and sloughed epithelial cells, also contribute to meeting microbial N requirements.

In many cases, diets fed to beef cattle contain sufficient RDP to meet microbial needs (NRC, 1996, 2000; Fu et al., 2001). Exceptions would be low-quality forage diets that are typically fed to beef cows and corn-based diets fed to feedlot cattle. Responses to RDP have been reported mainly with energy sources that are rapidly fermented (Chikunya et al., 1996). Formulating diets to ensure that microbial requirements for RDP are met permits maximal microbial growth and efficiency. If N is limiting, microbial yield and efficiency and the fermentability of the diet are decreased. On the other hand, formulating diets with excess RDP, such that NH_3 production exceeds bacterial needs, increases N excretion and the energy required for urea synthesis, decreasing the efficiency of use of RDP for ruminant production.

Nitrogen Recycling

A portion of the NH_3 produced in the rumen is absorbed across the ruminal epithelium into the portal vein and converted mostly to urea by the liver (Reynolds and Kristensen, 2008), as described in Chapter 6. Urea produced by the liver is partly excreted in the urine, with the remainder recycled back to the gut through either direct transfer from blood across the epithelial tissue or via saliva. In ruminants, 40 to 80% of the urea synthesized by the liver is returned to the gastrointestinal tract, and of this, 27 to 60% of the urea enters the rumen (Lapierre and Lobley, 2001). The recycled urea is converted to NH_3 in the rumen and provides a source of N for microbial protein synthesis, which is especially important in cases where RDP from the diet is insufficient. Thus, microbial growth and carbohydrate fermentation affect the extent of NH_3 absorption, and urea N recycling and excretion. Nitrogen recycling is considered to be an evolutionary advantage that allows the ruminant to survive when protein supplies are inadequate. Nitrogen recycling is an important aspect of N metabolism, and it is estimated that 10 to 30% of microbial N flow to the duodenum is derived from endogenous urea entry to the rumen (Reynolds and Kristensen, 2008). As the CP content of the diet increases, the fraction of urea that is produced and is returned to the gastrointestinal tract decreases and, correspondingly, the fraction of urea produced that is excreted in urine increases (Reynolds and Kristensen, 2008). In cattle fed low-protein diets (<12% CP), almost all the urea is transferred to the gastrointestinal tract and the fraction of urea used for microbial protein synthesis is greater (28 to 72%, although variable) than for higher-protein diets (17 to 26%).

The urea content of saliva is about 65% of that in the plasma, with plasma values ranging from 4 to 19 mg urea N/100 mL of plasma (Bailey and Balch, 1961b). In cattle, urea N represents 80 to 85% of the total N of parotid saliva (Phillipson and Mangan, 1959; Bailey and Balch 1961a). There does not seem to be a relationship between the rate of secretion of saliva and the total N or the urea N concentrations in saliva. Arterial urea N concentration, in addition to other factors, determines the amount of blood urea N transferred to the lumen of the gastrointestinal tract (Reynolds and Kristensen, 2008).

POSTRUMINAL DIGESTION AND ABSORPTION OF NUTRIENTS

Digestion in the SI is facilitated by secretions from accessory organs including the liver, pancreas, and small intestinal mucosa. In cattle, bile and pancreatic secretions enter the anterior duodenum through separate bile and pancreatic ducts, respectively. The secretion of HCO_3^- from both accessory organs is needed to neutralize the HCl of the abomasal contents entering the duodenum. In addition, the Brunner's glands of the duodenum produce a mucus-rich alkaline secretion, which serves to help neutralize the acidic stomach contents flowing from the abomasum through the pyloric sphincter. Digesta leaving the abomasum of the ruminant is acidic and pH increases at a relatively slow rate as digesta passes through the SI (Merchen, 1988). The increase in pH is important because amylolytic and proteolytic enzymes secreted by the pancreas and intestinal mucosa generally have neutral to slightly alkaline pH optima.

Bile is synthesized in the liver and stored and secreted by the gallbladder. Ruminant bile contains mucus, electrolytes, and the bile salts and pigments. Bile salts have an important role in digestion by emulsifying and solubilizing lipids entering the SI and enhancing the activity of pancreatic lipase. In a process known as enterohepatic circulation, more than 95% of the bile salts secreted into the duodenum are reabsorbed in the ileum and returned to the liver. Pigments of bile have no digestive function and consist largely of bilirubin, which is produced as a breakdown product of hemoglobin by hepatic cells. In cattle, bilirubin is oxidized to the green-colored biliverdin (Merchen, 1988).

Pancreatic juice secreted by ruminants contains amylolytic, lipolytic, and proteolytic enzymes in a solution of electrolytes, HCO_3^-, and water. Total volume of pancreatic juice secreted is 2.7 to 3.3 L/d in cattle (Walker and Harmon, 1995). Stable pancreatic juice secretion rate was also observed for food-deprived sheep (15.7 to 18 mL/h) and those with ad libitum access to feed (27.2 to 33.7 mL/h; Pierzynowski, 1986). Walker and Harmon (1995) reported that in steers fed 12 times daily, pancreatic juice secretion was consistent throughout the day. Entry of digesta into the duodenum is probably the most important factor that regulates the volume of juice secreted by the pancreas because the continuous pattern of secretion observed in ruminants corresponds to the continuous flow of digesta. Taylor (1962) reported that preventing digesta from flowing into the duodenum resulted in a 70% decrease in secretory rate.

Mucosa of the SI is the site of origin of enteric enzymes that are important in postruminal digestion of carbohydrates and proteins. The Brunner's glands secrete a neutral to slightly alkaline fluid that contains amylase and ribonuclease. Secretion rate of duodenal fluid from all sources was estimated at 13 mL/h in sheep fed once daily and 26 mL/h in sheep fed three times daily (Harrison and Hill, 1962).

In addition to enzymes secreted in the duodenum, the disaccharidases lactase, maltase, and isomaltase are present throughout the SI. These enzymes act on their substrate at the brush border of intact mucosal cells and are not secreted into the intestinal lumen. The age of the animal and diet influence the relative activities of different disaccharidases along the SI. Activities of these enzymes have been measured in dairy calves ranging from 1 to 44 days of age (Huber et al., 1961). Lactase activity was greatest at day 1 and declined thereafter, whereas maltase activity was not affected by age. The type of carbohydrate entering the intestine had no influence on relative activities of different dissacharidases. In contrast, calves fed diets containing increased levels of lactose had substantially greater levels of lactase activity in the intestine at 11 weeks of age than controls (Huber et al., 1964).

Lactase, maltase, and isomaltase activity seem to be greatest in the jejunum with least activity in both the duodenum and ileum (Huber et al., 1961). Ben-Ghedalia et al. (1974) observed that activities of the proteolytic enzymes trypsin, chymotrypsin, and carboxypeptidase A secreted by the pancreas increased gradually between the duodenum and a point located 7 m caudal to the pylorus and then decreased gradually throughout the remainder of the intestine. This pattern generally corresponds to the characteristic pattern of pH values reported in ruminant intestine. The pH of the mid-jejunum is 6 to 7, which approximates the pH optima of most of these enzymes. The duodenum and proximal jejunum (pH 2.6 to 5.1) are likely too acidic and the ileum too alkaline (pH 7.8 to 8.2) for optimal activity of many digestive enzymes (Ben-Ghedalia et al., 1974; Merchen, 1988).

Digestion and Absorption of Starch in the Small Intestine

Digestion and absorption of starch in the SI of ruminants occur in three distinct phases. These phases of starch digestion and absorption have been extensively reviewed (Owens et al., 1986; Harmon, 1992, 1993; Huntington, 1997; Harmon et al., 2004; Huntington et al., 2006). Digestion of starch begins in the duodenum via the action of α-amylase secreted from the pancreas. Digestion by α-amylase produces maltose and branched-chain products commonly referred to as limit dextrins. The ruminant pancreas lacks an adaptive response to increased dietary starch (Kreikemeier et al., 1990; Walker and Harmon, 1995), which remains a biological enigma. This has led to the speculation that pancreatic α-amylase is the limiting phase of intestinal starch assimilation (Huntington, 1997). In contrast, Oba and Allen (2003a,b), and Taylor and Allen (2005), have suggested that the physicochemical characteristics of grain particles limit intestinal starch digestion.

The second phase of intestinal starch digestion and absorption occurs at the brush border membrane through the action of the brush border carbohydrases (e.g., maltase and isomaltase). In general, these enzymes have been poorly characterized, and similar to pancreatic enzymes, their activities show little or no adaptive response to diet (Janes et al., 1985; Kreikemeier et al., 1990). Kreikemeier and Harmon (1995) analyzed the composition of ileal digesta in steers infused abomasally with glucose, corn dextrins, or corn starch. They observed an accumulation of α-glucosides in ileal digesta that were composed mainly of disaccharides with little free glucose. The authors concluded that under their experimental conditions, starch assimilation was limited by brush border α-glucosidase activity.

The third and final component of intestinal starch digestion and absorption is the transport of glucose out of the intestinal lumen and into portal circulation. Although it has been generally thought that glucose crosses the enterocyte brush border membrane via the action of the sodium-dependent glucose transporter SGLT1 (Harmon and McLeod, 2001), again, the activity of this transporter has shown no adaptive response to luminal substrate in cattle (Bauer et al., 2001; Rodriguez et al., 2004). Despite this lack of adaptive response in glucose transport, abomasally infused glucose disappears from the intestinal lumen (Kreikemeier et al., 1991) suggesting that mechanisms other than SGLT1 are responsible for glucose transport in cattle (Krehbiel et al., 1996; Au et al., 2002).

Digestion and Absorption of Lipids in the Small Intestine

Composition of lipids changes during microbial digestion in the reticulorumen before arrival at the SI. No appreciable degradation of long-chain fatty acids occurs in the rumen, and digestion and absorption of lipids other than the VFA occurs almost exclusively in the SI (Garton, 1965). Intestinal digestion of lipids depends on the presence of biliary and pancreatic secretions, and diverting the flow of these materials away from the SI results in only 15 to 20% of dietary lipids being absorbed (Caple and Heath, 1975).

The majority (70 to 80%) of the lipids arriving at the abomasum are in the form of non-esterified fatty acids (NEFA) produced by microbial lipolysis of dietary triglycerides in the rumen (Garton, 1965; Leat and Harrison, 1975). The balance of lipids is predominantly in the form of phospholipids, which are largely of microbial origin. As a result of ruminal hydrolysis of triglycerides and subsequent fermentation of glycerol, very little glycerol of dietary origin escapes the rumen. It is interesting to note that duodenal digesta often contains a larger fraction of mono-, di-, and triglycerides and phospholipids than abomasal digesta because of the presence of biliary secretions that contain these lipids. Bile salts secreted from the liver (gallbladder) act as potent emulsifying agents in the proximal duodenum.

Triglycerides that escape ruminal breakdown and esterified fatty acids of microbial origin are hydrolyzed by pancreatic lipase to release NEFA. Non-esterified fatty acids are subsequently taken into micellar solution by bile salts and lysolecithin. The micelles are disrupted on the surface of the microvilli on the intestinal mucosa and NEFA are taken up by the mucosal cells. Most absorption of lipids occurs in the proximal half of the SI, with the bile salts reabsorbed in the distal jejunum and in the ileum via enterohepatic circulation. In contrast to nonruminants, virtually no monoglycerides are absorbed from the intestine (Leat and Harrison, 1975).

Lipids absorbed by intestinal mucosa are transported from the intestine via the lymphatic circulation. Lipids derived from intestinal absorption appear in the lymph in chylomicron form. Lipid composition of chylomicrons is approximately 70 to 80% triglyceride and 15 to 20% phospholipid. Smaller proportions of NEFA, cholesterol, and cholesterol esters also exist (Leat and Harrison, 1975). Because lipids are taken up by the mucosa as NEFA, it is apparent that the absorbed fatty acids are reassembled into triglycerides within the intestinal mucosal cells. Glycerol required for the resynthesis of triglycerides is of endogenous origin because little glycerol is available for absorption from the SI. The intestinal mucosal cells rely largely on α-glycerol phosphate obtained from glycolytic intermediates as a source of glycerol. Triglycerides and phospholipids resynthesized in the mucosa are packaged into chylomicrons, which are delivered to lymph and later enter the venous circulation via the thoracic duct. The continuous digestion and flow of digesta to the duodenum in ruminants results in continuous absorption of lipid from the SI (Leat and Harrison, 1975).

Digestion and Absorption of Protein in the Small Intestine

Ruminant animals derive their AA supply from a mixture of feed protein that escapes ruminal degradation and microbial protein that is formed as a result of microbial fermentation of carbohydrates in the reticulorumen. Microbial protein is readily digested by the host animal and constitutes a well-balanced array of essential AA for ruminants. Similar to nonruminants, gastric digestion initiates protein digestion via the secretion of HCl by gastric parietal cells. Hydrochloric acid is required for the conversion of pepsinogens to pepsins and also for maintaining pepsin activity. Pepsins are secreted as inactive precursors (i.e., pepsinogens) by chief cells in the abomasum. Once secreted, pepsin increases the susceptibility of microbial and ruminally undegradable intake proteins to attack by pancreatic proteases by opening the tertiary and quaternary structure of the protein and exposing AA residues to the pancreatic endopeptidases (Guan and Green, 1996). Once pepsin is present in the lumen of the abomasum, the reaction becomes autocatalytic, which involves the splitting off of a peptide chain and peptide fragments. The pepsins are most active at pH less than 4.0 and become inactive at pH greater than 6.0, although the optimal pH for pepsin activity varies among species (Crevieu-Gabriel et al., 1999). In ruminants, lysozyme accompanies the acid and pepsinsogen secretions, which helps lyse bacteria and speed up the digestion of microbial protein (Van Soest, 1994).

Pepsins are most active at peptide bonds that include phenylalanine, tyrosine, leucine, valine, and glutamic acid (Ulshen, 1987). Their most pronounced effect is between leucine and valine, between leucine and tyrosine, or between the aromatic AA such as phenylalanine-phenylalanine or phenylalanine-tyrosine. The principal function of digestion in the abomasum is the transformation of protein into large polypeptides, although peptic digestion in acid pH can produce free or peptide-bound AA. As suggested above, these are good stimuli for release of hormones that stimulate pancreatic enzyme secretion (e.g., cholecystokinin). Ultimately, the object of abomasal proteolysis is to make available peptide molecules that are susceptible to further hydrolysis by proteolytic enzymes in the SI.

Digestion in the Small Intestinal Lumen

Products of HCl and pepsin digestion enter the duodenum through the pyloric sphincter. In the duodenum, proteins and polypeptides serve as substrate for enzymes secreted from the pancreas and SI. Protein and polypeptides entering the duodenum are broken down further in the intestine by the pancreatic exopeptidases trypsin, chymotrypsin, and elastase and the pancreatic endopeptidases carboxypeptidase A and B. The pancreas secretes proenzymes into the duodenum, which, when activated, hydrolyze peptide bonds. The conversion of inactive trypsinogen to active trypsin requires removal of an N-terminal peptide and is catalyzed by the brush-border enzyme enteropeptidase. Enteropeptidase selectively cleaves a hexapeptide (H_2N-Val-Asp-Asp-Asp-Asp-Lys) from the amino terminus of trypsinogen, resulting in trypsin (Kitamoto et al., 1994). Enteropeptidase activity seems to be regulated by pancreatic secretion (Kwong et al., 1978) and thus possibly by protein concentration in the duodenum. Following its conversion from trypsinogen, trypsin activates the other zymogens, and to a lesser degree, trypsinogen.

Trypsin, chymotrypsin, and elastase catalyze the breakdown of proteins, polypeptides, and peptides into smaller peptides and AA in the duodenum. Each pancreatic protease has a unique and complementary action (Krehbiel and Matthews, 2003). Trypsin catalyzes the breakdown of bonds that involve lysine and arginine, whereas linkages involving aromatic AA residues are susceptible to chymotrypsin catalysis (Alpers, 1994). Elastases in general catalyze the breakdown of peptide bonds containing aliphatic residues. The action of trypsin, chymotrypsin, and elastase releases numerous terminal peptide bonds, which in turn are further digested by aminopeptidases, carboxypeptidases, and other specific peptidases present in the lumen or mucosa of the SI. Pancreatic carboxypeptidase A and B are exopeptidases that catalyze the hydrolysis of the carboxy-terminal bonds in polypeptide chains, removing the AA in sequence. Proteolysis of an approximately 100-residue segment from the amino-terminal region results in the activation of procarboxypeptidases (Aviles et al., 1985). Carboxy-terminal aromatic or nonpolar AA exposed by the action of chymotrypsin and elastase are available to be cleaved by carboxypeptidase A, whereas carboxy-terminal basic AA exposed by trypsin can be cleaved by carboxypeptidase B. The products of pancreatic digestion are oligopeptides of up to six AA residues (approximately 60%) as well as free AA (approximately 40%; Alpers, 1994).

Mucosal Phase of Digestion

The final stages of protein digestion are carried out by a wide array of brush-border and cytosolic peptidases. A review of the brush-border and cytosolic peptidases of the SI has been conducted by Alpers (1994). The small intestinal peptidases are capable of splitting products of pancreatic digestion (i.e., oligopeptides of six or less AA). These enzymes are present in the intestinal mucosa in two groups that are associated with different cell fractions, the apical membrane and the cytosol (Kim et al., 1972, 1974). The apical-membrane enzymes are attached to the outer surface of the microvillus and extend out from the luminal surface of the enterocyte, whereas the cytosolic enzymes are found within the cell and do not make direct contact with the luminal contents. As such, these two groups of enzymes are distinct from one another, differing in location and physicochemical and immunochemical properties (Kim et al., 1972; Nóren et al., 1977; Tobey et al., 1985). Apical enzymes seem to be unique to the SI, whereas similar cytosolic peptidases have been found in a number of tissues.

Many di- and tripeptides are absorbed intact and split within the enterocyte by the action of the cytosolic peptidases. In mammals, as much as 90% of the total mucosal peptidase activity for dipeptides, 40% of activity for tripeptides, and 10% of the tetrapeptidase activity are associated with the cytosolic fraction (Sterchi and Woodley, 1980). It seems that the capability of cytosolic enzymes to hydrolyze oligopeptides with more than three AA is limited. Oligopeptides of four to six AA in length are split to shorter peptides and free AA by apical microvillus membrane peptidases, whereas many di- and tripeptides are potential substrates for either apical membrane or cytosolic peptidases (Alpers, 1986). Therefore, membrane hydrolysis of peptides with subsequent absorption of AA, and transport of peptides followed by intracellular hydrolysis into AA, can occur (Ugolev et al., 1990).

Characterization of mammalian peptide-bound (peptide) or free AA transport systems has revealed the presence of at least two H^+-dependent and one H^+-independent peptide transport activities, one H^+-dependent AA transporter, and at least eight free AA systems in intestinal tissue (Krehbiel and Matthews, 2003). Although data are limited, the intestinal epithelia of cattle seem to express a similar complement of transport activities and specific transport proteins as observed for other species. Information is lacking regarding the coordination of expression and function of individual peptide and AA transporters to account for the quantity and relative ratio of AA that are absorbed across the intestinal epithelium in ruminants.

Digestion in the Cecum and Large Intestine

In ruminants, digestion of fiber and starch occurs mainly in the rumen, but also in the cecum and proximal large intestine, especially when digestion in the reticulorumen is limited. Factors that increase the rate of particulate passage, such as increased DMI or feeding high-concentrate diets, tend to decrease ruminal starch and fiber digestibility, which enhances the contribution of hindgut digestibility to total-tract digestibility. Potentially digestible fiber and starch that escape ruminal degradation become available for digestion in the distal part of the digestive tract.

Similar to the rumen, digestion in the ruminant cecum and proximal colon occurs through anaerobic fermentation, and bacterial concentrations in the hindgut are similar to those found in the rumen (Gressley et al., 2011). Products of fermentation are similar to those in the rumen: VFA, NH_3, microbial cells, and gas. Methane produced in the hindgut is responsible for 6 to 14% of total CH_4 output by cattle (Immig, 1996). The VFA contribute as much as 17% of the total VFA absorbed (Hoover, 1978). Fermentation in the cecum also incorporates N into microbial cells with excess N contributing to the pool of N recycled to the gastrointestinal tract. For cattle fed high-grain diets, excessive flow of fermentable carbohydrates from the SI can result in hindgut acidosis, characterized by increased rates of production of VFA and lactic acid, decreased fecal pH, and appearance of mucin casts in feces resulting from damage to gut epithelium (Gressley et al., 2011).

REFERENCES

AFRC (Agricultural and Food Research Council). 1992. Technical Committee on Responses to Nutrients Report No. 9, Nutrient Requirements of Ruminant Animals: Protein. *Nutrition Abstracts and Reviews* 62 (Series B):787-835.

Allen, M. S. 1996. Physical constraints on voluntary intake of forages by ruminants. *Journal of Animal Science* 74:3063-3075.

Allen, M. S. 1997. Relationship between fermentation acid production in the rumen and the requirement for physically effective fiber. *Journal of Dairy Science* 80:1447-1462.

Allison, M. J. 1969. Biosynthesis of amino acids by ruminal microorganisms. *Journal of Dairy Science* 29:797-807.

Allison, M. J., J. A. Bucklin, and R. W. Dougherty. 1964. Ruminal changes after overfeeding with wheat and the effect of intraruminal inoculation on adaptation to a ration containing wheat. *Journal of Animal Science* 23:1164-1171.

Alpers, D. H. 1986. Uptake and fate of absorbed amino acids and peptides in the mammalian intestine. *Federation Proceedings* 45:2261-2267.

Alpers, D. H. 1994. Digestion and absorption of carbohydrates and proteins. Pp. 1723-1749 in *Physiology of the Gastrointestinal Tract*, 3rd Ed., L. R. Johnson, ed. New York: Raven Press.

Argenzio, R. A. 1993. General functions of the gastrointestinal tract and their control and integration. Pp. 325-335 in *Dukes Physiology of Domestic Animals*, 11th Ed., M. J. Swenson and W. O. Reece, eds. Ithaca, NY: Cornell University Press.

Argyle, J. L., and R. L. Baldwin. 1988. Modelling of rumen water kinetics and effects of rumen pH changes. *Journal of Dairy Science* 71:1178-1188.

Aschenbach, J. R., and G. Gäbel. 2000. Effect and absorption of histamine in sheep rumen: Significance of acidotic epithelial damage. *Journal of Animal Science* 78:464-470.

Aschenbach, J. R., S. Bilk, G. Tadesse, F. Stumpff, and G. Gäbel. 2009. Bicarbonate-dependent and bicarbonate-independent mechanisms contribute to nondiffusive uptake of acetate in the ruminal epithelium of sheep. *American Journal of Physiology—Gastrointestinal and Liver Physiology* 296:G1098-G1107.

Aschenbach, J. R., G. B. Penner, F. Stumpff, and G. Gäbel. 2011. Ruminant nutrition symposium: Role of fermentation acid absorption in the regulation of ruminal pH. *Journal of Animal Science* 89:1092-1107.

Ash, R. W. 1959. Inhibition and excitation of reticulo-rumen contractions following the introduction of acids into the rumen and abomasum. *Journal of Physiology* 147:58-73.

Ash, R. W., and A. Dobson. 1963. The effect of absorption on the acidity of rumen contents. *Journal of Physiology* 169:39-61.

Ash, R. W., and R. N. Kay. 1959. Stimulation and inhibition of reticulum contractions, rumination and parotid secretion from the forestomach of conscious sheep. *Journal of Physiology* 149:43-67.

Au, A., A. Gupta, P. Schembri, and C. I. Cheeseman. 2002. Rapid insertion of GLUT2 into the rat jejunal brush-border membrane promoted by glucagon-like peptide 2. *Biochemical Journal* 367:247-254.

Aviles, F. X., J. Vendrell, F. J., Burgos, F. Soriano, and E. Mendez. 1985. Sequential homologies between procarboxypeptidases A and B from porcine pancreas. *Biochemical and Biophysical Research Communications* 130:97-103.

Bach, A., S. Calsamiglia, and M. D. Stern. 2005. Nitrogen metabolism in the rumen. *Journal of Dairy Science* 88:(Suppl.):E9-E21.

Bailey, C. B. 1961. Saliva secretion and its relation to feeding in cattle. 3. The rate of secretion of mixed saliva in the cow during eating, with an estimate of the magnitude of the total daily secretion of mixed saliva. *British Journal of Nutrition* 15:443-451.

Bailey, C. B., and C. C. Balch, 1961a. Saliva secretion and its relation to feeding in cattle. 1. The composition and rate of secretion of parotid saliva in a small steer. *British Journal of Nutrition* 15:371-382.

Bailey, C. B., and C. C. Balch. 1961b. Saliva secretion and its relation to feeding in cattle. 2. The composition and rate of secretion of mixed saliva in the cow during rest. *British Journal of Nutrition* 15:383-402.

Balch, C. C. 1958. Observations on the act of eating in cattle. *British Journal of Nutrition* 12:330-345.

Balch, C. C., and R. C. Campling. 1965. Rate of passage of digesta through the ruminant digestive tract. Pp. 108-123 in *Physiology of Digestion in the Ruminant*, R. W. Dougherty, R. S. Allen, W. Burroughs, N. L. Jacobson, and A. D. McGilliard, eds. Washington, DC: Butterworths.

Baldwin, R. L. 1995. *Modeling Ruminant Digestion and Metabolism*. London: Chapman & Hall.

Baldwin, R. L., and R. S. Emery. 1960. The oxidation-reduction potential of rumen contents. *Journal of Dairy Science* 43:506-511.

Bannink, A., J. France, S. Lopez, W. J. Gerrits, E. Kebreab, S. Tamminga, and J. Dijkstra. 2008. Modelling the implications of feeding strategy on rumen fermentation and functioning of the rumen wall. *Animal Feed Science and Technology* 143:3-26.

Bartle, S. J., and R. L. Preston. 1992. Roughage level and limited maximum intake regimens for feedlot steers. *Journal of Animal Science* 70:3293-3303.

Bauer, M. L., D. L. Harmon, D. W. Bohnert, A. F. Branco, and G. B. Huntington. 2001. Influence of alpha-linked glucose on sodium-glucose cotransport activity along the small intestine in cattle. *Journal of Animal Science* 79:1917-1924.

Beauchemin, K. A. 1991. Ingestion and mastication of feed by dairy cattle. Pp. 439-463 in *The Veterinary Clinics of North America: Dairy Nutrition Management*, C. J. Sniffen and T. H. Herdt, eds. Philadelphia: W. B. Saunders Co.

Beauchemin, K. A. 1992. Effects of ingestive and ruminative mastication on digestion of forage by cattle. *Animal Feed Science and Technology* 40:41-56.

Beauchemin, K. A., and J. G. Buchanan-Smith. 1989. Evaluation of markers, sampling sites and models for estimating rates of passage of silage or hay in dairy cows. *Animal Feed Science and Technology* 27:59-75.

Beauchemin, K. A., T. A. McAllister, Y. Dong, B. I. Farr, and K.-J. Cheng. 1994. Effects of mastication on digestion of whole cereal grains. *Journal of Animal Science* 72:236-246.

Beauchemin, K. A., W. Z. Yang, and L. M. Rode. 2001. Effects of barley grain processing on the site and extent of digestion of beef feedlot finishing diets. *Journal of Animal Science* 79:1925-1936.

Beauchemin, K. A., L. Eriksen, P. Nørgaard, and L. M. Rode 2008. Salivary secretion during meals in lactating dairy cattle. *Journal of Dairy Science* 91:2077-2081.

Becker, R. B., S. P. Marshall, and P. T. Dix Arnold. 1963. Anatomy, development, and functions of the bovine omasum. *Journal of Dairy Science* 46:835-839.

Ben-Ghedalia, D., H. Tagari, and A. Bondi. 1974. Protein digestion in the intestine of sheep. *British Journal of Nutrition* 31:125-142.

Bevans, D. W., K. A. Beauchemin, K. S. Shwartzkopf-Genswein, J. J. McKinnon, and T. A. McAllister. 2005. Effect of rapid or gradual grain adaptation on subacute acidosis and feed intake by feedlot cattle. *Journal of Animal Science* 83:1116-1132.

Bilk, S., K. Huhn, K. U. Honscha, H. Pfannkuche, and G. Gäbel. 2005. Bicarbonate exporting transporters in the ovine ruminal epithelium. *Journal of Comparative Physiology B* 175:365-374.

Bohatier, J. 1991. The rumen protozoa: Taxonomy, cytology and feeding behaviour. Pp. 217-239 in *Rumen Microbial Metabolism and Ruminant Digestion*, J. P: Jouany, ed. Paris, France: Institut National de la Recherche Agronomique.

Bost, J. 1970. Omasal physiology. Pp. 52-65 in *Physiology of Digestion and Metabolism in the Ruminant: Proceedings of the Third International Symposium, August, 1969, Cambridge, England*, UK, A.T. Phillipson, ed. Newcastle upon Tyne, England: Oriel Press.

Bowman, G. R., K. A. Beauchemin, and J. A. Shelford. 2003. Fibrolytic enzymes and parity effects on feeding behavior, salivation, and ruminal pH of lactating dairy cows. *Journal of Dairy Science* 86:565-575.

Brent, B. E. 1976. Relationship of acidosis to other feedlot ailments. *Journal of Animal Science* 43:930-935.

Britton, R., and C. Krehbiel. 1993. Nutrient metabolism by gut tissues. *Journal of Dairy Science* 76:2125-2131.

Brockman, R. P. 2005. Glucose and short-chain fatty acid metabolism. Pp. 291-310 in *Quantitative Aspects of Ruminant Digestion and Metabolism*, 2nd Ed., J. Dijkstra, J. M. Forbes, and J. France, eds. Wallingford, Oxfordshire, UK: CABI Publishing.

Broderick, G. A., J. R. Wallace, and E. R. Ørskov. 1991. Control of rate and extent of protein degradation. Pp. 541-592 in *Physiological Aspects of Digestion and Metabolism in Ruminants: Proceedings of the Seventh International Symposium on Ruminant Physiology*, T. Tsuda, Y. Sasaki, and R. Kawashima, eds. San Diego, CA: Academic Press.

Brown, M. S., C. R. Krehbiel, M. L. Galyean, M. D. Remmenga, J. P. Peters, B. Hibbard, J. Robinson, and W. M. Moseley. 2000. Evaluation of acute and subacute acidosis on dry matter intake, ruminal fermentation, blood chemistry, and endocrine profiles of beef steers. *Journal of Animal Science* 78:3155-3168.

Brown, M. S., C. H. Ponce, and R. Pulikanti. 2006. Adaptation of beef cattle to high-concentrate diets: Performance and ruminal metabolism. *Journal of Animal Science* 84:E25-E33.

Bryant, M. P., and L. A. Burkey. 1953. Cultural methods and some characteristics of some of the more numerous groups of bacteria in the bovine rumen. *Journal of Dairy Science* 36:205-217.

Budras, K.-D., R. E. Habel, A. Wünsche, and S. Buda. 2003. *Bovine Anatomy, An Illustrated Text*, 1st Ed. Hanover, Germany: Schlütersche GmbH & Co. KG, Verlag und. Druckerei.

Burrin, D. G., R. A. Stock, and R. A. Britton. 1988. Monensin level during grain adaptation and finishing performance in cattle. *Journal of Animal Science* 66:513-521.

Caple, I. W., and T. J. Heath. 1975. Biliary and pancreatic secretions: Their regulation and roles. Pp. 91-100 in *Digestion and Metabolism in the Ruminant*, I. W. McDonald, and A. C. I. Warner, eds. Armidale, NSW, Australia: University of New England Publishing Unit.

Carro, M. D., and E. L. Miller. 1999. Effect of supplementing a fibre basal diet with different N forms on ruminal fermentation and microbial growth in an in vitro semi-continuous culture system (RUSITEC). *British Journal of Nutrition* 82:149-157.

Carter, R. R., and W. L. Grovum. 1990. A review of the physiological significance of hypertonic body fluids on feed intake and ruminal function: Salivation, motility, and microbes. *Journal of Animal Science* 68:2811-2832.

Cassida, K. A., and M. R. Stokes. 1986. Eating and resting salivation in early lactation dairy cows. *Journal of Dairy Science* 69:1282-1292.

Castillo-Lopez, E., B. I. Wiese, S. Hendrick, J. J. McKinnon, T. A. McAllister, K. A. Beauchemin, and G. B. Penner. 2014. Incidence, prevalence, severity, and risk factors for ruminal acidosis in feedlot steers during backgrounding, diet transition, and finishing. *Journal of Animal Science* 92:3053-3063.

Chacon, E., and T. H. Stobbs. 1976. Influence of progressive defoliation of a grass sward on the eating behaviour of cattle. *Australian Journal of Agricultural Research* 27:709-727.

Chen, G., and J. B. Russell. 1989. More monensin-sensitive, ammonia-producing bacteria from the rumen. *Applied and Environmental Microbiology* 55:1052-1057.

Cheng, K.-J., and T. A. McAllister. 1997. Compartmentation in the rumen. Pp. 492-522 in *The Rumen Microbial Ecosystem*, P. N. Hobson and C. S. Stewart, eds. London: Blackie Academic & Professional.

Cherney, D. J. R., J. H. Cherney, and R. F. Lucey. 1993. In vitro digestion kinetics and quality of perennial grasses as influenced by forage maturity. *Journal of Dairy Science* 76:790-797.

Chikunya, S., C. J. Newbold, L. Rode, X. B. Chen, and R. J. Wallace. 1996. Influence of dietary rumen-degradable protein on bacterial growth in the rumen of sheep receiving different energy sources. *Animal Feed Science and Technology* 63:333-340.

Chung, Y.-H., E. J. McGeough, S. Acharya, T. A. McAllister, S. M. McGinn, O. M. Harstad, and K. A. Beauchemin. 2013. Enteric methane emission, diet digestibility, and nitrogen excretion from beef heifers fed sainfoin or alfalfa. *Journal of Animal Science* 91:4861-4874.

Church, D. C. 1988. Salivary function and production. Pp. 117-124 in *The Ruminant Animal, Digestive Physiology and Nutrition*, D. C. Church, ed. Englewood Cliffs, NJ: Prentice Hall.

Clark, J. H., T. H. Klusmeyer, and M. R. Cameron. 1992. Microbial protein synthesis and flows of nitrogen fractions to the duodenum of dairy cows. *Journal of Dairy Science* 75:2304-2323.

Chase, L. E., P. J.Wangness, and B. R. Baumgardt. 1976. Feeding behavior of steers fed a complete mixed ration. *Journal of Dairy Science* 59:1923-1928.

Cochran, R. C., D. C. Adams, J. D. Wallace, and M. L. Galyean. 1986. Predicting digestibility of different diets with internal markers: Evaluation of four potential markers. *Journal of Animal Science* 63:1476-1483.

Coe, M. L., T. G. Nagaraja, Y. D. Sun, N. Wallace, E. G. Towne, K. E. Kemp, and J. P. Hutcheson. 1999. Effect of virginiamycin on ruminal fermentation in cattle during adaptation to a high concentrate diet and during an induced acidosis. *Journal of Animal Science* 77:2259-2268.

Cook, D. I. 1995. Salivary secretion in ruminants. Pp. 153-170 in *Ruminant Physiology: Digestion, Metabolism, Growth and Reproduction: Proceedings of the Eighth International Symposium on Ruminant Physiology*, W. von Engelhardt, S. Leonhard-Marek, G. Breves, and G. Giesecke, eds. Stuttgart, Germany: Ferdinand Enke Verlag.

Cooper, R. J., T. J. Klopfenstein, R. A. Stock, C. T. Milton, D. W. Herold, and J. C. Parrott. 1999. Effects of imposed feed intake variation on acidosis and performance of finishing steers. *Journal of Animal Science* 77:1093-1099.

Corona, L., F. N. Owens, and R. A. Zinn. 2006. Impact of corn vitreousness and processing on site and extent of digestion by feedlot cattle. *Journal of Animal Science* 84:3020-3031.

Counotte, G. H., A. T. van't Klooster, J. van der Kuilen, and R. A. Prins. 1979. An analysis of the buffer system in the rumen of dairy cattle. *Journal of Animal Science* 49:1536-1544.

Crevieu-Gabriel, I., J. Gomez, J. P. Caffin, and B. Carre. 1999. Comparison of pig and chicken pepsins for protein hydrolysis. *Reproduction, Nutrition, Development* 39:443-454.

Danielli, J. F., M. W. Hitchcock, R. A. Marshall, and A. T. Phillipson. 1945. The mechanism of absorption from the rumen as exemplified by the behaviour of acetic, propionic and butyric acids. *Journal of Experimental Biology* 22:75-84.

Dehority, B. A., and C. G. Orpin. 1988. Development of, and natural fluctuations in, rumen microbial populations. Pp. 196-245 in *The Rumen Microbial Ecosystem*, P. N. Hobson, ed. London, UK: Elsevier Applied Science.

Demeyer, D. I. 1991. Quantitative aspects of microbial metabolism in the rumen and hind gut. Pp. 217-239 in *Rumen Microbial Metabolism and Ruminant Digestion*, J. P. Jouany, ed. Paris, France: Institut National de la Recherche Agronomique.

DeVries, T. J., F. Dohme, and K. A. Beauchemin. 2008. Repeated ruminal acidosis challenges in lactating dairy cows at high and low risk for developing acidosis: Feed sorting. *Journal of Dairy Science* 91:3958-3967.

Dhanoa, M. S., R. C. Siddons, J. France, and D. L. Gale. 1985. A multicompartmental model to describe marker excretion patterns in ruminant faeces. *British Journal of Nutrition* 53:663-671.

Dhanoa, M. S., J. France, R. C. Siddons, S. Lopez, and J. B. Buchanan-Smith. 1995. A non-linear compartmental model to describe forage degradation kinetics during incubation in polyester bags in the rumen. *British Journal of Nutrition* 73:3-15.

Dijkstra, J., H. Boer, J. Van Bruchem, M. Bruining, and S. Tamminga. 1993. Absorption of volatile fatty acids from the rumen of lactating dairy cows as influenced by volatile fatty acid concentration, pH and rumen liquid volume. *British Journal of Nutrition* 69:385-396.

Dijkstra, J., J. L. Ellis, E. Kebreab, A. B. Strathe, S. López, J. France, and A. Bannink. 2012. Ruminal pH regulation and nutritional consequences of low pH. *Animal Feed Science and Technology* 172:22-33.

Dirksen, G. U., H. G. Liebich, and E. Mayer. 1985. Adaptive changes of the ruminal mucosa and their functional and clinical significance. *Bovine Practitioner* 20:116-120.

Dong, Y. 1990. The Significance of Chewing During Eating and Rumination on Forage Digestion in Cattle. M.S. Thesis. University of Guelph, Guelph, ON.

Dougherty, R. W., W. E. Stewart, M. M. Nold, I. L. Lindahl, C. H. Mullenax, and B. F. Leek. 1962. Pulmonary absorption of eructated gas in ruminants. *American Journal of Veterinary Research* 205-212.

Dougherty, R. W., M. J. Allison, and C. H. Mullenax. 1964. Physiological disposition of C^{14}-labeled rumen gases in sheep and goats. *American Journal of Physiology* 207:1181-1188.

Ellis, W. C., J. H. Matis, and C. Lascano. 1979. Quantitating ruminal turnover. *Federation Proceedings* 38:2702-2706.

Ellis, W. C., J. H. Matis, T. M. Hill, and M. R. Murphy. 1994. Methodology for estimating digestion and passage kinetics of forages. Pp. 682-756 in *Forage Quality, Evaluation, and Utilization*, G. C. Fahey, ed. Madison, WI: American Society of Agronomy, Crop Science Society of America, Soil Science Society of America.

Ellis, J. L., J. Dijkstra, E. Kebreab, A. Bannink, N. E. Odongo, B. W. McBride, and J. France. 2008. Aspects of rumen microbiology central to mechanistic modelling of methane production in cattle. *Journal of Agricultural Science* 146:213-233.

Engelhardt, W. V., and R. Hauffe, 1975. Role of the omasum in absorption and secretion of water and electrolytes in sheep and goats. Pp. 216-230 in *Digestion and Metabolism in the Ruminant: Proceedings of the IV International Symposium on Ruminant Physiology*, I. W. McDonald and A. C. I. Warner, eds. Armidale, NSW, Australia: University of New England Publishing Unit.

Erdman, R. A. 1988. Dietary buffering requirements of the lactating dairy cow: A review. *Journal of Dairy Science* 71:3246-3266.

Erickson, G. E., C. T. Milton, K. C. Fanning, R. J. Cooper, R. S. Swingle, J. C. Parrott, G. Vogel, and T. J. Klopfenstein. 2003. Interaction between bunk management and monensin concentration on finishing performance, feeding behavior, and ruminal metabolism during an acidosis challenge with feedlot cattle. *Journal of Animal Science* 81:2869-2879.

Fahey, G. C., Jr., and L. L. Berger. 1988. Carbohydrate nutrition of ruminants. Pp. 269-297 in *The Ruminant Animal—Digestive Physiology and Nutrition*, D. C. Church, ed. Englewood Cliffs, NJ: Prentice Hall.

Faichney, G. J. 1986. The kinetics of particulate matter in the rumen. Pp. 173-195 in *Control of Digestion and Metabolism in Ruminants: Proceedings of the Sixth International Symposium on Ruminant Physiology*, L. P. Milligan, W. L. Grovum, and A. Dobson, eds. Englewood Cliffs, NJ: Prentice Hall.

Faichney, G. J. 1993. Digesta flow. Pp. 53-85 in *Quantitative Aspects of Ruminant Digestion and Metabolism*, J. M. Forbes and J. France, eds. Wallingford, Oxon, UK: CAB International.

Fell, B. F., and T. E. C. Weekes. 1975. Food intake as a mediator of adaptation in the rumen epithelium. Pp. 101-118 in *Digestion and Metabolism in the Ruminant: Proceedings of the Fourth International Symposium on Ruminant Physiology*, I. W. McDonald and A. C. I. Warner, eds. Armidale, NSW, Australia: University of New England Publishing Unit.

Firkins, J. L., and Z. Yu. 2006. Characterization and quantification of the microbial populations of the rumen. Pp. 19-54 in *Ruminant Physiology. Digestion, Metabolism and Impact of Nutrition on Gene Expression, Immunology and Stress*, K. Sejrsen, T. Hvelplund, and M. O. Nielson, eds. Wageningen, The Netherlands: Wageningen Academic.

Firkins J. L., M. S. Allen, B. S. Oldick, and N. R. St-Pierre. 1998. Modeling ruminal digestibility of carbohydrates and microbial protein flow to the duodenum. *Journal of Dairy Science* 81:3350-3369.

Firkins, J. L., M. L. Eastridge, N. R. St-Pierre, and S. M. Noftsger. 2001. Effects of grain variability and processing on starch utilization by lactating dairy cattle. *Journal of Animal Science* 79(Suppl.):E218-E238.

Firkins, J. L., Z. Yu, and M. Morrison. 2007. Ruminal nitrogen metabolism: Perspectives for integration of microbiology and nutrition for dairy. *Journal of Dairy Science* 90(Suppl.):E1-E16.

Flint, H. J., and E. A. Bayer. 2008. Plant cell wall breakdown by anaerobic microorganisms from the mammalian digestive tract. *Annals of the New York Academy of Sciences* 1125:280-288.

France, J., and R. C. Siddons. 1993. Volatile fatty acid production. Pp. 107-121 in *Quantitative Aspects of Ruminant Digestion and Metabolism*, J. M. Forbes and J. France, eds. Wallingford, Oxon, UK: CAB International.

France, J., J. H. M. Thornley, M. S. Dhanoa, and R. C. Siddons. 1985. On the mathematics of digesta flow kinetics. *Journal of Theoretical Biology* 113:743-758.

Fu, C. J., E. E. Felton, J. W. Lehmkuhler, and M. S. Kerley. 2001. Ruminal peptide concentration required to optimize microbial growth and efficiency. *Journal of Animal Science* 79:1305-1312.

Fulton, W. R., T. J. Klopfenstein, and R. A. Britton. 1979a. Adaptation to high concentrate diets by beef cattle. I. Adaptation to corn and wheat diets. *Journal of Animal Science* 49:775-784.

Fulton, W. R., T. J. Klopfenstein, and R. A. Britton. 1979b. Adaptation to high concentrate diets by beef cattle. II. Effect of ruminal pH alteration on rumen fermentation and voluntary intake of wheat diets. *Journal of Animal Science* 49:785-789.

Gäbel, G., M. Bestmann, and H. Martens. 1991. Influences of diet, short-chain fatty acids, lactate and chloride on bicarbonate movement across the reticulo-rumen wall of sheep. *Journal of Veterinary Medicine Series A* 38:523-529.

Gäbel, G., J. R. Aschenbach, and F. Müller. 2002. Transfer of energy substrates across the ruminal epithelium: Implications and limitations. *Animal Health Research Reviews* 3:15-30.

Galyean, M. L., and L. O. Tedeschi. 2014. Predicting microbial protein synthesis in beef cattle: Relationship to intakes of total digestible nutrients and crude protein. *Journal of Animal Science* 92:5099-5111.

Garton, G. A. 1965. The digestion and assimilation of lipids. Pp. 390-398 in *Physiology of Digestion in the Ruminant*, R. W. Dougherty, ed. Washington, DC: Butterworths.

Gibb, D. J., T. A. McAllister, C. Huisma, and R. D. Wiedmeier. 1998. Bunk attendance of feedlot cattle monitored with radio frequency technology. *Canadian Journal of Animal Science* 78:707-710.

Goad, D. W., C. L. Goad, and T. G. Nagaraja. 1998. Ruminal microbial and fermentative changes associated with experimentally induced subacute acidosis in steers. *Journal of Animal Science* 76:234-241.

Gonyou, H., and W. Stricklin. 1981. Eating behavior of beef cattle groups fed from a single stall or trough. *Applied Animal Ethology* 7:123-133.

González, L. A., X. Manteca, S. Calsamiglia, K. S. Schwartzkopf-Genswein, and A. Ferret. 2012. Ruminal acidosis in feedlot cattle: Interplay between feed ingredients, rumen function and feeding behaviour (a review). *Animal Feed Science and Technology* 172:66-79.

Gottschalk, G. 1986. *Bacterial Metabolism*, 2nd Ed. New York: Springer-Verlag.

Graham, C., and N. L. Simmons. 2005. Functional organization of the bovine rumen epithelium. *American Journal of Physiology—Regulatory, Integrative, and Comparative Physiology* 288:R173-R181.

Grant, R. J., and S. J. Weidner. 1992. Digestion kinetics of fiber: Influence of in vitro buffer pH varied within observed physiological range. *Journal of Dairy Science* 75:1060-1068.

Gregory, P. C. 1987. Inhibition of reticulo-ruminal motility by volatile fatty acids and lactic acid in sheep. *Journal of Physiology* 382:355-371.

Gressley, T. F., M. B. Hall and L. E. Armentano. 2011. Productivity, digestion, and health responses to hindgut acidosis in ruminants. *Journal of Animal Science* 89:1120-1130.

Griswold, K. E., B. A. Apgar, J. Bouton, and J. L. Firkins. 2003. Effects of urea infusion and ruminal degradable protein concentration on microbial growth, digestibility, and fermentation in continuous culture. *Journal of Animal Science* 81:329-336.

Grovum, W. L., and V. J. Williams. 1973. Rate of passage of digesta in sheep. 4. Passage of marker through the alimentary tract and the biological relevance of rate-constants derived from the changes in concentration of marker in faeces. *British Journal of Nutrition* 30:313-329.

Guan, D., and G. M. Green. 1996. Significance of peptic digestion in rat pancreatic secretory response to dietary protein. *American Journal of Physiology* 271:G42-G47.

Harmeyer, J. 1986. Energy expenditures of rumen microbes for maintenance. *Archiv für Tierernaehrung* 36:143-148.

Harmon, D. L. 1992. Dietary influences on carbohydrases and small intestinal starch hydrolysis capacity in ruminants. *Journal of Nutrition* 122:203-210.

Harmon, D. L. 1993. Nutritional regulation of postruminal digestive enzymes in ruminants. *Journal of Dairy Science* 76:2102-2111.

Harmon, D. L., and K. R. McLeod. 2001. Glucose uptake and regulation by intestinal tissues: Implications and whole-body energetics. *Journal of Animal Science* 79(Suppl.):E59-E72.

Harmon, D. L., R. A. Britton, R. L. Prior, and R. A. Stock. 1985. Net portal absorption of lactate and volatile fatty acids in steers experiencing glucose-induced acidosis or fed a 70% concentrate diet ad libitum. *Journal of Animal Science* 60:560-569.

Harmon, D. L., R. M. Yamka, and N. A. Elam. 2004. Factors affecting intestinal starch digestion in ruminants: A review. *Canadian Journal of Animal Science* 84:309-318.

Harrison, F. A., and K. J. Hill. 1962. Digestive secretions and the flow of digesta along the duodenum of the sheep. *Journal of Physiology* 162:225-243.

Hespell, R. B., and M. P. Bryant. 1979. Efficiency of rumen microbial growth: Influence of some theoretical and experimental factors on Y_{ATP}. *Journal of Animal Science* 49:1640-1659.

Hibbard, B., J. P. Peters, S. T. Chester, J. A. Robinson, S. F. Kotarski, W. J. Croom, Jr., and W. M. Hagler, Jr. 1995. The effect of slaframine on salivary output and subacute and acute acidosis in growing beef steers. *Journal of Animal Science* 73:516-525.

Hironaka, R., K. A. Beauchemin, and T. J. Lysyk. 1992. The effect of thickness of steam-rolled barley on its utilization by beef cattle. *Canadian Journal of Animal Science* 72:279-286.

Hofmann, R. R. 1988. Anatomy of the gastro-intestinal tract. Pp. 14-43 in *The Ruminant Animal—Digestive Physiology and Nutrition*, D. C. Church, ed. Englewood Cliffs, NJ: Prentice Hall.

Holtshausen, L., K. S. Schwartzkopf-Genswein, and K. A. Beauchemin. 2013. Ruminal pH profile and feeding behaviour of feedlot cattle transitioning from a high-forage to a high concentrate diet. *Canadian Journal of Animal Science* 93:529-533.

Hoover, W. H. 1978. Digestion and absorption in the hindgut of ruminants. *Journal of Animal Science* 46:1789-1799.

Hoover, W. H., and S. R. Stokes. 1991. Balancing carbohydrates and protein for optimum rumen microbial yield. *Journal of Dairy Science* 74:3630-3644.

Horton, G. M. J. 1978. The intake and digestibility of ammoniated cereal straws by cattle. *Canadian Journal of Animal Science* 58:471-478.

Hristov, A. N., and J.-P. Jouany. 2005. Factors affecting the efficiency of nitrogen utilization in the rumen. Pp. 117-166 in *Nitrogen and Phosphorus Nutrition of Cattle: Reducing the Environmental Impact of Cattle Operations*, E. Pfeffer and A. N. Hristov, eds. Cambridge, MA: CAB International.

Huber, J. T., N. L. Jacobson, R. S. Allen, and P. A. Hartman. 1961. Digestive enzyme activities in the young calf. *Journal of Dairy Science* 44:1494-1501.

Huber, J. T., R. J. Rifkin, and J. M. Keith. 1964. Effect of level of lactose upon lactase concentrations in the small intestines of young calves. *Journal of Dairy Science* 47:789-792.

Huhtanen, P., S. Ahvenjärvi, M. R. Weisbjerg, and P. Nørgaard. 2006. Digestion and passage of fibre in ruminants. Pp. 87-138 in *Ruminant Physiology. Digestion, Metabolism and Impact of Nutrition on Gene Expression, Immunology and Stress*, K. Sejrsen, T. Hvelplund, and M. O. Nielson, eds. Wageningen, The Netherlands: Wageningen Academic.

Hungate, R. E. 1966. *The Rumen and Its Microbes*. New York: Academic Press.

Hünerberg, M., S. M. McGinn, K. A. Beauchemin, E. K. Okine, O. M. Harstad, and T. A. McAllister. 2013. Effect of dried distillers' grains with solubles on enteric methane emissions and nitrogen excretion from finishing beef cattle. *Canadian Journal of Animal Science* 93:373-385.

Huntington, G. B. 1997. Starch utilization by ruminants: From basics to the bunk. *Journal of Animal Science* 75:852-867.

Huntington, G. B., D. L. Harmon, and C. J. Richards. 2006. Sites, rates, and limits of starch digestion and glucose metabolism in growing cattle. *Journal of Animal Science* 84(Suppl.):E14-E24.

Immig, I. 1996. The rumen and hindgut as source of ruminant methanogenesis. *Environmental Monitoring and Assessment* 42:57-72.

Jami, E., and I. Mizrahi. 2012. Composition and similarity of bovine rumen microbiota across individual animals. *PLoS ONE* 7:e33306.

Jami, E., A. Israel, A. Kotser, and I. Mizrahi. 2013. Exploring the bovine rumen bacterial community from birth to adulthood. *ISME Journal* 7:1069-1079.

Janes, A. N., T. E. C. Weekes, and D. G. Armstrong. 1985. Carbohydrase activity in the pancreatic tissue and small intestine mucosa of sheep fed dried-grass or ground maize-based diets. *Journal of Agricultural Science* 104:435-443.

Janssen, P. H. 2010. Influence of hydrogen on rumen methane formation and fermentation balances through microbial growth kinetics and fermentation thermodynamics. *Animal Feed Science and Technology* 160:1-22.

Janssen, P. H., and M. Kirs. 2008. Structure of the Archaeal community of the rumen. *Applied and Environmental Microbiology* 74:3619-3625.

Jensen, R., H. M. Deane, L. J. Cooper, V. A. Miller, and W. R. Graham. 1954. The rumenitis-liver abscess complex in beef cattle. *American Journal of Veterinary Research* 55:202-216.

Kahn, C. M. 2010. *Merck Veterinary Manual*, 20th Ed. Whitehouse Station, NJ: Merck & Co.

Kim, Y. S., W. Birthwhistle, and Y. W. Kim. 1972. Peptide hydrolases in the brush border and soluble fractions of small intestinal mucosa of rat and man. *Journal of Clinical Investigation* 51:1419-1430.

Kim, Y. S., Y. W. Kim, and M. H. Sleisenger. 1974. Studies on the properties of peptide hydrolases in the brush border and soluble fractions of small intestinal mucosa of rat and man. *Biochimica et Biophysica Acta* 370:283-296.

Kitamoto, Y., X. Yuan, Q. Wu, D. W. McCourt, and J. E. Sadler. 1994. Enterokinase, the initiator of intestinal digestion, a mosaic protease composed of a distinctive assortment of domains. *Proceedings of the National Academy of Sciences of the United States of America* 91:7588-7592.

Kleen, J. L., G. A. Hooijer, J. Rehage, and J. P. T. M. Noordhuizen. 2003. Subacute ruminal acidosis (SARA): A review. *Journal of Veterinary Medicine Series A* 50:406-414.

Kohn, R. A., and T. F. Dunlap. 1998. Calculation of the buffering capacity of bicarbonate in the rumen and in vitro. *Journal of Animal Science* 76:1702-1709.

Kramer, T., T. Michelberger, H. Gurtler, and G. Gäbel. 1996. Absorption of short-chain fatty acids across ruminal epithelium of sheep. *Journal of Comparative Physiology B* 166:262-269.

Krause, D. O., S. E. Denman, R. I. Mackie, M. Morrison, A. L. Rae, G. T. Attwood, and C. S. McSweeney. 2003. Opportunities to improve fiber degradation in the rumen: Microbiology, ecology, and genomics. *FEMS Microbiology Reviews* 27:663-693.

Krause, D. O., T. G. Nagaraja, A. D. G. Wright, and T. R. Callaway. 2013. Rumen microbiology: Leading the way in microbial ecology. *Journal of Animal Science* 91:331-341.

Krause, M., K. A. Beauchemin, L. M. Rode, B. I. Farr, and P. Nørgaard. 1998. Fibrolytic enzyme treatment of barley grain and source of forage in high-grain diets fed to growing cattle. *Journal of Animal Science.* 76:2912-2920.

Krehbiel, C. R., and J. C. Matthews. 2003. Absorption of amino acids and peptides. Pp. 41-70 in *Amino Acids in Animal Nutrition*, 2nd Ed., J. P. F. D'Mello, ed. Wallingford, Oxfordshire, UK: CAB International.

Krehbiel, C. R., R. A. Britton, D. L. Harmon, J. P. Peters, R. A. Stock, and H. E. Grotjan. 1996. Effects of varying levels of duodenal or midjejunal glucose and 2-deoxyglucose infusion on small intestinal disappearance and net portal glucose flux in steers. *Journal of Animal Science* 74:693-700.

Kreikemeier, K. K., and D. L. Harmon. 1995. Abomasal glucose, maize starch and maize dextrin infusions in cattle: Small intestinal disappearance, net portal glucose flux and ileal oligosaccharide flow. *British Journal of Nutrition* 73:763-772.

Kreikemeier, K. K., D. L. Harmon, J. P. Peters, K. L. Gross, C. K. Armendariz, and C. R. Krehbiel. 1990. Influence of dietary forage and feed intake on carbohydrase activities and small intestinal morphology of calves. *Journal of Animal Science* 68:2916-2929.

Kreikemeier, K. K., D. L. Harmon, R. T. Brandt, Jr., T. B. Avery, and D. E. Johnson. 1991. Small intestinal starch digestion in steers: Effect of various levels of abomasal glucose, corn starch and corn dextrin infusion on small intestinal disappearance and net glucose absorption. *Journal of Animal Science* 69:328-338.

Kristensen, N. B., and D. L. Harmon. 2004. Splanchnic metabolism of volatile fatty acids absorbed from the washed reticulorumen of steers. *Journal of Animal Science* 82:2033-2042.

Kristensen, V. F., and P. Nørgaard. 1987. The effect of roughage quality and physical structure of the diet on feed intake and milk yield of the dairy cow. Pp 79-91 in *Cattle Production Research: Danish Status and Perspectives*. Copenhagen: Landhusholdingsselskabets Forlag.

Kwong, W. K., L. Seetharam, and D. H. Aplers. 1978. Effect of exocrine pancreatic insufficiency on small intestine in the mouse. *Gastroenterology* 74:1277-1282.

Lapierre, H., and G. E. Lobley. 2001. Nitrogen recycling in the ruminant: A review. *Journal of Dairy Science* 84(Suppl.):E223-E236.

Leat, W. M. F., and F. A. Harrison. 1975. Digestion, absorption, and transport of lipids in the sheep. Pp. 481-495 in *Digestion and Metabolism in the Ruminant: Proceedings of the Fourth International Symposium on Ruminant Physiology*, I. W. McDonald and A. C. I. Warner, eds. Armidale, NSW, Australia: University of New England Publishing Unit.

Lee, J. A., and G. R. Pearce. 1984. The effectiveness of chewing during eating on particle size reduction of roughages by cattle. *Australian Journal of Agricultural Research* 35:609-618.

Leek, B. F. 1986. Sensory receptors in the ruminant alimentary tract. Pp. 3-17 in *Control of Digestion and Metabolism in Ruminants: Proceedings of the Sixth International Symposium on Ruminant Physiology*, L. P. Milligan, W. L. Grovum, and A. Dobson, eds. Englewood Cliffs, NJ: Prentice Hall.

Leek, B. F. 2004. Digestion in the ruminant stomach. Pp. 438-474 in *Duke's Physiology of Domestic Animals*, 12th Ed., W. O. Reece, ed. Ithaca, NY: Cornell University Press.

Leek, B. F., and R. H. Harding. 1975. Sensory nervous receptors in the ruminant stomach and the reflex control of reticulo-ruminal motility. Pp. 60-76 in *Digestion and Metabolism in the Ruminant: Proceedings of the Fourth International Symposium on Ruminant Physiology*, I. W. McDonald and A. C. I. Warner, eds. Armidale, NSW, Australia: University of New England Publishing Unit.

Li, Y. L., T. A. McAllister, K. A. Beauchemin, M. L. He, J. J. McKinnon, and W. Z Yang. 2011. Substitution of wheat dried distillers grains with solubles for barley grain or barley silage in feedlot cattle diets: Intake, digestibility, and ruminal fermentation. *Journal of Animal Science* 89:2491-2501.

Lindberg, J. E. 1985. Estimation of rumen degradability of feed proteins with the in sacco technique and various in vitro methods: A review. *Acta Agriculturae Scandinavica* 25(Suppl.):64-97.

Lindberg, J. E., and T. Varvikko. 1982. The effect of bag pore size on the ruminal degration of dry matter, nitrogenous compounds and cell walls in nylon bags. *Swedish Journal of Agricultural Research* 12:163-171.

Lindsay, D. B. 1970. Carbohydrate metabolism in ruminants. Pp. 438-451 in *Physiology of Digestion and Metabolism in the Ruminant: Proceedings of the Third International Symposium, August 1969, Cambridge, England*, A. T. Phillipson, ed. Newcastle upon Tyne, England: Oriel Press.

López, S. 2005. In vitro and in situ techniques for estimating digestibility. Pp. 87-121 in *Quantitative Aspects of Ruminant Digestion and Metabolism*, J. Dijkstra, J. M. Forbes, and J. France, eds. Wallingford, UK: CAB International.

Lyford, S. J. 1988. Growth and development of the ruminant digestive system. Pp 44-63 in *The Ruminant Animal: Digestive Physiology and Nutrition*, D. C. Church, ed. Englewood Cliffs, NJ: Prentice Hall.

Lynd, L. R., P. J. Weimer, W. H. van Zyl, and I. S. Pretorius. 2002. Microbial cellulose utilization: Fundamentals and biotechnology. *Microbiology and Molecular Biology Reviews* 66:506-577.

Maekawa, M., K. A. Beauchemin, and D. A. Christensen. 2002a. Chewing activity, saliva production, and ruminal pH of primiparous and multiparous lactating dairy cows. *Journal of Dairy Science* 85:1176-1182.

Maekawa, M., K. A. Beauchemin, and D. A. Christensen. 2002b. Effect of concentrate level and feeding management on chewing activities, saliva production, and ruminal pH of lactating dairy cows. *Journal of Dairy Science* 85:1165-1175.

Marden, J. P., C. Bayourthe, F. Enjalbert, and R. Moncoulon. 2005. A new device for measuring kinetics of ruminal pH and redox potential in dairy cattle. *Journal of Dairy Science* 88:277-281.

Masson, F. M., and A. E. Oxford. 1951. The action of the ciliates of the sheep's rumen upon various water-soluble carbohydrates, including polysaccharides. *Journal of General Microbiology* 5:664-672.

Masson, M. J., and A. T. Phillipson. 1951. The absorption of acetate, propionate and butyrate from the rumen of sheep. *Journal of Physiology* 113:189-206.

Matis, J. H. 1972. Gamma time-dependency in Blaxter's compartment model. *Biometrics* 28:597-602.

Matis, J. H., T. E. Wehrly, and W. C. Ellis. 1989. Some generalized stochastic compartment models for digesta flow. *Biometrics* 45:703-720.

McAllister, T. A., H. D. Bae, G. A. Jones, and K. J. Cheng. 1994. Microbial attachment and feed digestion in the rumen. *Journal of Animal Science* 72:3004-3018.

McDonald, P., R. A. Edwards, J. F. D. Greenhalgh, C. A. Morgan, L. A. Sinclair, and R. G. Wilkinson, eds. 2010. *Animal Nutrition*, 7th Ed. Harlow, Essex, England: Pearson Education.

McDougall, E. I. 1948. Studies on ruminant saliva. 1. The composition and output of sheep's saliva. *Biochemical Journal* 43:99-109.

McLeod, M. N., and D. J. Minson. 1988. Breakdown of large particles in forage by simulated digestion and detrition. *Journal of Animal Science* 66:1000-1004.

Merchen, N. R. 1988. Digestion, absorption and excretion in ruminants. Pp. 172-201 in *The Ruminant Animal: Digestive Physiology and Nutrition*, D. C. Church, ed. Englewood Cliffs, NJ: Prentice Hall.

Mertens, D. R. 1993a. Kinetics of cell wall digestion and passage in ruminants. Pp. 535-570 in *Forage Cell Wall Structure and Digestibility*, H. G. Jung, D. R. Buxton, R. D. Hatfield, and J. Ralph, eds. Madison, WI: American Society of Agronomy, Crop Science Society of America, Soil Science Society of America.

Mertens, D. R. 1993b. Rate and extent of digestion. Pp.13-51 in *Quantitative Aspects of Ruminant Digestion and Metabolism*, J. M. Forbes and J. France, eds. Wallingford, Oxfordshire, UK: CAB International.

Miller, T. L. 1995. Ecology of methane production and hydrogen sinks in the rumen. Pp. 317-331 in *Ruminant Physiology: Digestion, Metabolism, Growth and Reproduction: Proceedings of the Eighth International Symposium on Ruminant Physiology*, W. von Engelhardt, S. Leonhard-Marek, G. Breves, and G. Giesecke, eds. Stuttgart, Germany: Ferdinand Enke Verlag.

Miron, J., D. Ben-Ghedalia, and M. Morrison. 2001. Invited review: Adhesion mechanisms of rumen cellulolytic bacteria. *Journal of Dairy Science* 84:1294-1309.

Moya, D., A. Mazzenga, L. Holtshausen, G. Cozzi, L. A. González, S. Calsamiglia, D. G. Gibb, T. A. McAllister, K. A. Beauchemin, and K. Schwartzkopf-Genswein. 2011. Feeding behavior and ruminal acidosis in beef cattle offered a total mixed ration or dietary components separately. *Journal of Animal Science* 89:520-530.

Murphy, M. R., and P. M. Kennedy. 1993. Particle dynamics. Pp. 88-105 in *Quantitative Aspects of Ruminant Digestion and Metabolism*, J. M. Forbes and J. France, eds. Wallingford, Oxfordshire, UK: CAB International.

Nagaraja, T. G., and M. M. Chengappa. 1998. Liver abscesses in feedlot cattle: A review. *Journal of Animal Science* 76:287-298.

Nagaraja, T. G., and E. C. Titgemeyer. 2007. Ruminal acidosis in beef cattle: The current microbiological and nutritional outlook. *Journal of Dairy Science* 90:E17-E38.

Nocek, J. 1985. Evaluation of specific variables affecting in situ estimates of ruminal dry matter and protein digestion. *Journal of Animal Science* 60:1347-1358.

Nocek, J. E. 1988. In situ and other methods to estimate ruminal protein and energy digestibility: A review. *Journal of Dairy Science* 71:2051-2069.

Nocek, J. 1997. Bovine acidosis: Implications on laminitis. *Journal of Dairy Science* 80:1005-1028.

Nocek, J. E., and J. E. English. 1986. In situ degradation kinetics: Evaluation of rate determination procedure. *Journal of Dairy Science* 69:77-87.

Nolan, J. V. 1993. Nitrogen kinetics. Pp. 123-143 in *Quantitative Aspects of Ruminant Digestion and Metabolism*, J. M. Forbes and J. France, eds. Wallingford, Oxfordshire, UK: CAB International.

Nóren, O., E. Dabelsteen, H. Sjostrom, and L. Josefsson. 1977. Histological localization of two dipeptidases in the pig small intestine and liver using immunofluorescence. *Gastroenterology* 72:87-92.

NRC (National Research Council). 1985. *Ruminant Nitrogen Usage*. Washington, DC: National Academy Press.

NRC. 1996. *Nutrient Requirements of Beef Cattle*, 7th Rev. Ed. Washington, DC: National Academy Press.

NRC. 2000. *Nutrient Requirements of Beef Cattle: Update 2000*, 7th Rev. Ed. Washington, DC: National Academy Press.

NRC. 2001. *Nutrient Requirements of Dairy Cattle*, 7th Rev. Ed. Washington, DC: National Academy Press.

Oba, M., and M. S. Allen. 2003a. Effects of corn grain conservation method on feeding behavior and productivity of lactating dairy cows at two dietary starch concentrations. *Journal of Dairy Science* 86:174-183.

Oba, M., and M. S. Allen. 2003b. Effects of corn grain conservation method on ruminal digestion kinetics for lactating dairy cows at two dietary starch concentrations. *Journal of Dairy Science* 86:184-194.

Ørskov, E. R. 2000. The in situ technique for the estimation of forage degradability in ruminants. Pp. 175-188 in *Forage Evaluation in Ruminant Nutrition*, D. I. Givens, E. Owen, E., R. F. E. Axford, and H. M. Omed, eds. Wallingford, UK: CAB International.

Ørskov, E. R., and I. McDonald. 1979. The estimation of protein degradability in the rumen from incubation measurements weighted according to rate of passage. *Journal of Agricultural Science* 92:499-503.

Owens, F. N., and A. L. Goetsch. 1986. Digesta passage and microbial protein synthesis. Pp. 196-223 in *Control of Digestion and Metabolism in Ruminants: Proceedings of the Sixth International Symposium on Ruminant Physiology*, L. P. Milligan, W. L. Grovum, and A. Dobson, eds. Englewood Cliffs, NJ: Prentice Hall.

Owens, F. N., and A. L. Goetsch. 1988. Ruminal fermentation. Pp. 145-171 in *The Ruminant Animal: Digestive Physiology and Nutrition*, D. C. Church, ed. Englewood Cliffs, NJ: Prentice Hall.

Owens, F. N., R. A. Zinn, and Y. K. Kim. 1986. Limits to starch digestion in the ruminant small intestine. *Journal of Animal Science* 63:1634-1648.

Owens, F. N., D. S. Secrist, W. J. Hill, and D. R. Gill. 1998. Acidosis in cattle: A review. *Journal of Animal Science* 76:275-286.

Paton, L. J., K. A. Beauchemin, D. M. Veira, and M. A. G. von Keyserlingk. 2006. Use of sodium bicarbonate, offered free choice or blended into the ration, to reduce the risk of ruminal acidosis in cattle. *Canadian Journal of Animal Science* 86:429-437.

Penner, G. B., K. A. Beauchemin, and T. Mutsvangwa. 2007. The severity of ruminal acidosis in primiparous Holstein cows during the periparturient period. *Journal of Dairy Science* 90:365-375.

Penner, G. B., J. R. Aschenbach, G. Gäbel, and M. Oba. 2009a. Epithelial capacity for the apical uptake of short-chain fatty acids is a key determinant for intra-ruminal pH and the susceptibility to sub-acute ruminal acidosis in sheep. *Journal of Nutrition* 139:1714-1720.

Penner, G. B., M. Taniguchi, L. L. Guan, K. A. Beauchemin, and M. Oba. 2009b. Effect of dietary forage to concentrate ratio on volatile fatty acid absorption and the expression of genes related to volatile fatty acid absorption and metabolism in ruminal tissue. *Journal of Dairy Science* 92:2767-2781.

Penner, G. B., M. Oba, G. Gäbel, and J. R. Aschenbach. 2010. A single mild episode of subacute ruminal acidosis does not affect ruminal barrier function in the short term. *Journal of Dairy Science* 93:4838-4845.

Petri, R. M., T. Schwaiger, G. B. Penner, K. A. Beauchemin, R. J. Forster, J. J. McKinnon, and T. A. McAllister. 2013a. Changes in the rumen epimural bacterial diversity of beef cattle as affected by diet and induced ruminal acidosis. *Applied and Environmental Microbiology* 79:3744-3755.

Petri, R. M., T. Schwaiger, G. B. Penner, K. A. Beauchemin, R. J. Forster, J. J. McKinnon, and T. A. McAllister. 2013b. Characterization of the core rumen microbiome in cattle during transition from forage to concentrate as well as during and after an acidotic challenge. *PLoS ONE* 8:e83424.

Phillipson, A. T., and J. L. Mangan. 1959. Bloat in cattle. XVI. Bovine saliva: The chemical composition of the parotid, submaxillary, and residual secretions. *New Zealand Journal of Agricultural Research* 2:990-1001.

Pierzynowski, S. G. 1986. The secretion of pancreatic juice in sheep in different fed treatments. *Polskie Archiwum Weterynaryjne* (Warszawa) 26:31-39.

Pitt, R. E., J. S. Van Kessel, D. G. Fox, A. N. Pell, M. C. Barry, and P. J. Van Soest. 1996. Prediction of ruminal volatile fatty acids and pH within the net carbohydrate and protein system. *Journal of Animal Science* 74:226-244.

Piveteau, P. 1999. Metabolism of lactate and sugars by dairy propionibacteria: A review. *Le Lait* 79:23-41.

Plaizier, J. C., D. O. Krause, G. N. Gozho, and B. W. McBride. 2008. Subacute ruminal acidosis in dairy cows: The physiological causes, incidence and consequences. *Veterinary Journal* 176:21-31.

Pond, K. R., W. C. Ellis, J. H. Matis, H. M. Ferreiro, and J. D. Sutton. 1988. Compartment models for estimating attributes of digesta flow in cattle. *British Journal of Nutrition* 60:571-595.

Putnam, P. A., R. Lehmann, and R. E. Davis. 1966. Feed intake and salivary secretion by steers. *Journal of Animal Science* 25:817-820.

Qi, M., K. D. Jakober, and T. A. McAllister. 2009. Rumen microbiology. Pp. 161-176 in *Animal and Plant Productivity—Volume One, Encyclopedia of Life Support Systems*, R. J. Hudson, ed. Oxford, UK: Eolss. Available online at http://www.eolss.net.login.ezproxy.library.ualberta.ca. Accessed on January 10, 2011.

Ranilla, M. J., M. D. Carro, S. López, C. J. Newbold, and R. J. Wallace. 2001. Influence of nitrogen source on the fermentation of fibre from barley straw and sugarbeet pulp by ruminal microorganisms in vitro. *British Journal of Nutrition* 86:717-724.

Reynolds, C. K., and N. B. Kristensen. 2008. Nitrogen recycling through the gut and the nitrogen economy of ruminants: An asynchronous symbiosis. *Journal of Animal Science* 86:E293-E305.

Robinson, T. J., D. E. Jasper, and H. R. Guilbert. 1951. The isolation of *Spherophorus necrophorus* from the rumen together with some feed lot data on abscess and telangiectasis. *Journal of Animal Science* 10:733-741.

Rodriguez, S. M., K. C. Guimaraes, J. C. Matthews, K. R. McLeod, R. L. Baldwin, and D. L. Harmon. 2004. Influence of abomasal carbohydrates on small intestinal sodium-dependent glucose co-transporter activity and abundance in steers. *Journal of Animal Science* 82:3015-3023.

Ruckebusch, Y. 1988. Motility of the gastro-intestinal tract. Pp. 64-107 in *The Ruminant Animal: Digestive Physiology and Nutrition*, D. C. Church, ed. Englewood Cliffs, NJ: Prentice Hall.

Rumsey, T. S., P. A. Putnam, and E. E. Williams. 1969. Salivary and ruminal characteristics, respiration rate and EKG patterns of steers fed a pelleted of ground roughage diet at two levels of intake. *Journal of Animal Science* 29:464-468.

Rumsey, T. S., P. A. Putnam, E. E. Williams, and G. Samuelson 1972. Effect of ruminal and esophageal fistulation on ruminal parameters, saliva flow, EKG patterns and respiratory rate of beef steers. *Journal of Animal Science* 35:1248-1256.

Russell, J. B. 1998. Strategies that ruminal bacteria use to handle excess carbohydrate. *Journal of Animal Science* 76:1955-1963.

Russell, J. B. 2002. *Rumen Microbiology and Its Role in Ruminant Nutrition*. Ithaca, NY: James B. Russell.

Russell, J. B., and R. J. Wallace.1988. Energy-yielding and energy-consuming reactions. Pp. 246-282 in *The Rumen Microbial Ecosystem*, P. N. Hobson, ed. London, UK: Elsevier Applied Science.

Russell, J. B., and D. B. Wilson. 1996. Why are ruminal cellulolytic bacteria unable to digest cellulose at low pH? *Journal of Dairy Science* 79:1503-1509.

Satter, L. D., T. J. Klopfenstein, and G. E. Erickson. 2002. The role of nutrition in reducing nutrient output from ruminants. *Journal of Animal Science* 80(Suppl. 2):E143-E156.

Schwab, C. G., P. Huhtanen, C. W. Hunt, and T. Hvelplund. 2005. Nitrogen requirements of cattle. Pp. 13-70 in *Nitrogen and Phosphorus Nutrition of Cattle: Reducing the Environmental Impact of Cattle Operations*, E. Pfeffer and A. N. Hristov, eds. Cambridge, MA: CAB International.

Schwaiger, T., K. A. Beauchemin, and G. B. Penner. 2013a. The duration of time that beef cattle are fed a high-grain diet affects the recovery from a bout of ruminal acidosis: Dry matter intake and ruminal fermentation. *Journal of Animal Science* 91:5729-5742.

Schwaiger, T., K. A. Beauchemin, and G. B. Penner. 2013b. Duration of time that beef cattle are fed a high-grain diet affects the recovery from a bout of ruminal acidosis: Short-chain fatty acid and lactate absorption, saliva production, and blood metabolites. *Journal of Animal Science* 91:5743-5753.

Schwartz, H. M., and F. M. C. Gilchrist. 1975. Microbial interactions with the diet and the host animal. Pp. 165-179 in *Digestion and Metabolism in the Ruminant: Proceedings of the Fourth International Symposium on Ruminant Physiology*, I. W. McDonald, and A. C. I. Warner, eds. Armidale, NSW, Australia: University of New England Publishing Unit.

Schwartzkopf-Genswein, K. S., K. A. Beauchemin, D. J. Gibb, D. H. Crews, Jr., D. D. Hickman, M. Streeter, and T. A. McAllister. 2003. Effect of bunk management on feeding behavior, ruminal acidosis and performance of feedlot cattle: A review. *Journal of Animal Science* 81(Suppl. 2):E149-E158.

Schwartzkopf-Genswein, K. S., K. A. Beauchemin, T. A. McAllister, D. J. Gibb, M. Streeter, and A. D. Kennedy. 2004. Effect of feed delivery fluctuations and feeding time on ruminal acidosis, growth performance, and feeding behavior of feedlot cattle. *Journal of Animal Science* 82:3357-3365.

Sehested, J., L. Diernaes, P. D. Moller, and E. Skadhauge. 1999. Transport of butyrate across the isolated bovine rumen epithelium: Interaction with sodium, chloride, and bicarbonate. *Comparative Biochemistry and Physiology Part A: Molecular and Integrative Physiology* 123:399-408.

Seo, S., L. O. Tedeschi, C. G. Schwab, and D. G. Fox. 2006. Development and evaluation of empirical equations to predict feed passage rate in cattle. *Animal Feed Science and Technology* 128:67-83.

Sniffen, C. J., and P. H. Robinson. 1987. Protein and fiber digestion, passage, and utilization in lactating cows. Microbial growth and flow as influenced by dietary manipulations. *Journal of Dairy Science* 70:425-441.

Sniffen, C. J., J. D. O'Connor, P. J. Van Soest, D. G. Fox, and J. B. Russell. 1992. A net carbohydrate and protein system for evaluating cattle diets: II. Carbohydrate and protein availability. *Journal of Animal Science* 70:3562-3577.

Sterchi, E. E., and J. F. Woodley. 1980. Peptide hydrolases of the human small intestinal mucosa: Distribution of activities between brush border membranes and cytosol. *Clinica Chimica Acta* 102:49-56.

Stern, M. D., and W. H. Hoover. 1979. Methods for determining and factors affecting rumen microbial protein synthesis: A review. *Journal of Animal Science* 49:1590-1603.

Steven, D. H., and A. B. Marshall. 1970. Organization of the rumen epithelium. Pp. 80-100 in *Physiology of Digestion and Metabolism in the Ruminant: Proceedings of the Third International Symposium, August, 1969, Cambridge, England*, A.T. Phillipson, ed. Newcastle upon Tyne, England: Oriel Press.

Stewart, C. S. 1991. The rumen bacteria. Pp. 15-26 in *Rumen Microbial Metabolism and Ruminant Digestion*, J. P. Jouany, ed. Paris, France: Institut National de la Recherche Agronomique.

Stewart, C. S., and M. P. Bryant. 1988. The rumen bacteria. Pp. 21-75 in *The Rumen Microbial Ecosystem*, P. N. Hobson, ed. London, UK: Elsevier Applied Science.

Stewart, C. S., M. Fèvre, and R. A. Prins. 1995. Factors affecting fermentation and polymer degradation by anaerobic fungi and the potential for manipulation of rumen function. Pp. 251-270 *Ruminant Physiology: Digestion, Metabolism, Growth and Reproduction: Proceedings of the Eighth International Symposium on Ruminant Physiology*, W. von Engelhardt, S. Leonhard-Marek, G. Breves, and G. Giesecke, eds. Stuttgart, Germany: Ferdinand Enke Verlag.

Stewart, C. S., H. J. Flynt, and M. P. Bryant. 1997. The rumen bacteria. Pp. 10-72 in *The Rumen Microbial Ecosystem*, P. N. Stewart and C. S. Ha, eds. New York: Blackie Academic and Professional.

Stouthamer, A. H. 1973. A theoretical study on the amount of ATP required for synthesis of microbial cell material. *Antonie van Leeuwenhoek* 39:545-565.

Sudweeks, E. M., M. E. McCullough, L. R. Sisk, and S. E. Law. 1975. Effects of concentrate type and level and forage type on chewing time of steers. *Journal of Animal Science* 41:219-224.

Sutherland, T. M. 1987. Particle separation in the forestomachs of sheep. Pp. 63-73 in *Aspects of Digestive Physiology in Ruminants*, A. Dobson and M. J. Dobson, eds. Ithaca, NY: Cornell University Press.

Sutton, J. D. 1979. Carbohydrate fermentation in the rumen—variations on a theme. *Proceedings of the Nutrition Society* 38:275-281.

Taylor, C. C., and M. S. Allen. 2005. Corn grain endosperm type and brown midrib 3 corn silage: Site of digestion and ruminant digestion kinetics in lactating cows. *Journal of Dairy Science* 88:1413-1424.

Taylor, R. B. 1962. Pancreatic secretion in the sheep. *Research in Veterinary Science* 3:63-77.

Tedeschi, L. O., D. G. Fox, R. D. Sainz, L. G. Barioni, S. R. Medeiros, and C. Boin. 2005. Using mathematical models in ruminant nutrition. *Scientia Agricola* 62:76-91.

Titgemeyer, E. C. 1997. Design and interpretation of nutrient digestion studies. *Journal of Animal Science* 75:2235-2247.

Tobey, N., R. Yeh, T. I. Huang, W. Heizer, and C. Hoffner. 1985. Human intestinal brush border peptidases. *Gastroenterology* 88:913-926.

Towne, G., T. G. Nagaraja, R. T. Brandt, Jr., and K. E. Kemp. 1990. Ruminal ciliated protozoa in cattle fed finishing diets with or without supplemental fat. *Journal of Animal Science* 68:2150-2155.

Turner, A. W., and V. E. Hodgetts. 1955. Buffer systems in the rumen of the sheep. II. Buffering properties in relationship to composition. *Australian Journal of Agricultural Research* 6:125-144.

Ugolev, A. M., N. M. Timofeeva, G. M. Roshchina, L. F. Smirnova, A. A. Gruzdkov, and S. A. Gusev. 1990. Localization of peptide hydrolysis in the enterocyte and the transport mechanisms of apical membrane. *Comparative Biochemistry and Physiology A, Comparative Physiology* 95:501-509.

Ulshen, M. H. 1987. Physiology of protein and amino acid absorption. Pp. 139-152 in *Pediatric Nutrition: Theory and Practice*, R. J. Grand, J. L. Sutphen, and W. H. Dietz, eds. Boston, MA: Butterworths.

Ulyatt, M. J., D. W. Dellow, A. John, C. S. W. Reid, and G. C. Waghorn. 1986. Contribution of chewing during eating and rumination to the clearance of digesta from the ruminoreticulum. Pp. 498-515 in *Control of Digestion and Metabolism in Ruminants: Proceedings of the Sixth International Symposium on Ruminant Physiology*, L. P. Milligan, W. L. Grovum, and A. Dobson, eds. Englewood Cliffs, NJ: Prentice Hall.

Ulyatt, M. J., S. K. Baker, G. J. McCrabb, and K. R. Lassey. 1999. Accuracy of SF_6 tracer technology and alternatives for field measurements. *Australian Journal of Agricultural Research* 50:1329-1334.

Ungerfeld, E. M., and Kohn, R. A. 2006. The role of thermodynamics in the control of ruminal fermentation. Pp. 55-85 in *Ruminant Physiology: Digestion, Metabolism and Impact of Nutrition on Gene Expression, Immunology and Stress*, K. Sejrsen, T. Hvelplund, and M. O. Nielson, eds. Wageningen, The Netherlands: Wageningen Academic.

University of Bristol. 1998. The Comparative Gross Anatomy of the Forestomachs of the Domestic Ruminants [tutorial]. Available online at http://137.222.110.150/Calnet/vetAB10/page2.htm. Accessed on July 6, 2015.

Van Soest, P. J. 1994. *Nutritional Ecology of the Ruminant*, 2nd Ed. Ithaca, NY: Cornell University Press.

Van Soest, P. J., M. E. Van Amburgh, J. B. Robertson, and W. F. Knaus. 2005. Validation of the 2.4 times lignin factor for ultimate extent of NDF digestion, and curve peeling rate of fermentation curves into pools. Pp. 150-166 in *Proceedings of Cornell Nutrition Conference for Feed Manufacturers, Rochester, NY.* Ithaca, NY: Cornell University.

Vanzant, E. S., R. C. Cochran, and E. C. Titgemeyer. 1998. Standardization of in situ techniques for ruminant feedstuff evaluation. *Journal of Animal Science* 76:2717-2729.

Vieira, R. A. M., L. O. Tedeschi, and A. Cannas. 2008. A generalized compartmental model to estimate the fibre mass in the ruminoreticulum: 2. Integrating digestion and passage. *Journal of Theoretical Biology* 255:357-368.

Voelker, J. A., and M. S. Allen. 2003. Pelleted beet pulp substituted for high-moisture corn: 2. Effects on digestion and ruminal digestion kinetics in lactating dairy cows. *Journal of Dairy Science* 86:3553-3561.

Waiss, A. C., Jr., J. Guggolz, G. O. Kohler, H. G. Walker, Jr., and W. N. Garrett. 1972. Improving digestibility of straws for ruminant feed by aqueous ammonia. *Journal of Animal Science* 35:109-112.

Waldo, D. R., L. W. Smith, and E. I. Cox. 1972. Model of cellulose disappearance from the rumen. *Journal of Dairy Science* 55:125-129.

Wallace, R. J. 1991. Rumen proteolysis and its control. Pp. 131-150 in *Rumen Microbial Metabolism and Ruminant Digestion*, J. P. Jouany, ed. Paris, France: Institut National de la Recherche Agronomique.

Walker, J. A., and D. L. Harmon. 1995. Influence of ruminal or abomasal starch hydrolysate infusion on pancreatic exocrine secretion and blood glucose and insulin concentrations in steers. *Journal of Animal Science* 73:3766-3774.

Walker, N. D., C. J. Newbold, and R. J. Wallace. 2000. Nitrogen metabolism in the rumen. Pp. 71-108 in *Nitrogen and Phosphorus Nutrition of Cattle: Reducing the Environmental Impact of Cattle Operations*, E. Pfeffer and A. N. Hristov, eds. Cambridge, MA: CAB International.

Walter, A., and J. Gutknecht. 1986. Permeability of small nonelectrolytes through lipid bilayer membranes. *Journal of Membrane Biology* 90:207-217.

Weakley, D. C., M. D. Stern, and L. D. Satter. 1983. Factors affecting disappearance of feedstuffs from bags suspended in the rumen. *Journal of Animal Science* 56:493-507.

Weimer, P. J. 1998. Manipulating ruminal fermentation: A microbial ecological perspective. *Journal of Animal Science* 76:3114-3122.

Welch, J. G. 1982. Rumination, particle size and passage from the rumen. *Journal of Animal Science* 54:885-894.

Wierenga, K. T., T. A. McAllister, D. J. Gibb, A. V. Chaves, E. K. Okine, K. A. Beauchemin, and M. Oba. 2010. Evaluation of triticale dried distillers grain as a substitute for barley silage in feedlot finishing diets. *Journal of Animal Science* 88:3018-3029.

Williams, A. G. 1986. Rumen holotrich ciliate protozoa. *Microbiological Reviews* 50:25-49.

Williams, A. G., and G. S. Coleman. 1988. The rumen protozoa. Pp. 77-128 in *The Rumen Microbial Ecosystem*, P. N. Hobson, ed. Barking, UK: Elsevier Science.

Williams, V. J., and D. D. Mackenzie. 1965. The absorption of lactic acid from the reticulo-rumen of the sheep. *Australian Journal of Biological Sciences* 18:917-934.

Wilson, R. K., and A. V. Flynn. 1975. A note on the eating behaviour of cattle offered grass silage ad libitum in troughs. *Irish Journal of Agricultural Research* 14:218-220.

Wilson, D. J., T. Mutsvangwa, and G. B. Penner. 2012. Supplemental butyrate does not enhance the absorptive or barrier functions of the isolated ovine ruminal epithelia. *Journal of Animal Science* 90:3153-3161.

Wolin, M. 1960. A theoretical rumen fermentation balance. *Journal of Dairy Science* 43:1452-1459.

Wolin, M. J., T. L. Miller, and C. S. Stewart. 1997. Microbe-microbe interactions. Pp. 467-491 in *The Rumen Microbial Ecosystem*, P. N. Hobson and C. S. Stewart, eds. London, UK: Blackie Academic and Professional.

Wyburn, R. S. 1980. The mixing and propulsion of the stomach contents of ruminants. Pp. 35-52 in *Digestive Physiology and Metabolism in Ruminants*, Y. Ruckebusch and P. Thivend, eds. Westport, CT: AVI.

Yarns, D. A., P. A. Putnam, and E. C. Leffel. 1965a. Daily salivary secretion by beef steers. *Journal of Animal Science* 24:173-176.

Yarns, D. A., P. A. Putnam, and E. C. Leffel. 1965b. Rumen constituents affecting salivary secretion. *Journal of Animal Science* 24:805-809.

Zhang, S., J. R. Aschenbach, D. R. Barreda, and G. B. Penner. 2013a. Recovery of absorptive function of the reticulo-rumen and total tract barrier function in beef cattle after short-term feed restriction. *Journal of Animal Science* 91:1696-1706.

Zhang, S., J. R. Aschenbach, D. R. Barreda, and G. B. Penner. 2013b. Short-term feed restriction impairs the absorptive function of the reticulo-rumen and total tract barrier function in beef cattle. *Journal of Animal Science* 91:1685-1695.

Zinn, R. A. 1990. Influence of flake density on the comparative feeding value of steam-flaked corn for feedlot cattle. *Journal of Animal Science* 68:767-775.

3

Energy Terms and Concepts

ENERGY UNITS

Energy is an abstraction and defined as the potential to do work. It can be measured only in reference to defined, standard conditions; thus, all defined units are equally absolute. The joule is the International System of Unit of expressing electrical, mechanical, and chemical energy. The joule can be converted to ergs, watt-hours, and calories; the converse is also true. Nutritionists now typically standardize their combustion calorimeters using specifically purified benzoic acid, the energy content of which has been determined in electrical units and computed in terms of joules/g mole (NRC, 1984, 1996, 2000). The calorie has been standardized to equal 4.184 joules and is approximately equal to the heat required to raise the temperature of 1 g of water from 16.5° to 17.5°C, which is referred to as the 17°C calorie. The 15°C calorie, the amount of energy to raise 1 g of water from 14.5° to 15.5°C, ranges from 4.1852 to 4.1858 joules when measured experimentally and averages 4.1855 joules. In practical terms, the decision to fix 1 calorie to equal 4.1840 joules results in a 17°C instead of a 15°C calorie (FAO/WHO, 1971). In practice, the calorie is a small amount of energy; thus, the kilocalorie (1 kcal = 1,000 calories; referred to in some literature as the "Large" calorie and designated as spelling with a capital C; Calorie) and megacalorie (1 Mcal = 1,000 kcal) are more convenient for use in conjunction with animal feeding standards.

A number of abbreviations have been used to describe energy fractions in the animal system. Many of the abbreviations used throughout this text are those recommended in the *Nutritional Energetics of Domestic Animals and Glossary of Energy Terms* (NRC, 1981). Gross energy (GE) or heat of combustion is the energy released as heat when an organic substance is completely oxidized to carbon dioxide and water. Gross energy is related to chemical composition, but it does not provide any information regarding energy availability to the animal. Thus, GE is of limited use for assessing the value of a particular diet or dietary ingredient as an energy source for the animal. Gross energy can be directly determined by bomb calorimetry or calculated as

$$\text{GEI(Mcal)} = (4.15 \times \text{Carbohydrate} + 5.65 \times \text{Protein} + 9.4 \times \text{Fat}) \times \text{DMI}/100$$

where gross energy intake (GEI), carbohydrate, protein, and fat are expressed as percent dry matter (DM), and DM intake (DMI) is in kilograms.

EXPRESSING ENERGY VALUES OF FEEDS

Gross energy of the food minus the energy lost in the feces is termed digestible energy (DE; Figure 3-1). Digestible energy as a proportion of GE can vary from 0.3 for a very mature, weathered forage to nearly 0.9 for processed, high-quality cereal grains. Digestible energy has some value for feed evaluation because it reflects diet digestibility and can be measured with relative ease; however, particularly for ruminants, DE fails to consider several major losses of energy associated with digestion and metabolism of food. As a result, DE overestimates the value of high-fiber feedstuffs such as hays or straws relative to low-fiber, highly digestible feedstuffs such as grains. Total digestible nutrients (TDN) are similar to DE. The use of TDN has no particular advantages or disadvantages over DE as the unit to describe feed values or to express the energy requirements of the animal. Total digestible nutrients can be converted to DE using the following relationship:

$$1 \text{ kg of TDN} = 4.4 \text{ Mcal of DE} \qquad \text{(Eq. 3-1)}$$

Metabolizable energy (ME) is defined as GE minus fecal energy (FE), urinary energy (UE), and gaseous energy (GASE) losses, or ME = DE – (UE + GASE). Metabolizable energy is an estimate of the energy available to the animal and represents an accounting progression to assess food energy values and animal requirements. Nonetheless, ME

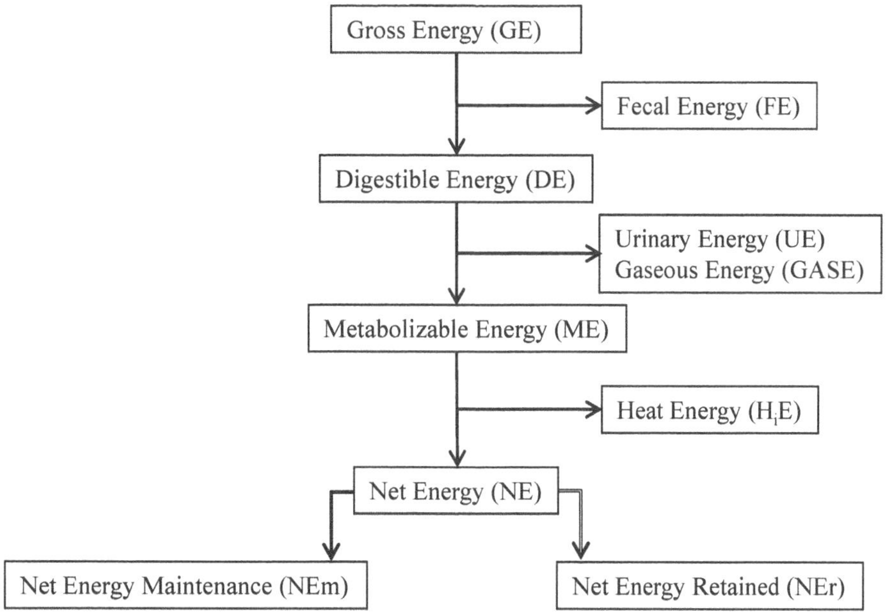

FIGURE 3-1 The idealized flow of energy through an animal. The net energy retained (NEr) indicated above includes body tissue, milk, and conceptus energy.

has many of the same weaknesses as DE; and because UE and GASE are highly predictable from DE, ME and DE are strongly correlated. In addition, the main source of GASE (primarily methane) is microbial fermentation, which also results in heat production. This heat is useful in helping to maintain body temperature in cold-stressed animals, but is otherwise an energy loss not accounted for by ME. The NRC (1984) suggested that ME = 0.82 DE (ARC, 1965; NRC, 1976). The NRC (1996, 2000) indicated that for most forages and mixtures of forages and cereal grains, the ratio of ME to DE is about 0.8, but can vary considerably (ARC, 1980; CSIRO, 1990) depending on intake, age of animal, and feed source.

The value of ME = 0.82 DE (NRC, 1976; Garrett, 1980) has been used for several decades; however, data summarized by Vermorel and Bickel (1980) clearly indicate that the ME:DE ratio ranged from 0.82 to 0.93 in growing cattle and was closer to 0.81 in adult sheep. In growing cattle (Vermorel and Bickel, 1980) the DE:ME was affected by cattle age (0.89 to 0.83, for 6- and 20-month-old cattle, respectively), dietary concentrate level, intake, and composition (ranges of 0.82 to 0.93 for 0 to 85% concentrate and low to high levels of intake). Differences in the data of Garrett (1980) and Vermorel and Bickel (1980) could be partially explained by differing cattle types and measurement techniques. Vermorel and Bickel (1980) indicated that higher ME:DE ratios in growing cattle compared with adults might reflect smaller methane and UE losses.

Recent data (Hales et al., 2012, 2013) indicate that Jersey steers fed high-concentrate diets containing wet distillers grains with solubles (WDGS) could have a higher ME:DE ratio than traditionally expected. For both experiments, Jersey steers, implanted with 36 mg of zeranol and consuming two times their maintenance requirement (5.71 kg/d) were fed 90% concentrate feeds (i.e., processed corn grain, WDGS, cottonseed meal, yellow grease, molasses, urea, and mineral supplement) and 10% alfalfa hay. In the Hales et al. (2012) study, 30% WDGS increased diet ME concentration from 2.97 to 3.08 Mcal/kg, whereas in the Hales et al. (2013) study, ME decreased from 3 to 2.8 Mcal/kg with the addition of WDGS. The difference in results is likely a result of ether extract concentrations greater than 8% of dietary DM in one of the Hales et al. (2013) treatments. Overall, however, their results suggest that Jersey steers fed a high-concentrate diet converted DE to ME with an efficiency of 95%. Hales et al. (2014) using MARC II composite steers (¼ Simmental, ¼ Angus, ¼ Hereford, and ¼ Gelbvieh) fed dry-rolled corn-based diets containing 25% WDGS and increasing concentrations of ground alfalfa hay (2, 6, 10, and 14% of dietary DM) demonstrated that the ME:DE ratio averaged 0.904 and ranged from 0.893 to 0.918 for steers fed diets containing 14 vs. 2% alfalfa, respectively. These values are consistent with data from growing lambs fed 85% concentrate diets (Vermorel and Bickel, 1980). These data also were approximated by using the equation ME = (1.01 × DE) – 0.45 (NRC, 2001), where ME and DE are in Mcal/kg. Corrections to ME for dietary ether extract content of greater than 3% of DM suggested for dairy cattle (NRC, 2001) were also helpful in predicting the ME:DE ratios observed by Hales et al. (2014). Equations for accurate prediction of ME from DE values need to be developed for growing and finishing beef cattle.

Based on the available data, the committee recognizes that a change is likely needed in the efficiency of DE to ME conversion. One option considered by the committee was to adopt the equation ME = (1.01 × DE) – 0.45 (NRC, 2001) as the method for determining the DE to ME conversion efficiency. This modification would adjust the calculated feed ingredient energy concentrations to more accurately reflect available data; however, this approach might not be optimal for limited feeding situations because the existing data indicate that intake level can affect ME:DE ratios. In addition, this equation was developed for dairy cows, and applying it to diets typical of those fed to beef cattle, particularly feedlot beef cattle fed high-grain, low-roughage diets, might not be valid. The relationship between the static ME:DE ratio of 0.82 used previously for beef cattle and the NRC (2001) approach is depicted in Figure 3-2.

When the NRC (2001) equation for calculating ME from DE was applied to practical beef cattle diets and observed vs. predicted average daily gain (ADG) was compared, results indicated a substantial overprediction of ADG, especially with high-grain, feedlot diets. In addition, the predicted ADG using the NRC (2001) equation was substantially greater than the predicted values using the NRC (1996, 2000) approach in which the ME:DE ratio was fixed at 0.82. Thus, despite a substantial amount of recent research data suggesting a conversion efficiency of greater than 0.82, application of higher values resulted in untenable overpredictions of ADG. The reason for the overprediction was the substantial increase in the ME concentration of the diet associated with application of the NRC (2001) equation.

As a result of these evaluations, the committee decided to retain the ME:DE ratio of 0.82 used in the previous edition and update of the *Nutrient Requirements of Beef Cattle* (NRC, 1996, 2000). Nonetheless, the option is presented within the Beef Cattle Nutrient Requirements Model (BCNRM; see Chapter 19, Model Equations and Sensitivity Analyses) for users to specify the ME:DE ratio (although the committee recommends this option be used with caution). Based on these observations and empirical evaluations, there is clearly a need for experiments defining relationships between diet composition, animal, and other factors, and the DE to ME conversion efficiency and concomitant predictions of net energy (NE) values of feed ingredients.

As defined by energy flows (Figure 3-1), ME can appear only as heat production (HE) or retained energy (RE), that is, ME = HE + RE. As indicated by this relationship, a major value of ME is its use as a reference unit and as a starting point for most systems based on the NE concept.

The value of feed energy for the promotion of energy retention is measured by determining the RE at two or more amounts of intake energy (IE). The NE of a feed or diet has classically been illustrated by the equation:

$$NE = \Delta RE/\Delta IE \quad \text{(Eq. 3-2)}$$

Determination of NE by this method assumes that the relationship between RE and feed intake is linear. Actually, the relationship is curvilinear and shows a diminishing-return effect (Figure 3-3; NRC, 1981; Garrett and Johnson, 1983; Ferrell and Oltjen, 2008); however, the relationship

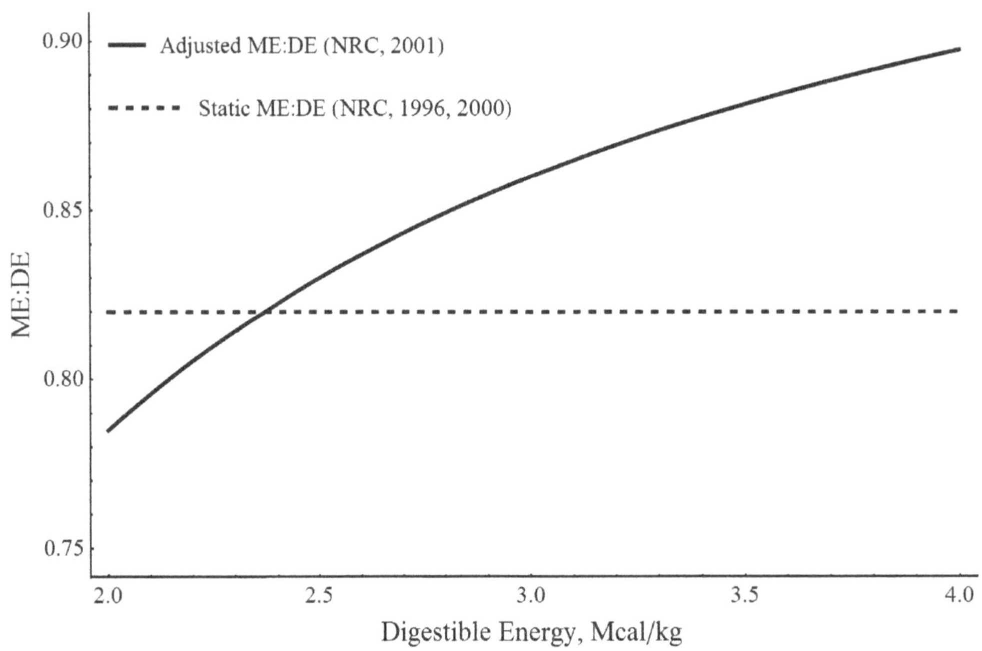

FIGURE 3-2 Representation of the relationship of ME to DE. The dashed line represents the use of a static value of 0.82 (NRC, 1996, 2000) and the solid line represents the NRC (2001) equation: ME = (1.01 × DE) – 0.45.

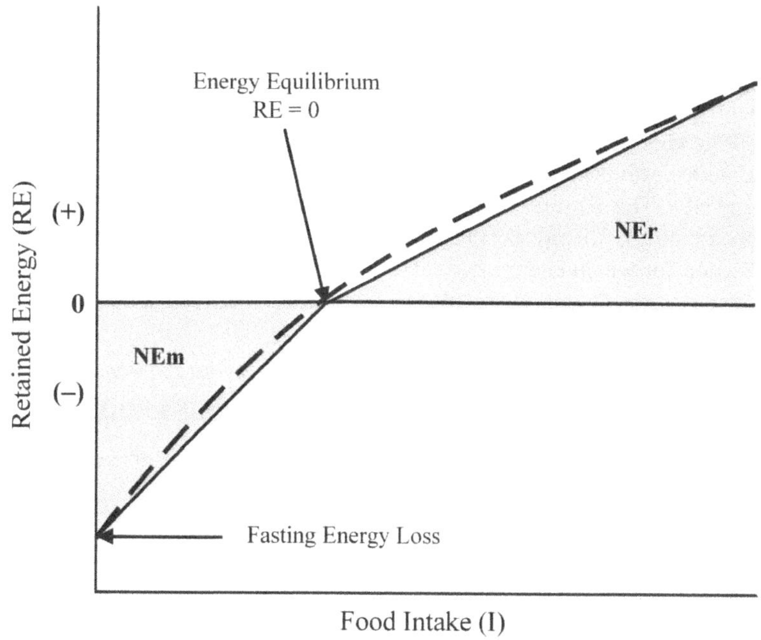

FIGURE 3-3 Representation of the relationship between RE and NE. The dashed line shows the curvilinearity between RE and food intake; the solid lines are linear approximations.

is conventionally approximated by two straight lines. The intersection of the two lines is the point at which RE = 0 and is defined as maintenance (M). Therefore, when RE = 0, ME = HE. The relationship between feed intake and body tissue loss (negative RE) comprises one portion of the curve and the relationship between body tissue gain (positive RE) comprises a second portion of the curve. Figure 3-3 clearly demonstrates that food intake is utilized more efficiently for energy maintenance than for energy retention. The heat production at zero feed intake (H_eE) is equivalent to the animal's NE requirement for maintenance. The ability of the food consumed to meet the NE required for maintenance is expressed as NEm and is represented by the following expression:

$$NEm = H_eE/I_m, \qquad \text{(Eq. 3-3)}$$

where I_m is the amount of feed consumed at RE = 0. Similarly, the value of feed consumed to promote energy retention is represented by the expression NEr and is determined as

$$NEr = RE/(I - I_m) \qquad \text{(Eq. 3-4)}$$

where $(I - I_m)$ represents the amount of feed consumed in excess of maintenance requirements.

The relationship ME = RE + HE can be rewritten in terms of NE. Thus, HE can be partitioned into H_eE, H_jE, and H_iE (heat increment of intake energy) as

$$ME = RE + H_eE + H_jE + H_iE, \qquad \text{(Eq. 3-5)}$$

where H_eE is equivalent to the NE requirement for maintenance; H_jE is heat of activity associated with obtaining feed; and H_iE is heat increment of intake energy.

Because in practical situations the heat of activity associated with obtaining feed (H_jE) is often included with H_eE, the expression becomes

$$ME = NEr + NEm + H_iE. \qquad \text{(Eq. 3-6)}$$

The NEr used in this expression does not distinguish among different forms in which energy can be retained, such as body tissue, milk, or tissues of the conceptus. Thus, the former expression might be expanded such that in a pregnant lactating heifer it becomes

$$RE = \text{Milk Energy} + \text{Conceptus Energy} + \text{Tissue Energy} \qquad \text{(Eq. 3-7)}$$

or

$$NEr = NEl + NEy + NEg, \qquad \text{(Eq. 3-8)}$$

where NEr, NEl, NEy, and NEg are equivalent to RE, milk energy, conceptus energy, and tissue energy, respectively.

Therefore,

$$ME = NEg + NEl + NEy + NEm + H_iE \qquad \text{(Eq. 3-9)}$$

In this expression, a portion of the heat increment (H_iE) is associated with the feed consumed for maintenance and each of the productive functions. The primary advantages of an NE system are that animal requirements stated as NE are independent of the diet, and the energy value of feeds for different physiological functions are estimated separately, for example, NEm, NEg, NEl, and NEy. This requires, however, that each feed must be assigned multiple NE values because the value varies with the function for which energy is used by the animal. Alternatively, the animal's energy requirement for various physiological functions may be expressed in terms of a single NE value, provided the relationships among efficiencies of utilization of ME for different functions are known.

Relationships for converting ME values to dietary available NEm and NEg (Mcal/kg DM) have been reported by Garrett (1980) and are shown below. For clarity, dietary available NEm and NEg will at times be designated as NEma and NEga. This will particularly be the case in Chapter 19 to distinguish NEm and NEg available in the diet (NEma and NEga) from animal NE requirements for maintenance and gain (NEm and NEg, respectively).

$$NEm = 1.37ME - 0.138ME^2 + 0.0105ME^3 - 1.12 \quad \text{(Eq. 3-10)}$$

$$NEg = 1.42ME + 0.174ME^2 + 0.0122ME^3 - 1.65 \quad \text{(Eq. 3-11)}$$

The NEm and NEg values used in the derivation of these equations were based on comparative slaughter studies involving 2,766 animals fed complete, mixed diets at or near ad libitum intake for 100 to 200 days (Garrett, 1980). Digestion trials were conducted on most diets fed at about 1.1 times the maintenance amount. The ME values were estimated as DE × 0.82. Data were not uniformly distributed across the range of ME concentrations encountered in practical situations (1%, <1.9 Mcal/kg; 22%, 1.9 to 2.6 Mcal/kg; 65%, 2.6 to 2.9 Mcal/kg; 12%, >2.9 Mcal/kg). *Caution should be exercised in use of these equations for predicting NEm or NEg values for individual feed ingredients or for feeds outside the ranges indicated above.* The relationship between DE and ME can vary considerably among feed ingredients or diets as a result of differences in intake, rate of digestion and passage, and composition (e.g., fiber vs. starch vs. fat). In addition, conversion of ME to NEm or NEg may vary beyond that associated with variation in dietary ME, in part because of differences in the composition of absorbed nutrients.

Available data, as discussed in Chapter 11 (Maintenance Consideration: Energy and Protein), indicate that efficiencies of ME use for lactation and maintenance are similar in beef cattle; thus, energy requirements for lactation have been expressed in NEm units. Efficiency of utilization of ME for accretion of energy in gravid uterine tissues is likewise discussed in a subsequent section of this report. Some evidence is available indicating that the efficiency of utilization of ME for maintenance (k_m) and pregnancy (k_y) vary similarly with changes in ME concentration in the diet (Robinson et al., 1980). For convenience, estimates of requirements for beef cows were converted to NEm equivalents. Conversion of requirements for lactation and pregnancy to NEm equivalents allow the energy value of feedstuffs to be adequately described by only two NE values (NEm and NEg). Additional information and equations for maintenance and growth are found in Chapters 11 and 12 (Growth), respectively.

SUMMARY RECOMMENDATIONS REGARDING ENERGY TERMS AND CONCEPTS

Overall, based on available data the committee recommends the following:

- Experiments need to be conducted to determine accurate relationships between dietary composition and other factors and ME:DE ratios in beef cattle.
- New equations need to be developed for determining the efficiency of conversion of DE to ME. Readers are referred to Galyean et al. (2016) for a recent reevaluation of the relationship between DE and ME.

REFERENCES

ARC (Agricultural Research Council). 1965. *The Nutrient Requirements of Farm Livestock. No. 2. Ruminants*. London, UK: Agricultural Research Council.

ARC. 1980. *The Nutrient Requirements of Ruminant Livestock: Technical Review.* Farnham Royal, UK: Commonwealth Agricultural Bureaux.

CSIRO (Commonwealth Scientific and Industrial Research Organisation). 1990. *Feeding Standards for Australian Livestock: Ruminants.* Melbourne, Australia: CSIRO Publishing.

FAO/WHO (Food and Agriculture Organization/World Health Organization). 1971. The Adoption of Joules as Units of Energy. FAO/WHO Ad Hoc Committee of Experts on Energy and Protein: Requirements and Recommended Intakes, March 22-April 2, 1971, Rome, Italy. Available online at http://www.fao.org/docrep/meeting/009/ae906e/ae906e17.htm. Accessed on May 31, 2013.

Ferrell, C. L., and J. W. Oltjen. 2008. ASAS Centennial Paper: Net energy systems for beef cattle: Concepts, application, and future models. *Journal of Animal Science* 86:2779-2794.

Galyean, M. L., N. A. Cole, L. O. Tedeschi, and M. E. Branine. 2016. Invited review: Efficiency of converting digestible energy to metabolizable energy and reevaluation of the California Net Energy System maintenance requirements and equations for predicting dietary net energy values for beef cattle. *Journal of Animal Science.* DOI:10.2527/jas.2015-0223.

Garrett, W. N. 1980. Energy utilization by growing cattle as determined in 72 comparative slaughter experiments. Pp. 3-8 in *Energy Metabolism: Proceedings, 8th Symposium on Energy Metabolism, September 1979, Cambridge, England*, L. E. Mount, ed. EAAP Publication No. 26. London: Butterworths.

Garrett, W. N., and D. E. Johnson. 1983. Nutritional energetics of ruminants. *Journal of Animal Science* 57(Suppl. 2):478-497.

Hales, K. E., N. A. Cole, and J. C. MacDonald. 2012. Effects of corn processing method and dietary inclusion of wet distillers grains with solubles on energy metabolism, carbon-nitrogen balance, and methane emissions of cattle. *Journal of Animal Science* 90:3174-3185.

Hales, K. E., N. A. Cole, and J. C. MacDonald. 2013. Effects of increasing concentrations of wet distillers grains with solubles in steam-flaked, corn-based diets on energy metabolism, carbon-nitrogen balance, and methane emissions of cattle. *Journal of Animal Science* 91:819-828.

Hales, K. E., T. M. Brown-Brandl, and H. C. Freetly. 2014. Effects of decreased dietary roughage concentration on energy metabolism and nutrient balance in finishing beef cattle. *Journal of Animal Science* 92:264-271.

NRC (National Research Council). 1976. *Nutrient Requirements of Beef Cattle*, 5th Rev. Ed. Washington, DC: National Academy Press.

NRC. 1981. *Nutritional Energetics of Domestic Animals and Glossary of Energy Terms.* Washington, DC: National Academy Press.

NRC. 1984. *Nutrient Requirements of Beef Cattle*, 6th Rev. Ed. Washington, DC: National Academy Press.

NRC. 1996. *Nutrient Requirements of Beef Cattle*, 7th Rev. Ed. Washington, DC: National Academy Press.

NRC. 2000. *Nutrient Requirements of Beef Cattle: Update 2000*, 7th Rev. Ed. Washington, DC: National Academy Press.

NRC. 2001. *Nutrient Requirements of Dairy Cattle*, 7th Rev. Ed. Washington, DC: National Academy Press.

Robinson, J. J., I. McDonald, C. Frazer, and J. G. Gordon. 1980. Studies on reproduction in prolific ewes. 6. The efficiency of energy utilization for conceptus growth. *Journal of Agricultural Science* 94:331-338.

Vermorel, M., and H. Bickel. 1980. Utilisation of feed energy by growing ruminants. *Annales de Zootechnie* 29:127-143.

4

Carbohydrates

INTRODUCTION

Carbohydrates are the major source of energy in beef cattle diets. In addition to supplying digestible energy (DE) to both the ruminal microorganisms and the animal itself, carbohydrates provide physical fiber needed to stimulate rumination and reticuloruminal motility. Various constituents contribute to the overall carbohydrate content of the diet, each with unique chemistry and availability. Based on their chemistry and functionality, carbohydrates can be broadly partitioned into nonfiber carbohydrates (NFC), also referred to as non-neutral detergent fiber (non-NDF) carbohydrates and nonstructural carbohydrates, and neutral detergent fiber (NDF) as shown in Figure 4-1. The NFC are generally found inside the cells of plants (except soluble fiber, which is found in the cell wall, and fructans found both in cell wall and cell contents) and is typically more digestible than NDF. Neutral detergent fiber is composed of hemicellulose, cellulose, and lignin, and is the slowly digestible carbohydrate fraction of feeds. Although lignin is not a carbohydrate, it is part of NDF and acid detergent fiber (ADF) and, as such, is included in the fiber carbohydrate fraction. Furthermore, lignin is resistant to digestion and negatively affects the digestibility of NDF and ADF. The organic acids (OA) are also not carbohydrates, but they are included in the NFC fraction because they are more closely related to carbohydrates than to fat or protein in terms of their digestion characteristics. Typical concentrations of NFC and NDF carbohydrates in forages are given in Table 4-1.

The current edition of the *Nutrient Requirements of Beef Cattle* has adopted a simplified carbohydrate fractionation approach based on analytical methodology, similarity of digestion rates among fractions, and availability of the fractions in the rumen. The mechanistic level of solution (MLS) of the Beef Cattle Nutrient Requirements Model (BCNRM; Chapter 19) separates the carbohydrates into six fractions: (1) OA, (2) water-soluble carbohydrates (WSC) including sugars and fructans (Fraction CA), (3) starch (Fraction CB1), (4) neutral detergent soluble fiber (Fraction CB2), (5) available NDF (Fraction CB3), and (6) unavailable NDF (Fraction CC). The C in the notation is used to distinguish carbohydrate fractions from protein fractions. The OA and WSC are completely available in the rumen and have no associated digestion rate constants; however, they are considered individually because, unlike the WSC, the OA from fermented feeds are not thought to provide substrate for microbial growth in the rumen. Although fructans are actually part of the soluble fiber fraction, in this classification they are considered part of the WSC fraction because they are soluble in water and are very rapidly available in the rumen (Hall et al., 2003a). Fractions CB1 and CB2 are assumed to be moderately available in the rumen, each with an associated digestion rate; however, starch and soluble fiber are considered separately within the CB fraction to reflect their inherent differences in rates of digestion and fermentation characteristics within the rumen. The CB3 fraction represents the slowly digestible fiber, with digestion rate varying among feed types, while the CC fraction is indigestible in the rumen. While this simplified approach might not adequately reflect the subtle differences in ruminal fermentability of all NFC fractions, its advantage is that the fractions and their digestion rates can be estimated from laboratory analysis, and it is thereby useful in practical feed formulation.

The aggregation used for the non-NDF fraction in the BCNRM differs slightly from the previous NRC (1996, 2000) model and the Cornell Net Carbohydrate and Protein System (CNCPS). Those models use a CA fraction that includes OA and sugars, and a CB1 fraction composed of neutral detergent soluble fiber and starch (Fox et al., 1992; Sniffen et al., 1992). Other ways of grouping the carbohydrates have been proposed based on their respective digestion properties. Hall et al. (1999) partitioned non-NDF as (1) OA, (2) sugars, (3) starch, and (4) neutral detergent soluble fiber (including fructans). Others have proposed more detailed fractionation (e.g., Lanzas et al., 2007) on the basis of ruminal degradation characteristics.

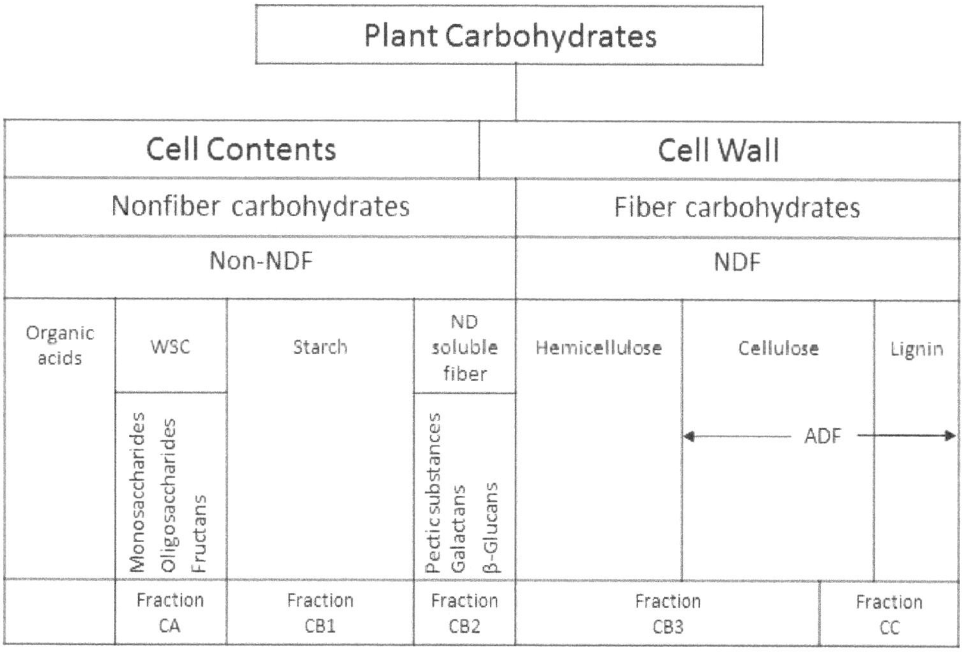

FIGURE 4-1 Simplified fractionation of plant carbohydrates. ADF = acid detergent fiber; ND = neutral detergent; NDF = neutral detergent fiber; WSC = water-soluble carbohydrates. Fiber carbohydrates represent cell wall recovered as NDF (fiber that is soluble in neutral detergent is not included). Nonfiber carbohydrates represent the carbohydrate fraction that is soluble in neutral detergent. Although fructans are actually part of the soluble fiber fraction, in this classification they are considered part of the WSC fraction because they are soluble in water and are very rapidly available in the rumen. Although not carbohydrates, organic acids are considered part of the non-NDF fraction. Organic acids are utilized as energy substrates by the animal, but they do not support appreciable microbial growth in the rumen and therefore are not included in Fraction CA. Lignin is considered part of the fiber carbohydrate fraction because it is included in ADF and NDF and affects their digestibility. Fractions CA, CB1, CB2, CB3, and CC are described in the text.

NON-NEUTRAL DETERGENT FIBER CARBOHYDRATES

Non-NDF carbohydrates (Figure 4-1) are soluble in neutral detergent solution and include OA, water-soluble carbohydrates (mono-, di-, and oligosaccharides, fructans), starch, and soluble fiber (pectic substances, mixed linkage β(1,3)(1,4)-glucans, galactans). The non-NDF fraction is determined by difference as % non-NDF = 100 – (% Crude Protein [CP] + % NDF + % Fat + % Ash). Because this fraction is calculated by difference, it comprises the cumulative errors associated with the analysis of CP, NDF, ash, and fat and, as such, should be considered an approximation of the highly digestible carbohydrate fraction (Hall, 2003a). The relative proportions of non-NDF and NDF, as well as the constituents within each, differ among feeds. This variation contributes to differences in the rate and extent of fermentation in the rumen, end products of fermentation, and contribution to microbial CP production (Lanzas et al., 2007). Thus, further analysis and consideration of the non-NDF fractions is recommended.

TABLE 4-1 Typical Concentrations of Some Carbohydrates (g/kg dry matter) in Forages[a]

Category	Temperate Legumes	Cool-Season Grasses	Warm-Season Grasses
Non-NDF			
Soluble sugars	20–50	30–60	10–50
Starch	10–110	0–20	10–50
Pectin	40–120	10–20	10–20
Fructans	—	30–100	—
NDF			
Cellulose	200–350	150–450	220–400
Hemicellulose	40–170	120–270	250–400

[a]Modified from Moore and Hatfield (1994).

Organic Acids

Organic acids are the fermentation acids found in silage and plant OA found in fresh forage and hay (Hall and Eastridge, 2014). Volatile fatty acids (VFA; mainly acetic, propionic, and butyric acids), lactic acid, and sometimes other organic acids are end products of the ensiling process, thus their concentrations are elevated in ensiled feeds (Cherney and Cherney, 2003). Most metabolic processes occur within the silo within 2 to 6 weeks of ensiling, resulting in a stable phase of ensiling as long as air is excluded during storage.

Lactic acid is the most abundant OA produced during an ideal fermentation ranging from 4 to 6% of dry matter (DM) in well-fermented corn silage and from 3 to 8% of DM in well-fermented cereal silage (McAllister et al.,1995) and alfalfa silage (Kung, 2008). Ingested lactic acid from silage is rapidly converted to propionate in the rumen, which is then absorbed and converted to glucose by the liver of the ruminant animal. Of the VFA in fermented feeds, acetic acid is greatest in concentration (1 to 3% of DM), whereas butyric (<0.5% of DM) and propionic acids (<0.5% of DM) are uncommon end products of normal silage fermentation (Kung, 2008). High levels of acetic (>3 to 4%) or butyric acid (>0.5%) in silage are indicators of less than desirable silage fermentation. Silages that have undergone heterolactic fermentation due to inoculation with *Lactobacillus buchneri* are an exception. These silages have elevated concentrations of acetic and propionic acids relative to silages that have not undergone heterolactic fermentation (Kung and Ranjit, 2001). These acids act as antifungal agents and improve aerobic stability of treated silage.

Ideally, the ratio of lactic acid to acetic acid in untreated silages should not be less than 3:1, and a higher ratio is better. Ethanol is sometimes found in silage, with higher concentrations indicating that the silage has undergone an extensive alcoholic fermentation caused by yeasts. These silages tend to heat up when exposed to air, thereby resulting in a short bunk life. In corn silage, ethanol is usually between 1 to 3% of the DM but can be much higher (6 to 7% of DM), whereas the ethanol content in grass and legume silages is usually <1.5% of DM (Kung, 2008). Ingested ethanol is converted to acetate in the rumen.

Nonvolatile OA found in unfermented forages, including shikimic, quinic, glyceric, glycolic, succinic, malic, citric, and fumaric acids, are usually found in relatively low concentrations (<5% of DM) in fresh grasses (Jones and Barnes, 1967), although greater concentrations (10% of DM) have been reported (Russell and Van Soest, 1984). Concentrations decrease markedly with harvesting and drying of forage, such that concentrations of nonvolatile OA are low in preserved hay. Acetate is the primary ruminal fermentation product of plant OA (Russell and Van Soest, 1984).

Although the OA from fermented feeds are used as energy substrates by the animal, they do not support appreciable microbial growth in the rumen (Hall and Eastridge, 2014). For this reason, the mechanistic solution of the BCNRM does not include OA in Fraction CA, which is used to estimate microbial yield in the rumen. Nonetheless, OA are assumed to be 100% digestible and are used in the calculation of apparent total digestible nutrients (TDN) of the total diet.

Water-Soluble Carbohydrates (Fraction CA)

The WSC fraction contains the wide range of sugars found in plants as monosaccharides (glucose, fructose, galactose, arabinose, and xylose) and oligosaccharides (sucrose, maltose, raffinose, and stachyose; Rooke and Hatfield, 2003). The naturally occurring oligosaccharides are not intermediates of polysaccharide degradation (Rooke and Hatfield, 2003). The most abundant monosaccharides are glucose and fructose, whereas the disaccharide sucrose is the most common oligosaccharide in plants (Moore and Hatfield, 1994). Sucrose is the main way that energy is transported, and in some cases, stored, within plants (Van Soest, 1994). Legumes, compared with grasses, typically have greater concentrations of sucrose and other soluble sugars, as do cool-season grasses compared with warm-season grasses (Moore and Hatfield, 1994). Most of the other monosaccharides and oligosaccharides are present in forages in relatively small concentrations.

Fructans are polymers of fructose molecules found both in the cell wall and cell content fraction of plants. Because most fructans are extractable in water (Pontis, 1990), they are considered WSC. Two main types of fructans are found in plants. The grass-type levan has β-(2,6) linkages and is found in cool-season grasses (e.g., timothy, orchardgrass, wheat, barley, and oats), whereas the composite-plant-type inulin has β-(2,1) linkages (Van Soest, 1994) and is found in dicotyledonous composites. Fructan content of plants generally increases with low growing temperatures. Fructans can contribute a large portion of the WSC when they occur. For example, fructan content can range from 2 to 17% of DM in barley, and from 0 to 20% in perennial ryegrass (Rooke and Hatfield, 2003).

The WSC represent a rapidly available energy source for the ruminal microorganisms and, from a practical standpoint, can be considered completely available in the rumen. For example, Weisbjerg et al. (1998) pulse-dosed various sugars (glucose, sucrose, and lactose) into the rumen and reported that the rate of hydrolysis varied from 248 to 1,404%/h. The rate of disappearance in the ruminal fluid of monosaccharides originating from disaccharide hydrolysis was very high, from approximately 300 to 700%/h. Likewise, fructans are rapidly degraded by ruminal microorganisms (Moore and Hatfield, 1994), but at a slower rate than monosaccharides (Thomas, 1960).

Certain feeds have high WSC concentrations (e.g., molasses, citrus pulp, beet pulp, almond hulls, some bakery waste), whereas sugar concentration in most forages is moderate (2 to 6% DM), except in high-sugar grasses, such as ryegrass (*Lolium perenne* L.) cultivars that have been selected for greater (up to 3 times greater) WSC content (Humphreys, 1989; Merry et al., 2006). In addition to effects of genotype, WSC content in forages is influenced by environmental conditions, with high light intensity and photosynthesis increasing WSC content, and high temperatures decreasing WSC content (Van Soest, 1994). Consequently, WSC accumulate in forage plants over the day as a result of photosynthesis, and this variation in WSC content can be exploited by harvesting or grazing forages later in the day (near sunset). In Europe, ryegrass cultivars bred for high WSC have increased nitrogen (N) utilization (Merry et al., 2006; Moorby et al.,

2006) and milk yield in dairy cows (Miller et al., 2001). Similarly, increasing the WSC content of grasses and legumes through harvest management has been shown to enhance animal preference (Fisher et al., 1999, 2002; Brito et al., 2008), dry matter intake (DMI; Burns et al., 2005; Brito et al., 2008), in vitro digestibility (Fisher et al., 1999), and N use efficiency (Brito et al., 2008, 2009). In silages, the WSC are fermented mainly to lactate during the ensiling process, so the WSC content of ensiled forage is usually low (<2% DM), although greater concentrations (10% DM) have been noted in high-sugar cultivars of grass silage (Merry et al., 2006).

The fermentation of WSC in the rumen tends to yield a greater butyrate concentration than other NFC sources and a greater lactate concentration than the fermentation of fiber components (Strobel and Russell, 1986; DeFrain et al., 2004). Not all sugars are fermented to VFA in the rumen, however, because some sugars are stored as glycogen (or microbial starch) by ruminal bacteria and protozoa (Hall, 2011). Glycogen synthesis can be an alternative route that allows the ruminal microorganisms to cope with captured substrate in excess of the energy needs of the cell (Hall and Eastridge, 2014).

The effects of sugar fermentation on ruminal pH have been inconsistent and depend on the conditions of the study. With sugar supplemented to the diet, in vivo pH decreased (Khalili and Huhtanen, 1991; Weisbjerg et al., 1998; Lee et al., 2003), remained unchanged (Broderick and Radloff, 2004; DeFrain et al., 2004), or increased (Heldt et al., 1999; Penner et al., 2009a,b).

Analytical Considerations. Sugars can be extracted with water or 80% ethanol (Hall, 2003a). Fructans are also mostly water soluble, while short-chain lower-molecular-weight fructans can be partially dissolved with ethanol (Pontis, 1990). Methods of carbohydrate detection used commercially are usually based on broad-spectrum colorimetric assays such as the reducing sugar assay (RSA) and the phenol-sulfuric acid (PSA) assay (Wylam, 1954; Gaillard, 1958). Chromatography can also be used to detect specific carbohydrates, but these methods are not yet used routinely for commercial feed analysis

The RSA (method 974.06; AOAC International, 2005) requires hydrolysis of oligo- and polysaccharides to allow measurement of released monosaccharides. The most common hydrolysis conditions used are those designed to hydrolyze sucrose, because it is often the major sugar in cattle diets. The PSA assay (condensation method; DuBois et al., 1956) does not require hydrolysis of the carbohydrates and detects mono-, oligo-, and polysaccharides.

A recent study by Hall (2013) evaluated the efficacy of the RSA and PSA assay as they are commonly used by feed analysis laboratories. These detection methods varied as to the recovery of purified carbohydrates. The RSA showed differing recoveries for the monosaccharides glucose, fructose, and galactose at 979, 1,042, and 706 g/kg, respectively. Carbohydrates such as maltose and lactose, which are not readily hydrolyzed, gave low values with RSA. In a direct comparison of the PSA and RSA methods on water extracts of feedstuffs relative to results determined using high-performance ion chromatography, PSA using a sucrose standard gave values more similar to chromatography than did RSA using a 50:50 glucose:fructose standard (Hall, 2014). The RSA gave values 54 g/kg greater than chromatography for cool-season grasses. The PSA gave values closer to chromatography than did RSA, and so is currently recommended as a detection method for WSC analysis. It was noted that the carbohydrate used to produce the standard curve should be representative of those measured.

Starch (Fraction CB1)

Starch is the main storage carbohydrate in plants and its concentration is abundant in cereal grains. Starch granules contain crystalline domains made of amylopectin and amorphous domains made of amylose and noncrystalline amylopectin (Ratnayake and Jackson, 2009). Amylose is primarily a linear glucose polymer with α-(1,4)glycosidic linkages and amylopectin is a highly branched glucose polymer with both α-(1,4) and α-(1,6) linkages (Svihus et al., 2005). Cereal starches are typically composed of 16 to 35% amylose and 65 to 84% amylopectin (Svihus et al., 2005; McAllister et al., 2006), but there are also waxy starches with elevated amylopectin concentration (>84%) and high amylose starches with high amylose content (>40% amylose; Giuberti et al., 2014). Corn starches are composed of 2% amylose in waxy types and 70% amylose in high-amylose types, with 24 to 30% of starch as amylose in typical dent corn hybrids (Zinn et al., 2011). The amylose:amylopectin ratio increases with corn kernel maturity and decreases with increasing growing temperatures (Zinn et al., 2011). The amylose:amylopectin ratio is usually negatively correlated with starch digestion in nonruminants (Svihus et al., 2005), but it may be less important for ruminants if the grain is adequately processed (Foley et al., 2006). Grain processing reduces particle size and in some cases, it affects the characteristics of the starch and the starch-protein matrix (i.e., gelatinization), thereby exposing the starch to microbial digestion in the rumen. Ruminal microorganisms produce a wide diversity of amylases and form complex microbial consortia that can overcome barriers to starch digestion (McAllister et al., 1994).

In cereal grains, starch is deposited in granules within the endosperm. Differences in the properties of the protein matrix within the endosperm of the various cereal grains affect their fractional digestion rate. Corn and sorghum are more-slowly fermented grains, whereas barley and wheat are more rapidly fermented (McAllister et al., 1993). Prolamins, endosperm storage proteins localized on the surface of starch granules, are the major endosperm storage proteins in most feed grains (Giuberti et al., 2014). Each feed grain has a specific type of prolamine: wheat (gliadin), barley (hordein), rye (secalin), corn (zein), sorghum (kafirin), and oats (avenin).

Prolamins generally comprise up to 50 to 60% of the total protein in feed grains, with lesser amounts found in oat and rice. Floury corn endosperm types have lower zein content compared with flint and normal dent corn endosperm types (Giuberti et al., 2014). The type and amount of storage prolamins in grains influence protein degradability in the rumen. In particular, zein is very insoluble in the rumen, affecting the degradability of protein and starch (van Barneveld, 1999).

Corn and sorghum have two types of endosperm: vitreous (also called hard, flinty, and horneous) and floury (also called soft or dent) endosperm. In floury (opaque) endosperm, starch granules are loosely associated with the protein matrix, whereas in vitreous (clear or glass-like) endosperm, the starch granules are densely compacted within the protein matrix (McAllister et al., 2006). Vitreous endosperm has starch granules that are polygonal in shape, whereas the granules in floury endosperm are spherical in shape and of small size with intracellular spaces that allow enzymes to diffuse easily (Giuberti et al., 2014). Greater kernel vitreousness is associated with lower ruminal starch availability (Correa et al., 2002) and less starch being digested in the total tract (Corona et al., 2006), although these negative effects can be partially overcome by extensive processing methods such as steam flaking (Corona et al., 2006). In barley and wheat, the starch granules are loosely associated with the protein matrix throughout the endosperm (McAllister et al., 2006). These differences in endosperm structure (e.g., protein matrix), combined with differences in kernel grain structure (e.g., resistant pericarp, fibrous hull for barley) determine the fractional digestion rate of the grain in the rumen (McAllister et al., 1993).

Grain processing can be used to partially or fully overcome the structural limitations to digestion imposed by the protein matrix and endosperm within the grain (Dehghan-banadaky et al., 2007). Effects of grain processing method differ depending on the grain type, processing conditions (final particle size or thickness of flakes, moisture, degree of starch gelatinization), type of animal (feedlot, cows), intake level, and diet (forage-to-grain ratio, byproduct inclusion). Processing alters the site and extent of digestion within the gastrointestinal tract of cattle, and consequently the DE supply to the animal. Generally, the intent of grain processing is to maximize total tract digestibility by maximizing ruminal fermentation, while avoiding digestive disturbances (e.g., acidosis, bloat; Beauchemin and McAllister, 2006). The effects of grain processing have been reviewed extensively for beef cattle (e.g., Hale, 1973; Theurer, 1986; Owens et al., 1997; Owens and Zinn, 2005).

Dry processing methods such as grinding with a hammer mill or dry rolling by passing the kernels between rollers mechanically compresses or cracks the fibrous hull and pericarp to enable access of ruminal microorganisms and enzymes to the internal endosperm and increases the surface area for microbial attachment (McAllister et al., 1994, 2006). Zinn et al. (2011) estimated that, based on three studies, feeding whole shelled maize decreased the average daily gain (ADG) by 2.5% and increased DMI by 3.2% compared with conventional dry processing methods.

Steam rolling and steam flaking of grains exposes the grain to steam at atmospheric, low, or high pressure before the grain is rolled. Steam flaking of grain is a more extensive process with longer steam conditioning times and thinner rolled flakes (steam flake density from 309 to 412 g/L [24 to 32 pounds/bushel]; Armbruster, 2006) than steam rolling. Typical retention times for steam flaking are 20 to 60 minutes for corn, 40 to 80 minutes for sorghum, and 20 to 30 minutes for wheat and barley (Armbruster, 2006), compared to less than 15 minutes for steam rolling.

Steam flaking is the method most commonly used to process corn grain in U.S. feedlots (Vasconcelos and Galyean, 2007). Most commercial feedlots using steam flaking apply 3 to 6% water to the incoming grain and the grain is steeped for 15 minutes to 24 hours. As a result, the steamed flakes range from 19 to 24% moisture as they exit the rolls (Armbruster, 2006). The combination of moisture, heat, and rolling causes gelatinization of the starch granules (i.e., swelling of granules as water is absorbed, disruption of the protein matrix and crystalline structure, dissolution of polysaccharides and diffusion from ruptured granules; Ratnayake and Jackson, 2009). With storage conditions that do not allow flakes to cool rapidly, starch in steam-flaked grains can become resistant to enzymatic hydrolysis (Ward and Galyean, 1999). Based on in vitro measurements (McMeniman and Galyean, 2007), this effect can lower the capacity of ruminal microorganisms to degrade DM, but its full effect on animal performance is unclear (Armbruster, 2006). Indices of starch availability, such as in vitro gas production, DM disappearance, birefringence, and enzymatic glucose release can be used to assess starch availability as a result of gelatinization when evaluating steam-flaked grains (Xiong et al., 1990; Ward and Galyean, 1999; Zinn et al., 2002). Based on 12 studies, Zinn et al. (2011) estimated that steam flaking increased ADG by 6.3% and decreased DMI by 5% compared with conventional dry processing.

High-moisture grain, achieved by harvesting grain at >24% moisture content followed by anaerobic fermentation, and reconstitution of grain also increase the ruminal starch availability of corn and sorghum (Defoor et al., 2006). Reconstituting grain is a process whereby water is added to grain to increase its moisture content up to 30%. The reconstituted grains are then allowed to ferment under oxygen-limiting conditions (Defoor et al., 2006). Averaged over a number of studies, high-moisture grain decreased ADG by 0.8% and DMI by 6.6% compared with conventional dry processing methods (Zinn et al., 2011).

For corn and sorghum, steam flaking or high-moisture storage increases the extent of starch digestion compared with dry rolling, both in the rumen and in the intestines

(Theurer et al., 1999; Zinn et al., 2011). Ruminal digestibility of corn and sorghum is usually greatest for high-moisture grain, followed by steam flaking, dry grinding, and then coarse cracking or dry rolling, and total tract digestibility of starch follows the same pattern (Firkins et al., 2001; Zinn et al., 2011).

Bulk density (test weight) of processed corn grain is often used to quantify extent of processing. Brown et al. (2000) suggested that the most favorable flaking density was 260 to 360 g/L (20 and 28 pounds/bushel) for optimal feedlot performance and efficiency, while Ponce et al. (2013) suggested that bulk density of steam-flaked corn can be increased up to 386 g/L (30 pounds/bushel) in diets with 25% of the DM as wet corn gluten feed without affecting ADG and gain:feed by finishing beef steers, although digestibility of starch was affected negatively by increasing bulk density. Zinn et al. (2002) reported that total tract starch digestion was maximized at a flake density of 310 g/L (24 pounds/bushel), with more extensive processing predisposing cattle to acidosis and bloat without significantly further increasing starch digestion. In a survey of U.S. feedlots, Vasconcelos and Galyean (2007) reported that the average bulk density of steam-flaked corn was 350 g/L (27 pounds/bushel), whereas for sorghum, the average was 330 g/L (26 pounds/bushel).

Steam-flaked, steam-rolled, and dry-rolled grains differ widely in the resulting size and thickness of kernels. Grain processing methods that maximize ruminal digestion of corn and sorghum generally maximize total tract digestibility (Ferraretto et al., 2013) and favor greater efficiency of feed use by beef cattle (Hale, 1973). Although low ruminal digestion of corn and sorghum starch can be partially compensated for by postruminal digestion, compensatory digestion is not always sufficient to avoid a decrease in total tract digestibility. Whole kernels and large kernel fragments that pass from the rumen are poorly digested in the intestines (Rowe et al., 1999; Huntington et al., 2006; Owens and Soderlund, 2006). When grain is the major source of starch in the diet, fecal starch concentration can be used as an indicator of total tract starch digestibility and the adequacy of grain processing (Zinn et al., 2007).

Barley is also typically processed to maximize its digestibility; digestibility of whole barley grain was 16% less than that of processed grain in beef studies summarized by Mathison (1996). Unlike corn and sorghum, few advantages of steam flaking over steam rolling have been demonstrated for barley and wheat (Fiems et al., 1990; Plascencia et al., 1998); however, it should be stated that few studies have been conducted using modern steam-flaking practices. Dry rolling, steam rolling, and temper rolling are generally the recommended methods of processing barley and wheat (Koenig et al., 2013). Most studies have shown little advantage of steam vs. dry processing of barley for beef cattle (Zinn, 1993; Grimson et al., 1987; Mathison et al., 1991; Engstrom et al., 1992), although further study using modern technology is needed to fully rule out any advantage of steam rolling. Many feedlots in western Canada feeding barley grain use a tempering process whereby moisture is added to grain before dry rolling. Water is added to increase the moisture content of grain to 14 to 24%, the moisture is allowed to penetrate the grain for a period of 8 to 24 h, and the grain is then subjected to rolling (Yang et al., 1996). Tempering grain before rolling produces larger, flatter particles, improves the uniformity of kernel thickness and particle size, and decreases the amount of fines relative to dry rolling (Beauchemin et al., 2001).

The extent to which barley grain is rolled can be adjusted to regulate the fractional degradation rate in the rumen, and it is related to growth performance in feedlot cattle (Koenig et al., 2013). The extent of barley grain processing can be assessed using a processing index (PI; Yang et al., 2000), which is determined as the weight of a given volume of processed grain divided by the weight of the same volume of the original whole grains (corrected for DM content when the grain is processed with added moisture). To maximize total tract digestibility and cattle performance, it is recommended that barley grain be rolled to a PI of 75 to 85% when diets are low in forage fiber (feedlot cattle), and more extensively processed (PI < 70%) when higher-forage diets (>20% forage) are fed (Yang et al., 2000; Beauchemin et al., 2001; Koenig et al., 2003; Bengochea et al., 2005; Koenig and Beauchemin, 2011).

Owens and Zinn (2005) summarized the results of 213 different treatments from 51 studies conducted from 1990 to 2004 using feedlot cattle (steers and heifers) and lactating dairy cows, with the results shown in Table 4-2. Feedlot cattle fed steam-flaked corn or sorghum grain had greater ruminal, postruminal, and total tract starch digestion, and site of digestion was shifted toward the rumen compared to cattle fed dry-rolled grain. Furthermore, high-moisture corn had higher ruminal digestion, but similar total tract digestion to steam-flaked corn. There were also important differences in the digestion kinetics of processed grains between feedlot beef cattle and dairy cattle. For feedlot cattle, a greater proportion of the total tract digestion of starch occurred in the rumen compared with dairy cows. Ruminal starch digestion is typically lower in dairy cows than feedlot cattle because of a faster rate of particulate passage from the rumen in dairy cows resulting from greater feed intake, increased fluid dilution rate resulting from greater proportion of forage in the diet, and increased size of the reticulo-omasal orifice.

With highly processed grains, postruminal digestibility of starch is extensive for both feedlot and dairy cows (Firkins et al., 2001). However, coarse processing of grains markedly decreases postruminal digestion of starch. Thus, whole and cracked grain kernels that escape ruminal digestion have very low postruminal digestibility and are visible in manure. The results from the meta-analysis by Owens and Zinn (2005) indicated that total tract digestibility of starch was maximized when diets were rich in N but low in NDF.

TABLE 4-2 Summary of Digestibility Studies with Beef Feedlot Cattle and Dairy Cows Fed Processed Grains[a]

	Feedlot Cattle				Dairy Cows			
Grain	Digested in the Rumen, % of Intake	Proportion of Total Tract Digestion in the Rumen, % of Total	Post-Ruminal Digestibility, % of Flow	Total Tract Digestibility, % of Intake	Digested in the Rumen, % of Intake	Proportion of Total Tract Digestion in the Rumen, % of Total	Postruminal Digestibility, % of Flow	Total Tract Digestibility, % of Intake
Corn								
Dry rolled	60.6	68.3	68.4	89.3	48.8	54.9	78.2	90.8
High moisture	91.0	91.8	90.4	99.2	76.3	79.4	82.9	96.0
Steam flaked	84.2	84.9	94.1	99.1	51.8	54.8	88.4	93.9
Steam rolled	—	—	—	—	57.9	61.2	87.6	93.9
Whole	74.3	89.5	31.4	83.6	—	—	—	—
Sorghum								
Dry rolled	66.8	69.4	89.0	96.5	49.3	55.8	77.2	87.8
Steam flaked	84.9	85.9	92.1	98.8	—	—	—	—
Barley								
Dry rolled	86.2	88.5	81.6	97.1	79.9	85.1	69.9	94.0
Steam flaked	89.2	90.2	90.5	99.1	—	—	—	—
Steam rolled	—	—	—	—	63.5	71.5	70.5	89.6
Wheat								
Dry rolled	86.0	87.8	84.6	97.9	—	—	—	—
Steam flaked	91.6	92.8	85.2	98.8	—	—	—	—
Steam rolled	—	—	—	—	69.2	69.3	97.7	99.3

[a]Summary of results for 213 treatment diets from 51 studies conducted from 1990 to 2004 using feedlot cattle (steers and heifers) and lactating dairy cows. Means are weighted by number of cattle per treatment. Adapted from Owens and Zinn (2005).

Ruminal starch fermentation was maximized in diets low in NDF and when fed at lower levels of intake. Postruminal digestion of starch was maximized in diets rich in N and low in NDF. Finally, diets low in N and NDF with a slower concentrate passage rate shifted the site of digestion toward the rumen.

The site of starch digestion is an important consideration for beef cattle. Extensive digestion in the rumen maximizes microbial protein synthesis, as it is largely driven by the quantity of carbohydrates fermented in the rumen once the requirement for ruminally degradable protein is met (Russell et al., 1992). Nonetheless, maximizing ruminal starch digestion can lead to acidosis as a result of rapid accumulation of VFA in ruminal fluid. Theoretically, fermentation in the rumen is less energetically efficient (~80%) than enzymatic digestion in the small intestine (~97%; Harmon and McLeod, 2001; Huntington et al., 2006). But, digestion in the large intestine is clearly less energetically efficient (40 to 45%) than digestion in either the rumen or small intestine. Thus, maximizing the proportion of starch digested in the small intestine is most efficient from an energy-efficiency perspective, but shifting the site of starch digestion from the rumen to the small intestine is only effective if starch digestibility in the small intestine is not limited (e.g., by particle size, shortage of digestive enzymes or other factors) and if additional microbial protein that could be produced from starch fermentation in the rumen is not needed to meet animal requirements. Starch not digested in the small intestine will be fermented in the large intestine or excreted in the feces, masking any gains in efficiency achieved by shifting the site of digestion from the rumen to the intestine. Such is the case for grains fed whole or coarsely cracked. Ground grains usually have high intestinal digestibility, although it is possible that some of this high digestibility arises from digestion in the large intestine. In addition to being energetically less efficient, fermentation of starch in the large intestine results in a loss of microbial N in feces, although VFA produced are available for absorption and use by the animal.

In summary, there are considerable differences in starch digestion depending on the grain and its method of processing, but cattle are efficient at and have high capacity for digesting starch from cereal grains (McLeod et al., 2006). Depending on the cereal grain and the processing method, total tract digestibility of starch in feedlot cattle is usually greater than 90% and often greater than 98%, because of the high extent of ruminal digestion of starch (greater than 75%) from processed grains. For beef cows and growing cattle fed high-forage diets for which the risk of digestive disturbances is low, it is recommended that grain be extensively processed to maximize digestibility in the rumen and in the total digestive tract.

Analytical Considerations. The starch content of feed can be determined enzymatically as described by Knudsen (1997), Hall (2003a, 2009), and Hall et al. (2001). The reader is encouraged to consult these references for a full description and understanding of the procedure. The starch concentration in feed is measured as the quantity of glucose released after gelatinization using heat (1 h at 90° to 100°C with intermittent agitation). Gelatinization involves the dissolution of hydrogen bonds among and within starch molecules to increase susceptibility to enzymatic hydrolysis. Samples are incubated in sealed tubes containing mildly acidic buffer (0.1 to 0.2 M acetate buffer) and a thermostable amylase. Further degradation of the released oligosaccharides to glucose monomers is achieved by incubation with amyloglucosidase (2 h at 100°C). Use of amyloglucosidase ensures that degradation of the released oligosaccharides to glucose monomers is achieved. The tubes are then centrifuged and the glucose monomers released in the supernatant after dilution are quantified with a glucose oxidase reagent.

Heat-stable endoamylases and exoamyloglucosidases that are specifically active only on α-(1,4) and α-(1,6) linkages are used to hydrolyze the α-(1 → 4) (linear chain) and α-(1 → 6) (branches) linkages present in starch (Hall, 2003a). Recovery should be 100 ± 2% for pure starch, confirming that the run conditions were as desired. Samples can be preextracted with aqueous ethanol (80:20 ethanol:water, vol/vol) to help decrease interference by removing low molecular-weight carbohydrates (Hall et al., 2001). Alternatively, free glucose can be determined on a second sample, and then subtracted from the free + enzymatically released glucose. Although this option requires running two samples, it avoids the potential errors (i.e., loss of sample) that can occur with ethanol preextraction.

There can be considerable variation in the accuracy of starch analyses because of differences in methodology, such as whether samples are preextracted, the gelatinization method and enzymes used, and the glucose detection method. As noted by Hall (2003a), the critical steps for accurate starch analysis are (1) complete gelatinization of starch, (2) specificity of enzymes, (3) complete hydrolysis of starch to glucose, (4) specific measurement of glucose yield from hydrolyzed starch, and (5) minimization of interference. Endogenous glucose in the sample and interference of other monosaccharides can cause overestimation of starch content.

Neutral Detergent Soluble Fiber (Fraction CB2)

Neutral detergent soluble fiber includes pectic substances, mixed-linkage β-glucans, galactans, and other nonstarch polysaccharides not recovered in NDF. These carbohydrates cannot be digested by mammalian enzymes and must be fermented by ruminal or intestinal microbes to be digested. Most sources of soluble fiber tend to ferment rapidly (20 to 40%/h), with some exceptions (e.g., soyhulls; Hall et al., 1998). The rate of fermentation of soluble fiber is reduced when ruminal pH is low, similar to the depression in fermentation of NDF with low pH (Hall, 2003b). Common dietary sources of soluble fiber include legume forages, citrus pulp, beet pulp, soyhulls, and soybean meal.

Pectins are the predominant type of soluble fiber in plants and are generally more abundant in legumes than in grasses. They are found within the cell wall but are soluble in neutral detergent. Pectic substances have a linear α-(1,4) linked galacturonic acid backbone with arabinan and galactan side chains. They are rapidly fermented in the rumen and tend to yield more acetate with little or no lactate (Hall and Eastridge, 2014) when fermented than the other NFC (Strobel and Russell, 1986). Hall et al. (1998) reported in vitro rates of digestion of 20 to 40%/h for neutral detergent soluble fiber from feeds (citrus pulp, beet pulp, and alfalfa) high in pectic substances.

Mixed-linkage β-(1,3)(1,4)glucans are found in very small quantities in the cell wall of grasses and in the bran of some grains, including oats and barley. Although they are part of the cell wall, β-glucans are soluble in neutral detergent and thus are considered part of the NFC fraction, more specifically in the soluble fiber portion. Although they are not digested by mammalian enzymes, they are highly fermentable in the rumen. Other types of glucans include xyloglucans, mannoglucans, and glucan gums.

NEUTRAL DETERGENT FIBER (FRACTIONS CB3 AND CC)

Neutral detergent fiber represents the fiber fraction of feeds, and comprises cellulose, hemicellulose and lignin. Lignin is considered part of this fraction because it is resistant to digestion, affects digestion of cell wall carbohydrates, and is insoluble in neutral and acid detergent. Forages, plant residues (straws, stover, stalks), and byproduct feeds are generally high in NDF content, ranging from 30 to 75% of DM. Some of the more common high-fiber byproduct feeds used in cattle diets are brewers spent grains, corn gluten feed, cottonseed hulls, distillers grains with solubles, soybean hulls, sugar beet pulp, and wheat middlings as discussed in Chapter 17 (Utilization of Byproduct Feeds by Cattle). While byproduct feeds are high in NDF content, considerable variation exists in their digestibility (Firkins, 1997; Bradford and Mullin, 2012).

Neutral detergent fiber is the major source of DE in most beef cattle diets, with the exception of high-concentrate diets fed to feedlot finishing cattle. In addition to supplying cattle with DE, NDF can provide a source of physically effective fiber that stimulates rumination, insalivation, and reticuloruminal motility, which help to elevate ruminal pH. Additionally, physically effective fiber helps form a mat or raft in the rumen that functions as a filtering system to prevent feed particles from passing from the rumen undigested.

Neutral detergent fiber comprises plant cell walls, the main structural constituent of plants (Figure 4-1). The cell walls provide support for the plant and create barriers against the environment and potentially pathogenic organisms (Scheller and Ulvskov, 2010). The primary cell wall is the outermost layer of the cell and is relatively digestible by ruminal microbes (Wilson, 1993). The secondary cell wall is formed inside the primary cell wall, and is often composed of up to three layers that are differentiated by the orientation of cellulose microfibrils (Wilson, 1993). The secondary cell wall layer can become thickened and highly lignified, making it resistant to microbial digestion. The tertiary cell wall layer is a very thin layer on the lumen side of the secondary cell wall and is of variable digestibility (Wilson, 1993). The digestibility of plant cell walls depends on the comprising tissues, with mesophyll being almost completely digestible and xylem being virtually indigestible (Akin, 1993).

Plant cell wall layers are composed of cellulose microfibrils and an interpenetrating matrix of hemicellulose, pectin, proteins, lignin, and phenolic compounds as shown in the conceptual diagram in Figure 4-2. These polymers are strongly bonded to one another and the makeup and quantity of each polymer varies according to the feedstuff, growing conditions, and part of the plant. The spatial orientation of cellulose and hemicellulose within the plant cell wall matrix and covalent linkages of lignin to the noncellulosic polysaccharides limit forage cell wall digestion in the rumen.

Cellulose consists of linear chains of D-glucopyranose residues linked by β-(1,4) bonds with alternate glucose residues in the same cellulose chain rotated by 180° (Martins et al., 2011). The chains exhibit a flat structure internally stabilized by hydrogen bonds. Parallel cellulose chains are linked via hydrogen bonding and van der Waals forces, resulting in microfibrils composed of about 40 glycan chains (Bidlack et al., 1992). Cellulose microfibrils are hydrogen-bonded

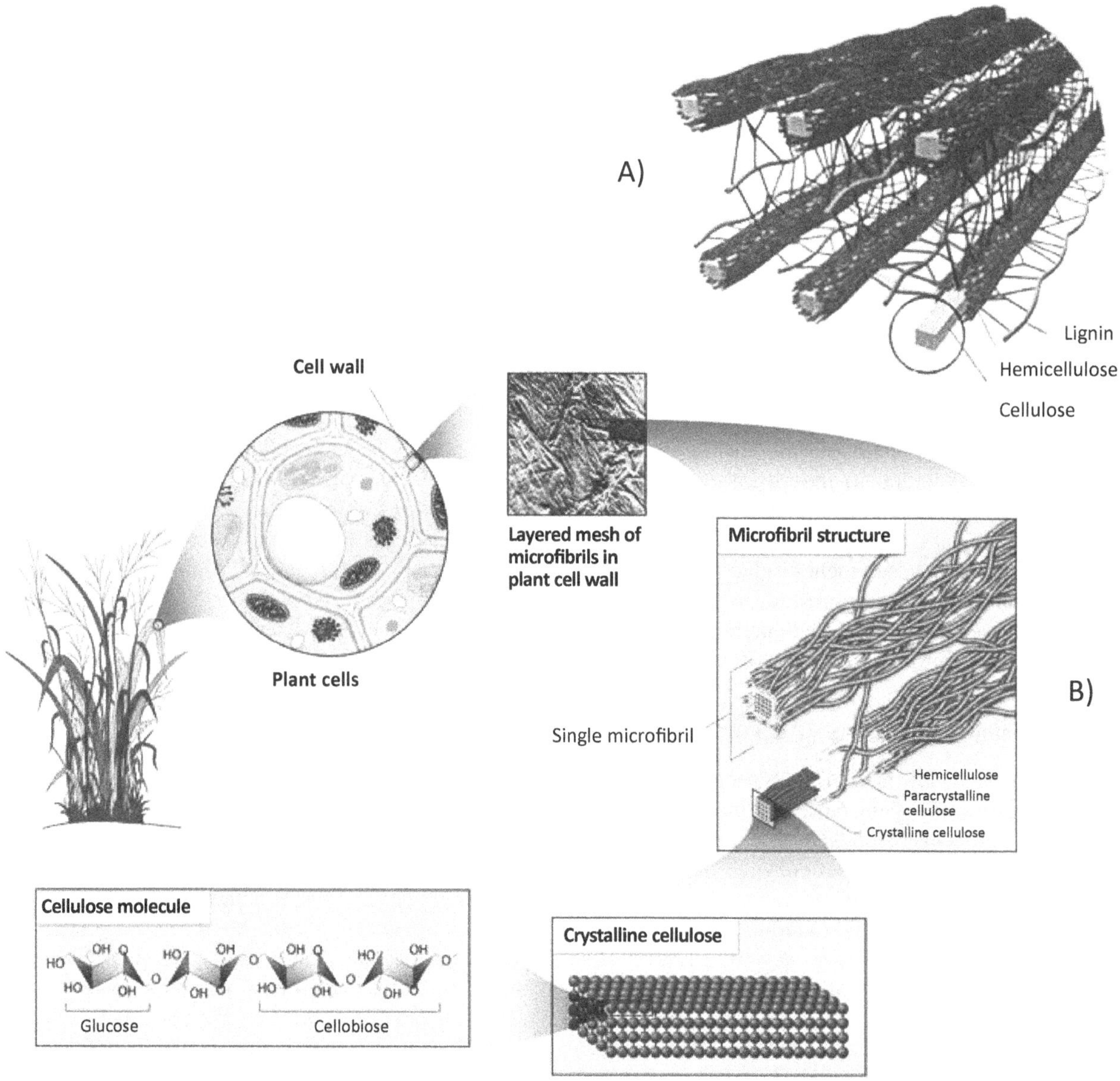

FIGURE 4-2 (A) Cellulose strands surrounded by hemicellulose and lignin (from Doherty et al., 2011); (B) Structural organization of the plant cell wall (from DOE, 2005, p. 204). Cellulose microfibrils and hemicellulose chains are embedded in lignin.

to hemicellulose, whereas ester and ether bonds connect hemicellulose to noncore lignin.

Hemicellulose is a heterogeneous group of polysaccharides, sometimes referred to as cross-linking glycans (Scheller and Ulvskov, 2010). They are characterized by having β-(1,4)-linked backbones of glucose, mannose, or xylose, the latter being a pentose. Four main groups of hemicelluloses can be defined according to their backbone chain composition: xylans, xyloglycans, mannans, and mixed-linkage β-glucans. Because the mixed-linkage β-glucans are soluble in neutral detergent, they are considered part of the neutral detergent soluble fiber fraction, as previously discussed. The backbone core polymer of hemicellulose is highly substituted with side chains of monomers including galactose, arabinose, glucoronic acid, and rhamnose (Hansen and Plackett, 2008). In grasses, hemicellulose in primary and secondary cell walls is composed mainly of arabinoxylan and glucuronoarabinoxylan (Nevins, 1993; Scheller and Ulvskov, 2010), whereas in dicots, the major hemicellulose is xyloglucan in primary cell walls and glucuronoxylan in secondary cell walls (Scheller and Ulvskov, 2010). Hemicellulose content is greater in grasses than in legumes when compared at similar stages of maturity. The most important biological role of hemicelluloses is their contribution to strengthening the cell wall through cross-linkages with cellulose and, in some cell walls, with lignin. The hemicellulosic xylans and arabinoxylans are hydrogen-bonded to cellulose and ester- or ether-bonded to noncore lignin. Hydrolysis of forage hemicellulose yields the neutral monosaccharides glucose, xylose, arabinose, mannose, galactose, rhamnose and fucose, and the uronic acids galacturonic, glucuronic, and 4-*O*-methylglucuronic (Moore and Hatfield, 1994).

As plants develop, the cell walls undergo lignification. Lignification of the cell wall causes a severe decrease in structural carbohydrate degradability in the rumen (Himmelsbach, 1993). Thus, there is a general relationship between lignification and digestibility within a plant species across stages of maturity but not across species (Moore and Hatfield, 1994). The amount, chemical composition, and distribution of lignin within plants differ widely depending on plant species, growing conditions, and physiological maturity of the plant. For example, at the same digestibility, lignin content is approximately double in legumes compared with grasses (Van Soest, 1994). Lignin itself is either indigestible or very low in digestibility (Van Soest, 1993), but it is the cross-linkage of lignin with carbohydrates, rather than concentration of lignin itself, that is detrimental to forage digestibility (Van Soest, 1994).

Lignin is a generic term for a large group of polymers (Vanholme et al., 2010). It is sometimes classified as core and noncore lignin based on susceptibility to hydrolysis (Jung and Deetz, 1993), although not all forage chemists accept this concept (e.g., Van Soest, 1993). Noncore lignin consists of low-molecular-weight phenolic compounds that are soluble by mild hydrolysis, mainly ρ-hydroxycinnamic acids including ρ-coumaric, ferulic, sinapinic, and cinnamic (Bidlack et al., 1992; Lapierre, 1993). These phenolic acids are present as ester- and ether-bound monomers, or as esterified dimers (Jung and Deetz, 1993). The ρ-hydroxycinnamic acids attach to arabinoxylans (hemicellulose) through ester bonds (Himmelsbach, 1993). Further bonding of the hydroxycinnamic moiety with the core lignin polymer occurs via ether linkages forming a bridge between carbohydrate and lignin. Ferulic and ρ-coumaric acids are major noncore lignin monomers that link hemicellulose and core lignin.

Core lignin is composed of highly condensed phenylpropanoid cell wall polymers that are largely resistant to degradation in the rumen. These complex polymers are composed of ρ-hydroxyphenyl, guaiacyl, and syringyl core units in various proportions (Lapierre, 1993). ρ-Hydroxyphenyl lignin is formed mainly from *p*-coumaryl alcohol, guaiacyl lignin is formed mainly from coniferyl alcohol, and syringyl lignin is formed mainly from sinapyl alcohol (Moore and Hatfield, 1994). Ether and carbon-carbon bonding interconnect these phenylpropanoid polymers. The syringyl:guaiacyl ratio in lignin has been shown to be negatively associated with forage fiber degradability, and generally the ratio increases with increasing maturity of perennial C3 grasses and legumes (Jung and Deetz, 1993). For further details on the chemical structures of lignin found in feed, the reader may refer to Van Soest (1994), Vanholme et al. (2010), and Samuel et al. (2014). Cell wall polysaccharides can be linked with lignin by glucosidic, ether, and ester cross-linkages, or by cinnamic acid bridges (White et al., 1993; Figure 4-3).

Analytical Considerations. Structural carbohydrates are typically measured as NDF and ADF. The detergent system of fiber analysis was originally developed by Peter J. Van Soest (1963) as a replacement for crude fiber with the methodology refined over the years (Van Soest and Wine, 1967; Goering and Van Soest, 1975; Van Soest et al., 1991) and approved as First Action by the AOAC International (method 2002.04; Mertens, 2002).

Neutral detergent fiber is the fraction of the cell wall that is insoluble after refluxing in a heated neutral (pH 7.0) detergent solution followed by filtering. The residue is composed of the three major components of the cell wall—hemicellulose, cellulose, and lignin—as well as some minor components including bound protein and N, minerals, and cuticle. There are several modifications of the NDF method for use with high-protein or high-starch feeds (Van Soest et al., 1991). Sodium sulfite is added during refluxing to remove contaminating proteins from NDF by cleaving disulfide bonds and dissolving many cross-linked proteins. The use of sodium sulfite is especially important for heat-treated feeds and protein supplements. Heat-stable amylase is used to eliminate contamination of starch, which is critical for high-starch feeds. The amylase-treated NDF modification (aNDF) uses both heat-stable amylase and sodium sulfite to obtain NDF with minimum contamination by either starch or protein. The aNDF method (AOAC method 2002.04; Mertens, 2002) is

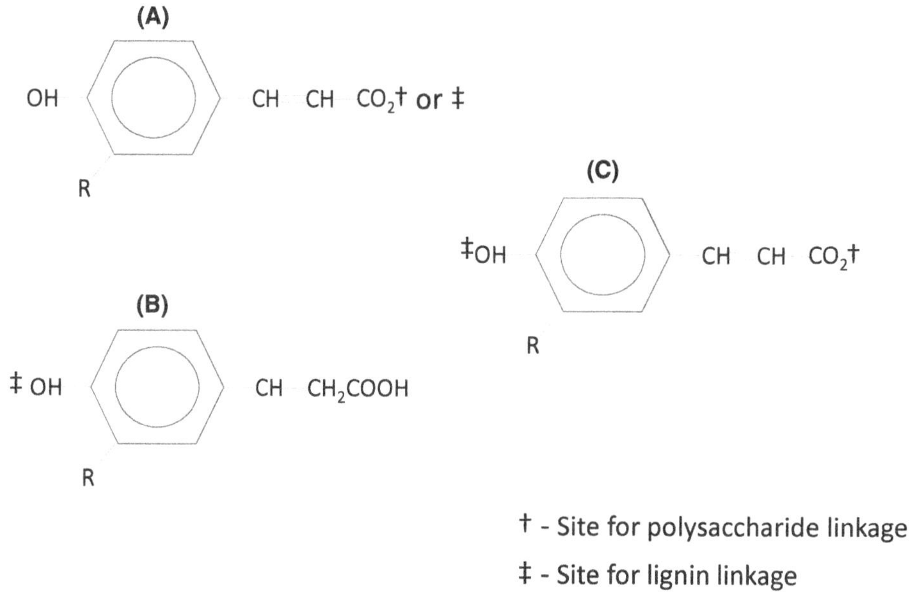

FIGURE 4-3 Linkages between lignin and polysaccharides: (A) Ester linkage with polysaccharide; (B) Ether linkage with lignin; (C) Ester-ether bridge between polysaccharide and lignin.

recommended in feed formulation by the current committee, as it provides an accurate measure of NDF carbohydrate with minimum contamination from protein and starch.

Acid detergent fiber is a further fractionation of the cell wall that is composed mainly of cellulose and lignin, with variable amounts of ash and N. The ADF fraction is analyzed according to AOAC (1990) method 973.18. The concentration of ADF is highly correlated with the concentration of NDF within a forage classification. If ADF analysis is not possible, an estimate can be obtained using regression equations from NRC (2001):

$$\text{Corn silage ADF, \%} = -1.15 + 0.62 \text{ NDF, \%}$$
$$(r^2 = 0.89, \text{RMSE} = 1.4, n = 2{,}425)$$

$$\text{Grass forage ADF, \%} = 6.89 + 0.50 \text{ NDF, \%}$$
$$(r^2 = 0.62, \text{RMSE} = 3.1, n = 722)$$

$$\text{Legume forage ADF, \%} = -0.73 + 0.82 \text{ NDF, \%}$$
$$(r^2 = 0.84, \text{RMSE} = 2.0, n = 2{,}899),$$

where RMSE = root mean square error and n = number of samples.

Lignin concentration estimates vary widely among methods. The current committee recommends using acid detergent lignin (ADL) to estimate core lignin. The ADL concentration is measured gravimetrically as the residue remaining after hydrolysis of the acid detergent residue using 72% sulfuric acid followed by ashing (Van Soest, 1963). Treatment of the ADF residue, which contains cellulose, lignin, cutin, and acid-insoluble ash (silica), with 72% sulfuric acid dissolves cellulose, and ashing the residue determines the crude lignin fraction including cutin. It is recognized that the ADL method underestimates lignin concentration as a result of loss of soluble lignin in the acid detergent step. Permanganate lignin, whereby lignin is oxidized from the ADF residue using a potassium permanganate solution, is an alternative to ADL (Goering and Van Soest, 1975). Although the reagents used for permanganate lignin are much less corrosive than sulfuric acid, large particles can be poorly penetrated by the reagents, yielding low values. Thus, sulfuric acid is the preferred method for lignin. Yet another method used to determine lignin is the Klason procedure, which determines lignin as the unhydrolyzed residue remaining after a two-stage sulfuric acid hydrolysis. Klason lignin concentrations are approximately 2 to 5 times higher for grasses and 30% higher for legumes compared with ADL (Jung et al., 1999). Although Klason lignin is a standard method of analysis for wood, it is not extensively used in animal nutrition.

Available NDF (Fraction CB3) and Unavailable NDF (Fraction CC)

The BCNRM assumes that fiber carbohydrates are represented by NDF. The proportion of cellulose, hemicellulose, and lignin within NDF will vary greatly among feeds, and consequently, digestibility of the NDF fraction is variable. The BCNRM has two levels of solution: empirical (ELS) and mechanistic (MLS). The MLS of the BCNRM incorporates a measure of either potentially fermentable or digestible (i.e., PD as defined in Chapter 2, Anatomy, Digestion, and Nutrient Utilization) NDF, which is available NDF (Fraction CB3), or unavailable NDF (Fraction CC). Available NDF is digested in a dynamic process at a fractional rate of degradation (kd) estimated using in situ or in vitro methods as discussed in Chapter 2. The CC fraction is the proportion of NDF remaining after a long incubation (preferably 240 h; Van Soest et al., 2005; Lopes et al., 2015). Available NDF is then estimated as NDF minus Fraction CC. Unavailable fiber accounts for the effects that lignin has on decreasing NDF degradability of forages. When it is not possible to conduct an in vitro or in situ assay, an alternative is to estimate Fraction CC using the equation from the CNCPS (Sniffen et al., 1992), as validated by Van Soest et al. (2005):

$$CC = 2.4 \times \text{lignin} (\% \text{ DM})$$

Available NDF is assumed to have a postruminal digestibility of 20% (Sniffen et al., 1992), whereas unavailable fiber is indigestible in the hindgut as well as in the rumen and small intestine.

FORAGES AND THEIR UTILIZATION

There are many factors that affect forage quality and utilization by cattle and this topic has received vast attention over the years and is the subject of numerous books (e.g., Heath et al., 1985; Jung et al., 1993; Fahey, 1994; Van Soest, 1994; Moser et al., 1996; Frame et al., 1998; Hopkins, 2000; Frame, 2005; Rayburn, 2007), book chapters (e.g., Galyean and Goetsch, 1993; Merchen and Bourquin, 1994; Moore and Hatfield, 1994; Buxton et al., 1996), and reviews (e.g., Reid and Klopfenstein, 1983; Leng, 1990; Rouquette et al., 2009). A brief discussion of some of the major factors affecting forage utilization of cattle follows with a summary given in Table 4-3. Grazed and harvested (hay, silage) forages comprise 90 to 100% of the diets fed to most beef cattle in North America, with the exception of feedlot finishing cattle (Galyean and Goetsch, 1993; refer to Chapter 1, Beef Production: Systems, Quality, and Safety, for a description of beef production systems used in North America).

Buxton et al. (1996) defined forage quality as the relative performance of animals when herbage is offered ad libitum. Forage quality determines the need to supplement the diet with grains and protein feeds to meet the animal's requirements for energy and CP to attain the desired level of production. Forages are also an important dietary source of minerals and vitamins. Galyean and Goetsch (1993) estimated that the diet TDN content of most beef cattle diets from fiber alone is only 30 to 40%; thus additional TDN must be supplemented to obtain the desired level of performance from forage-fed cattle (Paterson et al., 1994). Low-quality forages are often defined as <55% digestible, and in most cases deficient in CP (<8% of DM) and low in soluble sugars and starches (<10% DM) (Leng, 1990). In most production systems, excluding feedlot finishing cattle, the objective is to maximize the utilization of forages and minimize the need for supplementation; thus maximizing forage quality is critical.

As forage quality is defined in terms of animal performance, it represents the interactions between intake poten-

TABLE 4-3 Overview of Factors Affecting Forage Fiber Utilization by Ruminants[a]

Factor	Fiber Intake	Fiber Digestion
Animal species	Cattle > sheep = goats	Cattle = goats > sheep
Forage type	Legumes ≤ grass at ad libitum intake Warm-season grasses > cool-season grasses at similar maturity	Rate: legumes > grass Extent: grass > legumes; warm-season grasses > cool-season grasses at similar maturity
Forage maturity	Declines with maturity	Rate and extent declines with maturity
Supplements		
Grain/starch	Intake decreases with increasing grain Effects greater with high- than with low-quality forage	Decreases with >30% grain Effects greater with low- than with high-quality forage
Protein	Increases with low-N forages Little effect on high-quality forage	Rate and extent increases with low-N, high-fiber forages Variable effects on high-quality forages
Forage processing		
Pelleting/grinding	Increases via increased passage	Decreases extent
Conservation	Decreases with ensiling	Little effect, except decreases for heat-damaged silage
Chemical treatment	Increases intake of high-fiber forage	Extent increases if intake limited Rate possibly increases
Environmental factors		
Cold stress	Increases via increased rate of passage	Extent decreases
Heat stress	Decreases	Increases or no change

[a]Adapted from Galyean and Goetsch (1993).

tial, digestibility, and partitioning of metabolized products within the animal (Buxton et al., 1996). Intake accounts for a greater proportion of the variation in DE intake of cattle than digestibility, yet intake potential of forage is extremely difficult to estimate. The intake potential of forage is dictated by intrinsic properties of the forage including chemical composition, digestibility, and palatability, combined with animal characteristics (Minson and Wilson, 1994). The plant and animal factors interact such that the intake potential of a particular forage is not a constant, but rather varies depending upon the animal characteristics, as well as supplementation of additional feedstuffs. The complexity of these interactions makes it very difficult to estimate forage intake. Chapter 10 (Feed Intake) provides greater detail on the factors affecting DMI of cattle, and numerous publications have previously addressed this issue (e.g., Baile and McLaughlin, 1987; Mertens, 1994; Allen, 1996; Forbes, 1996). It is generally recognized that for low-quality forages, DMI is limited by rumen fill or limited physical capacity of the rumen, whereas for high-quality forages, the animal's energy requirement determines intake. As described by Mertens (1994), when cattle are fed diets that are palatable but low in energy, physical distention of the rumen limits intake, with the level of physical distention that signals satiety varying with animal body weight and physiological state. Mertens (1985) proposed that the NDF content of forages, adjusted for particle size, could be used as a proxy for this filling effect.

The digestibility of the feed largely determines its energy content, which is represented in the ELS of the BCNRM as tabular values of TDN (% DM), metabolizable energy (ME; Mcal/kg), net energy required for maintenance (NEm; Mcal/kg), and NE required for gain (NEg; Mcal/kg). These values can be adjusted by the user to represent the digestibility of the particular forage used. In the MLS, however, the TDN content of the forage is estimated from the apparent digestibilities of carbohydrates, CP, and fat. These apparent digestibilities are determined by simulating the degradation, passage, and digestion of the feed in the rumen and small intestine. As fiber carbohydrates are the main source of DE in forages, the proportion of potentially fermentable NDF (i.e., Fraction CB3) and its rate of degradation (kd_{CB3}) are the most important aspects affecting forage TDN content. Although the ruminal microbes are well-adapted to ferment a variety of carbohydrates, ruminal degradation of NDF is highly variable depending on plant species, morphological fraction (e.g., leaf versus stem), plant cell wall type (primary versus secondary), plant tissue (mesophyll, phloem, xylem), and degree of cell wall lignification, as well as differences among animals due to their rumen microbiome. Consequently, degradability of cellulose from forages ranges from 25 to 90%, whereas hemicellulose degradability ranges from 45 to 90% (Moore and Hatfield, 1994). The rate of degradation relative to the rate of passage from the rumen will determine the extent of ruminal digestion of the potentially fermentable NDF fraction.

The NDF content of forages and its digestibility are highly variable. With most forages, increasing physiological maturity causes a decrease in leaf-to-stem ratio (Merchen and Bourquin, 1994). With grasses, NDF and lignin contents of leaves and stems increase with advancing stage of maturity. Thus, the digestibility of leaves and stems decreases with maturity, although the rate of decrease is greater for stems than leaves. With most legumes (e.g., alfalfa), stems increase in NDF and lignin contents with plant maturity, but leaves remain relatively constant in composition. As a result, the stems decrease in digestibility with maturity, but the digestibility of the leaves remains stable.

Grasses usually contain greater NDF content than legumes at the same physiological stage of development because they contain a greater proportion of hemicellulose (Van Soest, 1994). Lignin content, however, is greater in legumes than in grasses at the same stage of maturity, and as a result, digestibility of legume NDF is usually lower than grass NDF. Likewise, warm-season forages (i.e., C4 species grow best at temperatures of 21° to 35°C) usually contain greater NDF and lignin concentrations than temperate cool-season forages (i.e., C3 species grow best when temperatures are 4° to 24°C; Van Soest, 1994). When compared at the same stage of maturity, the digestibility of cool-season grasses is usually greater than for warm-season grasses, with less difference in digestibility between cool-season and warm-season legumes (Merchen and Bourquin, 1994). Greater digestibility of temperate grasses is attributed to anatomical structure of C3 vs. C4 grasses, lower ambient temperatures during growth of cool- vs. warm-season grasses, and greater proportions of rapidly digestible mesophyll cells and lower proportions of the less-digestible parenchyma bundle sheath and epidermis.

Forage-based diets are frequently supplemented to improve animal productivity (Paterson et al., 1994). However, intake and digestibility of forage, when offered as part of a supplemented diet, may be greater or less than when the forage is fed alone because of digestive and metabolic interactions between forages and concentrates (i.e., associative effects). These associative effects result from changes in voluntary intake and digestibility of the forage, with the changes in intake usually greater than changes in digestibility (Merchen and Bourquin, 1994; Dixon and Stockdale, 1999). Increased forage intake is usually the result of increased rate of NDF degradability and consequently, reduction in ruminal fill. Positive associative effects when grains and protein supplements increase voluntary intake and/or digestion of forage are usually a result of the provision of a limiting nutrient, such as N for N-deficient forages or N and fermentable carbohydrates for low-quality roughages (e.g., straw; Leng, 1990; Paterson et al., 1994; Buxton et al., 1996), or in some cases, macrominerals, microminerals, or vitamins to meet the requirements of the ruminal microorganisms (Leng, 1990). Provision of the lacking nutrient stimulates rumen microbial growth and increases NDF degradation.

Negative associative effects when grains decrease voluntary intake and/or digestion of forage occur mainly with low- to medium-digestibility forage. The fermentable carbohydrates in grain decrease the rate of degradation of the forage NDF (Mould et al., 1983). Negative associative effects occur frequently in feedlot cattle fed high-grain diets mainly as a result of low ruminal pH (<6.0) and suboptimal conditions for fiber digestion (refer to Chapter 2). As discussed by Merchen and Bourquin (1994), the negative effect of NFC on NDF degradation is thought to be a result of (1) preference of ruminal microorganisms for NFC rather than NDF, (2) decreased ruminal pH caused by rapid degradation of NFC, and (3) preferential proliferation of NFC-digesting bacteria caused by competition for essential nutrients.

Associative effects of forages and grains can be accounted for in the ELS of the BCNRM by manually adjusting the TDN value of the forage. In the mechanistic level, this adjustment is made by changing NDF degradability or by decreasing the ruminal pH (negative associative effects only).

Forage preservation via ensiling affects forage utilization, and numerous reviews and books have been published on the subject (e.g., McDonald et al., 1991; Rotz and Muck, 1994; Charmley, 2001; Buxton et al., 2003). Changes in chemical and physical characteristics that occur during the ensiling process can affect DMI and digestibility by the animal. Dry matter intake is usually negatively affected when high-moisture silages or poorly fermented silages are fed, with little effect on DMI for well-preserved forages (Weiss et al., 2003). Losses of DM that occur pre-, during, and postensiling, and complex interactions between plant enzymes and microorganisms, affect forage quality primarily by altering the composition of the NFC fraction (particularly the CA fraction) and the nitrogenous compounds (proportion of nonprotein N, ruminal degradability of N, N bound to fiber fractions resulting from heating, etc.). During the anaerobic fermentation stage of ensiling, the NFC fraction, particularly mono- and oligosaccharides, provides fermentable substrate for the epiphytic lactic acid bacteria, resulting in the production of OA (see Organic Acids section) and a decrease in silage pH. The structural carbohydrates can also potentially be used to provide monosaccharides for the fermentation process, but in most cases, their contribution is low (Rooke and Hatfield, 2003). Thus, the ensiling process usually has relatively minor effects on NDF digestibility when sound harvest management is applied (Der Bedrosian et al., 2012); however, in the case of heat-damaged silage (usually high-DM silages) that has undergone Maillard reactions, concentrations of ADF and NDF increase and their digestibilities decrease (Weiss et al., 2003). Some metabolic processes continue after active fermentation of silage appears complete. For example, in vitro starch digestion, soluble protein, and ammonia N contents of corn silage have been shown to increase with prolonged storage in the silo (Der Bedrosian et al., 2012). The main nutritional differences between well-fermented silage and fresh or dried forage are lower concentrations of WSC, greater concentrations of OA and nonprotein N, and higher ruminal degradability of protein in silages (Weiss et al., 2003). In addition, pH of silage is lower and particle size is usually smaller than that of fresh forage or hay.

Decreasing particle size of forage by fine chopping, grinding, or pelleting affects the physical effectiveness of fiber (see Fiber and Prediction of Ruminal pH section), and can affect voluntary intake and digestibility. Significant reduction in the particle size of forage decreases its filling effect in the rumen, and consequently, intake is increased because of faster rate of passage from the rumen. The intake response to particle size reduction increases with forage maturity. Shortened retention time in the rumen causes a decrease in ruminal digestibility of NDF, and consequently, the hindgut makes a proportionally larger contribution to fiber digestion in the total tract. The net effect of grinding and pelleting, however, is a decrease in total tract digestibility of up to 15 percentage units for grasses and 3 to 6 percentage units for legumes (Merchen and Bourquin, 1994).

FIBER AND PREDICTION OF RUMINAL pH

In addition to providing cattle with digestible nutrients, long forage particles promote chewing during eating and rumination, which increases salivary secretion, thereby helping to elevate ruminal pH (see section on Ruminal pH and Acidosis in Chapter 2). Long particles also create a floating mat in the rumen, which stimulates reticuloruminal contractions and passage of digesta from the rumen. Without these mixing motions the rumen can become a stagnant pool, with declining removal of VFA absorption and fluid passage from the rumen, thereby decreasing ruminal pH. As NDF is more slowly digested than starch and sugar, inclusion of forage in the diet slows the rate of carbohydrate digestion in the rumen. Decreasing the rate of carbohydrate digestion decreases the rate of VFA production, thereby preventing large drops in ruminal pH. Feeding long-particle fiber can also shift the site of starch digestion from the rumen to the intestine, which decreases the potential for ruminal acidosis (Yang and Beauchemin, 2006).

Over the years, many ways of characterizing the physical properties of feeds and the physiological response to fiber have been proposed (Chenost, 1966; Balch, 1971; Sudweeks et al., 1981; Santini et al., 1983; Kristensen and Nørgaard, 1987; De Boever et al., 1990; Sauvant et al., 1990; Mertens, 1997). Various terms have been introduced to characterize the physiological aspects of fiber, but the most widely recognized terms in North America are effective fiber and physically effective fiber.

Effective fiber, or effective NDF (eNDF), is a term originally proposed for dairy diets (Mertens, 1986, 1997) to represent the ability of a feed to replace forage in a diet so that milk fat percentage is maintained. The eNDF value of a feed depends upon its particle length, buffering capacity, fermentation rate, and other inherent characteristics. In practice, the eNDF content of a feed is determined by measuring its NDF content, then measuring the percentage of the NDF remaining on a 1.18-mm screen after vertical shaking of the dry feed. The values are then adjusted for density, hydration, degree of lignification of the NDF, and other factors based on the judgment of the user (Mertens, 1997). The difficulty with the eNDF concept is the lack of a standardized method of assessment, and thus assignment of eNDF values to feeds is somewhat arbitrary. Furthermore, milk fat synthesis is affected by numerous factors other than fiber (Bauman et al., 2006) and not relevant for beef cattle.

The term physically effective fiber (peNDF) was introduced by Mertens (1997) to refine the concept of effective fiber. Physically effective fiber relates solely to the particle size of feed and is an indication of its potential to stimulate chewing. Thus, peNDF differs from eNDF in that peNDF is narrowly defined in terms of chewing, whereas eNDF encompasses more factors. The physical effectiveness factor (pef) of a feed ranges from 0 to 1.0; pef is multiplied by NDF content to determine peNDF content of the feed. Long-grass hay is used as a reference feed and has a pef of 1.0 (Mertens, 1997). The pef of other feeds are relative to this standard. For example, finely chopped alfalfa silage containing 42% NDF (DM basis) has a pef of 0.60, so its peNDF content would be 25.2% DM (42% × 0.60). A limitation to using chewing time to indicate the physical effectiveness of feeds is the need to rely on book values for pef. Thus, sieving methods that measure particle length of feeds are used to determine pef, based on the concept that long particles retained on sieves represent particles that require chewing.

The original concepts of eNDF and peNDF (Mertens, 1997) were based on measuring particle length using an oscillating dry-sieving technique (using a sieve with 1.18-mm openings). The assumption was that 1.18 mm is the critical size at which feed particles are considered physically effective. Because of ease of use, however, the Penn State Particle Separator (PSPS) has been widely adopted on-farm to measure particle length of feeds (Lammers et al., 1996; Kononoff et al., 2003). The 2013 PSPS consists of three sieves (19-, 8-, and 4-mm openings) and a bottom pan, with the 4-mm sieve replacing the 1.18-mm sieve in the 2002 version of the PSPS. The pef of a feed or diet is often calculated as the total proportion of material retained on the three sieves ($pef_{>4}$). These pef values are fairly similar to the previous eNDF values. However, caution should be used when determining the pef of total mixed rations containing whole grains and pelleted supplements, which can be trapped on the 4-mm sieve, thereby inflating the pef values of the ration. This limitation can be overcome by calculating dietary peNDF solely from the individual forages, omitting the manufactured feeds.

There is sometimes confusion between the terms eNDF and peNDF, and often they are incorrectly used interchangeably. While peNDF and eNDF are highly correlated, eNDF can be greater than peNDF for feeds that elevate ruminal pH but do not stimulate chewing activity (Mertens, 1997). It should be noted that the term eNDF is used in early versions of the CNCPS, the Cornell-Penn-Miner Dairy (CPM-Dairy) model, and the NRC (1996, 2000) model, whereas later versions of the CNCPS version 5 (Fox et al., 2003, 2004), CNCPS version 6 (Tylutki et al., 2008) and CPM-Dairy version 3 (Boston et al., 2000) use peNDF to predict ruminal pH using an equation from Pitt et al. (1996) as modified by Tylutki et al. (2008). Because peNDF and eNDF are indicators of chewing time, and mastication increases salivation, these terms are indirectly related to ruminal pH.

Fox and Tedeschi (2002) recommended 7 to 10% peNDF in the dietary DM of feedlot cattle to keep mean ruminal pH above 5.7 (based on the equation of Pitt et al., 1996), but commercial feedlot finishing diets are often below this recommendation for peNDF. For example, a diet (DM basis) containing 90% concentrate and 10% medium chopped grass hay (NDF content of 50%; pef of 0.9) would have a peNDF content from forage of only 4.5% (i.e., 10 × 0.5 × 0.9), although depending upon the type of grain and degree of processing the concentrate may also contribute peNDF. In commercial feedlots, diets often contain even less forage (<10%). In a survey of feedlots in the United States, Vasconcelos and Galyean (2007) reported that mean forage inclusion in corn-based diets averaged 8.3% of DM in summer (range: 4.5 to 13.5%) and 9.0% of DM in winter (range: 0 to 13.5%). Similarly, with barley-based diets used in western Canada, forage inclusion usually ranges from 4 to 10% of DM (Koenig and Beauchemin, 2011). Furthermore, silages with a pef of 0.6 to 0.9 (e.g., corn silage, alfalfa silage, barley silage) are used as the main roughage in feedlot finishing diets. Thus, the dietary peNDF content from forage of most feedlot finishing diets is often less than 5%, although additional peNDF may be supplied via the grain. However, prevalence of ruminal acidosis under commercial feedlot conditions has not been well established, and further research is needed. There is some limited evidence to suggest that even though the peNDF content of the diet remains unchanged throughout the finishing phase, the prevalence and severity of acidosis in feedlots increases with days on feed, possibly due to increased DMI (Castillo-Lopez et al., 2014).

Various authors have published empirical predictions of mean ruminal pH from peNDF or eNDF (Pitt et al., 1996; Mertens, 1997; Fox et al., 2004; Zebeli et al., 2006, 2008). It appears that the accuracy of these equations is low for beef cattle ($R^2 < 0.52$; Sarhan and Beauchemin, 2015). In level 2 of the previous model (NRC, 1996, 2000), peNDF was used to predict ruminal pH using the equation from Pitt et al. (1996; Table 4.4) and ruminal pH was used to adjust the degradation rate of available NDF. The peNDF content was also used directly to adjust bacterial yield (Pitt et al., 1996). In the ELS of the BCNRM, no adjustment is made for ruminal pH; however, in the MLS, degradation of available NDF (kd_{CB3}) and efficiency of microbial CP synthesis are adjusted for mean ruminal pH, as was done in the previous NRC (1996, 2000) model.

The current committee recognizes that ruminal pH in beef cattle is a complex function of VFA production and removal (absorption, neutralization, passage; see Chapter 2). As the concept of peNDF does not account for the fermentability of feed or absorption from the rumen, this term is limited as a sole predictor of ruminal pH, particularly for feedlot cattle diets. For dairy cows, Mertens (1997) found that 71% of the variation in ruminal pH was accounted for by peNDF, but peNDF accounted for considerably less (<50%) of the variation in ruminal pH for beef cattle (Sarhan and Beauchemin, 2015). Many additional factors other than peNDF can influence ruminal pH, including intake of NFC and its degradation rate, grain processing effects, use of ionophores, feeding management, and so forth. Furthermore, extant pH prediction equations based on peNDF (Mertens, 1997; Pitt et al., 1996; Fox et al., 2004; Zebeli et al., 2006, 2008) were developed using mainly data for dairy cows with only the equations from Pitt et al. (1996) and Fox et al. (2004) using limited data for beef cattle.

In the MLS of the BCNRM, mean ruminal pH is an input supplied by the user. For beef cows and growing cattle fed forage-based diets (>50% forage DM), mean ruminal pH is assumed to be >6.2. When the peNDF content of feeds is known, the equation from Pitt et al. (1996) can be used:

$$\text{Mean ruminal pH} = 5.46 + 0.038 \times \text{peNDF (\% DM)}, \text{peNDF} < 26.3\%, \text{pH} = 6.46 \text{ when peNDF} > 26.3\%; r^2 = 0.52; \text{n} = 55\ (n = 10 \text{ for beef cattle}, n = 1 \text{ for sheep, and } n = 44 \text{ for dairy cows}) \quad \text{(Eq. 4-1)}$$

An evaluation of Eq. 4-1 using an independent database of 65 published papers and 231 treatment means for beef cattle, using pef values for feeds from CPM-Dairy, resulted in a concordance correlation of 0.67, a coefficient of determination of 0.52, and a root mean square prediction error of 4.94% (Sarhan and Beauchemin, 2015). While this equation was the most reliable of the empirical equations evaluated in that study, the accuracy of prediction of mean ruminal pH from peNDF alone was relatively low.

When forage content of the diet is <50% of the DM and the peNDF content of the diet is not known, mean ruminal pH can be estimated using an unpublished equation (M. Sarhan and K. A. Beauchemin, Agriculture and Agri-Food Canada, Lethbridge, Alberta, Canada, personal communication, June 2014) where

$$\text{Mean ruminal pH} = 5.724 - 0.00963 \times \text{forage (\% of dietary DM)}; r^2 = 0.76; n = 97 \quad \text{(Eq. 4-2)}$$

This equation was developed using treatment means from 27 beef cattle studies published from 1998 to 2014 where mean ruminal pH was measured over 24-h periods, mostly using indwelling pH loggers. It should be noted that there is still considerable variability unaccounted for by this equation, particularly for diets containing <15% forage DM. Furthermore, the equation does not account for variability in starch content or particle length of forages, although including starch content into the equation did not further improve the accuracy of prediction.

The estimated mean ruminal pH using Eq. 4-1 or 4-2 should be increased or decreased by the user to reflect the risk factors for acidosis listed in Table 4-4. Numerous factors

TABLE 4-4 Some Factors Affecting Ruminal pH and the Risk of Acidosis in Feedlot Cattle

Factor	Increased pH	Decreased pH	References
Forage proportion and particle size	Greater proportion, longer particles	Lower proportion, shorter particles	White et al., 1969; Koenig et al., 2003; Crawford et al., 2008; Wierenga et al., 2010; Faleiro et al., 2011; Iraira et al., 2013
Grain type	Corn, sorghum	Barley, wheat	Zinn and Barajas, 1997; Beauchemin and McGinn, 2005
Grain processing	Whole, coarsely rolled	Finely ground, steam flaked; flat flakes, decreasing bulk density; high moisture	Lee et al., 1982; Zinn, 1990; Reinhardt et al., 1997; Beauchemin et al., 2001; Koenig and Beauchemin, 2011
Feed additives	Ionophores, buffers		Nicholson et al., 1963; Adams et al., 1981; Stroud et al., 1985; Boerner et al., 1987; Erickson et al., 2003; Farran et al., 2003
Feed management	Consistent daily allocation and delivery time	Inconstant daily allocation and delivery time	Soto-Navarro et al., 2000; Schwartzkopf-Genswein et al., 2003, 2004
Feeding frequency	More than once daily	Once daily	Soto-Navarro et al., 2000; Robles et al., 2007
Duration of feeding period	Shorter time on feed	Longer time on feed	Castillo-Lopez et al., 2014
Dietary transition and adaptation	Gradual dietary transition; longer adaptation to diets	Abrupt dietary transition; short adaptation to diet	Uhart and Carroll, 1967; Fulton et al., 1979; Bevans et al., 2005; Brown et al., 2006; Holtshausen et al., 2013; Schwaiger et al., 2013
Feed sorting by animal	Preferential selection of long forage particles	Preferential selection of grain	Provenza, 1995; Phy and Provenza, 1998; Keunen et al., 2002; Leonardi and Armentano, 2003; DeVries et al., 2014a,b

other than forage proportion and peNDF affect mean ruminal pH as listed in Table 4-4, and these are not accounted for in the current ruminal pH prediction equations. It should also be noted that area under the curve and time under an acidosis pH threshold (e.g., 5.6) is recognized as being more critical to ruminal health and function than low mean pH; however, prediction of these acidosis indices is difficult because of a lack of published data. Further discussion of the factors affecting ruminal pH is provided in Chapter 2.

REFERENCES

Adams, D. C., M. L. Galyean, H. E. Kiesling, J. D. Wallace, and M. D. Finkner. 1981. Influence of viable yeast culture, sodium bicarbonate and monensin on liquid dilution rate, rumen fermentation and feedlot performance of growing steers and digestibility in lambs. *Journal of Animal Science* 53:780-789.

Akin, D. E. 1993. Perspectives of cell wall biodegradation—Session synopsis. Pp. 73-82 in *Forage Cell Wall Structure and Digestibility*, H. G. Jung, D. R. Buxton, R. D. Hatfield, and J. Ralph, eds. Madison, WI: American Society of Agronomy, Crop Science Society of America, and Soil Science Society of America.

Allen, M. S. 1996. Physical constrains on voluntary intake of forages by ruminants. *Journal of Animal Science* 74:3063-3075.

AOAC (Association of Official Analytical Chemists). 1990. *Official Methods of Analysis of the Association of Official Analytical Chemists*, 15th Ed., K. Helrich, ed. Arlington, VA: Association of Official Analytical Chemists.

AOAC International. 2005. *Official Methods of Analysis*, 18th Rev. Ed., W. Horwitz and G. W. Latimer, Jr., eds. Gaithersburg, MD: AOAC International.

Armbruster, S. 2006. Steam flaking grains for feedlot cattle: A consultant's perspective. Pp. 46-55 in *Cattle Grain Processing Symposium, November 15-17, 2006, Tulsa, OK*. Stillwater: Oklahoma State University.

Baile, C. A., and C. L. McLaughlin. 1987. Mechanisms controlling feed intake in ruminants: A review. *Journal of Animal Science* 64:915-922.

Balch, C. C. 1971. Proposal to use time spent chewing as an index of the extent to which diets for ruminants possess the physical property of fibrousness characteristic of roughages. *British Journal of Nutrition* 26:383-392.

Bauman, D. E., I. H. Mather, R. J. Wall, and A. L. Lock. 2006. Major advances associated with the biosynthesis of milk. *Journal of Dairy Science* 89:1235-1243.

Beauchemin, K. A., and T. A. McAllister. 2006. Digestive disturbances: Acidosis, laminitis, and bloat. Pp. 221-231 in *Cattle Grain Processing Symposium November 15-17, 2006, Tulsa, OK*. Stillwater: Oklahoma State University.

Beauchemin, K. A., and S. M. McGinn. 2005. Methane emissions from feedlot cattle fed barley or corn diets. *Journal of Animal Science* 83:653-661.

Beauchemin, K. A., W. Z. Yang, and L. M. Rode. 2001. Effects of barley grain processing on the site and extent of digestion of beef feedlot finishing diets. *Journal of Animal Science* 79:1925-1936.

Bengochea, W. L., G. P. Lardy, M. L. Bauer, and S. A. Soto-Navarro. 2005. Effect of grain processing degree on intake, digestion, ruminal fermentation, and performance characteristics of steers fed medium-concentrate growing diets. *Journal of Animal Science* 83:2815-2825.

Bevans, D. W., K. A. Beauchemin, K. S. Schwartzkopf-Genswein, J. McKinnon, and T. A. McAllister. 2005. Effect of rapid or gradual grain adaptation on subacute acidosis and feed intake by feedlot cattle. *Journal of Animal Science* 83:1116-1132.

Bidlack, J., M. Malone, and R. Benson. 1992. Molecular structure and component integration of secondary cell walls in plants. *Proceedings of the Oklahoma Academy of Science* 72:51-56.

Boerner, B. J., F. M. Byers, G. T. Schelling, C. E. Coppock, and L. W. Greene. 1987. Trona and sodium bicarbonate in beef cattle diets: Effects on pH and volatile fatty acid concentrations. *Journal of Animal Science* 65:309-316.

Boston, R. C., D. G. Fox, C. Sniffen, E. Janczewski, R. Munson, and W. Chalupa. 2000. The conversion of a scientific model describing dairy cow nutrition and production to an industry tool: The CPM Dairy project. Pp. 361-378 in *Modelling Nutrient Utilization in Farm Animals*, J. P. McNamara, J. France, and D. E. Beever, eds. Wallingford, UK: CAB International.

Bradford, B. J., and C. R. Mullins. 2012. Invited review: Strategies for promoting productivity and health of dairy cattle by feeding nonforage fiber sources. *Journal of Dairy Science* 95:4735-4746.

Brito, A. F., G. F. Tremblay, A. Bertrand, Y. Castonguay, G. Bélanger, R. Michaud, H. Lapierre, C. Benchaar, H. V. Petit, D. R. Ouellet, and R. Berthiaume. 2008. Alfalfa cut at sundown and harvested as baleage improves milk yield of late-lactation dairy cows. *Journal of Dairy Science* 91:3968-3982.

Brito, A. F., G. F. Tremblay, H. Lapierre, A. Bertrand, Y. Castonguay, G. Bélanger, R. Michaud, C. Benchaar, D. R. Ouellet, and R. Berthiaume. 2009. Alfalfa cut at sundown and harvested as baleage increases bacterial protein synthesis in late-lactation dairy cows. *Journal of Dairy Science* 92:1092-1107.

Broderick, G. A., and W. J. Radloff. 2004. Effect of molasses supplementation on the production of lactating dairy cows fed diets based on alfalfa and corn silage. *Journal of Dairy Science* 87:2997-3009.

Brown, M. S., C. R. Krehbiel, G. C. Duff, M. L. Galyean, D. M. Hallford, and D. A. Walker. 2000. Effect of degree of corn processing on urinary nitrogen composition, serum metabolite and insulin profiles, and performance by finishing steers. *Journal of Animal Science* 78:2464-2474.

Brown, M. S., C. H. Ponce, and R. Pulikanti. 2006. Adaptation of beef cattle to high-concentrate diets: Performance and ruminal metabolism. *Journal of Animal Science* 84:E25-E33.

Burns, J. C., H. F. Mayland, and D. S. Fisher. 2005. Dry matter intake and digestion of alfalfa harvested at sunset and sunrise. *Journal of Animal Science* 83:262-270.

Buxton, D. R., D. R. Mertens, and D. S. Fisher. 1996. Forage quality and ruminant nutrition. Pp. 229-266 in *Cool-Season Forage Grasses*, L. E. Moser, D. R. Buxton, and M. D. Casler, eds. Agronomy Monograph 34, Madison, WI: American Society of Agronomy, Crop Science Society of America, and Soil Science Society of America.

Buxton, D. R., R. E. Muck, and J. H. Harrison, eds. 2003. *Silage Science and Technology*, Agronomy Monograph 42. Madison, WI: American Society of Agronomy, Crop Science Society of America, and Soil Science Society of America.

Castillo-Lopez, E., B. I. Wiese, S. Hendrick, J. J. McKinnon, T. A. McAllister, K. A. Beauchemin, and G. B. Penner. 2014. Incidence, prevalence, severity, and risk factors for ruminal acidosis in feedlot steers during backgrounding, diet transition, and finishing. *Journal of Animal Science* 92:3053-3063.

Charmley, E. 2001. Towards improved silage quality—A review. *Canadian Journal of Animal Science* 81:157-168.

Chenost, M. 1966. Fibrousness of forages: Its determination and its relation to feeding value. Pp. 406-411 in *Proceedings of the 10th International Grassland Congress, July 7-16, 1966, Helsinki, Finland*. Helsinki: Valtioneuvoston kirjapaino.

Cherney, J. H., and D. J. R. Cherney. 2003. Assessing silage quality. Pp. 141-198 in *Silage Science and Technology*, D. R. Buxton, R. E. Muck, and J. H. Harrison, eds. Madison, WI: American Society of Agronomy, Crop Science Society of America, and Soil Science Society of America.

Corona, L., F. N. Owens, and R. A. Zinn. 2006. Impact of corn vitreousness and processing on site and extent of digestion by feedlot cattle. *Journal of Animal Science* 84:3020-3031.

Correa, C. E. S., R. D. Shaver, M. N. Pereira, J. G. Lauer, and K. Kohn. 2002. Relationship between corn vitreousness and ruminal in situ starch degradability. *Journal of Dairy Science* 85:3008-3012.

Crawford, G. I., C. D. Keeler, J. J. Wagner, C. R. Krehbiel, G. E. Erickson, M. B. Crombie, and G. A. Nunnery. 2008. Effects of calcium magnesium carbonate and roughage level on feedlot performance, ruminal metabolism, and site and extent of digestion in steers fed high-grain diets. *Journal of Animal Science* 86:2998-3013.

De Boever, J., J. I. Andries, D. De Brabander, B. G. Cottyn, and F. X. Buysse. 1990. Chewing activity of ruminants as a measure of physical structure—A review of factors affecting it. *Animal Feed Science and Technology* 4:281-291.

Defoor, P. J., M. S. Brown, and F. N. Owens. 2006. Reconstitution of grain sorghum for ruminants. Pp. 93-98 in *Cattle Grain Processing Symposium, November 15-17, 2006, Tulsa, OK.* Stillwater: Oklahoma State University.

DeFrain, J. M., A. R. Hippen, K. F. Kalscheur, and D. J. Schingoethe. 2004. Feeding lactose increases ruminal butyrate and plasma β-hydroxybutyrate in lactating dairy cows. *Journal of Dairy Science* 87:2486-2494.

Dehghan-banadaky, M., R. Corbett, and M. Oba. 2007. Effects of barley grain processing on productivity of cattle. *Animal Feed Science and Technology* 137:1-24.

Der Bedrosian, M. C., K. E. Nestor, Jr., and L. Kung, Jr. 2012. The effects of hybrid, maturity, and length of storage on the composition and nutritive value of corn silage. *Journal of Dairy Science* 95:5115-5126.

DeVries, T. J., T. Schwaiger, K. A. Beauchemin, and G. B. Penner. 2014a. Impact of severity of ruminal acidosis on feed-sorting behaviour of beef cattle. *Animal Production Science* 54:1238-1242.

DeVries, T. J., T. Schwaiger, K. A. Beauchemin, and G. B. Penner. 2014b. The duration of time that beef cattle are fed a high-grain diet affects feed sorting behavior, both prior to and after acute ruminal acidosis. *Journal of Animal Science* 92:1728-1737.

Dixon, R. M., and C. R. Stockdale. 1999. Associative effects between forages and grains: Consequences for feed utilisation. *Australian Journal of Agricultural Research* 50:757-773.

DOE (U.S. Department of Energy). 2005. *Genomics: GTL Roadmap. Systems Biology for Energy and Environment.* DOE/SC-0090. U.S. Department of Energy Office of Science. Available online at http://genomicscience.energy.gov/roadmap/pdf/GenomicsGTL_Roadmap.pdf. Accessed May 7, 2015.

Doherty, W., M. Payam, and C. Fellows. 2011. Value-adding to cellulosic ethanol: Lignin polymers. *Industrial Crops and Products* 33(2):259-276.

DuBois, M., K. A. Gilles, J. K. Hamilton, P. A. Rebers, and F. Smith. 1956. Colorimetric method for determination of sugars and related substances. *Analytical Chemistry* 28:350-356.

Engstrom, D. F., G. W. Mathison, and L. A. Goonewardene. 1992. Effect of β-glucan, starch, and fibre content and steam vs. dry rolling of barley grain on its degradability and utilization by steers. *Animal Feed Science and Technology* 37:33-46.

Erickson, G. E., C. T. Milton, K. C. Fanning, R. J. Cooper, R. S. Swingle, J. C. Parrot, G. Vogel, and T. J. Klopfenstein. 2003. Interaction between bunk management and monensin concentration on finishing performance, feeding behavior, and ruminal metabolism during an acidosis challenge with feedlot cattle. *Journal of Animal Science* 81:2869-2879.

Fahey, G. C., ed. 1994. *Forage Quality, Evaluation, and Utilization.* Madison, WI: American Society of Agronomy, Crop Science Society of America, and Soil Science Society of America.

Faleiro, A. G., L. A. González, M. Blanch, S. Cavini, L. Castells, J. L. Ruiz de la Torre, X. Manteca, S. Calsamiglia, and A. Ferret. 2011. Performance, ruminal changes, behaviour and welfare of growing heifers fed a concentrate diet with or without barley straw. *Animal* 5:294-303.

Farran, T., G. E. Erickson, and T. J. Klopfenstein. 2003. Evaluation of buffering agents in feedlot diets for cattle. Pp. 35-38 in *2003 Beef Cattle Report* 1-1-2003. Lincoln: University of Nebraska.

Ferraretto, L. F., P. M. Crump, and R. D. Shaver. 2013. Effect of cereal grain type and corn grain harvesting and processing methods on intake, digestion, and milk production by dairy cows through a meta-analysis. *Journal of Dairy Science* 96:533-550.

Fiems, L. O., B. G. Cottyn, C. V. Boucque, J. M. Vanacker, and F. X. Buysse. 1990. Effect of grain processing on in sacco digestibility and degradability in the rumen. *Archiv für Tierernaehrung* 40:713-721.

Firkins, J. L. 1997. Effects of feeding nonforage fiber sources on site of fiber digestion. *Journal of Dairy Science* 80:1426-1437.

Firkins, J. L., M. L. Eastridge, N. R. St-Pierre, and S. M. Noftsger. 2001. Effects of grain variability and processing on starch utilization by lactating dairy cattle. *Journal of Animal Science* 79(Suppl.):E218-E238.

Fisher, D. S., H. F. Mayland, and J. C. Burns. 1999. Variation in ruminants' preference for tall fescue hays cut either at sundown or at sunup. *Journal of Animal Science* 77:762-768.

Fisher, D. S., H. F. Mayland, and J. C. Burns. 2002. Variation in ruminant preference for alfalfa hays cut at sunup and sundown. *Crop Science* 42:231-237.

Foley, A. E., A. N. Hristov, A. Melgar, J. K. Ropp, R. P. Etter, S. Zaman, C. W. Hunt, K. Huber, and W. J. Price. 2006. Effect of barley and its amylopectin content on ruminal fermentation and nitrogen utilization in lactating dairy cows. *Journal of Dairy Science* 89:4321-4335.

Forbes, J. M. 1996. Integration of regulatory signals controlling forage intake in ruminants. *Journal of Animal Science* 74:3029-3035.

Fox, D. G., and L. O. Tedeschi. 2002. Application of physically effective fiber in diets for feedlot cattle. Pp. 67-81 in *Proceedings of the Plains Nutrition Council Spring Conference, April 25-26, 2002, San Antonio, TX.* Publication AREC 02-20. Amarillo: Texas A&M Research and Extension Center. Available online at http://amarillo.tamu.edu/files/2010/10/2002PNC-Proceedings.pdf. Accessed on June 18, 2014.

Fox, D. G., C. J. Sniffen, J. D. O'Connor, J. B. Russell, and P. J. Van Soest. 1992. A net carbohydrate and protein system for evaluating cattle diets: III. Cattle requirements and diet adequacy. *Journal of Animal Science* 70:3578-3596.

Fox, D. G., T. P. Tylutki, L. O. Tedeschi, M. E. Van Amburgh, L. E. Chase, A. N. Pell, T. R. Overton, and J. B. Russell. 2003. The Cornell Net Carbohydrate and Protein System for Evaluating Herd Nutrition and Excretion. CNCPS Version 5.0. Model Documentation, Animal Science Mimeo 213. Ithaca, NY: Cornell University.

Fox, D., L. Tedeschi, T. Tylutki, J. Russell, M. Van Amburgh, L. Chase, A. Pell, and T. Overton. 2004. The Cornell Net Carbohydrate and Protein System model for evaluating herd nutrition and nutrient excretion. *Animal Feed Science and Technology* 112:29-78.

Frame, J. 2005. *Forage Legumes for Temperate Grasslands.* Wallingford, UK: CAB International.

Frame, J., J. F. L. Charlton, and, A. S. Laidlaw. 1998. *Temperate Forage Legumes.* Wallingford, UK: CAB International.

Fulton, W. R., T. J. Klopfenstein, and R. A. Britton. 1979. Adaptation to high concentrate diets by beef cattle. I. Adaptation to corn and wheat diets. *Journal of Animal Science* 49:775-784.

Gaillard, B. D. E. 1958. A detailed summative analysis of the crude fibre and nitrogen-free extractives fractions of roughages. I. Proposed scheme of analysis. *Journal of the Science of Food and Agriculture* 3:170-177.

Galyean, M. L., and A. L. Goetsch, 1993. Utilization of forage fiber by ruminants. Pp. 33-71 in *Forage Cell Wall Structure and Digestibility,* H. G. Jung, D. R. Buxton, R. D. Hatfield, and J. Ralph, eds. Madison, WI: American Society of Agronomy, Crop Science Society of America, and Soil Science Society of America.

Giuberti, G., A. Gallo, F. Masoero, L. F. Ferraretto, P. Hoffman, and R. Shaver. 2014. Factors affecting starch utilization in large animal food production system: A review. *Starch* 66:72-90.

Goering, H. K., and P. J. Van Soest. 1975. *Forage Fiber Analysis (Apparatus, Reagents, Procedures, and Some Applications).* Agriculture Handbook No. 379. Washington, DC: U.S. Department of Agriculture, Agricultural Research Service.

Grimson, R. E., R. D. Weisenburger, J. A. Basarab, and R. P. Stilborn. 1987. Effects of barley volume-weight and processing method on feedlot performance of finishing steers. *Canadian Journal of Animal Science* 67:43-53.

Hale, W. H. 1973. Influence of processing on the utilization of grains (starch) by ruminants. *Journal of Animal Science* 37:1075-1081.

Hall, M. B. 2003a. Challenges with nonfiber carbohydrate methods. *Journal of Animal Science* 81:3226-3232.

Hall, M. B. 2003b. Making sense of non-fibre carbohydrates in dairy rations. *Proceedings of the 24th Western Nutrition Conference*, September 2003, Winnipeg, Manitoba, Canada.

Hall, M. B. 2009. Determination of starch, including maltooligosaccharides, in animal feeds: Comparison of methods and a method recommended for AOAC collaborative study. *Journal of AOAC International* 92:42-49.

Hall, M. B. 2011. Isotrichid protozoa influence conversion of glucose to glycogen and other microbial products. *Journal of Dairy Science* 94:4589-4602.

Hall, M. B. 2013. Efficacy of reducing sugar and phenol-sulfuric acid assays for analysis of soluble carbohydrates in feedstuffs. *Animal Feed Science and Technology* 185:94-100.

Hall, M. B. 2014. Selection of an empirical detection method for determination of water-soluble carbohydrates in feedstuffs for application in ruminant nutrition. *Animal Feed Science and Technology* 198:28-37.

Hall, M. B., and M. L. Eastridge. 2014. Carbohydrate and fat: Considerations for energy and more. *The Professional Animal Scientist* 30:140-149.

Hall, M. B., A. N. Pell, and L. E. Chase. 1998. Characteristics of neutral detergent-soluble fiber fermentation by mixed ruminal microbes. *Animal Feed Science and Technology* 70:23-39.

Hall, M. B., W. H. Hoover, J. P. Jennings, and T. K. Miller Webster. 1999. A method for partitioning neutral detergent soluble carbohydrates. *Journal of the Science of Food and Agriculture* 79:2079-2086.

Hall, M. B., J. P. Jennings, B. A. Lewis, and J. B. Robertson. 2001. Evaluation of starch analysis methods for feed samples. *Journal of the Science of Food and Agriculture* 81:17-21.

Hansen, N. M., and D. Plackett. 2008. Sustainable films and coatings from hemicelluloses: A review. *Biomacromolecules* 6:1493-1505.

Harmon, D. L., and K. R. McLeod. 2001. Glucose uptake and regulation by intestinal tissues: Implications and whole-body energetics. *Journal of Animal Science* 79(Suppl.):E59-E72.

Heath, M. E., R. F. Barnes, and D. S. Metcalfe, eds. 1985. *Forages: The Science of Grassland Agriculture.* West Lafayette, IN: Purdue University.

Heldt, J. S., R. C. Cochran, G. K. Stokka, C. G. Farmer, C. P. Mathis, E. C. Tigemeyer, and T. G. Nagaraja. 1999. Effects of different supplemental sugars and starch fed in combination with degradable intake protein on low-quality forage use by beef steers. *Journal of Animal Science* 77:2793-2802.

Himmelsbach, D. S. 1993. Structure of forage cell walls—Session synopsis. Pp. 271-283 in *Forage Cell Wall Structure and Digestibility*, H. G. Jung, D. R. Buxton, R. D. Hatfield, and J. Ralph, eds. Madison, WI: American Society of Agronomy, Crop Science Society of America, and Soil Science Society of America.

Holtshausen, L., K. S. Schwartzkopf-Genswein, and K. A. Beauchemin. 2013. Short communication: Rumen fermentation, ruminal pH profile and feeding behaviour of feedlot cattle transitioning from a high-forage to a high-concentrate diet. *Canadian Journal of Animal Science* 93:529-533.

Hopkins, A., ed. 2000. *Grass: Its Production and Utilization*, 3rd Ed. Oxford, UK: Blackwell Science.

Humphreys, M. O. 1989. Water soluble carbohydrates in perennial ryegrass breeding. II. Cultivar and hybrid progeny performance in cut plots. *Grass and Forage Science* 44:237-244.

Huntington, G. B., D. L. Harmon, and C. J. Richards. 2006. Sites, rates, and limits of starch digestion and glucose metabolism in growing cattle. *Journal of Animal Science* 84(Suppl.):E14-E24.

Iraira, S. P., J. L. Ruíz de la Torre, M. Rodríguez-Prado, S. Calsamiglia, X. Manteca, and A. Ferret. 2013. Feed intake, ruminal fermentation, and animal behavior of beef heifers fed forage free diets containing nonforage fiber sources. *Journal of Animal Science* 91:3827-3835.

Jones, E. C., and R. J. Barnes. 1967. Non-volatile organic acids of grasses. *Journal of the Science of Food and Agriculture* 18:321-324.

Jung, H. G., and D. A. Deetz. 1993. Cell wall lignification and degradability. Pp. 315-346 in *Forage Cell Wall Structure and Digestibility*, H. G. Jung, D. R. Buxton, R. D. Hatfield and J. Ralph, eds. Madison, WI: American Society of Agronomy, Crop Science Society of America, and Soil Science Society of America.

Jung, H. G., D. R. Buxton, R. D. Hatfield, and J. Ralph, eds. 1993. *Forage Cell Wall Structure and Digestibility.* Madison, WI: American Society of Agronomy, Crop Science Society of America, and Soil Science Society of America.

Jung, H. J. G., V. H. Varel, P. J. Weimer, and J. Ralph. 1999. Accuracy of Klason lignin and acid detergent lignin methods as assessed by bomb calorimetry. *Journal of Agricultural and Food Chemistry* 47:2005-2008.

Keunen, J. E., J. C. Plaizier, I. Kyriazakis, T. F. Duffield, T. M. Widowski, M. I. Lindinger, and B. W. McBride. 2002. Effects of a subacute ruminal acidosis model on the diet selection of dairy cows. *Journal of Dairy Science* 85:3304-3313.

Khalili, H., and P. Huhtanen.1991. Sucrose supplements in cattle given grass silage-based diet. 1. Digestion of organic matter and nitrogen. *Animal Feed Science and Technology* 33:247-261.

Knudsen, K. E. B. 1997. Carbohydrate and lignin contents of plant materials used in animal feeding. *Animal Feed Science and Technology* 67:319-338.

Koenig, K. M., and K. A. Beauchemin. 2011. Optimum extent of barley grain processing and barley silage proportion in feedlot cattle diets: Growth, feed efficiency and fecal characteristics. *Canadian Journal of Animal Science* 91:411-422.

Koenig, K. M., K. A. Beauchemin, and L. M. Rode. 2003. Effect of grain processing and silage on microbial protein synthesis and nutrient digestibility in beef cattle fed barley-based diets. *Journal of Animal Science* 81:1057-1067.

Koenig, K. M., K. A. Beauchemin, and W. Z. Yang. 2013. Processing feed grains: Factors affecting the effectiveness of grain processing for beef and dairy cattle production. Pp. 62-73 in *Processing, Performance and Profit: Proceedings of the 34th Western Nutrition Conference, September 24-26*, Saskatoon, Saskatchewan, Canada. Available online at http://www.westernnutritionconference.ca/2013_WNC_Proceedings.pdf. Accessed on November 24, 2014.

Kononoff, P. J., A. J. Heinrichs, and D. R. Buckmaster. 2003. Modification of the Penn State Forage and Total Mixed Ration Particle Separator and the effects of moisture content on its measurements. *Journal of Dairy Science* 86:1858-1863.

Kristensen, V. F., and P. Nørgaard. 1987. Effect of roughage quality and physical structure of the diet on feed intake and milk yield of the dairy cow. Pp. 79-91 in *Cattle Production Research—Danish Status and Perspectives.* Copenhagen: Landhusholdningsselskabets Forlag.

Kung, L., Jr. 2008. Silage fermentation end products and microbial populations: Their relationships to silage quality and animal productivity. Annual Conference of the American Association of Bovine Practitioners, September 25-27, 2008, Charlotte, NC. Available online at http://ag.udel.edu/anfs/faculty/kung/documents/08SilageFermentationEndProductsandMicrobialPopulationsTheirRelationshipsto.pdf. Accessed on August 15, 2014.

Kung, L., Jr., and N. K. Ranjit. 2001. The effect of *Lactobacillus buchneri* and other additives on the fermentation and aerobic stability of barley silage. *Journal of Dairy Science* 84:1149-1155.

Lammers, B. P., D. R. Buckmaster, and A. J. Heinrichs. 1996. A simple method for the analysis of particle sizes of forage and total mixed rations. *Journal of Dairy Science* 79:922-928.

Lanzas, C., C. J. Sniffen, S. Seo, L. O. Tedeschi, and D. G. Fox. 2007. A revised CNCPS feed carbohydrate fractionation scheme for formulating rations for ruminants. *Animal Feed Science and Technology* 136:167-190.

Lapierre, C. 1993. Application of new methods for the investigation of lignin structure. Pp. 133-166 in *Forage Cell Wall Structure and Digestibility*, H. G. Jung, D. R. Buxton, R. D. Hatfield, and J. Ralph, eds. Madison,

WI: American Society of Agronomy, Crop Science Society of America, and Soil Science Society of America.

Lee, R. W., M. L. Galyean and G. P. Lofgreen. 1982. Effect of mixing whole shelled and steam flaked corn in finishing diets on feedlot performance and site and extent of digestion in beef steers. *Journal of Animal Science* 55:475-483.

Lee, M. R. F., R. J. Merry, D. R. Davies, J. M. Moorby, M. O. Humphreys, M. K. Theodorou, J. C. MacRae, and N. D. Scollan. 2003. Effect of increasing availability of water-soluble carbohydrates on in vitro rumen fermentation. *Animal Feed Science and Technology* 104:59-70.

Leng, R. A. 1990. Factors affecting the utilization of 'poor-quality' forages by ruminants particularly under tropical conditions. *Nutrition Research Reviews* 3:277-303.

Leonardi, C., and L. E. Armentano. 2003. Effect of quantity, quality, and length of alfalfa hay on selective consumption by dairy cows. *Journal of Dairy Science* 86:557-564.

Lopes, F., D. E. Cook, and D. K. Combs. 2015. Validation of an in vitro model for predicting rumen and total-tract fiber digestibility in dairy cows fed corn silages with different in vitro neutral detergent fiber digestibilities at 2 levels of dry matter intake. *Journal of Dairy Science* 98:574-585.

Martins, D. A. B., M. H. F. Alves do Prado, R. S. R. Leite, H. Ferreira, M. M. D. S. Moretti, R. da Silva, and E. Gomes. 2011. Agroindustrial wastes as substrates for microbial enzymes production and source of sugar for bioethanol production. Pp. 319-360 in *Integrated Waste Management*, Vol. II, S. Kumar, ed. Available online at http://www.intechopen.com/books/integrated-waste-management-volume-ii/agro-industrial-wastes-as-substrates-for-microbial-enzymes-production-and-source-of-sugar-for-bioetha. Accessed on August 15, 2014.

Mathison, G. W. 1996. Effects of processing on the utilization of grain by cattle. *Animal Feed Science and Technology* 58:113-125.

Mathison, G. W., R. Hironaka, B. K. Kerringan, I. Vlach, L. P. Milligan, and R. D. Weisenburger 1991. Rate of starch degradation, apparent digestibility and rate and efficiency of steer gain as influenced by barley grain volume-weight and processing method. *Canadian Journal of Animal Science* 71:867-878.

McAllister, T. A., R. C. Phillippe, L. M. Rode, and K. J. Cheng. 1993. Effect of the protein matrix on the digestion of cereal grains by ruminal microorganisms. *Journal of Animal Science* 71:205-212.

McAllister, T. A., H. D. Bae, G. A. Jones, and K. J. Cheng. 1994. Microbial attachment and feed digestion in the rumen. *Journal of Animal Science* 72:3004-3018.

McAllister, T. A., L. B. Selinger, R. McMahon, H. D. Bae, T. J. Lysyk, S. J. Oosting, and K.-J. Cheng. 1995. Intake, digestibility and aerobic stability of barley silage inoculated with mixtures of *Lactobacillus plantarum* and *Enterococcus faecium*. *Canadian Journal of Animal Science* 75:425-432.

McAllister, T. A., D. J. Gibb, K. A. Beauchemin, and Y. Wang. 2006. Starch type, structure and ruminal digestion. Pp. 30-41 in *Cattle Grain Processing Symposium, November 15-17, 2006, Tulsa, OK*. MP-177. Stillwater: Oklahoma State University. Available online at http://beefextension.com/proceedings/cattle_grains06/06-5.pdf. Accessed on November 25, 2014.

McDonald, P., A. R. Henderson, and S. J. E. Heron. 1991. *The Biochemistry of Silage*, 2nd Ed. Kingston, UK: Chalcombe Publications.

McLeod, K. R., R. L. Baldwin, S. W. El-Kadi, and D. L. Harmon. 2006. Site of starch digestion: Impact on energetic efficiency and glucose metabolism in beef and dairy cattle. Pp. 129-136 in *Cattle Grain Processing Symposium, November 15-17, 2006, Tulsa, OK*. MP-177. Stillwater: Oklahoma State University. Available online at http://beefextension.com/proceedings/cattle_grains06/06-18.pdf. Accessed on November 25, 2014.

McMeniman, J. P., and M. L. Galyean. 2007. Effect of simulated air-lift and conveyor leg takeaway systems on starch availability and *in vitro* dry matter disappearance of steam-flaked corn grain. *Animal Feed Science and Technology* 136:323-329.

Merchen, N. R., and L. D. Bourquin. 1994. Processes of digestion and factors influencing digestion of forage-based diets by ruminants. Pp. 564-612 in *Forage Quality, Evaluation, and Utilization*, G. C. Fahey, ed. Madison, WI: American Society of Agronomy, Crop Science Society of America, and Soil Science Society of America.

Merry, R. J., M. R. F. Lee, D. R. Davies, R. J. Dewhurst, J. M. Moorby, N. D. Scollan, and M. K. Theodorou. 2006. Effects of high-sugar ryegrass silage and mixtures with red clover silage on ruminant digestion. 1. In vitro and in vivo studies of nitrogen utilization. *Journal of Animal Science* 84:3049-3060.

Mertens, D. R. 1985. Factors influencing feed intake in lactating dairy cows: From theory to application using neutral detergent fiber. Pp. 1-18 in *Proceedings of the Georgia Nutrition Conference for the Feed Industry, February 13-15, 1985, Atlanta, GA*. Athens: University of Georgia.

Mertens, D. R. 1986. Effect of physical characteristics, forage particle size and density on forage utilization. Pp. 91-106 in *Proceedings from the AFIA Nutrition Symposium, November 12-13, 1986, St. Louis, MO*. Arlington, VA: American Feed Industry Association.

Mertens, D. R. 1994. Regulation of forage intake. Pp. 450-493 in *Forage Quality, Evaluation, and Utilization*, G. C. Fahey, ed. Madison, WI: American Society of Agronomy, Crop Science Society of America, and Soil Science Society of America.

Mertens, D. R. 1997. Creating a system for meeting the fiber requirements of dairy cows. *Journal of Dairy Science* 80:1463-1481.

Mertens, D. R. 2002. Gravimetric determination of amylase-treated neutral detergent fiber in feeds with refluxing in beakers or crucibles: Collaborative study. *Journal of AOAC International* 85:1217-1240.

Miller, L. A., J. M. Moorby, D. R. Davies, M. O. Humphreys, N. D. Scollan, J. C. MacRae, and M. K. Theodorou. 2001. Increased concentration of water-soluble carbohydrate in perennial ryegrass (*Lolium perenne* L.): Milk production from late-lactation dairy cows. *Grass and Forage Science* 56:383-394.

Minson, D. J., and J. R. Wilson. 1994. Prediction of intake as an element of forage quality. Pp. 533-563 in *Forage Quality, Evaluation, and Utilization*, G. C. Fahey, ed. Madison, WI: American Society of Agronomy, Crop Science Society of America, and Soil Science Society of America.

Moorby, J. M., R. T. Evans, N. D. Scollan, J. C. MacRae, and M. K. Theodorou. 2006. Increased concentration of water-soluble carbohydrate in perennial ryegrass (*Lolium perenne* L.): Evaluation in dairy cows in early lactation. *Grass and Forage Science* 61:52-59.

Moore, K. J., and R. D. Hatfield. 1994. Carbohydrates and forage quality. Pp. 229-280 in *Forage Quality, Evaluation, and Utilization*, G. C. Fahey, ed. Madison, WI: American Society of Agronomy, Crop Science Society of America, and Soil Science Society of America.

Moser, L. E., D. R. Buxton, and M. D. Casler, eds. 1996. *Cool-Season Forage Grasses*. Agronomy Monograph 34. Madison, WI: American Society of Agronomy, Crop Science Society of America, and Soil Science Society of America.

Mould, F. L., E. R. Ørskov, and S. O. Mann. 1983. Mixed feeds. I. Effects of type and level of supplementation and the influence of the rumen fluid pH on cellulolysis in vivo and dry matter digestion of various roughages. *Animal Feed Science and Technology* 10:15-30.

Nevins, D. J. 1993. Analysis of forage cell wall polysaccharides. Pp. 105-132 in *Forage Cell Wall Structure and Digestibility*, H. G. Jung, D. R. Buxton, R. D. Hatfield, and J. Ralph, eds. Madison, WI: American Society of Agronomy, Crop Science Society of America, and Soil Science Society of America.

Nicholson, J. W. G., H. M. Cunningham, and D. W. Friend. 1963. Effect of adding buffers to all-concentrate rations on feedlot performance of steers, ration digestibility and intra-rumen environment. *Journal of Animal Science* 22:368-373.

NRC (National Research Council). 1996. *Nutrient Requirements of Beef Cattle*, 7th Rev. Ed. Washington, DC: National Academy Press.

NRC. 2000. *Nutrient Requirements of Beef Cattle (Update 2000)*, 7th Rev. Ed. Washington, DC: National Academy Press.

NRC. 2001. *Nutrient Requirements of Dairy Cattle*, 7th Rev. Ed. Washington, DC: National Academy Press.

Owens, F., and S. Soderlund. 2006. Ruminal and postruminal starch digestion by cattle. Pp. 116-128 in *Cattle Grain Processing Symposium, November 15-17, 2006, Tulsa, OK.* Stillwater: Oklahoma State University. Available online at http://beefextension.com/proceedings/cattle_grains06/06-17.pdf. Accessed on November 25, 2014.

Owens, F., and R. A. Zinn. 2005. Corn grain for cattle: Influence of processing on site and extent of digestion. Pp. 86-112 in *Proceedings of the Southwest Nutrition and Management Conference, February 24-25, 2005, Tempe, AZ.* Tucson: University of Arizona.

Owens, F. N., D. S. Secrist, W. J. Hill, and D. R. Gill. 1997. The effect of grain source and grain processing on performance of feedlot cattle: A review. *Journal of Animal Science* 75:868-879.

Paterson, J. A., R. L. Belyea, J. P. Bowman, M. S. Kerley, and J. E. Williams. 1994. The impact of forage quality and supplementation regimen on ruminant animal intake and performance. Pp. 59-114 in *Forage Quality, Evaluation, and Utilization*, G. C. Fahey, ed. Madison, WI: American Society of Agronomy, Crop Science Society of America, and Soil Science Society of America.

Penner, G. B., L. L. Guan, and M. Oba. 2009a. Effect of feeding Fermenten and sugar to lactating cows on ruminal fermentation. *Journal of Dairy Science* 92:1725-1733.

Penner, G. B., L. L. Guan, and M. Oba. 2009b. Increasing dietary sugar concentration may improve dry matter intake, ruminal fermentation, and productivity of dairy cows in the postpartum phase of the transition period. *Journal of Dairy Science* 92:3341-3353.

Phy, T. S., and F. D. Provenza. 1998. Sheep fed grain prefer foods and solutions that attenuate acidosis. *Journal of Animal Science* 76:954-960.

Pitt, R. E., J. S. Van Kessel, D. G. Fox, A. N. Pell, M. C. Barry, and P. J. Van Soest. 1996. Prediction of ruminal volatile fatty acids and pH within the net carbohydrate and protein system. *Journal of Animal Science* 74:226-244.

Plascencia, A., J. F. Calderon, E. J. DePeters, M. A. Lopez-Soto, M. Vega, and R. A. Zinn. 1998. Influence of processing on the feeding value of barley for lactating cows. Pp. 257-263 in *Proceedings, Western Section, American Society of Animal Science*, Vol. 49. Champaign, IL: ASAS.

Ponce, C. H., E. M. Domby, U. Y. Anele, J. S. Schutz, K. K. Gautam, and M. L. Galyean. 2013. Effects of bulk density of steam-flaked corn in diets containing wet corn gluten feed on feedlot cattle performance, carcass characteristics, apparent total tract digestibility, and ruminal fermentation. *Journal of Animal Science* 91:3400-3407.

Pontis, H. G. 1990. Fructans. Pp. 353-370 in *Methods in Plant Biochemistry, Carbohydrates*, Vol. 2, P. M. Dey and J. B. Harborne, eds. San Diego, CA: Academic Press.

Provenza, F. D. 1995. Postingestive feedback as an elementary determinant of food preference and intake in ruminants. *Journal of Range Management* 48:2-17.

Ratnayake, W. S., and D. S. Jackson. 2009. Starch gelatinization. *Advances in Food and Nutrition Research* 55:221-268.

Rayburn, E. B., ed. 2007. *Forage Utilization for Pasture-Based Livestock Production.* NRAES Publication No. 173. Ithaca, NY: Plant and Life Sciences Publishing.

Reid, R. L., and T. J. Klopfenstein. 1983. Forages and crop residues: Quality evaluation and systems of utilization. *Journal of Animal Science* 57:534-562.

Reinhardt, C. D., R. T. Brandt, Jr., K. C. Behnke, A. S. Freeman, and T. P. Eck. 1997. Effect of steam-flaked sorghum grain density on performance, mill production rate, and subacute acidosis in feedlot steers. *Journal of Animal Science* 75:2852-2857.

Robles, V., L. A. González, A. Ferret, X. Manteca, and S. Calsamiglia. 2007. Effects of feeding frequency on intake, ruminal fermentation, and feeding behaviour in heifers fed high-concentrate diets. *Journal of Animal Science* 85:2538-2547.

Rooke, J. A., and R. D. Hatfield. 2003. Biochemistry of ensiling. Pp. 95-139 in *Silage Science and Technology*, D. R. Buxton, R. E. Muck, and J. H. Harrison, eds. Madison, WI: American Society of Agronomy, Crop Science Society of America, and Soil Science Society of America.

Rotz, C. A., and R. E. Muck. 1994. Changes in forage quality during harvest and storage. Pp. 828-868 in *Forage Quality, Evaluation, and Utilization*, G. C. Fahey, ed. Madison, WI: American Society of Agronomy, Crop Science Society of America, and Soil Science Society of America.

Rouquette, F. M., Jr., L. A. Redmon, G. E. Aiken, G. M. Hill, L. E. Sollenberger, and J. Andrae. 2009. ASAS centennial paper: Future needs of research and extension in forage utilization. *Journal of Animal Science* 87:438-446.

Rowe, J. B., M. Choct, and D. W. Pethick. 1999. Processing cereal grains for animal feeding. *Australian Journal of Agricultural Research* 50:721-736.

Russell, J. B., and P. J. Van Soest. 1984. In vitro ruminal fermentation of organic acids common in forage. *Applied and Environmental Microbiology* 47:155-159.

Russell, J. B., J. D. O'Connor, D. G. Fox, P. J. Van Soest, and C. J. Sniffen. 1992. A net carbohydrate and protein system for evaluating cattle diets: I. Ruminal fermentation. *Journal of Animal Science* 70:3551-3561.

Samuel, R., Y. Pu, N. Jiang, C. Fu, Z.-Y. Wang, and A. Ragauskas. 2014. Structural characterization of lignin in wild-type versus COMT down-regulated switchgrass. *Frontiers in Energy Research* 1:14.

Santini, F. J., A. R. Hardie, N. A. Jorgensen, and N. F. Finner. 1983. Proposed use of adjusted intake based on forage particle length for calculation of roughage indexes. *Journal of Dairy Science* 66:811-820.

Sarhan, M. A., and K. A. Beauchemin. 2015. Ruminal pH predictions for beef cattle: Comparative evaluation of current models. *Journal of Animal Science* 93:1741-1759.

Sauvant, D., J. P. Duphy, and B. Michalet-Doreau. 1990. Le concept d'indice de fibrosité des aliments des ruminants [The concept of a fibrosity index for ruminant feeds]. *INRA Productions Animales* 3:309-318.

Scheller, H. V., and P. Ulvskov. 2010. Hemicelluloses. *Annual Review of Plant Biology* 61:263-289.

Schwaiger, T., K. A. Beauchemin, and G. B. Penner. 2013. The duration of time that beef cattle are fed a high-grain diet affects the recovery from a bout of ruminal acidosis: Dry matter intake and ruminal fermentation. *Journal of Animal Science* 91:5729-5742.

Schwartzkopf-Genswein, K. S., K. A. Beauchemin, D. J. Gibb, D. H. Crews, Jr., D. D. Hickman, M. N. Streeter, and T. A. McAllister. 2003. Effect of bunk management on feeding behavior, ruminal acidosis and performance of feedlot cattle: A review. *Journal of Animal Science* 81(Suppl. 2):E149-E158.

Schwartzkopf-Genswein, K. S., K. A. Beauchemin, T. A. McAllister, D. J. Gibb, and M. Streeter. 2004. Effect of feed delivery fluctuations and feeding time on ruminal acidosis, growth performance and feeding behaviour of feedlot cattle. *Journal of Animal Science* 82:3357-3365.

Sniffen, C. J., J. D. O'Connor, P. J. Van Soest, D. G. Fox, and J. B. Russell. 1992. A net carbohydrate and protein system for evaluating cattle diets: II. Carbohydrate and protein availability. *Journal of Animal Science* 70:3562-3577.

Soto-Navarro, S. A., C. R. Krehbiel, G. C. Duff, M. L. Galyean, M. S. Brown, and R. L. Steiner. 2000. Influence of feed intake fluctuation and frequency of feeding on nutrient of digestion, digesta kinetics, and ruminal fermentation profiles in limit-fed steers. *Journal of Animal Science* 78:2215-2222.

Strobel, H. J., and J. B. Russell. 1986. Effect of pH and energy spilling on bacterial protein synthesis by carbohydrate-limited cultures of mixed rumen bacteria. *Journal of Dairy Science* 69:2941-2947.

Stroud, T. E., J. E. Williams, D. R. Ledoux, and J. A. Paterson. 1985. The influence of sodium bicarbonate and dehydrated alfalfa as buffers on steer performance and ruminal characteristics. *Journal of Animal Science* 60:551-559.

Sudweeks, E. M., L. O. Ely, D. R. Mertens, and L. R. Sisk. 1981. Assessing minimum amounts and form of roughages in ruminant diets: Roughage value index system. *Journal of Animal Science* 53:1406-1411.

Svihus, B., A. K. Uhlen, and O. M. Harstad. 2005. Effect of starch granule structure, associated components and processing on nutritive value of cereal starch: A review. *Animal Feed Science and Technology* 122:303-320.

Theurer, C. B. 1986. Grain processing effects on starch utilization by ruminants. *Journal of Animal Science* 63:1649-1662.

Theurer, C. B., J. T. Huber, A. Delgado-Elorduy, and R. Wanderley. 1999. Invited review: Summary of steam-flaking corn or sorghum grain for lactating dairy cows. *Journal of Dairy Science* 82:1950-1959.

Thomas, G. J. 1960. Metabolism of the soluble carbohydrates of grasses in the rumen of the sheep. *Journal of Agricultural Science* 54:360-372.

Tylutki, T. P., D. G. Fox, V. M. Durbal, L. O. Tedeschi, J. B. Russell, M. E. Van Amburgh, T. R. Overton, L. E. Chase, and A. N. Pell. 2008. Cornell Net Carbohydrate and Protein System: A model for precision feeding of dairy cattle. *Animal Feed Science and Technology* 143:174-202.

Uhart, B. A., and F. D. Carroll. 1967. Acidosis in beef steers. *Journal of Animal Science* 26:1195-1198.

van Barneveld, S, 1999. Chemical and physical characteristics of grains related to variability in energy and amino acid availability in ruminants: A review. *Australian Journal of Agricultural Research* 50:651-666.

Vanholme, R., K. Morreel, J. Ralph, and W. Boerjan, 2010. Lignin biosynthesis and structure. *Plant Physiology* 153:895-905.

Van Soest, P. J. 1963. Use of detergents in the analysis of fibrous feeds. II. A rapid method for the determination of fiber and lignin. *Journal of the Association of Official Analytical Chemists* 46:829-835.

Van Soest, P. J. 1993. Cell wall matrix interactions and degradation—Session synopsis. Pp. 377-395 in *Forage Cell Wall Structure and Digestibility*, H. G. Jung, D. R. Buxton, R. D. Hatfield and, J. Ralph, eds. Madison, WI: American Society of Agronomy, Crop Science Society of America, and Soil Science Society of America.

Van Soest, P. J. 1994. *Nutritional Ecology of the Ruminant,* 2nd Ed. Ithaca, NY: Cornell University Press.

Van Soest, P. J., and R H. Wine. 1967. Use of detergents in the analysis of fibrous feeds. IV. Determination of plant cell wall constituents. *Journal of the Association of Official Analytical Chemists* 50:50-55.

Van Soest, P. J., J. B. Robertson, and B. A. Lewis. 1991. Methods for dietary fiber, neutral detergent fiber, and nonstarch polysaccharides in relation to animal nutrition. *Journal of Dairy Science* 74:3583-3597.

Van Soest, P. J., M. E. Van Amburgh, J. B. Robertson, and W. F. Knaus. 2005. Validation of the 2.4 times lignin factor for ultimate extent of NDF digestion, and curve peeling rate of fermentation curves into pools. Pp. 139-149 in *Proceedings of Cornell Nutrition Conference for Feed Manufacturers, Rochester, NY.* Ithaca, NY: Cornell University Press. Available online at http://nutritionmodels.tamu.edu/papers/VanSoestetalCNC2005139.pdf. Accessed on May 6, 2015.

Vasconcelos, J. T., and M. L. Galyean. 2007. Nutritional recommendations of feedlot consulting nutritionists: The 2007 Texas Tech University survey. *Journal of Animal Science* 85:2772-2781.

Ward, C. F., and M. L. Galyean. 1999. The Relationship between Retrograde Starch as Measured by Starch Availability Estimates and In Vitro Dry Matter Disappearance of Steam-flaked Corn. Burnett Center Internet Progress Report No. 2, Texas Tech University. Available online at http://www.depts.ttu.edu/afs/burnett_center/progress_reports/bc2.pdf. Accessed on February 25, 2014.

Weisbjerg, M. R., T. Hvelplund, and B. M. Bibby. 1998. Hydrolysis and fermentation rate of glucose, sucrose and lactose in the rumen. *Acta Agriculturae Scandinavica, Animal Science* 48:12-18.

Weiss, W. P., D. G. Chamberlain, and C. W. Hunt. 2003. Feeding silages. Pp. 469-504 in *Silage Science and Technology*, D. R. Buxton, R. E. Muck, and J. H. Harrison, eds. Madison, WI: American Society of Agronomy, Crop Science Society of America, and Soil Science Society of America.

White, B. A., Mackie, R. I., and Doerner, K. C. 1993. Enzymatic hydrolysis of forage cell walls. Pp. 455-484 in *Forage Cell Wall Structure and Digestibility*, H. G. Jung, D. R. Buxton, R. D. Hatfield, and J. Ralph, eds. Madison, WI: American Society of Agronomy, Crop Science Society of America, and Soil Science Society of America.

White, T. W., W. L. Reynolds, and R. H. Klett. 1969. Roughage sources and levels in steer rations. *Journal of Animal Science* 29:1001-1005.

Wierenga, K. T., T. A. McAllister, D. J. Gibb, A. V. Chaves, E. K. Okine, K. A. Beauchemin, and M. Oba. 2010. Evaluation of triticale dried distillers grain as a substitute for barley silage in feedlot finishing diets. *Journal of Animal Science* 88:3018-3029.

Wilson, J. R. 1993. Organization of forage plant tissues. Pp. 1-32 in *Forage Cell Wall Structure and Digestibility*, H. G. Jung, D. R. Buxton, R. D. Hatfield, and J. Ralph, eds. Madison, WI: American Society of Agronomy, Crop Science Society of America, and Soil Science Society of America.

Wylam, C. B. 1954. Analytical studies on the carbohydrates of grasses and clovers. IV. Further developments in the methods of estimation of mono-, di- and oligo-saccharides and fructosan. *Journal of the Science of Food and Agriculture* 5:167-172.

Xiong, Y., S. J. Bartle and R. L. Preston. 1990. Improved enzymatic method to measure processing degree and availability of starch in sorghum grain. *Journal of Animal Science* 68:3861-3870.

Yang, W. Z., and K. A. Beauchemin. 2006. Increasing the physically effective fiber content of dairy cow diets may lower efficiency of feed use. *Journal of Dairy Science* 89:2694-2704.

Yang, W. Z., K. A. Beauchemin, and L. M. Rode. 1996. Ruminal digestion kinetics of temper-rolled hulless barley. *Canadian Journal of Animal Science* 76:629-632.

Yang, W. Z., K. A. Beauchemin, and L. M. Rode. 2000. Effects of barley grain processing on extent of digestion and milk production of lactating cows. *Journal of Dairy Science* 83:554-568.

Zebeli, Q., M. Tafaj, H. Steingass, B. Metzler, and W. Drochner. 2006. Effects of physically effective fiber on digestive processes and milk fat content in early lactating dairy cows fed total mixed rations. *Journal of Dairy Science* 89:651-668.

Zebeli, Q., J. Dijkstra, M. Tafaj, H. Steingass, B. N. Ametaj, and W. Drochner. 2008. Modeling the adequacy of dietary fiber in dairy cows based on the responses of ruminal pH and milk fat production to composition of the diet. *Journal of Dairy Science* 91:2046-2066.

Zinn, R. A. 1990. Influence of flake density on the comparative feeding value of steam-flaked corn for feedlot cattle. *Journal of Animal Science* 68:767-775.

Zinn, R. A. 1993. Influence of processing on the comparative feeding value of barley for feedlot cattle. *Journal of Animal Science* 71:3-10.

Zinn, R. A., and R. Barajas. 1997. Influence of flake density on the comparative feeding value of a barley-corn blend for feedlot cattle. *Journal of Animal Science* 75:904-909.

Zinn, R. A., F. N. Owens, and R. A. Ware. 2002. Flaking corn: Processing mechanics, quality standards, and impacts on energy availability and performance of feedlot cattle. *Journal of Animal Science* 80:1145-1156.

Zinn, R. A., A. Barreras, L. Corona, F. N. Owens, and R. A. Ware. 2007. Starch digestion by feedlot cattle: Predictions from analysis of feed and fecal starch and nitrogen. *Journal of Animal Science* 85:1727-1730.

Zinn, R. A., A. Barreras, L. Corona, F. N. Owens, and A. Plascencia. 2011. Comparative effects of processing methods on the feeding value of maize in feedlot cattle. *Nutrition Research Reviews* 24:183-190.

5

Lipids

INTRODUCTION

Lipids refer to all materials that are solubilized in fat-solubilizing solvents and include a broad array of compounds: mono-, di-, and triacylglycerols; phospholipids; glycolipids; free fatty acids (FFA); sterols; alcohols; fat-soluble vitamins; and other nonpolar products. Fats and oils both are triacylglycerols and differ in whether they are solids (fats) or liquids (oils) at room temperature (NRC, 2012).

Fats or oils and the fatty acids (FA) derived from them have important dietary roles. These include increasing energy density, providing essential FA, increasing palatability, decreasing dustiness of feed, improving feed efficiency, alleviating heat stress, influencing nutrient partitioning, enhancing intestinal solubility of fat-soluble nutrients, manipulating ruminal fermentation, and providing lubrication during feed mixing (Palmquist, 1988, Zinn and Jorquera, 2007).

Most natural plant feedstuffs have a small amount of lipid (Table 5-1). Oilseeds have a greater concentration, ranging from 18% in soybeans to 35% in sunflower seeds (Palmquist, 1988). Plant lipids can be grouped into storage lipids in seeds, which are predominantly triacylglycerols; leaf lipids, which are predominantly glycolipids and phospholipids; and additional compounds such as carotenoids, essential oils, chlorophyll, and waxes. Triacylglycerols are the main component of concentrate feedstuffs, whereas glycolipids predominate in forages (Van Soest, 1994). Fats used as feed supplements have FA primarily in the form of triacylglycerols (Palmquist, 1988; Zinn and Jorquera, 2007). Triacylglycerols from plant sources and glycolipids in forages have mostly unsaturated FA and contain high amounts of oleic (C18:1), linoleic (C18:2), or α-linolenic (C18:3) acids (Palmquist, 1988; Vernon and Flint, 1988; Table 5-1).

Beef cattle fed conventional diets consume relatively little lipids as a result of the small amount in natural feedstuffs. Supplemental fat, such as tallow and yellow grease (a combination of waste greases from the food industry and rendered animal fat), are the main sources of fat used in animal feeds. Supplemental fat in ruminant diets is limited because of the observed negative effects on some ruminal bacteria. Around 2% can be added without any adverse effects, but cattle require a period of adaptation when fat is added to the diet at concentrations of 3% or greater. The upper limit of supplementation in high-concentrate diets has been estimated at 6% (Zinn and Jorquera, 2007; Hess et al., 2008).

Historically, supplemental fats have been used as a concentrated energy source in ruminant diets. More recently, additional interest has focused on the role of specific FA to alter physiological processes in the animal and the desire to alter the composition of food products from ruminants to provide potential health benefits to the human diet. This recent interest has resulted in renewed attention to pathways of biohydrogenation (BH) and identification of pathway intermediates because of their links to animal metabolism and ruminant products (Jenkins et al., 2008). Activity in this area is reflected in the number of patents relating to ways to alter BH, the composition of absorbed FA, and the profile of FA in ruminant products (Ladeira et al., 2012).

DIGESTION AND ABSORPTION

Metabolism in the Rumen

Fatty acid metabolism in the rumen has a major influence on the FA composition of ruminant products. Dietary polyunsaturated fatty acids (PUFA), abundant in grass and other feedstuffs, do not result in high PUFA concentrations in meat and milk. Ruminal microorganisms can synthesize their own FA, but they cannot derive energy from β-oxidation of FA under anaerobic conditions (Lourenco et al., 2010).

Bacterial lipids originate from dietary long-chain fatty acids (LCFA) and de novo synthesis from short-chain FA. Polyunsaturated fatty acids are not synthesized by most bacteria (Jenkins, 1993). Protozoal lipids contain more unsaturated FA and products of BH than bacterial lipids, possibly as a consequence of incorporating bacterial lipids (Jenkins, 1993;

TABLE 5-1 Fatty Acid Composition of Some Typical Feedstuffs[a]

Feedstuff	Fatty Acids, %	Fatty Acids, Weight % of Total Fat							
		C14:0	C16:0	C16:1	C18:0	C18:1	C18:2	C18:3	C20:0
Cereals									
Barley	1.6	–	27.6	0.9	1.5	20.5	43.3	4.3	–
Corn	3.2	0.1	16.3	–	2.6	30.9	47.8	2.3	–
Milo	2.3	–	20.0	5.2	1.0	31.6	40.2	2.0	–
Oats	3.2	0.1	22.1	1.0	1.3	38.1	34.9	2.1	–
Wheat	1.0	0.1	20.0	0.7	1.3	17.5	55.8	4.5	–
Forages									
Dehydrated alfalfa meal (17%)	1.4	0.7	28.5	2.4	3.8	6.5	18.4	39.0	–
Perennial ryegrass	–	0.2	11.9	1.7	1.0	2.2	14.6	68.2	–
Pasture grasses	–	1.1	15.9	2.5	2.0	3.4	13.2	61.3	0.2
Red clover	–	1.5	14.2	–	3.7	–	5.6	72.3	–
White clover	–	1.1	6.5	2.5	0.5	6.6	18.5	60.7	2.0

[a]Representative values, adapted from Palmquist (1988).

Lourenco et al., 2010). Protozoa incorporate conjugated linoleic acid (CLA) and *trans*-11-18:1 FA formed by ingested bacteria and contribute to the flow of the respective FA and other PUFA to the duodenum and ultimately to the host animal (Jenkins et al., 2008). The FA composition of ruminal contents has much lower concentrations of PUFA and much greater concentrations of stearic acid when compared with the FA composition of the diet reflecting extensive ruminal BH (Harfoot, 1981; Vernon and Flint, 1988).

The main dietary lipids are triacylglycerols, phospholipids, and galactolipids. Lipids are transformed in the rumen by ruminal microbes as a result of lipolysis and BH. In general, hydrolysis of all three types of dietary lipids is rapid. Bacteria are primarily responsible for lipolysis with small contribution from plant lipases. Following hydrolysis, unsaturated FA are transformed to saturated FA by isomerization of *cis* to *trans* isomers, followed by hydrogenation (reviewed by Harfoot, 1981; Jenkins, 1993; Jenkins et al., 2008).

Hydrolysis of triacylglycerol by microbial lipases yields glycerol and FFA. Glycerol is metabolized by ruminal microorganisms to form propionate, which is absorbed from the rumen (Jenkins, 1993; Zinn and Jorquera, 2007) into capillaries and eventually to the portal vein (see Chapter 2, Anatomy, Digestion, and Nutrient Utilization). Hydrolysis of glycolipids, such as galactolipid, releases both glycerol and galactose, which are fermented (Van Soest, 1994). Hydrolysis of glycolipids and phospholipids also results in formation of FFA (Jenkins, 1993; Doreau and Chilliard, 1997).

Free fatty acids can be partially or completely biohydrogenated. In a meta-analysis of 25 published studies that included intake and duodenal flow of specific FA (Jenkins and Bridges, 2007), there was a linear relationship between the amount of the FA consumed and the ruminal loss of unsaturated FA, indicating BH. Biohydrogenation of oleic (C18:1) and α-linolenic (C18:3) acids was 86%, whereas BH of linoleic acid (C18:2) was 82%. These data suggest that the apparent BH ranking of α-linolenic acid is greater than that for linoleic acid. The treatments evaluated included FA present in feedstuffs, as well as those in added unprotected fat sources. Because the studies included in the meta-analysis differed in many factors, including diet composition and feeding level, the authors suggested that ruminal losses of dietary unsaturated FA could be predicted from dietary intake. Increasing the dietary concentration of unsaturated FA increased intestinal supply; however, supplementation is limited because of potential negative effects on ruminal fermentation and digestion (Jenkins and Bridges, 2007). Data from in vitro studies showed that BH of linoleic acid (C18:2) decreased with increased dietary concentration, suggesting that increasing the linoleic acid content of the diet will increase postruminal flow of the FA (Beam et al., 2000). Apart from dietary concentration of unsaturated FA, the extent of ruminal BH is altered by other factors including decreases associated with low ruminal pH, increased rate of passage, and ionophores or essential oils in the diet, while buffers in the diet can increase the extent of BH (Hall and Eastridge, 2014).

Specifics of Biohydrogenation

Biohydrogenation decreases the toxic effects of unsaturated FA on ruminal microorganisms, which results from detergent action of FA on the microbial cell membrane. The isomerase responsible for the initial step in the BH process requires a free carboxyl group. This indicates lipolysis precedes BH (Jenkins, 1993). Complete or partial BH by ruminal microorganisms results in formation of saturated FA, *trans* FA, and CLA. Conjugated linoleic acid refers to a series of positional and geometric isomers of linoleic acid characterized by the presence of a conjugated double bond.

Two major substrates for BH are α-linolenic acid (C18:3) for grazing animals, reflecting its prevalence in glycolipids and phospholipids of forages, and linoleic acid (C18:2), reflecting its prevalence in triacylglycerols for animals receiving concentrates or lipid supplements. Oleic acid (C18:1) is also present in significant amounts. There are both major and additional suggested pathways of BH based on identification of intermediates in the rumen; only the major pathways are described here (Figure 5-1).

Linoleic acid forms several isomers of CLA, with *cis*-9, *trans*-11-18:2 (rumenic acid) as the major form, which is then converted to *trans*-11-18:1 (*trans*-vaccenic acid) and finally to C18:0 (stearic acid) for complete BH (Lourenco et al., 2010). A total of 14 positional and geometric CLA isomers were identified in digesta contents from cattle or in vitro fermentation contents (Jenkins et al., 2008). Feeding high levels of starch or inclusion of supplemental oils (fish oil or vegetable oil) can result in *trans*-10,*cis*-12 CLA being a major CLA isomer which leads to formation of *trans*-10-18:1 (Bauman and Griinari, 2003; Lourenco et al., 2010). The BH pathway for α-linolenic acid is more complex because it has three double bonds. Partial BH of α-linolenic acid also results in the formation of *trans*-11-18:1 isomer (Jenkins et al., 2008; Lourenco et al., 2010). Intermediates of BH of FA having more than three double bonds (such as 20:5 and 22:6 in fish oil) are even less clearly defined (Jenkins et al., 2008).

Postruminal Digestion and Absorption

Lipids are released from the microorganisms in the abomasum. Microbial lipids and dietary lipids pass to the duodenum. Lipid flow to the small intestine is primarily composed of saturated LCFA and microbial lipids. Fatty acid flow to the small intestine can exceed intake because of microbial lipid synthesis; however, this is not always the case (Jenkins, 1993; Doreau and Ferlay, 1994; Doreau and Chilliard, 1997).

Microbial lipids and dietary lipids that were not hydrolyzed in the rumen are hydrolyzed by pancreatic lipase and colipase. Bile acids facilitate dissociation of FA from feed particles. Phospholipids in bile are converted to lysophospholipids by pancreatic phospholipase. These components promote solubilization and incorporation of FA into micelles (Vernon and Flint, 1988; Bauchart, 1993). Fatty acids are absorbed from micelles into the epithelial cells in the jejunum. Some stearate can be converted to oleate before being incorporated into triacylglycerols. The triacylglycerols are further incorporated into chylomicrons and very-low-density lipoproteins, which are released into the lymphatic system (Vernon and Flint, 1988; Doreau and Chilliard, 1997), eventually entering the vascular system at the thoracic duct, thereby bypassing the liver during absorption. Plasma lipids contain PUFA in phospholipids and cholesterol esters, but little is present in triacylglycerols. Tissues secrete lipoprotein lipase, which hydrolyzes triacylglycerols in lipoproteins, releasing the FA for use by tissues (Vernon and Flint, 1988).

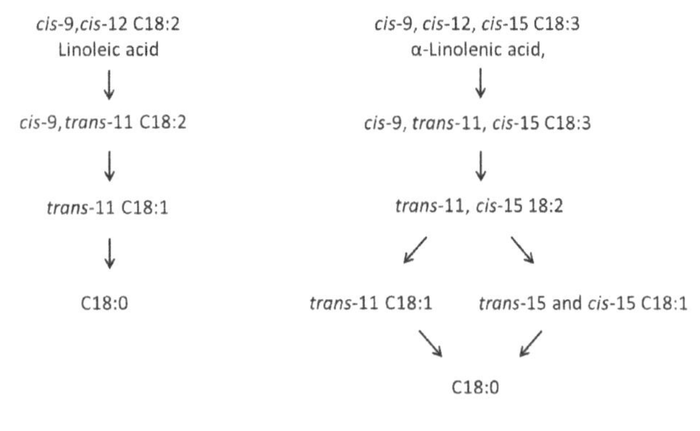

FIGURE 5-1 Major biohydrogenation pathways of linoleic acid and α-linolenic acid in the rumen (adapted from Jenkins et al., 2008).

DIGESTIBILITY AND ENERGY VALUE

Digestibility is the primary factor that determines the amount of energy provided by a fat source. Therefore, factors that influence digestion will influence energy value. Because dietary fat is extensively modified in the rumen, as described previously, measures of digestibility for the entire gastrointestinal tract might not provide a valid estimate. Estimates of digestibility between the duodenum and ileum or total intestinal digestibility are more reliable than total tract estimates.

Reported values for FA digestibility in the small intestine of ruminants ranged from 55 to 92%. The digestibility values were affected by chain length; with digestibility decreasing with increasing FA chain length above C18. Average intestinal digestibility for C16 and C18 FA was 70% and 77%, respectively (Doreau and Chilliard, 1997). In a series of studies wherein cattle were fed diets including yellow grease, tallow, or animal-vegetable blends, with the level of supplementation of 3 to 9% of dry matter (DM), the average intestinal digestibility of fat was 77% (Zinn and Jorquera, 2007). The average digestibility of C18 FA with varying number of double bonds was C18:0, 77%; C18:1, 85%; C18:2, 83%; and C18:3, 76% (Doreau and Chilliard, 1997) and the digestibility of saturated FA was about 80% of unsaturated FA (Zinn and Jorquera, 2007). Increased inclusion of soybean oil decreased small intestinal digestibility of total FA in sheep as a result of decreased digestibility of saturated FA (Kucuk et al., 2004). Summarizing data from seven studies, an inverse linear relationship was observed between FA intake per unit body weight (BW) in cattle and intestinal digestibility, demonstrating decreased digestibility as dietary intake increased. The decrease was primarily a result of decreased digestibility of saturated (C16:0 and C18:0) rather than unsaturated (C18:1, C18:2) FA (Plascencia et al., 2003). Based on decreased digestibility with increasing fatty acid intake, the optimal feeding value of supplemental fat was recommended at a total lipid intake of 1.0 g/kg BW or less or 7% of dietary DM (Zinn and Jorquera, 2007).

The energy value of fat can be estimated using various assumptions. For example, the gross energy value of FA varies with chain length and saturation and is about 9.33

Mcal/kg (Hall and Eastridge, 2014). If average digestibility of fat is 75%, the digestible energy (DE) value would be 7.0 Mcal/kg. If it is further assumed that there are no urinary or gaseous energy losses from fat (ME [metabolizable energy] = DE; NRC, 2001), and the efficiency of conversion of ME to net energy for lactation (NEl) or net energy required for maintenance (NEm) is 0.80, NEm for fat supplements can be estimated at 5.6 Mcal/kg (NRC, 2001; Hall and Eastridge, 2014). Using 0.55 as the efficiency of conversion of ME to net energy required for gain (NEg) for fat supplements, NEg of 3.85 is estimated (NRC, 2001). Empirical estimates of the efficiency of conversion of ME to NEg ranged from 61 to 74% in cattle fed protected lipid supplements (Garrett et al., 1976) and from 75 to 82% in sheep infused with unsaturated FA (Czerkawski et al., 1966). A value of 67% for the partial efficiency of ME from dietary fat for gain was used by Zinn and Jorquera (2007).

ESSENTIAL FATTY ACIDS

Early work on feeding fat-free diets to dairy calves demonstrated the essentiality of fat in the diet by showing that lack of fat resulted in poor growth, leg weakness, scaly dermatitis, and decreased plasma concentration of linoleic acid (Cunningham and Loosli, 1954; Lambert et al., 1954). In a more recent study, lipid supplements containing low or adequate essential fatty acids (EFA) were fed to 3-day-old calves in milk-replacer diets for 6 weeks. Calves fed milk replacer deficient in EFA had decreased plasma and tissue linoleic acid with concomitant increase in other FA metabolites, indicating low EFA status. Moreover, increased hemolysis of red blood cells was observed, suggesting a role of EFA in maintaining the integrity of red blood cell membrane structure. No external signs of EFA deficiency were visible in this time interval (Jenkins and Kramer, 1986). Other studies (Noble, 1981, 1984) have shown that the newborn ruminant has low linoleic acid stores at birth, which increase rapidly as the calf advances in age. These studies suggest the ruminant is very efficient at using and conserving dietary linoleic acid and might require less dietary linoleic acid, as a percentage of energy intake, than nonruminant species. In the adult ruminant, linoleic acid that escapes BH is preferentially incorporated into cholesterol esters and phospholipids (Noble, 1981, 1984).

Mammals lack the ability to synthesize linoleic acid (18:2 n-6) and α-linolenic acid (18:3 n-3) because they lack the Δ^{12} and Δ^{15} desaturase enzymes required to introduce a double bond at C12 or C15 from the carboxyl end of the FA making both linoleic acid and α-linolenic acid EFA (Palmquist, 2009). The designation, n-6 or n-3, refers to the location of the first double bond proximate to the methyl end of the FA. Both linoleic acid and α-linolenic acid are the precursors of two different series of long-chain PUFA, n-6 and n-3, respectively. They are metabolized by a common desaturase and elongase enzyme system to longer-chain PUFA (Figure 5-2).

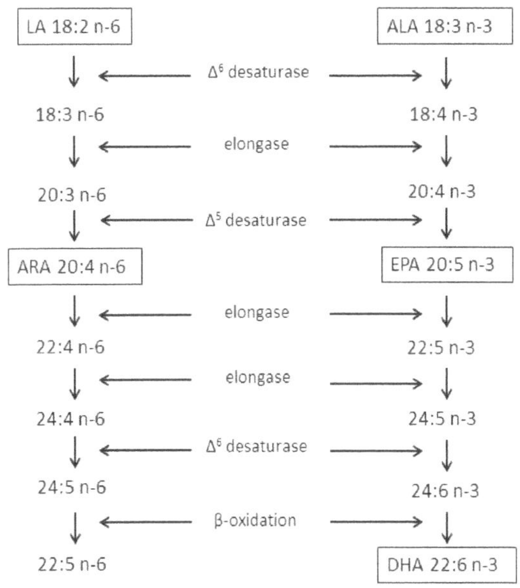

FIGURE 5-2 Synthesis of long-chain polyunsaturated fatty acids (PUFA) from the two essential FA (C18) precursors. The n-6 and n-3 pathways share desaturase and elongase enzymes. Long-chain PUFA of particular focus are identified. LA, linoleic acid; ARA, arachidonic acid; ALA, α-linolenic acid; EPA, eicosapentaenoic acid; DHA, docosahexaenoic acid (adapted from Leonard et al., 2004).

The elongase adds two carbon units to the carboxyl end of the FA, and the desaturase enzymes add a double bond at the carbon indicated (Leonard et al., 2004). The first step in the pathway, Δ^6 desaturase, adds a double bond at C6. This is rate limiting, and the 18C FA compete at this step. As a consequence, increasing dietary concentration of α-linolenic acid decreased metabolism of linoleic acid to arachidonic acid (ARA). The reciprocal relationship was also observed where increasing dietary linoleic acid decreased metabolism of α-linolenic acid to longer-chain PUFA (Holman, 1998).

The EFA are precursors of the eicosanoids, which include the C20 FA, ARA, and eicosapentaenoic acid (EPA), and their derivatives. In many species, EPA and docosahexaenoic acid (DHA) are considered EFA (reviewed by Palmquist, 2009). These PUFA are important components of membranes, providing fluidity and selective permeability. Fatty acid derivatives of the C20 FA include prostaglandins, thromboxanes, prostacyclins, leukotrienes, and other biologically active lipid metabolites that influence physiological processes including growth, reproduction, the immune response, inflammation, and vascular responses. These compounds are synthesized locally in response to the cellular environment (Palmquist, 2009).

The EFA are common in natural feedstuffs (Tables 5-1 and 5-2). Grains and soybean oil contain predominantly C18:2 n-6, whereas forages and flaxseed oil contain significant amounts of C18:3 n-3. Soybean and canola oils contain sig-

TABLE 5-2 Fatty Acid Composition of Selected Sources of Fats and Oils[a]

	Fatty Acids, Weight % of Total Fat														
Type of Lipid	I C10	C12:0	C14:0	C16:0	C16:1	C18:0	C18:1	C18:2	C18:3	C20:4	C20:5	C22:1	C22:5	C22:6	IV[b]
Animal fats															
Beef tallow	0	0.9	3.7	24.9	4.2	18.9	36.0	3.1	0.6	0	0	0	0	0	44
Choice white grease	0.2	0.2	1.9	21.5	5.7	14.9	41.1	11.6	0.4	0	0	0	0	0	60
Restaurant grease	–	–	1.9	16.2	2.5	10.5	47.5	17.5	1.9	0	0	0	0	0	75
Fish oils															
Menhaden	0	0	8.0	15.2	10.5	3.8	14.5	2.2	1.5	1.2	13.2	0.4	4.9	8.6	161
Salmon	0	0	3.3	9.8	4.8	4.3	17.0	1.5	1.1	0.7	13.0	3.4	3.0	18.2	195
Vegetable oils															
Canola	0	0	0	4.0	0.2	1.8	56.1	20.3	9.3	0	0	0.6	0	0	115
Corn	0	0	0	10.6	0.1	1.9	27.3	53.5	1.16	0	0	0	0	0	125
Cottonseed	0	0	0.8	22.7	0.8	2.3	17.0	51.5	0.2	0.1	0	0	0	0	110
Flaxseed	0	0	0	5.3	0	4.1	20.2	12.7	53.3	0	0	0	0	0	187
Palm kernel	3.7	47.0	16.4	8.1	0	2.8	11.4	1.6	0	0	0	0	0	0	13
Safflower	0	0	0	4.3	0	1.9	14.4	74.6	0	0	0	0	0	0	148
Soybean	0	0	0.1	10.3	0.2	3.8	22.8	51.0	6.8	0	0	0	0	0	132
Sunflower	0	0	0	5.4	0.2	3.5	45.3	39.8	0.2	0	0	0	0	0	114

[a]From NRC (2012); values represent the general fatty acid profile of each source; all fatty acids from each source are not included.
[b]IV, iodine value is a measure of the degree of unsaturation of the fatty acid.

nificant amounts of both C18:2 and C18:3. Long-chain n-6 and n-3 PUFA can be synthesized in the body from linoleic acid or α-linolenic acid, respectively, or can be provided from the diet. Some fish oils and algae contain the longer-chain (≥C20) n-3 PUFA and directly supply EPA and DHA.

FAT SUPPLEMENTATION

Fats are supplemented in the diet primarily to increase energy density of the ration. Moreover, fat supplements can increase the duodenal flow of specific FA present in the supplements. In addition, because of microbial metabolism in the rumen, fat supplements can increase duodenal flow of BH products. There is growing interest in increasing the flow of specific FA including n-3 PUFA, CLA, and *trans*-11-18:1, that are metabolically active and which enhance the value of ruminant products because of their potential beneficial effects on human health (Hess et al., 2008; Jenkins et al., 2008).

Fat supplements for livestock are co-products of other industries. The quality of feed-grade animal fats is determined by moisture, impurities, and unsaponifiables; total FA; FFA concentrations; and melting point (Palmquist, 1988; Zinn and Jorquera, 2007). Triglycerides contain approximately 90% FA and 10% glycerol. A pure fat source should have total FA in this range. High levels of FFA indicate hydrolysis of triglycerides. This could be associated with oxidative rancidity. The initial peroxide value gives information on the degree of oxidative rancidity and should be less than 5 mEq/kg fat (Zinn and Jorquera, 2007). The melting point (titer) depends on the chain length and the degree of unsaturation of FA. Tallow has a melting point of 38°C or higher; grease has a melting point below 38°C (Palmquist, 1988). Standards for various qualities of feed-grade fats are determined by the American Fats and Oils Association (Palmquist, 1988; Zinn and Jorquera, 2007).

Dietary lipids can disrupt ruminal fermentation, resulting in decreased digestibility of nonlipid energy sources, particularly structural carbohydrates. Ruminal changes observed with fat supplementation include decreased production of methane, hydrogen, and short-chain FA, with a lower acetate-to-propionate ratio (Jenkins, 1993; Doreau and Chilliard, 1997). Dietary lipids also disrupt ruminal digestion of proteins. Linseed oil infusion in the rumen of sheep decreased ruminal protein digestion and ammonia concentration and increased total nitrogen flow to the duodenum. Protozoal numbers also decreased following linseed oil infusions (Ikwuegbu and Sutton, 1982). On the contrary, dietary fat is less detrimental to nonstructural carbohydrates such as starch (Jenkins, 1993; Doreau and Chilliard, 1997).

Supplemental fat sources, depending on the degree of unsaturation, have variable effects on ruminal fermentation, with greater inhibition observed with unsaturated FA than saturated FA. Free fatty acids are more inhibitory to the fermentation than esterified FA, and free oils are more inhibitory than oils in whole seeds (Palmquist, 1988). Oils containing PUFA have a more toxic effect on biohydrogenating bacteria and ruminal fermentation than monoenoic FA. Fish oils that are rich in 20:5 n-3 (EPA) and 22:6 n-3 (DHA), resulting from consumption of marine algae, inhibit the last step of BH, conversion of *trans*-11-18:1 to 18:0, and inhibit the growth of biohydrogenating bacteria, resulting in increased *trans*-18:1 FA. Coconut oil, rich in lauric acid

(C12:0), inhibits methanogenesis. Some essential oils may alter BH because of their antimicrobial activity (Lourenco et al., 2010).

Feeding vegetable- and fish-based oil supplements increases BH intermediates available for tissue metabolism. A BH intermediate of primary interest is *trans*-11-18:1 (*trans*-vaccenic acid), formed from the partial BH of either linoleic or α-linolenic acids (Figure 5-1). This intermediate can be converted to *cis*-9,*trans*-11-18:2 (CLA, rumenic acid) by the Δ^9-desaturase enzyme in ruminant tissues. *Cis*-9,*trans*-11 CLA is the major isomer present in ruminant fat and dairy products and has been associated with a series of beneficial health effects in animal models. *Trans*-10,*cis*-12 CLA, another intermediate product of BH, is associated with decreased milk fat synthesis (Lock and Bauman, 2004). Mechanisms associated with inhibition of FA synthesis in the mammary gland, by *trans*-10,*cis*-12 CLA and other rumen-produced CLA isomers, have been described (see review by Bauman et al., 2011).

Fat supplementation increases the proportion of fat in carcasses. Despite the extensive BH of dietary C18 and other PUFA, the ratio of saturated to unsaturated FA tends to increase when animal fats are supplied and decrease when vegetable oils are supplied (Doreau and Chilliard, 1997; Hess et al., 2008).

Feedlot Diets

Grain feeding decreases BH and promotes increased unsaturation of fats in carcass and milk possibly because of low ruminal pH and decreased lipolysis. High-corn diets increased ruminal formation of *trans*-18:1 (Jenkins, 1993). Biohydrogenation values lower than 60 to 70% for linoleic and α-linolenic acids were associated with diets having greater than 70% concentrates (Doreau and Chilliard, 1997).

Tallow and yellow grease are the major sources of fat used in animal feeds. Choice white grease, poultry fats, and acidulated animal/vegetable soap stocks are also commonly used (Zinn and Jorquera, 2007). Extraction of oil, in a modification of the dry-grind processing of corn for ethanol production, results in a co-product that can be incorporated into feed (Berger and Singh, 2010). This or other co-products could be more important in the future. A survey of consulting feedlot nutritionists (Vasconcelos and Galyean, 2007) indicated that 71% of clients served by the nutritionists used added fat. The primary sources were tallow (47%) and yellow grease (24%). The average recommended level of supplemental fat was 3.1% and the maximum recommended level of added fat was 4.5%. Based on these inclusion rates of supplemental fat, the total fat recommended in feedlot diets averaged 7.6% with a maximum of 10%. Therefore, added fat contributes to high NEg concentration in finishing diets. Considering effects of fat on utilization of other dietary nutrients, ruminants fed high-concentrate diets and adapted to fat in the diet can receive up to 6% supplemental fat included in the diet without any negative effects on growth performance (Zinn and Jorquera, 2007; Hess et al., 2008).

Forage-Based Diets

A summary of a series of studies using soybean oil as the supplemental fat source (Hess et al., 2008) indicated that dietary soybean oil added to corn-based supplements in forage-based diets at greater than or equal to 6% of DM decreased the acetate-to-propionate ratio in the rumen and decreased ruminal organic matter (OM) digestion. Ruminal changes were also observed at lower inclusion level of soybean oil, which could have been associated with ruminal fermentation of the corn. Additional studies documented the effects of fat supplements on decreasing forage OM intake and diet digestibility. Fish oil added to grass silage diets increased duodenal flow of total *trans*-18:1 (mostly *trans*-11-18:1), 20:5 n-3 (EPA), and 22:6 n-3 (DHA) and decreased duodenal flow of 18:0. Addition of oil altered microbial species in the rumen (Kim et al., 2008). Duodenal flow of *trans*-11-18:1 increased regardless of type of supplemental fat added to forage-based diets. The optimal level of fat inclusion varies depending on the purpose of supplementing fat. To maximize the use of forage-based diets, the recommended optimal inclusion rate of supplemental fat was less than 3% of DM. To maintain intake of forage, supplemental fat should be limited to 2% of DM intake. To increase dietary DE by adding fat, supplementation should be limited to 4% of DM intake (Hess et al., 2008).

In diets containing a mix of forage and concentrate ranging from 72.9% forage to 18.4% forage, with added soybean oil to maintain similar amounts of fat in the DM, ruminal BH of C18 PUFA increased as the level of dietary forage increased. Despite greater BH, duodenal flow of C18:3 n-3 increased as forage intake increased (Kucuk et al., 2001).

Physiological Processes

Supplemental fats increase energy density of the diet and can have positive effects on reproduction when energy intake is inadequate. Supplemental fats can also directly affect reproductive function due to the presence of specific FA such as long-chain n-3 PUFA, including EPA (20:5 n-3) and DHA (22:6 n-3); and the long-chain n-6 PUFA ARA (20:4 n-6). The 20-C PUFA are precursors of various prostaglandins (PG), forming different PG series based on the n-6 and n-3 FA, respectively. Arachidonic acid is a precursor of the specific prostaglandin $PGF_{2\alpha}$, which has an important role in reproduction, including ovulation, estrus, embryo survival, and parturition (Funston, 2004; Hess et al., 2005; Gulliver et al., 2012).

Studies evaluating reproductive responses to dietary PUFA in cattle were summarized by Hess et al. (2005, 2008), Palmquist (2009), and Gulliver et al. (2012). Studies that evaluated the effect of diet on concentrations of $PGF_{2\alpha}$

or a metabolite of $PGF_{2\alpha}$ and reproduction in beef cattle included those of Filley et al. (1999, 2000), Grant et al. (2005), Wamsley et al. (2005), and Childs et al. (2008a,b,c). In general, results support the concept that dietary linoleic acid increases production of $PGF_{2\alpha}$; and long-chain n-3 PUFA (EPA and DHA) inhibit production of $PGF_{2\alpha}$, although this was not observed in all cases. The dietary ratio of n-6 to n-3, which determines the relative availability of the precursors for eicosanoid formation, could be a factor in varying results. Both n-3 and n-6 PUFA can also interact to control the synthesis and metabolism of steroid hormones. Further studies evaluating the effects of unsaturated fatty acid status and feeding n-3 PUFA on reproductive outcomes are needed (Gulliver et al., 2012).

Hess et al. (2008) recommended feeding supplemental fat to replacement heifers for 60 to 90 days before the breeding season to increase pregnancy rate; however, it is not clear whether responses were specifically associated with fat or were a response to energy supplementation (Funston, 2004; Hess et al., 2008). An effect of fat feeding on pregnancy rate was not observed in summarizing 324 observations with cows given fat for 4 to 90 days during the postpartum period (Hess et al., 2005). Consequently, postpartum supplementation was not recommended as a nutritional strategy to improve pregnancy rates in beef cows (Hess et al., 2008). Research is inconclusive on the role of supplemental fat and the influence it might have on reproduction beyond the energy contribution. Overall, dose-response studies are lacking, and quantitative responses are not described. There are many factors that contribute to the variation in responses observed, including the amount and source of fat added, length of supplementation, basal diet, timing of supplementation, and environmental conditions. Inclusion of supplemental fat will be determined in part by economic considerations (Funston, 2004).

Ruminant Products

Researchers have investigated the effectiveness of altering FA composition of ruminant milk and meat products for human consumption (reviewed by Shingfield et al., 2013; Scollan et al., 2014). Previous studies in cattle have documented significant changes in tissue FA profile in response to various dietary FA supplements. For example, Dunne et al. (2011) showed increased muscle EPA and DHA concentration in response to a ruminally protected source of these FA; Herdmann et al. (2010) showed increased C18:3 n-3, EPA, and DHA in muscle and adipose tissue in response to a diet enriched in sources of α-linolenic acid; and Noci et al. (2007) observed increased *trans*-11-18:1 and *cis*-9,*trans*-11-18:2 (CLA) in muscle adipose tissue and subcutaneous adipose tissue of heifers grazing grass pasture and receiving supplements of either sunflower oil or linseed oil.

Inclusion of 15% flaxseed in hay or silage diets resulted in increased n-3 FA, including C18:3 n-3, and decreased ratio of n-6 to n-3 FA in subcutaneous adipose tissue in cull beef cows. This approach with cull beef cows might be used to produce α-linolenic acid–enriched ground beef, but at a greater retail price to offset increased feed costs associated with supplementing flaxseed to the diet (He et al., 2012).

Although much of the dietary lipid is biohydrogenated in the rumen, there are differences in the tissue unsaturated FA profile between cattle fed grass-based and grain-based finishing diets without supplemental fat. Grass-fed cattle have a higher concentration of *trans*-11-18:1, which is a precursor for de novo synthesis of CLA (*cis*-9,*trans*-11-18:2). In many cases, grass-fed beef has greater concentrations of n-3 PUFA and a lower ratio of n-6 to n-3 FA. The specific n-3 FA C18:3, C20:5 (EPA), C22:5 (docosapentaenoic acid), and C22:6 (DHA) are often in greater concentration in grass-fed than grain-fed beef. In contrast, grain-fed beef has a higher concentration of monounsaturated FA (Daley et al., 2010). Differences in tissue FA composition in cattle from various production systems are addressed in greater detail in Chapter 1 (Beef Production: Systems, Quality, and Safety).

Protection of PUFA

It is well established that BH alters the profile of dietary FA. The need for rumen-protected fat sources implies the need for specific unsaturated FA in ruminal outflow. The main interests in this regard are the two EFA, linoleic acid and α-linolenic acid, and the two longer-chain n-3 PUFA, EPA and DHA.

Efforts to decrease the antimicrobial effects of dietary FA, prevent disruption of fermentation, and supply more PUFA to the duodenum include two main approaches: encapsulation of unsaturated FA, and altering the structure of FA to resist actions of microbial enzymes. Strategies using formaldehyde-treated fats or blocking the carboxyl group by forming calcium salts or fatty acyl amides have been extensively studied. Summarizing observations from 25 published studies (Jenkins and Bridges, 2007), ruminal loss as a percentage of intake was similar for linoleic acid and α-linolenic acid for unprotected fat, oilseeds, or protected fat. There was some benefit for oleic acid in feeding oilseed and protected fats. On average, about 15% of dietary PUFA will escape BH (Jenkins and Bridges, 2007). Other studies suggest that BH of calcium salts of unsaturated FA is about 20% less than that of conventional liquid fats (Zinn and Jorquera, 2007). A greater understanding of BH could lead to further developments in this area.

REFERENCES

Bauchart, D. 1993. Lipid absorption and transport in ruminants. *Journal of Dairy Science* 76:3864-3881.

Bauman, D. E., and J. M. Griinari. 2003. Nutritional regulation of milk fat synthesis. *Annual Review of Nutrition* 23:203-227.

Bauman, D. E., K. J. Harvatine, and A. L. Lock. 2011. Nutrigenomics, rumen-derived bioactive fatty acids and the regulation of milk fat synthesis. *Annual Review of Nutrition* 31:299-319.

Beam, T. M, T. C. Jenkins, P. J. Moate, R. A. Kohn, and D. L. Palmquist. 2000. Effects of amount and source of fat on the rates of lipolysis and biohydrogenation of fatty acids in ruminal contents. *Journal of Dairy Science* 83:2564-2573.

Berger, L., and V. Singh. 2010. Changes in evolution of corn coproducts for beef cattle. *Journal of Animal Science* 88(Suppl.):E143-E150.

Childs, S., F. Carter, C. O. Lynch, J. M. Sreenan, P. Lonergan, A. A. Hennessy, and D. A. Kenny. 2008a. Embryo yield and quality following dietary supplementation of beef heifers with n-3 polyunsaturated fatty acids (PUFA). *Theriogenology* 70:992-1003.

Childs, S., A. A. Hennessy, J. M. Sreenan, D. C. Wathes, Z. Cheng, C. Stanton, M. G. Diskin, and D. A. Kenny. 2008b. Effect of level of dietary n-3 supplementation on systemic and tissue fatty acid concentrations and on selected reproductive variables in cattle. *Theriogenology* 70:595-611.

Childs, S., C. O. Lynch, A. A. Hennessy, C. Stanton, D. C. Wathes, J. M. Sreenan, M. G. Diskin, and D. A. Kenny. 2008c. Effect of dietary enrichment with either n-3 or n-5 fatty acids on systemic metabolite and hormone concentration of ovarian function in heifers. *Animal* 2:883-893.

Cunningham, H. M., and J. K. Loosli. 1954. The effects of fat-free diets on young dairy calves with observations on metabolic fecal fat and digestion coefficients for lard and hydrogenated coconut oil. *Journal of Dairy Science* 37:453-461.

Czerkawski, J. W., K. L. Blaxter, and F. W. Wainman. 1966. The metabolism of oleic, linoleic and linolenic acids by sheep with reference to their effects on methane production. *British Journal of Nutrition* 20:349-362.

Daley, C. A., A. Abbott, P. S. Doyle, G. A. Nader, and S. Larson. 2010. A review of fatty acid profiles and antioxidant content in grass-fed and grain-fed beef. *Nutrition Journal* 9:10.

Doreau, M., and Y. Chilliard. 1997. Digestion and metabolism of dietary fat in farm animals. *British Journal of Nutrition* 78(Suppl. 1):S15-S35.

Doreau, M., and A. Ferlay. 1994. Digestion and utilization of fatty acids by ruminants. *Animal Feed Science and Technology* 45:379-396.

Dunne, P. G., J. Rogalski, S. Childs, F. J. Monahan, D. A. Kenny, and A. P. Moloney. 2011. Long chain n-3 polyunsaturated fatty acid concentration and color and lipid stability of muscle from heifers offered a ruminally protected fish oil supplement. *Journal of Agricultural and Food Chemistry* 59:5015-5025.

Filley, S. J., H. A. Turner, and F. Stormshak. 1999. Prostaglandin $F_{2\alpha}$ concentrations, fatty acid profiles, and fertility in lipid-infused postpartum beef heifers. *Biology of Reproduction* 61:1317-1323.

Filley, S. J., H. A. Turner, and F. Stormshak. 2000. Plasma fatty acids, prostaglandin $F_{2\alpha}$ metabolite, and reproductive response in postpartum heifers fed rumen bypass fat. *Journal of Animal Science* 78:139-144.

Funston, R. N. 2004. Fat supplementation and reproduction in beef females. *Journal of Animal Science* 82(Suppl.):E154-E161.

Garrett, W. N., Y. T. Yang, W. L. Dunkley, and L. M. Smith. 1976. Energy utilization, feedlot performance and fatty acid composition of beef steers fed protein encapsulated tallow or vegetable oils. *Journal of Animal Science* 42:1522-1533.

Grant, M. H., B. M. Alexander, B. W. Hess, J. D. Bottger, D. L. Hixon, E. A. Van Kirk, T. M. Nett, and G. E. Moss. 2005. Dietary supplementation with safflower seeds differing in fatty acid composition differentially influences serum concentrations of prostaglandin F metabolite in postpartum beef cows. *Reproduction Nutrition Development* 45:721-727.

Gulliver, C. E., M. A. Friend, B. J. King, and E. H. Clayton. 2012. The role of omega-3 polyunsaturated fatty acids in reproduction of sheep and cattle. *Animal Reproduction Science* 131:9-22.

Hall, M. B., and M. L. Eastridge. 2014. Invited review: Carbohydrate and fat: Considerations for energy and more. *The Professional Animal Scientist* 30:140-149.

Harfoot, C. G. 1981. Lipid metabolism in the rumen. Pp. 21-55 in *Lipid Metabolism in Ruminant Animals*, W. W. Christie, ed. Oxford, UK: Pergamon Press.

He, M. L., T. A. McAllister, J. P. Kastelic, P. S. Mir, J. L. Aalhus, M. E. R. Dugan, N. Aldai, and J. J. McKinnon. 2012. Feeding flaxseed in grass hay and barley silage diets to beef cows increases alpha-linolenic acid and its biohydrogenation intermediates in subcutaneous fat. *Journal of Animal Science* 90:592-604.

Herdmann, A., J. Martin, G. Nuernberg, D. Dannenberger, and K. Nuernberg. 2010. Effect of dietary n-3 and n-6 PUFA on lipid composition of different tissues of German Holstein bulls and the fate of bioactive fatty acids during processing. *Journal of Agricultural and Food Chemistry* 58:8314-8321.

Hess, B. W., S. L. Lake, E. J. Scholljegerdes, T. R. Weston, V. Nayigihugu, J. D. C. Molle, and G. E. Moss. 2005. Nutritional controls of beef cow reproduction. *Journal of Animal Science* 83:E90-E106.

Hess, B. W., G. E. Moss, and D. C. Rule. 2008. A decade of developments in the area of fat supplementation research with beef cattle and sheep. *Journal of Animal Science* 86:E188-E204.

Holman, R. T. 1998. The slow discovery of the importance of ω3 essential fatty acids in human health. *Journal of Nutrition* 128:427S-433S.

Ikwuegbu, O. A., and J. D. Sutton. 1982. The effect of varying the amount of linseed oil supplementation on rumen metabolism in sheep. *British Journal of Nutrition* 48:365-375.

Jenkins, K. J., and J. K. G. Kramer. 1986. Influence of low linoleic and linolenic acids in milk replacer on calf performance and lipids in blood plasma, heart, and liver. *Journal of Dairy Science* 69:1374-1386.

Jenkins, T. C. 1993. Lipid metabolism in the rumen. *Journal of Dairy Science* 76:3851-3863.

Jenkins, T. C., and W. C. Bridges, Jr. 2007. Protection of fatty acids against ruminal biohydrogenation in cattle. *European Journal of Lipid Science and Technology* 109:778-789.

Jenkins, T. C., R. J. Wallace, P. J. Moate, and E. E. Mosley. 2008. Board invited review: Recent advances in biohydrogenation of unsaturated fatty acids with the rumen microbial ecosystem. *Journal of Animal Science* 86:397-412.

Kim, E. J., S. A. Huws, M. R. F. Lee, J. D. Wood, S. M. Muetzel, R. J. Wallace, and N. D. Scollan. 2008. Fish oil increases the duodenal flow of long chain polyunsaturated fatty acids and *trans*-11 18:1 and decreases 18:0 in steers via changes in the rumen bacterial community. *Journal of Nutrition* 138:889-896.

Kucuk, O., B. W. Hess, P. A. Ludden, and D. C. Rule. 2001. Effect of forage:concentrate ratio on rumen digestion and duodenal flow of fatty acids in ewes. *Journal of Animal Science* 79:2233-2240.

Kucuk, O., B. W. Hess, and D. C. Rule. 2004. Soybean oil supplementation of a high-concentrate diet does not affect site and extent of organic matter, starch, neutral detergent fiber, or nitrogen digestion, but influences both ruminal metabolism and intestinal flow of fatty acids in limit-fed lambs. *Journal of Animal Science* 82:2985-2994.

Ladeira, M. M., O. R. M. Neto, M. L. Chizzotti, D. M. Oliveira, and A. C. Junior. 2012. Lipids in the diet and fatty acid profile in beef: A review and recent patents on the topic. *Recent Patents on Food, Nutrition & Agriculture* 4:123-133.

Lambert, M. R., N. L. Jacobson, R. S. Allen, and J. H. Zaletel. 1954. Lipid deficiency in the calf. *Journal of Nutrition* 52:259-272.

Leonard, A. E., S. L. Pereira, H. Sprecher, and Y.-S. Huang. 2004. Elongation of long-chain fatty acids. *Progress in Lipid Research* 43:36-54.

Lock, A. L., and D. E. Bauman. 2004. Modifying milk fat composition of dairy cows to enhance fatty acids beneficial to human health. *Lipids* 39:1197-1206.

Lourenco, M., E. Ramos-Morales, and R. J. Wallace. 2010. The role of microbes in rumen lipolysis and biohydrogenation and their manipulation. *Animal* 4:1008-1023.

Noble, R. C. 1981. Digestion, absorption and transport of lipids in ruminant animals. Pp. 57-93 in *Lipid Metabolism in Ruminant Animals*, W. W. Christie, ed. Oxford, UK: Pergamon Press.

Noble, R. C. 1984. Essential fatty acids in the ruminant. Pp. 185-200 in *Fats in Animal Nutrition*, J. Wiseman, ed. Boston: Butterworths.

Noci, F., P. French, F. J. Monahan, and A. P. Moloney. 2007. The fatty acid composition of muscle fat and subcutaneous adipose tissue of grazing heifers supplemented with plant oil-enriched concentrates. *Journal of Animal Science* 85:1062-1073.

NRC (National Research Council). 2001. *Nutrient Requirements of Dairy Cattle*, 7th Rev. Ed. Washington, DC: National Academy Press.

NRC. 2012. *Nutrient Requirements of Swine*, 11th Rev. Ed. Washington, DC: The National Academies Press.

Palmquist, D. L. 1988. The feeding value of fats. Pp. 293-311 in *Feed Science*, E. R. Ørskov, ed. New York: Elsevier.

Palmquist, D. L. 2009. Omega-3 fatty acids in metabolism, health, and nutrition and for modified animal product foods. *The Professional Animal Scientist* 25:207-249.

Plascencia, A., G. D. Mendoza, C. Vasquez, and R. A. Zinn. 2003. Relationship between body weight and level of fat supplementation on fatty acid digestion in feedlot cattle. *Journal of Animal Science* 81:2653-2659.

Scollan, N. D., D. Dannenberger, K. Nuernberg, I. Richardson, S. MacKintosh, J.-F. Hocquette, and A. P. Moloney. 2014. Enhancing the nutritional and health value of beef lipids and their relationship with meat quality. *Meat Science* 97:384-394.

Shingfield, K. J., M. Bonnet, and N. D. Scollan. 2013. Recent developments in altering the fatty acid composition of ruminant-derived foods. *Animal* 7(Suppl. 1):132-162.

Van Soest, P. J. 1994. *Nutritional Ecology of the Ruminant*, 2nd Ed. Ithaca, NY: Cornell University Press.

Vasconcelos, J. T., and M. L. Galyean. 2007. Nutritional recommendations of feedlot consulting nutritionists: The 2007 Texas Tech University survey. *Journal of Animal Science* 85:2772-2781.

Vernon, R. G., and D. J. Flint. 1988. Lipid metabolism in farm animals. *Proceedings of the Nutrition Society* 47:287-293.

Wamsley, N. E., P. D. Burns, T. E. Engle, and R. M. Enns. 2005. Fish meal supplementation alters uterine prostaglandin $F_{2\alpha}$ synthesis in beef heifers with low luteal-phase progesterone. *Journal of Animal Science* 83:1832-1838.

Zinn, R. A., and A. P. Jorquera. 2007. Feed value of supplemental fats used in feedlot cattle diets. *Veterinary Clinics of North America: Food Animal Practice* 23:247-268.

6

Protein and Amino Acids

INTRODUCTION

Digestion of dietary proteins and other nitrogenous compounds in ruminants is complex because of pregastric fermentation. Ruminal microorganisms degrade some of the dietary nitrogen (N) using the products for their own metabolism including protein synthesis. Both dietary protein and microbial protein contribute to the protein entering the duodenum, which will supply amino acids (AA) to the host animal. Chapter 2 (Anatomy, Digestion and Nutrient Utilization) provides background on protein and N digestion. This chapter considers the feed, processes in the gastrointestinal tract related to predicting the supply of AA to the host, and use of absorbed AA for productive functions.

The current edition of the *Nutrient Requirements of Beef Cattle* uses the terms ruminally degradable protein (RDP) and ruminally undegradable protein (RUP) to distinguish between dietary protein that is hydrolyzed and is not hydrolyzed in the rumen, respectively. This is consistent with major journals including the *Journal of Animal Science*. The previous edition and update of the *Nutrient Requirements of Beef Cattle* (NRC, 1996, 2000) used the terms degradable intake protein (DIP) and undegradable intake protein (UIP) to distinguish these two fractions.

METABOLIZABLE PROTEIN SYSTEM

The previous edition and update of the *Nutrient Requirements of Beef Cattle* (NRC, 1996, 2000) expressed protein requirements in terms of absorbed protein, a rationale presented by the Subcommittee on Nitrogen Usage in Ruminants (NRC, 1985). The change from the crude protein (CP) system to the metabolizable protein (MP) system was adopted in the *Nutrient Requirements of Dairy Cattle* (NRC, 1989) and by the Agricultural and Food Research Council (AFRC, 1992). The MP system accounts for rumen degradation of dietary protein and separates requirements into the needs of microorganisms and the needs of the animal. Incorporating the complexities of ruminal digestion into the system to describe protein requirements was a major step forward. Meeting microbial requirements with RDP maximizes the supply of microbial protein to the host. Ruminally degradable protein provides peptides, AA, and ammonia for microbial metabolism. Sources of nonprotein nitrogen (NPN), such as urea or ammonia, which provide a source of N for ruminal microorganisms, are also included in RDP. The amount of RDP required is based on predicted microbial protein synthesis. Accurate predictions of microbial protein synthesis as well as the RDP component of feeds are needed to meet requirements using the MP system. The predicted amount of RUP, which includes protein that is not degraded in the rumen as well as protected AA included in the diet, is also needed to determine the MP supply. Sources of MP entering the duodenum include three fractions: microbial protein, RUP, and endogenous protein. The endogenous fraction does not make a net contribution to the MP supply. The MP supply is the true protein digested in the intestine and the corresponding absorbed AA (NRC, 2001). Ultimately, AA are required by the animal for maintenance and productive functions.

Metabolizable protein is digested, and AA are absorbed into the portal vein. The ratio of portal absorption to small intestinal apparent absorption of AA varies with the AA as a result of endogenous secretions added to the digestive flow, non-reabsorption of endogenous secretions, and gut oxidation of AA. After absorption, AA first flow to the liver where differential net AA removal occurs (Lapierre et al., 2006). These processes alter the AA profile available for net tissue gain, conceptus growth, or milk production.

Supply and requirements should be expressed on an AA basis; however, the complexity of metabolism in the reticulorumen makes this difficult. More specific information on the AA supply from ruminal microbes and RUP will allow formulation of diets in which the AA supply from RUP provides a complementary AA pattern to ruminal microbes, to meet AA requirements and maintain productivity at lower

N intakes (NI) (Broderick and Colombini, 2010). Decreasing excess NI will decrease excretion of N and lessen the environmental impact of beef production (see Chapter 16, Environment).

MICROBIAL PROTEIN SYNTHESIS

Microbial CP (MCP) synthesized in the rumen can supply from 50% (NRC, 1985; Spicer et al., 1986) to essentially all the MP required by beef cattle, depending on the RUP content of the diet. Therefore, prediction of MCP synthesis is an important component of the MP system. Various equations have been used to predict MCP synthesis. Burroughs et al. (1974) proposed that MCP synthesis averaged 13.05% of total digestible nutrients (TDN). In *Ruminant Nitrogen Usage* (NRC, 1985), two equations were developed to predict MCP synthesis—one for diets containing more than 40% forage and one for diets containing less than 40% forage. Both equations are more complex than that of Burroughs et al. (1974). The use of these equations to predict MCP was discussed in the previous edition and update (NRC, 1996, 2000). The average MCP value for the data set used to develop the >40% forage equation (NRC, 1985) was 12.8% of TDN intake. The NRC (1996, 2000) suggested that diets with both high and low TDN concentrations would likely have efficiencies of less than 0.13. To simplify the NRC (1985) system, 13% of TDN intake for synthesis efficiency was used for diets containing more than 40% forage. For diets containing less than 40% forage, the equation of Russell et al. (1992) was used to apply a 2.2% decrease in MCP synthesis for every 1% decrease in the dietary concentration of physically effective neutral detergent fiber (peNDF) less than 20% peNDF (NRC, 1996, 2000). Based on studies in which the RDP supply from native forages was estimated, Lardy et al. (2004) recommended an efficiency of 0.1 for low-quality forages.

The requirement for RDP is related to microbial yield. In NRC (1996, 2000), the requirement for RDP (including NPN) was considered equal to MCP synthesis. It was assumed that recycled N could make up the deficit in RDP. However, the efficiency with which ruminally available N is trapped is less than 100% due to the dynamic nature of the ruminal environment. Processes contributing to efficiency less than 100% include both absorption of ammonia from the rumen and passage of fluid containing RDP from the rumen. A maximum trapping efficiency of 0.9 was suggested by NRC (1985) and adopted by the dairy NRC (1989). If N recycling is zero, based on an efficiency of 0.9, the net RDP required would be 1.11 × microbial N. The dairy NRC (2001) adopted a slightly lower efficiency value based on evaluation of a literature data set. The dairy estimate (NRC, 2001) of MCP assumes an efficiency of 85% for converting RDP to MCP, and therefore the requirement for RDP is 1.18 × microbial N. When RDP supply is less than 1.18 times the predicted MCP synthesis, MCP synthesis is calculated as the RDP supply multiplied by 0.85. While recognizing that the efficiency of converting RDP to MCP is less than 100%, there is uncertainty relative to the appropriate efficiency factor. Therefore the current Beef Cattle Nutrient Requirements Model (BCNRM; Chapter 19, Model Equations and Sensitivity Analyses) retains the concept that the requirement for RDP is equal to MCP synthesis adopted by the NRC (1996, 2000). In the mechanistic level of solution (MLS) of the BCNRM, the user has the option to change this value (see Chapter 19).

Many factors affect the efficiency of MCP synthesis (NRC, 1985; Russell et al., 1992). Russell et al. (1992) demonstrated the need for AA and peptides for optimal MCP synthesis. A lack of AA or peptides is unlikely to be a problem in most typical diets for beef cattle. Adequate RDP in finishing diets can be accomplished by adding urea (Sindt et al., 1993, 1994). Fiber-digesting bacteria use either ammonia or amino acids as a N source although the proportion of microbial N derived from ammonia is greater than that derived from AA (Atasoglu et al., 2001). Therefore, AA/peptides should not be limiting in the rumen; however, these fiber-digesting bacteria may require branched-chain volatile fatty acids, such as 2-methylbutyrate and isovalerate (Russell and Sniffen, 1984; NRC, 1985), which would be supplied by AA degradation. A need for RDP (other than NPN) might occur in diets containing mixtures of forage and grain such as "step-up" diets for finishing cattle (Sindt et al., 1993). The type of carbohydrate (structural vs. nonstructural) can also affect microbial maintenance requirements because of differences in rates of fermentation (microbial growth rate), rates of passage, and ruminal pH. Level of intake, as it changes rate of passage, and pH are also important. Lipids provide little, if any, energy for ruminal microorganisms, and the energy obtained from protein fermentation is minimal (Nocek and Russell, 1988). Grass silages provide less energy for ruminal fermentation than hay because of fermentation of nonstructural carbohydrates by microorganisms during silage fermentation. As a result, microbial production in the rumen is lower when cattle are fed grass silages rather than hay (Titgemeyer and Löest, 2001).

Total digestible nutrient intake (TDNI) has been used to determine microbial yield despite the fact that most ruminal bacteria are unable to use protein or ether-extractable components as an energy source (Russell et al., 1992). The AFRC (1992) observed that total tract digestible organic matter (OM) intake was the most precise indicator of MCP synthesis when NI was adequate. Digestible OM and TDN are roughly equivalent in feedstuffs and diets. The Commonwealth Scientific and Industrial Research Organisation (CSIRO, 2007) suggested energy supply for microbial synthesis was more usefully expressed as fermentable OM.

Recently, the NRC (1996, 2000) model was used to compare predicted values for duodenal flow of MCP with measured duodenal flow from 118 high-concentrate diets (Owens et al., 2014). Duodenal flow of microbial protein increased with TDNI or dry matter intake (DMI). The slope for microbial protein yield was 0.091 g/g TDNI. The use of TDNI accounted for 49% and the use of DMI accounted for 48% of the variation in MCP yield. There was no relationship between the percentage of OM digested in the rumen and the TDN concentration across all diets. In addition, there was also no relationship between MCP yield and TDN concentration of the diets. The amount of OM fermented in the rumen that yields energy for use by the microbes should be closely related to microbial yield. If OM disappearance in the rumen limits microbial growth, then ruminal OM disappearance should be used to predict microbial yield (Owens et al., 2014).

Despite widespread application of the NRC (1996, 2000, 2001) MP system, the validity of predicting MCP synthesis from TDNI using the factor of 0.13 and the peNDF adjustment has neither been extensively assessed experimentally nor evaluated by conducting analyses of published literature. Therefore, published data were used to evaluate the NRC (1996, 2000) method of predicting MCP synthesis as well as to develop and evaluate new equations based on intakes of dietary TDN and CP (Galyean and Tedeschi, 2014).

The data used represented 285 treatment means from 66 papers published in the *Journal of Animal Science* from 1983 to 2013. Selected papers represented a wide range in dietary TDN (1.37 to 12.25 kg/d) and CP (0.084 to 2.43 kg CP/d) intakes, and ether extract (EE) concentrations (1 to 11%). Potential independent variables for prediction of MCP synthesis were evaluated using the full database. After removing variation associated with study, data from the literature database were used to derive equations including TDNI, fat-free TDNI (FFTDNI), and CP intake (CPI) as independent variables. Mixed model regression procedures were used to fit the independent variable models.

There was a linear relationship between MCP yield (g/d), adjusted for study variation, and true ruminal OM digested (TROMD, g/d), with a slope of 0.082. Both TDNI and FFTDNI were strongly related to TROMD. The slopes for the regression of MCP synthesis on TDNI and FFTDNI were substantially less than the value of 0.13 recommended by the beef NRC (1996, 2000) and dairy NRC (2001); however, the inclusion of an intercept in the regression equations developed by Galyean and Tedeschi (2014) prevents direct comparison of the slope of TDNI (0.087) and FFTDNI (0.096) with the efficiency (0.13) developed by NRC (1996), which had the intercept set to zero. The NRC (1985) reported an intercept and slope of approximately 105 g/d and 0.05 g of MCP/g of TDNI, respectively.

The database was randomly divided into the Development and the Evaluation data sets at the citation level to evaluate the derived equations. The Evaluation data set was then used to test the validity of the equations developed from the Development data set. Using the equations developed, the r^2 for observed vs. predicted MCP synthesis ranged from 0.46 to 0.56. Inclusion of more than one independent variable only slightly improved the r^2.

None of the equations derived from the Development data set showed evidence of mean or linear bias, suggesting that although precision was less than desirable, accuracy was reasonably good. In contrast to the newly developed equations, the NRC method (1996, 2000) with or without adjustment for peNDF, showed significant mean and linear biases with a high degree of overprediction. Although there was no bias, standard errors of prediction averaged approximately 25 to 30% of mean values, indicating the need for further research to develop more accurate equations.

Based on the results of the methods used to assess the adequacy of the predictive equations, those based on TDNI, FFTNDI, and CPI are recommended for the prediction of MCP synthesis by beef cattle. These independent variables are readily available from feed composition databases or via analytical measurements on beef cattle diets. Practically, the simple linear regression equations based on TDNI and FFTDNI might have the most utility, as the additional precision and accuracy gained by adding CPI and quadratic terms for either TDNI or FFTDNI to these simple, one-variable models were relatively small. Between the TDNI- and FFTDNI-based equations, there seems to be an advantage to using FFTDNI, suggesting that when dietary EE values are known, correction of the TDN value is desirable.

Two equations are used to predict MCP in the BCNRM. For diets containing average amounts (<3.9%) of EE in the data set of Galyean and Tedeschi (2014), the equation in which TDNI predicts MCP is used (Eq. 6-1). For diets with EE greater than the average in the data set (≥3.9%), the equation based on FFTNDI is used (Eq. 6-2). Predicted MCP is equal to the required RDP supply. For users preferring to enter an efficiency value for MCP synthesis rather than use the BCNRM equations, that option is included in the BCNRM. This option would also allow users to test different scenarios. The default option is the BCNRM equations:

$$MCP = 0.087\ TDNI + 42.73, \qquad \text{(Eq. 6-1)}$$

$$MCP = 0.096\ FFTDNI + 53.33, \qquad \text{(Eq. 6-2)}$$

where MCP is microbial CP, g/d; TDNI is TDN intake, g/d; and FFTDNI is fat-free TDNI, g/d.

Comparison of values generated using Eq. 6-1 with those of the previous NRC (1996, 2000) reports showed similar prediction for MCP synthesis for a low-quality forage diet and a feedlot diet (Table 6-1). The value of 0.10 recommended by Lardy et al. (2004) for microbial efficiency was used for the low-quality forage diet, and the peNDF adjustment recommended by the NRC (1996, 2000) was used for

TABLE 6-1 Comparisons of Values Generated Using the New Equations to Predict MCP with Values Generated from the Previous NRC (1996, 2000) Equation[a]

Item	Low-Quality Forage (Cow)	Corn Silage-Based (Grower)	Steam-Flaked Corn-Based (Finishing)
Body weight, kg	500	250	450
DMI, kg/d	11.6	6.2	10.2
TDN, %	57.8	75.6	87.9
CP, %	12	13	13.5
peNDF, %	57.2	23.7	8.0
Predicted MCP synthesis BCNRM, g/d[b]	626	450	822
Predicted MCP synthesis NRC 1996/2000, g/d[c]	671	609	816

[a]Adapted from Galyean and Tedeschi (2014).

[b]Beef Cattle Nutrient Requirements Model, Equation 6-1.

[c]NRC (1996, 2000); Microbial efficiency, g MCP/g TDNI, for predicted MCP synthesis was 0.10 for the low-quality forage based on previous evaluation of the NRC (1996, 2000) model for beef cows (Lardy et al., 2004); 0.13 for the grower diet; and 0.091 for the feedlot diet accounting for peNDF.

the feedlot diet. In contrast, with a grower-phase diet, the BCNRM predicted an MCP of 450 g/d whereas the NRC (1996, 2000) predicted an MCP of 609 g/d. The equations used in the BCNRM show little evidence of bias, whereas those of the previous model (NRC, 1996, 2000) over-predicted MCP synthesis for a majority of the observations in the data set. Therefore, the current equations are recommended. However, the large standard errors of prediction emphasize the fact that microbial growth is measured with high variability using presently available techniques and depends on many other variables that cannot be accounted for by simple empirical equations, thereby highlighting the need for further research and data gathering to advance our ability to predict this important component of the MP system.

UREA RECYCLING

Ammonia absorption by the portal-drained viscera (PDV) is significant. Net ammonia absorption was 42% of dietary NI across 304 measurements in growing and lactating cattle (Firkins and Reynolds, 2005). Ammonia absorption by the mesenteric-drained viscera (MDV; small and large intestines) was 28 to 53% of total PDV ammonia absorption. The absorbed ammonia is extracted by the liver and most is converted to urea. Urea released by the liver is excreted in urine or transferred to the gastrointestinal tract (GIT). Transfer to the GIT is via saliva and across the epithelium of the rumen and other sections of the GIT. Urea N that is transferred to the gastrointestinal tract is hydrolyzed by microbial urease and provides ammonia that can supplement the RDP supply in the rumen. Urea N production increases with increased NI and accounted for 70% of NI across a series of studies, with NI ranging from approximately 60 to 200 g N/d, in growing cattle (Reynolds and Kristensen, 2008). Feeding excess N results in increased ammonia absorption, increased urea production, and increased urea excretion in urine (Reynolds and Kristensen, 2008).

Increasing the dietary N-to-fermentable carbohydrate ratio by increasing dietary N in diets of similar metabolizable energy (ME) and neutral detergent fiber (NDF) contents resulted in increased plasma and salivary urea N concentrations, and increased urinary N excretion (Marini and Van Amburgh, 2003). Estimates of urea transferred to the GIT in saliva were from 4 to 20% of gut entry and increased with increased dietary N (therefore, with increased plasma urea N). Urea entry rate (urea production, g N/d) and urinary urea N excretion (g N/d) increased linearly with NI (g N/d), but the quantity (g N/d) of urea recycled to the GIT was similar among diets. The percentage of urea entry that was recycled to the GIT decreased as dietary N increased and was 83.3% on the low-N diet, decreasing to 29% on the high-N diet (Marini and Van Amburgh, 2003).

Changing the dietary concentrate-to-forage ratio in isocaloric and isonitrogenous diets, using cracked corn and urea as the N source in the concentrate, resulted in increased plasma urea concentration and liver urea production with increased concentrate in the diet from approximately 25% up to 55% concentrate. The combination of limiting feed intake and using urea as the N source in the concentrate may have resulted in concentrations of ruminal ammonia that exceeded microbial needs (Huntington et al., 1996). At 20% or less dietary concentrate and low plasma urea concentration, 90% of urea produced was recycled to the GIT; this decreased to recycling 50% of urea produced to the GIT at 90% dietary concentrate. At dietary concentrate less than 20%, transfer of urea to MDV was 0. Transfer of urea to MDV increased to 9% of PDV transfer above 63% concentrate. Similar to the situation of increasing dietary NI, urea excretion in the urine increased with increased plasma urea concentration.

Urea transferred to the GIT is hydrolyzed to produce ammonia. The ammonia is used for microbial metabolism, synthesis of MCP and other products, reabsorbed as ammonia and recycled to the ornithine cycle in the liver, or excreted in the feces. The dual-labeled urea technique can be used to estimate urea kinetics as described by Lobley et al. (2000). This approach includes estimating urea transferred to the GIT through blood and saliva. Using this approach, ammonia produced from hydrolysis of urea transferred to the GIT is partitioned into ammonia returned to the ornithine cycle, excreted in feces, or used for anabolism. The fraction used for anabolism includes urea N used by the animal via microbial and mammalian metabolites of ammonia. Across a series of studies, the percentage of urea N transferred to the GIT that was reabsorbed as ammonia N varied from 20 to 50% and, when averaged across dietary N concentrations, was about 34%. Across dietary N concentrations, the percentage used for anabolic purposes averaged 54% (Reynolds and Kristensen, 2008).

At both low dietary N and low concentrate-to-forage ratio, plasma urea N concentration was low and a greater percentage of liver urea N production was recycled to the GIT. Nonetheless, the absolute transfer showed little change, suggesting that concentration is not the primary factor determining transfer of blood urea N to the GIT. Decreased dietary N has not resulted in increased urea recycling (i.e., low arterial urea N is not followed by increased PDV urea N uptake, g N/d); however, at low arterial urea N, PDV extraction of urea N increased from 1 to 7%. If flux remains the same with a decreasing arterial N supply, then extraction must increase. Gut permeability to urea could be increased when cattle are fed low-N diets. This may be a result of transport proteins; however, correlation between dietary NI and upregulation of the ruminal urea transporter (UT-B) has not been demonstrated (Calsamiglia et al., 2010).

Urea N produced by the liver and recycled to the GIT provides a source of N for microbial growth and also conserves N. The extent of urea N recycled to the GIT varies greatly from 20 to over 90% of urea N production. Recycling a higher percentage of urea N produced is associated with low intakes of dietary N. Factors that are related to urea N recycling include ruminal ammonia concentration, digestion of OM in the rumen, and plasma urea concentration (Kennedy and Milligan, 1980). The equation developed by the NRC (1985), and adopted by the NRC (1996, 2000), estimated the quantity of urea N recycled to the rumen based on relationships between ruminal ammonia and plasma urea N concentrations. In subsequent studies in growing beef cattle, urea N fluxes were measured using the dual-labeled urea isotopic approach. The flux data provide new information to predict both N recycled to the GIT and that component of recycled N that is used for anabolism. Urea N used for anabolism reflects retention of microbial products of ammonia, including amino acids, as well as anabolic reactions of ammonia within the body (Lobley et al., 2000).

Regression analysis of literature data based on the dual-labeled urea isotopic approach was used to derive equations to predict urea N used for anabolism (Eisemann and Tedeschi, in press). Data were obtained from nine studies (Archibeque et al., 2001, 2002; Marini and Van Amburgh, 2003; Wickersham et al., 2008a,b, 2009; Huntington et al., 2009; Brake et al., 2010; Bailey et al., 2012) in which the dual-labeled urea technique was used in growing cattle to estimate urea kinetics as described by Lobley et al. (2000). Treatments included in the database had a range in body weight (BW) of 204 to 366 kg, DMI of 4.10 to 8.86 kg/d, NI of 38.5 to 204 g N/d, and CP of 4.7 to 21.4% of DM. Most of the diets were predominantly forage-based, but two experiments included several high-concentrate diets. The basic relationships assumed to predict N metabolism were

$$\text{UER} = \text{UUE} + \text{GER} \qquad \text{(Eq. 6-3)}$$

and

$$\text{GER} = \text{ROC} + \text{UFE} + \text{UUA}, \qquad \text{(Eq. 6-4)}$$

where UER is urea N entry rate (urea N produced primarily in the liver), g N/d; UUE is urinary urea N excretion, g N/d; GER is gastrointestinal urea N entry rate, g N/d; ROC is urea N recycled to the ornithine cycle, g N/d; UFE is N from urea excreted in the feces, g N/d; and UUA is urea N used for anabolism, g N/d (Lobley et al., 2000).

The UER was linearly related to NI. The meta-regression analysis, after removing study variation, indicated that for each increment in NI, about 74.5% appeared as UER. This is similar to the value of 70% observed previously without adjusting for study variation (Reynolds and Kristensen, 2008). Urea potentially available for anabolism was estimated using the relationship between UUA:UER and NI, g N/d, or dietary CP%. The general pattern of the relationships of UUA:UER vs. NI or UUA:UER vs. CP (% DM) is that the fraction of UER used for anabolism decreases linearly as NI, g N/d, or CP (% DM) increases. These equations indicated, however, that the ratio of UUA to UER would become negative for animals consuming more than approximately 230 g of N/d or 25% CP (% DM). Although these values are extreme for the current database and practical feeding conditions, to prevent negative predictions of UUA:UER, nonlinear functions were fitted assuming the exponential decay form, as it does not yield negative values.

Based on the relationship between UUA:UER and CP, the equation was rearranged to predict UUA, and a decay form was used to prevent negative values:

$$\text{UUA}_{\text{CP}} = [-0.1113 + 0.996 \times e^{(-0.0616 \times \text{CP})}] \times [0.745 \times \text{NI} - 11.98], \qquad \text{(Eq. 6-5)}$$

where UUA is urea N used for anabolism, g N/d; *e* is the base of the natural (Naperian) logarithm (i.e., 2.718); CP is crude protein, % DM; and NI is nitrogen intake, g N/d.

Figure 6-1 depicts the relationship between N recycled to the rumen or urea-N used for anabolism (g N/d) and NI at DMI = 5 kg/d and 10 kg/d. The equation developed in this edition of the *Nutrient Requirements of Beef Cattle* for estimating UUA is compared with the equation of Reynolds and Kristensen (2008), which also estimates UUA, and the

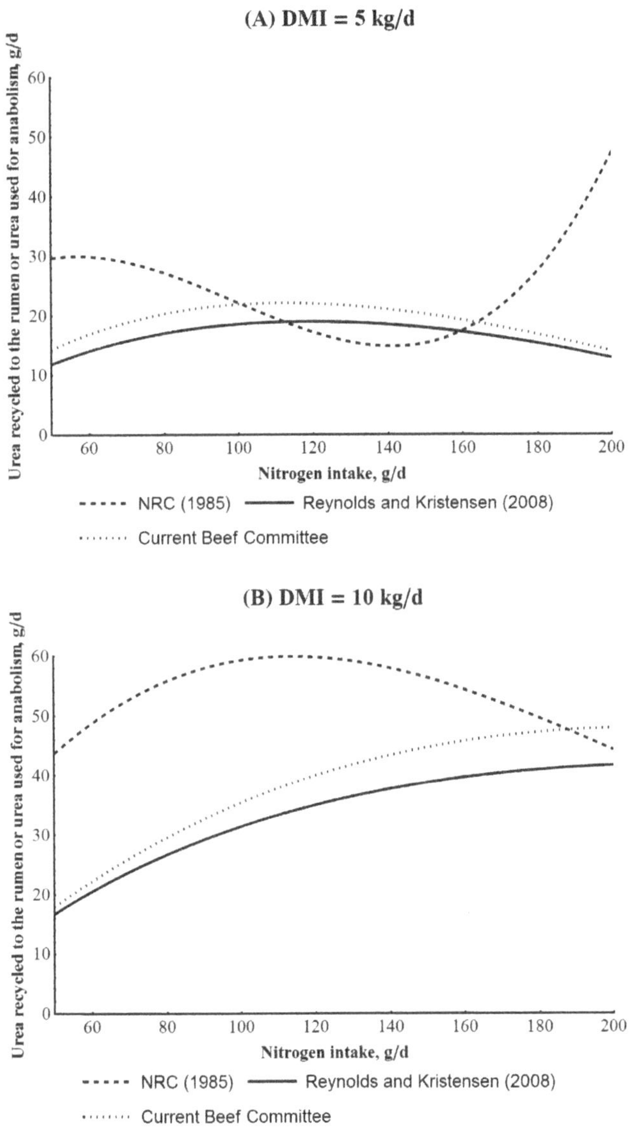

FIGURE 6-1 Relationship between N recycled to the rumen or urea N used for anabolism (g N/d) and nitrogen intake (g N/d) using the equations developed by the NRC (1985) (urea N recycled to the rumen, dashed lines), Reynolds and Kristensen (2008) (urea N used for anabolism, UUA, solid lines), and the current beef committee (UUA, dotted lines) for (A) DMI = 5 kg/d and (B) DMI = 10 kg/d.

previous NRC (1985) equation for recycled N. The current equation and that of Reynolds and Kristensen (2008) have similar results and show quadratic behavior. The NRC (1985) equation predicts greater recycled N at the lower and upper end of the NI range compared to predictions of UUA by the other two equations for the low-DMI scenario and greater recycled N throughout most of the NI range compared to the other two equations for the high-DMI scenario.

The isotopic approach determines the anabolic fate of urea N by difference. The estimate of UUA includes N from ammonia incorporated into amino acids of microbial protein that are digested, absorbed, and incorporated into body protein and the anabolic reactions of ammonia in the body. Errors associated with this method are likely to result in overestimation of the urea N used for anabolism (Lobley et al., 2000; Zuur et al., 2000). Our data set contains most of the available information for growing cattle consuming high-forage diets. The effect of high-concentrate diets on urea N kinetics and therefore the estimate of UUA is not certain. Implants or β-adrenergic agonists change postabsorptive use of N, increase N deposition, and may alter UUA (Titgemeyer et al., 2012). The relationships among components of urea N flux in other physiological states could also differ relative to growing animals.

The equation might be most applicable to growing cattle consuming forage-based diets, and further research is needed to verify the utility of the equation in other situations. This analysis confirms the significance of N recycling to the GIT, particularly in situations where low-protein diets are fed, to provide a source of N for microbial metabolism. More complex modeling that integrates the total RDP supply with absorption of ammonia from the rumen is needed to include an allowance for urea recycling in the model. Calculations using all three of the equations shown in Figure 6-1 are included as options in the MLS of the BCNRM (see Chapter 19). Further research is needed to improve our ability to estimate the components of urea recycling.

SYNCHRONY OF ENERGY AND PROTEIN

Optimal use of RDP (including NPN) would logically occur if protein and carbohydrate degradation in the rumen were occurring simultaneously, but this is not the case in many diets. Protein degradation of many of the forages, for example, is rapid, and degradation of energy-yielding components of NDF is much slower. With grains (e.g., corn and sorghum), the opposite is true—slow protein degradation and rapid starch degradation. This results in low ruminal ammonia concentrations from high-grain diets after feeding and high concentrations from forage diets, which is influenced by dietary CP concentrations.

The ruminant compensates for asynchrony of carbohydrate and protein degradation by recycling nitrogen. An excellent example of this is how cow performance is similar with protein supplementation either three times per week or once per day (Beaty et al., 1994). More basic studies with animals (Henning et al., 1993; Rihani et al., 1993) suggest little or no advantage to synchrony of energy availability and protein breakdown. Cattle also compensate by eating numerous meals per day such as in the feedlot. Several reviews support the concept that the value of nutrient synchrony in theory is not supported by experimental results (Cole and Todd, 2008; Hall and Huntington, 2008; Reynolds and Kristensen, 2008).

RUMINAL OUTFLOW OF INTACT DIETARY PROTEIN

The previous edition and update of the *Nutrient Requirements of Beef Cattle* (1996, 2000) listed values for RUP in individual feeds. The sum of RDP and RUP is equal to CP. These values were derived from several approaches including laboratory measurements of solubility, in situ incubations, in vivo measurements, and in vitro measurements for specific feeds (Tedeschi et al., 2005; Owens et al., 2014). Some of the values were those reported in the *Nutrient Requirements of Dairy Cattle* (NRC, 1989).

The extent of protein degradation in the rumen varies with a number of factors, including structural aspects of the protein, the ruminal dilution rate, forage:concentrate ratio in the diet, ruminal pH, nutrient interactions, and feed processing (NRC, 1985; Bach et al., 2005). As described in Chapter 2 and in previous reviews (NRC, 1985, 2001; Sniffen et al., 1992), feed CP is divided into three fractions based on rate of degradation. Fraction A (NPN) is degraded rapidly, fraction B (true protein) is potentially degradable, and fraction C is mostly unavailable protein (acid detergent insoluble crude protein) that is degraded very slowly or functionally not at all. The portion of fraction B that is degraded will depend on both the fractional degradation rate of its components and rate of passage, which are competing processes. Greater time in the rumen will increase degradation of this fraction.

There is no primary standard to validate estimates of RDP and RUP for individual feeds. Instead, methods based on different approaches can be compared for predicting RDP and RUP (Schwab et al., 2003). The Level 2 solution of the NRC (1996, 2000) model used protein fractions defined from in vitro chemical assays and digestion rates. The B fraction was subdivided into a B1, B2, and B3 to distinguish varying rates of degradation within the B fraction. These values were based on the Cornell Net Carbohydrate and Protein System (CNCPS) model (Sniffen et al., 1992). The *Nutrient Requirements of Dairy Cattle* (NRC, 2001) used values derived from the in situ procedure in which the A, B, and C protein fractions were quantified using in vitro enzymatic assays. There was a single estimate of rate of digestion (kd) for the B fraction.

The two models, NRC (2001) and CNCPS (version 4.0; Fox et al., 2004), were compared for predicted flows of RUP and total nonammonia and nonmicrobial N using diets from published studies that measured this flow. Total nonammonia and nonmicrobial N would include RUP and endogenous protein (Schwab et al., 2003). The two models predicted similar estimates for RUP flow to the duodenum. However, NRC (2001) predicted greater total nonammonia and nonmicrobial N flows than CNCPS because NRC (2001) includes an estimate for endogenous protein flow in addition to RUP.

The CNCPS model (Fox et al., 2004) was used to predict RUP values for individual feeds at different levels of DMI (Tedeschi et al., 2005). Predicted values were compared with tabular values for RUP in the previous reports (NRC, 1996, 2000). Estimates for forages were not affected by level of intake, and the overall r^2 relating tabular and predicted RUP values was low. In contrast, there was a stronger relationship between tabular and predicted values for concentrates. At lower intake levels, tabular values were greater than predicted values; however, at three times maintenance intake, there was no bias in the prediction. The tabular RUP values for concentrates were based on NRC (1989) and were developed using data from animals fed at three times maintenance intake (Tedeschi et al., 2005).

Based on data from 118 diets, a weak relationship was observed between measured RUP, calculated as the difference between total duodenal flow of N and the sum of microbial N and ammonia N, and predicted RUP using DMI, diet composition, and tabular RUP for feed components provided by the NRC (1996, 2000). Measured RUP tended to increase as predicted RUP increased. Measured RUP was also greater than predicted RUP. Part of the reason for the lack of agreement could be the inclusion of N from endogenous secretions in measured duodenal flows. Expressing the measured RUP values per unit of DMI resulted in a weaker relationship between measured and predicted RUP, which could reflect shorter ruminal residence time with higher feed intakes and decreased degradation of dietary protein (Owens et al., 2014). Using data from a series of experiments in dairy cattle, a similar observation of inadequate prediction was made (Bateman et al., 2005) between predicted RUP flow using tabular RUP values from the NRC (1985, 1989, 1996, 2000) and measured nonammonia and nonmicrobial N flow to the small intestine. Equations that adjusted tabular RUP values for increased ME intake above maintenance improved the accuracy of prediction (Bateman et al., 2005) supporting the effect of DMI on the apparent RUP value of feeds. In an evaluation of byproduct feeds (Winterholler et al., 2009), the estimate of RDP for cottonseed meal was greater when used to supplement a low-quality forage diet than the RDP values reported in the previous edition and update of the *Nutrient Requirements of Beef Cattle* (1996, 2001). The predicted passage rate (kp) was 0.05/h (NRC, 2001) and measured kp was 0.025/h. The difference in the estimated values for RDP is consistent with greater time in the rumen resulting in greater values for RDP. Passage rate of fluids is greater than kp of solids, and kp of both fractions increases as DMI increases (Broderick et al., 1991).

As suggested by Owens et al. (2014), a standardized method to estimate RUP is needed. Methodologies and considerations for determining forage protein degradation and estimating RUP for forages have been described (Klopfenstein et al., 2001; Haugen et al., 2006b; Edmunds et al., 2012; Buckner et al., 2013). Variation in the estimate of RUP of forages with rate of passage was addressed. Even if a standardized method is accepted, the RUP value of a given feed is not constant, and therefore predictions are needed to estimate RUP under specific conditions.

Tabular values were a good first step in estimating RUP for use in MP systems; however, tabular values relate to the situation in which they were determined and are not constants for a feed ingredient. The actual RUP is affected by the protein fractions and time in the rumen, particularly for the B fraction. Including a specific kp in the prediction of degradability would provide the adjustment for varying situations. The Feed Evaluation Unit of the UK Agricultural Development and Advisory Service (cited in CSIRO, 2007) published values from in situ measurements at several values of kp for a variety of feeds. For example, kp = 0.02 was used for cattle and sheep given completely ground diets or a very low level of feeding of a mixed diet (maintenance), and kp = 0.05 was used for calves, low-yielding dairy cows, beef cattle, and sheep given a high level of mixed diets.

PROTEIN SUPPLY TO THE SMALL INTESTINE

Protein flowing to the small intestine includes N from both exogenous and endogenous sources. Exogenous sources are from the diet and include both RDP and RUP. Endogenous nitrogen (EN) is added to the gastrointestinal tract in various forms, mainly protein and urea. Urea enters via saliva and is transferred across the gut wall. Proteins enter through a combination of secretions and sloughed cells (Lapierre et al., 2008). In the rumen, EN from both protein and urea contribute to microbial N in addition to RDP. Therefore, EN from protein entering the duodenum is free as well as a component of microbial N (Ouellet et al., 2002).

Endogenous N flow is difficult to measure (NRC, 2001; Ouellet et al., 2002; Marini et al., 2008). In other species, EN has been divided into basal and total EN (Stein et al., 2007). Basal secretions are related to DMI and are estimated by several techniques including feeding a protein-free diet. Total endogenous secretions include basal secretions plus secretions resulting from specific dietary ingredients and are estimated using isotopic techniques. Various techniques have been used to estimate endogenous flow in ruminants. For example, estimates of basal endogenous secretions were made by measuring nonammonia N flow from the rumen and abomasum using steers totally nourished with intragastric infusion of volatile fatty acids (Ørskov et al., 1986).

Infusion of ^{15}N-leucine to label body proteins can be used to estimate total endogenous flow and to model the contribution of urea and proteins to microbial N flow. Estimates of the contribution of nonurea EN to duodenal N flow suggest that it can be 15% (Ouellet et al., 2002) to 20% (Lapierre et al., 2006) of the total N flow in lactating dairy cows. Estimates of total EN were 3.4 to 6.0 g N/kg DMI and of free endogenous N were 1.4 to 2.4 g N/kg DMI (Ouellet et al., 2002; Lapierre et al., 2008). Values for free EN are similar to the value of 1.9 g of N/kg DMI, adopted by the NRC (2001) as an estimate of endogenous N flow.

Endogenous N was estimated in sections of the GIT using meta-analysis and a regression approach (Marini et al., 2008). The database included 65 studies with growing-finishing cattle and 42 studies with lactating cattle. Studies in growing-finishing cattle included a range in cattle weight and diets. Estimated EN in duodenal contents was 2.9 g N/kg DMI for free EN and 5.1 g N/kg DMI for bacterial N (Marini et al., 2008). Assuming protein and urea each contribute about half of the EN used by bacteria (Ouellet et al., 2002), then bacterial N from endogenous protein was 2.5 g N/kg DMI, and total nonurea EN flow was 5.4 g N/kg DMI. Net protein (NP) and AA supply is represented by RUP and MCP derived from RDP and urea. Crude protein from endogenous protein sources arriving at the duodenum does not represent a net increase in protein supply. Underestimating EN flow at the duodenum leads to an overestimation of net MP and AA supply (Lapierre et al., 2008).

Nonurea EN is a significant portion of N flow to the duodenum. Free EN ranges from 1.4 to 2.9 g N/kg DMI, and total endogenous flow ranges from 3.4 to 6.0 g N/kg DMI. Although much of this endogenous fraction is digested and absorbed in the small intestine, it does not contribute a net increase in MP supply. Endogenous secretions not absorbed in the small intestine contribute to endogenous losses in feces. Removing the nonurea endogenous fraction from measured duodenal flow would allow estimation of the actual net supply of MP. Not considering the contribution of EN, when determining the MP supply, results in overestimation of MP supply.

AMINO ACID ABSORPTION

Metabolizable protein is defined as the true protein digested in the intestine, supplied by microbial protein and RUP. True protein digested is equivalent to absorbed AA. Digestibility of protein is important for both MCP and RUP.

The NRC (1996, 2000) assumed that all RUP was 80% digestible. This reflected the lack of information on the digestibility of RUP to assign different values. The *Nutrient Requirements of Dairy Cattle* (NRC, 2001) includes variable estimates of RUP digestibility for different feed ingredients. Estimates were based on the mobile bag technique or in vitro procedures. Valid estimates of RUP digestibility are needed to predict MP supply.

Erasmus et al. (1994) showed the differences in digestibility of RUP across feed ingredients. Intestinal digestibility of RUP varied from 37.8% (hay) to 98% (soybean meal). Within feedstuffs, rumen degradation had a greater influence on postruminal provision of specific AA than postruminal digestion. Different proteins have differing AA composition, and therefore, one would expect the AA content of RUP and RDP to differ. The database of the AA content of the RUP fraction is limited, and research is needed to improve our knowledge in this area.

Microbial Crude Protein

The NRC (1996, 2000) used the value of 80% digestibility of microbial true protein (MTP; NRC, 1985). The NRC (1985, 1996, 2000) used 0.8 MCP = MTP because MCP contains approximately 20% nucleic acids. Considering both factors, digestibility of MTP and the AA content of ruminal bacteria, the MP from MCP equals 0.64 times MCP. The current committee maintained the use of these factors.

Forages

The structure of the forage cell is an important factor in the degradability of protein in the rumen and the digestibility of RUP in the intestine. Ruminal degradation of protein in leaves can result in a greater portion of cell wall N in RUP. Digestible RUP of forages has been estimated by correcting RUP for acid detergent insoluble protein (ADIP; Negi et al. 1988; NRC, 1996). Crude protein content of forages is greater in the vegetative state and decreases as forages mature (NRC, 1996, 2000, 2001). The RUP content of forages is relatively low: approximately 10 to 40% of CP (NRC, 1996, 2000, 2001).

Average RUP estimated for a diverse group of forages was 20% or less, while for warm-season grasses, the estimated average was approximately 39% (Buckner et al., 2013). Ruminally undegradable protein as a percentage of CP was lowest early in the growing season and increased as plants matured. Digestibility of RUP, based on the mobile bag technique, decreased as maturity increased. Across forage species and growing season, RUP digestibility ranged from 10 to 60%. Heat treatment of alfalfa increased RUP as a fraction of total CP and increased digestibility of RUP from 15 to 46% in freeze-dried alfalfa. Although ADIP is used to estimate unavailable N caused by heating, some of this N can be digestible (Haugen et al., 2006a). These results and others (Kononoff et al., 2007) suggest digestibility of RUP of forages is much lower than 80% (NRC, 1996, 2000) and much lower than tabular NRC (2001) values, which range from 60 to 75% for grasses and hays (Buckner et al., 2013).

Incubation time in the rumen is an important factor in estimation of both RUP and RUP digestibility. Shorter retention time will lead to estimation of a greater value for RUP. The overestimate of RUP leads to an overestimate of RUP digestibility and MP supply.

Concentrates and Byproduct Feeds

Digestibility of RUP in corn and several protein supplements was greater than that of forages and averaged 70 to 98% (Kononoff et al., 2007). Digestibility of RUP in the small intestine for several protein supplements varied from 55 to 80%. Amino acid digestibility in feedstuffs also varied (Titgemeyer et al., 1989). Feed ingredients investigated, with the exception of soybean meal, differed in these two publications. The approach used to estimate digestibility also differed.

Amino acids from proteins digested in the small intestine must be absorbed and also made available for productive function. This is a potential problem with heat-damaged feeds. Significant amounts of ADIP in heat-damaged corn gluten meal and distillers grains were digestible but the absorbed N from the heat-damaged proteins did not produce the same response in gain by lambs and cattle as the control feed ingredients (Nakamura et al., 1994). Decreased available lysine in the heat-damaged protein might explain the lower gain (Nakamura et al., 1994).

As pointed out earlier, the NRC (1996, 2000) assumed that all RUP was 80% digestible. The primary concern with digestibility values is for forages because of the number of studies that report intestinal digestibility less than 80%. An overestimate of RUP contributes to the problem of overestimating intestinal digestibility. Estimates of RUP digestibility of forages are variable but most estimates are less than 60%. The committee recommends that the RUP digestibility of forages be decreased to 60%.

METABOLIC PROTEIN REQUIREMENTS

The requirements of the NRC (1984, 1985) for MP were based on the factorial method. Factors included were metabolic fecal losses, urinary losses, scurf losses, growth, fetal growth, and lactation. Metabolic fecal, urinary, and scurf losses represent the requirement needed for maintenance. The previous NRC subcommittee (1996, 2000) estimated maintenance MP requirement based on metabolic body weight ($BW^{0.75}$). The factorial approach was used for maintenance, growth, fetal growth, and lactation. The use of this approach is continued by the current committee. The specific MP requirements are described in Chapter 11 (Maintenance), Chapter 12 (Growth), and Chapter 13 (Reproduction) of this publication.

Nitrogen Accretion

Nitrogen accumulation in body tissues can be measured directly using the comparative slaughter technique, which provides a single value over a relatively large weight range. Another approach is to measure N accumulation indirectly through N balance experiments, which are conducted over a shorter time frame. Conducting N balance experiments includes measurement of NI and all losses of N from the body. In growing animals, N is quantitated in feed, orts, urine, and feces. Nitrogen retained is determined by difference. In general, NI is overestimated and losses are underestimated, resulting in overestimating N balance. Several publications reviewed potential reasons for overestimating N balance (Owen, 1967; MacRae et al., 1993; Spanghero and Kowalski, 1997).

There are several assumptions underlying the N balance approach including that samples of feed, orts, and excreta are appropriate representative samples. Urine is collected under acidifying conditions to prevent microbial growth and volatilization of ammonia (Owen, 1967). Inadequate acidification of samples leads to ammonia loss (Spanghero and Kowalski, 1997), and at the same time, acidification of the urine can result in the formation of precipitates containing N (Owen, 1967). There are a number of nitrogenous compounds in the urine in addition to urea (Bristow et al., 1992). Filtering the sample would remove these precipitates but would decrease recovery of N (Owen, 1967). Evaporation of water or ammonia from feces during collections or loss of N from feces during sample drying also contributes to errors (Owen, 1967; Spanghero and Kowalski, 1997).

Nitrogen is also excreted via skin (epithelial cells, hair, and nails) that is lost from the body (Owen, 1967). These losses are difficult to measure and are often ignored. It is also assumed that no N is lost in volatile forms and no atmospheric N is fixed by the animal. There could be other sources of loss, such as N gas, that are not measured, or the formation of nitrate or nitrite compounds which are not measured by Kjeldahl analysis (Spanghero and Kowalski, 1997).

The current committee used data based on N accretion to estimate efficiency of conversion of MP to NP.

Conversion of Metabolizable Protein to Net Protein

Utilization of MP in beef cattle depends on MP digestibility in the small intestine and the efficiency of subsequent absorption and incorporation of AA into NP in muscle, fetus, or milk. Although N requirements are defined in terms of MP, true requirements are for the specific AA that are supplied by MCP, RUP, and endogenous protein. Studies by Armstrong and Hutton (1975) and Zinn and Owens (1983) reported that the average biological value of a protein, defined as the relative AA balance, was 66% (NRC, 1984). The NRC (1985) assumed a constant conversion of MP to NP for gain of 0.50 and to NP for milk of 0.65. These efficiency values were based on the biological value of the protein and the efficiency of use of an "ideal mixture of AA" (Oldham, 1987). Oldham (1987) suggested that the efficiency value for utilization of the "ideal mixture of AA" was 0.85 for all physiological functions. Biological values will vary with source(s) of RUP in the diet and for different biological functions (Oldham, 1987). For example, it is likely that the overall efficiency values for pregnancy and lactation are higher than for gain. Based on the data for lactation and pregnancy (NRC, 1985), the NRC (1996, 2000) chose to use 0.65 (0.85 × 0.76; efficiency × biological value for lactation and pregnancy) as the overall efficiency value for lactation and pregnancy.

In the 7th edition and update of the *Nutrient Requirements of Beef Cattle* (NRC, 1996, 2000), it was recognized that the efficiency of MP use for gain is likely not constant across BW (maturity) and rates of gain. Similarly, the Institut National de la Recherche Agronomique (INRA; Jarrige, 1988) system assumes a decreasing efficiency as BW increases. This was confirmed by Ainslie et al. (1993) and Wilkerson et al. (1993), who calculated incremental efficiency by determining daily gain response with increasing concentrations of MP. The equation developed by Ainslie et al. (1993) predicts a conversion efficiency of MP to NP of 66.3% for a 150-kg calf, whereas a 300-kg steer has an efficiency of 49.2%. The data of Ainslie et al. (1993) and Wilkerson et al. (1993) cover only the BW range from 150 to 300 kg, and these bounds were placed on the conversion efficiency equation used in the previous NRC (1996, 2000) reports. Thus, for cattle weighing more than 300 kg, protein requirements similar to those in older NRC publications (NRC, 1984, 1985) were retained in the 7th edition and update (NRC, 1996, 2000), and the low CP requirement of cattle weighing more than 400 kg was recognized (Preston, 1982).

As indicated, the Ainslie et al. (1993) equation predicts that the use of MP for growth decreases linearly from 72% for 100-kg calves to 49% for 300-kg cattle, and assumes a constant efficiency of 49% for cattle greater than 300 kg, which is the equivalent shrunk body weight. The equation assumes that all AA from MP are used with the same efficiency in growing calves, and that the efficiency is not affected by factors other than BW. In contrast, data from more recent studies described below suggest that efficiencies for utilization of postruminal AA for growth and lactation are affected by additional factors, which need consideration for future models of AA requirements in growing, reproducing, and lactating cattle. Thus, the current committee recognizes this as an area of needed research for improving subsequent models.

Since the publication of the previous edition and update (NRC, 1996, 2000), experiments have been conducted to study whole-body utilization of postruminal AA using ruminal and/or postruminal infusions of specific nutrients that do not influence rumen microbial synthesis (Titgemeyer, 2003; Löest and Titgemeyer, 2013). Diets that contained sufficient RDP and little RUP were fed so that the predominant source of MP was from a consistent supply of MCP. Calves were then provided with excess amounts of nutrients (e.g., AA, glucose, fat, volatile fatty acids) except the single AA selected to be limiting, and N balance was determined (Titgemeyer, 2003). Using this approach, Schroeder et al. (2006a) studied the effect of energy supply in growing steers that were receiving limiting amounts of methionine. Cattle received 0 or 3 g/d of supplemental L-methionine in combination with supplemental energy in the amounts of 0, 1.3, or 2.6 Mcal/d of gross energy. Nitrogen retention was increased in response to supplemental methionine and to increasing energy in the absence of methionine. It was concluded that the efficiency of methionine utilization for maintenance was improved as energy supply increased. The effects of energy and methionine supplementation were additive, such that the incremental efficiency of methionine utilization remained constant at 18%. In a companion study, Schroeder et al. (2006b) compared the effects of different energy sources on the efficiency of methionine utilization. Nitrogen retention was increased when 3 g/d of supplemental L-methionine was provided without supplemental energy. When cattle were supplemented with ruminal infusions of acetic or propionic acid or abomasal infusions of glucose or fat, the efficiency of methionine use increased from 11% for no supplemental energy to an average of 21% when any of the energy sources was provided. These data suggest that the efficiency of utilization of methionine by cattle is affected by concentration of energy. In contrast, when leucine was limiting (0 or 4 g/d supplemental L-leucine), increasing the energy concentration led to only small increases in N retention, suggesting that the efficiency of leucine utilization is not as responsive to energy as is that of methionine (Schroeder et al., 2007; Titgemeyer et al., 2012).

Similar to lean tissue, mammary tissue has the ability to change its removal of AA based on the supply of energy (Bequette et al., 2000; Hanigan et al., 2013). If mammary tissue is presented with an adequate supply of energy, milk production will reach its maximal potential and the mammary tissue will increase AA extraction efficiency. In contrast, if the mammary tissue is presented with inadequate energy, it will decrease extraction of AA from blood, and AA extraction efficiency is decreased. Therefore, efficiency of postabsorptive AA use for lean tissue growth or milk protein synthesis is likely not constant and depends on the supply of energy.

In addition to energy, quantities of AA in amounts greater than their requirement for maximal whole-body protein deposition have the potential to affect the conversion of MP to NP. First, N from excess AA catabolism must be eliminated from the body, largely as urea, which has an associated energy cost. In addition, AA supply can affect AA degradation pathways. For example, the rate-limiting enzyme in the degradation of branched-chain AA, branched-chain keto-acid dehydrogenase, has been shown to have increased activity when the supply of branched-chain AA is increased (Block, 1989). Therefore, antagonisms among branched-chain AA can occur with excesses of one AA increasing the degradation of another. Awawdeh et al. (2006) showed that supplementation of excess AA improved the efficiency of methionine use in steers maintained under conditions where only methionine was limiting. Supplementation with 200 g/d of a methionine-free mixture of essential AA led to large increases in N retention, even though methionine was first-limiting. The improvements in methionine utilization in response to excess AA supplementation were greater than could be attributed to the increase in energy supplied by the AA (Schroeder et al., 2006a). In a similar experiment, Awawdeh et al. (2006) observed that either a mixture of leucine-free essential AA (200 g/d) or a mixture of essential (100 g/d) and non-essential AA (100 g/d) could improve N retention when leucine was limiting. Nonetheless, improvements in the utilization of leucine in response to supplementation with excess amounts of AA were less than the improvements in methionine utilization.

The available data demonstrate that the efficiency of AA utilization is consistently less than that predicted by the NRC (1996, 2000). Across 11 experiments with efficiencies calculated across 22 treatments, incremental methionine utilization for growth averaged 26% (Campbell et al., 1996, 1997; Lambert et al., 2002, 2004; Löest et al., 2002; Awawdeh et al., 2004, 2006; Schroeder et al., 2006a,b; Löest and Titgemeyer, 2013). Across 4 experiments with efficiencies calculated across 10 treatments, efficiency of leucine utilization for growth averaged 38% (Awawdeh et al., 2005, 2006; Schroeder et al., 2007; Titgemeyer et al., 2012; Löest and Titgemeyer, 2013). The NRC (1996, 2000) model would predict efficiencies of AA utilization for growth to be 57 to 68% for cattle similar in BW to those used in these experiments (132 to 228 kg). In addition, the efficiency value of 0.65 for conversion of MP to milk protein, after subtraction of maintenance use, is likely too high. Hanigan et al. (1998) summarized publications reporting responses to postruminally infused casein and found a maximal efficiency of conversion of approximately 45% with an average conversion efficiency of 22%. In a more recent summary of literature data, Lapierre et al. (2007) reported that the highest efficiency was 43%, which decreased as MP supply and milk protein output increased. Some of the disparity might be attributable to differences in feed intake and source of RUP, with energy and AA supplies having an influence on the efficiencies of use, at least for some AA. It is unlikely, however, that the differences in nutrient supply account for all the difference between the observed efficiencies and those predicted by NRC (1996, 2000).

The current committee recognizes that the NRC (1996, 2000) equation used for converting MP to NP can result in over- or underprediction of MP allowable for gain or milk production; however, it is difficult to extrapolate recent data on individual AA efficiencies of utilization for N retention to conversion of total MP to NP used for gain. Incremental efficiencies of utilization of methionine and leucine from published values cited by Titgemeyer (2003) and Löest and Titgemeyer (2013) for cattle (mostly Holstein steers) ranging in BW from 135 to 370 kg were used to develop the following equation:

$$\text{MP to NP efficiency, \%} = 30 + 10{,}493.1 \times e^{(-0.0486 \times \text{BW})}, \quad \text{RMSE} = 13.2, \qquad \text{(Eq. 6-6)}$$

where e is the base of the natural (Naperian) logarithm (i.e., 2.718). This equation suggests that the efficiency of utilization of AA for N retention decreases at a decreasing rate as BW increases, and restricts efficiency from dropping below 0.30. The intercept (30%) was derived from the summary of values presented by Löest and Titgemeyer (2013). The conversion efficiency of 30% is less than the 49% value used in the previous NRC (1996, 2000) reports and is recommended primarily because of the limited number of observations for cattle above 250 kg.

Although data are limited, the current committee suggests a constant efficiency of 0.43 for converting MP to NP for milk. Consistent results from several experiments (Whitelaw et al., 1986; Hanigan et al., 1998; Lapierre et al., 2007) suggest that the greatest efficiency of conversion of MP to milk protein is approximately 0.43. Like growth, the efficiency for milk production is likely not constant, and probably decreases as MP supply increases; however, to improve the accuracy of estimating AA requirements in cattle in the future, factors that affect AA utilization, such as energy supply and excess MP supply, should be considered when developing and validating future models (Löest and Titgemeyer, 2013). Improving efficiency will require greater knowledge of AA supply and AA requirements so that diets can be formulated for AA as well as MP. We recommend this area as a priority area of research for subsequent models.

The current model was evaluated using Eq. 6-6 to estimate the efficiency of conversion of MP to NP for growing cattle. Incorporating this equation into a series of simulations resulted in estimates of dietary CP required that were well above those known to be adequate based on practice and empirical data. A similar result was obtained for simulations in lactating cattle when the estimate for the efficiency of conversion of MP to NP for milk was decreased from 0.65 to 0.43. The results of these simulations show that the new efficiency values do not fit the BCNRM, and therefore, Eq. 6-6 and a new value for the efficiency of conversion of MP to NP for milk were not included in the BCNRM. The BCNRM retains the equation adopted by the previous subcommittee (NRC, 1996, 2000). The current committee recommends evaluation of the whole system in the future and consideration of the proposed equation for growing animals (Eq. 6-6) and the proposed estimate of the efficiency of conversion of MP to NP for lactating animals as part of a revised system.

AMINO ACID REQUIREMENTS

Abomasal infusion of high-quality sources of AA significantly increased N balance in steers, despite the fact that they were fed diets balanced to optimize ruminal fermentation and to provide protein in excess of NRC requirements (Houseknecht et al., 1992). These results indicate that protein accretion was constrained by the quantity and/or proportionality of AA absorbed. Use of supplemental protein sources to supply increased quantities of specific amino acids is one approach to provide an optimal supply of amino acids to the small intestine (Merchen and Titgemeyer, 1992).

Amino acid requirements for tissue growth are a function of the percentage of each AA in the NP accretion and thereby depend on the accuracy of prediction of protein retained. Ainslie et al. (1993) summarized various studies that have determined essential AA content of tissue protein in selected muscles (Hogan, 1975; Evans and Patterson, 1985), in daily accretion (Early et al., 1990), or in the whole empty body (Williams, 1978; Rohr and Lebzien, 1991; Ainslie et al., 1993). In a sensitivity analysis with model-predicted vs. first-limiting AA allowable gain, the average of the three whole empty body studies gave the least bias (Fox et al., 1995). The average values of empty body protein (average of Williams, 1978; Rohr and Lebzien, 1991; Ainslie et al., 1993) are as follows (g/100 g empty body protein); arginine, 3.3; histidine, 2.5; isoleucine, 2.8; leucine, 6.7; lysine, 6.4; methionine, 2.0; phenylalanine, 3.5; threonine, 3.9; and valine, 4.0. Tryptophan values were not given because of limitations in assay procedures.

Several studies have evaluated tissue AA requirements by measuring net flux of essential AA across the hind limb of growing steers (Boisclair et al., 1994; Byrem et al., 1998). The proportionality of individual AA uptake did not markedly change when protein accretion was increased by infusing various compounds (cimaterol or bovine somatotropin). The proportions of the essential AA in the net flux in these studies followed the same trends as suggested by the tissue composition values listed above.

The above studies and the data previously cited in this section suggest that both quantity and proportionality of AA availability are important to achieve maximum energy allowable average daily gain. The NRC (1996, 2000) model Level 2 allowed the user to estimate both quantity and proportion of essential AA required by the animal and those supplied by the diet. The critical steps involved are the prediction of microbial growth and composition; amount and composition of dietary protein escaping ruminal degradation; intestinal digestion and absorption; and net flux of absorbed AA into tissue. Because of limitations in the ability to predict each of these components, these estimates of AA balances were recommended as a guide.

Net daily tissue synthesis of protein represents a balance between synthesis and degradation (Oltjen et al., 1986; Early et al., 1990; Lobley, 1992). Lobley (1992) indicated that a 500-kg steer with a net daily protein accretion of 150 g actually degrades and resynthesizes at least another 2,550 g of protein. Thus, balancing for daily net accretion accounts for only about 5.5% of the total daily protein synthesis. Protein metabolism is very dynamic, and a kinetic approach is needed to accurately predict AA requirements. Small changes in either the rate of synthesis or degradation can cause significant alterations in the rate of gain. Lobley (1992), however, concluded that the precision of kinetic methods is critical; a 2% change in synthesis rate would alter NP accretion 20 to 40%, and many of the procedures are not accurate within 4 to 5%. When combined with a system that has limitations in predicting absorbed AA from microbial and feed sources, errors could be greatly magnified with an inadequate mechanistic metabolism model.

EMPIRICAL AND MECHANISTIC LEVELS OF SOLUTION

Similar to the previous edition and update of the *Nutrient Requirements of Beef Cattle* (NRC, 1996, 2000), two levels of solution are provided to accommodate a broad range of users (see Chapter 19, Model Equations and Sensitivity Analyses). The empirical level of solution (ELS), described in this chapter, uses equations based on tabular values of TDN to determine TDNI and MCP synthesis. The specific equation used depends on the ether extract content of the diet. Undegraded protein from feed is calculated from CP intake and tabular values of RUP. In the MLS, ruminal degradation kinetics of protein is used to determine RDP and RUP; and ruminal degradation kinetics of carbohydrates is used to estimate MCP synthesized. For both levels of solution, the supply of MP is the sum of digested MTP and digested RUP.

The kinetics of predicting ruminal protein degradability using a three-pool, in situ-based ABC system is described in detail by the NRC (1985, 2001), and a similar approach was adopted herein (see Chapter 2, Anatomy, Digestion, and Nutrient Utilization, for further discussion). Briefly, ruminal protein degradation of feedstuffs is described using a first-order disappearance model (Ørskov and McDonald, 1979), with only one kd for the protein B fraction, which is a composite of the protein fractions B1 and B2 that were used in the in the previous edition and update of this publication (NRC, 1996, 2000). In the BCNRM, feed is incubated in situ and the weight of CP in the residue at various incubation times is used to calculate the potential degradability (P):

$$P = A + B\,(1 - e^{-kd \times t}) \qquad \text{(Eq. 6-7)}$$

where the A fraction is the percentage of CP that is instantaneously or completely degraded in the rumen (NPN and true protein that escapes from the in situ bag because it is highly soluble or very fine particle size); the B fraction is assumed to be potentially degraded CP, with the amount degraded at time *t* determined from kd.

The C fraction is the percentage of CP that is completely undegraded and considered to pass in its entirety to the small intestine, and is calculated as

$$C = 100 - (A + B). \qquad \text{(Eq. 6-8)}$$

The RDP and RUP values (% of CP) of a feedstuff can then be calculated as follows:

$$RDP = A + B\,[kd/(kd + kp)] \times 100, \qquad \text{(Eq. 6-9)}$$

$$RUP = B\,[kp/(kd + kp)] + C \times 100. \qquad \text{(Eq. 6-10)}$$

Thus, RDP consists of the entire A fraction, and the portion of B that is actually degraded in the rumen; RUP consists of the fraction of B that passes from the rumen before it is digested and the entire C fraction; and the sum of RDP and RUP is equal to 100%. The fractional rate of passage from the rumen (kp) can be determined directly, or values of 0.05 to 0.07 per hour are often assumed for protein feeds. As an example, if a rate of solids passage of 0.05 is assumed, with A = 0.25, B = 0.65, C = 0.10, and kd = 0.06, then RDP = 0.25 + [0.65 (0.06 / (0.06 + 0.05)] × 100 = 60.4% and RUP = 0.65 [0.05 / (0.06 + 0.05)] + 0.10 × 100 = 39.6%.

SUMMARY OF RECOMMENDED ADJUSTMENTS FOR PROTEIN

The committee recommends two new equations to predict MCP. For diets containing <3.9% EE, Eq. 6-1 in which TDNI predicts MCP is used. For diets with EE ≥3.9% of the dietary DM, Eq. 6-2 in which fat-free TDNI predicts MCP is used. Efficiency of conversion of RDP to MCP is assumed to be 100%, and therefore MCP is equal to the required RDP supply.

The committee also recommends that the RUP digestibility of forages be decreased to 60%.

REFERENCES

AFRC (Agricultural and Food Research Council). 1992. Technical Committee on Responses to Nutrients Report No. 9, Nutrient Requirements of Ruminant Animals: Protein. *Nutrition Abstracts and Reviews* 62(Series B):787-835.

Ainslie, S. J., D. G. Fox, T. C. Perry, D. J. Ketchen, and M. C. Barry. 1993. Predicting amino acid adequacy of diets fed to Holstein steers. *Journal of Animal Science* 71:1312-1319.

Archibeque, S. L., J. C. Burns, and G. B. Huntington. 2001. Urea flux in beef steers: Effects of forage species and nitrogen fertilization. *Journal of Animal Science* 79:1937-1943.

Archibeque, S. L., J. C. Burns, and G. B. Huntington. 2002. Nitrogen metabolism of beef steers fed endophyte-free tall fescue hay: Effects of ruminally protected methionine supplementation. *Journal of Animal Science* 80:1344-1351.

Armstrong, D. G., and K. Hutton. 1975. Fate of nitrogenous compounds entering the small intestine. Pp. 432-447 in *Digestion and Metabolism in the Ruminant*, I. W. McDonald and A. C. I. Warner, eds. Armidale, NSW, Australia: University of New England.

Atasoglu, C., C. J. Newbold, and R. J. Wallace. 2001. Incorporation of [^{15}N] ammonia by the cellulolytic ruminal bacteria *Fibrobacter succinogenes* BL2, *Ruminococcus albus* SY3, and *Ruminococcus flavefaciens* 17. *Applied and Environmental Microbiology* 67:2819-2822.

Awawdeh, M. S., E. C. Titgemeyer, K. C. McCuistion, and D. P. Gnad. 2004. Effects of ammonia load on methionine utilization by growing steers. *Journal of Animal Science* 82:3537-3542.

Awawdeh, M. S., E. C. Titgemeyer, K. C. McCuistion, and D. P. Gnad. 2005. Ruminal ammonia load affects leucine utilization by growing steers. *Journal of Animal Science* 83:2448-2454.

Awawdeh, M. S., E. C. Titgemeyer, G. F. Schroeder, and D. P. Gnad. 2006. Excess amino acid supply improves methionine and leucine utilization by growing steers. *Journal of Animal Science* 84:1801-1810.

Bach, A., S. Calsamiglia, and M. D. Stern. 2005. Nitrogen metabolism in the rumen. *Journal of Dairy Science* 88 (Suppl.):E9-E21.

Bailey, E. A., E. C. Titgemeyer, K. C. Olson, D. W. Brake, M. L. Jones, and D. E. Anderson. 2012. Effects of supplemental energy and protein on forage digestion and urea kinetics in growing beef cattle. *Journal of Animal Science* 90:3492-3504.

Bateman, H. G., II, J. H. Clark, and M. R. Murphy. 2005. Development of a system to predict feed protein flow to the small intestine of cattle. *Journal of Dairy Science* 88:282-295.

Beaty, J. L., R. C. Cochran, B. A. Lintzenich, E. S. Vanzant, J. L. Morrill, R. T. Brandt, Jr., and D. E. Johnson. 1994. Effect of frequency of supplementation and protein concentration in supplements on performance and digestion characteristics of beef cattle consuming low quality forages. *Journal of Animal Science* 72:2475-2486.

Bequette, B. J., M. D. Hanigan, A. G. Calder, C. K. Reynolds, G. E. Lobley, and J. C. MacRae. 2000. Amino acid exchange by the mammary gland of lactating goats when histidine limits milk production. *Journal of Dairy Science* 83:765-775.

Block, K. P. 1989. Interactions among leucine, isoleucine, and valine with special reference to the branched-chain amino acid antagonism. Pp. 229-244 in *Absorption and Utilization of Amino Acids*, M. Friedman, ed. Boca Raton, FL: CRC Press.

Boisclair, Y. R., D. E. Bauman, A. W. Bell, F. R. Dunshea, and M. Harkins. 1994. Nutrient utlization and protein turnover in the hindlimb of cattle treated with bovine somatotropin. *Journal of Nutrition* 124:664-673.

Brake, D. W., E. C. Titgemeyer, M. L. Jones, and D. E. Anderson. 2010. Effect of nitrogen supplementation on urea kinetics and microbial use of recycled urea in steers consuming corn-based diets. *Journal of Animal Science* 88:2729-2740.

Bristow, A. W., D. C. Whitehead, and J. E. Cockburn. 1992. Nitrogenous constituents in the urine of cattle, sheep and goats. *Journal of the Science of Food and Agriculture* 59:387-394.

Broderick, G. A., and S. Colombini. 2010. In vitro methods to determine rate and extent of ruminal protein degradation. Pp. 691-702 in *Energy and Protein Metabolism and Nutrition*, G. M. Crovetto, ed. EAAP Publication No. 127. Wageningen, The Netherlands: Wageningen Academic.

Broderick, G. A., R. J. Wallace, and E. R. Ørskov. 1991. Control of rate and extent of protein degradation. Pp. 541-591 in *Physiological Aspects of Digestion and Metabolism in Ruminants: Proceedings of the 7th International Symposium on Ruminant Physiology*, T. Tsuda, Y. Sasaki, and R. Kawashima, eds. New York: Academic Press.

Buckner, C. D., T. J. Klopfenstein, K. M. Rolfe, W. A. Griffin, M. J. Lamothe, A. K. Watson, J. C. MacDonald, W. H. Schacht, and P. Schroeder. 2013. Ruminally undegradable protein content and digestibility for forages using the mobile bag in situ technique. *Journal of Animal Science* 91:2812-2822.

Burroughs, W., A. H. Trenkle, and R. L. Vetter. 1974. A system of protein evaluation for cattle and sheep involving metabolizable protein (amino acids) and urea fermentation potential of feedstuffs. *Veterinary Clinics of North America: Small Animal Practice* 69:713-722.

Byrem, T. M., D. H. Beermann, and T. F. Robinson. 1998. The β-agonist cimaterol directly enhances chronic protein accretion in skeletal muscle. *Journal of Animal Science* 76:988-998.

Calsamiglia, S., A. Ferret, C. K. Reynolds, N. B. Kristensen, and A. M. van Vuuren. 2010. Strategies for optimizing nitrogen use by ruminants. *Animal* 4:1184-1196.

Campbell, C. G., E. C. Titgemeyer, and G. St-Jean. 1996. Efficiency of D- vs L-methionine utilization by growing steers. *Journal of Animal Science* 74:2482-2487.

Campbell, C. G., E. C. Titgemeyer, and G. St-Jean. 1997. Sulfur amino acid utilization by growing steers. *Journal of Animal Science* 75:230-238.

Cole, N. A., and R. W. Todd. 2008. Opportunities to enhance performance and efficiency through nutrient synchrony in concentrate-fed ruminants. *Journal of Animal Science* 86(14 Suppl.):E318-E333.

CSIRO (Commonwealth Scientific and Industrial Research Organisation). 2007. *Nutrient Requirements of Domesticated Ruminants*. Collingwood, Australia: CSIRO Publishing.

Early, R. J., B. W. McBride, and R. O. Ball. 1990. Growth and metabolism in somatrotropin-treated steers: III. Protein synthesis and tissue energy expenditures. *Journal of Animal Science* 68:4153-4166.

Edmunds, B., K.-H. Südekum, H. Spiekers, and F. J. Schwarz. 2012. Estimating ruminal crude protein degradation of forages using in situ and in vitro techniques. *Animal Feed Science and Technology* 175:95-105.

Eisemann, J. H., and L. O. Tedeschi. In press. Predicting the amount of urea nitrogen recycled and used for anabolism in growing cattle. *The Journal of Agricultural Science*.

Erasmus, L. J., P. M. Botha, C. W. Cruywagen, and H. H. Meissner. 1994. Amino acid profile and intestinal digestibility in dairy cows of rumen-undegradable protein from various feedstuffs. *Journal of Dairy Science* 77:541-551.

Evans, E. H., and R. J. Patterson. 1985. Use of dynamic modelling seen as good way to formulate crude protein, amino acid requirements for cattle diets. *Feedstuffs* 57(42):24-29.

Firkins, J. L., and C. K. Reynolds. 2005. Whole-animal nitrogen balance in cattle. Pp. 167-186 in *Nitrogen and Phosphorus Nutrition of Cattle*, E. Pfeffer and A. N. Hristov, eds. Wallingford, UK: CAB International.

Fox, D. G., M. C. Barry, R. E. Pitt, D. K. Roseler, and W. C. Stone. 1995. Application of the Cornell Net Carbohydrate and Protein Model for cattle consuming forages. *Journal of Animal Science* 73:267-277.

Fox, D. G., L. O. Tedeschi, T. P. Tylutki, J. B. Russell, M. E. Van Amburgh, L. E. Chase, A. N. Pell, and T. R. Overton. 2004. The Cornell Net Carbohydrate and Protein System model for evaluating herd nutrition and nutrient excretion. *Animal Feed Science and Technology* 112:29-78.

Galyean, M. L., and L. O. Tedeschi. 2014. Predicting microbial protein synthesis in beef cattle: Relationship to intakes of total digestible nutrients and crude protein. *Journal of Animal Science* 92:5099-5111.

Hall, M. B., and G. B. Huntington. 2008. Nutrient synchrony: Sound in theory, elusive in practice. *Journal of Animal Science* 86(14 Suppl.):E287-E292.

Hanigan, M. D., J. P. Cant, D. C. Weakley, and J. L. Beckett. 1998. An evaluation of post-absorptive protein and amino acid metabolism in the lactating dairy cow. *Journal of Dairy Science* 81:3385-3401.

Hanigan, M. D., S. I. Arriola-Apelo, and M. Aguilar. 2013. Predicting post-absorptive protein and amino acid metabolism. In *The Integration of Knowledge in Animal Production: Proceedings of the 50th Annual Meeting of the Brazilian Society of Animal Science, July 23-26, 2013, Campinas, São Paulo, Brazil* [in Portuguese]. Sociedade Brasileira de Zootecnia (SBZ).

Haugen, H. L., S. K. Ivan, J. C. MacDonald, and T. J. Klopfenstein. 2006a. Determination of undegradable intake protein digestibility of forages using the mobile nylon bag technique. *Journal of Animal Science* 84:886-893.

Haugen, H. L., M. J. Lamothe, T. J. Klopfenstein, D. C. Adams, and M. D. Ullerich. 2006b. Estimation of the undegradable intake protein in forages using neutral detergent insoluble nitrogen at a single in situ incubation time point. *Journal of Animal Science* 84:651-659.

Henning, P. H., D. G. Steyn, and H. H. Meissner. 1993. Effect of synchronization of energy and nitrogen supply on ruminal characteristics and microbial growth. *Journal of Animal Science* 71:2516-2528.

Hogan, J. P. 1975. Quantitative aspects of nitrogen utilization in ruminants. *Journal of Dairy Science* 58:1164-1177.

Houseknecht, K. L., D. E. Bauman, D. G. Fox, and D. F. Smith. 1992. Abomasal infusion of casein enhances nitrogen retention in somatotropin-treated steers. *Journal of Nutrition* 122:1717-1725.

Huntington, G. B., E. J. Zetina, J. M. Whitt, and W. Potts. 1996. Effects of dietary concentrate level on nutrient absorption, liver metabolism, and urea kinetics of beef steers fed isonitrogenous and isoenergetic diets. *Journal of Animal Science* 74:908-916.

Huntington, G. B., K. Magee, A. Matthews, M. Poore, and J. Burns. 2009. Urea metabolism in beef steers fed tall fescue, orchardgrass, or gamagrass hays. *Journal of Animal Science* 87:1346-1353.

Jarrige, R., ed 1988. *Alimentation des Bovins, Ovins, et Caprins*. Paris: Institut National de la Recherche Agronomique.

Kennedy, P. M., and L. P. Milligan. 1980. The degradation and utilization of endogenous urea in the gastrointestinal tract of ruminants: A review. *Canadian Journal of Animal Science* 60:205-221.

Klopfenstein, T. J., R. A. Mass, K. W. Creighton, and H. H. Patterson. 2001. Estimating forage protein degradation in the rumen. *Journal of Animal Science* 79(Suppl.):E208-E216.

Kononoff, P. J., S. K. Ivan, and T. J. Klopfenstein. 2007. Estimation of the proportion of feed protein digested in the small intestine of cattle consuming wet corn gluten feed. *Journal of Dairy Science* 90:2377-2385.

Lambert, B. D., E. C. Titgemeyer, G. L. Stokka, B. M. DeBey, and C. A. Löest. 2002. Methionine supply to growing steers affects hepatic activities of methionine synthase and betaine homocysteine methyltransferase, but not cystathionine synthase. *Journal of Nutrition* 132:2004-2009.

Lambert, B. D., E. C. Titgemeyer, C. A. Löest, and D. E. Johnson. 2004. Effect of glycine and vitamin supplementation on sulphur amino acid utilization by growing cattle. *Journal of Animal Physiology and Animal Nutrition* 88:288-300.

Lapierre, H., D. Pacheco, R. Berthiaume, D. R. Ouellet, C. G. Schwab, P. Dubreuil, G. Holtrop, and G. E. Lobley. 2006. What is the true supply of amino acids for a dairy cow? *Journal of Dairy Science* 89(Suppl.):E1-E14.

Lapierre, H., G. E. Lobley, D. R. Ouellet, L. Doepel, and D. Pacheco. 2007. Amino acid requirements for lactating dairy cows: Reconciling predictive models and biology. Pp. 39-59 in *Proceedings of Cornell Nutrition Conference for Feed Manufacturers*. New York: Cornell University.

Lapierre, H., D. R. Ouellet, R. Berthiaume, R. Martineau, G. Holtrop, and G. E. Lobley. 2008. Distribution of ^{15}N in amino acids during ^{15}N-leucine infusion: Impact on the estimation of endogenous flows in dairy cows. *Journal of Dairy Science* 91:2702-2714.

Lardy, G. P., D. C. Adams, T. J. Klopfenstein, and H. H. Patterson. 2004. Building beef cow nutritional programs with the 1996 NRC beef cattle requirements model. *Journal of Animal Science* 82(Suppl.):E83-E92.

Lobley, G. E. 1992. Control of the metabolic fate of amino acids in ruminants: A review. *Journal of Animal Science* 70:3264-3275.

Lobley, G. E., D. M. Bremner, and G. Zuur. 2000. Effects of diet quality on urea fates in sheep as assessed by refined, non-invasive [$^{15}N^{15}N$]urea kinetics. *British Journal of Nutrition* 84:459-468.

Löest, C. A., and E. C. Titgemeyer. 2013. Post-ruminal amino acid utilization by growing cattle. In *The Integration of Knowledge in Animal Production: Proceedings of the 50th Annual Meeting of the Brazilian Society of Animal Science, July 23-26, 2013, Campinas, São Paulo, Brazil* [in Portuguese]. Sociedade Brasileira de Zootecnia (SBZ).

Löest, C. A., E. C. Titgemeyer, G. St-Jean, D. C. Van Metre, and J. S. Smith. 2002. Methionine as a methyl group donor in growing cattle. *Journal of Animal Science* 80:2197-2206.

MacRae, J. C., A. Walker, D. Brown, and G. E. Lobley. 1993. Accretion of total protein and individual amino acids by organs and tissues of growing lambs and the ability of nitrogen balance techniques to quantitate protein retention. *Animal Science* 57:237-245.

Marini, J. C., and M. E. Van Amburgh. 2003. Nitrogen metabolism and recycling in Holstein heifers. *Journal of Animal Science* 81:545-552.

Marini, J. C., D. G. Fox, and M. R. Murphy. 2008. Nitrogen transactions along the gastrointestinal tract of cattle: A meta-analytical approach. *Journal of Animal Science* 86:660-679.

Merchen, N. R., and E. C. Titgemeyer. 1992. Manipulation of amino acid supply to the growing ruminant. *Journal of Animal Science* 70:3238-3247.

Nakamura, T., T. J. Klopfenstein, D. J. Gibb, and R. A. Britton. 1994. Growth efficiency and digestibility of heated protein fed to growing ruminants. *Journal of Animal Science* 72:774-782.

Negi, S. S., B. Singh, and H. P. S. Makkar. 1988. Rumen degradability of nitrogen in typical cultivated grasses and leguminous fodders. *Animal Feed Science and Technology* 22:79-89.

Nocek, J. E., and J. B. Russell. 1988. Protein and energy as an integrated system. Relationship of ruminal protein and carbohydrate availability to microbial synthesis and milk production. *Journal of Dairy Science* 71:2070-2107.

NRC (National Research Council). 1984. *Nutrient Requirements of Beef Cattle*, 6th Rev. Ed. Washington, DC: National Academy Press.

NRC. 1985. *Ruminant Nitrogen Usage*. Washington, DC: National Academy Press.

NRC. 1989. *Nutrient Requirements of Dairy Cattle*, 6th Rev. Ed. Washington, DC: National Academy Press.

NRC. 1996. *Nutrient Requirements of Beef Cattle*, 7th Rev. Ed. Washington, DC: National Academy Press.

NRC. 2000. *Nutrient Requirements of Beef Cattle: Update 2000*, 7th Rev. Ed. Washington, DC: National Academy Press.

NRC. 2001. *Nutrient Requirements of Dairy Cattle*, 7th Rev. Ed. Washington, DC: National Academy Press.

Oldham, J. D. 1987. Efficiencies of amino acid utilization. Pp. 171-186 in *Feed Evaluation and Protein Requirement Systems for Ruminants*, R. Jarrige, and G. Alderman, eds. Luxembourg: Commission of European Communities.

Oltjen, J. W., A. C. Bywater, R. L. Baldwin, and W. N. Garrett. 1986. Development of a dynamic model of beef cattle growth and composition. *Journal of Animal Science* 62:86-97.

Ørskov, E. R., and I. McDonald. 1979. The estimation of protein degradability in the rumen from incubation measurements weighted according to rate of passage. *Journal of Agricultural Science* 92:499-503.

Ørskov, E. R., N. A. MacLeod, and D. J. Kyle. 1986. Flow of nitrogen from the rumen and abomasum in cattle and sheep given protein-free nutrients by intragastric infusion. *British Journal of Nutrition* 56:241-248.

Ouellet, D. R., M. Demers, G. Zuur, G. E. Lobley, J. R. Seoane, J. V. Nolan, and H. Lapierre. 2002. Effect of dietary fiber on endogenous nitrogen flows in lactating cows. *Journal of Dairy Science* 85:3013-3025.

Owen, E. C. 1967. Nitrogen balances. *Proceedings of the Nutrition Society* 26:116-124.

Owens, F. N., S. Qi, and D. A. Sapienza. 2014. Invited review: Applied protein nutrition of ruminants—current status and future directions. *The Professional Animal Scientist* 30:150-179.

Preston, R. L. 1982. Empirical value of crude protein systems for feedlot cattle. Pp. 201-217 in *Protein Requirements of Cattle: Proceedings of an International Symposium*, F. N. Owens, ed. Stillwater: Oklahoma Agricultural Experiment Station.

Reynolds, C. K., and N. B. Kristensen. 2008. Nitrogen recycling through the gut and the nitrogen economy of ruminants: An asynchronous symbiosis. *Journal of Animal Science* 86(14 Suppl.):E293-E305.

Rihani, N., W. N. Garrett, and R. A. Zinn. 1993. Influence of level of urea and method of supplementation on characteristics of digestion of high-fiber diets by sheep. *Journal of Animal Science* 71:1657-1665.

Rohr, K., and P. Lebzien. 1991. Present knowledge of amino acid requirements for maintenance and production. Pp. 127-137 in *Protein Metabolism and Nutrition: Proceedings of the Sixth International Symposium on Protein Metabolism and Nutrition, June 9-14, 199, Herning, Denmark*, B. O. Eggum, ed. EAAP Publication No. 59. Tjele, Denmark: National Institute of Animal Science, Research Center Foulum.

Russell, J. B., and C. J. Sniffen. 1984. Effect of carbon-4 and carbon-5 volatile fatty acids on growth of mixed rumen bacteria in vitro. *Journal of Animal Science* 67:987-994.

Russell, J. B., J. D. O'Connor, D. G. Fox, P. J. Van Soest, and C. J. Sniffen. 1992. A net carbohydrate and protein system for evaluating cattle diets: I. Ruminal fermentation. *Journal of Animal Science* 70:3551-3561.

Schroeder, G. F., E. C. Titgemeyer, M. S. Awawdeh, J. S. Smith, and D. P. Gnad. 2006a. Effects of energy level on methionine utilization by growing steers. *Journal of Animal Science* 84:1497-1504.

Schroeder, G. F., E. C. Titgemeyer, M. S. Awawdeh, J. S. Smith, and D. P. Gnad. 2006b. Effects of energy source on methionine utilization by growing steers. *Journal of Animal Science* 84:1505-1511.

Schroeder, G. F., E. C. Titgemeyer, and E. S. Moore. 2007. Effects of energy supply on leucine utilization by growing steers at two body weights. *Journal of Animal Science* 85:3348-3354.

Schwab, C. G., T. P. Tylutki, R. S. Ordway, C. Sheaffer, and M. D. Stern. 2003. Characterization of proteins in feeds. *Journal of Dairy Science* 86(Suppl.):E88-E103.

Sindt, M. H., R. A. Stock, T. J. Klopfenstein, and D. H. Shain. 1993. Effect of protein source and grain type on finishing calf performance and ruminal metabolism. *Journal of Animal Science* 71:1047-1056.

Sindt, M. H., R. A. Stock, and T. J. Klopfenstein. 1994. Urea versus urea and escape protein for finishing calves and yearlings. *Animal Feed Science and Technology* 49:103-117.

Sniffen, C. J., J. D. O'Connor, P. J. Van Soest, D. G. Fox, and J. B. Russell. 1992. A net carbohydrate and protein system for evaluating cattle diets: II. Carbohydrate and protein availability. *Journal of Animal Science* 70:3562-3577.

Spanghero, M., and Z. M. Kowalski. 1997. Critical analysis of N balance experiments with lactating cows. *Livestock Production Science* 52:113-122.

Spicer, L. A., C. B. Theurer, J. Sowe, and T. H. Noon. 1986. Ruminal and post-ruminal utilization of nitrogen and starch from sorghum grain-, corn- and barley-based diets by beef steers. *Journal of Animal Science* 62:521-530.

Stein, H. H., B. Séve, M. F. Fuller, P. J. Moughan, and C. F. M. de Lange. 2007. Invited review: Amino acid bioavailability and digestibility in pig feed ingredients: Terminology and application. *Journal of Animal Science* 85:172-180.

Tedeschi, L. O., D. G. Fox, and P. H. Doane. 2005. Evaluation of the tabular feed energy and protein undegradability values of the National Research Council nutrient requirements of beef cattle. *The Professional Animal Scientist* 21:403-415.

Titgemeyer, E. C. 2003. Amino acid utilization by growing and finishing ruminants. Pp. 329-346 in *Amino Acids in Animal Nutrition*, J. P. F. D'Mello, ed. Wallingford, UK: CAB International.

Titgemeyer, E. C., and C. A. Löest. 2001. Amino acid nutrition: Demand and supply in forage fed ruminants. *Journal of Animal Science* 79(Suppl.):E180-E189.

Titgemeyer, E. C., N. R. Merchen, and L. L. Berger. 1989. Evaluation of soybean meal, corn gluten meal, blood meal and fish meal as sources of nitrogen and amino acids disappearing from the small intestine of steers. *Journal of Animal Science* 67:262-275.

Titgemeyer, E. C., K. S. Spivey, S. L. Parr, D. W. Brake, and M. L. Jones. 2012. Relationship of whole body nitrogen utilization to urea kinetics in growing steers. *Journal of Animal Science* 90:3515-3526.

Whitelaw, F. G., J. S. Milne, E. R. Ørskov, and J. S. Smith. 1986. The nitrogen and energy metabolism of lactating cows given abomasal infusions of casein. *British Journal of Nutrition* 55:537-556.

Wickersham, T. A., E. C. Titgemeyer, R. C. Cochran, E. E. Wickersham, and D. P. Gnad. 2008a. Effect of rumen-degradable intake protein supplementation on urea kinetics and microbial use of recycled urea in steers consuming low-quality forage. *Journal of Animal Science* 86:3079-3088.

Wickersham, T. A., E. C. Titgemeyer, R. C. Cochran, E. E. Wickersham, and E. S. Moore. 2008b. Effect of frequency and amount of rumen-degradable intake protein supplementation on urea kinetics and microbial use of recycled urea in steers consuming low-quality forage. *Journal of Animal Science* 86:3089-3099.

Wickersham, T. A., E. C. Titgemeyer, R. C. Cochran, and E. E. Wickersham. 2009. Effect of undegradable intake protein supplementation on urea kinetics and microbial use of recycled urea in steers consuming low-quality forage. *British Journal of Nutrition* 101:225-232.

Wilkerson, V. A., T. J. Klopfenstein, R. A. Britton, R. A. Stock, and P. S. Miller. 1993. Metabolizable protein and amino acid requirements of growing cattle. *Journal of Animal Science* 71:2777-2784.

Williams, A. P. 1978. The amino acid, collagen, and mineral composition of preruminant calves. *Journal of Agricultural Science* 90:617-624.

Winterholler, S. J., D. L. Lalman, T. K. Dye, C. P. McMurphy, and C. J. Richards. 2009. In situ ruminal degradation characteristics of by-product feedstuffs for beef cattle consuming low-quality forage. *Journal of Animal Science* 87:2996-3002.

Zinn, R. A., and F. N. Owens. 1983. Influence of feed intake level on site of digestion in steers fed a high concentrate diet. *Journal of Animal Science* 56:471-475.

Zuur, G., K. Russell, and G. E. Lobley. 2000. Multiple-entry urea kinetic model: Effects of incomplete data collection. Pp. 145-161 in *Modelling Nutrient Utilization in Farm Animals*, J. P. McNamara, J. France, and D. E. Beever, eds. Walllingford, UK: CAB International.

7

Minerals

INTRODUCTION

This chapter is an update to the National Research Council publications (NRC, 1996, 2000) on the mineral requirements for beef cattle and contains information about mineral requirements as well as functions, signs of deficiency, factors affecting requirements, toxicity, and sources of each essential mineral. At least 17 minerals are required by beef cattle. These minerals can be divided into two general categories: (1) macrominerals and (2) microminerals. Macrominerals are dietary minerals required in gram quantities and include calcium (Ca), magnesium (Mg), phosphorus (P), potassium (K), sodium (Na), chlorine (Cl), and sulfur (S). These minerals are important structural components of bone and other tissues and serve as important constituents of body fluids. They also play vital roles in the maintenance of acid-base balance, osmotic pressure, membrane electric potential, and nerve transmission. Microminerals or "trace minerals" are required in milligram or microgram amounts and include chromium (Cr), cobalt (Co), copper (Cu), iodine (I), iron (Fe), manganese (Mn), molybdenum (Mo), nickel (Ni), selenium (Se), and zinc (Zn). Trace minerals are present in body tissues in very low concentrations and often serve as components of metalloenzymes, enzyme cofactors, or as components of hormones of the endocrine system. Others, including arsenic, boron, lead, silicon, and vanadium have been shown to be essential for one or more animal species, but there is no evidence these minerals are of practical importance in beef cattle production, and therefore will not be discussed in detail in this chapter.

Calcium and P requirements discussed in the subsequent sections are included in the Beef Cattle Nutrient Requirements Model (BCNRM; Chapter 19, Model Equations and Sensitivity Analyses). Requirements and maximum tolerable concentrations for other minerals are shown in Table 7-1. For certain minerals, requirements are not listed because research data are inadequate to determine them.

Many of the essential minerals are usually found in sufficient concentrations in practical feedstuffs. Other minerals are frequently insufficient in diets fed to cattle, and supplementation is necessary to optimize animal performance or health. Supplementing diets at concentrations in excess of requirements greatly increases mineral loss in cattle waste. Thus, supplementation of minerals in excess of requirements should be avoided to prevent possible environmental problems associated with runoff from waste or application of cattle waste to soil.

For all minerals considered essential, detrimental effects on animal performance can be demonstrated from feeding excessive amounts. Generally, the dietary amount required for optimal performance is well below amounts found to be detrimental to performance; however, toxicity from several of the essential minerals, including Se, Mo, and Cu are problems that can occur under practical feeding conditions. The NRC (1980, 2005) described signs of toxicosis and the dietary concentrations of minerals that are considered excessive. Therefore, certain elements such as lead, cadmium, and mercury will be briefly discussed because they should always be considered toxic and are of practical concern because toxicosis from these elements unfortunately occasionally occurs.

A number of elements that are not required (or at least required only in very small amounts) can cause toxicity in beef cattle. Maximum tolerable concentrations of several elements known to be toxic to cattle are given in Table 7-2. The maximum tolerable concentration for a mineral has been defined as "the dietary concentration that, when fed for a limited period of time (as discussed in NRC, 2005), will not impair animal performance and should not produce unsafe residues in human food derived from the animal" (NRC, 1980, 2005).

Concentrations of mineral elements in both concentrate and forage feedstuffs vary greatly (Adams, 1975; Coppice and Fettman, 1977; Kertz, 1998). Reliable or typical analyses

TABLE 7-1 Mineral Requirements and Maximum Tolerable Concentrations (Dry Matter Basis)[a]

Requirement[b]					
			Cows		
Mineral	Unit	Growing and Finishing Cattle	Gestating	Early Lactation	Maximum Tolerable Concentration
Calcium	%	See Chapter 19			
Chlorine	%	–	–	–	–
Chromium	mg/kg	–	–	–	1,000.00[c]
Cobalt	mg/kg	0.15	0.15	0.15	25.00
Copper	mg/kg	10.00	10.00	10.00	40.00[d]
Iodine	mg/kg	0.50	0.50	0.50	50.00
Iron	mg/kg	50.00	50.00	50.00	500.00
Magnesium	%	0.10	0.12	0.20	0.40
Manganese	mg/kg	20.00	40.00	40.00	1,000.00
Molybdenum	mg/kg	–	–	–	5.00
Nickel	mg/kg	–	–	–	50.00
Phosphorus	%	See Chapter 19			
Potassium	%	0.60	0.60	0.70	2.00
Selenium	mg/kg	0.10	0.10	0.10	5.00
Sodium	%	0.06-0.08	0.06-0.08	0.10	–
Sulfur	%	0.15	0.15	0.15	0.30-0.50[d]
Zinc	mg/kg	30.00	30.00	30.00	500.00

[a]Adapted from NRC (1996, 2000).
[b]Dietary mineral requirements can be influenced by dietary antagonists.
[c]Highly dependent on source. See *Mineral Tolerance of Animals* (NRC, 2005).
[d]Dependent on animal age, diet type, and dietary antagonists. See *Mineral Tolerance of Animals* (NRC, 2005).

of concentrations of some mineral elements (e.g., chloride and various micromineral elements) in many feedstuffs are unavailable (Henry, 1995). Furthermore, the biological availability of macro- and microminerals from feedstuffs can vary depending on feedstuff type (Ammerman and Miller, 1972; Peeler, 1972; Spears, 2003; Suttle, 2010). Moreover, concentrations among samples of the same feed type may be quite variable depending on such factors as fertilization and manure application rates, soil type and conditions, and plant species (Butler and Jones, 1973). Concentrations in byproducts also are variable and can be influenced by the method of processing to produce the feedstuff. Therefore, laboratory analyses of feeds for macro- and micromineral element content are important for precise and accurate diet formulation to meet mineral requirements. Laboratory analyses using wet chemistry methods are critical for accurate determination.

The interactions between minerals and animal production are extremely complex. Many factors can affect an animal's response to mineral supplementation, such as the duration and concentration of mineral supplementation, physiological status of an animal (pregnant vs. nonpregnant), the absence or presence of dietary antagonists, environmental factors, and the influence of stress on mineral metabolism. Breed differences in mineral metabolism have also been documented (Wiener et al., 1978; Gooneratne, et al., 1994; Ward et al., 1995: Du et al., 1996; Mullis et al., 1997). These factors will be discussed in subsequent sections of this chapter where relevant.

TABLE 7-2 Maximum Tolerable Concentrations of Mineral Elements Toxic to Cattle[a]

Element	mg/kg
Aluminum	1,000.00
Arsenic	50.00 (100.00 for organic forms)
Bromide	200.00
Cadmium	0.50
Fluorine	40.00 to 100.00
Lead	30.00
Mercury	2.00
Strontium	2,000.00

[a]Adapted from NRC (1980).

MACROMINERALS

Calcium

Calcium (Ca) is the most abundant mineral in the body; approximately 98% functions as a structural component of bones and teeth. The remaining 2% is distributed in extracellular fluids and soft tissues and is involved in such vital functions as blood clotting, membrane permeability, muscle contraction, transmission of nerve impulses, cardiac regulation, secretion of certain hormones, and activation and stabilization of certain enzymes.

Calcium Requirements

Estimated requirements for Ca were calculated by summing the available Ca needed for maintenance, growth, pregnancy, and lactation and correcting for the percentage of dietary Ca absorbed. Calcium requirements are similar to those in the previous edition and update of this publication (NRC, 1996, 2000) because new information is not sufficient to justify a change at this time. The maintenance requirement was calculated as 15.4 mg Ca/kg body weight (BW; Hansard et al., 1954, 1957), which is equal to endogenous fecal loss of Ca. Retained Ca needs in excess of maintenance requirements were calculated as 7.1 g Ca/100 g protein gain. Calcium content of gain was calculated from slaughter data reported by Ellenberger et al. (1950). Recently, Watson et al. (2014) analyzed data from three separate serial slaughter experiments to calculate Ca retention/100 g of protein gain in feedlot cattle. There was no difference in Ca retention across treatments within each experiment. Calcium retention values expressed as grams of Ca/100 g of protein gained were 13.41, 8.24, and 14.44 for British crossbred steers in the first experiment, British crossbred steers in the second experiment, and Holstein steers in the third experiment, respectively. Because of the variation in the data reported by Watson et al. (2014), further research is needed to warrant changing the current recommendation of 7.1 g calcium/100 g protein.

The Ca requirement for lactation in excess of maintenance needs was calculated as 1.23 g Ca/kg milk produced. Fetal Ca content was assumed to be 13.7 g Ca/kg fetal weight. This requirement was distributed over the last 3 months of pregnancy.

Absolute Ca requirements were converted to dietary Ca requirements assuming a true absorption for dietary Ca of 50%. Lower absorption values have been obtained in older cattle, but in many instances, Ca intake might have exceeded dietary requirements in these animals (Hansard et al., 1954, 1957; Martz et al., 1990). Absorption of Ca is largely determined by the requirement relative to intake. True Ca absorption is decreased when intake exceeds the animal's need. The Agricultural and Food Research Council (AFRC, 1991) used a value of 68% absorption to calculate Ca requirements of cattle.

Factors Affecting Calcium Requirements

Calcium is absorbed primarily from the duodenum and jejunum by both active transport and passive diffusion (McDowell, 2003). It should be noted that diets high in fat can decrease Ca absorption through the formation of soaps (Oltjen, 1975). Vitamin D is required for active absorption of Ca (DeLuca, 1979, 2004). The amount of Ca absorbed is affected by the chemical form and source of the Ca, the interrelationships with other nutrients, and the animal's requirement. Requirement is influenced by such factors as age, weight, and type and stage of production. In natural feedstuffs, Ca occurs in oxalate or in phytate form. In alfalfa hay, 20 to 33% was present as insoluble calcium oxalate and apparently unavailable to the animal (Ward et al., 1979). True absorption of Ca from alfalfa hay was much less than Ca absorption from corn silage when fed to dairy cows (Martz et al., 1990). In cattle fed high-concentrate diets, dietary Ca in excess of requirements improved gain or feed efficiency in some studies (Huntington, 1983; Brink et al., 1984; Bock et al., 1991). Improvements in performance were likely the result of manipulation of digestive tract function (i.e., a possible buffering effect) and might not represent a specific metabolic Ca requirement. Increasing Ca from 0.25 to 0.40 or 1.11% decreased organic matter (OM) and starch digestion in the rumen but increased postruminal digestion of OM and starch (Goetsch and Owens, 1985). In finishing cattle fed a high-concentrate diet, increasing Ca more than 0.3% increased gain in one of two experiments but did not affect Ca status based on bone Ca, bone ash, and plasma ionizable Ca concentrations (Huntington, 1983).

Signs of Calcium Deficiency

The skeleton stores a large reserve of Ca that can be utilized to maintain critical blood Ca concentrations. Depending on the age and under certain conditions, older cattle can be fed Ca-deficient diets for extended periods without developing deficiency signs if previous Ca intake was adequate. Calcium deficiency in young, rapidly growing animals, however, prevents normal bone growth, thereby causing rickets and retarding growth and development. Rickets can be caused by a deficiency of Ca, P, or vitamin D. It is characterized by improper calcification of the organic matrix of bone, which results in weak, soft bones that may be easily fractured. Signs include swollen, tender joints, enlargement of the ends of bones, an arched back, stiffness of the legs, and development of beads on the ribs (NRC, 1996, 2000).

Osteomalacia is the result of demineralization of the bones of adult animals. Because Ca and P in bone are in a dynamic state, high demands on Ca and P stores, such as what occurs during pregnancy and lactation, can result in osteomalacia. This condition is characterized by weak, brittle bones that can break when stressed.

Blood Ca concentration is not a good indicator of Ca status because plasma Ca is maintained between 9 and 11 mg/dL by homeostatic mechanisms. Parathyroid hormone is released in response to a lowering of plasma Ca. It stimulates the production of 1,25-dihydroxy cholecalciferol (vitamin D_3), which increases Ca absorption from the intestine and, in conjunction with parathyroid hormone, increases Ca resorption from bone. If plasma Ca concentrations become elevated, calcitonin is produced and parathyroid hormone production is inhibited. Thus, Ca absorption and bone resorption are decreased (NRC, 1996, 2000).

Signs of Calcium Toxicity

High concentrations of dietary Ca are tolerated well by cattle. In cattle, dietary Ca concentrations greater than 1.0% have been associated with decreased dry matter intake (DMI) and lower performance (Miller, 1983). The amount of mineral added to experimental diets to achieve higher Ca concentrations is often unpalatable and also replaces the energy and protein that could be used by the animal for growth and other functions. Beede et al. (2001) fed 0.47, 0.98, 1.52, and 1.95% Ca diets to cows in late gestation that were receiving a high-chloride diet to prevent milk fever. Cows fed 1.52% Ca diets tended to eat less than cows offered 0.98 or 0.47% Ca diets, whereas those fed the 1.95% Ca diets had significantly lower feed intakes. Furthermore, protein and energy digestibility were decreased when beef steers were fed a diet containing 4.4% Ca (calcium carbonate; Ammerman et al., 1963). High concentrations of dietary Ca can affect metabolism of P, Mg, and other trace elements, but the changes are relatively small (NRC, 1980, 2005; Alfaro et al., 1988).

Calcium Sources

The Ca content in forage is affected by species, portion of plant consumed, maturity, quantity of exchangeable Ca in the soil, and climate (Minson, 1990). Forages are generally good sources of Ca, and legumes are higher in Ca content than grasses. Cereal grains are low in Ca; thus, high-grain diets require supplementation. Oilseed meals are much higher in Ca than grains. Sources of supplemental Ca include calcium carbonate, ground limestone, bone meal, dicalcium phosphate, defluorinated phosphate, monocalcium phosphate, and calcium sulfate. In young steers, true absorption of Ca from different sources ranged from 45% for ground limestone to 64% for dibasic calcium phosphate (Hansard et al., 1957).

Phosphorus

Phosphorus (P) is often discussed in conjunction with Ca because the two minerals function together in bone formation; however, the effect of the Ca:P ratio on ruminant performance has been overemphasized in the past. Several studies (Dowe et al., 1957; Wise et al., 1963; Ricketts et al., 1970; Alfaro et al., 1988) have shown that dietary Ca:P ratios of between 1:1 and 7:1 result in similar performance, provided that P intake is adequate to meet requirements.

Approximately 80% of P in the body is found in bones and teeth, with the remainder distributed in soft tissues. Phosphorus also functions in cell growth and differentiation as a component of DNA and RNA; energy utilization and transfer as a component of adenosine triphosphate (ATP), adenosine diphosphate (ADP), and adenosine monophosphate (AMP); phospholipid formation; and maintenance of acid-base and osmotic balance (for more information regarding minerals and acid-base balance, see *Mineral Tolerance of Animals,* NRC, 2005). Phosphorus is required by ruminal microorganisms for their growth and cellular metabolism.

Phosphorus Requirements

Since the release of the NRC (1996, 2000) publications, several experiments have been published examining P requirements of beef cattle, which are discussed later in this section. Phosphorus requirements in the NRC (1996, 2000) publications were calculated using the factorial method. Estimated requirements for maintenance, growth, pregnancy, and lactation were totaled and then corrected for the percentage of dietary P absorbed. The maintenance requirement for P was considered to be 16 mg P/kg BW. This value is similar to fecal endogenous losses observed in cattle fed P concentrations at or near requirements (Tillman and Brethour, 1958; Tillman et al., 1959; Challa and Braithwaite, 1988; Challa et al., 1989). Slightly lower fecal endogenous losses were observed for dairy cows in negative P balance (Martz et al., 1990). Retained P needs in excess of maintenance requirements were calculated as 3.9 g P/100 g protein gain. The P content of gain was calculated from data presented by Ellenberger et al. (1950). Phosphorus needs, during lactation, in excess of maintenance, were calculated as 0.95 g P/kg milk produced. Fetal P was assumed to be 7.6 g P/kg fetal weight, and this requirement was distributed over the last 3 months of pregnancy.

Work conducted by several researchers (Erickson et al. 1999, 2002; Block et al., 2004; Brokman et al., 2004) suggested that the NRC (1996, 2000) overestimated the P requirement of feedlot cattle. Erickson et al. (1999, 2002) reported no differences in performance, carcass characteristics, or indices of P status when finishing cattle were fed diets containing 0.14 to 0.34% P with yearling steers and 0.16 to 0.40% P with calf-fed steers. Therefore, the authors concluded that the dietary P requirement is less than 0.14% (dry matter [DM] basis) for yearling feedlot steers and less than 0.16% (DM basis) for calf-fed steers; however, as indicated by Geisert et al. (2010), in both experiments conducted by Erickson et al. (1999, 2002), the exact P requirements could not be determined because cattle performance was similar across all concentrations of P. Block et al. (2004) suggested the P requirement of finishing cattle is overestimated by the NRC (1996, 2000) because of incorrect estimates of P required for BW gain, maintenance P requirement, or the absorption coefficient assumption of 68%. Recently, Geisert et al. (2010) conducted two dose-response experiments with dietary P concentrations ranging from 0.10 to 0.38% P. Geisert et al. (2010) concluded that the P requirement for feedlot cattle is between 0.10 and 0.17% of dietary DM or 7.2 to 14.1 g/d and that the P requirement for finishing cattle is less than the P concentrations of typical feedlot diets (0.30 to 0.50%), as well as less than the NRC (1996, 2000) estimates. In contrast, Watson et al. (2014) analyzed data from three serial slaughter experiments to calculate P retention in feedlot cattle. There were no differences in P retention across treatments within each experiment. The authors indicated that P retention values from all three experiments were relatively consistent with the NRC (1996, 2000) values. Collectively, these data indicate that it is not necessary to supplement grain-based finishing diets with P because the P requirement for maximum performance is less than 0.17% of dietary DM (Geisert et al., 2010). There are currently no feasible methods for decreasing dietary P concentration less than that provided in basal ingredients in grain-based feedlot diets.

It should be noted that the diets fed by Erickson et al. (1999, 2002) and Geisert et al. (2010) were higher in fiber than traditional high-concentrate feedlot diets typically fed in the United States (Spears and Weiss, 2014). It has been documented that the major route of P excretion changes from feces to urine as the fiber content of the diet decreases, and that P plays a critical role in acid-base balance (Reed et al., 1965; Scott, 1972; Spears and Weiss, 2014). Furthermore, Scott (1972, 1975) reported that the increase in P excretion in the urine of sheep and cattle fed high-concentrate diets was not related to differences in P intake or renal acid excretion, but resulted from an increase in P absorption from the gut and a decrease in tubular P reabsorption by the kidney compared with animals fed a high-roughage-based diet. Further investigation as to how diet type and P concentration influence acid-base balance and P metabolism is warranted.

Factors Affecting Phosphorus Requirements

In the NRC (1996, 2000) publications, a true absorption of 68% was assumed in converting absolute P requirements to dietary requirements. Although this might be an underestimation for high-concentrate, grain-based diets, this value agrees well with most studies (Tillman and Brethour, 1958; Tillman et al., 1959; Challa et al., 1989; Martz et al., 1990) of cattle where true absorption has been measured. Absorption of P was much greater in young calves fed milk (Lofgreen et al., 1952). In their estimate of requirements, the AFRC (1991) assumed an absorption coefficient of 64% for P in forages and 70% for P in concentrates.

In young calves with an initial weight of 96 kg, 0.22% P was adequate for maximal weight gains, but increasing P to 0.30% increased bone ash (Wise et al., 1958). A more recent study with dairy calves weighing approximately 70 kg indicated that 0.26% P was not adequate for maximum growth or bone ash (Jackson et al., 1988). Call et al. (1978) fed Hereford heifers (165 kg initial weight), beginning at approximately 7 months of age, diets containing 0.14 or 0.36% P for 2 years. No differences between the two groups were detected in growth, rib bone morphology and P content, age at puberty, conception rate, or calving interval.

In a second study, Hereford heifers were fed low-P diets from weaning through their fifth gestation and lactation (Call et al., 1986). The low-P group received 6 to 12.1 g P/d, whereas controls received 20.6 to 38.1 g P/d, with P intake increased as the cattle grew larger. Females fed the low-P intake remained healthy, and growth and reproduction were similar to that observed in P-supplemented animals. When P intake of 6 to 12.1 g P/d was decreased to 5.1 to 6.6 g P/d, clinical signs of deficiency occurred within 6 months (Call et al., 1986). Reproduction was not impaired until cows were fed the very-low-P diet for more than 1 year. It was concluded that 12 g P/d throughout one production year was adequate for 450-kg Hereford cows (Call et al., 1986). No measurements of milk production or calf weaning weights were given in these papers (Call et al., 1978, 1986).

Signs of Phosphorus Deficiency

In grazing livestock, P deficiency has been described as the most prevalent mineral deficiency throughout the world (McDowell, 2003). Studies in South Africa and Texas of cattle that grazed forages low in P showed large improvements in fertility and calf weaning weights with P supplementation (Dunn and Moss, 1992). Phosphorus deficiency results in decreased growth and feed efficiency, decreased appetite, impaired reproduction, decreased milk production, and weak, fragile bones (Shupe et al., 1988; Underwood and Suttle, 1999). The skeleton provides a large reserve of P that can be drawn on during periods of inadequate P intake in mature animals. Skeletal reserves can subsequently be replaced during periods when P intake is high relative to requirements. Plasma P concentrations consistently below 4.5 mg/dL are indicative of a deficiency, but bone P is a more sensitive measure of P status (McDowell, 2003).

Phosphorus absorption occurs in the small intestine. The percentage absorbed is not greatly affected by the amount of P intake (AFRC, 1991). Varying endogenous fecal excretion is an important homeostatic mechanism for controlling P in cattle. Endogenous fecal losses consist largely of unabsorbed salivary P (Challa et al., 1989). Salivary P is affected by plasma P concentration, which depends on P intake as well as factors that affect salivary flow such as DMI and physical form of the diet (AFRC, 1991). Thus, fecal endogenous loss of P can vary depending on intake and other factors that affect salivary P. In estimating the maintenance requirement, it is important that endogenous fecal excretion of P be measured in cattle fed approximately their P requirement. Urinary losses of P are generally lower than in feces but can increase in cattle fed high-concentrate diets (Reed et al., 1965).

Signs of Phosphorus Toxicity

Toxicosis from P is rare in food-producing animals. Inorganic P supplements as feed ingredients are fairly expensive. Moreover, phosphate is readily excreted via urine, and thereby is well tolerated (Underwood and Suttle, 1999). Thus, animals can tolerate a wide range of dietary P intakes if their diets are balanced with Ca. In many cases, P toxicity is associated with metabolic disorders of Ca absorption and function, produced by a relative excess of P in relation to low Ca. Nevertheless, high levels of P can still be detrimental to animals even in the presence of adequate Ca (Laflamme and Jowsey, 1972; Carstairs et al., 1981; Matsuzaki et al., 1997). Studies have shown that cattle performance and health were not affected by feeding dietary P up to 0.7% DM basis (DeBore et al., 1981; Knowlton and Herbein, 2002).

Urolithiasis (urinary calculi) can be produced by excess P intake in relation to Ca in ruminants. The disease is caused by the formation of stones or calculi in the kidney or bladder, resulting in obstruction of urine excretion, particularly in males. The continuous buildup of urine in the bladder eventually ruptures the bladder or urethra, followed by abdominal distension, depression, and death as a result of uremia. The incidence was approximately 70% in male lambs fed a high level of P (0.8%) and low Ca (0.44%), and also occurred in lambs fed lower levels of P (Emerick and Embry, 1963, 1964). Different forms of sodium phosphates seemed to have equal ability to produce calculi, but elevating dietary Ca concentrations partially protected against the incidence in sheep (Bushman et al., 1965). Grazing sheep (45 kg BW) tolerated up to 1.5 g P from orthophosphoric acid and 3.0 g P from monosodium phosphate administered in water for 70 days (McMeniman, 1973). However, sheep that received 3.0 g P from orthophosphoric acid died between 7 and 10 weeks of supplementation.

Phosphorus Sources

Phosphorus-deficient soils are widespread, and forages produced on these soils are low in P. Drought conditions and increased forage maturity also can result in forage with low P concentrations. Cereal grains and oilseed meals contain moderate to high concentrations of P. Likewise, animal and fish products are high in P. In terms of availability, supplemental sources of P were ranked as follows: dicalcium phosphate, defluorinated phosphate, and bone meal (Peeler, 1972). More recent studies with calves have indicated that defluorinated phosphate (Miller et al., 1987) and monoammonium phosphate (Jackson et al., 1988) are equal in availability to dicalcium phosphate. Phytate phosphorus is not well utilized by nonruminants, but it seems to be utilized by ruminants as readily as P from inorganic sources (McGillivray, 1974).

Magnesium

More than 300 enzymes are known to be activated by Mg (Wacker, 1980). Magnesium is essential, as the complex Mg-ATP, for all biosynthetic processes including glycolysis, energy-dependent membrane transport, formation of cyclic-AMP, and transmission of the genetic code. Magnesium also is involved in the maintenance of electric potentials across nerve and muscle membranes and for nerve impulse transmission. Of the total percentage of Mg in the body, 65 to 70% is in bone, 15% in muscle, 15% in other soft tissues, and 1% in extracellular fluid (Mayland, 1988).

Magnesium Requirements

Dietary requirements for Mg vary depending on age, physiological state, and bioavailability from the diet. As a percentage of DM, recommended Mg requirements are as follows:

- growing and finishing cattle, 0.10%;
- gestating cows, 0.12%; and
- lactating cows, 0.20%.

Absolute requirements for Mg have been estimated as follows:

- replenishment of endogenous loss, 3 mg Mg/kg live weight;
- growth, 0.45 g Mg/kg gain;
- lactation, 0.12 g Mg/kg milk; and
- pregnancy, 0.12, 0.21, and 0.33 g Mg/d for early, mid, and late pregnancy, respectively (Grace, 1983).

O'Kelley and Fontenot (1969, 1973) found that beef cows required 7 to 9 g Mg/d during gestation and 18 to 21 g Mg/d during lactation to maintain serum Mg concentrations of 2.0 mg/dL. These daily quantities corresponded to 0.10 to 0.13% during gestation and 0.17 to 0.20% during lactation. In young calves fed milk, 12 to 16 mg Mg/kg body weight was adequate to maintain blood magnesium concentrations (Huffman et al., 1941; Blaxter and McGill, 1956).

Factors Affecting Magnesium Requirements

The rumen is the major site of Mg absorption in ruminants (Grace et al., 1974; Greene et al., 1983). Magnesium absorption is high in young calves fed milk but decreases with age (Peeler, 1972). True absorption values for Mg in mature ruminants fed hay and grass range from 10 to 37% (ARC, 1980). Magnesium in concentrates is more available than Mg in forages (Peeler, 1972). A number of studies have shown that high dietary potassium (K) decreases Mg absorption (Greene et al., 1983; Wylie et al., 1985). High dietary concentrations of nitrogen (N), organic acids (citric acid and *trans*-aconitate), long-chain fatty acids, Ca, and P also can decrease Mg absorption or utilization (Fontenot et al., 1989). High ruminal ammonia concentrations have been associated with hypomagnesemia in cows grazing spring pastures high in crude protein (Martens and Rayssiguier, 1980). Magnesium absorption has been enhanced by feeding soluble carbohydrates or carboxylic ionophores (Fontenot et al., 1989; Spears et al., 1989). Evidence suggests that Mg absorption from the rumen occurs by means of an active Na-linked process (Martens and Rayssiguier, 1980), and Na supplementation in a low-Na diet increases Mg absorption (Martens et al., 1987). It has also been reported that different breeds absorb Mg differently (Greene et al., 1989). Excess Mg absorbed is excreted primarily in the urine.

Signs of Magnesium Deficiency

Magnesium deficiency in calves results in excitability, anorexia, hyperemia, convulsions, frothing at the mouth, profuse salivation, and calcification of soft tissue (Moore et al., 1938; Blaxter et al., 1954). Grass tetany or hypomagnesemic tetany is characterized by low Mg concentrations in plasma and cerebrospinal fluid and is a problem in lactating beef cows. Initial signs of grass tetany are nervousness, decreased feed intake, and muscular twitching around the face and ears. Animals are uncoordinated and walk with a stiff gait. In the advanced stages, cows go down on their side with their head back and go into convulsions. Death usually occurs unless the animal is treated intravenously or subcutaneously with a magnesium-salt solution.

Grass tetany can occur in older (three or more lactations) cows grazing lush spring pastures or fed harvested forages low in Mg. With early spring pastures, the problem is more of insufficient availability rather than forage with low Mg concentrations per se. Fertilizing pastures with fertilizers high in N and K is associated with increased incidence of grass tetany. Cows depend on a frequent supply of Mg from the gastrointestinal tract to maintain normal blood Mg concentrations because homeostatic mechanisms are not sufficient to regulate blood Mg concentrations. Magnesium concentrations in bone are high, but mature animals lack the ability to mobilize large amounts of Mg from bone (Rook and Stony, 1962). In young calves, at least 30% of the skeletal Mg can be mobilized during Mg deficiency (Blaxter et al., 1954). For information regarding nutritional management to help prevent grass tetany in beef cows see NRC (2001) and Kvasnicka and Krysl (2014).

Signs of Magnesium Toxicity

Magnesium toxicity is not a problem in beef cattle. Maximum tolerable concentrations have been estimated at 0.4% (NRC, 1980, 2005). Cows fed 0.39% Mg showed no adverse effects (O'Kelley and Fontenot, 1969). Young calves fed 1.3% Mg had lower feed intake and weight gain and diarrhea with mucus in feces (Gentry et al., 1978). Steers fed 2.5 or 4.7% Mg exhibited severe diarrhea and a lethargic appearance, whereas 1.4% Mg decreased DM digestibility (Chester-Jones et al., 1990).

Magnesium Sources

Cereal grains generally contain 0.11 to 0.17% Mg, whereas plant protein sources contain approximately twice this concentration (Underwood and Suttle, 1999). Magnesium concentration in forages varies greatly depending on plant species, soil Mg, stage of growth, season, and environmental temperature (Minson, 1990). Legumes are usually higher in Mg than are grasses. Magnesium oxide and magnesium sulfate are good sources of supplemental Mg, but Mg in magnesite and dolomitic limestone is poorly available (Gerken and Fontenot, 1967; Ammerman et al., 1972).

Potassium

Potassium (K) is the third most abundant mineral in the body and the major cation in intracellular fluid. Potassium is important in acid-base balance, regulation of osmotic pressure, water balance, muscle contractions, nerve impulse transmission, and certain enzymatic reactions.

Potassium Requirements

Feedlot cattle require approximately 0.6% K in the dietary DM. Studies conducted with K in cattle receiving no ionophore have been inconsistent. Roberts and St. Omer (1965) observed a response in gain with K supplementation of steer diets containing 0.50 to 0.56% K in only one of three trials. Devlin et al. (1969) noted improvements in steers gain and feed intake when K was added to diets already containing 0.5% K. In contrast, Kelley and Preston (1984) observed no improvement in steer performance when K was supplemented to a basal diet containing 0.4% K. Studies with feedlot cattle fed lasalocid (Ferrell et al., 1983; Spears and Harvey, 1987) or monensin (Brink et al., 1984) indicate that the K requirement does not exceed 0.55%. Potassium requirements in young dairy calves not fed an ionophore also do not exceed 0.55% (Weil et al., 1988; Tucker et al., 1991). Because of the lower rates of gain observed in growing cattle in range conditions, K requirements for range cattle might be lower than those for feedlot cattle. Clanton (1980) concluded that growing cattle in range conditions require 0.3 to 0.4% K.

Potassium requirements of beef cows are not well defined. Clanton (1980) suggested that gestating beef cows require 0.5 to 0.7% K. Because of the relatively high secretion of K in milk (1.5 g/kg), requirements for K could be slightly greater in beef cows during lactation—for example, for cows producing 9 kg milk/d, approximately 13.5 g K/d or 0.13% of DMI would be needed for milk production.

Signs of Potassium Deficiency

A deficiency of K results in decreased feed intake and weight gain, pica, rough hair coat, and muscular weakness (Devlin et al., 1969). In beef cattle, a severe deficiency of K is unlikely. A marginal K deficiency results in decreased feed intake and retarded weight gain. Dietary K concentration is the best indicator of K status, and serum or plasma K are not reliable indicators of K status. Decreased feed consumption seems to be an early indicator of marginal K deficiency, but the depression in feed intake is usually of relatively small magnitude, making it difficult to detect in field conditions.

Potassium is absorbed from the rumen and omasum as well as the intestine, and absorption is very high. The major route of K excretion is the urine. Body stores of potassium are small; therefore, a deficiency can occur rapidly (Ward, 1966).

Signs of Potassium Toxicity

Increasing the K content of a liquid diet from 1.2 to 5.8% on a DM basis resulted in the deaths of three of eight calves as a result of cardiac insufficiency (Blaxter et al., 1960). In calves, increasing dietary K from 2.77 to 6.77% (as a percentage of the total air-dried feed) decreased feed intake and retarded weight gain (Neathery et al., 1980). A conservative maximum tolerable concentration of K for ruminants has been set at 1 to 2% for cattle (NRC, 2005). Nonetheless, cattle grazing lush, spring pastures often consume more than 3% K, and other than decreased absorption of Mg, no adverse effects have been reported.

Potassium Sources

Forages are excellent sources of K, usually containing between 1 and 4% K. As noted previously, high K content in lush spring pastures seems to be a major factor associated with the occurrence of grass tetany in beef cows (Mayland, 1988).

As forages mature, the K content decreases, and low concentrations of K have been observed in range forage and in accumulated tall fescue during the winter (Clanton, 1980). Cereal grains are often deficient (<0.5%) in K, and high-concentrate diets might require K supplementation unless a high-K forage or protein supplement is included in the diet. Oilseed meals are good sources of K. Potassium can be supplemented to cattle diets as potassium chloride, potassium bicarbonate, potassium sulfate, or potassium carbonate, and all forms are readily available.

Sodium and Chlorine

Sodium (Na) is the major cation, whereas chlorine (Cl) is the major anion in extracellular fluid. Both Na and Cl are involved in maintaining osmotic pressure, controlling water balance, and regulating acid-base balance. Sodium also functions in muscle contractions, nerve impulse transmission, and glucose and amino acid transport. Chlorine is necessary for the formation of hydrochloric acid in gastric juice and for the activation of amylase.

Sodium and Chlorine Requirements

Requirements for Na in nonlactating beef cattle do not exceed 0.06 to 0.08%, whereas lactating beef cows require approximately 0.10% Na (Morris, 1980). Ruminants have an appetite for Na, and if it is provided ad libitum, they will consume more salt than they actually require. In a 2-year study with beef cows grazing forage containing from 0.012 and 0.055% Na, providing salt ad libitum did not affect calf weaning weights or cow BW (Morris et al., 1980). Chlorine requirements are not well defined, but a deficiency of Cl does not seem likely in practical conditions (Neathery et al., 1981). Young calves fed 0.038% Cl performed similar to those fed 0.5% Cl (Burkhaltor et al., 1979).

Signs of Sodium Deficiency

Signs of deficiency of Na are rather nonspecific and include pica and decreased feed intake, growth, and milk production (Underwood and Suttle, 1999). When Na intake is low, the body conserves Na by increasing reabsorption of Na from the kidney in response to aldosterone (McDowell, 2003). The Na:K ratio in saliva has been used as an indicator of Na status. This ratio is normally 20:1, and a production response to Na supplementation is likely when the Na:K ratio is less than 10:1 (Morris, 1980). Serum or plasma Na concentration is not a reliable indicator of Na status, but dietary Na concentration is a good measure of Na adequacy.

Signs of Sodium Toxicity

High concentrations of salt have been used to regulate feed intake by beef cattle. Berger and Rasby (2011) assembled guidelines regarding level of salt inclusion for a desired intake of a particular feedstuff or supplement. Cattle can tolerate high dietary concentrations of salt provided that an adequate supply of clean drinking water is available at all times. Growing cattle were able to tolerate 9.3% salt for 84 days without adverse effects (Meyer et al., 1955); however, Leibholz et al. (1980) reported that 6.5% salt decreased OM intake and growth in calves. The maximum tolerable concentration for dietary salt in cattle was estimated at 9.0% in the *Mineral Tolerance of Domestic Animals* (NRC, 1980). In contrast, the NRC (2005) *Mineral Tolerance of Animals* suggests about 1 g salt/kg BW can be consumed by ruminants without affecting feed intake. Assuming 1 g salt/kg BW is the maximum total load the body can adapt to without adversely affecting feed intake, then a 410-kg beef feedlot steer might be expected to consume 9 kg DM (2.2% of BW) each day (NRC, 1996, 2000). Its voluntary salt intake would be about 410 g. Thus, any diet containing more than 410 g of salt/9 kg DM or 4.55% salt would be expected to limit intake of the steer.

Salt is much more toxic when present in the drinking water of cattle than in the feed. Growing cattle were able to tolerate 1.0% added salt in drinking water without adverse effects (Weeth et al., 1960; Weeth and Haverland, 1961); however, the addition of 1.25 to 2.0% salt resulted in anorexia, decreased weight gain or weight loss, decreased water intake, and physical collapse (Weeth et al., 1960). In some areas of the western United States, soils are high in salt, resulting in groundwater that can cause saline water intoxication. Consumption of water with more than 7,000 mg Na/L resulted in decreased feed and water intake, decreased growth, mild digestive disturbances, and diarrhea (Jensen and Mackey, 1979).

Sodium and Chlorine Sources

Cereal grains and oilseed meals usually provide inadequate amounts of Na for beef cattle. Animal products are much higher in Na and Cl than plant products (Meyer et al., 1950). The Na content of forages varies considerably (Minson, 1990). Sodium can be supplemented as sodium chloride or sodium bicarbonate, and both forms are highly available.

Sulfur

Sulfur (S) is a component of methionine, cysteine, and cystine, and the B-vitamins, thiamin and biotin, as well as a number of other organic compounds. Sulfate is a component of sulfated mucopolysaccharides and also functions in certain detoxification reactions in the body. All S-containing compounds with the exception of biotin and thiamin can be synthesized from methionine. Ruminal microorganisms are capable of synthesizing all organic S-containing compounds required by mammalian tissue from inorganic S (Block et al., 1951; Thomas et al., 1951). Sulfur is required by ruminal microorganisms for their growth and normal cellular metabolism.

Sulfur Requirements

The requirements of beef cattle for S are not well defined. The recommended concentration in beef cattle diets is 0.15%. Sulfur supplementation increased gain by steers fed corn silage/corn-urea-based diets containing 0.10 to 0.11% S (Hill, 1985). In steers fed high-concentrate diets containing 0.14% S, increasing dietary S tended to decrease ruminal lactic acid accumulation and improve feed efficiency (Rumsey, 1978). Other studies have indicated that 0.11 to 0.12% S was adequate for growing cattle (Bolsen et al., 1973; Pendlum et al., 1976). Zinn et al. (1997) reported that S addition (as ammonium sulfate) to a steam-flaked corn-based finishing diet in excess of 0.20% decreased performance and longissimus muscle area in crossbred feedlot heifers. These data are similar to those reported by Qi et al. (1993) in goats. Others (Thompson et al., 1972; Pendlum et al., 1976; Rumsey, 1978) have reported that supplementing S well above 0.15 % (range 0.26 to 0.44% S) to finishing cattle diets either had no effect on cattle performance or decreased DMI. As discussed by Zinn et al. (1997), different supplemental sources of S were used in the aforementioned experiments, and ruminal and total tract availability of supplemental S sources can differ (Johnson et al., 1971; Kahlon et al., 1975; Fron et al., 1990). In a more recent series of experiments, Sarturi et al. (2013a,b) reported that the availability of S for ruminal reduction might be a useful tool when formulating diets than total S in the diet, as evidenced by comparing ruminal S reduction, total tract digestibility, and feedlot performance in finishing cattle fed wet or dry distillers grains. The authors concluded that S in wet distillers grains with solubles increased ruminal hydrogen sulfide concentrations to a greater extent than dried distillers grains when both were fed to supply 1.16% dietary S on a dry matter basis. This concept of using ruminal and total tract digestibility of S needs to be expanded to other feedstuffs (including water) to more accurately predict the impact of total dietary S on animal performance.

In Australia, S supplementation increased gain by 12% in steers grazing sorghum × sudangrass containing 0.08 to 0.12% S (Archer and Wheeler, 1978). The S requirement of ruminants grazing sorghum × sudangrass might be increased because of the need for S in the detoxification of cyanogenic glucoside found in sorghum forages.

Signs of Sulfur Deficiency

Severe S deficiency results in anorexia, weight loss, weakness, dullness, emaciation, excessive salivation, and death (Thomas et al., 1951; Starks et al., 1953). Marginal deficiencies of S can decrease feed intake, digestibility, and microbial protein synthesis. A dietary limitation of S can dramatically decrease microbial numbers, as well as microbial digestion, and protein synthesis. Supplementation to increase the S content of hay from 0.04 to 0.075% increased counts of ruminal bacteria, protozoa, and sporangia of anaerobic fungi in sheep (Morrison et al., 1990). Impaired utilization of lactate by ruminal microorganisms, resulting in lactate accumulation in the rumen and blood, also can occur as a result of S deficiency (Whanger and Matrone, 1966).

Factors Affecting Sulfur Requirements

Most rumen bacteria are able to synthesize the S-containing amino acids from sulfide (Goodrich et al., 1978). Ruminal sulfide is derived from the reduction of inorganic S sources and from the degradation of S-containing amino acids. Sulfide can be absorbed from the rumen and oxidized by tissues to sulfate, a less toxic form of S. Sulfur is found in feedstuffs largely as a component of protein. Dietary S requirements might be greater when diets high in rumen escape protein are fed because of a limitation of S for optimal ruminal fermentation; however, most practical diets are adequate in S. When urea or other nonprotein N sources replace preformed protein, S supplementation might be needed. Mature forages, forages grown in S-deficient soils, corn silage, and sorghum × sudangrass can be low in S. Sorghum forages seem inherently low in S relative to most forages, and the S content of sorghum × sudangrass did not increase in response to S fertilization (Wheeler et al., 1980). For a thorough review of the influence of high S in beef cattle diets on animal performance, see Drewnoski et al. (2014).

Signs of Sulfur Toxicity

Ruminants are more susceptible to S toxicity than nonruminants. Acute S toxicity is characterized by diarrhea, muscular twitching, dyspnea, and, in prolonged cases, inactivity followed by death (Coghlin, 1944). Initially, S was considered a nontoxic mineral; however, in 1956, S toxicity was documented to cause polioencephalomalacia (PEM) in ruminants (Adams et al., 1956). Concentrations of S lower than those needed to cause clinical signs of toxicity can decrease feed intake and retard growth rate (Kandylis, 1984; Spears et al., 2011; Uwituze et al., 2011; Richter et al., 2012) and decrease copper (Smart et al., 1986; Pogge et al., 2014; Drewnoski et al., 2014) and selenium status (Ivancic and Weiss, 2001; Arthington, 2008) in cattle. Increasing dietary S from 0.12 to 0.41% using ammonium sulfate decreased feed intake by 32% in steers fed high-concentrate diets containing urea (Bolsen et al., 1973). Consumption of water high in sulfate decreased feed and/or water intake (Weeth and Hunter, 1971; Loneragan et al., 2001; Patterson and Johnson, 2003; Patterson et al., 2003; Cammack et al., 2010; Kessler et al., 2012).

The maximum tolerable concentration of dietary S was estimated at 0.40% (NRC, 1980); however, the NRC (2005) *Mineral Tolerance of Animals* suggested two maximum tolerable concentrations for S depending on the diet the animal is being fed. Cattle and sheep fed diets with less than 15% forage are at risk of PEM when the diet S is as low as 0.35%. Therefore, the NRC (2005) set the maximum tolerable concentration of S in diets that are more than 85% concentrate at 0.30%. Drinking water for cattle fed these high-concentrate diets should ideally contain less than 600 mg sulfate/L (200 mg S/L).

The maximum tolerable dietary S level based on the avoidance of PEM for cattle with at least 40% forage is 0.50% S. Cattle on these higher-forage diets can tolerate greater concentrations of S in the water. A series of studies (Weeth and Hunter, 1971; Weeth and Capps, 1972) examined the effect of drinking water with added sodium sulfate on feed intake and water intake of growing beef heifers fed predominantly hay rations, culminating in the conclusion that 2,500 mg sulfate/L (or 834 mg S/L) represents the maximal safe concentration of sulfate in drinking water (Digesti and Weeth, 1976) in cattle consuming a forage diet. However, a study of Loneragan et al. (2001) demonstrated that significant decreases in BW and carcass yield occurred when drinking water for feedlot steers contained 1,219 mg sulfate/L but not when water contained 582 mg sulfate/L. It should be remembered that other factors such as diet type, antagonists, S source, and availability of S from feedstuffs can influence the maximum dietary S concentration tolerated by cattle. Furthermore, data on dietary S interactions with Cu and Se availability in ruminants (Smart et al., 1986; Ivancic and Weiss, 2001; Arthington, 2008; Drewnoski, et al., 2014; Pogge et al., 2014) suggest that S content of cattle diets be limited to the requirement of the animal, which is 0.15% for beef cattle.

The majority of the PEM cases have been reported in feedlot and stocker cattle; however, PEM has been noted in most classes of cattle and sheep. Rumen pH has been shown to influence hydrogen sulfide (H_2S) production. As ruminal pH decreases, greater amounts of sulfide are converted to H_2S (Beauchamp et al., 1984; Morine et al., 2014a,b), suggesting that cattle receiving high-concentrate diets are more susceptible to PEM. Furthermore, increasing the roughage neutral detergent fiber (NDF) in finishing diets high in S (>0.40% S) increased ruminal pH and decreased ruminal H_2S concentration (Morine et al., 2014a,b). Prediction equations developed by Nichols et al. (2013) indicate that when ruminally available S is accounted for, the prevalence of PEM decreased by 19% for every 1% of roughage NDF added to the diet. Collectively, these data indicate that rumen pH can influence H_2S production; however, pH only explained approximately 12% of the variation of ruminal H_2S concentration (Sarturi et al., 2011), indicating that forage NDF could be influencing H_2S production through other associative mechanisms.

Many of the early recorded cases of PEM were a result of cattle consuming water containing 1,000 mg of sulfate/L or greater. Monitoring water sulfate concentrations and limiting access to high-sulfate water are the primary management strategies used by beef cattle producers to decrease the incidence of PEM; however, the availability of ethanol byproducts for cattle feeding has increased. Unfortunately, ethanol co-products can contain elevated S concentrations. Feeding of ethanol byproducts has led to incidences of PEM in locations where PEM has not been documented in the past. Currently, the majority of confirmed cases of PEM are in feedlot cattle that are consuming high-concentrate diets. Limited information is known about the true cause of PEM (Goetsch and Owens, 1987) or the factors that affect H_2S production in vivo. Furthermore, limited information is available regarding subclinical cases of PEM.

The National Animal Health Monitoring System (NAHMS; USDA APHIS, 2000) survey reported that 22.6% of the feedlots involved in the survey, had water that contained 300 or greater mg/L sulfate. The NAHMS survey considered less than 300 mg sulfate/L safe for livestock consumption. A steer weighing 454 kg with an ambient temperature of 14.4°C has a water requirement of approximately 41 L/d (NRC, 1996, 2000). If the water contained 300 mg sulfate/L the steer would be consuming 12.3 g sulfate/d, or 4.1 g elemental S/d. Assuming a DMI of 9.5 kg/d, that animal's S requirement would be 14.3 g S/d; therefore, the daily water intake alone would account for approximately 29% of

TABLE 7-3 Sulfur Concentration (%) in Common Feedstuffs[a]

Feedstuff	Sulfur	±SD
Cracked corn	0.11	0.12
Corn silage	0.10	0.05
Alfalfa hay	0.26	0.07
Distillers grains plus solubles	0.62	0.08
Soybean meal	0.41	0.03
Molasses	0.68	0.29

[a]Adapted from Neuhold (2012).

the total daily S requirement. If that same animal consumed water containing 1,000 mg/L sulfate at an elevated ambient temperature of 26.6°C, its S intake would be 18.1 g or 127% of its daily requirement without any S intake from feed.

Sulfur content can vary greatly both within and between feedstuffs. Table 7-3 demonstrates the variability of S content in common feedstuffs. Corn contributes the majority of S content of a typical finishing feedlot diet, even though corn only contains, on average, 0.11% S. This is a result of the fact that corn makes up between 70 to 90% of a typical feedlot diet. Table 7-4 shows the effect of replacing corn with distillers grains in a typical feedlot finishing diet and the addition of high-sulfate water on the S intake of a feedlot steer. This table demonstrates that the requirement of 0.15% and that the maximum tolerable S concentration can be rapidly exceeded with the combination of distillers grains and water sulfate.

In January 2011, the University of Nebraska and Iowa State University hosted a webinar on the subject of S concentrations in feedlot diets. This webinar questioned whether the maximum concentration set by the NRC (1996, 2000) is appropriate or if this limit should be increased. At the end of the webinar presentation, it was suggested that the maximum limit should be increased to 0.45% based on the data presented. Other work from the University of Nebraska suggests that not only the total concentration of S should be considered but also the type of S. Sarturi et al. (2011) suggested there are two types of S in a ruminant diet, rumen degradable S and rumen undegradable S, indicating the source and type of S should be considered when formulating diets.

TABLE 7-4 Calculated Influence of Water Sulfate Concentration and Distillers Grains Addition to Finishing Diets on Total Dietary Sulfur Content (%)[a]

	Water Sulfate, mg/L[b]			
Items[c]	0	300	1,000	1,500
Finishing diet 0% DG[d] (0.0% S)	0.15	0.20	0.31	0.39
Finishing diet 10% DG (0.62% S)	0.21	0.25	0.36	0.44
Finishing diet 20% DG (0.62% S)	0.24	0.29	0.40	0.48
Finishing diet 30% DG (0.62% S)	0.29	0.34	0.45	0.53
Finishing diet 40% DG (0.62% S)	0.34	0.39	0.50	0.58

[a]Adapted from Neuhold (2012).

[b]Based on 48 L/animal daily water and 10.0 kg/animal daily DMI.

[c]Based on a finishing diet containing steam-flaked corn, corn silage, alfalfa hay, and soybean meal formulated to contain 0.15% sulfur. The diet was formulated to provide a NEg of 1.4 Mcal/kg DM and no less than 13.5% CP.

[d]Distillers grains plus solubles.

Sulfur Sources

Sulfur can be supplemented in ruminant diets as sodium sulfate, ammonium sulfate, calcium sulfate, potassium sulfate, magnesium sulfate, or elemental sulfur. Based on in vitro microbial protein synthesis, the availability of sulfur to ruminal microorganisms from different sources has been ranked from most to least available as L-methionine, calcium sulfate, ammonium sulfate, D,L-methionine, sodium sulfate, sodium sulfide, elemental sulfur, and methionine hydroxy analog (Kahlon et al., 1975). High-sulfate water, high-sulfur-containing forages (alfalfa or Canada thistle), ethanol byproducts (wet distillers grain, dry distillers grain, corn gluten feed, and corn distillers solubles), dried whey, molasses-based feeds, fertilizers, blood meal, and feather meal, and sulfur salts are common beef cattle feed ingredients that can contribute to the S concentration in the diet. Forages receiving ammonium sulfate fertilizers as a source of N may have also increased forage S concentrations. Cows grazing these forages were shown to have lesser liver Cu concentrations vs. cows grazing nonfertilized pastures or pastures receiving an equivalent amount of N from ammonium nitrate (Arthington et al., 2002). In addition, this fertilizer-affected increase in forage S will also impact plant Se uptake. Plant Se uptake, particularly from the selenate form, is impacted by sulfate (Parker et al., 1992; Wu and Huang, 1992). This is an important consideration, particularly in regions of the United States that historically grow forages that are Se deficient.

MICROMINERALS

Chromium

Chromium (Cr) was first shown to be essential for mammals by Schwarz and Mertz (1959). Since then, Cr has been shown to influence carbohydrate metabolism (Mertz, 1993), lipid metabolism (Abraham et al., 1991; Bernhard et al., 2012a), and protein absorption and metabolism (Okada et al., 1983; Kornegay et al., 1997). Recently, Cr supplementation has been reported to decrease the effect of heat stress in lactating Holstein dairy cows through anti-inflammatory mechanisms (Zhang et al., 2014).

Chromium functions as a component of the glucose tolerance factor, which serves to potentiate the action of insulin (Mertz, 1992). The addition of 0.4 mg Cr/kg dietary DM to cattle diets (Bunting et al., 1994; Kegley and Spears, 1995) or 0.0, 5.0, 10.0, or 15.0 mg/animal daily (Sumner et al., 2007) increased glucose clearance rate following intravenous glucose administration. Furthermore, a decrease in insulin release following intravenous glucose administration has been reported in calves (Kegley et al., 1997b) and in grazing pre- and postpartum cows receiving Cr in free-choice mineral feeders (calculated Cr intake based on free-choice mineral intake was 3.5 mg/animal daily or 0.35 mg Cr/kg dietary DM assuming a DMI of 10 kg per cow; Stahlhut et al., 2006a,b) suggesting that Cr potentiates the action of insulin. Based on a Cr dose titration experiment measuring insulin sensitivity, Spears et al. (2012) estimated the dietary Cr requirement for growing heifers to be 0.47 mg of Cr/kg dietary DM. However, variable responses to Cr supplementation on certain aspects of glucose metabolism have also been reported in ruminants as discussed by Spears et al. (2012).

Variable responses to Cr supplementation have made it difficult to determine the specific effect of Cr on the immune system (Spears, 2000). Burton et al. (1994) reported that in newly weaned, stressed feedlot calves, supplementation at 0.5 mg of Cr/kg dietary DM for 30 days after transit to the feedlot increased the magnitude of peak antibody titer response to infectious bovine rhinotracheitis (IBR) vaccination but had no effect on antibody titers to bovine parainfluenza-3 virus (PI-3) vaccination relative to the unsupplemented controls. The addition of 0.4 mg of Cr/kg diet did not affect antibody titer responses to porcine erythrocyte immunization (Kegley et al., 1997a) or bovine herpesvirus-1 (Arthington. et al., 1997) in stressed cattle. In contrast, Cr supplementation at 0.0, 0.1, 0.2, and 0.3 mg Cr/kg DM increased performance (linear response) and helped to alleviate weight loss during an immune challenge in feedlot cattle (Bernhard et al., 2012b). Other experiments have shown that the addition of dietary Cr (0.2 to 1.0 mg/kg) improved growth and certain immune response measures and other indices of immunity in stressed cattle, with the more recent Cr research focusing on glucose metabolism in cattle (Chang and Mowat, 1992; Burton et al., 1993; Moonsie-Shageer and Mowat, 1993; Hayirli et al., 2001; Sumner et al., 2007; Bernhard et al., 2012b; Spears et al., 2012). Dairy cows supplemented with 0.5 mg of Cr/kg dietary DM had greater primary and secondary antibody responses to immunization of an ovalbumin antigen than control cows, but both groups had similar antibody responses to human erythrocytes antigen immunization (Burton et al., 1993). It is unclear as to why the effects of Cr were observed with one antigen and not the other. Inconsistent immune responses to Cr supplementation have also been observed in other livestock species (van Heugten and Spears, 1997; Gentry et al., 1999).

Spears et al. (2012) reported that in 2009, the U.S. Food and Drug Administration (FDA) Center for Veterinary Medicine issued a regulatory discretion letter permitting Cr addition, in the form of chromium propionate, to cattle diets at concentrations not to exceed 0.5 mg Cr/kg dietary DM. As reported by Spears et al. (2012), safety experiments conducted by Lloyd et al. (2010) in lactating dairy cows indicate that Cr supplementation at 4 times the FDA-allowable concentration (2.0 mg of Cr/kg dietary DM) for 120 days did not increase milk, muscle, or fat concentrations of Cr. Based on the review of the literature published since the NRC (1996, 2000) publications, further research is warranted to substantiate setting a dietary Cr requirement.

Based on studies with humans and laboratory animals, organically bound Cr is much more bioavailable than inorganic forms of Cr. The maximum tolerable concentration of trivalent Cr in the chloride form was estimated to be 1,000 mg of Cr/kg diet for cattle (NRC, 1980). No adverse effects were observed in steers fed 4.0 mg chromium polynicotinate complex/kg diet for 70 days (Claeys and Spears, unpublished data, 1999). Hexavalent Cr is much more toxic than the trivalent form (NRC, 1980, 2005).

Cobalt

Cobalt (Co) functions as a component of vitamin B_{12} (cobalamin). Cattle are not dependent on a dietary source of vitamin B_{12} because ruminal microorganisms can synthesize B_{12} from dietary Co. Measurements of the amount of dietary Co converted to vitamin B_{12} in the rumen have ranged from 3 to 13% of Co intake (Smith, 1987). Ruminal bacteria also produce a number of B_{12} analogues that are active in bacteria but apparently inactive in animal tissues (Bigger et al., 1976). Two vitamin B_{12}-dependent enzymes occur in mammalian tissues (Smith, 1987): (1) methylmalonyl CoA mutase is essential for the metabolism of propionate to succinate, as it catalyzes the conversion of L-methylmalonyl CoA to succinyl CoA; and (2) 5-methyltetrahydrofolate homocysteine methyltransferase (methionine synthase) catalyzes the transfer of methyl groups from 5-methyltetrahydrofolate to homocysteine to form methionine and tetrahydrofolate. This reaction is important in the recycling of methionine following transfer of its methyl group.

Cobalt Requirements

The NRC (1996, 2000) set the Co requirement of cattle at 0.10 mg/kg dietary DM based on research by Smith (1987). Cobalt concentrations between 0.07 and 0.11 mg Co/kg DM have been reported to be adequate in various studies (Smith, 1987). Young, rapidly growing cattle seem more sensitive to Co deficiency than older cattle. Feeding a high-concentrate diet could depress ruminal synthesis of vitamin B_{12} and increase production of B_{12} analogues (Walker and Elliot, 1972; Halpin et al., 1984); however, MacPherson and Chalmers (1985) found no evidence that Co requirements were increased when high-concentrate diets were consumed.

Since the NRC (1996, 2000) set the Co requirement of 0.1 mg/kg DM, several experiments have been conducted investigating the influence of Co supplementation on performance and vitamin B_{12} status in cattle. Schwarz et al. (2000) and Stangl et al. (2000) estimated the dietary Co requirement for intact male Simmental cattle consuming corn silage-based diets supplemented with a concentrate containing soybean meal, ground corn, barley, vitamin-mineral premix to be 0.26, 0.24, 0.12, and 0.17 mg Co/kg dietary DM for maximal plasma and liver vitamin B_{12} concentrations, average daily gain (ADG), and DMI, respectively, and suggested that a concentration of 0.20 mg cobalt/kg DM as the recommend requirements for growing cattle after taking into account live animal performance and biochemical data. Furthermore, results of a series of in vitro and in vivo experiments conducted by Tiffany (2003) and Tiffany et al. (2006) indicated that finishing Angus steers require approximately 0.15 mg Co/kg dietary DM based on vitamin B_{12} status, as well as vitamin B_{12} production in a continuous culture fermentation system. Nonetheless, in a separate finishing experiment, Tiffany and Spears (2005) reported no differences in plasma and liver vitamin B_{12} concentrations or overall ADG and DMI when supplemental Co was increased from 0.05 to 0.15. Based on the review of the current literature, the total dietary requirements of finishing cattle fed corn-based diets is 0.15 mg Co/kg dietary DM and limited research indicates that Co requirements may be higher (0.20 mg Co/kg DM) in cattle fed barley-based diets (Schwarz et al., 2000; Stangl et al., 2000; Tiffany, 2003; Tiffany et al., 2006; Spears and Weiss, 2014). Therefore, the recommended Co requirement for beef cattle is 0.15 mg Co/kg DM. Further research is warranted examining the effects of diet type on Co requirements of beef cattle.

Signs of Cobalt Deficiency

Decreased appetite and failure to grow or moderate weight loss are early signs of Co deficiency (Smith, 1987). If the deficiency is allowed to become severe, animals exhibit unthriftiness, rapid weight loss, fatty degeneration of the liver, and pale skin and mucous membranes as a result of anemia. Cobalt deficiency also has been reported to impair the ability of neutrophils to kill yeast and decrease disease resistance (MacPherson et al., 1988). Findings indicate that an inability by ruminal microorganisms to convert succinate to propionate is an early manifestation of Co deficiency (Kennedy et al., 1991). Ruminal and plasma succinate concentrations were greatly elevated in lambs fed Co-deficient diets. Liver vitamin B_{12} or Co concentrations can be used to assess Co status (Smith, 1987). Vitamin B_{12} concentrations in liver of 0.10 µg/g wet weight or less are indicative of Co deficiency. Measurement of serum B_{12} in cattle may be of limited value because of the presence of B_{12} analogues in bovine serum (Halpin et al., 1984).

Signs of Cobalt Toxicity

Cobalt toxicity is not likely to occur unless an error is made in formulating a mineral supplement. Cattle can tolerate approximately 100 times the dietary requirement for Co (NRC, 1980). Signs of chronic Co toxicity, with the exception of elevated liver Co, are similar to those of Co deficiency and include decreased feed intake and BW gain, anemia, emaciation, hyperchromia, debility, and increased liver Co (NRC, 1980). Ely et al. (1948) suggested that calves can tolerate up to 0.86 mg Co/kg BW, which would be approximately 34 mg Co/kg diet assuming a DMI of 2.5% of BW. Young dairy calves given up to 66 mg Co/kg BW for up to 28 weeks showed no adverse effects (Keener et al., 1949). The sulfate, carbonate, and chloride forms of Co were similar in terms of toxicity (Keener et al., 1949). Controlled experiments to access Co tolerance have not been reported in older cattle. Based on the available literature, the maximum tolerable concentrations for Co were set at 25 mg/kg diet for cattle (NRC, 2005).

Cobalt Sources

Soils deficient in Co occur in many areas of the world, including the southeastern Atlantic coast of the United States (Ammerman, 1970). Legumes are generally higher in Co than grasses, and availability of Co in soil is highly dependent on soil pH (Underwood and Suttle, 1999). Increasing soil pH from 5.4 to 6.4 decreased the Co content of ryegrass from 0.35 to 0.12 mg/kg (Mills, 1981). Cobalt can be supplemented to the diet in free-choice mineral mixtures. Feed-grade sources of Co include cobalt sulfate and cobalt carbonate. It is unclear how these two forms of Co compare in terms of relative bioavailability for vitamin B_{12} synthesis. Pellets containing cobalt oxide and finely divided iron and controlled-release glass pellets containing Co have been used in grazing ruminants. Both types of pellets remain in the reticulorumen and release cobalt over an extended period (NRC, 1996, 2000).

Copper

Copper (Cu) functions as an essential component of a number of enzymes including lysyl oxidase, cytochrome oxidase, superoxide dismutase, ceruloplasmin, and tyrosinase (McDowell, 2003).

Copper Requirements

Requirements of Cu can vary from 4 to more than 15 mg/kg dietary DM, depending largely on the concentration of dietary molybdenum (Mo) and S. The recommended concentration of Cu in beef cattle diets is 10 mg Cu/kg diet. This amount should provide adequate Cu if the diet does not exceed 0.25% S and 2 mg Mo/kg. Less than 10 mg Cu/kg diet might meet requirements of feedlot cattle because Cu is more available in concentrate diets than in forage diets. Conflicting results in terms of cattle performance have been noted in previous experiments evaluating the effect of dietary Cu supplementation to corn silage-based growing diets (Ward et al., 1993; Ward and Spears, 1997; 5.0 mg of supplemental

Cu added to basal diets containing 5.2 or 6.2 mg Cu/kg DM; respectively) and high-concentrate finishing diets (10 to 40 mg of supplemental Cu/kg DM; Engle and Spears, 2000a,b; Engle et al., 2000; basal diets contained 4.9 mg Cu/kg DM), but reasons for these variable results are not clear. Differences among experiments in the magnitude of response in liver and plasma Cu to Cu supplementation cannot explain the variable performance responses to increasing dietary Cu. Certain environmental, genetic, and dietary factors can also influence animal performance response to dietary Cu, which might explain some of the variability in experimental results.

Factors Affecting Copper Requirements

Copper requirements for ruminants are greatly increased by Mo and S in excess of 2 mg Mo/kg dietary DM and 0.25% S in most situations (Smart et al., 1986; Underwood and Suttle, 1999; Spears, 2003; Hansen et al., 2008; Kessler et al., 2012; Drewnoski, et al., 2014; Pogge et al., 2014). The antagonistic action of Mo on Cu metabolism is exacerbated when S is also high. Considerable evidence suggests that molybdate and sulfide interact to form thiomolybdates in the rumen (Suttle, 1991). Copper is believed to react with thiomolybdates in the rumen to form insoluble complexes that are poorly absorbed. Moreover, some thiomolybdates are absorbed and affect systemic metabolism of Cu (Gooneratne et al., 1989). Thiomolybdates can result in Cu being tightly bound to plasma albumin and not available for biochemical functions, and they might directly inhibit certain Cu-dependent enzymes. In cattle grazing pastures containing 3 to 20 mg Mo/kg, Cu concentrations in the range of 7 to 14 mg/kg were inadequate (Thornton et al., 1972).

Sulfur decreases Cu absorption, perhaps via formation of copper sulfide in the gut, which can be independent from its role in the Mo-Cu interaction (Suttle, 1974). Decreasing the sulfate content of drinking water high in sulfate from 500 to 42 mg/L by reverse osmosis increased the Cu status of cattle (Smart et al., 1986). A Cu concentration of 10 mg/kg was not adequate in cows receiving sulfated water, which resulted in total dietary S of 0.35% (Smart et al., 1986). High concentrations of Fe (Standish et al., 1969; Campbell et al., 1974; Humphries et al., 1983; Phillippo et al., 1987b; Mullis et al., 2003a) and Zn (Davis and Mertz, 1987) also decrease Cu status and can increase Cu requirements.

As discussed in the dairy report (NRC, 2001), it is difficult to predict the negative impact that S and Mo can have on Cu absorption and metabolism. As summarized and discussed by Underwood and Suttle (1999), the relationship between Cu availability and dietary Cu and Mo is more difficult to predict than the equation of Suttle and McLauchlan (1976; log [Copper Absorbable] = –1.153 – 0.076 [S, g/kg] – 0.013 [S, g/kg × Mo, mg/kg]) would suggest. The effect of S and Mo varies depending on the feedstuff serving as a source of Cu. Underwood and Suttle (1999) concluded that the absorbable amount of Cu in ensiled grass as not being greatly impaired by an increase in dietary Mo but was greatly depressed by the addition of S to the ration. When the diet contained 0.2% S, about 5.5% of the Cu was available, but when the diet contained 0.4% S, the absorbable Cu was decreased to about 1.5%. In hays, the inhibitory effect of Mo is present but relatively small. As S increases from 0.2 to 0.4% in hays, the percentage of absorbable Cu decreases by 20 to 30%. The percentage absorbable Cu in fresh grasses is lower than that of hays or ensiled grasses at any given S or Mo concentration and the addition of S or Mo drastically decreases Cu absorbability. The inhibitory effect of increasing diet Mo on Cu absorption is greatest when dietary Mo is low and seems to reach a plateau once diet Mo is about 4 to 5 mg/kg DM. Beyond this point, higher dietary Mo concentrations do not impair Cu absorbability significantly further (Gengelbach et al., 1994; Underwood and Suttle,1999).

Copper requirements can also be affected by breed. Simmental cattle excrete more Cu in their bile than Angus cattle (Gooneratne et al., 1994) and could differ in duodenal transporters involved in Cu absorption (Fry et al., 2013). Ward et al. (1995) reported that Simmental and Charolais cows and their calves were more susceptible to Cu deficiency than Angus cattle when fed the same diet. Mullis et al. (2003b) reported that Simmental heifers fed a corn silage-based diet containing no supplemental Cu (basal diet contained 4.4 mg of Cu/kg DM) had lower ($P < 0.05$) plasma Cu concentrations than Angus heifers at the end of the 112-d experiment. Furthermore, day-112 plasma Cu concentrations for Simmental heifers (0.37 mg Cu/L) were below concentrations indicative of Cu deficiency (plasma <0.60 mg Cu/L; Underwood, 1971; Mills, 1987), whereas plasma Cu concentrations for Angus heifers remained above deficient concentrations (0.80 mg Cu/L) throughout the entire experiment. However, it is difficult to predict overall Cu status from plasma Cu alone (Claypool et al., 1975).

Signs of Copper Deficiency

Copper deficiency is a widespread problem in many areas of the United States and Canada. Signs that have been attributed to Cu deficiency include

- anemia,
- decreased growth,
- depigmentation and changes in the growth and physical appearance of hair,
- cardiac failure,
- bones that are fragile and easily fractured,
- diarrhea, and
- low reproduction characterized by delayed or depressed estrus (Underwood and Suttle, 1999).

Achromotrichia or lack of hair pigmentation is generally the earliest clinical sign of Cu deficiency. Copper deficiency also decreases the ability of isolated neutrophils to kill yeast (Boyne and Arthur, 1981); and Cu deficiency in grazing lambs increased susceptibility to bacterial infections (Woolliams et al., 1986). As discussed in the Molybdenum section, some of the abnormalities that have been attributed to Cu deficiency might be caused by molybdenosis rather than Cu per se. Copper is poorly absorbed in ruminants with a developed rumen. Absorbed Cu is excreted primarily via the bile with small amounts lost in the urine (Gooneratne et al., 1989). Considerable storage of Cu can occur in the liver.

Signs of Copper Toxicity

Copper toxicity can occur in cattle as a result of excessive supplementation of Cu or the use of feeds that have been contaminated with Cu from agricultural or industrial sources. The liver can accumulate large amounts of Cu before signs of toxicity are observed. When Cu is released from the liver in large amounts (hemolytic crisis), hemolysis, methemoglobinemia, hemoglobinuria, jaundice, icterus, widespread necrosis, and death often occur (NRC, 1980, 2005). The maximum tolerable concentration of Cu for cattle has been estimated at 40 mg Cu/kg diet (NRC, 2005). Higher concentrations of Cu can be tolerated by cattle for several weeks of even months; however, long-term feeding of diets with slightly less than 40 mg Cu/kg DM have been reported to result in Cu toxicosis (Bradley, 1993).

The concentration of Cu needed to cause toxicity will depend on the concentration of Cu antagonists (i.e., Mo, S, Fe, etc.) in the diet as well as animal age and breed. Preruminant calves are much more sensitive to Cu toxicity than older cattle, particularly forage-fed cattle with a functioning rumen. In young preruminant calves, feeding 115 mg Cu/kg DM for 91 days resulted in signs of toxicity (Shand and Lewis, 1957). Reductions in performance have also been reported in Holstein and Ayrshire-Holstein calves fed milk replacer containing 50 or 200 mg Cu/kg DM (Jenkins and Hidiroglou, 1989). Lactating Holstein dairy cows tolerated 40 (Engle et al., 2001) or 80 mg (Du et al., 1996) supplemental Cu/kg DM for 60 days. However, over the 60-day period, liver Cu concentrations increased daily by over 7 mg Cu/kg DM in both experiments. It is unknown if Cu toxicity would have developed if the cattle were fed for a longer period of time. Furthermore, forage-fed cattle consuming high-S forages and supplements can tolerate and may require Cu-fortified diets that approach this suggested maximum concentration (Arthington and Pate, 2002; Arthington et al., 2002; Arthington, 2005).

Copper Sources

Forage Cu concentrations are of limited value in assessing Cu adequacy unless forage concentrations of Cu antagonists such as Mo, S, and Fe are also considered. Liver Cu concentrations less than 20 mg/kg on a DM basis or plasma concentrations less than 0.6 μg Cu/mL are indicative of deficiency (Underwood, 1971). Nonetheless, in the presence of high dietary Mo and S, Cu in liver and plasma might not accurately reflect Cu status because the Cu can exist in tightly bound forms unavailable for biochemical functions (Suttle, 1991). Forages vary greatly in Cu content, depending on plant species and available Cu in the soil (Minson, 1990). Legumes are usually higher in Cu than grasses. Milk and milk products are low in Cu. Cereal grains generally contain 4 to 8 mg Cu/kg, and oilseed meals and leguminous seeds contain 15 to 30 mg Cu/kg.

Copper is usually supplemented to diets or free-choice mineral mixtures in the sulfate, carbonate, or oxide forms. Studies indicate that copper oxide is very poorly available relative to copper sulfate (Langlands et al., 1989b; Kegley and Spears, 1994). In early studies, copper carbonate was at least equal to copper sulfate (Chapman and Bell, 1963). Various organic forms of Cu also are available. In calves fed diets high in Mo, copper proteinate was more available than copper sulfate (Kincaid et al., 1986); however, Wittenberg et al. (1990) found similar availability of Cu from copper proteinate and copper sulfate in steers fed high-Mo diets. Studies comparing copper lysine to copper sulfate have yielded inconsistent results. Ward et al. (1993) reported that copper lysine and copper sulfate were of similar bioavailability when fed to cattle; however, Nockels et al. (1993) found that copper lysine was more available than copper sulfate. Since the 1990s, numerous experiments have been conducted evaluating the influence of Cu source on feedlot and grazing cattle Cu tissue concentrations, animal performance, immune function, and other economically relevant parameters and have yielded variable results. Injectable forms of Cu such as copper glycinate or copper ethylenediaminetetraacetic acid (EDTA) have been given at 3- to 6-month intervals to prevent Cu deficiency (Underwood and Suttle, 1999). Although feed-grade copper oxide is largely unavailable, copper oxide needles, which remain in the gastrointestinal tract and slowly release Cu over a period of months, have been used as a Cu source for cattle (Cameron et al., 1989). In crossbred steers (Brahman × British dams and Brangus sires), copper oxide needle administration increased liver Cu concentrations and had no impact on OM intake or total tract OM digestibility. However, apparent digestibility of NDF and crude protein were greater for control cattle compared to cattle receiving a Cu bolus. These findings are difficult to explain because of the lack of a treatment effect on total OM digestibility (Arthington, 2005).

Iodine

Iodine (I) functions as an essential component of the thyroid hormones thyroxine (T4) and triiodothyronine (T3), which regulate the rate of energy metabolism in the body. Between 70 and 80% of dietary I is absorbed as iodide from the rumen with considerable resecretion occurring in the abomasum (Miller et al., 1988). Iodide that is secreted into the abomasum is largely reabsorbed from the small and large intestines. Absorbed iodide is mostly taken up by the thyroid gland for thyroid hormone synthesis or is excreted in the urine. In lactating cows, approximately 8% of dietary I is secreted in milk (Miller et al., 1988). When the thyroid hormones are catabolized, much of the I is reused by the thyroid gland.

Iodine Requirements

Iodine requirements of beef cattle are not well established; 0.5 mg I/kg diet should be adequate unless the diet contains goitrogenic substances that interfere with I metabolism. Iodine requirements have been estimated by measuring thyroid hormone secretion rate (ARC, 1980). Miller et al. (1988) calculated the theoretical I requirement to be 0.6 mg/100 kg BW assuming

- a daily thyroxine secretion rate of 0.2 to 0.3 mg I/100 kg BW,
- 30% uptake of dietary I by the thyroid, and
- 15% recycling of thyroxine iodine.

This would correspond to 0.2 to 0.3 mg I/kg in the total diet, depending on feed intake.

Factors Affecting Iodine Requirements

Goitrogenic substances in the feed can increase I requirements substantially (two- to fourfold) depending on the amount and type of goitrogens present. The cyanogenetic goitrogens include the thiocyanate derived from cyanide in white clover and the glucosinolates found in some *Brassica* forages such as kale, turnips, and rape. They impair I uptake by the thyroid, and their effect can be overcome by increasing dietary I. Soybean meal and cottonseed meal also have a goitrogenic effect (Miller et al., 1975). The thiouracil goitrogens are found in *Brassica* seeds and inhibit iodination of tyrosine residues in the thyroid gland. The action of thiouracil goitrogens is more difficult to reverse with I supplementation.

Signs of Iodine Deficiency

The first sign of I deficiency is usually enlargement of the thyroid (goiter) in the newborn (Miller et al., 1988). Iodine deficiency may result in calves born hairless, weak, or dead; decreased reproduction in cows characterized by irregular cycling, low conception rate, and retained placenta; and decreased libido and semen quality in males (McDowell, 2003). Deficiency signs may not appear for more than a year after cattle are fed an I-deficient diet. Protein-bound I, thyroid weight in newborns, and milk I have been used to assess iodine status (Underwood and Suttle, 1999).

Signs of Iodine Toxicity

The maximum tolerable level of I is 50 mg/kg diet (NRC, 1980, 2005). The degree to which I at this concentration elicits signs of toxicity might depend on the source. In calves, 50 mg/kg I as calcium iodate decreased weight gain and feed intake, and caused coughing and excessive nasal discharge (Newton et al., 1974). Iodine in the form of ethylenediamine dihydroiodide (EDDI) has been fed at concentrations exceeding 50 mg/kg without adverse effects in calves and lactating cows (NRC, 1980, 2005).

Iodine Sources

The I content of feeds depends on the I available in the soil. In the United States, much of the Northeast, the Great Lakes, and Rocky Mountain regions are deficient in I (Underwood and Suttle, 1999). Iodine is usually supplemented in diets or in free-choice minerals as calcium iodate or EDDI, an organically bound form of I. Both forms are highly available and stable in mineral supplements and diets. Forms of iodide such as potassium or sodium iodide are less stable, and considerable losses can occur as a result of exposure to heat, moisture, light, and other minerals. Ethylenediamine dihydroiodide has been widely used in cattle to prevent foot rot, but the amount of EDDI fed to prevent foot rot is much higher than dietary requirements. At present, 49.9 mg of I (or less than 50 mg/animal daily) from EDDI is the maximum concentration that can be fed per animal daily as stated in the U.S. FDA Compliance Policy Guide (CPG § 651.100).

Iron

Iron (Fe) is an essential component of a number of proteins involved in oxygen transport or utilization. These proteins include hemoglobin, myoglobin, and a number of cytochromes and Fe-S proteins involved in the electron transport chain. Several mammalian enzymes also either contain Fe or are activated by Fe (McDowell, 2003). More than 50% of the Fe in the body is present in hemoglobin, with smaller amounts present in other Fe-requiring proteins and enzymes, and in protein-bound stored Fe.

Iron Requirements

The Fe requirement is approximately 50 mg/kg diet in beef cattle. Studies with young calves fed milk diets have indicated that 40 to 50 mg Fe/kg is adequate to support growth and prevent anemia (Bremner and Dalgarno, 1973; Bernier et al., 1984). Iron requirements of older cattle are not well defined. Requirements in older cattle are probably less than in young calves because considerable recycling of Fe occurs when red blood cells turn over (Underwood, 1977), and in older animals blood volume is not increasing, or at least not to the extent that it is in young animals.

Signs of Iron Deficiency

A deficiency of Fe results in anemia (hypochromic microcytic), listlessness, decreased feed intake and weight gain, pale mucus membranes, and atrophy of the papillae of the tongue (Blaxter et al., 1957; Bremner and Dalgarno, 1973). Iron deficiency can occur in young calves fed exclusively milk, especially if they are housed in confinement. Most practical feedstuffs are more than adequate in Fe, and Fe deficiency is unlikely in other classes of cattle unless parasite infestations or diseases exist that cause chronic blood loss. In the absence of blood loss, only small amounts of iron are lost in the urine and feces (McDowell, 2003).

Signs of Iron Toxicity

Iron toxicity causes diarrhea, metabolic acidosis, hypothermia, and decreased gain and feed intake (NRC, 1980, 2005). The maximum tolerable concentration of Fe for cattle has been estimated at 500 mg Fe/kg DM (NRC, 2005). Dietary Fe concentrations as low as 250 to 500 mg/kg have caused Cu depletion in cattle (Bremner et al., 1987; Phillippo et al., 1987b). In areas where drinking water or forages are high in Fe, dietary Cu might need to be increased to prevent Cu deficiency. The tolerable concentrations were set based on the animal having normal Fe status and fed an Fe source that is highly digestible. Therefore, higher dietary concentrations can probably be fed when Fe is supplied from sources with low bioavailability or if the animal is in a deficient state (NRC, 2005).

Iron Sources

Cereal grains normally contain 30 to 60 mg Fe/kg; oilseed meals contain 100 to 200 mg Fe/kg (Underwood and Suttle, 1999). With the exception of milk and milk products, feeds of animal origin are high in Fe, with meat and fish meal containing 400 to 500 mg Fe/kg. Blood meal usually has more than 3,000 mg Fe/kg. The Fe content of forages is highly variable but most forages contain from 70 to 500 mg Fe/kg. Much of the variation in forage Fe is probably caused by soil contamination. Water and soil ingestion also can be significant sources of Fe for beef cattle. Availability of Fe from forages seems to be less than from most supplemental Fe sources (Raven and Thompson, 1959; Thompson and Raven, 1959). Iron from soil is probably of low availability; however, research by Healy (1972) indicated that a significant amount of Fe from various soil types was soluble in ruminal fluid. Furthermore, an in vitro bioaccessibility experiment performed by Hansen and Spears (2009) indicated that Fe from soil-contaminated corn silage was more soluble after silage fermentation.

Iron is generally supplemented in diets as ferrous sulfate, ferrous carbonate, or ferric oxide. The availability of Fe is greatest for ferrous sulfate, with ferrous carbonate being intermediate (Ammerman et al., 1967; McGuire et al., 1985). Ferric oxide is basically unavailable (Ammerman et al., 1967).

Manganese

Manganese (Mn) functions as a component of the enzymes pyruvate carboxylase, arginase, and superoxide dismutase and as an activator for a number of enzymes (Hurley and Keen, 1987), including a number of hydrolases, kinases, transferases, and decarboxylases. Of the many enzymes that can be activated by Mn, only the glycosyltransferases are known to specifically require Mn.

Manganese Requirements

The Mn requirement for growing and finishing cattle is approximately 20 mg of Mn/kg diet. Skeletal abnormalities were noted in calves from cows fed diets containing 15.8 mg Mn/kg, but these were not observed in calves from cows fed diets supplemented to contain 25 mg Mn/kg (Rojas et al., 1965). The quantity of Mn needed for maximal growth is less than that required for normal skeletal development. Manganese requirements for reproduction are greater than for growth and skeletal development, and the recommended concentration for breeding cattle is 40 mg/kg. Cows fed a diet containing 15.8 mg Mn/kg had lower conception rates than cows fed 25 mg Mn/kg (Rojas et al., 1965). Heifers fed 10 mg Mn/kg exhibited impaired reproduction (delayed cycling and decreased conception rate) compared with those fed 30 mg Mn/kg, but growth was similar for the two groups (Bentley and Phillips, 1951). Hansen et al. (2006a) reported that a diet containing 16.6 mg Mn/kg dietary DM fed to gestating heifers was not adequate for proper fetal development; however, supplementing 50 mg Mn/kg dietary DM was adequate (Hansen et al., 2006a). However, no other doses of Mn were fed in this experiment, making it impossible to estimate an Mn requirement from the data collected.

Supplementing a corn silage-based diet containing 32 mg Mn/kg with 14 mg Mn/kg from a manganese polysaccharide complex decreased services per conception from 1.6 to 1.1 but did not affect overall conception rate in beef cows (DiCostanzo et al., 1986). Since the publication of the NRC (1996, 2000) reports, research conducted by Legleiter et al. (2005) and Hansen et al. (2006b) would support the continued recommendation of 20 mg Mn/kg for growing and finishing cattle.

Signs of Manganese Deficiency

Inadequate intake of Mn in young animals results in skeletal abnormalities that can include stiffness, twisted legs, enlarged joints, and decreased bone strength (Hurley and Keen, 1987). In older cattle, Mn deficiency causes low reproductive performance characterized by depressed or irregular estrus, low conception rate, abortion, stillbirths, and low birth weights.

Factors Affecting Manganese Requirements

Absorption of Mn from $^{54}MnCl$ in lactating dairy cows was less than 1% (Van Bruwaene et al., 1984), and little is known concerning dietary factors that influence Mn absorption. Some evidence suggests that high dietary Ca and P can increase Mn requirements (Hawkins et al., 1955; Dyer et al., 1964; Lassiter et al., 1972). Biliary excretion of Mn plays an important role in Mn homeostasis, but little excretion of Mn occurs via the urine (Hidiroglou, 1979). High dietary Fe (Hansen et al., 2010), and possibly high dietary Ca and P (Spears, 2003), might also interfere with Mn availability.

Signs of Manganese Toxicity

In the *Mineral Tolerances of Domestic Animals* (NRC, 1980, 2005), the maximum tolerable concentration of Mn was set at 1,000 mg/kg, at least on a short-term basis. Calves fed 1,000 mg Mn/kg from manganese sulfate for 100 days showed no adverse effects (Cunningham et al., 1966); >2,000 mg Mn/kg was required in this study to decrease growth and feed intake. In young calves fed milk replacer, 500 mg Mn/kg decreased weight gain and feed efficiency (Jenkins and Hidiroglou, 1991).

Manganese Sources

The concentration of Mn in forages varies greatly depending on plant species, soil pH, and soil drainage (Minson, 1990). Forages generally contain adequate Mn, assuming that the Mn is available from the forage for absorption. Corn silage can be low, or at best marginal, in Mn content (Buchanan-Smith et al., 1974). Cereal grains usually contain between 5 and 40 mg Mn/kg, with corn being especially low (Underwood and Suttle, 1999). Plant protein sources normally contain 30 to 50 mg Mn, whereas animal protein sources contain only 5 to 15 mg Mn/kg. Manganese can be supplemented to ruminant diets as manganese sulfate, manganese oxide, or various organic forms (manganese methionine, manganese proteinate, manganese polysaccharide complex, or manganese amino acid chelate). Manganese sulfate is more available than manganese oxide (Wong-Ville et al., 1989; Henry et al., 1992). Compared with manganese sulfate, relative availability of Mn from manganese methionine is approximately 120% (Henry et al., 1992).

Molybdenum

Molybdenum (Mo) functions as a component of the enzymes xanthine oxidase, sulfite oxidase, and aldehyde oxidase (Mills and Davis, 1987). Requirements for Mo, however, are not established. There is no evidence that Mo deficiency occurs in cattle under practical conditions, but Mo might enhance microbial activity in the rumen in some instances. The addition of 10 mg Mo/kg to a high-roughage diet containing 1.7 mg Mo/kg increased the rate of in situ DM disappearance from the rumen of cattle (Shariff et al., 1990). In situ DM disappearance was not improved by Mo supplementation when steers were fed a ground barley-based diet containing 1.0 mg Mo/kg (Shariff et al., 1990). Molybdenum added to a semi-purified diet containing 0.36 mg Mo/kg increased growth and cellulose digestion in lambs (Ellis et al., 1958). In contrast, in three subsequent studies with lambs fed semi-purified or practical diets, no responses to added Mo were observed (Ellis and Pfander, 1960). Therefore, there is insufficient information to determine an Mo requirement for beef cattle.

Factors Affecting Molybdenum Utilization

Metabolism of Mo is greatly affected by Cu and S, with both minerals acting antagonistically. Sulfide and molybdate interact in the rumen to form thiomolybdates, resulting in decreased absorption and altered postabsorptive metabolism of Mo (Mills and Davis, 1987). Sulfate shares common transport systems with molybdate in the intestine and kidney, thus decreasing intestinal absorption and increasing urinary excretion of molybdate (Mills and Davis, 1987). It is well documented that relatively low dietary Mo can cause Cu deficiency and that increasing dietary Cu can overcome Mo toxicity.

Signs of Molybdenum Toxicity

In cattle, high concentrations of Mo (20 mg Mo/kg or greater) can cause toxicity characterized by diarrhea, anorexia, loss of weight, stiffness, and changes in hair color (Ward, 1978; Kessler et al., 2012). Providing large amounts of Cu will usually overcome molybdenosis. The maximum tolerable concentration of Mo for cattle has been estimated to be between 5 and 10 mg Mo/kg dietary DM (NRC, 2005). Molybdenum concentrations of less than 10 mg/kg can result in Cu deficiency, depending on the length of time the cattle are exposed to it and the concentration of dietary copper. Experimental results suggest that a relatively low concentration of Mo can exert direct effects on certain metabolic processes independent of alterations in Cu status. The addition of 5 mg Mo/kg to diets containing 0.1 mg Mo/kg caused Cu depletion associated with decreased growth and feed efficiency, loss of hair pigmentation, changes in hair texture, and infertility in heifers (Bremner et al., 1987; Phillippo et al., 1987a,b). In these same studies, cattle fed high dietary Fe had similar Cu status—based on plasma Cu, liver Cu, and ceruloplasmin and superoxide dismutase activity—as heifers fed Mo, but they did not show clinical signs of Cu deficiency. Supplementation with 5 mg Mo/kg starting at 13 to 19 weeks of age increased age at puberty, decreased live weight of heifers at puberty, and decreased conception rate (Phillippo et al., 1987a). Feeding beef cows and their calves an additional 5 mg Mo/kg decreased calf gains from birth to weaning by 28%, whereas calf gains were not affected by the addition of 500 mg Fe/kg (Gengelbach et al., 1994). Based on the responses of growth, liver Cu concentration, and plasma Cu distribution, Kincaid (1980) suggested the minimal toxic concentration of Mo in drinking water for calves was between 10 and 50 mg/L, and the critical Cu:Mo ratio is <0.5 when animals were given diets containing 13 mg Cu/kg and 0.29% S.

Molybdenum Sources

Forages vary greatly in Mo concentration depending on soil type and soil pH. Neutral or alkaline soils coupled with high moisture and OM favor Mo uptake by forages (McDowell, 2003). Cereal grains and protein supplements are less variable in Mo than forages.

Nickel

Nickel (Ni) deficiency has been produced experimentally in a number of animals (Nielson, 1987); however, the function of Ni in mammalian metabolism is unknown. Nickel is an essential component of urease in ureolytic bacteria (Spears, 1984). Supplementation of Ni to ruminant diets has increased ruminal urease activity in a number of studies (Spears, 1984; Oscar and Spears, 1988).

Research data are not sufficient to determine Ni requirements of beef cattle. The maximum tolerable concentration of Ni was estimated to be 50 mg/kg diet (NRC, 1980, 2005). Nonetheless, growing steers fed diets supplemented with 50 mg Ni/kg in the Cl form for 84 days showed no adverse effects (Oscar and Spears, 1988).

Selenium

Selenium (Se) was discovered in 1817 by Jons Berzelius and was recognized as a toxic substance for decades (Sunde, 1997). In the 1950s, Se was identified as an essential nutrient (Schwarz and Foltz, 1957; Patterson et al., 1957). In 1973, glutathione peroxidase was identified as the first known Se metalloenzyme (Rotruck et al., 1973). Glutathione peroxidase catalyzes the reduction of hydrogen peroxide and lipid hydroperoxides, thereby preventing oxidative damage to body tissues (Hoekstra, 1974). A second selenometalloenzyme, iodothyronine 5′-deiodinase, was identified (Arthur et al., 1990). This enzyme catalyzes the deiodination of T4 to the more metabolically active T3 in tissues. Selenium is also found in other proteins but the functions of those proteins remain unclear (Deagen et al., 1987; Yeh et al., 1997).

In mammalian tissues, organically bound Se is found in two large protein pools. The pools are represented by the specific amino acids that contain Se: (1) selenocysteine (SeCys), and (2) selenomethionine (SeMet). The Se-containing proteins glutathione peroxidase and deiodinase mentioned above are two of 25 known selenoproteins currently identified by genomic techniques (Beckett and Arthur, 2005; Suttle, 2010). Selenoproteins are synthesized by direct insertion of SeCys by a specific tRNA, which is coded for by the UGA codon that also has traditionally been identified with stop functions (Hatfield, 2001). The insertion of SeCys during translation to make specific selenoproteins is highly regulated (Sunde, 2001a). This unique amino acid (SeCys), having Se in place of the S molecule represents the 21st amino acid that is necessarily incorporated into mammalian proteins (Sunde 2001a,b; Reeves and Hoffmann, 2009). The inclusion of SeCys into selenoproteins is needed for functionality in all cases where the function of the respective selenoprotein is known.

The second protein pool is based on SeMet, which is methionine with Se instead of S. Consumption of SeMet is common because most plant and animal proteins will nondiscriminately incorporate SeMet in the place of Met into proteins during protein synthesis. Consequently, dietary SeMet will increase muscle Se concentrations to a much greater extent than SeCys. When SeMet is catabolized, the resulting Se molecule can then be incorporated into SeCys and then into selenoproteins as needed. Therefore, both organically bound forms of Se and inorganic salt forms can be used to meet the Se requirement of beef cattle. Interestingly, Se from organically bound forms (predominantly SeMet) seems to cross the ruminant placenta with greater efficiency than dietary inorganic forms (Taylor et al., 2009).

Selenium Requirements

Based on the available research data, the Se requirement of beef cattle can be met by 0.1 mg Se/kg dietary DM. Clinical or subclinical signs of Se deficiency have been reported in beef cows and calves receiving forages containing 0.02 to 0.05 mg Se/kg (Morris et al., 1984; Hidiroglou et al., 1985; Spears et al., 1986); however, calves housed in confinement have been fed semi-purified diets containing 0.02 to 0.03 mg Se/kg for months without showing clinical signs of deficiency, despite very low activities of glutathione peroxidase (Boyne and Arthur, 1981; Siddons and Mills, 1981; Reffett et al., 1988). Even in the absence of clinical deficiency signs, calves have decreased neutrophil activity (Boyne and Arthur, 1981) and humoral immune response (Reffett et al., 1988).

Cattle grazing areas of elevated Se concentration in the soil and forages or fed diets with elevated Se will have increased muscle Se concentrations (Hintze et al., 2001, 2002; Lawler et al., 2004). In these cases, increased muscle Se is predominantly in the form of SeMet. Muscle enrichment of Se can be used by ruminants as a nutrient reserve with the Se associated with SeMet being released over time at the rate of muscle protein turnover (Taylor et al., 2014). Elevated dietary levels of Se have also been shown to increase maternal intestinal (Soto-Navarro et al., 2004; Reed et al., 2007) and mammary gland vascularity (Vonnahme et al., 2011; Neville et al., 2013), which could have implications for nutrient transport to both the developing and neonatal offspring.

Factors Affecting Selenium Requirements

Factors that affect Se requirements are not well defined. The function of vitamin E and Se are interrelated, and a diet low in vitamin E can increase the amount of Se needed to prevent certain abnormalities such as nutritional muscular dystrophy (white muscle disease; Miller et al., 1988). High dietary S has resulted in an increased incidence of white muscle disease in some but not all studies (Miller et al., 1988) and high-S byproduct feeds decreased spermatozoa concentrations in young rams (Van Emon et al., 2013). In sheep, the occurrence of white muscle disease is greater when legume hay, rather than nonlegume hay, is consumed, even when Se contents are similar (Whanger et al., 1972). Harrison and Conrad (1984) reported that Se absorption in dairy cows was minimal at low (0.4%) and high (1.4%) Ca intakes and maximal when dietary Ca was 0.8%. In young calves, varying dietary Ca from 0.17 to 2.35% did not significantly affect Se absorption (Alfaro et al., 1987). High concentrations of unsaturated fatty acids in the diet or various stressors (environmental or dietary) also can increase the requirement for Se.

The form of Se also can affect dietary requirements. Selenium is generally supplemented in animal diets as sodium selenite, whereas selenomethionine is the predominant form of Se in most feedstuffs. Selenium from selenomethionine or an Se-containing yeast was approximately twice as available as sodium selenite or cobalt selenite in growing heifers (Pehrson et al., 1989). Availability of Se from sodium selenate was similar to sodium selenite (Podoll et al., 1992).

Selenium is absorbed primarily from the duodenum, with little or no absorption from the rumen or abomasum. Absorption of Se in ruminants is much lower than in nonruminants (Wright and Bell, 1966). The lower absorption of Se is believed to be related to the reduction of selenite to insoluble forms in the rumen. Absorption of SeMet seems to follow normal absorption mechanisms for methionine. Fecal excretion is greater than urinary excretion in mature ruminants. Pulmonary excretion of Se is important when intakes of selenium are high (Ganther et al., 1966).

Signs of Selenium Deficiency

White muscle disease in young ruminants is a common clinical sign of Se deficiency that results in degeneration and necrosis in both skeletal and cardiac muscle (Underwood and Suttle, 1999). Affected animals may show stiffness, lameness, or even cardiac failure. Other signs of Se deficiency that have been observed include unthriftiness (often times with weight loss and diarrhea; Underwood and Suttle, 1999), anemia with presence of Heinz bodies (Morris et al., 1984), and increased mortality and decreased calf weaning weights (Spears et al., 1986). Selenium-depleted cattle have shown decreased immune responses in a number of studies (Stabel and Spears, 1993). Arthur et al. (1988) reported that Se-deficient cattle had increased T4 and decreased T3 concentrations in plasma relative to Se-supplemented cattle. Depressed activity of iodothyronine 5′-deiodinase could explain the unthriftiness and poor growth often observed in Se deficiency. Decreases in glutathione peroxidase activity associated with Se deficiency can explain the occurrence of white muscle disease, Heinz body anemia, and possibly other signs of Se deficiency.

Selenium concentrations in plasma, serum, and whole blood, and glutathione peroxidase activities in plasma, whole blood, and erythrocytes, have been used to assess Se status. Glutathione peroxidase activities indicative of an Se deficiency can vary from one laboratory to another depending on assay conditions. Langlands et al. (1989a) concluded from a number of on-farm studies with cattle in Australia that Se concentrations in whole blood and plasma were poor indicators of responsiveness to Se supplementation unless unthriftiness was apparent.

Prevention and control of selenosis can be more challenging than trying to treat for an Se deficit. The best way to treat selenosis is to remove the animals from the area that is affected (McDowell, 2003). Other methods that exist include treatment of the soil to decrease the uptake in plants, treatment of the animals to decrease Se absorption or increase Se excretion, and modification of the diet (Underwood and Suttle, 1999). Feedstuffs that are high in Se can be blended with feedstuffs that are rather low in Se (McDowell, 2003), but all these options have practical limitations.

Signs of Selenium Toxicity

The maximum tolerable concentration of Se was set at 2 mg/kg dietary DM by NRC (1980). This recommendation has been challenged (Underwood and Suttle 1999; McDowell, 2003; NRC, 2005) as an underestimate of Se tolerance in ruminants under practical dietary conditions. In the NRC (2005) publication, the committee discussed the evidence in ruminants and adjusted the recommended maximum tolerable concentrations upward to 5 mg Se/kg DM. Since that time, additional data have supported that recommendation (Davis et al., 2006a,b, 2008; Neville et al., 2008; Juniper et al., 2008; Taylor et al., 2009). No signs of Se toxicosis were noted in controlled feedlot experiments where beef cattle were fed (for greater than 100 days) feedlot diets containing high seleniferous feedstuffs (wheat, hay, or alfalfa/grass hay) where the total mixed diets exceeded 2.0 mg Se/kg DM (Hintze et al., 2002; Lawler et al., 2004). Similar results have been reported in ruminants fed dietary supplements fortified with Se (Cristaldi et al., 2005; Taylor, 2005; Davis et al., 2006a,b; Juniper et al., 2008). Collectively, these experiments suggest that the toxicity of Se depends on how Se is incorporated into feedstuffs and suggests the need for further evaluation of the maximum tolerable concentration of Se in beef cattle under practical dietary conditions.

Selenium toxicity can occur as a result of excessive Se supplementation or consumption of plants naturally high in Se. Many species in the plant genera *Astragalus* and *Stanleya* grow primarily on seleniferous areas and can accumulate up to 3,000 mg Se/kg. Consumption of forages containing 5 to 40 mg Se/kg results in chronic toxicosis (alkali disease). Chronic toxicity signs include lameness; anorexia; emaciation; loss of vitality; sore feet; cracked, deformed, and elongated hoofs; liver cirrhosis; nephritis; and loss of hair from the tail (Rosenfeld and Beath, 1964). Acute Se toxicity was previously referred to as "blind staggers" as animals thought to be suffering from Se toxicity exhibited labored breathing, diarrhea, ataxia, abnormal posture, and death from respiratory failure (NRC, 1980, 2005). However, the neuropathology of blind staggers cannot be reproduced by elevated concentrations of pure Se alone. This indicates that other factors including alkaloid poisoning, starvation, and polioencephalomalacia might be responsible for or confounded with the disorder (O'Tool and Raisbeck, 1995; NRC, 2005).

If an animal is suffering from acute signs of Se toxicity, it can be treated by increasing Se excretion from the body (McDowell, 2003). For cattle that are suffering from chronic selenosis, 4 to 5 g of naphthalene for 5 days can be used to increase Se excretion (McDowell, 2003). The removal of the Se source from the diet will also help with the removal of Se from the body and help restore kidney function if there is not too much damage (McDowell, 2003).

In managing selenosis, the best control for Se is the rotation of pastures and the use of supplemental feeds that are low in Se concentration (McDowell, 2003). The dilution of high-Se feeds with low-Se feeds is a common practice in poultry and swine operations and is also used in diets for finishing cattle (McDowell, 2003). Feeding livestock feeds that are from low-Se areas is the best way to control selenosis (McDowell, 2003).

Selenium Sources

Feedstuffs grown in many areas of the United States and Canada are deficient or at least marginally deficient in Se. Selenium-deficient areas are located in the northwestern, northeastern, and southeastern parts of the United States. The Se content of forages and other feedstuffs varies greatly depending on plant species, agronomic biofortification (Whelan, 1989; Whelan et al., 1994a,b; Hall et al., 2011, 2013), and particularly the Se content of the soil. Selenium can legally be supplemented in beef cattle diets to provide 3 mg/animal daily or 0.3 mg/kg in the complete diet as stated in the Code of Federal Regulations (21 CFR § 573.920 [2013]). Alternate methods of supplementing Se include injecting Se every 3 to 4 months or at critical production stages and using boluses retained in the rumen that release Se over a period of months (Hidiroglou et al., 1985; Campbell et al., 1990).

Zinc

Zinc (Zn) functions as an essential component of a number of important metalloenzymes such as Cu-Zn superoxide dismutase, carbonic anhydrase, alcohol dehydrogenase, carboxypeptidase, alkaline phosphatase, and RNA polymerase, which affect metabolism of carbohydrates, proteins, lipids, and nucleic acids (Hambidge et al., 1986). In addition, other enzymes are activated by Zn. Enzymes that require Zn are involved in nucleic acid, protein, and carbohydrate metabolism (Hambidge et al., 1986) and reproduction (Graham, 1991). Zinc also is important for normal development and functioning of the immune system.

Zinc Requirements

The recommended requirement of Zn in beef cattle diets is 30 mg of Zn/kg dietary DM. This concentration will satisfy requirements in most situations. Pond and Oltjen (1988) reported no growth responses to Zn supplementation in medium- or large-framed steers fed silage-corn-based diets containing 22 to 26 mg Zn/kg. Growth responses to Zn supplementation were observed in two of four studies with finishing steers fed diets containing 18 to 29 mg Zn/kg (Perry et al., 1968). In later studies, Zn added to diets containing 17 to 21 mg Zn/kg improved gain in only one of seven experiments (Beeson et al., 1977). Other studies with growing and finishing cattle have indicated no response to Zn supplementation when diets contained 22 to 32 mg Zn/kg (Pringle et al., 1973; Spears and Samsell, 1984). Spears and Kegley (2002) reported that Zn supplementation (25 mg Zn/kg DM) to growing and finishing steers (basal finishing diet contained 26 mg Zn/kg dietary DM) had no effect on overall ADG, intake, and feed efficiency, and Galyean et al. (1995) reported that supplementation of 35 or 70 mg Zn/kg DM to diets that already had 30 mg Zn/kg dietary DM did not affect carcass characteristics. Collectively, these experiments indicate that 30 mg Zn/kg dietary DM is adequate to meet the needs of growing and finishing cattle.

Zinc requirements of beef cattle fed forage-based diets and requirements for reproduction and milk production are less well defined. Zinc supplementation increased gain by nursing calves grazing mature forages that contained 7 to 17 mg Zn/kg (Mayland et al., 1980).

Signs of Zinc Deficiency

Severe Zn deficiency in cattle results in decreased growth, feed intake, and feed efficiency; listlessness; excessive salivation; decreased testicular growth; swollen feet with open, scaly lesions; parakeratotic lesions that are most severe on the legs, neck, head, and around the nostrils; failure of wounds to heal; and alopecia (Miller and Miller, 1962; Miller et al., 1965; Ott et al., 1965; Mills et al., 1967). Thymus atrophy and impaired immune responses have been observed in calves with a genetic disorder that causes impaired absorption of Zn, resulting in Zn deficiency (Perryman et al., 1989). Subclinical deficiencies of Zn can decrease weight gain (Mayland et al., 1980) and perhaps reproductive performance. Plasma or liver Zn concentrations can be used to diagnose severe Zn deficiencies, but plasma Zn determination is of little value in detecting marginal deficiencies. Stress or disease causes a redistribution of Zn in the body that can temporarily result in low plasma concentrations characteristic of a severe deficiency (Hambidge et al., 1986).

Factors Affecting Zinc Requirements

Absorption of Zn occurs primarily from the abomasum and small intestine (Miller and Cragle, 1965). Zinc absorption is homeostatically controlled, and cattle adjust the percentage of dietary Zn absorbed based on their need for growth or lactation (Miller, 1975). Milk contains 3 to 5 mg Zn/L, but the increased demand for milk production is likely met by increased absorption, provided that dietary Zn is present in a form that can be absorbed. Dietary factors that affect Zn requirements in ruminants are not understood fully. Using a perfused rat-intestine system, Oestreicher and Cousins (1985) reported that high luminal Cu concentrations can attenuate Zn movement out of the mucosal cell. In contrast to nonruminants, high dietary Ca does not seem to increase Zn requirements greatly in ruminants (Pond, 1983; Pond and Wallace, 1986). Phytate also does not affect Zn absorption in ruminants with a functional rumen. A relatively large portion of the Zn in forages is associated with the plant cell wall (Whitehead et al., 1985), but it is not known whether zinc's association with fiber decreases absorption.

Signs of Zinc Toxicity

The amount of Zn necessary to cause toxicity is much greater than requirements. The maximum tolerable concentration of Zn is 500 mg/kg (NRC, 1980, 2005). Decreased weight gain was reported in calves fed 900 mg Zn/kg for 12 weeks (Ott et al., 1966). Young calves fed milk replacer tolerated 500 mg Zn/kg for 5 weeks without adverse effects, but 700 mg/kg decreased gain, feed intake, and feed efficiency (Jenkins and Hidiroglou, 1991).

Zinc Sources

The Zn content of forages is affected by a number of factors including plant species, maturity, and soil Zn (Minson, 1990). Legumes are generally higher in Zn than grasses. Cereal grains usually contain between 20 and 30 mg Zn/kg, whereas plant protein sources contain 50 to 70 mg Zn/kg. Feed-grade sources of bioavailable Zn include zinc oxide, zinc sulfate, zinc methionine, and zinc proteinate. Based on available data, Zn in the sulfate and oxide form are of similar bioavailability in ruminants (Kincaid, 1979; Kegley and Spears, 1992). Absorption of Zn from zinc methionine is similar to zinc oxide, but zinc methionine seems to be metabolized differently following absorption (Spears, 1989).

REFERENCES

Abraham, A. B., B. A. Brooks, and U. Eylath. 1991. Chromium and cholesterol-induced arteriosclerosis in rabbits. *Annals of Nutrition and Metabolism* 35:203-207.

Adams, O. R., L. A. Griner, and R. Jensen. 1956. Polioencephalomalacia of cattle and sheep. *Journal of the American Veterinary Medical Association* 129:311-321.

Adams, R. S. 1975. Variability in mineral and trace element content of dairy cattle feeds. *Journal of Dairy Science* 58:1538-1548.

AFRC (Agriculture and Food Research Council). 1991. A Reappraisal of the Calcium and Phosphorus Requirements of Sheep and Cattle. Technical Committee on Responses to Nutrients Report No. 6, CAB International. Reprinted in *Nutrition Abstracts and Reviews Series B* 61:573-612.

Alfaro, E., M. W. Neathery, W. J. Miller, R. P. Gentry, C. T. Crowe, A. S. Fielding, R. E. Etheridge, D. G. Pugh, and D. M. Blackmon. 1987. Effects of varying the amounts of dietary calcium on selenium metabolism in dairy calves. *Journal of Dairy Science* 70:831-836.

Alfaro, E., M. W. Neathery, W. J. Miller, C. T. Crowe, R. P. Gentry, A. S. Fielding, D. G. Pugh, and D. M. Blackmon. 1988. Influence of a wide range of calcium intakes on tissue distribution of macroelements and microelements in dairy calves. *Journal of Dairy Science* 71:1295-1300.

Ammerman, C. B. 1970. Recent developments in cobalt and copper in ruminant nutrition: A review. *Journal of Dairy Science* 53:1097-1106.

Ammerman, C. B., and S. M. Miller. 1972. Biological availability of minor mineral ions: A review. *Journal of Animal Science* 35:681-694.

Ammerman, C. B., L. R. Arrington, M. C. Jayaswal, R. L. Shirley, and G. K. Davis. 1963. Effect of dietary calcium and phosphorus levels on nutrient digestibility by steers. *Journal of Animal Science* 22:248 (Abstract 52).

Ammerman, C. B., J. M. Wing, B. G. Dunavant, W. K. Robertson, J. P. Feaster, and L. R. Arrington. 1967. Utilization of inorganic iron by ruminants as influenced by form of iron and iron status of the animal. *Journal of Animal Science* 26:404-410.

Ammerman, C. B., C. F. Chicco, P. E. Loggins, and L. R. Arrington. 1972. Availability of different inorganic salts of magnesium to sheep. *Journal of Animal Science* 34:122-126.

ARC (Agricultural Research Council). 1980. *The Nutrient Requirements of Ruminant Livestock*. Slough, UK: Commonwealth Agricultural Bureaux.

Archer, K. A., and J. L. Wheeler. 1978. Response by cattle grazing sorghum to salt-sulfur supplements. *Australian Journal of Experimental Agriculture and Animal Husbandry* 18:741-744.

Arthington, J. D. 2005. Effects of copper oxide bolus administration of high-level copper supplementations on forage utilization and copper status in beef cattle. *Journal of Animal Science* 83:2894-2900.

Arthington, J. D. 2008. Effects of supplement type and selenium source on measures of growth and selenium status in yearling beef steers. *Journal of Animal Science* 86:1472-1477.

Arthington, J. D., and F. M. Pate. 2002. Effect of corn- vs molasses-based supplements on trace mineral status in beef heifers. *Journal of Animal Science* 80:2787-2791.

Arthington, J. D., L. R. Corah, J. E. Minton, T. H. Elsasser, and F. Blecha. 1997. Supplemental dietary chromium does not influence ACTH, cortisol, or immune responses in young calves inoculated with bovine herpesvirus-1. *Journal of Animal Science* 75:217-223.

Arthington, J. D., J. E. Rechcigl, G. P. Yost, L. R. McDowell, and M. D. Fanning. 2002. Effect of ammonium sulfate fertilization on bahiagrass quality and copper metabolism in grazing beef cattle. *Journal of Animal Science* 80:2507-2512.

Arthur, J. R., P. C. Morrice, and G. J. Becket. 1988. Thyroid hormone concentrations in selenium-deficient and selenium-sufficient cattle. *Research in Veterinary Science* 45:122-123.

Arthur, J. R., F. Nicol, and G. J. Becket. 1990. Hepatic iodothyronine 5′-deiodinase. *Biochemical Journal* 272:537-540.

Beauchamp, R. O., Jr., J. S. Bus, J. A. Popp, C. J. Boreiko, and D. A. Andjelkovich. 1984. A critical review of the literature on hydrogen sulfide toxicity. *Critical Reviews in Toxicology* 13:25-97.

Beckett, G. J., and J. R. Arthur. 2005. Selenium and endocrine systems. *Journal of Endocrinology* 184:455-465.

Beede, D. K., T. E. Pilbeam, S. M. Puffenbarger, and R. J. Tempelman. 2001. Peripartum responses of Holstein cows and heifers fed graded concentrations of calcium (calcium carbonate) and anion (chloride) three weeks before calving. *Journal of Dairy Science* 84(Suppl. 1):83(Abstract 244).

Beeson, W. M., T. W. Perry, and T. D. Zurcher. 1977. Effect of supplemental zinc on growth and on hair and blood serum levels of beef cattle. *Journal of Animal Science* 45:160-165.

Bentley, O. G., and P. H. Phillips. 1951. The effect of low manganese rations upon dairy cattle. *Journal of Dairy Science* 34:396-403.

Berger, A. L., and R. J. Rasby. 2011. Limiting Feed Intake with Salt in Beef Cattle Diets. NebGuide. G2046. University of Nebraska–Lincoln Extension, Institute of Agriculture and Natural Resources.

Bernhard, B. C., N. C. Burdick, R. J. Rathmann, J. A. Carroll, D. N. Finck, M. A. Jennings, T. R. Young, and B. J. Johnson. 2012a. Chromium supplementation alters glucose and lipid metabolism in feedlot cattle during the receiving period. *Journal of Animal Science* 90:4857-4865.

Bernhard, B. C., N. C. Burdick, W. Rounds, R. J. Rathmann, J. A. Carroll, D. N. Finck, M. A. Jennings, T. R. Young, and B. J. Johnson. 2012b. Chromium supplementation alters the performance and health of feedlot cattle during the receiving period and enhances their metabolic response to a lipopolysaccharide challenge. *Journal of Animal Science* 90:3879-3888.

Bernier, J. F., F. J. Fillion, and G. J. Brisson. 1984. Dietary fibers and supplemental iron in a milk replacer for veal calves. *Journal of Dairy Science* 67:2369-2379.

Bigger, G. W., J. M. Elliot, and T. R. Richard. 1976. Estimated ruminal production of pseudovitamin B_{12}, factor A and factor B in sheep. *Journal of Animal Science* 43:1077-1081.

Blaxter, K. L., and R. F. McGill. 1956. Magnesium metabolism in cattle. *Veterinary Reviews and Annotations* 2:35-55.

Blaxter, K. L., J. A. F. Rook, and A. M. MacDonald. 1954. Experimental magnesium deficiency in calves: Clinical and pathological observations. *Journal of Comparative Pathology and Therapeutics* 64:157-175.

Blaxter, K. L., G. A. M. Sarman, and A. M. MacDonald. 1957. Iron-deficiency anemia in calves. *British Journal of Nutrition* 11:234-246.

Blaxter, K. L., B. Cowlishaw, and J. A. Rook. 1960. Potassium and hypomagnesaemic tetany in calves. *Animal Production* 2:1-10.

Block, H. C., G. E. Erickson, and T. J. Klopfenstein. 2004. Re-evaluation of phosphorus requirements and phosphorus retention of feedlot cattle: A review. *The Professional Animal Scientist* 20:319-329.

Block, R. J., J. A. Stekol, and J. K. Loosli. 1951. Synthesis of sulfur amino acids from inorganic sulfate by ruminants. II. Synthesis of cystine and methionine from sodium sulfate by the goat and by the microorganisms of the rumen of the ewe. *Archives of Biochemistry and Biophysics* 33:353-363.

Bock, B. J., D. L. Harmon, R. T. Brandt, Jr., and J. E. Schneider. 1991. Fat source and calcium level effects on finishing steer performance, digestion, and metabolism. *Journal of Animal Science* 69:2211-2224.

Bolsen, K. K., W. Woods, and T. Klopfenstein. 1973. Effect of methionine and ammonium sulfate upon performance of ruminants fed high corn rations. *Journal of Animal Science* 36:1186-1190.

Boyne, R., and J. R. Arthur. 1981. Effects of selenium and copper deficiency on neutrophil function in cattle. *Journal of Comparative Pathology* 91:271-276.

Bradley, C. H. 1993. Copper poisoning in a dairy herd fed a mineral supplement. *Canadian Veterinary Journal* 34:287-292.

Bremner, I., and A. C. Dalgarno. 1973. Iron metabolism in the veal calf. 2. Iron requirements and the effect of copper supplementation. *British Journal of Nutrition* 30:61-76.

Bremner, I., W. R. Humphries, M. Phillippo, M. J. Walker, and P. C. Morrice. 1987. Iron-induced copper deficiency in calves: Dose-response relationships and interactions with molybdenum and sulfur. *Animal Production* 45:403-414.

Brink, D. R., O. A. Turgeon, Jr., D. L. Harmon, R. T. Steele, T. L. Mader, and R. A. Britton. 1984. Effects of additional limestone of various types on feedlot performance of beef cattle fed high corn diets differing in processing method and potassium level. *Journal of Animal Science* 59:791-797.

Brokman, A. M., J. W. Lehmkuhler, and D. J. Undersander. 2004. Supplemental phosphorus removal for finishing yearling Holstein steers. *Journal of Animal Science* 82(Suppl. 1):44 (Abstract M176).

Buchanan-Smith, J. G., E. Evans, and S. O. Poluch. 1974. Mineral analysis of corn silage produced in Ontario. *Canadian Journal of Animal Science* 54:253-256.

Bunting, L. D., J. M. Fernandez, D. L. Thompson, and L. L. Southern. 1994. Influence of chromium picolinate on glucose usage and metabolic criteria in growing Holstein calves. *Journal of Animal Science* 72:1591-1599.

Burkhaltor, D. L., M. W. Neathery, W. J. Miller, R. H. Whitlock, and J. C. Allen. 1979. Effects of low chloride intake on performance, clinical characteristics, and chloride, sodium, potassium and nitrogen metabolism in dairy calves. *Journal of Dairy Science* 62:1895-1901.

Burton, J. L., B. A. Mallard, and D. N. Mowat. 1993. Effects of supplemental chromium on immune response of periparturient and early lactation dairy cows. *Journal of Animal Science* 71:1532-1539.

Burton, J. L., B. A. Mallard, and D. N. Mowat. 1994. Effects of supplemental chromium on antibody responses of newly weaned feedlot calves to immunization with infectious bovine rhinotracheitis and parainfluenza 3 virus. *Canadian Journal of Veterinary Research* 58:148-151.

Bushman, D. H., R. J. Emerick, and L. B. Embry. 1965. Incidence of urinary calculi in sheep as affected by various dietary phosphates. *Journal of Animal Science* 24:671-675.

Butler, G. W., and D. I. H. Jones. 1973. Mineral biochemistry of herbage. Pp. 127-162 in *Chemistry and Biochemistry of Herbage,* Vol. 2. G. W. Butler and R. W. Bailey, eds. London: Academic Press.

Call, J. W., J. E. Butcher, J. T. Blake, R. A. Smart, and J. L. Shupe. 1978. Phosphorus influence on growth and reproduction of beef cattle. *Journal of Animal Science* 47:216-225.

Call, J. W., J. E. Butcher, J. L. Shupe, J. T. Blake, and A. E. Olson. 1986. Dietary phosphorus for beef cows. *American Journal of Veterinary Research* 47:475-481.

Cameron, H. J., R. J. Boila, L. W. McNichol, and N. E. Stranger. 1989. Cupric oxide needles for grazing cattle consuming low-copper, high-molybdenum forage and high-sulfate water. *Journal of Animal Science* 67:252-261.

Cammack, K. M., C. L. Wright, K. J. Austin, P. S. Johnson, R. R. Cockrum, K. L. Kessler, and K. C. Olsen. 2010. Effects of high sulfate water and clinoptilolite on health and growth performance of steers fed forage-based diets. *Journal of Animal Science* 88:1777-1785.

Campbell, A. G., M. R. Coup, W. H. Bishop, and D. E. Wright. 1974. Effect of elevated iron intake on the copper status of grazing cattle. *New Zealand Journal of Agricultural Research* 17:393-399.

Campbell, D. T., J. Maas, D. W. Weber, O. R. Hedstrom, and B. B. Norman. 1990. Safety and efficacy of two sustained-release intracellular selenium supplements and the associated placental and colostral transfer of selenium in beef cattle. *American Journal of Veterinary Research* 51:813-817.

Carstairs, J. A., R. R. Neitzel, and R. S. Emery. 1981. Energy and phosphorus status as factors affecting postpartum performance and health of dairy cows. *Journal of Dairy Science* 64:34-41.

Challa, J., and G. D. Braithwaite. 1988. Phosphorus and calcium metabolism in growing calves with special emphasis on phosphorus homeostasis.

1. Studies on the effect of changes in the dietary phosphorus intake on phosphorus and calcium metabolism. *Journal of Agricultural Science* 110:573-581.

Challa, J., G. D. Braithwaite, and M. S. Dhanoa. 1989. Phosphorus homeostasis in growing calves. *Journal of Agricultural Science* 12:217-226.

Chang, X., and D. N. Mowat. 1992. Supplemental chromium for stressed and growing feeder calves. *Journal of Animal Science* 70:559-565.

Chapman, H. L., Jr., and M. C. Bell. 1963. Relative absorption and excretion by beef cattle of copper from various sources. *Journal of Animal Science* 22:82-85.

Chester-Jones, H., J. P. Fontenot, and H. P. Veit. 1990. Physiological and pathological effects of feeding high levels of magnesium to steers. *Journal of Animal Science* 68:4400-4413.

Clanton, D. C. 1980. Applied potassium nutrition in beef cattle. Pp. 17-32 in *Proceedings, Third International Mineral Conference, January 17-18, 1980, Miami, FL.*

Claypool, D. W., F. W. Adams, H. W. Pendell, N. A. Hartmann, and J. F. Bone. 1975. Relationship between the level of copper in the blood plasma and liver of cattle. *Journal of Animal Science* 41:911-914.

Coghlin, C. L. 1944. Hydrogen sulfide poisoning in cattle. *Canadian Journal of Comparative Medicine* 8:111-113.

Coppice, C. E., and M. J. Fettman. 1977. Chloride as a required nutrient for lactating dairy cows. P. 43 in *Proceedings of the Cornell Nutrition Conference*. Ithaca, NY: Cornell University Press.

Cristaldi, L. A., L. R. McDowell, C. D. Buergelt, P. A. Davis, N. S. Wilkinson, and F. G. Martin. 2005. Tolerance of inorganic selenium in wether sheep. *Small Ruminant Research* 56:205-213.

Cunningham, G. N., M. B. Wise, and E. R. Barrick. 1966. Effect of high dietary levels of manganese on the performance and blood constituents of calves. *Journal of Animal Science* 25:532-538.

Davis, G. K., and W. Mertz. 1987. Copper. Pp. 301-364 in *Trace Elements in Human and Animal Nutrition,* Vol. 1, W. Mertz, ed. New York: Academic Press.

Davis, P. A., L. R. McDowell, N. S. Wilkinson, C. D. Buergelt, R. Van Alstyne, R. N. Weldon, and T. T. Marshall. 2006a. Effects of selenium levels in ewe diets on selenium in milk and the plasma and tissue selenium concentrations in lambs. *Small Ruminant Research* 65:14-23.

Davis, P. A., L. R. McDowell, N. S. Wilkinson, C. D. Buergelt, R. Van Alstyne, R. N. Weldon, and T. T. Marshall. 2006b. Tolerance of inorganic selenium by range-type ewes during gestation and lactation. *Journal of Animal Science* 84:660-668.

Davis, P. A., L. R. McDowell, N. S. Wilkinson, C. D. Buergelt, R. Van Alstyne, R. N. Weldon, T. T. Marshall, and E. Y. Matsuda-Fugisaki. 2008. Comparative effects of various dietary levels of Se as sodium selenite or Se yeast on blood, wool, and tissue Se concentrations of wether sheep. *Small Ruminant Research* 74:149-158.

Deagen, J. T., J. A. Butler, M. A. Beilstein, and P. D. Whanger. 1987. Effects of dietary selenite, selenocystine, and selenomethionine on selenocysteine lyase and glutathione peroxidase activities and on selenium levels in rat tissues. *Journal of Nutrition* 117:91-98.

DeBore, G. J., G. Buchanan-Smith, G. K. MacLeod, and J. S. Walton. 1981. Responses of dairy cows fed alfalfa silage supplemented with phosphorus, copper, zinc, and manganese. *Journal of Dairy Science* 64:2370-2377.

DeLuca, H. F. 1979. The vitamin D system in the regulation of calcium and phosphorus metabolism. *Nutrition Reviews* 37:161-193.

DeLuca, H. F. 2004. Overview of general physiologic features and functions of vitamin D. *American Journal of Clinical Nutrition* 80(Suppl.):1689S-1696S.

Devlin, T. J., W. K. Roberts, and V. V. E. St. Omer. 1969. Effects of dietary potassium upon growth, serum electrolytes and intrarumen environment of finishing beef steers. *Journal of Animal Science* 28:557-562.

DiCostanzo, A., J. C. Meiske, S. D. Plegge, D. L. Haggard, and K. M. Chaloner. 1986. Influence of manganese copper and zinc on reproductive performance of beef cows. *Nutrition Reports International* 34:287-293.

Digesti, R. D., and H. J. Weeth. 1976. A defensible maximum for inorganic sulfate in drinking water of cattle. *Journal of Animal Science* 42:1498-1502.

Dowe, T. W., J. Matsushima, and V. H. Arthaud. 1957. The effects of adequate and excessive calcium when fed with adequate phosphorus in growing rations for beef calves. *Journal of Animal Science* 16:811-820.

Drewnoski, M. E., D. J. Pogge, and S. L. Hansen. 2014. High sulfur in beef cattle diets: A review. *Journal of Animal Science* 92:3763-3780.

Du, Z., R. W. Hemken, and R. J. Harmon. 1996. Copper metabolism of Holstein and Jersey cows and heifers fed diets high in cupric sulfate or copper proteinate. *Journal of Dairy Science* 79:1873-1880.

Dunn, T. G., and G. E. Moss. 1992. Effects of nutrient deficiencies and excesses on reproductive efficiency of livestock. *Journal of Animal Science* 70:1580-1593.

Dyer, I. A., W. A. Cassatt, and R. R. Rao. 1964. Manganese deficiency in the etiology of deformed calves. *BioScience* 14:31-32.

Ellenberger, H. G., J. A. Newlander, and C. H. Jones. 1950. Composition of the Bodies of Dairy Cattle. Vermont Agricultural Experiment Station Bulletin 558. Burlington: University of Vermont.

Ellis, W. C., and W. H. Pfander. 1960. Further studies on molybdenum as a possible component of the "alfalfa ash factor" for sheep. *Journal of Animal Science* 19:1260 (Abstract 116).

Ellis, W. C., W. H. Pfander, M. E. Muhrer, and E. E. Pickett. 1958. Molybdenum as a dietary essential for lambs. *Journal of Animal Science* 17:180-188.

Ely, R. E., K. M. Dunn, and C. F. Huffman. 1948. Cobalt toxicity in calves resulting from high oral administration. *Journal of Animal Science* 7:239-246.

Emerick, R. J., and L. B. Embry. 1963. Calcium and phosphorus levels related to the development of phosphate urinary calculi in sheep. *Journal of Animal Science* 22:510-513.

Emerick, R. J., and L. B. Embry. 1964. Effects of calcium and phosphorus levels and diethylstilbestrol on urinary calculi incidence and feedlot performance of lambs. *Journal of Animal Science* 23:1079-1083.

Engle, T. E., and J. W. Spears. 2000a. Dietary copper effects on lipid metabolism, performance, and ruminal fermentation in finishing steers. *Journal of Animal Science* 78:2452-2458.

Engle, T. E., and J. W. Spears. 2000b. Effects of copper concentration and source on performance and copper status of growing and finishing steers. *Journal of Animal Science* 78:2446-2451.

Engle, T. E., J. W. Spears, V. Fellner, and J. Odle. 2000. Effects of soybean oil and dietary copper on ruminal and tissue lipid metabolism in finishing steers. *Journal of Animal Science* 78:2713-2721.

Engle, T. E., V. Fellner, and J. W. Spears. 2001. Copper status, serum cholesterol, and milk fatty acid profile in Holstein cows fed varying concentrations of copper. *Journal of Dairy Science* 84:2308-2313.

Erickson, G. E., T. J. Klopfenstein, C. T. Milton, D. Hanson, and C. Calkins. 1999. Effect of dietary phosphorus on finishing steer performance, bone status, and carcass maturity. *Journal of Animal Science* 77:2832-2836.

Erickson, G. E., T. J. Klopfenstein, C. T. Milton, D. Brink, M. W. Orth, and K. M. Whittet. 2002. Phosphorus requirement of finishing feedlot calves. *Journal of Animal Science* 80:1690-1695.

Ferrell, M. C., F. N. Owens, and D. R. Gill. 1983. Potassium levels and ionophores for feedlot steers. Pp. 54-59 in *Animal Science Research Report* MP-114. Stillwater: Oklahoma State University, Agricultural Experiment Station. Available online at http://beefextension.com/research_reports/1983rr/83-10.pdf. Accessed on April 27, 2015.

Fontenot, J. P., V. G. Allen, G. E. Bunce, and J. P. Goff. 1989. Factors influencing magnesium absorption and metabolism in ruminants. *Journal of Animal Science* 67:3445-3455.

Fron, M. J., J. A. Boling, L. P. Bush, and K. A. Dawson. 1990. Sulfur and nitrogen metabolism in the bovine fed different forms of supplemental sulfur. *Journal of Animal Science* 68:543-552.

Fry, R. S., J. W. Spears, K. E. Lloyd, A. T. O'Nan, and M. S. Ashwell. 2013. Effect of dietary copper and breed on gene products involved in copper acquisition, distribution, and use in Angus and Simmental cows and fetuses. *Journal of Animal Science* 91:861-871.

Galyean, M. L., K. J. Malcolm-Callis, S. A. Gunter, and R. A. Berrie. 1995. Effect of zinc source and level and added copper lysine in the receiving diet on performance by growing and finishing steers. *The Professional Animal Scientist* 11:139-148.

Ganther, H. E., O. A. Levander, and C. A. Baumann. 1966. Dietary control of selenium volatilization in the rat. *Journal of Nutrition* 88:55-60.

Geisert, B. G., G. E. Erickson, T. J. Kolpfenstein, C. N. Macken, M. K. Lubbe, and J. C. MacDonald. 2010. Phosphorus requirement and excretion of finishing beef cattle fed different concentrations of phosphorus. *Journal of Animal Science* 88:2393-2402.

Gengelbach, G. P., J. D. Ward, and J. W. Spears. 1994. Effect of dietary copper, iron, and molybdenum on growth and copper status of beef cows and calves. *Journal of Animal Science* 72:2722-2727.

Gentry, L. R., J. M. Fernandez, T. L. Ward, T. W. White, L. L. Southern, T. D. Binder, D. L. Thompson, Jr., D. W. Horohov, A. M. Chapa, and T. Sahlu. 1999. Dietary protein and chromium tripicolinate in Suffolk wether lambs: Effects on production characteristics, metabolic and hormonal responses, and immune status. *Journal of Animal Science* 77:1284-1294.

Gentry, R. P., W. J. Miller, D. G. Pugh, M. W. Neathery, and J. B. Bynoum. 1978. Effects of feeding high magnesium to young dairy calves. *Journal of Dairy Science* 61:1750-1754.

Gerken, H. J., and J. P. Fontenot. 1967. Availability and utilization of magnesium from dolomitic and magnesium oxide in steers. *Journal of Animal Science* 26:1404-1408.

Goetsch, A. L., and F. N. Owens. 1985. Effects of calcium source and level on site of digestion and calcium levels in the digestive tract of cattle fed high-concentrate diets. *Journal of Animal Science* 61:995-1003.

Goetsch, A. L., F. N. Owens. 1987. Effects of supplemental sulfate (Dynamate) and thiamin-HCl on passage of thiamin to the duodenum and site of digestion in steers. *Archives of Animal Nutrition* 12:1075-1083.

Goodrich, R. D., T. S. Kahlon, D. E. Pamp, and D. P. Cooper. 1978. *Sulfur in Ruminant Nutrition.* Des Moines: National Feed Ingredient Association.

Gooneratne, S. R., W. T. Buckley, and D. A. Christensen. 1989. Review of copper deficiency and metabolism in ruminants. *Canadian Journal of Animal Science* 69:819-845.

Gooneratne, S. R., H. W. Symonds, J. V. Bailey, and D. A. Christensen. 1994. Effects of dietary copper, molybdenum and sulfur on biliary copper and zinc excretion in Simmental and Angus cattle. *Canadian Journal of Animal Science* 74:315-325.

Grace, N. D. 1983. Manganese. Pp. 80-83 in *The Mineral Requirements of Grazing Ruminants*, N. D. Grace, ed. Occasional Publication No. 9. New Zealand: New Zealand Society of Animal Producers.

Grace, N. D., M. J. Wyatt, and J. C. MacRae. 1974. Quantitative digestion of fresh herbage by sheep. III. The movement of Mg, Ca, P, K, and Na in the digestive tract. *Journal of Agricultural Science* 82:321-330.

Graham, T. W. 1991.Trace element deficiencies in cattle. *Veterinary Clinics of North America: Food Animal Practice* 7:153-215.

Greene, L. W., K. E. Webbe, Jr., and J. P. Fontenot. 1983. Effect of potassium level on site of absorption of magnesium and other macroelements in sheep. *Journal of Animal Science* 56:1214-1221.

Greene, L. W., J. F. Baker, and P. F. Hardt. 1989. Use of animal breeds and breeding to overcome the incidence of grass tetany: A review. *Journal of Animal Science* 67:3463-3469.

Hall, J. A., A. M. Harwell, R. J. Van Saun, W. R. Vorachek, W. C. Stewart, M. L. Galbraith, K. J. Hooper, J. K. Hunter, W. D. Mosher, and G. J. Pirelli. 2011. Agronomic biofortification with selenium: Effects on whole blood selenium and humoral immunity in beef cattle. *Animal Feed Science and Technology* 164:184-190.

Hall, J. A., G. Bobe, J. K. Hunter, W. R. Vorachek W. C. Stewart, J. A. Vanegas, C. T. Estill, W. D. Mosher, and G. J. Pirelli. 2013. Effect of feeding selenium-fertilized alfalfa hay on performance of weaned beef calves. *PLoS ONE* 8(3):e58188.

Halpin, C. G., D. J. Harris, I. W. Caple, and D. S. Patterson. 1984. Contribution of cobalamin analogues to plasma vitamin B_{12} concentrations in cattle. *Research in Veterinary Science* 37:249-253.

Hambidge, K. M., C. C. Casey, and N. F. Krebs. 1986. Zinc. Pp. 1-137 in *Trace Elements in Human and Animal Nutrition*, Vol. 2, W. Mertz, ed. New York: Academic Press.

Hansard, S. L., C. L. Comar, and M. P. Plumlee. 1954. The effects of age upon calcium utilization and maintenance requirements in the bovine. *Journal of Animal Science* 13:25-36.

Hansard, S. L., H. M. Crowder, and W. A. Lyke. 1957. The biological availability of calcium in feeds for cattle. *Journal of Animal Science* 16:437-443.

Hansen, S. L., and J. W. Spears. 2009. Bioaccessibility of iron from soils is increased by silage fermentation. *Journal of Dairy Science* 92:2896-2905.

Hansen, S. L., J. W. Spears, K. E. Lloyd, and C. S. Whisnant. 2006a. Feeding a low manganese diet to heifers during gestation impairs fetal growth and development. *Journal of Dairy Science* 89:4305-4311.

Hansen, S. L., J. W. Spears, K. E. Lloyd, and C. S. Whisnant. 2006b. Growth, reproductive performance, and manganese status of heifers fed varying concentrations of manganese. *Journal of Animal Science* 84:3375-3380.

Hansen, S. L., P. Schlegel, L. R. Legliter, K. E. Lloyd, and J. W. Spears. 2008. Bioavailability of copper from copper glycinate in steers fed high dietary sulfur and molybdenum. *Journal of Animal Science* 86:173-179.

Hansen, S. L., M. S. Ashwell, A. J. Moeser, R. S. Fry, M. D. Knutson, and J. W. Spears. 2010. High dietary iron reduces transporters involved in iron and manganese metabolism and increases intestinal permeability in calves. *Journal of Dairy Science* 93:656-665.

Harrison, J. H., and H. R. Conrad. 1984. Effect of dietary calcium on selenium absorption by the nonlactating dairy cow. *Journal of Dairy Science* 67:1860-1864.

Hatfield, D. L. 2001. *Selenium: Its Molecular Biology and Role in Human Health.* Boston, MA: Kluwer.

Hawkins, G. E., Jr., G. H. Wise, G. Matrone, R. K. Waugh, and W. L. Lott. 1955. Manganese in the nutrition of young dairy cattle fed different levels of calcium and phosphorus. *Journal of Dairy Science* 38:536-547.

Hayirli, A., D. R. Bremmer, S. J. Bertics, M. T. Socha, and R. R. Grummer. 2001. Effect of chromium supplementation on production and metabolic parameters in periparturient dairy cows. *Journal of Dairy Science* 84:1218-1230.

Healy, W. B. 1972. In vitro studies on the effects of soil on elements in ruminal, duodenal and ileal liquors from sheep. *New Zealand Journal of Agricultural Research* 15:289-305.

Henry, P. R. 1995. Sodium and chlorine bioavailability. Pp. 337-348 in *Bioavailability of Nutrients for Animals*, C. B. Ammerman, D. H. Baker, and A. J. Lewis, eds. New York: Academic Press.

Henry, P. R., C. B. Ammerman, and R. C. Littell. 1992. Relative bioavailability of manganese from a manganese-methionine complex and inorganic sources for ruminants. *Journal of Dairy Science* 75:3473-3478.

Hidiroglou, M. 1979. Manganese in ruminant nutrition. *Canadian Journal of Animal Science* 59:217-236.

Hidiroglou, M., J. Proulx, and J. Jolette. 1985. Intraruminal selenium pellet for control of nutritional muscular dystrophy in cattle. *Journal of Dairy Science* 68:57-66.

Hill, G. M. 1985. The relationship between dietary sulfur and nitrogen metabolism in the ruminant. Pp. 37-42 in *Proceedings of the Georgia Nutrition Conference*. Athens: University of Georgia.

Hintze, K. J., G. P. Lardy, M. J. Marchello, and J. W. Finley. 2001. Areas with high concentrations of selenium in the soil and forage produce beef with enhanced concentrations of selenium. *Journal of Agricultural and Food Chemistry* 49:1062-1067.

Hintze, K. J., G. P. Lardy, M. J. Marchello, and J. W. Finley. 2002. Selenium accumulation in beef: Effect of dietary selenium and geographical area of animal origin. *Journal of Agricultural and Food Chemistry* 50:3938-3942.

Hoekstra, W. G. 1974. Biochemical role of selenium. Pp. 61-77 in *Trace Element Metabolism in Animals*, No. 2, W. G. Hoekstra, J. W. Suttie, H. E. Ganther, and W. Mertz, eds. Baltimore: University Park Press.

Huffman, D. F., C. L. Conley, C. C. Lightfoot, and C. W. Duncan. 1941. Magnesium studies in calves. II. The effect of magnesium salts and various natural feeds upon the magnesium content of the blood plasma. *Journal of Nutrition* 22:609-620.

Humphries, W. R., M. Phillippo, B. W. Young, and I. Bremner. 1983. The influence of dietary iron and molybdenum on copper metabolism in calves. *British Journal of Nutrition* 49:77-86.

Huntington, G. B. 1983. Feedlot performance, blood metabolic profile and calcium status of steers fed high concentrate diets containing several levels of calcium. *Journal of Animal Science* 56:1003-1011.

Hurley, L. S., and C. L. Keen. 1987. Manganese. Pp. 185-223 in *Trace Elements in Human and Animal Nutrition*, Vol. 1, W. Mertz, ed. New York: Academic Press.

Ivancic, J., and W. P. Weiss. 2001. Effect of dietary sulfur and selenium concentrations on selenium valance in lactating Holstein cows. *Journal of Dairy Science* 84:225-232.

Jackson, J. A., Jr., D. L. Langer, and R. W. Hemken. 1988. Evaluation of content and source of phosphorus fed to dairy calves. *Journal of Dairy Science* 71:2187-2192.

Jenkins, K. J., and M. Hidiroglou. 1989. Tolerance of the calf for excess copper in milk replacer. *Journal of Dairy Science* 72:150-156.

Jenkins, K. J., and M. Hidiroglou. 1991. Tolerance of the preruminant calf for excess manganese or zinc in milk replacer. *Journal of Dairy Science* 74:1047-1053.

Jensen, R., and D. R. Mackey. 1979. *Diseases of Feedlot Cattle,* 3rd Ed. Philadelphia: Lea and Febiger.

Johnson, W. H., R. D. Goodrich, and J. C. Meiske. 1971. Metabolism of radioactive sulfur from elemental sulfur, sodium sulfate and methionine by lambs. *Journal of Animal Science* 32:778-783.

Juniper, D. T., R. H. Phipps, D. I. Givens, A. K. Jones, and G. Bertin. 2008. Tolerance of ruminant animals to high dose in-feed administration of a selenium-enriched yeast. *Journal of Animal Science* 86:197-204.

Kahlon, T. S., J. C. Meiske, and R. D. Goodrich. 1975. Sulfur metabolism in ruminants. 1. In vitro availability of various chemical forms of sulfur. *Journal of Animal Science* 41:1147-1153.

Kandylis, K. 1984. Toxicology of sulfur in ruminants: Review. *Journal of Dairy Science* 67:2179-2187.

Keener, H. A., G. P. Percival, K. S. Morrow, and G. H. Ellis. 1949. Cobalt tolerance in young dairy cattle. *Journal of Dairy Science* 32:527-533.

Kegley, E. B., and J. W. Spears. 1992. Performance and mineral metabolism of lambs as affected by source (oxide, sulfate or methionine) and level of zinc. *Journal of Animal Science* 70(Suppl. 1):302.

Kegley, E. B., and J. W. Spears. 1994. Bioavailability of feed grade copper sources (oxide, sulfate or lysine) in growing cattle. *Journal of Animal Science* 72:2728-2734.

Kegley, E. B., and J. W. Spears. 1995. Immune response, glucose metabolism, and performance of stressed feeder calves fed inorganic or organic chromium. *Journal of Animal Science* 73:2721-2726.

Kegley, E. B., J. W. Spears, and T. T. Brown. 1997a. Effects of shipping and chromium supplementation on performance, immune response, and disease resistance of steers. *Journal of Animal Science* 75:1956-1964.

Kegley, E. B., J. W. Spears, and J. H. Eisemann. 1997b. Performance and glucose metabolism in calves fed chromium-nicotinic acid complex or chromium chloride. *Journal of Dairy Science* 80:1744-1750.

Kelley, W. K., and R. L. Preston. 1984. Effect of dietary sodium, potassium and the anion form of these cations on the performance of feedlot steers. *Journal of Animal Science* 59(Suppl. 1):450.

Kennedy, D. G., P. B. Young, W. J. McCaugley, S. Kennedy, and W. J. Blanchflower. 1991. Rumen succinate production may ameliorate the effects of cobalt-vitamin B_{12} deficiency on methylmalonyl CoA mutase in sheep. *Journal of Nutrition* 121:1236-1242.

Kertz, A. 1998. Variability in delivery of nutrients to lactating dairy cows. *Journal of Dairy Science* 81:3075-3084.

Kessler, K. L., K. C. Olsen, C. L. Wright, K. J. Austin, P. S. Johnson, and K. M. Cammack. 2012. Effects of supplemental molybdenum on animal performance, liver copper concentrations, ruminal hydrogen sulfide concentrations, and the appearance of sulfur and molybdenum toxicity in steers receiving fiber-based diets. *Journal of Animal Science* 90:5005-5012.

Kincaid, R. L. 1979. Biological availability of zinc from inorganic sources with excess calcium. *Journal of Dairy Science* 62:1081-1085.

Kincaid, R. L. 1980. Toxicity of ammonium molybdate added to drinking water of calves. *Journal of Dairy Science* 63:608-610.

Kincaid, R. L., R. M. Blauwiekel, and J. D. Cronrath. 1986. Supplementation of copper as copper sulfate or copper proteinate for growing calves fed forages containing molybdenum. *Journal of Dairy Science* 69:160-163.

Knowlton, K. F., and J. H. Herbein. 2002. Phosphorus partitioning during early lactation in dairy cows fed diets varying in phosphorus content. *Journal of Dairy Science* 85:1227-1236.

Kornegay, E. T., Z. Wang, C. M. Wood, and N. D. Lindemann. 1997. Supplemental chromium picolinate influences nitrogen balance, dry matter digestibility, and carcass traits in growing and finishing pigs. *Journal of Animal Science* 75:1319-1323.

Kvasnicka, B., and L. J. Krysl. 2014. Grass tetany in beef cattle. BCH-3110. *Beef Cattle Handbook.* Iowa Beef Center, University of Iowa. Available online at http://www.iowabeefcenter.org/Beef%20Cattle%20Handbook/Grass-Tetany.pdf. Accessed on April 2, 2015.

Laflamme, G. H., and J. Jowsey. 1972. Bone and soft tissue changes with oral phosphate supplements. *Journal of Clinical Investigation* 51:2834-2840.

Langlands, J. P., G. E. Donald, J. E. Bowles, and A. J. Smith. 1989a. Selenium concentrations in the blood of ruminants grazing in northern New South Wales. III. Relationship between blood concentration and the response in live weight of grazing cattle given a selenium supplement. *Australian Journal of Agricultural Research* 40:1075-1083.

Langlands, J. P., G. E. Donald, J. E. Bowles, and A. J. Smith. 1989b. Trace element nutrition of grazing ruminants. III. Copper oxide powder as a copper supplement. *Australian Journal of Agricultural Research* 40:187-193.

Lassiter, J. W., W. J. Miller, F. M. Pate, and R. P. Gentry. 1972. Effect of dietary calcium and phosphorus on ^{54}Mn metabolism following a single tracer intraperitoneal and oral doses in rats. *Proceedings of the Society for Experimental Biology and Medicine* 139:345-348.

Lawler, T. L., J. B. Taylor, J. W. Finley, and J. S. Caton. 2004. Effect of supranutritional and organically bound selenium on performance, carcass characteristics, and selenium distribution in finishing beef steers. *Journal of Animal Science* 82:1488-1493.

Legleiter, L. R., J. W. Spears, and K. E. Lloyd. 2005. Influence of dietary manganese on performance, lipid metabolism, and carcass composition of growing and finishing cattle. *Journal of Animal Science* 83:2434-2439.

Leibholz, J., R. C. Kellaway, and G. T. Hargreave. 1980. Effects of sodium chloride and sodium bicarbonate in the diet on the performance of calves. *Animal Feed Science and Technology* 5:309-314.

Lloyd, K. E., V. Fellner, S. J. McLeod, R. S. Fry, K. Krafka, A. Lamptey, and J. W. Spears. 2010. Effects of supplementing dairy cows with chromium propionate on milk and tissue chromium concentrations. *Journal of Dairy Science* 93:4774-4780.

Lofgreen, G. P., M. Kleiber, and J. R. Luick. 1952. The metabolic fecal phosphorus excretion of the young calf. *Journal of Nutrition* 47:571-581.

Loneragan, G. H., J. J. Wagner, D. H. Gould, F. B. Garry, and M. A. Thoren. 2001. Effects of water sulfate concentration on performance, water intake, and carcass characteristics of feedlot steers. *Journal of Animal Science* 79:2941-2948.

MacPherson, A., and J. S. Chalmers. 1985. Effect of dietary energy concentration on the cobalt/vitamin B12 requirements of growing calves. Pp. 145-147 in *Trace Elements in Man and Animals (TEMA 5): Proceedings of the Fifth International Symposium on Trace Elements in Man and*

Animals, C. F. Mills, I. Bremner, and J. K. Chesters, eds. Slough, UK: Commonwealth Agricultural Bureaux.

MacPherson, A., G. Fisher, and J. E. Paterson. 1988. Effect of cobalt deficiency on the immune function of ruminants. Pp. 397-398 in *Trace Elements in Man and Animals 6*, L. S. Hurley, C. L. Keen, B. Lonnerdal, and R. B. Rucker, eds. New York: Plenum Press.

Martens, H., and Y. Rayssiguier. 1980. Magnesium metabolism and hypomagnesaemia. Pp. 447-466 in *Digestive Physiology and Metabolism in Ruminants*, Y. Ruckebusch and P. Thivend, eds. Lancaster, UK: MTP Press.

Martens, H., O.W. Kubel, G. Gabel, and H. Honig. 1987. Effects of low sodium intake on magnesium metabolism of sheep. *Journal of Agricultural Science* 108:237-243.

Martz, F. A., A. T. Belo, M. F. Weiss, R. L. Belyea, and J. P. Gaff. 1990. True absorption of calcium and phosphorus from alfalfa and corn silage when fed to lactating cows. *Journal of Dairy Science* 73:1288-1295.

Matsuzaki, H., M. Uehara, K. Suzuki, Q. L. Liu, S. Sato, Y. Kanke, and S. Goto. 1997. High-phosphorus diet rapidly induces nephrocalcinosis and proximal tubular injury in rats. *Journal of Nutritional Science and Vitaminology* 43:627-641.

Mayland, H. F. 1988. Grass tetany. Pp. 511-522 in *The Ruminant Animal—Digestive Physiology and Nutrition*, D. C. Church, ed. Englewood Cliffs, NJ: Prentice Hall.

Mayland, H. F., R. C. Rosenau, and A. R. Florence. 1980. Grazing cow and calf responses to zinc supplementation. *Journal of Animal Science* 51:966-974.

McDowell, L. R. 2003. *Minerals in Animal and Human Nutrition*, 2nd Ed. Amsterdam: Elsevier.

McGillivray, J. J. 1974. Biological availability of phosphorus in feed ingredients. Pp. 15-24 in *Proceedings of the 35th Minnesota Nutrition Conference*. St. Paul: University of Minnesota.

McGuire, S. O., W. J. Miller, R. D. Gentry, N. W. Neathery, S. Y. Ho, and D. M. Blackmon. 1985. Influence of high dietary iron as ferrous carbonate and ferrous sulfate on iron metabolism in young calves. *Journal of Dairy Science* 68:2621-2628.

McMeniman, N. P. 1973. The toxic effect of some phosphate supplements fed to sheep. *Australian Veterinary Journal* 49:150-152.

Mertz, W. 1992. Chromium: History and nutritional importance. *Biological Trace Element Research* 32:3-8.

Mertz, W. 1993. Chromium in human nutrition: A review. *Journal of Nutrition* 123:626-633.

Meyer, J. H., R. R. Grunert, R. H. Grummer, P. H. Phillips, and G. Bohstedt. 1950. Sodium, potassium, and chlorine content of feedstuffs. *Journal of Animal Science* 9:153-156.

Meyer, J. H., W. C. Weir, N. R. Ittner, and J. D. Smith. 1955. The influence of high sodium chloride intakes by fattening sheep and cattle. *Journal of Animal Science* 14:412-418.

Miller, J. K., and R. G. Cragle. 1965. Gastrointestinal sites of absorption and endogenous secretion of zinc in dairy cattle. *Journal of Dairy Science* 48:370-373.

Miller, J. K., and W. J. Miller. 1962. Experimental zinc deficiency and recovery of calves. *Journal of Nutrition* 76:467-474.

Miller, J. K., E. W. Swanson, and G. E. Spalding. 1975. Iodine absorption, excretion, recycling, and tissue distributions in the dairy cow. *Journal of Dairy Science* 58:1578-1593.

Miller, J. K., N. Ramsey, and F. C. Madsen. 1988. The trace elements. Pp. 342-401 in *The Ruminant Animal—Digestive Physiology and Nutrition*, D. C. Church, ed. Englewood Cliffs, NJ: Prentice Hall.

Miller, W. J. 1975. New concepts and developments in metabolism and homeostasis of inorganic elements in dairy cattle: A review. *Journal of Dairy Science* 58:1549-1560.

Miller, W. 1983. Using mineral requirement standards in cattle feeding programs and feed formulations. Pp. 69-74 in *Proceedings of the Georgia Nutrition Conference for the Feed Industry*. Athens: University of Georgia.

Miller, W. J., W. J. Pitts, C. M. Clifton, and J. D. Morton. 1965. Effects of zinc deficiency per se on feed efficiency, serum alkaline phosphatase, zinc in skin, behavior, greying, and other measurements in the Holstein calf. *Journal of Dairy Science* 48:1329-1334.

Miller, W. J., M. W. Neathery, R. P. Gentry, D. M. Blackmon, C. T. Crowe, G. O. Ward, and A. S. Fielding. 1987. Bioavailability of phosphorus from defluorinated and dicalcium phosphates and phosphorus requirements of calves. *Journal of Dairy Science* 70:1885-1892.

Mills, C. F. 1981. Cobalt deficiency and cobalt requirements of ruminants. Pp. 129-141 in *Recent Advances in Animal Nutrition*, W. Haresign, ed. Boston: Butterworths.

Mills, C. F. 1987. Biochemical and physiological indicators of mineral status in animals: Copper, cobalt, and zinc. *Journal of Animal Science* 65:1702-1711.

Mills, C. F., and G. K. Davis. 1987. Molybdenum. Pp. 429-463 in *Trace Elements in Human and Animal Nutrition*, Vol. 1, W. Mertz, ed. New York: Academic Press.

Mills, C. F., A. C. Dalgarno, R. B. Williams, and J. Quarterman. 1967. Zinc deficiency and zinc requirements of calves and lambs. *British Journal of Nutrition* 21:751-768.

Minson, D. J. 1990. *Forages in Ruminant Nutrition*. New York: Academic Press.

Moonsie-Shageer, S., and D. N. Mowat. 1993. Effect of level of supplemental chromium on performance, serum constituents, and immune status of stressed feeder calves. *Journal of Animal Science* 71:232-238.

Moore, L. A., E. T. Hallman, and L. B. Sholl. 1938. Cardiovascular and other lesions in calves fed diets low in magnesium. *Archives of Pathology* 26:820-838.

Morine, S. J., M. E. Drewnoski, and S. L. Hansen. 2014a. Increasing dietary NDF concentration decreases ruminal hydrogen sulfide concentrations in steers fed high-sulfur diets based on ethanol co-products. *Journal of Animal Science* 92:3035-3041.

Morine, S. J., M. E. Drewnoski, A. K. Johnson, and S. L. Hansen. 2014b. Determining the influence of dietary roughage concentration and source on ruminal parameters related to sulfur toxicity. *Journal of Animal Science* 92:4068-4076.

Morris, J. G. 1980. Assessment of sodium requirements of grazing beef cattle: A review. *Journal of Animal Science* 50:145-152.

Morris, J. G., R. E. Delmas, and J. L. Hull. 1980. Salt (sodium) supplementation of range beef cows in California. *Journal of Animal Science* 51:722-731.

Morris, J. G., W. S. Cripe, H. L. Chapman, Jr., D. F. Walker, J. B. Armstrong, J. D. Alexander, Jr., R. Miranda, A. Sanchez, Jr., B. Sanchez, J. R. Blair-West, and D. A. Denton. 1984. Selenium deficiency in cattle associated with Heinz bodies and anemia. *Science* 223:491-493.

Morrison, M., R. M. Murray, and A. N. Boniface. 1990. Nutrition metabolism and rumen microorganisms in sheep fed a poor-quality tropical grass hay supplemented with sulfate. *Journal of Agricultural Science* 115:269-275.

Mullis, L. A., J. W. Spears, and R. L. McCraw. 1997. Copper requirements of Angus and Simmental heifers. *Journal of Animal Science* 75(Suppl. 1):265.

Mullis, L. A., J. W. Spears, and R. L. McCraw. 2003a. Effect of breed (Angus vs Simmental) and copper and zinc source on mineral status of steers fed high dietary iron. *Journal of Animal Science* 81:318-322.

Mullis, L. A., J. W. Spears, and R. L. McCraw. 2003b. Estimated copper requirements of Angus and Simmental heifers. *Journal of Animal Science* 81:865-873.

Neathery, N. W., D. G. Pugh, W. J. Miller, R. P. Gentry, and R. H. Whitlock. 1980. Effects of sources and amounts of potassium on feed palatability and on potassium toxicity in dairy calves. *Journal of Dairy Science* 63:82-85.

Neathery, N. W., D. M. Blackmon, W. J. Miller, S. Heinmiller, S. McGuire, J. M. Tarabula, R. P. Gentry, and J. C. Allen. 1981. Chloride deficiency in Holstein calves from a low chloride diet and removal of abomasal contents. *Journal of Dairy Science* 64:2220-2233.

Neuhold, K. E. 2012. Sulfur Metabolism in Beef Cattle and Management Strategies to Improve Performance and Health in Newly Weaned Beef Cattle. Ph.D Dissertation. Colorado State University, Fort Collins, CO.

Neville, T. L., M. A. Ward, J. J. Reed, S. A. Soto-Navarro, S. L. Julius, P. P. Borowicz, J. B. Taylor, D. A. Redmer, L. P. Reynolds, and J. S. Caton. 2008. Effects of level and source of dietary selenium on maternal and fetal body weight, visceral organ mass, cellularity estimates, and jejunal vascularity in pregnant ewe lambs. *Journal of Animal Science* 86:890-901.

Neville, T. L., A. M. Meyer, A. Reyaz, P. P. Borowicz, D. A. Redmer, L. P. Reynolds, J. S. Caton, and K. A. Vonnahme. 2013. Mammary gland growth and vascularity at parturition and during lactation in primiparous ewes fed differing levels of selenium and nutritional plane during gestation. *Journal of Animal Science and Biotechnology* 4(1):6.

Newton, G. L., E. R. Barrick, R. W. Harvey, and M. B. Wise. 1974. Iodine toxicity: Physiological effects of elevated dietary iodine on calves. *Journal of Animal Science* 38:449-455.

Nichols, C. A., V. R. Bremer, A. K. Watson, C. D. Buckner, J. L. Harding, G. E. Erickson, T. J. Klopfenstein, and D. R. Smith. 2013. The effect of sulfur and use of ruminal available sulfur as a model to predict incidence of polioencephalomalacia in feedlot cattle. *Bovine Practitioner* 47:47-53.

Nielson, F. H. 1987. Nickel. Pp. 245-274 in *Trace Elements in Human and Animal Nutrition*, Vol. 1, W. Mertz, ed. New York: Academic Press.

Nockels, C. F., J. DeBonis, and J. Torrent. 1993. Stress induction affects copper and zinc balance in calves fed organic and inorganic copper and zinc sources. *Journal of Animal Science* 71:2539-2545.

NRC (National Research Council). 1980. *Mineral Tolerance of Domestic Animals*. Washington, DC: National Academy Press.

NRC. 1996. *Nutrient Requirements of Beef Cattle*, 7th Rev. Ed. Washington, DC: National Academy Press.

NRC. 2000. *Nutrient Requirements of Beef Cattle: Update 2000*, 7th Rev. Ed. Washington, DC: National Academy Press.

NRC. 2001. *Nutrient Requirements of Dairy Cattle*. Washington, DC: National Academy Press.

NRC. 2005. *Mineral Tolerance of Domestic Animals*. Washington, DC: The National Academies Press.

Oestreicher, P., and R. J. Cousins. 1985. Copper and zinc absorption in the rat: Mechanisms of mutual antagonism. *Journal of Nutrition* 115:159-166.

Okada, S., M. Suzuki, and H. Ohba. 1983. Enhancement of ribonucleic acids synthesis by chromium (III) in mouse liver. *Journal of Inorganic Biochemistry* 19:95-103.

O'Kelley, R. E., and J. P. Fontenot. 1969. Effects of feeding different magnesium levels to drylot-fed lactating beef cows. *Journal of Animal Science* 29:959-966.

O'Kelley, R. E., and J. P. Fontenot. 1973. Effects of feeding different magnesium levels to drylot-fed gestating beef cows. *Journal of Animal Science* 36:994-1000.

Oltjen, R. R. 1975. Fats for ruminants—utilization and limitations, including value of protected fats. Pp. 31-40 in *Proceedings of the Georgia Nutrition Conference*. Athens: University of Georgia.

Oscar, T. P., and J. W. Spears. 1988. Nickel-induced alterations of in vitro and in vivo ruminal fermentation. *Journal of Animal Science* 66:2313-2324.

O'Tool, D. and M. F. Raisbeck. 1995. Pathology of experimentally induced chronic selenosis (alkali disease) in yearling cattle. *Journal of Veterinary Diagnostic Investigation* 7:364-373.

Ott, E. A., W. H. Smith, M. Stab, H. E. Parker, and W. M. Beeson. 1965. Zinc deficiency syndrome in the young calf. *Journal of Animal Science* 24:735-741.

Ott, E. A., W. H. Smith, R. B. Harrington, and W. M. Beeson. 1966. Zinc toxicity in ruminants. II. Effects of high levels of dietary zinc on gains, feed consumption and feed efficiency of beef cattle. *Journal of Animal Science* 25:419-423.

Parker, D. R., A. L. Page, and P. F. Bell. 1992. Contrasting selenite-sulfate interactions in selenium-accumulating and nonaccumulating plant species. *Soil Science Society of America Journal* 56:1818-1824.

Patterson, E. L., R. Milstrey, and E. L. R. Stokstad. 1957. Effect of selenium in preventing exudative diathesis in chicks. *Proceedings of the Society for Experimental Biology and Medicine* 95:617-620.

Patterson, H. H., P. S. Johnson, and W. B. Epperson. 2003. Effect of total dissolved solids and sulfates in drinking water for growing steers. Pp. 378-380 in *Proceedings, Western Section, American Society of Animal Science, June 22-2, 2003, Phoenix, AZ*, Vol. 54. Champaign, IL: ASAS.

Patterson, T., and P. Johnson. 2003. Effects of water quality on beef cattle. *Proceedings, The Range Beef Cow Symposium XVIII, December 9-11, 2003, Mitchell, NE*. Available online at http://rangebeefcow.com/speakers/presentations/Patterson.pdf. Accessed on November 24, 2014.

Peeler, H. T. 1972. Biological availability of nutrients in feeds: Availability of major mineral ions. *Journal of Animal Science* 35:695-712.

Pehrson, B., M. Knutson, and M. Gyllensward. 1989. Glutathionine peroxidase activity in heifers fed diets supplemented with organic and inorganic selenium compounds. *Swedish Journal of Agricultural Research* 19:53-56.

Pendlum, L. C., J. A. Boling, and N. W. Bradley. 1976. Plasma and ruminal constituents and performance of steers fed different nitrogen sources and levels of sulfur. *Journal of Animal Science* 43:1307-1314.

Perry, T. W., W. M. Beeson, W. H. Smith, and M. T. Mohler. 1968. Value of zinc supplementation of natural rations for fattening beef cattle. *Journal of Animal Science* 27:1674-1677.

Perryman, L. E., D. R. Leach, W. C. Davis, W. D. Mickelson, S. R. Heller, H. D. Ochs, J. A. Ellis, and E. Brummerstedt. 1989. Lymphocyte alterations in zinc-deficient calves with lethal trait A46. *Veterinary Immunology and Immunopathology* 21:239-245.

Phillippo, M., W. R. Humphries, T. Atkinson, G. D. Henderson, and P. H. Garthwaite. 1987a. The effect of dietary molybdenum and iron on copper status, puberty, fertility and oestrous cycles in cattle. *Journal of Agricultural Science* 109:321-336.

Phillippo, M., W. R. Humphries, and P. H. Garthwaite. 1987b. The effect of dietary molybdenum and iron on copper status and growth in cattle. *Journal of Agricultural Science* 109:315-320.

Podoll, K. L., J. B. Bernard, D. E. Ullrey, S. R. DeBar, P. K. Ku, and W. T. Magee. 1992. Dietary selenium versus selenite for cattle, sheep, and horses. *Journal of Animal Science* 70:1965-1970.

Pogge, D. J., M. E. Drewnoski, and S. L. Hansen. 2014. High dietary sulfur decreased the retention of copper, manganese, and zinc in steers. *Journal of Animal Science* 92:2182-2191.

Pond, W. G. 1983. Effect of dietary calcium and zinc levels on weight gain and blood and tissue mineral concentrations of growing Columbia- and Suffolk-sired lambs. *Journal of Animal Science* 56:952-959.

Pond, W. G., and R. R. Oltjen. 1988. Response of large and medium frame beef steers to protein and zinc supplementation of a corn silage-corn finishing diet. *Nutrition Reports International* 38:737-743.

Pond, W. G., and M. H. Wallace. 1986. Effects of gestation-lactation diet calcium and zinc levels and of parenteral vitamin A, D and E during gestation on ewe body weight and on lamb weight and survival. *Journal of Animal Science* 63:1019-1025.

Pringle, W. L., W. K. Dawley, and J. E. Miltimore. 1973. Sufficiency of Cu and Zn in barley, forage, and corn silage rations as measured by response to supplements by beef cattle. *Canadian Journal of Animal Science* 53:497-502.

Qi, K., C. D. Lu, and F. N. Owens. 1993. Sulfate supplementation of growing goats: Effects on performance, acid-base balance, and nutrient digestibilities. *Journal of Animal Science* 71:1579-1587.

Raven, A. M., and A. Thompson. 1959. The availability of iron in certain grasses, clover and herb species. I. Perennial ryegrass, cocksfoot and timothy. *Journal of Agricultural Science* 52:177-186.

Reed, J. J., M. A. Ward, K. A. Vonnahme, T. L. Neville, S. L. Julius, P. P. Borowicz, J. B. Taylor, D. A. Redmer, A. T. Grazul-Bilska, L. P. Reynolds, and J. S. Caton. 2007. Effects of selenium supply and dietary restriction on maternal and fetal body weight, visceral organ mass, cellularity estimates, and jejunal vascularity in pregnant ewe lambs. *Journal of Animal Science* 85:2721-2733.

Reed, W. D. C., R. C. Elliott, and J. H. Topps. 1965. Phosphorus excretion of cattle fed on high-energy diets. *Nature* 208:953-954.

Reeves, M. A., and P. R. Hoffmann. 2009. The human selenoproteome: Recent insights into functions and regulation. *Cellular and Molecular Life Sciences* 66:2457-2478.

Reffett, J. K., J. W. Spears, and T. T. Brown, Jr. 1988. Effect of dietary selenium on the primary and secondary immune response in calves challenged with infectious bovine rhinotracheitis virus. *Journal of Nutrition* 118:229-235.

Richter, E. L., M. E. Drewnoski, and S. L. Hansen. 2012. The effect of dietary sulfur on performance, mineral status, rumen hydrogen sulfide, and rumen microbial populations in yearling beef steers. *Journal of Animal Science* 90:3945-3953.

Ricketts, R. E., J. R. Campbell, D. E. Weinman, and M. E. Tumbleson. 1970. Effect of three calcium:phosphorus ratios on performance of growing Holstein steers. *Journal of Dairy Science* 53:898-903.

Roberts, W. K., and V. V. E. St. Omer. 1965. Dietary potassium requirements of fattening steers. *Journal of Animal Science* 24:902 (Abstract 234).

Rojas, M. A., I. A. Dryer, and W. A. Cassatt. 1965. Manganese deficiency in the bovine. *Journal of Animal Science* 24:664-667.

Rook, J. A. F., and J. E. Stony. 1962. Magnesium in the nutrition of farm animals. *Nutrition Abstracts & Reviews* 32:1055-1077.

Rosenfeld, I., and O.A. Beath, eds. 1964. *Selenium.* New York: Academic Press.

Rotruck, J. T., A. L. Pope, H. E. Ganther, A. B. Swanson, D. G. Hafeman, and W. G. Hoekstra. 1973. Selenium: Biochemical role as a component of glutathione peroxidase. *Science* 179:588-590.

Rumsey, T. S. 1978. Effects of dietary sulfur addition and Synovex-S ear implants on feedlot steers fed an all-concentrate finishing diet. *Journal of Animal Science* 46:463-477.

Sarturi, J. O., G. E. Erickson, T. J. Klopfenstein, and C. D. Buckner. 2011. Ruminal degradable sulfur from organic and inorganic sources in beef cattle finishing diets. *Journal of Animal Science* 89 (E-Suppl. 1):450 (Abstract 417).

Sarturi, J. O., G. E. Erickson, T. J. Klopfenstein, K. M. Rolfe, C. D. Buckner, and M. K. Luebbe. 2013a. Impact of source of sulfur on ruminal hydrogen sulfide and logic for the ruminal available sulfur for reduction concept. *Journal of Animal Science* 91:3352-3359.

Sarturi, J. O., G. E. Erickson, T. J. Klopfenstein, J. T. Vasconcelos, W. A. Griffin, K. M. Rolfe, J. R. Benton, and V. R. Bremer. 2013b. Effect of sulfur content in wet or dry distillers grains fed at several inclusions on cattle growth performance, ruminal parameters, and hydrogen sulfide. *Journal of Animal Science* 91:4849-4860.

Schwarz, F. J., M. Kirchgessner, and G. I. Stangl. 2000. Cobalt requirement of beef cattle—feed intake and growth at different levels of cobalt supply. *Journal of Animal Physiology and Animal Nutrition* 83:121-131.

Schwarz, K., and C. M. Foltz. 1957. Selenium as an integral part of factor 3 against dietary necrotic liver degeneration. *Journal of the American Chemical Society* 79:3292-3293.

Schwarz, K., and W. Mertz. 1959. Chromium (III) and the glucose tolerance factor. *Archives of Biochemistry and Biophysics* 85:292-295.

Scott, D. 1972. Excretion of phosphorus and acid in the urine of sheep and calves fed either roughage or concentrate diets. *Quarterly Journal of Experimental Physiology* 57:379-392.

Scott, D. 1975. Changes in mineral, water and acid–base balance associated with feeding and diet. Pp. 205-215 in *Digestion and Metabolism in the Ruminant*, I. W. McDonald and A.C.I. Warner, eds. Armidale, NSW, Australia: University of New England Publishing Unit.

Shand, A., and G. Lewis. 1957. Chronic copper poisoning in young calves. *Veterinary Record* 69:618-619.

Shariff, M. A., R. J. Boila, and K. M. Wittenberg. 1990. Effect of dietary molybdenum on rumen dry matter disappearance in cattle. *Canadian Journal of Animal Science* 70:319-323.

Shupe, J. L., J. E. Butcher, J. W. Call, A. E. Olson, and J. T. Blake. 1988. Clinical signs and bone changes associated with phosphorus deficiency in beef cattle. *American Journal of Veterinary Research* 49:1629-1636.

Siddons, R. C., and C. F. Mills. 1981. Glutathione peroxidase activity and erythrocyte stability in calves differing in selenium and vitamin E status. *British Journal of Nutrition* 46:345-355.

Smart, M. E., R. Cohen, D. A. Christensen, and C. M. Williams. 1986. The effects of sulphate removal from the drinking water on the plasma and liver copper and zinc concentrations of beef cows and their calves. *Canadian Journal of Animal Science* 66:669-680.

Smith, R. M. 1987. Cobalt. Pp. 143-183 in *Trace Elements in Human and Animal Nutrition*, W. Mertz, ed. New York: Academic Press.

Soto-Navarro, S. A., T. L. Lawler, J. B. Taylor, L. P. Reynolds, J. J. Reed, J. W. Finley, and J. S. Caton. 2004. Effect of high-selenium wheat on visceral organ mass, and intestinal cellularity and vascularity in finishing beef steers. *Journal of Animal Science* 82:1788-1793.

Spears, J. W. 1984. Nickel as a "newer trace element" in the nutrition of domestic animals. *Journal of Animal Science* 59:823-835.

Spears, J. W. 1989. Zinc methionine for ruminants: relative bioavailability of zinc in lambs and effects of growth and performance of growing heifers. *Journal of Animal Science* 67:835-843.

Spears, J. W. 2000. Micronutrients and immune function in cattle. *Proceedings of the Nutrition Society* 59:587-594.

Spears, J. W. 2003. Trace mineral bioavailability in ruminants. *Journal of Nutrition* 133:1506S-1509S.

Spears, J. W., and R. W. Harvey. 1987. Lasalocid and dietary sodium and potassium effects on mineral metabolism, ruminal volatile fatty acids and performance of finishing steers. *Journal of Animal Science* 65:830- 840.

Spears, J. W., and E. B. Kegley. 2002. Effect of zinc source (zinc oxide vs. zinc proteinate) and level on performance, carcass characteristics, and immune response of growing and finishing steers. *Journal of Animal Science* 80:2747-2752.

Spears, J. W., and L. J. Samsell. 1984. Effect of zinc supplementation on performance and zinc status of growing heifers. *Journal of Animal Science* 49(Suppl. 1):407.

Spears, J. W., and W. P. Weiss. 2014. Mineral and vitamin nutrition in ruminants. *The Professional Animal Scientist* 30:180-191.

Spears, J. W., R. W. Harvey, and E. C. Segerson. 1986. Effect of marginal selenium deficiency on growth, reproduction and selenium status of beef cattle. *Journal of Animal Science* 63:586-594.

Spears, J. W., B. R. Schricker, and J. C. Burns. 1989. Influence of lysocellin and monensin on mineral metabolism in steers fed forage-based diets. *Journal of Animal Science* 67:2140-2149.

Spears, J. W., K. E. Lloyd, and R. S. Fry. 2011. Tolerance of cattle to increased dietary sulfur and effect of dietary cation-anion balance. *Journal of Animal Science* 89:2502-2509.

Spears, J. W., C. S. Whisnant, G. B. Huntington, K. E. Lloyd, R. S. Fry, K. Krafka, and A. Lamptey. 2012. Chromium propionate enhances insulin sensitivity in growing cattle. *Journal of Dairy Science* 95:2037-2045.

Stabel, J. R., and J. W. Spears. 1993. Role of selenium in immune responsiveness and disease resistance. Pp. 333-356 in *Human Nutrition—A Comprehensive Treatise, Vol. 8, Nutrition and Immunology*, D. M. Klurfeld, ed. New York: Plenum Press.

Stahlhut, H. S., C. S. Whisnant, K. E. Lloyd, E. J. Baird, L. R. Legleiter, S. L. Hansen, and J. W. Spears. 2006a. Effect of chromium supplementation and copper status on glucose and lipid metabolism in Angus and Simmental beef cows. *Animal Feed Science and Technology* 128:253-265.

Stahlhut, H. S., C. S. Whisnant, and J. W. Spears. 2006b. Effect of chromium supplementation and copper status on performance and reproduction of beef cows. *Animal Feed Science and Technology* 128:266-275.

Standish, J. F., C. B. Ammerman, C. F. Simpson, F. C. Neal, and A. Z. Palmer. 1969. Influence of graded levels of dietary iron as ferrous sulfate, on performance and tissue mineral composition of steers. *Journal of Animal Science* 29:496-503.

Stangl, G. I., F. J. Schwarz, H. Muller, and M. Kirchgessner. 2000. Evaluation of the cobalt requirement of beef cattle based on vitamin B_{12}, folate, homocysteine and methylmalonic acid. *British Journal of Nutrition* 84:645-653.

Starks, P. B., W. H. Hale, U. S. Garrigus, and R. M. Forbes. 1953. The utilization of feed nitrogen by lambs as affected by elemental sulfur. *Journal of Animal Science* 12:480-491.

Sumner, J. M., F. Valdez, and J. P. McNamara. 2007. Effects of chromium propionate on response to an intravenous glucose tolerance test in growing Holstein heifers. *Journal of Dairy Science* 90:3467-3474.

Sunde, R. A. 1997. Selenium. Pp 493-556 in *Handbook of Nutritionally Essential Mineral Elements*, F. L. O'Dell, and R. A. Sunde, eds. New York: Marcel Dekker.

Sunde, R. A. 2001a. Regulation of selenoprotein expression. Pp. 81-98 in *Selenium: Its Molecular Biology and Role in Human Health*, D. L. Hatfield, ed. Boston, MA: Kluwer.

Sunde, R. A. 2001b. Selenium. Pp 354-365 in *Present Knowledge in Nutrition*, B. A. Bowman and R. M. Russell, eds. Washington, DC: ILSI Press.

Suttle, N. F. 1974. Effects of organic and inorganic sulfur on the availability of dietary copper to sheep. *British Journal of Nutrition* 32:559-568.

Suttle, N. F. 1991. The interactions between copper, molybdenum and sulfur in ruminant nutrition. *Annual Review of Nutrition* 11:121-140.

Suttle, N. F. 2010. *Mineral Nutrition of Livestock,* 4th Ed. Cambridge, MA: CAB International.

Suttle, N. F., and M. McLauchlin. 1976. Predicting the effects of dietary molybdenum and sulphur on the availability of copper to ruminants. *Proceedings of the Nutrition Society* 35:22A-23A.

Taylor, J. B. 2005. Time-dependent influence of supranutritional organically bound selenium on selenium accumulation in growing wether lambs. *Journal of Animal Science* 83:1186-1193.

Taylor, J. B., L. P. Reynolds, D. A. Redmer, and J. S. Caton. 2009. Maternal and fetal tissue selenium loads in nulliparous ewes fed supranutritional and excessive selenium during mid- to late pregnancy. *Journal of Animal Science* 87:1828-1834.

Taylor, J. B, J. S. Caton, and R. Larsen. 2014. Selenium biofortification in North America: Using naturally selenium-rich feeds for livestock. Pp. 155-156 in *Selenium in the Environment and Human Health*, G. S. Banuelos, Z.-Q. Lin, and X. Yin, eds. London: CRC Press.

Thomas, W. E., J. K. Loosli, H. H. Williams, and L. A. Maynard. 1951. The utilization of inorganic sulfates and urea nitrogen by lambs. *Journal of Nutrition* 43:515-523.

Thompson, A., and A. M. Raven. 1959. The availability of iron in certain grasses, clover and herb species. II. Alsike, broad red clover, Kent wild white clover, trefoil and lucerne. *Journal of Agricultural Science* 53:224-229.

Thompson, L. H., M. B. Wise, R. W. Harvey, and E. R. Barrick. 1972. Starea, urea and sulfur in beef cattle rations. *Journal of Animal Science* 35:474-480.

Thornton, I., G. F. Kershaw, and M. K. Davies. 1972. An investigation into copper deficiency in cattle in the Southern Pennines. II. Response to copper supplementation. *Journal of Agricultural Science* 78:165-171.

Tiffany, M. E. 2003. Cobalt Requirements in Growing and Finishing Cattle Based on Performance, Vitamin B_{12} Status, and Metabolite Concentrations. Ph.D Dissertation. North Carolina State University, Raleigh.

Tiffany, M. E., and J. W. Spears. 2005. Differential responses to dietary cobalt in finishing steers fed corn- versus barley-based diets. *Journal of Animal Science* 83:2580-2589.

Tiffany, M. E., V. Fellner, and J. W. Spears. 2006. Influence of cobalt concentration on vitamin B_{12} production and fermentation of mixed ruminal microorganisms in continuous culture-flow through fermenters. *Journal of Animal Science* 84:635-640.

Tillman, A. D., and J. R. Brethour. 1958. Dicalcium phosphate and phosphoric acid as phosphorus sources for beef cattle. *Journal of Animal Science* 17:100-103.

Tillman, A. D., J. R. Brethour, and S. L. Hansard. 1959. Comparative procedures for measuring the phosphorus requirement of cattle. *Journal of Animal Science* 18:249-255.

Tucker, W. B., J. A. Jackson, D. M. Hopkins, and J. F. Hogue. 1991. Influence of dietary sodium bicarbonate on the potassium metabolism of growing dairy calves. *Journal of Dairy Science* 74:2296-2302.

Underwood, E. J. 1971. *Trace Elements in Human and Animal Nutrition,* 3rd Ed. New York: Academic Press.

Underwood, E. J. 1977. *Trace Elements in Human and Animal Nutrition,* 4th Ed. New York: Academic Press.

Underwood, E. J., and N. F. Suttle. 1999. *The Mineral Nutrition of Livestock,* 3rd Ed. New York: CAB International.

USDA APHIS (U.S. Department of Agriculture Animal and Plant Health Inspection Service). 2000. National Animal Health Monitoring System Feedlot 1999: Water Quality in U.S. Feedlots. Info Sheet No. 341.1200. Available online at http://www.aphis.usda.gov/animal_health/nahms/feedlot/downloads/feedlot99/Feedlot99_is_WaterQuality.pdf. Accessed on April 28, 2015.

Uwituze, S., G. L. Parsons, K. K. Karges, M. L. Gibson, L. C. Hollis, J. J. Higgins, and J. S. Drouillard. 2011. Effects of distillers grains with high sulfur concentration on ruminal fermentation and digestibility of finishing diets. *Journal of Animal Science* 89:2817-2828.

Van Bruwaene, R., G. B. Gerber, R. Kirchmann, J. Colard, and J. Van Kerkorn. 1984. Metabolism of Cr, Mn, Fe and Co in lactating dairy cows. *Health Physics* 46:1069-1082.

Van Emon, M. L., K. A. Vonnahme, P. T. Berg, R. R. Redden, M. M. Thompson, J. D. Kirsch, and C. S. Schauer. 2103. Influence of level of dried distillers grains with solubles on feedlot performance, carcass characteristics, serum testosterone concentrations, and spermatozoa motility and concentration of growing rams. *Journal of Animal Science* 91:5821-5828.

Van Heugten, E., and J. W. Spears. 1997. Immune response and growth of stressed weanling pigs fed diets supplemented with organic or inorganic chromium. *Journal of Animal Science* 75:409-416.

Vonnahme, K. A., C. M. Wienhold, P. P. Borowicz, T. L. Neville, D. A. Redmer, L. P. Reynolds, and J. S. Caton. 2011. Supranutritional selenium increases mammary gland vascularity in postpartum ewe lambs. *Journal of Dairy Science* 94:2850-2858.

Wacker, W. E. C. 1980. *Magnesium and Man.* Cambridge, MA: Harvard University Press.

Walker, C. K., and J. M. Elliot. 1972. Lactational trends in vitamin B12 status on conventional and restricted-roughage rations. *Journal of Dairy Science* 55:474-479.

Ward, G. M. 1966. Potassium metabolism of domestic ruminants: A review. *Journal of Dairy Science* 49:268-276.

Ward, G. M. 1978. Molybdenum toxicity and hypocuprosis in ruminants: A review. *Journal of Animal Science* 46:1078-1085.

Ward, G., L. H. Harbors, and J. J. Blaha. 1979. Calcium-containing crystals in alfalfa: Their fate in cattle. *Journal of Dairy Science* 62:715-722.

Ward, J. D., and J. W. Spears. 1997. Long-term effects of consumption of low-copper diets with or without supplemental molybdenum on copper status, performance, and carcass characteristics of cattle. *Journal of Animal Science* 75:3057-3065.

Ward, J. D., J. W. Spears, and E. B. Kegley. 1993. Effect of copper level and source (copper lysine versus copper sulfate) on copper status, performance, and immune response in growing steers fed diets with or without supplemental molybdenum and sulfur. *Journal of Animal Science* 71:2748-2755.

Ward, J. D., J. W. Spears, and G. P. Gengelbach. 1995. Differences in copper status and copper metabolism among Angus, Simmental, and Charolais cattle. *Journal of Animal Science* 73:571-577.

Watson, A. K., T. J. McEvers, M. P. McCurdy, M. J. Hersom, L. J. Walter, N. D. May, J. A. Reed, N. A. Cole, K. E. Hales, G. W. Horn, J. P. Hutcheson, C. R. Krehbiel, T. E. Lawrence, J. C. MacDonald, and G. E. Erickson. 2014. Phosphorus and calcium retention in serially harvested cattle. Pp. 167-168 in *Plains Nutrition Council Spring*

Conference Proceedings. Available online at http://amarillo.tamu.edu/files/2010/10/2014-Proceedings-final.pdf. Accessed on November 24, 2014.

Weeth, H. J., and D. L. Capps. 1972. Tolerance of growing cattle for sulfate-water. *Journal of Animal Science* 34:256-260.

Weeth, H. J., and L. H. Haverland. 1961. Tolerance of growing cattle for drinking water containing sodium chloride. *Journal of Animal Science* 20:518-521.

Weeth, H. J., and J. E. Hunter. 1971. Drinking of sulfate-water by cattle. *Journal of Animal Science* 32:277-281.

Weeth, H. J., L. H. Haverland, and D. W. Cassard. 1960. Consumption of sodium chloride water by heifers. *Journal of Animal Science* 19:845-851.

Weil, A. B., W. B. Tucker, and R. W. Hemken. 1988. Potassium requirements of dairy calves. *Journal of Dairy Science* 71:1868-1872.

Whanger, P. D., and G. Matrone. 1966. Effect of dietary sulfur upon the production and absorption of lactate in sheep. *Biochimica et Biophysica Acta* 124:273-279.

Whanger, P. D., P. H. Weswig, J. E. Oldfield, P. R. Cheeke, and O. H. Muth. 1972. Factors influencing selenium and white muscle disease: Forage types, salts, amino acids, and dimethyl sulfoxide. *Nutrition Reports International* 6:21-37.

Wheeler, J. L., D. A. Hedges, K. A. Archer, and B. A. Hamilton. 1980. Effect of nitrogen, sulfur and phosphorus fertilizer on the production, mineral content and cyanide potential of forage sorghum. *Australian Journal of Experimental Agriculture and Animal Husbandry* 20:330-338.

Whelan, B. R. 1989. Uptake of selenite fertilizer by subterranean clover pasture in Western Australia. *Australian Journal of Experimental Agriculture* 29:517-522.

Whelan, B. R., N. J. Barrow, and D. W. Peter. 1994a. Selenium fertilizers for pastures grazed by sheep. 1. Selenium concentrations in whole-blood and plasma. *Australian Journal of Agricultural Research* 45:863-875.

Whelan, B. R., N. J. Barrow, and D. W. Peter. 1994b. Selenium fertilizers for pastures grazed by sheep. 2. Wool and liveweight responses to selenium. *Australian Journal of Agricultural Research* 45:877-887.

Whitehead, D. C., K. M. Goulden, and R. D. Hartley. 1985. The distribution of nutrient elements in cell wall and other fractions of the herbage of some grasses and legumes. *Journal of the Science of Food and Agriculture* 36:311-318.

Wiener, G., N. F. Suttle, A. C. Field, J. G. Herbert, and J. A. Woolliams. 1978. Breed differences in copper metabolism in sheep. *Journal of Agricultural Science* 91:433-441.

Wise, M. B., S. E. Smith, and L. L. Barnes. 1958. The phosphorus requirement of calves. *Journal of Animal Science* 17:89-99.

Wise, M. B., A. L. Ordoreza, and E. R. Barrick. 1963. Influence of variations in dietary calcium:phosphorus ratio on performance and blood constituents of calves. *Journal of Nutrition* 79:79-84.

Wittenberg, K. M., R. J. Boila, and M. A. Shariff. 1990. Comparison of copper sulfate and copper proteinate as copper sources for copper-depleted steers fed high molybdenum diets. *Canadian Journal of Animal Science* 70:895-904.

Wong-Ville, J., P. R. Henry, C. B. Ammerman, and P. V. Rao. 1989. Estimation of the relative bioavailability of manganese sources for sheep. *Journal of Animal Science* 67:2409-2414.

Woolliams, C., N. F. Suttle, J. A. Woolliams, D. G. Jones, and G. Wiener. 1986. Studies on lambs from lines genetically selected for low and high copper status. *Animal Production* 43:293-301.

Wright, P. L., and M. C. Bell. 1966. Comparative metabolism of selenium and tellurium in sheep and swine. *American Journal of Physiology* 211:6-10.

Wu, L., and Z. Z. Huang. 1992. Selenium assimilation and nutrient element uptake in white clover and tall fescue under the influence of sulphate concentration and selenium tolerance of the plants. *Journal of Experimental Botany* 43:549-555.

Wylie, M. J., J. P. Fontenot, and L. W. Greene. 1985. Absorption of magnesium and other macrominerals in sheep infused with potassium in different parts of the digestive tract. *Journal of Animal Science* 61:1219-1229.

Yeh, J. Y., S. C. Vendeland, Q. Gu, J. A. Butler, B. R. Ou, and P. D. Whanger. 1997. Dietary selenium increases selenoprotein W levels in rat tissues. *Journal of Nutrition* 127:2165-2172.

Zhang, F. J., X. G. Weng, J. F. Wang, D. Zhou, C. C. Zhai, Y. X. Hou, and Y. H. Zhu. 2014. Effects of temperature-humidity index and chromium supplementation on antioxidant capacity, heat shock protein 72, and cytokine response of lactating cows. *Journal of Animal Science* 92:3026-3034.

Zinn, R. A., E. Alvarez, M. Mendez, M. Montano, E. Ramirez, and Y. Shen. 1997. Influence of dietary sulfur level on growth performance and digestive function in feedlot cattle. *Journal of Animal Science* 75:1723-1728.

8

Vitamins

INTRODUCTION

Vitamins are unique among dietary nutrient classes fed to ruminants. Unlike other nutrient classes, the vitamins are not necessarily similar to one another except for the fact that they are organic by nature. Vitamins are grouped for classification into lipid- and water-soluble categories. The lipid (fat)-soluble vitamins are A, D, E, and K, whereas the water-soluble vitamins include biotin, choline, vitamin B_{12} (cyanocobalamine), folic acid, niacin, pantothenic acid, pyridoxine (B_6), riboflavin, thiamin, and vitamin C. Vitamin C is synthesized by ruminant animals and is not required in the diet. Likewise, the ability of ruminal bacteria to synthesize the water-soluble vitamins and supply them to the host animal after passage to the small intestine has been known for decades. Consequently, minimum daily dietary requirements for water-soluble vitamins are either not needed or have not been established for beef cattle. Nevertheless, research on specific vitamins and potential needs have continued in ruminants. In this chapter, relevant aspects of each vitamin will be discussed, and recommendations or requirements will be provided as appropriate. The focus will be on data published since the last revision (NRC, 1996, 2000), with specific emphasis on beef cattle applications. For broader and more detailed reviews of the vitamins in livestock and human nutrition, metabolism, and health, readers are referred to the numerous books and other publications (NRC, 1987, 2007a,b, 2012; McDowell, 2000, 2013; DSM Nutrition Company, 2012; Erdman et al., 2012; Zempleni et al., 2014).

Vitamins are required metabolically by beef cattle to sustain normal body function and life processes. In addition, they are required in adequate amounts to enable animals to efficiently utilize other nutrients. Many metabolic processes are initiated and controlled by specific vitamins throughout life. Minimum dietary requirements are amounts needed in the diet to prevent the appearance of a deficiency disease or a metabolic syndrome associated with specific vitamin deficiencies and to provide for normal life and production processes. It is with this classical view of vitamin requirements in mind that dietary requirements for vitamins have been traditionally established.

Calves from adequately fed mothers have minimal stores of vitamins at birth. Unlike the adult ruminant, a young calf does not have a fully functional rumen and active microflora, which typically contribute to vitamin synthesis. Colostrum is rich in vitamins, particularly vitamin A, provided that vitamins have been adequately supplied to the dam. Thus, a dietary supply of vitamins is typically provided to the newborn calf through colostrum; however, deficiencies of the B vitamins have been produced experimentally in calves before ruminal development (Miller, 1979).

Intensive production systems have placed an increased emphasis on the importance of supplying adequate vitamin concentrations to meet animal requirements. Vitamin requirements might be increased in confinement feeding situations where greater levels of production increase metabolic requirements. Nonetheless, confinement settings should also allow for more precise formulation and delivery of diets that will meet animal requirements.

FAT-SOLUBLE VITAMINS

Vitamin A

Vitamin A Requirements

Vitamin A requirements for livestock are reported in either International Units (IU; NRC, 1996, 2000, 2001, 2012) or Retinol Equivalents (RE; NRC, 1989b, 2007b). Definitions and relationships between IU and RE are shown below in Table 8-1.

Beef cattle requirements for vitamin A were previously reported (NRC, 1984, 1996) at 2,200 IU/kg dry feed for beef feedlot cattle; 2,800 IU/kg dry feed for pregnant beef heifers

TABLE 8-1 Definitions and Relationships of International Units (IU) and Retinol Equivalents (RE) as Related to Vitamin A[a]

1 IU	0.3 μg of all-*trans* retinol
1 IU	0.344 μg of retinyl acetate
1 IU	0.55 μg of retinyl palmitate
1 RE	1 μg of all-*trans* retinol, 2 μg of synthetic supplemented all-*trans* β-carotene, 6 μg of dietary all-*trans* β-carotene, or 12 μg of other dietary provitamin A carotenoids[b]

[a]NRC (1989b, 2001, 2007b; IOM, 2000).
[b]NRC (1989b, 2007b, IOM, 2000).

and cows; and 3,900 IU/kg dry feed for lactating cows and breeding bulls. Readers are referred to the previous edition and update (NRC, 1996, 2000) for references that were used to establish those requirements. These requirements are approximately 47, 60, and 84 IU/kg body weight (BW) for feedlot cattle, dry pregnant heifers and cows, and lactating cows and bulls, respectively. The NRC (2001) reported a vitamin A requirement for lactating dairy cows of 110 IU/kg BW, which is a substantial increase from the 76 IU/kg BW reported in the previous NRC (1989a) report for dairy cattle. The recent report for small ruminants (NRC, 2007b) expressed vitamin A requirements in RE (1 RE = 3.33 IU of vitamin A; Table 8-1). The Committee on Nutrient Requirements of Small Ruminants (NRC, 2007b) increased vitamin A requirements to provide a greater margin of safety and to ensure liver storage. They set the requirement for small ruminants during late gestation at 45.5 RE/kg BW and late pregnant and lactating small ruminants at 53.5 RE/kg BW, which equates to 151 and 178 IU/kg BW, respectively. Likewise, in setting requirements for ruminant livestock, the Commonwealth Scientific and Industrial Research Organisation (CSIRO, 2007) included a significant margin above the available data and set requirements at 30, 45, and 67 RE/kg BW for feedlot cattle, pregnant beef heifers and cows, and lactating beef cows, respectively (or 100, 150, and 223 IU/kg BW).

After the 1970s, little research on vitamin A in U.S. beef cattle was published until the 1990s. Hill et al. (1995) reported that feedlot steers receiving 2,134 IU supplemental vitamin A/kg dry matter (DM) had greater gains and feed efficiencies than those receiving approximately 6,274 IU supplemental vitamin A/kg DM. In a receiving experiment, Zinn et al. (1996) also observed greater average daily gain (ADG) and feed efficiency in crossbred calves consuming 2,200 IU supplemental vitamin A/kg DM than in those consuming 11,000 IU supplemental vitamin A/kg DM. However, these differences were only noted during the first 28 d of the 56-d experiment and no carcass data were presented for either of these experiments. Pyatt et al. (2005) reported that ADG and feed efficiency were improved in cattle fed 2,300 IU/kg compared with those that received 7,250 IU/kg. Similarly, in Gorocica-Buenfil et al. (2007a,b,c, 2008) conducted four experiments (three of which were with beef cattle) comparing 2,200 or 2,700 IU supplemental vitamin A/kg vs. no supplemental vitamin A during the finishing period. Compared with the nonsupplemented cattle, the vitamin A-supplemented cattle had improved gains in one experiment (Gorocica-Buenfil et al., 2007b) and improved feed efficiencies in two of the experiments (Gorocica-Buenfil et al., 2007c, 2008). Recently, Bryant et al. (2010) utilized 360 single-source black yearling steers fed a 91% concentrate (steam-flaked corn base) diet to evaluate the effects of supplemental vitamin A (0, 1,103, 2,205, 4,410, or 8,820 IU/kg dietary DM) on performance and carcass quality. Final BW, feed efficiency, ADG, and daily dry matter intake (DMI) did not differ among treatments within each 28-d period or for the overall experiment. However, from day 57 to slaughter, average DMI (10.33, 10.28, 10.57, 9.75, and 10.22 kg/steer daily for 0, 1,103, 2,205, 4,410, and 8,820 IU of vitamin A/kg DM, respectively) was less by steers receiving 4,410 IU of supplemental vitamin A/kg dietary DM than for steers in the other treatments. Collectively these data support the NRC (1996, 2000) recommendations of 2,200 IU supplemental vitamin A/kg DM for beef feedlot cattle.

Limited recent data are available investigating vitamin A requirements in pregnant beef heifers and cows, lactating cows, preweaned calves, and breeding bulls. Definitive studies with modern breeds and beef production practices have not been conducted or are not available in the literature for the above-mentioned physiological categories of beef cattle. Minimal dietary requirements need to be established on new data, but the existing data indicate that requirements set by the previous NRC committees (NRC, 1984, 1996, 2000) are reasonable, even though in practice many nutritionists include an additional margin of safety (Vasconcelos and Galyean, 2007). The current committee recognizes that the majority of the existing data were mostly collected with breed types and production levels that may not be completely representative of current production practices. Nonetheless, insufficient evidence exists to elevate the vitamin A requirements for beef cattle above those recommended in the previous NRC (1996, 2000) publications. Therefore, the committee has retained the previous requirements of 2,200 IU/kg dry feed for beef feedlot cattle; 2,800 IU/kg dry feed for pregnant beef heifers and cows; and 3,900 IU/kg dry feed for lactating cows and breeding bulls, which are approximately equal to 47, 60, and 84 IU/kg BW for feedlot, dry pregnant heifers and cows, and lactating cows and bulls, respectively. The committee has also opted to express requirements in both IU/kg BW and IU/kg DMI, which should provide more flexibility for end users. It should be noted that these requirements are exclusive of any contributions from precursors (Table 8-1) and that at lower or higher levels of intake, vitamin concentrations in the diet may need to be adjusted.

Special Considerations

Vitamin A deficiency symptoms have plagued both humans and livestock since records have been kept. In the ancient writings, there are numerous examples that can be traced to vitamin A deficiency issues. For example, in Eber's Papyrus, an Egyptian medical document written between 1520 and1600 BC, it is reported that ingesting or topical application of cattle or poultry liver extracts will cure poor night vision (Aykroyd, 1958; Wolf, 1978, as cited by McDowell, 2013). In addition, events recorded in religious texts indicate observed Vitamin A deficiencies. For example, Jeremiah 14:6 (New International Version) states "Wild donkeys stand on the barren heights and pant like jackals; their eyesight fails for the lack of pasture." Thus, vitamin A deficiency has been with us since the dawn of time, and it continues to be a real and relevant issue for both livestock and humans today. McLaren (1986) indicated that at least 73 countries and territories could have potential problems with vitamin A deficiency. In 1993, livestock producers in eastern Russia reported that vitamin A sources during periods when grass was not green were inadequate for their children and nonexistent for their livestock (J. S. Caton, Animal Sciences Department, North Dakota State University, Fargo, ND, personal communication, February 10, 2014). More recently it was reported (WHO, 2009) that globally, low serum retinol concentrations affect an estimated 190 million preschool-age children and 19.1 million pregnant women. At the time, this was approximately 33.3% of preschool-age children and 15.3% of pregnant women in world populations that were at risk for vitamin A deficiency.

The world's population of beef cattle suffers from considerable vitamin A deficiency issues, and North American herds are not exempt. After scientists identified the structure of vitamin A and later succeeded in synthesizing it, vitamin A was used in livestock production and human nutrition and health practices during the 1950s and 1960s. Without adequate provision in the diet of ruminants, deficiencies will occur. Indeed, vitamin A deficiency is still a relevant concern in beef cattle production systems, particularly following drought or in late winter and early spring, with late-term abortions, retained placenta, or stillborn calves being the most likely observable symptoms (Fyksen, 2013).

Vitamin A (retinol) does not occur, as such, in plant material; however, its precursors, carotenes or carotenoids, are present in plants in various forms. The prototypic vitamin A is all-*trans* retinol, but various isomers, including all-*trans* retinal and all-*trans* retinoic acid, are metabolically important forms of the vitamin. Many of the carotenoids have vitamin A activity, but β-carotene, α-carotene, γ-carotene, and cryptoxanthine (in corn grain) are of particular dietary importance because of their high vitamin A activity. Efficiency of conversion of carotenoids to retinol is variable in beef cattle and is generally less than that for nonruminants (Ullrey, 1972). McDowell (2000) reported that ruminal degradation of active vitamin A was 67% for cattle on concentrate diets and about 18% for cattle on forage-based diets.

Diet type and rate of ruminal fermentation may affect vitamin A supply in ruminants. For example, retinyl acetate was degraded by ruminal fluid from concentrate-fed cattle more rapidly than from animals fed hay or straw (Rode et al., 1990). In practice, conversion rates of dietary β-carotene to active vitamin A are considered to be substantially less that the theoretical maximums because of ruminal degradation and inefficiencies in digestion, absorption, and enzymatic activity (McDowell, 2000). Additional information on chemistry and molecular interconversions of compounds with vitamin A activity are available elsewhere (McDowell, 2000; Olson, 2001; Solomons, 2012; Zempleni et al., 2014).

Few grains, except for yellow corn, contain appreciable amounts of carotenoids, which are rapidly destroyed by exposure to sunlight, oxygen, high temperatures, and pressure. The availability of carotene from corn silage can be low (Jordan et al., 1963; Smith et al., 1964; Miller et al., 1967) and without analytical verification, vitamin A concentrations from corn silage should not be trusted or assumed. High-quality forages provide carotenoids in large amounts but availability tends to be seasonal. Vitamin A is lipid-soluble, and absorption is associated with intestinal lipid absorption in mammals. Consequently, normal processes of fat digestion and absorption are necessary to allow for absorption of vitamin A.

The liver can store vitamin A, and it contains approximately 90% of the body's vitamin A (McDowell, 2000). Long-term liver vitamin A stores are primarily within the liver stellate cells in the form of retinyl esters. These storage forms can serve to prevent vitamin A deficiency, and they play a critical role in normal life cycle vitamin A supply in extensive rangeland production systems. Unfortunately, liver stores are highly variable and cannot be assessed accurately without taking biopsy samples. On a practical basis, no more than 2 to 4 months of protection from stored vitamin A can be expected, and cattle should be observed carefully for signs of deficiency whenever the diet is low in vitamin A concentrations.

Vitamin D

Vitamin D Requirements

Unlike aquatic species that store appreciable amounts of vitamin D in the liver, most land mammals, including ruminants, do not maintain extensive body stores of vitamin D. Because vitamin D is synthesized by beef cattle exposed to sunlight or fed sun-cured forages, these animals rarely require vitamin D supplementation. The existence of limited body stores of vitamin D, difficulty in determining amount of in vivo synthesis, limited information on Ca and P intake, and potential interactions with other nutrients make it difficult to set vitamin D requirements with certainty. For beef cattle housed outdoors and exposed to regular sunlight, it is likely that there is no dietary requirement for vitamin D. In these situations, recommended requirements should be viewed as extra safety margins. In cases where beef cattle are housed indoors away from sunlight or in locations where they experience long periods of darkness and have no sun-cured forages in their diet, vitamin D should be included in the diet at recommended levels.

The vitamin D requirement of beef cattle was previously reported to be 275 IU/kg dry diet (NRC, 1984, 1996, 2000). An IU of Vitamin D is defined as 0.025 µg of cholecalciferol (D_3) or its equivalent. Ergocalciferol (D_2) also is active in cattle. These previously recommended levels equate to approximately 5.7 IU/kg BW. These values are similar to NRC (2007b) recommendations of 5.6 IU/kg BW for maintenance and ewes in early pregnancy and CSIRO (2007) recommendations for grazing ewes (6 IU/kg BW). However, the small ruminant publication (NRC, 2007b) recommended greater dietary vitamin D levels for ewes in late pregnancy and/or lactating ewes, and CSIRO (2007) recommended 10 IU/kg BW for these classes of ewes. For dairy cattle, the NRC (2001) set a requirement of 30 IU/kg BW, which, admittedly, provided a wide margin of safety. It is important to note that little recent data exist regarding vitamin D requirements for beef cattle. Consequently, previous vitamin D recommendations for beef cattle of 275 IU/kg DMI or approximately 5.7 IU/kg BW have been retained in this publication.

Special Considerations

Vasconcelos and Galyean (2007) reported that mean vitamin D levels included in finishing cattle diets was 329 IU/kg DMI, which is slightly greater than current recommendations. It is important to note, however, that in their extensive survey of consulting feedlot nutritionists (Vasconcelos and Galyean, 2007) the mode value for vitamin D dietary inclusion was 0, indicating that most of survey respondents did not include additional vitamin D.

Swanek et al. (1999) reported that providing 5 to 7.5×10^6 IU of vitamin D daily to finishing cattle for 7 to 10 d before slaughter increased serum Ca, lowered Warner-Bratzler shear force, and increased sensory tenderness ratings compared with steers given no additional vitamin D. This response is likely mediated through calpains because of their Ca dependency. In further investigations with vitamin D and carcass quality, Montgomery et al. (2004a,b,c) reported that doses of 0.5×10^6 IU of vitamin D daily for 8 d before slaughter would also decrease shear force, increase tenderness, and decrease performance for the last few days before slaughter; these authors noted only small effects on vitamin D tissue residues of treated cattle. Carnagey et al. (2007) indicated that high levels of vitamin D and E supplementation independently before slaughter increased tenderness but that the response was not observed when vitamins D and E were supplemented at high levels in combination. The effects of vitamin D on tenderness seem to be less pronounced in mature beef cows fed for slaughter (Carnagey et al., 2008). Although the data using supranutritional levels of vitamin D to improve carcass tenderness and quality are intriguing, intakes and performance were decreased, and these practices are certainly outside the realm of minimum daily dietary requirements to prevent deficiency disease and promote normal health and growth. Additional work in this area should provide more insight into the usefulness of vitamin D supplementation on carcass quality of finishing beef cattle.

Vitamin E

Vitamin E Requirements

Vitamin E requirements have not been clearly established for beef cattle. Requirements are difficult to establish without extensive knowledge of dietary selenium and vitamin E concentrations and the status of several other dietary and metabolic factors affecting overall oxidative stress and whole-animal antioxidant capacity. The vitamin E requirement for beef cattle was estimated to be between 15 and 60 IU/ kg dry diet for young calves (NRC, 1984, 1996, 2000). For vitamin E, an IU is defined as 1 mg of all racemic α-tocopherol acetate (IOM, 2000; NRC, 1989b, 2001; Table 8-2).

Early work by Gullickson and Calverley (1946) demonstrated that even in cattle diets that were consistently very low in vitamin E, no effects on growth, reproduction, or lactation were observed across four generations. Galyean et al. (1999) indicated that adding vitamin E to receiving diets at ≥400 IU/animal daily seemed to be useful for increasing gain and decreasing bovine respiratory disease (BRD) morbidity. Others (Rivera et al., 2002; Carter et al., 2005) reported no effect of supplementing 2,000 IU/animal daily on performance and overall health status of receiving calves during the first 28 d on feed. It is possible that differences among studies could be associated with the overall degree of stress being experienced by calves used in these experiments. In their review, Duff and Galyean (2007) indicated that adding vitamin E to the diet of stressed receiving calves at levels of more than 1,000 IU/animal daily would likely be beneficial in terms of decreasing BRD morbidity. Horn et al. (2010a,b) provided crossbred cows with 1,000 IU of vitamin E/cow daily and reported increases in circulating α-tocopherol concentrations but no effects on reproductive performance, calf immune function, or calf performance. Interestingly, Vasconcelos and Galyean (2007) reported in their feedlot survey that mean vitamin E concentrations in feedlot diets was 25.7 IU/kg DMI, which is well within the range of previously reported recommendations of 15 to 60 IU/kg DMI (NRC, 1996, 2000). It could be debated that responses

TABLE 8-2 Factors for Converting International Units (IU) of Vitamins E to α-Tocopherol in Milligrams[a]

Form of Vitamin E	USP Conversion Factors[b] IU/mg	USP Conversion Factors[b] mg/IU	Molar Conversion Factors[c] μmol/IU	α-Tocopherol Conversion Factors[d] mg/IU
Synthetic Vitamin E and Esters				
DL-α-Tocopheryl acetate	1.00	1.00	2.12	0.45
DL-α-Tocopheryl succinate	0.89	1.12	2.12	0.45
DL-α-Tocopherol[e]	1.10	0.91	2.12	0.45
Natural Vitamin E and Esters				
D-α-Tocopheryl acetate	1.36	0.74	1.56	0.67
D-α-Tocopheryl succinate	1.21	0.83	1.56	0.67
D-α-*Tocopherol*[f]	1.49	0.67	1.56	0.67

[a] Adapted from IOM (2000).

[b]Official United States Pharmacopeia (USP) conversions where 1 IU is defined as 1 mg of *all-rac*-α-tocopheryl acetate (USP, 1979, 1999).

[c]To convert mg to μmol divide the mg by the molecular weight of the vitamin E compound (α-tocopheryl acetate = 472; α-tocopheryl succinate = 530; α-tocopherol = 430) and multiply by 1,000. Because the amount of free and succinate compounds are adjusted for their different molecular weights relative to α-tocopheryl acetate, these forms have the same conversion factors as the corresponding tocopherol compounds.

[d]To convert the μmol of the vitamin E compound to mg of α-tocopherol, multiply the μmol by the molecular weight of α-tocopherol (430) and divide by 1,000. The activities of the three synthetic α-tocopherol compounds have been divided by 2 because the 2*S* stereoisomers contained in synthetic α-tocopherol are not maintained in the blood.

[e]DL-α-Tocopherol = *all-rac*-(racemic) α-tocopherol = synthetic vitamin E; *all rac*-α-tocopherol = *RRR*-, *RRS*-, *RSR*-, *RSS*-, *SSS*-, *SRS*-, *SSR*-, and *SRR*-α-tocopherol isomers.

[f]D-α-Tocopherol = *RRR*-α-tocopherol = natural vitamin E.

to elevated levels of vitamin E recommended by Duff and Galyean (2007) for stressed receiving calves are either pharmacological or resulting from offsetting yet uncharacterized conditional or transitory elevated requirements. It is much more likely, however, that responses to elevated dietary levels of vitamin E in stressed receiving cattle occur because increasing the nutrient density of receiving diets offsets the effect of decreased intake in terms of nutrient demands and satisfying nutrient requirements.

Inadequate data currently exist to set vitamin E requirements for most classes of beef cattle. Consequently, previous recommendations for beef cattle still seem reasonable. Moreover, Cusack et al. (2009) concluded that, based a meta-analysis of existing data sets, supplemental dietary vitamin E should be fed within the NRC (1996, 2000) guidelines for beef cattle. One exception to this would be calves that have undergone stress associated with marketing and transportation. Recent work summarized in Chapter 15 (Effects of Stress on Beef Cattle Nutrient Requirements) indicates that stressed receiving calves should be provided from 400 to 500 IU/calf daily during the receiving period for a 250-kg calf. This level of supplemental vitamin E would equate to approximately 1.6 to 2.0 IU/kg BW. The NRC (1996, 2000) recommended providing vitamin E at 15 to 60 IU/kg DMI (or approximately 0.31 to 1.25 IU/kg BW), which is a wide range that was deemed applicable to young calves, and was unchanged from the previous published beef nutrient requirements (NRC, 1984). It is likely that in nonstressed mature beef cattle, vitamin E requirements are low, and that vitamin E needs are likely often met through normal dietary ingredients. For growing cattle, evidence exists for dietary recommendations of vitamin E. As a result, the current committee has set the recommendation for newly received stress calves at 400 to 500 IU/d or 1.6 to 2.0 IU/kg BW during the receiving period. Once received and adapted to conditions, vitamin E concentrations in the diet of normal healthy finishing cattle should fall within previously recommended ranges of 25 to 35 IU/kg DMI or 0.52 to 0.73 IU/kg BW. It should be noted that at higher or lower DMI, dietary levels of vitamins may need to be adjusted. The data are limited for other classes of beef cattle, and additional research during periods of stress would provide greater insight into vitamin E requirements for other classes of beef cattle.

Special Considerations

Arnold et al. (1992) published data indicating that elevated levels of supplemental vitamin E could increase the color stability of displayed beef cuts and consequently decrease economic losses during retail phases of beef sales. These responses were associated with improved oxidative stability (Schaefer et al., 1995). Liu et al. (1995) reported that supplying 500 mg α-tocopherol acetate/animal daily for 126 d was adequate to increase color stability for beef aged 14 d, which is supported by the work of others (Zerby et al., 1999), although length and dose of vitamin E supply to optimize the effect might vary (Roeber et al., 2001). Results with inclusion of elevated dietary vitamin E concentrations have been mixed when diets containing 35% wet distillers grains have been fed. Bloomberg et al. (2011) reported that finishing steers fed 35% distillers grains should be provided 500 IU of vitamin E/animal daily for 97 d to improve retail shelf life of beef cuts. In contrast, others (Burken et al., 2012) have indicated that providing 500 IU of vitamin E/animal daily had no effect on animal performance or carcass characteristics. Data evaluating effects of supranutritional vitamin E supplementation during the finishing period to increase shelf life and consequently economic returns associated with retail beef cuts continue to build a case for this practice. Nonetheless, this use of vitamin E is not associated with minimal daily requirements or recommendations to offset deficiencies and promote normal health and production. In contrast, the use of elevated levels of vitamin E during periods of high stress, as discussed previously with newly received beef calves, could have merit in numerous production scenarios to promote animal health. Research in this direction could lead to improved production practices, health outcomes, and additional insight into yet uncharacterized conditional or transitory requirements.

Summary of New Recommendations Regarding Vitamin E

Overall, based on available data the committee recommends the following:

- It is likely that for nonstressed mature beef cattle, vitamin E requirements are low, and that vitamin E needs are likely most often met through normal dietary ingredients.
- For growing cattle, evidence exists for dietary recommendations of vitamin E. For newly received stressed calves, the committee recommends 400 to 500 IU/d or 1.6 to 2.0 IU/kg BW during the receiving period. Once received and adapted to conditions, for finishing beef cattle, dietary vitamin E should fall within previously recommended ranges of between 25 to 35 IU/kg DMI or 0.52 to 0.73 IU/kg BW.
- The data are limited for other classes of beef cattle and additional research during periods of stress would provide insight into vitamin E requirements for beef cattle.

Vitamin K

Vitamin K Requirements

The term vitamin K is used to describe a group of quinone fat-soluble compounds that have characteristic antihemorrhagic effects. Vitamin K was proposed and discovered by Henrik Dam in Denmark and derives its name from first letter of the Danish word "koagulation," which translates to coagulation in English (McDowell, 2000). Vitamin K is required for the synthesis of several plasma clotting factors including prothrombin (factor II), proconvertin (factor VII), Christmas factor (factor IX), and Stuart-Prower factor (factor X). Two major natural sources of vitamin K are the phylloquinones (vitamin K_1), found in plant sources, and the menaquinones (vitamin K_2), which are produced by bacteria. For ruminants, vitamin K_2 is the most significant source of vitamin K because it is synthesized in large quantities by ruminal bacteria. In addition, vitamin K_1 is abundant in pasture and green roughages. Both forms possess similar biological activity and function in blood clotting. Vitamin K is supplied to beef cattle by both dietary forms and by bacterial synthesis in the rumen. No requirement has been set for beef cattle, as bacterial synthesis and naturally occurring dietary forms are deemed adequate (NRC, 1996, 2000, 2001, 2007b; McDowell, 2000; CSIRO, 2007).

Special Considerations

Vitamin K deficiency has been reported in cattle that have accidentally ingested the rodenticide warfarin or in cattle with the "sweet clover disease" syndrome. Sweet clover disease results from the metabolic antagonistic action of dicoumarol that occurs when an animal consumes moldy or improperly cured sweet clover hay. Consumption of dicoumarol, a fungal metabolite produced from substrates in sweet clover hay, leads to prolonged blood clotting times and has caused death from uncontrolled hemorrhaging. It is important to note that dicoumarol passes through the placenta, and thus, the fetus of pregnant animals can be affected.

The initial appearance and severity of signs associated with dicoumarol poisoning are directly related to the dicoumarol content of the hay consumed. If low levels are consumed, clinical signs may not be evident for several months. Mild cases can be treated effectively with vitamin K (McElroy and Goss, 1940a; Link, 1959).

Few systematic studies of the effects of excess vitamin K have been conducted in ruminant animals. Toxicity associated with excessive oral intake of phylloquinone or menadione has not been demonstrated in beef cattle. The toxic dietary level of menadione is at least 1,000 times the dietary requirement (NRC, 1987).

WATER-SOLUBLE VITAMINS

Preruminant calves have limited stores of water-soluble vitamins at birth and consequently depend on supplies in milk until the rumen becomes functional and begins to supply vitamins via bacterial synthesis. Requirements for the water-soluble vitamins have not been set for beef cattle, primarily because bacterial synthesis in the rumen provides adequate amounts for most normally occurring situations. Likewise, ruminants are capable of synthesizing their own vitamin C (ascorbic acid) because the enzyme L-gulonolactone oxidase is present in their tissues (NRC, 2007b). Likely as a result of the long-standing acceptance of the adequacy of water-soluble vitamins for ruminants via dietary forms and bacterial synthesis, limited research has been conducted in this area.

The B vitamins are abundant in milk and many other feeds, and synthesis of B vitamins by ruminal microorganisms is extensive (McElroy and Goss, 1940a,b, 1941a,b;

Wegner et al., 1940, 1941; Hunt et al., 1943) and begins very soon after the introduction of dry feed into the diet (Conrad and Hibbs, 1954). As concentrate in the diet increases, thiamin declines, whereas niacin increases substantially in the rumen, and the duodenal concentration of thiamin, niacin, riboflavin, and biotin does not change (Miller et al., 1986a,b). Niacin decreases in the duodenum and ileum when monensin is added (22 mg/kg diet), whereas thiamin, riboflavin, and biotin are not affected.

Signs of insufficient intake of B-complex vitamins have been clearly demonstrated for thiamin (Johnson et al., 1948), riboflavin (Wiese et al., 1947), pyridoxine (Johnson et al., 1950), pantothenic acid (Sheppard and Johnson, 1957), biotin (Wiese et al., 1946), nicotinic acid (Hopper and Johnson, 1955), vitamin B_{12} (Draper et al., 1952; Lassiter et al., 1953), and choline (Johnson et al., 1951) in young calves. Although B vitamin synthesis is altered by diet type, considerable change is possible without producing signs of deficiency (Hayes et al., 1966; Clifford et al., 1967). The established metabolic functions of B vitamins are important, and consequently, a physiological need for most B vitamins can be assumed for cattle of all ages.

Supplemental riboflavin, niacin, folic acid, B_{12}, and ascorbic acid are degraded or absorbed anterior to the small intestine, whereas biotin and pantothenic acid primarily escape the rumen (Zinn et al., 1987). As a result, practical vitamin B deficiency is limited to young animals with immature rumen development and to situations in which an antagonist is present or ruminal synthesis is limited by lack of precursors.

Some considerations for each of the water-soluble vitamins will be briefly discussed below. For extensive details on chemistry, function, deficiencies, sources, toxicities, and other items associated with vitamin nutrition in ruminants and other species, readers are referred to some of the excellent works in this area (NRC, 1987, 2007a,b, 2012; McDowell, 2000, 2013; DSM Nutrition Company, 2012; Erdman et al., 2012; Zempleni et al., 2014).

Biotin

Biotin is needed as a coenzyme in the metabolism of carbohydrates, proteins, and lipids. Specifically, biotin is a cofactor in pyruvate carboxylase and propionyl-CoA-carboxylase, which are important for gluconeogenesis; acetyl-CoA-carboxylase, the first and rate-limiting step in de novo fatty acid synthesis; and methylcrotonyl-CoA carboxylase (McDowell, 2000). A significant body of work with biotin supplementation in dairy cattle has been generated during the past 15 to 20 years (NRC, 2001; Chen et al., 2011; Lean and Rabiee, 2011). Generally, this work demonstrated improvements in milk production and hoof health in response to supplemental biotin. It is likely that research will continue with biotin supplementation in dairy cattle because of potential economic benefits associated with increased milk production and improved herd health. Nonetheless, little research has been conducted with biotin supplementation in beef cattle. One report indicated that biotin supplementation improved hoof health in Canadian beef herds (Campbell et al., 2000). Earlier work indicated that dietary biotin largely escaped the rumen and that a mixture of vitamins supplemented at elevated levels in feedlot cattle did not alter performance but tended to decrease morbidity (Zinn et al., 1987). In light of the existing dairy cattle data, additional research with biotin supplementation in beef cattle might be warranted.

Choline

Choline is needed for maintaining cell structure as a component of phosphatidylcholine. It also plays a critical role in hepatic lipid metabolism and is an integral component of the neurotransmitter acetylcholine. Choline also participates in one-carbon metabolism, such as methyl transfer and is, consequently, thought to spare methionine for protein synthesis (NRC, 2007b). Lobley et al. (1996) reported that intravenous infusions of choline in sheep would spare irreversible loss of methionine by 18 to 26%. Choline from dietary sources is only of value to adult cattle if it can escape ruminal degradation. Rumsey (1975) determined that for choline-supplemented steers fed an all-concentrate diet, supplementation did not affect feedlot performance, carcass measurements, acidosis, or products of ruminal fermentation. Most dietary choline is degraded in the rumen, and hence, rumen-protected choline has been developed for research and production applications. Increasing levels of dietary rumen-protected choline produced a linear increase in milk production for lactating dairy cows (Erdman and Sharma, 1991). Additional research with rumen-protected choline in finishing beef steers and lambs (Bryant et al., 1999) and with lactating dairy cows (Sales et al., 2010) has produced inconsistent results. Bindel et al. (2000) reported that optimal performance of finishing heifers was achieved with 20 g/d of rumen-protected choline.

Because ruminants synthesize choline, a requirement has not been determined; however, it has been suggested that milk-fed calves receive supplementation of 0.26% choline in milk replacers (NRC, 1996, 2000). Huber (1988) suggested a requirement of 21 mg/kg BW; however, data do not exist to set requirements for ruminants under normal production practices.

Vitamin B_{12}

Vitamin B_{12} is a generic descriptor for a group of compounds that have vitamin B_{12} activity. One of the unique features of vitamin B_{12} is that it contains 4.5% cobalt. The naturally occurring forms of vitamin B_{12} are adenosylcobalamin and methylcobalamin. Cyanocobalamin, an artificially produced form of vitamin B_{12}, is used extensively in animal feeding because it is relatively stable and readily available.

The primary functions of vitamin B_{12} involve metabolism of nucleic acids and proteins, in addition to metabolism of fats and carbohydrates. Specifically, this vitamin plays a role in purine and pyrimidine synthesis, transfer of methyl groups, protein formation, and metabolism of lipids and carbohydrates. Vitamin B_{12} is of special interest in ruminant nutrition because of its role in propionate metabolism (Marston et al., 1961) and the practical incidence of vitamin B_{12} deficiency as a secondary result of cobalt deficiency.

A vitamin B_{12} deficiency is similar to a cobalt deficiency. The signs of deficiency are not always specific and can include poor appetite, retarded growth, and poor condition. In severe deficiencies, muscular weakness and demyelination of peripheral nerves occurs. In young ruminant animals, vitamin B_{12} deficiency can occur when rumen microbial populations are not yet fully developed.

The ruminant's absolute metabolic requirement for vitamin B_{12} is greater than the nonruminant's requirement and is associated with the requirement for cobalt because this trace mineral is a component of vitamin B_{12}. Cobalt content of the diet is the primary limiting factor for ruminal microbial synthesis of vitamin B_{12}. Substantial areas of the United States, Australia, and New Zealand have soils without sufficient cobalt to produce adequate concentrations in plants to support optimal vitamin B_{12} synthesis in the rumen (Ammerman, 1970). For additional information on cobalt and the interrelationship with vitamin B_{12}, see Chapter 7 (Minerals).

Folic Acid

Folic acid (vitamin B_9) is metabolically essential for the transfer of one-carbon units. One-carbon unit transfer is an essential process during amino acid metabolism, biosynthesis of purine and pyrimidines, and DNA methylation and demethylation associated with epigenetic events. Consequently, folic acid is closely linked to cell division and rate of cellular growth. Folic acid works in concert with several other vitamins, minerals, amino acids, and metabolites to maintain a functional one-carbon metabolic pool; these include vitamin B_{12}, choline, betaine, methionine, serine, glycine, vitamin B_6, cobalt, sulfur, and others. One-carbon metabolism is critical for normal growth and developmental processes and is likely one of the primary underlying mechanisms associated with developmental programming events during fetal and perinatal development (Wu et al., 2006; Ikeda et al., 2012; Meyer et al., 2012). Additional information can be found within some of the previously mentioned excellent reviews on vitamins in nutrition and growth.

Folic acid synthesis is very active in the ruminal microbial population and, consequently, animals with functional rumens should have adequate supplies (McDowell, 2000). Preruminant animals need to receive their folic acid supply through the milk. Calves raised on milk replacer should receive supplemental folic acid and other B-complex vitamins.

Early work with folic acid supplementation (in conjunction with other factors related to one-carbon metabolism) to beef cattle fed finishing diets indicated increased weight gains and carcass yield grade (Smith et al., 1964). Most of the recent work with folic acid in ruminant diets has centered on dairy cattle (NRC, 2001; Girard et al., 2005; Ragaller et al., 2009). Existing data are insufficient to determine folic acid requirements for beef cattle; however, additional research in one-carbon metabolism, folic acid, and other factors affecting the metabolic one-carbon pool would likely provide needed insight into developmental and epigenetic events during the fetal and perinatal periods of growth.

Niacin

Niacin is supplied to the ruminant from three primary sources: dietary niacin, conversion of tryptophan to niacin, and ruminal synthesis. Although niacin is normally synthesized in adequate quantities in the rumen, there are several factors that can influence ruminant niacin requirements (Olentine, 1984). These factors include protein (amino acid) balance, dietary energy supply, dietary rancidity, de novo synthesis, and availability of niacin in feeds. Excess leucine, arginine, and glycine increase the niacin requirement, whereas increasing dietary tryptophan decreases the niacin requirement (NRC, 1996, 2000). High-energy diets and the use of particular antibiotics can increase the requirement for niacin.

Niacin functions in carbohydrate, protein, and lipid metabolism as a component of the coenzyme forms of nicotinamide, nicotinamide adenine dinucleotide, and nicotinamide adenine dinucleotide phosphate. Niacin is particularly important in ruminants because it is required for liver detoxification of portal blood ammonia to urea and liver metabolism of ketones.

Niacin has been reported to enhance protein synthesis by ruminal microorganisms (Riddell et al., 1980, 1981). Niacin synthesis in the rumen seemed adequate when no niacin was added to the diet; however, when 6 g was added per day, an increase in niacin flow from the rumen occurred (Riddell et al., 1985). In addition, supplemental niacin was more effective in increasing microbial protein synthesis with urea than with soybean meal (Brent and Bartley, 1984). Responses to supplemental niacin in feedlot cattle have been variable.

Young ruminants are most susceptible to niacin deficiencies, and a dietary source of niacin or tryptophan is required until the rumen is fully developed. The first signs of niacin deficiency in most species are loss of appetite, decreased growth, general muscular weakness, digestive disorders, and diarrhea. The skin can also be affected with a scaly dermatitis. Often, these signs are followed by a microcytic anemia.

In a summary of 25 research trials investigating niacin supplementation (NRC, 2001), no clear support was found for the use of dietary niacin to enhance milk production in dairy cattle. In was also reported (NRC, 2001) that supplemental niacin was not effective at decreasing lipid-related metabolic disorders in dairy cattle. Niacin requirements have not been set for beef cattle or other ruminants as dietary supply and ruminal synthesis seem more than adequate.

Pantothenic Acid

Pantothenic acid is a component of two enzymes critical to carbohydrate, fat, and protein metabolism (McDowell, 2000; NRC, 2001, 2007b). These enzymes are coenzyme A (CoA) and acyl carrier protein (ACP). Biochemically, CoA stands at a pivotal crossroads in metabolism of carbohydrates, lipids, and proteins. The enzyme ACP is a primary acyl group carrier in the formation of fatty acids. Pantothenic acid is widespread in nature and abundant in most dietary components fed to ruminants. More importantly, ruminal microbial synthesis of pantothenic acid has been estimated to be 20 to 30 times more than normal dietary amounts experienced by cattle (NRC, 2001). Pantothenic acid requirements have not been set for beef cattle or any animal with a functional rumen.

Pyridoxine

Pyridoxine is also known as vitamin B_6 and represents a group of three compounds, pyridoxol (primary found in plants), pyridoxal, and pyridoxamine, with the latter two being found predominantly in animal products. Vitamin activities of the three compounds are similar. Other chemical forms associated with this vitamin are pyridoxal phosphate and pyridoxamine phosphate. Pyridoxal phosphate is a cofactor in over 60 enzymatic reactions associated with carbohydrate, protein, and lipid metabolism (McDowell, 2000); however, it is best known for its major role in most reactions associated with amino acid metabolism. As summarized by the NRC (2007b), many cellulolytic bacteria require pyridoxine, and in vitro studies have suggested that pyridoxine may stimulate lysine production by ruminal protozoa, thereby indicating a potential mechanism for enhancing postruminal lysine supply. Nonetheless, considerable research is needed to further test this hypothesis. Pyridoxine requirements have not been set for beef cattle because it seems that bacterial synthesis and dietary supply are adequate to prevent deficiency symptoms and allow for normal growth and production.

Riboflavin

Riboflavin is a component of flavoproteins and functions in numerous reactions associated with metabolism of carbohydrates, proteins, and lipids. Two common phosphorylated forms (McDowell, 2000; NRC, 2007b) of riboflavin are flavin mononucleotide and flavin adenine dinucleotide. Synthesis by ruminal bacteria is approximately 1.5 times the estimated metabolic needs for riboflavin (NRC, 2001), and consequently, dietary requirements have not been established for beef cattle.

Thiamin

In all species, a thiamin deficiency results in central nervous system disorders because thiamin is an important component of the biochemical reactions that break down glucose to supply energy to the brain. Signs of thiamin deficiency include weakness, retracted head, and cardiac arrhythmia. As with other water-soluble vitamins, deficiencies can result in slowed growth, anorexia, and diarrhea.

Thiamin functions in all cells as a coenzyme cocarboxylase. Thiamin is the coenzyme responsible for all enzymatic carboxylations of keto acids in the tricarboxylic acid cycle, which provides energy to the body. Thiamin also plays a key role in glucose metabolism, as a coenzyme in the pentose phosphate pathway.

Synthesis of thiamin by ruminal bacteria makes it difficult to establish a ruminant requirement. Animals with a functional rumen can generally synthesize an adequate amount of thiamin; however, the synthesis of thiamin is subject to dietary factors, including levels of carbohydrate and nitrogen. In addition, high-sulfur diets have been associated with thiamin deficiency and polioencephalomalacia (PEM), a laminar softening or degeneration of brain gray matter in cattle (Gould et al., 1991). Animal size, genetic factors, and physiological status also influence thiamin requirements.

Thiamin antimetabolites have been found in raw fish products and bracken fern (Somogyi, 1973). Occurrence of PEM, a central nervous system disorder, in grain-fed cattle and sheep has been linked to thiaminase activity or production of a thiamin antimetabolite in the rumen (Loew and Dunlop, 1972; Sapienza and Brent, 1974). Early research with PEM indicated an association with acidosis (Brent, 1976). Brent and Bartley (1984) discussed several potential causes of PEM, including the possibility of a thiaminase activity that could be linked to ruminal sulfur metabolism. Brent and Bartley (1984) also suggested that supplemental thiamin at the rate of 1 g/animal daily could be used as a means of protecting against PEM. Others have reported that affected animals have responded to intravenous administration of thiamin (2.2 mg/kg BW; NRC, 1996, 2000). Thiamin analogs produced in the rumen by thiaminase I in the presence of a cosubstrate seemed to be responsible for PEM (Brent and Bartley, 1984); however, supplementation of high-concentrate diets with thiamin has yielded inconsistent results (Grigat and Mathison, 1982, 1983). In a review of PEM, Gould (1998) discussed S-associated PEM and considered implications of H_2S production in the gut. High sulfate in feed or drinking water has also been reported to decrease animal performance and is implicated in incidence of PEM (Zinn et al., 1997; Loneragan et al., 2001). Galyean and Eng (1998) indicated that the case for PEM being linked to acidosis was worth additional consideration and that more information was needed in regard to the role of dietary and water sulfur levels and the relationship to H_2S production and PEM. Grout et al. (2006) suggest that the chemical form of sulfur in drinking water was likely a consideration in the formation of PEM. Vasconcelos and Galyean (2008) indicated that additional research investigating the effects of dietary (especially diets containing distillers byproducts) and water S concentrations on incidence of PEM, as well as methodologies to mitigate negative consequences of PEM, would be beneficial. Recent work (Neville et al., 2012) demonstrated that increasing levels of distillers grains (up to 60% of the diet) increased ruminal hydrogen sulfide (H_2S) but not incidence of PEM. Clearly, additional work is needed in this area.

Vitamin C

Vitamin C or L-ascorbic acid is a water-soluble molecule that is synthesized from glucose in most mammalian species except humans, primates, and guinea pigs. Vitamin C is a strong water-soluble antioxidant and is linked to many of the major antioxidant pathways. Vitamin C also acts as a reducing agent in hydroxylation reactions (McDowell, 2000). Dietary vitamin C is most likely destroyed in the rumen, and so, provision to ruminants would need to be as injectable or rumen-protected products (NRC, 2001, 2007b). Alternatively, Padilla et al. (2007) reported that with increasing concentrations of dietary vitamin C, ruminal escape could occur. Recent data (Pogge and Hansen, 2013) indicated that rumen-protected vitamin C can improve marbling in feedlot cattle fed high-sulfur diets. Additional work in this direction might likely prove fruitful. Inadequate data, as well as tissue synthesis preclude the setting of vitamin C requirements for beef cattle.

REFERENCES

Ammerman, C. B. 1970. Recent developments in cobalt and copper in ruminant nutrition: A review. *Journal of Dairy Science* 53:1097-1107.

Arnold, R. N., K. K. Scheller, S. C. Arp, S. N. Williams, D. R. Buege, and D. M. Schaefer. 1992. Effect of long- or short-term feeding of alpha-tocopherol acetate to Holstein and crossbred beef steers on performance, carcass characteristics and beef stability. *Journal of Animal Science* 70:3055-3065.

Aykroyd, W. R. 1958. Hypovitaminosis A. *Federation Proceedings* 17:103-143.

Bindel, D. J., J. S. Drouillard, E. C. Titgemeyer, R. H. Wessels, and C. A. Löest. 2000. Effects of ruminally protected choline and dietary fat on performance and blood metabolites of finishing heifers. *Journal of Animal Science* 78:2497-2503.

Bloomberg, B. D., G. G. Hilton, K. G. Hanger, C. J. Richards, J. B. Morgan, and D. L. VanOverbeke. 2011. Effects of vitamin E on color stability and palatability of strip loin steaks from cattle fed distillers grains. *Journal of Animal Science* 89:3769-3782.

Brent, B. E. 1976. Relationship of acidosis to other feedlot ailments. *Journal of Animal Science* 43:930-935.

Brent, B. E., and E. E. Bartley. 1984. Thiamin and niacin in the rumen. *Journal of Animal Science* 59:813-822.

Bryant, T. C., J. D. Rivera, M. L. Galyean, G. C. Duff, D. M. Hallford, and T. H. Montgomery. 1999. Effects of dietary level of ruminally protected choline on performance and carcass characteristics of finishing beef steers and on growth and serum metabolites in lambs. *Journal of Animal Science* 77:2893-2903.

Bryant, T. C., J. J. Wagner, J. D. Tatum, M. L. Galyean, R. V. Anthony, and T. E. Engle. 2010. Effect of dietary supplemental vitamin concentration on performance, carcass merit, serum metabolites, and lipogenic enzyme activity in yearling beef steers. *Journal of Animal Science* 88:1465-1478.

Burken, D. B., R. B. Hicks, D. L. VanOverbeke, G. G. Hilton, J. L. Wahrmund, B. P. Holland, C. R. Krehbiel, P. K. Camfield, and C. J. Richards. 2012. Vitamin E supplementation in beef finishing diets containing 35% wet distillers grains with solubles: Feedlot performance and carcass characteristics. *Journal of Animal Science* 90:1349-1355.

Campbell, J. R., P. R. Greenough, and L. Petrie. 2000. The effects of dietary biotin supplementation on vertical fissures of the claw wall in beef cattle. *Canadian Veterinary Journal* 41:690-694.

Carnagey, K. M., E. J. Huff-Lonergan, A. Trenkle, A. E. Wertz-Lutz, R. L. Horst, and D. C. Beitz. 2007. Use of 25-hydroxyvitamin D_3 and vitamin E to improve tenderness of beef from the longissimus dorsi of heifers. *Journal of Animal Science* 86:1649-1657.

Carnagey, K. M., E. J. Huff-Lonergan, S. M. Lonergan, A. Trenkle, R. L. Horst, and D. C. Beitz. 2008. Use of 25-hydroxyvitamin D_3 and dietary calcium to improve tenderness of beef from the round of beef cows. *Journal of Animal Science* 86:1637-1648.

Carter, J. N., D. R. Gill, C. R. Krehbiel, A. W. Confer, R. A. Smith, D. L. Lalman, P. L. Claypool, and L. R. McDowell. 2005. Vitamin E supplementation of newly arrived feedlot calves. *Journal of Animal Science* 83:1924-1932.

Chen, R., C. Wang, Y. M. Wang, and J. X. Liu. 2011. Effect of biotin on milk performance of dairy cattle: A meta-analysis. *Journal of Dairy Science* 94:3537-3546.

Clifford, A. J., R. D. Goodrich, and A. D. Tillman. 1967. Effects of supplementing ruminant all-concentrate and purified diets with vitamins of the B complex. *Journal of Animal Science* 26:400-403.

Conrad, H. R., and J. W. Hibbs. 1954. A high roughage system for raising calves based on early rumen development. IV. Synthesis of thiamin and riboflavin in the rumen as influenced by ratio of hay to grain fed and initiation of dry feed consumption. *Journal of Dairy Science* 37:512-522.

CSIRO (Commonwealth Scientific and Industrial Research Organisation). 2007. *Nutrient Requirements of Domesticated Ruminants*. Collingwood, Australia: CSIRO Publishing.

Cusack, P., N. McMeniman, A. Rabiee, and I. Lean. 2009. Assessment of the effects of supplementation with vitamin E on health and production of feedlot cattle using meta-analysis. *Preventive Veterinary Medicine* 88:229-246.

Draper, H. H., J. T. Sime, and B. C. Johnson. 1952. A study of vitamin B12 deficiency in the calf. *Journal of Animal Science* 11:332-340.

DSM Nutrition Company. 2012. The Vitamin Nutrition Compendium. Ruminants: Vitamin Nutrition for Ruminants. Available online at http://dsm.agnow.net/. Accessed on September 8, 2013.

Duff, G. C., and M. L. Galyean. 2007. Board-invited review: Recent advances in management of highly stressed, newly received feedlot cattle. *Journal of Animal Science* 85:823-840.

Erdman, J. W., Jr., I. A. MacDonald, and S. H. Zeisel, eds. 2012. *Present Knowledge in Nutrition*, 10th Rev. Ed. Ames, IA: Wiley-Blackwell.

Erdman, R. A., and B. K. Sharma. 1991. Effect of dietary rumen-protected choline in lactating dairy cows. *Journal of Dairy Science* 74:1641-1647.

Fyksen, J. 2013. Vitamin A deficiency in beef herds: Lingering effect of drought. Agri-View, April 25. Available online at http://www.agriview.com/news/livestock/vitamin-a-deficiency-in-beef-herds-lingering-effect-of-drought/article_1cf91f9a-add0-11e2-97b2-0019bb2963f4.html. Accessed on August 31, 2013.

Galyean, M. L., and K. S. Eng. 1998. Application of research findings and summary of research needs: Bud Britton Memorial Symposium on Metabolic Disorders of Feedlot Cattle. *Journal of Animal Science* 76:323-327.

Galyean, M. L., L. J. Perino, and G. C. Duff. 1999. Interaction of cattle health/immunity and nutrition. *Journal of Animal Science* 77:1120-1134.

Girard, C. L., C. Benchaar, J. Chiquette, and A. Desrochers. 2005. Net flux of nutrients across the rumen wall of lactating dairy cows as influenced by dietary supplements of folic acid. *Journal of Dairy Science* 92:6116-6122.

Gorocica-Buenfil, M. A., F. L. Fluharty, T. Bohn, S. J. Schwartz, and S. C. Loerch. 2007a. Effect of low vitamin A diets with high-moisture or dry corn on marbling and adipose tissue fatty acid composition of beef steers. *Journal of Animal Science* 85:3355-3366.

Gorocica-Buenfil, M. A., F. L. Fluharty, C. K. Reynolds, and S. C. Loerch. 2007b. Effect of dietary vitamin A concentration and roasted soybean inclusion on marbling, adipose cellularity, and fatty acid composition of beef. *Journal of Animal Science* 85:2230-2242.

Gorocica-Buenfil, M. A., F. L. Fluharty, C. K. Reynolds, and S. C. Loerch. 2007c. Effect of dietary vitamin A restriction on marbling and conjugated linoleic acid content in Holstein steers. *Journal of Animal Science* 85:2243-2255.

Gorocica-Buenfil, M. A., F. L. Fluharty, and S. C. Loerch. 2008. Effect of vitamin A restriction on carcass characteristics and immune status of beef steers. *Journal of Animal Science* 86:1609-1616.

Gould, D. H. 1998. Polioencephalomalacia. *Journal of Animal Science* 76:309-314.

Gould, D. H., M. M. McAllister, J. C. Savage, and D. W. Hamar. 1991. High sulfide concentrations in rumen fluid associated with nutritionally induced polioencephalomalacia in calves. *American Journal of Veterinary Research* 52:1164-1167.

Grigat, G. A., and G. W. Mathison. 1982. Thiamin supplementation of all-concentrate diet for feedlot steers. *Canadian Journal of Animal Science* 62:807-819.

Grigat, G. A., and G. W. Mathison. 1983. Thiamin and magnesium supplementation diets for feedlot steers. *Canadian Journal of Animal Science* 63:117-131.

Grout, A. S., D. M. Veira, D. M. Weary, M. A. G. von Keyserlingk, and D. Fraser. 2006. Differential effects of sodium and magnesium sulfate on water consumption by beef cattle. *Journal of Animal Science* 84:1252-1258.

Gullickson, T. W., and C. W. Calverley. 1946. Cardiac failure in cattle on the vitamin E-free rations as revealed by electrocardiograms. *Science* 104:312-313.

Hayes, B. W., G. E. Mitchell, Jr., C. O. Little, and N. W. Bradley. 1966. Concentrations of B vitamins in ruminal fluid of steers fed different levels and physical forms of hay and grain. *Journal of Animal Science* 25:539-542.

Hill, G. M., S. E. Williams, S. N. Williams, L. R. McDowell, N. Wilkinson, and B. G. Mullinix. 1995. Vitamin A and vitamin E effects on performance and tissue α-tocopherol concentrations of feedlot steers. *Journal of Animal Science* 73(Suppl. 1):95 (Abstract).

Hopper, J. H., and B. C. Johnson. 1955. The production and study of an acute nicotinic acid deficiency in the calf. *Journal of Nutrition* 56:303-310.

Horn, M., P. Gunn, M. Van Emon, R. Lemenager, J. Burgess, N. A. Pyatt, and S. L. Lake. 2010a. Effects of natural (RRR α-tocopherol acetate) or synthetic (all-rac α-tocopherol acetate) vitamin E supplementation on reproductive efficiency in beef cows. *Journal of Animal Science* 88:3121-3127.

Horn, M. J., M. L. Van Emon, P. J. Gunn, S. D. Eicher, R. P. Lemenager, J. Burgess, N. Pyatt, and S. L. Lake. 2010b. Effects of maternal natural (RRR α-tocopherol acetate) or synthetic (all-rac α-tocopherol acetate) vitamin E supplementation on suckling calf performance, colostrum immunoglobulin G, and immune function. *Journal of Animal Science* 88:3128-3135.

Huber, J. T. 1988. Vitamins in ruminant nutrition. Pp. 313-325 in *The Ruminant Animal: Digestive Physiology and Nutrition*, D. C. Church, ed. Long Grove, IL: Waveland Press.

Hunt, C. H., E. W. Burroughs, R. M. Bethke, A. F. Schalk, and P. Gerlaugh. 1943. Further studies on riboflavin and thiamin in the rumen contents of cattle. II. *Journal of Nutrition* 25:207-216.

Ikeda, S., H. Koyama, M. Sugimoto, and S. Kume. 2012. Roles of one-carbon metabolism in preimplantation period—effects on short-term development and long-term programming. *Journal of Reproduction and Development* 58:38-43.

IOM (Institute of Medicine). 2000. *Dietary Reference Intakes for Vitamin A, Vitamin K, Arsenic, Boron, Chromium, Copper, Iodine, Iron, Manganese, Molybdenum, Nickel, Silicon, Vanadium, and Zinc.* Washington, DC: National Academy Press.

Johnson, B. C., T. S. Hamilton, W. B. Nevens, and L. E. Boley. 1948. Thiamin deficiency in the calf. *Journal of Nutrition* 35:137-145.

Johnson, B. C., J. A. Pinkos, and K. A. Burke. 1950. Pyridoxine deficiency in the calf. *Journal of Nutrition* 40:309-322.

Johnson, B. C., H. H. Mitchell, and J. A. Pinkos. 1951. Choline deficiency in the calf. *Journal of Nutrition* 43:37-48.

Jordan, H. A., G. S. Smith, A. L. Neumann, J. E. Zimmerman, and G. W. Breniman. 1963. Vitamin A nutrition of beef cattle fed corn silage. *Journal of Animal Science* 22:738-745.

Lassiter, C. A., G. M. Ward, C. F. Huffman, C. W. Duncan, and H. D. Webster. 1953. Crystalline vitamin B12 requirement of the young dairy calf. *Journal of Dairy Science* 36:997-1005.

Lean, I. J., and A. R. Rabiee. 2011. Effect of feeding biotin on milk production and hoof health in lactating dairy cows: A quantitative assessment. *Journal of Dairy Science* 94:1465-1476.

Link, K. P. 1959. The discovery of dicumarol and its sequels. *Circulation* 19:97-107.

Liu, Q., M. C. Lanari, and D. M. Schaefer. 1995. A review of dietary vitamin E supplementation for improvement of beef quality. *Journal of Animal Science* 73:3131-3140.

Lobley, G. E., A. Connell, and D. Revell. 1996. The importance of transmethylation reactions to methionine metabolism in sheep: Effects of supplementation with creatine and choline. *British Journal of Nutrition* 75:47-56.

Loew, F. M., and R. H. Dunlop. 1972. Induction of thiamin inadequacy and polioencephalomalacia in adult sheep with amprolium. *American Journal of Veterinary Research* 33:2195-2205.

Loneragan, G. H., J. J. Wagner, D. H. Gould, F. B. Garry, and M. A. Thoren. 2001. Effects of water sulfate concentration on performance, water intake, and carcass characteristics of feedlot steers. *Journal of Animal Science* 79:2941-2948.

Marston, H. R., S. H. Allen, and R. M. Smith. 1961. Primary metabolic defect supervening on vitamin B_{12} deficiency in sheep. *Nature* 190:1085-1091.

McDowell, L. R. 2000. *Vitamins in Animal Nutrition: Comparative Aspects to Human Nutrition.* San Diego, CA: Academic Press.

McDowell, L. R. 2013. *Vitamin History: The Early Years*, 1st Ed. Sarasota, FL: Design Publishing, Inc.

McElroy, L. W., and H. Goss. 1940a. A quantitative study of vitamins in the rumen contents of sheep and goats fed vitamin-low diets. I. Riboflavin and vitamin K. *Journal of Nutrition* 20:527-540.

McElroy, L. W., and H. Goss. 1940b. A quantitative study of vitamins in the rumen contents of sheep and goats fed vitamin-low diets. II. Vitamin B6 (pyridoxine). *Journal of Nutrition* 20:541-550.

McElroy, L. W., and H. Goss. 1941a. A quantitative study of vitamins in the rumen contents of sheep and cows fed vitamin-low diets. III. Thiamin. *Journal of Nutrition* 21:163-173.

McElroy, L. W., and H. Goss. 1941b. A quantitative study of vitamins in the rumen contents of sheep and cows fed vitamin-low diets. IV. Pantothenic acid. *Journal of Nutrition* 21:405-409.

McLaren, D. S. 1986. Pathogenesis of vitamin A deficiency. Pp. 153-176 in *Vitamin A Deficiency and Its Control*, J. C. Bauernfeind, ed. Orlando, FL: Academic Press.

Meyer, A. M., J. S. Caton, B. W. Hess, S. P. Ford, and L. P. Reynolds. 2012. Epigenetics and effects on the neonate that may impact feed efficiency. Pp. 195-224 in *Feed Efficiency in the Beef Industry*, R. Hill, ed. Hoboken, NJ: Wiley-Blackwell.

Miller, B. L., J. C. Meiske, and R. D. Goodrich. 1986a. Effects of dietary additives on B-vitamin production and absorption in steers. *Journal of Animal Science* 62:484-496.

Miller, B. L., J. C. Meiske, and R. D. Goodrich. 1986b. Effects of grain source and concentrate level on B-vitamin production and absorption in steers. *Journal of Animal Science* 62:473-483.

Miller, R. W., R. W. Hemken, D. R. Waldo, and L. H. Moore. 1967. Vitamin A metabolism in steers fed corn silage or alfalfa hay pellets. *Journal of Dairy Science* 50:997(Abstract P133).

Miller, W. J. 1979. *Dairy Cattle Feeding and Nutrition.* New York: Academic Press.

Montgomery, J. L., J. R. Blanton, Jr., R. L. Horst, M. L. Galyean, K. J. Morrow, Jr., D. B. Wester, and M. F. Miller. 2004a. Effects of biological type of beef steers on vitamin D, calcium, and phosphorus status. *Journal of Animal Science* 82:2043-2049.

Montgomery, J. L., M. L. Galyean, R. L. Horst, K. J. Morrow, Jr., J. R. Blanton, Jr., D. B. Wester, and M. F. Miller. 2004b. Supplemental vitamin D3 concentration and biological type of beef steers. I. Feedlot performance and carcass traits. *Journal of Animal Science* 82:2050-2058.

Montgomery, J. L., M. B. King, J. G. Gentry, A. R. Barham, B. L. Barham, G. G. Hilton, J. R. Blanton, Jr., R. L. Horst, M. L. Galyean, K. J. Morrow, Jr., D. B. Wester, and M. F. Miller. 2004c. Supplemental vitamin D3 concentration and biological type of steers. II. Tenderness, quality, and residues of beef. *Journal of Animal Science* 82:2092-2104.

Neville, B. W., G. P. Lardy, K. K. Karges, S. R. Eckerman, P. T. Berg, and C. S. Schauer. 2012. Interaction of corn processing and distillers dried grains with solubles on health and performance of steers. *Journal of Animal Science* 90:560-567.

NRC (National Research Council). 1984. *Nutrient Requirements of Beef Cattle*, 6th Rev. Ed. Washington, DC: National Academy Press.

NRC. 1987. *Vitamin Tolerance of Animals.* Washington, DC: National Academy Press.

NRC. 1989a. *Nutrient Requirements of Dairy Cattle*, 6th Rev. Ed. Washington, DC: National Academy Press.

NRC. 1989b. *Recommended Dietary Allowances*, 10th Ed. Washington, DC: National Academy Press.

NRC. 1996. *Nutrient Requirements of Beef Cattle*, 7th Rev. Ed. Washington, DC: National Academy Press.

NRC. 2000. *Nutrient Requirements of Beef Cattle: Update 2000*, 7th Rev. Ed. Washington, DC: National Academy Press.

NRC. 2001. *Nutrient Requirements of Dairy Cattle*, 7th Rev. Ed. Washington, DC: National Academy Press.

NRC. 2007a. *Nutrient Requirements of Horses*, 6th Rev. Ed. Washington, DC: The National Academies Press.

NRC. 2007b. *Nutrient Requirements of Small Ruminants, Sheep, Goats, Cervids, and New World Camelids.* Washington, DC: The National Academies Press.

NRC. 2012. *Nutrient Requirements of Swine*, 11th Rev. Ed. Washington, DC: The National Academies Press.

Olentine, C. 1984. B-vitamins for ruminants. *Feed Management* 35:18-24.

Olson, J. A. 2001. Vitamin A. Pp. 1-50 in *Handbook of Vitamins*, 3rd Ed., R. B. Rucker, J. W. Suttie, D. B. McCormick, and L. J. Machlin, eds. New York: Marcel Dekker.

Padilla, L., T. Matsui, S. Ikeda, M. Kitagawa, and H. Yano. 2007. The effect of vitamin C supplementation on plasma concentration and urinary excretion of vitamin C in cattle. *Journal of Animal Science* 85:3367-3370.

Pogge, D. J., and S. L. Hansen. 2013. Supplemental vitamin C improves marbling in feedlot cattle consuming high sulfur diets. *Journal of Animal Science* 91:4303-4314.

Pyatt, N. A., L. L. Berger, and T. G. Nash. 2005. Effects of vitamin A and restricted intake on performance, carcass characteristics, and serum retinol status in Angus × Simmental feedlot cattle. *The Professional Animal Scientist* 21:318-331.

Ragaller, V., L. Huther, and P. Lebzien. 2009. Folic acid in ruminant nutrition: A review. *British Journal of Nutrition* 101:153-164.

Riddell, D. O., E. E. Bartley, and A. D. Dayton. 1980. Effect of nicotinic acid on rumen fermentation in vitro and in vivo. *Journal of Dairy Science* 63:1429-1436.

Riddell, D. O., E. E. Bartley, and A. D. Dayton. 1981. Effect of nicotinic acid on microbial protein synthesis in vitro and on dairy cattle growth and milk production. *Journal of Dairy Science* 64:782-791.

Riddell, D. O., E. E. Bartley, J. J. Arambel, T. G. Nagaraja, A. D. Dayton, and G. W. Miller. 1985. Effect of niacin supplementation on ruminal niacin synthesis and degradation in cattle. *Nutrition Reports International* 31:407-413.

Rivera, J. D., G. C. Duff, M. L. Galyean, D. A. Walker, and G. A. Nunnery. 2002. Effects of supplemental vitamin E on performance, health, and humoral immune response of beef cattle. *Journal of Animal Science* 80:933-941.

Rode, L. M., T. A. McAllister, and K. J. Cheng. 1990. Microbial degradation of vitamin A in rumen fluid from steers fed concentrate, hay, or straw diets. *Canadian Journal of Animal Science* 70:227-233.

Roeber, D. L., K. E. Belk, J. D. Tatum, J. W. Wilson, and G. C. Smith. 2001. Effects of three levels of α-tocopheryl acetate supplementation to feedlot cattle on performance of beef cuts during retail display. *Journal of Animal Science* 79:1814-1820.

Rumsey, T. S. 1975. Vitamin requirements for ruminants. *Feedstuffs* 47:30-34.

Sales, J., P. Homolka, and V. Koukolova. 2010. Effect of dietary rumen-protected choline on milk production of dairy cows: A meta-analysis. *Journal of Dairy Science* 93:3746-3754.

Sapienza, D. A., and B. E. Brent. 1974. Ruminal thiaminase vs. concentrate adaptation. *Journal of Animal Science* 39:251 (Abstract 434).

Schaefer, D. M., Q. Liu, C. Faustman, and M. C. Yin. 1995. Supranutritional administration of vitamins E and C improves oxidative stability of beef. *Journal of Nutrition* 125(6 Suppl.):1792S-1798S.

Sheppard, J. J., and B. C. Johnson. 1957. Pantothenic acid deficiency in the growing calf. *Journal of Nutrition* 61:195-205.

Smith, G. S., W. M. Durdle, J. E. Zimmerman, and A. L. Neumann. 1964. Relationships of carotene intake, thyro-active substances and soil fertility to vitamin A depletion of feeder cattle fed corn silages. *Journal of Animal Science* 23:625-632.

Solomons, N. W. 2012. Vitamin A. Pp. 259-320 in *Present Knowledge in Nutrition*, 10th Rev. Ed., J. W. Erdman Jr., I. A. Macdonald, and S. H. Zeisel, eds. Ames, IA: Wiley-Blackwwell.

Somogyi, J. C. 1973. Antivitamins. Pp. 254-275 in *Toxicants Occurring Naturally in Foods.* Washington, DC: National Academy Press.

Swanek, S. S., J. B. Morgan, F. N. Owens, D. R. Gill, C. A. Strasia, H. G. Dolezal, and F. K. Ray. 1999. Vitamin D_3 supplementation of beef steers increases longissimus tenderness. *Journal of Animal Science* 77:874-881.

Ullrey, D. E. 1972. Biological availability of fat-soluble vitamins: Vitamin A and carotene. *Journal of Animal Science* 35:648-657.

USP (United States Pharmacopeia). 1979. The United States Pharmacopeia—National Formulary. Rockville, MD: USP.

USP. 1999. The United States Pharmacopeia 24—National Formulary 19. Rockville, MD: USP.

Vasconcelos, J. T., and M. L. Galyean. 2007. Nutritional recommendations of feedlot consulting nutritionists: The 2007 Texas Tech University survey. *Journal of Animal Science* 85:2772-2781.

Vasconcelos J. T., and M. L. Galyean. 2008. ASAS Centennial Paper: Contributions in the Journal of Animal Science to understanding cattle metabolic and digestive disorders. *Journal of Animal Science* 86:1711-1721.

Wegner, M. I., A. N. Booth, C. A. Elvehjem, and E. B. Hart. 1940. Rumen synthesis of the vitamin B complex. *Proceedings of the Society of Experimental Biology and. Medicine* 45:769-771.

Wegner, M. I., A. N. Booth, C. A. Elvehjem, and E. B. Hart. 1941. Rumen synthesis of the vitamin B complex on natural rations. *Proceedings of the Society of Experimental Biology and Medicine* 47:90-94.

WHO (World Health Organization). 2009. *Global Prevalence of Vitamin A Deficiency in Populations at Risk 1995-2005: WHO Global Database on Vitamin A Deficiency.* Geneva: WHO.

Wiese, A. C., B. C. Johnson, and W. B. Nevens. 1946. Biotin deficiency in the dairy calf. *Proceedings of the Society of Experimental Biology and Medicine* 63:521-522.

Wiese, A. C., B. C. Johnson, H. H. Mitchell, and W. B. Nevens. 1947. Riboflavin deficiency in the dairy calf. *Journal of Nutrition* 33:263-270.

Wolf, G. 1978. A historical note on the mode of administration of vitamin A for the cure of night blindness. *American Journal of Clinical Nutrition* 31:290-292.

Wu, G., F., W. Bazer, J. M. Wallace, and T. E. Spencer, 2006. Board invited review: Intrauterine growth retardation: Implications for the animal sciences. *Journal of Animal Science* 84:2316-2337.

Zempleni, J., J. W. Suttie, J. F. Gregory III, and P. J. Stover, eds. 2014. *Handbook of Vitamins,* 5th Ed. New York: CRC Press.

Zerby, H. N., K. E. Belk, J. N. Sofos, L. R. McDowell, and G. C. Smith. 1999. Case life of seven retail products from beef cattle supplemented with alpha-tocopheryl acetate. *Journal of Animal Science* 77:2458-2463.

Zinn, R. A., F. N. Owens, R. L. Stuart, J. R. Dunbar, and B. B. Norman. 1987. B-vitamins supplementation of diets for feedlot calves. *Journal of Animal Science* 65:267-277.

Zinn, R. A., E. Alvarez, and R. L. Stuart. 1996. Interaction of supplemental vitamin A and E on health and performance of crossbred and Holstein calves during the receiving period. *The Professional Animal Scientist* 12:17-23.

Zinn, R. A., E. Alvarez, M. Mendez, M. Montano, E. Ramirez, and Y. Shen. 1997. Influence of dietary sulfur level on growth performance and digestive function in feedlot cattle. *Journal of Animal Science* 75:1723-1728.

9

Water

INTRODUCTION

Water is an essential nutrient for beef cattle. Water constitutes approximately 99% of all molecules in the body (Macfarlane and Howard, 1972) and contributes to approximately 70% of body mass. It is needed for regulation of body temperature as well as for growth; reproduction; lactation; digestion; metabolism; excretion; hydrolysis of protein, fat, and carbohydrates; regulation of mineral homeostasis; lubrication of joints; nervous system cushioning; sound transport; and eyesight. Water is also an excellent solvent for glucose, amino acids, mineral ions, and water-soluble vitamins, and plays a vital role in cellular metabolism and waste transport in the body.

BODY WATER DISTRIBUTION

Water is distributed throughout the physiological body pools of the animal. The total amount is termed total body water (TBW) and is determined by gravimetric analysis at slaughter. An estimate of TBW can be made by dilution of water markers in vivo after appropriate correction factors are applied. The physiological pools through which TBW is distributed include (1) the extracellular fluid (ECF), amounting to 31 to 38% of TBW; and (2) the intracellular fluid (ICF), amounting to 62 to 69% of TBW (Macfarlane et al., 1956; Purohit et al., 1972; NRC, 2001, 2003). Water along with potassium (K^+) and other inorganic ions, and proteins within cell types of the body, constitute the ICF pool. Measurement of ICF-water can be made by bioelectrical impedance spectroscopy or as the difference between an estimation of nonalimentary water and the ECF water space using thiocyanate as a marker (Gad and Preston, 1990). The ECF includes blood plasma (25% of extracellular water) plus interstitial fluid (75% of extracellular water). Measurements of ECF sometimes include transcellular water in the gut.

The ICF pool is highly regulated in volume; this is achieved by alterations in the ECF, which in turn, is regulated to maintain a constant concentration of sodium. Volume sensors, hormones, and water transfer mechanisms involving the hepatic portal system, heart, and kidneys are responsible for maintaining water homeostasis (Macfarlane and Howard, 1972). Understanding the mechanisms that control the shift in water between these pools is of interest for managing free-ranging and wild animals (Macfarlane, 1964; Cameron and Luick, 1972; Macfarlane and Howard, 1972; King, 1982). Shifts in water between the various pools are also important during dehydration and rehydration in dry as well as humid environments (Macfarlane et al., 1956; Macfarlane and Howard, 1972; Purohit et al., 1972; Shkolnik et al., 1975, 1980; Degen, 1977a), during long-term undernutrition involving drought feeding (Morris et al., 1962; McGregor, 2003), and during gestation (Cameron et al., 1975) and lactation (Degen 1977b; Woodford et al., 1984; Odwongo et al., 1985).

REQUIREMENTS

Water intake from feedstuffs and intake of free water approximates the water requirement of cattle. Small amounts of metabolic water are produced by oxidation of organic nutrients (NRC, 1981); however, the contribution of metabolic water to the animal's requirement is minimal. Water requirement is influenced by several factors, including rate and composition of gain, pregnancy, lactation, activity, type of diet, feed intake, and environmental temperature. Restriction of water intake decreases feed intake (Utley et al., 1970), which decreases animal performance, but water restriction also tends to increase apparent digestibility and nitrogen retention. The minimum requirement of cattle for water is a reflection of that needed for body maintenance and growth, fetal growth, reproduction, lactation, and that lost by excretion in the urine, feces, or sweat or by evaporation from the lungs or skin. Anything influencing these needs or losses will influence the minimum requirement. The antidiuretic hormone vasopressin controls reabsorption of water from the kidney tubules and ducts; thus, it affects excretion of urine. Under conditions of restricted water intake, the body may reabsorb a greater amount of water than usual, thereby concentrating urine. Although this capacity to concentrate urine solutes is limited, it can decrease water requirements by a small amount. The amount of urine produced daily also varies with the activity of the animal, air temperature, and water consumption, as well as other factors. Water requirements can increase when a diet is high in protein, salt, minerals, or diuretic substances. The quantity of water lost in the feces depends largely on the diet. Succulent diets and diets with high mineral content contribute to more water in the feces.

The quantity of water lost through evaporation from the skin or lungs is important and can even exceed that lost in the urine. If temperature or physical activity increases, water loss through evaporation and sweating increases. Because feeds themselves contain some water, and the oxidation of certain nutrients in feeds produces metabolic water, not all water must be provided by drinking. Feeds such as silage, green chop, or growing pasture forage are usually very high in moisture, whereas grains, hays, and dormant pasture forage are low in moisture. High-energy feeds produce more metabolic water; low-energy feeds produce a lesser amount, which is an obvious complication in the matter of assessing water requirements. Nonetheless, when compared with evaporative losses through respiration and sweating, metabolic water has a minimal contribution to total body water content. Fasting animals or those fed a low-protein diet can form water from the catabolism of body protein or fat, but this is of minor importance to overall water balance.

The results of water requirement studies conducted under various conditions imply that thirst is a result of need, and that animals drink to meet this need. The need results from an increase in the electrolyte concentration in the body fluids, which activates the thirst mechanism. As this discussion suggests, water requirements are affected by many factors, and it is impossible to list specific requirements with accuracy.

A water requirement equation for feedlot steers was developed by Hicks et al. (1988):

$$\text{Water intake (L/d)} = -6.0716 + (0.70866 \times \text{MT}) + (2.432 \times \text{DMI}) - (3.87 \times \text{PP}) - (4.437 \times \text{DS})$$

where MT is the maximum temperature, °C; DMI is dry matter intake, kg/d; PP is the daily precipitation, cm; and DS is dietary salt, %.

The major influences on water intake in beef cattle fed typical diets are physical access to water, DMI, environmental temperature, as well as stage and type of production. Table 9-1 has been designed as a guide only, and it should be used with consideration for the various factors that can influence water intake.

TABLE 9-1 Approximate Total Daily Water Intake (Liters) of Beef Cattle[a]

Weight, kg	Temperature, °C[b]					
	4.4	10.0	14.4	21.1	26.6	32.2
Growing heifers, steers, and bulls						
182	15.1	16.3	18.9	22.0	25.4	36.0
273	20.1	22.0	25.0	29.5	33.7	48.1
364	23.0	25.7	29.9	34.8	40.1	56.8
Finishing cattle						
273	22.7	24.6	28.0	32.9	37.9	54.1
364	27.6	29.9	34.4	40.5	46.6	65.9
454	32.9	35.6	40.9	47.7	54.9	78.0
Wintering pregnant cows[c]						
409	25.4	27.3	31.4	36.7	—	—
500	22.7	24.6	28.0	32.9	—	—
Lactating cows[d]						
409	43.1	47.7	54.9	64.0	67.8	61.3
Mature bulls						
636	30.3	32.6	37.5	44.3	50.7	71.9
727	32.9	35.6	40.9	47.7	54.9	78.0

[a]Winchester and Morris (1956).

[b]Water intake of a given class of cattle in a specific management regime is a function of dry matter intake and ambient temperature. Water intake is quite constant up to 4.4°C.

[c]Dry matter intake has a major influence on water intake. Heavier cows are assumed to be higher in body condition and to require less dry matter and, thus, less water intake.

[d]Cows larger than 409 kg are included in this recommendation.

Factors Affecting Requirements

Earlier research by Winchester and Morris (1956) suggests a constant relationship between water intake and DMI for cattle at thermal neutral conditions. Estimated water intake was calculated from DMI projected from body weight (BW) and average daily gain (ADG). Since water intake generally increases and DMI generally decreases in warmer months of the year, with the opposite relationship occurring in the cooler months of the year, the prediction of water intake from DMI is not consistent. However, other researchers have reported that water intake increases as DMI increases (Murphy et al., 1983; Hicks et al., 1988; Loneragan et al., 2001). Utilizing records from seven separate feedlot experiments across several years and seasons to examine that impact of environmental factors on daily water intake of finishing cattle, Arias and Mader (2011) concluded that mean ambient temperature, minimum temperature, and temperature-humidity index were the primary factors that influenced daily water intake. These researchers reported that across season, the largest R^2 (0.65) values obtained were from the following prediction equations:

$$\text{DWI} = 5.92 + 1.03\ \text{DMI} + 0.04\ \text{SR} + 0.45\ \text{Tmin}$$

and

$$DWI = -7.31 + 1.00\ DMI + 0.04\ SR + 0.30\ THI,$$

where

DWI is daily water intake (L/d);
DMI is dry matter intake (kg/d);
SR is solar radiation (W/m^2);
Tmin is the daily minimum ambient temperature (°C); and
THI is the temperature-humidity index, where: THI = 0.8 Ta + [(RH/100) (Ta – 14.4)] + 46.4 (Thom, 1959; NOAA, 1976); and Ta is the mean ambient temperature and RH is the relative humidity.

Furthermore, Arias and Mader (2011) did report significant regression coefficients for predicting water intake from DMI. However, the partial R^2 values were low, leading these researchers to question the strength of the relationship between water intake and DMI.

Sexson et al. (2012) summarized water intake data from four separate beef cattle feedlot experiments conducted in the High Plains region of the United States over several years. Univariate analysis demonstrated that the continuous variables of body weight, humidity, and sea level pressure were negatively related to water intake, whereas DMI, temperature the previous day, daily temperature, change in temperature from the previous day, average wind speed, and the temperature-humidity index were positively related to daily water intake. Using a multivariate, parsimonious model approach, prediction of average daily water intake included average humidity, average humidity squared, high temperature squared, high humidity squared, low temperature, low temperature squared, low humidity, average sea level pressure, average wind speed, average daily BW, high sea level pressure, low sea level pressure, high humidity, and low humidity. The generalized R^2 of the parsimonious, multivariate model was 0.32. These results indicate that BW and numerous environmental factors are related to water intake, but DMI had a minimal impact on water intake by yearling feedlot steers.

Although the strength of the relationship between water intake and DMI appears to be low, the roughage content of the diet may influence this relationship. Water intakes of 2.99, 2.11, and 3.20 L/kg were observed for cattle consuming diets containing 0, 33, or 88% forage, respectively (Bond et al., 1975). Depriving cattle of feed resulted in a 13% increase in water intake for cattle receiving a diet containing no forage (high-concentrate diet) whereas cattle consuming a diet composed of almost all forage (88%) decreased water intake by 92% (Bond et al., 1975). When water was removed for up to 48 h, cattle receiving a diet containing no forage (high-concentrate diet) had a 44% decrease in feed intake, whereas cattle consuming an 88% forage-based diet had a 55% decrease in DMI (Bond et al., 1975). Collectively, the aforementioned data indicate that water intake may be influenced by dry matter composition and that water intake, particularly deprivation, affects DMI.

Other factors that can influence water intake from troughs are water temperature, physical access to water, and intake of water from environmental precipitation (rain and snow). Early experiments examining cooling drinking water (18.3°C) improved ADG in beef cattle (of British origin) fed during the summer months (Ittner et al., 1951; Bond, 1975). However, Brahman × British crossbred cattle had similar performance when consuming cooled (18.3°C) or warmed (32.2°C) water (Lofgreen et al., 1975). Although consuming cold water (0.0°C) has been reported to alter rumen fermentation patterns and decrease rumen temperature in sheep, nutrient digestibility was not altered (Brod et al., 1982). Similar results have been reported in nonlactating dairy cows (Cunningham et al., 1964). Collectively, these data indicate that water temperature under certain situations might influence animal performance. In dairy cattle, water temperature has only a slight effect on drinking behavior and animal performance (Milam et al., 1986; Stermer et al.,1986; Baker et al., 1988; Wilks et al., 1990) and under most conditions would not warrant the additional cost of cooling water (NRC, 2001).

The thermoregulatory benefit of water during times of high ambient temperatures is primarily through evaporation (skin and respiration) and not a result of the physical intake of cold water (CSIRO, 2007). Thus, preventing animal dehydration during times of heat stress is critical. In times of elevated ambient temperatures, Mader et al. (1997) reported that the linear space at the water trough for feedlot cattle should be increased from 2.5 cm to 7.5 cm. For cows, the *Beef Housing Handbook* (MWPS, 1987) indicates that for every 16 cows, 30.5 cm of space is needed for adequate access to water. For water supply management for grazing cattle, see Nelson (2012) and MWPS (1987).

Heating water to prevent freezing during times of low ambient temperatures helps to maintain drinking water availability. Cattle prefer liquid water but can consume snow and ice when liquid water is not available in adequate amounts to replace drinking water (Butcher, 1973; Young and Degen, 1980; Degen and Young, 1984, 1990a,b; CSIRO, 2007). Cattle can tolerate the thermal stress from rapid ingestion of snow and ice by drawing from stored body heat and an immediate increase in metabolic rate to compensate for the heat required to melt frozen water and bring it to body temperature (Degen and Young, 1984).

Water Quality

Water quality is important in maintaining water consumption of cattle. Cattle consume water from surface water sources such as ponds, lakes, and streams, as well as from groundwater sources such as wells. Numerous components within water can influence water intake and overall animal performance. Beef cattle requirements for water are a function of different metabolic priorities, and the restriction of water intake to less than the animal's requirement will decrease cattle performance.

The five criteria most often considered in assessing water quality for both humans and livestock are organoleptic properties (odor and taste); physiochemical properties (pH, total dissolved solids, total dissolved oxygen, and hardness); presence of toxic compounds (heavy metals, toxic minerals, organophosphates, and hydrocarbons); presence of excess minerals or compounds (nitrates, sodium, sulfates, and iron); and presence of bacteria. Research information on water contaminants and their effects on cattle performance are sparse. However, the following information is an attempt to define some common water quality problems in relation to cattle performance (adapted from NRC, 2001).

- *Salinity, total dissolved solids (TDS), and total soluble salts (TSS)* are measures of constituents soluble in water. Sodium chloride is the first consideration in this category, but other components such as bicarbonate, sulfate, calcium, magnesium, and silica are associated with salinity, TDS, or TSS (NRC, 1974). A secondary group of constituents, found in lower concentrations than the major constituents, consists of iron, nitrate, strontium, potassium, carbonate, phosphorus, boron, and fluoride. Guidelines for TSS in water for cattle are shown in Table 9-2.

Research by Ray (1986) evaluated the effects of saline water on feedlot cattle. Cattle drinking saline water (TDS, 6,000 mg/L) had lower weight gains than those drinking normal water (1,300 mg/L) when energy content of the diet was low and during heat stress. In contrast, high-energy diets and the lower temperature during the winter months negated the detrimental effects of high-saline water consumption.

TABLE 9-2 Guidelines for Total Soluble Salts (TSS) in Water for Cattle[a]

TSS (mg/L)	Comments
<1,000	Safe and should pose no health problems.
1,000-2,999	Generally safe but may cause a mild temporary diarrhea in animals not accustomed to the water.
3,000-4,999	Water may be refused when first offered to animals or cause temporary diarrhea. Animal performance may be less than optimum because water intake is not maximized.
5,000-6,999	Avoid these waters for pregnant or lactating animals. May be offered with reasonable safety to animals where maximum performance is not required.
7,000	These waters should not be fed to cattle. Health problems and/or poor production will result.

[a]NRC (1974, 2001).

- *Hardness* is generally expressed as the sum of calcium and magnesium reported in equivalent amounts of calcium carbonate. Other cations in water—such as zinc, iron, strontium, aluminum, and manganese—can contribute to hardness but are usually in very low concentrations compared with calcium and magnesium. Hardness categories are listed in Table 9-3.
- *Nitrate* can be used in the rumen as a source of nitrogen for synthesis of bacterial protein, but reduction to nitrite also occurs. When absorbed into the body, nitrite decreases the oxygen-carrying capacity of hemoglobin and in severe cases results in respiratory distress. Symptoms of acute nitrate or nitrite poisoning are asphyxiation and labored breathing, rapid pulse, frothing at the mouth, convulsions, blue muzzle and bluish tint around eyes, and chocolate-brown blood. More moderate effects of nitrate poisoning are poor growth, infertility, abortions, vitamin deficiencies, and general unhealthiness, but research has not always supported these claims (Stuart and Oehme, 1982). The general safe concentration of nitrate-nitrogen (NO_3-N) in water is less than 10 mg/L and of nitrate less than 44 mg/L (Table 9-4). In evaluating potential nitrate problems, feeds also should be analyzed for nitrate because the effects of feed and water nitrate are additive.
- *Sulfate* guidelines for water are not well defined, but general recommendations are less than 500 mg/L for calves and less than 1,000 mg/L for adult cattle. When sulfate exceeds 500 mg/L, the specific salt form of sulfate or sulfur should be identified. The form of sulfur is an important determinant of toxicity (NRC, 1980). Hydrogen sulfide is the most toxic form and concentrations as low as 0.1 mg/L can decrease water intake. Common forms of sulfate in water are calcium, iron, magnesium, and sodium salts. All act as laxatives, but sodium sulfate is the most potent in this regard. Cattle fed water that is high in sulfates (2,000 to 2,500 mg/L) show diarrhea initially but seem to become resistant to the laxative effect. Iron sulfate was reported by Horvath (1985) to be a more potent depressor of water intake than other forms of sulfate. Research findings from Digesti and Weeth (1976), Weeth and Capps (1972), and Weeth and Hunter (1971) have shown that cattle can tolerate sulfate at up to 2,500 mg/L in

TABLE 9-3 Water Hardness Guidelines[a]

Category	Hardness, mg/L[b]
Soft	0-60
Moderately hard	61-120
Hard	121-180
Very hard	≥181

[a]NRC (1980, 2001).
[b]1 grain/gal. = 17.1 mg/L.

TABLE 9-4 Nitrate in Water[a]

Nitrate (NO_3), mg/L	Nitrate Nitrogen (NO_3 N), mg/L	Guidelines
0-44	0-10	Safe for consumption by ruminants
45-132	11-20	Generally safe in balanced diets with low-nitrate feeds
133-220	21-40	Could be harmful if consumed over long periods
221-660	41-100	Cattle at risk and possible death
≥661	≥101	Unsafe—possible death; should not be used as a source of water

[a]NRC (1974, 2001).

water for short periods (less than 90 days) with no major metabolic problems. At 2,500 mg/L, heifers increased renal filtration of sulfate by 37% compared with heifers drinking water that contained only 110 mg/L. Heifers also rejected water that contained 2,500 mg/L if lower-sulfate water was available. Smart et al. (1986) reported that beef cows that consumed water that contained sulfate at 500 mg/L had lower concentrations of copper in plasma and liver than cows that consumed water that contained 42 mg/L. No significant differences in health, reproduction, cow weight change, or birth weight of calves were reported, but calves of cows that received the high-sulfate water had lower weaning weights than calves of cows that received low-sulfate water. Water and feed with high sulfate contents have been linked to polioencephalomalacia in cattle (Hibbs and Thilsted, 1983; Gould, 1998; Cammack et al., 2010; Kessler et al., 2012) and decreased feed and water intake (Weeth and Hunter, 1971; Loneragan et al., 2001; Patterson and Johnson, 2003; Patterson et al., 2003; Kessler et al., 2012).

- *pH* guidelines for water for cattle have not been established. However, the Environmental Protection Agency (EPA, 1997) pH recommendation for human drinking water is between 6.5 and 8.5.
- *Other nutrients and contaminants* are sometimes found in water and can pose a health hazard to cattle. For safe consumption, water contaminants should not exceed the guidelines in Table 9-5. Nonetheless, many dietary, physiological, and environmental factors affect these guidelines and make it impossible to determine precisely the concentrations at which problems will occur.

Microbiologic analysis of water for coliform bacteria and other microorganisms is necessary to determine sanitary quality. A common microbiologic analysis is for total coliforms, not for specific coliforms. Results from the assay are usually reported as a most probable number (MPN), which is an index of the number of coliforms present (0 MPN = satisfactory; 1 to 8 MPN = unsatisfactory; over 9 MPN = unsafe). A more specific analysis for contamination is a fecal coliform test. Coliforms found in human and animal feces can be determined directly, and information as to the source of contamination can be obtained. The effect of coliforms in water on the health of cattle or ruminal microorganisms is unknown at this time.

TABLE 9-5 Generally Considered Safe Concentrations of Some Potentially Toxic Nutrients and Contaminants in Water for Cattle[a]

Item	Upper-Limit Guideline, mg/L
Aluminum	0.5
Arsenic	0.05
Boron	5.0
Cadmium	0.005
Chromium	0.1
Cobalt	1.0
Copper	1.0
Fluorine	2.0
Lead	0.015
Manganese	0.05
Mercury	0.01
Nickel	0.25
Selenium	0.05
Vanadium	0.1
Zinc	5.0

[a]NRC (1974, 1980, 2001); EPA (1997).

SUMMARY

Water availability and quality are extremely important for animal health and productivity. Limiting water availability to cattle will severely depress animal performance. Some water contaminants such as nitrates, sodium chloride, and sulfates have been reported to affect animal performance and health; however, most water contaminants have an unknown effect on animal performance. This is particularly true for water that has low concentrations of contaminants and is consumed over a long period. Water quality might cause poor production or nonspecific diseases and should be tested as part of the procedures used to investigate such problems. For more detailed information on factors that influence water intake, refer to the National Research Council publications *Nutrients and Toxic Substances in Water for Livestock and Poultry* (NRC, 1974), *Mineral Tolerance of Animals* (NRC, 2005), and *Nutrient Requirements of Small Ruminants* (NRC, 2007).

REFERENCES

Arias, R. A., and T. L. Mader. 2011. Environmental factors affecting daily water intake on cattle finished in feedlots. *Journal of Animal Science* 89:245-251.

Baker, C. C., C. E. Coppock, J. K. Lanham, D. H. Nave, J. M. Labore, C. F. Brasington, and R. A. Stermer. 1988. Chilled drinking water effects on lactating Holstein cows in summer. *Journal of Dairy Science* 71:2699-2708.

Bond, J., T. S. Rumsey, and B. T. Weinland. 1975. Effect of deprivation and reintroduction of feed and water on the feed and water intake behavior of beef cattle. *Journal of Animal Science* 43:873-878.

Brod, D. L., K. K. Bolsen, and B. E. Brent. 1982. Effect of water temperature on rumen temperature, digestion, and rumen fermentation in sheep. *Journal of Animal Science* 54:179-182.

Butcher, J. E. 1973. Snow as the only source of water for sheep. Pp. 205-209 in *Proceedings of Symposium on Water-Animal Relations*, H. F. Mayland, ed. Kimberly: University of Idaho Press.

Cameron, R. D., and J. R. Luick. 1972. Seasonal changes in total body water, extracellular fluid, and blood volume in grazing reindeer. *Canadian Journal of Zoology* 50:107-116.

Cameron, R. D., R. G. White, and J. R. Luick. 1975. The accumulation of water in reindeer during winter. Pp. 374-378 in *Proceedings of the First International Reindeer and Caribou Symposium, University of Alaska, Fairbanks*. J. R. Luick, P. C. Lent, D. R. Klein, and R. G. White, eds.

Cammack, K. M., C. L. Wright, K. J. Austin, P. S. Johnson, R. R. Cockrum, K. L. Kessler, and K. C. Olsen. 2010. Effects of high-sulfur water and clinoptilolite on health and growth performance of steers fed forage-based diets. *Journal of Animal Science* 88:1777-1785.

CSIRO (Commonwealth Science and Industry Research Organization). 2007. *Nutrient Requirements of Domesticated Ruminants*. Melbourne, Australia: CSIRO Publishing.

Cunningham, M. D., F. A. Martz, and C. P Merilan. 1964. Effect of drinking-water temperature upon ruminant digestion, intraruminal temperature, and water consumption on nonlactating dairy cows. *Journal of Dairy Science* 47:382-385.

Degen, A. A. 1977a. Fat-tailed Awassi and German Mutton Merino sheep under semi-arid conditions. 1. Total body water, its distribution and water turnover. *Journal of Agricultural Science* 88:693-698.

Degen, A. A. 1977b. Fat-tailed Awassi and German Mutton Merino sheep under semi-arid conditions. 2. Total body water and water turnover during pregnancy and lactation. *Journal of Agricultural Science* 88:699-704.

Degen, A. A., and B. A. Young. 1984. Effects of ingestions of warm, cold and frozen water on heat balance in cattle. *Canadian Journal of Animal Science* 64:73-80.

Degen, A. A., and B. A. Young. 1990a. Average daily gain and water intake in growing beef calves offered snow as a water source. *Canadian Journal of Animal Science* 70:711-714.

Degen, A. A., and B. A. Young. 1990b. The performance of pregnant beef cows relying on snow as a water source. *Canadian Journal of Animal Science* 70:507-515.

Digesti, R. D., and H. J. Weeth. 1976. A defensible maximum for inorganic sulfate in drinking water of cattle. *Journal of Animal Science* 42:1498-1502.

EPA (U.S. Environmental Protection Agency). 1997. National primary drinking water standards Pp. 18-19 in *Water On Tap: A Consumer's Guide to the Nation's Drinking Water.* EPA815-K-97-002. Washington, DC: Office of Ground Water and Drinking Water, EPA.

Gad, S. M., and R. L. Preston. 1990. In vivo prediction of extracellular and intracellular water in cattle and sheep using thiocyanate and urea. *Journal of Animal Science* 68:3649-3653.

Gould, D. H. 1998. Polioencephalomalacia. *Journal of Animal Science* 76:309-314.

Hibbs, C. M., and J. P. Thilsted. 1983. Toxicosis in cattle from contaminated well water. *Veterinary & Human Toxicology* 25:253-254.

Hicks, R. B., F. N. Owens, D. R. Gill, J. J. Martin, and C. A. Strasia. 1988. Water intake by feedlot steers. Pp. 208-212 in *Animal Science Research Report* MP125. Stillwater: Oklahoma Agricultural Experiment Station. Available online at http://beefextension.com/research_reports/1988rr/88-45.pdf. Accessed May 1, 2015.

Horvath, D. J. 1985. Consumption of simulated acid mine water by sheep. *Journal of Animal Science* 61:474-479.

Ittner, N. R., C. F. Kelly, and H. R. Guilbert. 1951. Water consumption of Hereford and Brahman cattle and the effect of cooled drinking water in a hot climate. *Journal of Animal Science* 10:742-751.

Kessler, K. L., K. C. Olsen, C. L. Wright, K. J. Austin, P. S. Johnson, and K. M. Cammack. 2012. Effects of supplemental molybdenum on animal performance, liver copper concentrations, ruminal hydrogen sulfide concentrations, and the appearance sulfur and molybdenum toxicity in steers receiving fiber-based diets. *Journal of Animal Science* 90:5005-5012.

King, J. M. 1982. Field studies of game and livestock under African ranching conditions. Pp. 179-188 in *Use of Tritiated Water in Studies of Production and Adaptation in Ruminants.* Vienna: International Atomic Energy Agency.

Lofgreen, G. P., R. L. Givens, S. R. Morrison, and T. E. Bond. 1975. Effect of drinking water temperature on beef cattle performance. *Journal of Animal Science* 40:223-229.

Loneragan, G. H., J. J. Wagner, D. H. Gould, F. B. Garry, and M. A. Thoren. 2001. Effects of water sulfate concentration on performance, water intake, and carcass characteristics of feedlot steers. *Journal of Animal Science* 79:2941-2948.

Macfarlane, W. V. 1964. Terrestrial animals in dry heat: Ungulates. Pp. 509-531 in *Handbook of Physiology: Section 4, Adaptation to the Environment*, D. B. Dill, E. F. Adolph, and C. G. Wilber, eds. Washington, DC: American Physiological Society.

Macfarlane, W. V., and B. Howard. 1972. Comparative water and energy economy of wild and domestic animals. *Symposia of the Zoological Society of London* 31:261-296.

Macfarlane, W. V., R. J. Morris, and B. Howard. 1956. Water economy of tropical Merino sheep. *Nature* 178:304-305.

Mader, T. L., L. R. Fell, and M. J. McPhee. 1997. Behavior response of non-Brahman cattle to shade in commercial feedlots. Pp. 795-802 in *Livestock Environment V: Proceedings, of the 5th International Symposium, May 29-31, 1997*, R. W. Bottcher and S. J. Hoff, eds. St. Joseph, MI: American Society of Agricultural Engineers.

McGregor, B. A. 2003. *Nutrition of Goats During Drought.* Rural Industries Research and Development Corporation (RIRDC) Publication No. 03/016. Barton, Australia: RIRDC.

Milam, K. Z., C. E. Coppock, J. W. West, J. K. Lanham, D. H. Nave, J. M. Labore, R.A. Stermer, and C. F. Brasington. 1986. Effects of drinking water temperature on production responses in lactating Holstein cows in summer. *Journal of Dairy Science* 69:1013-1019.

Morris, R. J. H., B. Howard, and W. V. Macfarlane. 1962. Interaction of nutrition and air temperature with water metabolism of Merino wethers shorn in winter. *Australian Journal of Agricultural Research* 13:320-334.

Murphy, M. R., C. L. Davis, and G. C. McCoy. 1983. Factors affecting water consumption by Holstein cows in early lactation. *Journal of Dairy Science* 66:35-38.

MWPS (MidWest Plan Service). 1987. *Beef Housing and Equipment Handbook*, 4th Ed. MidWest Plan Service. Ames: Iowa State University.

Nelson, C. J., ed. 2012. *Conservation Outcomes from Pastureland and Hayland Practices: Assessment, Recommendations, and Knowledge Gaps.* Lawrence, KS: Allen Press.

NOAA (National Oceanic and Atmospheric Administration). 1976. Livestock Hot Weather Stress. Operations Manual Letter C-31-76. Kansas City, MO: NOAA.

NRC (National Research Council). 1974. *Nutrients and Toxic Substances in Water for Livestock and Poultry.* Washington, DC: National Academy Press.

NRC. 1980. *Mineral Tolerance of Domestic Animals.* Washington, DC: National Academy Press.

NRC. 1981. *Effect of Environment on Nutrient Requirements of Domestic Animals.* Washington, DC: National Academy Press.

NRC. 2001. *Nutrient Requirements of Dairy Cattle*, 7th Rev. Ed. Washington, DC: National Academy Press.

NRC. 2003. *Nutrient Requirements of Nonhuman Primates*, 2nd Rev. Ed. Washington, DC: The National Academies Press.

NRC. 2005. *Mineral Tolerance of Animals*. Washington, DC: The National Academies Press.

NRC. 2007. *Nutrient Requirements of Small Ruminants: Sheep, Goats, Cervids, and New World Camelids*. Washington, DC: The National Academies Press.

Odwongo, W. O., H. R. Conrad, A. E. Staubus, and J. H. Harrison. 1985. Measurement of water kinetics with deuterium oxide in lactating dairy cows. *Journal of Dairy Science* 68:1155-1164.

Patterson, H. H., P. S. Johnson, and W. B. Epperson. 2003. Effect of total dissolved solids and sulfates in drinking water for growing steers. Pp. 378-380 in *Proceedings, Western Section, American Society of Animal Science*, Vol. 54. Champaign, IL: ASAS.

Patterson, T., and P. Johnson. 2003. Effects of water quality on beef cattle. Paper 63. *Proceedings, The Range Beef Cow Symposium XVIII, December 9-11, 2003, Mitchell, NE.* Lincoln: University of Nebraska. Available online at http://digitalcommons.unl.edu/rangebeefcowsymp/63/. Accessed on June 9, 2014.

Purohit, G. R., P. K. Ghosh, and G. C. Taneja. 1972. Water metabolism in desert sheep. Effects of various degrees of water restriction on the distribution of body water in Marwari sheep. *Australian Journal of Agricultural Research* 23:685-691.

Ray, D. E. 1986. Water salinity and performance of feedlot cattle. *Proceedings of the Southwest Nutrition Conference.* Tucson: University of Arizona.

Sexson, J. L., J. J. Wagner, T. E. Engle, and J. Eickhoff. 2012. Predicting water intake by yearling feedlot steers. *Journal of Animal Science* 90:1920-1928.

Shkolnik, A., A. Borut, I. Choshniak, and E. Maltz. 1975. The black Bedouin goat—a desert ruminant that "stores" water. *Federation of American Societies for Experimental Biology Journal* 34:472.

Shkolnik, A., E. Maltz, and I. Choshniak. 1980. The role of the ruminant's digestive tract as a water reservoir. Pp. 731-742 in *Digestive Physiology and Metabolism in Ruminants*, Y. Ruckebusch and P. Thivend, eds. Lancaster, UK: MTP Press.

Smart, M. E., R. Cohen, D. A. Christensen, and C. M. Williams. 1986. The effects of sulphate removal from drinking water on the plasma and liver copper and zinc concentrations of beef cows and their calves. *Canadian Journal of Animal Science* 66:669-680.

Stermer, R. A., C. F. Brasington, C. E. Coppock, J. K. Lanham, and K. Z. Milam. 1986. Effect of drinking water temperature on heat stress of dairy cows. *Journal of Dairy Science.* 69:546-551.

Stuart, L. D., and F. W. Oehme. 1982. Environmental factors in bovine and porcine abortion. *Veterinary & Human Toxicology* 24:435-441.

Thom, E. C. 1959. The discomfort index. *Weatherwise* 12:57-59.

Utley, P. R., N. W. Bradley, and J. A. Boling. 1970. Effect of restricting water intake on feed intake, nutrient digestibility and nitrogen metabolism in steers. *Journal of Animal Science* 31:130-135.

Weeth, H. J., and D. L. Capps. 1972. Tolerance of growing cattle for sulfate-water. *Journal of Animal Science* 34:256-260.

Weeth, H. J., and J. E. Hunter. 1971. Drinking of sulfate-water by cattle. *Journal of Animal Science* 32:277-281.

Wilks, D. L., C. E. Coppock, J. K. Lanham, K.N. Brooks, C. C. Baker, W. L. Bryson, R. G. Elmore, and R. A. Stermer. 1990. Responses of lactating Holstein cows to chilled drinking water in high ambient temperatures. *Journal of Dairy Science* 73:1091-1099.

Winchester, C. F., and M. J. Morris. 1956. Water intake rates of cattle. *Journal of Animal Science* 15:722-740.

Woodford, S. T., M. R. Murphy, and C. L. Davis. 1984. Water dynamics of dairy cattle as affected by initiation of lactation and feed intake. *Journal of Dairy Science* 67:2336-2343.

Young, B. A., and A. A. Degan. 1980. Ingestion of snow by cattle. *Journal of Animal Science* 51:811-815.

10

Feed Intake

INTRODUCTION

Accurate estimates of feed intake are essential for predicting the extent to which diets meet requirements for maintenance and production of various classes of beef cattle. Nonetheless, because of the multifactorial nature of its control in ruminants (Forbes, 2003), predicting intake accurately is a significant practical challenge. Previous editions of the National Research Council (NRC) series for beef cattle established empirical regression relationships between dietary energy concentration (specifically dietary net energy required for maintenance [NEm] concentration) and dry matter intake (DMI) by beef cattle. The conceptual framework for these equations is that consumption of less digestible, low-energy (often high-fiber) diets is controlled by physical factors like ruminal fill and digesta passage, whereas consumption of highly digestible, high-energy (often low-fiber, high-concentrate) diets is controlled by energy demands of the animal and by metabolic factors (NRC, 1987). Mertens (1987) expressed these concepts mathematically, suggesting that the neutral detergent fiber (NDF) content of the diet was a useful indicator of limitations associated with fill, whereas animal net energy (NE) requirements and dietary concentration of NE should predict DMI in the absence of fill limitations. Although this model of intake regulation is functional in many cases, it is not fully compatible with existing data. For example, Ketelaars and Tolkamp (1992) used voluntary intake and digestibility data from 831 roughages to evaluate the relationship between organic matter digestibility (OMD) and organic matter intake (OMI). Across a range of 30 to 84% OMD, OMI and OMD were related linearly. Based on the physical/physiological control theory, if intake of highly digestible feeds is regulated by energy demand, OMI (or digestible OMI) would be expected to plateau with increasing OMD, as animals would meet their energy demands with less OMI. In addition, substantial variation in NDF intake across diets varying in OMD, large increases in intake with lactation and cold stress, and decreases that are often observed with advancing pregnancy, are difficult to reconcile with the physical/physiological control theory.

These disparities between observed data and the physical/physiological control theory of intake control have led scientists around the world to offer other explanations for how intake is controlled in ruminants. Tolkamp and Ketelaars (1992) and Ketelaars and Tolkamp (1996) hypothesized that ruminants do not strive to achieve maximum intake, but seek to optimize the cost and benefits of oxygen consumption. In effect, ad libitum intake in their model corresponds to the point at which NE intake per unit of oxygen consumption is maximized; however, finding appropriate data to test this theory has proved challenging. Forbes (2003) highlighted the observations of Ketelaars and Tolkamp (1992) that pointed out the limitations of the physical/physiological control theory, and further noted that intake by individual cattle is naturally variable from day to day, even with forage diets. Forbes (2003) attributed this variability to a hypothetical intake control mechanism, in which he proposed that cattle adjust intake from day to day in response to discomfort signals. For example, overeating might cause distension or an acid load that resulted in discomfort, causing the animal to subsequently decrease feed intake to minimize discomfort. Another alternative intake model was suggested by Allen et al. (2009), who applied the hepatic oxidation theory to ruminants, suggesting that oxidation of various fuel sources in the liver might provide insight into control of feeding behavior. Oxidation of propionate and fatty acids in the liver were suggested as probable hypophagic signals.

There is likely significant truth in all the conceptual models suggested to explain how intake is controlled by ruminants. Indeed, the same sets of data are often used by different authors to justify seemingly disparate theories. Given the complexity of intake control in ruminants, considerable experimentation and model development will be required before we are able to arrive at a unifying theory of intake control. Readers are referred to reviews by Mertens (1994),

Fisher (2002), and Roche et al. (2008) for further discussion on intake regulation theories.

Because our understanding of factors regulating intake by ruminants is incomplete, Fisher (2002) noted that empirical models for predicting intake have considerable value, ultimately contributing to our understanding of ruminant intake control. Intake prediction equations of the NRC (1984) and the Agricultural Research Council (ARC, 1980) relate feed intake to dietary energy concentration (NEm and metabolizable energy, respectively) and body weight (BW), whereas CSIRO (2007) included an adjustment for the relative size of the animal (ratio of the animal's current or "normal" BW to its mature or "standard reference" BW). Use of energy concentration in such equations probably accounts, in part, for effects on feed intake attributable to gastrointestinal fill, energy demands, and potential effects of absorbed nutrients, all of which are components of various intake control theories. Nonetheless, these equations do not account fully or directly for the numerous physiological, environmental, and management factors that alter feed intake. As a result, the methods of predicting feed intake described subsequently are intended to provide general guidelines. No single, general equation should be expected to apply in all production situations. Optimally, beef cattle producers should use general equations as a foundation for developing intake predictions that are specific for production situations and reflect knowledge of the animals and management circumstances in question that cannot be achieved through a generalized equation. Examples of potential physiological, environmental, and management adjustments to predictions of DMI obtained from general equations are discussed briefly in the following sections.

FACTORS AFFECTING FEED INTAKE BY BEEF CATTLE

Physiological Factors Affecting Feed Intake

Body composition, especially degree of fatness, seems to affect feed intake (NRC, 1987). As animals mature, adipose tissue could generate feedback signals (e.g., leptin or circulating unsaturated fatty acids) that affect centrally controlled regulation of feed intake (NRC, 1987). Regardless of the mechanism, degree of fatness has been considered in equations to predict feed intake by beef cattle. Fox et al. (1988) suggested that DMI decreases 2.7% for each 1% increase in body fat content over the range of 21.3 to 31.5% body fat. As a result of the relationship between feed intake and body fat, monitoring of feed intake can sometimes be a useful management tool to determine when cattle have reached appropriate slaughter condition.

Sex (steer vs. heifer) seems to have limited effects on feed intake (ARC, 1980; NRC, 1987). Intake differences attributable to sex can be evident at certain times, as Ingvartsen et al. (1992a) reported that at BW less than 250 kg, heifers had greater intake capacity than steers or bulls. The 6th edition of the *Nutrient Requirements of Beef Cattle* (NRC, 1984) recommended that predicted DMI should be decreased by 10% for medium-framed heifers. At a given BW, heifers are proportionally more mature (fatter) than steers; hence, Fox et al. (1988) suggested the use of a frame-equivalent weight adjustment to their equation for predicting DMI instead of a direct adjustment for sex. In the 7th edition and update of the *Nutrient Requirements of Beef Cattle* (NRC, 1996, 2000) equations for feed intake, adjustments for sex were not statistically justified from the database used to generate the DMI prediction equations for growing-finishing beef cattle. A recent reevaluation of the equations from the previous edition with a larger database (Anele et al., 2014) also yielded no evidence that DMI predicted from average BW and dietary NEm concentration should be adjusted for sex.

Age when an animal is placed on feed for finishing can affect feed intake. Older animals (e.g., yearlings vs. calves) typically consume more feed per unit BW than younger ones. Presumably, the greater ratio of age to BW (age relative to proportion of mature body composition) for yearling cattle drives greater feed intake. This effect has been likened to increased feed intake by cattle experiencing compensatory growth (NRC, 1987). Assuming that cattle started on feed at heavier BW are generally older cattle, age-related effects on feed intake are partly responsible for the positive relationship between initial BW on feed and DMI as noted by NRC (1987) and in the 7th edition and update of the *Nutrient Requirements of Beef Cattle* (NRC, 1996, 2000). Designed studies are needed in which independent effects of age and BW (body composition) on feed intake can be quantified. The NRC (1984) and Fox et al. (1988) suggested a 10% increase in predicted DMI by cattle started on feed as yearlings compared with cattle started on feed as calves, whereas the NRC (1996, 2000) equation for predicting DMI by growing-finishing beef cattle included an intercept adjustment for yearlings vs. calves. It should be noted that the designation of calf vs. yearling in the database used to derive the DMI prediction equation for the NRC (1996, 2000) publication was somewhat arbitrary, and reevaluation of a larger literature database (Anele et al., 2014) did not include an age adjustment, with no loss in predictive value compared with the NRC (1996, 2000) equation. Similar to the NRC (1996) equation, Hicks et al. (1990) noted an increase in DMI by yearlings vs. calves, and these authors also reported a different shape of the intake curve over time for yearlings.

Physiological state can markedly alter feed intake. Increases in feed intake of 35 to 50% for lactating vs. nonlactating animals of the same BW fed the same diet have been noted (ARC, 1980). For forages, Minson (1990) reported a mean increase in DMI of 30% with lactation. Based on data from dairy cows, ARC (1980) and NRC (1987) suggested that DMI increases by 0.2 kg/kg fat-corrected milk. Hence, beef cows bred for greater milk-producing ability would be expected to have greater feed intakes per unit BW during lactation. Advancing pregnancy affects feed intake negatively, most notably during the last month (ARC, 1980; NRC, 1987). Ingvartsen et al. (1992a) reported a 1.5% decrease per week during the last 14 weeks of pregnancy in Danish Black and White heifers fed diets of predominantly roughage, which agrees fairly well with the decrease of 2% per week during the last month of pregnancy suggested by NRC (1987).

Considerable diversity in frame size exists in beef cattle. The NRC (1984) adjusted intake predictions for differences in frame size with multiplicative adjustment factors, whereas Fox et al. (1988) suggested predictions could be adjusted by scaling animals of different frame sizes to an equivalent mature weight (frame-equivalent weight). Holstein and Holstein × beef crosses can consume more feed relative to BW than beef breeds (NRC, 1987), leading Fox et al. (1988) to suggest that intake predictions should be increased 8% for Holsteins and 4% for Holstein × British breed crosses relative to British-breed cattle. The DMI prediction equations in the NRC (1996, 2000) publication did not include specific breed adjustments; however, adjustments suggested by NRC (1987) were recommended and delineated in Table 10-4 of the NRC (1996, 2000) publications.

Environmental Factors Affecting Feed Intake

Considerable research has been conducted to evaluate the effects of ambient temperature on feed intake and digestive function, and the topic has been reviewed extensively (Kennedy et al., 1986; Minton, 1987; Young, 1987; Young et al., 1989). Thermal stress can markedly alter energetic efficiency of ruminants, as evidenced by the effects of cold stress on energy utilization by beef cattle (Delfino and Mathison, 1991). In experimental situations, feed intake increases as the temperature falls below the thermoneutral zone and decreases with elevated ambient temperatures (NRC, 1987). With cold stress, ruminal motility and digesta passage increase before changes in intake, prompting Kennedy et al. (1986) to conclude that the digestive tract response might be essential to accommodate greater feed intake. Nonetheless, Young (1987) noted that this general response to temperature can vary with thermal susceptibility of the animal, acclimation, and diet. Behavioral responses (e.g., decreased grazing time) to thermal stress are restricted by some experimental conditions, which could heighten the effects of thermal stress on feed intake by ruminants. For example, acute cold stress decreased forage intake by as much as 47% in grazing cattle (Adams, 1987), but in thermally adapted grazing cows, Beverlin et al. (1989) reported only small changes in forage intake with temperature deviations of 8° to −16°C. Feed intake by beef cattle confined in outdoor lots and fed finishing diets does not generally increase during cold stress, and is often less during winter than in other seasons (Stanton, 1995).

Other adverse environmental conditions (wind, precipitation, mud, and so on) can accentuate the effects of ambient temperature. Fox et al. (1988) suggested multiplicative correction factors to adjust intake predictions for various environmental effects. Duration of adverse conditions seems important, and because environmental effects are variable, feed intake is difficult to predict under such conditions (NRC, 1987). Thus, caution should be used in applying these multiplicative adjustment factors (e.g., Table 10-4 in the 7th edition and update of the *Nutrient Requirements of Beef Cattle*) without some appreciation for the duration of effects and a historical perspective in specific production settings to justify the magnitude of these corrections.

Seasonal or photoperiod (day length) effects on feed intake are understood less fully than are thermal effects. Hicks et al. (1990) grouped intake data into four seasons and thereby accounted for much of the seasonal pattern in feed intake. Nonetheless, temperature, photoperiod, animal, and perhaps management differences contribute to seasonal patterns in feed intake, and separate effects are difficult to delineate. Photoperiod has been suggested as a potentially important factor influencing feed intake by beef cattle (NRC, 1987). Ingvartsen et al. (1992b) evaluated the effects of day length on voluntary DMI capacity of Danish Black and White bulls, steers, and heifers. Voluntary DMI increased 0.32%/h increase in day length, with the range in the literature reviewed by the authors reflecting a change of −0.6 to 1.5%. Based on the deviation from the voluntary intake at 12 h of daylight, intake would be expected to be 1.5 to 2% greater in long-day months (summer season; July in the Northern Hemisphere) and 1.5 to 2% less in short-day months (winter season; January in the northern hemisphere). No adjustment for photoperiod was suggested for the DMI prediction equations in the NRC (1996, 2000) publications, and again, caution should be used in applying such corrections without supporting historical data for particular locations and production settings.

Management and Dietary Factors Affecting Feed Intake

With grazing cattle, quantity of forage available can affect feed intake. The NRC (1987) reviewed data summarized by Rayburn (1986) and concluded that grazed forage intake was maximized when forage availability was approximately 2,250 kg dry matter (DM)/ha or a forage allowance of 40 g organic matter (OM)/kg BW. Intake decreased rapidly to 60% of maximum when forage allowance was 20 g OM/kg BW (450 kg/ha; NRC, 1987). Minson (1990) noted that bite size decreases with forage mass less than 2,000 kg DM/ha, and that this decrease is only partially compensated for by increased grazing time, thereby resulting in decreased forage intake. The break point at which intake of grazed forage was decreased with decreasing forage allowance seemed to be between 30 and 50 g OM/kg BW. Effects of forage allowance on intake could vary with forage type and sward structure. McCollum et al. (1992) evaluated the effects of forage availability on cattle grazing annual winter wheat pasture and noted that peak intake of digestible OM was predicted at 1,247 kg DM/ha or an allowance of approximately 300 g DM/kg BW. The database for determining the relationship between forage availability and forage intake is derived largely from studies with actively growing pastures. As noted by Minson (1990), gain by sheep is related more closely to green (growing) forage allowance than to total forage DM offered. Similarly, Bird et al. (1989) reported that BW gain by grazing cattle could be modeled more effectively from green pasture mass than from total pasture mass. Cattle eat only small amounts of senescent forage when growing forage is available (Minson, 1990), so selective grazing of growing forage might increase in pastures with both growing and senescent material. Hence, the effects of forage availability on intake should be considered in light of pasture composition and the potential for selective grazing.

Growth-promoting implants tend to increase feed intake. In two trials with beef steers fed a 60% concentrate diet, administering an estradiol benzoate/progesterone implant increased DMI from 4 to 16%, depending on when the implant was given relative to slaughter (Rumsey et al., 1992). Fox et al. (1988) suggested that the predicted feed intake should be decreased by 8% for nonimplanted cattle. Approximately half the studies in the database used to develop the NRC (1996, 2000) equation to predict DMI by growing-finishing beef cattle involved implanted cattle. Thus, caution should be used in arbitrarily applying an adjustment for nonuse of implant to the DMI predicted by this equation.

Monensin, an ionophore feed additive, typically decreases feed intake, especially with diets with a high proportion of forage or silage. Fox et al. (1988) suggested that feed intake decreases by 10 and 6% with 33 and 22 mg/kg monensin, respectively, in the diet. With beef steers fed a 90% concentrate diet, Galyean et al. (1992) noted a 4% decrease in feed intake with monensin at 31 mg/kg dietary DM. Lasalocid, another ionophore approved for use in beef cattle, seems to have limited effects on feed intake. Fox et al. (1988) suggested that feed intake is decreased 2% by lasalocid, regardless of dietary level. Malcolm et al. (1992) observed that feed intake was increased by approximately 4% with 85% concentrate diets that contained 33 mg lasalocid/kg compared with a non-ionophore, control diet. Less data are available on the effects of laidlomycin propionate, an ionophore approved for confined growing and finishing cattle, on feed intake; however, a summary of data (Vogel, 1995) suggested that laidlomycin propionate has little effect on feed intake. Approximately half the cattle in the database used for development of the NRC (1996, 2000) equation were fed an ionophore, again suggesting the need for caution in applying multiplicative adjustments to DMI predicted by this equation.

Dietary nutrient deficiencies, particularly of protein, can decrease feed intake. With low-nitrogen, high-fiber forage, ruminal nitrogen deficiency is common, and provision of supplemental nitrogen often increases DMI substantially (Galyean and Goetsch, 1993). Forage intake responses to added nitrogen (protein) are most typical when forage crude protein content is less than 6 to 8% (NRC, 1987). Supplementation of forages with grain-based concentrates often decreases forage intake, such effects typically being greater with high- than with low-quality forages (Galyean and Goetsch, 1993).

Grinding feeds can affect intake, but effects depend on the type of feed. With forages, fine grinding can increase intake, presumably through effects on digesta passage (Galyean and Goetsch, 1993), but with concentrates, fine grinding often decreases feed intake. Adjustments to intake predictions for finely processed diets as a function of dietary NEm concentration were suggested by NRC (1987). Fermentation of feeds by ensiling generally has little effect on DMI unless the silage is unusually wet or dry and undesirable fermentation has occurred (NRC, 1987), and intake of wilted grass silage is usually greater than that of direct-cut silage (Minson, 1990).

PREDICTION OF FEED INTAKE BY BEEF CATTLE

General

Equations for growing-finishing beef cattle and beef cows were presented in the 7th revised edition and update of the *Nutrient Requirements of Beef Cattle* (NRC, 1996, 2000), with substantial description of the databases used in their development, as well as the methods and databases used for validation of the developed equations. Equations presented in the 7th edition and update have been used extensively in practice; however, concerns related to over- and underprediction have been expressed (e.g., McMeniman et al., 2009; Anele et al., 2014; Coleman et al., 2014). The approach used by the committee in its preparation of the current edition was to reevaluate equations suggested in the 7th revised edition and update (NRC, 1996, 2000), with the intention of developing new, more robust equations that used similar inputs. For growing-finishing beef cattle, these efforts were described in detail by Galyean et al. (2011) and Anele et al. (2014). For beef cows, efforts to reevaluate the equation proposed in the 7th revised edition and update (NRC, 1996, 2000) will be described in a subsequent section. It also should be noted that the equations developed by NRC (1996, 2000) for growing-finishing beef cattle and beef cows were based on data for average DMI over an extended feeding period. Although prediction of DMI for shorter periods might be desirable for some applications, the databases that would enable the committee to develop such prediction equations for the wide variety of production situations and feeds available to beef cattle producers do not seem to exist. As in the 7th revised edition and update of the *Nutrient Requirements of Beef Cattle* (NRC, 1996, 2000), no attempt was made to develop prediction equations for intake by nursing calves, and readers are referred to NRC (1987) and Tedeschi and Fox (2009) for proposed methods of predicting intake by nursing calves.

Growing-Finishing Cattle—Dietary Energy Concentration

As noted before, the 7th edition and update (NRC, 1996, 2000) provided an equation to predict DMI by growing-finishing beef cattle. To develop that equation, data were obtained from experiments conducted with growing and finishing beef cattle, the results of which were published in the *Journal of Animal Science* from 1980 to 1992. The 185 data points in this literature database represented treatment means for average DMI throughout feeding periods that varied in length from 56 to 212 days. Approximately 48% of the cattle were implanted with a growth-promoting implant, and approximately 50% were fed an ionophore. Information on frame size (small, medium, or large), sex (steer, heifer, or bull), age (calf or yearling), and initial and final BW was recorded. Dietary NEm concentration was calculated from tabular values (NRC, 1984), but actually determined NEm values were used when reported in the study. Total NEm intake was calculated as the product of DMI and dietary NEm concentration. Total NEm intake was then divided by the average metabolic BW (average $BW^{0.75}$ in kg; calculated as: $[(\text{initial shrunk BW} + \text{final shrunk BW})/2]^{0.75}$). The relationship between NEm intake per unit $BW^{0.75}$ and dietary NEm concentration was established by stepwise regression analysis, which accounted for 69.9% of the variation in NEm intake per unit average $BW^{0.75}$ in the data set and included an intercept adjustment for calves vs. yearlings, as follows:

$$\text{Calves: NEm intake, Mcal/day} = BW^{0.75} \times (0.2435 \times \text{NEm} - 0.0466 \times \text{NEm}^2 - 0.1128)$$

$$\text{Yearlings: NEm intake, Mcal/day} = BW^{0.75} \times (0.2435 \times \text{NEm} - 0.0466 \times \text{NEm}^2 - 0.0869) \quad \text{(Eq. 10-1)}$$

where BW is the average BW for a feeding period, and NEm is the dietary NEm concentration in Mcal/kg of DM.

Dry matter intake (kg/d) is calculated from Eq. 10-1 by dividing total NEm intake (Mcal/d) by dietary NEm concentration (Mcal/kg of DM). For diets with a NEm concentration I 0.95 Mcal/kg DM, the committee recommended that the divisor to determine DMI from Eq. 10-1 be set at 0.95.

Equation 10-1 has been reported to over- and under-predict DMI depending on dietary and animal conditions (e.g., Patterson et al., 2000; Block et al., 2001; McMeniman et al., 2009). Patterson et al. (2000) evaluated the ability of Eq. 10-1 to predict DMI using data from 7 studies with growing beef cattle in which 54 diets were fed. These authors noted that the equation overpredicted DMI with low-quality diets and underpredicted DMI with high-quality diets, with the overall conclusion that it did not accurately predict intake by cattle on low-quality roughage diets. Block et al. (2001) reported that Eq. 10-1 gave accurate predictions in one of two trials and when data from the two trials were combined with growing-finishing beef cattle in Western Canada; however, predicted DMI had a relatively high standard error, indicating lack of precision. McMeniman et al. (2009) evaluated Eq. 10-1 with a commercial feedlot database consisting of 3,363 pen means collected from three feedlots over a 4-year period and observed significant mean and linear biases across the range of observed DMI.

To address these concerns, Anele et al. (2014) undertook a reevaluation of the NRC (1996, 2000) equation with the objectives of developing and validating more robust equations for predicting average DMI for a feeding period from the same independent variables used in Eq. 10-1 (i.e., BW and dietary NEm concentration). These authors also considered the alternative approach of calculating the DMI required (DMIR) to achieve a specified level of performance. This method, which is similar to the approach used by Tedeschi et al. (2004), uses the NRC (1996, 2000) equations to determine the Mcal of NEm and NE for gain (NEg) required for a given BW and average daily gain (ADG), with the DMIR calculated by dividing the NE requirements by their respective NE concentrations. Anele et al. (2014) developed two new prediction equations from a literature data set, which represented 531 treatment means covering a wide range in dietary NEm concentrations (1.11 to 2.31 Mcal/kg of DM), and which included the observations in the original NRC (1996, 2000) equation plus additional data from experiments published in the *Journal of Animal Science* from 1993 to 2011. One of the new equations (End BW equation) predicted DMI (kg/d) directly, whereas the other new equation predicted DMI as a percentage of average shrunk BW for the feeding period (DMI/BW equation). Both equations included linear and quadratic terms for dietary NEm concentration, and the End BW equation used the ending shrunk BW for a feeding period. Neither of the two equations included an adjustment for age (e.g., calf vs. yearling). When applied to the database from which the equations were developed, 61 and 58% of the variation in observed DMI was accounted for by the End BW and DMI/BW equations, respectively, vs. 48% for Eq. 10-1.

Anele et al. (2014) used four independent data sets that included 7,751 pen and individual observations of DMI by animals of varying BW and feeding periods of varying length, to test DMI predictions based on Eq. 10-1, the End BW and DMI/BW equations, and the DMIR method. Variance in observed DMI accounted for by the various prediction methods ranged from 13.1 to 82.9%, with higher r^2 values for two feedlot pen data sets and lower values for pen and individual data sets that included animals fed lower energy growing diets, as well as those in feedlot settings. Across the four data sets, the r^2 values were consistently greatest and prediction errors consistently least for the DMIR method, but mean biases ($P < 0.01$) were evident for all the equations, ranging from 1.01 kg for the DMIR method to –1.03 kg for Eq. 10-1. Negative linear bias was evident for virtually all equations/data set combinations, suggesting that prediction errors changed as DMI increased. Comparing the two new equations, there was a strong trend for lower standard errors of prediction for the DMI/BW equation than for the End BW equation. Overall, despite a larger literature database for equation development, the two new equations did not offer major advantages over Eq. 10-1 when applied to the four evaluation data sets. Anele et al. (2014) recommended that because the DMIR approach accounted for the greatest percentage of variation in observed DMI and had the least standard errors of prediction in all data sets evaluated, this approach should be considered as a means of predicting DMI by growing-finishing beef cattle.

Based on the results presented by Anele et al. (2014), the committee recommends that Eq. 10-1 should continue to be used as a means of predicting DMI by growing-finishing beef cattle. Nonetheless, given the results observed with the evaluation data sets reported by Anele et al. (2014), the committee recommends that the DMI/BW equation and the DMIR method reported by Anele et al. (2014) also be used as methods of predicting DMI by growing-finishing beef cattle. The DMI/BW and DMIR equations from Anele et al. (2014) are as follows:

$$\text{DMI, \% BW} = 1.2425 + 1.9218 \times \text{NEm} - 0.7259 \times \text{NEm}^2, \qquad \text{(Eq. 10-2)}$$

where BW is the average BW for a feeding period, and NEm is the dietary NEm concentration in Mcal/kg of DM.

$$\text{DMIR, kg/d} = [(\text{NEm required, Mcal/d})/\text{NEm, Mcal/kg dietary DM}] + [(\text{NEg required, Mcal/d})/\text{NEg, Mcal/kg dietary DM}] \qquad \text{(Eq. 10-3)}$$

Although it might useful in some respects to recommend a single equation to apply in all situations, the statistical evaluation of DMI predictions from the various equations and approaches (e.g., DMIR) suggests that there are no compelling reasons to "pick" a favored equation. Users are encouraged to assess the various equations in their own production settings to determine which equation(s) might provide a consistently superior fit to observed intake data.

Feedlot Cattle—Initial Body Weight on Feed

For cattle fed mostly high-energy diets in feedlots, initial BW when cattle are started on feed seems to have considerable value for predicting DMI (NRC, 1987). Based on an evaluation of the relationship between initial BW and DMI using data from commercial feedlot cattle operations, the 7th edition and update (NRC, 1996, 2000) recommended the following equation: DMI, kg/d = 4.54 + 0.0125 × ISBW, where ISBW is initial shrunk BW in kilograms. This equation was based on two commercial data sets—one from feedlots in Texas, Arizona, and California that included 929 pen means for DMI by crossbred steers and heifers, and the other collected from one feedlot in Kansas that included 732 pen means for DMI by crossbred steers. Average initial shrunk BW in these data sets ranged from approximately 76 to 528 kg.

McMeniman et al. (2010) reevaluated the relationship between DMI by feedlot cattle using a database of 3,363 pen means for DMI by feedlot cattle collected at three commercial feedlots over a 4-year period. The resulting equations, which accounted for 76% of the variation in feeding period-average DMI, with a standard error of prediction of 0.46 kg (approximately 5.5% of the mean DMI in the data set) were similar to the equation noted above that was suggested in the NRC (1996, 2000) publication; however, the intercept terms were less, and the slopes were slightly greater. Galyean et al. (2011) used an independent data set collected from two commercial feedlots during a 1-year period to validate the initial BW equations reported by McMeniman et al. (2010). Applied to the validation data set, the McMeniman et al. (2010) equations accounted for 69% of the variation in observed DMI, with a standard error of prediction of 0.47 kg (approximately 5.2% of the mean). A mean bias of 0.42 kg was noted for the equations, but no linear bias was evident ($P > 0.20$).

Based on the validation work reported by Galyean et al. (2011), the committee recommends that the equations reported by McMeniman et al. (2010), with intercept and slope adjustments for sex, be used as a means for predicting DMI by feedlot cattle fed high-grain diets (e.g., generally ≥2.06 Mcal/kg NEm and ≥1.4 Mcal/kg NEg):

$$\text{Steers: DMI, kg/d} = 3.830 + 0.0143 \times \text{ISBW},$$

$$\text{Heifers: DMI, kg/d} = 3.184 + 0.01536 \times \text{ISBW}, \qquad \text{(Eq. 10-4)}$$

where ISBW is initial shrunk BW in kilograms.

One advantage of the approach of predicting DMI from ISBW is that feedlot producers can typically access historical data and develop regression equations of a similar form that are site- and management-specific. Because the calculations are straightforward, involving only simple linear regression (within sex), equations can be readily determined using statistical tools commonly available in spreadsheet programs.

Adjustments to Predictions for Growing-Finishing Beef Cattle

Changes in feed intake in response to various factors suggested by Fox et al. (1988, 1992) could be used to adjust intake predicted from the equations derived above. The NRC (1996, 2000) recommended several of these corrections (see Table 10-4 in the 7th edition and update), but the previous committee also noted that caution should be applied in using these adjustments because of the possibility of double accounting. For example, the literature databases used to derive Eqs. 10-1 and 10-2 recommended in this edition included intake data from cattle under a variety of management systems and an array of environmental conditions. Hence, the equations derived from this database might reflect partial adjustments for many of the factors suggested by Fox et al. (1988, 1992). In addition, most of the adjustments recommended in Table 10-4 of the NRC (1996, 2000) publications were designed to account for relatively short-term changes in feed intake. As the prediction equations recommended in the current edition are designed to predict the average DMI over an extended feeding period, application of short-term adjustments might be inappropriate. As a result of the uncertainty of applying many of these corrections, we recommend only the following three adjustments to prediction equations.

1. *Use of Ionophores.* The majority of the data points used to develop Eqs. 10-1, 10-2, and 10-4 were from experiments in which cattle were fed ionophores. In addition, cattle in virtually all the evaluation data were fed an ionophore. Nonetheless, statistical evaluation was not conducted to determine whether feeding an ionophore altered DMI in these databases. Based on field experience, there is considerable evidence that monensin can decrease feed intake, particularly with high-forage or high-silage diets, but this effect on DMI seems to decrease in magnitude with high-grain feedlot diets. A recent meta-analysis by Duffield et al. (2012) of 151 trials conducted over the last four decades that included diets over a range of dietary energy concentrations indicated that monensin decreased DMI by 3.1%. In contrast to monensin, feeding lasalocid and laidlomycin propionate seems to have little effect on DMI by cattle in feedlot settings. Considering these factors, the committee suggests that DMI predicted from Eqs. 10-1, 10-2, and 10-4 should be adjusted for use of monensin. Based on the meta-analysis of Duffield et al. (2012), when monensin is fed at typical concentrations of 27.5 to 33 mg/kg dietary DM, predicted DMI should be decreased by 3%. This adjustment is for all diets, but it is particularly applicable to beef cattle fed high-forage or high-silage and "grower" diets with a lesser grain or greater roughage concentration than typically included in current feedlot diets (e.g., ϑ10% roughage). Across diet types, historical data or field observations of the effect of monensin on DMI also should be considered in applying the adjustment. When the ionophores lasalocid and laidlomycin propionate are fed, no adjustment is recommended.
2. *Nonuse of Growth-Promoting Implants.* As with ionophores, a statistical evaluation was not conducted to determine the need to adjust DMI predictions for nonuse of implants. The vast majority of the cattle in the databases used to generate Eqs. 10-1, 10-2, and 10-4 were implanted, but the implant regimens used were variable among data points, particularly in the literature databases. Notwithstanding, considerable research and field evidence suggest nonimplanted cattle have lower DMI than implanted cattle. Hence, the committee recommends that values suggested by Fox et al. (1992) be used as a guideline for adjustments to predicted DMI in cases where implants are not used (e.g., 6% decrease in predicted DMI for nonuse of implants).
3. *Effects of Forage Allowance.* The committee recommends that the adjustment to predicted DMI for low forage availability suggested in the 7th edition and update (NRC, 1996, 2000) continue to be used. This adjustment would apply to Eqs. 10-1 and 10-2, as Eq. 10-4 should be used only for feedlot cattle. Data presented in NRC (1987) relative to forage availability were reevaluated by Rayburn (1992) to develop an adjustment. In brief, this adjustment is a two-step process in which the daily forage allowance (FORA) should be determined [FORA = (FM × 1,000 × grazing unit)/(BW × days of grazing)], where FM = available forage mass when FM ≤ 1,150 kg/ha, grazing unit is the pasture area per animal in hectares, and BW is in kilograms. If FM is ≥1,150 kg/ha or FORA is four times the predicted DMI (expressed as g/kg BW), no adjustment should be made to the predicted DMI. If neither condition is true, relative DMI should be calculated from the following equation: Relative DMI, % = $0.17 \times FM - 0.000074 \times FM^2 + 2.4$. The predicted DMI should be multiplied by the relative DMI (expressed as a decimal) to adjust predicted DMI for the effects of limited FM. This adjustment procedure should be applied to all types of grazing systems; however, rotational or other intensive grazing systems with heavy stocking rates will result in more rapid changes in FM than continuous systems with lower stocking rates. This necessitates careful attention to FM in intensive grazing systems and frequent reevaluation of relative DMI. Caution should be used in applying adjustments for available forage to pastures with less dense sods, such as sparse native rangelands (Coleman et al., 2014). Readers are referred to CSIRO (2007) for a discussion of an alternative approach to adjust intake predictions for limited forage allowance.
4. *Potential Adjustments with β-Agonists.* Two β-agonists (ractopamine hydrochloride and zilpaterol hydrochloride) are currently approved for use with cattle confined in feedlots. These two compounds are fed for a short period (28 to 42 days for ractopamine hydrochloride and 20 to 40 days for zilpaterol hydrochloride) before slaughter. Because Eqs. 10-1, 10-2, 10-4, and the DMIR methods (Eq. 10-3) are designed to predict average DMI over an extended period, adjustments for the use of β-agonists would only be required if the short-term effect on DMI associated with feeding these compounds is sufficiently large to affect the average DMI for the entire feeding period. Based on the results

of studies with feedlot steers (Abney et al., 2007; 100- and 200-mg/d dose) and heifers (Quinn et al., 2008; 200-mg/d dose), feeding ractopamine hydrochloride for up to 42 d did not significantly affect DMI for the period in which the product was fed. For zilpaterol hydrochloride, results of some studies indicate a decrease in feed intake when the product is fed for 20 to 40 d (Montgomery et al., 2009). In addition, considerable field-level anecdotal evidence suggests that DMI decreases for a period of time after zilpaterol hydrochloride is fed. Nonetheless, over the entire feeding period in four experiments conducted at commercial feedlots, Elam et al. (2009) did not observe a significant change in DMI when cattle were fed zilpaterol hydrochloride for 20, 30, or 40 d. Thus, the committee recommends that no adjustment to DMI predicted from Eqs. 10-1, 10-2, 10-4, and the DMIR method (Eq. 10-3) be made for use of ractopamine hydrochloride or zilpaterol hydrochloride in feedlot cattle.

Beef Cows—Dietary Energy Concentration

Using treatment means from articles in the *Journal of Animal Science* (1979 to 1993), unpublished theses, and unpublished data from individual scientists, the committee responsible for 7th edition and update (NRC, 1996, 2000) developed a database of 153 observations for feeding-period average DMI by nonpregnant beef cows or by cows during the middle and last third of pregnancy. Dietary NEm concentration in the data set ranged from 0.76 to 2.08 Mcal/kg. Total NEm intake per unit of $BW^{0.75}$ was predicted from dietary NEm concentration, resulting in the following equation:

$$\text{NEm intake, Mcal/d} = BW^{0.75} \times (0.04997 \times NEm^2 + 0.04631);$$
$$\text{Intercept for nonpregnant cows} = 0.03840. \quad \text{(Eq. 10-5)}$$

As with Eq. 10-1 for growing-finishing beef cattle, DMI (kg/d) is calculated from Eq. 10-5 by dividing total NEm intake (Mcal/d) by dietary NEm concentration (Mcal/kg DM). The NRC (1996, 2000) also noted that with low-quality forages where the NEm concentration is ≤1 Mcal/kg (approximately 50% total digestible nutrients [TDN]), division of the predicted NEm intake from Eq.10-5 by the observed dietary NEm concentration to determine DMI results in biologically unrealistic estimates of DMI. Thus, for feeds with dietary NEm concentrations I 0.95 Mcal/kg, the NRC (1996, 2000) recommended that the divisor be set at 0.95. In addition, for lactating cows, the NRC (1996, 2000) suggested that DMI be increased by a factor of 0.2 × the daily milk production (kg/d).

The NRC (1996, 2000) further noted that data points were not included in the regression analysis when nutrient deficiency situations existed, which is a particularly significant issue with beef cows fed low-quality forages that are deficient in crude protein (specifically ruminally degraded protein). Thus, the NRC (1996, 2000) cautioned users that the equation was not likely applicable for predicting DMI with protein-deficient forages.

In preparing the 8th revised edition of the *Nutrient Requirements of Beef Cattle*, the committee evaluated the possibility of updating Eq. 10-5. Similar to the approach used to derive the updated database used to reevaluate Eq. 10-1 and to create Eq. 10-2, an effort was made to prepare a database from the published literature of direct estimates of DMI (i.e., marker-based estimates of DMI were not considered) by beef cows fed diets that varied substantially in energy concentration. Despite an expanded scope of journals that were searched for data, this effort was ultimately abandoned because of the inability to create a sufficiently large database for cows in various physiological stages across a range of dietary energy values. To some extent, this failure to create a new database reflects the fact that, except for grazing situations in which forage availability is not limiting, beef cows fed mixed diets, silages, and harvested forages are often fed allowances that are designed to meet production goals (e.g., to achieve a particular BW or body condition score over a period of time) rather than given ad libitum access to feed. Thus, the committee recommends that Eq. 10-5 continue to be used to predict DMI by beef cows, with the same adjustments to predictions (e.g., change in the intercept term for nonpregnant cows, and increase DMI by 0.2 × kg milk produced daily) suggested by the NRC (1996, 2000).

Notwithstanding this recommendation, the committee recognizes that Eq. 10-5 has been criticized for not predicting DMI accurately at the extremes of energy concentration (Coleman et al., 2014), with particular concerns for cattle grazing forages. These authors suggested that the basic premise of predicting DMI from BW and energy concentration of the diet (i.e., diet digestibility) as used in Eq. 10-5 is sound. Nonetheless, Coleman et al. (2014) proposed that predictions could be improved by including adjustments for additional factors such as milk production, calf performance, and environmental factors.

Alternatives to using Eq. 10-5, especially for cattle grazing forages of medium to low quality, are available. Guidelines or "thumb rules" for DMI as a percentage of BW have been used extensively in practical feeding of beef cows, and values suggested by Lalman (2004) are shown in Table 10-1. Estimates across a range of dietary TDN concentrations based on these thumb rules are shown in Figures 10-1 and 10-2 along with those from Eq. 10-5 and the equations based on BW and OM digestibility reported by Coleman et al. (2014). The equations for predicting DMI from BW and OM digestibility reported by Coleman et al. (2014) yield remarkably similar patterns across energy concentrations for gestating and lactating beef cows to those from Eq. 10-5, with the values predicted from guidelines in Table 10-1 reflecting a plateau in DMI at TDN concentrations greater than 60%.

TABLE 10-1 Forage Intake Guidelines for Beef Cows[a]

Forage Type	TDN, %	Example Forages	Forage DMI Capacity, % BW	
			Dry	Lactating
Low quality	<52	Dry winter forage, mature legume and grass hay, straw	1.8	2.2
Average quality	52-59	Dry summer pasture, dry pasture during fall, late-bloom legume hay, boot-stage and early-bloom grass hay	2.2	2.5
High quality	>59	Mid-bloom, early-bloom, and pre-bloom legume hay; pre-boot-stage grass hay; lush, growing pasture; silages	2.5	2.7

[a]Adapted from Lalman (2004).

Another approach to predicting intake of forages by beef cows is to use NDF concentration of the forage. Mertens (1987) reported that NDF intake in dairy cows averaged 1.2 ± 0.1% of BW, leading some to suggest that intake of forage diets can be predicted from the NDF concentration of the diet by using factors of 1.1 to 1.3% of BW for NDF intake and then dividing this value by the forage NDF concentration to determine DMI. For low- to medium-quality forages, a value of 1.1% of BW is often suggested. The committee cautions that neither the NDF approach nor the guidelines in Table 10-1 have been evaluated for predicting DMI by beef cows. Nonetheless, the guidelines in Table 10-1 have been used extensively in practice for many years and are likely to provide reasonable estimates of DMI by beef cows grazing forages in the range of TDN concentrations noted in the table. The NDF-based approach to predicting DMI has been used less widely than the guidelines in Table 10-1, but it might prove useful for beef cow producers who have reliable NDF data for the forages used in their operation.

It is important to note that the BW estimate used for guidelines in Table 10-1 and for the 1.1% BW/NDF % DM approach should be a nonpregnant (or at least first trimester) BW. Using a BW obtained later in gestation would inappropriately inflate the estimated DMI as a result the weight gain or loss associated with the conceptus. The nonpregnant BW also would be the basis for calculations using these methods with lactating cows.

Finally, using the DMIR approach described for growing-finishing beef cattle is another alternative for beef cows. Tedeschi et al. (2006) used a similar approach to determine the energy efficiency index of beef cows. The DMIR method would be particularly useful when one desires to limit feed cows to achieve production goals rather than provide feed ad libitum. Similar to growing-finishing cattle, to use this method, one simply determines the NEm requirement of the cow for a given production setting and divides this requirement by the dietary NEm concentration to determine the DMI required to achieve the specified performance. As with the guidelines in Table 10-1 and the NDF per unit BW approach, the committee cautions that this method has not been evaluated with independent data. Nonetheless, as long as the dietary NEm concentration is known with a reasonable degree of accuracy, this approach should provide a reliable

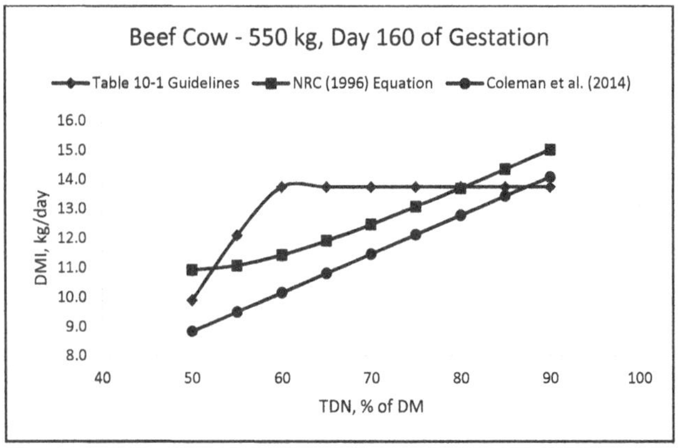

FIGURE 10-1 Dry matter intake predicted for a 550-kg beef cow at day 160 of gestation using the guidelines in Table 10-1, the NRC (1996, 2000) equation (Eq. 10-5), and the equation for dry cows of Coleman et al. (2014; OM intake, kg/d = 12.5 – 0.0299 × BW + 0.00002 × BW^2 – 0.05531 × D + 0.00032 × (BW × D), where BW = body weight and D = OM digestibility; for this calculation, OM was assumed to be 92% of DM, and TDN concentration was considered equivalent to OM digestibility).

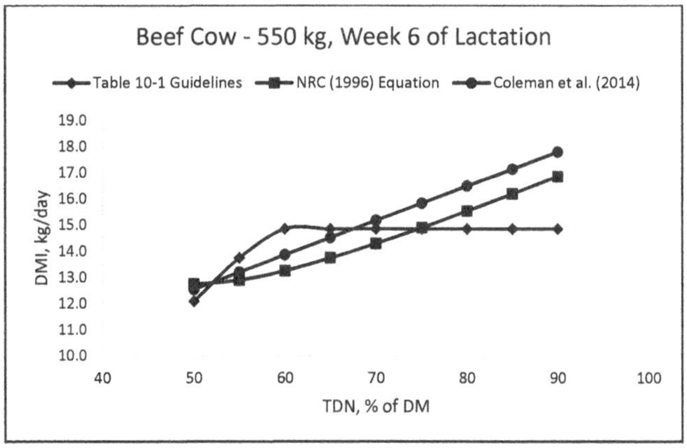

FIGURE 10-2 Dry matter intake predicted for a 550-kg beef cow at week 6 of lactation using the guidelines in Table 10-1, the NRC (1996, 2000) equation (Eq. 10-5 with adjustment for milk yield), and the equation for lactating cows of Coleman et al. (2014; OM intake, kg/day = 27.9 – 0.0902 × BW + 0.00009 × BW^2 – 0.05531 × D + 0.00032 × (BW × D), where BW = body weight, and D = OM digestibility; for this calculation, OM was assumed to be 92% of DM, and TDN concentration was considered equivalent to OM digestibility).

estimate of the quantity of feed required to achieve production goals.

Summary of Recommendations for Predicting Dry Matter Intake

Overall, the committee recommends the following approaches for predicting DMI.

- *Growing-Finishing Beef Cattle:* Use Eqs. 10-1, 10-2, or the DMIR method (Eq. 10-3) to predict feeding period average DMI. In addition to these methods, Eq. 10-4 may be used to predict DMI by cattle consuming high-grain diets (generally ≥2.06 Mcal/kg NEm and ≥1.4 Mcal/kg NEg/kg dietary DM). Adjustments to predictions for use of monensin (3% decrease) are recommended, particularly for beef cattle fed high-forage/high-silage and "grower" diets with a lesser grain or greater roughage concentration than typically included in current feedlot diets (e.g., ≥10% roughage). For high-grain diets, historical data or field observations should be used to confirm a need for this adjustment. No adjustment is recommended for use of lasalocid and laidlomycin propionate. Predicted intake should be decreased by 6% when cattle are not given a growth-promoting implant. When using Eqs. 10-1 and 10-2 to predict DMI by grazing beef cattle, adjustments for forage availability should be considered but might not be applicable to pastures with less dense sods, such as sparse native rangelands.
- *Beef Cows:* For mixed diets, silages, and harvested forages of medium to high quality, use Eq. 10-5, and increase the predicted DMI by a factor of 0.2 × kg milk produced/d with lactating cows. Guidelines in Table 10-1 may be used for gestating and lactating beef cows grazing forages ranging in TDN concentration from 50 to ≥60% of DM. If forage NDF concentration is known, forage DMI may be estimated by multiplying BW × 0.011 and dividing the result by the fractional value of the NDF concentration, with adjustments for lactation as described for Eq. 10-5. The BW used for predictions from Table 10-1 and for the NDF approach should be the nonpregnant (or first trimester) BW of the mature cow. Adjustments for inadequate forage availability recommended for growing-finishing beef cattle may be applied to DMI predictions with beef cows. For beef cows that are fed limited quantities of feed to achieve defined production goals, calculating the DMI required to meet NEm requirements for these goals (e.g., DMIR = (NEm required, Mcal/d)/(NEm, Mcal/kg dietary DM)) should be an appropriate method to determine feed requirements.
- *Nursing Beef Calves:* The approaches suggested by Tedeschi and Fox (2009) are recommended for estimating intake by nursing calves.

REFERENCES

Abney, C. S., J. T. Vasconcelos, J. P. McMeniman, S. A. Keyser, K. R. Wilson, G. J. Vogel, and M. L. Galyean. 2007. Effects of ractopamine hydrochloride on performance, rate and variation in feed intake, and acid-base balance in feedlot cattle. *Journal of Animal Science* 85:3090-3098.

Adams, D. C. 1987. Influence of winter weather on range livestock. Pp. 23-29 in *Proceedings of the Grazing Livestock Nutrition Conference, July, 1987, Jackson, WY.* Laramie: University of Wyoming.

Allen, M. S., B. J. Bradford, and M. Oba. 2009. Board-invited review: The hepatic oxidation theory of the control of feed intake and its application to ruminants. *Journal of Animal Science* 87:3317-3334.

Anele, U. Y., E. M. Domby, and M. L. Galyean. 2014. Predicting dry matter intake by growing and finishing beef cattle: Evaluation of current methods and equation development. *Journal of Animal Science* 92:2660-2667.

ARC (Agricultural Research Council). 1980. *The Nutrient Requirements of Ruminant Livestock.* Surrey, UK: Commonwealth Agricultural Bureaux.

Beverlin, S. K., K. M. Havstad, E. L. Ayers, and M. K. Petersen. 1989. Forage intake responses to winter cold exposure of free-ranging beef cows. *Applied Animal Behavior Science* 23:75-85.

Bird, P. R., M. J. Watson, and J. W. D. Cayley. 1989. Effect of stocking rate, season and pasture characteristics on liveweight gain of beef steers grazing perennial pastures. *Australian Journal of Agricultural Research* 40:1277-1291.

Block, H. C., J. J. McKinnon, A. F. Mustafa, and D. A. Christensen. 2001. Evaluation of the 1996 NRC beef model under western Canadian environmental conditions. *Journal of Animal Science* 79:267-275.

Coleman, S. W., S. A. Gunter, J. E. Sprinkle, and J. P. S. Neel. 2014. Difficulties associated with predicting forage intake by grazing beef cows. *Journal of Animal Science* 92:2775-2784.

CSIRO (Commonwealth Scientific and Industrial Research Organisation). 2007. *Nutrient Requirements of Domesticated Ruminants.* Collingwood, Australia: CSIRO Publishing.

Delfino, J. G., and G. W. Mathison. 1991. Effects of cold environment and intake level on the energetic efficiency of feedlot steers. *Journal of Animal Science* 69:4577-4587.

Duffield, T. F., J. K. Merrill, and R. N. Bagg. 2012. Meta-analysis of the effects of monensin in beef cattle on feed efficiency, body weight gain, and dry matter intake. *Journal of Animal Science* 90:4583-4592.

Elam, N. A., J. T. Vasconcelos, G. Hilton, D. L. VanOverbeke, T. E. Lawrence, T. H Montgomery, W. T. Nichols, M. N. Streeter, J. P. Hutcheson, D. A. Yates, and M. L. Galyean. 2009. Effect of zilpaterol hydrochloride duration of feeding on performance and carcass characteristics of feedlot cattle. *Journal of Animal Science* 87:2133-2141.

Fisher, D. S. 2002. A review of a few key factors regulating voluntary feed intake in ruminants. *Crop Science* 42:1651-1655.

Forbes, J. M. 2003. The multifactorial nature of food intake control. *Journal of Animal Science* 81(Suppl):E139-E144.

Fox, D. G., C. J. Sniffen, and J. D. O'Connor. 1988. Adjusting nutrient requirements of beef cattle for animal and environmental variations. *Journal of Animal Science* 66:1475-1495.

Fox, D. G., C. J. Sniffen, J. D. O'Connor, J. B. Russell, and P. J. Van Soest. 1992. A net carbohydrate and protein system for evaluating cattle diets: III. Cattle requirements and diet adequacy. *Journal of Animal Science* 70:3578-3596.

Galyean, M. L., and A. L. Goetsch. 1993. Utilization of forage fiber by ruminants. Pp. 33-71 in *Forage Cell Wall Structure and Digestibility*, H. G. Jung, D. R. Buxton, R. D. Hatfield, and J. Ralph, eds. Madison, WI: American Society of Agronomy, Crop Science Society of America, and Soil Science Society of America.

Galyean, M. L., K. J. Malcolm, and G. C. Duff. 1992. Performance of feedlot steers fed diets containing laidlomycin propionate or monensin plus tylosin, and effects of laidlomycin propionate concentration on intake patterns and ruminal fermentation in beefs steers during adaptation to a high-concentrate diet. *Journal of Animal Science* 70:2950-2958.

Galyean, M. L., N. DiLorenzo, J. P. McMeniman, and P. J. Defoor. 2011. Alpharma Beef Cattle Nutrition Symposium: Predictability of feedlot cattle growth performance. *Journal of Animal Science* 89:1865-1872.

Hicks, R. B., F. N. Owens, D R. Gill, J. W. Oltjen, and R. P. Lake. 1990. Dry matter intake by feedlot beef steers: Influence of initial weight, time on feed and season of the year received in yard. *Journal of Animal Science* 68:254-265.

Ingvartsen, K. L., H. R. Andersen, and J. Foldager. 1992a. Effect of sex and pregnancy on feed intake capacity of growing cattle. *Acta Agriculturae Scandinavica A. Animal Science* 42:40-46.

Ingvartsen, K. L., H. R. Andersen, and J. Foldager. 1992b. Random variation in voluntary dry matter intake and the effect of day length on feed intake capacity in growing cattle. *Acta Agriculturae Scandinavica A. Animal Science* 42:121-126.

Kennedy, P. M., R. J. Christopherson, and L. P. Milligan. 1986. Digestive responses to cold. Pp. 285-306 in *Control of Digestion and Metabolism in Ruminants*, L. P. Milligan, W. L. Grovum, and A. Dobson, eds. Englewood Cliffs, NJ: Prentice Hall.

Ketelaars, J. J. M. H., and B. J. Tolkamp. 1992. Toward a new theory of feed intake regulation in ruminants. 1. Causes of differences in voluntary feed intake: Critique of current views. *Livestock Production Science* 30:269-296.

Ketelaars, J. J., and B. J. Tolkamp. 1996. Oxygen efficiency and the control of energy flow in animals and humans. *Journal of Animal Science* 74:3036-3051.

Lalman, D. 2004. Supplementing Beef Cows. Oklahoma Cooperative Extension Fact Sheet ANSI-3010. Stillwater: Oklahoma State University. Available online at: http://osufacts.okstate.edu. Accessed on November 21, 2014.

Malcolm, K. J., M. E. Branine, and M. L. Galyean. 1992. Effects of ionophore management programs on performance of feedlot cattle. *Agri-Practice* 13:7-16.

McCollum, F. T., M. D. Cravey, S. A. Gunter, J. M. Mieres, P. B. Beck, R. San Julian, and G. W. Horn. 1992. Forage availability affects wheat forage intake by stocker cattle. Pp. 312-318 in *Animal Science Research Report*. MP-136. Stillwater: Oklahoma Agricultural Experiment Station. Available online at http://www.beefextension.com/research_reports/1992rr/92-47.pdf. Accessed on May 4, 2015.

McMeniman, J. P., P. J. Defoor, and M. L. Galyean. 2009. Evaluation of the National Research Council (1996) dry matter intake prediction equations and relationships between intake and performance by feedlot cattle. *Journal of Animal Science* 87:1138-1146.

McMeniman, J. P., L. O. Tedeschi, P. J. Defoor, and M. L. Galyean. 2010. Development and evaluation of feeding-period average dry matter intake prediction equations from a commercial feedlot database. *Journal of Animal Science* 88:3009-3017.

Mertens, D. R. 1987. Predicting intake and digestibility using mathematical models. *Journal of Animal Science* 64:1548-1558.

Mertens, D. R. 1994. Regulation of forage intake. Pp. 450-493 in *Forage Quality, Evaluation, and Utilization*, G. C. Fahey, ed. Madison, WI: American Society of Agronomy, Crop Science Society of America, and Soil Science Society of America.

Minson, D. J. 1990. *Forage in Ruminant Nutrition.* San Diego, CA: Academic Press.

Minton, J. E. 1987. Effects of heat stress on feed intake of beef cattle. Pp. 325-327 in *Symposium Proceedings: Feed Intake by Beef Cattle, November 20-22, 1986*, F. N. Owens, ed. MP-121. Stillwater: Oklahoma Agricultural Experiment Station. Available online at http://beeflextension.com/proceedings/feed_intake86.pdf. Accessed on May 4, 2015.

Montgomery, J. L., C. R. Krehbiel, J. J. Cranston, D. A. Yates, J. P. Hutcheson, W. T. Nichols, M. N. Streeter, D. T. Bechtol, E. Johnson, T. TerHune, and T. H. Montgomery. 2009. Dietary zilpaterol hydrochloride. I. Feedlot performance and carcass traits of steers and heifers. *Journal of Animal Science* 87:1374-1383.

NRC (National Research Council). 1984. *Nutrient Requirements of Beef Cattle*, 6th Rev. Ed. Washington, DC: National Academy Press.

NRC. 1987. *Predicting Feed Intake of Food-Producing Animals.* Washington, DC: National Academy Press.

NRC. 1996. *Nutrient Requirements of Beef Cattle*, 7th Rev. Ed. Washington, DC: National Academy Press.

NRC. 2000. *Nutrient Requirements of Beef Cattle: Update 2000*, 7th Rev. Ed. Washington, DC: National Academy Press.

Patterson, T., T. J. Klopfenstein, T. Milton, and D. R. Brink. 2000. Evaluation of the 1996 beef cattle NRC model predictions of intake and gain for calves fed low or medium energy density diets. Pp. 26-29 in 2000 *Nebraska Beef Cattle Report.* Paper 362. Lincoln: University of Nebraska.

Quinn, M. J., C. D. Reinhardt, E. R. Loe, B. E. Depenbusch, M. E. Corrigan, M. L. May, and J. S. Drouillard. 2008. The effects of ractopamine-hydrogen chloride (Optaflexx) on performance, carcass characteristics, and meat quality of finishing feedlot heifers. *Journal of Animal Science* 86:902-908.

Rayburn, E. B. 1986. *Quantitative Aspects of Pasture Management. Technical Manual.* Franklinville, NY: Seneca Trail RC&D.

Rayburn, E. B. 1992. Modeling the effect of forage availability on the forage intake of grazing cattle. Paper presented at the American Society of Agronomy Northeastern Branch Meetings, June 28-July 1, 1992, Storrs, CT.

Roche, J. R., D. Blache, J. K. Kay, D. R. Miller, A. J. Sheahan, and D. W. Miller. 2008. Neuroendocrine and physiological regulation of intake with particular reference to domesticated ruminant animals. *Nutrition Research Reviews* 21:207-234.

Rumsey, T. S., A. C. Hammond, and J. P. McMurtry. 1992. Response to reimplanting beef steers with estradiol benzoate and progesterone: Performance, implant absorption pattern, and thyroxine status. *Journal of Animal Science* 70:995-1001.

Stanton, T. L. 1995. Damage control strategies for cattle exposed to cold stress. Pp. 289-298 in *Symposium Proceedings: Intake by Feedlot Cattle, P-942, July 1995*, F. N. Owens, ed. Stillwater: Oklahoma Agricultural Experiment Station.

Tedeschi, L. O., and D. G. Fox. 2009. Predicting milk and forage intake of nursing calves. *Journal of Animal Science* 87:3380-3391.

Tedeschi, L. O., D. G. Fox, and P. J. Guiroy. 2004. A decision support system to improve individual cattle management. 1. A mechanistic, dynamic model for animal growth. *Agricultural Systems* 79:171-204.

Tedeschi, L. O., D. G. Fox, M. J. Baker, and K. Long. 2006. A model to evaluate beef cow efficiency. Pp. 84-98 in *Nutrient Digestion and Utilization in Farm Animals: Modeling Approaches*, E. Kebreab, J. Dijkstra, A. Bannink, W. J. J. Gerrits, and J. France, eds. Cambridge, MA: CABI International.

Tolkamp, B. J., and J. J. M. H. Ketelaars. 1992. Toward a new theory of feed intake regulation in ruminants. 2. Costs and benefits of feed consumption: An optimization approach. *Livestock Production Science* 30:297-317.

Vogel, G. 1995. The effect of ionophores on feed intake by feedlot cattle. Pp. 281-288 in *Symposium Proceedings: Intake by Feedlot Cattle, P-942, July 1995*, F. N. Owens, ed. Stillwater: Oklahoma Agricultural Experiment Station.

Young, B. A. 1987. Food intake of cattle in cold climates. Pp. 328-340 in *Symposium Proceedings: Feed Intake by Beef Cattle, MP-121, November 20-22, 1986*, F. N. Owens, ed. Stillwater: Oklahoma Agricultural Experiment Station. Available online at http://beefextension.com/proceedings/feed_intake86.pdf. Accessed on May 4, 2015.

Young, B. A., B. Walker, A. E. Dixon, and V. A. Walker. 1989. Physiological adaptation to the environment. *Journal of Animal Science* 67:2426-2432.

11

Maintenance Considerations: Energy and Protein

REQUIREMENTS FOR ENERGY

Measurement of Maintenance Requirements

The maintenance requirement for energy has been defined as the quantity of feed energy intake that will result in no net loss or gain of energy from the tissues of the animal's body. Processes or functions comprising maintenance energy requirements include body temperature regulation, essential metabolic processes, and physical activity. Energy maintenance does not necessarily equate to maintenance of body fat, body protein, or body fluid balance. Although for many practical situations maintenance might be considered a theoretical condition, it is useful and appropriate to consider maintenance energy requirements separate from energy requirements for production. The metabolizable energy (ME) required for maintenance functions represents approximately 70% of the total ME required by mature, producing beef cows (Ferrell and Jenkins, 1987) and more than 90% of the energy required by breeding bulls. Moreover, the fraction of total ME intake that growing cattle use for maintenance functions is rarely less than 40%, even at maximum intake. Successful management of beef cattle, whether for survival and production in poor nutritive environments or for maximal production, depends on the knowledge and understanding of their maintenance requirements.

Basically, three methods have been used to measure maintenance energy requirements. These include the use of

- long-term feeding trials to determine the quantity of feed required to maintain body weight (BW) or, conversely, determine BW maintained after feeding a predetermined amount of feed for an extended period of time (Taylor et al., 1981, 1986);
- calorimetric methods (ARC, 1965, 1980); or
- comparative slaughter (Lofgreen, 1965; Lofgreen and Garrett, 1968).

Each approach has advantages as well as limitations.

Long-Term Feeding Trial

Estimates of feed required for maintenance of BW, usually measured in long-term feeding trials, are obtainable with relative ease and can be determined with large numbers of cattle. Values obtained generally correlate well with energy for maintenance in mature, nonpregnant, nonlactating cattle (Jenkins and Ferrell, 1983; Ferrell and Jenkins, 1985b; Solis et al., 1988; Laurenz et al., 1991). Nonetheless, changes in body composition and composition of weight change in growing, pregnant, or lactating cattle are problematic with this approach. Expression of the results in terms of ME or net energy (NE) requirements depends on the use of information from other approaches.

Calorimetric Methods

The energy feeding systems of the Agricultural Research Council (ARC, 1965, 1980), the Ministry of Agriculture, Fisheries, and Food (MAFF, 1976, 1984), the Commonwealth Scientific and Industrial Research Organisation (CSIRO, 1990, 2007), and the Agricultural and Food Research Council (AFRC, 1993), and the energy requirements of dairy cows (NRC, 1989, 2001), sheep (NRC, 1985a), and small ruminants (NRC, 2007b) are primarily based on calorimetric methods. Fasting heat production (FHP) measured by calorimetry plus urinary energy lost during the same period provide measures of fasting metabolism (FM), which, by definition, equates to required net energy for maintenance (NEm). Measurement conditions are standardized such that animals are fed a specified diet at approximately maintenance for 3 weeks before measurement. Animals are trained to the calorimeter and kept in a thermoneutral environment. Measurements are usually made during the third and fourth day after withdrawal of feed. For practical use, FM values are adjusted for the difference between fasted weight of an animal and its liveweight when fed. In addition, recognizing that fasted animals are less physically active than fed animals, ARC (1980) adjusted FM by adding an activity allowance of 1 kcal/kg liveweight for cattle, and CSIRO (1990, 2007) incorporated additional corrections for breed, sex, proportional contribution of milk to the diet, energy intake, grazing activity, and cold stress.

Because of the complexity and cost of measurements, the numbers of animals that can be used is limited. With this approach, measurements are basically acute in that they are made over one or, at the most, a few days. Practical limitations of these systems stem largely from the difficulties in adjusting data obtained in well-controlled laboratory environments to the practical feeding situation.

Comparative Slaughter

The California Net Energy System, proposed by Lofgreen and Garrett (1968) and adopted in preceding editions of this publication (NRC, 1976, 1984, 1996, 2000), is based on comparative slaughter methods. In contrast to calorimetry, in which ME intake and heat energy (HE) are measured and retained energy (RE) is determined by difference, comparative slaughter procedures measure ME and RE directly and HE by difference. The RE is measured as the change in body energy content of animals fed at two or more levels of intake (one of which approximates maintenance) during a feeding period. The RE equates, by definition, to net energy required for gain (NEg) in a growing animal. As shown in Figure 3-3 (Chapter 3, Energy Terms and Concepts), the slope of the linear regression of RE on ME intake provides an estimate of the efficiency of utilization of ME for RE and in growing animals (k_g). The ME intake at which RE = 0 provides an estimate of ME required for maintenance (MEm). By convention, the intercept of the regression of log HE on ME intake is used to calculate an estimate of FHP, which equates to NEm. The efficiency of utilization of ME for maintenance (k_m) is calculated as the ratio of NEm to MEm. These approaches have an advantage over calorimetric methods because they allow experiments to be conducted under situations more similar to those found in the beef cattle industry. However, comparative slaughter measurements must be conducted over extended time periods to allow accurate assessment of body energy changes, and accurate assessment of body composition at the beginning and end of the feeding period is required.

In the current edition of the *Nutrient Requirements of Beef Cattle*, the NEm requirements of beef cattle expressed in Mcal are estimated as

$$\text{NEm} = 0.077\ \text{SBW}^{0.75}, \qquad \text{(Eq. 11-1)}$$

where shrunk body weight (SBW) is in kilograms (Lofgreen and Garrett, 1968; Garrett, 1980), and the expression indicates that 0.077 Mcal of NEm is needed for each kg of metabolic BW. This expression was derived using data from, primarily, growing steers and heifers of British ancestry that were penned in generally nonstressful environments. Effects of activity and environment are implicitly incorporated into NEm in this system. Similarly, influences of increased feed during the feeding period, altered activity, or environmental effects differing from those at maintenance are implicitly incorporated into estimates of NEg. Application to differing situations requires appropriate adjustments.

Variation in Energy Requirements for Maintenance

Maintenance energy expenditures vary with BW, breed or genotype, sex, age, season, temperature, physiological state, and previous nutrition. Metabolic modifiers can also affect maintenance requirements; this is discussed in detail in Chapter 14 (Compounds That Modify Digestion and Metabolism). Fasting heat production or NEm is more closely related to a fractional power of BW (either SBW or empty body weight [EBW] to account for fill) than to $BW^{1.0}$ (Brody, 1945; Kleiber, 1961); the correct power adjustment has been the subject of much debate. The expression $BW^{0.75}$, often referred to as metabolic BW, was originally used to confer proportionality on measurements of H_eE (the heat production at zero feed intake, which is equivalent to the animal's NEm requirement; Chapter 3) made in species differing considerably in mature weight (e.g., mice to elephants). The convention generally adopted for beef cattle is to use $SBW^{0.75}$ to scale energy requirements for BW, even though other functions could be more appropriate for specific applications. In the absence of actual SBW, which is 18 h without food but with water, SBW can be estimated as BW × 0.96.

Breed Differences in Maintenance

Armsby and Fries (1911) reported that "scrub" steers utilized energy less efficiently than "good" beef animals. Subsequently, numerous researchers noted differences in energy requirements or efficiencies of energy utilization among breeds of cattle; however, because of differences in procedures and approaches as well as diversity of breeds compared, direct comparisons among available data are difficult. Blaxter et al. (1966), using calorimetry, noted that Ayrshire steers had 20% greater FHP (kcal/$BW^{0.75}$) than black (Angus type) steers and 6% greater than crosses of those breeds. Results of Garrett (1971), using comparative slaughter, indicated that Holstein steers required 23% more feed to maintain body energy than Hereford steers. Similarly, Jenkins and Ferrell (1984) and Ferrell and Jenkins (1985b) indicated that feed required for weight or energy stasis in young bulls and heifers was greater for cattle in the Simmental breed than in those of the Hereford breed. Those data indicated that MEm was, averaged across sexes, 19% (126 vs. 106 kcal/$BW^{0.75}$) greater for Simmental than for Hereford cattle. Estimates reported for Simmental bulls were equal to those reported by Stetter et al. (1989). Values reported by Andersen (1980) and Byers (1982) indicated that Simmental cattle had 6 and 3% greater requirements than Herefords, respectively. Conversely, Old and Garrett (1987) and Andersen (1980) reported that maintenance requirements of Charolais and Hereford steers were similar. Estimates for growing Friesian cattle averaged approximately 13% greater (5 to 20%) than for Charolais (Robelin and Geay, 1976; Vermorel et al., 1976, 1982; Geay et al., 1980). Webster et al. (1976, 1982) reported that the predicted basal metabolism rates of Friesian cattle were greater than those of Angus (10%), Hereford (31%), or Friesian × Hereford (8%). Chestnutt et al. (1975) estimated maintenance requirements of Friesian cattle to be 20% greater than Friesian × Hereford and 14% greater than Angus steers, whereas estimates of Truscott et al. (1983) were 7% greater for Friesian than for Hereford steers. Wurgler and Bickel (1987) reported no consistent difference in estimates of maintenance requirements among Angus × Braunvieh, Braunvieh, or Friesian steers. Estimates of maintenance requirements of Limousin have been similar to those of Angus (Byers, 1982), Hereford, and Charolais (Andersen, 1980). Results of Webster et al. (1982) and Andersen (1980) indicated that Chianina cattle had about 30% greater energy expenditures than Angus and Hereford cattle. Several other reports (Vercoe, 1970; Vercoe and Frisch, 1974; Patle and Mudgal, 1975; Frisch and Vercoe 1976, 1977, 1982; van der Merwe and van Rooyen, 1980; Carstens et al., 1989b) indicate that maintenance energy requirements of *Bos indicus* breeds of cattle, including Africander, Barzona, Brahman, and Sahiwal, are about 10% less, and that British crosses with those breeds had about 5% lower maintenance energy requirements than British breeds. In contrast, data of Ledger (1977) and Ledger and Sayers (1977) suggest that maintenance requirements of Boran cattle could be about 5% greater than for Herefords; however, those results seem to conflict with the findings of Rogerson et al. (1968). It is likely that a generalization of lower NEm for a specific group of animals (e.g., *B. indicus*, British crossbreds) might not be appropriate if the genetics of breeds change over time. For instance, some experiments have shown that *B. indicus* crossbreds may have lower required NEm compared with *B. taurus*, varying from 71.2 (Chizzotti et al., 2007) to 74.5 kcal/kg $EBW^{0.75}$ (Ferrell and Jenkins, 1998), whereas other experimental results (Tedeschi et al., 2002) do not support the concept of lower required NEm for Nellore cattle (a *B. indicus* breed), reporting an average NEm for Nellore bulls and steers of 77.2 kcal/kg $EBW^{0.75}$. Similarly, in a meta-regression analysis of comparative slaughter of 389 cattle containing bulls, steers, and heifers of Nellore purebreds and Nellore × *B. taurus* crossbreds, Chizzotti et al. (2008) observed no difference for sex or breed and reported an average NEm of 75 kcal/kg $EBW^{0.75}$.

Data of Jenkins and Ferrell (1983) and Ferrell and Jenkins (1984a,b,c) indicated that maintenance requirements differed among genotypes of mature crossbred cows. The ME required for energy stasis (kcal/kg $BW^{0.75}$) of nonpregnant, nonlactating Jersey-, Simmental-, and Charolais-sired cows (from Angus or Hereford dams) was 112, 123, and 99% that of Angus-Hereford (130 kcal/kg $BW^{0.75}$) crossbred cows. Similarly, the data of Lemenager et al. (1980) suggested that the energy needs of Simmental × Hereford cows were about 25% greater than Hereford cows during gestation, whereas Angus × Hereford and Charolais × Hereford cows required about 5 and 7% more ME than Herefords. Laurenz et al. (1991) reported that Simmental cows required 21% more ME (kcal/kg $BW^{0.75}$) than Angus cows. Thornton et al. (2012) concluded that selection for more energetically efficient beef cattle via sire maintenance energy expected progeny difference (EPD; Williams et al., 2009) altered offspring muscle fiber type and meat color in a manner that improved meat quality. Klosterman et al. (1968) observed no difference in estimated energy requirements to maintain weight of mature nonpregnant, nonlactating Hereford and Charolais cows when adjusted for body condition. Similarly, when adjusted for body condition, Hereford × Friesian and White Shorthorn × Galloway cows were found to require similar amounts of energy to maintain liveweight (Russel and Wright, 1983). Estimates of ME (kcal/kg $BW^{0.75}$) for energy stasis of nonpregnant, nonlactating Red Poll-, Brown Swiss-, Gelbvieh-, Maine Anjou-, and Chianina-sired cows (C.L. Ferrell and T.G. Jenkins, U.S. Meat Animal Research Center (MARC), unpublished data, as cited by NRC, 1996, 2000) were 112, 122, 117, 113, and 108% of values for Angus-Hereford (126 kcal/kg $BW^{0.75}$) crossbred cows. Similar values were reported for weight stasis of those cows, with the exception of Gelbvieh and Chianina, which were higher (Ferrell and Jenkins, 1987). In that study, ME (kcal/kg $BW^{0.75}$) required for weight stasis of purebred Angus, Hereford, and Brown Swiss cows was 116, 115, and 155% of that estimated for Angus-Hereford crossbreds (119 kcal/kg $BW^{0.75}$). Results of Taylor and Young (1968) and Taylor et al. (1986) indicated that the energy required (recalculated as kcal/kg $BW^{0.75}$) for long-term weight equilibrium of British Friesian, Jersey, and Ayrshire cows was 20% greater than that of Angus and Hereford cows. Energy required by Dexter cows was 9% greater than the average of Angus and Hereford cows. Thompson et al. (1983) reported estimated values indicating that ME required for energy stasis was 9% greater in Angus × Holstein than in Angus × Hereford cows. Ritzman and Benedict (1938) observed no difference between energy required by Jersey and Holstein cows, whereas Brody (1945) observed slightly greater requirements for Holstein cows than Jersey cows. Solis et al. (1988) reported estimates of ME required for weight and energy stasis for 15 breed or breed crosses from a five-breed diallel. Simple correlation between the two estimates was 0.84 and the slope of the linear regression was 0.99, indicating good agreement between the two estimates. When pooled, estimates of ME required for energy stasis were 104, 96, 96, 112, and 106 kcal/kg $BW^{0.75}$/d for ½

Angus, ½ Brahman, ½ Hereford, ½ Holstein, and ½ Jersey cows, respectively.

Most of these reports observed differences between or among breeds compared and serve to document that considerable variation exists in maintenance requirements among cattle germplasm resources. Nonetheless, because of the diversity of breeds, methodologies, conditions, and other factors, direct comparisons are often tenuous. As a result, the NRC (1996, 2000) selected studies in which British breeds or British breed crosses were compared with other breeds or breed crosses and expressed the results as relative values. Based on the data reviewed in the preceding paragraphs, these generalizations can be made in growing cattle, *B. indicus* breeds of cattle excluding the Nellore data (e.g., Africander, Barzona, Brahman, and Sahiwal) require up to about 10% less energy than beef breeds of *B. taurus* cattle (e.g., Angus, Hereford, Shorthorn, Charolais, and Limousin) for maintenance, with crossbreds being intermediate. Under some circumstances (e.g., genetic improvement, high rates of gain), even *B. indicus* cattle may not need the 10% adjustment. For example, the analyses reported by Tedeschi et al. (2002) and Chizzotti et al. (2007) evaluated this premise. Thus, at this time, the Nellore breed, specifically those raised in Brazil, were removed from the 10% group until more data are available for all other breeds. Conversely, dairy or dual-purpose breeds of *B. taurus* cattle (e.g., Ayrshire, Brown Swiss, Braunvieh, Friesian, Holstein, and Simmental) seem to require about 20% more energy than beef breeds, with crosses being intermediate. Data on purebred, mature cows are more limited; however, available data with purebreds combined with those of crossbreds, indicate that relative differences between breeds in mature cows are similar to those observed in growing animals. This can be generalized further to indicate, in both adult and growing cattle, that a positive relationship exists between maintenance requirement and genetic potential for measures of productivity (e.g., rate of growth or milk production; Webster et al., 1977; Taylor et al., 1986; Ferrell and Jenkins, 1987; Montano-Bermudez et al., 1990).

Consistent with this concept, available data also suggest that animals having genetic potential for high productivity might have less advantage or be at a disadvantage in nutritionally or environmentally restrictive settings (Kennedy and Chirchir, 1971; Baker et al., 1973; Frisch, 1973; Moran, 1976; Peacock et al., 1976; Frisch and Vercoe, 1977; Ledger and Sayers, 1977; O'Donovan et al., 1978; Jenkins and Ferrell, 1984; Ferrell and Jenkins, 1985a,b; Jenkins et al., 1986). Results from these and other studies show that correlated responses to selection can result in a genotype– environment interaction. Selection may result in a population of animals that are highly adapted to a specific environment but less adapted to different environments and are less able to adapt to environmental changes (Frisch and Vercoe, 1977; Taylor et al., 1986; Jenkins et al., 1991).

Sex Differences in Maintenance

Garrett (1970) observed little difference in estimated fasting HE or ME required for maintenance between steers and heifers, which is supported by the findings of Chizzotti et al. (2007) using Nellore × Red Angus cattle. In a study based on comparative slaughter experiments involving 341 heifers and 708 steers, Garrett (1980) concluded that FHP (i.e., required NEm) of steers and heifers is similar. The ARC (1980) and CSIRO (1990, 2007) similarly concluded that fasting metabolism is similar between castrated males and heifers.

Ferrell and Jenkins (1985b) estimated similar FHP (kcal/kg $BW^{0.75}$ daily) for Hereford bulls (70.4) and heifers (69.3), but estimates for Simmental bulls (80.8) were 9% greater than for Simmental heifers (74.1). When expressed as ME required for maintenance, Hereford bulls and heifers differed by only 2%, but estimates for Simmental bulls were 16.5% greater than for Simmental heifers. Pooled across breeds, estimated ME required for energy stasis was 12% greater for intact males than for females (123 vs. 110 kcal ME/kg $BW^{0.75}$ daily). Maintenance energy requirements of steers (65.8 kcal/kg $EBW^{0.75}$), heifers (75.8 kcal/kg $EBW^{0.75}$), and bulls (72.4 kcal/kg $EBW^{0.75}$) of Nellore × Red Angus were statistically similar (Chizzotti et al., 2007). Differences between the NE estimates of Ferrell and Jenkins (1985b) and Chizzotti et al. (2007) may be partially explained by differences in relative gastrointestinal fill because of the difference in the way the values were calculated. Furthermore, in a meta-regression analysis by Chizzotti et al. (2008), which included 262 bulls, 103 steers, and 24 heifers of Nellore purebred and crossbred cattle, there was no difference among bulls (73 kcal/kg $EBW^{0.75}$), steers (76.6 kcal/kg $EBW^{0.75}$), and heifers (75.4 kcal/kg $EBW^{0.75}$) despite the sample size discrepancy. Webster et al. (1977) reported that Hereford × Friesian bulls had predicted basal metabolism values about 20% greater than steers of the same breed cross. In a subsequent report by Webster et al. (1982), the values presented indicated that bulls had 13 to 15% greater predicted basal metabolism than steers. Geay et al. (1980) also suggested greater maintenance requirements of bulls than heifers. The ARC (1980) and CSIRO (1990, 2007) cited the report of Graham (1968) as indicating that rams had 15% greater fasting metabolism than wethers and ewes. However, Bull et al. (1976) and Ferrell et al. (1979) estimated the MEm of rams to be only 2 to 3% greater than for ewe lambs. The average of available data, if the sheep data of Bull et al. (1976) and Ferrell et al. (1979) and the Nellore data of Chizzotti et al. (2008) are excluded, support the conclusion of ARC (1980) and CSIRO (1990, 2007) that the maintenance requirements of bulls are approximately 15% greater than that of steers or heifers of the same genotype. Discrepancies in the data sets could be attributed to differing physical activity and behavioral aspects, and not energetics per se. Nonetheless, bulls may have more muscle than steers, which would require more energy/kg to maintain.

Age Effects on Maintenance

The concept that maintenance per unit of size decreases with age in cattle and sheep (Blaxter, 1962; Graham et al., 1974; CSIRO, 1990, 2007) has been generally accepted. Data from sheep, predominantly castrate males, generally support this view (Graham and Searle, 1972a,b; Graham, 1980). The equation of Graham et al. (1974) indicated that maintenance decreased exponentially and was related to age by the relationship $e^{-0.08 \times age}$, which indicates a fractional decrease rate of 8% per year. Publications by CSIRO (1990, 2007) indicated a minimum of 84% of initial values to be attained at about 6 years of age. Young et al. (1989) noted that metabolic rate deviated substantially from allometric relationships; deviations were greatest during times of highest relative growth rate. They further suggested that significant deviations might also occur in association with other productive functions. Freetly et al. (1995) reported that, in sheep, age effects were confounded with variations in the exponent to express metabolic BW. They concluded that the exponent 0.75 was inappropriate for growing sheep and that the predicted exponent was different among growing Suffolk and Texel breeds. Their data indicated a need for additional research into the most appropriate exponent scaling factor in growing animals and its interrelationship with maturity.

In contrast to sheep, data reported from cattle are less consistent with respect to the effect of age on maintenance requirements. Blaxter et al. (1966) observed little influence of age (15 to 81 weeks), other than that associated with weight, on maintenance of steers. Results of Blaxter et al. (1966), Taylor et al. (1981), and Birkelo et al. (1989) were consistent with those findings. The data of Vermorel et al. (1980) indicated that maintenance requirements of cattle changed little between 5 and 34 weeks of age, but the data of Carstens et al. (1989b) indicated a 6% decrease in FHP and an 8% decrease in ME required for maintenance between 9 and 20 months. Conversely, data reported by Tyrrell and Reynolds (1989) indicated that ME required for maintenance (kcal/kg $SBW^{0.75}$) increased 14% in beef heifers as weight increased from 275 to 475 kg. To our knowledge, direct comparisons of mature, productive females to younger or nonreproducing animals are not available. Indirect evidence (see above) suggests that maintenance of mature, productive cows is not less than that of younger, growing animals after weaning. Therefore, the committee does not recommend adjusting NEm requirement for age of cattle.

Seasonal Effects on Maintenance

Although effects of season have typically been associated with effects of temperature, it has become increasingly evident that season per se can have significant effects on maintenance requirements of cattle and sheep. Christopherson et al. (1979), Blaxter and Boyne (1982), and Webster et al. (1982) noted lower maintenance requirements of sheep, cattle, and bison during the fall. Predicted basal metabolism of cattle in Scotland was 90.3, 92.0, 78.9, and 86.3 kcal/kg $BW^{0.75}$ during weeks 0 to 16, 17 to 32, 33 to 48, and 49 to 52, respectively (Webster et al., 1982). Data reported from Colorado by Birkelo et al. (1991) indicated that FHP measurements during fall, winter, and spring were 92.1, 96.9, and 97.8% of FHP measured during the summer, but MEm followed this pattern only when steers were fed a low versus a high plane of nutrition. Estimates of energy required for weight stasis of mature cows by Byers et al. (1987) for fall, winter, and spring were 86, 86, and 92%, and those for energy stasis were 94, 102, and 100% of estimates made during the summer. Laurenz et al. (1991) reported similar effects of season on energy required for weight stasis of Angus and Simmental cows and for energy stasis of Angus cows but a dissimilar pattern for energy stasis of Simmental cows. Byers and Carstens (1991) reported that as cow fatness increased, maintenance requirements increased during the spring and summer but decreased during the fall and winter. Walker et al. (1991) clearly demonstrated that seasonal effects in ewes are related to photoperiod, which suggests that both photoperiod and seasonal effects might be highly related to activity. Possible season/genotype or latitude effects have not been quantified.

Seasonal effects on maintenance have often been attributed to temperatures outside of the thermal neutral zone. However, energetic responses to seasonal cold stress, for example, have often been evaluated in controlled settings. Cattle extensively grazing western rangelands have opportunities to mitigate cold stress through modifying behavior to decrease wind exposure, increase solar radiation exposure, and optimize the effects of animal grouping. These behaviors can potentially alter maintenance energy demands associated with cold environments. Opposite behavioral activities are possible and probable for mitigating seasonal heat stress in extensive grazing environments. Keren and Olson (2006a) evaluated the effect of cold temperatures and high winds on grazing cattle using a thermal balance model (Keren and Olson, 2006b). Their model basically incorporated metabolic heat production with known pathways of heat gain and losses (Campbell and Norman, 1998) and followed the efforts of others working in hot environments (Brosh et al., 1998; Silva, 2000; Berman, 2004). They concluded that cattle respond to cold stress with behavioral patterns (e.g., changing their orientation to the wind and sun) that conserve energy and mitigate the energetic demands of cold stress. Their data are taken to indicate that the NRC (1996, 2000) model might overpredict energy demands in cold environments if irradiative environment and cattle acclimatization are not taken into account. Additional research in this area could be fruitful in terms of refining the energetic adjustments associated with seasonal temperature extremes.

Temperature Effects on Maintenance

Heat production in cattle arises from tissue metabolism and from fermentation in the digestive tract. Animals dissipate heat by evaporation, radiation, convection, and conduction. Both heat production and dissipation are regulated to maintain a nearly constant body temperature. Within the zone of thermoneutrality, HE (heat energy or heat production) is essentially independent of temperature and is determined by feed intake and the efficiency of energy use; body temperature control is primarily via regulation of heat dissipation. When effective ambient temperature increases above the zone of thermoneutrality that is higher than the upper critical temperature (UCT), productivity decreases, which was traditionally thought to be primarily as a result of decreased feed intake. Rhoads et al. (2010) observed that heat stress decreases intake in dairy cows, but that the decrease in intake only accounts for 35% of the associated decrease in milk production. Elevated body temperature increases tissue metabolic rate and increases "work" of dissipating heat (e.g., increased respiration and heart rates); consequently, energy requirements for maintenance increase. Data from Baumgard and Rhoads (2013) indicated that heat stress altered carbohydrate, lipid, and protein metabolism independent of observed reductions in feed intake. In addition, Tao and Dahl (2013) reported that maternal heat stress during late gestation in dairy cows resulted in fetal hypoxia and fetal growth restriction. Calves from heat-stressed dairy cows seem to have suppressed and compromised immune function (Tao and Dahl, 2013; Monteiro et al., 2014). These recent findings clearly demonstrate the need for additional work in the area of heat stress, maternal energetics, and offspring outcomes in beef cattle.

When effective ambient temperature decreases below the zone of thermoneutrality that is below the lower critical temperature (LCT), HE produced from "normal" tissue metabolism and fermentation is inadequate to maintain body temperature. As a result, animal metabolism (NRC, 1981) and (or) behavior (Keren and Olson, 2006a) must change to provide adequate heat to maintain body temperature, thereby affecting energy requirements for maintenance. Both UCT and LCT vary with the rate of heat production in thermoneutral conditions, as well as the animal's ability to dissipate or conserve heat. As noted in other sections of this report (Chapter 3), heat production of animals in thermoneutral conditions can differ substantially as functions of feed intake, physiological state, genotype, sex, and activity.

The word acclimatization is used to describe adaptive changes in response to changes in the climatic conditions and include behavioral as well as physiological changes. Behavioral modification in cattle includes using the variation in terrain or other topographical features such as windbreaks, huddling in groups or changing posture to minimize heat loss in cold weather and during decreased activity, seeking shade to decrease exposure to radiant heat, standing instead of lying down, seeking a hill to increase exposure to wind, or wading in water to increase heat dissipation in high temperatures. Physiological adaptations include changes in basal metabolism, respiration rate, distribution of blood flow to the skin and lungs, feed and water consumption, rate of passage of feed through the digestive tract, hair coat, and body composition. Physiological changes usually associated with acute temperature changes include shivering and sweating, as well as acute changes in feed and water consumption, respiration rate, heart rate, and activity. It should also be noted that animals differ greatly in their behavioral responses and in their ability to physiologically adapt to the thermal environment. Genotype differences are particularly evident in this regard.

Recognizing the importance of adaptation, the NRC (1981) committee concluded that required NEm of cattle adapted to the thermal environment is related to the previous ambient (air) temperature (T_p, °C) in the following manner:

$$NEm = [0.0007(20 - T_p) + 0.077]SBW^{0.75}. \quad \text{(Eq. 11-2)}$$

This equation indicates that the NEm requirement of cattle changes by 0.0007 Mcal/kg $SBW^{0.75}$ for each degree that previous ambient temperature differed from 20°C. It should be noted that these corrections for previous temperature are largely opposite to the photoperiod effect discussed previously in ewes.

Heat or cold stresses occur when effective ambient temperature is greater than UCT or less than LCT, respectively. Both UCT and LCT are functions of how much heat the animal produces and how much heat is lost to the environment. The HE of the animal can be calculated as shown previously:

$$HE = ME - RE \quad \text{(Eq. 11-3)}$$

or

$$HE = NEm/k_m + RE\ (1 - k_g), \quad \text{(Eq. 11-4)}$$

where ME is ME intake and RE is retained energy, which would include NEg, NE requirement for lactation (NEl), NE requirement for pregnancy (NEy), etc. As shown in Figure 3-3 (Chapter 3), the slope of the linear regression of RE on ME intake provides an estimate of the efficiency of utilization of ME for RE and in growing animals (k_g). The ME intake at which RE = 0 provides an estimate of MEm. By convention, the intercept of the regression of log HE on ME intake is used to calculate an estimate of FHP, which equates to NEm. The efficiency of utilization of ME for maintenance (k_m) is calculated as the ratio of NEm to MEm.

Cold Stress

Both environmental and animal factors contribute to differences in heat loss from the animal. Environmental factors include air movement, precipitation, humidity, contact surfaces, and thermal radiation. Although results are not completely satisfactory, numerous efforts have been made to integrate these effects with animal responses.

Factors contributing to differences in animal heat loss from conduction, convection, and radiation are surface area (SA), which includes surface or external insulation (EI), and internal or tissue insulation (TI). Evaporative losses are affected by respiration volume as well as SA, EI, and TI. Respiratory losses, although not quantified by NRC (1981), represent 5 to 25% and total evaporative heat losses represent 20 to 80% of total heat losses (Ehrlemark, 1991).

Surface area (m^2) is related to body weight by the equation:

$$SA = 0.09\ BW^{0.67}; \qquad \text{(Eq. 11-5)}$$

hence,

$$HE/SA = (HE/BW^{0.75}) \times (BW^{0.75}/SA). \qquad \text{(Eq. 11-6)}$$

Tissue insulation (TI, °C × m^2 × d/Mcal) is primarily a function of subcutaneous fat and skin thicknesses. Typical values for TI are 2.5 for a newborn calf, 6.5 for a 1-month-old calf, 5.5 to 8.0 for yearling cattle, and 6.0 to 12 for adult cattle. External insulation (EI) is provided by hair coat plus the layer of air surrounding the body. Thus, external insulation is related to hair depth; however, the effectiveness of hair as external insulation is influenced by wind, precipitation, mud, and hide thickness. These effects have been described as follows:

$$EI = (6.1816 - 0.5575\ WS + 0.0152\ WS^2 + 5.298\ HD - 0.4297\ HD^2 - 0.1029\ WS \times HD) \times MUD \times HIDE \qquad \text{(Eq. 11-7)}$$

where EI is expressed as °C × m^2 × d/Mcal, WS is wind speed (km/h), HD is effective hair depth (cm), MUD is hair coat, dimensionless (1 for no mud, 0.8 for some mud on lower body, 0.5 for mud on lower body and sides, and 0.2 for heavily covered with mud), and HIDE is dimensionless (0.8 for thin hide thickness, 1 for average hide thickness, and 1.2 for thick hide thickness). The equation for predicting EI (Eq. 11-7) was re-derived using the data from Table 5 of Fox et al. (1988), which is based on the NRC (1981) publication, except that instead of 1.5 cm for HD, 2 cm was used as indicated in the original NRC (1981) publication. Linear, quadratic, and interaction of the linear factors were added to the re-derived equation, resulting in an R^2 of 0.976 and a root of the MSE of 0.625°C × m^2 × d/Mcal. The variables MUD and HIDE are adjustments for mud and hide thickness. Total insulation (IN) is calculated as follows:

$$IN = TI + EI, \qquad \text{(Eq. 11-8)}$$

and LCT may be calculated as (NRC, 1981)

$$LCT = 39 - IN \times (HE/SA - H_e) \qquad \text{(Eq. 11-9)}$$

where LCT, IN, and HE/SA are as described previously. The term H_e represents the minimal total evaporative heat loss and is estimated (Ehrlemark, 1991) as:

$$H_e = 0.15\ HE/SA. \qquad \text{(Eq. 11-10)}$$

The animal can receive or lose heat by solar or long-wave radiation. The net effect of thermal radiation on the animal depends on the difference between the combined solar and long-wave radiation received by the animal and the long-wave radiation emitted by the animal. For animals in bright sunlight, a net gain of heat by thermal radiation usually exists (Keren and Olson 2006a,b; see discussion in the section above), resulting in an increased effective ambient temperature (EAT) of 3° to 5°C (NRC, 1981). In bright sunlight, this effect lowers LCT by 3° to 5°C. Conversely, CSIRO (1990, 2007) indicated that the rate of heat loss by long-wave radiation increases on cold clear nights, resulting in an increase in the LCT. Within the temperature range of −10° to 10°C, this effect is about 5°C.

The increase in energy required to maintain productivity in an environment colder than the animal's LCT can be estimated as

$$MEcs = SA \times (LCT - EAT)/IN, \qquad \text{(Eq. 11-11)}$$

where

MEcs is the increase in maintenance energy requirement because of cold stress (Mcal/d);
SA is surface area (m^2);
LCT is lower critical temperature (°C);
EAT is effective ambient temperature (°C) adjusted for thermal radiation; and
IN is total insulation (°C × m^2 × d/Mcal).

It follows that

$$NEcs = k_m \times MEcs. \qquad \text{(Eq. 11-12)}$$

As per NRC (1996, 2000), k_m is diet NEm/diet ME (assumed 0.576 in derivation) and NEcs is the increase in net energy requirement associated with cold stress. Total net energy for maintenance under conditions of cold stress (NEmcs) becomes

$$NEmcs = NEm + NEcs \qquad \text{(Eq. 11-13)}$$

Heat Stress

If ambient temperature and thermal radiation exceed the temperature of the skin surface, the animal cannot lose heat by sensible means (i.e., conduction, convection, and radiation) and will gain heat by these routes. Evaporative heat loss occurs from the skin (cutaneous) or through respiration. The effectiveness of both cutaneous and respiratory evaporative heat loss diminishes as relative humidity (RH) of the air increases and is totally ineffective when RH = 100. Animals can store some heat in their bodies during the day and dissipate the stored heat during cooler daytime periods or at night, if the animal's heat production exceeds its ability to dissipate heat; but if hyperthermia persists, animals cannot survive.

There has been much study of the various aspects of heat stress on animal performance, but there are no established bases for quantitative description of effects; consequently, additional research is needed to assess the effects of heat stress. Ehrlemark (1991), for example, developed a regression of respiratory heat loss on the ratio of ambient temperature minus LCT to body temperature minus LCT but did not include cutaneous evaporative heat loss or the influence of RH. It is generally agreed that adjustments to maintenance energy requirement for heat stress should be based on the severity of heat stress; however, severity can vary considerably among animals, depending on animal behavior, acclimatization, diet, level of productivity, radiant heat load, or genotype. The color and thickness of hair coat can also alter susceptibility to heat stress. The type and intensity of panting by an animal can provide an index for the appropriate adjustment in maintenance requirement, an increase of 7% when there is rapid shallow breathing and 11 to 25% when there is deep, open-mouth panting (NRC, 1981). With severe heat, feed consumption is decreased and consequently metabolic heat production and productivity are decreased.

Effects of Physiological State on Maintenance

Total heat production increases during gestation (Brody, 1945). Although indirect evidence is available to suggest that maintenance requirements of cows increase during gestation (Brody, 1945; Kleiber, 1961; Ferrell and Reynolds, 1987), an increase has not been directly measurable by comparative slaughter evaluations (Ferrell et al., 1976). Combined data from Brody (1945) and Ferrell et al. (1976) indicate that heat production can increase by nearly 50% near the end of pregnancy in cattle. Scheaffer et al. (2003), Meyer et al. (2010), and others have indicated that maternal visceral tissues, particularly the liver and gut, are responsive to the stage of pregnancy and can either conserve or increase energy use associated with pregnancy, depending on plane of nutrition. Increased heat production associated with pregnancy, for the purpose of estimating energy requirements, can be assumed to be attributable to the productive process of pregnancy independent of specific tissue use. Moe et al. (1970) estimated ME requirements for maintenance to be 22% greater in lactating than in nonlactating cows (primarily Holstein). A similar difference (23%) was reported by Flatt et al. (1969), whereas Ritzman and Benedict (1938) reported a larger (49%) difference. Neville and McCullough (1969) and Neville (1974), using Hereford cows and different approaches, estimated the maintenance requirement of lactating cows to be more than 30% greater than nonlactating cows. The reports of Patle and Mudgal (1975, 1977) agree with those observations, whereas data of Ferrell and Jenkins (1985a, 1987) and Montano-Bermudez et al. (1990) suggest a difference of 10 to 20%. Reynolds and Tyrrell (2000) reported that when expressed on a $BW^{0.75}$ basis, efficiency of ME use above maintenance for milk energy was similar among cattle breeds and that breeds differ in their propensity for milk yield and partitioning of ME between synthesis of milk and tissue, but not efficiency of ME use for milk production. On the whole, the available data indicate that the maintenance requirements of lactating cows are about 20% greater than those of nonlactating cows across beef breeds.

Effects of Activity on Maintenance

Few data are available regarding the efficiency of ME use for muscular work. In addition, it is debatable whether activity is a maintenance or productive function. It is highly probable that grazing cattle walk considerably farther than penned animals and therefore expend more energy for work; however, the extent to which grazing animals expend more energy standing, changing positions, eating, or ruminating than penned cattle is not well documented. It is recognized that energy expenditure for work by grazing cattle is influenced by numerous factors, including herbage quality and availability, topography, weather, distribution of water, genotype, or interactions among these factors; and variation among individuals can be substantial. In a review of available literature, CSIRO (1990) estimated that the increase in maintenance energy requirements of grazing compared with penned cattle was 10 to 20% in best grazing conditions and about 50% for cattle on extensive, hilly pastures where animals walk considerable distances to preferred grazing areas and water. The NRC (1996, 2000) used the CSIRO (1990) approach to estimate the NE required for activity (NEm_{pa}, Mcal/day) and the equation was devised as follows:

$$NEm_{pa} = [0.006\ DMI \times (0.9 - D) + 0.05T/(GF + 3)] \times BW/4.184, \quad \text{(Eq. 11-14)}$$

where

DMI is dry matter intake from pasture (kg/d);
D is digestibility of dry matter (as a decimal);
T is terrain (level, 1.0; undulating, 1.5; or hilly, 2.0);
BW is body weight (kg); and
GF is green forage availability (1,000 kg/ha).

If no green forage is available, replacement of GF with total forage available (TF) was suggested on the premise that selectivity, and thereby distance walked, decreases when no green forage is not available.

The updated CSIRO (2007) publication presented a revised equation that is somewhat more defined and adaptable to both sheep and cattle. When comparing the two equations, the more recent one (CSIRO, 2007) can, in some circumstances, give fairly high NEm_{pa} values. Moreover, the CSIRO (2007) equation has a greater number of input variables, which increases the complexity of application.

Lardy et al. (2004), in their assessment of the NRC (1996, 2000) model for use in grazing cows, concluded that "our recommendation is that the 'On Pasture' feature in the model not be used because it unrealistically increases energy requirements. In almost all cases, if the feature is not used, model output data more closely resemble biological data gathered at field stations." Their work is supported by recent efforts (Waterman et al., 2014) using data sets with extensively grazing beef cows from Montana and New Mexico. Consequently, readers are cautioned with respect to the use of the existing equations for NEm_{pa} in mature cows that are well adapted to extensive native range conditions associated with beef production systems in the western United States and Canada. Nonetheless, the equation might be more applicable to naive cows that are not adapted to extensive grazing conditions. Original data associated with estimates of energy costs of grazing and/or muscular activity of walking most likely used animals (sheep or cattle) adapted to pen-based management systems. Consequently, they were not adapted to extensive grazing environments where light to moderate physical (muscular) activities are the normal conditions. This is a particularly relevant observation in the light of data cited in the *Nutrient Requirements of Horses* (NRC, 2007a) indicating that unconditioned (not adapted) horses experiencing muscular work would have heart rates of 120 to 150 beats/min, whereas conditioned (adapted) horses performing the same level of work would have heart beats of 70 to 110 beats/min. Length of time to adapt horses to light or moderate workloads has been estimated to range from 4 to 8 weeks (T. J. Swanson and C. J. Hammer, North Dakota State University, Fargo, ND, personal communication, May 2013). Therefore, it is likely that existing equations for NEm_{pa} for beef cows are best suited for cows during the adaptation to their environment and that estimates of NEm_{pa} expenditure for grazing activities of mature well-adapted cows are substantially less. Additional work is needed in this area before more accurate prediction equations associated with the muscular activity of grazing can be developed. Users may consider the various options that have been proposed by the CSIRO (1990, 2007) and the NRC (1996, 2000) to account for the effects of activity on maintenance requirements; however, until additional data are available, the committee has chosen not to include an adjustment to the maintenance requirements for activity associated with grazing in the Beef Cattle Nutrient Requirements Model (BCNRM; Chapter 19, Model Equations and Sensitivity Analyses).

Effects of Previous Nutrition/Compensatory Gain

The phenomenon of compensatory gain is described as a period of faster or more efficient rate of growth following a period of slower or less efficient rate of growth that could result from nutritional or environmental stress or planned management strategies. Numerous reports are available to document this phenomenon in cattle and other species (Wilson and Osbourn, 1960; Carrol et al., 1963; Lawrence and Pierce, 1964; Hironaka and Kozub, 1973; Lopez-Sanbidet and Verde, 1976; O'Donovan, 1984; Hovell et al., 1987; Abdalla et al., 1988; Drouillard et al., 1991). The response to previous nutritional deprivation is highly variable. Data are available, for example, that show that at similar BW, body fat is decreased (Smith et al., 1977; Mader et al., 1989; Carstens et al., 1991), not changed (Fox et al., 1972; Burton et al., 1974; Rompala et al., 1985), or increased (Searle and Graham, 1975; Tudor et al., 1980; Abdalla et al., 1988) after a period of realimentation. Many factors contribute to these differences, including: animal genotype; variation in the severity, nature, and duration of restriction; the overall nutritional regimen; and the measurement interval during realimentation.

A major component of compensatory growth by animals given abundant feed after a period of restriction is increased feed intake. This response will cause increased gut fill and body weight, but there is also evidence for higher efficiency of energy use. Several reports (Graham and Searle, 1979; Thomson et al., 1980; Carstens et al., 1991) have provided evidence to suggest greater net efficiency of ME use for body energy gain; however, the duration of these effects is subject to debate (Butler-Hogg, 1984; Ryan et al., 1993a,b).

Results of studies by Marston (1948) have contributed to an understanding of the other possible mechanisms involved in compensatory growth. Those results showed that the level of feed intake can affect the metabolic rate of sheep and cattle. These and other reports (Graham and Searle, 1972a,b, 1975; Graham et al., 1974; Thomson et al., 1980; Ferrell et al., 1986; Ferrell and Koong, 1987) have shown that FHP decreases in response to decreased feed intake. Similarly, several reports (Wilson and Osbourn, 1960; Walker and Garrett, 1970; Foot and Tulloh, 1977; Ledger, 1977; Ledger and Sayers, 1977; Andersen, 1980; Gray and McCracken, 1980; Corbett et al., 1982) have shown that maintenance requirements in rats, swine, cattle, and sheep are decreased after periods of decreased nutritional intake. Some of the possible explanations for altered metabolism associated with different planes of nutrition have been discussed by Milligan and Summers (1986), Ferrell (1988), and Johnson et al. (1990). Briefly, metabolic bases for changes include altered rates of ion pumping and metabolite cycling (Milligan and Summers, 1986; Summers et al., 1988; Harris et al., 1989; McBride and Kelly, 1990; Lobley et al., 1992), altered size and metabolic rate of visceral organs (Cañas et al., 1982; Koong et al., 1982, 1985; Burrin et al., 1989), and enhanced mitochondrial function and decreased liver size (Connor et al., 2010).

There is much, although not total, support for the general conclusion that maintenance is decreased during and for some time after a period of feed restriction (Graham and Searle, 1972a; Thorbek and Henckel, 1976; Ledger and Sayers, 1977; Andersen, 1980; Schnyder et al., 1982; Stetter et al., 1989); however, reports on the extent of this decrease have been variable, and range from about 10% to more than 50%. Little definitive information is available regarding the duration of the decreased maintenance. Stated another way, the length of time that an animal exhibits compensatory gain after it has access to abundant feed is not well defined. Further, critical description of animals such that the expected degree of compensation can be predicted with confidence, without knowing their genotype and history (the nature and severity of restriction, etc.), is lacking. Because of these types of problems, generalizations are difficult to make, although several mathematical descriptions have been proposed (Frisch and Vercoe, 1977; Baldwin et al., 1980; Koong et al., 1985; Corbett et al., 1987). A decrease in maintenance of 20% for a compensating animal seems a reasonable generalization (Crabtree et al., 1976; Thorbek and Henckel, 1976; Frisch and Vercoe, 1977; Andersen, 1980; Baldwin et al., 1980; Thomson et al., 1980; Koong et al., 1982, 1985; Schnyder et al., 1982; Vermorel et al., 1982; Webster et al., 1982; Ferrell et al., 1986; Koong and Nienaber, 1987; Wurgler and Bickel, 1987; Birkelo et al., 1989; Burrin et al., 1989; Carstens et al., 1989a). The duration of decreased maintenance is subject to the extent and duration of restricted growth and to nutritional regimen during the recovery periods; typically, 60 to 90 d of compensation is expected.

Use of Energy from Weight Loss

Animals, particularly in a pasture or range situation, intermittently lose BW when feed quantity or quality is inadequate to meet the animal's nutrient requirements. Available data indicate that the composition of liveweight loss is approximately equal to the composition of liveweight gain in animals (ARC, 1980; CSIRO, 1990, 2007). Thus, the energy content of liveweight loss and gain is similar. Energy content and composition of weight gain are discussed in subsequent sections of this report.

Buskirk et al. (1992) argued that the energy content of EBW gain in mature cows varies depending on cow body condition. They estimated energy content of EBW change in cows with body condition scores (1 to 5 scale) of 1, 2, 3, 4, and 5 to be 2.57, 3.82, 5.06, 6.32, and 7.57 Mcal/kg, respectively. Similarly, CSIRO (1990) adopted relationships established by Hulme et al. (1986) indicating that the energy content of liveweight change in dairy cattle increases linearly from 3.0 to 7.1 Mcal/kg as condition score increases from 1 to 8 (on a scale of 1 to 8). Composition of weight change in mature cows is discussed in greater detail in Chapters 3 and 13 (Reproduction).

Although limited data are available, data from sheep (Marston, 1948), dairy cows (Flatt et al., 1965; Moe et al., 1970), and beef cows (Russel and Wright, 1983) indicate the efficiency of use of energy from body tissue loss for maintenance or milk production to be 77 to 84%, with the mean being approximately 80%.

Summary of New Recommendations Regarding Maintenance Energy Concepts

Overall, based upon available data the committee recommends the following:

- The data indicate that the NRC (1996, 2000) model may overpredict energy demands in cold environments if irradiative environment and cattle acclimatization are not taken into account. Additional research is needed to refine the energetic adjustments associated with seasonal temperature extremes.
- Readers are cautioned on the use of the existing equations for NEm_{pa} in mature cows that are well adapted to extensive native range conditions associated with beef production systems in the western United States and Canada. Conversely, the equation might be applicable to naive cows that are not adapted to extensive grazing conditions. Existing equations for NEm_{pa} for beef cows are best suited for cows during adaptation to their environment and predicted NEm_{pa} expenditure for grazing activities of mature well adapted cows are substantially less. Additional research is needed in this area before more accurate prediction equations associated with the muscular activity of grazing can be developed.

MAINTENANCE REQUIREMENTS FOR PROTEIN

The previous edition and update of the *Nutrient Requirements of Beef Cattle* (NRC, 1996, 2000) expressed protein requirements in terms of metabolizable protein (MP) rather than crude protein (CP). The adoption of an MP or absorbable protein system for ruminants in various forms (INRA, 1988; NRC, 1989, 1996, 2000, 2001, 2007b; AFRC, 1993; CSIRO, 2007) has been a steady transition since the publication of the *Nitrogen Usage in Ruminants* (NRC, 1985b) and has been recently reviewed (see Owens et al., 2014; Waterman et al., 2014). In 1985, the Subcommittee on Nitrogen Usage in Ruminants (NRC, 1985b) presented a rationale for expressing protein requirements in terms of absorbed protein. There are basically two reasons for using the MP system rather than the CP system. The first is that there is now extensive information about the two components of the MP system, bacterial (microbial) crude protein (BCP) synthesis and ruminally undegradable protein (RUP), which allows more accurate prediction of BCP and RUP than was previously possible. The second reason is that the CP system is based on an invalid assumption that all feedstuffs have an equal extent of protein degradation in the rumen, with CP being converted to MP with equal efficiency in all diets. The change from the CP system to the MP system was also adopted in the *Nutrient Requirements of Dairy Cattle* (NRC, 1989) and by the AFRC (1992). For a detailed discussion of protein, readers are referred to Chapter 6 (Protein and Amino Acids).

Metabolizable Protein Requirements for Maintenance

The NRC requirements (1984, 1985b) for CP and MP were based on the factorial methods. Factors included were metabolic fecal losses, urinary losses, scurf losses, growth, fetal growth, and milk. Metabolic fecal, urinary, and scurf losses represent the requirement needed for maintenance. It is difficult, however, to measure fecal and urinary losses independent of each other. It also is difficult to separate microbial (complete cells or cell walls) losses in the feces from true metabolic fecal losses. In a previous edition of this publication (NRC, 1984), metabolic fecal loss was calculated as a percentage of dry matter intake (DMI), whereas the NRC (1985b) calculated metabolic fecal loss as a percentage of indigestible DMI. Diet digestibility obviously affects the resulting calculated metabolic fecal losses. Beef cows are often fed diets containing 50 to 60% total digestible nutrients (TDN) during gestation. Consequently, for most beef cows, MP and CP requirements, using the calculation based on indigestible DMI (NRC, 1985b), are unrealistically high. The high requirement can be attributed to the fact that nitrogen is being excreted in the feces as microbial protein rather than as urea in the urine (NRC, 1985b) as a result of microbial growth in the postruminal digestive tract.

The Institute National de la Recherche Agronomique (INRA, 1988), using nitrogen balance studies that included scurf, urinary, and metabolic fecal losses, determined that the maintenance requirement was 3.25 g MP/kg $SBW^{0.75}$. This system simplifies calculations and is based on metabolic body weight ($BW^{0.75}$), as are the maintenance energy requirements, and is similar to the concept and value proposed by Smuts (1935). Assuming CP × 0.64 (CP converted to BCP: 80% true protein × 80% digestibility) = MP, Smuts (1935) calculated the requirement to be 3.52 g MP/kg $BW^{0.75}$. Wilkerson et al. (1993) estimated the maintenance requirement of 253-kg growing calves was 3.8 g MP/kg $BW^{0.75}$ using growth as the basis for the estimate. Their diets were high in roughage and were based on the assumption that 0.13 TDN = BCP. If actual BCP synthesis efficiency was less than 0.13, the estimate of the maintenance would be less than 3.8 g MP/kg $BW^{0.75}$. In NRC (1996, 2000), 3.8 g MP/kg $SBW^{0.75}$ was used because the maintenance requirement estimated was based on animal growth rather than on nitrogen balance. Nonetheless, nitrogen balance data reported by Susmel et al. (1993) support the 3.8-g MP/kg $SBW^{0.75}$ value, as do recent assessments by Owens et al. (2014).

The current committee has opted to retain the previous method of calculating MP for maintenance (MPm):

$$MPm = 3.8 \times SBW^{0.75} \qquad \text{(Eq. 11-17)}$$

where MPm = MP in g/d and SBW = body weight in kg. It should be noted that the BW data used to derive the above equation was based on the average of three consecutive days of BW measurements taken before morning feeding. It is likely that these methodologies produce BW that are more akin to a partial SBW than an actual full animal BW; however, the original manuscript of Wilkerson et al. (1993) referred to the animal weights as BW and not SBW. The NRC (1996, 2000) used SBW in their equation to predict MPm as noted in the NRC 1996/2000 equations chapter. It is interesting to note that in the recent assessment of protein requirements for maintenance (Owens et al., 2014) the above equation underpredicted actual protein losses, being off by approximately 4%. Thus, the Owens et al. (2014) analysis provides confidence that the current expression of MP requirements for maintenance approximates the MP maintenance requirement.

REFERENCES

Abdalla, H. O., D. G. Fox, and M. L. Thonney. 1988. Compensatory gain by Holstein calves after underfeeding protein. *Journal of Animal Science* 66:2687-2695.

AFRC (Agricultural and Food Research Council). 1992. Nutritive requirements of ruminant animals: Protein. *Nutrition Abstracts and Reviews Series B* 62:787-835.

AFRC. 1993. *Energy and Protein Requirements of Ruminants*. Wallingford, UK: CAB International.

Andersen, B. B. 1980. Feeding trials describing net requirements for maintenance as dependent on weight, feeding level, sex and genotype. *Annales de Zootechnie* 29:85-92.

ARC (Agricultural Research Council). 1965. *The Nutrient Requirements of Farm Livestock. No. 2. Ruminants*. London, UK: Agricultural Research Council.

ARC. 1980. *The Nutrient Requirements of Ruminant Livestock: Technical Review*. Farnham Royal, UK: Commonwealth Agricultural Bureaux.

Armsby, H. P., and J. A. Fries. 1911. The Influence of Type and of Age upon the Utilization of Feed by Cattle. Bulletin No. 128. Washington, DC: U.S. Department of Agriculture, Bureau of Animal Industry.

Baker, F. S., Jr., A. Z. Palmer, and J. W. Carpenter. 1973. Brahman × European crosses vs. British breeds. Pp. 227-284 in *Crossbreeding Beef Cattle, Series 2*, M. Koger, T. J. Cunha, and P. C. Warnick, eds. Gainesville: University of Florida Press.

Baldwin, R. L., N. E. Smith, J. Taylor, and M. Sharp. 1980. Manipulating metabolic parameters to improve growth rate and milk secretion. *Journal of Animal Science* 51:1416-1428.

Baumgard, L. H., and R. P. Rhoads. 2013. Effects of heat stress on postabsorptive metabolism and energetics. *Annual Review of Animal Biosciences* 1:311-337.

Berman, A. 2004. Tissue and external insulation estimates and their effects on prediction of energy requirements and of heat stress. *Journal of Dairy Science* 87:1400-1412.

Birkelo, C. P., D. E. Johnson, and H. W. Phetteplace. 1989. Plane of nutrition and season effects on energy maintenance requirement of beef cattle.

Pp. 263-266 in *Energy Metabolism of Farm Animals: Proceedings of the 11th Symposium, September 18-24, 1988, Lunteren, Netherlands*, Y. van der Honing, ed. EAAP Publication No. 43. Wageningen, The Netherlands: Pudoc.

Birkelo, C. P., D. E. Johnson, and H. W. Phetteplace. 1991. Maintenance requirements of beef cattle as affected by season on different planes of nutrition. *Journal of Animal Science* 69:1214-1222.

Blaxter, K. L. 1962. *The Energy Metabolism of Ruminants*. Springfield, IL: Charles C Thomas.

Blaxter, K. L., and A. W. Boyne. 1982. Fasting and maintenance metabolism of sheep. *Journal of Agricultural Science* 99:611-620.

Blaxter, K. L., J. L. Clapperton, and F. W. Wainman. 1966. Utilization of the energy and protein of the same diet by cattle of different ages. *Journal of Agricultural Science* 67:67-75.

Brody, S. 1945. *Bioenergetics and Growth*. New York: Hafner.

Brosh, A., Y. Aharoni, A. A. Degen, D. Wright, and B. A. Young. 1998. Effects of solar radiation, dietary energy, and time of feeding on thermoregulatory responses and energy balance in cattle in a hot environment. *Journal of Animal Science* 76:2671-2677.

Bull, L. S., H. F. Tyrrell, and J. T. Reid. 1976. Energy utilization by growing male and female sheep and rats, by comparative slaughter and respiration techniques. Pp. 137-140 in *Energy Metabolism of Farm Animals: Proceedings of the 7th Symposium, September 1976, Vichy, France*. EAAP Publication No. 19. Newcastle upon Tyne, England: Oriel Press.

Burrin, D. G., C. L. Ferrell, and R. A. Britton. 1989. Effect of feed intake of lambs on visceral organ growth and metabolism. Pp. 103-106 in *Energy Metabolism of Farm Animals: Proceedings of the 11th Symposium, September 18-24, 1988, Lunteren, Netherlands*, Y. van der Honing, ed. EAAP Publication No. 43. Wageningen, The Netherlands: Pudoc.

Burton, J. H., M. Anderson, and J. T. Reid. 1974. Some biological aspects of partial starvation: The effect of weight loss and regrowth on body composition in sheep. *British Journal of Nutrition* 32:515-527.

Buskirk, D. D., R. P. Lemenager, and L. A. Horstman. 1992. Estimation of net energy requirements (NEm and NE change) of lactating beef cows. *Journal of Animal Science* 70:3867-3876.

Butler-Hogg, B. W. 1984. Growth patterns in sheep: Changes in the chemical composition of the empty body and its constituent parts during weight loss and compensatory growth. *Journal of Agricultural Science* 103:17-24.

Byers, F. M. 1982. Patterns of energetic efficiency of tissue growth in beef cattle of four breeds. Pp. 92-95 in *Energy Metabolism of Farm Animals: Proceedings of the 9th Symposium, September 1982, Lillehammer, Norway*, F. Sundstol and A. Ekern, eds. EAAP Publication No. 29. Aas, Norway: Department of Animal Nutrition, Agricultural University of Norway.

Byers, F. M., and G. E. Carstens. 1991. Seasonality of maintenance requirements in beef cows. Pp. 450-453 in *Energy Metabolism of Farm Animals: Proceedings of the 12th Symposium, September 1-7, 1991, Kartause Ittingen, Switzerland*, C. Wenk and M. Boessinger, eds. EAAP Publication No. 58. London: Butterworths.

Byers, F. M., G. T. Schelling, and R. D. Goodrich. 1987. Maintenance requirements of beef cows with respect to genotype and environment. Pp. 230-233 in *Energy Metabolism of Farm Animals: Proceedings of the 10th Symposium, September 1985, Airlie*, VA, P. W. Moe, H. F. Tyrrell, and P. J. Reynolds, eds. EAAP Publication No. 32. New York: Rowman & Littlefield.

Campbell, G. S., and J. M. Norman. 1998. *An Introduction to Environmental Biophysics*, 2nd Ed. New York: Springer.

Cañas, R., J. J. Romero, and R. L. Baldwin. 1982. Maintenance energy requirements during lactation in rats. *Journal of Nutrition* 112:1876-1880.

Carrol, F. D., J. E. Ellsworth, and D. Kroger. 1963. Compensatory carcass growth in steers following protein and energy restriction. *Journal of Animal Science* 22:197-201.

Carstens, G. E., D. E. Johnson, and M. A. Ellenberger. 1989a. Energy metabolism and composition of gain in beef steers exhibiting normal and compensatory growth. Pp. 131-134 in *Energy Metabolism of Farm Animals: Proceedings of the 11th Symposium, September 18-24, 1988, Lunteren, Netherlands*, Y. van der Honing, ed. EAAP Publication No. 43. Wageningen, The Netherlands: Pudoc.

Carstens, G.E., D. E. Johnson, K. A. Johnson, S. K. Hotovy, and T. J. Szymanski. 1989b. Genetic variation in energy expenditures of monozygous twin beef cattle at 9 and 20 months of age. Pp. 312-315 in *Energy Metabolism of Farm Animals: Proceedings of the 11th Symposium, September 18-24, 1988, Lunteren, Netherlands*, Y. van der Honing, ed. EAAP Publication No. 43. Wageningen, The Netherlands: Pudoc.

Carstens, G. E., D. E. Johnson, M. A. Ellenberger, and J. D. Tatam. 1991. Physical and chemical components of the empty body during compensatory growth in beef steers. *Journal of Animal Science* 69:3251-3264.

Chestnutt, D. M. B., R. Marsh, J. G. Wilson, T. A. Stewart, T. A. McCullough, and T. McCallion. 1975. Effects of breed of cattle on energy requirements for growth. *Animal Production* 21:109-119.

Chizzotti, M. L., S. C. Valadares Filho, L. O. Tedeschi, F. H. M. Chizzotti, and G. E. Carstens. 2007. Energy and protein requirements for growth and maintenance of F1 Nellore × Red Angus bulls, steers, and heifers. *Journal of Animal Science* 85:1971-1981.

Chizzotti, M. L., L. O. Tedeschi, and S. C. Valadares Filho. 2008. A meta-analysis of energy and protein requirements for maintenance and growth of Nellore cattle. *Journal of Animal Science* 86:1588-1597.

Christopherson, R. J., R. J. Hudson, and M. K. Christopherson. 1979. Seasonal energy expenditures and thermoregulatory response of bison and cattle. *Canadian Journal of Animal Science* 59:611-617.

Connor, E. E., S. Kahl, T. H. Elsasser, J. S. Parker, R. W. Li, C. P. Van Tassell, R. L. Baldwin VI, and S. M. Barao. 2010. Enhanced mitochondrial complex gene function and reduced liver size may mediate improved feed efficiency of beef cattle during compensatory growth. *Functional and Integrative Genomics* 10:39-51.

Corbett, J. L., E. P. Furnival, and R. S. Pickering. 1982. Energy expenditure at pasture of shorn and unshorn border Leichester ewes during late pregnancy and lactation. Pp. 34-37 in *Energy Metabolism of Farm Animals: Proceedings of the 9th Symposium, September 1982, Lilehammer, Norway*, F. Sundstol and A. Ekern, eds. EAAP Publication No. 29. Aas, Norway: Department of Animal Nutrition, Agricultural University of Norway.

Corbett, J. L., M. Freer, and N. M. Graham. 1987. A generalized equation to predict the varying maintenance metabolism of sheep and cattle. Pp. 62-67 in *Energy Metabolism of Farm Animals: Proceedings of the 10th Symposium, September 1985, Airlie, VA*, P. W. Moe, H. F. Tyrrell, and P. J. Reynolds, eds. EAAP Publication No. 32. New York: Rowman & Littlefield.

Crabtree, R. M., M. Kay, and A. J. F. Webster. 1976. The net availabilities of ME for body gain of two pelleted diets offered to Hereford × Friesian castrate males over different live-weight ranges. *Animal Production* 22:156-157.

CSIRO (Commonwealth Scientific and Industrial Research Organisation). 1990. *Feeding Standards for Australian Livestock: Ruminants*. Melbourne, Australia: CSIRO Publishing.

CSIRO. 2007. *Nutrient Requirements of Domesticated Ruminants*. Collingwood, Australia: CSIRO Publishing.

Drouillard, J. S., C. L. Ferrell, T. J. Klopfenstein, and R. A. Britton. 1991. Compensatory growth following metabolizable protein or energy restrictions in beef steers. *Journal of Animal Science* 69:811-818.

Ehrlemark, A. 1991. Heat and Moisture Dissipation from Cattle: Measurements and Simulation Model. Ph.D. Dissertation. Swedish University of Agricultural Sciences, Uppsala, Sweden.

Ferrell, C. L. 1988. Contribution of visceral organs to animal energy expenditures. *Journal of Animal Science* 66(Suppl. 3):23-34.

Ferrell, C. L., and T. G. Jenkins. 1984a. A note on energy requirements for maintenance of lean and fat Angus, Hereford and Simmental cows. *Animal Production* 35:305-309.

Ferrell, C. L., and T. G. Jenkins. 1984b. Energy utilization by mature, nonpregnant, nonlactating cows of different breeds. *Journal of Animal Science* 58:234-243.

Ferrell, C. L., and T. G. Jenkins. 1984c. Relationships among various body components of mature cows. *Journal of Animal Science* 58:222-233.

Ferrell, C. L., and T. G. Jenkins. 1985a. Cow type and the nutritional environment: Nutritional aspects. *Journal of Animal Science* 61:725-741.

Ferrell, C. L., and T. G. Jenkins. 1985b. Energy utilization by Hereford and Simmental males and females. *Animal Production* 41:53-61.

Ferrell, C. L., and T. G. Jenkins. 1987. Influence of biological type on energy requirements. Pp. 1-7 in *Proceedings of the Grazing Livestock Nutrition Conference*. Stillwater: Oklahoma Agricultural Experiment Station, Oklahoma State University.

Ferrell, C. L., and T. G. Jenkins. 1998. Body composition and energy utilization by steers of diverse genotypes fed a high-concentrate diet during the finishing period: II. Angus, Boran, Brahman, Hereford, and Tuli sires. *Journal of Animal Science* 76:647-657.

Ferrell, C. L., and L. J. Koong. 1987. Response of body organs of lambs to differing nutritional treatments. Pp. 26-29 in *Energy Metabolism of Farm Animals: Proceedings of the 10th Symposium, September 1985, Airlie, VA*, P. W. Moe, H. F. Tyrrell, and P. J. Reynolds, eds. EAAP Publication No. 32. New York: Rowman & Littlefield.

Ferrell, C. L., and L. P. Reynolds. 1987. Oxidative metabolism of gravid uterine tissues of the cow. Pp. 298-301 in *Energy Metabolism of Farm Animals: Proceedings of the 10th Symposium, September 1985, Airlie, VA*, P. W. Moe, H. F. Tyrrell, and P. J. Reynolds, eds. EAAP Publication No. 32. New York: Rowman & Littlefield.

Ferrell, C. L., W. N. Garrett, N. Hinman, and G. Gritchting. 1976. Energy utilization by pregnant and nonpregnant heifers. *Journal of Animal Science* 42:937-950.

Ferrell, C. L., J. D. Crouse, R. A. Field, and J. L. Chant. 1979. Effects of sex, diet and stage of growth upon energy utilization by lambs. *Journal of Animal Science* 49:790-801.

Ferrell, C. L., L. J. Koong, and J. A. Nienaber. 1986. Effect of previous nutrition on body composition and maintenance energy costs of growing lambs. *British Journal of Nutrition* 56:595-605.

Flatt, W. P., L. A. Moore, N. W. Hooven, and R. D. Plowman. 1965. Energy metabolism studies with a high-producing lactating dairy cow. *Journal of Dairy Science* 48:797-798.

Flatt, W. P., P. W. Moe, A. W. Munson, and T. Cooper. 1969. Energy utilization by high-producing dairy cows. II. Summary of energy balance experiments with lactating Holstein cows. Pp. 235-239 in *Energy Metabolism of Farm Animals: Proceedings of the 4th Symposium, September 1967, Warsaw, Poland*, K. I. Blaxter, J. Kielanowski, and G. Thorbeck, eds. EAAP Publication No. 12. Newcastle upon Tyne, England: Oriel Press.

Foot, J. Z., and N. M. Tulloh. 1977. Effects of two paths of liveweight change on efficiency of feed use and on body composition of Angus steers. *Journal of Agricultural Science* 88:135-142.

Fox, D. G., R. R. Johnson, R. L. Preston, T. R. Dockerty, and E.W. Klosterman. 1972. Protein and energy utilization during compensatory growth in beef cattle. *Journal of Animal Science* 34:310-318.

Fox, D. G., C. J. Sniffen, and J. D. O'Connor. 1988. Adjusting nutrient requirements of beef cattle for animal and environmental variations. *Journal of Animal Science* 66:1475-1495.

Freetly, H. C., J. Nienaber, K. A. Leymaster, and T. G. Jenkins. 1995. Relationships among heat production, body weight, and age in Suffolk and Texel ewes. *Journal of Animal Science* 73:1030-1037.

Frisch, J. E. 1973. Comparative drought resistance of *Bos indicus* and *Bos taurus* crossbred herds in Central Queensland. 2. Relative mortality rates, calf birth weights and weights and weight changes of breeding cows. *Australian Journal of Experimental Agriculture and Animal Husbandry* 13:117-126.

Frisch, J. E., and J. E. Vercoe. 1976. Maintenance requirements, fasting metabolism and body composition in different cattle breeds. Pp. 209-212 in *Energy Metabolism of Farm Animals: Proceedings of the 7th Symposium, September 1976, Vichy, France*. EAAP Publication No. 19. Newcastle upon Tyne, England: Oriel Press.

Frisch, J. E., and J. E. Vercoe. 1977. Feed intake, eating rate, weight gains, metabolic rate and efficiency of feed utilization in *Bos taurus* and *Bos indicus* crossbred cattle. *Animal Production* 25:343-358.

Frisch, J. E., and J. E. Vercoe. 1982. The effect of previous exposure to parasites on the fasting metabolism and feed intake of three cattle breeds. Pp. 100-103 in *Energy Metabolism of Farm Animals: Proceedings of the 9th Symposium, September 1982, Lillehammer, Norway*, F. Sundstol and A. Ekern, eds. EAAP Publication No. 29. Aas, Norway: Department of Animal Nutrition, Agricultural University of Norway.

Garrett, W. N. 1970. The influence of sex on the energy requirements of cattle for maintenance and growth. Pp. 101-104 in *Energy Metabolism of Farm Animals: Proceedings of 5th Symposium, September 1970, Vitznau, Switzerland*. EAAP Publication No. 13. Zurich: Juris Verlag.

Garrett, W. N. 1971. Energetic efficiency of beef and dairy steers. *Journal of Animal Science* 32:451-456.

Garrett, W. N. 1980. Energy utilization by growing cattle as determined in 72 comparative slaughter experiments. Pp. 3-8 in *Energy Metabolism: Proceedings of the 8th Symposium, September 1979, Cambridge, England*, L. E. Mount, ed. EAAP Publication No. 26. London: Butterworths.

Geay, Y., J. Robelin, and M. Vermorel. 1980. Influence of the metabolizable energy content of the diet on energy utilization for growth in bulls and heifers. Pp. 9-12 in *Energy Metabolism: Proceedings of the 8th Symposium, September 1979, Cambridge, England*, L. E. Mount, ed. EAAP Publication No. 26. London: Butterworths.

Graham, N. M. 1968. Effects of undernutrition in late pregnancy on the nitrogen and energy metabolism of ewes. *Australian Journal of Agricultural Research* 19:555-565.

Graham, N. M. 1980. Variation in energy and nitrogen utilization by sheep between weaning and maturity. *Australian Journal of Agricultural Research* 31:335-345.

Graham, N. M., and T. W. Searle. 1972a. Balance of energy and matter in growing sheep at several ages, body weights and planes of nutrition. *Australian Journal of Agricultural Research* 23:97-108.

Graham, N. M., and T. W. Searle. 1972b. Growth in sheep. II. Efficiency of energy and nitrogen utilization from birth to 2 years. *Journal of Agricultural Science* 79:383-389.

Graham, N. M., and T. W. Searle. 1975. Studies on weaner sheep during and after a period of weight stasis. I. Energy and nitrogen utilization. *Australian Journal of Agricultural Research* 26:343-353.

Graham, N. M., and T. W. Searle. 1979. Studies of weaned lambs before, during and after weight loss. I. Energy and nitrogen utilization. *Australian Journal of Agricultural Research* 30:513-523.

Graham, N. M., T. W. Searle, and D. A. Griffiths. 1974. Basal metabolic rate in lambs and young sheep. *Australian Journal of Agricultural Research* 25:957-971.

Gray, R., and K. J. McCracken. 1980. Plane of nutrition and the maintenance requirement. Pp. 163-167 in *Energy Metabolism: Proceedings of the 8th Symposium, September 1979, Cambridge, England*, L. E. Mount, ed. EAAP Publication No. 26. London: Butterworths.

Harris, P. M., P. J. Garlick, and G. E. Lobley. 1989. Interactions between energy and protein metabolism in the whole body and hind limb of sheep in response to intake. Pp. 167-170 in *Energy Metabolism of Farm Animals: Proceedings of the 11th Symposium, September 18-24, 1988, Lunteren, Netherlands*, Y. van der Honing, ed. EAAP Publication No. 43. Wageningen, The Netherlands: Pudoc.

Hironaka, R., and G. C. Kozub. 1973. Compensatory growth of beef cattle restricted at two energy levels for two periods. *Canadian Journal of Animal Science* 53:709-715.

Hovell, F. D. DeB., E. R. Ørskov, D. J. Kyle, and N. A. MacLeod. 1987. Undernutrition in sheep: Nitrogen repletion by N-depleted sheep. *British Journal of Nutrition* 57:77-88.

Hulme, D. J., R. C. Kellaway, and P. J. Booth. 1986. The CAMDAIRY model for formulating and analyzing dairy cow rations. *Agricultural Systems* 22:81-108.

INRA (Institut National de la Recherche Agronomique). 1988. *Alimentation des Bovins, Ovins, et Caprins*, R. Jarrige, ed. Paris: Institut National de la Recherche Agronomique.

Jenkins, T. G., and C. L. Ferrell. 1983. Nutrient requirements to maintain weight of mature, nonlactating, nonpregnant cows of four diverse breed types. *Journal of Animal Science* 56:761-770.

Jenkins, T. G., and C. L. Ferrell. 1984. Characterization of postweaning traits of Simmental and Hereford bulls and heifers. *Animal Production* 39:355-364.

Jenkins, T. G., C. L. Ferrell, and L. V. Cundiff. 1986. Relationship of components of the body among mature cows as related to size, lactation potential and possible effects on productivity. *Animal Production* 43:245-254.

Jenkins, T. G., J. A. Nienaber, and C. L. Ferrell. 1991. Heat production of mature Hereford and Simmental cows. Pp. 296-299 in *Energy Metabolism of Farm Animals: Proceedings of the 12th Symposium, September 1-7, 1991, Kartause Ittingen, Switzerland*, C. Wenk and M. Boessinger, eds. EAAP Publication No. 58. London: Butterworths.

Johnson, D. E., K. A. Johnson, and R. L. Baldwin. 1990. Changes in liver and gastrointestinal tract energy demands in response to physiological workload in ruminants. *Journal of Nutrition* 120:649-655.

Kennedy, J. F., and G. I. K. Chirchir. 1971. A study of the growth rates of F1 and F2 Africander cross, Brahman cross and British cross cattle from birth to 18 months old in a tropical environment. *Australian Journal of Experimental Agriculture and Animal Husbandry* 11:593-598.

Keren, E. N., and B. E. Olson. 2006a. Thermal balance of cattle grazing winter range: Model application. *Journal of Animal Science* 84:1238-1247.

Keren, E. N., and B. E. Olson. 2006b. Thermal balance of cattle grazing winter range: Model development. *Journal of Thermal Biology* 31:371-377.

Kleiber, M. 1961. *The Fire of Life: An Introduction to Animal Energetics*. New York: John Wiley & Sons.

Klosterman, E. W., L. G. Sanford, and C. F. Parker. 1968. Effect of cow size and condition and ration protein content upon maintenance requirements of mature beef cows. *Journal of Animal Science* 27:242-246.

Koong, L. J., and J. A. Nienaber. 1987. Changes in fasting heat production and organ size of pigs during prolonged weight maintenance. Pp. 46-49 in *Energy Metabolism of Farm Animals: Proceedings of the 10th Symposium, September 1985, Airlie, VA*, P. W. Moe, H. F. Tyrrell, and P. J. Reynolds, eds. EAAP Publication No. 32. New York: Rowman & Littlefield.

Koong, L. J., C. L. Ferrell, and J. A. Nienaber. 1982. Effects of plane of nutrition on organ size and fasting heat production in swine and sheep. Pp. 245-248 in *Energy Metabolism of Farm Animals: Proceedings of the 9th Symposium, September 1982, Lillehammer, Norway*, F. Sundstol and A. Ekern, eds. EAAP Publication No. 29. Aas, Norway: Department of Animal Nutrition, Agricultural University of Norway.

Koong, L. J., C. L. Ferrell, and J. A. Nienaber. 1985. Assessment of interrelationships among level of intake and production, organ size and fasting heat production in growing animals. *Journal of Nutrition* 115:1383-1390.

Lardy, G. P., D. C. Adams, T. J. Klopfenstein, and H. H. Patterson. 2004. Building beef cow nutritional programs with the 1996 NRC beef cattle requirements model. *Journal of Animal Science* 82(Suppl.):E83-E92.

Laurenz, J. C., F. M. Byers, G. T. Schelling, and L. W. Green. 1991. Effects of seasonal environment on the maintenance requirement of mature beef cows. *Journal of Animal Science* 69:2168-2176.

Lawrence, T. L. J., and J. Pierce. 1964. Some effects of wintering yearling cattle on different planes of nutrition. II. Slaughter data and carcass evaluation. *Journal of Agricultural Science* 63:23-34.

Ledger, H. P. 1977. The utilization of dietary energy by steers during periods of restricted feed intake and subsequent realimentation. 2. The comparative energy requirements of penned and exercised steers for long term maintenance at constant liveweight. *Journal of Agricultural Science* 88:27-33.

Ledger, H. P., and A. R. Sayers. 1977. The utilization of dietary energy by steers during periods of restricted feed intake and subsequent realimentation. 1. The effect of time on the maintenance requirements of steers held at constant liveweight. *Journal of Agricultural Science* 88:11-26.

Lemenager, R. P., L. A. Nelson, and K. S. Hendrix. 1980. Influence of cow size and breed type on energy requirements. *Journal of Animal Science* 51:566-576.

Lobley, G. E., P. M. Harris, P. A. Skene, D. Brown, E. Milne, A. G. Calder, S. E. Anderson, P. J. Garlick, I. Nevison, and A. Connell. 1992. Responses in tissue protein synthesis to sub- and supra-maintenance intake in young, growing sheep: Comparison of large-dose and continuous-infusion techniques. *British Journal of Nutrition* 68:373-388.

Lofgreen, G. P. 1965. A comparative slaughter technique for determining net energy values with beef cattle. Pp. 309-317 in *Energy Metabolism: Proceedings of the 3rd Symposium, May 1964, Troon, Scotland*, K. L. Blaxter, ed. EAAP Publication No. 11. London: Academic Press.

Lofgreen, G. P., and W. N. Garrett. 1968. A system for expressing net energy requirements and feed values for growing and finishing cattle. *Journal of Animal Science* 27:793-806.

Lopez-Sanbidet, C., and L. S. Verde. 1976. Relationship between liveweight, age and dry matter intake for beef cattle after different levels of feed restriction. *Animal Production* 22:61-69.

Mader, T. L., O. A. Turgeon, Jr., T. J. Klopfenstein, D. R. Brink, and R. R. Oltjen. 1989. Effects of previous nutrition, feedlot regimen and protein level on feedlot performance of beef cattle. *Journal of Animal Science* 67:318-328.

MAFF (Ministry of Agriculture, Fisheries and Food). 1976. Energy Allowances and Feeding Systems for Ruminants. Technical Bulletin No. 33. London: Her Majesty's Stationery Office.

MAFF. 1984. *Energy Allowances and Feeding Systems for Ruminants*. ADAS Reference Book 433. London: Her Majesty's Stationery Office.

Marston, H. R. 1948. Energy transactions in sheep. I. The basal heat production and heat increment. *Australian Journal of Scientific Research* B1:93-129.

McBride, B. W., and J. M. Kelly. 1990. Energy cost of absorption and metabolism in the ruminant gastrointestinal tract and liver: A review. *Journal of Animal Science* 68:2997-3010.

Meyer, A. M., J. J. Reed, K. A. Vonnahme, S. A. Soto-Navarro, L. P. Reynolds, S. P. Ford, B. W. Hess, and J. S. Caton. 2010. Effects of stage of gestation and nutrient restriction during early to mid-gestation on maternal and fetal visceral organ mass and indices of jejunal growth and vascularity in beef cows. *Journal of Animal Science* 88:2410-2424.

Milligan, L. P., and M. Summers. 1986. The biological basis for maintenance and its relevance to assessing responses to nutrients. *Proceedings of the Nutrition Society* 45:185-193.

Moe, P. W., H. F. Tyrrell, and W. P. Flatt. 1970. Partial efficiency of energy use for maintenance, lactation, body gain and gestation in the dairy cow. Pp. 65-68 in *Energy Metabolism of Farm Animals: Proceedings of the 5th Symposium, September 1970, Vitznau, Switzerland*. EAAP Publication No. 13. Zurich: Juris Verlag.

Montano-Bermudez, M., M. K.Nielsen, and G. Deutscher. 1990. Energy requirements for maintenance of crossbred beef cattle with different genetic potential for milk. *Journal of Animal Science* 68:2279-2288.

Monteiro, A. P. A., S. Tao, I. M. Thompson, and G. E. Dahl. 2014. Effect of heat stress during late gestation on immune function and growth performance of calves: Isolation of altered colostral and calf factors. *Journal of Dairy Science* 97:6426-6439.

Moran, J. B. 1976. The grazing feed intake of Hereford and Brahman cross cattle in a cool temperate environment. *Journal of Agricultural Science* 86:131-134.

Neville, W. E., Jr. 1974. Comparison of energy requirements of nonlactating and lactating Hereford cows and estimates of energetic efficiency of milk production. *Journal of Animal Science* 38:681-686.

Neville, W. E., Jr., and M. E. McCullough. 1969. Calculated net energy requirements of lactating and nonlactating Hereford cows. *Journal of Animal Science* 29:823-829.

NRC (National Research Council). 1976. *Nutrient Requirements of Beef Cattle*. Washington, DC: National Academy of Sciences.

NRC. 1981. *Effect of Environment on Nutrient Requirements of Domestic Animals.* Washington, DC: National Academy Press.

NRC. 1984. *Nutrient Requirements of Beef Cattle*, 6th Rev. Ed. Washington, DC: National Academy Press.

NRC. 1985a. *Nutrient Requirements of Sheep*, 6th Rev. Ed. Washington, DC: National Academy Press.

NRC. 1985b. *Ruminant Nitrogen Usage.* Washington, DC: National Academy Press.

NRC. 1989. *Nutrient Requirements of Dairy Cattle*, 6th Rev. Ed. Washington, DC: National Academy Press.

NRC. 1996. *Nutrient Requirements of Beef Cattle*, 7th Rev. Ed. Washington, DC: National Academy Press.

NRC. 2000. *Nutrient Requirements of Beef Cattle*, 7th Rev. Ed.: Update 2000. Washington, DC: National Academy Press.

NRC. 2001. *Nutrient Requirements of Dairy Cattle*, 7th Rev. Ed. Washington, DC: National Academy Press.

NRC. 2007a. *Nutrient Requirements of Horses*, 6th Rev. Ed. Washington, DC: The National Academies Press.

NRC. 2007b. *Nutrient Requirements of Small Ruminants.* Washington, DC: The National Academies Press.

O'Donovan, P. B. 1984. Compensatory gain in cattle and sheep. *Nutrition Abstracts and Reviews* 54:389-410.

O'Donovan, P. B., A. Gebrewolde, B. Kebede, and E. S. E. Galal. 1978. Fattening studies with crossbred (European × Zebu) bulls. 1. Performance on diets of native hay and concentrate. *Journal of Agricultural Science* 90:425-429.

Old, C. A., and W. N. Garrett. 1987. Effects of energy intake on energetic efficiency and body composition of beef steers differing in size at maturity. *Journal of Animal Science* 65:1371-1380.

Owens, F. N., S. Qi, and D. A. Sapienza. 2014. Invited review: Applied protein nutrition of ruminants: Current status and future directions. *The Professional Animal Scientist* 30:150-179.

Patle, B. R., and V. D. Mudgal. 1975. Maintenance requirements for energy in crossbred cattle. *British Journal of Nutrition* 33:127-139.

Patle, B. R., and V. D. Mudgal. 1977. Utilization of dietary energy requirements for maintenance, milk production and lipogenesis by lactating crossbred cows during their midstage of lactation. *British Journal of Nutrition* 37:23-33.

Peacock, F. M., M. Koger, W. G. Kirk, E. M. Hodges, and J. R. Crockett. 1976. Beef Production of Brahman, Shorthorn, and Their Crosses on Different Pasture Programs. Technical Bulletin No. 780. Gainesville: University of Florida Agricultural Experiment Station.

Reynolds, C. K., and H. F. Tyrrell. 2000. Energy metabolism in lactating beef heifers. *Journal of Animal Science* 78:2696-2705.

Rhoads, M. L., J. W. Kim, R. J. Collier, B. A. Crooker, Y. R. Boisclair, L. H. Baumgard, and R. P. Rhoads. 2010. Effects of heat stress and nutrition on lactating Holstein cows: II. Aspects of hepatic growth hormone responsiveness. *Journal of Dairy Science* 93:170-179.

Ritzman, E. G., and F. G. Benedict. 1938. *Nutritional Physiology of the Adult Ruminant.* Washington, DC: Carnegie Institute.

Robelin, J., and Y. Geay. 1976. Changes with age (9, 13, 16, 19 months) of protein and energy retention, and energy utilization by growing Limousin bulls. Pp. 213-216 *in Energy Metabolism of Farm Animals: Proceedings of the 7th Symposium, September 1976, Vichy, France.* EAAP Publication No. 19. Newcastle upon Tyne, England: Oriel Press.

Rogerson, A., H. P. Ledger, and G. H. Freeman. 1968. Feed intake and liveweight gain comparisons of *Bos indicus* and *Bos taurus* steers on a high plane of nutrition. *Animal Production* 10:373-380.

Rompala, R. E., S. D. M. Jones, J. G. Buchanan-Smith, and H. S. Bayley. 1985. Feedlot performance and composition of gain in late maturing steers exhibiting normal and compensatory growth. *Journal of Animal Science* 61:637-646.

Russel, A. J. F., and I. A. Wright. 1983. Factors affecting maintenance requirements of beef cows. *Animal Production* 37:329-334.

Ryan, W. J., I. H. Williams, and R. J. Moir. 1993a. Compensatory growth in sheep and cattle. I. Growth pattern and feed intake. *Australian Journal of Agricultural Research* 44:1609-1621.

Ryan, W. J., I. H. Williams, and R. J. Moir. 1993b. Compensatory growth in sheep and cattle. II. Changes in body composition and tissue weights. *Australian Journal of Agricultural Research* 44:1623-1633.

Scheaffer, A. N., J. S. Caton, M. L. Bauer, and L. P. Reynolds. 2003. The impact of pregnancy on visceral growth and energy use in beef heifers. *Journal of Animal Science* 81:1853-1861.

Schnyder, W., H. Bickel, and A. Schürch. 1982. Energy metabolism during retarded and compensatory growth of Braunvieh steers. Pp. 96-99 in *Energy Metabolism of Farm Animals: Proceedings of the 9th Symposium, September 1982, Lillehammer, Norway*, F. Sundstol and A. Ekern, eds. EAAP Publication No. 29. Aas, Norway: Department of Animal Nutrition, Agricultural University of Norway.

Searle, T. W., and N. M. Graham. 1975. Studies of weaner sheep during and after a period of weight stasis. II. Body composition. *Australian Journal of Agricultural Research* 26:355-361.

Silva, R. G. 2000. A heat balance model for cattle in tropical environments. *Revista Brasileira de Zootecnia* 29:1244-1252.

Smith, G. M., J. D. Crouse, R. W. Mandigo, and K. L. Neer. 1977. Influence of feeding regime and biological type on growth, composition and palatability of steers. *Journal of Animal Science* 45:236-253.

Smuts, D. 1935. The relation between the basal metabolism and the endogenous nitrogen metabolism, with particular reference to the maintenance requirement of protein. *Journal of Nutrition* 9:403-433.

Solis, J. C., F. M. Byers, G. T. Schelling, C. R. Long, and L. W. Green. 1988. Maintenance requirements and energetic efficiency of cows of different breeds. *Journal of Animal Science* 66:764-773.

Stetter, R., A. Susenbeth, and K. H. Menke. 1989. Energy metabolism and protein retention of Simmental bulls fed four different feeding levels from maintenance to ad libitum at 250, 450 and 550 kg body mass. Pp. 21-24 in *Energy Metabolism of Farm Animals: Proceedings of the 11th Symposium, September 18-24, 1988, Lunteren, Netherlands*, Y. van der Honing, ed. EAAP Publication No. 43. Wageningen, The Netherlands: Pudoc.

Summers, M., B. W. McBride, and L. P. Milligan. 1988. Components of basal energy expenditure. Pp. 257-286 in *Aspects of Digestive Physiology in Ruminants*, A. Dobson and M. J. Dobson, eds. Ithaca, NY: Cornell University Press.

Susmel, P., M. Spanghero, B. Stefano, C. R. Mills, and E. Plazzotta. 1993. Digestibility and allantoin excretion in cows fed diets differing in nitrogen content. *Livestock Production Science* 36:213-222.

Tao, S., and G. E. Dahl. 2013. Invited review: Heat stress effects during late gestation on dry cows and their calves. *Journal of Dairy Science* 96:4079-4093.

Taylor, C. S., and G. B. Young. 1968. Equilibrium weight in relation to feed intake and genotype in twin cattle. *Animal Production* 10:393-412.

Taylor, C. S., H. G. Turner, and G. B. Young. 1981. Genetic control of equilibrium maintenance efficiency in cattle. *Animal Production* 33:179-194.

Taylor, C. S., R. B. Theissen, and J. Murray. 1986. Inter-breed relationship of maintenance efficiency to milk yield in cattle. *Animal Production* 43:37-61.

Tedeschi, L. O., C. Boin, D. G. Fox, P. R. Leme, G. F. Alleoni, and D. P. Lanna. 2002. Energy requirement for maintenance and growth of Nellore bulls and steers fed high-forage diets. *Journal of Animal Science* 80:1671-1682.

Thompson, W. R., D. H. Theuninck, J. C. Meiske, R. D. Goodrich, J. R. Rust, and F. M. Byers. 1983. Linear measurements and visual appraisal as estimators of percentage empty body fat of beef cows. *Journal of Animal Science* 56:755-760.

Thomson, E. F., M. Gingins, J. W. Blum, H. Bickel, and A. Schürch. 1980. Energy metabolism of sheep during nutritional limitation and realimentation. Pp. 427-430 in *Energy Metabolism: Proceedings of the 8th Symposium, September 1979, Cambridge, England*, L. E. Mount, ed. EAAP Publication No. 26. London: Butterworths.

Thorbek, G., and S. Henckel. 1976. Studies on energy requirements for maintenance in farm animals. Pp. 117-120 in *Energy Metabolism of Farm Animals: Proceedings of the 7th Symposium, September 1976, Vichy, France*. EAAP Publication No. 19. Newcastle upon Tyne, England: Oriel Press.

Thornton, K. J., C. M. Welch, L. C. Davis, M. E. Doumit, R. A. Hill, and G. K. Murdoch. 2012. Bovine sire selection based on maintenance energy affects muscle fiber type and meat color of F1 progeny. *Journal of Animal Science* 90:1617-1627.

Truscott, T. G., J. D. Wood, N. G. Gregory, and I. C. Hart. 1983. Fat deposition in Hereford and Friesian steers. 3. Growth efficiency and fat mobilization. *Journal of Agricultural Science* 100:277-284.

Tudor, G. D., D. W. Utling, and P. K. O'Rourke. 1980. The effect of pre- and post-natal nutrition on the growth of beef cattle. III. The effect of severe restriction in early postnatal life on the development of body components and chemical composition. *Australian Journal of Agricultural Research* 31:191-204.

Tyrrell, H. F., and C. K. Reynolds. 1989. Effect of stage of growth on utilization of energy by beef heifers. Pp. 17-20 in *Energy Metabolism of Farm Animals: Proceedings of the 11th Symposium, September 18-24, 1988, Lunteren, Netherlands*, Y. van der Honing, ed. EAAP Publication No. 43. Wageningen, The Netherlands: Pudoc.

van der Merwe, F. J., and P. van Rooyen. 1980. Estimates of ME needs for maintenance and gain in beef steers of four genotypes. Pp. 135-140 in *Energy Metabolism: Proceedings of the 8th Symposium, September 1979, Cambridge, England*, L. E. Mount, ed. EAAP Publication No. 26. London: Butterworths.

Vercoe, J. E. 1970. Fasting metabolism and heat increment of feeding in Brahman × British and British cross cattle. Pp. 85-88 in *Energy Metabolism of Farm Animals: Proceedings of the 5th Symposium, September 1970, Vitznau, Switzerland*. EAAP Publication No. 13. Zurich: Juris Verlag.

Vercoe, J. E., and J. E. Frisch. 1974. Fasting metabolism, liveweight and voluntary intake of different breeds of cattle. Pp. 131-134 in *Energy Metabolism of Farm Animals: Proceedings of the 6th Symposium, September 1973, Hohenheim, BDR*, K. H. Menke, H. J. Lantzsch, and J. R. Reichl, eds. EAAP Publication No. 14. Universitat Hohenheim Dokumentationsstelle.

Vermorel, M., J. C. Bouvier, and Y. Geay. 1976. The effect of genotype (normal and double muscled Charolais and Friesian) on energy utilization at 2 and 16 months of age. Pp. 217-220 in *Energy Metabolism of Farm Animals: Proceedings of the 7th Symposium, September 1976, Vichy, France*. EAAP Publication No. 19. Newcastle upon Tyne, England: Oriel Press.

Vermorel, M., C. Bouvier, and Y. Geay. 1980. Energy utilization by growing calves: Effects of age, milk intake and feed level. Pp. 49-53 in *Energy Metabolism: Proceedings of the 8th Symposium, September 1979, Cambridge, England*, L. E. Mount, ed. EAAP Publication No. 26. London: Butterworths.

Vermorel, M., Y. Geay, and J. Robelin. 1982. Energy utilization by growing bulls, variations with genotype, liveweight, feeding level and between animals. Pp. 88-91 in *Energy Metabolism of Farm Animals: Proceedings of the 9th Symposium, September 1982, Lillehammer, Norway*, F. Sundstol and A. Ekern, eds. EAAP Publication No. 29. Aas, Norway: Department of Animal Nutrition, Agricultural University of Norway.

Walker, J. J., and W. N. Garrett. 1970. Shifts in energy metabolism of male rats during adaptation to prolonged undernutrition and during their subsequent realimentation. Pp. 193-196 in *Energy Metabolism of Farm Animals: Proceedings of 5th Symposium, September 1970, Vitznau, Switzerland*. EAAP Publication No. 13. Zurich: Juris Verlag.

Walker, V. A., B. A. Young, and B. Walker. 1991. Does seasonal photoperiod directly influence energy metabolism. Pp. 372-375 in *Energy Metabolism of Farm Animals: Proceedings of the 12th Symposium, September 1-7, 1991, Kartause Ittingen, Switzerland*, C. Wenk and M. Boessinger, eds. EAAP Publication No. 58. London: Butterworths.

Waterman, R. C., J. S. Caton, C. A. Löest, M. K. Petersen, and A. J. Roberts. 2014. Beef Species Symposium: An assessment of the 1996 Beef NRC: Metabolizable protein supply and demand and effectiveness of model performance prediction of beef females within extensive grazing systems. *Journal of Animal Science* 92:2773-2774.

Webster, A. J. F., J. S. Smith, and G. Mollison. 1976. On the prediction of heat production in growing cattle. Pp. 221-224 in *Energy Metabolism of Farm Animals: Proceedings of the 7th Symposium, September 1976, Vichy, France*. EAAP Publication No. 19. Newcastle upon Tyne: Oriel Press.

Webster, A. J. F., J. S. Smith, and G. S. Mollison. 1977. Prediction of the energy requirements for growth in beef cattle. 3. Body weight and heat production in Hereford×British Friesian bulls and steers. *Animal Production* 24:237-244.

Webster, A. J. F., J. S. Smith, and G. S. Mollison. 1982. Energy requirements of growing cattle: Effects of sire breed, plane of nutrition, sex and season on predicted basal metabolism. Pp. 84-87 in *Energy Metabolism of Farm Animals: Proceedings of the 9th Symposium, September 1982, Lillehammer, Norway*, F. Sundstol and A. Ekern, eds. EAAP Publication No. 29. Aas, Norway: Department of Animal Nutrition, Agricultural University of Norway.

Wilkerson, V. A., T. J. Klopfenstein, R. A. Britton, R. A. Stock, and P. S. Miller. 1993. Metabolizable protein and amino acid requirements of growing beef cattle. *Journal of Animal Science* 71:2777-2784.

Wilson, P. N., and D. F. Osbourn. 1960. Compensatory growth after undernutrition in mammals and birds. *Biological Reviews* 35:324-363.

Williams, J. L., D. J. Garrick, and S. E. Speidel. 2009. Reducing bias in maintenance energy expected progeny difference by accounting for selection on weaning and yearling weights. *Journal of Animal Science* 87:1628-1637.

Wurgler, F., and H. Bickel. 1987. The partial efficiency of energy utilization in steers of different breeds. Pp. 90-93 in *Energy Metabolism of Farm Animals: Proceedings of the 10th Symposium, September 1985, Airlie, VA*, P. W. Moe, H. F. Tyrrell, and P. J. Reynolds, eds. EAAP Publication No. 32. New York: Rowman & Littlefield.

Young, B. A., A. W. Bell, and R. T. Hardin. 1989. Mass specific metabolic rate of sheep from fetal life to maturity. Pp. 155-158 in *Energy Metabolism of Farm Animals: Proceedings of the 11th Symposium, September 18-24, 1988, Lunteren, Netherlands*, Y. van der Honing, ed. EAAP Publication No. 43. Wageningen, The Netherlands: Pudoc.

12

Growth

INTRODUCTION

Animal growth depends on diets being formulated with adequate amounts and proportions of energy and essential nutrients. When cattle consume energy in excess of that required for body maintenance, growth, reproduction, and lactation can occur. Relative to growth, the work by Lofgreen and Garrett (1968) continues to stand as the basis of the net energy (NE) system incorporated into the previous editions and update of the *Nutrient Requirements of Beef Cattle* (NRC, 1976, 1984, 1996, 2000) recommendations as well as in the current edition of this publication. This system, like other systems currently in use, was rooted in the concepts developed by Armsby, Atwater, Blaxter, Brody, Kellner, and Kleiber, among others (Ferrell and Oltjen, 2008).

ENERGY AND PROTEIN REQUIREMENTS FOR GROWING AND FINISHING CATTLE

The system developed for predicting energy and protein requirements of growing cattle assumes that cattle have a similar body composition at the same degree of physiological maturity. Based on this assumption, an equivalent shrunk body weight (SBW) is implemented by adjusting the body weight (BW) of cattle of various body sizes and sexes to a BW at which they are equivalent in body composition to steers in the Garrett (1980) database. In equation form, this adjustment is given by EQSBW = SBW × (SRW/FSBW), where EQSBW = the SBW equivalent to the NRC (1984) medium-framed-size steer; SBW = actual SBW being evaluated; SRW = standard reference BW for the expected final body fat; and FSBW = final SBW at the expected final body fat of the actual animal. The components of this equation are further explained throughout this chapter. Inclusion of this concept (Fox and Black, 1984; Tylutki et al., 1994) allows the incorporation of various factors such as compensatory gain, feeding an ionophore, use of anabolic agents, or breed effects that are expected to affect mature BW or slaughter weight as continuous rather than discrete variables. Animal requirements are dynamic and they change as animals undergo normal growth and development or undergo changes in physiological state. Diverse genotypes, changing environmental conditions, energy intake, hormonal status, and tissue turnover also affect rate and composition of tissue accretion (Owens et al., 1995).

As with the previous NE systems, net energy for gain (NEg) is defined as retained energy (RE), and equations for this publication were taken from the National Research Council report (NRC, 1984) to convert RE to SBW and SBW gain. Net energy for gain is defined as the energy content of the tissue deposited, which is a function of the proportion of fat and protein in the empty body tissue gain (Garrett et al., 1959; fat contains 9.367 kcal/g and nonfat organic matter contains an average of 5.686 kcal/g). Simpfendorfer (1974) summarized data from steers of British beef breeds from birth to maturity and observed that within cattle of a similar mature size, 95.6 (ash, not shown) to 98.9% (water) of the variation in the chemical components and empty body energy content was associated with the variation in BW (Figure 12-1A and B). When energy does not limit growth, the empty body contains an increasingly smaller percentage of protein and an increasingly larger percentage of fat, and reaches chemical maturity when additional weight gain contains little additional protein. Figure 12-1B shows that steers contained little additional protein in the gain after an SBW of 750 kg (Simpfendorfer, 1974). At SBW in excess of 200 to 300 kg, there seemed to be an influence of the effect of plane of nutrition, as evidenced by the scatter of points on the plot of body fat content (Figure 12-1B). In addition, as discussed in Chapter 14 (Compounds That Modify Digestion and Metabolism), it should be noted that anabolic implants and β-adrenergic agonists change this relationship at a given SBW by increasing protein and decreasing fat in weight gain.

The energy content of weight gain across a wide range of metabolizable energy (ME) intakes and rates of gain was described in equation form by Garrett (1980). This data set

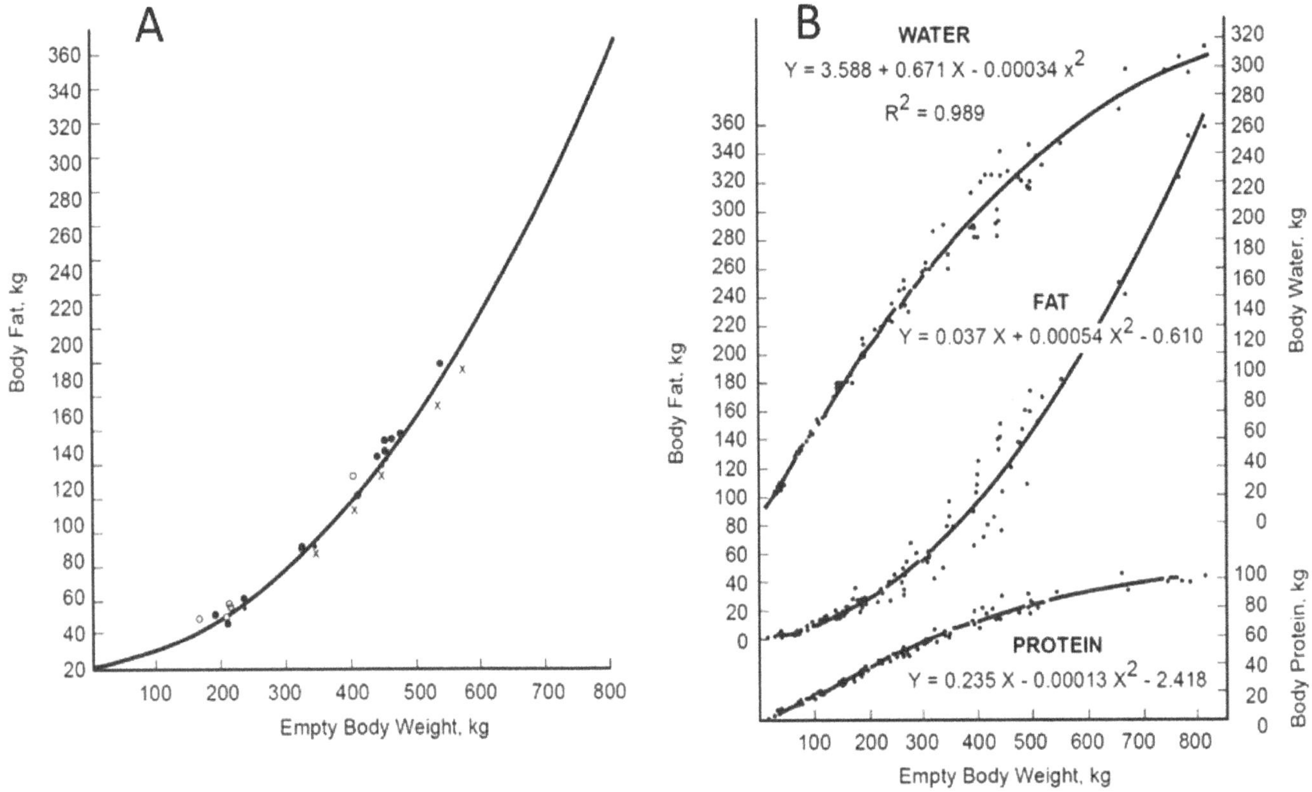

FIGURE 12-1 Relationship between empty body weight (kg) and body fat (kg) in male castrates of British beef breeds. (A): From Simpfendorfer (1974). (B): From Simpfendorfer (1974); superimposed points are from Lofgreen and Garrett (1968), Fox et al. (1972), Jesse et al. (1976), Crickenberger et al. (1978), Harpster (1978), Lomas et al. (1982), and Woody et al. (1983).

included 72 comparative slaughter experiments conducted at the University of California between 1960 and 1980 of approximately 3,500 cattle receiving various diets. The equation developed with British-breed steers describes the relationship between RE and empty body weight gain (EBG) for a given empty body weight (EBW):

$$RE = 0.0635 \times EBW^{0.75} \times EBG^{1.097} \quad \text{(Eq. 12-1)}$$

Because energy is retained as either protein or fat, the composition of the gain at different weights can be estimated from RE computed in Eq. 12-1 (Garrett, 1987):

$$\text{Proportion of fat} = 0.12 \times RE - 0.14 \quad \text{(Eq. 12-2)}$$

and

$$\text{Proportion of protein} = 0.253 - 0.027 \times RE \quad \text{(Eq. 12-3)}$$

Using these relationships, the relationship between stage of growth (percentage of mature weight), rate of gain, and composition of gain can be computed (Table 12-1). The resulting NEg requirement for various SBW and shrunk weight gains (SWG) are those presented in Table 12-1 for a medium-framed steer (note that SWG is in increments of 0.2 kg, except the last line shows requirements for 1.3 kg SWG/d rather than 1.2 kg SWG/d). These SWG represent ranges in SWG from cattle used in the Garrett (1987) database. Several relationships are shown in Table 12-1: first, for a medium-framed steer, energy content of the gain at a particular SWG increases with weight in a particular SBW; second, protein and fat content of the gain and expected body fat at a particular weight depend on rate of gain. Equations 12-1, 12-2, and 12-3 were used to compute the expected percentage of body fat at different SBW from the NE concentrations in the gain (Mcal/kg) when a medium-framed steer was grown from 200 kg SBW at 11.6% body fat at an SWG of 1 kg/d (1.01 Mcal NEg/kg diet) for the first 100 kg and 1.3 kg/d (1.35 Mcal NEg/kg diet) to various SBW (Table 12-1; NRC, 1984, 1996, 2000). Equations 12-1 and 12-2 were used to calculate protein and fat at various SWG, using constants of 0.891 and 0.956, respectively, for converting EBW and EBG to SBW and SWG. It should be noted that these constants used for short time periods at the beginning or end of the growth curve might result in incorrect estimates. Table 12-1 shows the percentage body fat expected at various weights for a medium-framed steer with typical two-phase feeding programs (grown on high-quality forage and finished on high-energy diets), indicating that even at low rates of gain and early stages of growth, some fat is deposited and both

TABLE 12-1 Relationship of Stage of Growth and Rate of Gain to Body Composition, Based on NRC (1984) Medium-Framed Steer

	Shrunk Body Weight, kg						
Shrunk ADG, kg	200	250	300	350	400	450	500
				NEg required, Mcal/d[a]			
0.6	1.68	1.99	2.28	2.56	2.83	3.09	3.34
0.8	2.31	2.73	3.13	3.51	3.88	4.24	4.59
1.0	2.95	3.48	4.00	4.49	4.96	5.42	5.86
1.3	3.93	4.65	5.33	5.98	6.61	7.22	7.81
				Protein in gain, %[b]			
0.6	20.4	19.5	18.8	18.0	17.3	16.6	16.0
0.8	18.7	17.6	16.5	15.5	14.6	13.6	12.7
1.0	17.0	15.6	14.2	13.0	11.7	10.5	9.3
1.3	14.4	12.5	10.7	9.0	7.3	5.7	4.2
				Fat in gain, %[b]			
0.6	5.9	9.7	13.2	16.6	19.9	23.1	26.2
0.8	13.6	18.7	23.6	28.2	32.8	37.1	41.4
1.0	21.4	27.9	34.1	40.1	45.6	51.5	56.9
1.3	22.3	29.0	35.4	41.5	47.4	53.2	58.7
				Body fat, %[c]			
0.6	11.6	10.8	10.9	11.5	12.3	13.4	14.5
0.8	11.6	12.5	13.9	15.6	17.5	19.4	21.4
1.0	11.6	14.2	17.0	19.9	22.8	25.6	28.5
1.3	11.6	14.4	17.4	20.4	23.4	26.4	29.3
1 then 1.3	11.6	14.2	17.0	20.1	23.1	26.1	29.1

[a]Computed from the NRC (1984) equation, which was determined from 72 comparative slaughter experiments (Garrett, 1980); retained energy (RE) = $0.0635 \times EBW^{0.75} \times EBG^{1.097}$, where EBW is 0.891 SBW and EBG is 0.956 SBG.

[b]Computed from the equations of Garrett (1987), which were determined from the NRC (1984) database; proportion of fat in the shrunk body weight gain = $0.12 \times RE - 0.14$, and proportion of protein = $0.253 - 0.027 \times RE$.

[c]Percentage body fat was determined when grown at 1 kg ADG to 300 kg and 1.3 kg ADG to each subsequent weight as described above.

protein and fat are synthesized as rate of gain increases. Lightweight (90 kg) Holstein calves restricted to 0.23 to 0.53 kg average daily gain (ADG) had 14.2 to 16.5% fat in the gain, respectively (Abdalla et al., 1988), which agrees with this concept. Phospholipids are required for cellular membrane growth (Murray et al., 1988). As energy intake above maintenance increases, protein synthesis rate becomes first-limiting, and excess energy is deposited as fat; this dilutes body content of protein, ash, and water, which are deposited in nearly constant ratios to each other at a particular physiological maturity (Garrett, 1987).

In a review of growth and development of feedlot cattle, Owens et al. (1995) evaluated the relationships of fat and protein accretion to RE for bulls and steers fed high-energy diets ad libitum. The derived equation for fat accretion (kg/day) = $0.15\ RE - 0.0057 \times RE^2 - 0.162$ ($R^2 = 0.94$) indicated that as RE increased, fat accretion increased at a decreasing rate. In contrast, the derived equation for protein accretion (kg/d) = $0.267 - 0.072\ RE + 0.0094\ RE^2$ ($R^2 = 0.34$) indicated that protein accretion decreased slightly as RE increased. These relationships suggest that as EBW increases, the accretion rate for fat is greater than the accretion rate for protein (Table 12-1; Figure 12-1). The relationships between RE, proportion of fat in gain, and proportion of protein in gain are influenced by dietary ME concentration as shown by Krehbiel et al. (2006). Quadratic functions were significant ($P < 0.001$) for RE, proportion of fat in gain, and proportion of protein in gain as dietary ME increased. Based on the first derivative of the derived equation, RE reached an asymptote when dietary ME was 3.27 Mcal/kg dry matter (DM). Similar to the data summarized by Owens et al. (1995), data presented by Krehbiel et al. (2006) indicated that the increase in RE with increasing dietary ME is associated with an increase in the proportion of fat in gain and a slight decrease in the proportion of protein in gain when high-energy diets are fed. As suggested by Owens et al. (1995), the relationships of weight gain and proportion of fat and protein in gain are confounded with BW because most studies feed cattle for constant days on feed and not to a constant final BW. The exponent for EBW gain (1.097) in the RE equation ($RE = 0.0635 \times EBW^{0.75} \times EBW\ gain^{1.097}$; NRC, 1996, 2000) indicates that the energy content of gain increases with increasing rate of gain at any given BW, although the effect is expected to be relatively small. Therefore, the main determinant of composition of gain seems to be BW, similar to the conclusions of Simpfendorfer (1974).

To predict NEg required for SBW and SWG, EBW and EBG were converted to 4.4% shrunk liveweight gain to develop the following equations from the Garrett (1980) body composition database:

$$EBW = 0.88 \times SBW + 14.6 \times NEm - 22.9; \quad (r = 0.98) \quad \text{(Eq. 12-4)}$$

and

$$EBG = 0.93 \times SWG + 0.174 \times NEm - 0.28; \quad (r = 0.96) \quad \text{(Eq. 12-5)}$$

or with constants of 0.891 × SBW and 0.956 × SWG. These equations were used to predict EBG and SWG:

$$EBG = 12.341 \times \left(\frac{RE}{EBW^{0.75}}\right)^{0.9116} = 12.341 \times EBW^{-0.6837} \times RE^{0.9116} \quad \text{(Eq. 12-6)}$$

$$SWG = 13.91 \times RE^{0.9116} \times SBW^{-0.6837} \quad \text{(Eq. 12-7)}$$

In these equations, RE is equivalent to NE available for gain (NEFG). Thus, if intake is known, NEFG can be calculated as (DM intake – feed required for maintenance) × dietary NEg. The NEFG can then be substituted into Eqs. 12-6 and 12-7 for RE to predict ADG.

Given the relationship between energy retained and protein content of gain, protein content of SWG is given as (NRC, 1984, 1996, 2000):

$$\text{Protein retained} = SWG \times \{268 - [29.4 \times (RE/SWG)]\}; \ (r^2 = 0.96). \quad \text{(Eq. 12-8)}$$

The BW at which cattle reach the same chemical composition differs depending on mature size and sex; hence, composition is different even when the weight is the same (Fortin et al., 1980; Figure 12-2A and B). Each type reached 28% body fat (equivalent body composition) at different weights (Figure 12-2A). Figure 12-2B shows a similar plot for empty body protein, with the end of the line corresponding to the weight at 28% body fat. Weight at the same 12th-rib lipid content varied by 170 kg among steers of different biological types (Cundiff et al., 1981).

The first NRC NE system (NRC, 1976) used the Lofgreen and Garrett (1968) system to predict energy requirements, which was based on British breed steers given an estrogenic implant. From 1970 to 1990, larger mature-size European breed sires were increasingly used with the North American-based British breed cow herd, resulting in the development of more diverse types of cows in the United States and Canada. This change, along with the use of sire evaluation programs that led to selection for larger body size to achieve greater absolute daily gain, has resulted in an increase in average steer slaughter weights. The NRC (1984) provided equations

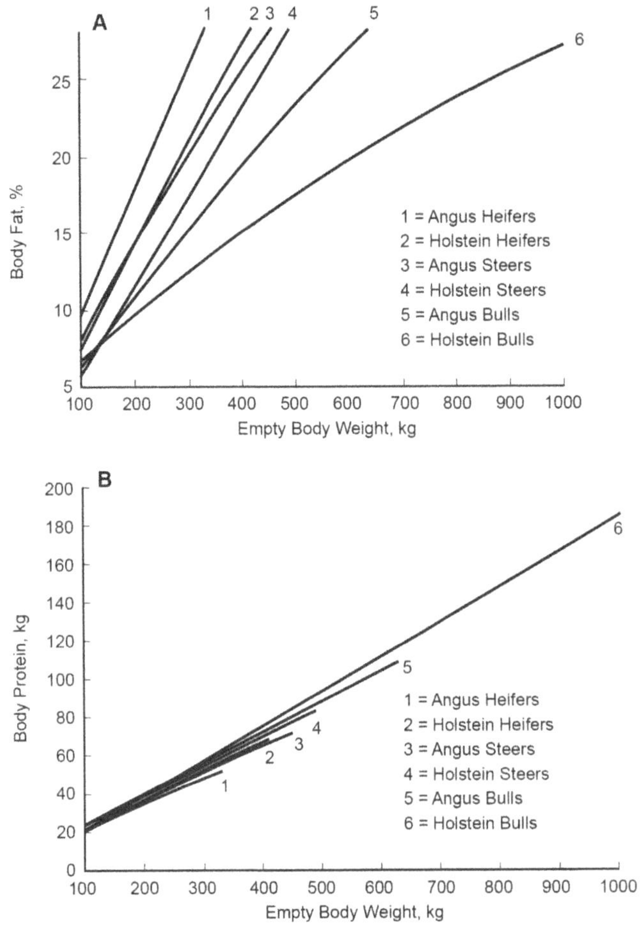

FIGURE 12-2 Relationship between empty body weight (kg) and body fat (%) in Angus and Holstein heifers, steers, and bulls; composition differs even when weight is the same. (A): Each type reached 28% body fat (equivalent body composition) at different weights. (B): A similar plot for empty body protein; the end of the line corresponds to the weight at 28% body fat (NRC, 1996, 2000).

for medium- and large-framed cattle to adjust requirements for these changes. The current population of beef cattle in the United States and Canada continues to vary in biological type and slaughter weight.

All systems developed since the NRC (1984) system use some type of size-scaling approach to adjust for differences in weight at a given composition. The Commonwealth Scientific and Industrial Research Organisation system (CSIRO, 1990, 2007) uses one table of energy requirements for a proportion of an SRW, and then gives a table of SRW for different breed types. This SRW is defined as the weight at which skeletal development is complete and the empty body contains 25% fat, which corresponds to a condition score (CS) 3 on a 0 to 5 scale. Oltjen et al. (1986) developed a mechanistic model to predict protein accretion from initial and mature DNA content, with the residual between NE available for gain and that required for protein synthesis assumed to be deposited as fat. The animal's current weight as a proportion of mature weight is used to adjust for differences in mature size and use of implants.

The Institut National de la Recherche Agronomique system (INRA, 1989) used allometric relationships between the EBW and SBW, the weight of the chemical components, and the weight of the fat-free body mass to predict energy and protein requirements. Coefficients in the equations are parameters from the Gompertz equation (Taylor, 1968), which represents changes in liveweight with time. Initial and final weights with growth curve coefficients were given for six classes of bulls, two classes of steers, and two classes of heifers for finishing cattle, and two classes each for male and female growing cattle. The amount of lipid deposited daily is proportional to the daily liveweight gain raised to the power 1.8. Daily gain of protein is calculated from the gain in the fat-free body mass because protein content of fat-free gain varies little with type of animal, growth rate, or feeding level (Garrett, 1987). Byers et al. (1989) developed an equation for steers similar to that of NRC (1984), except that weight was replaced by proportional weight (current weight/dam mature weight). A different exponent was used for "no growth regulator", and the growth regulator equation assumes use of an estrogenic implant.

Fox et al. (1992) developed a system to interrelate energy and protein requirements with the Beef Improvement Federation (BIF) frame-size system for describing breeding females and the USDA system for describing feeder cattle. Dam mature weight was predicted from the BIF (1986) frame sizes of 1 to 9, which is assumed to be the same as the weight at which a similar frame size steer is 28% body fat (USDA low-Choice grade). That weight was subsequently divided into the frame size of steer assumed to represent the NRC (1984), medium-framed steer equation, to obtain an adjustment factor to compute the weight at which other frame sizes and sexes are equivalent in body composition. This approach is similar to the CSIRO (1990, 2007) SRW system.

Because neither the actual composition nor mature weight is known, body composition and subsequent NE requirements must be predicted from the estimated mature cow weight for breeding cattle or final weight and grade of feeder cattle. Because of the large number of breed types used, the widespread use of crossbreeding, anabolic implants, β-adrenergic agonists, steers rather than bulls, feeding systems, and carcass grading systems used in North America, the European and CSIRO systems used to predict energy and protein requirements might not be readily adaptable to North American conditions. Other proposed systems (Oltjen et al., 1986; Byers et al., 1989; INRA, 1989; CSIRO, 1990, 2007; AFRC, 1993) either did not account for as much of the variation with the validation data set described later or are not sufficiently complete to allow prediction of requirements from common descriptions of cattle given all the conditions that must be taken into account such as gender (bulls, steers, heifers), various implant combinations, use of a β-adrenergic agonist, and wide variations in body size, feeding systems, and final weights and grades.

Based on the evaluations discussed previously, the system developed for predicting energy and protein requirements for growing cattle assumes that cattle have a similar body composition at the same degree of maturity. The NRC (1984) medium-framed steer equation (Eq. 12-1) is used as the standard reference base to compute the energy content of gain at various stages of growth and rates of gain for all cattle types. This is accomplished by adjusting the BW of cattle of various body sizes and sexes to a weight at which they are equivalent in body composition to the steers in the Garrett (1980) database, as described by Tylutki et al. (1994):

$$EQSBW = SBW \times (SRW/FSBW), \quad \text{(Eq. 12-9)}$$

where

EQSBW is weight equivalent to the NRC (1984) medium-framed steer;
SBW is shrunk BW being evaluated;
SRW is standard reference weight for the expected final body fat (Table 12-2); and
FSBW is final shrunk BW at the expected final body fat (Table 12-2).

TABLE 12-2 Standard Reference Weights for Different Final Body Compositions

	Average Marbling Score		
	Traces	Slight	Small
Body fat, % SE[a]	25.2 ± 2.9	26.8 ± 3.0	27.8 ± 3.4
Standard reference weight, kg[b]	435	462	478

[a]The means and standard errors (SE) shown for body fat in each marbling score category were determined by averaging the percentage of body fat across all cattle in each of three marbling categories in the energy and protein retained validation data (Harpster, 1978; Danner et al., 1980; Lomas et al., 1982; Woody et al., 1983). In a second comparison evaluation, the body fat data of Perry et al. (1991a,b) and Ainslie et al. (1992) averaged 28.4% (±4.1) for those in the small marbling category. These relate to the current USDA and Canadian grading standards, respectively, as follows: Traces, Standard or A; Slight, Select or AA; and Small, Choice or AAA.

[b]The standard reference body weights (SBW basis) were determined from the NEg concentrations in the gain (Mcal/kg) when the reference animal (NRC [1984] medium-framed steer) was grown from 200 kg SBW at 11.6% body fat at SWG of 1 kg/d (1.01 Mcal NEg/kg diet) for the first 100 kg and 1.3 kg/d (1.35 Mcal NEg/kg diet) until the percentage of body fat in Table 12-2 was reached. Equations 12-1 and 12-2 in the text were used for the computations, using constants of 0.891 and 0.956, respectively, for converting EBW and EWG to SBW and SWG. The SRW and FSBW (mature weight) of replacement heifers (18 to 22% fat) is assumed to be the same as the 28% fat weight as implanted steer mates, based on the data of Smith et al. (1976), Cundiff et al. (1981), Jenkins and Ferrell (1984), and Harpster (1978) and accumulated fat content when heifers are grown at replacement heifer rates (Table 12-1). Breeding bulls are assumed to be 67% greater than cows, giving an SRW of 800 kg.

These values were determined by averaging the percentage of body fat within all cattle in each of three marbling categories in the energy and protein retained validation data used in the NRC (1996, 2000), Harpster (1978), Danner et al. (1980), Lomas et al. (1982), and Woody et al. (1983). Body fat percentage averaged (±SD) 27.8 (±3.4), 26.8 (±3.0), and 25.2 (±2.9) for those pens in the small, slight, or trace marbling categories, respectively. In comparison, the body fat data of Perry et al. (1991a,b) and Ainslie et al. (1992) averaged 28.4% (±4.1) for those in the USDA Small marbling category. These steers had been selected to be a cross section of the current breed types and body sizes used in the United States. This variable SRW allows for adapting the system to both U.S. and Canadian grading systems and determining SRW for marketing cattle at different end points. For breeding herd replacement heifers, FSBW is the expected mature weight. When computed as shown in Table 12-1 for heifers grown at 0.6 to 0.8 kg/d, accumulated fat content was 18 to 22% at the 28% fat steer SRW. Therefore, the SRW for breeding herd replacement heifers was assumed to be the same as the 1984 medium-framed steer fed to 28% fat. This approach is supported by a summary of the U.S. Meat Animal Research Center (MARC) data (Smith et al., 1976; Cundiff et al., 1981; Jenkins and Ferrell, 1984) in which mature weights of heifer mates averaged 10% more than implanted steer mates finished on high-energy diets after weaning. Based on MARC data, breeding bulls are assumed to be 67% heavier at maturity than cows, giving an SRW of 800 kg, which is the mature weight of a bull with the same genotype as the NRC (1984) medium-framed steer.

The EQSBW computed from the SRW/FSBW multiplier can be used in Eq. 12-7 to compute the NEg requirement. If Eq. 12-1 or 12-6 is used, SBW is adjusted to EBW with Eq. 12-4. Alternatively, the equation of Williams et al. (1992; EBW = full BW × [1 − gut fill], where gut fill is 0.0534 + 0.329 × fraction of forage NDF) can be used to predict EBW from unshrunk liveweight. Predicted gut fill is then corrected with multipliers for full BW, physical form of forage, and fraction of concentrates.

Because a table of requirements can be generated for any body size using the Beef Cattle Nutrient Requirements Model (BCNRM; Chapter 19) software provided, only one example is shown in Box 12-1 (600-kg FSBW to represent the average steer in the United States). A similar table can be created and printed for any body size with the BCNRM. In this representative example, an FSBW change of 35 kg alters the NEg requirement by approximately 5%. Heifers and bulls with similar parents as the steers represented in this table have 18% greater and lesser, respectively, NEg requirements at the same weight as these steers. This system requires an accurate estimation of FSBW. Most cattle feeders are experienced with results expected with feedlot finishing on a high-energy diet of backgrounded calves or yearlings that have received an estrogenic implant. Guidelines for other conditions are

BOX 12-1
Example of Calculation of Requirements

For this example, a 320-kg steer with an FSBW of 600 kg (or herd replacement heifer with a mature weight of 600 kg) has an EQSBW of 320 × (478/600) = 255 kg. A 320-kg heifer with an FSBW of 480 has an EQSBW of 320 × (478/480) = 319 kg. The predicted SWG for the 320-kg steer consuming 5 Mcal NEg/d is (Eq. 12-7): $13.91 \times 5^{0.9116} \times 255^{-0.6837} = 13.91 \times 4.337 \times 0.02263 = 1.365$ kg/d. The SWG of the heifer consuming the same amount of energy will be $13.91 \times 5^{0.9116} \times 319^{-0.6837} = 1.17$ kg/d. To compute the NEg requirement in this example for a 320-kg steer using Eq. 12-1 (0.891 × SBW to compute EBW and 0.956 × SWG to compute EBG): 255 × 0.891 = 227 kg EBW; 1.365 × 0.956 = 1.305 kg EBG/d; RE = $0.0635 \times 227^{0.75} \times 1.305^{1.097} = 0.0635 \times 58.5 \times 1.339 = 4.97$ Mcal/d. Assuming NEm requirement is $0.077 \times SBW^{0.75}$, the NEm requirement is $0.077 \times 320^{0.75} = 5.83$ Mcal/d. For the heifer, the NEg requirement is also 4.97 Mcal/d and the NEm requirement is 5.83 Mcal/d. Net protein requirement for gain is then (Eq. 12-8): [268 − 29.4 × (5/1.365)] × 1.365 = 219 g/d for the 320-kg steer and 167 g/d for the 320-kg heifer. These values are then divided by the efficiency of use of absorbed protein to obtain the metabolizable protein (MP) required for gain [max (0.492, 0.834 − 0.00114 × EQSBW)], which is then added to the MP required for maintenance ($3.8 \times SBW^{0.75}$) to obtain the total MP required. For the 320-kg steer, MP = 219/max (0.492, 0.834 − 0.00114 × 255) + $3.8 \times 320^{0.75}$ = 691 g/d, and for the 320-kg heifer, MP = 167/max (0.492, 0.834 − 0.00114 × 319) + $3.8 \times 320^{0.75}$ = 623 g/d.

- decrease FSBW 25 to 45 kg for nonuse of an estrogenic implant (see Chapter 14);
- increase FSBW 25 to 45 kg for use of an implant containing trenbolone acetate (TBA) plus estrogen (see Chapter 14);
- increase FSBW 6 to 36 kg for use of a β-adrenergic agonist (see Chapter 14);
- increase FSBW 25 to 45 kg for extended periods at slow rates of gain; and
- decrease FSBW 25 to 45 kg for continuous use of a high-energy diet from weaning.

PREVIOUS PLANE OF NUTRITION EFFECTS

Energy intake above maintenance can vary considerably, depending on the diet fed during early growth in stocker and backgrounding programs. Table 12-1 indicates that a decreased intake above maintenance results in a greater proportion of protein in the gain at a particular weight, which is supported by several studies (Fox and Black, 1984; Abdalla et al., 1988; Byers et al., 1989; Sainz et al., 1995) and by the model of Keele et al. (1992); however, when thin cattle are placed on a high-energy diet, compensatory energy deposition occurs. In general, most of the improved efficiency of gain results from a decreased maintenance requirement and increased feed intake (Fox and Black, 1984; Ferrell et al., 1986; Carstens et al., 1987; Abdalla et al., 1988; Choat et al., 2003). Variation in response as a result of previous diet has been suggested, such that all diets that restrict gain might not result in a compensatory gain response (Sainz et al., 1995; Hersom et al., 2004; McCurdy et al., 2010). For example, Sainz et al. (1995) reported that increased feed DM intake (DMI) accounted for 60 to 104% of the increased growth rate during finishing of previously restricted steers, and that energy restriction decreased the maintenance energy requirements of steers limit-fed a high-concentrate diet but increased the maintenance energy requirements of steers fed an alfalfa/oat straw-based diet. In addition, Hersom et al. (2004) observed that fatter cattle, which had grazed winter wheat pasture, had similar rates of BW gain and feed efficiency compared with leaner cattle, which had grazed low-quality native range, when all cattle were fed to a common compositional end point. The NRC (1996, 2000) assumes that NEm requirement is 20% less in a very thin animal (CS 1), is increased by 20% in a very fleshy animal (CS 9), and changes 5% per CS relative to a base score of 5. The NEm adjustment for previous nutrition (COMP) is computed as

$$COMP = 0.8 + (BCS - 1) \times 0.05 \quad \text{(Eq. 12-10)}$$

where BCS is body condition score (range = 1 to 9). The effect of plane of nutrition is taken into account by the rate of gain function (increased fat deposition with increased rate of gain) and EQSBW in the primary equations. Therefore, the user determines the expected final weight and body fat, and the EQSBW to use in calculating the NEg required is determined as shown in Eq. 12-9. The change in efficiency of energy utilization is accounted for by a decreased NEm requirement and increased DMI above maintenance.

EFFECTS OF SPECIAL DIETARY FACTORS

Although the current committee recommends the use of Eq. 12-10, special dietary factors should be considered. Diet composition and level of intake differences will cause the composition of the ME (ruminal volatile fatty acids, intestinally digested carbohydrates, and fat) to vary (Ferrell, 1988; McCurdy et al., 2010), which can affect the composition of gain (Fox and Black, 1984). Most of these effects will alter rate of gain, which is taken into account by the primary equations; however, fat distribution can be altered, which could affect carcass grade (Fox and Black, 1984).

Sainz et al. (1995) reported that carcass and empty-body fat of steers fed concentrate ad libitum was greater than in steers limit-fed concentrate or forage-fed steers, but that the empty-body protein mass was greater in the limit-fed and forage-fed steers. Similar results were reported by Hersom et al. (2004): initial empty-body fat and energy content of steers grazing wheat pasture was increased and empty-body fat-free organic matter content was decreased compared with steers grazing restricted wheat pasture and steers grazing dormant winter range. In the study by Hersom et al. (2004), feed intake, expressed as a percentage of mean weight, of restricted wheat pasture and dormant winter range steers was greater than for steers that grazed winter wheat, but neither live- nor empty-BW ADG of restricted or winter-range steers were increased during finishing. The increased feed DMI and increased heat production of winter range steers were similar to the results of Sainz et al. (1995) for steers fed the alfalfa/oat straw-based diet. Calculated heat production during finishing for steers that had previously grazed dormant native range was about 15% greater than for steers grazing winter wheat. The authors suggested that consumption of low-quality, dormant native-range forage by steers during the winter growing program increased their maintenance energy requirements during the subsequent finishing phase. Blum et al. (1985) examined changes in heat production during and immediately after energy restriction of 2-year-old steers. At the end of a 151-d restriction phase, daily heat production by restricted steers was 100 kcal/kg $BW^{0.75}$; however, by the fifth day of refeeding, daily heat production by previously restricted steers had increased to 148 kcal/kg $BW^{0.75}$ and remained constant through day 24 of realimentation. Therefore, restricted feeding during the growing phase might not always result in compensatory gain during the subsequent finishing phase, and dietary factors should be considered.

UNIQUE BREED EFFECTS

Most of the unique breed effects on NEg requirements are accounted for by differences in the weight at which different breeds reach a given chemical composition (Harpster, 1978; Cundiff et al., 1986; INRA, 1989). Nonetheless, breeds can differ in fat distribution, which can influence carcass grade (Cundiff et al., 1986; Perry et al., 1991a). Ferrell and Jenkins (1998a,b) compared Angus-, Belgian Blue-, Hereford-, Piedmontese-, Boran-, Brahman-, and Tuli-sired steers and reported differences among sire breeds in BW, composition of gain, maintenance requirements, and efficiencies of energy use for gain among breeds. Their results indicated that for many traits, dam breed (Angus, Hereford, and MARC III) and the interactions between sire or dam and nutritional treatment were at least as important as sire breed effects. The slope and intercept of linear relationships between RE (energy gain and ME intake) were influenced by sire breed and dam breed. When analyzed by sire breed, the efficiency of use of ME for empty body energy gain (k_g) varied from 0.27 for Piedmontese- to 0.44 for Belgian Blue-sired steers. Similarly, when the data were analyzed by dam breed, k_g varied from 0.33 for steers from MARC III to 0.60 for steers from Hereford cows. Expected k_g for a diet having 3.13 Mcal ME/kg is 0.47 (NRC, 1984). The slope of linear regressions of energy gain on ME intake indicated that the efficiency of ME use for gain was least for Angus- and greatest for Brahman-sired steers (Ferrell and Jenkins, 1998b). With the

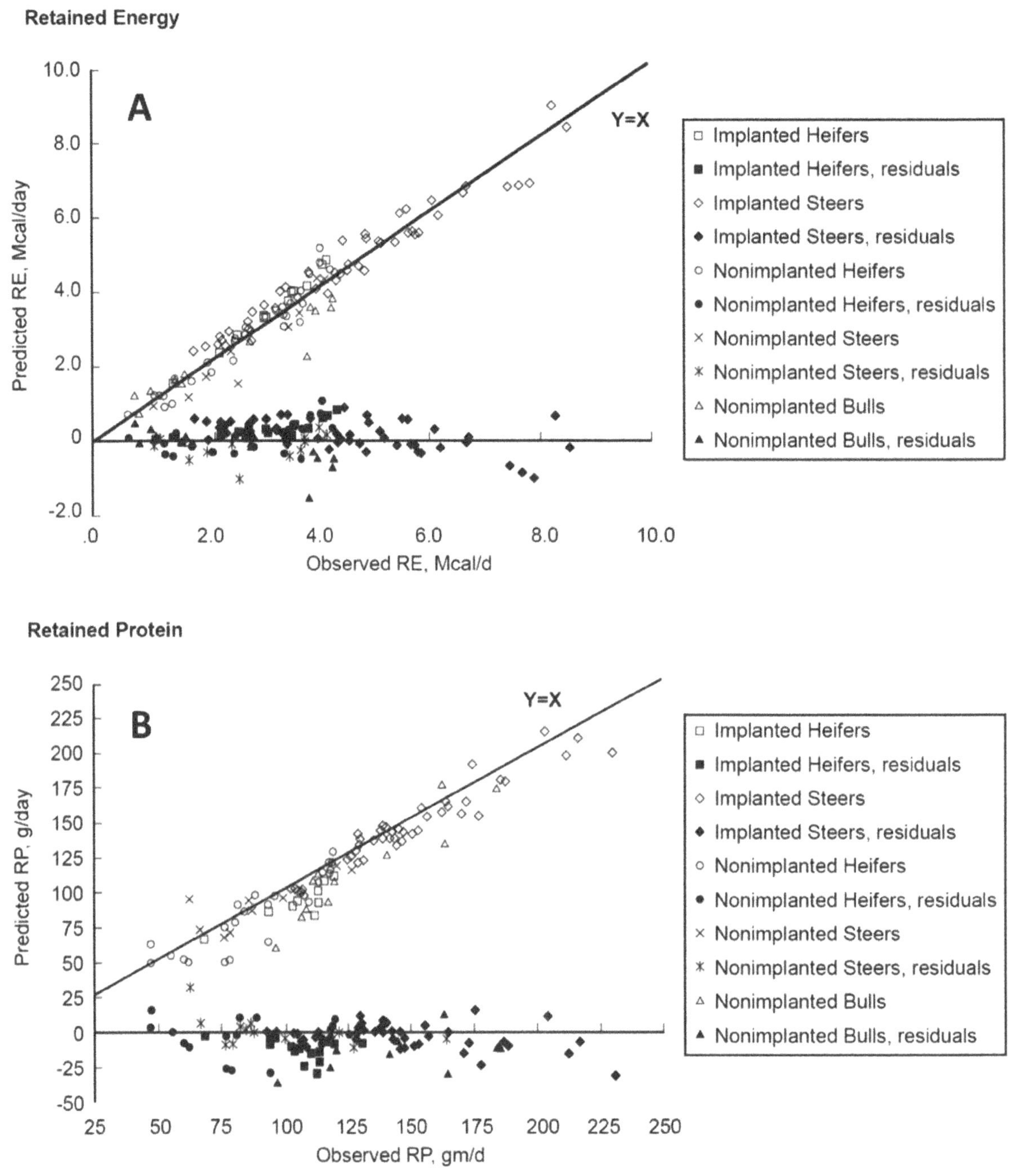

FIGURE 12-3 Use of the NRC (1984) medium-framed steer as a standard reference base to predict retained energy (A) and retained protein (B) in growing cattle across wide variations in cattle breed, body size, implant, and nutritional management systems. Adapted from Tylutki et al. (1994).

exception of Brahman-sired steers, estimates tended to be lower than expected (0.47) based on tabular values shown in Table 12-1 (NRC, 1984); however, the authors suggested that digestion or metabolism of the diet might have been lower than predicted because tabular values were used to calculate ME content of the diet.

EVALUATION OF THE ENERGY SYSTEM

In the previous edition and update of the *Nutrient Requirements of Beef Cattle* (1996, 2000), the SRW approach was evaluated and compared with the NRC (1984) equations using three distinctly different data sets that were completely independent of those used to develop the NRC (1984) and the NRC (1996, 2000) equations. The Oltjen et al. (1986) model was also compared with the NRC (1984) and the NRC (1996, 2000) equations for the first two data sets. For the NRC (1984) comparison, cattle with frame sizes larger than 6 were considered large-framed. For the NRC (1996, 2000) comparison, the SRW (478 kg) was divided by the pen mean weight at 28% body fat to obtain the body size adjustment factor, which was then applied to the actual weight for use

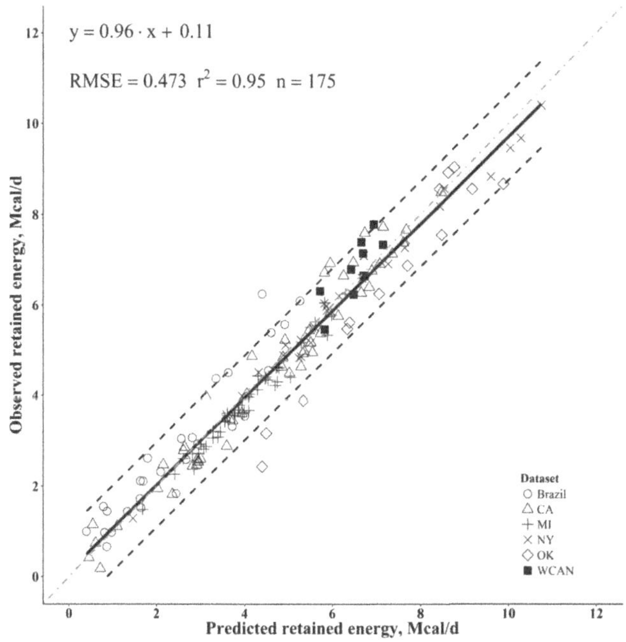

FIGURE 12-4 Use of the NRC (1984) medium-framed steer as a standard reference base to predict net energy requirements for growth across wide variations in cattle breed, body size, implant, and nutritional management systems. Data points were added to the second data set from Perry and Fox (1997; NY), Guiroy (2001; NY), Tedeschi (2001; Brazil), Basarab et al. (2003; Western Canada [WCAN]), Hersom et al. (2004; OK), and McCurdy et al. (2010; OK). The Beef Cattle Nutrient Requirements Model accounted for 95% of the variation for retained energy.

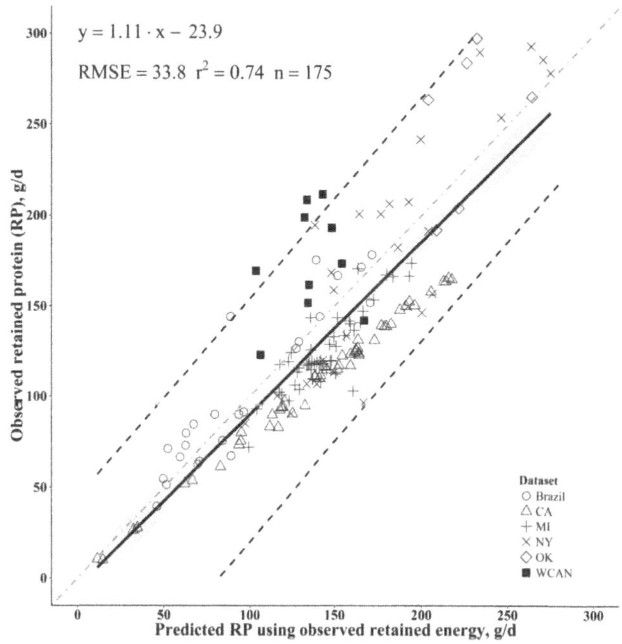

FIGURE 12-5 Use of the NRC (1984) medium-framed steer as a standard reference base to predict net energy requirements for growth across wide variations in cattle breed, body size, implant, and nutritional management systems. Data points were added to second data set from Perry and Fox (1997; NY), Guiroy (2001; NY), Tedeschi (2001; Brazil), Basarab et al. (2003; Western Canada [WCAN]), Hersom et al. (2004; OK), and McCurdy et al. (2010; OK). The Beef Cattle Nutrient Requirements Model accounted for 74% of the variation in retained protein when retained protein is predicted from observed retained energy.

in the standard reference equations to predict energy and protein retained.

The first data set (Harpster, 1978; Danner et al., 1980; Lomas et al., 1982; Woody et al., 1983) included 82 pen observations (65 pens of steers and 17 pens of heifers) with body composition determined by the same procedures used by Garrett (1980) in developing the NRC (1984) equations. Included were FSBW representative of the range in cattle fed in North America; all-silage- to all-corn-based diets; no anabolic implant, estrogen only or estrogen + TBA; and *Bos taurus* breed types representative of those fed in North America (British, European, Holstein, and their crosses).

The second data set included 142 serially slaughtered (whole body chemical analysis by component; Fortin et al., 1980; Anrique et al., 1990) nonimplanted steers, heifers, and bulls ranging widely in body size. A detailed description of these data sets, validation procedures, and results was published by Tylutki et al. (1994) for the Cornell Net Carbohydrate and Protein System (CNCPS), except that in the comparison the SRW was increased from 467 to 478 kg. In nearly every subclass, the system developed for NRC (1996, 2000) accounted for more of the variation and had less bias than did the other two systems. Nearly identical results were obtained between the NRC (1984) and NRC (1996, 2000) systems when energy retained was used to predict SWG in Eq. 12-7; this equation is the one most commonly used to predict ADG. Figure 12-3 shows the results when all subclasses were combined. The BCNRM accounted for 94% of the variation with a 2% overprediction bias for RE and 91% of the variation in retained protein with a 2% underprediction bias. Figure 12-3 shows that use of the NRC (1984) medium-framed steer as a standard reference base results in accurate prediction of NE requirements for growth across wide variations in cattle breed, body size, implant, and nutritional management systems.

The third data set included ADG predicted from independent trials with 96 different diets fed to a total of 943 *B. indicus* (Nellore breed) steers and bulls for which ME intake and body composition were determined (Lanna et al., 1996). The FSBW was determined from final EBW fat content. For NRC (1984) and NRC (1996, 2000), the r^2 was 0.58 and 0.72, and the bias was −20% and −2%, respectively.

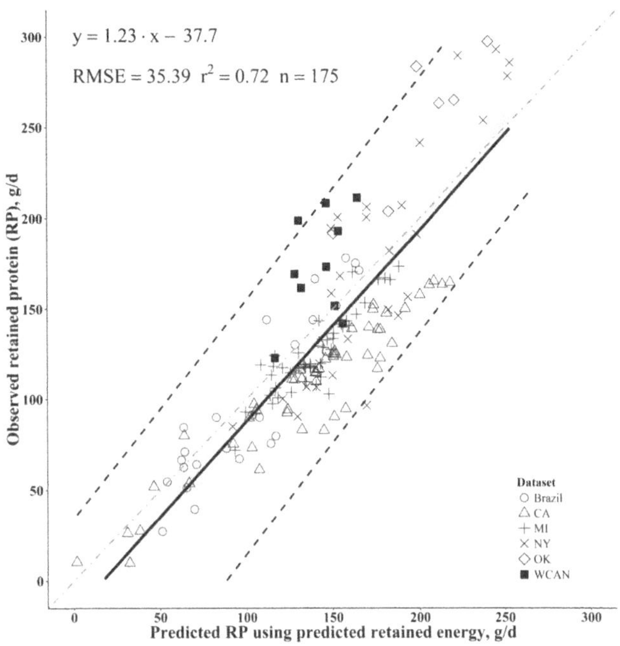

FIGURE 12-6 Use of the NRC (1984) medium-framed steer as a standard reference base to predict net energy requirements for growth across wide variations in cattle breed, body size, implant, and nutritional management systems. Data points were added to the second data set from Perry and Fox (1997; NY), Guiroy (2001; NY), Tedeschi (2001; Brazil), Basarab et al. (2003; Western Canada [WCAN]), Hersom et al. (2004; OK), and McCurdy et al. (2010; OK). The Beef Cattle Nutrient Requirements Model accounted for 72% of the variation when retained protein is predicted from predicted retained energy.

For the present publication, data points from Perry and Fox (1997), Guiroy (2001), Tedeschi (2001), Basarab et al. (2003), Hersom et al. (2004) and McCurdy et al. (2010) were added to the second data set to evaluate the RE and RP predictions by the current model. With the additional data points, the BCNRM accounted for 95% of the variation for RE (Figure 12-4) and 74% of the variation in retained protein when retained protein is predicted from observed RE (Figure 12-5), and 72% when retained protein is predicted from predicted RE (Figure 12-6).

These validations indicate that given the accuracies obtained, problems with predicting net energy and protein requirements and SWG are likely to include one of the following:

1. choosing the wrong FSBW;
2. short-term, transitory effects of previous nutrition, gut fill, or anabolic implants;
3. variation in NEm requirement;
4. variation in ME value assigned to the feed because of variations in feed composition and extent of ruminal or intestinal digestion;
5. variation in NE and NEg derived from the ME because of variation in end products of digestion and their metabolism; and
6. variations in gut fill.

Numerous additions have been incorporated in this edition of the *Nutrient Requirements of Beef Cattle* to allow for application of NE concepts to express nutrient requirements during the life cycle of beef cattle. Examples include energy and protein requirements for breeding herd replacements, which include management recommendations to achieve target BW and rates of gain for replacement heifers and young cows. Estimation of energy and protein reserves of mature beef cows, their relationships with body size and cow BCS, and NEm provided from body reserves or required to replace body reserves are included in Chapter 13 (Reproduction). Feed additives and modifiers were included in Chapter 14. In concept, the equations for expressing energy requirements of animals and values of feed resources to meet those requirements have not changed a great deal for many years. Primary changes have been to expand earlier equations to encompass more distinguishable segments of beef cattle production and include predictions relating to different physiological states (Ferrell and Oltjen, 2008).

REFERENCES

Abdalla, H. O., D. G. Fox, and M. L. Thonney. 1988. Compensatory gain by Holstein calves after underfeeding protein. *Journal of Animal Science* 66:2687-2695.

AFRC (Agricultural and Food Research Council). 1993. *Energy and Protein Requirements of Ruminants.* Wallingford, UK: CAB International.

Ainslie, S. J., D. G. Fox, and T. C. Perry. 1992. Management systems for Holstein steers that utilize alfalfa silage and improve carcass value. *Journal of Animal Science* 70:2643-2651.

Anrique, R. G., M. L. Thonney, and H. J. Ayala. 1990. Dietary energy losses of cattle influenced by body type, size, sex and intake. *Animal Production* 50:467-474.

Basarab, J. A., M. A. Price, J. L. Aalhus, E. K. Okine, W. M. Snelling, and K. L. Lyle. 2003. Residual feed intake and body composition in young growing cattle. *Canadian Journal of Animal Science* 83:189-204.

BIF (Beef Improvement Federation). 1986. *Guidelines for Uniform Beef Improvement Programs,* 5th Ed. Raleigh, NC: Beef Improvement Federation.

Blum, J. W., W. Schnyder, P. L. Kunz, A. K. Blom, H. Bickel, and A. Schürch. 1985. Reduced and compensatory growth: Endocrine and metabolic changes during food restriction and refeeding in steers. *Journal of Nutrition* 115:417-424.

Byers, F. M., G. T. Schelling, and L. W. Greene. 1989. Development of growth functions to describe energy density of growth in beef cattle. Pp. 195-198 in *Energy Metabolism of Farm Animals: Proceedings of the 11th Symposium, September 18-24, 1988, Lunteren, Netherlands,* Y. van der Honing, ed. EAAP Publication No. 43. Wageningen, The Netherlands: Pudoc.

Carstens, G. E., D. E. Johnson, and M. A. Ellenberger. 1987. The energetics of compensatory growth in beef cattle [abstract]. *Journal of Animal Science* 65(Suppl. 1):263-264.

Choat, W. T., C. R. Krehbiel, G. C. Duff, R. E. Kirksey, L. M. Lauriault, J. D. Rivera, B. M. Capitan, D. A. Walker, G. B. Donart, and C. L. Goad. 2003. Influence of previously grazed dormant native range or winter wheat pasture on finishing performance, carcass characteristics and ruminal metabolism. *Journal of Animal Science* 81:3191-3201.

Crickenberger, R. G., D. G. Fox, and W. T. Magee. 1978. Effect of cattle size and protein level on the utilization of high corn silage or high grain rations. *Journal of Animal Science* 46:1748-1758.

CSIRO (Commonwealth Scientific and Industrial Research Organisation). 1990. *Feeding Standards for Australian Livestock: Ruminants*. Melbourne, Australia: CSIRO Publishing.

CSIRO. 2007. *Nutrient Requirements of Domesticated Ruminants*. Collingwood, Australia: CSIRO Publishing.

Cundiff, L. V., R. W. Koch, and G. M. Smith. 1981. Characterization of biological types of cattle (cycle II). IV. Postweaning growth and feed efficiency of steers. *Journal of Animal Science* 53:332-346.

Cundiff, L. V., K. E. Gregory, R. H. Koch, and G. E. Dickerson. 1986. Genetic diversity among cattle breeds and its use to increase beef production efficiency in a temperate climate. Pp. 271-282 in *Proceedings of the Third World Congress on Genetics Applied to Beef Production*. University of Nebraska-Lincoln. Available online at http://digitalcommons.unl.edu/wcgalp/39. Accessed on April 10, 2015.

Danner, M. L., D. G. Fox, and J. R. Black. 1980. Effect of feeding system on performance and carcass characteristics of yearling steers, steer calves and heifer calves. *Journal of Animal Science* 50:394-404.

Ferrell, C. L. 1988. Energy metabolism. Pp. 251-268 in *The Ruminant Animal—Digestive Physiology and Nutrition*, D. C. Church, ed. Englewood Cliffs, NJ: Prentice Hall.

Ferrell, C. L., and T. G. Jenkins. 1998a. Body composition and energy utilization by steers of diverse genotypes fed a high-concentrate diet during the finishing period: I. Angus, Belgian Blue, Hereford, and Piedmontese sires. *Journal of Animal Science* 76:637-646.

Ferrell, C. L., and T. G. Jenkins. 1998b. Body composition and energy utilization by steers of diverse genotypes fed a high-concentrate diet during the finishing period: II. Angus, Boran, Brahman, Hereford, and Tuli sires. *Journal of Animal Science* 76:647-657.

Ferrell, C. L., and J. W. Oltjen. 2008. ASAS Centennial Paper: Net energy systems for beef cattle—Concepts, application, and future models. *Journal of Animal Science* 86:2779-2794.

Ferrell, C. L., L. J. Koong, and J. A. Nienaber. 1986. Effect of previous nutrition on body composition and maintenance energy costs of growing lambs. *British Journal of Nutrition* 56:595-605.

Fortin, A., S. Simpfendorfer, J. T. Reid, H. J. Ayala, R. Anrique, and A. F. Kertz. 1980. Effect of level of energy intake and influence of breed and sex on the chemical composition of cattle. *Journal of Animal Science* 51:604-614.

Fox, D. G., and J. R. Black. 1984. A system for predicting body composition and performance of growing cattle. *Journal of Animal Science* 58:725-739.

Fox, D. G., R. R. Johnson, R. L. Preston, T. R. Dockerty, and E. W. Klosterman. 1972. Protein and energy utilization during compensatory growth in beef cattle. *Journal of Animal Science* 34:310-318.

Fox, D. G., C. J. Sniffen, J. D. O'Connor, J. B. Russell, and P. J. Van Soest. 1992. A net carbohydrate and protein system for evaluating cattle diets. III. Cattle requirements and diet adequacy. *Journal of Animal Science* 70:3578-3596.

Garrett, W. N. 1980. Energy utilization by growing cattle as determined in 72 comparative slaughter experiments. Pp. 3-7 in *Energy Metabolism: Proceedings, 8th Symposium on Energy Metabolism, September 1979, Cambridge, England*, L. E. Mount, ed. EAAP Publication No. 26. London: Butterworths.

Garrett, W. N. 1987. Relationship between energy metabolism and the amounts of protein and fat deposited in growing cattle. Pp. 98-101 in *Energy Metabolism of Farm Animals: Proceedings of the 10th Symposium September 1985, Airlie, VA*, P. W. Moe, H. F. Tyrrell, and P. J. Reynolds, eds. EAAP Publication No. 32. New York: Rowman & Littlefield.

Garrett, W. N., J. H. Meyer, and G. P. Lofgreen. 1959. The comparative energy requirements of sheep and cattle for maintenance and gain. *Journal of Animal Science* 18:528-547.

Guiroy, P. J. 2001. A System to Improve Local Beef Production Efficiency and Consistency in Beef Quality and Its Implementation Through the Creation of a Strategic Alliance. Ph.D. Dissertation. Cornell University, Ithaca, NY.

Harpster, H. W. 1978. Energy Requirements of Cows and the Effect of Sex, Selection, Frame Size, and Energy Level on Performance of Calves of Four Genetic Types. Ph.D. Dissertation. Michigan State University, East Lansing, MI.

Hersom, M. J., G. W. Horn, C. R. Krehbiel and W. A. Phillips. 2004. Effect of live weight gain of steers during winter grazing: I. Feedlot performance, carcass characteristics, and body composition of beef steers. *Journal of Animal Science* 82:262-272.

INRA (Institut National de la Recherche Agronomique). 1989. *Ruminant Nutrition*. Montrouge, France: Libbey Eurotext.

Jenkins, T. G., and C. L. Ferrell. 1984. Output/input differences among biological types. Pp. 15-37 in *Proceedings: Beef Cow Efficiency Symposium*. East Lansing: Michigan State University.

Jesse, G. W., G. B. Thompson, J. L. Clark, H. B. Hedrick, and K. G. Weimer. 1976. Effects of ration energy and slaughter weight on composition of empty body and carcass gain of cattle. *Journal of Animal Science* 43:418-425.

Keele, J. W., C. B. Williams, and G. L. Bennett. 1992. A computer model to predict the effects of level of nutrition on composition of empty body gain in beef cattle: I. Theory and development. *Journal of Animal Science* 70:841-857.

Krehbiel, C. R., J. J. Cranston, and M. P. McCurdy. 2006. An upper limit for caloric density of finishing diets. *Journal of Animal Science* 84(Suppl.):E34-E49.

Lanna, D. P. D., D. G. Fox, C. Boin, M. J. Traxler, and M. C. Barry. 1996. Validation of the CNCPS estimates of nutrient requirements of growing and lactating zebu germplasm in tropical conditions. *Journal of Animal Science* 76(Suppl. 1):287.

Lofgreen, G. P., and W. N. Garrett. 1968. A system for expressing net energy requirements and feed values for growing and finishing cattle. *Journal of Animal Science* 27:793-806.

Lomas, L. W., D. G. Fox, and J. R. Black. 1982. Ammonia treatment of corn silage. I. Feedlot performance of growing and finishing cattle. *Journal of Animal Science* 55:909-923.

McCurdy, M. P., G. W. Horn, J. J. Wagner, P. A. Lancaster, and C. R. Krehbiel. 2010. Effects of winter growing programs on subsequent feedlot performance, carcass characteristics, body composition, and energy requirements of beef steers. *Journal of Animal Science* 88:1564-1576.

Murray, R. K., D. K. Granner, P. A. Mayes, and V. W. Rodwell. 1988. *Harper's Biochemistry.* Norwalk, CT: Appleton and Lange.

NRC (National Research Council). 1976. *Nutrient Requirements of Beef Cattle.* Washington, DC: National Academy of Sciences.

NRC. 1984. *Nutrient Requirements of Beef Cattle,* 6th Revised Ed. Washington, DC: National Academy Press.

NRC. 1996. *Nutrient Requirements of Beef Cattle,* 7th Rev. Ed. Washington, DC: National Academy Press.

NRC. 2000. *Nutrient Requirements of Beef Cattle: Update 2000,* 7th Rev. Ed. Washington, DC: National Academy Press.

Oltjen, J. W., A. C. Bywater, R. L. Baldwin, and W. N. Garrett. 1986. Development of a dynamic model of beef cattle growth and composition. *Journal of Animal Science* 62:86-97.

Owens, F. N., D. S. Secrist, and D. R. Gill. 1995. Impact of grain sources and grain processing on feed intake by and performance of feedlot cattle. Pp. 235-256 in *Symposium: Intake by Feedlot Cattle*. Research Report P-942. Stillwater: Oklahoma State University.

Perry, T. C., and D. G. Fox. 1997. Predicting carcass composition and individual feed requirement in live cattle widely varying in body size. *Journal of Animal Science* 75:300-307.

Perry, T. C., D. G. Fox, and D. H. Beermann. 1991a. Effect of an implant of trenbolone acetate and estradiol on growth, feed efficiency and carcass composition of Holstein and beef breed steers. *Journal of Animal Science* 69:4696-4702.

Perry, T. C., D. G. Fox, and D. H. Beermann. 1991b. Influence of final weight, implants, and frame size on performance and carcass quality of Holstein and beef breed steers. Pp. 142-159 in *Holstein Beef Production: Proceedings from the Holstein Beef Production Symposium, February 13-15, 1991, Harrisburg, PA*. NRAES 44. Ithaca, NY: Northeast Regional Agricutural Engineering Service.

Sainz, R. D., F. Del Torra, and J. W. Oltjen. 1995. Compensatory growth and carcass quality in growth-restricted and refed beef steers. *Journal of Animal Science* 73:2971-2979.

Simpfendorfer, S. 1974. Relationship of Body Type, Size, Sex, and Energy Intake to the Body Composition of Cattle. Ph.D. Dissertation. Cornell University, Ithaca, NY.

Smith, G. M., D. B. Laster, and L. V. Cundiff. 1976. Characterization of biological types of cattle. II. Postweaning growth and feed efficiency of steers. *Journal of Animal Science* 43:37-47.

Taylor, C. S. 1968. Time taken to mature in relation to mature weight for sexes, strains, and species for domesticated mammals and birds. *Animal Production* 10:157-169.

Tedeschi, L. O. 2001. Development and Evaluation of Models for the Cornell Net Carbohydrate and Protein System: 1. Feed Libraries, 2. Ruminal Nitrogen and Branched-Chain Volatile Fatty Acid Deficiencies, 3. Diet Optimization, 4. Energy Requirement for Maintenance and Growth. Ph.D. Dissertation, Cornell University, Ithaca, NY.

Tylutki, T. P., D. G. Fox, and R. G. Anrique. 1994. Predicting net energy and protein requirements for growth of implanted and nonimplanted heifers and steers and nonimplanted bulls varying in body size. *Journal of Animal Science* 72:1806-1813.

Williams, C. B., J. W. Keele, and D. R. Waldo. 1992. A computer model to predict empty body weight in cattle from diet and animal characteristics. *Journal of Animal Science* 70:3215-3222.

Woody, H. D., D. G. Fox, and J. R. Black. 1983. Effect of diet grain content on performance of growing and finishing cattle. *Journal of Animal Science* 57:717-728.

13

Reproduction

INTRODUCTION

Reproduction is the main factor that limits the production efficiency of beef cattle (Dickerson, 1970; Dziuk and Bellows, 1983; Koch and Algeo, 1983). Productive beef females in the U.S. and Canadian beef industries are expected to reach puberty by 13 to 15 months of age, conceive early in the breeding season, produce their first calf at 23 to 24 months of age, return to estrus, conceive within approximately 82 days after calving, consistently produce a calf every 365 days until they are 10 or more years of age, and have slaughter value at the end of their productive life. Many factors affect the cow's ability to accomplish these production demands, such as nutrition (nutrient availability, intake, metabolism, and utilization), genotype (genetic make-up, size, and production potential), management practices, environmental conditions (weather, terrain, altitude), and health status, as well as the complex interactions between these factors. In any given beef cow-calf operation, meeting the nutritional needs of the herd is the most significant variable affecting reproductive competency and health. The largest loss in potential calf crop occurs because cows fail to become pregnant (Wiltbank et al., 1961; Bellows et al., 1979), and the interval from parturition to first estrus (postpartum interval [PPI]) largely determines the likelihood of cows becoming pregnant during the breeding season (Wiltbank, 1970). Undernutrition is a major factor affecting reproductive performance, resulting in delayed attainment of puberty, extended periods of postpartum anestrus, as well as compromised conception rates, embryonic survival, sexual behavior, and, as emerging documentation suggests, compromised developmental programming of offspring.

Nutrient requirements of beef cows vary throughout the yearly production cycle. Cow energy requirements are lowest after weaning, increase during pregnancy, and peak during early lactation. In addition, nutrient availability of grazed and harvested forages fluctuates by forage species, stage of maturity, soil type, and environmental conditions. This variation in nutrient requirements and nutrient availability typically results in cows increasing and decreasing in body weight (BW) and body condition score (BCS) in a cyclic pattern throughout the year. Short et al. (1990), in their review of the literature, stated that allocation of nutrients to various body functions is commonly referred to as nutrient partitioning, and proposed that nutrients are partitioned by priority to first maintain life and then to propagate the species. They further proposed the approximate order of nutrient priority to be (1) basal metabolism, (2) activity to gather food, (3) growth, (4) basic energy reserves, (5) maintenance of pregnancy, (6) lactation to support an existing offspring, (7) accumulation of additional energy reserves, (8) estrous cycles and initiation of pregnancy, and (9) accumulation of excess energy reserves.

There is an energetic cost associated with both the synthesis and catabolism of body energy reserves and it has been assumed that maintaining weight is biologically more efficient. Nonetheless, Freetly and Nienaber (1998) conducted a study with nonpregnant, nonlactating cows and concluded that the efficiency of energy retention in cows that were feed-restricted, followed by realimentation, did not differ from cows fed to maintain energy balance. In a second study with pregnant beef cows, Freetly et al. (2008) did not observe a decrease in total amount of feed when pregnant cows were fed to gain the same amount of BW during the study using different patterns of BW gain. These data suggest that cows have the ability to adapt energy metabolism to periods of feed restriction and realimentation, which allows flexibility in the utilization of forage resources and timing of supplementation strategies.

Depending on the physiological status of the animal and the duration and severity of poor-feed conditions, cyclic loss and gain in weight and body condition can be biologically less efficient, but more economically efficient if total production system efficiencies are not compromised (Freetly et al., 2005). Although it might not be practical or economically advantageous in some cases to supplement all the nutrients necessary to maintain animal weight and body reserves throughout the year, emerging data in the area of maternal nutrition and the impact on developmental programming of offspring suggest that these traditional concepts might need to be reevaluated. The clear challenge is to make sure that nutrient requirements are met during critical stages of production and development that allow optimal performance.

BODY CONDITION SCORE

Optimal management of body energy reserves is critical for economic success within beef production systems. Cows at either extreme of body energy reserve (too fat or thin) are at risk from metabolic challenges, disease, decreased milk yield, low conception rates, dystocia, and compromised prenatal and postnatal outcomes (Meyer et al., 2010a,b; Funston et al., 2012b; Long et al., 2012). Considerable evidence in sheep supports observations in cattle regarding negative outcomes associated with nutrient restriction (too thin) or nutrient excess (too fat) on both maternal and offspring outcomes (Wu et al., 2006; Swanson et al., 2008; Neville et al., 2010; Meyer et al., 2011, 2013; Yunusova et al., 2013). Overconditioning of beef cows is expensive and often leads to calving problems and lower dry matter intake (DMI; Buskirk et al., 1992) during early lactation. Conversely, thin cows that do not have sufficient reserves for maximal milk production will have extended PPI, which will delay rebreeding and could deliver compromised calves that are predisposed to long-term health and production challenges.

Body energy reserves at calving and energy balance between calving and breeding are important factors affecting not only ovulation, but also the length of the postpartum anestrous interval. A plethora of literature indicates that the PPI is heavily influenced by BCS at calving and is optimized when cows calve in moderate, or near-moderate, body condition (Whitman, 1975; Dunn and Kaltenbach, 1980; Graham et al., 1984; Richards et al., 1986; Houghton et al., 1990). Houghton et al. (1990) reported that the postpartum anestrous interval decreased as BCS at calving increased, but this did not necessarily mean higher first-service or season-long pregnancy rates in mature beef cows. Thin mature cows at calving in the Houghton et al. (1990) study had an extended PPI, but "if" they cycled before termination of the breeding season, their fertility was very high—the challenge is getting thin mature cows to initiate cyclicity during a limited breeding season. In the Houghton et al. (1990) study, mature cows that were in moderate body condition at calving and maintained condition to breeding, cows that were moderately thin at calving and approached moderate condition by breeding, and moderately fleshy cows at calving and approached moderate condition by breeding all had shorter PPI and higher breeding-season pregnancy rates than either thin cows getting thinner or fleshy cows getting fatter between calving and breeding. In contrast, primiparous beef females seem to be more sensitive to nutritional and BCS changes than mature cows. Lalman et al. (1997) reported that thin primiparous females at calving that managed to approach moderate body condition by breeding have a lower pregnancy rate than those that calve in moderate body condition and either maintain or gain body condition before breeding. Collectively, results of these studies strongly suggest that cows need to be managed to obtain near-moderate BCS at calving, and then at least maintain condition to breeding to optimize PPI, first-service conception rate, breeding season pregnancy rate, and preweaning calf performance.

Body condition scores are a subjective measure by which we describe, with objective criteria, the body energy content of an animal. The BCS system attempts to define an array of tissue and whole animal responses that are both continuous and dynamic with a visual appraisal from a distance of 0 to 30 m. Consequently, the opportunity for errors in both BCS assessment and in expectations of what a BCS assessment actually means in terms of biology can be significant. While subjective, a visual BCS system has been useful in refining beef nutrient requirements and overall animal energy balance within their environment. The Commonwealth Scientific and Industrial Research Organisation (CSIRO, 1990, 2007) and the French National Institute for Agricultural Research (INRA, 1989) adopted the 0 to 5 BCS system of Wright and Russel (1984a,b) to describe body energy reserves and develop nutrient recommendations. Research from Oklahoma State University (Cantrell et al., 1982; Wagner, 1984; Selk et al., 1988) and Colorado State University (Whitman, 1975) developed a 9-point body condition scoring system, whereas the Purdue University researchers (Houghton et al., 1990; Buskirk et al., 1992; Graffam, 1992) used a 5-point scale with minus, average, and plus within each BCS. When the data from Houghton et al. (1990) were transformed to a 1 to 9 BCS scale, the resulting percent carcass lipid content was 4.2, 7.6, 10.9, 14.3, 17.7, 21.1, 24.4, 27.8, and 31.3, respectively. Herd and Sprott (1986) used a 9-point scale and reported 0, 4, 8, 12, 16, 20, 24, 28, and 32% body fat, for BCS 1 through 9, respectively. The National Research Council (NRC, 1996, 2000) publications adopted the 1 to 9 body condition scoring system to describe energy reserves and define energy requirements. Body fat percentages of 3.77, 7.54, 11.30, 15.07, 18.84, 22.61, 26.38, 30.15, and 33.91 for BCS 1 to 9, respectively, were developed from equations using the body composition data set provided by the U.S. Meat Animal Research Center (MARC; C. L. Ferrell, U.S. MARC, Clay Center, NE, personal communication, June 23, 1995). This data set included body composition of 105 mature cows of diverse breed types and body sizes. Characteristics of this data set were EBW = 0.851 × SBW; mean EBW = 546 (range 302 to 757) kg, percentage of empty body fat = 19.3% (range 4.03 to 31.2), percentage of empty body protein = 15.3% (range 13.2 to 18.0), and BCS = 5.56 (range 2.25 to 8.0), where EBW is empty body weight and SBW is shrunk body weight. The developed equations were validated using an independent set of data collected from 65 mature cows (C. L. Ferrell, U.S. MARC, Clay Center, NE, personal communication, June 23, 1995), which included 9-year old cows of diverse sire breeds and either Angus or Hereford dams. The resulting best-fit equations to describe relationships between BCS and empty body percentage of fat, protein, water, and ash were linear (Figure 13-1). A zero-intercept model was used to describe the relationship between percentage of empty body fat and BCS. In the validation of this model, BCS accounted for 67, 52, and 66% of the variation in body fat, body protein, and body energy, respectively. The following equations were used in the Beef Cattle Nutrient Requirements Model (BCNRM; Chapter 19) to predict body fat, body protein, water, ash, and body energy reserves:

Body fat, protein, water, and ash for each BCS (1 to 9) are computed as

$$AFat_{BCS} = 0.037683 \times BCS; \quad r^2 = 0.67, \qquad \text{(Eq. 13-1)}$$

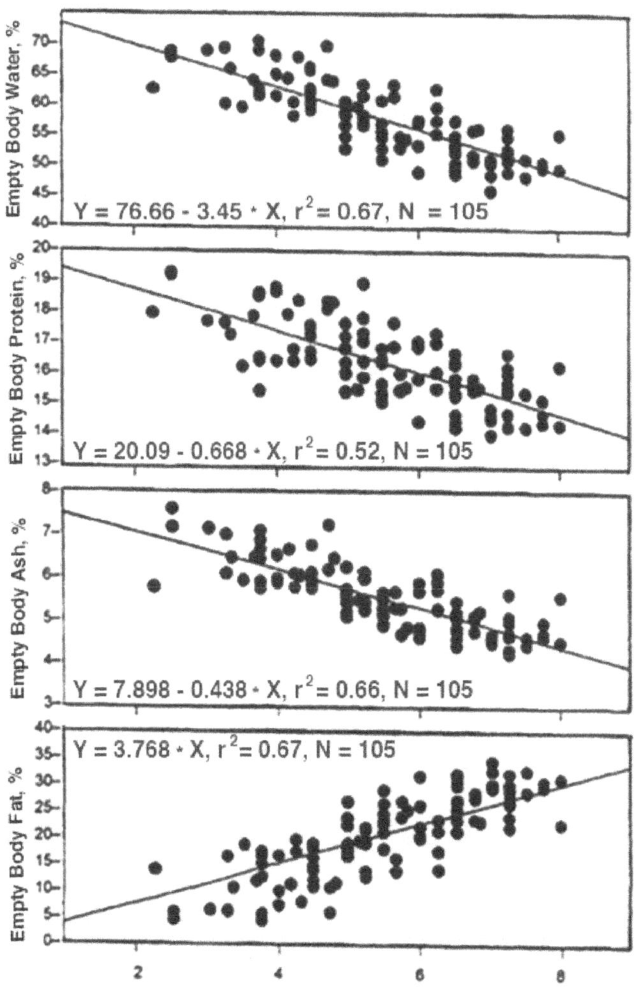

FIGURE 13-1 Relationships between body condition score (BCS) and empty body percentage of fat, protein, water, and ash (NRC, 1996, 2000).

$$AProt_{BCS} = 0.200886 - 0.0066762 \times BCS; \quad r^2 = 0.52, \qquad \text{(Eq. 13-2)}$$

$$AWaterBCS = 0.766637 - 0.034506 \times BCS; \quad r^2 = 0.67, \qquad \text{(Eq. 13-3)}$$

$$AAsh_{BCS} = 0.078982 - 0.00438 \times BCS; \quad r^2 = 0.66, \qquad \text{(Eq. 13-4)}$$

where

$AFat_{BCS}$ is the proportion of empty body fat at BCS, %;
$AProt_{BCS}$ is the proportion of empty body protein at BCS, %;
$AWater_{BCS}$ is the proportion of empty body water at BCS, %; and
$AAsh_{BCS}$ is the proportion of empty body ash at BCS, %.

Table 13-1 provides the description of the 9-point BCS system recommended by the current committee, and Table 13-2 provides a decision tree that can be used to help identify the BCS of beef cows.

Relationship Between Body Condition Score and Weight

The CSIRO (1990, 2007) nutrient requirements assume 83 kg of BW change per unit change in BCS (1 to 5 BCS system), which is equivalent to about 46 kg of BW change per unit BCS change on a 1 to 9 BCS scale. Shrunk body weights in the Houghton et al. (1990), Buskirk et al. (1992), Graffam (1992), and Ferrell and Jenkins (1996) studies with mature cows, as well as the Graffam (1992), and Lalman et al. (1997) studies with primiparous beef females were linearly related to BCS. When the Houghton et al. (1990), Buskirk et al. (1992), and Graffam (1992) studies were transformed from a 1^- to 5^+ BCS scale to a 1 to 9 scale, the equivalent weight changes were 33.3, 37.8, and 38.6 kg, respectively, per unit change of BCS in mature cows, whereas Wagner (1984) reported a 38-kg change, and Ferrell and Jenkins (1996) reported a 51-kg change. The NRC (2000) equations, based on the Cornell model (Fox et al., 1992), predict percent SBW changes from base weight (mature, BCS 5) for beef females in BCS 1 to 9 of 76.5, 81.3, 86.7, 92.9, 100.0, 108.3 118.1, 129.9, and 144.3%, respectively. The Fox et al. (1992) numbers suggest that fat cows require more weight, and therefore more energy, to change one BCS than thin cows. If BCS represents body energy reserves and there are differences in energy density between fat and protein, then one could argue that fat cows gaining almost exclusively fat should require a smaller weight change per BCS to equal the body energy change of thin cows that are gaining primarily lean muscle mass. On the other hand, the differential turnover rate between protein and fat and the associated differences in efficiency of metabolizable energy (ME) use for protein and fat could negate the difference in energy between lean and adipose deposition, which would

TABLE 13-1 Description of Cow Body Condition Score (BCS)

BCS	Description[a]
1	**Emaciated.** The cow is severely emaciated and physically weak. Bone structure of shoulder, ribs, spinous processes, transverse processes, hooks, and pins are sharp to the touch and easily visible. No evidence of fat deposits or normal muscling. Cattle in this score are extremely rare and are usually inflicted with a disease and/or parasitism. Cows in this condition would typically produce carcasses that qualify for the USDA market news category of "Lights." Empty body fat content would be approximately 3.8%.
2	**Very Thin.** The cow appears emaciated, similar to 1, above, but not weak. Muscle atrophy is significant in the shoulder, over the loin and rump, and through the hindquarters. The spinous and transverse processes, hooks, and pins feel sharp to the touch and are easily seen. Ribs can be easily identified. Cows in this condition would typically produce carcasses that qualify for the USDA market news category of either "Leans" or "Lights" depending on carcass size. Empty body fat content would be approximately 7.5%.
3	**Thin.** The cow appears thin with no fat in the brisket. There is evidence of muscle atrophy over the shoulder, loin, and rump and through the hindquarter, but there is the beginning of some fat deposition over the shoulder, loin and rump. The foreribs have slight evidence of fat deposition, but the last three or more ribs (11th, 12th, and 13th) are visible. Vertebrae of the spine are easily visible and can be identified individually by touch. Spaces between the processes have some evidence of some fat infiltration. Bone structure of the hooks and pins are evident under the hide. Cows in this condition would typically produce carcasses that qualify for the USDA market news category of either "Leans" or "Lights" depending on carcass size. Empty body fat content would be approximately 11.3%.
4	**Borderline.** The cow appears thin, with the last two ribs (12th and 13th) easily visible, especially in mature cows with a big spring of rib and broad, flat ribs. Muscle atrophy over the shoulders and loin and through the hindquarters is evident, but is approaching normal. Movement of the muscles in the shoulder is prominent just under the hide as the animal moves. Individual spinous processes along the topline, and transverse processes along the edge of the loin between the hooks and last rib, are not visible, but can be easily palpated with minimal pressure and feel rounded rather than sharp. The hooks and pins are covered in minimal fat and easily identified. Cows in this condition would typically produce carcasses that qualify for the USDA market news category of either "Leans" or "Lights" depending on carcass size. Empty body fat content would be approximately 15.1%.
5	**Moderate.** There is slight evidence of fat deposition in the brisket. Muscle expression in the shoulder, loin, rump, and hindquarters are normal. A thin layer of fat covers the muscles in the shoulder, and when the animal is in motion; the muscle and scapula movement under the hide are not prominent. The last two ribs (12th and 13th) can only be seen if the cow has less than normal gut fill. Individual spinous processes along the topline and transverse processes along the loin edge between the hooks and last rib appear smooth and are not visible, but they can be palpated with firm pressure. The hooks and pins are covered with a layer of fat, but still distinguishable. Areas on each side of the tail head are fairly smooth, but not mounded. Cows in this condition would typically produce carcasses that qualify for the USDA market news category of "Boners" (Boning Utility). Empty body fat content would be approximately 18.8%.
6	**Good.** Some fat deposition in the brisket is evident. The cow exhibits a smooth appearance throughout without any evidence of muscle atrophy. A layer of fat covers the muscles in the shoulder, and shoulder movement visually takes on a fluid motion. There is noticeable sponginess over the foreribs just behind the shoulder. The ribs on young cows are fully covered by fat and cannot be detected visually. Flat, wide ribs on older cows may be seen. The spinous processes along the topline and transverse processes between the hooks and last rib are embedded in muscle and fat, and can only be felt with firm pressure. The hooks are embedded in fat and appear round and smooth. The pins are embedded in fat and there is noticeable sponginess on each side of the tail head. Cows in this condition would typically produce carcasses that qualify for the USDA market news category of "Boners" (Boning Utility). Empty body fat content would be approximately 22.6%.
7	**Fleshy.** The brisket is full, but not distended. A layer of fat covers the muscles in the shoulder, and the movement appears fluid. The ribs appear smooth, and there is a uniform layer of fat from the forerib to the last rib. The spinous processes along the topline and transverse processes between the hooks and last rib are embedded in fat, and the topline is beginning to take on a square appearance. The pins are embedded in a layer of spongy fat, and there are pones of fat forming on either side of the tail head. Cows in this condition would typically produce carcasses that qualify for the USDA market news category of either "Breakers" or "Boners" (Boning Utility). Empty body fat content would be approximately 26.4%, which would be approaching a similar carcass fat content of a steer or heifer that grades USDA Choice.
8	**Obese.** The neck appears short and thick and the brisket is distended with fat. There is no evidence of bone structure in the shoulder, along the topline, hooks, or pins. A thick layer of fat covers the shoulder, and shoulder muscle movement under the hide is not evident. The animal appears very square and blocky over the topline and smooth along the sides. Fat cover over the shoulder, topline, and ribs appear spongy both visually and by touch. The pins are embedded in pones of fat on both sides of the tail head. There is evidence of fat deposition in the udder. Animal movement may be somewhat impaired due to excess fat. Cows in this condition would produce carcasses that typically qualify for the USDA market news category of "Breakers." Empty body fat content would be approximately 30.2%.
9	**Very Obese.** These cows are very obese and are rarely seen. The animal appears very square and blocky over the topline, smooth along the sides, and spongy throughout. Bone structures are difficult to identify and the animal appears short necked with a full, distended brisket. The pins are embedded in patchy pone fat on both sides and over the top of the tail head. There is evidence of significant fat deposition in the udder. Animal mobility may be seriously impaired due to excess fat. Cows in this condition would produce carcasses that typically qualify for the USDA market news category of "Breakers." Empty body fat content would be approximately 33.9%.

[a]Adapted from Whitman (1975); Cantrell et al. (1982); Wagner (1984); Selk et al. (1988); Herd and Sprott (1986); Houghton et al. (1990); Buskirk et al. (1992); Fox et al. (1992); NRC (1996, 2000); and Selk (2009).

TABLE 13-2 Cow Body Condition Score (BCS) Decision Tree[a]

Reference Points	Body Condition Score								
	1	2	3	4	5	6	7	8	9
Physically weak	Yes	No	No	No	No	No	No	No	No
Muscle atrophy[b]	Yes	Yes	Some	Slight	No	No	No	No	No
Fat in brisket	No	No	No	No	Slight	Some	Full	Distended	Extreme
Fat over shoulder	No	No	No	No	Slight	Some	Yes	Yes	Yes
Visible ribs, no.	All	All	3-5	1-2	Maybe 1-2	None	None	None	None
Visible spinous processes (vertebrae along topline)	Yes	Yes	Yes	Slight	No	No	No	No	No
Visible transverse processes between hooks and last rib	Yes	Yes	Yes	Slight	No	No	No	No	No
Visible hooks/pins	Yes	Yes	Yes	Yes	Yes	Some	Slight	No	No
Tail head fat pones	No	No	No	No	No	Slight	Some	Yes	Extreme
Fat in udder	No	No	No	No	No	No	Slight	Yes	Yes
Mobility	Poor	Marginal	Ok	Ok	Ok	Ok	Ok	Marginal	Poor

[a]Adapted from Whitman (1975); Cantrell et al. (1982); Wagner (1984); Herd and Sprott (1986); Selk et al. (1988); Houghton et al. (1990); Buskirk et al. (1992), Fox et al. (1992); NRC (1996, 2000), and Eversole et al., 2009.

[b]Muscles of loin, rump, and hindquarter appear concave, indicating a loss of muscle tissue.

make BW change across BCS similar. To address the issue of BW and energy changes associated with changes in BCS, data from the literature were used to reevaluate the SBW weight/BCS relationship. Using the average of nine mean postpartum SBW weights (mean = 503.7 kg, range = 467.2 to 547.8) and BCS (mean = 2.84, range = 2.61 to 3.27, 1^- to 5^+ scale) reported by Houghton et al. (1990), Buskirk et al. (1992), and Graffam (1992), an SBW of 514.66 kg was calculated when transformed to a BCS 5 cow base weight (1 to 9 scale). These same three studies reported an average SBW change/BCS of 65.8 kg (range = 60 to 69.5, 1^- to 5^+ scale), which is 36.57 kg/BCS when transformed to a 1 to 9 scale. An adjustment factor can then be calculated for BCS 1 to 9, respectively, as 71.58, 78.69, 85.79, 92.90, 100.00, 107.11, 114.21, 121.32, 128.42% of the value assigned to a mature BCS 5 cow. This represents a 7.105% change in SBW/BCS and results in a change in SBW/BCS of 28.4, 35.5, 42.6, 49.7 and 56.8 kg for 400-, 500-, 600-, 700-, and 800-kg mature cows, respectively. Similarly, mean cow weight, BCS, and SBW change/BCS in the U.S. MARC mature cow data set (C. L. Ferrell, U.S. MARC, Clay Center, NE, personal communication, June 13, 1995) reported in NRC (1996, 2000) were 546, 5.56, and 44 kg, respectively. When these data are converted to a mature, BCS 5 equivalent basis, the mean cow weight was calculated to be 521.4 kg with an 8.439% change in SBW/BCS. The 36.57-kg SBW weight change for 514.3-kg cows seems more realistic and consistent with past recommendations than 44 kg and, therefore, the 7.105% change in SBW/BCS was adopted as the default value in the BCNRM. Users are given the opportunity, however, to change the default value in the software for the BCNRM if desired. The committee believes that this approach is a significant improvement to the NRC (1996, 2000) model, but also recognizes that the change in SBW/BCS might be more appropriately described by a hyperbolic function, and not a constant, within a given mature cow size. Nonetheless, there are limited data in the literature with widely diverse breeds and cow sizes to define this hyperbolic function.

Studies conducted with primiparous beef females suggest that it takes between 33.0 (Lalman et al., 1997) and 38.6 (Graffam, 1992) kg of weight change to gain one BCS, but only 8.6 kg (Graffam, 1992) to lose one BCS. In contrast, Ripberger (1997) and Bradford (1998) reported that primiparous beef females required 70.2 kg and 61.7 kg of liveweight change, respectively, to increase one BCS. A change in SBW/BCS of 50.9 kg results when these four literature values are averaged. A difference in weight change per unit of BCS between multiparous and primiparous females makes sense because the primiparous female would need to gain weight (skeletal growth and muscle) to just maintain body condition. This difference in SBW/BCS for primiparous females versus multiparous females has not been accounted for in the NRC (1996, 2000) model, but the committee recognizes that change in SBW associated with growth should be accounted for. Literature values that define the change in SBW/BCS for primiparous females gaining and losing body energy reserves are limited; therefore, no adjustment was made in this edition of the *Nutrient Requirements of Beef Cattle*. Nonetheless, one could argue that the four studies previously cited warrant an adjustment of 0.4 and 1.6, respectively, be made to the mature cow SBW/BCS value of 7.105% for primiparous females that are losing and gaining weight, respectively. When primiparous females are losing weight, the adjustment factor in the BCNRM can be changed from 1.00 to 0.40 resulting in an 11.5-kg and 14.6-kg loss in SBW, respectively, for BCS 5, 400- and 500-kg primiparous females to lose one BCS. Similarly, changing the default value from 1.0 to 1.6 would result in a 46-kg and 58.6-kg gain in SBW, respectively, for

400- and 500-kg primiparous females to gain one BCS. In both cases, these default values can be changed by the user as more data become available. The revised weight adjustment factor and estimated SBW for cows in BCS 5 are calculated as follows, with results shown in Table 13-3.

Weight adjustment factor and SBW at BCS 5 are calculated as

$$WAF_{BCS} = 1 - 0.07105 \times (5 - BCS); \quad \text{(Eq. 13-5)}$$

$$SBW_5 = SBW/WAF_{BCS}, \quad \text{(Eq. 13-6)}$$

where

WAF_{BCS} is the weight adjustment factor at BCS;
BCS is the current body condition score (1 to 9);
SBW_5 is the shrunk body weight at BCS 5, kg; and
SBW is the current shrunk body weight, kg.

Relationship Between Body Condition Score and Net Energy Requirements

When Bradford (1998) estimated fasting heat production (FHP) in mature cows using the average of all linear, quadratic, and linear plus quadratic equations, estimates were 72.8, 119.1, and 77.8 kcal/kg$^{0.75}$ of SBW, with all R^2 values ranging between 0.81 and 0.85. The linear estimates of FHP obtained by all methods of analysis ranged from 70.3 to 74.5 kcal/kg$^{0.75}$ of SBW. Similarly, Buskirk et al. (1992) estimated FHP in mature cows at 72.5 kcal/kg$^{0.75}$ of SBW (R^2 = 0.94), which is not significantly different than the classical 77 kcal/BW$^{0.75}$ used to define the net energy required for maintenance (NEm). Graffam (1992) validated the Buskirk et al. (1992) FHP equations for mature cows and gestating primiparous beef females, but suggested that FHP of lactating primiparous beef females ranged between 111.5 and 121.3 kcal/kg$^{0.75}$ of SBW across all equations developed. At the time, these higher values seemed unrealistic and did not conform to the classical 77 kcal/kg$^{0.75}$ of SBW definition of maintenance. The young beef female is, however, the only component of the beef production system that is expected to simultaneously grow and lactate, and it is realistic to expect that lactating primiparous beef females have a higher NEm requirement. Ferrell and Jenkins (1985) reported that the variation in NEm requirements seems to be positively associated with an animal's genetic potential for growth rate and milk production. Research conducted by Graham (1980) also suggests that a higher NEm requirement could be expected in the young, lactating female. Ripberger (1997) reported an FHP of 97.2 kcal/kg$^{0.75}$ of SBW (standard error with limits of 96.8 and 97.6 kcal; 25% higher than 77 kcal/kg$^{0.75}$ of SBW) and 106.9 (standard error with limits of 106.3 and 107.5 kcal; 39% higher than 77 kcal/kg$^{0.75}$ of SBW), respectively, for Angus and Simmental primiparous females when the best-fit model (linear) was used. Ripberger (1997) speculated that a combination of growth, age, and lactation, as well as the rate of protein synthesis and degradation, contributed to an increased demand on the metabolically active organs (especially the liver and gastrointestinal tract; Ferrell and Jenkins, 1985). In contrast, Bradford (1998) did not observe a significant difference in FHP between lactating primiparous Angus vs. Simmental beef females and determined that a quadratic equation resulted in the highest R^2 (0.85) and yielded an FHP estimate of 111.5 kcal/kg$^{0.75}$ of SBW (standard error with limits of 107.5 and 115.6 kcal; 45% higher than 77 kcal/kg$^{0.75}$ of SBW). These results are similar to the range of 111.5 to 121.3 kcal/kg$^{0.75}$ of SBW reported by Buskirk et al. (1992). There was a significant metabolizable protein (MP) × ME interaction in the Bradford (1998) study, however, which might have inflated the FHP value because energy is a major factor that not only determines amino acid supply to the tissue, but also controls the nature and extent of tissue metabolism of amino acids (Asplund, 1994). In the absence of adequate energy in the form of glucose, glucogenic amino acids can be metabolized for energy, even when the supply of amino acids is limiting, or when nonglucose

TABLE 13-3 Estimated Shrunk Body Weights for Cows Differing in Body Condition Score (BCS) and Weight

		Mature Cow Shrunk Body Weight (BCS, kg)				
BCS	Weight Adjustment[a]	400	500	600	700	800
1	0.716	286	357	428	500	571
2	0.787	314	393	471	550	628
3	0.858	343	429	514	600	686
4	0.929	371	464	557	650	743
5	1.000	400	500	600	700	800
6	1.071	429	536	643	750	857
7	1.142	457	572	686	800	914
8	1.213	486	607	729	850	972
9	1.284	514	643	772	900	1029

[a]Adjustments are based on an SBW mean of 514.66 kg for a BCS = 5 cow and 36.57 kg of SBW change/BCS from transformed data reported by Houghton et al. (1990), Buskirk et al. (1992), and Graffam (1992).

energy is adequate (Asplund, 1994). Regardless, the NEm requirement of lactating primiparous beef females seems to be somewhat higher than the classical 77 kcal/kg$^{0.75}$ of SBW, and more research is needed to characterize true maintenance energy requirements. The BCNRM does not adjust NEm for lactating primiparous females, but one could argue that the NEm requirement might need to be increased at least 25% (97 kcal/kg$^{0.75}$ of SBW) as reported Ripberger (1997). The default adjustment in the BCNRM has been set to 1.00 × 77 kcal/kg$^{0.75}$ of SBW for lactating primiparous females, however; this number can be increased in the software for the BCNRM by the user as more data become available.

The CSIRO (1990, 2007) assumed the energy content of 1.0-kg weight change is 6.4 Mcal/kg for British breeds and 5.5 Mcal/kg for the larger European breeds. These are similar to the 6.0 Mcal/kg value used by INRA (1989). Buskirk et al. (1992) developed equations to predict energy lost and gained (data transformed from a BCS 1^- to 5^+ system to a 1 to 9 system) in mature beef cows that resulted in net energy for weight change (NE_Δ)/kg SBW values of 2.16, 2.88, 3.61, 4.33, 5.06, 5.79, 6.53, 7.26, and 7.96 Mcal, respectively, for BCS 1 through 9. These numbers fit the biological limits of 1.2 and 8.0 Mcal/kg energy content of fat-free lean (Reid et al., 1955; Garrett and Hinman, 1969) and adipose tissue (NRC, 1984). The Buskirk et al. (1992) equations were validated by Graffam (1992) for mature gestating and lactating cows. Fox et al. (1992) used the 1 through 9 BCS system and modeled results from 14 studies of body composition in cows to predict weight and energy lost or gained with changes in age, mature size, and condition score to develop the NRC (1996, 2000) requirements and concluded that the energy content was a constant of 5.82 Mcal of net energy (NE)/kg of weight change across all condition scores and cow weights. Biologically, however, it seems logical that the energy content of gain in a thin cow (primarily lean gain) should be lower than the energy content of gain in a fat cow (primarily adipose gain). Predicted cow weight change/day values are fairly similar for moderately sized cows in the mid-BCS range between the Buskirk et al. (1992) energy system, which uses a variable energy content of gain, and the NRC (1996, 2000) energy system, which uses a constant body energy content. For example, when cows are fed submaintenance energy diets, the two systems are fairly similar for cows in BCS 4 through 7. However, the Buskirk et al. (1992) system predicts more weight loss/d in thin, big cows and a smaller weight loss/day in thin, small cows compared with NRC (1996, 2000). Similarly, the Buskirk et al. (1992) equations predict less weight loss/d in fat cows compared with NRC (1996, 2000). When diets provide energy above maintenance, the Buskirk et al. (1992) equations predict more rapid gain/day in thin cows (BCS 1 to 4) across all mature sizes, similar gains for 600-kg cows in the BCS range of 5 to 9, and slower weight gain/d in smaller, fat cows (BCS 7 to 9) compared with NRC (1996, 2000). In an attempt to resolve this issue, the body energy reserves model described below is different from that in the previous editions of the *Nutrient Requirements for Beef Cattle*. It is based on the Fox et al. (1999) and Tedeschi et al. (2006) models, except for the percent BW change per unit change of BCS, which is currently assumed to be 7.105% instead of 6.85% as proposed by Fox et al. (1999). Body fat and body protein provided by the U.S. MARC data set (C. L. Ferrell, U.S. MARC, Clay Center, NE, personal communication, June 23, 1995) and the estimated SBW values (7.105% change in SBW/BCS) were used in the BCNRM to estimate body energy reserves for each BCS that is presented in Table 13-4. Because ash and water do not contribute to body energy reserves, total body energy can be calculated using Eq. 13-1 through Eq. 13-7 for each BCS, and empty body energy reserves can be calculated for cows differing in BCS and weight according to the following equations:

TABLE 13-4 Estimated Body Energy Reserves for Cows Differing in Body Condition Score (BCS) and Weight

	Empty Body Composition[a]		Mature Cow Empty Body Energy Reserves, Mcal						
BCS	Fat, %[b]	Protein, %[c]	400 kg	500 kg	600 kg	700 kg	800 kg	Mcal/kg EBW gain[d]	Mcal/kg EBW loss[e]
1	3.77	19.42	356	445	534	623	712	4.22	3.69
2	7.54	18.75	476	595	714	833	952	4.76	4.22
3	11.30	18.09	611	764	917	1,070	1,223	5.30	4.76
4	15.07	17.42	762	952	1,143	1,333	1,524	5.84	5.30
5	18.84	16.75	928	1,160	1,392	1,624	1,856	6.38	5.84
6	22.61	16.08	1,109	1,386	1,664	1,941	2,218	6.91	6.38
7	26.38	15.42	1,306	1,632	1,958	2,285	2,611	7.45	6.91
8	30.15	14.75	1,517	1,897	2,276	2,655	3,035	7.99	7.45
9	33.91	14.08	1,744	2,181	2,617	3,053	3,489		7.99

[a]EBW = 0.851 × SBW (NRC, 1996, 2000).
[b]EB Fat × 9.4 = Mcal of empty body energy from fat.
[c]EB Protein × 5.7 = Mcal empty body energy from protein.
[d]Mcal required to gain 1 kg of EBW.
[e]Mcal required to lose 1 kg of EBW.

$$EBW = 0.851 \times SBW, \quad \text{(Eq. 13-7)}$$

$$TFat_{BCS} = AFat_{BCS} \times EBW_{BCS}, \quad \text{(Eq. 13-8)}$$

$$TProt_{BCS} = AProt_{BCS} \times EBW_{BCS}, \quad \text{(Eq. 13-9)}$$

$$TE_{BCS} = 9.4 \times TFat_{BCS} + 5.7 \times TProt_{BCS}, \quad \text{(Eq. 13-10)}$$

where

$TFat_{BCS}$ is the empty body fat at BCS, kg;
$TProt_{BCS}$ is the empty body protein at BCS, kg; and
TE_{BCS} is the total body energy reserves at BCS, Mcal.

The amount of body energy reserves needed to gain or lose 1 BCS can then be computed as

$$\Delta TE_{BCS,Lose} = TE_{BCS} - TE_{BCS-1} \text{ when } 9 \geq BCS \geq 2, \quad \text{(Eq. 13-11)}$$

$$\Delta TE_{BCS,Gain} = TE_{BCS+1} - TE_{BCS} \text{ when } 1 \geq BCS \geq 8, \quad \text{(Eq. 13-12)}$$

where

$\Delta TE_{BCS,Lose}$ is the loss of total body energy reserves (mobilization) per BCS, Mcal; and
$\Delta TE_{BCS,Gain}$ is the gain of total body energy reserves (deposition) per BCS, Mcal.

During mobilization, 1 Mcal of retained body energy will be substituted for 0.80 Mcal of NE available for maintenance (NEma), and during repletion, 1 Mcal NEma is converted to an ME basis by dividing NE by 0.60. Using these coefficients, days to change 1 BCS can be computed by dividing this ME equivalent retained body energy by the ME balance (ΔME). The difference of NE mobilized or replenished per BCS change between these two models increases as BW or BCS increases. The amount of body energy reserve needed to gain or lose 1 BCS can be calculated as follows, with results shown in Table 13-5:

$$Days_{Lose} = \Delta TE_{BCS,Lose} \times 0.80/0.60 \times \Delta ME, \text{ when } \Delta ME < 0, \quad \text{(Eq. 13-13)}$$

$$Days_{Gain} = \Delta TE_{BCS,Gain}/0.60 \times \Delta ME, \text{ when } \Delta ME > 0, \quad \text{(Eq. 13-14)}$$

where

$Days_{Lose}$ is the days to lose 1 BCS, d; and
$Days_{Gain}$ is the days to gain 1 BCS, d.

PUBERTY AND HEIFER DEVELOPMENT

Heifer puberty has been extensively reviewed in the older literature, but Funston et al. (2012a) recently reviewed the nutritional aspects of heifer development, attainment of puberty, and subsequent reproductive competency, and concluded that the major goal in heifer development programs should be the production of a sound, functional, low-cost pregnant heifer. Heifer puberty and reproductive performance are directly influenced by energy balance and plane of nutrition (Short and Adams, 1988; Butler and Smith, 1989; Swanson, 1989; Randel, 1990; Robinson, 1990), while postweaning growth rate has been reported to have an inverse relationship with age at puberty (Wiltbank et al., 1966, 1969, 1985; Arije and Wiltbank, 1971; Short and Bellows, 1971; Ferrell, 1982). Additionally, heifers that reach puberty early, and have several estrous cycles before the breeding season, conceive earlier during the breeding season (Byerley et al., 1987) and can have a higher pregnancy rate than those that reach puberty after initiation of the breeding season (Short and Bellows, 1971; Byerley et al., 1987). Lesmeister et al. (1973) also reported that heifers calving early during their first calving season have greater lifetime calf production than those calving late and are more likely to conceive earlier during their second breeding season.

TABLE 13-5 Energy Reserves (Mcal) for Cows with Different Body Weights and Condition Scores to Change One Body Condition Score[a]

BCS		Current SBW at BCS 5, kg								
Gain	Lose	400	450	500	550	600	650	700	750	800
1 → 2	2 → 1	120	135	150	165	180	195	210	225	240
2 → 3	3 → 2	135	152	169	186	203	220	237	254	271
3 → 4	4 → 3	151	169	188	207	226	245	264	282	301
4 → 5	5 → 4	166	187	207	228	249	270	290	311	332
5 → 6	6 → 5	181	204	226	249	272	294	317	340	362
6 → 7	7 → 6	196	221	246	270	295	319	344	368	393
7 → 8	8 → 7	212	238	265	291	318	344	371	397	424
8 → 9	9 → 8	227	255	284	312	341	369	397	426	454

[a]Example: It takes 228 Mcal of body energy reserves for a 550-kg SBW cow to change from a BCS 4 to 5 and to change from a BCS 5 to 4.

Both age and weight at puberty differ substantially among breeds of cattle (Laster et al., 1972, 1976, 1979; Stewart et al., 1980; Sacco et al., 1987). Within beef breeds, those having larger mature size tend to reach puberty at a later age and at a heavier weight. *Bos indicus* heifers tend to reach puberty at an older age than *Bos taurus* heifers, and heifers from higher milk-producing breeds are generally younger at puberty than those from breeds having lower milk production. Some of these differences are likely the result of direct maternal effects expressed through higher preweaning gains of calves from higher milk-producing breeds. Evidence also indicates that maternal nutrient supply during gestation can affect the reproductive performance of offspring (Funston et al., 2010), suggesting that nutritional management of both the dam and the female offspring in utero is relevant in terms of long-term reproductive success.

Numerous data sets are available that indicate neither age nor weight is a reliable indicator of reproductive development, but that threshold values for both age and weight must be obtained before puberty can occur. Brody (1945) suggested that puberty occurred around the point where energy was shifting away from lean growth and toward energy reserves. This conclusion is similar to the "physiological maturity" concept proposed by Joubert (1963) and the "target weight" concept proposed by Lamond (1970). These concepts have been used by Spitzer et al. (1975), Dziuk and Bellows (1983), Wiltbank et al. (1985), Patterson et al. (1992), Larson (2007), and Funston et al. (2012a) to recommend that replacement heifers should be fed to reach a preselected "target" weight at a given age. Heifers of most *B. taurus* breeds are expected to reach puberty by 13 to 15 months of age when fed adequately to allow calving by 23 to 24 months of age; however, threshold ages of some heifers of *B. indicus* breeds at puberty can be older than 14 months. Generally, heifers of typical *B. taurus* beef breeds (e.g., Angus, Charolais, Hereford, Limousin) are expected to reach puberty at about 60% of mature weight. Heifers of dual-purpose or dairy breeds (e.g., Braunvieh, Brown Swiss, Friesian, Gelbvieh, Red Poll) tend to reach puberty at a younger age and at a lower weight, relative to mature weight (about 55% of mature weight) than those of beef breeds. Conversely, heifers of *B. indicus* breeds (e.g., Brahman, Nellore, Sahiwal) generally reach puberty at older ages and at heavier weights (about 65% of mature weight) than those of *B. taurus* beef breeds (Laster et al., 1972, 1976, 1979; Stewart et al., 1980; Ferrell, 1982; Sacco et al., 1987; Martin et al., 1992; Gregory et al., 1999; Vera et al., 1993).

Mature weight refers to weight reached by 5 years of age in cows of the same genotype in a nonrestrictive environment (i.e., mature weight as determined by genetic potential). In a restrictive environment (high environmental temperature, limited nutrition, parasite loads, etc.), mature weight of cows is often less than that of cows of similar genotype maintained in a less restrictive environment (Butts et al., 1971; Pahnish et al., 1983). Heifer weight at puberty is also decreased, but to a lesser extent than is mature weight. Thus, under a restrictive environment, weight at puberty is generally a greater percentage of observed mature weight than described above (Vera et al., 1993).

Clanton et al. (1983), Lynch et al. (1997), and Freetly et al. (2001) evaluated altering the rate and timing of heifer gain. In these studies, heifers were programmed for either continuous growth or slow early growth followed by rapid late growth between weaning and breeding to reach a similar target weight. It can be concluded from these studies that there are minimal differences in calving rate, age at calving, PPI, or second-year pregnancy rates, but total energy required to reach target weight can be decreased by using a slow followed by more rapid gain strategy. The decrease in total energy required to reach target weight is likely the result of maintaining less BW during the early postweaning period followed by compensatory growth during the period leading up to the breeding season. The challenge with relying on late rapid gain is that environmental conditions can affect programmed gains, which in turn could result in missing the targeted weight at breeding.

Excess BW and energy reserves can have a negative effect on heifer reproduction (Bagley, 1993). The older literature suggests that milk production is decreased when dairy heifers are developed to weights that create excess body energy reserves (Swanson, 1967). Similarly, Mangus and Brinks (1971) and Martin et al. (1981), respectively, reported decreased weaning weights in heifers raised by cows with higher milk production and in heifers that were creep-fed. In addition, Ferrell et al. (1976a) reported a difference in pregnant heifer udder weight between heifers fed a low vs. high nutrient intake. Collectively, these studies suggest that mammary development is negatively affected by the accumulation of excess body energy reserves. Moreover, Arnett et al. (1971) reported that heifers allowed to become obese require more services per conception, experience greater dystocia, produce less milk, and wean lighter calves than their leaner contemporaries, with differences persisting through the third lactation.

During the last several decades, the beef industry has had access to genetic tools such as expected progeny differences (EPD) and, more recently, genomic-enhanced EPDs, to increase selection pressure on economically important traits. For example, Brinks et al. (1978) identified a relationship between scrotal circumference in bulls and age at puberty in their heifer offspring, and scrotal circumference subsequently became an indicator trait for puberty (Funston et al., 2012a). Breed association websites show a significant increase in not only growth and carcass traits, but also in scrotal circumference since the mid-1980s, which demonstrates progress in modifying these economically important traits. Although age at puberty is not reported by the breed associations, one might expect that age at puberty has also decreased during this same time frame. The appropriate target weight at breeding was reevaluated in a review conducted by Funston et al. (2012a) who proposed that feed cost and cost per pregnant heifer could be decreased if a target weight of 55% (51 to 57%) of expected mature weight was used. This review acknowledged that developing heifers to a lower target weight could result in a reduced percentage of pubertal heifers at breeding season initiation, and a slightly older age at puberty. Nonetheless, pregnancy rates during a limited breeding season (48 to 60 days), calf production, and proportion of heifers retained as pregnant 2-year-olds were similar compared to heifers developed to the higher target weights (60 to 65%). The concept of earlier puberty and lighter heifer weights at puberty is supported by Gunn (2013), who reported that Simmental-Angus heifers obtained puberty at 46.8 to 50.8% of their estimated mature weight and that all heifers were cycling by the beginning of the breeding season when a target weight of 60 to 65% was programmed. The Funston et al. (2012a) review suggested that heifer age and growth rate from birth to weaning accounted for more variation in puberty status and artificial insemination (AI) pregnancy rate than postweaning growth rate.

There are pros and cons associated with both the higher (60 to 65%) and lower (51 to 57%) heifer target-weight systems. The higher target-weight system could be expected to result in higher feed cost/pregnant heifer, decreased age at puberty, increased percentage of pubertal heifers with more than one estrous cycle before breeding season initiation, a larger percentage of heifers calving early in their first calving season, and less risk of environmental (weather and nutrition) stresses. This system might be preferred when one or more of the following herd conditions exist: spring calving with larger cow size, *B. indicus* breed types, risk of environmental extremes (especially mud and cold stress) that could negatively affect attainment of the desired target weights, replacement heifers are developed as a group with large variation in heifer weights and/or ages, and either environmental or soil conditions that prevent compensatory growth before initiation of the breeding season. The lower target-weight system could be expected to result in decreased feed cost, decreased BW at puberty, the need to develop more heifers to maintain the desired number of pregnant females, and the potential for decreased mature cow weights with lower maintenance energy requirements. This system might be preferred when one or more of the following herd conditions exist: fall calving, small to moderate cow size, breed types with more milk production, low risk of environmental extremes, environmental and soil conditions that allow compensatory growth before initiation of the breeding season, and when a progestin is used in an estrous synchronization protocol to help "jump start" nonpubertal heifers.

Many spring calving herds develop replacement heifers in drylot and then abruptly move them to early grazing season pasture just before breeding or immediately following breeding. If this pasture at turnout is low in dry matter and nutrient profile, heifers could lose weight, thereby resulting in a greater incidence of reproductive failure. A series of studies (Arias et al., 2012, 2013; Lake et al., 2013) were designed to evaluate the impact of energy in the diet during the 21-d period immediately following a timed-AI (TAI) mating. Diets were formulated to continue the winter feeding period gain (~0.7 kg/d), maintain weight, or lose weight (80% of NEm requirement [NRC, 1996, 2000]). Heifers that continued to gain during the 21-d period had higher TAI conception rates than heifers programmed to either maintain or lose weight. In addition, the gaining heifers had 6-d embryos that were more advanced in their development, had more total cells, and had more live cells compared to heifers that lost weight. Collectively, these data suggest that nutrition during the period immediately after ovulation (not semen or oocyte quality or fertilization) plays a significant role in embryo quality and early embryo survival. These data could also relate to the previous target weight discussion. Heifers developed to a lighter target weight that are not provided a period of compensatory growth before breeding could have decreased reproductive success. Similarly, reproductive success could be compromised in heifers that are developed using the higher target-weight system if they abruptly go from a positive to a maintenance or negative energy balance immediately before or after breeding.

Metabolic modifiers are discussed in Chapter 14 (Compounds That Modify Digestion and Metabolism), but the topic of ionophores deserves brief mention in this chapter as it relates to heifer puberty. Increased concentrations of ruminal propionate have been associated with feeding diets high in nutrient density (starch) as well as dietary inclusion of an ionophore (Bagley, 1993). Rutter et al. (1983) reported that direct infusion of propionate into the abomasum resulted in hormonal alterations that would lead to an earlier age at puberty in heifers. This observation is supported by McCartor et al. (1979) and Moseley et al. (1977, 1982), who reported a decreased age at puberty when either corn or monensin were fed compared to a control diet formulated to produce similar weight gain. This concept has potential applications in heifer development programs where lighter target weights are used.

The following equations were developed in the BCNRM to compute target breeding weights and growth rates for replacement heifers using the data summarized above for default, target breeding weights of 55, 60, and 65% of expected mature weight for dual-purpose or dairy, *B. taurus*, and *B. indicus* breed types, respectively. The user has the option of modifying these targets in the BCNRM software since selecting the dual purpose or dairy breed type (target weight of 55%) will also result in a 20% increase in maintenance requirement automatically (based on Table 19-1, see Chapter 19, Model Equations and Sensitivity Analyses), which would not be appropriate for some *B. taurus* and *B. indicus* breed types. The equations described in Chapters 10 (Feed Intake), 11 (Maintenance), and 12 (Growth) are used to predict net energy and protein requirements for maintenance and growth. Based on the data summarized by Gregory et al. (1999), the BCNRM uses a target primiparous female weight that is 80% of the expected mature cow weight, which is the six-breed average for 2-year-old females vs. mature cow weight from the U.S. MARC database. Target calving weight coefficients used in the BCNRM for cows calving at 3 and 4 years of age are 0.92 and 0.96, respectively, based on the U.S. MARC database (Gregory et al., 1999). Predicted target weights and rates of gain for young beef females are computed as

$$\text{TPW} = \text{MW} \times \begin{cases} 0.55, \text{ for dual-purpose and dairy cattle,} \\ 0.60, \text{ for } \textit{Bos taurus}, \text{ and} \\ 0.65, \text{ for } \textit{Bos indicus}, \end{cases} \quad \text{(Eq. 13-15)}$$

$$\text{TPA} = \text{TCA} - 280, \quad \text{(Eq. 13-16)}$$

$$\text{ADGbp} = (\text{TPW} - \text{SBW})/(\text{TPA} - \text{Age}), \quad \text{(Eq. 13-17)}$$

$$\text{TCW1} = \text{MW} \times 0.80, \quad \text{(Eq. 13-18)}$$

$$\text{ADGap} = (\text{TCW1} - \text{TPW})/(280 - \text{DP}), \quad \text{(Eq. 13-19)}$$

$$\text{TCW2} = \text{MW x } 0.92, \quad \text{(Eq. 13-20)}$$

$$\text{ADGac1} = (\text{TCW2} - \text{TCW1})/\text{CI}, \quad \text{(Eq. 13-21)}$$

$$\text{TCW3} = \text{MW} \times 0.96, \quad \text{(Eq. 13-22)}$$

$$\text{ADGac2} = (\text{TCW3} - \text{TCW2})/\text{CI}, \quad \text{(Eq. 13-23)}$$

$$\text{TCW4} = \text{MW}, \quad \text{(Eq. 13-24)}$$

$$\text{ADGac3} = \text{TCW4} - \text{TCW3}/\text{CI}, \quad \text{(Eq. 13-25)}$$

where

TPW is the target first pregnancy weight, kg;
MW is the mature body weight, kg;
TPA is the target pregnancy age, d;
TCA is the target calving age, d;
ADGbp is the average daily gain before first pregnancy, kg/d;
SBW is the current shrunk body weight, kg;
Age is the current age, d;
TCW1 is the target first calving weight, kg;
ADGap is the average daily gain after first pregnancy, kg/d;
DP is days pregnant, d;
TCW2 is the target second calving weight, kg;
ADGac1 is the average daily gain after first calving, kg/d;
CI is calving interval, d;
TCW3 is the target third calving weight, kg;
ADGac2 is the average daily gain after second calving, kg/d;
TCW4 is the target fourth calving weight, kg; and
ADGac3 is the average daily gain after third calving, kg/d.

Equations in Chapter 12 are used to compute requirements for the targeted average daily gain (ADG), with adjustments based on previous nutrition, to determine ADG and net energy required for gain (NEg) requirements needed to achieve these target weights. For pregnant animals, gain associated with gravid uterus growth is added to the predicted daily SBW gain as follows:

$$\text{ADGpreg} = \text{CBW} \times 0.01828 \times (0.02 - 0.0000286 \times \text{DP}) \times e^{(0.02 \times \text{DP} - 0.0000143 \times \text{DP}^2)}, \quad \text{(Eq. 13-26)}$$

where

ADGpreg is the average daily gain associated with the gravid uterus, kg/d;
CBW is the calf birth weight, kg; and
DP is days pregnant, d.

For pregnant heifers, weight of fetal and associated uterine tissues should be deducted from SBW to compute growth requirements. The conceptus weight (CW) can be calculated as follows:

$$\text{CW} = \text{CBW} \times 0.01828 \times e^{(0.02 \times \text{DP} - 0.0000143 \times \text{DP}^2)}, \quad \text{(Eq. 13-27)}$$

where CW is the conceptus body weight, kg.

POSTPARTUM INTERVAL

The physiological mechanisms controlling postpartum anestrous and infertility in postpartum cows were reviewed by Short et al. (1990). Several factors have been reported to influence the length of the PPI which include pre- and postpartum nutrition/energy balance (Rutter and Randel, 1984), frequency of suckling (Randel, 1981), cow age (Dimmick et al., 1991), and body condition (Randel, 1990). From these data, one can make the following conclusions: to have a high probability of conception by approximately 80 days post-calving, cows need to be managed in a manner that will allow resumption of estrus and ovulation to occur approximately 40 days postpartum.

Return to estrus is orchestrated via an integration of multiple signals within the hypothalmo-hypophyseal-ovarian axis (Hess et al., 2005). Energy balance, which is perceived by the reproductive axis through a variety of nutritionally induced cues, has a profound effect on the duration of postpartum anestrous. Duration of the postpartum anestrous period is increased in cows fed low concentrations of dietary energy during either late gestation or early lactation. The reproductive response to either low dietary energy intake prepartum, or weight loss during gestation depends on body condition at calving. Cows that are in good body condition at calving (BCS ≥ 5) are minimally affected by moderate pre- or postpartum weight changes; however, postpartum anestrous intervals are significantly increased by weight loss in cows that are thin (BCS ≤ 4) at calving. This problem is exacerbated by insufficient energy intake and weight loss during the postpartum period in cows that are thin at calving. Negative effects of poor body condition at calving on the postpartum interval can be partly overcome by increased feeding of energy postpartum (Houghton et al., 1990; Lalman, et al., 1997), but the postpartum period is a time of high metabolic demand, and significantly increasing cow weight and BCS can be a challenge. In addition, it can be expensive to feed enough energy to cows during early lactation to compensate for poor body condition at calving. This problem is intensified in heifers because of the additional nutrient demands for growth during early lactation. Conversely, cows that are obese at calving, even if they lose weight postpartum, are at greater risk of metabolic, infectious, digestive, and reproductive disorders than cows in near-moderate body condition.

Feeding supplemental fat to beef cows during late gestation has been evaluated as a method to alleviate the negative effects of prepartum nutritional inadequacy on reproductive performance. Hess et al. (2005, 2008) in their reviews of the literature concluded that fat supplementation during late gestation is a nutritional strategy that can improve beef cow pregnancy rates. In three reviews (Funston et al., 2004; Hess et al., 2005, 2008) the reproductive response (postpartum anestrous interval, first-service conception rate, or overall pregnancy rate) was inconsistent, but reproduction was not compromised. Additionally, there is some evidence to suggest that other nutritional factors, such as lipids high in linoleic acid, can positively influence reproduction by affecting critical components of the reproductive axis (Hess et al., 2005). Hess et al. (2008) concluded in their review that the level of dietary fat inclusion on a high-forage cow diet should not exceed 2% (DM basis) to prevent negative associative effects on fiber digestion, while feeding more than 4% (DM basis) will not increase dietary energy intake.

Duration of the postpartum anestrous interval is longer in suckled cows than in either nonlactating cows or cows that are milked (Yavas and Walton, 2000). The delay in initiation of estrous cycles postpartum seems to result primarily from calf contact rather than suckling or lactation per se. In addition, the calf stimulus interacts with the nutritional status of the cow such that the postpartum anestrous interval is increased to a greater extent in cows with poor body condition than those with good condition. Early weaning of calves (Houghton et al., 1990) as well as short-term or partial weaning (such as once-per-day suckling; Short et al., 1990) have decreased the postpartum interval in anestrous beef cows, but successful use of these approaches requires more intensive management and additional inputs.

The physiological mechanism controlling postpartum anestrus is thought to be mediated at the level of the hypothalamic release of gonadotropin-releasing hormone (GnRH) and subsequent changes in the pulsatile patterns of the pituitary hormone LH (Short et al., 1990). Estrous synchronization is becoming a more widely adopted tool for managing reproduction in beef cattle. The hormone progesterone is used in several synchronization programs to suppress estrus and ovulation, mimicking the luteal phase of the cow's estrus cycle. Once the progesterone source is removed, the female can then come into estrus and ovulate. Inclusion of progesterone in the synchronization program has been shown to stimulate anestrus cows to come into estrus and ovulate (Fike et al., 1997). This phenomenon has been called "progesterone priming" whereby progesterone treatment to anestrus cows can elicit a GnRH-induced LH surge and ovulation with a corpus luteum forming on the ovary capable of sustaining a pregnancy (Troxel and Kessler, 1984; Smith et al., 1987).

PREGNANCY

Reproductive efficiency in the beef cow is ultimately measured as a binomial trait (i.e., pregnancy), derived by the success or failure of multiple biological steps. Understanding these biological steps allows for progress to be made toward alleviating reproductive failures, but progress toward this goal has been limited (Geary et al., 2013). Ultimately, pregnancy success is the summation of the following simplified steps: ovulation of a viable oocyte, sperm transport and fertilization, regular cell division and growth by the developing embryo, secretion and reception of pregnancy recognition signals that ultimately prevent luteolysis, and placentation within the uterus (Geary et al., 2013).

Multiple factors (Atkins et al., 2013), including follicular characteristics that have been summarized by Geary et al. (2013), and deficiencies in the uterine environment reviewed by Bridges et al. (2013), affect embryo survival and reproductive efficiency in beef cattle. However, the impact of nutrient availability and nutrient flux on hormonal regulation, follicular dynamics, embryo survival, and uterine function are not as well established as hormone regulation and thus deserve more attention.

Conception

It is well established that general undernutrition of the cow during the prepartum and postpartum periods negatively impacts pregnancy success and reproductive efficiency in beef cattle (Diskin et al., 2003). In cattle, there is a strong correlation between negative energy balance in early lactation and the resumption of ovulation postpartum (Canfield and Butler, 1990). Dietary restriction has been shown to alter follicular growth in cattle. Murphy et al. (1991) reported that heifers on a low dietary intake had decreased size and persistence of the dominant follicle compared with heifers offered higher energy intakes. Mackey et al. (1999) demonstrated that acute nutritional restriction (0.4 times maintenance) for about 12 d decreased dominant follicle growth rate, decreased maximum follicle diameter, and prevented ovulation after induced luteolysis with prostaglandin.

Several studies with ewes have shown that feeding dietary fat altered the growth pattern of follicles and that this effect is at least partially independent of energy intake (Martin et al., 1998). Supplemental fat in the dairy cow increased the number of follicles (Beam and Butler, 1997; Lucy et al., 1991) and increased the size of the preovulatory dominant follicle (Lucy et al., 1990), but this response has not been well established in beef females. An increase in follicle size, however, may have beneficial effects on bovine oocyte quality (Lonergan et al., 1994) and on corpus luteum function (Mattos et al., 2000), both of which can result in higher pregnancy rates.

Cows maintained in a negative energy state or fed excess protein can have decreased ovulated oocyte quality (Leroy et al., 2008). High dietary protein, resulting in high concentrations of urea nitrogen in plasma and milk (>19 mg/dL), has been associated with decreased fertility in dairy cattle (Jordan et al., 1983; Elrod et al., 1993; Butler et al. 1996). This is supported by a recent study in which cows were fed a high level of dried distillers grains with solubles (DDGS) as a primary energy source in a corn stover-based diet. In this study, cows that had plasma urea levels over 16 mg/dL failed to conceive (R. P. Lemenager, Purdue University, West Lafayette, IN, personal communication, May 10, 2014). Elrod and Butler (1993) suggested that plasma urea decreases fertility by altering the uterine environment. In contrast, several authors (Fahey et al., 1998; Boland et al., 2001) concluded that the deleterious effect of plasma urea on reproductive competence is likely to occur at the level of the oocyte and would affect embryo quality, rather than cause disruptions in the uterine environment. Regardless of the mechanism, this suggests that nutrition during the days and weeks leading up to ovulation and maternal recognition of pregnancy, especially protein intake that increases plasma urea beyond the threshold (16 to 19 mg/dL), can negatively affect female fertility.

A recent review by Bridges et al. (2013) suggests that pregnancy is not limited by fertilization rate, but is greatly affected by early embryonic loss. Early embryonic death can depend on many factors such as chromosomal abnormalities, stress (nutritional, environmental, and disease) affecting the dam, and either poor oocyte or semen quality. Fertilization and very early embryonic development occur in the oviduct of the cow. The embryo then travels to the uterus where it will continue to develop before attaching and implanting in the uterine epithelium (Bazer et al., 2009). Ayalon (1978) reported that fertilization failure ranged between 0.0 to 3.4% in virgin heifers and between 15.0 to 17.0% in normal-cycling cows. Several studies have since reported fertilization rates in beef cattle ranging from 76 to 94% (Diskin and Sreenan, 1980; Smith et al., 1982; Maurer and Chenault, 1983; Ahmad et al., 1995). When inseminated after an estrous and/or ovulation synchronization program, fertilization rates are approximately 90% in beef cows and between 75 and 85% in dairy cows (Bridges et al., 2013), but only 50 to 60% of beef embryos and only 30 to 40% of dairy embryos remain viable by day 30 of gestation.

Early embryo mortality and late embryo mortality, respectively, refer to mortalities occurring between fertilization and approximately day 24 of gestation and between days 25 and 45 of gestation (CBRN, 1972). Days 14 to19 have been recognized as the period of maternal recognition of pregnancy. Maternal progesterone concentrations and preovulatory estradiol concentrations can impact the ability of the uterus to maintain a viable embryo before implantation, as well as the ability of the embryo to provide an adequate signal for the dam to recognize pregnancy (Mann and Lamming, 2001; Bridges et al., 2013). Embryo mortalities in *B. taurus* beef heifers accounted for 25 to 30% (Diskin and Sreenan, 1980; Roche et al., 1981) of reproductive failure in heifers, with early embryo mortality accounting for about 75 to 80% of all embryo and fetal mortalities. The greatest mortalities occur near maternal recognition (between days 8 and 18), but after fertilization (Diskin and Sreenan, 1980; Roche et al., 1981). Of the remaining mortalities, most estimates suggest that about 10 to 15% of mortalities occur at or near the time of implantation (days 25 to 45) and 5 to 8% of the mortalities take place between placental attachment and birth (Sreenan and Diskin, 1986).

Early embryo mortality represents a significant source of reproductive failure and the controlling mechanisms are not well understood; however, an emerging area of interest revolves around the role of histotrophs. Since uterine histotrophs are composed of enzymes, cytokines, growth factors, ions, hormones, glucose, fructose, amino acids, transport proteins, and adhesion molecules (Bazer et al., 2012; Mullen et al., 2012) and involve secretions from the glandular epithelium, they have been identified as a critical factor for conceptus growth and survival (Bridges et al., 2013). In addition, uterine histotrophs are the sole supply of nutrients for the developing embryo before implantation. Recent reports indicate that total uterine protein content, as well as the quantities of different uterine proteins are increased in heifers with higher antral follicle counts, has a known association with increased fertility in heifers (Vallet et al., 2014). Although the mechanisms are not yet clear, nutrient intake can affect the synthesis and metabolism of hormones (Sangsritavong et al., 2002) and histotrophs associated with reproduction and early embryonic development. This area of nutrition by reproduction interaction deserves more attention.

Conception rates, a consequence of both fertilization rates and embryo survival, have been found to be significantly higher in heifers/cows with three follicular waves per estrous cycle compared to females with only two follicular waves (Ahmad et al., 1997; Inskeep, 2002; Townson et al., 2002). It has been reported that the proportion of animals with two or three waves per cycle varies between herds (Sirois and Fortune, 1988; Ginther et al., 1989a) and that this is repeatable for individual females (Ginther et al., 1989b; Rhodes et al., 1995; Ahmad et al., 1997), and varies with both nutrition (Murphy et al., 1991) and body condition (Burke et al., 1998). These results suggest that earlier conception rates could be achieved by developing targeted supplementation strategies that increase the frequency of three follicular waves per estrous cycle.

Heat stress is another condition that can compromise steroidogenesis (Zeron et al., 2001), decrease oocyte quality and the viability of oocytes (Hansen, 2002), cause a decrease in oocyte competence (Al-Katanani et al., 2002), decrease fertilization rate (Putney et al., 1989), and lower embryo survival in nontropically and tropically adapted *B. taurus*, *B. indicus*, and *B. indicus* cross heifers and cows (Turner, 1982; Putney et al., 1989; Holroyd et al., 1993; Smith and Stevenson, 1995). In beef females, both heat stress and ingestion of endophyte-infected tall fescue can work individually and collectively to decrease serum cholesterol and the diameter of the corpus lutea (Burke et al., 2001). They concluded that the combination of heat stress and the consumption of endophyte-infected tall fescue led to decreased serum estradiol, fewer large follicles during the estrous cycle, and decreased diameter of the preovulatory dominant follicle, which might explain reduced pregnancy rates commonly observed in heifers grazing infected tall fescue pasture during the heat of summer.

Role of the Uterus and Placenta

In contrast to pigs and horses that have a diffuse type of placentation, the morphology of the ruminant placenta is termed cotyledonary (Vonnahme et al., 2013). In cattle, the fetal placenta attaches to discrete sites on the uterine wall called caruncles, which are aglandular sites appearing as knobs along the uterine luminal surface of nonpregnant animals (Ford, 1999). Placental membranes attach at these sites by chorionic villi in cotyledonary areas. This caruncular-cotyledonary unit, called a placentome, is both physiologically and functionally important in the exchange of all types of compounds between the dam and fetus. Unlike the ewe (Stegeman, 1974), placental growth in the cow increases progressively throughout gestation (Reynolds et al., 1990; Vonnahme et al., 2007). During gestation, many changes occur in the maternal cardiovascular system, including dramatic growth and development of uterine and placental vascular beds. Compared with sheep, cow placenta exhibits relatively modest changes in capillary area density (blood vessel growth) from mid- to late gestation, with capillary area density and cotyledonary tissue increasing approximately 190%, whereas caruncular and cotyledonary tissue masses increase approximately 530% and 650%, respectively (Vonnahme et al., 2007, 2013).

The placenta is responsible for the exchange of metabolites, water, heat, and respiratory gases; synthesis and secretion of numerous hormones; and extensive interconversion of nutrients and other metabolites (Munro et al., 1983; Battaglia, 1992). Efficiency of placental nutrient transport is directly related to uteroplacental blood flow (Reynolds and Redmer, 1995; Reynolds et al., 2005, 2010), which increases dramatically during gestation to support the nutritional demands of the rapidly growing fetus. To support this demand, the dam increases its plasma volume by 30 to 40%, as well as cardiac output (i.e., 35% increase in stroke volume and 15% in heart rate; Rosenfeld, 1984). Placental transport of oxygen, glucose, amino acids, and urea and placental clearance of highly diffusible solutes increase during gestation as indicated by net fetal uptake or loss in both sheep and cattle (Bell et al., 1986; Reynolds et al., 1986). Because of the numerous metabolic functions of the uterus and placenta (uteroplacenta), there is extensive oxidative metabolism throughout gestation with approximately equal energy consumption between the uteroplacenta and fetus (Reynolds et al., 1986). Even in late gestation, uteroplacental net use of glucose is at least 70% of gravid uterine glucose uptake. These observations are consistent with previous research that suggested about half the total energy expenditure associated with pregnancy can be attributed to gravid uterus tissues and about a quarter can be attributed to the fetus per se (Kleiber, 1961; Ferrell and Reynolds, 1987). Likewise, a major proportion of the net amino acid uptake from the uterine circulation is metabolized by the uteroplacenta (Reynolds et al., 1986; Ferrell, 1991b).

Factors Affecting Birth Weight and Dystocia

It is generally accepted that nutrient needs for pregnancy are proportional to calf birth weight. Factors known to affect calf birth weight include breed of sire, breed of dam, heterosis, parity of the dam, number of fetuses, sex of the fetus, environmental temperature, and nutrition of the dam (Ferrell, 1991a). Of the factors affecting calf birth weight, breed or genotype of the sire and dam generally has the greatest influence (Andersen and Plum, 1965). Typical birth weights of calves of various breeds are listed in Table 13-6. Birth weights of calves in one study differed by as much as 18 kg (AFRC, 1990), and a 10-kg range in calf birth weights has been reported for the breeds typically used in the United States (Gregory et al., 1982; Cundiff et al., 1988; BIF, 1990). Heterosis results in a 6 to 7% increased birth weight in *B. taurus* breed crosses, less (0 or negative) when *B. taurus* sires are crossed on *B. indicus* dams, but considerably higher (20 to 25%) birth weights when the reciprocal mating is made (Ellis et al., 1965; Long, 1980; Gregory et al., 1992). Heifer calf birth weights average 7% less than bull calves at birth (AFRC, 1990; BIF, 1990), and weight of calves born to 2-, 3-, and 4-year-old cows, respectively, average 8, 5, and 2% less than those born to 5- to 10-year-old cows (BIF, 1990; Gregory et al., 1990). Birth weight of calves born as twins is

TABLE 13-6 Estimated Birth Weight of Calves of Different Breeds or Breed Crosses (kg)[a]

Breed	BIF	AFRC	U.S. MARC
Angus	31	26	35
Braford	36	–	–
Brahman	31	–	41
Brangus	33	–	–
Braunvieh	–	–	39
Charolais	39	43	40
Chianina	–	–	41
Devon	32	34	–
Galloway	–	–	36
Gelbvieh	39	–	39
Hereford	36	35	37
Holstein	–	43	–
Jersey	–	25	31
Limousin	37	38	39
Longhorn	–	–	33
Maine-Anjou	40	–	41
Nellore	–	–	32
Piedmontese	–	–	38
Pinzgaur	38	–	40
Polled Hereford	33	–	36
Red Poll	–	–	36
Sahiwal	–	–	38
Santa Gertrudis	33	–	–
Salers	35	–	38
Shorthorn	37	32	39
Simmental	39	43	40
South Devon	33	42	38
Tarentaise	33	–	38

NOTE: BIF, Beef Improvement Federation; AFRC, Agricultural and Food Research Council; U.S. MARC, Roman L. Hruska U.S. Meat Animal Research Center (USDA/ARS).

[a]BIF (1990), AFRC (1990), U.S. MARC, from data reported by Cundiff et al. (1988), and Gregory et al. (1982), which are from a particular sire breed on mature Angus and Hereford cows.

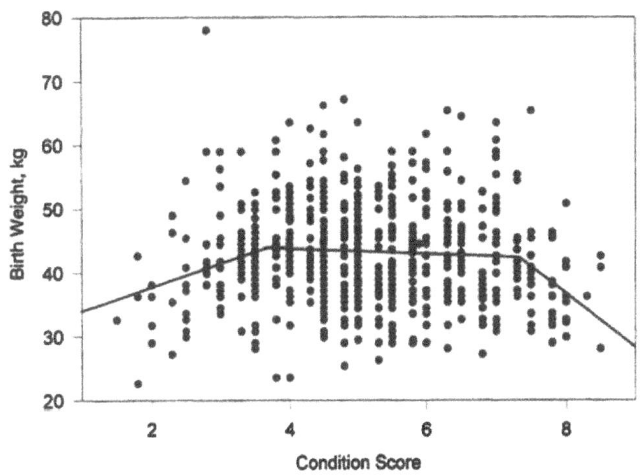

FIGURE 13-2 Relationship of calf birth weight to cow condition score in mature cows of nine breeds (NRC, 1996, 2000).

25% less, but the total weight of twins averages 150% of the birth weight of calves born as singles (Gregory et al., 1990).

Severe energy or protein underfeeding has resulted in marked decreases in calf birth weight (Hight, 1966, 1968a,b; Tudor, 1972). Inadequate food intake during late pregnancy is also associated with weak labor, increased dystocia, decreased milk production and growth of progeny, and decreased rebreeding performance of the dam (Bellows and Short, 1978; Kroker and Cummins, 1979). Conversely, gross overfeeding during pregnancy can also result in decreased birth weight and milk production, increased dystocia and neonatal death loss, and poor rebreeding performance (Arnett et al., 1971; Robinson, 1977). The relationship of calf birth weight to cow condition score is typified by data shown in Figure 13-2 from nine breeds (NRC, 1996, 2000). Calf birth weights are positively related to BCS when cows are below BCS 3.5, remain constant between BCS 3.5 to 7, and are inversely related to BCS when cows are above BCS 7. It is generally accepted that calf birth weight is not substantially influenced by cow nutritional status within a broad range, but it can be affected by extreme over- or underfeeding. Historically, the negative influences of nutrition have been on rebreeding performance, dystocia, etc., rather than on calf birth weight per se. Recently however, more research has been focused on the effect of over- and underfeeding on fetal programming, which is discussed in a later section of this chapter.

Weight gain during pregnancy and loss at parturition is about 1.7 times the calf birth weight and represents weight gain or loss of the fetus, fetal fluids, placenta, and altered uterus weight. A regression equation was developed by Ewing et al. (1966) to adjust cow weight change at calving: postcalving cow BW = precalving cow BW – [(calf birth BW × 1.9697) – 8.62]. This equation has been used by some researchers to adjust cow weights from a pregnant to nonpregnant basis and vice versa.

Environmental Temperature

Although this section is primarily concerned with factors affecting calf birth weight, it is important to note that high environmental temperature during or shortly after conception can significantly increase embryonic mortality in cattle, as well in as several other species (Bell, 1987). In addition, high environmental temperatures, particularly during early pregnancy, might result in a wide range of congenital defects. Limited data are available from well-controlled studies of cattle to characterize the influence of elevated temperatures on calf birth weight (Collier et al., 1982), and to the committee's knowledge, no data are available from controlled experiments to characterize the influence of chronic cold exposure in cattle, although these effects have been well documented in sheep (Alexander and Williams, 1971;

Rutter et al., 1971, 1972; Cartwright and Thwaites, 1976; Thompson et al., 1982; Bell, 1987). A number of studies have shown that calves born in the spring are heavier than those born in the fall (McCarter et al., 1991b), calves born in the northern areas of the United States are heavier than those born in southern areas, and genotype–environment interactions can have important influences on calf birth weight (Burns et al., 1979; Olson et al., 1991). It can be concluded from these sources that the magnitude of response of calf birth weight to environmental temperature is influenced by severity, duration, and timing of exposure, as well as the genotype of the dam.

Factors Affecting Fetal Growth

Considerable progress has been made toward understanding how various factors affect fetal growth and the ensuing birth weight. Normal fetal growth follows an exponential pattern (Figure 13-3), and in cattle, the weight of uterine and placental tissues increases exponentially (Ferrell et al., 1976a; Prior and Laster, 1979). Growth and development of the uterus and placental tissues precede fetal growth and are required to support subsequent fetal growth (Ferrell, 1991b,c). Growth of the fetus is a result of its genetic potential for growth, which is reflected in its demand for nutrients and constraints imposed by the maternal and placental systems in meeting that demand (Gluckman and Liggins, 1984; Ferrell, 1989). The potential of the maternal and placental systems to meet those demands is reflected in uterine blood flow, placental size, and functional capacity of the uterus, which is discussed in detail elsewhere in this chapter. The influence of maternal nutrition on fetal programming is complicated by the fact that the fetus can be undernourished in well-fed dams because of inadequate placental size or compromised function. Conversely, even though the dam is undernourished, the maternal and placental systems may compensate such that fetal malnutrition is minimal (Bassett, 1986, 1991), at least as measured historically by birth weight, calf survival, and weaning weight. Weight and perfusion of uterine and placental tissues are decreased with heat (Alexander and Williams, 1971; Cartwright and Thwaites, 1976; Reynolds et al., 1985; Bell et al., 1987) and with twins compared with single fetuses (Bellows et al., 1990; Ferrell and Reynolds, 1992). These variables are also influenced by genotype of sire, dam, or fetus (Ferrell, 1991c). Numerous reports indicate that perfusion of uterine and placental tissues and functional capacity of the placenta have central roles in fetal growth (Alexander, 1964a,b; Owens et al., 1986).

Death of calves perinatally represents a major production loss for beef cattle. Neonatal mortality is related to birth weight, with the greatest losses occurring at low and high birth weights and lower mortality associated with moderate birth weights. Because dystocia is positively associated with calf birth weight, it is a major cause of neonatal calf death (Laster and Gregory, 1973; Bellows et al., 1987). Some cattle producers have attempted to decrease calf birth weight, particularly in first-calf heifers, by underfeeding during the last trimester of pregnancy. As noted previously, malnutrition must be relatively severe to result in substantial decreases in calf birth weight. In nine studies reviewed by Dunn (1980), severe underfeeding of the dam resulted in decreased calf birth weights in all but one study, but dystocia was decreased in only one study (Dunn and Moss, 1992). It is important to note that underfeeding to decrease birth weight also resulted

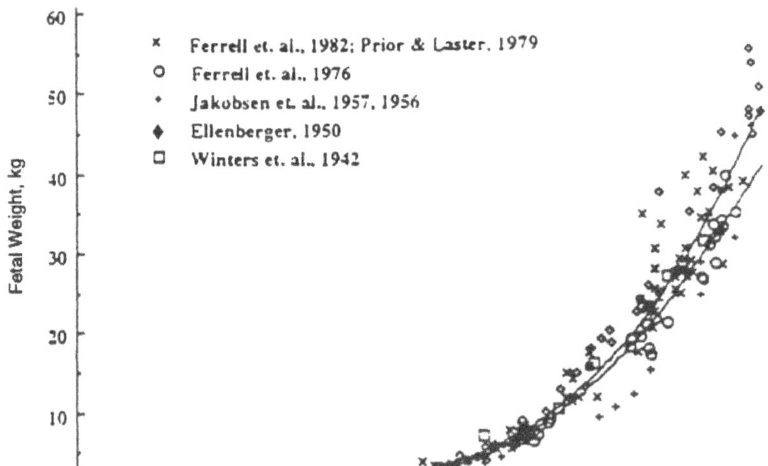

FIGURE 13-3 Relationship of fetal weight to day of gestation in cattle (NRC, 1996, 2000).

in decreased calf survival. In addition, underfeeding during late pregnancy results in a longer interval between calving and rebreeding (Short et al., 1990; Ferrell, 1991d; Dunn and Moss, 1992), lower milk production, and decreased calf weaning weight (Hight, 1968b; Corah et al., 1975; Bellows and Short, 1978), with the effects being more severe in primiparous females than in mature multiparous cows.

Developmental Programming

In production systems, there are many opportunities for dams to be exposed to suboptimal nutritional conditions; whether as a result of environmental temperatures (high or low), advancing forage maturity that affects nutrient availability and intake, competition between nutrient partitioning for developing heifers and their developing fetus, or increased nutrient requirements resulting from selection for greater milk yield (Wu et al., 2006; Caton and Hess, 2010). A majority of the nutrition–reproduction interaction research conducted historically has focused on the production phases of birth to weaning, weaning to puberty, puberty to first calf, first calf to second calf, and mature cow gestation through lactation with little emphasis on long-term effects on productivity. An increasing body of literature has recently focused on the effects of dam nutrition on prenatal and early neonatal development. A majority of this research has been devoted to undernutrition and strongly suggests that altering maternal nutrition during the production cycle can negatively or positively affect offspring development and productivity, depending on the timing and level of nutrition provided.

Developmental programming can be defined as a developmental insult, whether nutritional, environmental, or stress-related during pregnancy or early lactation, which can have metabolic and physiological consequences on growth, body composition, and reproductive capacity of the resulting offspring later in life (Barker et al., 2012; Reynolds and Caton, 2012; Casellas and Caja, 2014). The concept of developmental programming was first highlighted with malnourished women during the Dutch Famine of 1944 who had children with normal birth weights, but with increased incidence of obesity, dyslipidemia, insulin resistance, and coronary heart disease in later life (Barker et al., 2012). In his review, Bach (2012) concluded that there is a clear relationship between nutrient supply and hormonal signals at specific windows during pre- and early postnatal development. The interaction between nutrient supply and hormonal signals can exert permanent changes in the metabolism of humans (Fall, 2011) as well as changes in performance, body composition, and metabolic function of the offspring from livestock (Wu et al., 2006; Caton and Hess, 2010; Reynolds and Caton, 2012; Funston et al., 2012b) through processes generically referred to as developmental programming. Many terms have been used in the literature to categorize this developmental process: metabolic imprinting (Waterland and Garza, 1999), metabolic programming (Lucas, 2000), fetal programming (Barker and Clark, 1997), developmental programming (Reynolds et al., 2010), lactational programming (Hinde and Capitanio, 2010), and neonatal programming (Moura and Passos, 2005). Regardless of the terms used, long-term modifications of the metabolic function of the offspring during fetal development can occur through epigenetic changes (Wu et al., 2006; Bonasio et al., 2010; Anderson et al., 2012).

In beef cattle, undernutrition (75% of recommended allowance) during early stages of pregnancy can compromise placental angiogenesis, cotyledon weight, and therefore, fetal development (Vonnahme et al., 2007; Long et al., 2009). Maternal nutritional status has also been implicated in the programming of nutrient partitioning, which in turn ultimately affects not only fetal growth, but also the development and function of the major organ systems (Wallace, 1948; Godfrey and Barker, 2000; Wallace et al., 2001; Redmer et al., 2004; Luther et al., 2005; Wu et al., 2006; Caton et al., 2009; Long et al., 2009; Caton and Hess 2010). Prenatal growth trajectory is sensitive to the direct and indirect effects of maternal dietary nutrient intake in early embryonic development (Robinson et al., 2000).

The effect of maternal undernutrition on progeny growth is the best-studied example of developmental programming. Barker and Clark (1997) stated that nutrition is among the most influential intrauterine factors dictating placental and fetal growth. In addition, Waterland and Jirtle (2004) noted that fetal development is most vulnerable to maternal nutrition around the peri-implantation period and during rapid placental development. Freetly et al. (2000, 2005) reported that moderate nutrient intake restriction during the second trimester of gestation, followed by realimentation during the third trimester of pregnancy was a plausible management strategy that did not negatively affect calf production; however, negative nutrient balance during the third trimester of pregnancy can decrease calf birth weights (Freetly et al., 2005). This is supported by Corah et al. (1975) who demonstrated that feeding only 65% of the NRC (1970) requirements during the last 100 d of gestation decreased calf birth weight and weaning weight.

Heifer offspring from nutrient-restricted cows during gestation have a lower pregnancy rate (Martin et al., 2007), and steer offspring from feed-restricted cows have been reported to have decreased BW entering the finishing phase (Stalker et al., 2006). These studies suggest that maternal nutrient balance during gestation can impact fetal and developmental programming, and that the only logical time for nutrient restriction, as well as weight and BCS cycling, would be during the second trimester of pregnancy immediately following calf weaning. The impact of maternal nutrition during the second trimester, however, could also be a critical fetal development period in cattle, which could affect later-life performance and is an emerging area of research.

Most studies conclude that undernutrition of the dam not only decreases offspring birth weight, but also predisposes them to decreased glucose tolerance, increased insulin response, and increased adiposity. In a study by Gardner et al. (2005), feeding ewes 50% of the Agricultural and Food Research Council (AFRC, 1990) requirements of ME from day 1 to 30 of gestation had no effect on glucose, but a 50% restriction from 110 d until parturition led to decreased glucose tolerance and insulin resistance. In the Gardner et al. (2005) study, lambs from ewes that were feed-restricted during late gestation exhibited a decrease in GLUT 4 receptors, the major glucose transporter found on cell membranes. This decrease was seen in adipose, but not muscle tissue, and led to the suggestion by the authors that glucose intolerance was related to an inability of adipose tissue to absorb glucose, rather than the ability of muscle to consume glucose (Gardner et al., 2005). In a study by Ford et al. (2007), however, feeding ewes 50% of the NRC (1985) requirements from days 28 to 78 of gestation resulted in increased levels of glucose and insulin in progeny 63 d after lambing. By 250 days of age, lambs from nutrient-restricted dams had increased glucose, but decreased insulin response compared to control lambs. At slaughter, their carcasses were heavier and had more visceral adipose tissue (Ford et al., 2007). When ewes were fed 60% of maintenance ME requirements from days 1 to 39 of gestation (Muñoz et al., 2009), male offspring had increased adipose tissue and females had elevated levels of circulating leptin. It was concluded that offspring of dams that were nutrient-restricted during early pregnancy had elevated leptin concentrations, greater subcutaneous fat, and less muscle.

Some studies have tried to evaluate how maternal nutrition during gestation affects the reproductive success of the offspring. Female mammals are born with a finite number of ovarian follicles and oocytes in the ovaries, which develop in utero and decrease as the animal ages (Evans et al., 2012). In cattle, oogonia are formed from primordial germ cells in the first or second fetal month and continue to develop into primordial follicles, reaching a maximum concentration of approximately 2.1 million between 90 and 140 days of fetal age (Evans et al., 2012). Furthermore, apoptosis begins in late fetal life and decreases the number of primordial follicles in the ovary that would be available for ovulation in the adult animal. Factors such as maternal health and nutrition during gestation have been shown to play a role in determination of postnatal follicle numbers (Evans et al., 2012). During fetal folliculogenesis (days 65 to 110 of gestation) in sheep, undernutrition of the dam can delay fetal follicular development and subsequently reduce ovulation rate in the female progeny (Rae et al., 2001, 2002). Interestingly, Rae et al. (2002) found no effects of undernutriton of the dam on bull reproductive development and adult function; however, attainment of puberty might be affected. Mossa et al. (2009) presented preliminary evidence that energy restriction (i.e., 60% of recommendations) during the first trimester of pregnancy resulted in heifers with a 30% decrease in the number of high-quality follicles (i.e., ≥3 mm) at 250 d of age, without a decrease in BW. It was speculated that a reduction in follicle count could lead to negative consequences on future reproductive ability. Animal numbers in this study were limited, however, and more research is needed to corroborate these results and further understand the underlying physiological mechanisms.

Inadequate dam nutrition during the last stages of pregnancy may compromise the health of the offspring. It is well established that calves with poor passive transfer of immunoglobulins (i.e., <10 mg IgG/mL serum) have increased risk of diarrhea and respiratory problems and higher mortality rates (Davis and Drackley, 1998). Passive transfer of immunity depends primarily on the quality and quantity of colostrum offered and absorption capacity of the calf. Absorption capacity of the calf can be affected by the plane of nutrition of the dam during late gestation. Protein intake during the last 100 d of gestation was linearly related to serum IgG concentrations in the calf, despite the fact that IgG concentration in the colostrum and amount of colostrum consumed by the calf remained similar (Blecha et al., 1981). It is possible that colostrum from protein-restricted dams is either deficient in some fundamental constituents or that calves were affected and programmed by the protein restriction of the dam, impairing their ability to properly absorb IgG. Meyer et al. (2010b) support this concept since they observed that a moderate (i.e., ~80% of recommendations) nutrient restriction during early to mid-pregnancy of beef cattle altered the jejunal proliferation and total intestinal vascularity of the fetus, which could modify the capacity for IgG absorption.

Skeletal muscle is the primary tissue that uses glucose and fatty acids, so decreased muscle mass should conserve glucose and fatty acids, decrease insulin sensitivity and potentially predispose the offspring to endocrine disorders such as diabetes. Because myofiber number is established at birth and postnatal muscle growth occurs by hypertrophy, prenatal muscle fiber number has the potential to affect postnatal growth and development (Zhu et al., 2004). Organs grow and mature based on the stage of development and resources; and when nutrients are scarce, the fetus partitions nutrients based on a hierarchy of priorities that focus on short-term survival. The brain is at the apex of the hierarchy, whereas organs such as the kidneys and lungs are placed at the bottom of the hierarchy and receive fewer nutrients (Barker, 2012). The availability of nutrients can influence cell proliferation, because undernutrition during early gestation has been shown to decrease muscle fiber size and muscle fiber number (Bedi et al., 1982; Zhu et al., 2004; Quigley et al., 2005) and the secondary to primary fiber ratio (Zhu et al., 2004; Quigley et al., 2005).

Maternal overnutrition during gestation appears to result in similar effects in the adult offspring as undernutrition during gestation. Feeding 140% of nutritional requirements to growing ewes from day 40 of pregnancy to parturition resulted in a decrease in birth weights (Swanson et al., 2008). Fetuses of adolescent sheep fed twice the ME requirements of control ewes had smaller livers, spleens, kidneys, and hearts. These effects were magnified when the obese adolescent sheep were switched from the high-energy diet to half the ME during the last trimester of gestation (Redmer et al., 2012).

Feeding 150% of the NRC (1985) requirements to ewes from 60 d before conception until parturition led to greater adiposity and subsequently greater levels of circulating leptin (Long et al., 2010). In addition, progeny had increased feed intake despite elevated circulation of leptin, indicating leptin resistance, which is a classical symptom of diabetes, as well as decreased insulin sensitivity, another determinant of diabetes. In another study (Zhang et al., 2011), where ewes were fed 150% of the NRC (1985) recommendations from 60 d before conception until parturition, obesity resulted in decreased fetal pancreatic weights and fewer β-cells near parturition, but ewes were observed to have elevated circulating concentrations of leptin. At birth, however, circulating leptin levels were lower and glucose levels were higher in lambs whose dams were overfed during gestation. Because obesity is affiliated with impairment of β-cell function, the offspring of overnourished ewes could be prone to insulin resistance, glucose intolerance, and diabetes in later life (Zhang et al., 2011).

Several studies have looked at the effects of feeding specific nutrients or dietary ingredients during gestation on developmental programming. Feeding corn (Radunz et al., 2010) or DDGS (Radunz et al., 2010; Gunn, 2013) during late gestation resulted in heavier birth weights compared to calves from cows fed control diets. Heavier birth weights were also seen in lambs from ewes that were fed either corn or DDGS (Radunz et al., 2011a). In the Gunn (2013) study, birth weights were increased 4.4 kg, and dystocia was increased from 26.3 to 59.5% in first-parity cows when DDGS were used as an energy source to meet requirements in a corn stalk-based diet (DMI = 12.0 kg; crude protein [CP] intake = 1781 g/d; NEg intake = 2.13 Mcal/d) compared to an isocaloric control diet. The increased birth weights and resulting dystocia in these studies were attributed to excess glucose in the corn diet and protein (amino acid and/or ruminally undegradable protein [RUP]) intake in the DDGS diets during the last trimester of gestation. Interestingly, weaning weights were heavier in the Gunn (2013) study as a result of both increased birth weight and increased gain during the preweaning period, but no differences in weaning weight were seen in the Radunz et al. (2012) study. Radunz et al. (2012) reported no differences in insulin sensitivity, but an increase in marbling scores in steers from dams fed DDGS, and suggested that maternal diet during late gestation altered offspring adipose deposition. In the lamb study (Radunz et al., 2011b), lambs from dams fed DDGS had an increased initial insulin response as well as heavier BW and increased internal fat deposits compared to lambs from hay-fed, control dams. The increase in internal fat deposition was attributed to the altered insulin response. Similarly, supplementing protein during late gestation to cows fed winter range or crop residue tended to have increased weaning weights and increased quality grades at slaughter (Larson et al., 2009; Funston et al., 2012b). Feeding isonitrogenous and isocaloric supplements differing in percentage of RUP (59 vs. 34%) to cows also resulted in increased calf birth and weaning weights compared with nonsupplemented cows, but maternal RUP supplementation had no effects on feedlot performance (Summers et al., 2012).

Micronutrients can also affect fetal development. Bach (2012) reviewed the effect of dam nutrition on developmental programming and reported that amino acids (arginine, methionine) and B vitamins (folate, B_{12}) have the potential to affect fetal development and gene expression. Arginine is a precursor to nitric oxide, which has angiogenic properties and participates in regulation of blood flow to the uterus and placenta (Bird et al., 2003). Methionine is an essential amino acid that, when limiting in the dam diet, could lead to hypomethylation of DNA and dysregulation of gene expression and metabolism in the offspring (Petrie et al., 2002). Folate and vitamin B_{12}, while not typically deficient in ruminants, are important in methyl transfer and have improved milk production (Preynat et al., 2009) and have the potential of altering fetal development and gene expression (Bach, 2012).

Postnatal growth and development can be affected by maternal nutrient supply during gestation independent of milk nutrient supply postpartum (Meyer et al. 2010a; Neville et al., 2010; Yunusova et al., 2013), but there are relatively few beef studies that have investigated the effects of dam nutrition on lactational programming of the neonate. Similar to the fetus, neonates depend on their dam for nutrition to support growth and development, and thereby can be affected by milk yield and composition. The neonate has an abundance of totipotent stem cells, also known as satellite cells, which lie between the sarcolemma and basal lamina (Brameld et al., 2010; Du et al., 2010), that can proliferate into muscle, adipose, or connective tissue. Postnatal skeletal development depends on hypertrophy, and these satellite cells increase muscle fiber size by proliferating, differentiating, and fusing with existing muscle fibers (Brameld et al., 2010). Satellite cells decrease with age, however, and the window between birth and weaning can provide an opportunity to manipulate growth and development (Du et al., 2010).

Growth and development of the neonate are influenced by milk yield and composition, although most of the research in this area has been performed on neonatal dairy calves fed milk replacer, and few studies have looked at the effects of milk composition on beef calf performance. However, there is a positive correlation between dam postpartum energy status and calf gain (Perry et al., 1991; Lalman et al., 2000), milk yield and calf gain (Beal et al., 1990; Lake et al., 2005), and milk fat and calf gain (Beal et al., 1990; Brown and Brown, 2002). Several studies have shown that milk replacer composition, especially protein and energy concentration, affects calf growth, protein accumulation, and fat deposition (Donnelly and Hutton, 1976; Gerrits et al., 1996; Diaz et al.,

2001; Tikofsky et al., 2001; Bascom et al., 2007; Hill et al., 2011; Stamey et al., 2012). Feeding isocaloric milk replacer varying from 16 to 26% CP at 1.5% of BW has been reported to linearly increase BW and protein deposition as well as increase growth efficiency (Blome et al., 2003); however, Bartlett et al. (2006) concluded that increased protein concentration of milk replacer was only beneficial if the energy supply was adequate. In one beef study, feeding DDGS to dams during late gestation and early lactation decreased percentage of milk fat and increased milk protein, but calves from these dams were significantly heavier at birth and remained heavier until weaning (Winterholler et al., 2012). This observation is supported by studies (Gunn, 2013; Gunn et al., 2015), wherein DDGS (~1.2% of BW) were fed as a primary energy source in a corn stalk-based diet during both late gestation and early lactation to primiparous females. In these studies, they observed a decrease in milk fat and milk solids, but an increase was noted in milk urea, conjugated linoleic acid, monounsaturated fatty acid, and polyunsaturated fatty acid concentrations, as well as calf weaning weights. The steer offspring were followed into the feedlot and the heifer offspring were followed into their yearling breeding season. Muscle fatty acid composition of the steer offspring in later life was altered because of the difference in milk profile of dams fed DDGS. Heifer offspring from DDGS-fed cows had an increased frame size at weaning; obtained puberty at a heavier BW (298 vs. 324 kg), with similar BCS and age at puberty; and had a significantly higher timed-AI conception rate (70.6 vs. 33.3%), with similar overall breeding season pregnancy rate compared to heifer offspring from control cows. Because diets in these studies spanned both late gestation and early lactation, it is unclear whether these changes were the result of fetal and/or lactational programming of the heifer offspring. Regardless, it is reasonable to conclude from these studies that the observed developmental programming effects were likely mediated through either protein supply (RDP vs. RUP and/or amino acid supply) or fatty acid profile delivered to either the fetus in utero or neonate by milk consumption.

Reproductive Technologies

Marginally restricted diets that delay onset of puberty in heifers or extend the PPI in cows can be partially offset by the use of estrus synchronization protocols that use a progestin. Procedures that facilitate synchronization of estrus in cycling females and induction of an ovulatory estrus in peripubertal heifers or anestrous postpartum cows have been shown to increase reproductive rates and expedite genetic progress (Patterson et al., 2012). Estrus synchronization can be an effective means of increasing the percentage of females that become pregnant early in the breeding season, which can result in a shorter calving season and a more uniform calf crop (Dziuk and Bellows, 1983). Females that conceived to a synchronized estrus calved earlier in the calving season and calves were on average 13 d older and 9.5 kg heavier at weaning than calves from nonsynchronized females (Schafer et al., 2007).

LACTATION

Milk production in the beef cow is difficult to assess, but the nutritional balance of a lactating cow is important to the calculation of energy and protein requirements. In contrast to the dairy cow, which is generally milked by machine at least twice daily and milk production can be easily measured, the beef cow is generally in a pasture or range environment with milk consumed by the suckling calf. The primary methods of estimating milk production include the use of weaning weights as an indirect estimate, and making direct measurements through one of the following methods: teat cannulation following oxytocin injection (Lamond et al., 1969); milk removal by hand (Totusek et al., 1973); machine milking (Anthony et al., 1959); and determining differences in calf weights before and after nursing, which is commonly referred to as the weigh-suckle-weigh procedure, following a period of separation from the dam (Knapp and Black, 1941). Time of separation of the calf from the cow before and between milk production estimates has been reported to vary from 4 h (Williams et al., 1979a) to more than 19 h (Beal et al., 1988). Most data in the literature using the weigh-suckle-weigh procedures were based on multiple estimates of milk production at 4- to 8-h intervals to estimate 24-h milk production.

Numerous reports have included measures of milk yield of beef cows either to assess the relative yields for breed group comparisons or to estimate the relative influence of milk yield on calf preweaning growth (Drewry et al., 1959; Notter et al., 1978; Reynolds et al., 1978; Robinson et al., 1978; Williams et al., 1979b; Bartle et al., 1984; Marshall et al., 1984; Miller and Deutscher, 1985; Fiss and Wilton, 1989; Montano-Bermudez et al., 1990; Green et al., 1991; Freking and Marshall, 1992; Gregory et al., 1992, 1999). Only a limited number of studies have reported data that will allow the development of a lactation curve (Deutscher and Whiteman, 1971; Kropp et al., 1973; Totusek et al., 1973; Grainger and Wilhelm, 1979; Neidhardt et al., 1979; Gaskins and Anderson, 1980; Chenette and Frahm, 1981; Jenkins and Ferrell, 1984, 1992; Holloway et al., 1985; Jenkins et al., 1986; Clutter and Neilson, 1987; Sacco et al., 1987; Mezzadra et al., 1989; McCarter et al., 1991a; Buskirk et al., 1992; Hohenboken et al., 1992; Ripberger, 1997). These studies, unlike those with dairy cows, generally include a limited number of data points for a given cow during lactation, largely because of logistical problems described previously. Figure 13-4 provides a generalized lactation curve based on meeting the nutrient requirements for cows providing 5, 8, 11, and 14 kg of milk at peak lactation.

The most widely applied equation for describing the lactation curve of dairy cattle has been the one proposed and described by Wood (1980). Several other approaches have

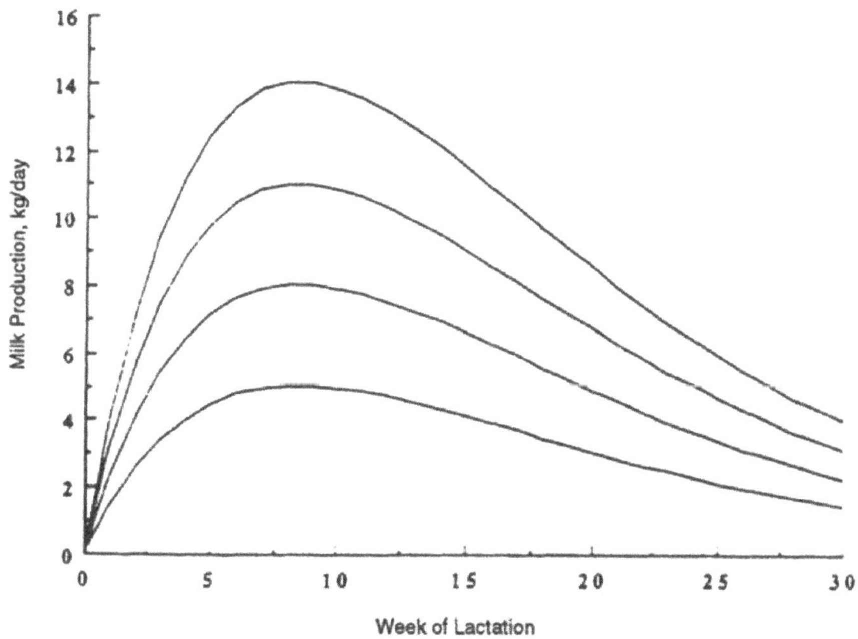

FIGURE 13-4 Generalized lactation curves for cows producing 5, 8, 11, or 14 kg of milk at peak milk production (NRC, 1996, 2000).

been proposed (Rowlands et al., 1982; Elston et al., 1989; Morant and Granaskthy, 1989) but, as with Wood's equation, their use with beef cattle milk production has been limited because of the relatively large number of data points required to fit the equation form. Jenkins and Ferrell (1984) proposed a similar equation form that requires fewer data points, which is used in the BCNRM:

$$Yn = n/(a \times e^{(k \times n)}) \times \text{age coefficient}, \quad \text{(Eq. 13-28)}$$

$$a = 1/(\text{aPKYD} \times k \times e^{1}), \quad \text{(Eq. 13-29)}$$

$$k = 1/\text{T}, \quad \text{(Eq. 13-30)}$$

$$n = \text{DIM}/7, \quad \text{(Eq. 13-31)}$$

$$\text{aPKYD} = (0.125 \times \text{RMY} + 0.375) \times \text{PKYD}, \quad \text{(Eq. 13-32)}$$

where

Yn is the daily milk yield (kg/d) at week n postpartum;
n is week of lactation;
a and k are intermediate variables;
e is the base of the natural (Naperian) logarithm (i.e., 2.718);
Age coefficient = 0.74 for age ≤ 2 yr; 0.88 for ages > 2, but ≤3 yr; and 1 for age ≥ 3 yr;
DIM is days in milk or days since calving, d;
aPKYD is the adjusted peak milk yield, kg/d;
RMY is the relative milk yield, kg/d;
PKYD is peak milk yield (Table 19-1), kg/d; and
T is the week of peak milk yield.

This equation form has been criticized (Hohenboken et al., 1992), but it has an advantage over Wood's equation in that it can be fitted with a minimal number of data points. In addition, curve parameters can be estimated from published data with minimal information. Data from Holloway et al. (1985), Jenkins et al. (1986), Clutter and Neilson (1987), Sacco et al. (1987), Lubritz et al. (1989), Mezzadra et al. (1989), McCarter et al. (1991a), Hohenboken et al. (1992), and Jenkins and Ferrell (1992) indicate that peak lactation occurred at approximately 8.5 weeks postpartum in cows with suckling calves. These data included a wide variety of breeds and breed crosses of cows and calves, milk yields, and sampling protocols. The peak is somewhat later than generally observed for dairy cows and could reflect the relationship between milk production and the calf's capacity to consume milk.

Total milk yield can be estimated using the following equation:

$$\text{TotalY} = -7/(a \times k) \times (n \times e^{(-k \times n)} + 1/k\, e^{(-k \times n)} - 1/k), \quad \text{(Eq. 13-33)}$$

where TotalY is the total milk yield for the n weeks of lactation, kg.

BULLS

There are limited new data on bull nutrition and the nutrition–reproduction interaction since the publication of the *Nutrient Requirements of Beef Cattle,* 7th Revised Edition (NRC, 1996, 2000); however, the current committee believes that the earlier published research remains applicable. Nutrient intakes that are below the requirements result in decreased growth rates and delayed puberty in the male in a manner similar to the female and, if severe enough, can permanently impair sperm output (Bratton et al., 1959; VanDemark et al., 1964; Nolan et al., 1990). Similarly, the reproductive potential of young males can also be impaired by overfeeding (Coulter and Kozub, 1984). Overfeeding has been associated with decreased scrotal circumference, epididymal sperm reserves, and seminal quality; however, in production, it seems more likely to underfeed, particularly bulls of large breeds, than to overfeed (Pruitt and Corah, 1985). Negative influences of specific nutrient deficiencies have been discussed in detail by Hurley and Doane (1989).

Mating behavior is an important aspect of male reproductive function as it has a direct bearing on the number of females mated. Specific nutrient deficiencies can result in lowered physical ability to mate, in addition to specific effects noted by Hurley and Doane (1989) such as decreased testicular weight, secretory output of the accessory sex glands, sperm motility, and sperm concentration. Moderate deficiencies or excesses in energy and/or protein, however, seem to have little effect on mating behavior, spermatogenesis, or semen quality. Severe deficiencies can result in diminished libido, depression of endocrine testicular function, and arrest of growth and secretory activity of accessory sex glands. Prolonged severe malnutrition, particularly insufficient intake of energy, protein, or water can lead to a decrease or cessation of spermatogenesis and are accompanied by decreased size of testes and accessory sex glands. Atrophy of the interstitial and Sertoli cell populations can accompany these changes. Overall, it is evident that unless males are moderately to severely deprived, there is minimal effect on the sexual responses and efficiency of the mating responses. Conversely, overfeeding and obesity may result in diminished libido and sexual activity.

It should be noted that the negative effects of malnutrition are more evident in the young male than in older animals. The mature male is remarkably resistant to nutritional stress, and infertility problems of nutritional origin are not often identified. Both young and mature males frequently lose weight during the breeding season, resulting from both decreased food consumption and substantially increased physical activity. Therefore, bulls should be in good body condition at the beginning of the breeding season to provide energy and protein reserves for use during breeding.

Both males and females are susceptible to decreases in fertility from increased environmental temperatures and humidity, typically associated with changes in season (Parkinson, 1987). Specifically in the male, decreases in sperm number and decreased testosterone concentrations can result from environmental heat stress (Rhynes and Ewing, 1973). As little as 12 h of exposure to an ambient temperature of 40°C can impair sperm function at the levels of the developing cells in the testicle (Gwazdauskas, 1985). Spermatogenesis is a continuous process in the male and is composed of events inside the testicle where young sperm cells divide, followed by other developmental and maturational events in the testicle and epididymis. The length of time required for a sperm cell to develop and mature in the bull's reproductive tract is approximately 61 d (Amann, 1970). The sperm cells in the early stages of development in the testicle are the most susceptible to negative effects of heat stress (Rahman et al., 2011). Thus, the detrimental effects on spermatogenesis can be expected to be seen in the ejaculates for 61 days following the last insult. This is supported by early work done by Meyerhoeffer et al. (1978), who placed bulls in a controlled environment of 35°C for 8 h and 30.5°C for 16 h daily for 8 weeks and compared them to bulls in a constant environment of 22.8°C. During the treatment period, heat-stressed bulls had elevated rectal temperatures, and sperm motility was significantly decreased within the first 2 weeks. Sperm motility did not return to normal until 8 weeks following the 8-week heat-stress period.

ENERGY AND PROTEIN REQUIREMENTS

Terms and concepts for nutrients are reported in Chapters 3 (Energy Terms and Concepts) through 9 (Water) of this publication. Chapter 10 (Intake), Chapter 11 (Maintenance), and Chapter 12 (Growth) provide specific information that relates directly to cow and bull energy and protein requirements. In brief, the efficiencies of ME use for lactation and maintenance are similar in beef cattle; thus, energy requirements for lactation have been expressed in NEm units. Efficiency of ME utilization for accretion of energy in gravid uterine tissues is likewise discussed in a subsequent section of this chapter. Some evidence is available to indicate that the efficiency of ME utilization for maintenance (k_m) and pregnancy (k_y) vary similarly with changes in ME concentration in the diet (Robinson et al., 1980). For convenience, estimates of requirements for beef cows were converted to NEm equivalents in the model. Conversion of requirements for lactation and pregnancy to NEm equivalents allows the energy value of feedstuffs to be adequately described by only two NE values (NEm and NEg).

Energy and Protein Requirements for Pregnancy

Energy accretion in the gravid uterus of Hereford heifers bred to Hereford bulls has been reported by Ferrell et al. (1976a). The equation used to describe the relationship of energy content of the gravid uterus is:

$$\text{NEu} = 69.73 \times e^{(0.03233 \times \text{DP} - 0.0000275 \times \text{DP}^2)}, \quad \text{(Eq. 13-34)}$$

where

NEu is the energy content of the gravid uterus, kcal; and
DP is days pregnant.

Similar values can be calculated from the data of Prior and Laster (1979), who used crossbred heifers bred to Brown Swiss bulls, and from the data of Jakobsen (1956) and Jakobsen et al. (1957), who used Red Danish cattle. Other information related to bovine fetal growth and weight change of the pregnant cow is available (Winters et al., 1942; Ellenberger et al., 1950; Eley et al., 1978; Silvey and Haycock, 1978). Eq. 13-34 was associated with a predicted calf birth weight of 38.5 kg and scaling it by birth weight yields the following equation (kcal):

$$NEy = CBW \times 1.811 \times e^{(0.03233 \times DP - 0.0000275 \times DP^2)}, \quad \text{(Eq. 13-35)}$$

where

NEy is the net energy required for pregnancy, Mcal/d; and
CBW is the calf birth weight, kg.

This equation may be differentiated with respect to DP to estimate daily energy accretion in the tissues of the gravid uterus, yielding (kcal/d)

$$NEy = [CBW \times (0.05855 - 0.0000996 \times DP) \times e^{(0.03233 \times DP - 0.0000275 \times DP^2)}]/1{,}000. \quad \text{(Eq. 13-36)}$$

The gross efficiency of ME use for accretion in the gravid uterus of cattle averaged 14% (Ferrell et al., 1976b). Other estimates with cattle and sheep average about 13% (Graham, 1964; Langlands and Southerland, 1968; Lodge and Heaney, 1970; Moe et al., 1970; Moe and Tyrrell, 1971; Sykes and Field, 1972; Rattray et al., 1974; Robinson et al., 1980). The use of the fixed partial efficiency (k_y of 0.13) results in the following equation to estimate the daily ME requirement for pregnancy in cattle:

$$MEy = NEy/0.0013, \quad \text{(Eq. 13-37)}$$

where MEy is the metabolizable energy requirement for pregnancy, Mcal/d.

Some evidence is available to indicate that efficiencies of ME use for maintenance and pregnancy vary similarly (Robinson et al., 1980). Values for efficiency of utilization of ME for maintenance (k_m) can be calculated from the equation of Garrett (1980b) as follows:

$$k_m = (1.37\ ME - 0.138\ ME^2 + 0.0105\ ME^3 - 1.12)/ME \quad \text{(Eq. 13-38)}$$

or

$$NEm = k_m \times MEy, \quad \text{(Eq. 13-39)}$$

where

NEm is the net energy requirement for maintenance of pregnancy, Mcal/d; and
k_m is the partial efficiency of use of ME to NE for maintenance, but otherwise an average of 0.576 can be used if a typical forage-based diet of 2.0 Mcal ME/kg, and then k_m is assumed.

Dry matter intake required to support pregnancy can be calculated as

$$FFP = MEy/ME, \quad \text{(Eq. 13-40)}$$

where FFP is the feed dry matter needed to support pregnancy, kg/d.

Estimates of the NEm required for pregnancy, from Eq. 13-39, are shown in Table 13-7. Current NEm requirements for pregnancy are the same as in NRC (1996, 2000). For comparison purposes, previous estimates from NRC (1984) and CSIRO (1990) are shown.

Protein requirements for pregnancy can be estimated using the approach used with energy. Estimates of nitrogen (N) content of gravid uterine tissues at various stages of gestation have been reported by Jakobsen (1956), Ferrell et al. (1976a), and Prior and Laster (1979). The equation derived by Ferrell et al. (1976a) to relate N (g) content of those tissues to day of gestation (*t*) was:

$$N = 2.310 \times e^{(0.0278 \times DP - 0.0000176 \times DP^2)} \quad \text{(Eq. 13-41)}$$

As with energy, this relationship can be scaled by predicted calf birth weight (38.5 kg) to derive the following equation:

$$N = CBW \times 0.06 \times e^{(0.0278 \times DP - 0.0000176 \times DP^2)} \quad \text{(Eq. 13-42)}$$

Daily accretion of protein in gravid uterine tissues can be calculated by differentiation of Eq. 13-42 with respect to DP as follows:

TABLE 13-7 Estimates of NEm (Mcal/d) Required for Pregnancy[a]

Day of Gestation	NRC (1996, 2000)	NRC (1984)	CSIRO (1990)
130	0.327	0.199	0.280
160	0.634	0.505	0.509
190	1.166	1.083	0.923
220	2.027	1.952	1.673
250	3.333	2.916	3.029
280	5.174	3.518	5.478

[a]Estimates are based on calf birth weight of 38.5 kg.

default value, but this number can be changed by the user if desired.

2. The model does not have a default adjustment factor for first-calf heifer SBW change/BCS, The default adjustment factor used in the model is set at 1.0 for these primiparous females when they are gaining or losing body energy reserves. Although the data are limited, one could logically argue that an adjustment factor of 0.40 should be used for primiparous females when they are losing weight and 1.6 when they are gaining weight, compared to the default value of 1.0 used for mature cows. The default adjustment factors can be changed by the user if desired.
3. The FHP and resulting NEm requirement of lactating primiparous females in the current model is set at 77 kcal/SBW$^{0.75}$. There are several data sets that suggest that the maintenance requirement is higher. The default adjustment in the model is set at 1.0, but limited data would suggest that the NEm requirement is 25% higher (adjustment factor of 1.25) when primiparous females enter lactation, independent of breed type. The default adjustment factors can be changed by the user if desired.
4. The energy value for 1 kg of weight change was changed from a constant of 5.82 Mcal of NE/kg to a variable number related to each BCS. Based on fat (9.4 Mcal/kg) and protein (5.7 Mcal/kg) percentages for each BCS from the U.S. MARC data set used in NRC (1996, 2000), and the revised SBW/BCS factor of 7.105%, the energy required to change 1 kg of weight can be calculated for any given cow weight and BCS.
5. The current model uses default puberty target weights of 65, 60, and 55% of expected mature weight, respectively, for *B. indicus*, *B. taurus*, and dual-purpose/dairy breeds to calculate requirements and gain. The default adjustment factors can be changed by the user if desired.

REFERENCES

AFRC (Agricultural and Food Research Council). 1990. Nutrient requirements of ruminant animals: Energy. *Nutrition Abstracts and Reviews Series B* 60:729-804.

Ahmad, N., F. N. Schrick, R. L. Butcher, and E. K. Inskeep. 1995. Effect of persistent follicles on early embryonic losses in beef cows. *Biology of Reproduction* 52:1129-1135.

Ahmad, N., E. C. Townsend, R. A. Dailey, and E. K. Inskeep. 1997. Relationships of hormonal patterns and fertility to occurrence of two or three waves of ovarian follicles, before and after breeding, in beef cows and heifers. *Animal Reproduction Science* 49:13-28.

Alexander, G. 1964a. Studies on the placenta of the sheep (*Ovis aries* L.): Effect of surgical reduction of the number of caruncles. *Journal of Reproduction and Fertility* 7:307-322.

Alexander, G. 1964b. Studies on the placenta of the sheep (*Ovis aries* L.): Placental size. *Journal of Reproduction and Fertility* 7:289-305.

Alexander, G., and D. Williams. 1971. Heat stress and development of the conceptus in domestic sheep. *Journal of Agricultural Science* 76:53-72.

Al-Katanani, Y. M., F. F. Paula-Lopes, and P. J. Hansen. 2002. Effect of season and exposure to heat stress on oocyte competence in Holstein cows. *Journal of Dairy Science* 85:390-396.

Amann, R. P. 1970. Sperm production rates. Pp. 433-482 in *The Testis* Vol. 1, A. D. Johnson, W. R. Gomes, and N. L. VanDemark, eds. New York: Academic Press.

Andersen, H., and M. Plum. 1965. Gestation length and birth weight in cattle and buffaloes: A review. *Journal of Dairy Science* 48:1224-1235.

Anderson, O. S., K. E. Sant, and D. C. Dolinoy. 2012. Nutrition and epigenetics: An interplay of dietary methyl donors, one-carbon metabolism and DNA methylation. *Journal of Nutritional Biochemistry* 23:853-859.

Anthony, W. B., P. F. Parks, E. L. Mayton, L. V. Brown, J. G. Starling, and T. B. Patterson. 1959. A new technique for securing milk production data for beef cows nursing calves in nutrition studies. *Journal of Animal Science* 18:1541 (Abstract 206).

ARC (Agricultural Research Council). 1980. *The Nutrient Requirements of Ruminant Livestock: Technical Review.* Farnham Royal, UK: Commonwealth Agricultural Bureaux.

Arias, R. P., P. J. Gunn, R. P. Lemenager, and S. L. Lake. 2012. Effects of post-AI nutrition on growth performance and fertility of yearling beef heifers. Pp. 117-121 in *Proceedings of Western Section, American Society of Animal Science, July 15-19, 2012, Phoenix, AZ,* Volume 63. Available online at http://www.jtmtg.org/JAM/2012/abstracts/WSASAS_2012.pdf. Accessed on April 8, 2015.

Arias, R. P., P. J. Gunn, R. P. Lemenager, G. A. Perry, G. A. Bridges, and S. L. Lake. 2013. Effects of post-AI nutrition in fertility of yearling beef heifers. Pp. 126-130 in *Proceedings of Western Section, American Society of Animal Science, June 18-20, 2013, Bozeman, MT,* Volume 64. Available online at https://www.asas.org/docs/default-source/western-section/proceedings.pdf?sfvrsn=0. Accessed on April 8, 2015.

Arije, G. F., and J. N. Wiltbank. 1971. Age and weight at puberty in Hereford heifers. *Journal of Animal Science* 33:401-406.

Arnett, D. W., G. L. Holland, and R. Totusek. 1971. Some effects of obesity in beef females. *Journal of Animal Science* 33:1129-1136.

Arnott, G., D. Roberts, J. A. Rooke, S. P. Turner, A. B. Lawrence, and K. M. D. Rutherford. 2012. Board invited review: The importance of the gestation period for welfare of calves: Maternal stressors and difficult births. *Journal of Animal Science* 90:5021-5034.

Asplund, J. M. 1994. *Principles of Protein Nutrition of Ruminants.* Boca Raton, FL: CRC Press.

Atkins, J. A., M. F. Smith, M. D. MacNeil, E. M. Jinks, F. M. Abreu, L. J. Alexander, and T. W. Geary. 2013. Pregnancy establishment and maintenance in cattle. *Journal of Animal Science* 91:722-733.

Ayalon, N. 1978. A review of embryonic mortality in cattle. *Journal of Reproduction & Fertility* 54:483-493.

Bach, A. 2012. Ruminant Nutrition Symposium: Optimizing Performance of the Offspring: Nourishing and managing the dam and postnatal calf for optimal lactation, reproduction, and immunity. *Journal of Animal Science* 90:1835-1845.

Bagley, C. P. 1993. Nutritional management of replacement beef heifers: A review. *Journal of Animal Science* 71:3155-3163.

Barker, D. J. P. 2012. Developmental origins of chronic disease. *Public Health* 126:185-189.

Barker, D. J. P., and P. M. Clark. 1997. Fetal undernutrition and disease in later life. *Reviews of Reproduction* 2:105-112.

Barker, D. J. P., M. Lamp, T. Roseboom, and N. Winder. 2012. Resource allocation and health in later life. *Placenta* 33(Suppl. 2):e30-e34.

Bartle, S. J., J. R. Males, and R. L. Preston. 1984. Effect of energy intake on the postpartum interval in beef cows and the adequacy of the cow's milk production for calf growth. *Journal of Animal Science* 58:1068-1074.

Bartlett, K. S., F. K. McKeith, M. J. VandeHaar, G. E. Dahl, and J. K. Drackley. 2006. Growth and body composition of dairy calves fed milk replacers containing different amounts of protein at two feeding rates. *Journal of Animal Science* 84:1454-1467.

Bartol, F. F., A. A. Wiley, D. J. Miller, A. J. Silva, K. E. Roberts, M. L. P. Davolt, J. C. Chen, A. L. Frankshun, M. E. Camp, K. M. Rahman, J. L. Vallet, and C. A. Bagnell. 2013. Lactation Biology Symposium: Lactocrine signaling and developmental programming. *Journal of Animal Science* 91:696-705.

Eversole, D. E., M. F. Browne, J. B. Hall, and R. E. Dietz. 2009. *Body Condition Scoring Beef Cows.* Virginia Tech Extension Fact Sheet 400-795. Available online at https://pubs.ext.vt.edu/400/400-795/400-795.html. Accessed on April 10, 2015.

Ewing, S. A., L. Smithson, D. Stephens, and D. McNutt. 1966. Weight Loss Patterns of Beef Cows at Calving. Oklahoma Agricultural Experiment Station Miscellaneous Publication MP 78:64.

Fahey, J., M. P. Boland, and D. O'Callaghan. 1998. Effects of dietary urea on embryo development in superovulated donor ewes and on embryo survival following transfer in recipient ewes. Abstract 182 in *Proceedings of the Annual Meeting of the British Society of Animal Science*, Penicuik: British Society of Animal Science.

Fall, C. H. D. 2011. Evidence for the intra-uterine programming of adiposity in later life. *Annals of Human Biology* 38:410-428.

Ferrell, C. L. 1982. Effects of postweaning rate of gain on onset of puberty and productive performance of heifers of different breeds. *Journal of Animal Science* 55:1272-1283.

Ferrell, C. L. 1989. Placental regulation of fetal growth. Pp. 1-19 in *Animal Growth Regulation*, D. R. Campion, G. J. Hausman, and R. J. Martin, eds. New York: Plenum Press.

Ferrell, C. L. 1991a. Maternal and fetal influences on uterine and conceptus development in the cow. I. Growth of tissues of the gravid uterus. *Journal of Animal Science* 69:1945-1953.

Ferrell, C. L. 1991b. Maternal and fetal influences on uterine and conceptus development in the cow. II. Blood flow and nutrient flux. *Journal of Animal Science* 69:1954-1965.

Ferrell, C. L. 1991c. Nutrient requirements and various factors affecting fetal growth and development. Pp. 76-92 in *52nd Minnesota Nutrition Conference, September 16-18, 1991, Bloomington, MN.*

Ferrell, C. L. 1991d. Nutritional influences on reproduction. Pp. 577-603 in *Reproduction in Domestic Animals*, 4th Ed., P. T. Cupps, ed. Los Angeles: Academic Press.

Ferrell, C. L., and T. G. Jenkins. 1985. Cow type and the nutritional environment: Nutritional aspects. *Journal of Animal Science* 61:725-741.

Ferrell, C. L., and T. G. Jenkins. 1996. Relationships between body condition score and empty body weight, water, fat, protein, and energy percentages in mature beef cows of diverse breeds (Abstract). *Journal of Animal Science* 74(Suppl. 1):245.

Ferrell, C. L., and L. P. Reynolds. 1987. Oxidative metabolism of gravid uterine tissues of the cow. Pp. 298-301 in *Energy Metabolism of Farm Animals: Proceedings of the 10th Symposium, September 1985, Airlie, VA*, P. W. Moe, H. F. Tyrrell, and P. J. Reynolds, eds. EAAP Publication No. 32. New York: Rowman & Littlefield.

Ferrell, C. L., and L. P. Reynolds. 1992. Uterine and umbilical blood flows and net nutrient uptake by fetuses and uteroplacental tissues of cows gravid with either single or twin fetuses. *Journal of Animal Science* 70:426-433.

Ferrell, C. L., W. N. Garrett, and N. Hinman. 1976a. Growth, development and composition of the udder and gravid uterus of beef heifers during pregnancy. *Journal of Animal Science* 42:1477-1489.

Ferrell, C. L., W. N. Garrett, N. Hinman, and G. Grichting. 1976b. Energy utilization by pregnant and non-pregnant heifers. *Journal of Animal Science* 42:937-950.

Ferrell, C. L., S. P. Ford, R. K. Christenson, and L. Prior. 1983. Blood flow, steroid secretion and nutrient uptake of the gravid bovine uterus and fetus. *Journal of Animal Science* 56:656-667.

Fike, K. E., M. L. Day, E. K. Inskeep, J. E. Kinder, P. L. Lewis, R. E. Short, and H. D. Hafs. 1997. Estrus and luteal function in suckled beef cows that were anestrous when treated with an intravaginal device containing progesterone with or without a subsequent injection of estradiol benzoate. *Journal of Animal Science* 75:2009-2015.

Fiss, C. F., and J. W. Wilton. 1989. Effects of breeding system, cow weight and milk yield on reproductive performance in beef cattle. *Journal of Animal Science* 67:1714-1721.

Ford, S. P. 1999. Cotyledonary placenta. Pp. 730-738 in *Encyclopedia of Reproduction*, E. Knobil and J. D. Neil, eds. San Diego, CA: Academic Press.

Ford, S. P., B. W. Hess, M. M. Schwope, M. J. Nijland, J. S. Gilbert, K. A. Vonnahme, W. J. Means, H. Han, and P. W. Nathanielsz. 2007. Maternal undernutrition during early to mid-gestation in the ewe results in altered growth, adiposity, and glucose tolerance in male offspring. *Journal of Animal Science* 85:1285-1294.

Fox, D. G., C. J. Sniffen, J. D. O'Connor, J. B. Russell, and P. J. Van Soest. 1992. A net carbohydrate and protein system for evaluating cattle diets: III. Cattle requirements and diet adequacy. *Journal of Animal Science* 70:3578-3596.

Fox, D. G., M. E. Van Amburgh, and T. P. Tylutki. 1999. Predicting requirements for growth, maturity, and body reserves in dairy cattle. *Journal of Dairy Science* 82:1968-1977.

Freetly, H. C., and J. A. Nienaber. 1998. Efficiency of energy and nitrogen loss and gain in mature cows. *Journal of Animal Science* 76:896-905.

Freetly, H. C., C. L. Ferrell, and T. G. Jenkins. 2000. Timing of realimentation of mature cows that were feed-restricted during pregnancy influences calf birth weights and growth rates. *Journal of Animal Science* 78:2790-2796.

Freetly, H. C., C. L. Ferrell, and T. G. Jenkins. 2001. Production performance of beef cows raised on three different nutritionally controlled heifer development programs. *Journal of Animal Science* 79:819-826.

Freetly, H. C., C. L. Ferrell, and T. G. Jenkins. 2005. Nutritionally altering weight gain patterns of pregnant heifers and young cows changes the time that feed resources are offered without any differences in production. *Journal of Animal Science* 83:916-926.

Freetly, H. C., J. A. Nienaber, and T. Brown-Brandl. 2008. Partitioning of energy in pregnant beef cows during nutritionally induced body weight fluctuation. *Journal of Animal Science* 86:370-377.

Freking, B. A., and D. M. Marshall. 1992. Interrelationships for heifer milk production and other biological traits with production efficiency to weaning. *Journal of Animal Science* 70:646-655.

Funston, R. N. 2004. Fat supplementation and reproduction in beef females. *Journal of Animal Science* 82:E154-E161.

Funston, R. N., D. M. Larson, and K. A. Vonnahme. 2010. Effects of maternal nutrition on conceptus growth and offspring performance: Implications for beef cattle production. *Journal of Animal Science* 88:E205-E215.

Funston, R. N., J. L. Martin, D. M. Larson, and A. J. Roberts. 2012a. Physiology and Endocrinology Symposium: Nutritional aspects of developing replacement heifers. *Journal of Animal Science* 90:1166-1171.

Funston, R. N., A. F. Summers, and A. J. Roberts. 2012b. Alpharma Beef Cattle Nutrition Symposium: Implications of nutritional management for beef cow-calf systems. *Journal of Animal Science* 90:2301-2307.

Gadeken, D., K. Rohr, and P. Lebzien. 1991. Effect of nitrogen fertilizing and of harvest season on net energy values of hay in dairy cows. Pp. 321-324 in *Energy Metabolism of Farm Animals: Proceedings of the 12th Symposium, September 1-7, 1991, Kartause Ittingen, Switzerland*, C. Wenk and M. Boessinger, eds. EAAP Publication No. 58. London: Butterworths.

Gardner, D. S., K. Tingey, B. W. M. Van Bon, S. E. Ozanne, V. Wilson, J. Dandrea, D. H. Keisler, T. Stephenson, and M. E. Symonds. 2005. Programming of glucose-insulin metabolism in adult sheep after maternal undernutrition. *American Journal of Physiology* 289:R947-R954.

Garrett, W. N. 1980a. Discussion Paper: Use of energy in reproduction and lactation. Pp. 383-389 in *Energy Metabolism: Proceedings, 8th Symposium on Energy Metabolism, September 1979, Cambridge, England*, L. E. Mount, ed. EAAP Publication No. 26. London: Butterworths.

Garrett, W. N. 1980b. Energy utilization by growing cattle as determined in 72 comparative slaughter experiments. Pp. 3-7 in *Energy Metabolism of Farm Animals: Proceedings of the 8th Symposium, September 1979, Cambridge, England*, L. E. Mount, ed. European Association for Animal Production Publication No. 26. London: Butterworths.

Garrett, W. N., and N. Hinman. 1969. Re-evaluation of the relationship between carcass density and body composition of beef steers. *Journal of Animal Science* 28:1-5.

Gaskins, C. T., and D. C. Anderson. 1980. Comparison of lactation curves in Angus-Hereford, Jersey-Angus and Simmental-Angus cows. *Journal of Animal Science* 50:828-832.

Gasser, C. L. 2013. Joint Alpharma-Beef Species Symposium: Considerations on puberty in replacement beef heifers. *Journal of Animal Science* 91:1336-1340.

Geary, T. W., M. F. Smith, M. D. MacNeil, M. L. Day, G. A. Bridges, G. A. Perry, F. M. Abreu, J. A. Atkins, K. G. Pohler, E. M. Jinks, and C. A. Madsen. 2013. Triennial Reproduction Symposium: Influence of follicular characteristics at ovulation on early embryonic survival. *Journal of Animal Science* 91:3014-3021.

Gerrits, W. J., G. H. Tolman, J. W. Schrama, S. Tamminga, M. W. Bosch, and M. W. Verstegen. 1996. Effect of protein and protein-free energy intake on protein and fat deposition rates in preruminant calves of 80 to 240 kg live weight. *Journal of Dairy Science* 74:2129-2139.

Ginther, O. J., J. P. Kastelic, and L. Knopf. 1989a. Composition and characteristics of follicular waves during the bovine oestrus cycle. *Animal Reproduction Science* 20:187-200.

Ginther, O. J., J. P. Kastelic, and L. Knopf. 1989b. Intraovarian relationships among dominant and subordinate follicles and the corpus luteum in heifers. *Theriogenology* 32:787-795.

Gleddie, V. M., and R. T. Berg. 1968. Milk production in range beef cows and its relationship to calf gains. *Canadian Journal of Animal Science* 48:323-333.

Gluckman, P. D., and G. C. Liggins. 1984. Regulation of fetal growth. Pp. 511-557 in *Fetal Physiology and Medicine,* R. W. Beard and P. W. Nathanielsz, eds. London: Butterworths.

Godfrey, K. M., and D. J. P. Barker. 2000. Fetal nutrition and adult disease. *American Journal of Clinical Nutrition* 71:1344S-1352S.

Graffam, W. S. 1992. Net Energy for Weight Change (NE_{Δ}) with Mature and First-Calf Heifers. M.S. Thesis. Purdue University, West Lafayette, IN.

Graham, N. M. 1964. Energy exchanges in pregnant and lactating ewes. *Australian Journal of Agricultural Research* 15:127-141.

Graham, N. M. 1980. Variation in energy and nitrogen utilization by sheep between weaning and maturity. *Australian Journal of Agricultural Research* 31:335-345.

Graham, J. F., A. J. Clark, and S. A. Spiker. 1984. The repeatability and accuracy of condition scoring beef cattle. P. 684 in *Proceedings of the 15th Biennial Conference of the Australian Society of Animal Production, Armidale, February 1984, New South Wales.* Sydney: Pergamon Press.

Grainger, C., and G. Wilhelm. 1979. Effect of pattern and duration of underfeeding in early lactation on milk production and reproduction of dairy cows. *Australian Journal of Experimental Agriculture and Animal Husbandry* 19:395-401.

Grainger, C., C. W. Holmes, A. M. Bryant, and Y. Moore. 1983. A note on the concentration of energy in the milk of Friesian and Jersey cows. *Animal Production* 36:307-308.

Green, R. D., L. V. Cundiff, G. E. Dickerson, and T. G. Jenkins. 1991. Output/input differences among nonpregnant, lactating *Bos indicus-Bos taurus* and *Bos taurus-Bos taurus* F1 cross cows. *Journal of Animal Science* 69:3156-3166.

Greenwood, P. L., and L. M. Cafe. 2007. Prenatal and pre-weaning growth and nutrition of cattle: Long-term consequences for beef production. *Animal* 1:1283-1296.

Gregory, K. E., L. V. Cundiff, and R. M. Koch. 1982. Characterization of breeds representing diverse biological types: Preweaning traits. Pp. 7-8 in *Beef Research Program, Progress Report No. 1*. ARM-NC-21:7. Peoria, IL: Agricultural Research North Central Region, Science and Education Administration, U.S. Department of Agriculture. Available online at http://www.ars.usda.gov/sp2UserFiles/Place/30400000/BeefResearchReports/BeefResearchProgressReportNo.1.pdf. Accessed on April 8, 2015.

Gregory, K. E., S. E. Echternkamp, G. E. Dickerson, L. V. Cundiff, R. M. Koch, and L. D. Van Vleck. 1990. Twinning in cattle: III. Effects of twinning on dystocia, reproductive traits, calf survival, calf growth and cow productivity. *Journal of Animal Science* 68:3133-3144.

Gregory, K. E., L. V. Cundiff, and R. M. Koch. 1992. Breed effects and heterosis on milk yield and 200-day weight in advanced generations of composite populations of beef cattle. *Journal of Animal Science* 70:2366-2372.

Gregory, K. E., L.V. Cundiff, and R. M. Koch. 1999. Composite Breeds to Use Heterosis and Breed Differences to Improve Efficiency of Beef Production. Technical Bulletin No. 1875. Available online at http://naldc.nal.usda.gov/download/CAT10879520/PDF. Accessed on April 8, 2015.

Gunn, P. J. 2013. Inclusion of Dried Distiller's Grains with Solubles in Beef Cow Diets and Impact on Reproduction and Subsequent Development of Progeny. Ph.D. Dissertation. Purdue University, West Lafayette, IN.

Gunn, P. J., J. P. Schoonmaker, R. P. Lemenager, and G. A. Bridges. 2015. Feeding distiller's grains as an energy source to gestating and lactating beef heifers: Impact on female progeny growth, puberty attainment and reproductive processes. *Journal of Animal Science* 93:746-757.

Gwazdauskas, F. C. 1985. Effects of climate on reproduction in cattle. *Journal of Dairy Science* 68:1568-1578.

Hansen, P. J. 2002. Embryonic mortality in cattle from the embryo's perspective. *Journal of Animal Science* 80(Suppl. 2):E33-E44.

Hansen, P. J., D. H. Baik, J. J. Rutledge, and E. R. Hauser. 1982. Genotype × environmental interactions on reproductive traits of bovine females. II. Postpartum reproduction as influenced by genotype, dietary regimen, level of milk production and parity. *Journal of Animal Science* 55:1458-1472.

Herd, D. B., and L. R. Sprott. 1986. Body Condition, Nutrition and Reproduction of Beef Cows. Texas A&M University Extension Bulletin 1526.

Hess, B. W., S. L. Lake, E. J. Scholljegerdes, T. R. Weston, V. Nayigihugu, J. D. C. Molle, and G. E. Moss. 2005. Nutritional controls of beef cow reproduction. *Journal of Animal Science* 83(Suppl.):E90-E106.

Hess, B. W., G. E. Moss, and D. C. Rule. 2008. A decade of developments in the area of fat supplementation research with beef cattle and sheep. *Journal of Animal Science* 86(Suppl.):E188-E204.

Hight, G. K. 1966. Effects of undernutrition in late pregnancy on beef cattle production. *New Zealand Journal of Agricultural Research* 9:479-490.

Hight, G. K. 1968a. Comparison of the effects of three nutritional levels in late pregnancy on beef cows and their calves to weaning. *New Zealand Journal of Agricultural Research* 11:477-486.

Hight, G. K. 1968b. Plane of nutrition effects in late pregnancy and during lactation on beef cows and their calves to weaning. *New Zealand Journal of Agricultural Research* 11:71-84.

Hill, T. M., M. J. VandeHaar, L. M. Sordillo, D. R. Catherman, H. G. Bateman II, and R. L. Schlotterbeck. 2011. Fatty acid intake alters growth and immunity in milk-fed calves. *Journal of Dairy Science* 94:3936-3948.

Hinde, K., and J. P. Capitanio. 2010. Lactational programming? Mother's milk energy predicts infant behavior and temperament in rhesus macaques (*Macaca mulatta*). *American Journal of Primatology* 72:522-529.

Hohenboken, W. D., A. Dudley, and D. E. Moddy. 1992. A comparison among equations to characterize lactation curves in beef cows. *Animal Production* 55:23-28.

Holloway, J. W., D. F. Stephens, J. V. Whiteman, and R. Totusek. 1975. Performance of 3-year-old Hereford, Hereford × Holstein and Holstein cows on range and in drylot. *Journal of Animal Science* 40:114-125.

Holloway, J. W., W. T. Butts, Jr., J. R. McCurley, E. E. Beaver, H. L. Peeler, and W. L. Backus. 1985. Breed × nutritional environmental interactions for beef female weight and fatness, milk production and calf growth. *Journal of Animal Science* 61:1354-1363.

Holroyd, R. G., K. W. Entwistle, and R. K. Shepherd. 1993. Effects on reproduction of estrous cycle variations, rectal temperatures and live weights in mated Brahman cross heifers. *Theriogenology* 40:453-464.

Houghton, P. L., R. P. Lemenager, L. A. Horstman, K. S. Hendrix, and G. E. Moss. 1990. Effects of body composition, pre- and postpartum energy level and early weaning on reproductive performance of beef cows and preweaning calf gain. *Journal of Animal Science* 68:1438-1446.

Hurley, W. L., and R. M. Doane. 1989. Recent developments in the roles of vitamins and minerals in reproduction. *Journal of Dairy Science* 72:784-804.

INRA (Institut National de la Recherche Agronomique). 1989. *Ruminant Nutrition.* Montrouge, France: Libbey Eurotext.

Inskeep, E. K. 2002. Factors that affect embryonic survival in the cow: Application of technology to improve calf crop. Pp. 255-279 in *Factors Affecting Calf Crop: Biotechnology of Reproduction*, M. J. Fields, R. S. Sand, and J. Y. Yelich, eds. Boca Raton, FL: CRC Press.

Jakobsen, P. E. 1956. Protein requirements for fetus formation in cattle. Pp. 115-126 in *Proceedings: 7th International Congress on Animal Husbandry, Madrid, Spain.*

Jakobsen, P. E., P. H. Sorensen, and H. Larsen. 1957. Energy investigations as related to fetus formation in cattle. *Acta Agriculturae Scandinavica* 7:103-112.

Jenkins, T. G., and C. L. Ferrell. 1984. A note on the lactation curves of crossbred cows. *Animal Production* 39:479-482.

Jenkins, T. G., and C. L. Ferrell. 1992. Lactation characteristics of nine breeds of cattle fed various quantities of dietary energy. *Journal of Animal Science* 70:1652-1660.

Jenkins, T. G., C. L. Ferrell, and L. V. Cundiff. 1986. Relationship of components of the body among mature cows as related to size, lactation potential and possible effects on productivity. *Animal Production* 43:245-254.

Jordan, E. R., T. E. Chapman, D. W. Holtan, and L. V. Swanson. 1983. Relationship of dietary crude protein to composition of uterine secretions and blood in high producing postpartum dairy cows. *Journal of Dairy Science* 66:1854-1862.

Joubert, D. M. 1963. Puberty in female farm animals. *Animal Breeding Abstracts* 31:295-306.

Kleiber, M. 1961. *The Fire of Life: An Introduction to Animal Energetics.* New York: Wiley & Sons.

Knapp, B., Jr., and W. H. Black. 1941. Factors influencing rate of gain of beef calves during the suckling period. *Journal of Agricultural Research* 63:249-254.

Koch, R. M., and J. W. Algeo. 1983. The beef cattle industry: Changes and challenges. *Journal of Animal Science* 57(Suppl. 2):28-43.

Kroker, G. A., and L. J. Cummins. 1979. The effect of nutritional restriction on Hereford heifers in late pregnancy. *Australian Veterinary Journal* 55:467-474.

Kropp, J. R., D. F. Stephens, J. W. Holloway, J. V. Whiteman, L. Knori, and R. Totusek. 1973. Performance on range and in drylot of two-year-old Hereford, Hereford × Holstein and Holstein females as influenced by level of winter supplementation. *Journal of Animal Science* 37:1222-1232.

Lake, S. L., E. J. Scholljegerdes, R. L. Atkinson, V. Nayigihugu, S. I. Paisley, D. C. Rule, G. E. Moss, T. J. Robinson, and B. W. Hess. 2005. Body condition score at parturition and postpartum supplemental fat effects on cow and calf performance. *Journal of Animal Science* 83:2908-2917.

Lake, S. L., R. Arias, P. Gunn, and G. A. Bridges. 2013. Nutritional management post-AI to enhance pregnancy outcome. Pp. 51-59 in *Proceedings, Range Beef Cow Symposium, December 3-5, Rapid City, SD.* Available online at http://www.rangebeefcow.com/2013/proceedings/08-lake-post-ai-nutrition.pdf. Accessed on December 3, 2014.

Lalman, D. L., D. H. Keisler, J. E. Williams, E. J. Scholljegerdes, and D. M. Mallett. 1997. The influence of postpartum weight and body condition change on duration of anestrus by undernourished suckled beef heifers. *Journal of Animal Science* 75:2003-2008.

Lalman, D. L., J. E. Williams, B. W. Hess, M. G. Thomas, and D. H. Keisler. 2000. Effect of dietary energy on milk production and metabolic hormones in thin, primiparous beef heifers. *Journal of Animal Science* 78:530-538.

Lamond, D. R. 1970. The influence of undernutrition on reproduction in the cow. *Animal Breeding Abstracts* 38:359-372.

Lamond, D. R., J. H. Holmes, and K. P. Haydock. 1969. Estimation of yield and composition of milk produced by grazing beef cows. *Journal of Animal Science* 29:606-611.

Langlands, J. P., and H. A. M. Southerland. 1968. An estimate of the nutrients utilized for pregnancy by Merino sheep. *British Journal of Nutrition* 22:217-227.

Larson, D. M., J. L. Martin, D. C. Adams, and R. N. Funston. 2009. Winter grazing system and supplementation during late gestation influence performance of beef cows and steer progeny. *Journal of Animal Science* 87:1147-1155.

Larson, R. L. 2007. Heifer development: Reproduction and nutrition. *Veterinary Clinics of North America: Food Animal Practice* 23:53-68.

Laster, D. B., and K. E. Gregory. 1973. Factors affecting peri- and early postnatal calf mortality. *Journal of Animal Science* 37:1092-1097.

Laster, D. B., H. A. Glimp, and K. E. Gregory. 1972. Age and weight at puberty and conception in different breeds and breed crosses of beef heifers. *Journal of Animal Science* 34:1031-1036.

Laster, D. B., G. M. Smith, and K. E. Gregory. 1976. Characterization of biological types of cattle. IV. Postweaning growth and puberty of heifers. *Journal of Animal Science* 43:63-70.

Laster, D. B., G. M. Smith, L. V. Cundiff, and K. E. Gregory. 1979. Characterization of biological types of cattle (Cycle II). II. Postweaning growth and puberty of heifers. *Journal of Animal Science* 48:500-508.

Leroy, J. L., A. Van Soom, G. Opsomer, I. G. Goovaerts, and P. E. Bols. 2008. Reduced fertility in high-yielding dairy cows: Are the oocyte and embryo in danger? Part II. Mechanisms linking nutrition and reduced oocyte and embryo quality in high-yielding dairy cows. *Reproduction in Domestic Animals* 43:623-632.

Lesmeister, J. L., P. J. Burfening, and R. L. Blackwell. 1973. Date of first calving in beef cows and subsequent calf production. *Journal of Animal Science* 36:1-6.

Lodge, G. A., and D. P. Heaney. 1970. Energy cost of pregnancy in the ewe. Pp. 109-112 in *Energy Metabolism of Farm Animals: Proceedings of the 5th Symposium on Energy Metabolism, September 1970, Vitznau, Switzerland.* EAAP Publication No. 13. Zurich: Juris Verlag.

Lonergan, P., P. Monaghan, D. Rizos, M. P. Boland, and I. Gordon. 1994. Effect of follicle size on bovine oocyte quality and developmental competence following maturation, fertilization and culture in vitro. *Molecular Reproduction and Development* 37:48-53.

Long, C. R. 1980. Crossbreeding for beef production: Experimental results. *Journal of Animal Science* 51:1197-1223.

Long, N. M. 2010. Effects of nutrient restriction of bovine dams during early gestation on postnatal growth, carcass and organ characteristics, and gene expression in adipose tissue and muscle. *Journal of Animal Science* 88:3251-3261.

Long, N. M., K. A. Vonnahme, B. W. Hess, P. W. Nathanielsz, and S. P. Ford. 2009. Effects of early gestational undernutrition on fetal growth, organ development, and placentomal composition in the bovine. *Journal of Animal Science* 87:1950-1959.

Long, N. M., L. A. George, A. B. Uthlaut, D. T. Smith, M. J. Nijland, P. W. Nathanielsz, and S. P. Ford. 2010. Maternal obesity and increased nutrient intake before and during gestation in the ewe results in altered growth, adiposity, and glucose tolerance in adult offspring. *Journal of Animal Science* 88:3546-3553.

Long, N. M., C. B. Tousley, K. R. Underwood, S. I. Paisley, W. J. Means, B. W. Hess, M. Du and S. P. Ford. 2012. Effects of early- to mid-gestational undernutrition with or without protein supplementation on offspring growth, carcass characteristics, and adipocyte size in beef cattle. *Journal of Animal Science* 90:197-206.

Lowman, B. G., R. A. Edwards, S. H. Somerville, and G. M. Jolly. 1979. The effect of plane of nutrition in early lactation on the performance of beef cows. *Animal Production* 29:293-303.

Lubritz, D. L., K. Forrest, and O. W. Robinson. 1989. Age of cow and age of dam effects on milk production of Hereford cows. *Journal of Animal Science* 67:2544-2549.

Lucas, A. 2000. Programming not metabolic imprinting. *American Journal of Clinical Nutrition* 71:602.

Lucy, M. C., T. S. Gross, and W. W. Thatcher. 1990. Effect of intravenous infusion of soybean oil emulsion on plasma concentrations of 15-keto-13,14-dihydro-prostaglandin F2 and ovarian function in cycling Holstein heifers. Pp. 119-132 in *Livestock Reproduction in Latin America: Proceedings of the Final Research Co-ordination Meeting of the Arcal III Programme, September 19-23, 1988, Bogota*. Vienna, Austria: International Atomic Energy Agency.

Lucy, M. C., C. R. Staples, F. M. Michel, and W. W. Thatcher. 1991. Feeding of calcium soaps to early postpartum dairy cows on plasma prostaglandin $F_{2\alpha}$, luteinizing hormone, and follicular growth. *Journal of Dairy Science* 74:483-489.

Luther, J. S., D. A. Redmer, L. P. Reynolds, and J. M. Wallace. 2005. Nutritional paradigms of ovine fetal growth restriction: Implications for human pregnancy. *Human Fertility* 8:179-187.

Lynch, J. M., G. C. Lamb, B. L. Miller, R. T. Brandt, Jr., R. C. Cochran, and J. E. Minton. 1997. Influence of timing of gain on growth and reproductive performance of beef replacement heifers. *Journal of Animal Science* 75:1715-1722.

Mackey, D. R., J. M. Sreenan, J. F. Roche, and M. G. Diskin. 1999. Effect of acute restriction on incidence of anovulation and periovulatory estradiol and gonadotropin concentrations in beef heifers. *Biology of Reproduction* 61:1601-1607.

Mangus, W. L., and J. S. Brinks. 1971. Relationships between direct and maternal effects on growth in Herefords: I. Environmental factors during preweaning growth. *Journal of Animal Science* 32:17-23.

Mann, G. E., and G. E. Lamming. 2001. Relationship between maternal endocrine environment, early embryo development and inhibition of the luteolytic mechanism in cows. *Reproduction* 121(1):175-180.

Marshall, D. M., R. R. Frahm, and G. W. Horn. 1984. Nutrient intake and efficiency of calf production by two-breed cross cows. *Journal of Animal Science* 59:317-328.

Martin, J. L., K. A. Vonnahme, D. C. Adams, G. P. Lardy, and R. N. Funston. 2007. Effects of dam nutrition on growth and reproductive performance of heifer calves. *Journal of Animal Science* 85:841-847.

Martin, L. C., J. S. Brinks, R. M. Bourdon, and L. V. Cundiff. 1992. Genetic effects on beef heifer puberty and subsequent reproduction. *Journal of Animal Science* 70:4006-4017.

Martin, R., G. Moscoso, R. J. Scaramuzzi, P. T. Loughna, P. Johnson, and A. J. Leigh. 1998. The effect of maternal hyperglycemia on embryo development in the ewe. *Journal of Reproduction and Fertility Abstract Series* 21:90.

Martin, T. G., R. P. Lemenager, G. Srinivasan, and R. Alenda. 1981. Creep feed as a factor influencing performance of cows and calves. *Journal of Animal Science* 53:33-39.

Masilo, B. S., J. S. Stevenson, R. R. Shalles, and J. E. Shirley. 1992. Influence of genotype and yield and composition of milk on interval to first partum ovulation in milked beef and dairy cows. *Journal of Animal Science* 70:379-385.

Mattos, R., C. R. Staples, and W. W. Thatcher. 2000. Effects of dietary fatty acids on reproduction in ruminants. *Reviews of Reproduction* 5:38-45.

Maurer, R. R., and J. R. Chenault. 1983. Fertilization failure and embryonic mortality in parous and nonparous beef cattle. *Journal of Animal Science* 56:1186-1189.

McCarter, M. N., D. S. Buchanan, and R. R. Frahm. 1991a. Comparison of crossbred cows containing various proportions of Brahman in spring or fall calving systems. II. Milk production. *Journal of Animal Science* 69:77-84.

McCarter, M. N., D. S. Buchanan, and R. R. Frahm. 1991b. Comparison of crossbred cows containing various proportions of Brahman in spring or fall calving systems: III. Productivity as three-, four-, and five-year olds. *Journal of Animal Science* 69:2754-2761.

McCartor, M. M., R. D. Randel, and L. H. Carroll. 1979. Dietary alteration of ruminal fermentation on efficiency of growth and onset of puberty in Brangus heifers. *Journal of Animal Science* 48:488-494.

McMorris, M. R., and J. W. Wilton. 1986. Breeding systems, cow weight and milk yield effects on various biological variables in beef production. *Journal of Animal Science* 63:1361-1372.

Melton, A. A., J. K. Riggs, L. A. Nelson, and T. C. Cartwright. 1967. Milk production, composition and calf gains of Angus, Charolais and Hereford cows. *Journal of Animal Science* 26:804-809.

Meyer, A. M., J. J. Reed, T. L. Neville, J. B. Taylor, C. J. Hammer, L. P. Reynolds, D. A. Redmer, K. A. Vonnahme, and J. S. Caton. 2010a. Effects of nutritional plane and selenium supply during gestation on ewe and neonatal offspring performance, body composition, and serum selenium. *Journal of Animal Science* 88:1786-1800.

Meyer, A. M., J. J. Reed, K. A. Vonnahme, S. A. Soto-Navarro, L. P. Reynolds, S. P. Ford, B. W. Hess, and J. S. Caton. 2010b. Effects of stage of gestation and nutrient restriction during early to mid-gestation on maternal and fetal visceral organ mass and indices of jejunal growth and vascularity in beef cows. *Journal of Animal Science* 88:2410-2424.

Meyer, A. M., J. J. Reed, T. L. Neville, J. F. Thorson, K. R. Maddock-Carlin, J. B. Taylor, L. P. Reynolds, D. A. Redmer, J. S. Luther, C. J. Hammer, K. A. Vonnahme, and J. S. Caton. 2011. Nutritional plane and selenium supply during gestation impact yield and nutrient composition of colostrum and milk in primiparous ewes. *Journal of Animal Science* 89:1627-1639.

Meyer, A. M., T. L. Neville, J. J. Reed, J. B. Taylor, L. P. Reynolds, D. A. Redmer, C. J. Hammer, K. A. Vonnahme, and J. S. Caton. 2013. Maternal nutritional plane and selenium supply during gestation impacts visceral organ mass and intestinal growth and vascularity of neonatal lamb offspring. *Journal of Animal Science* 91:2628-2639.

Meyerhoeffer, D. C., R. P. Wettemann, S. W. Coleman, and M. E. Wells. 1978. Reproductive criteria of beef bulls during and after exposure to increased ambient temperature. *Journal of Animal Science* 60:352-357.

Mezzadra, C., R. Paciaroni, S. Vulich, E. Villarreal, and L. Melucci. 1989. Estimation of milk consumption curve parameters for different genetic groups of bovine calves. *Animal Production* 49:83-87.

Micke, G. C., T. M. Sullivan, K. L. Gatford, J. A. Owens, and V. E. A. Perry. 2010. Nutrient intake in the bovine during early and mid-gestation causes sex-specific changes in progeny plasma IGF-1, live weight, height and carcass traits. *Animal Reproduction Science* 121:208-217.

Micke, G. C., T. M. Sullivan, I. C. McMillen, S. Gentili, and V. E. A. Perry. 2011. Protein intake during gestation affects postnatal bovine skeletal muscle growth and relative expression of IGF-1, IGF1R, IGF-2, and IGF-2R. *Molecular and Cellular Endocrinology* 332:234-241.

Miller, H. L., and G. H. Deutscher. 1985. Beef production of Simmental × Angus and Hereford × Angus cows under range conditions. *Journal of Animal Science* 61:1364-1369.

Moe, P. W. 1981. Energy metabolism of dairy cattle. *Journal of Dairy Science* 64:1120-1139.

Moe, P. W., and H. F Tyrrell. 1971. Metabolizable energy requirements of pregnant dairy cows. *Journal of Dairy Science* 55:480-483.

Moe, P. W., H. F. Tyrrell, and W. P. Flatt. 1970. Partial efficiency of energy use for maintenance, lactation, body gain and gestation in the dairy cow. Pp. 65-68 in *Energy Metabolism of Farm Animals: Proceedings of the 5th Symposium on Energy Metabolism, September 1970, Vitznau, Switzerland*. EAAP Publication No. 13. Zurich: Juris Verlag.

Moe, P. W., W. P. Flatt, and H. F. Tyrrell. 1972. Net energy value of feeds for lactation. *Journal of Dairy Science* 55:945-958.

Mondragon, I., J. W. Wilton, O. B. Allen, and H. Song. 1983. Stage of lactation effects: Repeatabilities and influences on weaning weights of yield and composition of milk in beef cattle. *Canadian Journal of Animal Science* 63:751-761.

Montano-Bermudez, M., M. K. Nielson, and G. H. Deutscher. 1990. Energy requirements for maintenance of crossbred beef cattle with different genetic potentials for milk. *Journal of Animal Science* 68:2279-2288.

Morant, S. V., and A. Granaskthy. 1989. A new approach to the mathematical formulation of lactation curves. *Animal Production* 49:151-162.

Moseley, W. M., M. M. McCartor, and R. D. Randel. 1977. Effects of monensin on growth and reproductive performance of beef heifers. *Journal of Animal Science* 45:961-968.

Moseley, W. M., T. G. Dunn, C. C. Kaltenbach, R. E. Short, and R. B. Staigmiller. 1982. Relationship of growth and puberty in beef heifers fed monensin. *Journal of Animal Science* 55:357-362.

Mossa, F., D. Kenny, F. Jimenez-Krassel, G. W. Smith, D. Berry, S. Butler, S., and A. C. Evans. 2009. Undernutrition of heifers during the first trimester of pregnancy diminishes size of the ovarian reserve in female offspring. *Biology of Reproduction* 81(Suppl. 1):Abstract 135.

Moura, E. G., and M. C. Passos. 2005. Neonatal programming of body weight regulation and energetic metabolism. *Bioscience Reports* 25:251-269.

Mullen, M., G. Elia, M. Hilliard, M. H. Parr, M. G. Diskin, A. C. O. Evans, and M. A. Crow. 2012. Proteomic characterization of histotroph during the preimplantation phase of the estrous cycle in cattle. *Journal of Proteome Research* 11:3004-3018.

Munger, A. 1991. Milk production efficiency in dairy cows of different breeds. Pp. 292-295 in *Energy Metabolism of Farm Animals: Proceedings of the 12th Symposium, September 1-7, 1991, Kartause Ittingen, Switzerland*, C. Wenk and M. Boessinger, eds. EAAP Publication No. 58. London: Butterworths.

Muñoz, C., A. F. Carson, M. A. McCoy, L. E. R. Dawson, A. R. G. Wylie, and A. W. Gordon. 2009. Effects of plane of nutrition of ewes in early and mid-pregnancy on performance of the offspring: Female reproduction and male carcass characteristics. *Journal of Animal Science* 87:3647-3655.

Munro, H. N., S. J. Philistine, and M. E. Fant. 1983. The placenta in nutrition. *Annual Review of Nutrition* 3:97-124.

Murphy, M. G., W. J. Enright, M. A. Crowe, K. McConnell, L. J. Spicer, M. P. Boland, and J. F. Roche. 1991. Effect of dietary intake on pattern of growth of dominant follicles during the oestrous cycle in beef heifers. *Journal of Reproduction and Fertility* 92:333-338.

Neidhardt, R., D. Plasse, J. H. Weniger, O. Verde, J. Beltran, and A. Benavides. 1979. Milk yield of Brahman cows in a tropical beef production system. *Journal of Animal Science* 48:1-6.

Neville, T. L., J. S. Caton, C. J. Hammer, J. J. Reed, J. S. Luther, J. B. Taylor, D. A. Redmer, L. P. Reynolds, and K. A. Vonnahme. 2010. Ovine offspring growth and diet digestibility are influenced by maternal Se supplementation and nutritional intake level during pregnancy despite a common postnatal diet. *Journal of Animal Science* 88:3645-3656.

Nolan, C. J., D. A. Nevendorff, R. W. Godfrey, P. G. Harms, T. H. Welsh, Jr., N. H. McArthur, and R. D. Randel. 1990. Influence of dietary energy intake on prepubertal development of Brahman bulls. *Journal of Animal Science* 68:1087-1096.

Notter, D. R., L. V. Cundiff, G. M. Smith, D. B. Laster, and K. E. Gregory. 1978. Characterization of biological types of cattle. VII. Milk production in young cows and transmitted and maternal effects on preweaning growth of progeny. *Journal of Animal Science* 46:908-921.

NRC (National Research Council). 1970. *Nutrient Requirements of Beef Cattle*, 4th Rev. Ed. Washington, DC: National Academy Press.

NRC. 1984. *Nutrient Requirements of Beef Cattle*, 6th Rev. Ed. Washington, DC: National Academy Press.

NRC. 1985. *Nutrient Requirements of Sheep*, 6th Rev. Ed. Washington, DC: National Academy Press.

NRC. 1996. *Nutrient Requirements of Beef Cattle*, 7th Rev. Ed. Washington, DC: National Academy Press.

NRC. 2000. *Nutrient Requirements of Beef Cattle: Update 2000*, 7th Rev. Ed. Washington, DC: National Academy Press.

Olson, T. A., K. Euclides Fiho, L. V. Cundiff, M. Koger, W. T. Butts, Jr., and K. E. Gregory. 1991. Effects of breed group by location interaction on crossbred cattle in Nebraska and Florida. *Journal of Animal Science* 69:104-114.

Owens, J. A., J. Falconer, and J. S. Robinson. 1986. Effects of restriction of placental growth on umbilical and uterine blood flows. *American Journal of Physiology* 250:R427-R434.

Pahnish, O. F., M. Koger, J. J. Urick, W. C. Burns, W. T. Butts, and G. V. Richardson. 1983. Genotype × environmental interaction in Hereford cattle. III. Postweaning traits of heifers. *Journal of Animal Science* 56:1039-1046.

Parkinson, T. J. 1987. Seasonal variations in semen quality of bulls: Correlations with environmental temperature. *Veterinary Record* 120:479-482.

Paten, A. M., P. R. Kenyon, N. Lopez-Villalobos, S. W. Peterson, C. M. C. Jenkinson, S. J. Pain, and H. T. Blair. 2013. Lactation Biology Symposium: Maternal nutrition during early and mid-to-late pregnancy: Comparative effects on milk production of twin-born ewe progeny during their first lactation. *Journal of Animal Science* 90:676-684.

Patle, B. R., and V. D. Mudgal. 1976. Utilization of diet energy for maintenance and milk production by lactating crossbred cows (Brown Swiss × Sahiwal) during their early stage of lactation. *Indian Journal of Animal Science* 46:1-7.

Patterson, D. J., R. C. Perry, G. H. Kiracofe, R. A. Bellows, R. B. Staigmiller, and L. R. Corah. 1992. Management considerations in heifer development and puberty. *Journal of Animal Science* 70:4018-4035.

Patterson, D. J., N. T. Martin, J. M. Thomas, and M. F. Smith. 2012. Control of estrus in heifers. Pp. 53-84 in *Proceedings of the Range Beef Cow Symposium XVII: Applied Reproductive Strategies in Beef Cattle, December 3-4, 2012, Sioux Falls, SD*. Available online at http://www.appliedreprostrategies.com/pdfs/2012ARSBC_06PattersonProceedings.pdf. Accessed on December 3, 2014.

Perry, R. C., L. R. Corah, R. C. Cochran, W. E. Beal, J. S. Stevenson, J. E. Minton, D. D. Simms, and J. R. Brethour. 1991. Influence of dietary energy on follicular development, serum gonadotropins, and first postpartum ovulation in suckled beef cows. *Journal of Animal Science* 69:3762-3773.

Petrie, L., S. J. Duthie, W. D. Rees, and J. M. L. McConnell. 2002. Serum concentrations of homocysteine are elevated during early pregnancy in rodent models of fetal programming. *British Journal of Nutrition* 88:471-477.

Preynat, A., H. Lapierre, M. C. Thivierge, M. F. Palin, J. J. Matte, A. Desrochers, and C. L. Girard. 2009. Influence of methionine supply on the response of lactational performance of dairy cows to supplementary folic acid and vitamin B_{12}. *Journal of Dairy Science* 92:1685-1695.

Prior, R. L., and D. B. Laster. 1979. Development of the bovine fetus. *Journal of Animal Science* 48:1546-1553.

Pruitt, R. J., and L. R. Corah. 1985. Effect of energy intake after weaning on sexual development of beef bulls. 1. Semen characteristics and serving capacity. *Journal of Animal Science* 61:1186-1193.

Putney, D. J., M. Drost, and W. W. Thatcher. 1989. Influence of summer heat stress on pregnancy rates of lactating dairy cattle following embryo transfer or artificial insemination. *Theriogenology* 31:765-778.

Quigley, S. P., D. O. Kleemann, M. A. Kakar, J. A. Owens, G. S. Nattrass, S. Maddocks, and S. K. Walker. 2005. Myogenesis in sheep is altered by maternal feed intake during the periconception period. *Animal Reproduction Science* 87:241-251.

Radunz, A. E., F. L. Fluharty, M. L. Day, H. N. Zerby, and S. C. Loerch. 2010. Prepartum dietary energy source fed to beef cows: I. Effects on pre- and postpartum cow performance. *Journal of Animal Science* 88:2717-2728.

Radunz, A. E., F. L. Fluharty, I. Susin, T. L. Felix, H. N. Zerby, and S. C. Loerch. 2011a. Winter-feeding systems for gestating sheep. II. Effects on feedlot performance, glucose tolerance, and carcass composition of lamb progeny. *Journal of Animal Science* 89:478-488.

Radunz, A. E., F. L. Fluharty, H. N. Zerby, and S. C. Loerch. 2011b. Winter-feeding systems for gestating sheep. I. Effects on pre- and postpartum ewe performance and lamb progeny preweaning performance. *Journal of Animal Science* 89:467-477.

Radunz, A. E., F. L. Fluharty, G. D. Lowe, and S. C. Loerch. 2012. Prepartum dietary energy source fed to beef cows: II. Effects on progeny postnatal growth, glucose tolerance, and carcass composition. *Journal of Animal Science* 90:4962-4974.

Rae, M. T., S. Palassio, C. E. Kyle, A. N. Brooks, R. G. Lea, D. W. Miller, and S. M. Rhind. 2001. Effect of maternal undernutrition during pregnancy on early ovarian development and subsequent follicular development in sheep fetuses. *Reproduction* 122:915-922.

Rae, M. T., C. E. Kyle, D. W. Miller, A. J. Hammond, A. N. Brooks, and S. M. Rhind. 2002. The effects of undernutrition, in utero, on reproductive function in adult male and female sheep. *Animal Reproduction Science* 72:63-71.

Rahman, M. B., L. Vandaele, T. Rijsselaere, D. Maes, M. Hoogewijs, A. Frijters, J. Noordman, A. Granados, E. Dernelle, M. Shamsuddin, J. Parrish, and A. Van Soom. 2011. Scrotal insulation and its relationship to abnormal morphology, chromatin protamination and nuclear shape of spermatozoa in Holstein-Friesian and Belgian Blue bulls. *Theriogenology* 76:1246-1257.

Randel, R. D. 1981. Effect of once-daily suckling on postpartum interval and cow-calf performance of first-calf Brahman × Hereford heifers. *Journal of Animal Science* 53:755-757.

Randel, R. D. 1990. Nutrition and postpartum rebreeding in cattle. *Journal of Animal Science* 68:853-862.

Rattray, P. V., W. N. Garrett, N. E. East, and N. Hinman. 1974. Efficiency of utilization of metabolizable energy during pregnancy in sheep. *Journal of Animal Science* 38:383-393.

Redmer, D. A., J. M. Wallace, and L. P. Reynolds. 2004. Effect of nutrient intake during pregnancy on fetal and placental growth and vascular development. *Domestic Animal Endocrinology* 27:199-217.

Redmer, D. A., J. S. Milne, R. P. Aitken, M. L. Johnson, P. P. Borowicz, L. P. Reynolds, J. S. Caton, and J. M. Wallace. 2012. Decreasing maternal nutrient intake during the final third of pregnancy in previously overnourished adolescent sheep: Effects on maternal nutrient partitioning and feto-placental development. *Placenta* 33:114-121.

Reid, J. T., G. H. Wellington, and H. O. Dunn. 1955. Some relationships among the major chemical components of the bovine body and their application to nutritional investigations. *Journal of Dairy Science* 38:1344-1359.

Reynolds, L. P., and J. S. Caton. 2012. Role of the pre- and post-natal environment in developmental programming of health and productivity. Invited review. *Molecular and Cellular Endocrinology* 354:54-59.

Reynolds, L. P., and D. A. Redmer. 1995. Utero-placental vascular development and placental function. *Journal of Animal Science* 73:1839-1851.

Reynolds, L. P., C. L. Ferrell, J. A. Nienaber, and S. P. Ford. 1985. Effects of chronic environmental heat stress on blood flow and nutrient uptake of the gravid bovine uterus and foetus. *Journal of Agricultural Science* 104:289-297.

Reynolds, L. P., C. L. Ferrell, D. A. Robertson, and S. P. Ford. 1986. Metabolism of the gravid uterus, foetus and utero-placenta at several stages of gestation in cows. *Journal of Agricultural Science* 106:437-444.

Reynolds, L. P., C. L. Ferrell, D. A. Robertson, and J. Klindt. 1990. Growth hormone, insulin and glucose concentrations in bovine fetal and maternal plasmas at several stages of gestation. *Journal of Animal Science* 68:725-733.

Reynolds, L. P., P. P. Broowicx, K. A. Vonnahme, M. L. Hohnson, A. T. Grazul-Bilska, J. M. Wallace, D. A. Redmer, and J. S. Caton, 2005. Placental angiogenesis in sheep models of compromised pregnancy. *Journal of Physiology* 565:43-58.

Reynolds, L. P., P. P Borowicz, J. S. Caton, K. A. Vonnahme, J. S. Luther, D. S. Buchanan, S. A. Hafez, A. T. Grazul-Bilska, and D. A. Redmer. 2010. Utero-placental vascular development and placental function: An update. *International Journal of Developmental Biology* 54:355-366.

Reynolds, W. L., T. M. DeRouen, and R. A. Bellows. 1978. Relationships of milk yield of dam to early growth rate of straightbred and crossbred calves. *Journal of Animal Science* 47:584-594.

Rhodes, F. M., G. De'Ath, and K. W. Entwistle. 1995. Animal and temporal effects on ovarian follicular dynamics in Brahman heifers. *Animal Reproduction Science* 38:265-277.

Rhynes, W. E., and L. L. Ewing. 1973. Testicular endocrine function in Hereford bulls exposed to high ambient temperature. *Endocrinology* 92:509-515.

Richards, M. W., J. C. Spitzer, and M. B. Warner. 1986. Effect of varying levels of postpartum nutrition and body condition at calving on subsequent reproductive performance in beef cattle. *Journal of Animal Science* 62:300-306.

Ripberger, T. J. 1997. Determination of Net Energy Requirements (NE_m and NE_Δ) of the Primiparous Heifer. M.S. Thesis. Purdue University, West Lafayette, IN.

Robinson, D. L., L. M. Café, and P. L. Greenwood. 2013. Meat Science and Muscle Biology Symposium: Developmental programming in cattle: Consequences for growth efficiency, carcass, muscle, and beef quality characteristics. *Journal of Animal Science* 91:1428-1442.

Robinson, J. J. 1977. The influence of maternal nutrition on ovine growth. *Proceedings of the Nutrition Society* 36:9-16.

Robinson, J. J. 1990. Nutrition in the reproduction of farm animals. *Nutrition Research Reviews* 3:253-276.

Robinson, J. J., I. McDonald, C. Fraser, and J. G. Gordon. 1980. Studies on reproduction in prolific ewes. 6. The efficiency of energy utilization for conceptus growth. *Journal of Agricultural Science* 94:331-338.

Robinson, J. S., V. M. Moore, J. A. Owens, and I. C. McMillen. 2000. Origins of fetal growth restriction. *European Journal of Obstetrics and Gynecology and Reproductive Biology* 92:13-19.

Robinson, O. W., M. K. M. Yusuff, and E. U. Dillard. 1978. Milk production in Hereford cows. I. Means and correlations. *Journal of Animal Science* 47:131-136.

Roche, J. F., M. P. Boland, and T. A. McGeady. 1981. Reproductive wastage following artificial insemination in cattle. *Veterinary Record* 109:95-97.

Rosenfeld, C. R. 1984. Consideration of the uteroplacental circulation in intrauterine growth. *Seminars in Perinatology* 8:42-51.

Rowlands, G. J., S. Lucey, and A. M. Russell. 1982. A comparison of different models of the lactation curve of dairy cattle. *Animal Production* 35:135-144.

Rutter, L. M., and R. D. Randel. 1984. Postpartum nutrient intake and body condition: Effect on pituitary function and onset of estrus in beef cattle. *Journal of Animal Science* 58:265-274.

Rutter, W., T. R. Laird, and P. J. Broadbent. 1971. The effects of clipping pregnant ewes at housing and of feeding different basal roughages. *Animal Production* 13:329-336.

Rutter, W., T. R. Laird, and P. J. Broadbent. 1972. A note on the effects of clipping pregnant ewes at housing. *Animal Production* 14:127-130.

Rutter, L. M., R. D. Randel, G. T. Schelling, and D. W. Forrest. 1983. Effect of abomasal infusion of propionate on the GnRH-induced luteinizing hormone release in prepuberal heifers. *Journal of Animal Science* 56:1167-1173.

Sacco, R. E., J. F. Baker, and T. C. Cartwright. 1987. Production characteristics of primiparous females of a five-breed diallel. *Journal of Animal Science* 64:1612-1618.

Sangsritavong, S., D. K. Combs, R. Sartori, L. E. Armentano, and M. C. Wiltbank. 2002. High feed intake increases liver blood flow and metabolism of progesterone and estradiol-17 beta in dairy cattle. *Journal of Dairy Science* 85:2831-2842.

Schafer, D.J., J. F. Bader, J. P. Meyer, J. K. Haden, M. R. Ellersieck, M. C. Lucy, M. F. Smith, and D. J. Patterson. 2007. Comparison of progestin-based protocols to synchronize estrus and ovulation before fixed-time artificial insemination in postpartum beef cows. *Journal of Animal Science* 85:1940-1945.

Schillo, K. K., J. B. Hall, and S. M. Hileman. 1992. Effect of nutrition and season on onset of puberty in the beef heifers. *Journal of Animal Science* 70:3994-4005.

Selk, G. 2009. Know USDA Cull Cow Grades before Marketing Culls. Beef Magazine, October 30. Available online at http://beefmagazine.com/cowcalfweekly/1030-know-usda-cull-cow-grades-market. Accessed on May 12, 2014.

Selk, G. E., R. P. Wettemann, K. S. Lusby, J. W. Oltjen, S. L. Mobley, R. J. Rasby, and J. C. Garmendia. 1988. Relationship among weight change, body condition and reproductive performance of range beef cows. *Journal of Animal Science* 66:3153-3159.

Short, R. E., and D. C. Adams. 1988. Nutritional and hormonal interrelationships in beef cattle reproduction. *Canadian Journal of Animal Science* 68:29-40.

Short, R. E., and R. A. Bellows. 1971. Relationships among weight gains, age at puberty and reproductive performance in heifers. *Journal of Animal Science* 32:127-131.

Short, R. E., R. A. Bellows, R. B. Staigmiller, J. G. Berardinelli, and E. E. Custer. 1990. Physiological mechanisms controlling anestrous and infertility in postpartum beef cattle. *Journal of Animal Science* 68:799-816.

Silvey, M. W., and K. P. Haycock. 1978. A note on live-weight adjustment for pregnancy in cows. *Animal Production* 27:113-116.

Sirois, J., and J. E. Fortune. 1988. Ovarian follicular dynamics during the oestrus cycle in heifers monitored by real-time ultrasonography. *Biology of Reproduction* 39:308-317.

Smith, M. F., K. J. Nix, D. C. Kraemer, M. S. Amoss, M. A. Herron, and J. N. Wiltbank. 1982. Fertilization rate and early embryonic loss in Brahman crossbred heifers. *Journal of Animal Science* 54:1005-1011.

Smith, M. W., and J. S. Stevenson. 1995. Fate of the dominant follicle, embryonal survival and pregnancy rates in dairy cattle treated with prostaglandin F2α and progestins in the absence or presence of a functional corpus luteum. *Journal of Animal Science* 73:3743-3751.

Smith, V. G., J. R. Chenault, J. F., McAllister, and J. W. Lauderdale. 1987. Response of postpartum beef cows to exogenous progestogens and gonadotropin releasing hormone. *Journal of Animal Science* 64:540-551.

Spitzer, J. C., J. N. Wiltbank, and D. C. Lefever. 1975. Increasing Beef Cow Productivity by Increasing Reproductive Performance. Colorado State University Experiment Station General Series No. 949. Fort Collins, CO: Colorado State University Experiment Station.

Sreenan, J. M., and M. G. Diskin. 1986. The extent and timing of embryonic mortality in the cow. Pp. 142-158 in *Embryonic Mortality in Farm Animals*, J. M. Sreenan and M. G. Diskin, eds. The Hague, The Netherlands: Martinus Nijhoff.

Stalker, L. A., D. C. Adams, T. J. Klopfenstein, D. M. Feuz, and R. N. Funston. 2006. Effects of pre- and postpartum nutrition on reproduction in spring calving cows and calf feedlot performance. *Journal of Animal Science* 84:2582-2589.

Stamey, J. A., N. A. Janovick, A. F. Kertz, and J. K. Drackley. 2012. Influence of starter protein content on growth of dairy calves in an enhanced early nutrition program. *Journal of Dairy Science* 95:3327-3336.

Stegeman, J. H. J. 1974. Placental development in the sheep and its relation to fetal development. *Bijdragen tot de Dierkunde* 44:3-72.

Stewart, T. S., C. R. Long, and T. C. Cartwright. 1980. Characterization of cattle of a five breed diallel. III. Puberty in bulls and heifers. *Journal of Animal Science* 50:808-820.

Summers, A. F., S. T. Weber, T. L. Meyer, and R. N. Funston. 2012. Late gestation supplementation impacts primiparous beef heifers and progeny. Pp. 22-23 in *Nebraska Beef Cattle Report*. Paper 701. Lincoln: University of Nebraska. Available online at http://digitalcommons.unl.edu/animalscinbcr/701. Accessed on April 10, 2015.

Swanson, E. W. 1967. Optimum growth patterns for dairy cattle. *Journal of Dairy Science* 50:244-252.

Swanson, L. V. 1989. Discussion—Interactions of nutrition and reproduction. *Journal of Dairy Science* 72:805-814.

Swanson, T. J., C. J. Hammer, J. S. Luther, D. B. Carlson, J. B. Taylor, D. A. Redmer, T. L. Neville, J. J. Reed, L. P. Reynolds, J. S. Caton, and K. A. Vonnahme. 2008. Effects of gestational plane of nutrition and selenium supplementation on mammary development and colostrum quality in pregnant ewe lambs. *Journal of Animal Science* 86:2415-2423.

Sykes, A. R., and A. C. Field. 1972. Effect of dietary deficiencies of energy, protein and calcium on the pregnant ewe. III. Some observations on the use of biochemical parameters in controlling undernutrition during pregnancy and on the efficiency of energy utilization of energy and protein for fetal growth. *Journal of Agricultural Science* 78:127-133.

Tedeschi, L. O., S. Seo, D. G. Fox, and R. Ruiz. 2006. Accounting for energy and protein reserve changes in predicting diet-allowable milk production in cattle. *Journal of Dairy Science* 89:4795-4807.

Thompson, G. E., J. M. Bassett, D. E. Samson, and J. Lee. 1982. The effects of cold exposure of pregnant sheep on foetal plasma nutrients, hormones and birth weight. *British Journal of Nutrition* 48:59-64.

Tikofsky, J. N., M. E. Van Amburgh, and D. A. Ross. 2001. Effect of varying carbohydrate and fat content of milk replacer on body composition of Holstein bull calves. *Journal of Animal Science* 79:2260-2267.

Totusek, R., D. W. Arnett, G. L. Holland, and J. V. Whiteman. 1973. Relation of estimation method, sampling interval and milk composition to milk yield of beef cows and calf gain. *Journal of Animal Science* 37:153-158.

Townson, D. H., P. C. W. Tsang, W. R. Butler, M. Frajblat, L. C. Griel, Jr., C. J. Johnson, R. A. Milvae, G. M. Niksic, and J. L. Pate. 2002. Relationship of fertility to ovarian follicular waves before breeding in dairy cows. *Journal of Animal Science* 80:1053-1058.

Troxel, T. R., and D. J. Kessler. 1984. The effect of progestin and GnRH treatments in ovarian function and reproductive hormone secretions of anoestrous postpartum suckled beef cows. *Theriogenology* 21:699-711.

Tudor, G. D. 1972. The effect of pre- and post-natal nutrition on the growth of beef cattle. I. The effect of nutrition and parity of the dam on calf birth weight. *Australian Journal of Agricultural Research* 23:389-395.

Turner, H. G. 1982. Genetic variation of rectal temperature in cows and its relationship to fertility. *Animal Production* 35:401-402.

Tyrrell, H. F., and J. T. Reid. 1965. Prediction of the energy value of cow's milk. *Journal of Dairy Science* 48:1215-1223.

Unsworth, E. F. 1991. The efficiency of utilization of metabolizable energy for lactation from grass silage-based diets. Pp. 329-332 in *Energy Metabolism of Farm Animals: Proceedings of the 12th Symposium, September 1-7, 1991, Kartause Ittingen, Switzerland*, C. Wenk and M. Boessinger, eds. EAAP Publication No. 58. London: Butterworths.

Vallet, J. L., R. A. Cushman, A. K. McNeell, E. C. Wright, E. Larimore, J. R. Miles, C. C. Chasel, C. A. Lents, J. R. Wood, A. S. Cupp, and G. A. Perry. 2014. Global changes in uterine protein secretion are associated with differences in the number of antral follicles in heifers. *Journal of Animal Science* 92(Suppl. 2):89 (Abstract 200).

VanDemark, N. L., G. R. Fritz, and R. E. Marger. 1964. Effect of energy intake on reproductive performance of dairy bulls. II. Semen production and replenishment. *Journal of Dairy Science* 47:798-802.

van der Honing, Y. 1980. The utilization by high-yielding cows of energy from animal tallow or soya bean oil added to a diet rich in concentrates. Pp. 315-318 in *Energy Metabolism: Proceedings, 8th Symposium on Energy Metabolism, September 1979, Cambridge, England*, L. E. Mount, ed. EAAP Publication No. 26. London: Butterworths.

Vera, R. R., C. A. Ramirez, and H. Ayala. 1993. Reproduction in continuously underfed Brahman cows. *Animal Production* 57:193-198.

Vonnahme, K. A., M. J. Zhu, P. P. Borowicz, T. W. Geary, B. W. Hess, L. P. Reynolds, J. S. Caton, W. J. Means, and S. P. Ford. 2007. Effect of early gestational undernutrition on angiogenic factor expression and vascularity in the bovine placentome. *Journal of Animal Science* 85:2464-2472.

Vonnahme, K. A., C. O. Lemley, P. Shukla, and S. T. O'Rourke. 2013. 2011 and 2012 Early Careers Achievement Awards: Placental programming: How the maternal environment can impact placental function. *Journal of Animal Science* 91:2467-2480.

Wagner, J. J. 1984. Carcass Composition in Mature Hereford Cows: Estimation and Influence on Metabolizable Energy Requirements for Maintenance During Winter. Ph.D. Dissertation, Oklahoma State University, Stillwater.

Wallace, J., D. Bourke, P. Da Silva, and R. Aitken. 2001. Nutrient partitioning during adolescent pregnancy. *Reproduction* 122:347-357.

Wallace, L. R. 1948. Growth of lambs before and after birth in relation to level of nutrition. *Journal of Agricultural Science* 38:367-401.

Waterland, R. A., and C. Garza. 1999. Potential mechanisms of metabolic imprinting that lead to chronic disease. *American Journal of Clinical Nutrition* 69:179-197.

Waterland, R. A., and R. L. Jirtle. 2004. Early nutrition, epigenetic changes at transposons and imprinted genes, and enhanced susceptibility to adult chronic diseases. *Nutrition* 20:63-68.

Whitman, R. W. 1975. Weight Change, Body Condition and Beef Cow Reproduction. Ph.D. Dissertation. Colorado State University, Fort Collins.

Williams, J. H., D. C. Anderson, and D. D. Kress. 1979a. Milk production in Hereford cattle. I. Effects of separation interval on weigh-suckle-weigh milk production estimates. *Journal of Animal Science* 49:1438-1442.

Williams, J. H., D. C. Anderson, and D. D. Kress. 1979b. Milk production in Hereford cattle. II. Physical measurements: Repeatabilities and relationship with milk production. *Journal of Animal Science* 49:1443-1448.

Wilson, L. L., J. E. Gillooly, M. C. Rugh, C. E. Thompson, and H. R. Purdy. 1969. Effects of energy intake, cow body size and calf sex on composition and yield of milk by Angus-Holstein cows and preweaning growth rate of progeny. *Journal of Animal Science* 28:789-795.

Wiltbank, J. N. 1970. Research needs in beef cattle reproduction. *Journal of Animal Science* 31:755-762.

Wiltbank, J. N., E. J. Warwick, E. H. Vernon, and B. M. Priode. 1961. Factors affecting net calf crop in beef cattle. *Journal of Animal Science* 20:409-415.

Wiltbank, J. N., K. E. Gregory, L. A. Swiger, J. E. Ingalls, J. A. Rothlisberger, and R. M. Koch. 1966. Effects of heterosis on age and weight at puberty in beef heifers. *Journal of Animal Science* 25:744–751.

Wiltbank, J. N., C. W. Kasson, and J. E. Ingalls. 1969. Puberty in crossbred and straightbred beef heifers. *Journal of Animal Science* 29:602-605.

Wiltbank, J. N., S. Roberts, J. Nix, and L. Rowden. 1985. Reproductive performance and profitability of heifers fed to weight 272 or 318 kg at the start of the first breeding season. *Journal of Animal Science* 60:25-34.

Windisch, W., M. Kirchgessner, and H. L. Muller. 1991. Effect of different energy supply on energy metabolism in lactating dairy cows after a period of energy restriction. Pp. 304-307 in *Energy Metabolism of Farm Animals: Proceedings of the 12th Symposium, September 1-7, 1991, Kartause Ittingen, Switzerland*, C. Wenk and M. Boessinger, eds. EAAP Publication No. 58. London: Butterworths.

Winterholler, S. J., C. P. McMurphy, G. L. Mourer, C. R. Krehbiel, G. W. Horn, and D. L. Lalman. 2012. Supplementation of dried distillers grains with solubles to beef cows consuming low-quality forage during late gestation and early lactation. *Journal of Animal Science* 90:2014-2025.

Winters, L. M., W. W. Green, and R. E. Comstock. 1942. The Prenatal Development of the Bovine. Technical Bulletin No. 151. University of Minnesota Agricultural Experiment Station.

Wood, P. D. P. 1980. Breed variation in the shape of the lactation curve of cattle and their implications for efficiency. *Animal Production* 31:133-141.

Wright, V. A., and A. J. Russel. 1984a. Estimation in vivo of the chemical composition of the bodies of mature cows. *Animal Production* 38:33-43.

Wright, V. A., and A. J. Russel. 1984b. Partition of fat, body composition, and body condition score in mature cows. *Animal Production* 38:23-32.

Wu, G., F. W. Bazer, J. M. Wallace, and T. E. Spencer. 2006. Board-invited review: Intrauterine growth retardation: Implications for the animal sciences. *Journal of Animal Science* 84:2316-2337.

Yavas, Y., and J. S. Walton. 2000. Postpartum acyclicity in suckled beef cows: A review. *Theriogenology* 54:25-55.

Yunusova, R., T. L. Neville, K. A. Vonnahme, C. J. Hammer, J. J. Reed, J. B. Taylor, D. A. Redmer, L. P. Reynolds, and J. S. Caton. 2013. Impacts of maternal selenium supply and nutritional plane on visceral tissues and intestinal biology in offspring. *Journal of Animal Science* 91:2229-2242.

Zeron, Y., A. Ocheretny, O. Kedar, A. Borochov, D. Sklan, and A. Arav. 2001. Seasonal changes in bovine fertility: Relation to developmental competence of oocytes, membrane properties and fatty acid composition of follicles. *Reproduction* 121:447-545.

Zhang, L., N. M. Long, S. M. Hein, Y. Ma, P. W. Nathanielsz, and S. P. Ford. 2010. Maternal obesity in ewes results in reduced fetal pancreatic β-cell numbers in late gestation and decreased circulating insulin concentration at term. *Domestic Animal Endocrinology* 40:30-39.

Zhu, M. J., S. P. Ford, P. W. Nathanielsz, and M. Du. 2004. Effect of maternal nutrient restriction in sheep on the development of fetal skeletal muscle. *Biology of Reproduction* 71:1968-1973.

14

Compounds That Modify Digestion and Metabolism

INTRODUCTION

This chapter includes a brief description of feed additives and other compounds used in beef cattle production systems to improve animal health and the efficiency of nutrient use, increase growth rate, and decrease the environmental impact of cattle (Tedeschi et al., 2011; Stackhouse et al., 2012; Stackhouse-Lawson et al., 2013). Some of these compounds alter digestive processes to improve the profile of nutrients available from the feed, whereas others have their effects on postabsorptive metabolism. In the United States, those compounds defined as drugs are approved for use by the U.S. Food and Drug Administration (FDA) based on data demonstrating both efficacy and safety. Their use, including appropriate dose and withdrawal (if any), is regulated by the FDA.

ADDITIVES THAT ALTER RUMINAL FERMENTATION

This group of additives in diets of beef cattle is used to alter microbial fermentation in order to prevent or treat digestive disorders or to improve feed efficiency. Two major digestive disorders are ruminal acidosis and bloat.

Ruminal acidosis is described in Chapter 2 (Anatomy, Digestion, and Nutrient Utilization). Normal fermentation in the reticulorumen produces organic acids as end products that are removed by absorption. Keeping production and removal of acids in balance maintains a stable ruminal pH. Ruminal pH below 5.8 and 5.0 to 5.2 are indicators of subacute and acute acidosis, respectively. Acidosis is most prevalent when cattle are consuming, or being introduced to, diets containing large amounts of rapidly fermentable carbohydrates, which produce organic acids including lactic acid. Excessive intake resulting from larger meal size increases the risk for acidosis (Owens et al., 1998; Vasconcelos and Galyean, 2008).

Bloat is a condition that results from the inability to expel ruminal gas. Distension of the rumen leads to pressure on the diaphragm and lungs. If severe and not alleviated, it can lead to death. Types of bloat include free gas bloat and frothy bloat. Free gas bloat is often caused by obstruction of the esophagus or impaired rumen motility. In frothy bloat, gas is entrapped within the fluid portion of ruminal contents. Management, microbial, animal, and plant factors all contribute to development of bloat. Primary agents responsible for the foam of frothy bloat seem to be derived from plant components in pasture bloat or microorganisms in the rumen in feedlot bloat (Cheng et al., 1998; Berg et al., 2000).

Pasture bloat is predominantly a result of grazing specific legume (primarily clover or alfalfa) or wheat pastures. Several other cereal grain pastures are categorized as having moderate risk of bloat (Wang et al., 2012). Other legumes such as sainfoin, birdsfoot trefoil, and cicer milkvetch do not cause bloat (Majak et al., 1995) and are considered low-risk forages (Wang et al., 2012). Production of both a large volume of gas and stable foam characterize the bloat-causing legumes (Fay et al., 1980) and annual winter wheat (Vasconcelos and Galyean, 2008). Legumes that are typically safe have leaves that contain low levels of soluble proteins or that resist cell disruption, thereby decreasing substrate to form proteinaceous foams. They can also have leaves that contain enough proanthocyanidins (PA; condensed tannins) to precipitate soluble proteins and prevent foam formation. Concentrations of PA from 1 to 5 mg/g dry matter (DM) can have positive effects to prevent bloat; however, they decrease intake at greater concentrations. Alfalfa and the clovers have very low concentrations of PA, high concentrations of soluble proteins, and rapid initial rates of digestion. High soluble protein and low lignin concentrations are important for their nutritive value. A slow initial rate of digestion can also be protective (Li et al., 1996). Winter wheat pasture has greater soluble protein and nonprotein nitrogen (NPN) in the vegetative growth stage than the reproductive growth stage. Bloat occurs primarily when consumed forage is in the vegetative growth stage (Min et al., 2005b).

Grazing management strategies that limit access to or

intake of bloat-producing legume pastures will decrease the risk of frothy bloat. The use of legume pastures can also be managed to decrease the occurrence of bloat by promoting rapid ruminal clearance and decreasing soluble protein intake. Soluble protein decreases as alfalfa matures; therefore, grazing more mature alfalfa that has begun to flower decreases bloat potential. Increasing intake and consequent ruminal clearance rate by continuous grazing was beneficial compared to grazing for 6 hours daily (Majak et al., 1995). Risk can be lessened by reducing intake of bloat-producing pastures through supplementing with fibrous roughages low in soluble protein and fermentable carbohydrate (Howarth, 1975; Wang et al., 2012).

Several feed additives can decrease the incidence of pasture bloat. Using poloxalene as a feed additive lowered risk of legume bloat and bloat associated with grazing winter wheat pastures (Majak et al., 1995). Poloxalene, ionophores, and condensed tannins all decreased wheat pasture bloat by reducing ruminal gas production, ruminal foam production, or both. Poloxalene and condensed tannins were more effective in decreasing foam production and strength than monensin and were also more effective in preventing frothy bloat (Min et al., 2005a, 2006).

As noted by Cheng et al. (1998), feedlot bloat is associated with feeding high-grain diets that lead to rapid production of organic acids from the fermentation of starch and production of bacterial mucopolysaccharides (slime). It frequently occurs during the transition of animals from forage-based diets to high-grain diets. Slower transition to allow microbial populations to adapt will typically decrease bloat. Processing grain increases the rate and extent of ruminal starch digestion and can increase occurrence of bloat. A decrease in particle size is associated with greater digestion of starch by microbial enzymes, resulting in a decline in ruminal pH and an increase in viscosity of ruminal fluid. Other processing methods, such as steam flaking, also increase ruminal starch digestion. As discussed in a subsequent section of this chapter, ionophores also decrease feedlot bloat (Cheng et al., 1998).

Ionophore Antibiotics

Ionophores are carboxylic polyether compounds included in diets of growing and finishing cattle to improve feed efficiency and animal health. Four products are currently licensed in North America, referred to by their chemical name as lasalocid, laidlomycin propionate, monensin, and salinomycin. Lasalocid and monensin are licensed in both the United States and Canada, laidlomycin propionate is licensed in the United States, and salinomycin is licensed in Canada. They are effective for cattle grazing pasture and confined cattle consuming high-concentrate diets.

Ionophores are toxic to many ruminal microorganisms. The mechanism of action of ionophores is initiated by binding to cations and facilitating their movement across cell membranes. The result is disruption in ionic gradients and decreased intracellular pH. Activity of the proton pump to expel protons results in depletion of ATP. The mechanisms of ionic transport are similar for monensin and lasalocid but not identical. Monensin has greater affinity for sodium ions (Na^+) than potassium ions (K^+) whereas lasalocid has higher affinity for K^+ and similar affinity for calcium ions (Ca^{+2}) and Na^+. Gram-positive bacteria are susceptible to their effects, whereas many gram-negative bacteria are resistant, which may be related to the impermeability of the bacterial outer membrane of gram-negative bacteria. As a result of susceptibility to ionophores, there are changes in bacterial populations and a shift in volatile fatty acids (VFA) produced in the rumen toward more propionate with corresponding decreases in acetate and butyrate (Bergen and Bates, 1984).

Several obligate amino acid-fermenting ruminal bacteria were inhibited by monensin in culture experiments (Russell and Strobel, 1989; Krause and Russell, 1996). Monensin decreased ammonia concentration in ruminal fluid and increased bacterial protein (Yang and Russell, 1993). By decreasing amino acid-fermenting bacteria, monensin has an amino acid-sparing effect, which results in decreased deamination of amino acids and ammonia concentration in the rumen, leading to increased flow of dietary amino acids to the abomasum (Bergen and Bates, 1984; Russell and Strobel, 1989; McGuffey et al., 2001). Methane production decreases also, although the decrease in methane seems to be transitory (4 to 6 weeks) and could be related to the initial effects on ciliated protozoa, which adapt to ionophores (Guan et al., 2006). The decrease in methane production was noted for a longer time interval in steers fed a low-concentrate diet (Guan et al., 2006).

Spears (1990) reported that ionophores tend to increase the digestible energy content of the diet and increase the apparent digestibility of nitrogen (N). In addition, ionophores affect the apparent absorption of some minerals. In general, they enhance absorption of magnesium, phosphorus, zinc, and selenium, with inconsistent effects on calcium, potassium, and sodium (Spears, 1990).

In a review summarizing 228 trials in which monensin was included in the diet of feedlot cattle at 31.8 mg/kg DM (standard deviation 7.5 mg/kg DM), Goodrich et al. (1984) concluded that monensin increased average daily gain (ADG) by 1.6% and decreased feed intake by 6.5%, resulting in a 7.5% improvement in feed conversion. Carcass characteristics were not altered by feeding monensin. In situations where ADG was low in control cattle, the proportional response in ADG to monensin inclusion was greater than in situations of greater ADG in control cattle (Goodrich et al., 1984). In a series of feedlot trials conducted in 1990 (Stock et al., 1995), dry matter intake (DMI) decreased by 1%, ADG increased by 3%, and feed conversion improved by 4% in cattle receiving monensin. The lesser response to monensin was consistent with changes in feedlot diets over time toward diets with greater net energy (NE) concentration. Effects of monensin in beef cattle were evaluated using a meta-analysis approach including data from 40 peer-reviewed papers and

24 trial reports containing a total of 169 trials (Duffield et al., 2012). The mean dose of monensin used was 28.1 mg/kg DM and ranged from 3 to 98 mg/kg DM. Across all studies with growing and finishing cattle that were included in the analysis, monensin improved feed conversion by 6.4%, decreased DMI by 3.1%, and increased ADG by 2.5%. The improvement in feed conversion was 8.1% in the 1970s decreasing to 2.3 to 3.5% in the 1990s and since 2000, respectively. Feed conversion in control cattle improved from 8.79 to 6.39 kg feed/kg gain over the same time frame of diminished response to monensin. Similar to the observations of Goodrich et al. (1984), the ADG response to monensin was less in trials in which the control cattle had greater ADG (Duffield et al., 2012). Results of recent studies, including three large feedlot trials in which corn wet distillers grain with solubles was fed, noted no response to monensin in DMI or ADG; however, feed efficiency improved by 3.1% in one experiment. Inclusion of both monensin (32 or 37 mg/kg DM) and tylosin (7.9 mg/kg DM) improved feed efficiency by 4.5% (Meyer et al., 2013).

In steers fed a high-energy diet with 4% added yellow grease, there was little production response to monensin (33 mg/kg dietary DM; Zinn and Borques, 1993). A similar observation was made for response to monensin and tylosin (25 and 10 mg/kg, respectively) in steers fed high-concentrate diets supplemented with 4% tallow; the combination decreased DMI and ADG with no difference in gain:feed relative to control cattle for the total trial. Monensin plus tylosin (MTY) decreased ruminal acetate and increased ruminal propionate concentration with no added tallow. With added tallow, concentrations of acetate and propionate did not differ for control and MTY and were similar to those of MTY cattle without tallow (Clary et al., 1993). Monensin decreased biohydrogenation of linoleic acid (C18:2, n-6) in ruminal cultures (Fellner et al., 1997). A lesser effect of monensin in diets with animal fat was summarized by Tedeschi et al. (2003). Performance response to monensin or MTY could be modified by addition of 4% fat.

The effects of ionophores in grazing systems or in cattle fed high-forage diets (ϑ75% of DM) have not been studied to the same degree as the effects in feedlot cattle. A summary of 41 publications included results from 46 comparisons of monensin and 15 comparisons of lasalocid (Bretschneider et al., 2008). In growing beef cattle consuming forage-based diets, ADG increased by 12.1% (0.66 to 0.74 kg/d) with monensin and 10.3% (0.78 to 0.86 kg/d) with lasalocid compared to cattle not receiving the respective ionophore. Dry matter intake was not changed by either monensin or lasalocid, although it tended to decrease (2.7%) with monensin. Therefore, the change in ADG resulted in improved feed conversion. The response in ADG of cattle receiving monensin decreased as the ADG of cattle fed the control diet increased in response to improved forage quality, whereas the response to lasalocid increased with increasing ADG of control cattle. The description of forages was limited in many studies and ADG of control cattle was used as a proxy for forage quality. Differences in response to forage quality could be related to differences in the mechanism of the two ionophores and differing Na^+/K^+ ratio in the rumen as a function of forage quality. The response in ADG to monensin was greater when cattle were fed forage-based diets than high-concentrate diets, consistent with lesser response to monensin as rate of gain increases. In contrast, there was a greater effect on DMI when cattle were fed high-concentrate diets (Bretschneider et al., 2008). Further studies might better characterize responses to monensin for cattle grazing cool- and warm-season grasses of varying quality.

Equations relating dose of monensin to alterations in VFA profile and thus potential energy substrates available for metabolism were developed for cattle fed diets containing at least 80% concentrate. Molar proportions of propionate increased and acetate and butyrate decreased with no change in total VFA as dose of monensin increased (Ellis et al., 2012). This response supports a significant improvement in the capture of feed energy during ruminal fermentation with less methane produced. Thus, metabolizable energy (ME) and NE values of feeds should increase when ionophores are consumed.

In a comparative slaughter trial, Byers (1980) assumed monensin increased the ME value of the diet by 5 to 10% to reflect changes in ruminal fermentation. Using this approach, the primary effect of monensin (30 to 33 mg/kg DM) was to improve the efficiency of energy use for maintenance or increase the net energy for maintenance (NEm) value of the diet by 5.7% with no effect on net energy for gain (NEg). Delfino et al. (1988) made a similar observation with respect to lasalocid increasing the ME concentration of the diet. The NEm value of the diet increased by 10% in response to lasalocid with little change in NEg.

No interactions were observed for anabolic implants (zeranol, progesterone-estradiol, or testosterone-estradiol) and dietary monensin in feedlot cattle fed high-concentrate diets (Goodrich et al., 1984), or implants containing estradiol and dietary monensin in forage-fed cattle (Bretschneider et al., 2008). Beck et al. (2014) demonstrated additive effects of monensin and an implant containing trenbolone acetate (TBA) and estradiol in steers grazing wheat pasture. Average daily gain was improved 7% with monensin, 12% with implants, and 21.7% (0.23 kg/d) with combined implants and monensin compared with nonimplanted cattle without monensin; using both technologies decreased cost of gain by 26% (Beck et al., 2014).

Additive responses are consistent with different mechanisms of action for monensin and implants. Monensin changes microbial populations in the rumen, whereas implants alter postabsorptive metabolism; however, overall response could depend on the diet and ADG. When cattle were grazing high-quality pastures and therefore had greater rates of gain, there was a response to estradiol but no additional response to monensin (Bretschneider et al., 2008). More information is needed on the use of monensin in implanted cattle grazing pastures to evaluate the combined response.

In the previous edition and update of the *Nutrient Requirements of Beef Cattle* (NRC, 1996, 2000), it was concluded that adjustments made to values for dietary NEm based on the use of ionophores were independent of anabolic agents. This conclusion is maintained in the current edition.

Ionophores have beneficial effects on cattle health by decreasing the risk of subclinical acidosis. Monensin decreased meal size and variation in DMI in cattle fed high-concentrate diets (Stock et al., 1995; McGuffey et al., 2001; Birkelo, 2003; Erickson et al., 2003). Ionophores decreased growth of gram-positive bacteria, including lactic acid and mucopolysaccharide-producing bacteria such as *Streptococcus bovis* and *Lactobacillus* spp., and thus decreased lactate production (Bergen and Bates, 1984; Goodrich et al., 1984; Cheng et al., 1998; Birkelo, 2003). Effects on specific microbial species as well as on eating behavior could contribute to the effect of ionophores on decreasing acidosis.

Monensin decreased frothy bloat in cattle grazing winter wheat pasture (Branine and Galyean, 1990). Monensin was more effective than lasalocid in decreasing legume bloat (Bartley et al., 1983; Katz et al., 1986) and bloat in cattle grazing winter wheat (Horn et al., 2005). Although both ionophores decreased frothy bloat, neither ionophore prevented legume or wheat bloat. In contrast, lasalocid was more effective than monensin in decreasing grain bloat (Bartley et al., 1983).

In the previous edition and update of the *Nutrient Requirements of Beef Cattle* (NRC, 1996, 2000), based on simulations using data for response of cattle to monensin in which diets averaged 28 mg/kg monensin in DM, a 12% increase in dietary NEm accounted for the improvements in performance, including a 4% decrease in DMI, 1.5% improvement in ADG, and 5.3% improvement in feed conversion in cattle fed a 90% concentrate diet (Tedeschi et al., 2003). The recommended 12% increase in NEm was for any ionophore. This adjustment is much greater than the observed estimate of 5.7% (Byers, 1980) and, based on field experience, seems to be too high. Considering observations of decreased response in DMI and feed efficiency to monensin concurrent with improved performance of cattle fed high-concentrate diets without monensin, the adjustment for monensin was reevaluated and changed as described below.

Based on the meta-analysis of Duffield et al. (2012), DMI was decreased by 3% when monensin was fed at 27.5 to 33.0 mg/kg DM. A similar trend was observed in DMI for cattle fed forage-based diets (Bretschneider et al., 2008). Therefore, predicted DMI was decreased by 3% when feeding monensin, regardless of the proportion of dietary concentrate. There was no decrease in predicted DMI when lasalocid or laidlomycin propionate was fed (see Chapter 10, Feed Intake). The feed conversion (feed:gain ratio) was improved by 3.5% for both monensin and lasalocid. Thus, to account for these changes in DMI and the feed:gain ratio, ADG had to be improved on average by 0.52% for monensin and 3.63% for lasalocid. Based on the mode of action of ionophores in the rumen, the committee decided to change dietary ME to account for the improvements in ADG and feed efficiency rather than adjust dietary NEm. The magnitude of the dietary ME change was obtained through a sensitivity analysis of combinations of dietary ME (2.3 to 3.2 Mcal/kg) and body weight (BW; 250 to 450 kg) using the Beef Cattle Nutrient Requirements Model (BCNRM) of the current edition of the *Nutrient Requirements of Beef Cattle*. The NEm was computed from the unadjusted ME to predict DMI. On average, the increase of dietary ME was 2.3% (range from 1.9 to 2.6%) for monensin and 1.5% (range from 1 to 1.8%) for lasalocid. Therefore, this committee recommends that dietary ME be increased by 2.3% for monensin and 1.5% for lasalocid. The recommendation for laidlomycin propionate is the same as for lasalocid.

Poloxalene

Nonionic surfactants prevent legume bloat (Bartley, 1965). Poloxalene, a nonionic surfactant that decreases foam stability, prevented legume bloat with no negative effects on feed intake, milk production, or animal health (Helmer et al., 1965). It was effective in cattle grazing Ladino clover pastures (Foote et al., 1968) and in cattle grazing wheat pasture that produced a foamy bloat (Bartley et al., 1975). It is used for prevention of legume (alfalfa, clover; Majak et al., 1995) and wheat pasture bloat, and it can be top dressed, incorporated into the feed, or provided in a molasses or salt block. Access to poloxalene should be provided several days before exposure to bloat-producing conditions. A concentrate form is available for preparation of a drench to treat animals showing signs of bloat (Code of Federal Regulations: 21 CFR §§ 520.1840, 558.464 [2014]).

Plant Secondary Metabolites

Plants produce a series of metabolites that are used for defense rather than growth or reproduction of the plant. Collectively, these compounds are termed plant secondary metabolites and include essential oils, saponins, tannins, and organosulfur compounds. They are of interest to ruminant nutritionists because of their potential to alter rumen microbial populations, resulting in improved digestive efficiency, animal productivity, and health. As such, they provide a potential alternative to inclusion of antibiotics. One specific interest is in their potential to decrease ruminal methane production. Effects of plant secondary metabolites on methane production vary with dose, chemical structure, molecular weight, and composition of the diets to which they are added. Depending on the dose and the type of compound, they can also adversely affect digestibility (reviewed by Patra and Saxena, 2009, 2010). A meta-analysis that included the effects of saponins, tannins, essential oils, and organosulfurs on ruminal fermentation and digestibility taken from 36 studies emphasized the difference in the potential mechanisms of methane suppression by different groups of phytochemicals (Patra, 2010).

The effect of essential oils on ruminal fermentation is a relatively recent focus compared to saponins and tannins. The essential oils are obtained by steam distillation and represent the volatile fraction of plant compounds derived from any portion of the plant. Much of the data on the potential of essential oils to alter ruminal fermentation was derived from in vitro systems, although there are some in vivo data (Calsamiglia et al., 2007).

Inclusion of two different essential oil mixtures in the diet of feedlot cattle had little effect on ruminal fermentation, ADG, DMI, feed efficiency, or carcass characteristics. When tylosin was included, there was a slight improvement in feed efficiency (Meyer et al., 2009). Short-term dose-response studies were conducted with eugenol and cinnamaldehyde in beef heifers fed a high-concentrate diet (Yang et al., 2010b,c). Eugenol supplementation resulted in decreased neutral detergent fiber (NDF) digestibility with increasing dose and trended toward decreased ruminal degradability of protein (Yang et al., 2010c). Cinnamaldehyde had variable effects depending on the dose; the low dose increased DMI and ruminal digestibility of organic matter (OM) and NDF, while the high dose caused negative effects, including decreased DMI, and decreased ruminal OM, NDF, and N digestibility (Yang et al., 2010b). However, in a longer-term feeding trial with more animals, using the same cinnamaldehyde doses, cinnamaldehyde had no effect on DMI, ADG, feed efficiency, or carcass characteristics (Yang et al., 2010a). The effect of cinnamaldehyde addition on in vitro fermentation of high-grain substrates showed decreased VFA production, nutrient disappearance, methane production, and bacterial protein synthesis and increased acetate:propionate ratio. These changes suggest that the feeding value of a high-grain diet might be decreased with cinnamaldehyde (Lia et al., 2012).

Whether plant secondary compounds will have application in beef cattle production systems is not clear at this point. A systematic evaluation of individual compounds including appropriate dose, potential for adaptation of ruminal microorganisms, and the fate of the compound in the animal are required to define the potential for dietary inclusion to improve digestive efficiency.

Buffers

Lactic acid accumulation in the rumen is a key feature of acidosis in feedlot cattle (Huber, 1976). Lactic acid is a stronger acid (p*Ka* 3.8) than acetic acid (p*Ka* 4.8) and therefore, more would exist in the ionized state at pH 5.0 during acidosis, resulting in increased hydrogen ion concentration. Addition of buffers to the diet has had little effect on subclinical acidosis in feedlot cattle, which could be a result of the lag time between a meal and minimal rumen pH (6 to 10 h) along with the outflow of soluble materials (Owens et al., 1998). Addition of 5% sodium bicarbonate to a mixed concentrate and corn silage diet increased ruminal fluid pH, dilution rate, acetate:propionate ratio, and decreased starch digestion (Rogers and Davis, 1982). Addition of 2.5 or 5% sodium bicarbonate to medium- or high-concentrate diets increased the fluid dilution rate and particulate rate of passage but did not alter degradation of soybean meal protein in the rumen. Both levels increased ruminal pH (Okeke et al., 1983).

Including buffers at lower concentrations could be beneficial during diet transitions and during the first 4 weeks of feeding, but longer-term responses are variable. Addition of 0.9% (Russell et al., 1980) or 1% (Peirce et al., 1983) sodium bicarbonate to high-energy diets or 1.2% sodium bicarbonate to silage diets (Worley et al., 1986) increased DMI or gain during initial feeding of the diets, respectively; however, improvements were not maintained for the total feeding period. Ruminal dilution rates were not changed with the lower buffer concentration. Sodium bicarbonate included at 0.75% increased DMI and tended to improve weight gain over the entire feeding trial in one study in which highly processed grain diets were fed (Zinn, 1991), but in another study it did not affect performance of cattle fed diets including 4% yellow grease (Zinn and Borques, 1993). Low levels of added sodium bicarbonate might maintain (Russell et al., 1980; Peirce et al., 1983) or slightly increase (Boerner et al., 1987; Zinn, 1991) ruminal pH. There were small negative effects of sodium bicarbonate on carcass characteristics in some studies (Zinn, 1991).

Responses to buffers in general were limited or variable. Buffers sometimes enhanced performance during the adaptation phase to high-concentrate or high-silage diets, but did not show consistent benefits throughout the feeding period. These responses do not justify the cost. As a result, buffers have not been widely adopted in the beef cattle industry.

ADDITIVES THAT ALTER RUMINAL FERMENTATION AND OTHER ASPECTS OF GASTROINTESTINAL TRACT FUNCTION

The goal of using these additives in diets of ruminants is to improve the utilization of feed resources and enhance production efficiency. Both direct-fed microbials (DFM) and feed enzymes have the potential to decrease the use of antibiotics in production systems. A comprehensive review of this topic was published by Beauchemin et al. (2006).

Bacterial and Fungal Direct-Fed Microbials

These products are potential alternatives to antibiotics for producing selective changes in microbial populations in the gastrointestinal tract. Direct-fed microbials contain live microorganisms (bacteria, yeasts, or molds). Products vary in genus and species and can contain multiple organisms. Common bacteria include species of *Lactobacillus*, *Propionibacterium*, *Streptococcus*, and *Megasphaera*. A common yeast is *Saccharomyces cerevisiae*, and a common mold is *Aspergillus oryzae* (Birkelo, 2003; McAllister et al., 2011). In a recent review, McAllister et al. (2011) summarized the types of microorganisms used in ruminant studies since 1991; there were 36 studies with lactic acid-producing bacteria, 7 studies with ruminal bacteria including lactic acid-utilizing bacteria, 18 studies with other bacteria, and 30 studies with yeast and fungi.

Direct-fed microbial is a more focused term than probiotic. Probiotics can include microbial culture extracts and enzymes in addition to the microorganisms (Krehbiel et al., 2003). The original concept of feeding probiotics was to improve the balance of microorganisms in the gastrointestinal tract to benefit the health of the animal (Fuller, 1989). The concept applied initially to the intestine. Direct-fed microbials could alter both ruminal fermentation and intestinal function to result in improved growth performance (McAllister et al., 2011).

Lactic acid-producing bacteria are included in most DFM products with the rationale that production of lactic acid will stimulate growth of lactic acid-utilizing bacteria. In addition to producing lactic acid, these bacteria also produce antimicrobial peptides known as bacteriocins. The peptides inhibit bacteria closely related to the species that produced the bacteriocins. Whether these peptides have a significant role in the efficacy of DFM is not known. Lactic acid-utilizing bacteria should prevent the accumulation of lactic acid by promoting metabolism of lactate to propionate (McAllister et al., 2011).

Animal Health and Production Efficiency

Krehbiel et al. (2003) reviewed the use of DFM in beef production. Studies were conducted using DFM cultures containing various *Lactobacillus* spp. and *Streptococcus faecium* in feed for calves entering the feedlot. In some studies, DFM resulted in improved daily gain, increased feed consumption, and improvement in feed:gain ratio, whereas in other studies no performance responses were observed. In some studies, feeding several lactic acid-producing bacteria improved recovery of morbid, newly received feedlot calves. Direct-fed microbials might benefit calves prone to health problems more than healthy animals. Specific species, strain composition within species, and dose all contribute to variation in response. Potential mechanisms of action of DFM include inhibiting pathogenic microorganisms from adhering to the intestinal epithelium, indirectly inhibiting the growth of pathogens, and augmenting the immune response to pathogens. These mechanisms relate to effects on postruminal microbial metabolism (Krehbiel et al., 2003). A recent meta-analysis of results from 28 studies, in which supplementation of a *Saccharomyces cerevisiae* fermentation product was evaluated in diets of feedlot cattle, showed improved ADG, feed efficiency, and carcass quality grade across all studies (Wagner et al., 2013).

Ruminal Acidosis

The high proportion of grains and, therefore, starch in the diet of feedlot cattle renders them susceptible to subacute ruminal acidosis. To date, various DFM have been evaluated for their efficacy in preventing ruminal acidosis. Across a number of trials, DFM demonstrated some potential to decrease the risk for subacute acidosis by lessening the time ruminal pH remained below 5.6 (Krehbiel et al., 2003). Addition of the lactic acid-producing bacterium *Enterococcus faecium*, with or without yeast, had a limited effect on decreasing acidosis and resulted in only minor changes in digestive function (Beauchemin et al., 2003b). The effect of yeast supplements was evaluated using a meta-analysis including data from 157 experiments in which yeast supplements, including at least one strain of *Saccharomyces cerevisiae*, were fed to ruminants (Desnoyers et al., 2009). The authors concluded that yeast supplementation resulted in increased ruminal pH, VFA concentration, and OM digestibility. Changes were in the range of 1% or less, and data were quite variable across papers.

Prevalence of *E. coli* O157:H7

Escherichia coli O157:H7 is a foodborne pathogen that is prevalent in feedlot cattle and their environment (Elder et al., 2000). The bacterium can cause hemorrhagic colitis and hemolytic uremic syndrome in humans (McAllister et al., 2011). One application of DFM is in preharvest intervention to decrease contamination of *E. coli* O157:H7 in beef products. Direct-fed microbials containing *Lactobacillus* spp. (Brashears et al., 2003) or combinations of *Lactobacillus* and *Propionibacterium* (Elam et al., 2003; Younts-Dahl et al., 2004; Tabe et al., 2008) decreased fecal shedding of *E. coli* O157:H7 with little influence on performance. Specific strains of *L. acidophilus* plus *P. freudenreichii* decreased both prevalence and concentration of *E. coli* O157:H7 (Stephens et al., 2007). In contrast, a DFM containing *Bacillus subtilis* did not alter prevalence or shedding of *E. coli* O157:H7 and had no effect on performance of feedlot cattle (Arthur et al., 2010). These results are encouraging for preharvest control using DFM to decrease the number of cattle that carry *E. coli* and the number of cells in carriers (Stephens et al., 2007), and show that both the species and strain of DFM are important in this response.

Direct-fed microbials might be a useful tool to promote health in cattle. More predictable biological responses and greater description of the manner in which specific DFM alter the microbial ecosystem in the gastrointestinal tract of the host will lead to development of more effective products (McAllister et al., 2011).

Feed Enzymes

Inclusion of exogenous enzymes in the diet might enhance hydrolysis of specific dietary fractions and increase the available nutrient supply, resulting in improved feed efficiency (Birkelo, 2003). The fiber fraction of forages has been a main focus of feed enzyme technology because of the large fraction of NDF in forages with a digestibility of 65% or less. Exogenous fibrolytic enzymes could increase forage utilization and improve production efficiency. They are expected to produce the greatest animal response under conditions in which fiber digestion is compromised and dietary energy is the first-limiting component of the diet (Beauchemin et al., 2003a, 2004; Beauchemin and Holtshausen, 2010). Positive responses in ADG were observed in cattle fed high-forage diets and in ADG or feed conversion in cattle fed diets high in barley grain but not corn grain when enzyme mixtures of xylanase and cellulase or endoglucanase were included in the diet (Beauchemin and Holtshausen, 2010).

Enzymes that degrade plant cell-wall structural polysaccharides include the general category of cellulases and hemicellulases. Some products include these enzymes as well as others (amylases, proteases). Xylanases are important for hydrolysis of the xylan core polymer of hemicelluloses. Available products are typically blends of cellulases and xylanases, but the components of the products are not well described. Improved diet digestibility was demonstrated in some studies; however, the effects were variable (Beauchemin et al., 2003a). Results of a meta-analysis of data from in vitro studies including data from 14 research studies with 144 enzyme treatments for alfalfa hay or corn silage (Eun and Beauchemin, 2008) suggested that exoglucanase was a key enzyme associated with enhanced in vitro NDF digestibility in both alfalfa and corn silage.

While addition of exogenous enzymes to the diet improves hydrolysis of dietary components in the rumen, better characterization of the factors responsible for variation in animal response could result in more effective products. These factors include the specific enzymes included, level of supplementation, appropriate enzyme for specific feed ingredients, method of application to the feed, and energy balance of the animals. In addition, newer technologies such as metatranscriptomics could identify specific carbohydrases to add to the ruminal environment to improve efficiency of digestion (Meale et al., 2014).

METABOLIC MODIFIERS THAT AFFECT POSTABSORPTIVE METABOLISM

Anabolic Implants

As noted in an extensive review by Preston (1999), a variety of anabolic implants has been used to increase the efficiency of beef cattle production since the 1950s. They are available for use in steers and heifers destined for slaughter to enhance growth rate, feed efficiency, and lean tissue accretion. Active ingredients include estrogens (estradiol, estradiol benzoate, and zeranol), androgens (testosterone propionate and TBA) and progesterone. Estradiol, progesterone, and testosterone are naturally occurring steroid hormones, whereas TBA is a synthetic steroid. Zeranol is a fungal compound with estrogenic activity that is not normally produced in animals. Implants can be composed of a single compound or a combination of an estrogen and androgen or an estrogen and progesterone (Preston, 1999).

Implant products are placed under the skin in the middle third of the ear. The active ingredients are released from the implant over a period of time (payout time) that is related to the carrier or the matrix of the implant (Preston, 1999). Trade names, active ingredients, and animal use for products currently available in North America are given in Table 14-1. The products currently available are a combination of original products and identical generic products. Some lower-dose implants have 50% of the dose of active ingredient(s) of similar implants. Lower concentrations are for immature cattle or situations where energy is limiting. Implants are approved for use by the FDA in the United States and by the Veterinary Drug Directorate in Canada, although not all of the products listed in Table 14-1 are approved in both countries.

Implant products are available for suckling calves, grazing cattle, and finishing cattle. The effect of implanting, in each production phase, was an increase in ADG of 5, 15, and 20% for suckling steer calves, stocker steers, and feedlot steers, respectively (Duckett and Andrae, 2001). In general, response is greater when cattle consume diets providing a higher plane of nutrition. In feedlot steers, feed conversion was improved by up to 13.5% and hot carcass weight increased up to 7.5% depending on the implant type and strategy (Duckett and Andrae, 2001). The rate of empty body protein gain was increased with implants, with little effect on the rate of fat gain; thus, cattle were less mature at a given BW (reviewed by Montgomery et al., 2001).

Several reviews describe implant strategies including the selection of implants and their sequence as determined by animal sex, weight, age, and feed costs. Strategies include one, two, or three implants. The optimal approach will depend on specific goals and the balance between improved production traits, such as ADG and feed conversion, and effects on carcass quality (reviewed by Webb et al., 2002; Reinhardt, 2007).

Implants enhance both rate of gain and feed intake; however, rate of gain is usually enhanced more than intake, and consequently, feed conversion is improved. It was concluded by the NRC (1996, 2000) that the effect of implants on nutrient utilization was minimal. Therefore, their effect on nutrient requirements was accounted for by their effect on protein, fat, and energy accretion, which was taken into account by adjusting slaughter weight at constant finish (NRC, 1984, 1996, 2000; Tylutki et al., 1994). Most of the increase in weight gain can be accounted for by an increased growth of lean tissue and skeleton (Trenkle, 1990). Previous studies (Trenkle, 1990; Perry et al., 1991; Bartle et al., 1992) indicated that compared with not using an implant, estrogenic implants increased protein content of gain equivalent to a 35-kg change in final shrunk body weight (FSBW), whereas estradiol and TBA combination implants altered the protein content of gain equivalent to a change of approximately 70 kg in FSBW. The NRC (1996, 2000) recommended decreasing the FSBW by 25 to 45 kg when no implant was given or increasing the FSBW by 25 to 45 kg when estradiol and TBA

TABLE 14-1 List of Anabolic Implants Available for Use in Beef Cattle[a]

Trade Name	Active Ingredients			Animal Use[d]
	Estrogenic, mg[b]	Progesterone, mg	Androgenic, mg[c]	
Ralgro	36 Z			Suckling calves—S, H; Stockers—S, H; Feedlot—S, H
Compudose	25.7 E			Suckling calves—S; Stockers—S; Feedlot—S, H
Encore	43.9 E			Suckling calves—S; Stockers—S; Feedlot—S, H
Synovex-C	10 EB	100		Suckling calves—S, H
Component E-C	10 EB	100		Suckling calves—S, H
Synovex-S	20 EB	200		Stockers—S; Feedlot—S
Component E-S	20 EB	200		Stockers—S; Feedlot—S
Synovex-H	20 EB		200 TP	Stockers—H; Feedlot—H
Component E-H	20 EB		200 TP	Stockers—H; Feedlot—H
Component T-S			140 TBA	Feedlot—S
Finaplix-H			200 TBA	Feedlot—H
Component T-H			200 TBA	Feedlot—H
Component TE-G	8 E		40 TBA	Stockers—S, H
Revalor-G	8 E		40 TBA	Stockers—S, H
Revalor-IH	8 E		80 TBA	Feedlot—H
Component TE-IH	8 E		80 TBA	Feedlot—H
Revalor-IS	16 E		80 TBA	Feedlot—S
Component TE-IS	16 E		80 TBA	Feedlot—S
Component TE-100	10 E		100 TBA	Feedlot—S, H
Synovex Choice	14 EB		100 TBA	Feedlot—S, H
Revalor-S	24 E		120 TBA	Feedlot—S
Component TE-S	24 E		120 TBA	Feedlot—S
Revalor-H	14 E		140 TBA	Feedlot—H
Component TE-H	14 E		140 TBA	Feedlot—H
Synovex ONE-Grass	21 EB		150 TBA	Stockers—S, H
Revalor-200	20 E		200 TBA	Feedlot—S, H
Component TE-200	20 E		200 TBA	Feedlot—S, H
Synovex Plus	28 EB		200 TBA	Feedlot—S, H
Synovex ONE-Feedlot	28 EB		200 TBA	Feedlot—S, H
Revalor-XS (Timed release)	40 E		200 TBA	Feedlot—S

[a]Adapted from Duckett and Andrae (2001), Montgomery et al. (2001), the North American TBA Implant Database (TTU, 2014), Animal Drugs@FDA (http://www.accessdata.fda.gov/scripts/animaldrugsatfda, accessed on December 1, 2014), and Health Canada's Veterinary Drug Directorate, Drug Product Database (http://hc-sc.gc.ca/dhp-mps/prodpharma/databasdon/index-eng.php, accessed on April 3, 2015).

[b]Estrogenic activity: Z, zeranol; E, estradiol-17β; or EB, estradiol benzoate. Estradiol benzoate contains 71.4% estradiol by weight (Montgomery et al., 2001).

[c]Androgenic activity: TP, testosterone propionate; or TBA, trenbolone acetate.

[d]Animal use abbreviations: S, steer; H, heifer.

were given. This adjustment changed the NEg requirement by 5%, decreasing it with implant use or increasing it with no use of implants, and was consistent with a 5% increase in NE requirements when estrogenic implants were not used (NRC, 1984) or 4% adjustment for nonuse of implants in the model of Oltjen et al. (1986).

Based on a summary of 13 implant trials conducted between 1997 and 2000, including various implant strategies, effects on BW and carcass characteristics were summarized (Guiroy et al., 2002). Implants increased the adjusted FSBW (AFSBW) at 28% empty body fat (EBF) by 30 to 39 kg in heifers and by 14 to 42 kg in steers, confirming the increase in potential mature size. The percentage of animals that graded USDA low Choice or above decreased in implanted cattle. The response in FSBW and grade depended on the implant strategy, with greater response as the number of implants and dose of anabolic agents increased. Guiroy et al. (2002) suggested that the response to anabolic implants included a decrease in the proportion of DMI required for maintenance, a decrease in the energy content of gain, and an increase in the efficiency of use of absorbed energy. They suggested an increase of dietary ME of up to 4.2% in steers and 3.1% in heifers given an anabolic implant. Increased efficiency of ME use would be consistent with increased protein deposited per unit protein synthesized.

The mode of action by which anabolic agents enhance the rate of protein accretion in the body is not completely understood. Studies support the role of increased endogenous secretion of growth hormone and insulin-like growth factor-1 (IGF-1) as part of the response. Additional studies support increased responsiveness of satellite cells to muscle growth factors, such as IGF-1, and increased proliferation of these cells in implanted steers, resulting in increased muscle growth (NRC, 1994; Preston, 1999). When compounds present in a combined TBA/estradiol-17β implant were administered individually, estradiol-17β, but not TBA, increased IGF-I mRNA in the longissimus muscle of implanted cattle (Pampusch et al., 2008). Studies including both the individual compounds and the combined compounds in implants might provide greater insight on mechanism.

Guiroy et al. (2002) adjusted shrunk BW (SBW) to 28% EBF and grouped implant strategies with similar AFSBW into five categories of incremental anabolic response compared with the no-implant treatment (Category 1). Category 2 had the lowest increment, with a change of 13.7 kg, and included an estrogenic implant or an intermediate dose of estradiol-17β plus TBA. Category 5 had the highest increment, with a change of 41.8 kg, and included varying concentrations of estradiol-17β plus TBA in both a first and second implant. Fewer implant strategies were evaluated for heifers and they were grouped into only three categories. The committee recommends continuing the approach of adjusting the FSBW as described in NRC (1996, 2000). It was recommended previously to decrease the FSBW by 25 to 45 kg when no implant was given or to increase the FSBW by 25 to 45 kg when estradiol and TBA were given. The summary of Guiroy et al. (2002) reported a similar range in responses and could be used for a more specific recommendation depending on implant strategy.

Melengesterol Acetate

Melengesterol acetate (MGA) is an orally active synthetic progestin. It was developed in 1962 and used initially for controlling the estrous cycle. It was included in diets of feedlot heifers to improve feed efficiency and rate of gain (Bloss et al., 1966) by allowing ovarian follicular development, while inhibiting estrus and ovulation (Zimbelman and Smith, 1966; Patterson et al., 1989). It prevents estrus and ovulation by blocking the preovulatory surge of luteinizing hormone (Imwalle et al., 2002). Inclusion of MGA in the diet to provide a daily dose of 0.25 to 0.50 mg/heifer daily suppresses estrus and improves rate of gain (10%) and feed efficiency (3%) and also minimizes changes in feed intake patterns associated with cycling (in a review by Birkelo, 2003). Dry matter intake increased in response to inclusion of MGA in some studies (Perrett et al., 2008) but not in others (Kreikemeier and Mader, 2004). Part of the response to MGA could be a consequence of elevated estradiol released from immature follicles (Zimbelman and Smith, 1966; Birkelo, 2003).

For management strategies that include MGA, the combination of MGA and implants should be integrated in the growth promotant program (Mader and Lechtenberg, 2000). Based on a meta-analysis, using data from 18 publications that included MGA and at least one TBA implant treatment resulting in 101 treatment means, MGA increased carcass weight and fat depth at the 12th rib and decreased longissimus muscle area. This analysis showed that the effects of MGA could be additive with an androgenic implant but not with an estrogen–androgen combination implant, which might reflect the lack of response to endogenous follicular estrogens when exogenous estrogens are present. Some responses could reflect direct effects of progestins on muscle satellite cells to decrease their proliferation (Wagner et al., 2007). At present, the committee does not recommend an adjustment to ME or FSBW for inclusion of MGA in the diet.

β-Adrenergic Agonists

Interest in the effects of this general class of compounds in growing cattle developed from the observation of increased lean tissue and decreased fat in response to feeding a β-adrenergic agonist to cattle (Ricks et al., 1984). β-adrenergic agonists are compounds that bind to β-adrenergic receptors in the plasma membrane, which are coupled to guanine nucleotide-binding regulatory proteins (G proteins). The coupling of the receptor and G proteins leads to the consequent activation of several intracellular signaling pathways associated with receptor binding. The activation of the signaling pathway associated with adenylate cyclase, cAMP, and protein kinase A is thought to be one component of the response leading to the anabolic response in muscle; however, additional signaling pathways independent of protein kinase A are activated in skeletal muscle in association with hypertrophy (Lynch and Ryall, 2008). There are at least three β-adrenergic receptor (AR) subtypes in mammalian tissues designated β_1-AR, β_2-AR, and β_3-AR. The distribution of receptor subtype varies with tissue. Expression of β_2-AR mRNA was much greater than β_1-AR mRNA in semimembranosus muscle from implanted steers (Winterholler et al., 2007; Baxa et al., 2010) and heifers (Sissom et al., 2007) suggesting that the β_2-AR is more abundant than β_1-AR in this tissue.

Two physiological compounds that stimulate these receptors are epinephrine and norepinephrine. Some structurally similar synthetic β-adrenergic agonists also stimulate these receptors. These compounds are orally active and increase skeletal muscle growth (Mersmann, 1998). Two of these synthetic compounds are approved by the FDA for use as feed additives in finishing diets for cattle: ractopamine hydrochloride was approved in 2003, and zilpaterol hydrochloride was approved in 2006. They are classified as having selectivity for the β_1-AR and β_2-AR, respectively (Moody et al., 2000). Further research is needed to define the interaction of each compound with specific cellular receptors and subsequent intracellular responses.

One of the characteristics of β-AR signaling is the attenuation of response over time to a continual stimulus, which is described as desensitization. This could be a result of changes in the receptor, such as phosphorylation, which targets the receptor for internalization where it can be degraded, resulting in a decline in the total number of β-AR (Mills, 2002; Lynch and Ryall, 2008). Thus, β-adrenergic agonists are relatively short-acting and are therefore included in the diet for a relatively short duration at the end of the feeding period.

Ractopamine

Ractopamine hydrochloride is intended for use during the last 28 to 42 days of feeding at a dose of 9 to 27 mg/kg DM to provide 70 to 430 mg/animal daily. Initial dose-response studies with ractopamine evaluated inclusion at a rate to result in consumption of approximately 100, 200, or 300 mg/animal daily for 28 to 42 days in finishing steers and heifers (Schroeder et al., 2004). However, the majority of studies reported included ractopamine at levels to result in consumption of approximately 200 mg/animal daily for 28 days. General responses included increased ADG and improved feed efficiency with no effect on DMI in steers and heifers fed ractopamine. Some combination of increased final BW, hot carcass weight, and longissimus muscle area was observed in steers (Gruber et al., 2007; Winterholler et al., 2007; Bryant et al., 2010; Boler et al., 2012) and heifers (Sissom et al., 2007) fed ractopamine for the last 28 days. Effects of ractopamine on fat content were less consistent than on muscle, but ractopamine decreased fat accumulation in studies where there was an effect. Steers of diverse biological type responded similarly to ractopamine (Gruber et al., 2007), and response was not influenced by days on feed in steers (Winterholler et al., 2007) or heifers (Sissom et al., 2007).

Recent meta-analyses of 32 studies with steers and 16 studies with heifers in which growth responses and carcass characteristics were measured following ractopamine inclusion were reported by Pyatt et al. (2013a,b). The meta-analysis showed that DMI was not affected by ractopamine in either gender, whereas ADG, feed efficiency, dressing percentage, and longissimus muscle area were improved. Liveweight gain increased by 3.4, 6.8, and 10.2 kg and carcass weight increased by 3.1, 6.1, and 9.2 kg in steers fed 100, 200, or 300 mg ractopamine daily, respectively, compared with controls. Similarly, liveweight gain increased by 2.7, 5.4, and 8.1 kg, and carcass weight increased by 2.1, 4.3, and 6.4 kg in heifers fed 100, 200, or 300 mg ractopamine daily, respectively, compared with controls. There was a shift toward lower-quality grade with increasing dose in both sexes. Another meta-analysis (Lean et al., 2014) indicated increased longissimus muscle area of 1.84 cm^2 in cattle (combined steers and heifers) receiving ractopamine.

The equations developed by Guiroy et al. (2001) were used by Guiroy et al. (2002) to evaluate the effect of different implant strategies on AFSBW. A similar approach was used to assess the effect of ractopamine hydrochloride on AFSBW, using data from Elanco Animal Health (2012, 2013). The AFSBW for steers increased by 17 kg, and for heifers it increased by 6 kg.

Because of the prevalence of implant use, several studies evaluated the response of the combination of implants, containing TBA and estradiol 17β, and ractopamine. The effects of the combination were additive with no interactions related to the growth response or most carcass variables in steers (Bryant et al., 2010) and no interactions in heifers (Sissom et al., 2007). These observations support differing mechanisms; however, the overall response to ractopamine could depend on the specific implant strategy (Sissom et al., 2007).

The mechanism by which ractopamine enhances muscle growth is not well established. Increased protein deposition could result from increased protein synthesis, decreased protein breakdown, or a combination of both. Increased protein synthesis is supported by observations of increased myosin mRNA (Smith et al., 1989). To maximize the response in protein deposition, either increased amino acid supply or increased efficiency of amino acid use is needed. Studies in pigs showed that inadequate protein supply limited the magnitude of the response to ractopamine (NRC, 1994). The most recent edition of the *Nutrient Requirements of Swine* (NRC, 2012) recommended an increase in the daily requirement for standardized ileal digestible lysine for maximal growth response in conjunction with feeding ractopamine. The publication also recommended a change in the amino acid profile to reflect a greater response in deposition of muscle protein than nonmuscle protein.

There are few data available to assess whether there is a change in the amino acid requirements of cattle consuming ractopamine. Most of the published studies that included ractopamine in the diet used a single corn-based diet for all treatments. Dietary crude protein (CP) ranged from around 13.5 to 16% with no estimate of metabolizable protein (MP) supply. Based on a recent survey of consulting feedlot nutritionists (Vasconcelos and Galyean, 2007), the average CP used by clients served by the nutritionists ranged from 12.5 to 14.0%. It is likely that the MP supply in many published studies exceeded the requirements in control animals and did not limit the response to ractopamine, which improved the efficiency of amino acid use. One study (Walker et al., 2006) in which dietary ingredients were varied to alter MP supply (NRC, 1996, 2000) suggested a possible interaction between diet and response to ractopamine. The ADG of control heifers improved with greater dietary ruminally undegradable protein (RUP) as a percentage of CP, whereas heifers fed ractopamine showed no response to changes in dietary N source. Heifers fed the diet containing the greatest RUP as a percentage of CP did not respond to ractopamine (Walker et al., 2006). Other research results demonstrated decreased ammonia and amino acid concentration when ractopamine was added to in vitro incubations of ruminal fluid, suggesting a decrease in proteolysis and deamination and the potential need for greater ruminally degraded protein, although effects varied with protein source and method of grain processing (Walker and Drouillard, 2010). Although specific β-adrenergic agonists might have different effects in various species (Mersmann, 1998), it would be expected that increased protein deposition in response to ractopamine in cattle would require increased amino acid supply for maximal response unless supply was already in excess. Enhanced ability to predict total MP supply and the contribution of dietary RUP and bacteria to the supply in response to feeding ractopamine, along with systematic studies to evaluate MP requirements in cattle, might clarify this area.

Ractopamine enhances rate of gain, deposition of carcass protein, and the efficiency of nutrient use. There are no data at this time to suggest a change in nutrient requirements when ractopamine is fed.

Zilpaterol

Zilpaterol hydrochloride is intended for use during the last 20 to 40 days of feeding at a dose of 8.33 mg/kg DM to provide 60 to 90 mg/animal daily with a 3-day withdrawal. Although recently approved in the United States, zilpaterol was approved earlier in South Africa and Mexico. A number of studies evaluated the effects of zilpaterol inclusion on growth and carcass variables under U.S. conditions. Inclusion of zilpaterol increased ADG and improved gain:feed ratio with variable effects on DMI. Zilpaterol increased hot carcass weight, dressing percentage, and longissimus muscle area, while producing less visceral fat, lower marbling, and less fat at the 12th rib in steers (Vasconcelos et al., 2008; Elam et al., 2009; Kellermeier et al., 2009; Baxa et al., 2010) and heifers (Rathmann et al., 2012). Warner-Bratzler shear force increased in longissimus muscle from steers (Kellermeier et al., 2009) and heifers (Rathmann et al., 2012) fed zilpaterol.

For studies including zilpaterol hydrochloride (8.3 mg/kg DM) in diets fed to steers for 20 days (Vasconcelos et al., 2008; Elam et al., 2009; Montgomery et al., 2009), feed intake was unchanged, ADG increased by 0.26 kg, and gain:feed increased by 0.03. There were increases in BW of 9.5 kg, carcass weight of 14.6 kg, longissimus muscle area of 8.3 cm^2, and dressing percentage of 1.5 units. Responses in heifers (Montgomery et al., 2009; Rathmann et al., 2012) were increased ADG of 0.18 kg, decreased DMI of 0.34 kg, and improved gain:feed ratio of 0.03. There were increases in BW of 5.2 kg, carcass weight of 11.1 kg, longissimus muscle area of 5.7 cm^2, and dressing percentage of 1.5 units. Using the equations of Guiroy et al. (2001), zilpaterol fed for 20 days increased the AFSBW at 28% EBF by 34 to 36 kg in steers (Elam et al., 2009; Leheska et al., 2009) and by 26 to 27 kg in heifers (Leheska et al., 2009; Rathmann et al., 2012), suggesting an increase in mature size.

The mechanisms responsible for muscle hypertrophy for zilpaterol are not well defined. The ratio of the increase in carcass weight to the increase in BW of 1.5 in steers and 2.1 in heifers supports the emphasis on muscle growth and increased dressing percentage, which are key attributes of the response to zilpaterol. Results from a series of studies in which zilpaterol was fed to cattle and meat quality was observed were reviewed by Delmore et al. (2010).

Zilpaterol alters rate and composition of gain at a lower dose and with a shorter feeding period than ractopamine. In a comparison study, in which both ractopamine and zilpaterol were fed, ractopamine resulted in greater improvement of ADG, and zilpaterol resulted in greater improvement of carcass weight, dressing percentage, and carcass leaness, whereas improvement in gain:feed ratio was similar between the two compounds (Scramlin et al., 2010).

The response to zilpaterol was similar at differing stages of compositional maturity in steers (Vasconcelos et al., 2008) and heifers (Rathmann et al., 2012). Therefore, cattle can be fed for a longer period before receiving zilpaterol to harvest at a specific EBF percentage (Rathmann et al., 2012). Studies focused on the response to zilpaterol in the last 30 days of feeding, with or without a terminal implant containing TBA and estradiol-17β (Kellermeier et al., 2009; Baxa et al., 2010) or using two single implants containing TBA and estradiol-17β with different doses and payout patterns (Parr et al., 2011), showed additive effects of zilpaterol and implants. Changes in nutrient requirements have not received much focus. Results of a recent study suggest that the response to zilpaterol might increase with the inclusion of rumen-protected amino acids in the diet (Hosford et al., 2015). Monitoring effects on beef quality and carcass weight is an important component of incorporating zilpaterol into a management program (Delmore et al., 2010).

Using Adjusted Final Body Weight

The effect of β-adrenergic agonists and implants on AFSBW (an estimate of the SBW at 28% body fat) is similar, with greatest change in AFSBW in response to the most aggressive implant strategies. However, in the case of zilpaterol, the prediction of carcass fat using carcass traits of Montgomery et al. (2009) and the equation of Guiroy et al. (2001) overestimated carcass fat by at least 35% and 20%

for steers and heifers, respectively, based on observed values (Leheska et al., 2009). This was the case for both control cattle and cattle fed zilpaterol. Guiroy et al. (2002) showed that the percentage of carcasses grading USDA Choice or greater decreased with increasing implant aggressiveness even though the average EBF categories were similar. In addition, Guiroy et al. (2002) reported that even at a 30% EBF, many carcasses did not grade USDA Choice or greater. Similarly, Schneider et al. (2007) reported that marbling and quality grade trended down with increasing doses of TBA and estradiol-17β in heifers with similar EBF. Therefore, cattle fed β-adrenergic agonists and/or implanted and fed to the same EBF endpoint may not grade similarly to cattle that did not receive growth modifiers.

Sensitivity analysis of the distributions of hot carcass weight and longissimus muscle area for control steers and steers receiving zilpaterol hydrochloride (Elam et al., 2009) showed that the treatment effect was significant; this would affect the calculation of AFSBW. The differences for control and treated heifers were less than those for steers. In addition to carcass traits, the value of 14.26 kg empty body weight (EBW) to increase EBF by one percentage unit (ΔEBW/ΔEBF) is assumed in calculating AFSBW. This value was derived from regressing EBW (kg) on EBF (% EBW) of different breeds of cattle. The value varied from 12.6 to 16.1% (Guiroy et. al., 2001). Sensitivity analysis indicated that a change in EBW per unit change in EBF has a lesser effect on the estimated AFSBW than changes in carcass traits, although it was still significant.

Using data from Elanco Animal Health (2012, 2013) the actual ΔEBW/ΔEBF was estimated at 20.0 kg/% for steers and 15.9 kg/% for heifers receiving ractopamine hydrochloride. Estimated values of ΔEBW/ΔEBF were 23 to 53 kg/% for steers (Elam et al., 2009; Montgomery et al., 2009) and 19.6 to 28.5 kg/% for heifers (Montgomery et al., 2009; Rathmann et al., 2012) fed zilpaterol hydrochloride. More data are needed to confirm these estimates.

The implication of a greater ΔEBW/ΔEBF is that predicted ΔAFSBW will be (1) smaller than initially predicted when predicted EBF of the treated animals is greater than the desired threshold (usually 28%) because the EBF will be adjusted downward and there is more EBW per % EBF; or (2) greater than initially predicted when predicted EBF of the treated animals is less than the desired threshold (usually 28%) because, as they deposit more EBW per % EBF, animals would have reached the desired EBF at a heavier EBW.

Based on these analyses, the variances in carcass traits are important in determining final EBF and are more important than the variation in the ΔEBW/ΔEBF coefficient to calculate AFSBW. The equation of Guiroy et al. (2001) might not accurately predict carcass fat and EBF for animals receiving zilpaterol hydrochloride. Using adjusted values for ΔEBW/ΔEBF to calculate ΔAFSBW in animals receiving β-adrenergic agonists decreased values significantly and suggested that there was little change in AFSBW as a result of feeding either ractopamine or zilpaterol. Calculation of AFSBW using standard equations might not appropriately address changes in protein utilization responsible for heavier and leaner carcasses observed in cattle treated with β-adrenergic agonists.

SUMMARY OF RECOMMENDED ADJUSTMENTS FOR INCLUSION OF MODIFIERS

Ionophores

The committee recommends reducing predicted DMI by 3% when monensin is fed, regardless of the proportion of dietary concentrate. No decrease in predicted DMI is recommended when lasalocid or laidlomycin propionate is fed. Based on the mode of action of ionophores in the rumen, the committee decided to change dietary ME to account for the improvements in ADG and feed efficiency rather than adjust dietary NEm. Therefore, the committee recommends that dietary ME be increased by 2.3% for monensin and 1.5% for lasalocid and laidlomycin propionate.

Implants

The committee recommends no changes from the previous NRC (1996, 2000) with the use of implants. The user should increase the AFSBW based on the type of implant used; the paper of Guiroy et al. (2002) provides estimates for different categories of implants.

β-Adrenergic Agonists

Ractopamine

The committee recommends no adjustment for DMI when ractopamine is fed. No adjustment for nutrient requirements is suggested with the available information.

Zilpaterol

Albeit some data suggest a decrease in DMI when zilpaterol is fed, the committee recommends no adjustment for DMI because of the inconsistent results. No adjustment for nutrient requirements is suggested with the available information.

Adjusted Final SBW

Variation in carcass traits has a significant effect on the value for final EBF. Both carcass traits and ΔEBW/ΔEBF play an important role in calculating AFSBW. More reliable equations to predict carcass fat and EBF for control and β-adrenergic agonist-treated cattle are needed. An increase in AFSBW at 28% EBF when ractopamine or zilpaterol is fed can be estimated using standard equations; however, including a variable estimate for ΔEBW/ΔEBF resulted in a predicted AFSBW that was significantly less than that obtained using the standard equations and suggested that there was no change in AFSBW as a result of feeding either ractopamine or zilpaterol. More data are needed to confirm these observations, as are data on composition of gain of the whole animal. Calculation of AFSBW using standard equations might not appropriately address changes in protein utilization responsible for observed responses in cattle treated with β-adrenergic agonists.

REFERENCES

Arthur, T. M., J. M. Bosilevac, N. Kalchayanand, J. E. Wells, S. D. Shackelford, T. L. Wheeler, and M. Koohmaraie. 2010. Prevalence and load of *Escherichia coli* O157 in feedlot cattle. *Journal of Food Protection* 73:366-371.

Bartle, S. J., R. L. Preston, R. E. Brown, and R. J. Grant. 1992. Trenbolone acetate/estradiol combinations in feedlot steers: Dose-response and implant carrier effects. *Journal of Animal Science* 70:1326-1332.

Bartley, E. E. 1965. Bloat in cattle. VI. Prevention of legume bloat with a nonionic surfactant. *Journal of Dairy Science* 48:102-104.

Bartley, E. E., G. W. Barr, and R. Mickelsen. 1975. Bloat in cattle. XVII. Wheat pasture bloat and its prevention with poloxalene. *Journal of Animal Science* 41:752-759.

Bartley, E. E., T. G. Nagaraja, E. S. Pressman, A. D. Dayton, M. P. Katx, and L. R. Fina. 1983. Effects of lasalocid or monensin on legume or grain (feedlot) bloat. *Journal of Animal Science* 56:1400-1406.

Baxa, T. J., J. P. Hutcheson, M. F. Miller, J. C. Brooks, W. T. Nichols, M. N. Streeter, D. A. Yates, and B. J. Johnson. 2010. Additive effects of a steroidal implant and zilpaterol hydrochloride on feedlot performance, carcass characteristics, and skeletal muscle messenger ribonucleic acid abundance in finishing steers. *Journal of Animal Science* 88:330-337.

Beauchemin, K. A., and L. Holtshausen. 2010. Developments in enzyme usage in ruminants. Pp. 206-230 in *Enzymes in Farm Animal Nutrition*, 2nd Ed., M. R. Bedford and G. G. Partridge, eds. Oxfordshire, UK: CAB International.

Beauchemin, K. A., D. Colombatto, D. P. Morgavi, and W. Z. Yang. 2003a. Use of exogenous fibrolytic enzymes to improve feed utilization in ruminants. *Journal of Animal Science* 81(Suppl. 2):E37-E47.

Beauchemin, K. A., W. Z. Yang, D. P. Morgavi, G. R. Ghorbani, W. Kautz, and J. A. Z. Leedle. 2003b. Effects of bacterial direct-fed microbials and yeast on site and extent of digestion, blood chemistry, and subclinical ruminal acidosis in feedlot cattle. *Journal of Animal Science* 81:1628-1640.

Beauchemin, K. A., D. Colombatto, and D. P. Morgavi. 2004. A rationale for the development of feed enzyme products for ruminants. *Canadian Journal of Animal Science* 84:23-36.

Beauchemin, K. A., C. R. Krehbiel, and C. J. Newbold. 2006. Enzymes, bacterial direct-fed microbials and yeast: Principles for use in ruminant nutrition. Pp. 251-284 in *Biology of Nutrition in Growing Animals*, R. Mosenthin, J. Zentek, and T. Zebrowska, eds. New York: Elsevier.

Beck, P., T. Hess, D. Hubbell, G. D. Hufstedler, B. Fieser, and J. Caldwell. 2014. Additive effects of growth promoting technologies on performance of grazing steers and economics of the wheat pasture enterprise. *Journal of Animal Science* 92:1219-1227.

Berg, B. P., W. Majak, T. A. McAllister, J. W. Hall, D. McCartney, B. E. Coulman, B. P. Goplen, S. N. Acharya, R. M. Tait, and K.-J. Cheng. 2000. Bloat in cattle grazing alfalfa cultivars selected for a low initial rate of digestion: A review. *Canadian Journal of Plant Science* 80:493-502.

Bergen, W. G., and D. B. Bates. 1984. Ionophores: Their effect on production efficiency and mode of action. *Journal of Animal Science* 58:1465-1483.

Birkelo, C. P. 2003. Pharmaceuticals, direct-fed microbials, and enzymes for enhancing growth and feed efficiency of beef. *Veterinary Clinics of North America: Food Animal Practice* 19:599-624.

Bloss, R. E., J. I. Northam, L. W. Smith, and R. G. Zimbelman. 1966. Effects of oral melengestrol acetate on the performance of feedlot cattle. *Journal of Animal Science* 25:1048-1053.

Boerner, B. J., F. M. Byers, G. T. Shelling, C. E. Coppock, and L. W. Greene. 1987. Trona and sodium bicarbonate in beef cattle diets: Effects on pH and volatile fatty acid concentrations. *Journal of Animal Science* 65:309-316.

Boler, D. D., A. L. Shreck, D. B. Faulkner, J. Killefer, F. K. McKeith, J. W. Homm, and J. A. Scanga. 2012. Effect of ractopamine hydrochloride (Optaflexx) dose on live performance, carcass characteristics and tenderness in early weaned beef steers. *Meat Science* 92:458-463.

Branine, M. E., and M. L. Galyean. 1990. Influence of grain and monensin supplementation on ruminal fermentation, intake, digesta kinetics and incidence and severity of frothy bloat in steers grazing winter wheat pastures. *Journal of Animal Science* 68:1139-1150.

Brashears, M. M., M. L. Galyean, G. H. Loneragan, J. E. Mann, and K. Killinger-Mann. 2003. Prevalence of *Escherichia coli* O157:H7 and performance by beef feedlot cattle given *Lactobacillus* direct-fed microbials. *Journal of Food Protection* 66:748-754.

Bretschneider, G., J. C. Elizalde, and F. A. Perez. 2008. The effect of feeding antibiotic growth promoters on the performance of beef cattle consuming forage-based diets: A review. *Livestock Science* 114:135-149.

Bryant, T. C., T. E. Engle, M. L. Galyean, J. J. Wagner, J. D. Tatum, R. V. Anthony, and S. B. Laudert. 2010. Effects of ractopamine and trenbolone acetate implants with or without estradiol on growth performance, carcass characteristics, adipogenic enzyme activity, and blood metabolites in feedlot steers and heifers. *Journal of Animal Science* 88:4102-4119.

Byers, F. M. 1980. Determining effects of monensin on energy value of corn silage diets for beef cattle by linear or semi-log methods. *Journal of Animal Science* 51:158-169.

Calsamiglia, S., M. Busquet, P. W. Cardozo, L. Castillejos, and A. Ferret. 2007. Invited review: Essential oils as modifiers of rumen microbial fermentation. *Journal of Dairy Science* 90:2580-2595.

Cheng, K.-J., T. A. McAllister, J. D. Popp, A. N. Hristov, Z. Mir, and H. T. Shin. 1998. A review of bloat in feedlot cattle. *Journal of Animal Science* 76:299-308.

Clary, E. M., R. T. Brandt, Jr., D. L. Harmon, and T. G. Nagaraja. 1993. Supplemental fat and ionophores in finishing diets: Feedlot performance and ruminal digesta kinetics in steers. *Journal of Animal Science* 71:3115-3123.

Delfino, J., G. W. Mathison, and M. W. Smith. 1988. Effect of lasalocid on feedlot performance and energy partitioning in cattle. *Journal of Animal Science* 66:136-150.

Delmore, R. J., J. M. Hodgen, and B. J. Johnson. 2010. Perspectives on the application of zilpaterol hydrochloride in the United States beef industry. *Journal of Animal Science* 88:2825-2828.

Desnoyers, M., S. Giger-Reverdin, G. Bertin, C. Duvaux-Ponter, and D. Sauvant. 2009. Meta-analysis of the influence of *Saccharomyces cerevisiae* supplementation on ruminal parameters and milk production of ruminants. *Journal of Dairy Science* 92:1620-1632.

Duckett, S. K., and J. G. Andrae. 2001. Implant strategies in an integrated beef production system. *Journal of Animal Science* 79:E110-E117.

Duffield, T. F., J. K. Merrill, and R. N. Bagg. 2012. Meta-analysis of the effects of monensin in beef cattle on feed efficiency, body weight gain, and dry matter intake. *Journal of Animal Science* 90:4583-4592.

Elam, N. A., J. F. Gleghorn, J. D. Rivera, M. L. Galyean, P. J. Defoor, M. M. Brashears, and S. M. Younts-Dahl. 2003. Effects of live cultures of *Lactobacillus acidophilus* (strains NP45 and NP51) and *Propionibacterium freudenreichii* on performance, carcass, and intestinal characteristics, and *Escherichia coli* strain O157 shedding in finishing steers. *Journal of Animal Science* 81:2686-2698.

Elam, N. A., J. T. Vasconcelos, G. Hilton, D. L. VanOverbeke, T. E. Lawrence, T. H. Montgomery, W. T. Nichols, M. N. Streeter, J. P. Hutcheson, D. A. Yates, and M. L. Galyean. 2009. Effect of zilpaterol hydrochloride duration of feeding on performance and carcass characteristics of feedlot cattle. *Journal of Animal Science* 87:2133-2141.

Elanco Animal Health. 2012. Effects of Optaflexx® on Performance and Carcass Characteristics in Finishing Steers: 32-Trial Summary. Optaflexx Research Brief 5. Available online at http://www.elanco.us/pdfs/usbbuopt00027_optaflexxrb5.pdf. Accessed on November 13, 2014.

Elanco Animal Health. 2013. Effects of Optaflexx® on Performance and Carcass Characteristics in Finishing Heifers: 16-Trial Summary. Optaflexx Research Brief 7. Available online at http://www.elanco.us/pdfs/Optaflexx-Research-Brief-7.pdf. Accessed on November 13, 2014.

Elder, R. O., J. E. Keen, G. R. Siragusa, G. A. Barkocy-Gallagher, M. Koohmaraie, and W. W. Laegreid. 2000. Correlation of enterohemorrhagic *Escherichia coli* O157 prevalence in feces, hides, and carcasses of beef cattle during processing. *Proceedings of the National Academy of Sciences of the United States of America* 97:2999-3003.

Ellis, J. L., J. Dijkstra, A. Bannink, E. Kebreab, S. E. Hook, S. Archibeque, and J. France. 2012. Quantifying the effect of monensin dose on the rumen volatile fatty acid profile in high-grain-fed beef cattle. *Journal of Animal Science* 90:2717-2726.

Erickson, G. E., C. T. Milton, K. C. Fanning, R. J. Cooper, R. S. Swingle, J. C. Parrott, G. Vogel, and T. J. Klopfenstein. 2003. Interaction between bunk management and monensin concentration on finishing performance, feeding behavior, and ruminal metabolism during an acidosis challenge with feedlot cattle. *Journal of Animal Science* 81:2869-2879.

Eun, J.-S., and K. A. Beauchemin. 2008. Relationship between enzymatic activities and in vitro degradation of alfalfa hay and corn silage. *Animal Feed Science and Technology* 145:53-67.

Fay, J. P., K. J. Cheng, M. R. Hanna, R. E. Howarth, and J. W. Costerton. 1980. In vitro digestion of bloat-safe and bloat-causing legumes by rumen microorganisms: Gas and foam production. *Journal of Dairy Science* 63:1273-1281.

Fellner, V., F. D. Sauer, and J. K. G. Kramer. 1997. Effect of nigericin, monensin, and tetronasin on biohydrogenation in continuous flow-through ruminal fermenters. *Journal of Dairy Science* 80:921-928.

Foote, L. E., R. E. Girouard, Jr., J. E. Johnston, J. Rainey, P. B. Brown, and W. H. Willis. 1968. Poloxalene for prevention of legume bloat. *Journal of Dairy Science* 51:584-590.

Fuller, R. 1989. Probiotics in man and animals. *Journal of Applied Bacteriology* 66:365-378.

Goodrich, R. D., J. E. Garrett, D. R. Ghast, M. A. Kirich, D. A. Larson, and J. C. Meiske. 1984. Influence of monensin on the performance of cattle. *Journal of Animal Science* 58:1484-1498.

Gruber, S. L., J. D. Tatum, T. E. Engle, M. A. Mitchell, S. B. Laudert, A. L. Schroeder, and W. J. Platter. 2007. Effects of ractopamine supplementation on growth performance and carcass characteristics of feedlot steers of differing biological type. *Journal of Animal Science* 85:1809-1815.

Guan, H., K. M. Wittenberg, K. H. Ominski, and D. O. Krause. 2006. Efficacy of ionophores in cattle diets for mitigation of enteric methane. *Journal of Animal Science* 84:1896-1906.

Guiroy, P. J., D. G. Fox, L. O. Tedeschi, M. J. Baker, and M. D. Cravey. 2001. Predicting individual feed requirements of cattle fed in groups. *Journal of Animal Science* 79:1983-1995.

Guiroy, P. J., L. O. Tedeschi, D. G. Fox, and J. P. Hutcheson. 2002. The effects of implant strategy on finished body weight of beef cattle. *Journal of Animal Science* 80:1791-1800.

Helmer, L. G., E. E. Bartley, and R. M. Meyer. 1965. Bloat in cattle. IX. Effect of poloxalene, used to prevent legume bloat, on milk production, feed intake, health, reproduction, and rumen fermentation. *Journal of Dairy Science* 48:576-579.

Horn, G. W., P. A. Beck, J. G. Andrae, and S. I. Paisley. 2005. Designing supplements for stocker cattle grazing wheat pasture. *Journal of Animal Science* 83(Suppl.):E69-E78.

Hosford, A. D., J. E. Hergenreder, J. K. Kim, J. O. Baggerman, F. R. B. Ribeiro, M. J. Anderson, K. S. Spivey, W. Rounds, and B. J. Johnson. 2015. Effects of supplemental lysine and methionine with zilpaterol hydrochloride on feedlot performance, carcass merit, and skeletal muscle fiber characteristics in finishing feedlot cattle. *Journal of Animal Science* 93:4532-4544.

Howarth, R. E. 1975. A review of bloat in cattle. *The Canadian Veterinary Journal* 16:281-294.

Huber, T. L. 1976. Physiological effects of acidosis on feedlot cattle. *Journal of Animal Science* 43:902-909.

Imwalle, D. B., D. L. Fernandez, and K. K. Schillo. 2002. Melengestrol acetate blocks the preovulatory surge of luteinizing hormone, the expression of behavioral estrus, and ovulation in beef heifers. *Journal of Animal Science* 80:1280-1284.

Katz, M. P., T. G. Nagaraja, and L. R. Fina. 1986. Ruminal changes in monensin- and lasalocid-fed cattle grazing bloat-provocative alfalfa pasture. *Journal of Animal Science* 63:1246-1257.

Kellermeier, J. D., A. W. Tittor, J. C. Brooks, M. L. Galyean, D. A. Yates, J. P. Hutcheson, W. T. Nichols, M. N. Streeter, B. J. Johnson, and M. F. Fuller. 2009. Effects of zilpaterol hydrochloride with or without an estrogen-trenbolone acetate terminal implant on carcass traits, retail cutout, tenderness, and muscle fiber diameter in finishing steers. *Journal of Animal Science* 87:3702-3711.

Krause, D. O., and J. B. Russell. 1996. An rRNA approach for assessing the role of obligate amino acid-fermenting bacteria in ruminal amino acid deamination. *Applied and Environmental Microbiology* 62:815-821.

Krehbiel, C. R., S. R. Rust, G. Zhang, and S. E. Gilliland. 2003. Bacterial direct-fed microbials in ruminant diets: Performance response and mode of action. *Journal of Animal Science* 81(Suppl. 2):E120-E132.

Kreikemeier, W. M., and T. L. Mader. 2004. Effects of growth-promoting agents and season on yearling feedlot heifer performance. *Journal of Animal Science* 82:2481-2488.

Lean, I. J., J. M. Thompson, and F. R. Dunshea. 2014. A meta-analysis of zilpaterol and ractopamine effects on feedlot prerformance, carcass traits and shear strength of meat in cattle. *PLOS ONE* 9(12):e115904.

Leheska, J. M., J. L. Montgomery, C. R. Krehbiel, D. A. Yates, J. P. Hutcheson, W. T. Nichols, M. Streeter, J. R. Blanton, Jr., and M. F. Miller. 2009. Dietary zilpaterol hydrochloride. II. Carcass composition and meat palatability of beef cattle. *Journal of Animal Science* 87:1384-1393.

Li, Y.-G., G. Tanner, and P. Larkin. 1996. The DMACA-HCl protocol and the threshold proanthocyanidin content for bloat safety in forage legumes. *Journal of the Science of Food and Agriculture* 70:89-101.

Lia, Y., M. He, C. Li, R. Forster, K. A. Beauchemin, and W. Yang. 2012. Effects of wheat dried distillers' grains with solubles and cinnamaldehyde on in vitro fermentation and protein degradation using the Rusitec technique. *Archives of Animal Nutrition* 66:131-148.

Lynch, G. S., and J. G. Ryall. 2008. Role of β-adrenoceptor signaling in skeletal muscle: Implications for muscle wasting and disease. *Physiological Reviews* 88:729-767.

Mader, T. L., and K. F. Lechtenberg. 2000. Growth-promoting systems for heifer calves and yearlings finished in the feedlot. *Journal of Animal Science* 78:2485-2496.

Majak, W., J. W. Hall, and W. P. McCaughey. 1995. Pasture management strategies for reducing the risk of legume bloat in cattle. *Journal of Animal Science* 73:1493-1498.

McAllister, T. A., K. A. Beauchemin, A. Y. Alazzeh, J. Baah, R. M. Teather, and K. Stanford. 2011. Review: The use of direct fed microbials to mitigate pathogens and enhance production in cattle. *Canadian Journal of Animal Science* 91:193-211.

McGuffey, R. K., L. F. Richardson, and J. I. D. Wilkinson. 2001. Ionophores for dairy cattle: Current status and future outlook. *Journal of Dairy Science* 84(Suppl.):E194-E203.

Meale, S. J., K. A. Beauchemin, A. N. Hristov, A. V. Chaves, and T. A. McAllister. 2014. Board-invited review: Opportunities and challenges in using exogenous enzymes to improve ruminant production. *Journal of Animal Science* 92:427-442.

Mersmann, H. J. 1998. Overview of the effects of beta-adrenergic receptor agonists on animal growth including mechanism of action. *Journal of Animal Science* 76:160-172.

Meyer, N. F., G. E. Erickson, T. J. Klopfenstein, M. A. Greenquist, M. K. Luebbe, P. Williams, and M. A. Engstrom. 2009. Effects of essential oils, tylosin, and monensin on finishing steer performance, carcass characteristics, liver abscesses, ruminal fermentation, and digestibility. *Journal of Animal Science* 87:2346-2354.

Meyer, N. F., G. E. Erickson, T. J. Klopfenstein, J. R. Benton, M. K. Luebbe, and S. B. Laudert. 2013. Effects of monensin and tylosin in finishing diets containing corn wet distillers grains with solubles with differing corn processing methods. *Journal of Animal Science* 91:2219-2228.

Mills, S. E. 2002. Implications of feedback regulation of beta-adrenergic signaling. *Journal of Animal Science* 80:E30-E35.

Min, B. R., W. E. Pinchak, J. D. Fulford, and R. Puchala. 2005a. Effect of feed additives on in vitro and in vivo rumen characteristics and frothy bloat dynamics in steers grazing wheat pasture. *Animal Feed Science and Technology* 123-124:615-629.

Min, B. R., W. E. Pinchak, J. D. Fulford, and R. Puchala. 2005b. Wheat pasture bloat dynamics, in vitro ruminal gas production, and potential bloat mitigation with condensed tannins. *Journal of Animal Science* 83:1322-1331.

Min, B. R., W. E. Pinchak, R. C. Anderson, J. D. Fulford, and R. Puchala. 2006. Effects of condensed tannins supplementation level on weight gain and in vitro and in vivo bloat precursors in steers grazing winter wheat. *Journal of Animal Science* 84:2546-2554.

Montgomery, J. L., C. R. Krehbiel, J. J. Cranston, D. A. Yates, J. P. Hutcheson, W. T. Nichols, M. N. Streeter, D. T. Bechtol, E. Johnson, T. TerHune, and T. H. Montgomery. 2009. Dietary zilpaterol hydrochloride. I. Feedlot performance and carcass traits of steers and heifers. *Journal of Animal Science* 87:1374-1383.

Montgomery, T. H., P. F. Dew, and M. S. Brown. 2001. Optimizing carcass value and the use of anabolic implants in beef cattle. *Journal of Animal Science* 79:E296-E306.

Moody, D. E., D. L. Hancock, and D. B. Anderson. 2000. Phenethanolamine repartitioning agents. Pp. 65-96 in *Farm Animal Metabolism and Nutrition*, J. P. F. D'Mello, ed. Wallingford, Oxon, UK: CAB International.

NRC (National Research Council). 1984. *Nutrient Requirements of Beef Cattle*, 6th Rev. Ed. Washington, DC: National Academy Press.

NRC. 1994. *Metabolic Modifiers: Effects on the Nutrient Requirements of Food-Producing Animals.* Washington, DC: National Academy Press.

NRC. 1996. *Nutrient Requirements of Beef Cattle*, 7th Rev. Ed. Washington, DC: National Academy Press.

NRC. 2000. *Nutrient Requirements of Beef Cattle: Update 2000*, 7th Rev. Ed. Washington, DC: National Academy Press.

NRC. 2012. *Nutrient Requirements of Swine*, 11th Rev. Ed. Washington, DC: The National Academies Press.

Okeke, G. C., J. G. Buchanan-Smith, and W. L. Grovum. 1983. Effects of buffers on ruminal rate of passage and degradation of soybean meal in steers. *Journal of Animal Science* 56:1393-1399.

Oltjen, J. W., A. C. Bywater, R. L. Baldwin, and W. N. Garrett. 1986. Development of a dynamic model of beef cattle growth and composition. *Journal of Animal Science* 62:86-97.

Owens, F. N., D. S. Secrist, W. J. Hill, and D. R. Gill. 1998. Acidosis in cattle: A review. *Journal of Animal Science* 76:275-286.

Pampusch, M. S., M. E. White, M. R. Hathaway, T. J. Baxa, K. Y. Chung, S. L. Parr, B. J. Johnson, W. J. Weber, and W. R. Dayton. 2008. Effects of implants of trenbolone acetate, estradiol, or both, on muscle insulin-like growth factor-I, insulin-like growth factor-I receptor, estrogen receptor-α, and androgen receptor messenger ribonucleic acid levels in feedlot steers. *Journal of Animal Science* 86:3418-3423.

Parr, S. L., K. Y. Chung, M. L. Galyean, J. P. Hutcheson, N. DiLorenzo, K. E. Hales, M. L. May, M. J. Quinn, D. R. Smith, and B. J. Johnson. 2011. Performance of finishing beef steers in response to anabolic implant and zilpaterol hydrochloride supplementation. *Journal of Animal Science* 89:560-570.

Patra, A. K. 2010. Meta-analyses of effects of phytochemicals on digestibility and rumen fermentation characteristics associated with methanogenesis. *Journal of the Science of Food and Agriculture* 90:2700-2708.

Patra, A. K., and J. Saxena. 2009. Dietary phytochemicals as rumen modifiers: A review of the effects on microbial populations. *Antonie van Leeuwenjoek Journal of Microbiology* 96:363-375.

Patra, A. K., and J. Saxena. 2010. A new perspective on the use of plant secondary metabolites to inhibit methanogenesis in the rumen. *Phytochemistry* 71:1198 1222.

Patterson, D. J., G. H. Kiracofe, J. S. Stevenson, and L. R. Corah. 1989. Control of the bovine estrous cycle with melengestrol acetate (MGA): A review. *Journal of Animal Science* 67:1895-1906.

Peirce, S. B., L. D. Muller, and H. W. Harpster. 1983. Influence of sodium bicarbonate and magnesium oxide on digestion and metabolism in yearling beef steers abruptly changed from high forage to high energy diets. *Journal of Animal Science* 57:1561-1567.

Perrett, T., B. K. Wildman, G. K. Jim, A. R. Vogstad, R. K. Fenton, S. J. Hannon, O. C. Schunicht, S. M. Abutarbush, and C. W. Booker. 2008. Evaluation of the efficacy and cost-effectiveness of melengestrol acetate in feedlot heifer calves in western Canada. *Veterinary Therapeutics* 9:223-240.

Perry, T. C., D. G. Fox, and D. H. Beermann. 1991. Effect of an implant of trenbolone acetate and estradiol on growth, feed efficiency and carcass composition of Holstein and beef breed steers. *Journal of Animal Science* 69:4696-4702.

Preston, R. L. 1999. Hormone containing growth promoting implants in farmed livestock. *Advanced Drug Delivery Reviews* 38:123-138.

Pyatt, N. A., G. J. Vogel, J. W. Homm, R. L. Botts, and C. D. Bokenkroger. 2013a. Effects of ractopamine hydrochloride on performance and carcass characteristics in finishing heifers: 16-trial summary. *Journal of Animal Science* 91(Suppl. 2):692-693 (Abstract 742).

Pyatt, N. A., G. J. Vogel, J. W. Homm, R. L. Botts, and C. D. Bokenkroger. 2013b. Effects of ractopamine hydrochloride on performance and carcass characteristics in finishing steers: 32-trial summary. *Journal of Animal Science* 91(Suppl. 2):79 (Abstract).

Rathmann, R. J., B. C. Bernhard, R. S. Swingle, T. E. Lawrence, W. T. Nichols, D. A. Yates, J. P. Hutcheson, M. N. Streeter, J. C. Brooks, M. F. Miller, and B. J. Johnson. 2012. Effects of zilpaterol hydrochloride and days on the finishing diet on feedlot performance, carcass characteristics, and tenderness in beef heifers. *Journal of Animal Science* 90:3301-3311.

Reinhardt, C. 2007. Growth-promotant implants: Managing the tools. *Veterinary Clinics of North America: Food Animal Practice* 23:309-319.

Ricks, C. A., R. H. Dalrymple, P. K. Baker, and D. L. Ingle. 1984. Use of a β-agonist to alter fat and muscle deposition in steers. *Journal of Animal Science* 59:1247-1255.

Rogers, J. A., and C. L. Davis. 1982. Rumen volatile fatty acid production and nutrient utilization in steers fed a diet supplemented with sodium bicarbonate and monensin. *Journal of Dairy Science* 65:944-952.

Russell, J. B., and H. J. Strobel. 1989. Effect of ionophores on ruminal fermentation. *Applied and Environmental Microbiology* 55:1-6.

Russell, J. R., A. W. Young, and N. A. Jorgensen. 1980. Effect of sodium bicarbonate and limestone additions to high grain diets on feedlot performance and ruminal and fecal parameters in finishing steers. *Journal of Animal Science* 51:996-1002.

Schneider, B. A., J. D. Tatum, T. E. Engle, and T. C. Bryant. 2007. Effects of heifer finishing implants on beef carcass traits and longissimus tenderness. *Journal of Animal Science* 85:2019-2030.

Schroeder, A. L., D. M. Polser, S. B. Laudert, G. J. Vogel, T. Ripberger, and M. T. Van Koevering. 2004. The effect of Optaflexx™ on growth performance and carcass traits of steers and heifers. Pp. 65-81 in *Proceedings of the 19th Southwest Nutrition and Management Conference, February 26-27, Tempe, AZ.* Tucson: University of Arizona.

Scramlin, S. M., W. J. Platter, R. A. Gomez, W. T. Choat, F. K. McKeith, and J. Killefer. 2010. Comparative effects of ractopamine hydrochloride and zilpaterol hydrochloride on growth performance, carcass traits, and longissimus tenderness of finishing steers. *Journal of Animal Science* 88:1823-1829.

Sissom, E. K., C. D. Reinhardt, J. P. Hutcheson, W. T. Nichols, D. A. Yates, R. S. Swingle, and B. J. Johnson. 2007. Response to ractopamine-HCl in heifers is altered by implant strategy across days on feed. *Journal of Animal Science* 85:2125-2132.

Smith, S. B., D. K. Garcia, and D. B. Anderson. 1989. Elevation of a specific mRNA in longissimus muscle of steers fed ractopamine. *Journal of Animal Science* 67:3495-3502.

Spears, J. W. 1990. Ionophores and nutrient digestion and absorption in ruminants. *Journal of Nutrition* 120:632-638.

Stackhouse, K. R., C. A. Rotz, J. W. Oltjen, and F. M. Mitloehner. 2012. Growth-promoting technologies decrease the carbon footprint, ammonia emissions, and costs of California beef production systems. *Journal of Animal Science* 90:4656-4665.

Stackhouse-Lawson, K. R., M. S. Calvo, S. E. Place, T. L. Armitage, Y. Pan, Y. Zhao, and F. M. Mitloehner. 2013. Growth promoting technologies reduce greenhouse gas, alcohol, and ammonia emissions from feedlot cattle. *Journal of Animal Science* 91:5438-5447.

Stephens, T. P., G. H. Loneragan, L. M. Chichester, and M. M. Brashears. 2007. Prevalence and enumeration of *Escherichia coli* O157 in steers receiving various strains of *Lactobacillus*-based direct-fed microbials. *Journal of Food Protection* 70:1252-1255.

Stock, R. A., S. B. Laudert, W. W. Stroup, E. M. Larson, J. C. Parrott, and R. A. Britton. 1995. Effect of monensin and monensin and tylosin combination on feed intake variation of feedlot steers. *Journal of Animal Science* 73:36-44.

Tabe, E. S., J. Oloya, D. K. Doetkott, M. L. Bauer, P. S. Gibbs, and M. L. Khaitsa. 2008. Comparative effect of direct-fed microbials on fecal shedding of *Escherichia coli* O157:H7 and *Salmonella* in naturally infected feedlot cattle. *Journal of Food Protection* 71:539-544.

Tedeschi, L. O., D. G. Fox, and T. P. Tylutki. 2003. Potential environmental benefits of ionophores in ruminant diets. *Journal of Environmental Quality* 32:1591-1602.

Tedeschi, L. O., T. R. Callaway, J. P. Muir, and R. C. Anderson. 2011. Potential environmental benefits of feed additives and other strategies for ruminant production. *Revista Brasileira de Zootecnia* 40 (Suppl.):291-309.

Trenkle, A. 1990. Impact of implant strategies on performance and carcass merit of feedlot cattle. P. 13 in *Proceedings of the 1990 Southwest Nutrition and Management Conference, Tempe, AZ.* Tucson: University of Arizona.

TTU (Texas Tech University). 2014. Trade Names, Composition, and Approved Uses of Currently Available Implants for Beef Cattle, 2004. Intervet and Texas Tech University North American TBA Implant Database. Available online at http://www.depts.ttu.edu/afs/implantDB/dbhome/Implant%20Chart_Intervet_2004.pdf. Accessed on December 1, 2014.

Tylutki, T. P., D. G. Fox, and R. G. Anrique. 1994. Predicting net energy and protein requirements for growth of implanted and nonimplanted heifers and steers and nonimplanted bulls varying in body size. *Journal of Animal Science* 72:1806-1813.

Vasconcelos, J. T., and M. L. Galyean. 2007. Nutritional recommendations of feedlot consulting nutritionists: The 2007 Texas Tech University survey. *Journal of Animal Science* 85:2772-2781.

Vasconcelos, J. T., and M. L. Galyean. 2008. ASAS Centennial Paper: Contributions in the *Journal of Animal Science* to understanding cattle metabolic and digestive diseases. *Journal of Animal Science* 86:1711-1721.

Vasconcelos, J. T., R. J. Rathmann, R. R. Reuter, J. Leibovich, J. P. McMeniman, K. E. Hales, T. L. Covery, M. F. Miller, W. T. Nichols, and M. L. Galyean. 2008. Effects of duration of zilpaterol hydrochloride feeding and days on the finishing diet on feedlot cattle performance and carcass traits. *Journal of Animal Science* 86:2005-2015.

Wagner, J. J., N. E. Davis, and C. D. Reinhardt. 2007. A meta-analysis evaluation of feeding melengestrol acetate to feedlot heifers implanted with estradiol, trenbolone acetate, or the combination of estradiol and trenbolone acetate. *The Professional Animal Scientist* 23:625-631.

Wagner, J. J., T. E. Engle, and C. R. Belknap. 2013. Meta-analysis examining the effects of *Saccharomyces cerevisiae* fermentation products on feedlot performance and carcass traits. *Journal of Animal Science* 91(Suppl. 2):79 (Abstract O241).

Walker, C. E., and J. S. Drouillard. 2010. Effects of ractopamine hydrochloride are not confined to mammalian tissue: Evidence for direct effects of ractopamine hydrochloride supplementation on fermentation by ruminal microorganisms. *Journal of Animal Science* 88:697-706.

Walker, D. K., E. C. Titgemeyer, J. S. Drouillard, E. R. Loe, B. E. Depenbusch, and A. S. Webb. 2006. Effect of ractopamine and protein source on growth performance and carcass characteristics of feedlot heifers. *Journal of Animal Science* 84:2795-2800.

Wang, Y., W. Majak, and T. A. McAllister. 2012. Frothy bloat in ruminants: Cause, occurrence, and mitigation strategies. *Animal Feed Science and Technology* 172:103-114.

Webb, A. S., R. W. Rogers, and B. J. Rude. 2002. Review: Androgenic, estrogenic, and combination implants: Production and Meat Quality in Beef. *The Professional Animal Scientist* 18:103-106.

Winterholler, S. J., G. L. Parsons, C. D. Reinhardt, J. P. Hutcheson, W. T. Nichols, D. A. Yates, R. S. Swingle, and B. J. Johnson. 2007. Response to ractopamine-hydrogen chloride is similar in yearling steers across days on feed. *Journal of Animal Science* 85:413-419.

Worley, R. R., J. A. Paterson, K. P. Coffey, D. K. Bowman, and J. E. Williams. 1986. The effects of corn silage dry matter content and sodium bicarbonate addition on nutrient digestion and growth by lambs and calves. *Journal of Animal Science* 63:1728-1736.

Yang, C.-M. J., and J. B. Russell. 1993. The effect of monensin supplementation on ruminal ammonia accumulation in vivo and the numbers of amino acid-fermenting bacteria. *Journal of Animal Science* 71:3470-3476.

Yang, W. Z., B. N. Ametaj, C. Benchaar, and K. A. Beauchemin. 2010a. Cinnamaldehyde in feedlot cattle diets: Intake, growth performance, carcass characteristics, and blood metabolites. *Journal of Animal Science* 88:1082-1092.

Yang, W. Z., B. N. Ametaj, C. Benchaar, and K. A. Beauchemin. 2010b. Dose response to cinnamaldehyde supplementation in growing beef heifers: Ruminal and intestinal digestion. *Journal of Animal Science* 88:680-688.

Yang, W. Z., C. Benchaar, B. N. Ametaj, and K. A. Beauchemin. 2010c. Dose response to eugenol supplementation in growing beef cattle: Ruminal fermentation and intestinal digestion. *Animal Feed Science and Technology* 158:57-64.

Younts-Dahl, S. M., M. L. Galyean, G. H. Loneragan, N. A. Elam, and M. M. Brashears. 2004. Dietary supplementation with *Lactobacillus*- and *Propionibacterium*-based direct-fed microbials and prevalence of *Escherichia coli* O157 in beef feedlot cattle and on hides at harvest. *Journal of Food Protection* 67:889-893.

Zimbelman, R. G., and L. W. Smith. 1966. Control of ovulation in cattle with melengestrol acetate. *Journal of Reproduction and Fertility* 11:193-201.

Zinn, R. A. 1991. Comparative feeding value of steam-flaked corn and sorghum in finishing diets supplemented with or without sodium bicarbonate. *Journal of Animal Science* 69:905-916.

Zinn, R. A., and J. L. Borques. 1993. Influence of sodium bicarbonate and monensin on utilization of a fat-supplemented, high-energy growing-finishing diet by feedlot steers. *Journal of Animal Science* 71:18-25.

15

Effects of Stress on Beef Cattle Nutrient Requirements

OVERVIEW

Definition and Types of Stressors

Stress in beef cattle has been defined as an adaptive change or nonspecific response initiated by an internal or external environmental stimulus (Frazer et al., 1975; Selye, 1976; Breazile, 1987). Stress is not inherently harmful to an animal, and in fact, it could be beneficial (Selye, 1976). For example, the beneficial effects of short-term stress are observed in the fight-or-flight stress response which is the fundamental survival mechanism. Eustress is brought on by stimuli that initiate responses beneficial to an animal's well-being and functions to maintain homeostasis. Neutral stress produces responses that are neither harmful nor beneficial to the animal. In contrast, distress can alter the steady state of the body by causing harmful responses that can challenge adaptive physiological processes or evoke pathological changes interfering with the health, comfort, or well-being of the animal. Nutrition and stress are interactive and consequential, in that stress can accentuate or aggravate nutritional deficiencies, and nutritional deficiencies can induce a stress response. The major stressors observed in beef cattle are weaning, crowding, transportation, feed and water deprivation during marketing or drought, exposure to an array of pathogens, introduction to new feedstuffs, and adaptation to unfamiliar surroundings. Other stressors encountered by cattle include extreme weather changes and procedures associated with processing (e.g., castration, dehorning, vaccination, deworming). All these stressors can influence nutrient requirements of beef cattle, and because nutrition and stress are interrelated, they should be considered as continuous processes. Management of stress in cattle has two major components: (1) management of the cause of stress and (2) management of the effects of stress as determined by the quantified changes observed in animals. Both could involve nutrition.

The weaning stage in mammals naturally occurs during the transition to adulthood, but stress associated with weaning is difficult to eliminate. Preweaning and preconditioning management strategies have been used to decrease weaning stress (Thrift and Thrift, 2011). Although no long-term negative effects of weaning have been detected, the distress of separating a calf from its dam could initiate an extended period of vocalization (Stookey et al., 1997). Excessive vocalization in response to distress could increase susceptibility of the respiratory tract to infection by pathogens, potentially leading to bovine respiratory disease (BRD; Loerch and Fluharty, 1999).

Calves are often subjected to the distress of weaning coupled with additional stressors such as transportation and marketing. During the marketing process, when the animal is deprived of feed and water, ruminal fermentation processes and capacity are decreased and remain so for a few days after refeeding (Cole and Hutcheson, 1985a; Fluharty et al., 1994). Changes associated with the stressors of feed deprivation and transport include increased ruminal pH, serum osmolality, and serum glucose and urea nitrogen concentrations. Once animals are refed, these response variables return to predeprivation levels within 24 h (Cole and Hutcheson, 1985b, 1987a). The number of ruminal protozoa (Galyean et al., 1981; Cole and Hutcheson, 1981; Fluharty et al., 1994, 1996) and bacteria (Cole and Hutcheson, 1981; Galyean et al., 1981) is lower in steers subjected to fasting and transit stress than in animals not fasted and transported, and the number increases more slowly when fasting occurs in conjunction with transit than when fasting is the only stressor. Baldwin (1967) suggested that the number of ruminal protozoa and bacteria decreases sharply following transportation stress. Fluharty et al. (1994) conducted an experiment with newly weaned, ruminally fistulated steers to determine the effects of energy density and protein source in receiving diets on ruminal pH, in situ dry matter (DM) disappearance, and concentrations of ruminal bacteria and protozoa. On the day that calves arrived at the feedlot, dry matter intake (DMI) was 62% of the DMI on day 7 after arrival; however, in situ

DM disappearance 48 hours following arrival at the feedlot was not different compared with values measured before weaning. In contrast with previous experiments, no differences were observed in the concentration of total bacteria or cellulolytic bacteria as a result of feed and water deprivation. This finding suggests that the ruminal microbial population was not inhibited in its ability to digest available substrate immediately following weaning, fasting, and trucking (Fluharty et al., 1994).

Fluharty et al. (1996) conducted two experiments to determine the effects of the duration of feed and water deprivation on ruminal microbes and ruminal characteristics of newly weaned and feedlot-adapted calves. With newly weaned calves, DMI, ruminal volume, and weight of ruminal contents decreased as duration of feed and water deprivation increased; however, on day 4 after arrival, there were no longer any differences in these variables. In both experiments, 48 and 72 h of feed and water deprivation decreased protozoal numbers on days 0 and 4 compared with the 0-h treatment group, but there were no decreases in the ruminal numbers of cellulolytic or total bacteria as a result of duration of feed and water deprivation. These results indicate that ruminal volume, DM, total weight of ruminal contents, and protozoal numbers decrease as the duration of the fasting period increases, and this decrease is related to a decrease in DMI. Decreased performance of stressed calves seems to be the result of low DMI, not decreased bacterial numbers and digestive capacity (Loerch and Fluharty, 1999). Therefore, reestablishing DMI following marketing and transport seems to be of greatest importance for the subsequent health and nutrition of stressed calves.

The cause of low feed intake in market-transport-stressed calves is not entirely clear. Using sheep as a model, Cole (1991a) noted that exchanging about 50% of ruminal contents between fed and fasted animals did not affect subsequent feed intake. Feed and water deprivation also affected animal metabolic and hormonal (insulin, growth hormone) responses to glucose or propionate loads (Cole et al., 1993), and postprandial shifts in body water compartments (Cole, 2000). Nonetheless, these effects seemed to be transitory and not clearly related to long-term feed intake depression.

Effects of Stressors on Inflammation and Growth

Calves are often removed from their dams when they are placed on the truck destined for marketing, a receiving lot, a stocker/backgrounding facility, or a feedlot. Once in their new environment, calves are typically supplied with unfamiliar water and feed sources, exposed to the processing regimen, and commingled with a new group of calves. In addition, calves can be marketed and transported during a time when the weather is not ideal, exacerbating an already stressful situation. When combined with the stress of weaning and marketing, exposure to respiratory pathogens can lead to a high incidence of BRD (Blecha et al., 1984). Pathogens responsible for BRD infection include both viruses and bacteria. Common viral pathogens include bovine viral diarrhea virus (BVDV), bovine herpes virus-1 (BHV-1), bovine respiratory syncytial virus, and bovine coronavirus. Although some of these pathogens, such as BVDV, might not initially seem to have direct effects on the respiratory system, they cause respiratory problems by affecting the whole body's immune status (Ellis, 2001). The decrease in immune status then allows bacterial infections to gain hold in the respiratory tract. In addition, more than one virus is usually isolated from cattle diagnosed with BRD, once again indicating the complexity of the disease (Duff and Galyean, 2007).

Bacterial pathogens usually include a combination of *Mannheimia* (*Pasteurella*) *haemolytica*, *Pasteurella multocida*, and *Histophilus somni* (*Haemophilus somnus*; Ellis, 2001; Apley, 2006). These gram-negative bacteria all produce lipopolysaccharides (LPS) and/or lipooligosaccharides as pathogenic factors (Corbeil, 2007; Dabo et al., 2007; Rice et al., 2007). Of the major bacterial pathogens, *M. haemolytica* is generally considered to be the most prevalent and pathogenic. Interestingly, *M. haemolytica* is naturally found in the upper respiratory tract, but it acts as an opportunistic pathogen. It does not become pathogenic unless the immune system is compromised, such as during times of stress or infection. *M. haemolytica* produces not only an LPS virulence factor, but also a ruminant-specific leukotoxin that targets ruminant leukocytes, which greatly enhances its virulence (Zecchinon et al., 2005).

Another less-known bacterial pathogen is *Mycoplasma bovis*. In recent histopathological evaluations of cattle diagnosed with BRD in Canadian feedlots, *M. bovis* was second only to *M. haemolytica* in prevalence, and actually was identified more often than *M. haemolytica* in cattle that were classified as chronically infected with BRD (Booker et al., 2008). Viral and bacterial pathogens involved in the BRD complex work in a synergistic fashion, with detrimental results in the host animal. It is hypothesized that management or environmental stress factors along with viral infection alter the upper respiratory tract epithelium, thereby allowing bacterial pathogens to colonize (Lafleur et al., 2001; Jeyaseelan et al., 2002).

Following breach of the animal's defense mechanisms by viral and bacterial pathogens, trauma, or stress, the innate immune system initiates a rapid and potentially systemic acute inflammatory reaction. This acute-phase response usually occurs within the first 48 h of infection or injury. The acute-phase response is initiated by the production of proinflammatory cytokines (e.g., tumor-necrosis factor-α, interleukin-1, interleukin-6) by phagocytic immune tissues and leads to fever, anorexia, muscle catabolism, coagulation, increased glucocorticoid hormones, changes in liver protein synthesis, and leukocytosis (Gruys et al., 2005; Cray et al., 2009; Eckersall and Bell, 2010). Although such a response is valuable to the health of the animal, it can have severe growth consequences, especially in the short term. In addition, sustained stimulation of the inflammatory response impairs normal growth and development and can limit productivity by preventing an animal from attaining its full genetic potential for growth and carcass quality.

The acute-phase response provides an early nonspecific defense against pathogen challenge through a dynamic process that involves both systemic and metabolic changes in the body (Peterson, 2004). For many years, it has been known that the sedimentation rate in stabilized blood from ill patients is greater than in healthy individuals. This is caused by plasma proteins called acute-phase proteins. Acute-phase proteins are produced in the liver as part of an early defense mechanism in response to cellular injury (Eckersall and Conner, 1988), which can result from infection or inflammation (Saini et al., 1998). During the acute-phase response, pro-inflammatory cytokines promote skeletal muscle catabolism to supply amino acids and energy substrates for immune tissues. Further, during this early immune response, the liver changes its metabolic priorities to the production of acute-phase proteins for use in host defense. Although the long-term consequences of an acute-phase response are not understood, shifts in systemic metabolism might explain the detrimental effects on performance and carcass traits commonly associated with BRD in feedlot calves (Cole et al., 1984b; Schneider et al., 2009; Garcia et al., 2010; Holland et al., 2010; Brooks et al., 2011). Inflammation decreases DMI, average daily gain (ADG), and gain:feed in feedlot calves, decreasing growth rate and increasing days on feed, which results in economic losses during the feeding period. Additional data to help explain the physiological effects of inflammation on cattle growth and carcass merit will likely benefit the beef cattle industry.

Body Shrink and Shrink Recovery

Shrink or body weight (BW) loss during transport is a physiological process associated with loss of rumen fill, urine, and feces (fill shrink) and of fluid (both extracellular and intracellular) and tissue (tissue shrink; Coffey et al., 2001). Fill is typically recovered in a short period of time after feed and water intake returns to normal; however, cattle take longer to recover from tissue shrink than from fill shrink (Self and Gay, 1972). Several factors can contribute to shrink, including type of feed consumed, gathering and sorting cattle, weaning, length of time of food and water deprivation, length of haul, excessive handling, ambient temperature, temperament of the animal, preshipping diet and preconditioning before transportation (Phillips et al., 1985; Cole et al., 1986b; Pritchard and Mendez, 1990; Grandin, 1997; Coffey et al., 2001). The stress of transportation can produce several negative consequences such as a decrease in BW and DMI, as well as impairment of the immune system and increased morbidity and mortality (Grandin, 1997; Coffey et al., 2001). Relationships between the distances traveled with health and performance measures of feedlot cattle have been quantified (Cole and Hutcheson, 1988; Cernicchiaro et al., 2012). Camp et al. (1983) showed a positive correlation between average transit BW loss and incidence of BRD. Cole (1991b) noted that the incidence of BRD in loads of calves was positively related to average shrink between arrival at the auction barn and arrival at the feedlot. Nonetheless, the relationship between shrink of individual calves hauled approximately 24 h and incidence of BRD was quadratic, with highest morbidity in calves that shrink more than 8% or less than 3%. These quadratic results are most likely a result of the nutritional condition of the individual calf on arrival at the auction barn (i.e., individual calves with little feed in the gut at the time of arrival at the auction barn shrink little during marketing and transport, but are highly prone to becoming sick).

Research to determine whether preshipment diet and management can influence shrink has produced variable results (Hutcheson et al., 1984; Cole and Hutcheson, 1985a, 1987a; Phillips et al., 1985). Phillips et al. (1985) reported that steers fed hay had greater BW loss and fecal excretions during both 13 and 46 h of trucking than steers fed a 50% concentrate diet. In two experiments, Hutcheson et al. (1984) fed steers diets of hay or 55% concentrate before shipment from east Tennessee to the Texas Panhandle. In the first experiment, little difference in shrink was observed as a result of preshipment diet. In the second experiment, steers fed hay had less shrink than those fed the 55% concentrate diet. In contrast, steers fed the 55% concentrate diet had a lower percentage of morbidity in the second experiment than those steers fed hay before shipment. Results from the Hutcheson et al. (1984) studies are difficult to interpret because the concentrate diet contained an antibiotic, whereas steers fed hay received no antibiotic. Cole and Hutcheson (1985a, 1987a) fed diets varying in roughage concentration or fed different levels of a 65% concentrate diet before fasting. They reported that cattle fed a 65% concentrate diet tended to shrink less than those cattle fed a high-roughage diet. In addition, feeding a 65% concentrate diet ad libitum resulted in greater shrink than feeding restricted levels of the 65% concentrate diet. Although cattle on lower-quality forage shrink less than cattle on lush green grass during transport, it is questionable whether feeding concentrate before shipment will decrease shrink compared with cattle fed roughage because of variable results.

Coffey et al. (2001) suggested that providing key nutrients rather than specific types of feed can affect shrink and decrease the overall effect of fasting and transportation. Excess losses of body water and energy result from excess losses of urine and feces during transport. The application of electrolyte solutions to minimize transport stress in cattle has been investigated (Schaefer et al., 1997; Parker et al., 2003). There is a tendency for increases in extracellular fluid, carcass weight, and BW of cattle when electrolyte solutions are fed vs. when no electrolyte solutions are offered. The effects of the electrolyte solutions fed were to replenish lost total body water in the animals involved (Schaefer et al., 1997; Parker et al., 2003). Similar effects have been observed when cattle were offered water post-transport (Wythes et al., 1980, 1983). Cole et al. (1986b) showed that total nitrogen losses during transportation were highly correlated with nonevaporative water losses ($r = 0.95$; $P < 0.001$); therefore, protein is also an important consideration. Hutcheson et al. (1984) reported a positive response to supplemental potassium (K) in the receiving diet and recommended that K concentrations of diets for receiving calves be increased 20% to offset K loss during transport. Hutcheson et al. (1984) suggested that 1.2 to 1.4% K in the diet for 2 weeks is the optimal concentration for newly arrived, stressed calves. Additional K might not increase gain response if cattle shrink 2 to 4%, but with shrinkage of 7% or more, a significant effect could be observed with added K by allowing the electrolyte and water balance to return to normal. When K is added as potassium chloride, care should be taken to limit salt to 0.25% of dietary DM so as not to increase chloride intake. In addition, Chang and Mowat (1992) reported a response to supplemental chromium (Cr) in receiving diets, indicating that the body might be eliminating this element during transportation. Further research is needed to determine which nutrients are lost during transport and whether providing those nutrients to cattle before shipping would decrease the negative effects of shrink.

Preconditioning

Preconditioning programs are designed to decrease the stress associated with weaning, enhance the immune systems of calves, and ease the transition to the next phase of production. In general, preconditioning programs ensure that the animals have been weaned for 30 to 45 days, vaccinated (with clostridial and viral vaccines), treated with an anthelmintic, castrated, dehorned, and accustomed to feed bunks and water troughs before being transported (Duff and Galyean, 2007; Thrift and Thrift, 2011). Preconditioning periods less than 30 d generally do not allow calves time to produce enough weight gain to offset fixed costs, immunization might not be complete, and calves might not have fully recovered from the stress of weaning (Thrift and Thrift, 2011). Therefore, the optimal period for preconditioning seems to be about 45 d (TAMU, 2005). Most feedlot producers believe that preconditioning cattle is beneficial in decreasing morbidity and mortality in calves weighing less than 318 kg (USDA-APHIS, 2000a; Duff and Galyean, 2007); however, only 32.4% of all feedlots surveyed received information about the previous history of the calves "always or most of the time" (USDA-APHIS, 2000b; Duff and Galyean, 2007). According to Cravey (1996), preconditioned calves were more profitable in the feedlot, primarily because of greater feedlot daily gain, improved feed conversion, decreased medicine cost, and decreased mortality compared with non-preconditioned calves.

In a review of preconditioning studies, Cole (1985) observed that preconditioning decreased feedlot morbidity by 6% and mortality by 0.7%; however, feed conversion in the feedlot was actually negatively affected by preconditioning. Arthington et al. (2013) reported that delivery of a commercial vaccine to overtly healthy, weaned beef calves resulted in the activation of the acute-phase reaction and decreased ADG and feed efficiency, which may explain this response. Step et al. (2008) reported that calves weaned and held on the ranch of origin 45 d before shipment had lower BRD rates, regardless of vaccination status at weaning, than calves assembled at auction markets or transported from the same ranch at weaning. Nutrition, including deficiencies caused by previous management and low DMI by stressed calves, can also affect the susceptibility of cattle to BRD (Galyean et al., 1999). Therefore, allowing calves to go through the stress of the weaning process before transport is beneficial to beef cattle production. During the initial weaning process, long-stem hay and a concentrate supplement placed in feed bunks positioned at a right angle to the fence line will promote feed intake because calves will generally walk the outside fence of the pen and thereby locate the feed (Edwards, 2010; Thrift and Thrift, 2011). It has also been suggested that calves will learn to drink from a fountain or water trough quicker if the water is initially set to trickle, creating a sound that attracts calves to the water source (Sowell et al., 1999; Buhman et al., 2000; Thrift and Thrift, 2011).

The effects of early calf weaning (80 to 90 d of age) on subsequent measures of stress and performance have been determined (Arthington et al., 2005; Carroll et al., 2009). Arthington et al. (2005) reported that calves weaned at 89 d of age, kept on the ranch of origin, provided supplement (1% of BW), and grazed on annual and perennial pastures had decreased acute-phase protein concentrations and increased feed efficiency following transport and feedlot entry compared to normal-weaned (300 d of age) contemporaries. In addition, early-weaned calves had decreased concentrations of proinflammatory cytokines following an endotoxin challenge (Carroll et al., 2009).

Nutrient requirements of calves during preconditioning can be met by a variety of successful programs. After the initial weaning phase, calves may remain in drylot or be placed back on pasture. As the fall season approaches for calves weaned in the fall, availability of high-quality forage from most warm-season grasses decreases, especially after first frost (Mathis et al., 2008, 2009). Therefore, supplemental feed is required for calves being preconditioned on pasture to achieve gain conducive to promoting continued growth during the 30- to 45-d preconditioning period. Lalman (2004a,b) provided several examples of preconditioning supplements for calves grazing pastures or fed free-choice, high-quality grass hay in drylot. In addition, byproduct feeds such as corn dried distillers grains plus solubles can be used with hay as an effective preconditioning diet (Winterholler et al., 2009). The optimal marginal improvement in BW gain per unit of feed consumed was estimated at an intake of corn dried distillers grains plus solubles of approximately 1.4% of BW for both steers and heifers when fed with native tall grass prairie hay (Winterholler et al., 2009). Controlled experiments comparing effects of pasture-based preconditioning methods on calf performance and profitability through slaughter are limited. Preconditioning calves on ryegrass pastures resulted in greater ADG and decreased feed costs compared with preconditioning in a drylot (St. Louis et al., 2003); however, whether the improved performance on grass was a result of decreased morbidity or other factors was not determined. Mathis et al. (2009) determined that grazing calves on native rangelands at a higher rate of BW gain can better prepare calves to remain healthy after shipping. Nonetheless, if calves are sold after the preconditioning period, the increased supplemental feed input costs required to achieve the higher rate of BW gain on pasture might not be cost-effective relative to a lower-cost approach (Mathis et al., 2009).

Effects on Feed Intake

Whether one purchases preconditioned calves or market-sourced calves, newly received calves will have been subjected to stress and limited consumption of feed and water during transport. As the length of time and number of commingling events increases between leaving the home farm or ranch and arrival, there is an increased disease risk for lightweight, recently weaned calves. Objectives of the receiving health and nutrition program are to assist the calf in recovering from stress, optimize the immune response, and shorten the time to begin productive weight gain. These goals can be met by a variety of health and nutrition programs.

Dry matter intake by lightweight, stressed calves averages only 1.5% of BW during the first 2 weeks after arrival at a feeding facility (Hutcheson and Cole, 1986; Galyean and Hubbert, 1995). In a summary of 18 experiments involving transit-stressed calves, only 83.4% of morbid calves and 94.6% of healthy calves had consumed any feed by day 7 following arrival at the feedlot (Hutcheson and Cole, 1986). In addition, measured DMI of morbid calves was 58, 68, and 88% of healthy calves across days 1 to 7, 1 to 14, and 1 to 56, respectively (Table 15-1). Similarly, Sowell et al. (1999) recorded the frequency and timing of visits made to the feed bunk by newly received calves during receiving and growing periods. In the first of two experiments, 94% of calves identified as healthy and 87% of morbid calves visited the feed bunk on the day of arrival, but 100% of healthy calves and only 91% of morbid calves had visited the bunk by day 3. In the second experiment, only 13 and 10% of healthy and morbid calves, respectively, visited the feed bunk on day 1. Again, all healthy calves had visited the bunk by day 4, but only 76% of morbid calves were observed at the feed bunk. In both experiments, healthy calves had more overall feeding events per day and spent more time at the bunk daily than morbid animals, both during the first 4 d and throughout the 32-d experiment. In the first experiment, 52% of calves were identified as morbid, whereas 82% were classified as morbid in the second experiment. In both experiments, 80% of morbid calves were identified within 10 d of arrival.

TABLE 15-1 Dry Matter Intake of Newly Arrived Calves (% of BW)[a]

Age, d	Healthy (SD)[b]	Diseased (SD)[b]
0 to 7	1.55 (0.51)	0.90 (0.75)
0 to 14	1.90 (0.50)	1.43 (0.70)
0 to 28	2.71 (0.50)	1.84 (0.66)
0 to 56	3.03 (0.43)	2.68 (0.68)

[a]Adapted from Hutcheson and Cole (1986).
[b]SD = standard deviation.

It is evident that newly received, highly stressed calves consume less feed than healthy calves exposed to fewer stress factors. As such, current recommendations are to increase the density of nutrients in diets of stressed calves so that animal requirements for nutrients are met even when DMI is low (Table 15-2). It is unclear, in commercial settings, whether disease causes decreased intake, or decreased intake is responsible for disease incidence. After recovery, DMI can remain low or be similar to that of untreated animals. In a combined viral and bacterial challenge experiment, Burciaga-Robles et al. (2010) observed the largest decreases in DMI during and immediately after the challenge. Nevertheless, DMI continued to be slightly decreased in challenged steers compared with controls throughout the finishing period (112 d). In contrast, there is evidence that, after recovery, morbid animals experience compensatory gain compared with untreated animals. This compensation may be a result of recovering gastrointestinal fill or decreased competition for nutrients when cattle are moved from preconditioning pens to pasture (Montgomery et al., 2009) or are adapted to a finishing diet (McBeth et al., 2001; Holland et al., 2010).

Duff and Galyean (2007) concluded that, with the possible exception of K, the stresses of weaning, marketing, transport, and disease do not seem to increase the total nutrient requirements of calves. Because of low DMI, however, the concentrations of nutrients in the diet need to be increased to meet the nutrient requirements of the animals (Table 15-2).

DIETARY ENERGY

Energy deficiency in cattle can severely depress the immune system (Nockles, 1988); however, excess dietary energy can also have detrimental effects. Calves newly arrived at a feedlot and fed a high-energy diet (75% concentrate) experienced increased performance, but the incidence of disease was 57% compared with 47% when a 25% concentrate diet was fed (Preston and Kunkle, 1974; Preston and Smith, 1974). Supplementing high-energy diets with hay for 3 to 7

TABLE 15-2 Suggested Nutrient Concentrations for Stressed Calves (Dry Matter Basis)

				Daily Nutrient Intake for 250-kg Calf[a]	
Nutrient	Unit	Suggested Range	Unit/d	0-7 d	0-14 d
Dry matter	%	80.0-85.0	kg	3.88	4.75
Crude protein	%	12.5-14.5	kg	0.48-0.56	0.59-0.69
Net energy of maintenance	Mcal/kg	1.3-1.6	Mcal	5.04-6.21	6.18-7.60
Net energy of gain	Mcal/kg	0.8-0.9	Mcal	3.10-3.49	3.80-4.28
Calcium	%	0.6-0.8	g	23.0-31.0	28.5-38.0
Phosphorus	%	0.4-0.5	g	16.0-19.4	19.0-23.8
Potassium	%	1.2-1.4	g	47.0-54.3	57.0-66.5
Magnesium	%	0.2-0.3	g	8.0-11.6	9.5-14.3
Sodium	%	0.2-0.3	g	8.0-11.6	9.5-14.3
Copper	mg/kg	10.0-15.0	mg	38.8-58.2	47.5-71.3
Iron	mg/kg	100.0-200.0	mg	388.0-776.0	475.0-950.0
Manganese	mg/kg	40.0-70.0	mg	155.2-271.6	190.0-332.5
Zinc	mg/kg	75.0-100.0	mg	291.0-388.0	356.3-475.0
Cobalt	mg/kg	0.1-0.2	mg	0.39-0.78	0.48-0.95
Selenium	mg/kg	0.1-0.2	mg	0.39-0.78	0.48-0.95
Iodine	mg/kg	0.3-0.6	mg	1.16-2.33	1.43-2.85
Vitamin E	IU/d	400.0-500.0	IU	400.0-500.0	400.0-500.0

[a]Dry matter intake levels are based on 1.55% for days 0 through 7 and 1.90% for days 0 through 14 from Table 15-1. Cattle need ad libitum access to fresh, clean drinking water.

days can overcome the adverse health effects of the high-energy diet (Lofgreen et al., 1981; Lofgreen, 1983, 1988), and is a recommended management practice.

In many experiments, the effects of energy intake are confounded with changes in dietary ingredients, particularly roughage (Duff and Galyean, 2007). In summarizing the classical research on dietary preferences of lightweight, stressed cattle, Lofgreen (1983) noted that such cattle have: (1) an abnormally low DMI relative to BW; and (2) a preference for and greater consumption of a high-concentrate than a high-roughage diet. Stressed calves selected diets with 72% concentrate when given a choice among feed mixtures varying in concentrate level during the first week after arrival at the feedlot (Lofgreen, 1983). Therefore, intake and performance by lightweight, newly received calves seems to be optimized with greater- compared with lower-concentrate diets.

In addition to the studies by Lofgreen, others have examined the effects of dietary energy concentration on performance and health of receiving calves. Corn silage-based diets with 1.15, 1.21, 1.25, or 1.30 Mcal/kg net energy required for gain (NEg) were fed to individually housed steers in a 28-d receiving experiment (Fluharty and Loerch, 1996). There was a linear increase in DMI with increasing dietary energy concentration, but there was no difference in daily gain, feed efficiency, or health status for the 28-d period. Similarly, DMI was improved and daily gain was not different between high-energy (1.17 Mcal/kg NEg) and low-energy (1.01 Mcal/kg NEg) diets in a 28-d preconditioning study conducted by Pritchard and Mendez (1990). Berry et al. (2004a,b) attempted to sort out the confounding effects of roughage and energy concentrations by feeding high- and low-starch concentrations within each of two dietary roughage concentrations. They observed that energy concentration did not influence performance or overall morbidity, but morbid calves fed diets with the greater concentration had less shedding of *P. multocida* and *H. somnus* than those fed the lower-energy diets. Dietary roughage concentration varied over a narrow range of 35 to 45% in the Berry et al. (2004a,b) studies, and therefore, comparison with results from experiments with greater variation in roughage/energy concentration is not possible.

Whitney et al. (2006) fed early-weaned beef steers with bermudagrass hay alone, hay with 0.175 or 0.35% BW of supplemental soybean meal, or a 70% concentrate diet for an 84-d backgrounding phase and observed that calves fed the concentrate diet had greater DMI and ADG than those fed the hay-based diets. Rivera et al. (2005) analyzed data collected at the New Mexico State University Clayton Livestock Research Center to evaluate the relationships between BRD and dietary roughage concentration in lightweight, stressed cattle. Diets ranged from all-hay to 75% concentrate. The analysis showed that morbidity (i.e., percentage of calves treated for BRD using visual observation and rectal temperature as a means of diagnosis) decreased slightly as dietary roughage concentration increased (morbidity, % = 49.59 – (0.0675 × roughage, %); $P = 0.003$) and that ADG and DMI were decreased by increasing the dietary roughage concentration. In addition, economic analysis indicated that the slightly lesser morbidity noted with greater roughage concentrations would not offset the loss in profit resulting from decreased ADG. Rivera et al. (2005) concluded that greater-concentrate, milled diets would likely provide the optimal receiving diet for lightweight, highly stressed, newly received cattle, with limited effects on BRD.

Grain type used in receiving diets does not seem to affect calf health or performance. For example, Smith (1998) reported that grain type—corn, grain sorghum, barley, or wheat—used in starter and receiving diets did not affect calf health or performance. Faster rates of gain have been obtained with a mixture of grains (Addis et al., 1975, 1978). Although more research is needed, highly stressed calves could have low tolerance for added fat, leading Cole and Hutcheson (1987b) to suggest that fat should probably not exceed 4% of dietary DM in receiving diets. Stressed calves prefer a dry diet compared to a diet high in corn silage, but they adapt to high amounts of corn silage in the diet after 7 to 14 days (Preston and Smith, 1973, 1974; Preston and Kunkle, 1974; Koers et al., 1975; Davis and Caley, 1977).

DIETARY PROTEIN

Dietary protein requirements for beef cattle are calculated from equations that are integrated with BW and energy intake (NRC, 1984, 1996, 2000). Energy intake seems to be the first-limiting factor involved with weight gain; therefore, protein deposited in gain is largely dependent on energy intake (Galyean, 1996). Because newly received stressed calves often have very low DMI during the first several days after arrival, protein requirements might be low; however, requirements would increase as energy intake increases. In general, protein requirements of stressed calves do not seem to be different from those of nonstressed calves (Duff and Galyean, 2007), but because of the low initial DMI, the concentration of protein in the diet should be increased for stressed or diseased calves (Cole and Hutcheson, 1990; Hutcheson et al., 1993; Duff and Galyean, 2007). Protein concentrations of 13.5 to 14.5% on a DM basis in receiving diets typically meet the protein requirements of stressed calves (Embry, 1977; Bartle et al., 1988; Cole and Hutcheson, 1988, 1990; Eck et al., 1988). Diseased calves exhibit a hypermetabolic response with increased excretion of nitrogen (N; Cole et al., 1986a), and the N kinetics of virus-infected calves are affected by shifts in the rates of protein metabolism (Orr et al., 1989). When fed increased protein, hyperurinary excretion of N during disease is partially alleviated (Boyles et al., 1989).

The effects of various protein levels and sources for newly received calves have been characterized; Galyean et al. (1993) fed three levels (12, 14, or 16%) of supplemental crude protein (CP) from soybean meal to 120 calves (185 kg) in a 42-d receiving experiment and observed that daily gain increased and DMI tended to increase linearly with increasing CP concentration. They also observed that morbidity was greater for calves fed the 16% CP diet compared with the 14% CP diet. Fluharty and Loerch (1995) conducted a series of experiments to assess protein requirements of newly arrived cattle. In the first of two experiments, newly weaned Simmental × Angus crossbred (243 kg) steers were fed increasing CP concentrations (12, 14, 16, or 18%) from two sources: spray-dried blood meal or soybean meal. In this experiment, feed efficiency improved linearly with increasing CP concentration for the first 7 d and for the entire 42-d feeding period and daily gain increased linearly with increasing CP concentration during the first week after arrival. For the entire receiving period, calves fed the blood meal diets had a 7.4% greater gain compared with calves fed the soybean meal diets. Similar to results reported by Galyean et al. (1993), morbidity also increased linearly with increasing CP concentration. In the second experiment, 246-kg Simmental × Angus steers were fed 11, 14, 17, 20, 23, or 26% CP diets with protein supplied by spray-dried blood meal or soybean meal. In this experiment, they observed that DMI was not affected by CP concentration; daily gain and feed efficiency both responded quadratically, with the 20% CP diet yielding the greatest performance. There were no differences in health status between treatment groups. In a summary of several experiments, Galyean et al. (1999) reported that as the protein concentration in receiving diets increased up to approximately 20% of DM, animal performance improved, but the incidence of BRD increased slightly.

A study at South Dakota State (Pritchard and Boggs, 2006) indicated that dried distillers grains could effectively replace soybean meal as a protein supplement for incoming feedlot cattle. Nonetheless, morbidity rates in their study were very low (<3%), so the effects of feeding corn byproducts in heavily stressed calves is not known. Van Koevering et al. (1991) reported that replacing soybean meal with dried distillers grains in a receiving supplement decreased performance but did not affect the incidence or severity of BRD.

Some experiments suggest that stressed calves have a lower tolerance for nonprotein nitrogen (urea) than do nonstressed calves. Urea intakes of 30 g/d or less seem to be tolerated by newly arrived or stressed calves during the first 2 weeks of feeding (Preston and Kunkle, 1974; Gates and Embry, 1975; Cole et al., 1984a). In contrast, Duff et al. (2000) reported that including urea at 1% of the dietary DM in a 70% concentrate diet for newly received calves did not affect performance or BRD morbidity during a 28-d receiving period. Feeding ruminally undegradable protein (RUP) to stressed calves has resulted in increased performance (Preston and Kunkle, 1974; Preston and Smith, 1974; Grigsby, 1981; Phillips, 1984). Feeding RUP at 5.4% dietary DM, or 45% total protein, resulted in increased daily gains and DMI (Preston and Bartle, 1990; Gunter et al., 1993; Hutcheson et al., 1993; Fluharty and Loerch, 1995). It seems that RUP concentrations of 5.4% dietary DM are generally adequate for stressed calves. Zinn and Shen (1998) evaluated ruminally degradable protein (RDP) and metabolizable indispensable amino acid requirements of feedlot calves during the early receiving period. Four Holstein steer calves (249 kg) cannulated in the rumen and proximal duodenum were used in an experiment to study effects on organic matter (OM) and N digestion and amino acid supply to the small intestine. Treatment diets were steam-flaked corn-based diets supplemented with urea, soybean meal, or fish meal at three concentrations (1.5, 3.0, or 4.5% DM). Digestibility of feed N, OM in the rumen, and microbial N flow to the small intestine decreased in diets without urea; however, postruminal OM digestibility was greater in calves fed fish meal compared with calves fed urea. As a result, there were no total tract treatment effects on OM digestibility. Ruminal microbial N efficiency (g microbial N/kg OM fermented) increased linearly with increasing concentration of fish meal. Supply of metabolizable indispensable amino acids to the small intestine increased linearly as level of fish meal in the diet increased. Daily gain and DMI increased and feed efficiency improved linearly as additional fish meal was added to the diet; however, there were no differences compared with soybean meal.

Protein metabolism is affected by disease as demonstrated by Waggoner et al. (2009a,b). In their experiment, cattle that were injected with a dose of LPS had decreases in plasma concentrations of methionine, threonine, leucine, isoleucine, phenylalanine, tryptophan, glycine, serine, asparagine, and tyrosine, whereas alanine increased (Waggoner et al., 2009b). In a similar experiment, decreases were noted in plasma threonine, lysine, leucine, phenylalanine, tryptophan, asparagine, ornithine, and glutamate, although alanine again increased (Waggoner et al., 2009a). Increased valine was also interpreted as a possible biomarker of concurrent BHV-1 and *M. haemolytica* infection (Aich et al., 2009).

In general, the activation of the immune system during an infection causes a decrease in most plasma amino acid concentrations. This results from the increased need for production of immune system cells, such as leukocytes, which can require high levels of specific amino acids (Colditz, 2002). In addition, the activation of the acute-phase response and the subsequent production of acute-phase proteins heighten the need for amino acids and could decrease available plasma amino acid levels (Sandberg et al., 2007). Amino acids are often transported from the muscle to the liver for this specific purpose. Degradation of these amino acids from muscle can also be induced by disease, and not only are these amino acids used for production of proteins or immune cells, they can be excreted, resulting in further N loss and muscle wasting (Powanda and Beisel, 2003). Moreover, as decreased intake is considered a clinical sign of disease (Duff and Galyean, 2007), less protein intake via feedstuffs exacerbates the effects of increased N usage by the body in order to mount an immune response. To meet the demands of the acute-phase response and immune system, specific amino acid requirements could increase. Unless specific amino acids are supplemented in the diet, which is an obvious challenge for ruminant diets, muscle protein will be catabolized to synthesize plasma proteins (Obled, 2003).

MINERALS

Beneficial effects of supplemental nutrients on immunity and the incidence of BRD in beef cattle would be most likely seen in animals with a marginal or deficient status of the nutrient; however, it is highly unusual to know the nutrient status of cattle used in most applied receiving studies. In general, it does not seem that mineral requirements for stressed cattle are different than those of nonstressed cattle (Orr et al., 1990; Duff and Galyean, 2007). When intracellular water is lost as a result of marketing and shipping stress, cellular deficiencies of K and sodium (Na) can occur. Therefore, the K requirement of stressed calves is 20% greater than that of nonstressed calves (Hutcheson et al., 1984). Many factors affect immune system response (Nockels, 1988). During disease states, trace mineral requirements can be affected by immune system response. High concentrations of zinc (Zn) have been shown to be beneficial to the animal's health during disease (Chirase et al., 1991), and Zn, copper (Cu), selenium (Se), and iron (Fe) seem to be necessary for immunocompetence (Chandra and Dayton, 1982; Droke and Loerch, 1989; Erskine et al., 1989, 1990). Galyean et al. (1999) and Duff and Galyean (2007) reported on the potential effects of Cr, Cu, Se, and Zn supplementation on immune function, BRD morbidity, and performance by newly received calves. A summary is provided below.

Chromium

Chromium supplementation has been shown to have positive effects on health and production of animals and humans. Chromium has been studied over the past two decades because of its involvement with glucose and insulin regulation. Chromium potentiates the action of insulin binding to insulin receptor sites (Kegley et al., 2000). Cefalu et al. (2002) postulated that Cr picolinate augmented insulin-sensitive glucose uptake by binding with chromodulin (Vincent, 2001), which enhanced the translocation of glucose transporters to the cell membrane. This mineral has been shown to increase glucose clearance and insulin sensitivity (Kegley et al., 2000) and enhance performance (Moonsie-Shageer and Mowat, 1993) and immune responses (Chang and Mowat, 1992) of feeder calves.

Performance results in beef cattle supplemented with Cr have varied from a 30% increase in ADG (Chang and Mowat, 1992) to no improvements (Kegley and Spears, 1995), but organic sources of Cr, such as Cr propionate or Cr methionine, have shown more consistent results. Bernhard et al. (2012) fed 221-kg steers 0, 0.1, 0.2, or 0.3 mg Cr/kg (DM basis) from Cr propionate. Chromium-supplemented steers had improvements in DMI, feed efficiency, and performance within the first 28 d of the experiment, when cattle would have been under significant nutritional and physiological stress. Through the remainder of the experiment, Cr-supplemented steers maintained these advantages. In the study by Bernhard et al. (2012) the addition of 0.3 mg Cr/kg resulted in the greatest improvement in feed efficiency and daily weight gain. There are other literature reports that either support or contradict these results (Chang and Mowat, 1992;

Mowat et al., 1993; Kegley and Spears, 1995; Kegley et al., 1997, 2000).

Moonsie-Shageer and Mowat (1993) reported that Cr supplementation (0.2, 0.5, and 1.0 mg Cr/kg) decreased morbidity, with 0.2 mg/kg resulting in a significant improvement compared with other treatments through a 30-d trial. Similarly, Mowat et al. (1993) reported that high-Cr yeast and chelated Cr decreased morbidity during a 35-d study. On the other hand, Cr supplementation at 0.2 mg/kg or 0.4 mg/kg of a high-Cr yeast did not affect morbidity in a study by Chang and Mowat (1992). Bernhard et al. (2012) reported no differences for cattle treated at least twice; however, supplementation of 0.3 mg/kg Cr resulted in a lower number of morbid steers, such that the number of steers treated at least once tended to decrease linearly with increasing Cr concentrations.

Variations in Cr results can be attributed to disparities in stress levels of the cattle (which may result in temporary Cr deficiencies), basal Cr concentration of the diet, level of Cr supplementation, and the source of Cr supplementation (Spears, 2000).

Copper

Results of experiments do not provide compelling evidence for consistent effects of Cu supplementation on immunity and provided only limited evidence of responses to Cu supplementation in field studies with stressed calves (Galyean et al., 1999). Ward and Spears (1999) injected Angus bull calves with 90 mg of Cu glycinate 28 d before weaning and subsequently supplemented these calves with Cu from copper sulfate ($CuSO_4$; 7.5 and 5 mg/kg DM during the receiving and growing phases, respectively). Control steers did not receive supplemental Cu during the study. During the growing phase, half the steers in each Cu treatment group were supplemented with 5 mg of Mo/kg DM, and at day 168, half the steers were stressed by loading and transport. Specific immunity of stressed cattle was not greatly affected by Cu deficiency and supplemental Mo (Ward and Spears, 1999).

Bailey et al. (2001) fed Angus × Hereford heifers a basal grass hay-based diet that contained 6 mg Cu/kg DM or supplemented the basal diet with 49 mg/kg DM as $CuSO_4$; 22 mg/kg from $CuSO_4$; 22 mg/kg from a 50:50 combination of $CuSO_4$ and Cu-amino acid complex; or 22 mg Cu/kg from a 25:50:25 mixture of $CuSO_4$, Cu-amino acid complex, and copper oxide (CuO). All diets contained added Mo, S, and Fe. It was observed that the treatments did not affect ADG or cell-mediated immune function.

Selenium

If organic complexes of minerals are more bioavailable than inorganic sources, then an organically complexed mineral might be beneficial in preventing or correcting deficiencies during stress and periods of low feed intake (Greene, 1995). Beck et al. (2005) evaluated lymphocyte blastogenesis and the skin-swelling response to phytohaemagglutinin (PHA) in calves after weaning from cows fed mineral supplements containing no supplemental Se or 26 mg Se/kg DM from sodium selenite (Na_2SeO_3) or Se-yeast. They reported that macrophage phagocytosis was increased in calves supplemented with Se-yeast compared with control and Na_2SeO_3 calves. Although not affected by Se source, skin swelling response after injection of PHA tended to be increased by Se supplementation (Beck et al., 2005). Fry et al. (2005) fed Angus-crossbred calves fescue hay plus a corn-soybean meal supplement that provided no supplemental Se or 1.7 mg supplemental Se/d from Na_2SeO_3 or Se-yeast. In this study, it was observed that both sources increased whole-blood Se, with the effects being more rapid in the Se-yeast group, but lymphocyte proliferation and phagocytosis did not differ among treatments. Even though dietary Se is considered vital for practically all components of the immune system (Arthur et al., 2003), Suttle and Jones (1989) concluded that the evidence was not strong that Se affected resistance to infection in ruminants.

Zinc

Gunter et al. (2001) supplemented steers with 103 mg Zn/d from zinc sulfate ($ZnSO_4$), Zn-amino acid complex, or Zn-polysaccharide during a 116-d grazing period on Bermudagrass pastures, after which, the steers were shipped (14 h) to a research feedlot where they continued on the same Zn sources that were fed during grazing. In this experiment, grazing and feedlot performance, serum Zn concentrations, and the number of steers treated for BRD were not affected by Zn source. Spears and Kegley (2002) fed Angus steers (246 kg initial BW) growing (silage-based) and finishing (corn-based) diets with no added Zn or 25 mg supplemental Zn/kg DM from zinc oxide (ZnO) and two different zinc proteinates and observed that all three sources of supplemental Zn increased ADG during the growing period, as well as carcass quality grade, whereas the two proteinates tended to increase finishing period ADG and improve gain:feed compared with ZnO. They also noted that lymphocyte blastogenesis and humoral antibody titers after IBR vaccination during the growing period did not differ among treatments.

In summary, Galyean et al. (1999) and Duff and Galyean (2007) concluded that formulation of receiving diets should account for decreased DMI by highly stressed, newly received beef cattle and for any known nutrient deficiencies, but that adding trace minerals to such diets beyond the concentrations needed to compensate for low feed intake is difficult to justify. In addition, although there is some evidence that organically complexed mineral sources might occasionally have different effects on performance and immune function, the effects seem too variable to recommend feeding particular sources.

VITAMINS

Adding B vitamins to receiving diets for stressed calves increased their performance and feed intake in one (Overfield et al., 1976) but not all (Cole et al., 1979, 1982) experiments. Niacin added at 125 ppm seemed to increase ADG by healthy calves (Hutcheson and Cummins, 1984); however, diseased calves receiving niacin at 250 ppm seemed to have the best ADG. The most significant gains were observed when the cattle received 271 mg/45.4 kg BW daily (Hutcheson and Cummins, 1984). Results with supplemental B vitamins in the diets of newly weaned/received cattle have been variable (Galyean et al., 1999), and there seems to be little justification for B vitamin supplementation of nutritionally balanced receiving diets.

Vitamin E has been shown to be involved in immune system response; lymphocyte-stimulation indices were highest for calves fed 227.5 mg (250 IU) all-*rac*-α-tocopheral compared to controls (Cipriano et al., 1982). Increasing vitamin E intake during disease or infection produced varying results, but in general the data indicate that vitamin E is necessary for optimal functioning of the immune system. Han et al. (1999) observed a marked decrease in plasma α-tocopherol concentrations of supplemented cattle in response to transit stress. Vitamin E fed at 400 IU/d in receiving and starting diets of newly arrived feeder calves decreased disease and number of sick days and increased gain (Hicks, 1985). Additional studies (Gill et al., 1986; Hays et al., 1987) evaluating the effects of vitamin E-supplemented cattle reported improvements in animal performance and improved immune system function when cattle received 800 IU of supplemental dietary vitamin E.

Vitamin E fed at 450 IU/d to cattle that experienced more than 10% shrink increased gain (Lee et al., 1985). Calves receiving 125 mg/d (125 IU/d) of all-*rac*-α-tocopheral acetate consumed more than calves that did not receive additional vitamin E or 500 mg/d (500 IU/d; Reddy et al., 1985). Elam (2006) analyzed data from experiments done by Secrist et al. (1997), Rivera et al. (2003), Carter et al. (2005), and Cusack et al. (2005) using mixed-model statistical methods to evaluate the effects of vitamin E intake on performance and health of newly received cattle. In these experiments, supplemental vitamin E intake ranged from 0 to 2,000 IU/animal daily and ADG, DMI, and gain:feed were not significantly related to intake of vitamin E; however, BRD morbidity decreased by 0.35% for every 100-IU increase in daily vitamin E intake (Elam, 2006). Galyean et al. (1999) concluded that vitamin E at doses of greater than 400 IU/animal daily seemed beneficial for increasing ADG and decreasing BRD morbidity. Therefore, the committee recommends that vitamin E should be fed at between 400 and 500 IU/animal daily during the receiving and starting periods.

FEED ADDITIVES

Because ionophores can negatively affect DMI, their concentrations are often limited in receiving diets to avoid these negative effects. Duff et al. (1995) compared a control diet (no ionophore) with diets containing lasalocid at 33 mg/kg and monensin at 22 or 33 mg/kg dietary DM. They noted that ionophores decreased DMI compared with the control diet for the 28-d experiment, but ADG and the efficiency of gain were not altered by treatments. Numerically, monensin at 33 mg/kg resulted in a lower DMI than lasalocid at the same concentration. All three ionophore treatments decreased the presence of coccidial oocysts (Duff et al., 1995). These findings indicate that ionophores might have negative effects on DMI in newly received cattle, but these effects can be moderated by the choice of ionophore or, with monensin, by decreasing the dietary concentration.

SUMMARY

Table 15-2 gives the suggested nutrient concentrations for receiving diets of stressed cattle. Many of the nutrients are based on the committee's calculations. Decreased intake during disease stress is the single most common observation. Nutrient quantities recommended in Table 15-2 are for the first 2 weeks after arrival or until the cattle are consuming feed, on a DM basis, of 2% of BW or more. Table 15-2 also gives the nutrient amounts that would be consumed per day when suggested amounts are calculated: 1.55% of BW, the average amount of feed consumed during the first week; and 1.90% of BW, the average amount of feed consumed during the first 2 weeks.

REFERENCES

Addis, D. G., G. P. Lofgreen, J. G. Clark, J. R. Dunbar, and C. Adams. 1975. Barley vs milo in receiving rations. Pp. 53-58 in *California Feeders Day Report 14*. University of California, Davis.

Addis, D. G., G. P. Lofgreen, J. G. Clark, C. Adams, F. Prigge, J. R. Dunbar, and B. Norman. 1978. Barley vs wheat in receiving rations for new calves. Pp. 54-60 in *California Feeders Day Report 17*. University of California, Davis.

Aich, P., L. A. Babiuk, A. A. Potter, and P. Griebel. 2009. Biomarkers for prediction of bovine respiratory disease outcome. *OMICS: A Journal of Integrative Biology* 13:199-209.

Apley, M. 2006. Bovine respiratory disease: Pathogenesis, clinical signs, and treatment in lightweight calves. *Veterinary Clinics of North America: Food Animal Practice* 22:399-411.

Arthington, J. D., J. W. Spears, and D. C. Miller. 2005. The effect of early weaning on feedlot performance and measures of stress in beef calves. *Journal of Animal Science* 83:933-939.

Arthington, J. D., R. F. Cooke, T. D. Maddock, D. B. Araujo, P. Moriel, N. DiLorenzo, and G. C. Lamb. 2013. Effects of vaccination on the acute-phase protein response and measures of performance in growing beef calves. *Journal of Animal Science* 91:1831-1837.

Arthur, J. R., R. C. McKenzie, and G. J. Beckett. 2003. Selenium in the immune system. *Journal of Nutrition* 133:1457S-1459S.

Bailey, J. D., R. P. Ansotegui, J. A. Paterson, C. K. Swenson, and A. B. Johnson. 2001. Effects of supplementing combinations of inorganic and complexed copper on performance and liver mineral status of beef heifers consuming antagonists. *Journal of Animal Science* 79:2926-2934.

Baldwin, R. L. 1967. Effect of starvation and refeeding upon rumen fermentation. Pp. 7-12 in *California Feeders Day Report 7*. University of California, Davis.

Bartle, S. J., K. J. Karr, J. G. Ross, D. L. Hancock, and R. L. Preston. 1988. Supplemental protein sources and dietary crude protein level for receiving feedlot cattle. Pp. 61-62 in *Texas Tech University Animal Science Research Report T-5-251*. Lubbock: Texas Tech University.

Beck, P. A., T. J. Wistuba, M. E. Dais, and S. A. Gunter. 2005. Case study: Effects of feeding supplemental organic or inorganic selenium to cow-calf pairs on selenium status and immune responses of weaned beef calves. *The Professional Animal Scientist* 21:114-120.

Bernhard, B. C., N. C. Burdick, W. Rounds, R. J. Rathmann, J. A. Carroll, D. N. Finck, M. A. Jennings, T. R. Young, and B. J. Johnson. 2012. Chromium supplementation alters the performance and health of feedlot cattle during the receiving period and enhances their metabolic response to a lipopolysaccharide (LPS) challenge. *Journal of Animal Science* 90:3879-3888.

Berry, B. A., A. W. Confer, C. R. Krehbiel, D. R. Gill, R. A. Smith, and M. Montelongo. 2004a. Effects of dietary energy and starch concentrations for newly received feedlot calves: II. Acute phase protein response. *Journal of Animal Science* 82:845-850.

Berry, B. A., C. R. Krehbiel, A. W. Confer, D. R. Gill, R. A. Smith, and M. Montelongo. 2004b. Effects of dietary energy and starch concentrations for newly received feedlot calves: I. Growth performance and health. *Journal of Animal Science* 82:837-844.

Blecha, F., S. L. Boyles, and J. G. Riley. 1984. Shipping suppresses lymphocyte blastogenic responses in Angus and Brahman × Angus feeder calves. *Journal of Animal Science* 59:576-583.

Booker, C. W., S. M. Abutarbush, P. S. Morley, P. T. Guichon, B. K. Wildman, G. K. Jim, O. C. Schunict, T. J. Pittman, T. Perret, J. A. Ellis, G. Appleyard, and D. M. Haines. 2008. The effect of bovine viral diarrhea virus infections on health and performance of feedlot cattle. *Canadian Veterinary Journal* 49:253-260.

Boyles, D. W., C. H. Richardson, C. S. Cobb, M. D. Miller, and N. A. Cole. 1989. Influence of protein status on the severity of the hypermetabolic response of calves with infectious bovine rhinotracheitis virus. Pp. 14-15 in *Texas Tech University Animal Science Research Report T-5-263*. Lubbock: Texas Tech University.

Breazile, J. E. 1987. Physiological basis and consequences of distress in animals. *Journal of the American Veterinary Medical Association* 191:1212-1215.

Brooks, K. R., K. C. Raper, C. E. Ward, B. P. Holland, C. R. Krehbiel, and D. L. Step. 2011. Economic effects of bovine respiratory disease on feedlot cattle during backgrounding and finishing phases. *The Professional Animal Scientist* 27:195-203.

Buhman, M. J., L. J. Perino, M. L. Galyean, and R. S. Swingle. 2000. Eating and drinking behaviors of newly received feedlot calves. *The Professional Animal Scientist* 16:241-246.

Burciaga-Robles, L. O., D. L. Step, C. R. Krehbiel, B. P. Holland, C. J. Richards, M. A. Montelongo, A. W. Confer, and R. W. Fulton. 2010. Effects of exposure to calves persistently infected with bovine viral diarrhea virus type 1b and subsequent infection with *Mannheima haemolytica* on clinical signs and immune parameters: Model for bovine respiratory disease via viral and bacterial interaction. *Journal of Animal Science* 88:2166-2178.

Camp, T. H., R. A. Stermer, D. G. Stevens, and J. P. Anthony. 1983. Shrink of feeder calves as an indicator of incidence of bovine respiratory disease and time to return to purchase weight. *Journal of Animal Science* 57 (Suppl. 1):66 (Abstract).

Carroll, J. A., J. D. Arthington, and C. C. Chase. 2009. Early weaning alters the acute-phase reaction to an endotoxin challenge in beef calves. *Journal of Animal Science* 87:4167-4172.

Carter, J. N., D. R. Gill, C. R. Krehbiel, A. W. Confer, R. A. Smith, D. L. Lalman, P. L. Claypool, and L. R. McDowell. 2005. Vitamin E supplementation of newly arrived feedlot calves. *Journal of Animal Science* 83:1924-1932.

Cefalu, W. T., Z. Q. Wang, X. H. Zhang, L. C. Baldor, and J. C. Russell. 2002. Oral chromium picolinate improves carbohydrate and lipid metabolism and enhances skeletal muscle GLUT-4 translocation in obese, hyperinsulinemic (JCR-LA corpulent) rats. *Journal of Nutrition* 132:1107-1114.

Cernicchiaro, N., B. J. White, D. G. Renter, A. H. Babcock, L. Kelly, R. Slattery. 2012. Associations between the distance traveled from sale barns to commercial feedlots in the United States and overall performance, risk of respiratory disease, and cumulative mortality in feeder cattle during 1997 to 2009. *Journal of Animal Science* 90:1929-1939.

Chandra, R. K., and D. H. Dayton. 1982. Trace element regulation of immunity and infection. *Nutrition Research* 2:721-733.

Chang, X., and D. N. Mowat. 1992. Supplemental chromium for stressed and growing feeder calves. *Journal of Animal Science* 70:559-565.

Chirase, N. K., D. P. Hutcheson, and G. B. Thompson. 1991. Feed intake, rectal temperature, and serum mineral concentrations of feedlot cattle fed zinc oxide or zinc methionine and challenged with infectious bovine rhinotracheitis virus. *Journal of Animal Science* 69:4137-4145.

Cipriano, J. E., J. L. Morrill, and N. V. Anderson. 1982. Effect of dietary vitamin E on immune responses of calves. *Journal of Dairy Science* 65:2357-2365.

Coffey, K. P., W. K. Coblentz, J. B. Humphry, and F. K. Brazle. 2001. Basic principles and economics of transportation shrink in beef cattle. *The Professional Animal Scientist* 17:247-255.

Colditz, I. G. 2002. Effects of the immune system on metabolism: Implications for production and disease resistance in livestock. *Livestock Production Science* 75:257-268.

Cole, N. A. 1985. Preconditioning calves for the feedlot. *Veterinary Clinics of North America, Food Animal Practice* 1:401-411.

Cole, N. A. 1991a. Effects of animal-to-animal exchange of ruminal contents on the feed intake and ruminal characteristics of fed and fasted lambs. *Journal of Animal Science* 69:1795-1803.

Cole, N. A. 1991b. Nutritional and management procedures for marketing/transport stressed calves. Pp. 75-116 in *Proceedings of the Symposium on Receiving and Management of Stressed Cattle*. Great Plains Veterinary Education Center, Clay Center, NE.

Cole, N. A. 2000. Changes in postprandial plasma and extracellular and ruminal fluid volumes in wethers fed or unfed for 72 hours. *Journal of Animal Science* 78:216-223.

Cole, N. A., and D. P. Hutcheson. 1981. Influence on beef steers of two sequential short periods of feed and water deprivation. *Journal of Animal Science* 53:907-915.

Cole, N. A., and D. P. Hutcheson. 1985a. Influence of prefast feed intake on recovery from feed and water deprivation by beef steers. *Journal of Animal Science* 60:772-780.

Cole, N. A., and D. P. Hutcheson. 1985b. Influence of realimentation diet on recovery of rumen activity and feed intake in beef steers. *Journal of Animal Science* 61:692-701.

Cole, N. A., and D. P. Hutcheson. 1987a. Influence of pre-fast dietary roughage content on recovery from feed and water deprivation in beef steers. *Journal of Animal Science* 65:1049-1057.

Cole, N. A., and D. P. Hutcheson. 1987b. Influence of receiving diet fat level on the health and performance of feeder calves. *Nutrition Reports International* 36:965-970.

Cole, N. A., and D. P. Hutcheson. 1988. Influence of protein concentration in prefast and postfast diets on feed intake of steers and nitrogen and phosphorus metabolism of lambs. *Journal of Animal Science* 66:1764-1777.

Cole, N. A., and D. P. Hutcheson. 1990. Influence of dietary protein concentrations on performance and nitrogen and repletion in stressed calves. *Journal of Animal Science* 68:3488-3497.

Cole, N. A., M. R. Irwin, and J. B. McLaren. 1979. Influence of pretransit feeding regimen and posttransit B-vitamin supplementation on stressed feeder steers. *Journal of Animal Science* 49:310-317.

Cole, N. A., J. B. McLaren, and D. P. Hutcheson. 1982. Influence of preweaning and B-vitamin supplementation of the feedlot receiving diet on calves subjected to marketing and transit stress. *Journal of Animal Science* 54:911-917.

Cole, N. A., D. P. Hutcheson, J. B. McLaren, and W. A. Phillips. 1984a. Influence of pretransit zeranol implant and receiving diet protein and urea levels on performance of yearling steers. *Journal of Animal Science* 58:527-534.

Cole, N. A., T. H. Montgomery, R. Adams, D. P. Hutcheson, and J. B. McLaren. 1984b. Influence of feeder calf management and bovine respiratory disease on carcass traits of beef steers. Pp. 319-322 in *Proceedings, Western Section, American Society of Animal Science,* Volume 35.

Cole, N. A., D. D. Delaney, J. M. Cummins, and D.P. Hutcheson. 1986a. Nitrogen metabolism of calves inoculated with bovine adenovirus-3 or with infectious bovine rhinotracheitis virus. *American Journal of Veterinary Research* 47:1160-1164.

Cole, N. A., W. A. Phillips, and D. P. Hutcheson. 1986b. The effect of prefast diet and transport on nitrogen metabolism of calves. *Journal of Animal Science* 62:1719-1731.

Cole, N. A., D. M. Hallford, and R. Gallavan. 1993. Influence of a glucose load in fed or unfed lambs on blood metabolites and hormone patterns. *Journal of Animal Science* 71:765-773.

Corbeil, L. B. 2007. *Histophilus somni* host-parasite relationships. *Animal Health Research Reviews* 8:151-160.

Cravey, M. D. 1996. Preconditioning effect on feedlot performance. Pp. 33-37 in *Proceedings of the Southwest Nutrition and Management Conference, University of Arizona, Tucson.*

Cray, C., J. Zaias, and N. H. Altman. 2009. Acute phase response in animals: A review. *Comparative Medicine* 59:517-526.

Cusack, P. M. V., N. P. McMeniman, and I. J. Lean. 2005. The physiological and production effects of increased dietary intake of vitamins E and C in feedlot cattle challenged with bovine herpesvirus 1. *Journal of Animal Science* 83:2423-2433.

Dabo, S. M., J. D. Taylor, and A. W. Confer. 2007. *Pasteurella multocida* and bovine respiratory disease. *Animal Health Research Reviews* 8:129-150.

Davis, G. V., and H. K. Caley. 1977. Influence of management and rations on the performance of stress calves. Pp. 16-29 in *Cattle Feeders' Day. Report of Progress 288.* Manhattan: Kansas State University, Agricultural Experiment Station.

Droke, E. A., and S. C. Loerch. 1989. Effects of parenteral selenium and vitamin E on performance, health and humoral immune response of steers new to the feedlot environment. *Journal of Animal Science* 67:1350-1359.

Duff, G. C., and M. L. Galyean. 2007. Board-invited review: Recent advances in management of highly stressed, newly received feedlot cattle. *Journal of Animal Science* 85:823-840.

Duff, G. C., M. L. Galyean, K. J. Malcom-Callis, and D. R. Garcia. 1995. *Effects of Ionophore Type and Level on Performance of Newly Received Beef Calves.* Clayton Livestock Research Center Progress Report No. 96. Clayton: New Mexico State University.

Duff, G. C., K. J. Malcolm-Callis, M. L. Galyean, S. A. Soto-Navarro, and M. W. Wiseman. 2000. *Effects of Urea Concentration in Receiving Diet on Health and Performance of Newly Received Beef Steers.* Clayton Livestock Research Center Progress Report No. 105. Las Cruces: New Mexico Agricultural Experiment Station.

Eck, T. P., S. J. Bartle, R. L. Preston, R. T. Brandt, and C. R. Richardson, 1988. Protein source and level for incoming feedlot cattle. *Journal of Animal Science* 66:1871-1876.

Eckersall, P. D., and R. Bell. 2010. Acute phase proteins: Biomarkers of infection and inflammation in veterinary medicine. *Veterinary Journal* 185:23-27.

Eckersall, P. D., and J. G. Conner. 1988. Bovine and canine acute phase proteins. *Veterinary Research Communications* 12:169-178.

Edwards, T. A. 2010. Control methods for bovine respiratory disease for feedlot cattle. *Veterinary Clinics of North America: Food Animal Practice* 26:273-284.

Elam, N. A. 2006. Impact of vitamin E supplementation on newly received calves. *Proceedings, Colorado Nutrition Roundtable.* Ft. Collins: Colorado State University.

Ellis, J. A. 2001. The immunology of the bovine respiratory disease complex. *Veterinary Clinics of North America: Food Animal Practice* 17:535-550, vi-vii.

Embry, L. B. 1977. Feeding and management of new feedlot cattle. Pp. 47-52 in *South Dakota Cattle Feeders Day Report 1977.* Brookings: South Dakota State University.

Erskine, J. R., R. J. Eberhart, P. J. Grasso, and R. W. Scholz. 1989. Induction of *Escherichia coli* mastitis in cows fed selenium-deficient or selenium-supplemented diets. *American Journal of Veterinary Research* 50:2093-2100.

Erskine, J. R., R. J. Eberhart, and R. W. Scholz. 1990. Experimentally induced *Staphylococcus aureus* mastitis in selenium-deficient and selenium-supplemented dairy cows. *American Journal of Veterinary Research* 51:1107-1111.

Fluharty, F. L., and S. C. Loerch. 1995. Effects of protein concentration and protein source on performance of newly arrived feedlot steers. *Journal of Animal Science* 73:1585-1594.

Fluharty, F. L., and S. C. Loerch. 1996. Effect of dietary energy source and level on performance of newly arrived feedlot cattle. *Journal of Animal Science* 74:504-513.

Fluharty, F. L., S. C. Loerch, and B. A. Dehority. 1994. Ruminal characteristics, microbial populations, and digestive capabilities of newly weaned, stressed calves. *Journal of Animal Science* 72:2969-2979.

Fluharty, F. L., S. C. Loerch, and B. A. Dehority. 1996. Effects of feed and water deprivation on ruminal characteristics and microbial population of newly weaned and feedlot-adapted calves. *Journal of Animal Science* 74:465-474.

Frazer, D., J. S. D. Ritchie, and A. F. Frazer. 1975. The term "stress" in a veterinary context. *British Veterinary Journal* 131:653-658.

Fry, R. S., E. B. Kegley, M. E. Davis, M. D. Ratcliff, D. L. Galloway, and R. A. Dvorak. 2005. Level and source of supplemental selenium for beef steers. Pp. 105-108 in *Arkansas Animal Science Department Report*, Z. B. Johnson and D. W. Kellogg, eds. Research Series 535. Fayetteville: Arkansas Agricultural Experiment Station. Available online at http://arkansasagnews.uark.edu/535-26.pdf. Accessed on May 8, 2015.

Galyean, M. L. 1996. Protein levels in beef cattle finishing diets: Industry application, university research, and systems results. *Journal of Animal Science* 74:2860-2870.

Galyean, M. L., and M. E. Hubbert. 1995. Effects of season, health, and management on feed intake by beef cattle. Pp. 226-234 in *Symposium: Intake by Feedlot Cattle.* Oklahoma Agricultural Experiment Station P-942. Stillwater: Oklahoma State University.

Galyean, M. L., R. W. Lee, and M. E. Hubbert. 1981. Influence of fasting and transit stress on rumen fermentation in beef steers. *Journal of Animal Science* 53:7-18.

Galyean, M. L., S. A. Gunter, K. J. Malcolm-Callis, and D. R. Garcia. 1993. *Effects of Crude Protein Concentration in the Receiving Diet on Performance and Health of Newly Received Beef Calves.* Clayton Livestock Research Center Progress Report No. 88. Las Cruces: New Mexico Agricultural Experiment Station.

Galyean, M. L., L. J. Perino, and G. C. Duff. 1999. Interaction of cattle health/immunity and nutrition. *Journal of Animal Science* 77:1120-1134.

Garcia, M. D., R. M. Thallman, T. L. Wheeler, S. D. Shackelford, and E. Casas. 2010. Effect of bovine respiratory disease and overall pathogenic disease incidence on carcass traits. *Journal of Animal Science* 88:491-496.

Gates, R. N., and L. B. Embry. 1975. Soybean meal or urea during feedlot adaptation and growing calves. Pp. 27-33 in *South Dakota. Cattle Feeders Day Report AS 7525.*

Gill, D. R., R. A. Smith, R. B. Hicks, and R. L. Ball. 1986. The effect of vitamin E supplementation on health and performance of newly arrived stocker cattle. Pp. 240-243 in *Animal Science Research Report MP-118.* Stillwater: University of Oklahoma.

Grandin, T. 1997. Assessment of stress during handling and transport. *Journal of Animal Science* 75:249-257.

Greene, L. W. 1995. The nutritional value of inorganic and organic mineral sources. Pp. 23-31 in *Proceedings of the Plains Nutrition Council Symposium.* Lubbock: Texas Tech University.

Grigsby, M. E. 1981. *Protein Supplementation for Stressed Calves.* Clayton Livestock Research Center Progress Report No. 25. Las Cruces: New Mexico Agricultural Experiment Station.

Gruys, E., M. J. M. Toussaint, T. A. Niewold, and S. J. Koopmans. 2005. Acute phase reaction and acute phase proteins. *Journal of Zhejiang University-Science B* 6(11):1045-1056.

Gunter, S. A., M. L. Galyean, K. J. Malcolm-Callis, and D. R. Garcia. 1993. *Effects of Origin of Cattle and Supplemental Protein Source on Performance of Newly Received Feeder Cattle.* Clayton Livestock Research Center Progress Report No. 85. Las Cruces: New Mexico Agricultural Experiment Station.

Gunter, S. A., K. J. Malcolm-Callis, G. C. Duff, and E. B. Kegley. 2001. Performance of steers supplemented with zinc during grazing and receiving at the feedlot. *The Professional Animal Scientist* 17:280-286.

Han, H., T. C. Stovall, J. T. Wagner, and D. R. Gill. 1999. Impact of Agrado on tocopherol metabolism by transport-stressed heifers. Pp. 119-125 in *Animal Science Research Report MP-973.* Stillwater: University of Oklahoma.

Hays, V. S., D. R. Gill, R. A. Smith, and R. L. Ball. 1987. The effect of vitamin E supplementation on performance of newly arrived stocker cattle. Pp. 198-201 in *Oklahoma Agricultural Experiment Station Report MP-119.* Stillwater: University of Oklahoma.

Hicks, R. B. 1985. Effects of Nutrition, Medical Treatments, and Management Practices of Newly Received Stocker Cattle. M.S. Thesis. Oklahoma State University, Stillwater.

Holland, B. P., L. O. Burciaga-Robles, J. N. Shook, D. L. VanOverbeke, D. L. Step, C. J. Richards, and C. R. Krehbiel. 2010. Effects of bovine respiratory disease during preconditioning on subsequent feedlot growth performance, carcass characteristics, and meat quality of heifers. *Journal of Animal Science* 88:2486-2499.

Hutcheson, D. P., and N. A. Cole. 1986. Management of transit-stress syndrome in cattle: Nutritional and environmental effects. *Journal of Animal Science* 62:555-560.

Hutcheson, D. P., and J. M. Cummins. 1984. The use of niacin in receiving diets for feeder calves. Pp. 120-121 in *Proceedings of the Western Section, American Society of Animal Science.* Champaign, IL: American Society of Animal Science.

Hutcheson, D. P., N. A. Cole, and J. B. McLaren. 1984. Effects of pretransit diets and post-transit potassium levels for feeder calves. *Journal of Animal Science* 58:700-707.

Hutcheson, J. P., T. L. Stanton, C. F. Nockels, and O. Robertson. 1993. The effect of three protein sources on feedlot performance of receiving calves. Pp. 67-73 in *Beef Progress Report.* Fort Collins: Colorado State University.

Jeyaseelan, S., S. Sreevatsan, and S. K. Maheswaran. 2002. Role of *Mannheimia haemolytica* leukotoxin in the pathogenesis of bovine pneumonic pasteurellosis. *Animal Health Research Reviews* 3:69-82.

Kegley, E. B., and J. W. Spears. 1995. Immune response, glucose metabolism, and performance of stressed feeder calves fed inorganic and organic chromium. *Journal of Animal Science* 73:2721-2726.

Kegley, E. B., J. W. Spears, and T. T. Brown, Jr. 1997. Effects of shipping and chromium supplementation on performance, immune response, and disease resistance of steers. *Journal of Animal Science* 75:1956-1964.

Kegley, E. B., D. L. Galloway, and T. M. Fakler. 2000. Effect of dietary chromium-L-methionine on glucose metabolism of beef steers. *Journal of Animal Science* 78:3177-3183.

Koers, W. C., J. C. Parrott, L. B. Sherrod, R. H. Klett, and W. G. Sheldon. 1975. Receiving and sick pen rations for stressed calves. Pp. 73-76 in *Texas Tech Research Report 25.* Lubbock: Texas Tech University.

Lafleur, R. L., C. Malazdrewich, S. Jeyaseelan, E. Bleifield, M. S. Abrahamsen, and S. K. Maheswaran. 2001. Lipopolysaccharide enhances cytolysis and inflammatory cytokine induction in bovine alveolar macrophages exposed to *Pasteurella (Mannheimia) haemolytica* leukotoxin. *Microbial Pathogenesis* 30:347-357.

Lalman, D. L. 2004a. Preconditioning nutrition and management. P. 165 in *Oklahoma Beef Cattle Manual,* 4th Ed. Oklahoma Cooperative Extension Service MP E-913. Stillwater: Oklahoma State University.

Lalman, D. L. 2004b. Supplementing and feeding calves and stocker cattle. P. 144 in *Oklahoma Beef Cattle Manual,* 4th Ed. Oklahoma Cooperative Extension Service MP E-913. Stillwater: Oklahoma State University.

Lee, R. W., R. L. Stuart, K. R. Perryman, and K. R. Ridenhour. 1985. Effect of vitamin supplementation on the performance of stressed beef calves. *Journal of Animal Science* 6l(Suppl.1):425.

Loerch, S. C., and F. L. Fluharty. 1999. Physiological changes and digestive capabilities of newly received feedlot cattle. *Journal of Animal Science* 77:1113-1119.

Lofgreen, G. P. 1983. Nutrition and management of stressed beef calves. *Veterinary Clinics of North America: Large Animal Practice* 5:87-101.

Lofgreen, G. P. 1988. Nutrition and management of stressed beef calves: An update. *Veterinary Clinics of North America: Food Animal Practice* 4:509-522.

Lofgreen, G. P., A. E. Eltayeb, and H. E. Kiesling. 1981. Millet and alfalfa hays alone and in combination with a high-energy diet for receiving stressed calves. *Journal of Animal Science* 52:959-968.

Mathis, C. P., S. H. Cox, C. A. Loest, M. K. Petersen, R. L. Endecott, A. M. Encinias, and J. C. Wenzel. 2008. Comparison of low input pasture to high-input drylot backgrounding on performance and profitability of beef calves through harvest. *The Professional Animal Scientist* 24:169-174.

Mathis, C. P., S. H. Cox, C. A. Loest, M. K. Petersen, and J. T. Mulliniks. 2009. Pasture preconditioning calves at a higher rate of gain improves feedlot health but not postweaning profit. *The Professional Animal Scientist* 25:475-480.

McBeth, L. J., D. R. Gill, C. R. Krehbiel, R. L. Ball, S. S. Swaneck, W. T. Choat, and C. E. Markham. 2001. Effect of health status during the receiving period on subsequent feedlot performance and carcass characteristics. In *2001 Animal Science Research Reports: Beef and Dairy Cattle, Swine, Poultry, Sheep, Horses and Animal Products.* Oklahoma Agricultural Experiment Station Research Report P-986. Stillwater: Oklahoma Agricultural Experiment Station, Division of Agricultural Science and Natural Resources, Oklahoma State University. Available online at http://www.ansi.okstate.edu/research/research-reports-1/2001. Accessed on April 15, 2015.

Montgomery, S. P., J. J. Sindt, M. A. Greenquist, W. F. Miller, J. N. Pike, E. R. Loe, M. J. Sulpizio, and J. S. Drouillard. 2009. Plasma metabolites of receiving heifers and the relationship between apparent bovine respiratory disease, body weight gain, and carcass characteristics. *Journal of Animal Science* 87:328-333.

Moonsie-Shageer, S., and D. N. Mowat. 1993. Effect of level of supplemental chromium on performance, serum constituents, and immune response of stressed feeder steer calves. *Journal of Animal Science* 71:232-238.

Mowat, D. N., X. Chang, and W. Z. Yang. 1993. Chelated chromium for stressed feeder calves. *Canadian Journal of Animal Science* 73:49-55.

Nockels, C. F. 1988. The role of vitamins in modulating disease resistance. *Veterinary Clinics of North America: Food Animal Practice* 4:531-542.

NRC (National Research Council). 1984. *Nutrient Requirements of Beef Cattle,* 6th Revised Ed. Washington, DC: National Academy Press.

NRC. 1996. *Nutrient Requirements of Beef Cattle,* 7th Revised Ed. Washington, DC: National Academy Press.

NRC. 2000. *Nutrient Requirements of Beef Cattle: Update 2000,* 7th Rev. Ed. Washington, DC: National Academy Press.

Obled, C. 2003. Amino acid requirements in inflammatory states. *Canadian Journal of Animal Science* 83:365-373.

Orr, C., D. P. Hutcheson, J. M. Cummins, and G. B. Thompson. 1989. Nitrogen kinetics of infectious bovine rhinotracheitis stressed calves. *Journal of Animal Science* 66:1982-1989.

Orr, C. L., D. P. Hutcheson, R. B. Grainger, J. M. Cummins, and R. E. Mock. 1990. Serum copper, zinc, calcium, and phosphorus concentrations of calves stressed by bovine respiratory disease and infectious bovine rhinotracheitis. *Journal of Animal Science* 68:2893-2900.

Overfield, J. R., D. L. Hixon, and E. E. Hatfield. 1976. Effect of nutritional treatment of stressed feeder calves. Pp. 28-33 in *University of Illinois Cattle Feeders Day Report AS-6721*. Champaign: University of Illinois.

Parker, A. J., G. P. Hamlin, C. J. Coleman, and L. A. Fitzpatrick. 2003. Quantitative analysis of acid-base balance in *Bos indicus* steers subjected to transportation of long duration. *Journal of Animal Science* 81:1434-1439.

Peterson, H. H. 2004. Application of acute phase protein measurements in veterinary clinical chemistry. *Veterinary Research* 35:163-187.

Phillips, W. A. 1984. The effect of protein source on the poststress performance of steer and heifer calves. *Nutrition Reports International* 30:853-858.

Phillips, W. A., N. A. Cole, and D. P. Hutcheson. 1985. The effect of diet on the amount and source of weight lost by beef steers during transit or fasting. *Nutrition Reports International* 32:765-776.

Powanda, M. C., and W. R. Beisel. 2003. Metabolic effects of infection on protein and energy status. *Journal of Nutrition* 133:322S-327S.

Preston, R. L., and S. J. Bartle. 1990. Quantification of ruminal escape protein and amino acids for new feedlot cattle. Pp. 17-19 in *Texas Tech Animal Science Research Report T-5-283*. Lubbock: Texas Tech University.

Preston, R. L., and W. E. Kunkle. 1974. Role of roughage source in the receiving ration of yearling feeder steers. Pp. 51-53 in *Ohio Agricultural Research and Development Center, Research Summary 77*. Wooster: Ohio State University.

Preston, R. L., and C. K. Smith. 1973. Feedlot response of new feeder calves following a creep feeding period on the same or different protein source as that fed immediately after shipment. Pp. 29-31 in *Ohio Agricultural Research and Development Center, Research Summary 68*. Wooster: Ohio State University.

Preston, R. L., and C. K. Smith. 1974. Role of protein level, protected soybean protein, and roughage on the performance of new feeder calves. Pp. 47-50 in *Ohio Agricultural Research and Development Center, Research Summary 77*. Wooster: Ohio State University.

Pritchard, R. H., and D. L. Boggs. 2006. Use of DDGS as the primary source of supplemental crude protein in calf receiving diets. Pp. 6-9 in *2006 South Dakota Beef Report*. Brookings: South Dakota State University.

Pritchard, R. H., and J. K. Mendez. 1990. Effects of preconditioning on pre- and post-shipment performance of feeder calves. *Journal of Animal Science* 68:28-34.

Reddy, P. G., J. L. Morrill, R. A. Frey, M. B. Morrill, H. C. Minocha, S. J. Galitzer, and A. D. Dayton. 1985. Effects of supplemental vitamin E on the performance and metabolic profiles of dairy calves. *Journal of Dairy Science* 68:2259-2270.

Rice, J. A., L. Carrasco-Medina, D. C. Hodgins, and P. E. Shewen. 2007. *Mannheimia haemolytica* and bovine respiratory disease. *Animal Health Research Reviews* 8:117-128.

Rivera, J. D., G. C. Duff, M. L. Galyean, D. M. Hallford, and T. T. Ross. 2003. Effects of graded levels of vitamin E on inflammatory response and evaluation of methods of delivering supplemental vitamin E on performance and health of beef steers. *The Professional Animal Scientist* 19:171-177.

Rivera, J. D., M. L. Galyean, and W. T. Nichols. 2005. Review: Dietary roughage concentration and health of newly received cattle. *The Professional Animal Scientist* 21:345-351.

Saini, P. K., M. Raiz, D. W. Webert, P. D. Eckersall, C. R. Young, L. H. Stanker, E. Chakrabarti, and J. C. Judkins. 1998. Development of a simple enzyme immunoassay for blood haptoglobin concentration in cattle and its application in improving food safety. *American Journal of Veterinary Research* 59:1101-1107.

Sandberg, F. B., G. C. Emmans, and I. Kyriazakis. 2007. The effects of pathogen challenges on the performance of naïve and immune animals: The problem of prediction. *Animal* 1:67-86.

Schaefer, A. L., S. D. M. Jones, and R. W. Stanley. 1997. The use of electrolyte solutions for reducing transport stress. *Journal of Animal Science* 75:258-265.

Schneider, M. J., R. G. Tait, Jr., W. D. Busby, and J. M. Reecy. 2009. An evaluation of bovine respiratory disease complex in feedlot cattle: Impact on performance and carcass traits using treatment records and lung scores. *Journal of Animal Science* 87:1821-1827.

Secrist, D. S., F. N. Owens, and D. R. Gill. 1997. Effects of vitamin E on performance of feedlot cattle: A review. *The Professional Animal Scientist* 13:47-54.

Self, H. L., and N. Gay. 1972. Shrink during shipment of feeder cattle. *Journal of Animal Science* 35:489-494.

Selye, H. 1976. *The Stress of Life*. New York: McGraw-Hill.

Smith, R. A. 1998. Impact of disease on feedlot performance: A review. *Journal of Animal Science* 76:272-274.

Sowell, B. F., M. E. Branine, J. G. P. Bowman, M. E. Hubbert, H. E. Sherwood, and W. Quimby. 1999. Feeding and watering behavior of healthy and morbid steers in a commercial feedlot. *Journal of Animal Science* 77:1105-1112.

Spears, J. W. 2000. Micronutrients and immune function in cattle. *Proceedings of the Nutrition Society* 59:587-594.

Spears, J. W., and E. B. Kegley. 2002. Effect of zinc source (zinc oxide vs zinc proteinate) and level on performance, carcass characteristics, and immune response of growing and finishing steers. *Journal of Animal Science* 80:2747-2752.

St. Louis, D. G., T. J. Engelken, R. D. Little, and N. C. Edwards. 2003. Case study: Systems to reduce the cost of preconditioning calves. *The Professional Animal Scientist* 19:357-361.

Step, D. L., C. R. Krehbiel, H. A. DePra, J. J. Cranston, R. W. Fulton, J. G. Kirkpatrick, D. R. Gill, M. E. Payton, M. A. Montelongo, and A. W. Confer. 2008. Effects of commingling beef calves from different sources and weaning protocols during a forty-two-day receiving period on performance and bovine respiratory disease. *Journal of Animal Science* 86:3146-3158.

Stookey, J. M., K. S. Schwartzkopf-Genswein, C. S. Waltz, and J. M. Watts. 1997. Effects of remote and contact weaning on behavior and weight gain of beef calves. *Journal of Animal Science* 75(Suppl. 1):157 (Abstract).

Suttle, N. F., and D. G. Jones. 1989. Recent developments in trace element metabolism and function: Trace elements, disease resistance and immune responsiveness in ruminants. *Journal of Nutrition* 119:1055-1061.

TAMU (Texas A&M University System). 2005. Value Added Calf (VAC)-Management Program. ASWeb-120. Texas AgriLIFE Extension Service. Available online at http://animalscience-old.tamu.edu/beef-skillathon/pdf/beef-vac-mgmt.pdf. Accessed on September 6, 2013.

Thrift, F. A., and T. A. Thrift. 2011. Review: Update on preconditioning beef calves prior to sale by cow-calf producers. *The Professional Animal Scientist* 27:73-82.

USDA-APHIS (U. S. Department of Agriculture Animal and Plant Health Inspection Service). 2000a. Attitudes Towards Pre-Arrival Processing in U.S. Feedlots. Info Sheet No. N340.1100. Fort Collins, CO: APHIS. Available online at http://www.aphis.usda.gov/animal_health/nahms/feedlot/downloads/feedlot99/Feedlot99_is_Prearrival.pdf. Accessed on December 1, 2014.

USDA-APHIS. 2000b. *Part II: Baseline Reference of Feedlot Health Management*, 1999. Feedlot National Animal Health Monitoring System Report No. N335.1000. Fort Collins, CO: Centers for Epidemiology and Animal Health. Available online at http://www.aphis.usda.gov/animal_health/nahms/feedlot/downloads/feedlot99/Feedlot99_dr_PartII.pdf. Accessed on December 1, 2014.

Van Koevering, M. T., D. R. Gill, and F. N. Owens. 1991. The effects of escape protein on health and performance of shipping stressed calves. Pp. 156-162 in *Oklahoma State University Animal Science Research*

Report MP-134. Stillwater: Oklahoma Agricultural Experiment Station, Division of Agricultural Science and Natural Resources, Oklahoma State University.

Vincent, J. B. 2001. The bioinorganic chemistry of chromium(III). *Polyhedron* 20:1-26.

Waggoner, J. W., C. A. Loest, C. P. Mathis, D. M. Hallford, and M. K. Petersen. 2009a. Effects of rumen-protected methionine supplementation and bacterial lipopolysaccharide infusion on nitrogen metabolism and hormonal responses of growing beef steers. *Journal of Animal Science* 87:681-692.

Waggoner, J. W., C. A. Loest, J. L. Turner, C. P. Mathis, and D. M. Hallford. 2009b. Effects of dietary protein and bacterial lipopolysaccharide infusion on nitrogen metabolism and hormonal responses of growing beef steers. *Journal of Animal Science* 87:3656-3668.

Ward, J. D., and J. W. Spears. 1999. The effects of low-copper diets with or without supplemental molybdenum on specific immune response of stressed cattle. *Journal of Animal Science* 77:230-237.

Whitney, T. R., G. C. Duff, J. K. Collins, D. W. Schafer, and D. M. Hallford. 2006. Effects of diet for early-weaned crossbred beef steers on metabolic profiles and febrile response to an infectious bovine herpesvirus-1 challenge. *Livestock Science* 101:1-9.

Winterholler, S. J., B. P. Holland, C. P. McMurphy, C. R. Krehbiel, G. W. Horn, and D. L. Lalman. 2009. Use of dried distillers grains in preconditioning programs for weaned beef calves and subsequent impact on wheat pasture, feedlot, and carcass performance. *The Professional Animal Scientist* 25:722-730.

Wythes, J. R., W. R. Shorthose, P. J. Schmidt, and C. B. Davis. 1980. The effect of various rehydration procedures after a long journey on the liveweight, carcasses and muscle properties of cattle. *Australian Journal of Agricultural Research* 31:849-855.

Wythes, J. R., M. J. Brown, W. R. Shorthose, and M. R. Clarke. 1983. Effect of method of sale and various water regimens at saleyards on liveweight, carcass traits and muscle properties of cattle. *Australian Journal of Experimental Agriculture and Animal Husbandry* 23:235-242.

Zecchinon, L., T. Fett, and D. Desmecht. 2005. How *Mannheimia haemolytica* defeats host defence through a kiss of death mechanism. *Veterinary Research* 36:133-156.

Zinn, R. A., and Y. Shen. 1998. An evaluation of ruminally degradable intake protein and metabolizable amino acid requirements of feedlot calves. *Journal of Animal Science* 76:1280-1289.

16

Environment

BACKGROUND

Since the publication of the *Nutrient Requirements of Beef Cattle*, 7th Revised Edition and Update (NRC, 1996, 2000) there has been growing concern among producers, regulators, and the general public about the effects of livestock operations on the environment (Doering, 1994; NRC, 2003, 2012; 40 CFR Parts 122, 123, and 412 [2003]; EPA, 2003; NRCS, 2004; USDA, 2011, 2014). Beef cattle retain only a portion of the nutrients they consume, often less than 20% (Chapter 6, Protein and Amino Acids). The remainder is lost in feces or urine, or via respiration, eructation, and flatulence. The effects of these excreted nutrients, pharmacologically active compounds (PAC), and pathogens on groundwater, surface waters, air quality, global climate change, environmental sustainability, land use, biodiversity, and quality of life are potentially affected by nutritional and management programs used by producers (Steinfield, 2006; USDA, 2014). Although environmental concerns normally revolve around concentrated animal feeding operations (CAFO), some effects (greenhouse gas [GHG] emissions, surface water contamination) can also be a concern in extensive systems such as pasture-based cow-calf and stocker operations, especially when pastures are overgrazed or overfertilized (40 CFR Parts 122, 123, and 412 [2003]; EPA, 1998a,b, 2003).

In North America, most beef cattle are finished in open pens with native soil surfaces. In some regions, cattle are fed in semi-confinement and/or bedding is provided during the winter. Manure (a mixture of feces and urine that sometimes includes soil, bedding, or other materials) is normally removed from feeding pens after each lot of cattle is finished, and the manure is immediately applied to fields, stockpiled for later use, or composted. Manure management can represent up to 10% of fixed costs of a CAFO (Sweeten, 1978). Because several months can elapse between manure excretion and manure collection/utilization, some of the nitrogen (N) and carbon (C) is normally lost to the atmosphere before the manure is used as a fertilizer, resulting in a decrease in its fertilizer value. The quantity and source (runoff, volatilization) of nutrient losses will vary with diet, length of storage, and manure management (Hao, 2007; Hao et al., 2001, 2004, 2005, 2011a,b,c).

Despite concerns about negative effects, cattle can also have beneficial effects on the environment. Cattle are capable of converting forages and industrial and agricultural byproducts that are not suitable as human food, or are potential waste products, into a high-quality food source. In addition, manure from grazing cattle or cattle in CAFO is a potentially valuable fertilizer that can provide important nutrients and organic matter (OM) to soil (Whalen et al., 2000).

Adverse environmental effects are potentially decreased when nutrient inputs are similar to managed nutrient outputs and the whole-farm nutrient balance approaches zero. Under native range conditions, the balance between nutrient inputs and managed nutrient outputs can approach zero (Heitschmidt et al., 1996; Liebig et al., 2010). However, under intensive production systems, such as grazing of fertilized improved forages (Liebig et al., 2010) or confined feeding (Cole et al., 2005b; Cole and Todd, 2009; Erickson and Klopfenstein, 2010), nutrient inputs (as feed or fertilizer, etc.) generally exceed managed outputs for a variety of biological and management reasons.

By more precisely feeding and supplementing livestock to meet their nutrient requirements, excess nutrient losses can be decreased. Under practical conditions, however, the use of precision feeding systems to manage environmental impacts may be limited by factors such as (1) inherent biological inefficiencies in the animal, (2) variability in animal performance and/or nutrient requirements, (3) variability in composition of feed ingredients, (4) seasonal/climatic effects, (5) logistics, (6) economics, and (7) use of safety margins (Galyean, 1999, 2000; Cole, 2003; Vasconcelos and Galyean, 2007).

Water Quality and Soil Concerns

When livestock are concentrated into relatively small geographic areas, excreted feed nutrients are also concentrated into those areas. This can occur in the vicinity of CAFO or in areas of pastures where animals congregate to rest, feed, or water. Soils have variable capacities to store and bind nutrients and chemicals. In extreme cases, nutrient accumulations in soils can be toxic to plants (Van der Watt et al., 1994; Raven and Loeppert, 1997); however, soil nutrients (usually N, C, and phosphorus [P] are of greatest concern) and PAC can potentially run off into surface waters or percolate into groundwater (Boxall et al., 2003; 40 CFR Parts 122, 123, and 412 [2003]; EPA, 2003; Hanselman et al., 2003; Orlando et al., 2004). Nitrogen and P losses to surface waters can potentially cause eutrophication, and losses of C (primarily as OM) to surface waters can potentially affect the biological oxygen demand of the waters, resulting in oxygen depletion and fish kills.

In North America, regulations have been instituted to protect groundwater and surface water (i.e., Clean Water Act, Total Maximum Daily Loads [TMDL], etc.) from excess nutrients and other pollutants. The U.S. Clean Water Act (33 U.S.C. §1251 et seq. [1972]) established the basic structure for regulating discharges of pollutants into surface waters, the permitting process (National Pollutant Discharge Elimination System), and the Effluent Limitation Guidelines and Standards for CAFO (40 CFR Parts 122, 123, and 412 [2003]; EPA, 2003). In the United States, CAFO are required to develop Comprehensive Nutrient Management Plans that address factors such as feed management, manure handling, and land application of manure (NRCS, 2004). Similar regulations are in place in Canada and many other nations. Retention ponds, lagoons, or other runoff control structures (RCS) are usually required to control and capture runoff from open-lot feedyards and manure storage areas to prevent contamination of surface waters and groundwaters. In some regions of North America, naturally ventilated, covered housing, such as monoslope barns and covered barns with slatted floors, are used to prevent runoff from pen areas entering surface waters. Livestock operations with properly designed and maintained RCS normally have little, if any, effect on surface water or groundwater quality.

Because of strict regulations to limit runoff from CAFO, the most likely location for cattle manure to run off into surface waters is from fields that have been fertilized with manure; this is especially the case if the manure is applied to frozen ground and/or in excess of crop needs. Because manure collected from feedyard pens tends to have a lower N:P ratio than most crops require (1:1 to 3:1 vs. 5:1 to 8:1, respectively), manure applied to cropland or pastures to meet the crop's N requirements results in a quantity of manure P that is 5- to 10-fold greater than the P needs of the crop and potentially leads to excess runoff of both soluble and insoluble P (Lander et al., 1998; Kellogg and Lander, 1999; Kellogg, 2000; Kellogg et al., 2000; Andraski et al., 2003).

In areas with impaired surface water quality, agricultural operations may be required to adhere to TMDLs to limit runoff from fields or pastures to lakes or streams. This could require producers to provide filter strips and buffer zones around waterways and to limit animal access to riparian areas. Areas where animals congregate (water or feed sources, shade, wind breaks, etc.) in pastures could be isolated from riparian areas to decrease runoff of manure to surface waters. By providing water and supplements in multiple areas, potential effects of animal congregation on the environment can be decreased (Bowman and Sowell, 1997, 2003; Sowell et al., 1999; White et al., 2001; Tate et al., 2003). Nutrients in supplemental feeds can also be excreted and thus help decrease fertilizer requirements on pastures (Greenquist et al., 2009, 2011).

Atmospheric Concerns

The atmospheric emissions from cattle operations of greatest concern vary with geographic location and proximity to neighbors. Particulate matter (PM; dust), odors (volatile organic compounds [VOC]), and hydrogen sulfide (H_2S) are usually a local challenge related to quality-of-life or human health concerns. In contrast, emissions of ammonia (NH_3) and GHG (methane [CH_4], nitrous oxide [N_2O], and carbon dioxide [CO_2]) are usually regional or global in scope (NRC, 2003). Ammonia emissions are related to regional haze, eutrophication of surface waters (Hutchinson and Viets, 1969), adverse effects on sensitive ecosystems (Van der Eerden et al., 1998; Krupa, 2003; Todd et al., 2004, 2008b), and formation of small particulates ($PM_{2.5}$), whereas, GHG are related to global climate change. Although not a GHG, ammonia may deposit onto soils downwind of its source and be converted to the GHG N_2O via nitrification and denitrification. In some regions of North America, where large agricultural operations, cities, and automobile traffic are in close proximity to each other, reactive VOC are a concern because they can contribute to the formation of ground-level ozone and smog (Hobbs et al., 2004). There are a variety of federal and state regulations for protecting air quality in North America, among these are the Clean Air Act, and the Emergency Planning and Community Right to Know Act.

Air pollutants can originate from many sources at livestock operations, including the cattle, pen or pasture surfaces, manure stockpiles, feed or cattle alleys, feed mills, feed storage areas, and RCS. The quantity, composition, and timing of emissions depend on factors such as environmental conditions (i.e., air temperature, wind, atmospheric stability, precipitation, humidity), animal stocking density, animal weight and/or age, stage of production, the diet fed, manure management techniques, and pasture/pen surface characteristics.

DIETARY EFFECTS ON NUTRIENT EXCRETION

Assuming an average dry matter intake (DMI) of 9 kg, a dry matter (DM) digestibility of 75%, and that the diet is balanced for crude protein (CP; 12 to 15% DM) and minerals, during an average 150-d feeding period, 1,000 finishing beef cattle will excrete approximately the following quantities of nutrients: 338,000 kg DM; 243,000 kg C; 24,000 kg N; 6,000 kg calcium (Ca); 3,600 kg P; 1,800 kg magnesium (Mg); 1,500 kg sulfur (S); 6,300 kg chlorine (Cl); 8,300 kg potassium (K); 2,800 kg sodium (Na); 180 g cobalt (Co); 17 kg copper (Cu); 860 g iodine (I); 100 kg Fe (Fe); 40 kg manganese (Mn); 200 g selenium (Se); and 50 kg zinc (Zn). Factors such as stage of production, diet composition, forage quality, grain and forage processing, use of bedding, feed additives, supplementation strategies, etc. can influence the route of excretion (feces, urine, respiration), the quantity of nutrients excreted, and the movements and transformations of those nutrients.

The chemical composition of manure collected from feedlot pens is highly variable. This variability is due to factors such as the diet fed, the length and type of manure storage, type of housing, timing and method of manure collection, pen surface, location in pen, bedding used, season of year, soil contamination, etc. (Erickson et al., 2003a; Cole et al., 2009; ASABE, 2010; Buttrey et al., 2012; Luebbe et al., 2012a,b).

Factors Affecting Manure Output

As previously noted, manure is a mixture of feces and urine that can also contain soil, bedding, or other materials. The quantity of feces excreted is normally determined by the quantity of feed consumed and the digestibility of that feed. Diet digestibility is determined by factors such as forage/fiber quantity and quality, starch intake and gelatinization, rate of passage, and grain and forage processing. Dietary effects on urine excretion are more complex. According to the American Society of Agricultural and Biological Engineers standards (ASABE, 2010), beef cattle fed high-concentrate diets excrete approximately 28 kg wet manure daily (85% moisture) per 454 kg body weight (BW). In general, as the fiber content of finishing diets increases, the quantity of fecal OM excreted also increases because of greater feed intake and/or lower diet digestibility (Hales et al., 2014). In addition, the digestibility of the concentrate portion of the diet can also affect manure production. For example, compared with feeding dry-rolled corn (DRC) or dry-rolled grain sorghum, feeding finishing diets based on steam-flaked corn (SFC) or steam-flaked grain sorghum decreases total manure DM and OM excretion and manure OM collected from feedlot pens by 10 to 30% (Tucker and Watts, 1993; Corrigan et al., 2009; Buttrey et al., 2012; Hales et al., 2012a; Luebbe et al., 2012a,b).

The replacement of feed grains with high-fiber byproducts such as corn gluten feed (CGF) or distillers grains with solubles (DGS) in finishing diets can decrease overall diet digestibility, and thereby affect manure output and composition. Corrigan et al. (2009) reported that feeding 40% wet DGS (WDGS) in DRC-, SFC-, and high-moisture corn (HMC)-based finishing diets decreased DM and OM digestion by about 5 percentage points but did not significantly affect neutral detergent fiber (NDF), fat, or starch digestion. In other studies, feeding low concentrations (5 to 15%) of WDGS did not affect DM or OM excretion from cattle fed SFC-based diets (Cole et al., 2011); however, at greater concentrations, replacing SFC with WDGS in finishing diets linearly decreased digestion of OM and starch and linearly increased digestibility of NDF (Luebbe et al., 2012b). Hales et al. (2013) reported that when SFC-based finishing diets were fed to cattle at equal energy intake, digestion of NDF and energy decreased linearly with increasing dietary WDGS concentration (0, 15, 30, and 45% WDGS) and excretion of OM increased 10 to 25% when the concentration of WDGS was increased from 0 to 30% of dietary DM (Hales et al., 2012a, 2013). In contrast, Spiehs and Varel (2009) reported no effect of WDGS concentration (0, 20, 40, or 60%) on the digestibility of DRC-based diets—in part a result of lower DMI by cattle fed WDGS. The quantity of manure OM collected from pen surfaces increased about 42% when WDGS concentration in SFC- and DRC-based diets were increased from 0 to 20% of DM (Buttrey et al., 2012). Similarly, Luebbe et al. (2012a) reported that the quantity of wet manure and manure OM removed from feedlot pens increased by 30 to 94% as the concentration of WDGS in DRC-based diets was increased from 0 to 30% of dietary DM.

With DRC-based diets, feeding 35% wet corn gluten feed (WCGF) increased the total manure DM collected by about 35% and manure OM by about 53%, probably because of the 33% increase in DM excreted (Farran et al., 2006). Sindt et al. (2003) with SFC-based finishing diets and Montgomery et al. (2004) with SFC-based growing diets both reported that increasing WCGF in the diet decreased DM digestibility but increased NDF digestibility.

Factors Affecting Nitrogen Excretion

The primary factors affecting N excretion are the quantity and form of N consumed in the diet (see Chapter 2, Anatomy, Digestion, and Nutrient Utilization, and Chapter 6). In general, as dietary N intake increases, excretion of N increases. Once the dietary ruminally degradable protein (RDP) and metabolizable protein (MP) requirements are met, the increased N excretion occurs primarily in the urine (Vasconcelos et al., 2009; Koenig and Beauchemin, 2013a,b). Vasconcelos et al. (2009) noted that the daily N retention of cattle fed a finishing diet decreased (43, 44, and 31 g/d) and daily N excretion increased (77, 76, and 103 g/d) as days on feed increased from approximately 30 to 85 to 155 d.

With DRC-based diets, excretion of N increased linearly as the concentration of WDGS in the diet was increased from 0 to 60% (Spiehs and Varel, 2009; Luebbe et al., 2012a). Buttrey et al. (2012) similarly noted that feeding 20% WDGS (compared with 0% WDGS) increased total N collected in manure, but did not affect the proportion of consumed N recovered in manure or the concentration of N in the manure. Hales et al. (2013) reported that feeding WDGS in iso-fat, SFC-based diets increased excretion of N by 28 to 46% (95 to 140 g/d for 0 and 45% WDGS, respectively). Buttrey et al. (2012) noted that recovery of consumed feed N in manure tended to be greater in cattle fed DRC-based diets (40%) than in cattle fed SFC-based diets (29%). Feeding WCGF increased N intake and excretion, but did not affect the quantity of N apparently lost via volatilization (Farran et al., 2006)

Factors Affecting Phosphorus Excretion/Losses

Erickson et al. (1999, 2002) and Geisert et al. (2010) noted that performance of yearling steers and calves was not adversely affected by feeding finishing diets with P intakes as low as 14.1 g/animal daily (0.17% of dietary DM). Similarly, in growing steers, Greene et al. (2001) noted that decreasing the dietary P concentration from 0.33 to 0.22% decreased P intake and P excretion by 50% or more, but it did not affect average daily gain (ADG) or gain:feed. Vasconcelos et al. (2009) noted that replacing cottonseed meal with urea as the supplemental CP source decreased dietary P concentration by 29% and P excretion by 20%. The quantity of P excreted increased with days on feed (59, 74, and 88% of P intake on days 30, 85, and 155, respectively). Spiehs and Varel (2009) and Luebbe et al. (2012a) found that excretion of P increased as the concentration of WDGS in the diet increased from 0 to 60% of diet DM. The effect of WDGS on the recovery of consumed feed P in manure was somewhat inconsistent. Buttrey et al. (2012) reported that recovery of fed P in pen manure tended to be lower in cattle fed DRC-based (71% of intake) than cattle fed SFC-based diets (77%). When WDGS replaced 20% of corn in the diet, P intake and P recovered in manure increased, and recovery of fed P in manure tended to decrease. However, the P concentration (DM basis) in the collected manure was not changed (Buttrey et al., 2012).

It may be possible to decrease the quantities of minerals required in the diet or supplement by feeding more biologically available mineral sources. This would subsequently lead to a decrease in the quantities that are excreted to the environment. Nonetheless, the bioavailability of different mineral sources for ruminants has not been thoroughly studied, and values in the literature vary such that is it is not possible to make recommendations (Henry et al., 1988; Greene, 1995; see Chapter 7, Minerals).

Estimating the Quantity and Route of Nutrient Excretion

There are a number of potential methods to estimate nutrient excretion by beef cattle. The ASABE and the American Society of Animal Science (Erickson et al., 2003b; ASABE, 2010) jointly developed the ASABE Standard 384.2 MAR2005 (R2010) to estimate excretion of DM, OM, N, P, and Ca and to estimate the as-excreted composition of livestock manures. Most of the calculations were based on the fundamental models:

$$\text{Nutrient excretion} = (\text{Feed nutrient intake}) - (\text{Nutrient retained})$$

or

$$\text{Nutrient excretion} = (\text{Feed nutrient intake}) \times (1 - \text{Apparent nutrient digestibility})$$

Retention of N, P, and Ca were estimated using NRC (1996, 2000) equations. Fecal excretion was estimated from typical nutrient digestibilities and urinary excretion was based on BW, weight gain, relative BW, and/or days on feed. The equations were developed from 300 observations of urine excretion by beef cattle ranging in weight from 100 to 620 kg. Values for beef cows and growing calves were taken from the MidWest Plan Service-18 Section 1 (MWPS, 2004), the Natural Resources Conservation Service Agricultural Waste Management Field Handbook (NRCS, 2004), and an earlier version of the American Society of Agricultural Engineers D384.1 (ASAE, 1999). The Beef Cattle Nutrient Requirements Model (BCNRM; Chapter 19, Model Equations and Sensitivity Analyses) used in this publication is based on a similar system to estimate nutrient excretion.

Waldrip et al. (2013) and Dong et al. (2014) used data from independent nutrient balance studies to develop regression equations to predict urine N excretion, fecal N excretion, and urine N excretion as a proportion of total N excretion of beef cattle fed growing-finishing diets using dietary CP concentration or total N intake. Many of the equations were very similar; however, the equations of Dong et al. (2014) tended to have smaller root mean square errors (RMSE), possibly because they used a larger database than Waldrip et al. (2013). Equations based on daily N intake tended to be more accurate than models based on dietary CP concentration. Some of the simplest but most accurate equations developed by Dong et al. (2014) were as follows:

$$\text{Urine N} = 2.39\ (\pm 4.05) + 0.55\ (\pm 0.03)\ \text{N intake} - 3.36\ (\pm 0.70)\text{DMI}$$
$$r^2 = 0.78,\ \text{RMSE} = 3.88\ \text{g/d}, \qquad \text{(Eq. 16-1)}$$

$$\text{Fecal N} = 66.16\ (\pm 2.59) + 0.29\ (\pm 0.01)\ \text{N intake} - 0.94\ (\pm 0.04)\ \text{N digestibility}$$
$$r^2 = 0.92,\ \text{RMSE} = 1.58\ \text{g/d}, \qquad \text{(Eq. 16-2)}$$

$$\text{Fecal N} = 15.82\ (\pm 1.88) + 0.20\ (\pm 0.01)\ \text{N intake}$$
$$r^2 = 0.65,\ \text{RMSE} = 2.68\ \text{g/d}, \qquad \text{(Eq. 16-3)}$$

$$\text{Urine N/Total N} = -0.162\ (\pm 0.04) + 0.010\ (\pm 0.0006)\ \text{N digestibility}$$
$$r^2 = 0.70,\ \text{RMSE} = 0.022, \qquad \text{(Eq. 16-4)}$$

$$\text{Urine N/Total N} = 0.328\ (\pm 0.03) + 0.016\ (\pm 0.002)\ \text{Dietary CP}$$
$$r^2 = 0.26,\ \text{RMSE} = 0.027, \quad \text{(Eq. 16-5)}$$

where

Urine N is urinary N excretion, g/d;
N intake is N intake, g/d;
Fecal N is fecal N excretion, g/d;
Urine/Total N is urinary N as a proportion of total N excreted;
DMI is dry mater intake, kg/d;
N digestibility is digestibility of N in the total tract; and
Dietary CP is dietary CP as % DM.

In the studies by Waldrip et al. (2013) and Dong et al. (2014), the equations seemed to work best within the N intake range of about 100 to 300 g/d, which would correspond to dietary CP concentrations ranging from approximately 6 to 23% of DM. In addition, in both studies, if the calculated urine and fecal N excretion are used to estimate N retention by difference, the calculated N retention ranged from approximately 25 to 28% of N intake. These are values typically seen in N balance trials, but they tend to be greater than N retentions typically calculated using animal performance or the NRC (1996, 2000) model. It also should be noted that if fecal and urinary N excretion (g/d) are calculated separately, the sum of fecal and urine N excretion typically does not equal the total N excreted (g/d). H. Waldrip (Conservation and Production Research Laboratory, U.S. Department of Agriculture-Agricultural Research Service [USDA-ARS], Bushland, TX, personal communication, Sept., 2014) subsequently developed a nonlinear equation to calculate the proportion of total N excreted that was excreted in urine. This non-linear equation had a greater r^2 than the linear and multiple regression equations based on N intake presented by Waldrip et al. (2013) and Dong et al. (2014) and is more biologically explainable. Therefore, the committee opted to calculate urinary (and thus fecal) N excretion using this single nonlinear equation:

$$\text{Urine N/Total N} = 0.2331\ (\ln \text{N intake}) - 0.6244$$
$$r^2 = 0.548,\ \text{RMSE} = 0.068. \quad \text{(Eq. 16-6)}$$

Geisert et al. (2010) reported that total P excretion of beef cattle increased linearly as P intake increased from about 8 g/animal daily to about 55 g/animal daily and thus P excretion could be estimated from P intake using the following equation ($r^2 = 0.49$):

$$\text{P excretion, g/d} = 0.82 + 0.57\ \text{P intake, g/d} \quad \text{(Eq. 16-7)}$$

Most feed grains contain at least 0.25% P and many supplemental protein sources and byproducts contain more than 0.5% P. Thus, supplemental P is not needed in most grain-based or byproduct-based feedlot growing or finishing diets.

GREENHOUSE GAS EMISSIONS

Cattle operations are a potential source of GHG emissions. The GHG of greatest concern from beef cattle are CO_2, CH_4, and N_2O. The global warming potential (GWP) of these GHG differs considerably and is based on CO_2 equivalents (CO_{2e}). The GWP of CH_4 is normally 21 to 25 and of N_2O is 298 to 310 times that of CO_{2e} (IPCC, 2006; EPA, 2011; USDA, 2011). Methane emissions come from the animal directly (via enteric fermentation, see Chapter 2) and by fermentation of manure on pasture, on the surface of pens, in stockpiles, or in RCS. Nitrous oxide emissions occur via nitrification and denitrification of manure N on pastures or pen surfaces or from nitrification and denitrification of ammonia that is deposited onto soils. Carbon dioxide emissions occur via normal animal metabolism and from energy used on-farm (e.g., cropping, feed processing, feeding cattle) and to supply inputs for feed production (e.g., feed, fertilizer, herbicide manufacture, etc.). Biogenic CO_2 emissions from the animal are usually assumed to be equivalent to CO_2 captured via photosynthesis by the pastures and crops consumed by the animal and are thus not considered in calculating the C footprint (i.e., the sum of the GHG emissions expressed as CO_{2e}). Carbon dioxide emissions can also be affected by changes in land use that induce gain or loss of soil C. For example, decreasing tillage intensity, planting perennial forages in rotation, or establishing permanent grass on cropland will increase soil C sequestration, leading to removal of atmospheric CO_2, whereas increased tillage intensity leads to soil C losses and atmospheric CO_2 emissions.

Published C footprints (sum of the GHG emissions expressed as CO_{2e}) of typical cattle production systems can be variable and are affected by nutritional and management practices (Peters et al., 2010). Five studies recently estimated the C footprint of various North American beef cattle production systems: Vergé et al. (2008) and Beauchemin et al. (2010) in Canada, Pelletier et al. (2010) and Lupo et al. (2013) in the midwestern United States, and Stackhouse et al. (2012) and Stackhouse-Lawson et al. (2012) in California. In these studies, the C footprint of various beef cattle production scenarios ranged from 10.4 to 19.2 kg CO_{2e}/kg final BW or 16.7 to 32.5 kg CO_{2e}/kg HCW. Results of the five studies were remarkably similar. From 55 to 63% of the total CO_{2e} was via enteric CH_4, 18 to 23% was via manure N_2O, and 14 to 24% was from secondary emissions and fossil energy use. The cow-calf phase accounted for 64 to 80%, the stocker-grazing phase accounted for 8 to 20%, and the feedlot phase accounted for 12 to 16% of total CO_{2e} emitted by the beef cattle industry.

In these five studies, the daily C footprint was consistently greater during the stocker-grazing phase than during the feedlot-finishing phase. In addition, two of the studies (Pelletier et al., 2010; Stackhouse et al., 2012) reported that the C footprint was slightly less for calves weaned on pasture and moved directly to the feedlot (calf-fed; 21.1 and 23.0 kg CO_{2e}/kg HCW, respectively) than for cattle that went through a stocker-grazing phase before entering the feedlot (22.6 and 26.1 kg CO_{2e}/kg HCW, respectively). Likewise, both Pelletier et al. (2010) and Lupo et al. (2013) noted that the C footprint of cattle finished on grass was greater than calf-fed cattle finished on typical grain- and byproduct-based diets, but was similar to cattle that went through a stocker-grazing phase before entering the feedlot. Factors such as animal performance, final BW, HCW, pasture C sequestration (Phetteplace et al., 2001; Liebig et al., 2010), and carcass dressing percent had major effects on the C footprint.

Enteric Methane—"Typical" Emissions: Grazing Cattle

Enteric CH_4 production of grazing cattle is highly dependent on forage quality. For example, Westberg et al. (2001) noted a large seasonal variation in enteric CH_4 emissions of beef cows grazing stockpiled native range in October and the same pasture during lush spring growth in May. In October, when cows were losing BW, they produced an average of 87 g CH_4/animal daily, whereas, on the same pasture in May, they produced approximately 252 g CH_4/animal daily. Lassey (2007) summarized much of the CH_4 emissions data from grazing cattle determined using the SF_6 tracer technique (Johnson et al., 1994). Forage intake was calculated from either a requirements model or from the use of external markers. In vitro forage digestibility ranged from 49 to 83%. Calculated CH_4 conversion factors (Ym) ranged from 3.7 to 9.5% of gross energy intake (GEI). The mean Ym from all of the studies was 6.25 of GEI, which agrees well with the Ym of 6.5% for grazing beef cattle recommended by the International Panel on Climate Change (IPCC, 2006).

Enteric Methane—"Typical" Emissions: Feedlot Finishing Cattle

Published enteric CH_4 losses from finishing beef cattle range from 36 to 145 g/animal daily (Johnson and Johnson, 1995; Harper et al., 1999, 2013; McGinn et al., 2004, 2009; Beauchemin et al., 2008; Loh et al., 2008; Vergé et al., 2008; Hales et al., 2012a, 2013; Todd et al., 2014). The IPCC (2006) Tier 2 CH_4 Ym for feedlot cattle is 3 ± 1% of GEI, which is similar to values reported for cattle fed corn-based high-concentrate finishing diets (Harper et al., 1999, 2013; Beauchemin and McGinn, 2005; Hales et al., 2012a, 2013, 2014; Todd et al., 2014) but less than diets based on barley or with moderate (15 to 35%) roughage concentrations (McGinn et al., 2004, 2009; Beauchemin and McGinn, 2005, 2006a; Beauchemin et al., 2007b, 2008).

McGinn et al. (2008) used a backward Lagrangian stochastic model to estimate CH_4 emissions (enteric + pen surface) from feedlots in Australia and Canada. In their paper, the values determined in Australian feedyards were less than the emissions measured in Canadian feedyards (166 vs. 124 g/animal daily) possibly because of a difference in feed grain source (sorghum in Australia, barley in Canada). The CH_4 emissions had a significant diel pattern. Todd et al. (2014) and Harper et al. (2013) also noted a diel pattern, possibly related to feeding, in CH_4 emissions from two Texas feedyards. In their studies, total CH_4 emissions ranged from 79 to 183 g/animal daily and Ym ranged from 2.4 to 3.8% of GEI.

Factors Affecting Enteric Methane Losses

The dietary and management factors that can affect enteric CH_4 production have been reviewed over the past several years (Pelchen and Peters, 1998; Kurihara et al., 1999; Benchaar et al., 2001; Monteny and Chadwick, 2006; Beauchemin et al., 2007b, 2008; Molano and Clark, 2008; Hegarty et al., 2010; Cottle et al., 2011; Hristov et al., 2013a,b). In general, many of the effects seem to be greater in cattle fed high-forage diets than in cattle fed high-concentrate diets, most probably because of the greater enteric CH_4 production on high-forage diets.

The following is a brief review of the literature on the factors that influence enteric CH_4 production in beef cattle. Benchaar et al. (2001) used the model of Dijkstra et al. (1992) to estimate the effects of dietary regimens on enteric CH_4 production of beef and dairy cattle. In brief, enteric CH_4 losses (g/d) decreased when (1) DMI decreased, (2) the dietary forage:concentrate ratio decreased, (3) fibrous concentrate (beet pulp) was replaced with starchy concentrate (barley), (4) corn replaced barley, (5) silage replaced hay, and (6) forage particle size was decreased via processing. Enteric CH_4 losses, as a percent of GEI, decreased when (1) DMI increased, (2) forage:concentrate ratio decreased, (3) fibrous concentrate (beet pulp) was replaced with starchy concentrate (barley), (4) corn replaced barley, (5) silage replaced hay, and (6) forage particle size was decreased via processing.

The effects of animal genetics on enteric CH_4 production are not clear. Hegarty et al. (2007) reported that cattle selected for greater feed efficiency (i.e., with lower residual feed intake [RFI]) had lower daily enteric CH_4 production (g/d), but CH_4 production standardized for intake (g/kg DMI) was similar. In contrast, Freetly and Brown-Brandl (2013) reported that cattle selected for lower RFI actually had greater enteric CH_4 production (measured over 6 h) when adjusted for DMI.

Forage Concentration and Quality

In general, as the proportion of forage in the diet increases, the proportion of GEI lost as CH_4 (i.e., Ym) also increases (Blaxter and Wainman, 1964; Harper et al., 1999; Yan et al., 2000, 2009; IPCC, 2006; Hales et al., 2014). Nonetheless, total daily CH_4 emissions can decrease because of lower GEI (Johnson and Johnson, 1995). In addition, total CH_4 emissions over the animal's lifetime, or over the feeding period, can also be greater with increased forage feeding because more days are required to reach final market weight. Based on in vitro studies, Dijkstra et al. (2007) suggested that enteric CH_4 losses from starch, soluble sugars, hemicellulose, and cellulose were about 2.11, 3.18, 3.38, and 3.10 g/kg substrate, respectively, at a ruminal pH of 6.5. Hales et al. (2014) reported that enteric CH_4 emissions (Mcal/d and % of GEI) increased linearly (3.1, 3.4, 3.8, and 4.2% of GEI, respectively) as the concentration of alfalfa hay increased (2, 6, 10, or 14%) in DRC-based finishing diets containing 25% WDGS. Similarly, Doreau et al. (2011) found that enteric CH_4 production was greater (6.9% of GEI) in hay- and silage-based diets than in grain-based diets (3.2% of GEI). Total (enteric + manure) CH_4 production was least for a corn-based diet, whereas N_2O and CO_2 emissions from fossil fuel use and feed production were greatest with the corn-based diet; however, total GHG emissions were least for the corn-based diet.

Grain Source and Grain Processing

Grain source and grain processing method can affect enteric CH_4 losses, most likely by their effects on ruminal starch digestion. In general, the greater the ruminal starch digestibility, the lower the enteric CH_4 Ym. At constant energy intake (2× maintenance), Hales et al. (2012a) reported approximately 20% lower (2.5 vs. 3.0% of GEI) enteric CH_4 emission in cattle fed SFC-based diets than in steers fed DRC-based diets. Archibeque et al. (2006) noted similar responses with the feeding of HMC-based compared with DRC-based diets. Beauchemin and McGinn (2005) noted significantly lower enteric CH_4 in cattle fed DRC-based diets (2.8% of GEI) than for diets based on steam-rolled barley (4% of GEI), possibly as a result of the higher fiber content of barley.

Dietary Lipids

A number of studies have reported that dietary fat affects enteric CH_4 emissions. In two reviews, Beauchemin et al. (2007b, 2008) calculated that for beef cattle, dairy cattle, and sheep the enteric CH_4 production (g/kg DMI) decreased approximately 5.6% for each 1% increase in dietary added fat, and Martin et al. (2010) calculated for ruminants an average decrease of 3.8% (as g/kg DMI) with each 1% addition of fat. Moraes et al. (2014) reported that enteric CH_4 production of lactating and nonlactating dairy cows decreased 0.19 to 0.38 MJ/d for each 1% increase in dietary ether extract, but noted no significant effect of dietary fat on enteric methane production of growing steers and heifers. Using data for beef cattle, P. Escobar-Bahamondes and K. Beauchemin (Agriculture and Agri-Food Canada Research Centre, Lethbridge, Alberta, Canada, personal communication, June 2014) observed that increasing dietary fat concentration decreased CH_4 production (g/kg DMI) only for cattle consuming low levels of intake, and that at higher DMI (>8.2 kg/d), CH_4 production (g/kg DMI) actually increased with increasing dietary fat concentration.

Distillers Grains and Other Co-products

McGinn et al. (2009) reported that feeding corn-based DGS decreased enteric CH_4 emissions and attributed the effect to an increase in dietary fat concentration. This hypothesis was later substantiated in a study by Hunerberg et al. (2013) with steers fed growing diets. Hunerberg et al. (2013) noted that supplementing steers with corn-based dried DGS (DDGS) decreased enteric CH_4 production, whereas supplementing with wheat-based DDGS did not affect enteric CH_4. When corn oil was added to the wheat-based DDGS diet so that dietary fat content was similar to the corn-based DDGS diet, enteric CH_4 was decreased similarly to that with corn-based DDGS.

Hales et al. (2013) reported that when control (no DGS) and treatment (containing DGS) diets had similar fat content, the feeding of 15% WDGS did not affect enteric CH_4 production (2.4 vs. 2.5% of GEI), but feeding higher (30 to 45% of DM) concentrations of WDGS increased enteric CH_4 (2.9 and 3.8% of GEI, respectively). The increased enteric CH_4 emissions at higher WDGS inclusion rates were probably the result of greater fiber and lower starch intakes. Effects of 30% WDGS (DM basis) on enteric CH_4 production were similar for SFC- and DRC-based diets (Hales et al., 2012a).

Using lambs, Avila-Stagno et al. (2013) reported that feeding glycerol, a byproduct of the biodiesel industry, at concentrations of 0, 7, 14, or 21% dietary DM did not affect ruminal CH_4 production.

Dry Matter Intake

The classical work of Blaxter and Wainman (1964) and Blaxter and Clapperton (1965) indicated that enteric CH_4 emission, as a percent of GEI, was 23% greater in steers fed at maintenance energy than in steers fed at 2× maintenance energy (average 8.1 vs. 6.6% of GEI, respectively). Similarly, Beauchemin and McGinn (2006a) noted that enteric CH_4 Ym decreased approximately 0.773 percentage points for each unit (expressed as level of intake above maintenance) increase in DMI. This finding was similar to that of Yan et al. (2000; 0.78%) but less than the estimate using the Blaxter and Clapperton (1965) equation (0.93 to 1.28 percentage points) or the value suggested by Johnson and Johnson (1995; 1.6 percentage points). In contrast, with ryegrass diets, Molano and Clark (2008) reported that as DMI increased from 0.75× of maintenance to 2× maintenance, CH_4 Ym (range of 4.9 to 9.5% of GEI or 15.9 to 30.4 g/kg DMI) was not affected by DMI, although total daily enteric CH_4 emissions (g/d) increased linearly (r^2 = 0.80 to 0.84) with increasing DMI. Mills et al. (2010) reported that the enteric CH_4 Ym for beef and dairy cattle decreased approximately 0.103% for each kilogram increase in DMI; however, much of this effect seemed to occur at DMI above 12 kg/d, values appreciably greater than under most beef production scenarios. In addition, the variability was greater at intakes below 12 kg than above 12 kg/d.

Ionophores

Tedeschi et al. (2003, 2011) and McGinn et al. (2004) reported that the ionophore monensin decreased enteric CH_4 emissions by 10 to 25% in short-term finishing cattle studies. In contrast, longer-term studies suggest that in feedlot cattle, the inhibitory effect might be transitory, lasting for only 30 days or less (Guan et al., 2006). In agreement with Guan et al. (2006), Cooprider et al. (2011) and Stackhouse et al. (2012) reported that the daily CH_4 production (g/d) of cattle fed through a "natural" program, with no growth-promoting implants or dietary antibiotics or ionophores, was similar to cattle fed in more traditional systems that used anabolic implants, dietary ionophores, and β-agonists. Nonetheless, because "traditional" cattle had greater ADG and took fewer days to reach the same BW end point, cattle fed using available growth-promoting technologies had 31% lower GHG emissions per animal and 28% lower CH_4 emissions per unit of BW gain (3.92 vs. 5.02 CO_{2e}/kg BW gain, respectively) than cattle finished on a "natural" program.

Other Potential Feed Additives

Various chemical compounds with a specific inhibitory effect on rumen methanogens have been evaluated, including bromochloromethane, 2-bromoethane sulfonate, chloroform, and cyclodextrin (Johnson, 1972, 1974; Trei et al., 1972; Cole and McCroskey, 1975; Tomkins and Hunter, 2004; Tomkins et al., 2009; Abecia et al., 2013; Hristov et al., 2013a). Although these compounds have been very effective in decreasing CH_4 (up to 50% decreases), their safety, environmental impact, and long-term efficacy are barriers to their use. As research into the development of CH_4 inhibitors continues, it could be possible in the future to use other compounds to decrease CH_4 emissions from cattle, provided safety is established and consumers accept such an approach. One such promising compound is 3-nitrooxypropanol, which is noncarcinogenic and nonmutagenic and has been shown in initial short-term studies to be safe and effective in cattle. Recent studies have shown up to a 60% decrease in enteric CH_4 production in beef and dairy cattle fed moderate- to high-roughage diets, with no effects on diet digestibility (Haisan et al., 2013, 2014; Martinez-Fernandez et al., 2013; Romero-Perez et al., 2014). Similar to bromoethane sulfonate, 3-nitrooxypropanol is a structural analog of methyl-coenzyme M, a cofactor involved in the terminal step of methanogenesis that transfers a methyl group to methyl-coenzyme M reductase.

In addition to inhibitors, compounds such as nitrate (Weichenthal et al., 1963; Lichtenwalner et al., 1973; van Zijderveld et al., 2010; Hulshof et al., 2012; Hegarty et al., 2013; Velazco et al., 2013), sulfates (van Zijderveld et al., 2010), and nitroethane (Anderson and Rasmussen, 1998; Anderson et al., 2003) that act as electron acceptors have been shown to decrease CH_4 emissions from cattle. Results from a range of studies suggest that feeding calcium nitrate to cattle decreases CH_4 by approximately 10% per 1% supplementary nitrate. Because nitrate is potentially toxic, a gradual acclimation of animals to nitrate is required as poisoning can be triggered by higher blood methemoglobin levels in nonadapted cattle. At this time, supplemental nitrate is not permitted in cattle feed in Canada or the United States.

Other compounds such as plant secondary compounds (saponins, tannins, phenolics, essential oils), dicarboxylic acids (fumarate, malate, acrylate), direct-fed microbials (yeast cultures, bacteria), and enzymes have decreased enteric CH_4 emissions. Nonetheless, results have been inconsistent over a range of studies (McGinn et al., 2004; Beauchemin and McGinn, 2006b; Beauchemin et al., 2007b; Ungerfeld et al., 2007; Eckard et al., 2010; Martin et al., 2010; Soliva et al., 2011; Chung et al., 2013; Jayanegara et al., 2013).

Predicting Enteric Methane Production

Over the past 80+ years a variety of empirical and mechanistic models have been developed to estimate enteric CH_4 production from ruminants (Table 16-1, modified from USDA, 2014). These range from simple linear regression equations based on DMI to mechanistic models based on H ion balance within the rumen. Most published models were developed from data sets that used sheep or dairy cattle fed high-forage diets. Few experiments have used beef cows or stocker cattle grazing native range or cattle fed high-concentrate finishing diets typical of today's North American cattle industry. Therefore, the ability of many models to accurately predict enteric CH_4 production under these conditions could be limited.

Dietary carbohydrate composition and DMI seem to be the two factors with the greatest effect on enteric CH_4 emissions; therefore, most prediction equations and models include these factors. Yan et al. (2000, 2009) suggested that if the level of feed intake is not considered in the model, the model will generally underpredict enteric CH_4 production when DMI is low and overpredict enteric CH_4 production when DMI is high. Nonlinear models usually have the advantage of being able to theoretically determine the minimum and maximum enteric CH_4 production rates, and thus are more biologically explainable than linear equations, which can give unrealistically high enteric CH_4 production rates when DMI is high. In addition, the use of linear empirical models is often limited to the range of data used to develop the model (Kebreab et al., 2006). In contrast, nonlinear models can be used to extrapolate beyond the range of data used to develop the equation and can be used (as well as mechanistic models) to evaluate CH_4 mitigation strategies (Kebreab et al., 2006).

The most frequently cited mechanistic models used to predict enteric CH_4 production in cattle are MOLLY (Baldwin, 1995) and COWPOLL (Mills et al., 2001). Both estimate ruminal CH_4 production based on hydrogen balance within

TABLE 16-1 Empirical and Mechanistic Models Developed to Estimate Enteric CH_4 Emissions from Cattle[a]

Reference	Model Inputs and Comments
Kriss (1930)	Dry matter intake (DMI)
Bratzler and Forbes (1940)	Digested carbohydrates
Axelsson (1949)	DMI
Blaxter and Clapperton (1965)	Digestible energy (DE), % at the maintenance level of feeding, gross energy intake (GEI), and feeding level (multiple of maintenance)
Moe and Tyrrell (1979)	Digestible soluble carbohydrates, digestible hemicellulose, digestible cellulose
Holter and Young (1992)	Digestible soluble carbohydrates, cellulose, hemicellulose, fat intake
COWPOLL (Dijkstra et al., 1992; Mills et al., 2003; Bannink et al., 2006; Kebreab et al., 2008)	DMI, neutral detergent fiber (NDF), degradable NDF, total starch, degradable starch, soluble sugars in diet, dietary N, total ammoniacal nitrogen (NH_x-N) in diet, indigestible protein, rate of degradation of starch and crude protein (CP)
MOLLY (Baldwin, 1995)	Similar to COWPOLL
Mills et al. (2001)	DMI
Phetteplace et al. (2001)	Animal class, animal age and body weight (BW), quantity of meat/milk produced, feed type, feed intake, manure management
Mills et al. (2003)	Metabolizable energy intake (MEI), starch and acid detergent fiber (ADF) intake
Kebreab et al. (2004, 2009)	DMI, NDF, degradable NDF, total starch, degradable starch, soluble sugars in diet, diet N, NH_x-N in diet, indigestible protein, rate of degradation of starch and protein
Yan et al. (2000)	Grass silage diets: DE intake, feed level (× maintenance), silage portion of diet
IPCC (2006)	No. of animals, animal species, animal type; emission factor for each animal type (Tier 2 CH_4 conversion factor; Ym)
HOLOS (Little et al., 2008)	Based on IPCC (2006)
Ellis et al. (2007)	Metabolizable energy (ME), ADF, acid detergent lignin (ADL) intake
Hunter (2007)	Correction to Kurihara et al. (1999): DMI of tropical forages
Charmley et al. (2008)	Forage quality, estimated DMI, biomass (tropical forages)
Yan et al. (2009)	DE, silage, and total DM, intake, silage and dietary ADF
Ellis et al. (2009)	ME, cellulose, hemicellulose, and fat intake; nonfiber carbohydrate (NFC), NDF, and DMI
Cornell Net Carbohydrates and Protein System (Van Amburgh et al., 2010)	Uses equation of Mills et al. (2003) for dairy and Ellis et al. (2007) for beef; animal characteristics, dietary nutrient composition, protein fractions, in situ degradability of feeds
Integrated Farm System Model (Rotz et al., 2012)	Uses the Mits3 equation of Mills et al. (2003) for enteric CH_4, IPCC (2006) for manure CH_4 and either DAYCENT (Chianese et al., 2009c) or IPCC (2006) for manure N_2O
Ricci et al. (2013)	GEI, DE, ME, DM digestibility (DMD), DMI, forage, lactation status
Moraes et al. (2014)	Equations developed from Beltsville Energy Lab data: GEI, NDF, fat, BW, lactation status
Escobar-Bahamondes and Beauchemin[b]	BW, DMI, CP, NDF, NFC, ADL, fat

[a]Adapted from USDA (2014).
[b]P. Escobar-Bahamondes and K. Beauchemin, Agriculture and Agri-Food Canada Research Centre, Lethbridge, Alberta, Canada, personal communication, June 2014.

the rumen. They also both require a significant number of sophisticated inputs that are generally beyond the knowledge of most cattle producers.

The Cornell Net Carbohydrate and Protein System model (CNCPS) and the Integrated Farm System Model (IFSM; Rotz et al., 2005, 2012; Chianese et al., 2009a,b,c,d) are mixed empirical and mechanistic farm nutrient management models. The IFSM uses a nonlinear (Mits3) equation developed by Mills et al. (2003) to estimate enteric CH_4 emissions from beef cattle. Similarly, the CNCPS model estimates enteric CH_4 production with a submodel (Van Amburgh et al., 2010) based on an empirical equation of Ellis et al. (2007).

A number of studies have evaluated the predictive ability of empirical and mechanistic enteric CH_4 models using independent data sets (Benchaar et al., 1998; Kurihara et al., 1999; Yan et al., 2000, 2009; Mills et al., 2003; Ellis et al., 2007, 2009; Hunter, 2007; Kebreab et al., 2008; McGinn et al., 2008; Legesse et al., 2011; Tomkins et al., 2011; Ricci et al., 2013; Moraes et al., 2014). Unfortunately, the results and conclusions have been highly variable and the predictions can vary substantially among equations, even among equations developed using the same database (USDA, 2014; Table 16-1). Some of these prediction equations were developed using a limited data set or diets not typically fed to beef cattle. There is now a substantial body of literature that suggests different prediction equations should be used for beef cattle vs. dairy cows and sheep, as well as for beef cattle fed high-forage vs. high-grain diets. Ellis et al. (2009) used a number of Canadian studies to develop and/or test linear and nonlinear equations to predict enteric CH_4 emissions from beef cattle. Tomkins et al. (2011) noted that measured (using the tracer gas SF_6 method) enteric CH_4 emissions of steers on Australian tropical pastures (114 to 136 g/d) were similar to estimates using an Ellis et al. (2009) equation (113 g/d), and Ellis et al. (2009) noted that several of the equations developed by Mills et al. (2003) and Ellis et al. (2007, 2009) seemed to effectively predict enteric CH_4 losses from feedlot cattle fed barley-based diets. Nonetheless, those same equations tended to overestimate enteric losses from cattle fed corn- and DGS-based finishing diets (Hales et al., 2012a, 2013; Todd et al., 2014), possibly because the nonlinear equations of Mills et al. (2003) were developed using predominately grazing and/or dairy data. In contrast, three equations of Ellis et al. (2007, Eq. 3b, 10b, and 12b) based on forage intake, DMI and forage intake, and DMI and fat intake, respectively, accurately predicted enteric CH_4 emissions from feedlot cattle in a Texas feedyard (Todd et al., 2014) and in cattle fed SFC-based diets in respiration chambers (K. H. Hales and N. A. Cole, USDA-ARS, unpublished data). Unfortunately, at the present time, many extant equations have not been truly validated using high-concentrate, corn-based, or DGS-based diets, and because of a very limited data set, none consider grain processing or starch availability of the grain in the calculations, despite the fact that grain processing is known to have a significant effect on enteric CH_4 production (Hales et al., 2012a).

Rather than empirical or mechanistic models, the USDA (2014) proposed to use the IPCC (2006) Tier II Ym value of 6.5% of GE intake to estimate CH_4 emissions from cattle on forage-based diets. A modified IPCC (2006) method was recommended to estimate emissions from cattle on high-concentrate diets. For cattle fed high-concentrate diets, the IPCC (2006) Tier II Ym value of 3% of GE intake was adjusted down for use of a dietary ionophore (Tedeschi et al., 2003; Guan et al., 2006), increased dietary fat concentration (Zinn and Shen, 1996; Beauchemin et al., 2008; Martin et al., 2010; Hales et al., 2013), and more intensive grain processing (Archibeque et al., 2006; Hales et al., 2012a), and adjusted up for use of barley in place of corn or sorghum (Beauchemin and McGinn, 2005) and for increasing forage content (USDA, 2014).

To evaluate existing equations for use with beef cattle fed diets ranging in forage content and ingredient composition, P. Escobar-Bahamondes and K. Beauchemin (Agriculture and Agri-Food Canada Research Centre, Lethbridge, Alberta, Canada, personal communication, June 2014) constructed a database from 58 beef cattle studies published between 2000 and 2014 containing 197 treatment means. Over 54 equations from the literature were evaluated for precision and accuracy using the concordance correlation coefficient and root mean square prediction error (g/d), respectively. Accuracy and precision of extant prediction equations were evaluated specifically for high- and low-forage diets. Several extant equations of Mills et al. (2003), Ellis et al. (2007, 2009), IPCC (2006), and Moraes et al. (2014) seemed to accurately predict enteric CH_4 production. Nonetheless, because of the relatively low accuracy and precision of most extant equations, as well as their lack of consistency across diets ranging in forage proportion, new equations specific to high-forage and low-forage diets were developed by P. Escobar-Bahamondes and K. Beauchemin (Agriculture and Agri-Food Canada Research Centre, Lethbridge, Alberta, Canada, personal communication, June 2014) using appropriate statistical and modeling techniques (Manly, 1997; Tedeschi, 2006).

The best model developed for high-forage diets (≥40% dietary DM) was (n = 123; R^2 = 0.96; P < 0.0001):

$$CH_4\text{, g/d} = 71.5\ (\pm\ 11.45) + 0.12\ (\pm\ 0.03) \times BW + 0.10\ (\pm\ 0.01) \times DMI^3 - 244.8\ (\pm\ 56.44) \times Fat^3 \qquad \text{(Eq. 16-8)}$$

where

BW is body weight, kg; minimum BW = 100 kg and maximum BW = 700 kg;
DMI is dry matter intake, kg/d; and
Fat is crude fat intake, kg/d.

For low-forage diets (≤20% dietary DM), the best model was (n = 34; R^2 = 0.85; P < 0.0001):

$$CH_4\text{, g/d} = -10.1\ (\pm\ 0.62) + 0.21\ (\pm\ 0.001) \times BW + 0.36\ (\pm\ 0.003) \times DMI^3 - 69.2\ (\pm\ 1.65) \times Fat^3 + 13.0\ (\pm\ 0.45) \times (CP/NDF) - 4.9\ (\pm\ 0.07) \times (Starch/NDF), \qquad \text{(Eq. 16-9)}$$

where

Minimum BW = 300 kg and maximum BW = 600 kg;
CP is crude protein intake, kg/d;
NDF is neutral detergent fiber intake, kg/d; and
Starch is starch intake, kg/d.

At the present time, these new equations, developed to account for dietary forage concentration, could be preferable to extant equations for calculating CH_4 emissions from beef cattle. Because of the large variation in the results obtained by using existing equations, the committee opted to use multiple equations (Table 16-2) in the BCNRM to provide a median, an average, and a range in estimated enteric CH_4 production. Different sets of equations are used for diets containing less than 20% forage, diets containing 20 to 40% forage, and diets containing more than 40% forage. The equations used are presented in Table 16-2 with a brief explanation of their use.

Greenhouse Gas Emissions from Manure and Compost

There have been few studies that measured GHG flux from beef cattle manure on pasture (Flessa et al., 1996) or in feedlots (Lodman et al., 1993; Muir et al., 2010). Factors such as diet composition, grain and forage processing, bedding, residual antibiotics, air temperature, manure moisture content, N excretion, and manure oxidation-reduction potential seem to affect GHG emissions from manure systems (Jarvis et al., 1995; Yamulki et al., 1999; Hao et al., 2001, 2004, 2005, 2011a,b,c; Hao, 2007; Xu et al., 2007a,b; de Klein and Eckard, 2008; Montes et al., 2013).

The standard method used by regulators to estimate CH_4 emissions from manure is multiplying the maximum CH_4 production potential of manure under ideal conditions (Bo; usually determined from in vitro fermentations) by a CH_4 conversion factor (MCF; IPCC, 2006; EPA, 2011). The MCF is an estimate of the percentage of the Bo that is produced (range of 0 to 5%) and is based on "typical" environmental conditions and manure handling procedures. The values typically used for Bo of beef cattle manure range from 0.17 to 0.33 m^3/kg volatile solids (i.e., OM) excreted (Hashimoto et al., 1981; IPCC, 2006; EPA, 2011). Similarly, emissions of N_2O from manure are typically estimated as a percentage of total N excretion (range of 0.2 to 10%), and depend upon manure management (IPCC, 2006; EPA, 2011). Emissions of N_2O from feedlot pen surfaces have not been adequately measured because of technical challenges (low emission vs. high background concentrations) and spatial and temporal variability; but they probably range from 0 to 2% of total N excretion (IPCC, 2006; EPA, 2011; USDA, 2011). Based on these methods, it is estimated that N_2O comprises about 74% (7.8 Tg CO_{2e}/y) and CH_4 comprises about 26% (2.7 Tg CO_{2e}/y) of GHG emissions from beef cattle manure in the United States (EPA, 2011).

The quantity of CH_4 emitted from manure is highly dependent on the quantity of fermentable substrate present. Therefore, the quantity of CH_4 emitted from cattle manure will theoretically decrease as the digestibility of the diet increases. Thus, factors such as grain processing could affect GHG emissions from manure via its effects on nutrient excretion and fecal starch content (Barajas and Zinn, 1998; Hales et al., 2012a). Similarly, the concentration and digestibility of forages in the diet will affect fecal excretion of volatile solids and CH_4 production.

Cooprider et al. (2011) reported that the daily manure N_2O production of cattle fed through a traditional feeding system that used anabolic implants, dietary ionophores, and β-agonists was similar to cattle fed in a "natural" program with no antibiotics, ionophores, or growth promoters. Because of improved performance, however, cattle fed using common growth technologies had 31% lower N_2O emissions per animal than cattle fed through a "natural" program.

AMMONIA EMISSIONS

Ammonia (NH_3) from cattle operations is primarily produced via hydrolysis of urinary urea to ammonium (NH_4^+) and CO_2 (Monteny et al., 2002; Stratton et al., 2013). This reaction occurs rapidly on feedlot pen surfaces (Cole et al., 2009) because of the abundance of soil and fecal microbes. Ammonia emissions from pastures are typically less than those from feedlots because some urine may enter the soil and some NH_4^+ can be used by plants for growth (Petersen, et al., 1998). Atmospheric NH_3 concentrations and emission rates at feedyards follow temperature-dependent seasonal and diel patterns with lowest NH_3 emissions normally occurring at night and highest emissions occurring during the day (Hutchinson et al., 1982; Todd et al., 2008a, 2011; Rhoades et al., 2010). Measured NH_3 emissions from feedlots in the southern Great Plains of the United States have ranged from 66 to 283 g/animal daily with an annual average of approximately 119 g/animal daily (139 g/animal daily in summer and 68 g/animal daily in winter; Todd et al., 2008a, 2011; Hristov et al., 2011). As a proportion of the N consumed, feedlot NH_3 emissions average about 30% in winter, 57% in summer, and 48% annually (Todd et al., 2008a, 2011; Hristov et al., 2011). Temperature, precipitation, and bedding management significantly affect NH_3 emissions. For example, under winter conditions in western Canada (mean temperature −8° to 4°C), NH_3 emissions from cattle in a research feedlot during a backgrounding phase (CP content of diets 12 to 14%) from December to February ranged from 5 to 31 g/steer daily, corresponding to 3.8 to 16.3% of N intake (Koenig et al., 2013). For the same cattle, emissions during the finishing phase from April to July (mean temperature 5° to 19°C, 11 mm total cumulative precipitation) ranged from 12 to 93 g/steer daily or 4.4 to 26.7% of N intake. In a study conducted between June and October at a large commercial feedlot located in southern Alberta (McGinn et al., 2007), NH_3 emissions were similar to those reported by Todd et al. (2008a, 2011) in Texas, averaging 140 g/animal daily.

Factors Affecting Ammonia Emissions and Potential Control Methods

Ammonia emissions can be decreased by four potential methods: (1) decreasing urinary urea-N excretion, (2) decreasing hydrolysis of urinary urea to NH_3, (3) binding NH_3 in the pen surface, and/or (4) converting volatile NH_3 to less volatile NH_4^+.

TABLE 16-2 Equations Used in the Beef Cattle Nutrient Requirements Model to Predict Enteric Methane Production from Beef Cattle

Reference	Equation	Comments
	High-forage diets (≥40% dietary DM)	
Escobar-Bahamondes and Beauchemin,[a] Eq. 16-8	CH_4, g/d = 71.5 (± 11.45) + 0.12 (± 0.03) × BW, kg + 0.10 (±0.01) × DMI^3, kg/d – 244.8 (± 56.44) × Fat^3, kg/d	Use when intake and detailed diet composition are known
IPCC (2006, Tier II)[b]	CH_4, MJ/d = DMI, kg/d × GE, MJ/kg × 0.065, % of GE	Use when there is limited information on diet composition
Ellis et al. (2009, Eq. G)	CH_4, MJ/d = –1.01 + 2.76 × NDF, kg/d + 0.722 × Starch, kg/d	Use when only NDF and starch intake are known and when forage content is ≤75% dietary DM; BW range of 180 to 630 kg
Mills et al. (2003, Eq. NL2)	CH_4, MJ/d = 45.98 – (45.98 × $e^{-0.003 \times \text{MEI, MJ/d}}$)	Mitscherlich nonlinear equation developed using dairy data in which 45.98 = maximum CH_4 production, 0 = minimum CH_4 production, and 0.003 = shape parameter; use when forage content is ≤75% dietary DM
Ellis et al. (2009, Eq. N)	CH_4, MJ/d = 2.68 – 1.14 × (Starch/NDF) + 0.786 × DMI, kg/d	Use when only NDF and starch intake are known and where forage content is ≤75%, dietary DM; BW range of 180 to 630 kg
Moraes et al. (2014)	CH_4, g/d = (–1.487 + 0.046 × GEI × 4.184 + 0.032 × NDF, % DMI + 0.006 × BW) × 1,000/55.65	Animal level: Equation for heifers; BW range of 195 to 540 kg
Moraes et al. (2014)	CH_4, g/d = (–0.221 + 0.048 × GEI × 4.184 + 0.005 × BW) × 1,000/55.65	Animal level: Equation for steers; BW range of 170 to 630 kg
Moraes et al. (2014)	CH_4, g/d = (–0.163 + 0.051 × GEI × 4.184 + 0.038 × NDF, % DMI) × 1,000/55.65	Diet level: Equation for heifers; BW range of 195 to 540 kg
Moraes et al. (2014)	CH_4, g/d = (0.743 + 0.054 × GEI × 4.184) × 1,000/55.65	Diet level: Equation for steers; BW range of 170 to 630 kg
	Intermediate-forage diets (20 to 40% dietary DM)	
Ellis et al. (2007, Eq. 12b)	CH_4, MJ/d = 2.7 + 1.16 × DMI, kg/d – 15.8 × EE, kg/d	May underestimate CH_4 production for cattle fed high-fat diets (>5% fat); BW range of 200 to 660 kg
IPCC (2006, Tier II)[b]	CH_4, MJ/d = DMI, kg/d × GE, MJ/kg × 0.065, % of GE	Use when there is limited information on diet composition.
	Low-forage diets (≤20% dietary DM)	
Escobar-Bahamondes and Beauchemin,[a] Eq. 16-9	CH_4, g/d = 10.1 (± 0.62) + 0.21 (± 0.001) × BW, kg + 0.36 (± 0.003) × DMI^2, kg/d – 69.2 (± 1.65) × Fat, kg/d + 13.0 (± 0.45) × (CP/NDF) – 4.9 (± 0.07) × (Starch/NDF)	Use when intake and detailed diet composition are known; CP, NDF, and starch all kg/d
Ellis et al. (2009, Eq. G)	CH_4, MJ/d = –1.01 + 2.76 × NDF, kg/d + 0.722 × Starch, kg/d	Use when only NDF and starch intake are known and when forage content is ≥9% dietary DM; BW range of 180 to 630 kg
IPCC (2006, Tier II)[b]	CH_4, MJ/d = DMI, kg/d × GE, MJ/kg × 0.03, % of GE	Use when limited data on diet composition are available
Ellis et al. (2007, Eq. 9b)	CH_4, MJ/d = 0.357 + 0.0591 × MEI, MJ/d + 0.05 × Forage, %	Seems to overestimate CH_4 for corn-based diets but to work well with barley-based diets; BW range of 200 to 660 kg
Ellis et al. (2007, Eq. 10b)	CH_4, MJ/d = –1.02 + 0.681 × DMI, kg/d + 0.0481 × Forage, %	Use when DMI and dietary forage are known. Seems to work well with diets based on highly processed corn; BW range of 200 to 660 kg

NOTE: BW = body weight; CH_4 = enteric methane; CP = crude protein; DMI = dry matter intake; GE = gross energy; MEI = metabolizable energy intake; NDF = neutral detergent fiber; NFC = nonfiber carbohydrates. The equations above estimate daily enteric methane production. Values can be converted to MJ/d, Mcal/d, and g/d using the following conversion factors: The complete combustion of methane yields 891 kJ/mol or 55.5486 MJ/kg or 13.2764 Mcal/kg, assuming 16.04 g/mol. GE (Mcal/kg) = (0.0415 × carbohydrate, %) + (0.094 × fat, %) + (0.057 × protein, %) (NRC, 2001) or use 18.45 MJ/kg DM when dietary composition is not available.

[a] P. Escobar-Bahamondes and K. Beauchemin, Agriculture and Agri-Food Canada Research Centre, Lethbridge, Alberta, Canada, personal communication, June 2014.

[b] Use CH_4, MJ/d = DMI, kg/d × GE, MJ/kg × 0.065, % of GE for diets with >10% forage DM.

In beef cattle, 40 to 80% of nonretained N is excreted in the urine and this quantity typically increases as dietary CP and/or RDP concentration increases in the diet (Cole et al., 2005a; Archibeque et al., 2007; Vasconcelos et al., 2009; Erickson and Klopfenstein, 2010; Koenig and Beauchemin, 2013a,b). Because of increased urinary urea N excretion, NH_3 emissions increase with increased N intake. In SFC-based diets, N excretion in urine and subsequent NH_3 losses were 24 to 50% greater when a 13% CP diet was fed, than when an 11.5% CP diet was fed to finishing beef cattle (Cole et al., 2005a: Todd et al., 2006). Similarly, with barley-based growing and finishing diets, Koenig et al. (2013) noted a 42% decrease in NH_3 emissions when dietary CP concentrations were decreased from about 14 to 12%. Cole and Todd (2009) noted that when MP intake was below animal requirements, the N volatilization losses at a feedyard averaged approximately 40 g/animal daily. Thus, this might be the minimum NH_3 emission rate that can be obtained through dietary manipulations of finishing diets.

There are few published measurements of NH_3 emissions from pastures grazed by cattle and most of them used methods (i.e., closed or flow-through chambers) that tend to underestimate emissions and also have limited or inadequate spatial footprints. Using wind tunnels, Petersen et al. (1998) reported than NH_3 emissions from fecal patties on high-quality pastures were minimal, whereas NH_3 emissions from urine patches ranged from 3 to 52% of urinary N and increased with dietary CP content.

The primary factors controlling NH_3 emissions are temperature and N excretion; therefore, Todd et al. (2013) used NH_3 emissions measured semi-continuously for 2 years at two commercial feedyards in Texas to develop the following equation to predict feedlot NH_3 emissions based on dietary CP concentration and environmental temperature ($R^2 = 0.80$):

$$\ln(NH_3) = 8.82 - 1{,}629\ (1/T) + 0.108\ (CP) \quad \text{(Eq. 16-10)}$$

where

NH_3 is the average monthly ammonia emission rate, g/animal daily;
T is the average monthly temperature, in kelvin (K); and
CP is the % dietary crude protein (on a DM basis).

It should be noted, however, that because this equation was developed using feedlot cattle at ad libitum intake (9 to 11 kg DM/d), it may overestimate ammonia emissions in animals at lower feed intakes. To convert NH_3 to NH_3-N, multiply by 0.823.

Other factors such as dietary fat content (Machmuller et al., 2006; Cole et al., 2007) and pen surface OM content (Bierman et al., 1999; Adams et al., 2004; Erickson and Klopfenstein, 2010; Sayer et al., 2013) can affect NH_3 emissions; however, these effects are less pronounced, and less consistent than the effects of dietary N.

Although decreasing the CP concentration of the diet can potentially decrease NH_3 emissions, it can also potentially decrease animal performance; however, dietary protein requirements change with the stage of growth, which could make decreasing CP a feasible option. In three sequential metabolism trials, Vasconcelos et al. (2009) noted that approximately 75, 85, and 100% of dietary N intake was excreted at 30 (350 kg average BW), 85 (446 kg), and 155 (515 kg) days on feed. Similarly, Koenig et al. (2013) noted an increase in N excretion (80% and 92% of N intake for days 1 to 21 and days 85 to 106, respectively) and decrease in N retention as days on feed increased. Several studies have shown that it is possible to decrease N excretion and NH_3 emissions by decreasing dietary CP (or RDP or ruminally undegradable protein) concentrations of beef cattle finishing diets as time on feed increases (or phase feeding), with little or no effect on animal performance (Cole et al., 2006; Vasconcelos et al., 2006; Erickson and Klopfenstein, 2010). The effects of phase feeding on animal performance and NH_3 emissions could be affected by factors such as implanting strategy and grain processing method (Cole et al., 2006; Vasconcelos et al., 2006; Erickson and Klopfenstein 2010). Despite these potential effects, Cole et al. (2006), with SFC-based diets, and Erickson and Klopfenstein (2010), with DRC-based diets, both noted that cumulative N volatilization losses for the entire feeding period were decreased by about 25% by decreasing the dietary CP concentration during the final 56 d in the feedlot. The practicality of phase feeding under current feeding and management situations is equivocal because of potential logistic (additional supplements, diets, and feed truck scheduling), economic, and animal health obstacles. For example, blood NH_3 could serve as a renal/systemic buffer in cattle fed high-concentrate diets (Trenkle, 1978). Thus, decreasing dietary CP concentrations could potentially increase the incidence of acidosis in phase-fed cattle. The feeding of a β-agonist during the last 20 to 45 d on feed might make it more practical to institute phase feeding; however, the effects of β-agonists on protein requirements late in the feeding period are not clear. Moreover, the increased use of high-protein byproducts such as DGS in finishing diets limits the nutritionist's ability to lower dietary CP concentrations to less than 12% of dietary DM.

The effects of feeding low (<20%) concentrations of WDGS on feedlot NH_3 emissions have been somewhat inconsistent, probably as a result of differences in forage concentration and composition, dietary RDP, and/or dietary CP content. Feeding relatively low concentrations (<15% of DM) of DGS in finishing diets usually causes little or no increase in dietary N intake, but can result in a decrease in manure pH and an increase in carbohydrate OM on the pen surface (Cole et al., 2010). In contrast, Luebbe et al. (2012a) and Buttrey et al. (2013) noted increased (25%) daily N volatilization losses when dietary WDGS concentrations were increased from 0 to 15 or 20% of dietary DM. At higher (>25%) dietary concentrations, feeding of DGS has routinely increased NH_3 emissions by 20 to 53% because of an increase in dietary N intake and subsequent increase in urinary N excretion (Cole et al., 2010; Todd et al., 2011; Luebbe et al., 2012a,b; Hunerberg et al., 2013).

Kellems et al. (1979) reported that grain source (corn vs. sorghum vs. barley) and grain concentration (25, 50, or 75% of dietary DM) affected NH_3 emissions, primarily via their effects on fecal pH. Buttrey et al. (2012) noted that N volatilization losses, as a percentage of N intake and as a percentage of N excreted, were less in cattle fed a DRC-based diet than in cattle fed an SFC-based diet. This could be the result of lower manure pH in cattle fed the DRC-based diet caused by greater fecal starch excretion (Owens et al., 1997).

Theoretically, urinary and/or fecal pH can be lowered, and thus more highly volatile NH_3 can be converted to less volatile NH_4^+, by altering the dietary cation-anion balance (DCAB). Luebbe et al. (2011) altered urinary pH by altering DCAB; however, N volatilization losses were not affected.

HYDROGEN SULFIDE AND REDUCED SULFUR

Average atmospheric concentrations of hydrogen sulfide (H_2S) at feedyards are relatively low but tend to increase with higher temperatures and increased S in the diet (Rhoades et al., 2003; Koelsch et al., 2004). Emissions of H_2S are highly episodic, occurring in short bursts, typically after rains (K. Casey, Texas A&M AgriLife Research, Amarillo, TX, personal communication, Sept. 20, 2013). The atmospheric concentrations of H_2S at open-lot feedlots are typically too low to be a health hazard to cattle or humans.

Distillers grains with solubles are typically high in S and are thereby a primary source of S in many diets. The high S content of DGS occurs via the concentration of S normally present in the grain (amino acids, etc.) and additions of sulfuric acid used to modify fermentation or clean the system during ethanol production. Some forms of S seem to be more ruminally available (Sarturi et al., 2013a,b) and thus produce higher ruminal H_2S concentrations; however, the effects on urinary and fecal S excretion and subsequent emissions of H_2S are not yet clear.

PARTICULATE MATTER AND DUST

Emissions of particulate matter (PM) from livestock operations can affect visibility, quality of life, and possibly, human and cattle health (MacVean et al., 1986; Loneragan and Brashears, 2005). Dust particles also can carry other pollutants such as odorants, pathogens, and endotoxins (Wilson et al., 2002; Purdy et al., 2004). Most feedyard PM is organic, originating from manure, feed, and animal dander (Sweeten et al., 1988, 1998) and has an average aerodynamic diameter greater than 10 microns (i.e., PM-10). Dust emissions from feed mills can vary depending on the commodities used; how they are unloaded, transported, and processed; and how feeds are mixed (EPA, 1998a). The concentrations of PM in feedyard air are highly variable and are affected by the environment (i.e., temperature, precipitation, wind, atmospheric stability), animal activity, pen surface conditions, (i.e., moisture, manure depth; Auvermann and Romanillos, 2000; Auvermann et al., 2002; Miller and Berry, 2005), and possibly dietary factors such as fat content of diet.

ODORS/VOLATILE ORGANIC COMPOUNDS

Odors from cattle facilities can originate from the pen surface manure, manure stockpiles, RCS, or fields fertilized with manure (Sweeten et al., 1977; Watts et al., 1994). Feedyard odors are normally comprised of VOC such as amines, volatile fatty acids, alcohols, carbonyls, and reduced sulfur compounds (Mackie et al., 1998; Parker et al., 2005, 2010). Some odorants chemically react with other atmospheric compounds rapidly and thus travel only short distances, whereas, others, such as *p*-cresol, can travel long distances (Parker et al., 2005). The primary source of feedlot odors seems to be the microbial fermentation of fecal starch, although fermentation of proteins and S-containing compounds are also involved (Miller and Varel, 2001; Miller and Berry, 2005; Miller et al., 2006). Thus, decreasing starch in feces can decrease odorant production. Archibeque et al. (2006) noted that cattle fed diets based on HMC produced feces with less starch and less odorant production potential than cattle fed DRC-based diets. Adding DGS to finishing diets tended to increase potential odor production, most likely because of its high N and S content (Varel et al., 2008; Spiehs and Varel, 2009; Hales et al., 2012b,c).

REFERENCES

Abecia, L., A. I. Martin-Garcia, G. Martinez, C. J. Newbold, and D. R. Yañez-Ruiz. 2013. Nutritional intervention in early life to manipulate rumen microbial colonization and methane output by kid goats postweaning. *Journal of Animal Science* 91:4832-4840.

Adams, J. R., T. B. Farran, G. E. Erickson, T. J. Klopfenstein, C. N. Macken, and C. B. Wilson. 2004. Effect of organic matter addition to the pen surface and pen cleaning frequency on nitrogen mass balance in open feedlots. *Journal of Animal Science* 82:2153-2163.

Anderson, R. C., and M. A. Rasmussen. 1998. Use of a novel nitrotoxin-metabolizing bacterium to reduce ruminal methane production. *Bioresource Technology* 64:89-95.

Anderson, R. C., T. R. Callaway, J. A. S. VanKessel, Y. S. Jung, T. S. Edrington, and D. J. Nisbet. 2003. Effect of select nitrocompounds on ruminal fermentation; An initial look at their potential to reduce economic and environmental costs associated with ruminal methanogenesis. *Bioresource Technology* 90:59-63.

Andraski, T. W., L. G. Bundy, and K. C. Kilian. 2003. Manure history and long-term tillage effects on soil properties and phosphorus losses in runoff. *Journal of Environmental Quality* 32:1782-1789.

Archibeque, S. L., D. N. Miller, H. C. Freetly, and C. L. Ferrell. 2006. Feeding high-moisture corn instead of dry-rolled corn reduces odorous compound production in manure of finishing beef cattle without decreasing performance. *Journal of Animal Science* 84:1767-1777.

Archibeque, S. L., D. N. Miller, H. C. Freetly, E. D. Berry, and C. L. Ferrell. 2007. The influence of oscillating dietary protein concentrations on finishing cattle. II. Nutrient retention and ammonia emissions. *Journal of Animal Science* 85:1496-1503.

ASABE (American Society of Agricultural and Biological Engineers). 2010. Manure Production and Characteristics. Standard D384.2 MAR2005 (R2010). St. Joseph, MI: American Society of Agricultural and Biological Engineers.

ASAE (American Association of Agricultural Engineers). 1999. Manure Production and Characteristics. Standard D348.1. St. Joseph, MI: American Society of Agricultural Engineers.

Auvermann, B. W., and A. Romanillos. 2000. Effect of stocking density manipulation on fugitive PM_{10} emissions from cattle feedyards. In *Proceedings of the Society of Engineering in Agriculture Conference, April 3, 2000, Adelaide, South Australia.*

Auvermann, B. W., D. B. Parker, and J. M. Sweeten. 2002. Manure Harvesting Frequency—The Key to Feedyard Dust Control in a Summer Drought. Texas Cooperative Extension Publication E-52, 11-00. Available online at http://texaserc.tamu.edu. Accessed on June 12, 2013.

Avila-Stagno, J., A. V. Chaves, M. L. He, O. M. Harstad, K. A. Beauchemin, S. M. McGinn, and T. A. McAllister. 2013. Effects of increasing concentrations of glycerol in concentrate diets on nutrient digestibility, methane emissions, growth, fatty acid profiles, and carcass traits of lambs. *Journal of Animal Science* 91:829-837.

Axelsson, J. 1949. The amount of produced methane energy in the European metabolic experiments with adult cattle. *Annals of the Royal Agricultural College of Sweden* 16:404-419.

Baldwin, R. L. 1995. *Modeling Ruminant Digestion and Metabolism.* London, UK: Chapman and Hall.

Bannick, A., J. Kogut, J. Dijkstra, and J. France. 2006. Estimation of stoichiometry of volatile fatty acid production in the rumen of the lactating cow. *Journal of Theoretical Biology* 238:36-51.

Barajas, R., and R. A. Zinn. 1998. The feeding value of dry-rolled and steam-flaked corn in finishing diets for feedlot cattle: Influence of protein supplementation. *Journal of Animal Science* 76:1744-1752.

Beauchemin, K. A., and S. M. McGinn. 2005. Methane emissions from feedlot cattle fed barley or corn diets. *Journal of Animal Science* 83:653-661.

Beauchemin, K. A., and S. M. McGinn. 2006a. Enteric methane emissions from growing beef cattle as affected by diet and level of intake. *Canadian Journal of Animal Science* 86:401-408.

Beauchemin, K. A., and S. M. McGinn. 2006b. Methane emission from beef cattle: Effects of fumaric acid, essential oil, and canola oil. *Journal of Animal Science* 84:1489-1496.

Beauchemin, K. A., S. M. McGinn, T. F. Martinez, and T. A. McAllister. 2007a. Use of condensed tannin extract from quebracho trees to reduce methane emissions from cattle. *Journal of Animal Science* 85:1900-1906.

Beauchemin, K. A., S. M. McGinn, and H. Petit. 2007b. Methane abatement strategies for cattle: Lipid supplementation of diets. *Canadian Journal of Animal Science* 87:431-440.

Beauchemin, K. A., M. Kreuzer, F. O'Mara, and T. A. McAllister. 2008. Nutritional management for enteric methane abatement: A review. *Australian Journal of Experimental Agriculture* 48:21-27.

Beauchemin, K. A., H. H. Janzen, S. M. Little, T. A. McAllister, and S. M. McGinn. 2010. Life cycle assessment of greenhouse gas emissions from beef production in western Canada: A case study. *Agricultural Systems* 103:371-379.

Benchaar, C., J. Rivest, C. Pomar, and J. Chiquette. 1998. Prediction of methane production from dairy cows using existing mechanistic models and regression equations. *Journal of Animal Science* 76:617-627.

Benchaar, C., C. Pomar, and J. Chiquette. 2001. Evaluation of dietary strategies to reduce methane production in ruminants: A modeling approach. *Canadian Journal of Animal Science* 81:563-574.

Bierman, S., G. E. Erickson, T. J. Klopfenstein, R. A. Stock, and D. H. Shain. 1999. Evaluation of nitrogen and organic matter balance in the feedlot as affected by level and source of dietary fiber. *Journal of Animal Science* 77:1645-1653.

Blaxter, K. L., and J. L. Clapperton. 1965. Prediction of the amount of methane produced by ruminants. *British Journal of Nutrition* 19:511-521.

Blaxter, K. L., and F. W. Wainman. 1964. The utilization of the energy of different rations by sheep and cattle for maintenance and fattening. *Journal of Agricultural Science Cambridge* 63:113-128.

Bowman, J. G. P., and B. F. Sowell. 1997. Delivery method and supplement consumption by grazing ruminants: A review. *Journal of Animal Science* 75:543-550.

Bowman, J. G. P., and B. F. Sowell. 2003. Technology to complement forage-based beef production systems in the West. *Journal of Animal Science* 81(Suppl. 1):E18-E26.

Boxall, A. B. A., D. W. Kolpin, B. Halling-Sorensen, and J. Tolls. 2003. Are veterinary medicines causing environmental risks? *Environmental Science and Technology* 37:287A-294A.

Bratzler, J. W., and E. B. Forbes. 1940. The estimation of methane production by cattle. *Journal of Nutrition* 19:611-613.

Buttrey, E. K., N. A. Cole, K. H. Jenkins, B. E. Meyer, F. T. McCollum III, S. L. M. Preece, B. W. Auvermann, K. R. Heflin, and J. C. MacDonald. 2012. Effects of twenty percent corn wet distiller's grains plus solubles in steam-flaked and dry-rolled corn-based finishing diets on heifer performance, carcass characteristics, and manure characteristics. *Journal of Animal Science* 90:5086-5098.

Buttrey, E. K., K. H. Jenkins, J. B. Lewis, S. B. Smith, R. K. Miller, T. E. Lawrence, F. T. McCollum III, P. J. Pinedo, N. A. Cole, and J. C. MacDonald. 2013. Effects of 35% corn wet distiller's grain plus solubles in steam-flaked and dry-rolled corn-based finishing diets on animal performance, carcass characteristics, beef fatty acid composition, and sensory attributes. *Journal of Animal Science* 91:1850-1865.

Charmley, E., M. L. Stephens, and P. M. Kennedy. 2008. Predicting livestock productivity and methane emissions in northern Australia: Development of a bio-economic modeling approach. *Australian Journal of Experimental Agriculture* 48:109-113.

Chianese, D. S., C. A. Rotz, and T. L. Richard. 2009a. Simulation of carbon dioxide emissions from dairy farms to assess greenhouse gas reduction strategies. *Transactions of the ASABE* 52:1301-1312.

Chianese, D. S., C. A. Rotz, and T. L. Richard. 2009b. Simulation of methane emissions from dairy farms to assess greenhouse gas reduction strategies. *Transactions of the ASABE* 52:1313-1323.

Chianese, D. S., C. A. Rotz, and T. L. Richard. 2009c. Simulation of nitrous oxide emissions from dairy farms to assess greenhouse gas reduction strategies. *Transactions of the ASABE* 52:1325-1335.

Chianese, D. S., C. A. Rotz, and T. L. Richard. 2009d. Whole-farm greenhouse gas emissions: A review with application to a Pennsylvania dairy farm. *Applied Engineering in Agriculture* 25:431-444.

Chung, Y.-H., E. J. McGeough, S. Acharya, T. A. McAllister, S. M. McGinn, O. M. Harstad, and K. A. Beauchemin. 2013. Enteric methane emission, diet digestibility, and nitrogen excretion from beef heifers fed sainfoin or alfalfa. *Journal of Animal Science* 91:4861-4874.

Cole, N. A. 2003. Precision feeding: Opportunities and limitations. In *Proceedings of the Plains Nutrition Council, April 3-4, 2003, San Antonio, TX.* TAES Publication AREC 03-13. Amarillo: Texas A&M Research and Extension Center.

Cole, N. A., and J. E. McCroskey. 1975. Effects of hemiacetal of chloral and starch on the performance of beef steers. *Journal of Animal Science* 41:1735-1741.

Cole, N. A., and R. W. Todd. 2009. Nitrogen and phosphorus balance of beef cattle feedyards. Pp. 17-24 in *Proceedings of the Texas Manure Management Conference, September 29-30, 2009, Round Rock, TX.* College Station: Texas A&M AgriLife Extension Service.

Cole, N. A., R. N. Clark, R. W. Todd, C. R. Richardson, A. Gueye, L. W. Greene, and K. McBride. 2005a. Influence of dietary protein concentration and source on potential ammonia emissions from beef cattle manure. *Journal of Animal Science* 83:722-731.

Cole, N. A., R. C. Schwartz, and R. W. Todd. 2005b. Assimilation versus accumulation of macro- and micronutrients in soils: Relations to livestock and poultry feeding operations. *Journal of Applied Poultry Research* 14:393-405.

Cole, N. A., P. J. Defoor, M. L. Galyean, G. C. Duff, and J. F. Gleghorn. 2006. Effects of phase feeding crude protein on performance, carcass characteristics, serum urea nitrogen concentrations, and manure nitrogen of finishing beef steers. *Journal of Animal Science* 84:3421-3432.

Cole, N. A., R. W. Todd, and D. B. Parker. 2007. Use of fat and zeolite to reduce ammonia emission from beef cattle feedyards. In *Proceedings of the International Symposium on Air Quality and Waste Management for Agriculture, September 16-17, 2007, Broomfield, CO.* ASABE Publication No. 701P0907cd. St. Joseph, MI: ASABE.

Cole, N. A., A. M. Mason, R. W. Todd, M. Rhoades, and D. B. Parker. 2009. Chemical composition of pen surface layers of beef cattle feedyards. *The Professional Animal Scientist* 25:541-552.

Cole, N. A., K. E. Hales, and R. W. Todd. 2010. Environmental issues facing the beef cattle industry: Challenges and some possible solutions. In *Proceedings of Husker Beef Nutrition Conference, November 12, 2010, Lincoln, NE.* University of Nebraska.

Cole, N. A., K. McCuistion, L. W. Greene, and F. T. McCollum. 2011. Effects of concentration and source of set distillers grains on digestibility of steam-flaked corn-based diets fed to finishing steers. *The Professional Animal Scientist* 27:302-311.

Cooprider, K. L., F. M. Mitloehner, T. R. Famula, E. Kebreab, Y. Zhao, and A. L. Van Eenennaam. 2011. Feedlot efficiency implications on greenhouse gas emission and sustainability. *Journal of Animal Science* 89:2643-2656.

Corrigan, M. E., G. E. Erickson, T. J. Klopfenstein, M. K. Luebbe, K. J. Vander Pol, N. F. Meyer, C. D. Buckner, S. J. Vanness, and K. J. Hanford. 2009. Effect of corn processing method and corn wet distillers grains plus solubles inclusion level in finishing steers. *Journal of Animal Science* 87:3351-3360.

Cottle, D. J., J. V. Nolan, and S. G. Wiedemann. 2011. Ruminant enteric methane mitigation: A review. *Animal Production Science* 51:491-514.

de Klein, C. A. M., and R. J. Eckard. 2008. Targeted technologies for nitrous oxide abatement from animal agriculture. *Australian Journal of Experimental Agriculture* 48:14-20.

Dijkstra, J., H. D. St. C. Neal, D. E. Beever, and J. France. 1992. Simulation of nutrient digestion, absorption and outflow in the rumen: Model description. *Journal of Nutrition* 122:2239-2256.

Dijkstra, J., E. Kebreab, J. A. N. Mills, W. F. Pellikaan, S. Lopez, A. Bannink, and J. France. 2007. Predicting the profile of nutrients available for absorption: From nutrient requirement to animal response and environmental impact. *Animal* 1:99-111.

Doering, O. 1994. Public perceptions and policy imperatives: Animal agriculture and the environment. *Journal of Dairy Science* 78:469-475.

Dong, R. L., K. A. Beauchemin, G. Y. Zhao, and L. L. Chai. 2014. Prediction of urinary and fecal nitrogen excretion by beef cattle. *Journal of Animal Science* 92:4669-4681.

Doreau, M., H. M. G. van der Werf, D. Micol, H. Dubroeucq, J. Agabriel, Y. Rochette, and C. Martin. 2011. Enteric methane production and greenhouse gases balance of diets differing in concentrate in the fattening phase of a beef production system. *Journal of Animal Science* 89:2518-2528.

Eckard, R. J., C. Grainger, and C. A. M. de Klein. 2010. Options for the abatement of methane and nitrous oxide from ruminant production: A review. *Livestock Science* 130:47-56.

Ellis, J. L., E. Kebreab, N. E. Odongo, B. W. McBride, E. K. Okine, and J. France. 2007. Prediction of methane production from dairy and beef cattle. *Journal of Dairy Science* 90:3456-3466.

Ellis, J. L., E. Kebreab, N. E. Odongo, K. Beauchemin, S. McGinn, J. D. Nkrumah, S. S. Moore, R. Christopherson, G. K. Murdoch, B. W. McBride, E. K. Okine, and J. France. 2009. Modeling methane production from beef cattle using linear and nonlinear approaches. *Journal of Animal Science* 87:1334-1345.

EPA (U.S. Environmental Protection Agency). 1998a. *Emission Factor Documentation for AP-42 Section 9.9.1. Grain Elevators and Grain Processing Plants.* Final Report. Washington, DC: U.S. Environmental Protection Agency.

EPA. 1998b. *Strategy for Addressing Environmental and Public Health Impacts from Animal Feeding Operations.* Draft Report. Washington, DC: U.S. Environmental Protection Agency. Available online at http://www.epa.gov/npdes/pubs/astrat.pdf. Accessed on October 21, 2014.

EPA. 2003. *Producers' Compliance Guide for CAFOs: Revised Clean Water Act Regulation for Concentrated Animal Feeding Operations.* EPA 821-R-03-010. Washington, DC: U.S. Environmental Protection Agency. Available online at http://www.epa.gov/npdes/cafo/producersguide. Accessed on March 12, 2014.

EPA. 2011. Inventory of U.S. Greenhouse Emissions and Sinks. Washington, DC: U.S. Environmental Protection Agency. Available online at http://www.epa.gov/climatechange/emissions/downloads11/US-GHG-Inventory-2011-Chapter-6-Agriculture.pdf. Accessed on January 28, 2013.

Erickson, G. E., and T. Klopfenstein. 2010. Nutritional and management methods to decrease nitrogen losses from beef feedlots. *Journal of Animal Science* 88 (Suppl.):E172-E180.

Erickson, G., T. Klopfenstein, C. T. Milton, D. Hanson, and C. Calkins. 1999. Effect of dietary phosphorus on finishing steer performance, bone status, and carcass maturity. *Journal of Animal Science* 77:2832-2836.

Erickson, G., T. Klopfenstein, C. T. Milton, D. Brink, M. W. Orth, and K. M. Whittet. 2002. Phosphorus requirement of finishing feedlot calves. *Journal of Animal Science* 80:1690-1695.

Erickson, G. E., J. R. Adams, T. B. Farran, C. B. Wilson, C. N. Macken, and T. J. Klopfenstein. 2003a. Impact of cleaning frequency of pens and carbon to nitrogen (C:N) ratio as influenced by the diet or pen management on N losses from outdoor beef feedlots. Pp. 397-404 in *Proceedings of the Ninth International Symposium on Animal, Agricultural and Food Processing Wastes, October 12-15, 2003, Raleigh, NC.* ASAE Publication 701P1203. St. Joseph, MI: American Society of Agricultural Engineers.

Erickson, G., E. B. Auvermann, R. Eigenberg, L. W. Greene, T. Klopfenstein and R. Koelsch. 2003b. Proposed beef cattle manure excretion and characteristics standard for ASAE. Pp. 269-276 in *Proceedings of the Ninth International Symposium on Animal, Agricultural and Food Processing Wastes, October 12-15, 2003. Raleigh, NC.* ASAE Publication 701P1203. St. Joseph, MI: American Society of Agricultural Engineers.

Farran, T. B., G. E. Erickson, T. J. Klopfenstein, C. N. Macken, and R. U. Lindquist. 2006. Wet corn gluten feed and alfalfa hay levels in dry-rolled corn finishing diets: Effects on finishing performance and feedlot nitrogen mass balance. *Journal of Animal Science* 84:1205-1214.

Flessa, H., P. Dorsch, R. Beese, H. Konig, and A. F. Bouwman. 1996. Influence of cattle wastes on nitrous oxide and methane fluxes in pasture land. *Journal of Environmental Quality* 25:1366-1370.

Freetly, H. C., and T. M. Brown-Brandl. 2013. Enteric methane production from beef cattle that vary in feed efficiency. *Journal of Animal Science* 91:4826-4831.

Galyean, M. L. 1999. Review: Restricted and programmed feeding of beef cattle—definitions, application, and research results. *The Professional Animal Scientist* 15:1-6.

Galyean, M. L. 2000. Environmental stewardship in the future: Nutrient management issues and options for beef cattle feeding operations. *Journal of Animal Science* 79(Suppl. 1):1-9.

Geisert, B. G., G. E. Erickson, T. J. Klopfenstein, C. N. Macken, M. K. Luebbe, and J. C. MacDonald. 2010. Phosphorus requirement and excretion of finishing beef cattle fed different concentrations of phosphorus. *Journal of Animal Science* 88:2393-2402.

Greene, L. W. 1995. The nutritional value of inorganic and organic mineral sources. Pp. 23-31 in *Update on Mineral Nutrition of Beef Cattle. Proceedings of the Plains Nutrition Council Symposium, November 10, 1995. Lubbock, TX.* Publication TAMUS-AREC-95-2. Amarillo: Texas A&M University Research and Extension Center.

Greene, L. W., F. T. McCollum, N. K. Chirase, and T. M. Montgomery. 2001. Performance and conservation of phosphorus in growing cattle. *Journal of Animal Science* 79(Suppl. 1):293.

Greenquist, M. A., T. J. Klopfenstein, W. H. Schacht, G. E. Erickson, K. J. Vander Pol, M. K. Luebbe, K. R. Brink, A. K. Schwarz, and L. B.

Baleseng. 2009. Effects of nitrogen fertilization and dried distillers grains supplementation: Forage use and performance of yearling steers. *Journal of Animal Science* 87:3639-3646.

Greenquist, M. A., A. K. Schwarz, T. J. Klopfenstein, W. H. Schacht, G. E. Erickson, K. J. Vander Pol, M. K. Luebbe, K. R. Brink, and L. B. Baleseng. 2011. Effects of nitrogen fertilization and dried distillers grains supplementation: Nitrogen use efficiency. *Journal of Animal Science* 89:1146-1152.

Guan, H., K. M. Wittenberg, K. H. Ominski, and D. O. Krause. 2006. Efficiency of ionophores in cattle diets for mitigation of enteric methane. *Journal of Animal Science* 84:1896-1906.

Haisan J., Y. Sun, K. Beauchemin, L. Guan, S. Duval, D. R. Barreda, and M. Oba. 2013. Effect of feeding 3-nitrooxypropanol on methane emissions and productivity of lactating dairy cows. *Advances in Animal Biosciences* 4:260.

Haisan J., Y. Sun, L. L. Guan, K. A. Beauchemin, A. Iwaasa, S. Duval, D. R. Barreda, and M. Oba. 2014. The effects of feeding 3-nitrooxypropanol on methane emissions and productivity of Holstein cows in mid lactation. *Journal of Dairy Science* 97:3110-3119.

Hales, K. E., N. A. Cole, and J. C. MacDonald. 2012a. Effect of corn processing method and dietary inclusion of wet distiller's grains with solubles on energy metabolism carbon-nitrogen balance, and enteric methane emissions of finishing cattle. *Journal of Animal Science* 90:3174-3185.

Hales, K. E., N. A. Cole, and V. H. Varel. 2012b. Effects of corn processing method and dietary inclusion of corn wet distillers grains with solubles on odor and gas production in cattle manure. *Journal of Animal Science* 90:3988-4000.

Hales, K. E., D. B. Parker, and N. A. Cole. 2012c. Potential odorous volatile organic compound emission from feces and urine from cattle fed corn-based diets with wet distiller's grains and solubles. *Atmospheric Environment* 60:292-297.

Hales, K. E., N. A. Cole, and J. C. MacDonald. 2013. Effects of increasing concentrations of wet distillers grains with solubles in steam-flaked, corn-based diets on energy metabolism carbon-nitrogen balance, and methane emissions of cattle. *Journal of Animal Science* 91:819-828.

Hales, K. E., H. C. Freetly, and T. M. Brown-Brandl. 2014. Effects of decreased dietary roughage concentration on energy metabolism and nutrient balance in finishing beef cattle. *Journal of Animal Science* 92:264-271.

Hanselman, T. A., D. A. Graetz, and A. C. Wilkie. 2003. Manure-borne estrogens as potential environmental contaminants: A review. *Environmental Science & Technology* 37:5471-5479.

Hao, X. 2007. Nitrate accumulation and greenhouse gas emissions during compost storage. *Nutrient Cycling in Agroecosystems* 78:189-195.

Hao, X., C. Chang, F. J. Larney, and G. R. Travis. 2001. Greenhouse gas emissions during cattle feedlot manure composting. *Journal of Environmental Quality* 30:376-386.

Hao, X., C. Chang, and F. J. Larney. 2004. Carbon, nitrogen balances and greenhouse gas emission during cattle feedlot manure composting. *Journal of Environmental Quality* 33:37-44.

Hao, X., F. J. Larney, C. Chang, G. R. Travis, C. K. Nichol, and E. Bremer. 2005. The effect of phosphogypsum on greenhouse gas emissions during cattle manure composting. *Journal of Environmental Quality* 34:774-781.

Hao, X., M. B. Benke, C. Li, F. J. Larney, K. A. Beauchemin, and T. A. McAllister. 2011a. Nitrogen transformations and greenhouse gas emission during composting of manure from cattle fed diets containing corn dried distillers grains with solubles and condensed tannins. *Animal Feed Science and Technology* 166-167:539-549.

Hao, X., M. Benke, F. J. Larney, and T. A. McAllister. 2011b. Greenhouse gas emissions when composting manure from cattle fed wheat dried distillers' grains with solubles. *Nutrient Cycling in Agroecosystems* 89:105-114.

Hao, X., S. Xu, F. J. Larney, K. Stanford, A. J. Cessna, and T. A. McAllister. 2011c. Inclusion of antibiotics in feed alters greenhouse gas emissions from feedlot manure during composting. *Nutrient Cycling in Agroecosystems* 89:257-267.

Harper, L. A., O. T. Denmead, J. R. Freney, and F. M. Byers. 1999. Direct measurements of methane emissions from grazing and feedlot cattle. *Journal of Animal Science* 77:1391-1401.

Harper, L. A., T. K. Flesch, R. Todd, N. A. Cole, J. D. Wilson, and H. Adam. 2013. Methane and ammonia emissions from beef feeding in a U.S. southern High Plains region. *Advances in Animal Biosciences* 4:469.

Hashimoto, A. G., V. H. Varel, and Y. R. Chen. 1981. Ultimate methane yield from beef cattle manure: Effect of temperature, ration constituents, antibiotics and manure age. *Agricultural Wastes* 3:241-256.

Hegarty, R. S., J. P. Goopy, R. M. Herd, and B. McCorkell. 2007. Cattle selected for lower residual feed intake have reduced daily methane production. *Journal of Animal Science* 85:1479-1486.

Hegarty, R. S., D. Alcock, D. L. Robinson, J. P. Goopy, and P. E. Vercoe. 2010. Nutritional and flock management options to reduce methane output and methane per unit product from sheep enterprises. *Animal Production Science* 50:1026-1033.

Hegarty, R. S., J. Miller, D. L. Robinson, L. Li, N. Oelbrandt, K. Luijben, J. McGrath, G. Bremner, and H. B. Perdok. 2013. Growth, efficiency, and carcass attributes of feedlot cattle supplemented with calcium nitrate or urea. *Advances in Animal Biosciences* 4:440.

Heitschmidt, R. K., R. E. Short, and E. E. Grings. 1996. Ecosystems, sustainability, and animal agriculture. *Journal of Animal Science* 74:1395-1405.

Henry, P., M. Echevarria, C. Ammerman, and P. Rao. 1988. Estimation of the relative biological availability of inorganic selenium sources for ruminants using tissue uptake of selenium. *Journal of Animal Science* 66:2306-2314.

Hobbs, P. J., J. Webb, T. T. Mottram, B. Grant, and T. M. Misselbrook. 2004. Emissions of volatile organic compounds originating from UK livestock agriculture. *Journal of the Science of Food and Agriculture* 84:1414-1420.

Holter, J. B., and A. J. Young. 1992. Methane prediction in dry and lactating Holstein cows. *Journal of Dairy Science* 75:2165-2175.

Hristov, A., M. Hanigan, N. A. Cole, R. Todd, T. A. McAllister, P. M. Ndegwa, and A. Rotz. 2011. Review: Ammonia emissions from dairy farms and beef feedlots. *Canadian Journal of Animal Science* 91:1-35.

Hristov, A. N., T. Oh, J. L. Firkins, J. Dijkstra, E. Kebreab, G. Waghorn, H. P. S. Makkar, A. T. Adesogan, W. Yang, C. Lee, P. J. Gerber, B. Henderson, and J. M. Tricarico. 2013a. Special Topics: Mitigation of methane and nitrous oxide emissions from animal operations: I. A review of enteric methane mitigation options. *Journal of Animal Science* 91:5045-5069.

Hristov, A. N., T. Ott, J. Tricarico, A. Rotz, G. Waghorn, A. Adesogan, J. Dijkstra, F. Montes, J. Oh, E. Kebreab, S. J. Oosting, P. J. Gerber, B. Henderson, H. P. S. Makkar, and J. L. Firkins. 2013b. Special Topics: Mitigation of methane and nitrous oxide emissions from animal operations: III. A review of animal management mitigation options. *Journal of Animal Science* 91:5095-5113.

Hulshof, R. B. A., A. Berndt, W. J. J. Gerrits, J. Dijkstra, S. M. van Zijderveld, J. R. Newbold, and H. B. Perdok. 2012. Dietary nitrate supplementation reduces methane emission in beef cattle fed sugarcane-based diets. *Journal of Animal Science* 90:2317-2323.

Hunerberg, M., S. M. McGinn, K. A. Beauchemin, E. K. Okine, O. M. Harstad, and T. A. McAllister. 2013. Effect of dried distillers grains plus solubles on enteric methane emissions and nitrogen excretion from growing beef cattle. *Journal of Animal Science* 91:2846-2857.

Hunter, R. A. 2007. Methane production by cattle in the tropics. *British Journal of Nutrition* 98:657.

Hutchinson, G. L., and F. G. Viets, Jr. 1969. Nitrogen enrichment of surface water by absorption of ammonia volatilized from cattle feedlots. *Science* 166:514-515.

Hutchinson, G. L., A. R. Mosier, and C. E. Andre. 1982. Ammonia and amine emissions from a large cattle feedlot. *Journal of Environmental Quality* 11:288-293.

IPCC (Intergovernmental Panel of Climate Change) 2006. *2006 IPCC Guidelines for National Greenhouse Gas Inventories*, H. S. Eggleston, L. Buendia, K. Miwa, T. Ngara, and K. Tanabe, eds. Hayama, Japan: Institute for Global Environmental Strategies. Available online at http://www.ipcc-nggip.iges.or.jp/public/2006gl/index.html. Accessed on January 28, 2013.

Jarvis, S. C., R. D. Lovell, and R. Panayides. 1995. Patterns of methane emission from excreta of grazing animals. *Soil Biology and Biochemistry* 27(12):1581-1588.

Jayanegara, A., S. Marquardt, E. Wina, M. Kreuzer, and F. Leiber. 2013. In vitro indications for favourable non-additive effects on ruminal methane mitigation between high-phenolic and high-quality forages. *British Journal of Nutrition* 109:615-622.

Johnson, D. E. 1972. Effects of hemiacetal of chloral and starch on methane production and energy balance of sheep fed a pelleted diet. *Journal of Animal Science* 35:1064-1068.

Johnson, D. E. 1974. Adaptational responses in nitrogen and energy balance of lambs fed a methane inhibitor. *Journal of Animal Science* 38:154-157.

Johnson, K. A., and D. E. Johnson. 1995. Methane emissions from cattle. *Journal of Animal Science* 73:2483-2492.

Johnson, K., M. Huyler, H. Westberg, B. Lamb, and P. Zimmerman. 1994. Measurement of methane emissions from ruminant livestock using a sulfur hexafluoride tracer technique. *Environmental Science & Technology* 28:359-362.

Kebreab, E., J. A. N. Mills, L. A. Crompton, A. Bannink, J. Dijkstra, W. J. J. Gerrits, and J. France. 2004. An integrated mathematical model to evaluate nutrient partition in dairy cattle between the animal and its environment. *Animal Feed Science and Technology* 112:131-154.

Kebreab, E., K. Clark, C. Wagner-Riddle, and J. France. 2006. Methane and nitrous oxide emissions from Canadian animal agriculture: A review. *Canadian Journal of Animal Science* 86:135-137.

Kebreab, E., K. A. Johnson, S. L. Archibeque, D. Pape, and T. Wirth. 2008. Model for estimating enteric methane emissions from United States dairy and feedlot cattle. *Journal of Animal Science* 86:2738-2748.

Kebreab, E., J. Dijkstra, A. Bannink, and J. France. 2009. Recent advances in modeling nutrient utilization in ruminants. *Journal of Animal Science* 87(Suppl.):E111-E122.

Kellems, R. O., J. R. Miner, and D. C. Church. 1979. Effect of ration, waste composition and length of storage on the volatilization of ammonia, hydrogen sulfide and odors from cattle waste. *Journal of Animal Science* 48:436-445.

Kellogg, R. L. 2000. Potential priority watersheds for protection of water quality from contamination by manure nutrients. P. 20 in *Animal Residuals Management Conference November 12-14, 2000, Kansas City, MO.* Available online at http://www.nrcs.usda.gov/technical/land/pubs/. Accessed on June 20, 2005.

Kellogg, R. L., and C. H. Lander. 1999. Trends in the Potential for Nutrient Loading from Confined Livestock Operations. Poster presentation at The State of North America's Private Land Conference, January 19-21, 1999, Chicago, IL. Available online at http://www.nrcs.usda.gov/wps/portal/nrcs/detail/national/technical/nra/?&cid=nrcs143_014183. Accessed on October 23, 2014.

Kellogg, R. L., C. H. Lander, D. E. Moffitt, and N. Gollehon. 2000. *Manure Nutrients Relative to the Capacity of Cropland and Pastureland to Assimilate Nutrients: Spatial and Temporal Trends for the United States.* Publication No. NPS 00-0579. Washington, DC: USDA-NRCS and ERS. Available online at http://www.nrcs.usda.gov/wps/portal/nrcs/detail/national/technical/?cid=nrcs143_014126. Accessed on June 20, 2005.

Koelsch, R. K., D. D. Schulte, B. L. Woodbury, D. N. Miller, and D. E. Stenberg. 2004. Total reduced sulfur concentrations in vicinity of beef cattle feedlots. *Applied Engineering in Agriculture* 20:77-85.

Koenig, K. M., and K. A. Beauchemin. 2013a. Nitrogen metabolism and route of excretion in beef feedlot cattle fed barley-based backgrounding diets varying in protein concentration and rumen degradability. *Journal of Animal Science* 91:2295-2309.

Koenig, K. M., and K. A. Beauchemin. 2013b. Nitrogen metabolism and route of excretion in beef feedlot cattle fed barley-based finishing diets varying in protein concentration and rumen degradability. *Journal of Animal Science* 91:2310-2320.

Koenig, K. M., S. M. McGinn, and K. A. Beauchemin. 2013. Ammonia emissions and performance of backgrounding and finishing beef feedlot cattle fed barley-based diets varying in dietary crude protein concentration and rumen degradability. *Journal of Animal Science* 91:2278-2294.

Kriss, M. 1930. Quantitative relations of the dry matter of the food consumed, the heat production, the gaseous outgo, and the insensible loss of body weight of cattle. *Journal of Agricultural Research* 40:283-295.

Krupa, S. V. 2003. Effects of atmospheric ammonia (NH_3) on terrestrial vegetation: A review. *Environmental Pollution* 124:179-221.

Kurihara, M., T. Magner, R. A. Hunter, and G. J. McCrabb. 1999. Methane production and energy partition of cattle in the tropics. *British Journal of Nutrition* 82:227-234.

Lander, C. H., D. Moffitt, and K. Alt. 1998. *Nutrients Available from Livestock Manure Relative to Crop Growth Requirements.* Resources Assessment and Strategic Planning Working Paper 98-1. Washington, DC: U.S. Department of Agriculture, Natural Resources Conservation Service. Available online at http://www.nrcs.usda.gov/wps/portal/nrcs/detail/national/technical/nra/?cid=nrcs143_014175. Accessed on February 24, 2014.

Lassey, K. R. 2007. Livestock methane emission: From the individual grazing animal through national inventories to the global methane cycle. *Agricultural and Forest Meteorology* 142:120-132.

Legesse, G., J. A. Small, S. L. Scott, G. H. Crow, H. C. Block, A. W. Alemu, C. D. Robins, and E. Kebreab. 2011. Predictions of enteric methane emissions for various summer pasture and winter feeding strategies for cow calf production. *Animal Feed Science and Technology* 166-167:678-687.

Lichtenwalner, R. E., J. P. Fontenot, and R. E. Tucker. 1973. Effect of source of supplemental nitrogen and level of nitrate on feedlot performance and vitamin A metabolism of fattening beef calves. *Journal of Animal Science* 37:837-847.

Liebig, M. A., J. R. Gross, S. L. Kronberg, R. L. Phillips, and J. D. Hanson. 2010. Grazing management contributions to net global warming potential: A long-term evaluation in the Northern Great Plains. *Journal of Environmental Quality* 39:799-809.

Little, S., J. Linderman, K. MacLean, and H. Janzen. 2008. *HOLOS—A Tool to Estimate and Reduce Greenhouse Gases from Farms: Methodology and Algorithms for Versions 1.1.x.* Publication No. A52-136/2008E-PDF. Agriculture and Agri-Food Canada. Available online at http://publications.gc.ca/collections/collection_2009/agr/A52-136-2008E.pdf. Accessed on December 1, 2014.

Lodman, D. W., M. E. Branine, B. R. Carmean, P. Zimmerman, G. M. Ward, and D. E. Johnson. 1993. Estimates of methane emissions from manure of U.S. cattle. *Chemosphere* 26:189-199.

Loh, Z., D. Chen, M. Bai, T. Naylor, D. Griffith, J. Hill, T. Denmead, S. McGinn, and R. Edis. 2008. Measurement of greenhouse gas emissions from Australian feedlot beef production using open-path spectroscopy and atmospheric dispersion modelling. *Australian Journal of Experimental Agriculture* 48:244-247.

Loneragan, G. H., and M. M. Brashears. 2005. Effects of using retention-pond water for dust abatement on performance of feedlot steers and carriage of *Escherichia coli* O157 and *Salmonella* spp. *Journal of the American Veterinary Medical Association* 226:1378-1383.

Luebbe, M. K., G. E. Erickson, T. J. Klopfenstein, M. A. Greenquist, and J. R. Benton. 2011. Effect of dietary cation-anion difference on urinary pH, feedlot performance, nitrogen mass balance, and manure pH in open feedlot pens. *Journal of Animal Science* 89:489-500.

Luebbe, M. K., G. E. Erickson, T. J. Klopfenstein, and M. A. Greenquist. 2012a. Nutrient mass balance and performance of feedlot cattle fed corn wet distillers grains plus solubles. *Journal of Animal Science* 90:296-306.

Luebbe, M. K., J. M. Patterson, K. H. Jenkins, E. K. Buttrey, T. C. Davis, B. E. Clark, F. T. McCollum, N. A. Cole, and J. C. MacDonald. 2012b. Wet distillers grains plus solubles concentration in steam-flaked corn-based diets: Effects on feedlot cattle performance, carcass characteristics, nutrient digestibility, and ruminal fermentation characteristics. *Journal of Animal Science* 90:1589-1602.

Lupo, C. D., D. E. Clay, J. L. Benning, and J. J. Stone. 2013. Life-cycle assessment of the beef cattle production systems for the northern Great Plains, USA. *Journal of Environmental Quality* 42:1386-1394.

Machmuller, A., D. A. Ossowski, and M. Kreuzer. 2006. Effect of fat supplementation on nitrogen utilization of lambs and nitrogen emission from their manure. *Livestock Science* 101:159-168.

Mackie, R. I., P. G. Stroot, and V. H. Varel. 1998. Biochemical identification and biological origin of key odor components in livestock waste. *Journal of Animal Science* 76:1331-1342.

MacVean, D. W., D. K. Franzen, T. J. Keefe, and B. W. Bennett. 1986. Airborne particle concentration and meteorological conditions associated with pneumonia incidence in feedlot cattle. *American Journal of Veterinary Research* 47:2676-2682.

Manly, B. F. J. 1997. *Randomization, Bootstrap and Monte Carlo Methods in Biology*, 2nd Ed. London, UK: Chapman & Hall.

Martin, C., D. P. Morgavi, and M. Doreau. 2010. Methane mitigation in ruminants: From microbe to the farm scale. *Animal* 4:351-365.

Martinez-Fernandez, G., A. Arco, L. Abecia, G. Cantalapiedra-Hijar, E. Molina-Alcaide, A. I. Martin-Garcia, M. Kindermann, S. Duval, and D. R. Yanez-Ruiz. 2013. The addition of ethyl-3-nitrooxy propionate and 3-nitrooxypropanol in the diet of sheep sustainably reduces methane emissions and the effect persists over a month. *Advances in Animal Biosciences* 4(Pt. 2):368.

McGinn, S. M., K. A. Beauchemin, T. Coates, and D. Colombatto. 2004. Methane emissions from beef cattle: Effects of monensin, sunflower oil, enzymes, yeast, and fumaric acid. *Journal of Animal Science* 82:3346-3356.

McGinn, S. M., T. K. Flesch, B. P. Crenna, K. A. Beauchemin, and T. Coates. 2007. Quantifying ammonia emissions from a cattle feedlot using a dispersion model. *Journal of Environmental Quality* 36:1585-1590.

McGinn, S. M., D. Chen, A. Loh, J. Hill, K. A. Beauchemin, and O. T. Denmead. 2008. Methane emissions from feedlot cattle in Australia and Canada. *Australian Journal of Experimental Agriculture* 48:183-185.

McGinn, S. M., Y. H. Chung, K. A. Beauchemin, A. D. Iwaasa, and C. Grainger. 2009. Use of corn distillers' dried grains to reduce enteric methane loss from beef cattle. *Canadian Journal of Animal Science* 89:409-413.

Miller, D. N., and E. D. Berry. 2005. Cattle feedlot soil moisture and manure content: I. Impacts on greenhouse gases, odor compounds, nitrogen losses, and dust. *Journal of Environmental Quality* 34:644-655.

Miller, D. N., and V. H. Varel. 2001. In vitro study of the biochemical origin and production limits of odorous compounds in cattle feedlots. *Journal of Animal Science* 79:2949-2956.

Miller, D. N., E. D. Berry, J. E. Wells, C. L. Ferrell, S. L. Archibeque, and H. C. Freetly. 2006. Influence of genotype and diet on steer performance, manure odor, and carriage of pathogenic and other fecal bacteria. III. Odorous compound production. *Journal of Animal Science* 84:2533-2545.

Mills, J. A., J. Dijkstra, A. Banning, S. B. Cammell, E. Kebreab, and J. France. 2001. A mechanistic model of whole-tract digestion and methanogenesis in the lactating dairy cow: Model development, evaluation, and application. *Journal of Animal Science* 79:1584-1597.

Mills, J. A. N., E. Kebreab, C. M. Yates, L. A. Crompton, S. B. Cammell, M. S. Dhanoa, R. E. Agnew, and J. France. 2003. Alternative approaches to predicting methane emissions from dairy cows. *Journal of Animal Science* 81:3141-3150.

Mills, J. A. N., L. A. Crompton, A. Bannink, and C. K. Reynolds. 2010. Estimating methane emissions from enteric fermentation for the UK greenhouse gas inventory. *Proceedings of the Nutrition Society* 69(OCE4):E336.

Moe, P. W., and H. F. Tyrrell. 1979. Methane production in dairy cows. *Journal of Dairy Science* 62:1583-1586.

Molano, G., and H. Clark. 2008. The effect of level of intake and forage quality on methane production by sheep. *Australian Journal of Experimental Agriculture* 48:219-222.

Monteny, G. L., and D. Chadwick. 2006. Greenhouse gas abatement strategies for animal husbandry. *Agriculture, Ecosystems & Environment* 112:163-170.

Monteny, G. L., M. C. J. Smits, G. van Duinkerken, H. Mollenhorst, and I. J. M. De Boer. 2002. Prediction of ammonia emissions from dairy barns using feed characteristics. Part II: Relationships between urinary urea concentration and ammonia emission. *Journal of Dairy Science* 85:3389-3394.

Montes, F., R. Meinen, C. Dell, A. Rotz, A. N. Hristov, J. Oh, G. Waghorn, P. J. Gerber, B. Henderson, H. P. S. Makkar, and J. Dijkstra. 2013. Special Topics: Mitigation of methane and nitrous oxide emissions from animal operations: II. A review of manure management mitigation options. *Journal of Animal Science* 91:5070-5094.

Montgomery, S. P., J. S. Drouillard, E. C. Titgemeyer, J. J. Sindt, T. B. Farran, J. N. Pike, C. M. Coetzer, A. M. Trater, and J. J. Higgins. 2004. Effects of wet corn gluten feed and intake level on diet digestibility and ruminal passage rate in steers. *Journal of Animal Science* 82:3526-3536.

Moraes, L. E., A. B. Strathe, J. G. Fadel, D. P. Casper, and E. Kebreab. 2014. Prediction of enteric methane emission from cattle. *Global Change Biology* 20:2140-2148.

Muir, S., D. Chen, D. Rowell, and J. Hill. 2010. Manure pad composition and greenhouse gas emissions from Australian beef feedlots. Pp. 66-67 in *Proceedings of the 4th International Conference on Greenhouse Gases in Animal Agriculture, October 3-8, 2010, Banff, Canada.* Lethbridge, Alberta: Agri-Food Canada.

MWPS (MidWest Plan Service). 2004. Manure Management Systems Series. Section 1: Manure Characteristics. MidWest Plan Service. Available online at https://www-mwps.sws.iastate.edu/catalog/manure-management/sect-1. Accessed on May 15, 2013.

NRC (National Research Council). 1996. *Nutrient Requirements of Beef Cattle,* 7th Rev. Ed. Washington, DC: National Academy Press.

NRC. 2001. *Nutrient Requirements of Dairy Cattle,* 7th Rev. Ed. Washington, DC: National Academy Press.

NRC. 2003. *Air Emissions from Animal Feeding Operations: Current Knowledge, Future Needs.* Washington, DC: The National Academies Press.

NRC. 2012. *Climate Change: Evidence, Impacts, and Choices* (PDF Booklet). Available online at http://www.nap.edu/catalog.php?record_id=14673. Accessed on October 20, 2013.

NRCS (Natural Resources Conservation Service). 2004. *Agricultural Waste Management Field Handbook.* USDA-Natural Resources Conservation Service. Available online at http://www.nrcs.usda.gov/wps/portal/nrcs/detailfull/national/technical/ecoscience/mnm/. Accessed on May 15, 2013.

Orlando, E. F., A. S. Kolok, G. A. Binzcik, J. L. Gates, M. K. Horton, C. S. Lambright, L. E. Gray, Jr., A. M. Soto, and L. J. Guillette, Jr. 2004. Endocrine-disrupting effects of cattle feedlot effluent on an aquatic sentinel species, the fathead minnow. *Environmental Health Perspectives* 112:353-358.

Owens, F. N., D. S. Secrist, W. J. Hill, and D. R. Gill. 1997. The effect of grain source and grain processing on performance of feedlot cattle: A review. *Journal of Animal Science* 75:868-879.

Parker, D. B., M. B. Rhoades, G. L. Schuster, J. A. Koziel, and Z. L. Perschbacher. 2005. Odor characterization at open-lot beef cattle feedyards using triangular forced-choice olfactometry. *Transactions of the ASABE* 48:1527-1535.

Parker, D. B., Z. L. Perschbacker-Buser, N. A. Cole, and J. A. Koziel. 2010. Recovery of agricultural odors and odorous compounds from polyvinyl fluoride film bags. *Sensors* 10:8536-8552.

Pelchen, A., and K. J. Peters. 1998. Methane emissions from sheep. *Small Ruminant Research* 27:137-150.

Pelletier, N., R. Pirog, and R. Rasmussen. 2010. Comparative life cycle environmental impacts of three beef production strategies in the upper midwestern United States. *Agricultural Systems* 103:380-389.

Peters, G. M., H. V. Rowley, S. Wiedemann, R. Tucker, M. D. Short, and M. Schulz. 2010. Red meat production in Australia: Life cycle assessment and comparison with overseas studies. *Environmental Science & Technology* 44:1327-1332.

Petersen, S. O., S. G. Sommer, O. Aaes, and K. Soegaard. 1998. Ammonia losses from urine and dung of grazing cattle: Effect of N intake. *Atmospheric Environment* 32:295-300.

Phetteplace, H. W., D. E. Johnson, and A. F. Seidl. 2001. Greenhouse gas emissions from simulated beef and dairy livestock systems in the United States. *Nutrient Cycling in Agroecosystems* 60:99-102.

Purdy, C. W., D. C. Straus, and R. N. Clark. 2004. Diversity of *Salmonella* serovars in feedyard and nonfeedyard playas of the southern High Plains in the summer and winter. *American Journal of Veterinary Research* 65:45-52.

Raven, K. P., and R. H. Loeppert. 1997. Trace element composition of fertilizers and soil amendments. *Journal of Environmental Quality* 26:551-557.

Rhoades, M. B., D. B. Parker, and B. Dye. 2003. Measurement of hydrogen sulfide in beef cattle feedlots in the Texas High Plains. Paper No. 034108 in *Proceedings of the ASABE Annual Meeting, July 27-30, 2003, Las Vegas, NV*. St. Joseph, MI: American Society of Agricultural and Biological Engineers.

Rhoades, M. B., D. B. Parker, N. A. Cole, R. W. Todd, E. A. Caraway, B. W. Auvermann, D. R. Topliff, and G. L. Schuster. 2010. Continuous ammonia emission measurements from a commercial beef feedyard in Texas. *Transactions of the ASABE* 53:1823-1831.

Ricci, P., J. A. Rooke, I. Nevison, and A. Waterhouse. 2013. Methane emissions from beef and dairy cattle: Quantifying the effect of physiological stage and diet characteristics. *Journal of Animal Science* 91:5379-5389.

Romero-Perez, A., E. K. Okine, S. M. McGinn, L. L. Guan, M. Oba, S. M. Duval, and K. A. Beauchemin. 2014. The potential of 3-nitrooxypropanol to lower enteric methane emissions from beef cattle. *Journal of Animal Science* 92:4682-4693.

Rotz, C. A., D. R. Buckmaster, and J. W. Comerford. 2005. A beef herd model for simulating feed intake, animal performance, and manure excretion in farm systems. *Journal of Animal Science* 83:231-242.

Rotz, A. C., M. S. Corson, D. S. Chianese, F. Montes, S. D. Hafner, and C. U. Coiner. 2012. The Integrated Farm System Model: Reference Manual, Version 3.6. University Park, PA: USDA-ARS. Available online at http://ars.usda.gov/SP2UserFiles/Place/19020000/ifsmreference.pdf. Accessed on October 2, 2014.

Sarturi, J. O., G. E. Erickson, T. J. Klopfenstein, K. M. Rolfe, C. D. Buckner, and M. K. Luebbe. 2013a. Impact of source of sulfur on ruminal hydrogen sulfide and logic for the ruminal available sulfur for reduction concept. *Journal of Animal Science* 91:3352-3359.

Sarturi, J. O., B. E. Erickson, T. J. Klopfenstein, J. T. Vasconcelos, W. A. Griffin, K. M. Rolfe, J. R. Benton, and V. R. Bremer. 2013b. Effect of sulfur content in wet or dry distillers grains fed at several inclusions on cattle growth performance, ruminal parameters, and hydrogen sulfide. *Journal of Animal Science* 91:4849-4860

Sayer, K. M., C. D. Buckner, G. E. Erickson, T. J. Klopfenstein, C. N. Macken, and T. W. Loy. 2013. Effect of corn bran and steep inclusion in finishing diets on diet digestibility, cattle performance, and nutrient mass balance. *Journal of Animal Science* 91:3847-3858.

Sindt, J. J., J. S. Drouillard, E. C. Titgemeyer, S. P. Montgomery, C. M. Coetzer, T. B. Farran, J. N. Pike, J. J. Higgins, and R. T. Ethington. 2003. Wet corn gluten feed and alfalfa hay combinations in steam-flaked corn finishing cattle diets. *Journal of Animal Science* 81:3121-3129.

Soliva, C. R., S. L. Amelchanka, S. M. Duval, and M. Kreuzer. 2011. Ruminal methane inhibition potential of various pure compounds in comparison with garlic oil as determined with a rumen simulation technique (Rusitec). *British Journal of Nutrition* 106:114-122.

Sowell, B. F., J. C. Moslety, and J. G. P. Bowman. 1999. Social behavior of grazing beef cattle: Implications for management. *Journal of Animal Science* 77(Suppl.):1-6.

Spiehs, M. J., and V. H. Varel. 2009. Nutrient excretion and odorant production in manure from cattle fed corn wet distillers grains with solubles. *Journal of Animal Science* 87:2977-2984.

Stackhouse, K. R., C. A. Rotz, J. W. Oltjen, and F. M. Mitloehner. 2012. Growth promoting technologies reduce the carbon footprint, ammonia emissions, and cost of California beef production systems. *Journal of Animal Science* 90:4656-4665.

Stackhouse-Lawson, K. R., C. A. Rotz, J. W. Oltjen, and F. M. Mitloehner. 2012. Carbon footprint and ammonia emissions of California beef production systems. *Journal of Animal Science* 90:4641-4655.

Steinfield, H. 2006. *Livestock's Long Shadow: Environmental Issues and Options*. Rome: Food and Agriculture Organization. Available online at http://www.fao.org/docrep/010/a0701e/a0701e00.htm. Accessed on January 28, 2013.

Stratton, J., J. M. Ham, and T. Borch. 2013. Activation energy of urea hydrolysis and ammonia Henry constant effects on ammonia release from confined animal feeding operations (CAFOs). In *Proceedings of the Waste to Worth: "Spreading" Science and Solution, April 1- 5, 2013, Denver, CO*. Available online at http://www.extension.org/pages/67805/abstracts-for-waste-to-worth. Accessed on February 16, 2014.

Sweeten, J. M. 1978. Manure management for cattle feedlots. *Great Plains Beef Cattle Handbook L-1094*. College Station: Texas Agricultural Experiment Station.

Sweeten, J. M., D. L. Reddell, L. Schake, and B. Garner. 1977. Odor intensities at cattle feedlots. *Transactions of the ASAE* 20(3):502-508.

Sweeten, J. M., C. B. Parnell, R. S. Etheredge, and D. Osborne. 1988. Dust emissions in cattle feedlots. *Veterinary Clinics of North America Food Animal Practice* 4(3):557-578.

Sweeten, J. M., C. B. Parnell, B. W. Shaw, and B. W. Auvermann. 1998. Particle size distribution of cattle feedlot dust emission. *Transactions of the ASAE* 41(5):1477-1481.

Tate, K. W., E. R. Atwill, N. K. McDougald, and M. R. George. 2003. Spatial and temporal patterns of cattle feces deposition on rangeland. *Journal of Range Management* 56:432-438.

Tedeschi, L. O. 2006. Assessment of the adequacy of mathematical models. *Agricultural Systems* 89:225-247.

Tedeschi, L. O., D. G. Fox, and T. P. Tylutki. 2003. Potential environmental benefits of ionophores in ruminant diets. *Journal of Environmental Quality* 32:1591-1602.

Tedeschi, L. O., T. R. Callaway, J. P. Muir., R. C. Anderson. 2011. Potential environmental benefits of feed additives and other strategies for ruminant production. *Revista Brasileira de Zootecnia* 40(Suppl.):291-309.

Todd, R. W., W. Guo, B. A. Stewart, and C. Robinson. 2004. Vegetation, phosphorus, and dust gradients downwind from a cattle feedyard. *Journal of Range Management* 57:291-299.

Todd, R. W., N. A. Cole, and R. N. Clark. 2006. Reducing crude protein in beef cattle diets reduces ammonia emissions from artificial feedyard surfaces. *Journal of Environmental Quality* 35:404-411.

Todd, R. W., N. A. Cole, R. N. Clark, T. K. Flesch, L. A. Harper, and B. Baek. 2008a. Ammonia emissions from a beef cattle feedyard on the southern High Plains. *Atmospheric Environment* 42:6797-6805.

Todd, R. W., N. A. Cole, R. N. Clark, W. C. Rice, and W. X. Guo. 2008b. Soil nitrogen distribution and deposition on shortgrass prairie adjacent to a beef cattle feedyard. *Biology and Fertility of Soils* 44:1099-1102.

Todd, R. W., N. A. Cole, M. B. Rhoades, D. B. Parker, and K. D. Casey. 2011. Daily, monthly, seasonal, and annual ammonia emissions from southern High Plains cattle feedyards. *Journal of Environmental Quality* 40:1090-1095.

Todd, R. W., N. A. Cole, H. M. Waldrip, and R. M. Aiken. 2013. Arrhenius equation for modeling feedyard ammonia emissions using temperature and diet crude protein. *Journal of Environmental Quality* 42:666-671.

Todd, R. W., M. B. Altman, N. A. Cole, and H. M. Waldrip. 2014. Methane emissions from a beef cattle feedyard during the winter and summer on the southern High Plains of Texas. *Journal of Environmental Quality* 43:1125-1130.

Tomkins, N., and R. A. Hunter. 2004. Methane mitigation in beef cattle using a patented anti-methanogen. In *Proceedings of the 2nd Joint Australia and New Zealand Forum on Non-CO_2 Greenhouse Gas Emissions from Agriculture, October 2-9, 2003, Canberra*, R. J. Eckard and W. Slattery, eds. Canberra: CRC for Greenhouse Accounting.

Tomkins, N., S.M. Colegate, and R. A. Hunter. 2009. A bromochloromethane formulation reduces enteric methanogenesis in cattle fed grain-based diets. *Animal Production Science* 49:1053-1058.

Tomkins, N., S. M. McGinn, D. A. Turner, and E. Charmley. 2011. Comparison of open-circuit respiration chambers with a micrometeorological method for determining methane emissions from beef cattle grazing a tropical pasture. *Animal Feed Science and Technology* 166-167:240-247.

Trei, J. E., G. C. Scott, and R. C. Parish. 1972. Influence of methane inhibition on energetic efficiency of lambs. *Journal of Animal Science* 34:510-515.

Trenkle, A. 1978. The relationship between acid-base balance and protein metabolism in ruminants. In *Proceedings of the Regulation of Acid-Base Balance Symposium, November 8-9, 1978, Tucson, AZ.* University of Arizona and Church and Dwight Co., Inc.

Tucker, R. W., and P. J. Watts. 1993. Waste minimization in cattle feedlots by ration modification. *Transactions of the ASAE* 36:1461-1466.

Ungerfeld, E. M., R. A. Kohn, R. J. Wallace, and C. J. Newbold. 2007. A meta-analysis of fumarate effects on methane production in ruminal batch culture. *Journal of Animal Science* 85:2556-2563.

USDA (U.S. Department of Agriculture). 2011. U.S. Agriculture and Forestry Greenhouse Gas Inventory: 1990-2008. Technical Bulletin No. 1930. Climate Change Program Office, Office of the Chief Economist, USDA, Washington, DC. Available online at http://www.usda.gov/oce/climate_change/AFGG_Inventory/USDA_GHG_Inv_1990-2008_June2011.pdf. Accessed on October 21, 2014.

USDA. 2014. Quantifying greenhouse gas sources and sinks in animal production systems. Pp. 257-416 in *Quantifying Greenhouse Gas Fluxes in Agriculture and Forestry: Methods for Entity-Scale Inventory*, M. Eve, D. Pape, M. Flugge, R. Steele, D. Man, M. Riley-Gilbert, and S. Biggar, eds. Technical Bulletin 1939, Climate Change Program Office, Office of the Chief Economist, USDA, Washington, DC. Available online at http://www.usda.gov/oce/climate_change/Quantifying_GHG/USDATB1939_07072014.pdf. Accessed on October 21, 2014.

Van Amburgh, M. E., L. E. Chase, T. R. Overton, D. A. Ross, E. B. Recktenwald, R. J. Higgs, and T. P. Tylutki. 2010. Updates to the Cornell Net Carbohydrate and Protein System v6.1 and implications for ration formulation. Pp. 144-159 in *Proceedings of the 72nd Cornell Nutrition Conference for Feed Manufacturers, October 19-21, 2010, Syracuse, NY.* Available online at www.ansci.cornell.edu/cnconf/proceedings.html. Accessed on May 23, 2013.

Van der Eerden, L. J. M., P. H. B. de Visser, and C. J. van Dijk. 1998. Risk of damage to crops in the direct neighborhood of ammonia sources. *Environmental Pollution* 102(Suppl. 1):49-53.

Van der Watt, H. H. M. E, Summer, and M. L. Cabrera. 1994. Bioavailability of copper, manganese, and zinc in poultry litter. *Journal of Environmental Quality* 23:43-49.

van Zijderveld, S. M., W. J. J. Gerrits, J. A. Apajalahti, J. R. Newbold, J. Dijkstra, R. A. Leng, and H. B. Perdok. 2010. Nitrate and sulfate: Effective alternative hydrogen sinks for mitigation of ruminal methane production in sheep. *Journal of Dairy Science* 93:5856-5866.

Varel, V. H., J. E. Wells, E. D. Berry, M. J. Spiehs, D. N. Miller, C. L. Ferrell, S. D. Shackelford, and M. Koohmaraie. 2008. Odorant production and persistence of *Escherichia coli* in manure slurries from cattle fed zero, twenty, forty, or sixty percent wet distillers grain with solubles. *Journal of Animal Science* 86:3617-3627.

Vasconcelos, J. T., and M. L. Galyean. 2007. Nutritional recommendations of feedlot consulting nutritionists: The 2007 Texas Tech University survey. *Journal of Animal Science* 85:2772-2781.

Vasconcelos, J., L. W. Greene, N. A. Cole, M. S. Brown, F. T. McCollum, and L Tedeschi. 2006. Effects of phase feeding of protein on performance, blood urea nitrogen concentration, manure N:P ratio, and carcass characteristics of feedlot cattle. *Journal of Animal Science* 84:3032-3038.

Vasconcelos, J. T., N. A. Cole, K. W. McBride, A. Gueye, M. L. Galyean, C. R. Richardson, and L. W. Greene. 2009. Effects of dietary crude protein and supplemental urea concentrations on nitrogen and phosphorus utilization by feedlot cattle. *Journal of Animal Science* 87:1174-1183.

Velazco, J. I., G. Bremner, L. Li, K. Luijben, R. S. Hegarty, and H. Perdok. 2013. Short-term emission measurements in beef feedlot cattle to demonstrate enteric methane mitigation from dietary nitrate. *Advances in Animal Biosciences* 4:579.

Vergé, X. P. C., J. A. Dyer, R. L. Desjardins, and D. Worth. 2008. Greenhouse gas emissions from the Canadian beef industry. *Agricultural Systems* 98:126-134.

Waldrip, H. M., R. W. Todd, and N. A. Cole. 2013. Prediction of nitrogen excretion by beef cattle: A meta-analysis. *Journal of Animal Science* 91:4290-4302.

Watts, P. J., M. Jones, S. C. Lott, R. W. Tucker, and R. J. Smith. 1994. Feedlot odor emissions following heavy rainfall. *Transactions of the ASAE* 37(2):629-636.

Weichenthal, B. A., L. B. Embry, R. J. Emerick, and F. W. Whetzal. 1963. Influence of sodium nitrate, vitamin A and protein level on feedlot performance and vitamin A status of fattening cattle. *Journal of Animal Science* 22:979-984.

Westberg, H., B. Lamb, K. A. Johnson, and M. Huyler. 2001. Inventory of methane emissions from U. S. cattle. *Journal of Geophysical Research* 106 (12):12633-12642.

Whalen, J. K., C. Chang, G. W. Clayton, and J. P. Carefoot. 2000. Cattle manure amendments can increase the pH of acid soils. *Soil Science Society of America Journal* 64:962-966.

White, S. L., R. E. Sheffield, S. P. Washburn, L. D. King, and J. T. Green. 2001. Spatial and time distribution of dairy cattle excreta in an intensive pasture system. *Journal of Environmental Quality* 30:2180-2187.

Wilson, S. C., J. Morrow-Tesch, D. C. Straus, J. D. Cooley, W. C. Wong, F. M. Mitlohner, and J. J. McGlone. 2002. Airborne microbial flora in a cattle feedlot. *Applied and Environmental Microbiology* 68:3238-3242.

Xu, S., X. Hao, K. Stanford, T. A. McAllister, F. J. Larney, and J. Wang. 2007a. Greenhouse gas emissions during co-composting of calf mortalities with manure. *Journal of Environmental Quality* 36:1914-1919.

Xu, S., X. Hao, K. Stanford, T. A. McAllister, F. J. Larney, and J. Wang. 2007b. Greenhouse gas emissions during co-composting of cattle mortalities with manure. *Nutrient Cycling in Agroecosystems* 78:177-187.

Yamulki, S., S. C. Jarvis, and P. Owen. 1999. Methane emission and uptake from soils as influenced by excreta deposition from grazing animals. *Journal of Environmental Quality* 28:676-682.

Yan, T., R. E. Agnew, F. J. Gordon, and M. G. Porter. 2000. Prediction of methane energy output in dairy and beef cattle offered grass-silage based diets. *Livestock Production Science* 64:253-263.

Yan, T., M. G. Porter, and C. S. Mayne. 2009. Prediction of methane emission from beef cattle using data measured in indirect open-circuit respiration calorimeters. *Animal* 3(10):1455-1462.

Zinn, R. A., and Y. Shen. 1996. Interaction on dietary calcium and supplemental fat on digestive function and growth performance in feedlot steers. *Journal of Animal Science* 74:2303-2309.

17

Utilization of Byproduct Feeds by Beef Cattle

AVAILABILITY AND OVERVIEW

Numerous alternative feeds are used within the beef industry. These alternative feeds are produced as a result of manufacturing a product or a number of products, so they are often referred to as byproducts and sometimes as co-products; however, the focus of processing plants is usually to manufacture other products with the byproduct as a secondary product. Many times, these byproduct feeds are greater in fiber, protein, or minerals than the original feed source. Byproducts can be defined as a feed produced from the manufacturing of products which are often intended for human consumption or uses. This chapter focuses on the most widely used (i.e., produced) feed byproducts, which generally include byproducts made from corn or soybeans because those two crops comprise the most acres and result in the greatest cash receipts in the United States (NASS, 2014). Many other byproduct feeds are available, however, and will only be commented on here, with greater detail provided in Chapter 18 (Composition of Selected Feeds for Beef Cattle). Table 17-1 provides a summary of various feed byproducts that are typically used in beef nutrition, in addition to those discussed in this chapter. These byproduct feeds are often very advantageous to use in the locale of production and can be economical feeds to use for beef cattle.

Another important concept when evaluating feed byproducts is that energy values determined from laboratory assays and reported from chemical analysis are often incorrect. For example, their energy value is frequently "calculated" from fiber content using equations that were developed for silages and forages. Although these equations are very useful for the forages from which they were developed, the calculated energy values for nonforage, fibrous byproducts are meaningless. For this reason, experiments comparing byproduct feeds to traditional energy sources (either forage or grain) are necessary to accurately predict performance and energy content.

CORN BYPRODUCTS

Corn Dry Grinding

The most common feed byproduct produced today in North America is distillers grains plus solubles (DGS), which is produced from the dry-grind ethanol process (Figure 17-1) and is usually derived from corn. Despite widespread use today, feeding distillers grains (and brewers byproducts) is not a new concept, as shown by early reports from the late 1800s in Kentucky and a boom in cattle feeding in eastern Kentucky from 1932 to 1957. Early reports indicated that cattle were fed wet "slop" to fatten with feed piped to cattle troughs (Ball and Cornett, 1996). While plants producing alcohol for human consumption produce DGS, the majority of DGS is produced as a byproduct of fuel ethanol.

During ethanol production, grain (corn or other starch source) is ground, converted to sugar enzymatically, and then the sugar is fermented to ethanol and CO_2 by yeast. After the ethanol distillation step, whole stillage is centrifuged to separate the distillers grains from the distillers solubles. Distillers solubles are evaporated and typically added back to the distillers grains to produce DGS. The nutrient composition of DGS depends on the relative ratios of distillers grains to distillers solubles in the final feed product. If all the solubles are added back to the grains, DGS are approximately 80% distillers grains and 20% distillers solubles on a DM basis (Corrigan et al., 2009b). Most distillers grains contain some solubles, but this can vary from plant to plant.

The supply of DGS has increased dramatically over the past 10 years because of the dramatic increase in ethanol production from corn dry milling (Figure 17-2). Ethanol production capacity from grain is approximately 50 billion L (13.2 billion gal.) and is projected to remain relatively constant in future years (RFA, 2014). Therefore, the supply of DGS from dry-mill ethanol should remain relatively constant going forward; however, there is increased interest in removing components that were traditionally included in distillers

TABLE 17-1 Byproduct Energy and Protein Feeds Used in Beef Cattle Diets

Ingredient	IFN[a]	AAFCO[b]	Comments
Blood meal	5-26-005	9.61	High in rumen undegradable protein; amount depends on heating temperature
Brewer's grains	5-00-516	15.1	May be wet or dried
Canola meal, solvent extracted	5-05-146	71.77	Similar in value to soybean meal
Citrus pulp, dried	4-01-237	21.1	High in fiber and pectin; high levels may cause ruminal parakeratosis
Cottonseed, full fat	5-01-609	24.4	High in both protein and oil; calves may be sensitive to the gossypol content
Cottonseed meal	5-01-632	24.12	Value depends on processing method; can be a replacement for soybean meal; calves may be sensitive to the gossypol content
Fats			Fat is high in energy and addition increases the energy value of the diet, increases palatability, decreases dustiness, and decreases wear on pelleting dies
Feather meal	5-03-795	9.15	Value may depend on cooking and processing method; sometimes fed in combination with blood meal
Fish meal	5-01-977	51.14	High in protein with an excellent amino acid profile; can be a source of rumen undegradable protein depending on how it is cooked
Oat groats	—	69.1	Generally too expensive to feed to ruminants
Oats	4-03-309	—	Crushed oats are very palatable to cattle
Peanut meal, expelled	5-03-649	71.9	High in protein; can replace soybean meal
Peanut meal, extracted	5-03-650	71.9	High in protein; can replace soybean meal
Potato byproducts			Raw potatoes are not very palatable and have a laxative effect
Poultry byproduct meal	5-03-798	9.71	Quality depends on the components that go into the meal; can be a source of rumen undegradable protein, depending on how it is cooked
Rice bran	4-03-928	75.7	A high-fiber ingredient
Sugar beet pulp	4-00-669	63.36	Palatable, bulky, and slightly laxative
Wheat bran	4-05-190	93.1	A good source of protein, minerals, and fiber; slight laxative effect
Wheat middlings	4-05-201	93.6	Also known as wheat shorts
Whey powder	4-01-182	54.7	Generally too expensive except for calf diets

[a]International Feed Number.
[b]Association of American Feed Control Officials Number.

byproducts such as corn oil and perhaps corn fiber and corn protein, which could dramatically change the chemical composition and thus change the use of DGS by beef cattle producers and nutritionists.

Effect of Grain Used for Fermentation

Because DGS are produced from starch, the remaining feed product(s) originate from the grain components minus starch. As a result, many nutrients or components are concentrated roughly threefold because approximately two-thirds of the grain is starch. For example, corn contains approximately 0.3% phosphorus (P), whereas DGS contains approximately 0.9% (dry matter [DM] basis). Similar results are observed for crude protein (CP), fiber, fat, and many other minerals. Fat and sulfur (S) are not necessarily concentrated threefold, which will be discussed later. Other grain sources can be used for conversion of starch to ethanol. The second most common grain in the United States used for ethanol production is grain sorghum (Wang et al., 2008), whereas wheat is a common grain source used in Canada and internationally (Paz et al., 2013).

Sorghum DGS. The grain that is used for feedstock in the ethanol process has an effect on the nutrient composition of the resulting DGS. Few studies have compared DGS that originated from different grains. More research is needed to elucidate whether sorghum as a feedstock is different than corn in terms of cattle performance. Sorghum contains more fiber and protein and less fat than corn grain, which results in DGS that are greater in fiber and protein and lower in fat than corn DGS. When corn- or sorghum-based DGS were fed at 30% of the diet in place of corn, no statistically significant differences were detected in the gain-to-feed ratio or feed efficiency (G:F; Al-Suwaiegh et al., 2002); however,

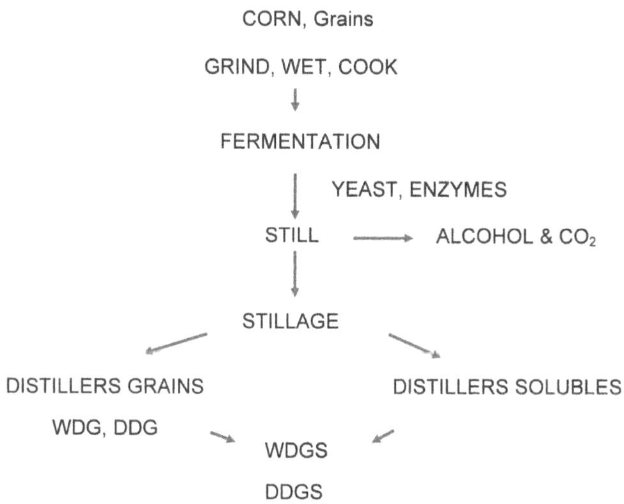

FIGURE 17-1 Schematic of the dry milling industry with the feed products produced. WDG = wet distillers grains; DDG = dry distillers grains; WDGS = wet distillers grains plus solubles; DDGS = dry distillers grains plus solubles.

cattle fed corn-based DGS were 3% more efficient than those fed sorghum DGS, which calculates to a 10% greater feeding value for corn over sorghum DGS. In steam-flaked corn (SFC) diets, cattle fed 10% corn wet DGS (WDGS) were more efficient that cattle fed 10% sorghum WDGS, but the difference was not statistically significant (Vasconcelos et al., 2007). Nonetheless, the feeding value of corn DGS was about 40% greater when comparing the two DGS. May et al. (2010) compared sorghum or corn WDGS included at 15 or 30% of dietary DM and observed that cattle fed corn-based WDGS had greater G:F than cattle fed sorghum-based WDGS. Similarly, Depenbusch et al. (2009) fed 15% WDGS or dry distillers grains plus solubles (DDGS) from corn or sorghum in SFC diets and observed no significant difference

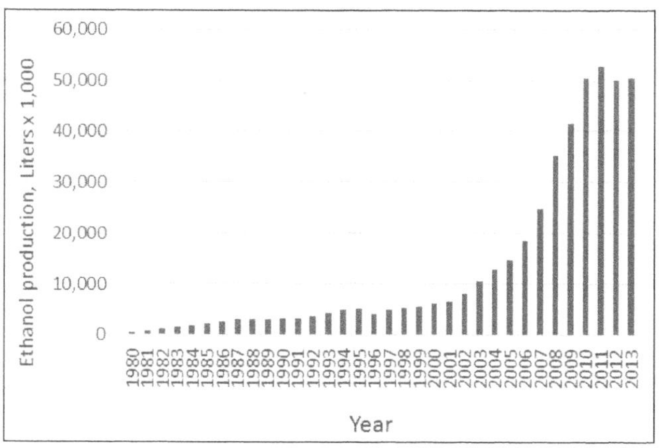

FIGURE 17-2 Ethanol production by year in the United States. SOURCE: Renewable Fuels Association (RFA, 2014).

as a result of the grain used to produce DGS, but again these authors calculated a 25% difference in feeding value between the two grain sources.

Wheat DGS. As with sorghum, wheat DGS has a different chemical composition than corn DGS. Paz et al. (2013) summarized results from three studies (Widyaratne and Zijlstra, 2007; Nuez-Ortin and Yu, 2009; Abdelqader and Oba, 2012) that evaluated wheat distillers and concluded that wheat DGS contains more CP, less fat, and similar neutral detergent fiber (NDF) compared to corn DGS. Numerous studies have evaluated wheat DGS, primarily as a replacement for barley. Cattle have similar gain and G:F when wheat WDGS replaced barley at very low inclusions in the diet (Ojowi et al., 1997). More recently, when DDGS from wheat replaced half or all the barley in growing diets, performance was either similar to that of cattle fed barley (Gibb et al., 2008) or gains and G:F were improved (McKinnon and Walker, 2008). When finishing cattle were fed 0, 20, 40, or 60% DDGS from wheat in place of barley, intakes increased linearly and G:F decreased linearly (Gibb et al., 2008). Maximal gain was observed at 32% inclusion when growing cattle were fed wheat DDGS (Beliveau and McKinnon, 2008). When those cattle were finished and fed 6, 12, 18, or 23% wheat DDGS, no differences in performance were detected (Beliveau and McKinnon, 2008). These data suggest that wheat-based DDGS contain energy similar to or greater than that of barley grain in growing and finishing diets. As with corn-based DGS, experimental results suggest that some control of ruminal pH is provided when wheat DDGS are included in the diet, but roughage should not be replaced by wheat DDGS because rumen pH and intake are decreased (Li et al., 2011). In one study that compared wheat DDGS to corn DDGS at 20 or 40% in place of barley grain, corn DDGS increased dietary energy compared with barley, whereas no difference was observed between wheat DDGS and barley (Walter et al., 2010). Gains were all fairly similar, but days on feed were adjusted so cattle were slaughtered at similar weights in this study. Feed efficiencies were similar when 0, 20, or 40% wheat DDGS were fed, whereas feeding 20 or 40% corn DDGS in place of barley quadratically improved G:F (Walter et al., 2010). These data suggest that wheat DDGS is very similar to barley grain when used as an energy source in growing or finishing diets and slightly lower in energy than corn DDGS, which is probably related to fiber and fat composition of DDGS.

Corn DGS. Based on several published articles summarized by Benton (2010), the average nutrient composition for corn DGS is approximately 31.5% CP, 10.5% fat, 6% starch, 43.2% NDF, 0.80% P, and 0.71% S. Relatively low variation was observed for CP, NDF, P, and S with coefficients of variation (CV) of 10.7, 10.5, 8.4, and 6.3%, respectively. Greater variation was observed for fat and starch with CV of 31.4 and 36.3%, respectively. Although DM variation is probably of greatest importance with wet byproducts, both fat and S concentrations can vary in DGS. Variability in nutrient content of byproducts is perceived to be a big challenge. More variation is likely across ethanol plants than within plants (Spiehs et al., 2002; Buckner et al., 2011b). Buckner et al. (2011b) collected samples of WDGS and modified DGS (MDGS, partially dried, 42 to 50% DM) for 5 consecutive days, across 4 different months, and within 6 dry milling plants and analyzed them for DM, CP, fat, P, and S. The variation in DM content within each plant was minimal (CV was less than 3%), but average DM content was different across plants. On average, DGS contained 31.0% CP, 11.9% fat, 0.84% P, and 0.77% S on a DM basis. Variation within days, across days, and within the same plants remained small for CP and P (CV less than 4%), but P varied some across plants. Variation in fat content was slightly higher but remained relatively small (CV less than 5%) within plants and within days, but greater variation was observed among ethanol plants. Fat content varied from 10.9 to 13.0% by plant, likely a result of varying amounts of distillers solubles that the plants added back to distillers grains. Fractionation of the corn kernel in dry milling plants and centrifuging (removal) of a portion of the oil has been recently adopted by many ethanol plants, which will lead to greater variation in oil or fat content across plants. Variation in S content was the largest for all nutrients tested as CV within days and across days (within the same ethanol plants) ranged from 3 to 13%. These data suggest that S values should be routinely monitored as this can lead to nutritional challenges (see Chapter 7, Minerals).

All plants produce wet distillers grains (WDG) and distillers solubles. Plants can combine the two to produce WDGS, or market them separately as WDG and distillers solubles, or sometimes referred to as syrup. Plants can also dry a portion (MDGS) or completely dry distillers grains to produce either DDG or DDGS.

Numerous reviews have focused on feeding DGS to dairy (Kalscheur et al., 2012; Paz et al., 2013) or beef cattle (Klopfenstein et al., 2007; Erickson et al., 2012) or focused on feeding DGS to finishing cattle (Klopfenstein et al., 2008; Galyean et al., 2012). Readers are referred to these papers for additional information on the feeding value of DGS.

Value of Distillers Grains with Solubles as a Protein Supplement

Historically, DGS was used in beef cattle diets as a protein supplement (Klopfenstein et al., 1978). Using the conventional ethanol process, corn DGS average 30 to 32% CP (Spiehs et al., 2002; Buckner et al., 2011b; see Chapter 18), consisting of the original zein protein, which is about 40% ruminally degradable protein or RDP (% CP; McDonald, 1954; Little et al., 1968). The ruminally undegradable protein (RUP) of DDGS is variable, which is likely a result of measurement technique (Aines et al., 1986; Klopfenstein et al., 2008). A number of studies that compared the RUP of DDGS relative to soybean meal (30% RUP) showed 2.3 to 2.6 times the RUP of soybean meal for DDGS, or about 70% RUP as a percentage of CP in DGS (Klopfenstein et al., 1978; Aines et al., 1986; Ham et al., 1994). Using the in situ technique often inflates ruminal disappearance of protein and shows that the RUP is 30 to 35% (Corrigan et al., 2009b); however, using other techniques, such as ammonia release, shows that the RUP is 69 to 83% and decreases as the proportion of solubles included in DGS increases. Using duodenally fistulated steers, Castillo-Lopez et al. (2013) recently measured 63% RUP (% CP) for DGS when accounting for bacterial protein using two separate techniques and accounting for protein from yeast in DGS. These measurements were made with cattle fed a forage-based diet containing sorghum silage and brome hay.

Digestibility of RUP in DGS has been evaluated and found to be relatively high. Apparent digestibility is dramatically less than true digestibility of protein from the feed when total tract digestion studies are used to evaluate protein digestibility. These incorrect, low-digestibility values are a result of hindgut fermentation and microbial protein synthesis in the large intestine. Corrigan et al. (2009b) measured the digestibility of the RUP fraction of DDGS that varied in the amount of solubles from 0 to 110% of what the plant was producing (approximately 22% of the DDGS, DM basis). The digestibility of RUP from DGS in the small intestine ranged from 90 to 97%, expressed as a percentage of CP entering the intestine using the mobile bag technique. Similar results have been observed using primarily dairy diets consisting of 40 to 60% concentrate (Kononoff et al., 2007; Janicek et al., 2008; Kelzer et al., 2010; Paz et al., 2013). Similar or slightly lower digestibilities of 86 to 95% were measured by Boucher et al. (2009); Li et al. (2012); and Mjoun et al. (2010) using a modified three-step in vitro procedure developed by Gargallo et al. (2006).

Excessive heating of protein is often measured as acid detergent insoluble nitrogen (ADIN; or protein). As ADIN increases in forages, protein availability or digestibility in the small intestine is decreased (Yu and Thomas, 1976; Van Soest et al., 1991; McNiven et al., 2002). Drying of distillers grains to produce DDGS can result in overheating and increase ADIN content of DDGS. Nonetheless, no evidence exists that protein digestion is hindered in the intestine or that performance is negatively affected when DDGS with increasing ADIN were compared (Ham et al., 1994).

Because of a greater proportion of RUP in DGS than in other protein sources, the supply of RDP in finishing and forage-based diets is thought to be limiting. When using prediction models, diets that include DGS are sometimes inadequate in RDP, suggesting that urea or other sources of RDP are necessary despite quite large amounts of dietary protein. The previous NRC committee (1996, 2000) did not adequately account for recycling of nitrogen (N) that can occur when metabolizable protein (MP) is in excess because of dietary RUP. The current committee recognizes that when dietary supply of MP is greater than the requirements, N is recycled to the rumen, which may provide sufficient RDP when DGS are used for protein supplementation. A few studies have evaluated RDP supply in cattle fed finishing diets supplemented with DGS. In diets with 25% WDGS, dietary CP was 14.2%, yet when performance was predicted using the NRC (1996, 2000) model, it was found that cattle were 83 g/d short on RDP. When cattle were supplemented with 0, 0.5, or 1.0% urea to increase the RDP balance from –83 to 36 or 153 g/d positive balance, cattle performance was not affected (Jenkins et al., 2011). In another experiment, individually fed steers received either 10 or 20% DDGS with or without supplemental urea to ensure RDP requirements were met (with added urea). Numerically, cattle gained more when supplemented with urea with 10% DDGS, but no statistical difference was observed statistically when urea was added to diets with either 10 or 20% DDGS. Similar results have been observed in forage-based diets. Heifers fed a forage-based diet with 30% DDGS had similar intake, gain, and G:F when fed 0, 0.4, 0.8, 1.2, or 1.6% added urea despite a calculated deficiency of RDP, assuming that no N recycling occurred (Stalker et al., 2007). Based on calculations using the NRC (1996, 2000) model, heifers supplemented with 1.2% urea met the predicted RDP requirement. In a second experiment by the same authors, heifers fed meadow hay along with 1.4 kg DDGS had similar intake, gain, and G:F whether fed 0 or 45 g/d of urea (100% of RDP predicted requirements). These data suggest that when cattle are fed sufficient DGS that provide excess MP, the N recycling to the rumen meets the RDP requirements of growing and finishing cattle.

Somewhat in contrast, Ponce et al (2009) reported that performance of cattle fed SFC-based diets containing 15% WDGS was improved by the addition of urea (0.53 to 1.06%), although urea additions did not improve the performance of cattle fed 30% WDGS diets. In a second study (Ponce et al., 2010), performance of cattle fed 15% WDGS diets was not affected by dietary urea additions.

The protein in DGS is comprised of the same amino acid (AA) profile as the grain from which it originates. The AA profile of DGS from different feedstocks is reported in Table 17-2 along with that of corn grain. As with many other nutrients, the AA are generally concentrated approximately threefold when expressed on a DM basis. Readers are referred to Paz et al. (2013) for additional information on the AA content of dry corn milling byproducts.

Use of Distillers Grains plus Solubles as an Energy Supplement

At low inclusions, DGS provides protein in diets and substitutes for traditional protein supplements; however, formulating diets to provide adequate RDP and MP is critical to ensure optimal performance of cattle. As dietary inclusion of DGS increases, the protein supply may far exceed requirements for RDP and MP, especially if recycled MP

TABLE 17-2 Amino Acid Composition of Corn Grain and Dried Distillers Grains plus Solubles (DDGS) from Corn, Sorghum, and Wheat

	Stein and Shurson (2009)[a]			Stein et al. (2006)	NRC (2012)	
	Corn DDGS	Sorghum DDGS	Wheat DDGS	Corn DDGS	Corn DDGS	Corn Grain
CP	30.42	35.00	45.19	30.89	27.33	8.24
Indispensable AA						
Arg	1.27	1.18	1.70	1.24	1.16	0.37
His	0.79	0.76	1.02	0.87	0.71	0.24
Ile	1.11	1.46	1.50	1.15	1.02	0.28
Leu	3.46	4.47	2.96	3.51	3.13	0.96
Lys	0.84	0.73	0.72	0.89	0.77	0.25
Met	0.60	0.57	0.59	0.69	0.55	0.18
Phe	1.47	1.80	2.13	1.50	1.34	0.39
Thr	1.16	1.14	1.34	1.12	0.99	0.28
Trp	0.23	0.38	0.44	0.19	0.21	0.06
Val	1.49	1.77	1.89	1.59	1.35	0.38
Dispensable AA						
Ala	2.10	3.10	1.64	1.99	1.93	0.60
Asp	2.00	2.32	2.13	2.19	1.82	0.54
Cys	0.58	0.52	0.81	0.79	0.51	0.19
Glu	4.71	6.76	10.90	3.92	4.35	1.48
Gly	1.13	1.10	1.80	1.10	1.04	0.31
Pro	2.29	2.68	4.57	2.19	2.09	0.71
Ser	1.27	1.50	2.09	1.21	1.18	0.38
Tyr	1.11	–	–	1.04	1.04	0.26

[a]Adapted from Stein and Shurson, 2009; data were divided by 0.90 to convert from a perceived as-is basis for DDGS to a DM basis. In addition, data from Stein et al. (2006) are presented as an average of 10 different DDGS samples.

is providing additional RDP to the rumen. Excess MP is deaminated and used as an energy source, along with the other components in DGS. With the expansion of fuel ethanol production since 2006, the abundant supply of DGS changed the pricing structure, making DGS competitive relative to traditional energy sources such as corn. As a result, a large amount of research has focused on using DGS as an energy source, substituting for corn or forages in beef cattle diets.

Feedlot Diets. Buckner et al. (2008) evaluated DDGS inclusions in finishing diets and determined that performance improved as DDGS replaced dry-rolled corn (DRC). Similar experiments have been performed evaluating WDGS (Farlin, 1981; Firkins et al., 1985; Larson et al., 1993a; Watson et al., 2014) and MDGS (Watson et al., 2014). Bremer et al. (2011) summarized numerous experiments that evaluated multiple inclusions of DDGS, MDGS, and WDGS. In that summary, performance improved relative to diets based on DRC and high-moisture corn (HMC) that did not contain DGS (Table 17-3; Bremer et al., 2011). The authors attributed this increase in performance to feeding value rather than the energy content of distillers grains. An energy value equivalent to 130 to 140% that of corn is difficult to explain as the fat and RUP contents of DGS that can be used as energy only accounts for about 20% greater energy than corn (Klopfenstein et al., 2008). Moreover, these values are difficult to explain nutritionally because the digestibility (Corrigan et al., 2009a; Vander Pol et al., 2009; May et al., 2010; Luebbe et al., 2012; Hales et al., 2012, 2013b) of diets containing DGS were lower than the corn it replaced, with partial digestibility coefficients that ranged from 70 to 90% compared with corn. Digestibility was decreased with DGS diets in al these studies, but the decrease was not always statistically significant.

Combined results from many experiments comparing WDGS, MDGS, and DDGS relative to DRC or HMC indicate that WDGS have a greater feeding value than MDGS, with both being greater than DDGS (Bremer et al., 2011). In experiments where WDGS, MDGS, and DDGS were compared directly on the same cattle within an experiment, WDGS or MDGS had a greater feeding value than DDGS and by a similar magnitude of difference (Ham et al., 1994; Nuttelman et al., 2011, 2013), which supports the meta-analysis data of Bremer et al. (2011).

The feeding value of DGS relative to corn in feedlot diets interacts with grain processing when DGS are fed as an energy substitute at greater than 20% of dietary DM (Cole et al., 2006; Corrigan et al., 2009a). The effects of feeding DGS on animal performance seem to be less in diets based on

TABLE 17-3 Meta-Analysis of Finishing Steer Performance When Fed Different Dietary Inclusions of Corn Wet Distillers Grains Plus Solubles (WDGS), Modified Distillers Grains Plus Solubles (MDGS), or Dried Distillers Grains Plus Solubles (DDGS) Replacing Dry-Rolled and High-Moisture Corn[a]

	DGS Inclusion[b]					P^c	
	0	10	20	30	40	L	Q
				WDGS[d]			
DMI, kg/d	10.4	10.6	10.6	10.4	10.2	0.01	<0.01
ADG, kg/d	1.60	1.71	1.77	1.78	1.75	<0.01	<0.01
G:F	0.155	0.162	0.168	0.171	0.173	<0.01	<0.01
Feeding value, %[e]		150	143	136	130		
				MDGS[f]			
DMI, kg/d	10.4	10.8	10.9	10.9	10.6	0.95	<0.01
ADG, kg/d	1.60	1.71	1.77	1.78	1.74	<0.01	<0.01
G:F	0.155	0.159	0.162	0.164	0.165	<0.01	0.05
Feeding value, %[e]		128	124	120	117		
				DDGS[f]			
DMI, kg/d	10.4	10.9	11.2	11.3	11.3	<0.01	0.03
ADG, kg/d	1.60	1.66	1.72	1.77	1.83	<0.01	0.50
G:F	0.155	0.156	0.158	0.160	0.162	<0.01	0.45
Feeding value, %[e]		112	112	112	112		

[a]WDGS = wet distillers grains with solubles, MDGS = modified wet distillers grains with solubles, DDGS = dried distillers grains with solubles. References are cited in Bremer et al. (2011).

[b]Dietary treatment levels (DM basis) of distillers grains plus solubles (DGS), 0 DGS = 0% DGS, 10 DGS = 10% DGS, 20 DGS = 20% DGS, 30 DGS = 30% DGS, 40 DGS = 40% DGS.

[c]Estimation equation linear (L) and quadratic (Q) term *t*-statistic for variable of interest response to DGS level.

[d]WDGS data presented are from Bremer et al., (2011).

[e]Percentage of corn feeding value, calculated from DGS inclusion level G:F relative to 0 WDGS G:F, divided by DGS inclusion.

[f]MDGS and DDGS steer performance was scaled to the WDGS intercept for equal comparison across byproduct types. This process was validated by Nuttelman et al. (2010).

SFC than in diets based on DRC or HMC. Vander Pol et al. (2008) evaluated corn processing methods in diets containing 30% WDGS and concluded that in terms of G:F, steam-flaked corn was equal to or lower than DRC, whereas whole corn was 11% lower and HMC was 4% greater than DRC. In a follow-up experiment, Corrigan et al. (2009a) replaced either DRC, HMC, or SFC with 0, 15, 27.5, or 40% WDGS. They found that as WDGS increased in the diet, G:F increased linearly when it replaced DRC and quadratically when it replaced HMC, with 27.5% WDGS giving the greatest G:F, or was not changed in diets containing SFC. Comparing across experiments, Meyer et al. (2013) observed similar to slightly poorer G:F when SFC and supplemental fat were replaced with WDGS, but performance was improved when a blend of DRC and HMC were replaced with WDGS. When corn-based WDGS replaced SFC and tallow, the value of WDGS was about equal to SFC when 25% or less of the diet was comprised of DGS (May et al., 2010; Luebbe et al., 2012; Meyer et al., 2013). Buttrey et al. (2012a, 2013) reported no significant DGS × grain processing interaction in finishing cattle fed diets containing 20 or 35% WDGS; however, in both trials feeding WDGS tended to improve performance of cattle fed a DRC-based diet, but not cattle fed an SFC-based diet.

Interaction with Roughage. Distillers grains byproducts are a very palatable feed. In addition, feeding WDGS or MDGS also minimizes sorting in total mixed rations. When using DGS, particularly WDGS or MDGS in finishing diets, low-quality, fibrous, often unpalatable forages can be fed as a roughage source for finishing cattle. Likewise, because DGS contain a relatively large proportion of NDF and very little starch, research has focused on the amount of roughage required to minimize ruminal acidosis and maintain intakes as DGS use increases. When cattle were fed DRC-based diets containing 25% WDGS with either 2, 6, 10, or 14% alfalfa hay as a roughage source, optimal performance was observed at 6% hay, with maximal average daily gain (ADG) and G:F observed at that inclusion (Hales et al., 2013c). Intakes increased linearly as alfalfa hay increased, which is a typical roughage response observed in diets without DGS (Defoor et al., 2002). Similarly, Depenbusch et al. (2009) fed 6% alfalfa hay in SFC-based diets with 0 or 15% distillers grains (dry or wet sorghum distillers grains) and reported increased intake and gain, but no effect on G:F compared to the same diets without hay. Thus, feeding conventional concentrations of roughage is necessary in diets containing distillers grains. This conclusion is supported by the results of May et al. (2011), where feeding 7.5, 10, or 12.5% alfalfa hay with either 15 or 30% WDGS increased intakes linearly and decreased G:F linearly, suggesting that 7.5% hay was sufficient to optimize intake, gain, and G:F.

Because of the palatability and high CP content of distillers grains, one possibility when using DGS is to feed lower-quality forages as a roughage source. Some studies have evaluated different forages as roughages in finishing diets containing DGS. Little differences were observed between feeding 11% corn silage or 6% alfalfa hay when diets contained 0 or 25% DDGS in SFC-based diets (Uwituze et al., 2010). Cattle fed corn silage had slightly greater dry matter intake (DMI), but similar gain and G:F as those fed alfalfa hay, and there were no interactions between roughage source and DDGS. In contrast, Quinn et al. (2011) observed slightly greater G:F when bermudagrass hay was fed as a roughage source compared with sorghum silage with either 15 or 30% WDGS and cattle fed either of the lower-quality sources gained faster than cattle fed alfalfa hay. In that study, despite formulating for equal NDF content from the different sources of roughages, cattle tended to respond better to larger particle, lower-quality hay as roughage vs. smaller-particle roughages; however, the response was similar whether cattle were fed 15 or 30% WDGS. In contrast, Benton et al. (2015) fed 30% WDGS to all cattle, but fed 0% roughage compared with 3, 4, or 6% corn stalks, alfalfa, or corn silage as well as 6, 8, or 12% corn stalks, alfalfa, or corn silage, respectively. The rationale for different inclusions was to provide equal NDF content from the different roughage sources. In their experiment, cattle gained more, ate more, but had similar or poorer G:F as roughage concentration increased in the diet. No differences were observed among roughage sources, suggesting that feeding 6% cornstalks gave similar gain and G:F compared to higher-quality forages such as alfalfa hay or corn silage. From these data, it can be concluded that feeding DGS provides greater dietary NDF, but it should not be used as a substitute for roughages. Nonetheless, lower-quality forage can be used when distillers are included in feedlot diets.

Sulfur. Whereas most nutrients are concentrated threefold in DGS relative to corn, S concentration is often greater than three times the concentration in corn. Sulfur content of corn ranges from 0.1 to 0.15% of DM, whereas S content of DGS is usually greater than 0.6% of DM. The elevated S in DGS is a result of added sulfuric acid during the fermentation process for pH control, which elevates S concentration in DGS. A large body of research has been conducted on the relationship between S in water (Loneragan et al., 2005), plants (Gould et al., 2002), and DGS (Nichols et al., 2013; Drewnoski et al., 2014; Ponce et al., 2014) and its relationship to sulfur-induced polioencephalomalacia (PEM). Ruminally available S can be converted to hydrogen sulfide gas by rumen microbes, which in turn increases the incidence of PEM (see Chapter 7). Not all dietary S is available to rumen microbes, particularly the sulfur-containing amino acids that comprise a portion of RUP in DGS. Nonetheless, much of the excess S in DGS is available, which makes PEM a risk when inclusions are greater than 50% of the dietary DM or when the percentage of S in DGS is greater than 0.8% (Nichols et al., 2013; Ponce et al., 2014). To more clearly address the risk of sulfur-induced PEM, Sarturi et al. (2013a) proposed that a concept of ruminal available S should be used to account for both inorganic (available) and organic (portions unavailable) forms of S. The portion of organic S that is not available is from S amino acids contained in RUP, which can be quite large in DGS. This concept is supported by research in dairy cattle where different proteins varying in degradability of S amino acids affected hydrogen sulfide concentration in the rumen gas cap (Fonseca et al., 2013). In a recent review article, Drewnoski et al. (2014) summarized the results of experiments wherein the effect of inorganic S was compared with that of S added through DGS. The authors concluded that increased inorganic S decreased intake, gain, and carcass weight; however, increasing S by increasing DGS inclusion (and thus other components changing) resulted in a smaller decrease in intake and gain, and no effect on carcass weight.

Nichols et al. (2013) suggested that with low-fiber diets, a dietary S concentration of approximately 0.5% DM (DGS as the primary source) can lead to increased risk of PEM, which corresponds to approximately 0.3% DM ruminally available S. Interestingly, previous work using either dietary or water sulfates (i.e., 100% rumen available S) suggests that about 0.3% dietary S is the upper limit for risk of PEM (Zinn et al., 1997; NRC, 2005). As roughage or fiber increases in cattle diets, the risk of PEM from S decreases (Nichols et al., 2013) and feeding more NDF to finishing cattle decreases ruminal hydrogen sulfide concentration (Morine et al., 2014a,b). Ruminal pH is likely an important predictor of risk associated with hydrogen sulfide production, and therefore sulfur-induced PEM, because ruminal pH and hydrogen sulfide concentration are negatively correlated (Morine et al., 2014b).

There is evidence that before dietary S becomes toxic and induces PEM in a subset of the population, elevated dietary S decreases intake and growth. Uwituze et al. (2011) evaluated feeding cattle DDGS at 30% DM inclusion in DRC or SFC finishing diets with or without sulfuric acid added to the DDGS to mimic S addition during ethanol production. Diets that contained 0.65% S had 8.9% lower DMI and 12.9% less ADG, resulting in 4.3% lighter carcass weights compared to diets that contained 0.42% sulfur. These cattle also had greater concentrations of ruminal hydrogen sulfide gas concentration. Sarturi et al. (2013b) fed 20, 30, or 40% wet or dry DGS that contained either 0.82 or 1.16% S to individually fed finishing steers. When DGS contained 1.16% S, and particularly when fed as wet DGS, cattle intake and gain decreased, with little effect on G:F. These data suggest that although cattle might not exhibit clinical signs of PEM, cattle may consume less feed to offset high S intakes, and weight gain is decreased. Likewise, hydrogen sulfide concentration was greater when measured in ruminally fistulated steers fed DGS that contained 1.16% S compared to 0.82% S.

Effect of Fat in Distillers Grain and Fat Metabolism. One plausible reason why DGS can replace corn in finishing diets is that DGS contain concentrated oil that originated from the grain, in most cases corn. Historically, the fat content (i.e., oil) of corn DGS measured as ether extract is between 11 and 13% (DM basis; Spiehs et al., 2002; Buckner et al., 2011b; see Chapter 18). Fatty acid (FA) profiles of the ether extract in DGS are similar to the grain that was used as a feedstock for ethanol production. For example, FA profiles of corn oil (Pavan et al., 2007) were similar to distillers grains from corn (Leonardi et al., 2005). As a result, the FA within DGS are similar to corn, only concentrated as a percentage of DM (Vander Pol et al., 2009). Readers are referred to Dose et al. (2011) for details on the FA profile of corn DDGS.

Diets fed to cattle that are abundant in unsaturated FA are normally biohydrogenated to some extent, which results in a different FA profile leaving the rumen than that provided in the diet (Jenkins et al., 2008). Biohydrogenation also ensures that unsaturated FA do not become toxic to the rumen microbiota (Jenkins et al., 2008). Vander Pol et al. (2009) evaluated different fat sources including WDGS in both feeding and metabolism studies. The ratio of unsaturated FA relative to saturated FA increased in the small intestine of steers fed WDGS compared with corn-based diets, or corn-based diets with either added tallow (saturated fat) or added corn oil (unsaturated FA). These data suggest that a portion of the FA in DGS are "protected" from biohydrogenation in the rumen and thus enter the small intestine intact. Similar results were observed by Bremer et al. (2010) where the unsaturated:saturated FA ratio of duodenal contents increased from approximately 0.40 to 0.50 for corn, corn oil, tallow, and distillers solubles to 0.83 for WDGS. All diets fed by Bremer et al. (2010) contained approximately 8.5% fat, except the corn control (3.6%). Fatty acid digestibility was greater than 93% for all diets. Vander Pol et al. (2009) measured the FA profile of duodenal contents and observed increased concentrations of 18:1, 18:2, and additional polyunsaturated FA in duodenal digesta of steers fed WDGS compared to all other diets.

These data suggest that the corn oil found in DGS is "protected" from ruminal biohydrogenation and that all fat sources are quite digestible. In addition, the increase in flow/supply of unsaturated and polyunsaturated FA is observed in the FA in meat from cattle fed DGS diets compared with corn-based diets (Gill et al., 2008; Koger et al., 2010; Mello et al., 2012b; Buttrey et al., 2013). This change in FA profile could have positive and negative effects on beef, but it particularly affects oxidation and discoloration, which can negatively affect shelf life (Mello et al., 2012a).

Forage-Based Diets

The high energy content and palatability of the combination of wet DGS and low-quality forages has led to research with supplementing DGS to beef cows or growing calves as a source of both protein and energy. Likewise, the concentrated P originating from the grain (DGS range: 0.75 to 0.95% P) provides supplemental P that might be needed by the cow or growing calf fed forage diets.

Supplementing DGS to growing beef cattle in place of forages results in improved gains and G:F because DGS have greater energy content than the forages they replace. A summary of results from 13 pasture grazing experiments (35 treatment means) where cattle were supplemented with DGS at a rate of 0 to >1% of body weight (BW) indicated that ending BW and gain increased linearly as DGS supplementation increased (Griffin et al., 2012). In this summary, daily gain increased to 1.05 kg when DGS was supplemented at 1% of BW compared to 0.67 kg/d for nonsupplemented cattle (Figure 17-3). One difficulty encountered in grazing experiments is measuring feed intake. Griffin et al. (2012) looked at results from seven growing experiments (28 treatment means) where DDGS was supplemented up to 1.3% of BW to growing cattle fed forage-based diets in confinement and found that in these studies, nonsupplemented cattle gained 0.54 kg/d, and gain increased quadratically to 1.13 kg/d when DDGS was supplemented at 1% of BW (Figure 17-3). When forage intake was measured in confinement, forage intake decreased quadratically as DDGS supplementation increased while total DMI increased (Figure 17-4). The first kilogram of DDGS fed replaced 0.23 kg of forage, whereas the second kilogram of DDGS replaced an additional 0.13 kg of forage. In a limited number of studies evaluating pasture forage intake with growing cattle, 2.3 kg/d of DDGS reduced forage DM intake 0.7 to 1 kg/d (MacDonald et al., 2007; Watson et al., 2012).

Loy et al. (2008) measured the total digestible nutrient (TDN) concentration of DDGS relative to corn and they concluded that DDGS had 130% of the energy of corn when fed at low levels, but when fed at high levels it was only about 118%. This decrease could be a result of the fat content of the DDGS and the subsequent effect on the ability of ruminal microorganisms to ferment fiber. Total dietary fat concentrations greater than 5% have been shown to decrease fiber digestion through a variety of proposed mechanisms. In

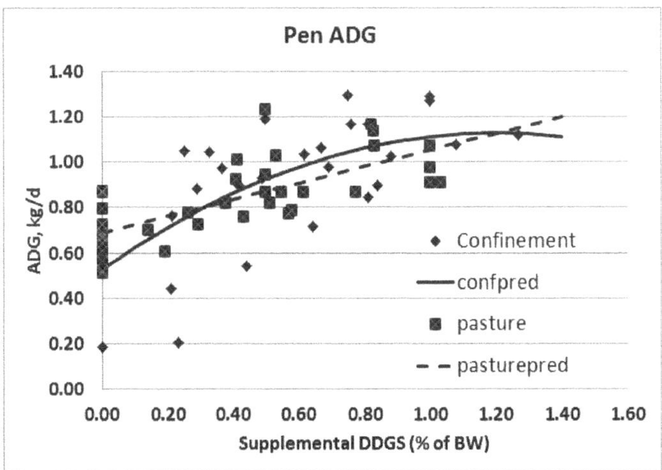

FIGURE 17-3 Effect of dried distillers grains plus solubles (DDGS) supplementation on ADG for growing cattle supplemented DDGS. Pasture ADG (pasturepred) = 0.6895 + 0.3638x (Linear < 0.01; Quadratic = 0.21). Confinement ADG (confpred) = 0.5289 + 0.9966x – 0.4152 × 2 (Linear < 0.01; Quadratic < 0.01). Adapted from Griffin et al. (2012).

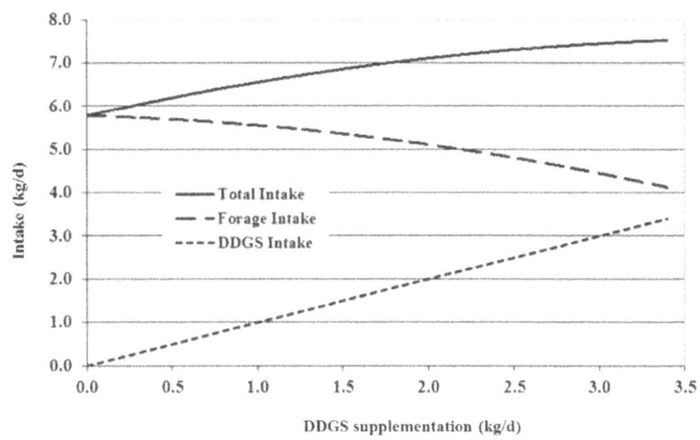

FIGURE 17-4 Effect of dried distillers grains plus solubles (DDGS) supplementation on intake for cattle fed in confinement studies. DDG Supplementation = x. Forage intake = 5.7781 – 0.1238x – 0.1069x^2 (Linear = 0.31; Quadratic < 0.01). Total Intake = 5.7781 + 0.8762x – 0.1069x^2 (Linear < 0.01; Quadratic < 0.01). Adapted from Griffin et al. (2012).

the Loy et al. (2008) study, the fat concentration of the high level of DDGS diet was about 5.2%.

When compared with corn supplementation for energy in growing diets, DGS contains 115 to 130% of the energy of corn (Loy et al., 2008; Nuttelman et al., 2010; Ahern et al., 2015). The energy content was similar for DDGS, MDGS, and WDGS relative to corn across experiments when supplemented in forage-based diets (Ahern et al., 2015). Likewise, the response to supplementing DGS appears similar in both high-quality forages (TDN > 55) and in low-quality forage-based diets (TDN < 55). Morris et al. (2005) evaluated response to feeding 0 to 2.7 kg of DDGS supplemented to growing cattle individually fed either a low-quality forage (TDN = 53) or a high-quality forage (TDN = 65). Gains increased linearly when DDGS was fed in either forage-based diet, but the slope was greater for increased ADG when DDGS replaced the lower-quality forage. Supplementing DGS to cattle grazing corn residue increases ADG and is dose dependent (Jones et al., 2014, 2015).

Distillers Solubles

Distillers solubles are often marketed as a separate feed by some plants and can vary in nutrient content from plant to plant, which is likely related to how solubles and grains are centrifuged in whole stillage. Distillers solubles are sometimes referred to as syrup that is low in fiber concentration, but a good source of protein, fat, P, and S (Corrigan et al., 2009b). Solubles contain approximately 25% CP, which is estimated to be 25 to 30% RUP, 20% fat, 1.57% P, 0.92% S, and 2.3% NDF. Distillers solubles have become a popular base for liquid feed supplements. As molasses prices have increased, liquid supplement companies use steep from the wet milling industry and distillers solubles from the dry milling industry to replace a portion of molasses in liquid supplements. In addition, solubles can replace corn and protein in finishing diets. In two trials, Trenkle (2003, 2004) observed decreased or unchanged performance in finishing cattle consuming up to 8 or 12% condensed distillers solubles. Pesta et al. (2015) fed 0, 9, 18, 27, or 36% distillers solubles in DRC and HMC diets (1:1 ratio, DM basis). In their study, a quadratic response to adding solubles was observed for ADG, with maximum gain for cattle fed 20.8% of the dietary DM calculated from the quadratic equation. Pesta et al. (2015) also observed a quadratic improvement in G:F as solubles inclusion increased, with maximum G:F calculated at 32.5% inclusion of solubles. Similar treatments were imposed in finishing diets based on SFC (Harris et al., 2014). Gain and G:F increased quadratically with increasing solubles in SFC diets, with cattle fed 9, 18, or 27% solubles having greater ADG than cattle fed 0 or 36% solubles. The greatest G:F was observed with 27% inclusion of solubles in SFC diets (Harris et al., 2014).

Fractionation of Distillers Grains

Changes in the dry milling process have occurred recently, and further changes in processing will likely continue in the future. There is interest in fractionating portions of the kernel (prefractionation) before entering the dry grinding ethanol process and changes during or after the dry grind milling process (postfractionation). Examples include removal of a portion of the oil from solubles, isolating a portion of the bran and producing a new feed byproduct, increasing the protein content of distillers grains, and so forth (Berger and Singh, 2010). Some data are available evaluating distillers-type byproducts that have been produced with these different process changes implemented. One example is prefractionation and removal of a portion of the bran that is then mixed with distillers solubles. When a product (Dakota bran®) was fed up to 45% of the diet, ADG and G:F were increased and were similar to DDGS at 30% inclusion, even though the composition was slightly different for Dakota bran® compared to DDGS, with Dakota bran® containing 14.7% CP, 32% NDF, and 10.9% fat (Buckner et al., 2011a).

A new process of partially removing the oil from the distillers solubles has been rapidly adopted by the industry. This process results in DGS being lower in fat (7 to 9% ether extract compared with 11 to 13%). No peer-reviewed literature evaluating this process is currently available. Based on experiment station reports, however, removing a portion of the oil from the solubles stream decreases the fat content from approximately 12% to 8%, with minimal effect on G:F of finishing cattle. Results of three experiments that compared partially deoiled DGS with normal fat DGS suggest that energy from DGS was either not affected (Jolly et al., 2013) or decreased about 2% for every 1% fat removed (Bremer et al., 2014; Jolly et al., 2014). Within experiments, the two DGS were not significantly different. These data suggest that the improved performance as a result of feeding WDGS compared with corn grain is at least partially associated with the higher fat content in the WDGS. Similar observations were made by Lodge et al. (1997) and Vander Pol et al. (2008).

Corn Wet Milling

Numerous publications have outlined the wet milling process (Blanchard, 1992; Stock et al., 2000). Corn wet milling produces a variety of products for human consumption; therefore, #2 grade corn or better is used in the United States wet milling industry. Wet milling begins with a kernel steeping process where a dilute sulfurous acid solution softens the kernel. After steeping, the wet kernel goes through a series of grinding and separation processes. The starch can then be dried and sold as unmodified starch or processed into several products, which include conversion to dextrose to be used in the fermentation process to produce ethanol or sweetener. Water enters the process converse to the entry of whole kernels, so the flow of starch and liquid is opposite. After steeping, the liquid not absorbed by the corn is separated as steep liquor, which contains 5 to 10% solids (Blanchard, 1992). Another liquid stream in the wet milling process is the remaining liquefied solution after production of ethanol (if plants produce ethanol from starch) referred to as distillers solubles, which contains yeast cells and unfermented sugars. The distillers solubles from wet milling are dramatically different than those from dry milling as few feed particles remain in the wet milling process. The steep liquor and distillers solubles go through an evaporation process separately or together to produce a 40 to 45% DM product, and the combined products are called steep (Stock et al., 2000). Steep can be marketed several ways, including as a liquid protein source or combined with other feed streams to produce corn gluten feed (CGF). The next byproduct produced from wet milling is the germ meal, left over from oil extraction. Because the germ is dried to increase corn oil yield, germ meal is typically 90% DM.

The primary fiber component of the corn kernel separated during wet milling is called bran and is the outer coating of the kernel. It is typically pressed after separation to remove water and contains approximately 40% DM (Stock et al., 2000). Bran can be dried to 85% DM, but bran is seldom marketed as an individual ingredient and is typically combined with steep to produce CGF (Blanchard, 1992).

In the last separation step of wet milling, gluten is removed from the starch. The gluten protein is marketed as corn gluten meal. Corn gluten meal contains greater than 70% CP and is also relatively high in RUP (60% or more). Corn gluten meal is a good source of methionine and xanthophylls, which makes it appealing for the poultry and pet food industries. By mixing some of the byproducts of the wet milling process, CGF is produced. Corn gluten feed generally contains bran and steep but can also include germ meal. Corn gluten feed can be marketed in a wet (WCGF) or dry (DCGF) form (Stock et al., 2000).

Feedlot Diets

In a review by Stock et al. (2000), they divided WCGF into two categories based on the amount of steep added to the bran/germ. In that review, WCGF with less steep had a feeding value equal to corn (seven experiments), and WCGF with more steep had a feeding value of 109% the feeding value of DRC (five experiments). The net energy of the two types of WCGF were 1% more than corn for WCGF with less steep and 15% more than corn for WCGF with more steep. In a summary of experiments with WCGF included at up to 40% of the diet to replace DRC or HMC, Bremer et al. (2008) found that WCGF with less steep yielded gain and G:F similar to corn control diets, with a feeding value that was 99% of corn. The summary also showed that feeding WCGF that contained more steep resulted in better performance than feeding corn, with a linear improvement in gain and G:F and a feeding value of 112% of corn (Bremer et al., 2008). Studies evaluating the bran and steep separately indicate that steep contains greater energy than corn, whereas bran was approximately 80% the feeding value of corn (Scott et al., 1997; Sayer et al., 2013). When evaluated within the same cattle experiment, as the proportion of steep increased from 37.5 to 50% relative to corn bran/germ meal, cattle become more efficient and the feed had a greater energy content (Macken et al., 2004b).

Although some WCGF contains bran that is dried, drying the bran/germ portion does not affect the energy value when fed to finishing cattle (Macken et al., 2004a); however, similar to DGS, drying CGF decreases its apparent energy content relative to corn in finishing diets. When fed at 25 to 30% of the diet, DCGF had a feeding value of 88% of DRC, whereas WCGF had 100% (Green et al., 1987) to 110% (Ham et al., 1995) the value of corn. The majority of the CGF fed to finishing cattle consists of WCGF, including Cargill's branded WCGF product (Sweet Bran®), which is fed across many cattle feeding regions in the United States.

Feeding WCGF helps control ruminal acidosis more than does DGS likely as a result of a larger particle size. Rumen pH was decreased, but returned to prefeeding pH faster for steers fed WCGF compared to those fed corn (Krehbiel et al., 1995). Numerous experiments have measured greater average ruminal pH and decreased duration of low pH when corn bran or WCGF are fed in feedlot diets (Adams et al., 2004; Farran et al., 2006; Sayer et al., 2013).

In contrast to DGS in finishing diets, where the relative value compared to corn changes based on inclusion rate, the response to feeding WCGF in finishing diets seems to be more consistent across inclusion rates. When 0, 10, 20, 25, 30, or 35% Sweet Bran® was fed in SFC-based diets, no differences were observed for ADG, G:F, or calculated dietary net energy required for gain (NEg) content (Macken et al., 2004b). Nonetheless, the greatest numeric ADG and G:F values were observed between 20 and 25% inclusion. Block et al. (2005) evaluated 20, 30, or 40% inclusion in SFC-based diets. Feed efficiency varied from 0.173 kg gain/kg DMI for the control cattle fed 0% Sweet Bran® to between 0.168 and 0.180 kg gain/kg DMI for cattle fed different inclusions, which was a quadratic response. Others have reported similar quadratic responses to inclusion of Sweet Bran® in SFC based-diets, although differences are quite small for G:F and calculated dietary energy (Domby et al., 2014).

As a result of an improved ruminal environment and acidosis control, more intense processing of corn might be beneficial in diets containing CGF. A few studies have directly compared different corn processing methods within diets containing CGF. When whole corn was compared with DRC in diets with 32% Sweet Bran® WCGF, dietary NEg was improved 12% for DRC vs. whole corn. According to the results of Scott et al. (2003), if the feeding value is calculated based on corn inclusion in the diet (52.5%), then DRC had a 13.6% greater feeding value than whole corn. Likewise, in this study, feeding SFC improved G:F, with a 12.7% greater feeding value than DRC. The feeding value for HMC was slightly less than SFC, but it still had a 9.5% greater feeding value than DRC, as cattle were significantly more efficient. In a second study, Scott et al. (2003) observed a greater response to SFC, but a lesser response to HMC relative to DRC in diets with 22% Sweet Bran®. Macken et al. (2006) fed cattle diets with 25% Sweet Bran® and observed improved performance for cattle fed HMC and SFC compared with cattle fed DRC. In their study, net energy of the diets was improved by 10% for HMC and 14.9% for SFC compared with DRC. These data clearly illustrate that corn processing is an important factor in diets with WCGF and large improvements are observed, especially for HMC, in diets containing WCGF compared with the response expected in diets based on primarily corn grain without WCGF.

Interaction with Roughage. Similar to feeding DGS, feeding CGF increases dietary NDF content, which dilutes starch in finishing diets. Corn gluten feed offers greater protection for ruminal acidosis than many other byproducts, likely because of the particle size and fermentation pattern associated with corn bran. The benefit of feeding CGF has prompted research on optimizing roughage inclusion in diets with WCGF or Sweet Bran®. Farran et al. (2006) found that in diets with 35% WCGF, DM intake increased as alfalfa hay inclusion increased from 0 to 7.5% without an improvement in ADG, which led to the optimal G:F for diets containing no alfalfa hay, DRC and 35% WCGF. Nonetheless, in the same study with diets containing 0% WCGF, optimal gain and intakes were observed with 3.75% alfalfa hay inclusion, which was similar in performance to feeding 7.5% alfalfa. The response has been slightly different in diets based on SFC in combination with WCGF. In diets with 40% WCGF, gain and intake increased linearly as alfalfa hay increased from 0 to 9%, with no effect on G:F (Parsons et al., 2007).

In a commercial study conducted by the same authors, they noted that cattle fed 40% Sweet Bran® with 4.5% alfalfa hay gained the same, consumed less, and were slightly more efficient than cattle fed 9% alfalfa, suggesting that 4.5% hay was sufficient with 40% Sweet Bran® in diets based on SFC (Parsons et al., 2007). In SFC diets with 25, 35, or 45% WCGF, cattle fed either 2 or 6% alfalfa hay performed similarly within each inclusion of WCGF, suggesting that 2% hay was sufficient. Overall, it seems that feeding WCGF might allow for less roughage to be included in finishing diets, which could depend on other factors such as grain processing and inclusion rate of WCGF.

As a result of improved control of ruminal acidosis and lack of starch in corn gluten feed, the use of WCGF and Sweet Bran® for receiving diets and for grain adaptation is now a common practice in the beef industry. A commercial product (RAMP®) is commonly used as a starter diet for receiving cattle and used during grain adaptation programs. Research has documented improved performance, even over the entire feeding period, by replacing traditional forages used for grain adaptation by either Sweet Bran® or RAMP® (Huls et al., 2009; Buttrey et al., 2012b; MacDonald and Luebbe, 2012; Schneider et al., 2012a,b, 2013).

Forage-Based Diets

Corn gluten feed is an excellent source of nutrients for forage-based diets. There is little to no starch in CGF, which results in no negative associative effects on fiber digestion. Corn gluten feed contains highly digestible fiber and degradable protein, which are good sources of energy and protein for ruminal microbes, especially in forage-based diets (DeHaan et al., 1983; Kampman and Loerch, 1989). A comparison of wet and dry CGF with DRC at 40% dietary inclusion for growing calves fed grass hay, wheat straw, and corn stalklage by Oliveros et al. (1987) showed that the supplements nearly doubled gains and improved feed conversion. Wet and dry CGF had better feed conversions than corn, and WCGF resulted in improved feed conversion compared with DCGF. The apparent feeding value of DCGF was 10% greater than corn, whereas WCGF was 31% greater than DCGF and 42% greater than corn in these forage-based diets.

Hominy Feed

Dry milling of corn grain includes dry separation of the kernel components into flour, grits, and meal. During this process, endosperm is partially separated from germ and pericarp by dry grinding. The resulting byproduct, hominy feed, is slightly concentrated in fiber, oil, and protein relative to the original grain when the endosperm is removed. Although less common than dry grinding for ethanol production, this form of dry corn milling results in a byproduct that is normally fed to finishing cattle. Less research is available on the use of hominy, but data have primarily focused on corn replacement in feedlot diets. Three trials were conducted by Larson et al. (1993b) to evaluate digestibility and performance when corn was replaced by hominy. The hominy feed in these studies contained, on the average, 57% starch, 25 to 30% NDF, 11 to 11.5% CP, and 5.3 to 7.7% fat, suggesting it was slightly concentrating all the nonstarch components of corn grain, but still contained over half the starch of corn. Another consideration with hominy is its relatively small particle size, with geometric mean diameters that are 8 to 10 times smaller (0.5 to 0.6 mm) than corn. The authors concluded that hominy contained 87% of the energy of corn grain when fed up to 40% of dietary DM. Although hominy contains less starch, starch digestibility was similar to ground or rolled corn. The lower energy content of hominy feed compared to corn is likely a result of its greater NDF concentration.

SOYBEAN BYPRODUCTS

Soybeans are the second largest annual cash crop produced in the United States (USDA ERS, 2012). The process of crushing results in the production of soybean meal and the extraction of soybean oil for human consumption. Soybean meal is a high-protein (44 and 48% CP) feed primarily used in nonruminant diets in the United States and is also a major export commodity. The higher-CP soybean meal is produced by dehulling the bean, which concentrates protein and provides an additional feed byproduct referred to as soybean hulls or soyhulls. Besides human consumption as oil, soy oil and other fats/oils also provide esterified FA for production of biodiesel. Biodiesel production is rapidly increasing in North America, which yields an additional feed byproduct, termed glycerin or glycerol.

Soybean Hulls (Soyhulls)

Soyhulls, the outer covering of the soybean, are separated during soybean processing into soybean meal and soy oil. Soyhulls contain a large proportion of NDF (60 to 70%; see Chapter 18) and are relatively low in protein (10 to 15%). When measured in situ, fiber digestibility is quite high because of low lignin content (Hsu et al., 1987). Because of effects of particle size and passage rate, however, in vivo digestibility is lower than estimates of digestibility using in situ or in vitro measurements (Hsu et al., 1987). Soyhulls are generally used to either supplement energy in a forage-based diet or as a replacement for a portion of the grain in finishing diets for beef cattle.

Supplementing soyhulls in forage-based diets generally increases gain with little effect on forage intake until supplementation exceeds 0.5% of BW. When cows were supplemented with 0, 1, 2, or 3 kg/d of soyhulls, intake of low-quality native hay decreased only 0.64 kg for 0 vs. 3 kg of soyhulls (Martin and Hibberd, 1990). As expected, supplementation resulted in a linear increase in organic matter (OM) and acid detergent fiber digestibilities in that study. When compared with corn supplementation, soyhulls resulted in equal gains when fed at 25% of energy intake and greater gains when fed at 50% of energy intake, without affecting digestibility (Garces-Yepez et al., 1997). Others have reported improved fiber digestibility when soyhulls were supplemented in a forage-based diet compared with corn supplementation, especially at high rates of supplementation (Garces-Yepez et al., 1997), suggesting no negative associative effect of soyhulls supplementation on fiber digestion. When soyhulls were compared with corn supplementation with cattle fed bermudagrass, NDF digestibility improved, whereas decreased digestibility was reported when corn was supplemented (Galloway et al., 1993). Similar decreases in fiber digestibility were observed when corn was supplemented in a more digestible forage diet consisting of orchardgrass hay, but no effect on digestion was evident when soyhulls were added (Galloway et al., 1993). With grazing cattle on brome or on corn residue, supplementing soyhulls increased gain compared with nonsupplemented controls across five trials, which was numerically greater than supplementing corn in three of the four trials that compared the two supplements (Anderson et al., 1988).

When soyhulls or corn were supplemented to steers fed at approximately 50% of their intake and the remaining diet was bermudagrass, OM and especially fiber digestibility was dramatically improved for steers fed soyhulls compared with those supplemented with corn (Orr et al., 2008). When low-quality brome hay was replaced with 0, 15, 30, 45, or 60% soyhulls, ruminal and total-tract digestibility of NDF increased linearly from 42 to 64% (Grigsby et al., 1992). Thus, as large amounts of soyhulls are added to forage-based diets, digestibility is enhanced. Because soyhulls are a fibrous byproduct, they are frequently used as a creep feed for calves. However, no performance differences were observed in calves provided a soyhull-based or corn-based creep feed when creep feed was limit-fed or offered ad libitum, despite greater fiber digestibility in calves provided by soyhull-based creep (Faulkner et al., 1994). Caution should be exercised to prevent bloat; anecdotal evidence suggests that less than 1% of BW (DM basis) is appropriate.

Data on the use of soyhulls in finishing diets are limited in refereed publications, even though many producers in the Northern plains include soyhulls at a relatively low concentration (less than 20% inclusion) in finishing diets. In finishing diets, soyhulls have mostly been used to replace varying proportions of corn. Although soyhulls contain large amounts of fiber, using soyhulls as a replacement for fiber from traditional forages is not advised. Soyhulls were compared to corn at 50 or 70% of the diet, with the remainder of the diet composed of forage (Hsu et al., 1987). When soyhulls were the main energy source for finishing lambs, lambs fed soyhulls had lower G:F than lambs fed corn; however, when corn was compared to soyhulls in grass- or corn silage-based finishing diets, steers fed soyhulls tended to eat more, gain less, and were significantly less efficient (Hsu et al., 1987). Based on G:F, soyhulls were 17 to 28% lower in feeding value than corn in these diets; however, these finishing diets would be more similar to growing diets used today. When soyhulls were fed at 0, 10.5, 21, or 31.4% of the diet in place of ground corn to finishing lambs for 56 days, lambs increased DM intake linearly but gained the same, which linearly decreased G:F (Ferreira et al., 2011). When soyhulls replaced corn in a high-concentrate (5% roughage as cobs) diet, intakes increased, gain decreased, and G:F decreased linearly (Ludden et al., 1995). The decrease in G:F resulted in a calculated feeding value for soyhulls of 50 to 60% the value of corn. In more recent studies evaluating soyhulls as a replacement for corn, but in diets with 40% WDGS, feeding 12.5 or 25% soyhulls in place of corn increased intake, gain, and improved G:F (Bittner et al., 2013b), whereas feeding 37.5% soyhulls decreased performance. The calculated feeding value was 95% of corn at 37.5%, but 107 to 119% of corn at 25 and 12.5% inclusion, respectively. In contrast, a similarly designed study by the same authors, but with 40% MDGS, showed that ADG and G:F were linearly decreased, resulting in calculated soyhull feeding values of only 60 to 70% of corn (Bittner et al., 2013a). Based on the feedlot studies comparing soyhulls to corn, which are limited, soyhulls seem to have approximately 70% the feeding value of corn for replacement.

Glycerin (Glycerol)

Plants synthesize FA and store them in triacylglycerides. One of the more recent biofuel byproducts that is receiving interest as a cattle feed is glycerin (glycerol), which is produced during the manufacture of biodiesel. Depending on the FA that are linked to the glycerol, glycerin normally comprises approximately 10% of the oil weight entering biodiesel production. The process uses methanol (and potassium hydroxide) during esterification, so some methanol can remain in the glycerin following separation (Drouillard, 2012), which is not likely a major concern for ruminants because of microbial metabolism of methanol, which minimizes safety concerns related to methanol content (Elam et al., 2008).

Few data are available on the effect of feeding glycerin on beef cattle production but Drouillard (2012) provided an overview of the issues related to feeding glycerin. Based on ruminal fermentation studies, glycerin is converted to proprionate in the rumen and decreases the acetate:proprionate ratio (El-Nor et al., 2010; Krueger et al., 2010; Drouillard, 2012). The energy content of glycerin, calculated from feeding studies has been somewhat variable. Most comparisons with beef cattle have been relative to corn in high-concentrate diets, with glycerin inclusions from 0 to 20% of dietary DM. Results of these studies have generally indicated a slight decrease in intake for steers (Elam et al., 2008; Parsons et al., 2009) and lambs (Avila-Stagno et al., 2013), or resulted in no change in intake in steers (Mach et al., 2009) or lambs (Gunn et al., 2010). In these studies, when intake decreased, so did gain (Elam et al., 2008; Avila-Stagno et al., 2013); however, G:F has been either not affected (Elam et al., 2008; Mach et al., 2009; Gunn et al., 2010; Avila-Stagno et al., 2013) or improved (Parsons et al., 2009). Many of these studies replaced dietary increments of 4 to 7% with inclusions as large as 12, 15, or 21% of dietary DM. Parsons et al. (2009) fed cattle 0, 2, 4, 8, 12, or 16% glycerin and observed quadratic effects on ADG, with increases when 2 to 4% glycerin was fed and a gradual decrease to a gain similar to the control (0% glycerin) when feeding glycerin at 16% of dietary DM. As a result of decreasing intake in their study, G:F was improved by feeding glycerin in place of SFC. More data are needed to evaluate glycerin as an energy source for cattle fed less than 20% glycerin as a replacement for corn in finishing diets.

Limited data are available on adding glycerin to forage-based diets. When glycerin was fed at 0, 2.5, 5, 7.5, or 10% of the diet in place of SFC in a 50% forage-based diet, growing cattle tended to eat and gain well when fed up to 7.5% glycerin (Hales et al., 2013a). Once inclusion increased from 7.5 to 10% of dietary DM, gain and G:F were decreased (Hales et al., 2013a). The same authors completed a second experiment where 7.5% glycerin replaced either alfalfa hay or SFC in a growing diet. Gain, intake, and G:F did not differ between feeding 0% glycerin or 7.5% glycerin in place of SFC in growing diets; however, gain was increased when 7.5% glycerin replaced alfalfa hay, but intakes and G:F were not statistically different (Hales et al., 2013a). These data suggest that glycerin can be supplemented in forage-based diets, which increases dietary energy concentration.

Overall, glycerin seems to have a similar energy content to corn whether added to finishing or forage-based diets. Nonetheless, the energy value seems to depend on the inclusion rate, a situation somewhat similar to DGS. Inclusions of less than 7% (DM basis) seem more effective than inclusion levels above 10%. More data are needed to evaluate glycerin as an energy source when replacing forage for growing cattle and beef cows.

REFERENCES

Abdelqader, M. M., and M. Oba. 2012. Lactation performance of dairy cows fed increasing concentrations of wheat dried distillers grains with solubles. *Journal of Dairy Science* 95:3894-3904.

Adams, J. R., T. B. Farran, G. E. Erickson, T. J. Klopfenstein, C. N. Macken, and C. B. Wilson. 2004. Effect of organic matter addition to the pen surface and pen cleaning frequency on nitrogen balance in open feedlots. *Journal of Animal Science* 82:2153-2163.

Ahern, N. A., B. L. Nuttelman, T. J. Klopfenstein, J. C. MacDonald, and G. E. Erickson. 2015. Comparison of wet or dry distillers grains plus solubles to corn as an energy source in forage-based diets. Pp. 34-35 in *Nebraska Beef Cattle Report MP 101*. Lincoln: University of Nebraska Agricultural Research Division. Available online at http://ianrpubs.unl.edu/live/mp101/build/mp101.pdf. Accessed on April 14, 2015.

Aines, G., T. Klopfenstein, and R. Stock. 1986. *Distillers Grains. Nebraska Beef Cattle Report* MP51. Lincoln: University of Nebraska Agricultural Research Division. Available online at http://digitalcommons.unl.edu/cgi/viewcontent.cgi?article=1777&context=extensionhist. Accessed on April 16, 2015.

Al-Suwaiegh, S., K. C. Fanning, R. J. Grant, C. T. Milton, and T. J. Klopfenstein. 2002. Utilization of distillers grains from the fermentation of sorghum or corn in diets for finishing beef and lactating dairy cattle. *Journal of Animal Science* 80:1105-1111.

Anderson, S. J., J. K. Merrill, and T. J. Klopfenstein. 1988. Soybean hulls as an energy supplement for the grazing ruminant. *Journal of Animal Science* 66:2959-2964.

Avila-Stagno, J., A. V. Chaves, M. L. He, O. M. Harstad, K. A. Beauchemin, S. M. McGinn, and T. A. McAllister. 2013. Effects of increasing concentrations of glycerol in concentrate diets on nutrient digestibility, methane emissions, growth, fatty acid profiles, and carcass traits of lambs. *Journal of Animal Science* 91:829-837.

Ball, C. E., and S. Cornett. 1996. Cattle feeding in the United States: The first 300 years. Pp. 3-15 in *Cattle Feeding: A Guide to Management*, 2nd Ed., R. C. Albin and G. B. Thompson, eds. Amarillo, TX: Trafton Printing.

Beliveau, R. M., and J. J. McKinnon. 2008. Effect of graded levels of wheat-based dried distillers grains with solubles on performance and carcass characteristics of feedlot steers. *Canadian Journal of Animal Science* 88:677-684.

Benton, J. R. 2010. Interaction between Roughages and Corn Milling Byproducts in Finishing Cattle Diets. Ph.D. Dissertation. University of Nebraska, Lincoln.

Benton, J. R., A. K. Watson, G. E. Erickson, T. J. Klopfenstein, K. J. Vander Pol, N. F. Meyer, and M. A. Greenquist. 2015. Effects of roughage source and inclusion in beef finishing diets containing corn wet distillers grains plus solubles. *Journal of Animal Science* 9:4358-4367.

Berger, L., and V. Singh. 2010. Changes and evolution of corn coproducts for beef cattle. *Journal of Animal Science* 88(Suppl.):E143-E150.

Bittner, C. J., G. E. Erickson, T. L. Mader, and L. J. Johnson. 2013a. Utilization of soybean hulls when fed in combination with MDGS in finishing diets. Pp. 86-87 in *Nebraska Beef Cattle Report MP98*. Lincoln: University of Nebraska Agricultural Research Division.

Bittner, C. J., B. L. Nuttelman, C. J. Schneider, D. B. Burken, T. J. Klopfenstein, and G. E. Erickson. 2013b. Effects of feeding increasing levels of soyhulls in finishing diets with WDGS. Pp. 88-89 in *Nebraska Beef Cattle Report MP98*. Lincoln: University of Nebraska Agricultural Research Division.

Blanchard, P. H. 1992. *Technology of Corn Wet Milling and Associated Processes*. Industrial Chemistry Library Vol. 4. New York: Elsevier.

Block, H. C., C. N. Macken, T. J. Klopfenstein, G. E. Erickson, and R. A. Stock. 2005. Optimal wet corn gluten and protein levels in steam-flaked corn-based finishing diets for steer calves. *Journal of Animal Science* 83:2826-2834.

Boucher, S. E., S. Calsamiglia, C. M. Parsons, M. D. Stern, M. Ruiz Mereno, M. Vazquez-Anon, and C. G. Schwab. 2009. In vitro digestibility of individual amino acids in rumen-undegraded protein: The modified three-step procedure and the immobilized digestive enzyme assay. *Journal of Dairy Science* 92:3939-3950.

Bremer, V. R., G. E. Erickson, and T. J. Klopfenstein. 2008. Meta-analysis of UNL feedlot trials replacing corn with WCGF. Pp. 37-38 in *Nebraska Beef Cattle Report MP 91*. Lincoln: University of Nebraska Agricultural Research Division.

Bremer, V. R., K. M. Rolfe, C. D. Buckner, G. E. Erickson, and T. J. Klopfenstein. 2010. Metabolism characteristics of feedlot diets containing different fat sources. Pp. 80-82 in *Nebraska Beef Cattle Report MP93*. Lincoln: University of Nebraska Agricultural Research Division.

Bremer, V. R., A. K. Watson, A. J. Liska, G. E. Erickson, K. G. Cassman, K. J. Hanford, and T. J. Klopfenstein. 2011. Effect of distillers' grains moisture and inclusion level in livestock diets on greenhouse gas emissions in the corn-ethanol-livestock life cycle. *The Professional Animal Scientist* 27:449-455.

Bremer, M. L., A. K. Watson, D. B. Burken, J. C. MacDonald, and G. E. Erickson. 2014. Energy value of de-oiled modified distillers grains plus solubles in a forage-based diet. Pp. 32-33 in *Nebraska Beef Cattle Report MP99*. Lincoln: University of Nebraska Agricultural Research Division.

Buckner, C. D., T. L. Mader, G. E. Erickson, S. L. Colgan, D. R. Mark, V. R. Bremer, K. K. Karges, and M. L. Gibson. 2008. Evaluation of dry distillers grains plus solubles inclusion on performance and economics of finishing beef steers. *The Professional Animal Scientist* 24:404-410.

Buckner, C. D., V. R. Bremer, T. J. Klopfenstein, G. E. Erickson, K. J. Vander Pol, K. K. Karges, and M. L. Gibson. 2011a. Evaluation of a prefermentation-fractionated by-product in growing and finishing cattle diets. *The Professional Animal Scientist* 27:295-301.

Buckner, C. D., M. F. Wilken, J. R. Benton, S. J. Vanness, V. R. Bremer, T. J. Klopfenstein, P. J. Kononoff, and G. E. Erickson. 2011b. Nutrient variability for distillers grains plus soluble and dry matter determination of ethanol by-products. *The Professional Animal Scientist* 27:57-64.

Buttrey, E. K., N. A. Cole, K. H. Jenkins, B. E. Meyer, F. T. McCollum, S. L. M. Preece, B. W. Auvermann, K. R. Heflin, and J. C. MacDonald. 2012a. Effects of twenty percent wet distillers grains plus solubles in steam-flaked and dry-rolled corn-based finishing diets on heifer performance, carcass characteristics, and manure characteristics. *Journal of Animal Science* 90:5086-5098.

Buttrey, E. K., M. K. Luebbe, R. G. Bondurant, and J. C. MacDonald. 2012b. Case study: Grain adaptation of yearling steers to steam-flaked corn-based diets using a complete starter feed. *The Professional Animal Scientist* 28:482-488.

Buttrey, E. K., K. H. Jenkins, J. B. Lewis, S. B. Smith, R. K. Miller, T. E. Lawrence, F. T. McCollum, P. J. Pinedo, N. A. Cole, and J. C. MacDonald. 2013. Effects of 35% corn wet distillers grains plus solubles in steam-flaked and dry-rolled corn-based finishing diets on animal performance, carcass characteristics, beef fatty acid composition, and sensory attributes. *Journal of Animal Science* 91:1850-1865.

Castillo-Lopez, E., T. J. Klopfenstein, S. C. Fernando, and P. J. Kononoff. 2013. In vivo determination of rumen undegradable protein of dried distillers grains with solubles and evaluation of duodenal microbial crude protein flow. *Journal of Animal Science* 91:924-934.

Cole, N. A., M. L. Galyean, J. MacDonald, and M. S. Brown. 2006. Interaction of grain co-products with grain processing: Associative effects and management. Pp. 192-204 in *Cattle Grain Processing Symposium, November 15-17, 2006, Tulsa, OK.* Oklahoma State University, Stillwater. Available online at http://beefextension.com/files/Proceedings%20final.pdf. Accessed on December 8, 2014.

Corrigan, M. E., G. E. Erickson, T. J. Klopfenstein, M. K. Luebbe, K. J. Vander Pol, N. F. Meyer, C. D. Buckner, S. J. Vanness, and K. J. Hanford. 2009a. Effect of corn processing method and corn wet distillers grains plus solubles inclusion level in finishing steers. *Journal of Animal Science* 87:3351-3362.

Corrigan, M. E., T. J. Klopfenstein, G. E. Erickson, N. F. Meyer, K. J. Vander Pol, M. A. Greenquist, M. K. Luebbe, K. K. Karges, and M. L. Gibson. 2009b. Effects of level of condensed distillers solubles in corn dried distillers grains on intake, daily body weight gain, and digestibility in growing steers fed forage diets. *Journal of Animal Science* 87:4073-4081.

DeHaan, K., T. Klopfenstein, and R. Stock. 1983. Corn gluten feed as a protein and energy source for ruminants. Pp. 19-21 in *Nebraska Beef Cattle Report MP44.* Lincoln: University of Nebraska Agricultural Research Division.

Defoor, P. J., M. L. Galyean, G. B. Salyer, G. A. Nunnery, and C. H. Parsons. 2002. Effects of roughage source and concentration on intake and performance by finishing heifers. *Journal of Animal Science* 80:1395-1404.

Depenbusch, B. E., E. R. Loe, J. J. Sindt, N. A. Cole, J. J. Higgins, and J. S. Drouillard. 2009. Optimizing use of distillers grains in finishing diets containing steam-flaked corn. *Journal of Animal Science* 87:2644-2652.

Domby, E. M., U. Y. Anele, K. K. Gautam, J. E. Hergenreder, A. R. Pepper-Yowell, and M. L. Galyean. 2014. Interactive effects of bulk density of steam-flaked corn and concentration of Sweet Bran on feedlot cattle performance, carcass characteristics, and apparent total tract nutrient digestibility. *Journal of Animal Science* 92:1133-1143.

Dose, C. S., T. C. Jenkins, L. O. Tedeschi, K. Karges, and P. J. Kononoff. 2011. A chemical evaluation of the chemical composition of four corn milling co-products with focus on fatty acids. *Journal of Dairy Science* 94 (Suppl. 1):507 (Abstract 566).

Drewnoski, M. E., D. J. Pogge, and S. L. Hansen. 2014. High-sulfur in beef cattle diets: A review. *Journal of Animal Science* 92:3763-3780.

Drouillard, J. S. 2012. Utilization of crude glycerin in beef cattle. Pp. 155-161 in *Biofuel Co-products as Livestock Feed—Opportunities and Challenges*, H. P. S. Makkar, ed. Rome, Italy: FAO. Available online at http://www.fao.org/docrep/016/i3009e/i3009e.pdf. Accessed on December 8, 2014.

Elam, N. A., K. S. Eng, B. Bechtel, J. M. Harris, and R. Crocker. 2008. Glycerol from biodiesel production: Considerations for feedlot diets. In *Proceedings of the 23rd Southwest Nutrition and Management Conference, February 21-22, 2008, Tempe, AZ.* Tucson: University of Arizona.

El-Nor, S. A., A. A. AbuGhazaleh, R. B. Potu, D. Hastings, and M. S. A. Khattab. 2010. Effects of differing levels of glycerol on rumen fermentation and bacteria. *Animal Feed Science and Technology* 162:99-105.

Erickson, G. E., T. J. Klopfenstein, and A. K. Watson. 2012. Utilization of feed byproducts from the wet and dry milling industry for beef cattle. Pp. 77-99 in *Biofuel Co-products as Livestock Feed—Opportunities and Challenges*, H. P. S. Makkar, ed. Rome, Italy: FAO. Available online at http://www.fao.org/docrep/016/i3009e/i3009e.pdf. Accessed on December 8, 2014.

Farlin, S. D. 1981. Wet distillers grains for finishing cattle. *Animal Nutrition and Health* 36:35.

Farran, T. B., G. E. Erickson, T. J. Klopfenstein, C. N. Macken, and R. U. Lindquist. 2006. Wet corn gluten feed and alfalfa hay levels in dry-rolled corn finishing diets: Effects on finishing performance and feedlot nitrogen balance. *Journal of Animal Science* 84:1205-1214.

Faulkner, D. B., D. F. Hummel, D. D. Buskirk, L. L. Berger, D. F. Parrett, and G. F. Cmarik. 1994. Performance and nutrient metabolism by nursing calves supplemented with limited or unlimited corn or soyhulls. *Journal of Animal Science* 72:470-477.

Ferreira, E. M., A. V. Pires, I. Susin, C. Q. Mendes, R. S. Gentil, R. C. Araujo, R. C. Amaral, and S. C. Loerch. 2011. Growth, feed intake, carcass characteristics, and eating behavior of feedlot lambs fed high-concentrate diets containing soybean hulls. *Journal of Animal Science* 89:4120-4126.

Firkins, J. L., L. L. Berger, and G. C. Fahey, Jr. 1985. Evaluation of wet and dry distillers grains and wet and dry corn gluten feeds for ruminants. *Journal of Animal Science* 60:847-860.

Fonseca, A. J. M., A. R. J. Cabrita, L. A. O. Pinho, E. J. Kim, and R. J. Dewhurst. 2013. Effects of protein sources on concentrations of hydrogen sulphide in the rumen headspace gas of dairy cows. *Animal* 7:75-81.

Galloway, D. L., Sr., A. L. Goetcsh, L. A. Forster, Jr., A. R. Patil, W. Sun, and Z. B. Johnson. 1993. Feed intake and digestibility by cattle consuming bermudagrass or orchardgrass hay supplemented with soybean hulls and(or) corn. *Journal of Animal Science* 71:3087-3095.

Galyean, M. L., N. A. Cole, M. S. Brown, J. C. MacDonald, C. H. Ponce, and J. S. Schutz. 2012. Utilization of wet distillers grains in high-energy beef cattle diets based on processed grain. Pp. 61-76 in *Biofuel Co-products as Livestock Feed—Opportunities and Challenges*, H. P. S. Makkar, ed. Rome, Italy: FAO. Available online at http://www.fao.org/docrep/016/i3009e/i3009e.pdf. Accessed on December 8, 2014.

Garces-Yepez, P., W. E. Kunkle, D. B. Bates, J. E. Moore, W. W. Thatcher, and L. E. Sollenberger. 1997. Effects of supplemental energy source and amount on forage intake and performance by steers and intake and diet digestibility by sheep. *Journal of Animal Science* 75:1918-1925.

Gargallo, S., S. Calsamiglia, and A. Ferret. 2006. Technical note: A modified three-step in vitro procedure to determine intestinal digestion of protein. *Journal of Animal Science* 84:2163-2167.

Gibb, D. J., X. Hao, and T. A. McAllister. 2008. Effect of dried distillers grains from wheat on diet digestibility and performance of feedlot cattle. *Canadian Journal of Animal Science* 88:659-665.

Gill, R. K., D. L. VanOverbeke, B. Depenbusch, J. S. Drouillard, and A. DiCostanzo. 2008. Impact of beef cattle diets containing corn or sorghum distillers grains on beef color, fatty acid profiles, and sensory attributes. *Journal of Animal Science* 86:923-935.

Gould, D. H., D. A. Dargatz, F. B. Garry, D. W. Hamer, and P. F. Ross. 2002. Potentially hazardous sulfur conditions on beef cattle ranches in the United States. *Journal of the American Veterinary Medical Association* 221:673-677.

Green, D. A., R. A. Stock, F. K. Goedeken, and T. J. Klopfenstein. 1987. Energy value of corn wet milling by-product feeds for finishing ruminants. *Journal of Animal Science* 65:1655-1666.

Griffin, W. A., V. R. Bremer, T. J. Klopfenstein, L. A. Stalker, L. W. Lomas, J. L. Moyer, and G. E. Erickson. 2012. A meta-analysis evaluation of supplementing dried distillers grains plus solubles to cattle consuming forage-based diets. *The Professional Animal Scientist* 28:306-312.

Grigsby, K. N., M. S. Kerley, J. A. Paterson, and J. C. Weigel. 1992. Site and extent of nutrient digestion by steers fed a low-quality bromegrass hay diet with incremental levels of soybean hull substitution. *Journal of Animal Science* 70:1941-1949.

Gunn, P. J., M. K. Neary, R. P. Lemenager, and S. L. Lake. 2010. Effects of crude glycerin on performance and carcass characteristics of finishing wether lambs. *Journal of Animal Science* 88:1771-1776.

Hales, K. E., N. A. Cole, and J. C. MacDonald. 2012. Effects of corn processing method and dietary inclusion of wet distillers grains with solubles on energy metabolism, carbon-nitrogen balance, and methane emissions of cattle. *Journal of Animal Science* 90:3174-3185.

Hales, K. E., R. G. Bondurant, M. K. Luebbe, N. A. Cole, and J. C. MacDonald. 2013a. Effects of crude glycerin in steam-flaked corn-based diets fed to growing feedlot cattle. *Journal of Animal Science* 91:3875-3880.

Hales, K. E., N. A. Cole, and J. C. MacDonald. 2013b. Effects of increasing concentration of wet distillers grains with solubles in steam-flaked corn-based diets on energy metabolism, carbon-nitrogen balance, and methane emissions of cattle. *Journal of Animal Science* 91:819-828.

Hales, K. E., H. C. Freetly, S. D. Shackelford, and D. A. King. 2013c. Effects of roughage concentration in dry-rolled corn-based diets containing wet distillers grains with solubles on performance and carcass characteristics of finishing beef steers. *Journal of Animal Science* 91:3315-3321.

Ham, G. A., R. A. Stock, T. J. Klopfenstein, E. M. Larson, D. H. Shain, and R. P. Huffman. 1994. Wet corn distillers byproducts compared with dried corn distillers grains with solubles as a source of protein and energy for ruminants. *Journal of Animal Science* 72:3246-3257.

Ham, G. A., R. A. Stock, T. J. Klopfenstein, and R. P. Huffman. 1995. Determining the net energy value of wet and dry corn gluten feed in beef growing and finishing diets. *Journal of Animal Science* 73:353-359.

Harris, M. E., G. E. Erickson, K. H. Jenkins, and M. K. Luebbe. 2014. Evaluating corn condensed distillers solubles concentration in steam-flaked corn finishing diets on cattle performance and carcass characteristics. Pp. 86-87 in *Nebraska Beef Cattle Report MP 99*. Lincoln: University of Nebraska Agricultural Research Division.

Hsu, J. T., D. B. Faulkner, K. A. Garleb, R. A. Barclay, G. C. Fahey, Jr., and L. L. Berger. 1987. Evaluation of corn fiber, cottonseed hulls, oat hulls and soybean hulls as roughage sources for ruminants. *Journal of Animal Science* 65:244-255.

Huls, T. J., M. K. Luebbe, W. A. Griffin, G. E. Erickson, T. J. Klopfenstein, and R. A. Stock. 2009. Using wet corn gluten feed to adapt cattle to finishing diets. Pp. 53-55 in *Nebraska Beef Cattle Report MP 92*. Lincoln: University of Nebraska Agricultural Research Division.

Janicek, B. N., P. J. Kononoff, A. M. Gehman, and P. H. Doane. 2008. The effect of feeding dried distillers grains plus solubles on milk production and excretion of urinary purine derivatives. *Journal of Dairy Science* 91:3544-3553.

Jenkins, K. H., K. J. Vander Pol, J. T. Vasconcelos, S. A. Furman, C. T. Milton, G. E. Erickson, and T. J. Klopfenstein. 2011. Effect of degradable intake protein supplementation in finishing diets containing dried distillers grains or wet distillers grains plus solubles on performance and carcass characteristics. *The Professional Animal Scientist* 27:312-318.

Jenkins, T. C., R. J. Wallace, P. J. Moate, and E. E. Mosley. 2008. Board-invited review: Recent advances in biohydrogenation of unsaturated fatty acids within the rumen microbial ecosystem. *Journal of Animal Science* 86:397-412.

Jolly, M. L., B. L. Nuttelman, D. Burken, C. J. Schneider, T. J. Klopfenstein, and G. E. Erickson. 2013. Effects of modified distillers grains plus solubles and condensed distillers solubles with and without oil extraction on finishing performance. Pp. 64-65 in *Nebraska Beef Cattle Report MP98*. Lincoln: University of Nebraska Agricultural Research Division.

Jolly, M. L., B. L. Nuttelman, D. Burken, C. J. Schneider, T. J. Klopfenstein, and G. E. Erickson. 2014. Effects of increasing inclusion of wet distillers grains plus solubles with and without oil extraction on finishing performance. Pp. 81-82 in *Nebraska Beef Cattle Report MP99*. Lincoln: University of Nebraska Agricultural Research Division.

Jones, M., J. C. MacDonald, G. Erickson, T. J. Klopfenstein, and A. K. Watson. 2014. Effect of distillers grains supplementation on calves grazing irrigated or non-irrigated corn residue. Pp. 48-49 in *Nebraska Beef Cattle Report MP99*. Lincoln: University of Nebraska Agricultural Research Division.

Jones, M., J. C. MacDonald, G. Erickson, T. J. Klopfenstein, and R. Bondurant. 2015. Dried distillers grains supplementation of calves grazing irrigated corn residue. Pp. 25-26 in *Nebraska Beef Cattle Report MP101*. Lincoln: University of Nebraska Agricultural Research Division.

Kalscheur, K. F., A. D. Garcia, D. J. Schingoethe, F. D. Royon, and A. R. Hippen. 2012. Feeding biofuel co-products to dairy cattle. Pp. 115-153 in *Biofuel Co-products as Livestock Feed—Opportunities and Challenges*, H. P. S. Makkar, ed. Rome, Italy: FAO. Available: http://www.fao.org/docrep/016/i3009e/i3009e.pdf. Accessed on December 8, 2014.

Kampman, K. A., and S. C. Loerch. 1989. Effects of dry corn gluten feed on feedlot cattle performance and fiber digestibility. *Journal of Animal Science* 67:501-512.

Kelzer, J. M., P. J. Kononoff, L. O. Tedeschi, T. C. Jenkins, K. Karges, and M. L. Gibson. 2010. Evaluation of protein fractionation and ruminal and intestinal digestibility of corn milling co-products. *Journal of Dairy Science* 93:2803-2815.

Klopfenstein, T., J. Waller, N. Merchen, and L. Petersen. 1978. Distillers grains as a naturally protected protein for ruminants. Pp. 38-46 in *Distillers Feed Conference Proceedings*, Vol. 33. Cincinnatti, OH: Distillers Feed Research Council.

Klopfenstein, T. J., G. E. Erickson, and V. R. Bremer. 2007. Feeding corn milling byproducts to beef cattle. *Veterinary Clinics of North America: Food Animal Practice* 23:223-245.

Klopfenstein, T. J., G. E. Erickson, and V. R. Bremer. 2008. Use of distillers co-products in diets fed to beef cattle. Pp. 5-56 in *Using Distillers Grains in the U.S. and International Livestock and Poultry Industries*, B. A. Babcock, D. J. Hayes, and J. D. Lawrence, eds. Ames, IA: MATRIC. Available online at http://www.matric.iastate.edu/DGbook/. Accessed on November 23, 2014.

Koger, T. J., D. M. Wulf, A. D. Weaver, C. L. Wright, K. E. Tjardes, K. S. Mateo, T. E. Engle, R. J. Maddock, and A. J. Smart. 2010. Influence of feeding various quantities of wet and dry distillers grains to finishing steers on carcass characteristics, meat quality, retail-case life of ground beef, and fatty acid profile of longissimus muscle. *Journal of Animal Science* 88:3399-3408.

Kononoff, P. J., S. K. Ivan, and T. J. Klopfenstein. 2007. Estimation of the proportion of feed protein digested in the small intestine of cattle consuming wet corn gluten feed. *Journal of Dairy Science* 90:2377-2385.

Krehbiel, C. R., R. A. Stock, D. W. Herold, D. H. Shain, G. A. Ham, and J. E. Carulla. 1995. Feeding wet corn gluten feed to reduce subacute acidosis in cattle. *Journal of Animal Science* 73:2931-2939.

Krueger, N. A., R. C. Anderson, L. O. Tedeschi, T. R. Callaway, T. S. Edrington, and D. J. Nisbert. 2010. Evaluation of feeding glycerol on free-fatty acid production and fermentation kinetics of mixed ruminal microbes in vitro. *Bioresource Technology* 101:8469-8472.

Larson, E. M., R. A. Stock, T. J. Klopfenstein, M. H. Sindt, and R. P. Huffman. 1993a. Feeding value of wet distillers byproducts for finishing ruminants. *Journal of Animal Science* 71:2228-2236.

Larson, E. M., R. A. Stock, T. J. Klopfenstein, M. H. Sindt, and D. H. Shain. 1993b. Energy value of hominy feed for finishing ruminants. *Journal of Animal Science* 71:1092-1099.

Leonardi, C., S. Bertics, and L. E. Armentano. 2005. Effect of increasing oil from distillers grains or corn oil on lactation performance. *Journal of Dairy Science* 88:2820-2827.

Li, C., J. Q. Li, W. Z. Yang, and K. A. Beauchemin. 2012. Ruminal and intestinal amino acid digestion of distillers grain vary with grain source and milling process. *Animal Feed Science and Technology* 175:121-130.

Li, Y. L., T. A. McAllister, K. A. Beauchemin, M. L. He, J. J. McKinnon, and W. Z. Yang. 2011. Substitution of wheat dried distillers grains with solubles for barley grain or barley silage in feedlot cattle diets: Intake, digestibility, and ruminal fermentation. *Journal of Animal Science* 89:2491-2501.

Little, C. O., G. E. Mitchell, Jr., and G. D. Potter. 1968. Nitrogen in the abomasums of wethers fed different protein sources. *Journal of Animal Science* 27:1722-1726.

Lodge, S. L., R. A. Stock, T. J. Klopfenstein, D. H. Shain, and D. W. Herold. 1997. Evaluation of wet distillers composite for finishing ruminants. *Journal of Animal Science* 75:44-50.

Loneragan, G., D. Gould, J. Wagner, F. Garry, and M. Thoren. 2005. The magnitude and patterns of ruminal hydrogen sulfide production, blood thiamin concentration, and mean pulmonary arterial pressure in feedlot steers consuming water of different sulfate concentrations. *Bovine Practitioner* 39:16-22.

Loy, T. W., T. J. Klopfenstein, G. E. Erickson, C. N. Macken, and J. C. MacDonald. 2008. Effect of supplemental energy source and frequency on growing calf performance. *Journal of Animal Science* 86:3504-3510.

Ludden, P. A., M. J. Cecava, and K. S. Hendrix. 1995. The value of soybean hulls as a replacement for corn in beef cattle diets formulated with or without added fat. *Journal of Animal Science* 73:2706-2711.

Luebbe, M. K., J. M. Patterson, K. H. Jenkins, E. K. Buttrey, T. C. Davis, B. E. Clark, F. T. McCollum, N. A. Cole, and J. C. MacDonald. 2012. Wet distillers grains plus solubles concentration in steam-flaked corn-based diets: Effects on feedlot cattle performance, carcass characteristics, nutrient digestibility, and ruminal fermentation characteristics. *Journal of Animal Science* 90:1589-1602.

MacDonald, J. C., and M. K. Luebbe. 2012. Case study: Grain adaptation for feedlot cattle using high concentrations of wet corn gluten feed in steam-flaked corn-based diets. *The Professional Animal Scientist* 28:131-139.

MacDonald, J. C., T. J. Klopfenstein, G. E. Erickson, and W. A. Griffin. 2007. Effects of dried distillers grains and equivalent undegradable intake protein or ether extract on performance and forage intake of heifers grazing smooth bromegrass pastures. *Journal of Animal Science* 85:2614-2624.

Mach, N., A. Bach, and M. Devant. 2009. Effects of crude glycerin supplementation on performance and meat quality of Holstein bulls fed high-concentrate diets. *Journal of Animal Science* 87:632-638.

Macken, C. N., G. E. Erickson, T. J. Klopfenstein, C. T. Milton, and R. A. Stock. 2004a. Effects of dry, wet, and rehydrated corn bran and corn processing method in beef finishing diets. *Journal of Animal Science* 82:3543-3548.

Macken, C. N., G. E. Erickson, T. J. Klopfenstein, and R. A. Stock. 2004b. Effects of concentration and composition of wet corn gluten feed in steam-flaked corn-based finishing diets. *Journal of Animal Science* 82:2718-2723.

Macken, C. N., G. E. Erickson, T. J. Klopfenstein, and R. A. Stock. 2006. Effects of corn processing method and protein concentration in finishing diets containing wet corn gluten feed on cattle performance. *The Professional Animal Scientist* 22:14-22.

Martin, S. K., and C. A. Hibberd. 1990. Intake and digestibility of low-quality native grass hay by beef cows supplemented with graded levels of soybean hulls. *Journal of Animal Science* 68:4319-4325.

May, M. L., J. C. DeClerck, M. J. Quinn, N. DiLorenzo, J. Leibovich, D. R. Smith, K. E. Hales, and M. L. Galyean. 2010. Corn or sorghum wet distillers grains with solubles in combination with steam-flaked corn: Feedlot performance, carcass characteristics, and apparent total tract digestibility. *Journal of Animal Science* 88:2433-2443.

May, M. L., M. J. Quinn, N. DiLorenzo, D. R. Smith, and M. L. Galyean. 2011. Effects of roughage concentration in steam-flaked corn-based diets containing wet distillers grains with solubles on feedlot cattle performance, carcass characteristics, and in vitro fermentation. *Journal of Animal Science* 89:549-559.

McDonald, I. W. 1954. The extent of conversion of food protein to microbial protein in the rumen of the sheep. *Biochemical Journal* 56:120-125.

McKinnon, J. J., and A. M. Walker. 2008. Comparison of wheat-based dried distillers grain with solubles to barley as an energy source for backgrounding cattle. *Canadian Journal of Animal Science* 88:721-724.

McNiven, M. A., E. Prestlokken, L. T. Mydland, and A. W. Mitchell. 2002. Laboratory procedure to determine protein digestibility of heat-treated feedstuffs for dairy cattle. *Animal Feed Science and Technology* 96:1-13.

Mello, A. S., Jr., C. R. Calkins, B. E. Jenschke, T. P. Carr, M. E. R. Dugan, and G. E. Erickson. 2012a. Beef quality of calf-fed steers finished on varying levels of corn-based wet distillers grains plus solubles. *Journal of Animal Science* 90:4625-4633.

Mello, A. S., Jr., B. E. Jenschke, L. S. Senaratne, T. P. Carr, G. E. Erickson, and C. R. Calkins. 2012b. Effects of feeding modified distillers grains plus solubles on marbling attributes, proximate composition, and fatty acid profile of beef. *Journal of Animal Science* 90:4634-4640.

Meyer, N. F., G. E. Erickson, T. J. Klopfenstein, J. R. Benton, M. K. Luebbe, and S. B. Laudert. 2013. Effects of monensin and tylosin in finishing diets containing corn wet distillers grains with solubles with differing corn processing methods. *Journal of Animal Science* 91:2219-2228.

Mjoun, K., K. F. Kalscheur, A. R. Hippen, and D. J. Schingoethe. 2010. Ruminal degradability and intestinal digestibility of protein and amino acids in soybean and corn distillers grains products. *Journal of Dairy Science* 93:4144-4154.

Morine, S. J., M. E. Drewnoski, and S. L. Hansen. 2014a. Determining the influence of dietary roughage concentration and source on ruminal parameters related to sulfur toxicity. *Journal of Animal Science* 92:4068-4076.

Morine, S. J., M. E. Drewnoski, and S. L. Hansen. 2014b. Increasing dietary NDF concentration decreases ruminal hydrogen sulfide concentrations in steers fed high-sulfur diets based on ethanol co-products. *Journal of Animal Science* 92:3035-3041.

Morris, S. E., T. J. Klopfenstein, D. C. Adams, G. E. Erickson, and K. J. Vander Pol. 2005. The effects of dried distillers grains on heifers consuming low or high quality forage. Pp. 18-20 in *Nebraska Beef Cattle Report MP83-A*. Lincoln: University of Nebraska Agricultural Research Division.

NASS (National Agricultural Statistics Service). 2014. USDA Crop Production 2013 Summary. USDA National Agricultural Statistics Service. Available online at http://usda.mannlib.cornell.edu/usda/current/CropProdSu/CropProdSu-01-10-2014.pdf. Accessed on November 23, 2014.

Nichols, C. A., V. R. Bremer, A. K. Watson, C. D. Buckner, J. L. Harding, G. E. Erickson, T. J. Klopfenstein, and D. R. Smith. 2013. The effect of sulfur and use of ruminal available sulfur as a model to predict incidence of polioencephalomalacia in feedlot cattle. *The Bovine Practitioner* 47(1):47-53.

NRC (National Research Council). 1996. *Nutrient Requirements of Beef Cattle*, 7th Rev. Ed. Washington, DC: National Academy Press.

NRC. 2000. *Nutrient Requirements of Beef Cattle: Update 2000*, 7th Rev. Ed. Washington, DC: National Academy Press.

NRC. 2005. *Mineral Tolerance of Animals*, 2nd Ed. Washington, DC: The National Academies Press.

NRC. 2012. *Nutrient Requirements of Swine*, 11th Rev. Ed. Washington, DC: The National Academies Press.

Nuez-Ortin, W. G., and P. Q. Yu. 2009. Nutrient variation and availability of wheat DDGS, corn DDGS, and blend DDGS from bioethanol plants. *Journal of the Science of Food and Agriculture* 89:1754-1761.

Nuttelman, B. L., M. K. Lubbe, T. J. Klopfenstein, J. R. Benton, and G. E. Erickson. 2010. Comparing the energy value of wet distillers grains to dry rolled corn in high forage diets. Pp. 49-50 in *Nebraska Beef Cattle Report MP93*. Lincoln: University of Nebraska Agricultural Research Division.

Nuttelman, B. L., W. A. Griffin, J. R. Benton, G. E. Erickson, and T. J. Klopfenstein. 2011. Comparing dry, wet, or modified distillers grains plus solubles on feedlot cattle performance. Pp. 50-52 in *Nebraska Beef Report MP94*. Lincoln: University of Nebraska Agricultural Research Division.

Nuttelman, B. L., D. B. Burken, C. J. Schneider, G. E. Erickson, and T. J. Klopfenstein. 2013. Comparing wet or dry distillers grains plus solubles for yearling finishing cattle. Pp. 62-63 in *Nebraska Beef Report MP98*. Lincoln: University of Nebraska Agricultural Research Division.

Ojowi, M., J. J. McKinnon, A. Mustafa, and D. A. Christensen. 1997. Evaluation of wheat-based distillers grains for feedlot cattle. *Canadian Journal of Animal Science* 77:447-454.

Oliveros, B., F. Goedeken, E. Hawkins, and T. Klopfenstein. 1987. Dry or wet bran or gluten feed for ruminants. Pp. 14-16 in *Nebraska Beef Cattle Report MP52*. Lincoln: University of Nebraska Agricultural Research Division.

Orr, A. I., J. C. Henley, and B. J. Rude. 2008. The substitution of corn with soybean hulls and subsequent impact on digestibility of a forage-based diet offered to beef cattle. *The Professional Animal Scientist* 24:566-571.

Parsons, C. H., J. T. Vasconcelos, R. S. Swingle, P. J. Defoor, G. A. Nunnery, G. B. Salyer, and M. L. Galyean. 2007. Effects of wet corn gluten feed and roughage levels on performance, carcass characteristics, and feeding behavior of feedlot cattle. *Journal of Animal Science* 85:3079-3089.

Parsons, G. L., M. K. Shelor, and J. S. Drouillard. 2009. Performance and carcass traits of finishing heifers fed crude glycerin. *Journal of Animal Science* 87:653-657.

Pavan, E., S. K. Duckett, and J. G. Andrae. 2007. Corn oil supplementation to steers grazing endophyte-free tall fescue. I. Effects on in vivo digestibility, performance, and carcass traits. *Journal of Animal Science* 85:1330-1339.

Paz, H. A., E. Castillo-Lopez, H. A. Ramirez-Ramirez, D. A. Christensen, T. J. Klopfenstein, and P. J. Kononoff. 2013. Invited review. Ethanol co-products for dairy cows: There goes our starch… now what? *Canadian Journal of Animal Science* 93:407-524.

Pesta, A. C., B. L. Nuttelman, A. L. Shreck, W. A. Griffin, T. J. Klopfenstein, and G. E. Erickson. 2015. Finishing performance of feedlot cattle fed condensed distillers solubles. *Journal of Animal Science* 93:4350-4357.

Ponce, C. H., M. S. Brown, N. A. Cole, C. L. Maxwell, and J. C. Silva. 2009. Effects of ruminally degradable N in diets containing wet distiller's grains and steam-flaked corn on feedlot cattle performance and carcass characteristics. P. 101 in *Plains Nutrition Council Spring Conference, April 9-10, 2009, San Antonio, TX*. Publication AREC 09-18. Amarillo: Texas Agrilife Research and Extension Center. Available online at http://amarillo.tamu.edu/files/2010/11/2009PNC-Proceedings1.pdf. Accessed on December 8, 2014.

Ponce, C. H., M. S. Brown, J. Osterstock, N. A. Cole, T. Lawrence, J. MacDonald, C. Maxwell, J. Wallace, and B. Coufal. 2010. Effects of non-protein nitrogen and wet distillers grains with solubles on growth performance, carcass merit, mineral status, tissue enzyme activities, and visceral organ mass by feedlot cattle. P. 114 in *Plains Nutrition Council Spring Conference, April 22-23, 2010, San Antonio, TX*. Publication AREC 10-57. Available online at http://amarillo.tamu.edu/files/2010/11/2010PNC-Proceedings1.pdf. Accessed on December 8, 2014.

Ponce, C. H., M. S. Brown, J. Osterstock, N. A. Cole, T. Lawrence, S. Soto-Navarro, J. MacDonald, B. D. Lambert, and C. Maxwell. 2014. Effects of wet distillers grains with solubles on visceral organ mass, trace mineral status and polioencephalomalacia biomarkers of individually-fed cattle. *Journal of Animal Science* 92:4034-4046.

Quinn, M. J., M. L. May, N. DiLorenzo, C. H. Ponce, D. R. Smith, S. L. Parr, and M. L. Galyean. 2011. Effects of roughage source and distillers grain concentration on beef cattle finishing performance, carcass characteristics, and in vitro fermentation. *Journal of Animal Science* 89:2631-2642.

RFA (Renewable Fuels Association). 2014. Statistics: Ethanol. Available online at http://www.ethanolrfa.org/pages/statistics#A. Accessed on December 10, 2014.

Sarturi, J. O., G. E. Erickson, T. J. Klopfenstein, K. M. Rolfe, C. D. Buckner, and M. K. Luebbe. 2013a. Impact of source of sulfur on ruminal hydrogen sulfide and logic for the ruminal available sulfur for reduction concept. *Journal of Animal Science* 91:3352-3359.

Sarturi, J. O., G. E. Erickson, T. J. Klopfenstein, J. T. Vasconcelos, W. A. Griffin, K. M. Rolfe, J. R. Benton, and V. R. Bremer. 2013b. Sulfur content in wet or dry distillers grains at several inclusions on cattle growth performance, ruminal parameters, and hydrogen sulfide. *Journal of Animal Science* 91:4849-4860.

Sayer, K. M., C. D. Buckner, G. E. Erickson, T. J. Klopfenstein, C. N. Macken, and T. W. Loy. 2013. Effect of corn bran and steep inclusion in finishing diets on diet digestibility, cattle performance, and nutrient mass balance. *Journal of Animal Science* 91:3847-3858.

Schneider, C. J., B. L. Nuttelman, K. M. Rolfe, W. A. Griffin, G. E. Erickson, and T. J. Klopfenstein. 2012a. Complete-feed diet RAMP in grain adaptation programs. Pp. 85-86 in *Nebraska Beef Cattle Report MP95*. Lincoln: University of Nebraska Agricultural Research Division.

Schneider, C. J., B. L. Nuttelman, K. M. Rolfe, W. A. Griffin, D. R. Smith, T. J. Klopfenstein, and G. E. Erickson. 2012b. Use of complete-feed diets RAMP and test starter for receiving cattle. Pp. 87-88 in *Nebraska Beef Cattle Report MP95*. Lincoln: University of Nebraska Agricultural Research Division.

Schneider, C. J., B. L. Nuttelman, W. A. Griffin, D. B. Burken, D. R. Smith, T. J. Klopfenstein, and G. E. Erickson. 2013. Using RAMP for receiving cattle compared to traditional receiving diets. Pp. 84-85 in *Nebraska Beef Cattle Report MP98*. Lincoln: University of Nebraska Agricultural Research Division.

Scott, T., T. Klopfenstein, R. Stock, and M. Klemesrud. 1997. Evaluation of corn bran and corn steep liquor for finishing steers. Pp. 72-76 in *Nebraska Beef Report MP67-A*. Lincoln: University of Nebraska Agricultural Research Division.

Scott, T. L., C. T. Milton, G.E. Erickson, T. J. Klopfenstein, and R. A. Stock. 2003. Corn processing method in finishing diets containing wet corn gluten feed. *Journal of Animal Science* 81:3182-3190.

Spiehs, M. J., M. H. Whitney, and G. C. Shurson. 2002. Nutrient database for distiller's dried grains with solubles produced from new ethanol plants in Minnesota and South Dakota. *Journal of Animal Science* 80:2639-2645.

Stalker, L. A., D. C. Adams, and T. J. Klopfenstein. 2007. Urea inclusion in distillers dried grains supplements. *The Professional Animal Scientist* 23:390-394.

Stein, H. H., and G. C. Shurson. 2009. Board-invited review: The use and application of distillers dried grains with solubles in swine diets. *Journal of Animal Science* 87:1292-1303.

Stein, H. H., C. Pedersen, M. L. Gibson, and M. G. Boersma. 2006. Amino acid and energy digestibility in ten samples of distillers dried grain with solubles by growing pigs. *Journal of Animal Science* 84:853-860.

Stock, R. A., J. M. Lewis, T. J. Klopfenstein, and C. T. Milton. 2000. Review of new information on the use of wet and dry milling feed by-products in feedlot diets. *Journal of Animal Science* 77(Suppl.):E1-E12.

Trenkle, A. H. 2003. Relative feeding value of wet corn distillers solubles as a feed for finishing cattle. AS Leaflet R1772 in *2002 Beef Research Report*. Iowa State University. Available online at http://www.extension.iastate.edu/Pages/ansci/beefreports/asl1772.pdf. Accessed on December 8, 2014.

Trenkle, A. H. 2004. Effects of replacing corn grain and urea with condensed corn distillers solubles on performance and carcass value of finishing steers. AS Leaflet R1884 in *Animal Industry Report*. Iowa State University. Available online at http://lib.dr.iastate.edu/cgi/viewcontent.cgi?article=1043&context=ans_air. Accessed on December 8, 2014.

USDA ERS (U.S. Department of Agriculture Economic Research Service). 2012. Soybeans and Oil Crops: Background. Available online at http://www.ers.usda.gov/topics/crops/soybeans-oil-crops/background.aspx#.UV3SYZMX-5Q. Accessed on March 13, 2015.

Uwituze, S., G. L. Parsons, M. K. Shelor, B. E. Depenbusch, K. K. Karges, M. L. Gibson, C. D. Reinhardt, J. J. Higgins, and J. S. Drouillard. 2010. Evaluation of dried distillers grains and roughage source in steam-flaked corn finishing diets. *Journal of Animal Science* 88:258-274.

Uwituze, S., G. L. Parsons, C. J. Schneider, K. K. Karges, M. L. Gibson, L. C. Hollis, J. J. Higgins, and J. S. Drouillard. 2011. Evaluation of sulfur content of dried distillers grains with solubles in finishing diets based on steam-flaked corn or dry-rolled corn. *Journal of Animal Science* 89:2582-2591.

Van Soest, P. J., J. B. Robertson, and B. A. Lewis. 1991. Methods for dietary fiber, neutral detergent fiber, and nonstarch polysaccharides in relation to animal nutrition. *Journal of Dairy Science* 74:3583-3597.

Vander Pol, K. J., M. A. Greenquist, G. E. Erickson, T. J. Klopfenstein, and T. Robb. 2008. Effect of corn processing in finishing diets containing wet distillers grains on feedlot performance and carcass characteristics of finishing steers. *The Professional Animal Scientist* 24:439-444.

Vander Pol, K. J., M. K. Luebbe, G. I. Crawford, G. E. Erickson, and T. J. Klopfenstein. 2009. Performance and digestibility characteristics of finishing diets containing distillers grains, composites of corn processing coproducts, or supplemental corn oil. *Journal of Animal Science* 87:639-652.

Vasconcelos, J. T., L M. Shaw, K. A. Lemon, N. A. Cole, and M. L. Galyean. 2007. Effects of graded levels of sorghum wet distillers grains and degraded intake protein supply on performance and carcass characteristics of feedlot cattle fed steam-flaked corn-based diets. *The Professional Animal Scientist* 23:467-475.

Walter, L. J., J. L. Aalhus, W. M. Robertson, T. A. McAllister, D. J. Gibb, M. E. R. Dugan, N. Aldai, and J. J. McKinnon. 2010. Evaluation of wheat or corn dried distillers grains with solubles on performance and carcass characteristics of feedlot steers. *Canadian Journal of Animal Science* 90:259-269.

Wang, D., S. Bean, J. McLaren, P. Seib, R. Madl, M. Tuinstra, Y. Shi, M. Lenz, X. Wu, and R. Zhao. 2008. Grain sorghum is a viable feedstock for ethanol production. *Journal of Industrial Microbiology and Biotechnology* 35:313-320.

Watson, A. K., T. J. Klopfenstein, W. H. Schacht, G. E. Erickson, D. R. Mark, M. K. Luebbe, K. R. Brink, and M. A. Greenquist. 2012. Smooth bromegrass pasture beef growing systems: Fertilization strategies and economic analysis. *The Professional Animal Scientist* 28:443-451.

Watson, A. K., K. J. Vander Pol, T. J. Huls, M. K. Luebbe, G. E. Erickson, T. J. Klopfenstein, and M. A. Greenquist. 2014. Effect of dietary inclusion of wet or modified distillers grains plus solubles on performance of finishing cattle. *The Professional Animal Scientist* 30:585-596.

Widyaratne, G. P., and R. T. Zijlstra. 2007. Nutritional value of wheat and corn distiller's dried grain with solubles: Digestibility and digestible contents of energy, amino acids and phosphorus, nutrient excretion and growth performance of grower-finisher pigs. *Canadian Journal of Animal Science* 87:103-114.

Yu, Y., and J. W. Thomas. 1976. Estimation of the extent of heat damage in alfalfa haylage by laboratory measurements. *Journal of Animal Science* 42:766-774.

Zinn, R. A., E. Alvarez, M. Mendez, M. Montaño, E. Ramirez, and Y. Shen. 1997. Influence of dietary sulfur concentration on growth performance and digestive function in feedlot cattle. *Journal of Animal Science* 75:1723-1728.

18

Composition of Selected Feeds for Beef Cattle

INTRODUCTION

The average composition data for feeds are provided in this chapter to aid in the formulation and evaluation of diets in the absence of nutrient analyses. Similar to the *Nutrient Requirements of Beef Cattle,* 7th Revised Edition and Update (NRC, 1996, 2000), the mean, standard deviation (SD), and sample size are provided by nutrient within feeds. Not all feeds are available, and forages, particularly grazed forages, are difficult to summarize given the variation associated with forage type, maturity, month, and whether they originated from mixed pastures or monocultures. A major challenge with grazed forages is whether samples were clipped or if they originated from diet sampling with fistulated cattle. Evidence suggests that cattle select for greater energy and protein concentration when clipped samples are compared with samples collected via esophagealy or ruminally fistulated steers (Lesperance et al., 1960). An attempt was made to briefly provide data on fiber, protein, and digestibility of forages across seasons and regions.

DATA BACKGROUND

Data are provided by feed in Table 18-1. These data originated from three commercial laboratories (DairyOne, Ithaca, NY; Servi-Tech Laboratories, Hastings, NE; and Ward Laboratories, Kearney, NE).

Data were removed if feeds could not be properly identified, contained more than one ingredient, or were diet samples of mixed ingredients. In addition, any feeds with less than 20 entries were not summarized. Initially, data were summarized by source to generate a summary by nutrient that included minimum, maximum, average, and standard deviation. Histograms were generated for each feed, and each nutrient was analyzed within the feed to visually evaluate distribution curves and anomalies. In some cases, such as dry matter (DM) for distillers grains, some feeds were likely mislabeled and included dry and wet versions of the same feeds. When possible, feeds were separated using DM to ensure accuracy of reporting. All nutrients besides DM were evaluated, but assessment of quality was based on deviation from the mean. Any values falling outside of 3.5 SD above or below the mean were removed. The distributions were evaluated by nutrient tested and data were removed only for that nutrient. For example, if the phosphorus content of corn fell outside of 3.5 SD, those values were removed for phosphorus, but the entire feed entry was not removed from consideration of other nutrients.

USE OF THE DATA

Feed composition tables should not replace nutrient analysis by nutritionists or producers; however, these summary data can be used to evaluate whether the nutrient analysis is within a "normal" range when assessed with the mean and SD. The sample size is provided to offer confidence for each individual feed for a specific nutrient. For example, for corn silage, 209,858 samples were used to summarize DM. The mean was 33.07, and the SD was 5.59. This means that approximately two-thirds of the samples were within 1 SD of the mean or range from 27.48 to 38.66. A description of a normal distribution is provided in Figure 18-1.

In most cases, the distribution of values for given nutrients appeared to have a normal distribution; however, tests of normality were not conducted. For many microminerals across most feeds, the SD is often as large as or similar to the mean. In these cases, data were not normally distributed and analysis is recommended. If an individual feed has a relatively large SD relative to its mean, caution is warranted when using a book value or the means provided in Table 18-1. If individual nutrients are variable, then sample analysis is critical to ensuring dietary supplies meet requirements. Likewise, if very few observations are available for an individual nutrient and a given feed, sample analysis is warranted as variation is poorly understood. Because crop varieties, weather, soil fertility and type, processing method,

TABLE 18-1 Means and Standard Deviations for the Composition Data of Feeds Commonly Used in Beef Diets[a]

	Feed Name					
Component	Alfalfa Cubes	Alfalfa Dehy	Alfalfa, Fresh	Alfalfa Greenchop	Alfalfa Hay	Alfalfa Haylage
DM (% AF)	91.04 ± 1.37 (826)	93.83 ± 1.22 (1,530)	30.73 ± 12.16 (276)	40.50 ± 13.03 (139)	87.03 ± 3.68 (50,308)	41.04 ± 11.00 (3,565)
Ash (% DM)	11.98 ± 2.36 (448)	10.29 ± 2.88 (10)	6.23 ± 3.99 (3)	11.09 ± 1.99 (2)	11.87 ± 2.54 (7,751)	12.09 ± 2.58 (125)
TDN (% DM)	56.0 ± 3.03 (757)	62.4 ± 3.60 (10)	63.0	59.0	55.2 ± 5.86 (1,135)	63.0 ± 7.67 (109)
DE (Mcal/kg)	2.47 ± 0.15 (756)					
ME (Mcal/kg)	2.02	2.25	2.28	2.13	1.99	2.28
NEm (Mcal/kg)	1.18	1.39	1.41	1.28	1.15	1.41
NEg (Mcal/kg)	0.61	0.81	0.83	0.71	0.59	0.83
Sugar (% DM)				7.10	8.67 ± 3.02 (37)	1.93 ± 1.11 (24)
Starch (% DM)	1.35 ± 0.84 (606)	0.93 ± 0.15 (3)		2.10	2.97 ± 1.37 (95)	1.88 ± 1.09 (20)
Fat (% DM)	2.13 ± 0.51 (575)	3.99 ± 3.41 (16)	1.47 ± 0.31 (3)	2.47 ± 0.12 (3)	1.55 ± 0.42 (2,315)	2.01 ± 1.11 (164)
NDF (% DM)	45.46 ± 6.13 (774)	40.37 ± 1.66 (10)	37.86 ± 6.64 (224)	34.39 ± 5.11 (142)	41.73 ± 8.53 (33,696)	42.49 ± 6.97 (1,828)
ADF (% DM)	35.41 ± 4.21 (770)	31.23 ± 3.45 (15)	31.40 ± 5.95 (243)	29.43 ± 4.44 (146)	33.25 ± 5.91 (35,791)	36.06 ± 5.92 (2,171)
Lignin (% DM)	7.57 ± 1.22 (424)			6.21 ± 3.46 (2)	6.77 ± 1.30 (1,726)	6.37 ± 1.97 (35)
CP (% DM)	18.08 ± 2.75 (801)	18.49 ± 1.99 (1,753)	23.08 ± 3.68 (279)	23.14 ± 2.81 (147)	19.81 ± 3.18 (50,015)	20.11 ± 3.23 (2,707)
RDP (% CP)	68.82 ± 4.61 (267)					
RUP (% CP)	31.01 ± 4.55 (267)					
Soluble CP (% CP)	39.30 ± 5.32 (362)			12.40	25.19 ± 5.58 (140)	25.18 ± 3.83 (32)
ADICP (% DM)	8.22 ± 1.80 (337)		8.67 ± 3.80 (64)	6.47 ± 3.23 (2)	12.21 ± 6.10 (603)	13.74 ± 6.59 (116)
Ca (% DM)	1.49 ± 0.37 (751)	2.23 ± 3.30 (12)	1.77 ± 0.54 (67)	1.51 ± 0.33 (13)	1.47 ± 0.26 (17,180)	1.56 ± 0.31 (1,469)
P (% DM)	0.23 ± 0.05 (639)	0.40 ± 0.27 (12)	0.34 ± 0.08 (67)	0.29 ± 0.05 (13)	0.26 ± 0.05 (17,310)	0.30 ± 0.07 (1,462)
Mg (% DM)	0.28 ± 0.07 (623)	0.32 ± 0.06 (5)	0.37 ± 0.10 (64)	0.35 ± 0.07 (11)	0.30 ± 0.07 (16,034)	0.33 ± 0.07 (1,278)
K (% DM)	2.05 ± 0.56 (744)	2.32 ± 0.70 (5)	2.92 ± 0.68 (65)	2.88 ± 0.58 (10)	2.55 ± 0.53 (16,221)	2.75 ± 0.60 (1,307)
Na (% DM)	0.16 ± 0.05 (171)	0.05 ± 0.01 (4)	0.12 ± 0.08 (45)	0.07	0.10 ± 0.07 (3,484)	0.13 ± 0.08 (512)
Cl (% DM)	0.70 ± 0.24 (308)	0.27 ± 0.01 (3)		0.66	0.52 ± 0.24 (47)	0.63 ± 0.45 (7)
S (% DM)	0.25 ± 0.05 (335)	0.28 ± 0.01 (6)	0.33 ± 0.08 (57)		0.26 ± 0.07 (6,124)	0.31 ± 0.07 (718)
Co (mg/kg)	0.77 ± 0.26 (5)					
Cu (mg/kg)	8.54 ± 2.64 (268)	11.17 ± 3.33 (4)	8.31 ± 2.26 (26)		7.82 ± 1.89 (3,196)	7.84 ± 1.99 (387)
I (mg/kg)						
Fe (mg/kg)	648.52 ± 678.17 (268)					
Mn (mg/kg)	44.07 ± 16.90 (265)	65.25 ± 21.47 (4)	39.70 ± 15.38 (27)		44.55 ± 15.26 (2,998)	42.40 ± 14.23 (381)
Mo (mg/kg)		3.25 ± 0.84 (3)	4.54 ± 2.66 (5)		1.78 ± 0.94 (851)	1.43 ± 0.87 (19)
Se (mg/kg)	0.81 ± 0.59 (2)					
Zn (mg/kg)	24.34 ± 4.18 (266)	50.20 ± 40.76 (4)	24.69 ± 9.25 (26)		23.47 ± 5.84 (3,183)	25.01 ± 5.77 (385)

continued

TABLE 18-1 Continued

Component	Feed Name					
	Almond Hulls	Apple Pomace	Bakery Products	Barley Grain	Barley Grain, Steam Flaked	Barley Hay
DM (% AF)	89.21 ± 2.10 (288)	18.49 ± 4.76 (122)	88.86 ± 4.24 (2,338)	89.69 ± 2.40 (4,038)	81.12 ± 4.60 (160)	87.99 ± 3.08 (2,473)
Ash (% DM)	8.29 ± 2.98 (112)	3.14 ± 1.65 (36)	4.08 ± 1.27 (1,394)	2.77 ± 0.65 (1,825)	2.87 ± 0.94 (63)	8.36 ± 2.47 (789)
TDN (% DM)	59.1 ± 6.62 (184)	70.9 ± 6.56 (123)	91.9 ± 4.10 (874)	84.1 ± 2.13 (1,755)	84.0	60.2 ± 5.21 (1,002)
DE (Mcal/kg)	2.61 ± 0.30 (185)	3.13 ± 0.27 (123)	4.05 ± 0.20 (948)	3.71 ± 0.09 (1,746)		2.65 ± 0.25 (991)
ME (Mcal/kg)	2.14	2.56	3.32	3.04	3.04	2.17
NEm (Mcal/kg)	1.28	1.66	2.29	2.06	2.06	1.31
NEg (Mcal/kg)	0.71	1.05	1.59	1.40	1.40	0.74
Sugar (% DM)	15.05 ± 6.58 (2)		6.32 ± 8.71 (23)	10.65 ± 0.07 (2)		
Starch (% DM)	2.50 ± 1.46 (47)	3.98 ± 4.90 (41)	37.61 ± 11.60 (670)	56.74 ± 4.54 (2,283)	59.27 ± 2.91 (65)	5.66 ± 5.52 (757)
Fat (% DM)	2.80 ± 1.02 (115)	6.10 ± 2.47 (36)	10.04 ± 3.48 (2,059)	2.20 ± 0.39 (2,126)	2.05 ± 0.55 (15)	2.41 ± 0.88 (795)
NDF (% DM)	38.96 ± 12.61 (247)	45.56 ± 8.60 (122)	13.87 ± 7.29 (999)	18.29 ± 3.96 (1,778)	26.30	56.88 ± 8.59 (1,549)
ADF (% DM)	32.73 ± 10.50 (223)	38.72 ± 8.27 (111)	7.37 ± 4.89 (925)	7.09 ± 2.11 (1,846)	8.38 ± 1.06 (27)	33.88 ± 6.47 (2,349)
Lignin (% DM)	11.06 ± 5.29 (62)	14.85 ± 5.23 (22)	1.99 ± 1.24 (115)	1.75 ± 0.54 (1,281)		4.32 ± 1.53 (776)
CP (% DM)	5.47 ± 1.21 (235)	6.37 ± 1.68 (123)	13.14 ± 2.31 (2,230)	12.78 ± 2.83 (3,877)	12.53 ± 1.29 (121)	10.95 ± 3.84 (2,424)
RDP (% CP)	53.12 ± 27.51 (2)	43.07 ± 7.98 (6)	66.36 ± 9.91 (22)	49.14 ± 8.62 (1,160)		67.10 ± 8.37 (719)
RUP (% CP)	46.89 ± 27.51 (2)	56.76 ± 7.83 (6)	33.42 ± 9.81 (22)	50.77 ± 8.59 (1,160)		32.77 ± 8.39 (719)
Soluble CP (% CP)	40.92 ± 14.32 (94)	20.18 ± 8.15 (73)	23.23 ± 8.96 (444)	27.58 ± 8.16 (1,368)		44.73 ± 9.80 (773)
ADICP (% DM)	11.46 ± 4.84 (18)	10.63 ± 3.14 (12)	5.74 ± 3.41 (70)	2.21 ± 1.24 (1,129)		4.12 ± 1.76 (649)
Ca (% DM)	0.30 ± 0.32 (101)	0.15 ± 0.05 (83)	0.26 ± 0.26 (718)	0.08 ± 0.05 (1,636)	0.11 ± 0.03 (17)	0.37 ± 0.18 (1,926)
P (% DM)	0.11 ± 0.03 (98)	0.14 ± 0.04 (84)	0.39 ± 0.18 (720)	0.38 ± 0.07 (1,734)	0.35 ± 0.04 (44)	0.24 ± 0.08 (1,960)
Mg (% DM)	0.11 ± 0.03 (94)	0.09 ± 0.03 (85)	0.15 ± 0.09 (583)	0.13 ± 0.02 (1,604)	0.15 ± 0.01 (17)	0.18 ± 0.06 (1,874)
K (% DM)	2.38 ± 0.64 (117)	0.83 ± 0.29 (88)	0.53 ± 0.26 (533)	0.53 ± 0.09 (1,583)	0.56 ± 0.09 (17)	1.92 ± 0.68 (1,993)
Na (% DM)	0.02 ± 0.01 (93)	0.02 ± 0.02 (65)	0.59 ± 0.26 (757)	0.02 ± 0.02 (483)	0.03(3)	0.23 ± 0.25 (318)
Cl (% DM)	0.09 ± 0.06 (17)	0.04 ± 0.03 (13)	0.80 ± 0.31 (238)	0.17 ± 0.05 (156)		0.88 ± 0.53 (660)
S (% DM)	0.04 ± 0.02 (107)	0.08 ± 0.03 (67)	0.16 ± 0.04 (454)	0.14 ± 0.03 (1,288)	0.14 ± 0.01 (36)	0.18 ± 0.06 (1,348)
Co (mg/kg)						1.67
Cu (mg/kg)	7.06 ± 3.31 (50)	8.36 ± 2.30 (49)	6.75 ± 4.18 (312)	6.12 ± 2.00 (254)	5.43 ± 0.48 (4)	7.05 ± 2.69 (375)
I (mg/kg)						
Fe (mg/kg)	368.12 ± 222.14 (46)	105.61 ± 97.91 (48)	192.36 ± 133.20 (296)	99.40 ± 68.29 (219)		310.18 ± 305.58 (144)
Mn (mg/kg)	20.94 ± 11.05 (50)	10.79 ± 5.14 (19)	44.41 ± 20.85 (312)	21.85 ± 10.81 (256)	14.67 ± 1.53 (3)	35.05 ± 15.50 (369)
Mo (mg/kg)				1.37 ± 0.26 (4)	0.51 ± 0.09 (3)	1.60 ± 0.90 (194)
Se (mg/kg)	0.04			1.00		
Zn (mg/kg)	16.26 ± 10.06 (50)	7.84 ± 3.74 (50)	43.06 ± 19.68 (313)	30.64 ± 10.17 (256)	31.20 ± 2.89 (4)	25.35 ± 9.66 (373)

TABLE 18-1 Continued

Component	Feed Name					
	Barley Silage	Barley Straw	Beet Pulp, Dry	Beet Pulp, Wet	Bermudagrass, Fresh	Bermudagrass Hay
DM (% AF)	33.63 ± 7.85 (2,334)	85.07 ± 10.35 (19)	91.49 ± 1.56 (1,189)	21.95 ± 3.65 (901)	34.94 ± 10.37 (1,024)	92.99 ± 1.58 (14,528)
Ash (% DM)	8.65 ± 2.46 (1,457)	12.08	6.84 ± 2.09 (526)	8.59 ± 3.86 (187)	8.63 ± 1.84 (492)	7.94 ± 1.34 (9,364)
TDN (% DM)	60.6 ± 4.59 (1,761)	48.3 ± 1.36 (7)	66.6 ± 3.08 (856)	66.6 ± 3.43 (517)	57.3 ± 2.85 (873)	56.3 ± 2.68 (12,499)
DE (Mcal/kg)	2.67 ± 0.19 (1,758)		2.94 ± 0.13 (850)	2.94 ± 0.14 (442)	2.53 ± 0.17 (877)	2.48 ± 0.13 (12,496)
ME (Mcal/kg)	2.19	1.75	2.41	2.41	2.07	2.04
NEm (Mcal/kg)	1.33	0.91	1.52	1.52	1.22	1.19
NEg (Mcal/kg)	0.75	0.36	0.93	0.93	0.65	0.62
Sugar (% DM)			8.55 ± 5.05 (4)	23.21 ± 27.46 (26)		5.80
Starch (% DM)	9.17 ± 8.16 (1,513)		0.93 ± 0.72 (444)	1.65 ± 2.04 (163)	1.79 ± 1.34 (534)	4.78 ± 3.04 (10,345)
Fat (% DM)	3.47 ± 0.72 (1,472)	1.00	1.14 ± 0.35 (515)	0.86 ± 0.30 (167)	2.76 ± 0.61 (494)	1.86 ± 0.39 (9,391)
NDF (% DM)	54.77 ± 7.18 (1,819)	71.63 ± 3.60 (7)	41.33 ± 4.58 (872)	48.23 ± 7.14 (679)	66.60 ± 5.97 (984)	66.98 ± 4.89 (13,303)
ADF (% DM)	34.73 ± 5.07 (2,211)	50.09 ± 2.44 (7)	26.35 ± 3.43 (855)	28.06 ± 3.19 (755)	36.14 ± 4.27 (1,016)	35.65 ± 3.82 (13,727)
Lignin (% DM)	4.77 ± 1.31 (1,481)	5.16	3.94 ± 2.08 (338)	4.37 ± 2.63 (72)	5.03 ± 1.20 (492)	5.41 ± 1.27 (9,782)
CP (% DM)	12.05 ± 2.99 (2,259)	6.08 ± 1.53 (11)	9.07 ± 1.31 (1,002)	9.55 ± 1.25 (862)	15.16 ± 4.41 (1,032)	11.11 ± 2.90 (13,844)
RDP (% CP)	79.13 ± 7.22 (1,386)		46.71 ± 8.73 (54)	44.11 ± 8.95 (26)	67.49 ± 4.56 (477)	58.33 ± 4.89 (8,299)
RUP (% CP)	20.79 ± 7.22 (1,386)		53.15 ± 8.72 (54)	55.66 ± 8.99 (26)	32.40 ± 4.47 (476)	41.58 ± 4.91 (8,296)
Soluble CP (% CP)	65.32 ± 9.34 (1,517)		21.89 ± 10.95 (588)	20.78 ± 10.20 (292)	42.43 ± 6.79 (489)	32.73 ± 4.84 (8,466)
ADICP (% DM)	4.92 ± 1.79 (1,635)	0.60	7.04 ± 3.83 (282)	4.53 ± 2.40 (49)	6.31 ± 1.93 (440)	5.35 ± 1.37 (7,731)
Ca (% DM)	0.41 ± 0.16 (2,110)	0.52 ± 0.22 (10)	0.96 ± 0.29 (733)	0.99 ± 0.37 (620)	0.48 ± 0.15 (640)	0.49 ± 0.10 (10,794)
P (% DM)	0.30 ± 0.07 (2,128)	0.21 ± 0.20 (10)	0.08 ± 0.02 (688)	0.10 ± 0.03 (602)	0.30 ± 0.07 (721)	0.20 ± 0.05 (10,824)
Mg (% DM)	0.18 ± 0.07 (2,096)	0.18 ± 0.04 (8)	0.23 ± 0.05 (686)	0.25 ± 0.05 (531)	0.22 ± 0.05 (637)	0.20 ± 0.05 (10,335)
K (% DM)	2.06 ± 0.69 (2,252)	2.29 ± 0.88 (8)	0.71 ± 0.50 (686)	0.57 ± 0.25 (531)	2.21 ± 0.60 (764)	1.65 ± 0.44 (11,588)
Na (% DM)	0.19 ± 0.19 (1,005)	0.52 ± 0.14 (6)	0.19 ± 0.20 (612)	0.10 ± 0.08 (309)	0.03 ± 0.03 (242)	0.13 ± 0.11 (4,928)
Cl (% DM)	0.86 ± 0.42 (1,294)		0.12 ± 0.14 (261)	0.09 ± 0.07 (71)	0.76 ± 0.23 (446)	0.74 ± 0.24 (7,278)
S (% DM)	0.19 ± 0.30 (1,570)	0.21± 0.03 (6)	0.30 ± 0.12 (558)	0.20 ± 0.11 (454)	0.27 ± 0.07 (493)	0.40 ± 0.12 (8,165)
Co (mg/kg)	1.41 ± 0.57 (2)		0.99 ± 0.31 (9)			0.60 ± 0.46 (73)
Cu (mg/kg)	6.94 ± 2.57 (418)	5.10 ± 1.45 (6)	8.40 ± 3.47 (243)	8.79 ± 3.89 (132)	10.75 ± 3.06 (101)	10.65 ± 3.07 (2,221)
I (mg/kg)						
Fe (mg/kg)	353.03 ± 376.65 (314)		640.72 ± 333.11 (249)	748.18 ± 462.69 (117)	254.68 ± 116.03 (99)	217.79 ± 94.15 (2,164)
Mn (mg/kg)	38.54 ± 20.83 (417)	42.67 ± 9.83 (6)	61.59 ± 16.70 (243)	60.79 ± 21.76 (130)	76.97 ± 47.51 (100)	63.11 ± 30.79 (2,202)
Mo (mg/kg)	1.20 ± 0.86 (7)	1.63				
Se (mg/kg)			0.12		0.20 ± 0.17 (5)	0.27 ± 0.10 (48)
Zn (mg/kg)	28.34 ± 14.38 (406)	25.40 ± 4.20 (6)	29.91 ± 59.71 (249)	22.56 ± 7.23 (133)	41.38 ± 15.81 (100)	34.84 ± 9.95 (2,232)

continued

TABLE 18-1 Continued

	Feed Name					
Component	Bermudagrass Silage	Blood Meal	Bluegrass Hay	Bluestem, Fresh	Bluestem Hay	Brewers Dried Grains
DM (% AF)	39.00 ± 9.78 (1,024)	89.56 ± 2.79 (1,203)	89.32 ± 2.27 (37)	51.23 ± 12.16 (14)	89.19 ± 5.40 (173)	93.16 ± 2.69 (189)
Ash (% DM)	8.68 ± 2.44 (746)	2.80 ± 1.76 (233)	9.33 ± 1.20 (6)		9.70 ± 0.57 (2)	4.57 ± 1.33 (148)
TDN (% DM)	55.4 ± 3.69 (1,035)	74.6 ± 2.65 (877)	51.0	53.0	50.0	72.0 ± 5.17 (169)
DE (Mcal/kg)	2.44 ± 0.18 (1,039)	3.29 ± 0.17 (880)				3.17 ± 0.23 (168)
ME (Mcal/kg)	2.00	2.70	1.84	1.92	1.81	2.60
NEm (Mcal/kg)	1.15	1.78	1.00	1.07	0.97	1.70
NEg (Mcal/kg)	0.59	1.15	0.45	0.52	0.42	1.08
Sugar (% DM)						
Starch (% DM)	2.60 ± 1.44 (668)	0.62 ± 0.25 (5)				5.77 ± 5.52 (127)
Fat (% DM)	3.15 ± 0.88 (774)	1.21 ± 1.34 (304)	1.40 ± 0.40 (5)		1.35 ± 0.33 (4)	8.52 ± 2.58 (159)
NDF (% DM)	66.59 ± 5.58 (1,037)	4.53 ± 8.41 (9)	68.83 ± 4.22 (6)		69.71 ± 5.53 (31)	52.12 ± 9.75 (170)
ADF (% DM)	40.32 ± 3.85 (1,030)	0.70 ± 0.43 (5)	40.40 ± 3.33 (25)	43.51 ± 2.83 (11)	43.32 ± 4.41 (123)	25.39 ± 5.28 (170)
Lignin (% DM)	6.40 ± 1.43 (686)	2.00 ± 0.96 (3)				6.65 ± 2.03 (143)
CP (% DM)	13.48 ± 3.20 (1,047)	95.05 ± 4.88 (411)	7.50 ± 2.18 (28)	6.52 ± 2.29 (14)	6.02 ± 2.32 (181)	25.02 ± 5.09 (179)
RDP (% CP)	70.31 ± 7.92 (749)	25.23 ± 10.99 (54)				40.31 ± 13.32 (87)
RUP (% CP)	29.67 ± 7.96 (750)	74.64 ± 10.86 (54)				59.14 ± 13.86 (88)
Soluble CP (% CP)	53.57 ± 9.94 (817)	14.18 ± 8.40 (176)			43.00	17.91 ± 13.60 (144)
ADICP (% DM)	7.99 ± 2.43 (1,014)	21.41 ± 24.65 (128)				19.25 ± 11.12 (133)
Ca (% DM)	0.51 ± 0.13 (898)	0.14 ± 0.40 (183)	0.28 ± 0.09 (25)	0.44 (2)	0.41 ± 0.12 (26)	0.32 ± 0.25 (132)
P (% DM)	0.29 ± 0.07 (905)	0.23 ± 0.21 (187)	0.19 ± 0.05 (25)	0.15 ± 0.02 (2)	0.12 ± 0.07 (27)	0.65 ± 0.16 (132)
Mg (% DM)	0.23 ± 0.06 (901)	0.04 ± 0.04 (155)	0.13 ± 0.04 (22)		0.21 ± 0.19 (17)	0.22 ± 0.06 (127)
K (% DM)	2.16 ± 0.67 (950)	0.41 ± 0.23 (157)	1.94 ± 0.47 (22)		1.11 ± 0.31 (16)	0.34 ± 0.47 (134)
Na (% DM)	0.05 ± 0.03 (357)	0.26 ± 0.14 (150)			0.06 ± 0.05 (2)	0.02 ± 0.01 (70)
Cl (% DM)	0.71 ± 0.28 (705)	0.32 ± 0.16 (5)				0.10 ± 0.08 (35)
S (% DM)	0.24 ± 0.06 (770)	0.65 ± 0.26 (152)	0.14 ± 0.03 (19)		0.87 ± 1.77 (5)	0.29 ± 0.06 (124)
Co (mg/kg)						1.13
Cu (mg/kg)	13.32 ± 6.47 (68)	4.35 ± 2.00 (96)	3.83 ± 1.19 (12)		9.57 ± 12.66 (6)	17.64 ± 6.60 (48)
I (mg/kg)						
Fe (mg/kg)	536.86 ± 538.96 (67)	2,378.00 ± 513.70 (96)				327.41 ± 249.85 (46)
Mn (mg/kg)	75.31 ± 34.14 (72)	5.33 ± 4.01 (78)	75.55 ± 18.35 (11)		117.80 ± 139.59 (5)	56.09 ± 19.61 (49)
Mo (mg/kg)			2.88 ± 3.04 (4)		0.02	
Se (mg/kg)						
Zn (mg/kg)	43.87 ± 13.15 (69)	36.59 ± 18.98 (96)	14.64 ± 5.52 (11)		20.68 ± 14.53 (6)	214.62 ± 209.01 (50)

TABLE 18-1 Continued

	Feed Name					
Component	Brewers Wet Grains	Bromegrass Hay	Bromegrass Silage	Buffalo Grass	Cane, Fresh	Cane Hay
DM (% AF)	25.96 ± 6.24 (3,921)	88.27 ± 2.92 (1,010)	42.05 ± 14.28 (8)	93.38 ± 1.76 (384)	28.04 ± 10.71 (330)	82.79 ± 5.14 (6,904)
Ash (% DM)	4.38 ± 0.61 (1,654)	8.84 ± 1.12 (12)				11.14 ± 3.49 (14)
TDN (% DM)	73.9 ± 2.68 (2,389)	52.0 ± 4.29 (67)	55.0	54.0	63.0	63.0 ± 7.73 (290)
DE (Mcal/kg)	3.26 ± 0.15 (2,394)					
ME (Mcal/kg)	2.67	1.88	1.99	1.95	2.28	2.28
NEm (Mcal/kg)	1.75	1.04	1.14	1.11	1.41	1.41
NEg (Mcal/kg)	1.13	0.49	0.58	0.55	0.83	0.83
Sugar (% DM)	0.50	9.85 ± 3.46 (8)				18.06 ± 2.72 (5)
Starch (% DM)	4.81 ± 3.50 (1,487)	2.64 ± 0.35 (7)		5.49 ± 2.44 (383)		71.40
Fat (% DM)	9.51 ± 1.37 (1,899)	1.64 ± 0.25 (16)				1.34 ± 0.29 (38)
NDF (% DM)	49.99 ± 5.46 (2,500)	65.92 ± 4.64 (316)	71.05 ± 4.45 (2)		60.00 ± 9.17 (11)	63.33 ± 6.52 (418)
ADF (% DM)	24.32 ± 3.11 (2,484)	40.29 ± 3.74 (856)	43.05 ± 4.70 (8)		33.27 ± 5.28 (69)	39.26 ± 6.13 (2,590)
Lignin (% DM)	6.74 ± 0.92 (1,544)					4.20 ± 1.42 (7)
CP (% DM)	28.52 ± 4.47 (2,896)	8.34 ± 2.33 (965)	9.03 ± 2.12 (9)	7.60	11.29 ± 4.01 (106)	8.02 ± 2.87 (4,696)
RDP (% CP)	36.15 ± 5.16 (1,420)					
RUP (% CP)	63.77 ± 5.17 (1,420)					
Soluble CP (% CP)	11.40 ± 4.88 (1,726)	24.55 (2)				36.45 ± 8.22 (45)
ADICP (% DM)	21.75 ± 6.56 (1,443)	7.44 ± 3.18 (79)	11.88		10.63	5.88 ± 3.52 (136)
Ca (% DM)	0.35 ± 0.11 (1,971)	0.55 ± 0.10 (186)	0.47 ± 0.05 (5)		0.39 ± 0.11 (24)	0.42 ± 0.13 (813)
P (% DM)	0.68 ± 0.10 (1,966)	0.18+ 0.05 (431)	0.22 ± 0.08 (5)		0.30 ± 0.14 (24)	0.18 ± 0.06 (824)
Mg (% DM)	0.23 ± 0.03 (1,932)	0.18 ± 0.15 (252)	0.17 ± 0.03 (4)		0.24 ± 0.04 (12)	0.23 ± 0.06 (334)
K (% DM)	0.12 ± 0.09 (1,884)	1.59 ± 0.48 (298)	1.67 ± 0.81 (5)		3.61 ± 0.99 (13)	2.11 ± 0.67 (416)
Na (% DM)	0.02 ± 0.02 (696)	0.04 ± 0.05 (6)	0.04 ± 0.03 (2)			0.11 ± 0.21 (12)
Cl (% DM)	0.09 ± 0.08 (139)	0.19 ± 0.07 (4)				
S (% DM)	0.32 ± 0.06 (1,727)	0.17 ± 0.39 (89)	0.13 ± 0.02 (5)		0.14 ± 0.04 (3)	0.11 ± 0.03 (151)
Co (mg/kg)	1.50 ± 0.71 (2)					
Cu (mg/kg)	17.41 ± 5.33 (238)	5.96 ± 7.04 (27)	5.00 ± 0.82 (4)			7.73 ± 9.83 (35)
I (mg/kg)						
Fe (mg/kg)	222.05 ± 86.41 (218)					
Mn (mg/kg)	51.26 ± 11.83 (240)	67.38 ± 39.77 (26)	69.00 ± 6.08 (3)			69.12 ± 71.05 (32)
Mo (mg/kg)	2.00 ± 0.28 (2)	2.23 ± 2.43 (7)				0.71 ± 0.369 (12)
Se (mg/kg)						
Zn (mg/kg)	93.01 ± 16.801 (237)	16.24 ± 4.85 (27)	16.25 ± 6.24 (4)			21.01 ± 8.26 (35)

continued

TABLE 18-1 Continued

Component	Feed Name					
	Cane Silage	Canola Grain	Canola Meal	Citrus Pulp, Dry	Citrus Pulp, Wet	Corn Cobs
DM (% AF)	29.33 ± 7.06 (570)	94.72 ± 1.77 (171)	90.43 ± 2.13 (3,137)	87.69 ± 1.94 (561)	19.41 ± 6.64 (177)	89.26 ± 11.11 (674)
Ash (% DM)		4.33 ± 1.28 (16)	7.41 ± 0.92 (664)	7.42 ± 1.17 (176)	6.78 ± 1.85 (75)	2.56 ± 0.99 (157)
TDN (% DM)	65.0 ± 8.43 (18)	109.2 ± 5.54 (5)	71.1 ± 5.46 (1,446)	70.0 ± 2.75 (422)	70.2 ± 3.76 (146)	58.3 ± 4.89 (250)
DE (Mcal/kg)		4.81 ± 0.24 (5)	3.13 ± 0.23 (1,474)	3.09 ± 0.11 (422)	3.10 ± 0.15 (145)	2.57 ± 0.23 (237)
ME (Mcal/kg)	2.35	3.95	2.57	2.53	2.54	2.11
NEm (Mcal/kg)	1.47	2.78	1.67	1.63	1.64	1.25
NEg (Mcal/kg)	0.88	1.99	1.06	1.03	1.03	0.68
Sugar (% DM)			8.75 ± 0.78 (2)		0.90	
Starch (% DM)	11.40	1.40 ± 1.65 (11)	1.29 ± 0.93 (271)	1.00 ± 1.06 (127)	1.70 ± 0.82 (57)	1.38 ± 1.26 (121)
Fat (% DM)	2.40 ± 0.14 (2)	39.79 ± 4.24 (150)	7.32 ± 5.26 (1,290)	2.44 ± 0.64 (202)	3.17 ± 1.68 (83)	0.92 ± 1.00 (123)
NDF (% DM)	52.99 ± 9.06 (114)	28.25 ± 8.82 (31)	30.16 ± 3.67 (1,523)	24.02 ± 3.09 (426)	26.29 ± 4.32 (148)	78.26 ± 13.60 (259)
ADF (% DM)	35.26 ± 5.65 (474)	21.99 ± 7.14 (34)	21.42 ± 2.99 (1,497)	20.43 ± 3.33 (383)	23.16 ± 5.08 (163)	42.03 ± 9.63 (284)
Lignin (% DM)	19.70	6.40 ± 2.90 (10)	8.83 ± 1.87 (325)	2.45 ± 1.41 (75)	3.21 ± 2.25 (35)	4.05 ± 1.81 (210)
CP (% DM)	7.70 ± 2.14 (536)	23.90 ± 3.54 (159)	40.86 ± 2.84 (3,014)	6.91 ± 0.91 (529)	8.58 ± 1.57 (167)	4.18 ± 2.21 (613)
RDP (% CP)		62.86 ± 15.95 (7)	57.46 ± 9.62 (134)	59.49 ± 10.25 (5)	61.64	69.53 ± 6.59 (42)
RUP (% CP)		36.99 ± 15.97 (7)	42.25 ± 9.75 (135)	40.31 ± 10.20 (5)	38.06	30.13 ± 6.29 (41)
Soluble CP (% CP)	35.18 ± 6.43 (6)	41.48 ± 17.89 (20)	32.17 ± 11.78 (786)	41.06 ± 10.34 (149)	56.65 ± 9.00 (82)	37.03 ± 11.81 (147)
ADICP (% DM)	10.44 ± 4.54 (27)	11.11 ± 3.86 (9)	17.36 ± 6.64 (285)	5.63 ± 2.00 (52)	4.48 ± 2.52 (18)	3.40 ± 1.71 (110)
Ca (% DM)	0.35 ± 0.10 (289)	0.53 ± 0.18 (23)	0.69 ± 0.13 (932)	1.84 ± 0.45 (237)	1.28 ± 0.65 (147)	0.06 ± 0.06 (174)
P (% DM)	0.19 ± 0.05 (289)	0.70 ± 0.22 (23)	1.10 ± 0.14 (1,042)	0.11 ± 0.02 (215)	0.16 ± 0.04 (116)	0.09 ± 0.07 (182)
Mg (% DM)	0.29 ± 0.34 (185)	0.35 ± 0.07 (20)	0.56 ± 0.09 (881)	0.13 ± 0.02 (204)	0.13 ± 0.03 (115)	0.05 ± 0.04 (155)
K (% DM)	1.91 ± 0.63 (218)	0.85 ± 0.16 (22)	1.26 ± 0.18 (969)	0.98 ± 0.15 (217)	1.22 ± 0.33 (124)	0.89 ± 0.31 (162)
Na (% DM)	0.10 ± 0.06 (8)	0.02 ± 0.01 (13)	0.06 ± 0.06 (635)	0.05 ± 0.04 (204)	0.04 ± 0.04 (97)	0.03 ± 0.03 (58)
Cl (% DM)		0.07 ± 0.02 (8)	0.12 ± 0.12 (171)	0.12 ± 0.05 (49)	0.11 ± 0.06 (27)	0.25 ± 0.08 (106)
S (% DM)	0.35 ± 1.25 (147)	0.41 ± 0.08 (16)	0.71 ± 0.10 (723)	0.11 ± 0.04 (135)	0.11 ± 0.03 (93)	0.05 ± 0.03 (147)
Co (mg/kg)			0.14			
Cu (mg/kg)	21.93 ± 22.16 (15)	5.31 ± 3.29 (9)	6.46 ± 4.34 (418)	6.40 ± 2.63 (99)	5.48 ± 2.11 (67)	7.31 ± 6.27 (26)
I (mg/kg)						
Fe (mg/kg)		98.09 ± 37.41 (7)	230.51 ± 101.35 (396)	124.58 ± 121.22 (94)	120.95 ± 97.85 (62)	245.68 ± 193.22 (26)
Mn (mg/kg)	129.87 ± 112.29 (15)	74.51 ± 69.23 (9)	68.40 ± 12.28 (416)	9.92 ± 4.40 (97)	11.89 ± 8.01 (62)	13.93 ± 9.73 (26)
Mo (mg/kg)	0.30 ± 0.39 (8)		1.00 ± 0.32 (8)			
Se (mg/kg)			1.10			
Zn (mg/kg)	14.94 ± 14.40 (15)	46.52 ± 14.47 (9)	59.84 ± 10.69 (411)	12.09 ± 13.48 (98)	12.11 ± 4.18 (67)	26.47 ± 13.03 (26)

TABLE 18-1 Continued

	Feed Name					
Component	Corn, Ear Corn	Corn Earlage	Corn Germ Meal	Corn Gluten Feed and Distillers; Golden Synergy	Corn Gluten Feed, Wet	Corn Gluten Feed, Wet; Sweet Bran
DM (% AF)	83.28 ± 6.25 (1,221)	62.54 ± 6.89 (6,642)	90.59 ± 2.12 (562)	50.44 ± 4.65 (69)	43.76 ± 8.18 (767)	60.07 ± 2.16 (75)
Ash (% DM)	1.69 ± 0.39 (603)	1.86 ± 0.41 (3,540)	4.31 ± 1.71 (237)	4.40 ± 0.42 (11)	6.40 ± 1.94 (816)	
TDN (% DM)	84.6 ± 3.15 (686)	84.3 ± 3.12 (3,893)	78.6 ± 11.70 (301)	89.0 ± 7.17 (22)	86.0	89.0 ± 3.01 (63)
DE (Mcal/kg)	3.73 ± 0.13 (663)	3.72 ± 0.12 (3,747)	3.46 ± 0.42 (297)		3.79	3.92
ME (Mcal/kg)	3.06	3.05	2.84	3.22	3.11	3.22
NEm (Mcal/kg)	2.08	2.07	1.90	2.21	2.12	2.21
NEg (Mcal/kg)	1.41	1.41	1.26	1.52	1.45	1.52
Sugar (% DM)		1.29 ± 1.26 (7)			3.40 ± 2.53 (11)	
Starch (% DM)	60.68 ± 6.69 (773)	60.16 ± 6.03 (4,308)	19.67 ± 6.74 (76)	11.50	15.23 ± 4.03 (774)	
Fat (% DM)	3.60 ± 0.64 (618)	3.54 ± 0.49 (3,725)	11.50 ± 8.81 (431)	8.70 ± 0.78 (49)	4.29 ± 2.19 (1,608)	4.65 ± 1.83 (27)
NDF (% DM)	19.41 ± 5.52 (750)	21.04 ± 5.61 (4,342)	39.41 ± 9.39 (313)	29.28 ± 4.07 (4)	38.53 ± 6.39 (833)	26.75 ± 3.61 (2)
ADF (% DM)	9.04 ± 2.88 (932)	9.89 ± 2.88 (5,885)	12.27 ± 3.20 (402)	10.27 ± 1.56 (22)	11.78 ± 2.17 (995)	9.79 ± 1.19 (64)
Lignin (% DM)	1.80 ± 0.50 (601)	1.74 ± 0.46 (3,507)	2.44 ± 1.04 (130)		1.60 ± 0.77 (112)	
CP (% DM)	8.28 ± 0.93 (1,077)	8.09 ± 0.82 (6,524)	22.14 ± 4.90 (525)	24.55 ± 1.66 (64)	21.70 ± 3.83 (2,359)	23.76 ± 1.48 (71)
RDP (% CP)	36.27 ± 5.89 (585)	51.28 ± 13.80 (3,490)	52.21 ± 14.94 (11)		65.70 ± 7.01 (47)	
RUP (% CP)	63.59 ± 5.95 (585)	48.63 ± 13.82 (3,490)	47.80 ± 14.94 (11)		34.11 ± 6.97 (47)	
Soluble CP (% CP)	24.68 ± 9.20 (691)	38.91 ± 14.34 (3,655)	29.53 ± 12.88 (77)	16.30	59.12 ± 15.48 (412)	50.47 ± 9.41 (3)
ADICP (% DM)	3.24 ± 1.12 (520)	2.29 ± 1.04 (3,193)	6.42 ± 5.48 (148)		6.95 ± 5.02 (67)	
Ca (% DM)	0.04 ± 0.03 (765)	0.05 ± 0.02 (5,112)	0.04 ± 0.05 (145)	0.09 ± 0.04 (29)	0.05 ± 0.03 (1,397)	0.06 ± 0.03 (66)
P (% DM)	0.28 ± 0.04 (773)	0.27 ± 0.03 (5,132)	1.07 ± 0.27 (249)	0.82 ± 0.09 (29)	0.90 ± 0.29 (1,449)	0.97 ± 0.08 (67)
Mg (% DM)	0.12 ± 0.02 (675)	0.12 ± 0.01 (4,127)	0.30 ± 0.16 (139)	0.34 ± 0.03 (16)	0.41 ± 0.14 (979)	0.40 ± 0.03 (58)
K (% DM)	0.44 ± 0.07 (671)	0.47 ± 0.07 (4,159)	0.69 ± 0.48 (145)	1.17 ± 0.15 (16)	1.32 ± 0.45 (1,236)	1.39 ± 0.17 (59)
Na (% DM)	0.01 ± 0.00 (39)	0.02 ± 0.03 (86)	0.04 ± 0.04 (114)	0.21 ± 0.07 (4)	0.16 ± 0.10 (533)	
Cl (% DM)	0.09 ± 0.03 (14)	0.11 ± 0.03 (41)	0.10 ± 0.12 (21)	0.14	0.19 ± 0.12 (191)	
S (% DM)	0.10 ± 0.01 (591)	0.10 ± 0.02 (3,593)	0.06 ± 0.24 (192)	0.62 ± 0.08 (55)	0.52 ± 0.14 (2,040)	0.49 ± 0.05 (57)
Co (mg/kg)						
Cu (mg/kg)	4.44 ± 2.50 (29)	3.75 ± 2.40 (133)	7.18 ± 1.48 (119)	5.35 ± 0.42 (4)	5.41 ± 1.15 (230)	
I (mg/kg)						
Fe (mg/kg)	91.62 ± 63.54 (27)	85.17 ± 52.79 (97)	129.72 ± 29.67 (117)		152.45 ± 66.07 (69)	
Mn (mg/kg)	10.24 ± 5.20 (29)	11.16 ± 6.40 (124)	16.40 ± 5.61 (118)	17.75 ± 2.63 (4)	17.42 ± 3.82 (232)	
Mo (mg/kg)	0.42	0.34 ± 0.17 (12)		0.96 ± 0.09 (4)	1.39 ± 0.11 (3)	
Se (mg/kg)						
Zn (mg/kg)	24.95 ± 8.17 (29)	21.56 ± 5.65 (134)	106.31 ± 30.31 (120)	62.45 ± 6.62 (4)	62.41 ± 12.63 (232)	

continued

TABLE 18-1 Continued

Component	Feed Name					
	Corn Gluten Feed, Dry	Corn Gluten Meal	Corn Grain, Dry Rolled	Corn Grain, High Moisture	Corn Grain, Steamed Flaked	Corn Greenchop
DM (% AF)	88.92 ± 1.54 (3,654)	90.40 ± 1.54 (934)	87.22 ± 3.25 (31,123)	70.54 ± 4.79 (25,343)	80.70 ± 3.21 (53,435)	34.29 ± 5.85 (61,863)
Ash (% DM)	8.20 ± 1.53 (1,356)	2.91 ± 0.95 (550)	1.44 ± 0.29 (7,166)	1.53 ± 0.26 (14,152)	1.26 ± 0.34 (1,108)	4.04 ± 1.08 (56,953)
TDN (% DM)	80.0 ± 2.01 (1,224)	87.8 ± 3.14 (202)	87.6 ± 1.83 (6,452)	90.4 ± 1.23 (14,617)	95.0 ± 1.15 (539)	67.3 ± 2.33 (60,687)
DE (Mcal/kg)	3.53 ± 0.09 (1,169)	3.87 ± 0.25 (201)	3.86 ± 0.07 (5,967)	3.98 ± 0.05 (14,421)	4.19 ± 0.05 (540)	2.97 ± 0.10 (60,634)
ME (Mcal/kg)	2.89	3.17	3.17	3.27	3.43	2.43
NEm (Mcal/kg)	1.94	2.17	2.17	2.25	2.38	1.55
NEg (Mcal/kg)	1.30	1.49	1.49	1.56	1.67	0.95
Sugar (% DM)	2.68 ± 1.99 (34)	0.23 ± 0.06 (3)	1.81 ± 1.11 (20)	2.16 ± 0.67 (5)		
Starch (% DM)	16.92 ± 3.05 (1,029)	15.42 ± 3.47 (56)	72.07 ± 3.18 (14,066)	71.30 ± 2.98 (15,103)	76.24 ± 2.43 (51,310)	33.44 ± 7.34 (59,234)
Fat (% DM)	3.32 ± 0.86 (1,852)	2.40 ± 0.93 (583)	3.81 ± 0.52 (15,057)	3.93 ± 0.58 (14,699)	3.19 ± 0.92 (1,355)	2.87 ± 0.47 (57,027)
NDF (% DM)	35.05 ± 4.78 (1,575)	8.07 ± 6.16 (291)	9.72 ± 1.83 (6,999)	9.86 ± 1.76 (14,639)	8.97 ± 1.41 (605)	42.07 ± 5.83 (62,114)
ADF (% DM)	11.18 ± 2.14 (1,443)	4.81 ± 2.80 (293)	3.56 ± 0.88 (7,582)	3.69 ± 0.86 (15,957)	3.59 ± 0.77 (600)	24.21 ± 4.03 (61,341)
Lignin (% DM)	1.86 ± 0.96 (210)	2.26 ± 1.38 (34)	1.18 ± 0.29 (4,633)	1.15 ± 0.50 (13,715)	1.25 ± 0.35 (450)	3.36 ± 0.67 (58,129)
CP (% DM)	22.64 ± 2.64 (3,368)	68.21 ± 9.92 (863)	8.79 ± 0.97 (22,868)	8.75 ± 0.77 (23,493)	8.48 ± 0.62 (10,737)	7.82 ± 0.97 (58,015)
RDP (% CP)	63.69 ± 10.47 (159)	29.99 ± 21.81 (21)	34.60 ± 5.25 (4,253)	44.62 ± 11.36 (13,719)	29.50 ± 6.15 (337)	59.27 ± 8.35 (56,859)
RUP (% CP)	37.10 ± 11.99 (147)	69.72 ± 21.66 (21)	65.31 ± 5.29 (4,254)	55.30 ± 11.37 (13,719)	70.36 ± 6.18 (337)	40.68 ± 8.34 (56,860)
Soluble CP (% CP)	51.70 ± 12.95 (782)	14.74 ± 14.05 (127)	21.08 ± 5.17 (4,967)	29.98 ± 10.99 (16,551)	8.20 ± 4.32 (690)	33.58 ± 7.74 (57,638)
ADICP (% DM)	8.85 ± 5.99 (269)	15.08 ± 11.53 (56)	3.09 ± 1.30 (4,953)	2.52 ± 0.80 (12,637)	4.11 ± 1.44 (394)	3.84 ± 0.79 (53,292)
Ca (% DM)	0.10 ± 0.10 (1,346)	0.04 ± 0.04 (151)	0.03 ± 0.06 (6,655)	0.02 ± 0.02 (15,433)	0.02 ± 0.02 (713)	0.20 ± 0.05 (56,200)
P (% DM)	1.01 ± 0.18 (1,455)	0.55 ± 0.15 (230)	0.29 ± 0.05 (10,980)	0.30 ± 0.03 (19,015)	0.25 ± 0.05 (2,048)	0.22 ± 0.03 (56,422)
Mg (% DM)	0.43 ± 0.09 (1,062)	0.09 ± 0.07 (197)	0.11 ± 0.02 (8,130)	0.12 ± 0.02 (16,106)	0.09 ± 0.02 (616)	0.15 ± 0.03 (55,583)
K (% DM)	1.41 ± 0.37 (1,154)	0.26 ± 0.23 (197)	0.37 ± 0.05 (8,362)	0.39 ± 0.04 (16,525)	0.33 ± 0.06 (625)	1.05 ± 0.24 (57,514)
Na (% DM)	0.30 ± 0.23 (879)	0.05 ± 0.06 (137)	0.03 ± 0.03 (910)	0.02 ± 0.01 (381)	0.01 ± 0.00 (66)	0.01 ± 0.01 (1,606)
Cl (% DM)	0.26 ± 0.07 (129)	0.11 ± 0.04 (34)	0.15 ± 0.65 (274)	0.08 ± 0.03 (126)	0.08 ± 0.02 (97)	0.27 ± 0.11 (49,157)
S (% DM)	0.58 ± 0.14 (2,048)	0.82 ± 0.11 (470)	0.11 ± 0.12 (8,171)	0.10 ± 0.01 (13,531)	0.10 ± 0.01 (738)	0.09 ± 0.01 (54,045)
Co (mg/kg)	0.11 ± 0.03 (3)		0.51 ± 0.35 (4)			0.73 ± 0.51 (10)
Cu (mg/kg)	6.30 ± 1.89 (456)	12.70 ± 2.80 (141)	2.63 ± 2.07 (3,004)	2.22 ± 0.74 (818)	2.69 ± 0.94 (121)	6.23 ± 1.81 (1,370)
I (mg/kg)						
Fe (mg/kg)	133.33 ± 49.88 (280)	74.17	49.97 ± 38.22 (689)	49.47 ± 23.94 (359)	33.76 ± 13.45 (69)	144.28 ± 90.85 (1,344)
Mn (mg/kg)	20.07 ± 6.39 (453)	5.70 ± 5.23 (136)	7.58 ± 6.86 (3,111)	6.31 ± 2.55 (833)	7.14 ± 3.37 (120)	31.04 ± 17.56 (1,369)
Mo (mg/kg)	1.21 ± 0.18 (8)	0.84 ± 0.03 (3)	0.17 ± 0.15 (110)	0.30 ± 0.17 (29)		
Se (mg/kg)	0.22 ± 0.08 (2)		0.61 ± 0.47 (8)			0.08
Zn (mg/kg)	68.57 ± 18.51 (462)	29.08 ± 8.20 (139)	20.49 ± 8.21 (3,110)	19.44 ± 3.46 (846)	19.76 ± 5.77 (121)	26.06 ± 7.96 (1,380)

TABLE 18-1 Continued

Component	Feed Name					
	Corn Screenings	Corn Silage	Corn Snaplage	Corn Stalklage	Cornstalks	Corn Steep
DM (% AF)	86.15 ± 1.78 (538)	33.07 ± 5.59 (209,858)	58.94 ± 7.95 (2,817)	40.74 ± 15.99 (6,061)	85.81 ± 5.57 (10,601)	46.41 ± 11.66 (14)
Ash (% DM)	1.87 ± 0.76 (132)	4.24 ± 1.10 (158,269)	1.99 ± 0.39 (2,486)	12.15 ± 6.28 (1,325)	11.10 ± 4.53 (3,800)	11.29 ± 1.08 (101)
TDN (% DM)	72.8 ± 2.92 (50)	67.7 ± 2.41 (174,228)	82.0 ± 2.11 (2,504)	53.6 ± 6.73 (1,584)	52.7 ± 7.64 (1,332)	98.0 ± 5.94 (4)
DE (Mcal/kg)		2.98 ± 0.10 (173,568)	3.61 ± 0.12 (2,740)	2.36 ± 0.32 (1,589)	2.32 ± 0.35 (1,218)	
ME (Mcal/kg)	2.63	2.45	2.96	1.94	1.90	3.54
NEm (Mcal/kg)	1.72	1.56	2.00	1.09	1.06	2.47
NEg (Mcal/kg)	1.10	0.96	1.35	0.54	0.51	1.74
Sugar (% DM)	2.80	4.26 ± 3.69 (77)			3.10 ± 1.85 (5)	15.03 ± 6.68 (307)
Starch (% DM)	67.74 ± 4.50 (297)	32.58 ± 6.95 (169,620)	57.02 ± 6.13 (2,536)	5.98 ± 4.78 (643)	10.80 ± 12.36 (577)	11.40 ± 3.04 (4)
Fat (% DM)	3.32 ± 0.43 (187)	3.25 ± 0.48 (158,803)	3.46 ± 0.49 (2,500)	1.99 ± 0.93 (723)	1.44 ± 0.78 (689)	4.51 ± 2.13 (1,760)
NDF (% DM)	14.76 ± 2.83 (110)	42.98 ± 5.49 (193,210)	23.28 ± 5.33 (2,739)	63.78 ± 10.25 (2,357)	70.83 ± 11.05 (2,242)	3.55 ± 1.49 (58)
ADF (% DM)	5.32 ± 1.36 (242)	25.46 ± 3.85 (193,013)	11.24 ± 2.96 (2,739)	45.61 ± 8.56 (2,309)	46.75 ± 8.19 (3,164)	2.72 ± 1.74 (34)
Lignin (% DM)		3.17 ± 0.60 (161,976)	1.95 ± 0.47 (2,487)	6.12 ± 1.73 (1,130)	6.31 ± 2.45 (1,184)	
CP (% DM)	8.84 ± 0.97 (500)	8.24 ± 1.07 (203,902)	8.08 ± 0.81 (2,772)	6.81 ± 3.24 (2,256)	6.07 ± 2.30 (4,252)	31.78 ± 5.08 (2,779)
RDP (% CP)		74.52 ± 6.65 (154,000)	55.18 ± 15.50 (2,452)	70.91 ± 8.64 (661)	63.40 ± 11.50 (595)	
RUP (% CP)		25.38 ± 6.64 (153,970)	44.67 ± 15.52 (2,452)	29.01 ± 8.57 (660)	36.30 ± 11.05 (590)	
Soluble CP (% CP)	3.10 (2)	55.75 ± 9.74 (165,018)	42.90 ± 16.07 (2,558)	49.94 ± 10.26 (729)	42.15 ± 11.99 (717)	21.17 ± 8.20 (3)
ADICP (% DM)	7.03 ± 3.20 (4)	3.69 ± 0.88 (141,026)	2.22 ± 1.15 (2,075)	5.30 ± 1.56 (659)	6.09 ± 2.52 (501)	8.75 ± 9.51 (5)
Ca (% DM)	0.06 ± 0.08 (249)	0.24 ± 0.06 (166,171)	0.05 ± 0.02 (2,234)	1.76 ± 1.76 (1,554)	0.55 ± 0.59 (1,526)	0.10 ± 0.09 (1,205)
P (% DM)	0.25 ± 0.04 (284)	0.23 ± 0.03 (166,698)	0.27 ± 0.03 (2,266)	0.16 ± 0.08 (1,077)	0.11 ± 0.06 (1,417)	2.05 ± 0.47 (1,878)
Mg (% DM)	0.12 ± 0.02 (101)	0.17 ± 0.04 (163,010)	0.11 ± 0.01 (2,272)	0.22 ± 0.07 (850)	0.20 ± 0.08 (984)	0.90 ± 0.15 (1,268)
K (% DM)	0.39 ± 0.11 (114)	1.07 ± 0.25 (172,353)	0.51 ± 0.07 (2,250)	1.62 ± 0.65 (1,011)	1.25 ± 0.53 (1,397)	3.19 ± 0.68 (1,592)
Na (% DM)	0.03 ± 0.02 (3)	0.02 ± 0.01 (18,522)	0.01 ± 0.01 (156)	0.24 ± 0.56 (220)	0.07 ± 0.20 (392)	0.59 ± 0.25 (1,242)
Cl (% DM)	0.15 ± 0.04 (2)	0.25 ± 0.11 (127,790)	0.13 ± 0.05 (48)	0.53 ± 0.30 (584)	0.42 ± 0.25 (301)	0.40 ± 0.08 (97)
S (% DM)	0.10 ± 0.02 (35)	0.10 ± 0.05 (150,876)	0.10 ± 0.01 (2,135)	0.10 ± 0.04 (682)	0.09 ± 0.03 (832)	1.19 ± 0.45 (1,516)
Co (mg/kg)		0.71 ± 0.48 (105)			0.11	0.13
Cu (mg/kg)	3.08 ± 0.97 (16)	6.21 ± 2.52 (12,053)	3.79 ± 1.26 (247)	7.81 ± 3.08 (282)	6.47 ± 3.15 (524)	10.19 ± 3.77 (593)
I (mg/kg)						
Fe (mg/kg)		187.10 ± 121.92 (10,059)	107.92 ± 76.53 (241)	1,021.00 ± 829.18 (258)	984.15 ± 880.09 (285)	
Mn (mg/kg)	13.94 ± 2.92 (16)	32.80 ± 16.94 (12,052)	11.98 ± 4.99 (246)	63.89 ± 32.62 (275)	62.48 ± 46.70 (530)	42.78 ± 11.35 (608)
Mo (mg/kg)		0.54 ± 0.43 (132)		1.05 ± 0.99 (4)	0.46 ± 0.29 (15)	1.85 ± 0.07 (2)
Se (mg/kg)		0.16 ± 0.01 (7)			0.08	
Zn (mg/kg)	24.41 ± 4.96 (16)	26.41 ± 10.89 (12,174)	22.48 ± 4.99 (249)	30.02 ± 14.23 (278)	24.16 ± 13.50 (538)	121.04 ± 42.00 (611)

continued

TABLE 18-1 Continued

Component	Feed Name					
	Cotton Burrs	Cotton Gin Trash	Cottonseed Hulls	Cottonseed Meal	Cottonseed, Whole	Distillers Grain Solubles Dry, Corn
DM (% AF)	90.55 ± 3.73 (395)	90.87 ± 3.96 (780)	91.43 ± 1.87 (649)	88.59 ± 1.67 (2,887)	92.63 ± 2.10 (529)	89.99 ± 2.07 (28,793)
Ash (% DM)	15.34 ± 5.53 (280)	12.05 ± 6.92 (444)	3.62 ± 0.71 (286)	7.53 ± 1.76 (549)	4.12 ± 0.36 (95)	5.32 ± 0.88 (9,665)
TDN (% DM)	45.2 ± 4.04 (5)	48.5 ± 7.32 (388)	42.0	69.6 ± 6.59 (454)	93.0	89.0 ± 4.48 (6,335)
DE (Mcal/kg)		2.14 ± 0.33 (371)	1.85 ± 0.21 (152)	3.07 ± 0.36 (434)		3.92 ± 0.19 (5,976)
ME (Mcal/kg)	1.63	1.75	1.52	2.52	3.36	3.22
NEm (Mcal/kg)	0.80	0.91	0.68	1.62	2.33	2.21
NEg (Mcal/kg)	0.26	0.37	0.15	1.02	1.62	1.52
Sugar (% DM)	2.70 ± 1.21 (5)					1.16 ± 1.26 (2,137)
Starch (% DM)	6.03 ± 6.32 (9)	1.06 ± 0.77 (30)	1.13 ± 1.01 (41)	1.70 ± 4.02 (105)	2.20 (2)	5.88 ± 2.43 (7,954)
Fat (% DM)	2.48 ± 2.12 (101)	3.64 ± 2.07 (120)	2.71 ± 1.40 (291)	3.93 ± 2.74 (961)	19.45 ± 2.59 (534)	10.73 ± 2.05 (24,960)
NDF (% DM)	60.90 ± 9.03 (232)	60.57 ± 12.57 (384)	81.07 ± 6.56 (421)	33.60 ± 11.66 (481)	47.82 ± 6.96 (192)	33.66 ± 3.51 (12,700)
ADF (% DM)	55.93 ± 8.96 (139)	52.26 ± 12.63 (394)	65.10 ± 7.02 (267)	23.67 ± 9.72 (494)	42.85 ± 5.80 (90)	16.17 ± 3.15 (11,655)
Lignin (% DM)	16.60 ± 0.92 (12)	15.85 ± 5.26 (65)	19.29 ± 4.69 (47)	8.51 ± 4.18 (223)		4.96 ± 1.52 (3,226)
CP (% DM)	8.66 ± 2.03 (361)	12.29 ± 4.09 (406)	6.68 ± 2.33 (551)	44.98 ± 4.73 (2,833)	22.87 ± 2.53 (536)	30.79 ± 2.67 (24,403)
RDP (% CP)		35.57 ± 8.62 (10)	29.88 ± 10.07 (7)	57.19 ± 7.35 (63)		32.00 ± 6.26 (1,933)
RUP (% CP)		63.93 ± 8.49 (10)	70.12 ± 10.07 (7)	42.70 ± 7.32 (63)		67.93 ± 6.27 (1,933)
Soluble CP (% CP)		29.00 ± 12.06 (190)	21.51 ± 12.76 (67)	16.90 ± 6.10 (306)	32.77 ± 2.66 (3)	16.53 ± 4.73 (3,532)
ADICP (% DM)	32.08 ± 15.14 (6)	19.69 ± 9.49 (32)	17.03 ± 4.87 (16)	14.31 ± 9.55 (127)	22.87 ± 18.48 (165)	27.85 ± 12.58 (7,641)
Ca (% DM)	1.40 ± 0.48 (266)	1.57 ± 0.85 (291)	0.22 ± 0.09 (259)	0.28 ± 0.09 (675)	0.22 ± 0.12 (92)	0.05 ± 0.04 (18,286)
P (% DM)	0.17 ± 0.04 (263)	0.24 ± 0.12 (259)	0.16 ± 0.10 (263)	1.13 ± 0.21 (736)	0.53 ± 0.09 (94)	0.86 ± 0.11 (21,246)
Mg (% DM)	0.31 ± 0.08 (205)	0.31 ± 0.10 (255)	0.22 ± 0.05 (240)	0.63 ± 0.11 (464)	0.38 ± 0.13 (54)	0.32 ± 0.06 (14,340)
K (% DM)	2.26 ± 0.43 (205)	1.86 ± 0.63 (296)	1.20 ± 0.17 (255)	1.56 ± 0.26 (508)	1.12 ± 0.14 (57)	1.05 ± 0.18 (14,843)
Na (% DM)	0.04 ± 0.03 (164)	0.05 ± 0.05 (228)	0.02 ± 0.02 (157)	0.14 ± 0.12 (301)	0.08 ± 0.10 (4)	0.18 ± 0.10 (10,380)
Cl (% DM)	0.51 ± 0.44 (7)	0.47 ± 0.34 (30)	0.10 ± 0.07 (14)	0.08 ± 0.03 (78)		0.27 ± 0.26 (1,103)
S (% DM)	0.32 ± 0.10 (180)	0.39 ± 0.23 (174)	0.11 ± 0.05 (163)	0.43 ± 0.10 (320)	0.39 ± 1.15 (51)	0.66 ± 0.16 (18,646)
Co (mg/kg)	0.45	0.03				0.72 ± 0.90 (4)
Cu (mg/kg)	5.66 ± 2.29 (154)	11.28 ± 8.79 (129)	5.50 ± 2.68 (85)	12.44 ± 4.13 (161)	8.95 ± 3.72 (28)	6.69 ± 4.10 (7,055)
I (mg/kg)						
Fe (mg/kg)		818.76 ± 624.99 (124)	101.37 ± 107.57 (85)	149.92 ± 80.36 (156)		103.41 ± 37.64 (1,054)
Mn (mg/kg)	44.64 ± 18.62 (160)	73.15 ± 36.22 (131)	20.70 ± 4.93 (83)	22.30 ± 5.67 (162)	21.70 ± 31.64 (23)	22.04 ± 18.32 (6,921)
Mo (mg/kg)	1.30 ± 0.35 (3)	0.36		0.88 ± 0.13 (4)	0.23 ± 0.33 (3)	1.33 ± 0.48 (68)
Se (mg/kg)		0.49		1.29		1.14 (2)
Zn (mg/kg)	22.17 ± 21.24 (157)	28.27 ± 15.75 (130)	17.70 ± 8.36 (84)	63.90 ± 19.69 (179)	29.44 ± 8.62 (29)	63.34 ± 19.24 (7,022)

TABLE 18-1 Continued

Component	Feed Name					
	Distillers Grain Solubles Modified, Corn	Distillers Grain Solubles Wet, Corn	Distillers Solubles, Corn	Feather Meal	Fescue Hay	Field Peas, High CP
DM (% AF)	47.83 ± 4.09 (1,198)	31.44 ± 8.02 (6,616)	30.89 ± 6.02 (584)	92.00 ± 2.35 (906)	88.93 ± 3.58 (96)	88.17 ± 1.49 (111)
Ash (% DM)	6.65 ± 0.72 (16)	5.13 ± 1.13 (10,280)	9.11 ± 1.72 (2,126)	2.75 ± 1.46 (326)	8.35 ± 0.70 (4)	
TDN (% DM)	93.0 ± 5.71 (798)	98.0 ± 10.88 (3,672)	98.0 ± 7.79 (112)	79.1 ± 6.00 (605)	58.3 ± 2.52 (5)	75.8 ± 4.23 (5)
DE (Mcal/kg)	4.10	4.32 ± 0.23 (2,347)	4.32	3.49 ± 0.24 (607)		
ME (Mcal/kg)	3.36	3.54 ± 0.24 (2,347)	3.54	2.86 ± 0.25 (606)	2.11	2.74
NEm (Mcal/kg)	2.33	2.47 ± 0.19 (2,353)	2.47	1.92 ± 0.22 (601)	1.25	1.81
NEg (Mcal/kg)	1.62	1.74 ± 0.16 (2,350)	1.74	1.27 ± 0.19 (603)	0.68	1.19
Sugar (% DM)		0.90 ± 1.04 (8,662)	4.02 ± 4.41 (1,823)	17.70		
Starch (% DM)	3.36 ± 1.07 (51)	6.06 ± 2.61 (13,429)	10.68 ± 5.46 (3,220)	0.74 ± 0.43 (55)		43.90 ± 3.46 (5)
Fat (% DM)	10.22 ± 2.21 (813)	10.84 ± 1.75 (32,368)	16.85 ± 5.00 (9,764)	9.67 ± 3.04 (338)	2.10 ± 0.80 (6)	0.95 ± 0.41 (4)
NDF (% DM)	28.73 ± 3.67 (176)	31.52 ± 5.57 (8,052)	4.71 ± 2.74 (99)	25.80 (2)	64.99 ± 4.12 (14)	13.10 ± 2.83 (3)
ADF (% DM)	14.81 ± 3.06 (810)	15.27 ± 4.46 (10,162)	3.81 ± 2.14 (325)	14.25 (2)	40.30 ± 4.40 (88)	7.16 ± 1.97 (12)
Lignin (% DM)		4.70 ± 1.39 (1,304)			3.10 ± 0.57 (5)	
CP (% DM)	29.08 ± 2.45 (1,101)	30.63 ± 3.22 (34,175)	18.94 ± 4.92 (9,791)	91.07 ± 4.56 (879)	9.22 ± 3.02 (95)	25.17 ± 2.29 (112)
RDP (% CP)		29.92 ± 5.18 (593)		29.16 ± 7.24 (4)		
RUP (% CP)		69.93 ± 5.30 (594)		70.84 ± 7.24 (4)		
Soluble CP (% CP)	6.69 ± 7.03 (24)	15.58 ± 10.01 (3,448)	43.64 ± 28.99 (66)	10.31 ± 5.44 (70)		
ADICP (% DM)	2.10 ± 1.27 (2)	26.41 ± 12.79 (4,891)	10.40 ± 6.39 (451)	113.62 ± 47.71 (53)	8.65 ± 4.01 (19)	
Ca (% DM)	0.08 ± 0.05 (775)	0.05 ± 0.05 (18,972)	0.11 ± 0.07 (4,477)	0.76 ± 0.83 (73)	0.48 ± 0.18 (45)	0.14 ± 0.08 (16)
P (% DM)	0.94 ± 0.14 (798)	0.81 ± 0.18 (20,116)	1.52 ± 0.35 (4,759)	0.44 ± 0.37 (69)	0.22 ± 0.08 (45)	0.45 ± 0.09 (16)
Mg (% DM)	0.38 ± 0.07 (475)	0.35 ± 0.16 (10,941)	0.69 ± 0.16 (2,114)	0.08 ± 0.18 (66)	0.17 ± 0.06 (19)	0.15 ± 0.01 (11)
K (% DM)	1.27 ± 0.23 (492)	1.00 ± 0.26 (14,018)	2.34 ± 0.58 (4,194)	0.24 ± 0.27 (69)	1.73 ± 0.50 (38)	1.04 ± 0.13 (11)
Na (% DM)	0.39 ± 0.19 (49)	0.20 ± 0.13 (7,429)	0.42 ± 0.20 (2,238)	0.21 ± 0.14 (67)	0.10 ± 0.14 (8)	0.02
Cl (% DM)	0.16 ± 0.03 (3)	0.17 ± 0.04 (504)	0.37 ± 0.06 (522)	0.27 ± 0.13 (54)	1.45	
S (% DM)	0.67 ± 0.16 (856)	0.65 ± 0.54 (29,648)	0.82 ± 0.30 (8,715)	1.69 ± 0.35 (69)	0.16 ± 0.04 (32)	0.18 ± 0.03 (5)
Co (mg/kg)		0.11				
Cu (mg/kg)	7.03 ± 1.64 (43)	7.02 ± 2.90 (5,594)	7.41 ± 3.67 (1,865)	12.20 ± 5.07 (8)	4.75 ± 1.83 (13)	5.90 (2)
I (mg/kg)						
Fe (mg/kg)		154.92 ± 90.32 (185)		283.16 ± 115.93 (4)		
Mn (mg/kg)	19.31 ± 3.75 (42)	30.47 ± 24.85 (5,506)	35.81 ± 13.02 (1,878)	56.69 ± 69.53 (7)	90.31 ± 40.8 (13)	15.50 (2)
Mo (mg/kg)	0.97 ± 0.39 (26)	0.37 ± 0.31 (526)	1.10 ± 0.57 (13)		1.10 ± 0.35 (3)	2.10
Se (mg/kg)		0.99 (2)	6.28 ± 0.03 (2)			
Zn (mg/kg)	72.94 ± 8.07 (41)	55.06 ± 16.53 (5,510)	101.61 ± 38.93 (1,906)	171.11 ± 134.72 (7)	21.32 ± 5.26 (13)	32.30 (2)

continued

TABLE 18-1 Continued

	Feed Name					
Component	Field Peas, Low CP	Fish Meal	Flaxseed	Forage Sorghum Silage	Forage Sorghum Sudan	Forage Sorghum Sudan Hay
DM (% AF)	89.86 ± 1.70 (40)	92.30 ± 1.98 (331)	91.63 ± 2.05 (402)	28.86 ± 6.07 (12,964)	31.82 ± 15.22 (1,737)	88.77 ± 3.34 (2,606)
Ash (% DM)		20.02 ± 4.82 (220)	5.12 ± 1.30 (98)	8.89 ± 2.91 (6,273)	10.38 ± 3.00 (446)	8.83 ± 2.27 (904)
TDN (% DM)	59.0	81.9 ± 8.87 (279)	81.6 ± 8.91 (49)	57.4 ± 4.72 (8,053)	57.1 ± 5.26 (602)	56.8 ± 5.62 (1,043)
DE (Mcal/kg)		3.61 ± 0.35 (276)	3.60 ± 0.73 (186)	2.53 ± 0.21 (7,979)	2.52 ± 0.24 (585)	2.50 ± 0.26 (1,006)
ME (Mcal/kg)	2.13	2.96	2.95	2.08	2.06	2.05
NEm (Mcal/kg)	1.28	2.00	1.99	1.22	1.21	1.20
NEg (Mcal/kg)	0.71	1.35	1.34	0.66	0.65	0.64
Sugar (% DM)				1.44 ± 0.89 (50)		12.60 ± 0.99 (2)
Starch (% DM)	46.28 ± 2.35 (4)	5.82 ± 9.07 (29)	1.98 ± 1.51 (108)	9.79 ± 8.71 (6,341)	2.08 ± 1.77 (427)	2.86 ± 2.83 (826)
Fat (% DM)	1.90	11.89 ± 4.26 (255)	27.67 ± 10.19 (272)	2.90 ± 0.78 (6,469)	2.98 ± 0.85 (451)	2.02 ± 0.61 (897)
NDF (% DM)		13.60 ± 11.47 (80)	31.84 ± 7.27 (191)	57.71 ± 7.88 (9,891)	59.85 ± 6.70 (689)	62.70 ± 7.14 (1,204)
ADF (% DM)	41.30 ± 14.22 (3)	3.14 ± 2.07 (76)	18.94 ± 5.31 (194)	37.02 ± 5.51 (11,726)	35.84 ± 5.69 (820)	38.38 ± 6.05 (1,571)
Lignin (% DM)			5.75 ± 1.98 (31)	5.34 ± 1.32 (6,402)	4.24 ± 1.23 (496)	4.79 ± 1.59 (925)
CP (% DM)	14.66 ± 2.66 (31)	66.24 ± 8.07 (316)	28.68 ± 7.19 (389)	9.03 ± 2.53 (13,474)	12.77 ± 4.55 (1,007)	11.03 ± 4.26 (2,275)
RDP (% CP)		54.96 ± 12.20 (42)	65.08 ± 9.37 (38)	69.68 ± 8.58 (5,724)	70.56 ± 6.50 (445)	64.77 ± 8.11 (884)
RUP (% CP)		45.04 ± 12.20 (42)	34.77 ± 9.21 (38)	30.18 ± 8.56 (5,724)	29.28 ± 6.51 (445)	35.19 ± 8.09 (884)
Soluble CP (% CP)		25.24 ± 10.13 (116)	46.33 ± 12.30 (95)	49.56 ± 9.95 (6,336)	44.44 ± 9.17 (461)	39.32 ± 8.66 (907)
ADICP (% DM)		13.20 ± 15.00 (24)	7.44 ± 3.66 (22)	6.31 ± 2.23 (7,452)	5.43 ± 2.14 (389)	4.53 ± 1.50 (864)
Ca (% DM)	0.44 ± 0.12 (3)	5.20 ± 2.41 (143)	0.31 ± 0.12 (142)	0.44 ± 0.15 (8,536)	0.47 ± 0.13 (525)	0.41 ± 0.12 (1,107)
P (% DM)	0.26 ± 0.19 (3)	2.90 ± 1.07 (146)	0.70 ± 0.20 (144)	0.24 ± 0.07 (8,617)	0.30 ± 0.17 (538)	0.23 ± 0.07 (1,113)
Mg (% DM)	0.28 ± 0.03 (3)	0.20 ± 0.07 (75)	0.45 ± 0.11 (135)	0.26 ± 0.09 (7,857)	0.30 ± 0.17 (492)	0.29 ± 0.08 (1,002)
K (% DM)	0.95 ± 0.44 (3)	0.92 ± 0.34 (89)	0.94 ± 0.25 (135)	1.89 ± 0.66 (8,900)	2.51 ± 0.81 (572)	2.21 ± 0.63 (1,079)
Na (% DM)		0.65 ± 0.31 (100)	0.06 ± 0.03 (122)	0.02 ± 0.03 (2,151)	0.03 ± 0.03 (133)	0.05 ± 0.10 (78)
Cl (% DM)		1.15 ± 0.58 (23)	0.09 ± 0.04 (25)	0.70 ± 0.32 (5,151)	0.93 ± 0.38 (379)	0.98 ± 0.39 (558)
S (% DM)		0.82 ± 0.23 (75)	0.29 ± 0.07 (80)	0.15 ± 0.17 (6,358)	0.25 ± 0.51 (414)	0.16 ± 0.05 (941)
Co (mg/kg)		1.24 ± 0.68 (3)	0.71	1.98 ± 0.86 (2)		
Cu (mg/kg)		9.00 ± 7.12 (25)	14.93 ± 4.14 (66)	8.85 ± 3.54 (1,388)	13.29 ± 7.80 (97)	8.88 ± 3.34 (52)
I (mg/kg)						
Fe (mg/kg)		931.75 ± 827.49 (25)	109.60 ± 82.04 (63)	422.95 ± 397.98 (941)	398.01 ± 295.76 (48)	324.81 ± 271.15 (29)
Mn (mg/kg)		79.97 ± 78.73 (26)	32.92 ± 9.36 (65)	53.37 ± 23.07 (1,389)	79.01 ± 69.24 (100)	66.99 ± 29.33 (51)
Mo (mg/kg)		0.44	0.60	0.56 ± 0.56 (9)	0.70 ± 0.32 (6)	0.98 ± 0.37 (5)
Se (mg/kg)		1.37 ± 0.58 (6)				
Zn (mg/kg)		128.67 ± 69.24 (26)	52.72 ± 14.72 (66)	34.10 ± 18.56 (1,399)	38.89 ± 17.75 (99)	34.92 ± 12.58 (51)

TABLE 18-1 Continued

	Feed Name					
Component	Forage Sorghum Sudan Silage	Forage Sorghum, Fresh	Forage Sorghum Hay	Foxtail Millet	Fresh Soybean Forage	Glycerin
DM (% AF)	31.28 ± 9.80 (1,676)	30.55 ± 8.89 (6,910)	88.66 ± 4.15 (6,622)	76.01 ± 17.59 (21)	33.35 ± 11.80 (66)	80.25 ± 9.84 (121)
Ash (% DM)	10.19 ± 2.90 (1,422)	7.79 ± 2.98 (3,323)	7.59 ± 2.74 (2,928)		8.56 ± 2.35 (50)	6.69 ± 1.06 (20)
TDN (% DM)	56.1 ± 4.24 (1,649)	60.1 ± 5.49 (4,358)	58.8 ± 5.02 (3,330)	47.6	60.3 ± 4.82 (67)	69.0 ± 8.72 (7)
DE (Mcal/kg)	2.47 ± 0.19 (1,651)	2.65 ± 0.24 (4,355)	2.59 ± 0.21 (3,263)		2.66 ± 0.23 (67)	
ME (Mcal/kg)	2.03	2.17	2.13	1.72	2.18	2.49
NEm (Mcal/kg)	1.18	1.31	1.27	0.88	1.32	1.60
NEg (Mcal/kg)	0.62	0.74	0.70	0.34	0.75	1.00
Sugar (% DM)	5.77 ± 7.82 (3)					1.40(2)
Starch (% DM)	2.90 ± 2.95 (1,330)	11.98 ± 9.60 (3,340)	9.52 ± 10.40 (2,254)		5.89 ± 3.58 (51)	0.35
Fat (% DM)	3.28 ± 0.83 (1,445)	2.21 ± 0.60 (3,345)	2.09 ± 0.57 (2,263)		3.54 ± 1.72 (52)	6.24 ± 8.17 (34)
NDF (% DM)	61.13 ± 5.75 (1,705)	56.04 ± 9.24 (4,455)	58.52 ± 9.39 (3,933)	60.30	45.34 ± 7.78 (67)	0.30
ADF (% DM)	39.36 ± 4.20 (1,733)	34.93 ± 6.65 (4,687)	36.852 ± 6.45 (4,930)	42.03 ± 3.42 (6)	33.11 ± 5.88 (67)	0.15 ± 0.08 (8)
Lignin (% DM)	5.55 ± 1.20 (1,409)	4.00 ± 1.37 (3,437)	5.02 ± 1.52 (2,980)		7.07 ± 1.29 (51)	
CP (% DM)	12.27 ± 3.51 (1,739)	8.90 ± 3.32 (4,911)	9.05 ± 3.46 (5,602)	9.81 ± 3.25 (12)	19.59 ± 5.01 (67)	0.84 ± 0.84 (39)
RDP (% CP)	69.45 ± 7.08 (1,414)	65.57 ± 10.53 (3,352)	61.95 ± 9.16 (2,191)		67.07 ± 5.57 (52)	
RUP (% CP)	30.44 ± 7.01 (1,413)	34.38 ± 10.53 (3,352)	38.02 ± 9.17 (2,191)		32.79 ± 5.63 (52)	
Soluble CP (% CP)	50.46 ± 8.61 (1,499)	40.55 ± 9.94 (3,405)	36.88 ± 10.90 (2,229)		39.61 ± 7.02 (53)	100.0
ADICP (% DM)	6.45 ± 1.94 (1,542)	5.06 ± 1.96 (3,228)	4.98 ± 1.90 (2,143)		8.72 ± 2.32 (45)	
Ca (% DM)	0.55 ± 0.20 (1,466)	0.33 ± 0.11 (4,102)	0.38 ± 0.15 (3,241)	0.62 ± 0.46 (3)	1.27 ± 0.30 (60)	0.08 ± 0.13 (24)
P (% DM)	0.29 ± 0.08 (1,462)	0.22 ± 0.06 (4,116)	0.21 ± 0.08 (3,284)	0.16 ± 0.01 (3)	0.31 ± 0.08 (54)	0.19 ± 0.16 (44)
Mg (% DM)	0.27 ± 0.07 (1,428)	0.22 ± 0.08 (4,068)	0.27 ± 0.10 (2,880)	0.20	0.38 ± 0.11 (54)	0.07 ± 0.10 (7)
K (% DM)	2.48 ± 0.77 (1,585)	1.67 ± 0.58 (4,223)	1.77 ± 0.78 (3,044)	2.65	1.79 ± 0.67 (61)	0.53 ± 1.96 (36)
Na (% DM)	0.03 ± 0.03 (306)	0.02 ± 0.01 (505)	0.02 ± 0.04 (413)		0.02 ± 0.01 (6)	2.48 ± 0.66 (23)
Cl (% DM)	0.76 ± 0.32 (1,182)	0.60 ± 0.26 (2,828)	0.74 ± 0.40 (2,172)		0.29 ± 0.20 (40)	4.49 ± 0.79 (26)
S (% DM)	0.18 ± 0.04 (1,347)	0.14 ± 0.05 (3,256)	0.13 ± 0.06 (2,305)		0.21 ± 0.05 (46)	1.18 ± 0.78 (53)
Co (mg/kg)						
Cu (mg/kg)	10.57 ± 3.58 (90)	10.46 ± 7.09 (171)	8.60 ± 4.26 (257)		10.52 ± 1.87 (5)	5.08 ± 2.78 (4)
I (mg/kg)						
Fe (mg/kg)	747.05 ± 839.98 (88)	431.55 ± 420.75 (154)	258.66 ± 279.901 (206)		285.02 ± 152.71 (5)	
Mn (mg/kg)	61.34 ± 34.31 (90)	61.24 ± 27.93 (174)	57.29 ± 28.16 (252)		51.70 ± 23.85 (5)	22.25 ± 13.67 (4)
Mo (mg/kg)		1.27 ± 0.81 (3)	1.36 ± 0.53 (5)			
Se (mg/kg)	0.15		0.13			
Zn (mg/kg)	37.10 ± 11.99 (92)	35.52 ± 12.35 (171)	30.49 ± 14.13 (253)		38.22 ± 9.57 (5)	7.22 ± 8.61 (16)

continued

TABLE 18-1 Continued

Component	Feed Name					
	Grain Sorghum, Grain	Grain Sorghum Hay	Grain Sorghum Silage	Grain Sorghum Stalks	Grain Sorghum, Steam Flaked	Grain Sorghum, High Moisture
DM (% AF)	88.70 ± 1.40 (3,703)	84.79 ± 6.42 (1,366)	36.47 ± 9.37 (312)	82.63 ± 6.02 (630)	81.02 ± 5.02 (1,356)	69.91 ± 8.31 (286)
Ash (% DM)	2.09 ± 0.48 (1,482)	10.97 ± 2.70 (134)	10.08 ± 1.95 (33)	19.53 ± 5.63 (7)	1.36 ± 0.73 (7)	2.70 ± 1.41 (2)
TDN (% DM)	86.0 ± 1.86 (1,213)	54.5 ± 13.91 (212)	59.0 ± 7.65 (16)	39.5 ± 6.36 (23)	93.0	86.0
DE (Mcal/kg)	3.79 ± 0.09 (1,002)					
ME (Mcal/kg)	3.11	1.97	2.13	1.43	3.36	3.11
NEm (Mcal/kg)	2.12	1.12	1.28	0.58	2.33	2.12
NEg (Mcal/kg)	1.45	0.56	0.71	0.06	1.62	1.45
Sugar (% DM)	0.10		0.19 ± 0.08 (9)			
Starch (% DM)	71.16 ± 5.55 (2,651)	7.33 ± 5.11 (40)	4.63 ± 2.52 (22)		75.18 ± 2.73 (1,327)	72.89 ± 3.30 (60)
Fat (% DM)	3.50 ± 0.45 (2,195)	1.59 ± 0.37 (31)	2.40 ± 0.50 (42)	1.52 ± 1.26 (11)	2.62 ± 0.47 (46)	3.45 ± 0.35 (4)
NDF (% DM)	7.20 ± 3.58 (1,224)	56.44 ± 7.08 (259)	49.17 ± 7.00 (105)	64.82 ± 9.23 (77)	9.70	9.28 ± 0.31 (4)
ADF (% DM)	4.57 ± 1.81 (1,379)	36.49 ± 6.41 (797)	31.08 ± 6.20 (211)	40.11 ± 7.79 (446)	6.26 ± 0.62 (9)	5.52 ± 1.55 (64)
Lignin (% DM)	1.15 ± 0.43 (967)	2.90 ± 1.05 (167)	5.64 ± 0.70 (21)			
CP (% DM)	11.64 ± 1.83 (2,961)	8.95 ± 3.13 (1,346)	9.20 ± 2.43 (292)	7.67 ± 3.09 (791)	10.09 ± 1.12 (1,007)	10.41 ± 1.83 (224)
RDP (% CP)	28.61 ± 4.25 (855)					
RUP (% CP)	71.11 ± 4.27 (856)					
Soluble CP (% CP)	19.50 ± 5.12 (978)	38.50 ± 6.85 (6)	3.84 ± 1.47 (21)	31.95 ± 7.67 (8)		3.50 ± 0.71 (2)
ADICP (% DM)	8.66 ± 2.17 (1,172)	8.40 ± 3.06 (16)	8.95 ± 4.73 (30)	7.31 ± 2.00 (29)	82.19 ± 94.97 (6)	9.25 ± 3.17 (5)
Ca (% DM)	0.06 ± 0.05 (1,579)	0.44 ± .013 (161)	0.36 ± 0.15 (130)	0.48 ± 0.14 (130)	0.03 ± 0.03 (10)	0.04 ± 0.04 (105)
P (% DM)	0.34 ± 0.05 (1,613)	0.17 ± 0.06 (163)	0.23 ± 0.05 (133)	0.13 ± 0.06 (129)	0.26 ± 0.04 (20)	0.31 ± 0.05 (108)
Mg (% DM)	0.15 ± 0.02 (1,187)	0.26 ± 0.10 (54)	0.19 ± 0.05 (80)	0.27 ± 0.11 (30)	0.12 ± 0.01 (7)	0.14 ± 0.02 (37)
K (% DM)	0.39 ± 0.07 (1,226)	1.71 ± 0.54 (69)	1.66 ± 0.47 (107)	1.79 ± 0.49 (61)	0.38 ± 0.14 (7)	0.41 ± 0.22 (48)
Na (% DM)	0.12 ± 0.54 (102)	0.03 ± 0.00 (7)	0.02 ± 0.01 (6)	0.03 (9)		0.06 (2)
Cl (% DM)	0.32 ± 0.64 (87)			0.36)	0.46	0.08
S (% DM)	0.11 ± 0.01 (1,508)	0.10 ± 0.02 (32)	0.10 ± 0.02 (78)	0.11 ± 0.02 (45)	0.11	0.10 ± 0.02 (22)
Co (mg/kg)	0.65					
Cu (mg/kg)	4.95 ± 8.66 (546)	8.47 ± 1.94 (11)	4.75 ± 1.25 (48)	6.53 ± 1.16 (12)	3.00	8.23 ± 6.93 (3)
I (mg/kg)						
Fe (mg/kg)	42.94 ± 16.73 (704)					
Mn (mg/kg)	20.11 ± 11.98 (542)	97.64 ± 22.79 (11)	43.34 ± 15.36 (47)	86.17 ± 17.21 (12)	18.00	28.67 ± 19.63 (3)
Mo (mg/kg)	0.65	0.76 ± 0.41 (3)		1.10 ± 0.42 (8)		0.44
Se (mg/kg)						
Zn (mg/kg)	19.90 ± 6.84 (1,133)	30.44 ± 12.18 (11)	20.83 ± 6.60 (48)	21.29 ± 7.90 (12)	19.00	38.40 ± 16.17 (3)

TABLE 18-1 Continued

	Feed Name					
Component	Grape Pomace, Dry	Grape Pomace, Wet	Hominy	Johnsongrass	Kochia	Kochia Silage
DM (% AF)	91.81 ± 2.59 (67)	41.88 ± 6.68 (203)	88.74 ± 1.62 (1,725)	50.05 ± 24.36 (411)	75.00 ± 23.92 (202)	37.10 ± 7.51 (28)
Ash (% DM)	9.65 ± 4.72 (42)	15.11 ± 12.02 (72)	2.64 ± 0.73 (321)		11.82 ± 2.65 (11)	14.90
TDN (% DM)	47.0 ± 7.22 (24)	56.0 ± 10.84 (65)	87.2 ± 2.60 (820)	54.0	50.9 ± 12.89 (8)	58.0
DE (Mcal/kg)	2.07 ± 0.66 (52)	2.47 ± 0.49 (188)	3.85 ± 0.10 (791)			
ME (Mcal/kg)	1.70	2.02	3.15	1.95	1.84	2.10
NEm (Mcal/kg)	0.86	1.17	2.16	1.11	1.00	1.24
NEg (Mcal/kg)	0.32	0.61	1.48	0.55	0.45	0. 68
Sugar (% DM)			1.10 ± 1.13 (2)			
Starch (% DM)	0.97 ± 1.70 (20)	0.99 ± 0.71 (54)	56.77 ± 9.54 (689)			
Fat (% DM)	8.87 ± 2.49 (50)	8.95 ± 1.97 (91)	7.15 ± 2.12 (999)		1.10 ± 0.25 (9)	1.20
NDF (% DM)	51.78 ± 8.42 (53)	50.06 ± 10.20 (187)	16.79 ± 5.60 (874)	62.28 ± 6.01 (5)	45.98 ± 12.77 (37)	45.70
ADF (% DM)	46.28 ± 7.95 (51)	43.43 ± 10.85 (182)	5.62 ± 1.87 (1,036)	41.18 ± 5.63 (61)	33.22 ± 8.20 (96)	31.69 ± 2.94 (7)
Lignin (% DM)	31.91 ± 7.02 (23)	27.60 ± 9.87 (30)	1.48 ± 0.55 (985)			
CP (% DM)	12.27 ± 2.62 (65)	11.69 ± 2.22 (186)	10.27 ± 1.46 (1,587)	7.83 ± 3.49 (86)	13.49 ± 4.99 (150)	15.41 ± 4.11 (10)
RDP (% CP)	25.00 ± 3.46 (8)	24.47 ± 9.54 (25)	33.40 ± 7.81 (43)			
RUP (% CP)	74.87 ± 3.40 (8)	75.41 ± 9.50 (25)	66.54 ± 7.87 (43)			
Soluble CP (% CP)	20.57 ± 4.33 (26)	21.86 ± 7.06 (117)	28.02 ± 7.55 (374)			
ADICP (% DM)	35.47 ± 13.66 (20)	23.41 ± 8.62 (13)	3.25 ± 1.66 (83)		5.94 ± 1.33 (2)	8.13
Ca (% DM)	0.62 ± 0.21 (32)	0.51 ± 0.13 (122)	0.04 ± 0.03 (894)	0.59 ± 0.22 (8)	1.48 ± 0.37 (34)	1.10 ± 0.26 (7)
P (% DM)	0.27 ± 0.08 (32)	0.25 ± 0.05 (122)	0.54 ± 0.15 (1,108)	0.18 ± 0.06 (8)	0.19 ± 0.08 (34)	0.25 ± 0.03 (7)
Mg (% DM)	0.13 ± 0.04 (30)	0.12 ± 0.03 (123)	0.20 ± 0.07 (615)	0.15 ± 0.04 (2)	0.73 ± 0.26 (27)	0.59 ± 0.13 (5)
K (% DM)		1.61 ± 0.79 (141)	0.62 ± 0.18 (628)	1.30 ± 0.03 (2)	3.24 ± 1.38 (28)	3.08 ± 0.78 (6)
Na (% DM)	0.03 ± 0.02 (28)	0.05 ± 0.07 (114)	0.02 ± 0.02 (304)		0.09 ± 0.05 (7)	0.16 ± 0.085 (3)
Cl (% DM)	0.16 ± 0.21 (18)	0.05 ± 0.02 (9)	0.11 ± 0.03 (55)			
S (% DM)	0.16 ± 0.04 (24)	0.15 ± 0.04 (77)	0.12 ± 0.02 (667)	0.09	0.40 ± 0.15 (18)	0.37 ± 0.11 (6)
Co (mg/kg)						
Cu (mg/kg)	48.73 ± 71.82 (13)	20.87 ± 9.85 (77)	3.83 ± 1.29 (237)	4.00	25.06 ± 45.39 (7)	9.50 ± 0.58 (4)
I (mg/kg)						
Fe (mg/kg)	543.79 ± 495.55 (12)	601.25 ± 459.44 (77)	65.26 ± 22.52 (199)			
Mn (mg/kg)	31.45 ± 22.73 (13)	28.45 ± 9.83 (79)	10.65 ± 3.64 (232)	26.00	108.86 ± 66.11 (7)	117.00 ± 30.34 (4)
Mo (mg/kg)			0.45 ± 0.14 (26)			
Se (mg/kg)						
Zn (mg/kg)	20.24 ± 9.65 (13)	18.16 ± 5.59 (78)	37.51 ± 9.82 (240)	24.00	28.33 ± 15.54 (7)	29.75 ± 3.95 (4)

continued

TABLE 18-1 Continued

	Feed Name					
Component	Linseed Meal	Meadow Hay	Millet Forage, Fresh	Millet Forage Hay	Millet Grain	Millet Silage
DM (% AF)	90.47 ± 1.99 (165)	88.76 ± 2.86 (355)	36.19 ± 15.88 (751)	86.27 ± 4.62 (3,299)	86.70 ± 6.40 (85)	34.86 ± 10.93 (271)
Ash (% DM)	6.18 ± 1.07 (42)	9.94 ± 1.65 (65)		11.38 ± 2.65 (126)	5.34 ± 2.72 (58)	11.79 ± 3.15 (250)
TDN (% DM)	73.6 ± 7.57 (80)	52.9 ± 4.31 (50)	52.5 ± 9.51 (10)	52.5 ± 4.90 (253)	76.2 ± 7.07 (39)	53.0 ± 5.62 (315)
DE (Mcal/kg)	3.24 ± 0.32 (80)			2.32 ± 0.21 (191)	3.36 ± 0.32 (39)	2.34 ± 0.25 (313)
ME (Mcal/kg)	2.66	1.91	1.90	1.90	2.75	1.92
NEm (Mcal/kg)	1.74	1.07	1.05	1.06	1.83	1.07
NEg (Mcal/kg)	1.12	0.51	0.50	0.50	1.20	0.52
Sugar (% DM)		13.95 ± 4.45 (2)		51.60 ± 5.47 (62)		
Starch (% DM)	2.54 ± 2.39 (28)			2.91 ± 2.06 (116)	49.24 ± 9.92 (34)	3.60 ± 3.04 (236)
Fat (% DM)	11.96 ± 12.10 (95)	1.84 ± 0.35 (70)		1.72 ± 1.23 (134)	3.46 ± 1.10 (62)	2.33 ± 0.93 (257)
NDF (% DM)	32.10 ± 5.30 (80)	60.85 ± 4.57 (86)	65.28 ± 7.92 (33)	62.48 ± 6.02 (975)	21.61 ± 6.33 (36)	61.94 ± 4.96 (325)
ADF (% DM)	17.28 ± 3.95 (78)	35.79 ± 3.31 (336)	34.53 ± 7.06 (142)	37.69 ± 5.26 (1,765)	13.88 ± 5.13 (40)	40.28 ± 5.08 (357)
Lignin (% DM)	5.72 ± 2.33 (20)			6.50 ± 2.03 (120)	3.21 ± 1.15 (19)	5.99 ± 1.82 (254)
CP (% DM)	36.93 ± 5.12 (170)	8.79 ± 2.64 (368)	12.16 ± 5.32 (247)	9.53 ± 3.29 (2,369)	11.27 ± 2.04 (80)	12.28 ± 4.37 (366)
RDP (% CP)	68.06 ± 6.25 (12)			60.21 ± 7.55 (114)	28.59 ± 11.41 (2)	68.00 ± 8.33 (241)
RUP (% CP)	31.94 ± 6.25 (12)			39.63 ± 7.57 (114)	71.42 ± 11.41 (2)	31.89 ± 8.32 (241)
Soluble CP (% CP)	44.00 ± 12.42 (47)		43.65 ± 6.01 (2)	36.81 ± 9.52 (145)	19.87 ± 8.82 (24)	47.73 ± 12.43 (266)
ADICP (% DM)	8.59 ± 2.82 (23)	14.79 ± 9.57 (3)	8.03 ± 3.12 (13)	5.67 ± 1.74 (144)	18.45 ± 5.31 (21)	6.34 ± 3.03 (295)
Ca (% DM)	0.39 ± 0.09 (65)	0.50 ± 0.16 (291)	0.47 ± 0.19 (89)	0.50 ± 0.19 (1,080)	0.68 ± 0.78 (41)	0.51 ± 0.16 (281)
P (% DM)	0.88 ± 0.18 (67)	0.18 ± 0.05 (295)	0.26 ± 0.09 (90)	0.21 ± 0.06 (1,084)	0.30 ± 0.07 (37)	0.32 ± 0.08 (280)
Mg (% DM)	0.58 ± 0.10 (53)	0.19 ± 0.05 (254)	0.39 ± 0.62 (66)	0.32 ± 0.11 (862)	0.15 ± 0.04 (37)	0.37 ± 0.13 (280)
K (% DM)	1.18 ± 0.22 (56)	1.75 ± 0.50 (256)	2.84 ± 0.94 (77)	2.58 ± 0.73 (911)	0.38 ± 0.18 (40)	2.84 ± 0.93 (311)
Na (% DM)	0.09 ± 0.06 (51)	0.04 ± 0.01 (22)	0.16 (2)	0.03 ± 0.03 (63)	0.03 ± 0.03 (24)	0.03 ± 0.04 (64)
Cl (% DM)	0.08 ± 0.03 (8)			0.90 ± 0.58 (91)	0.09 ± 0.04 (21)	1.21 ± 0.66 (223)
S (% DM)	0.38 ± 0.04 (45)	0.18 ± 0.05 (204)	0.27 ± 0.65 (48)	0.20 ± 0.56 (306)	0.35 ± 0.23 (30)	0.20 ± 0.06 (249)
Co (mg/kg)						
Cu (mg/kg)	21.39 ± 3.33 (19)	4.48 ± 1.25 (184)	38.00 (2)	8.98 ± 10.56 (61)	10.26 ± 5.78 (9)	9.32 ± 2.40 (16)
I (mg/kg)						
Fe (mg/kg)	280.49 ± 249.34 (17)			450.23 ± 426.81 (23)	116.31 ± 80.73 (9)	618.24 ± 525.76 (30)
Mn (mg/kg)	50.73 ± 18.41 (19)	103.86 ± 52.52 (136)	90.00 (2)	72.73 ± 73.90 (60)	20.36 ± 4.88 (9)	97.66 ± 55.72 (31)
Mo (mg/kg)		1.81 ± 0.80 (165)	0.03	1.14 ± 0.71 (14)		
Se (mg/kg)					0.55	
Zn (mg/kg)	73.91 ± 11.60 (19)	19.62 ± 6.15 (182)	31.64 (2)	29.79 ± 13.58 (61)	39.39 ± 13.52 (9)	46.52 ± 15.83 (30)

TABLE 18-1 Continued

Component	Feed Name					
	Molasses, Beet	Molasses, Cane	Native Prairie Hay	Oat Forage	Oat Grain	Oat Hay
DM (% AF)		66.04 ± 15.09 (55)	88.01 ± 3.42 (1,647)	29.58 ± 12.64 (1,279)	89.86 ± 1.86 (1,638)	89.61 ± 2.47 (16,085)
Ash (% DM)	27.20	12.20 ± 5.34 (82)	10.26 ± 2.21 (18)	9.73 ± 2.63 (819)	3.06 ± 0.82 (845)	7.07 ± 2.12 (9,954)
TDN (% DM)	75.0	72.0	48.4 ± (4.77 (130)	61.1 ± 4.75 (1,076)	83.0 ± 4.23 (1,029)	59.9 ± 4.22 (11,168)
DE (Mcal/kg)				2.70 ± 0.25 (973)	3.66 ± 0.18 (995)	2.64 ± 0.18 (11,010)
ME (Mcal/kg)	2.71	2.60	1.75	2.21	3.00	2.17
NEm (Mcal/kg)	1.79	1.70	0.91	1.35	2.03	1.31
NEg (Mcal/kg)	1.16	1.08	0.37	0.77	1.37	0.73
Sugar (% DM)	70.62 ± 6.39 (222)	60.04 ± 8.86 (551)				
Starch (% DM)		11.98 ± 12.45 (5)		2.67 ± 1.90 (665)	44.09 ± 7.94 (782)	3.97 ± 2.57 (9,567)
Fat (% DM)		1.86 ± 1.58 (15)	1.80 ± 0.61 (24)	3.66 ± 0.92 (727)	6.16 ± 1.57 (942)	2.22 ± 0.59 (9,921)
NDF (% DM)			66.58 (4.82 (462)	52.71 ± 8.59 (990)	26.65 ± 8.62 (1,004)	59.13 ± 6.40 (12,345)
ADF (% DM)			41.45 ± 3.93 (1,380)	34.02 ± 5.49 (1,028)	13.30 ± 4.77 (1,109)	37.08 ± 4.66 (13,620)
Lignin (% DM)			2.05	4.21 ± 1.33 (763)	3.00 ± 0.77 (576)	4.69 ± 1.51 (9,988)
CP (% DM)	10.86 ± 2.92 (410)	8.59 ± 3.54 (820)	6.76 ± 2.02 (1,555)	16.46 ± 5.76 (1,152)	12.55 ± 1.89 (1,493)	8.73 ± 2.56 (14,209)
RDP (% CP)				73.35 ± 6.43 (717)	43.54 ± 6.21 (470)	65.92 ± 7.39 (9,305)
RUP (% CP)				26.40 ± 6.48 (717)	56.47 ± 5.90 (467)	33.91 ± 7.48 (9,306)
Soluble CP (% CP)			22.95 ± 8.83 (11)	50.91 ± 8.76 (751)	27.35 ± 6.08 (537)	40.84 ± 6.31 (9,972)
ADICP (% DM)			6.53 ± 1.60 (78)	5.55 ± 2.06 (584)	4.27 ± 1.95 (439)	3.73 ± 1.36 (8,084)
Ca (% DM)	0.11 ± 0.13 (155)	0.88 ± 0.29 (382)	0.49 ± 0.13 (798)	0.48 ± 0.17 (742)	0.10 ± 0.05 (876)	0.29 ± 0.13 (10,533)
P (% DM)	0.14 ± 0.24 (39)	0.22 ± 0.33 (197)	0.13 ± 0.05 (807)	0.33 ± 0.08 (750)	0.38 ± 0.09 (894)	0.21 ± 0.06 (10,808)
Mg (% DM)	0.09 ± 0.16 (29)	0.40 ± 0.17 (295)	0.17 ± 0.08 (632)	0.20 ± 0.05 (621)	0.14 ± 0.03 (821)	0.14 ± 0.05 (10,309)
K (% DM)	3.45 ± 0.68 (229)	4.46 ± 1.52 (415)	1.04 ± 0.50 (675)	2.77 ± 0.89 (812)	0.50 ± 0.11 (822)	1.65 ± 0.61 (12,061)
Na (% DM)	1.03 ± 0.50 (6)	1.04 ± 0.92 (50)	0.05 ± 0.07 (97)	0.21 ± 0.23 (143)	0.02 ± 0.01 (317)	0.41 ± 0.26 (3,000)
Cl (% DM)	1.09 ± 1.87 (32)	2.21 ± 1.07 (64)	0.51 ± 0.44 (5)	1.02 ± 0.50 (586)	0.14 ± 0.05 (83)	0.92 ± 0.47 (8,507)
S (% DM)	0.41 ± 0.11 (166)	0.68 ± 0.29 (342)	0.12 ± 0.05 (321)	0.22 ± 0.07 (580)	0.17 ± 0.05 (530)	0.13 ± 0.05 (8,778)
Co (mg/kg)					0.52 ± 0.30 (29)	0.88 ± 0.55 (19)
Cu (mg/kg)	3.00	6.55 ± 4.44 (34)	3.75 ± 1.45 (205)	8.96 ± 3.61 (73)	6.18 ± 2.30 (209)	6.91 ± 3.08 (1,197)
I (mg/kg)						
Fe (mg/kg)				603.40 ± 713.08 (67)	105.03 ± 57.06 (205)	249.09 ± 204.60 (1,017)
Mn (mg/kg)	3.00 ± 1.00 (3)	23.19 ± 22.45 (36)	65.31 ± 38.35 (206)	63.16 ± 39.86 (73)	50.29 ± 14.40 (207)	65.28 ± 28.66 (1,184)
Mo (mg/kg)		1.36	1.66 ± 1.20 (99)	2.97 ± 0.67 (3)	1.70	1.24 ± 0.72 (69)
Se (mg/kg)		5.33		1.15	0.28	0.14 ± 0.16 (3)
Zn (mg/kg)	14.00 ± 6.71 (5)	17.36 ± 10.62 (43)	20.61 ± 15.08 (210)	27.86 ± 8.56 (73)	31.07 ± 9.12 (211)	21.19 ± 9.80 (1,190)

continued

TABLE 18-1 Continued

	Feed Name					
Component	Oat Hulls	Oat Straw	Oatlage	Onions	Orchardgrass Hay	Pea Hulls
DM (% AF)	91.60 ± 1.65 (290)	84.19 ± 9.18 (51)	33.84 ± 10.17 (6,018)	23.17 ± 9.18 (97)	91.47 ± 2.16 (177)	88.25 ± 1.88 (60)
Ash (% DM)	5.24 ± 1.92 (39)	6.92 ± 1.27 (3)	9.80 ± 2.46 (4,978)	6.16 ± 3.03 (17)	10.54 ± 2.74 (6)	4.20 ± 1.12 (4)
TDN (% DM)	56.5 ± 5.12 (110)	44.3 ± 5.04 (3)	58.0 ± 4.32 (5,384)	68.9	56.2 ± 4.27 (103)	71.0
DE (Mcal/kg)	2.49 ± 0.25 (111)		2.56 ± 0.19 (5,384)			
ME (Mcal/kg)	2.04	1.60	2.10	2.49	2.03	2.57
NEm (Mcal/kg)	1.19	0.76	1.24	1.60	1.18	1.66
NEg (Mcal/kg)	0.63	0.23	0.67	1.00	0.62	1.05
Sugar (% DM)						
Starch (% DM)	15.83 ± 8.37 (40)	1.35 ± 0.35 (2)	3.11 ± 2.41 (4,782)	54.53 ± 5.92 (3)		40.00
Fat (% DM)	2.80 ± 1.44 (43)	1.33 ± 0.35 (3)	3.67 ± 0.85 (5,060)	13.63 ± 7.53 (88)	2.30 ± 0.62 (7)	1.33 ± 0.64 (7)
NDF (% DM)	64.44 ± 11.93 (116)	73.75 ± 5.12 (22)	58.88 ± 5.90 (5,623)	10.54 ± 7.93 (19)	57.35 ± 5.82 (39)	54.93 ± 15.60 (4)
ADF (% DM)	35.87 ± 6.90 (134)	49.29 ± 5.55 (35)	38.49 ± 4.27 (5,951)	8.84 ± 8.11 (19)	36.73 ± 4.09 (167)	18.82 ± 16.81 (32)
Lignin (% DM)	5.54 ± 1.82 (25)	7.07	5.33 ± 1.27 (4,989)		6.02	
CP (% DM)	6.10 ± 2.12 (284)	4.83 ± 2.55 (39)	12.70 ± 3.23 (6,065)	10.14 ± 1.75 (93)	13.77 ± 3.60 (173)	20.10 ± 6.16 (61)
RDP (% CP)	53.62		77.47 ± 7.67 (4,803)			
RUP (% CP)	46.38		22.38 ± 7.68 (4,802)			
Soluble CP (% CP)	31.57 ± 15.58 (31)	53.00	63.41 ± 9.86 (5,124)			
ADICP (% DM)	7.02 ± 0.81 (13)	1.90 (2)	5.48 ± 1.83 (4,775)		8.13	
Ca (% DM)	0.14 ± 0.07 (80)	0.30 ± 0.12 (27)	0.46 ± 0.18 (4,921)	0.51 ± 0.51 (30)	0.53 ± 0.16 (127)	0.20 ± 0.16 (34)
P (% DM)	0.20 ± 0.09 (80)	0.14 ± 0.06 (28)	0.32 ± 0.07 (4,947)	0.32 ± 0.14 (30)	0.24 ± 0.06 (128)	0.39 ± 0.13 (34)
Mg (% DM)	0.13 ± 0.05(79)	0.13 ± 0.06 (26)	0.18 ± 0.05 (4,812)	0.17 ± 0.08 (7)	0.26 ± 0.07 (38)	0.20 ± 0.08 (29)
K (% DM)	0.60 ± 0.12 (77)	2.23 ± 0.59 (26)	2.53 ± 0.77 (5,630)	1.26 ± 0.71 (7)	2.60 ± 0.56 (40)	0.97 ± 0.25 (29)
Na (% DM)	0.05 ± 0.14 (69)	0.22 (2)	0.23 ± 0.24 (1,066)		0.05 ± 0.02 (5)	
Cl (% DM)	0.15 ± 0.04 (11)	0.27	0.91 ± 0.42 (4,240)			
S (% DM)	0.09 ± 0.02(31)	0.13 ± 0.02 (6)	0.19 ± 0.13 (4,521)	0.34 ± 0.19 (7)	0.22 ± 0.07 (13)	0.17 ± 0.06 (8)
Co (mg/kg)			0.49			
Cu (mg/kg)	7.52 ± 3.21 (45)	8.88 ± 11.53 (4)	8.16 ± 3.23 (353)		6.62 ± 1.37 (13)	6.50 ± 3.32 (4)
I (mg/kg)						
Fe (mg/kg)	241.58 ± 125.71 (36)		561.68 ± 553.43 (327)			
Mn (mg/kg)	54.11 ± 18.17 (45)	70.33 ± 40.99 (3)	71.29 ± 38.72 (351)		157.73 ± 38.14 (11)	13.25 ± 3.77 (4)
Mo (mg/kg)		0.01	1.07 ± 0.73 (4)		0.76 ± 0.15 (5)	10.10 ± 0.71 (2)
Se (mg/kg)						
Zn (mg/kg)	25.20 ± 8.11 (44)	23.50 ± 10.39 (3)	29.79 ± 10.15 (355)		20.87 ± 6.43 (13)	41.25 ± 10.50 (4)

TABLE 18-1 Continued

Component	Feed Name					
	Peanut Hay	Peanut Hulls	Peanut Meal	Peanut Silage	Peas	Peavine Forage
DM (% AF)	90.78 ± 1.91 (575)	93.41 ± 1.92 (200)	94.11 ± 0.93 (551)	38.03 ± 10.76 (16)	87.28 ± 10.06 (298)	18.18 ± 9.67 (37)
Ash (% DM)	10.91 ± 3.89 (274)	3.77 ± 0.81 (36)	5.55 ± 1.26 (44)	9.63 ± 1.88 (5)	3.75 ± 1.06 (62)	9.64 ± 1.83 (33)
TDN (% DM)	57.5 ± 4.80 (448)	42.8 ± 3.55 (172)	80.8 ± 7.72 (54)	56.3 ± 3.87 (18)	80.0 ± 4.41 (199)	62.1 ± 3.74 (37)
DE (Mcal/kg)	2.54 ± 0.22 (448)	1.89 ± 0.17 (163)	3.56 ± 0.60 (58)	2.48 ± 0.18 (18)	3.53 ± 0.15 (183)	2.74 ± 0.18 (37)
ME (Mcal/kg)	2.08	1.55	2.92	2.04	2.89	2.25
NEm (Mcal/kg)	1.23	0.71	1.96	1.19	1.94	1.38
NEg (Mcal/kg)	0.66	0.17	1.32	0.62	1.30	0.80
Sugar (% DM)						
Starch (% DM)	4.00 ± 2.87 (236)	1.24 ± 0.77 (13)	6.93 ± 3.56 (6)	2.25 ± 0.77 (4)	42.66 ± 11.04 (95)	3.47 ± 1.86 (31)
Fat (% DM)	2.04 ± 0.90 (253)	1.16 ± 1.49 (137)	7.73 ± 2.40 (193)	3.74 ± 0.72 (5)	1.86 ± 1.04 (99)	3.81 ± 1.31 (34)
NDF (% DM)	47.40 ± 7.35 (502)	68.46 ± 5.20 (173)	19.89 ± 4.71 (55)	53.59 ± 6.22 (18)	13.67 ± 6.02 (193)	44.31 ± 7.78 (37)
ADF (% DM)	39.13 ± 6.47 (509)	58.87 ± 5.58 (180)	13.15 ± 4.08 (54)	40.33 ± 3.72 (18)	9.23 ± 6.26 (219)	31.96 ± 5.39 (37)
Lignin (% DM)	8.45 ± 2.09 (256)	23.02 ± 3.40 (25)	3.30 ± 1.33 (9)	9.13 ± 3.09 (6)	1.06 ± 0.71 (28)	5.75 ± 0.99 (33)
CP (% DM)	11.04 ± 3.22 (560)	9.46 ± 1.92 (190)	44.96 ± 3.80	13.69 ± 2.42 (19)	23.91 ± 3.04 (278)	20.13 ± 4.01 (35)
RDP (% CP)	66.13 ± 4.31 (227)	47.01 ± 8.49 (2)	71.80	65.57 ± 12.90 (5)	84.40 ± 2.89 (9)	73.67 ± 5.05 (32)
RUP (% CP)	33.51 ± 4.33 (227)	52.50 ± 7.79 (2)	28.20	34.43 ± 12.90 (5)	15.49 ± 3.02 (9)	26.33 ± 5.05 (32)
Soluble CP (% CP)	35.22 ± 8.10 (251)	24.45 ± 10.16 (122)	29.83 ± 10.66 (22)	45.36 ± 18.11 (6)	69.73 ± 10.27 (73)	51.48 ± 10.38 (35)
ADICP (% DM)	9.01 ± 2.65 (231)	19.48 ± 7.55 (6)	12.53 ± 8.60 (18)	10.42 ± 3.51 (18)	4.38 ± 5.16 (13)	6.27 ± 2.07 (32)
Ca (% DM)	1.24 ± 0.34 (423)	0.29 ± 0.27 (159)	0.21 ± 0.12 (32)	1.33 ± 0.31 (15)	0.13 ± 0.08 (153)	1.07 ± 0.31 (37)
P (% DM)	0.15 ± 0.06 (324)	0.10 ± 0.05 (161)	0.59 ± 0.19 (32)	0.27 ± 0.06 (15)	0.42 ± 0.10 (160)	0.35 ± 0.08 (37)
Mg (% DM)	0.55 ± 0.23 (299)	0.14 ± 0.10 (154)	0.32 ± 0.06 (32)	0.39 ± 0.09 (15)	0.18 ± 0.14 (127)	0.26 ± 0.06 (37)
K (% DM)	1.52 ± 0.49 (405)	0.70 ± 0.21 (157)	1.11 ± 0.20 (40)	1.84 ± 0.58 (16)	1.07 ± 0.23 (135)	2.81 ± 0.91 (37)
Na (% DM)	0.05 ± 0.04 (148)	0.03 ± 0.02 (90)	0.03 ± 0.03 (29)	0.03 ± 0.02 (10)	0.03 ± 0.03 (69)	0.02 ± 0.01 (2)
Cl (% DM)	0.68 ± 0.37 (144)	0.10 ± 0.07 (7)	0.21 ± 0.25 (6)	0.74 ± 0.44 (5)	0.13 ± 0.03 (17)	0.54 ± 0.30 (32)
S (% DM)	0.14 ± 0.04 (205)	0.10 ± 0.03 (132)	0.27 ± 0.08 (20)	0.17 ± 0.07 (6)	0.57 ± 1.16 (78)	0.21 ± 0.05 (33)
Co (mg/kg)			1.19			
Cu (mg/kg)	8.36 ± 4.85 (182)	12.52 ± 3.90 (95)	14.10 ± 4.24 (30)	27.75 ± 13.72 (6)	8.80 ± 2.20 (69)	
I (mg/kg)						
Fe (mg/kg)	695.35 ± 676.09 (140)	634.00 ± 284.55 (89)	462.90 ± 391.68 (29)	396.26 ± 174.55 (6)	112.70 ± 79.51 (60)	
Mn (mg/kg)	60.34 ± 34.48 (184)	50.00 ± 11.62 (95)	32.11 ± 12.76 (31)	37.53 ± 29.69 (6)	21.47 ± 11.37 (67)	
Mo (mg/kg)					0.81 (5)	
Se (mg/kg)	0.10					
Zn (mg/kg)	24.28 ± 10.41 (183)	16.09 ± 4.58 (93)	53.45 ± 10.35 (31)	79.59 ± 45.33 (6)	36.26 ± 8.65 (69)	

continued

TABLE 18-1 Continued

Component	Feed Name					
	Peavine Hay	Peavine Silage	Popcorn Grain	Potato Byproduct, Dry	Potato Byproduct, Wet	Potato Peels
DM (% AF)	90.47 ± 2.09 (130)	34.11 ± 12.91 (119)	87.81 ± 1.67 (153)	90.90 ± 3.80 (121)	23.38 ± 9.54 (394)	43.37 ± 36.74 (62)
Ash (% DM)	9.48 ± 2.37 (56)	10.75 ± 3.39 (138)		5.71 ± 4.16 (76)	7.69 ± 5.47 (199)	6.55 ± 2.19 (2)
TDN (% DM)	58.6 ± 4.74 (99)	59.2 ± 6.32 (161)	72.7 ± 0.85 (2)	79.2 ± 9.90 (91)	79.1 ± 8.73 (308)	78.0
DE (Mcal/kg)	2.58 ± 0.27 (100)	2.61 ± 0.28 (161)		3.49 ± 0.62 (97)	3.49 ± 0.43 (330)	
ME (Mcal/kg)	2.12	2.14	2.63	2.86	2.86	2.82
NEm (Mcal/kg)	1.26	1.28	1.72	1.92	1.91	1.88
NEg (Mcal/kg)	0.69	0.71	1.10	1.28	1.27	1.24
Sugar (% DM)						
Starch (% DM)	6.50 ± 4.15 (75)	5.58 ± 5.02 (137)	65.50 ± 1.10 (3)	44.34 ± 23.79 (57)	44.27 ± 19.34 (223)	
Fat (% DM)	2.04 ± 0.62 (56)	3.76 ± 0.87 (140)	3.50 ± 0.42 (2)	3.87 ± 6.99 (85)	7.95 ± 8.82 (240)	0.55 ± 0.35 (2)
NDF (% DM)	45.87 ± 9.10 (100)	49.19 ± 8.58 (164)		18.38 ±12.02 (95)	21.66 ± 11.84 (329)	16.00
ADF (% DM)	34.88 ± 6.37 (100)	35.27 ± 6.11 (163)	4.50 ± 0.59 (9)	13.31 ± 11.11 (96)	15.46 ± 9.57 (319)	10.00
Lignin (% DM)	6.65 ± 1.99 (57)	6.09 ± 1.67 (140)		3.20 ± 2.34 (54)	3.25 ± 2.03 (87)	
CP (% DM)	15.60 ± 5.46 (100)	19.68 ± 5.18 (238)	11.39 ± 0.99 (12)	8.13 ± 4.20 (98)	12.08 ± 6.80 (381)	13.80 ± 4.38 (2)
RDP (% CP)	70.07 ± 8.97 (73)	75.95 ± 7.89 (205)		51.47 ± 21.48 (13)	46.48 ± 13.14 (50)	
RUP (% CP)	29.85 ± 8.95 (73)	24.01 ± 7.89 (205)		48.38 ± 21.24 (13)	53.32 ± 13.16 (50)	
Soluble CP (% CP)	42.93 ± 9.56 (76)	59.80 ± 11.11 (224)		50.12 ± 20.48 (69)	40.64 ± 19.11 (207)	
ADICP (% DM)	6.35 ± 2.47 (71)	6.42 ± 2.38 (228)		4.49 ± 3.69 (44)	6.22 ± 5.52 (57)	
Ca (% DM)	1.21 ± 0.54 (97)	1.00 ± 0.35 (154)	0.04 ± 0.02 (3)	0.17 ± 0.17 (88)	0.41 ± 0.38 (207)	0.29 ± 0.07 (60)
P (% DM)	0.27 ± 0.11 (97)	0.32 ± 0.07 (153)	0.27 ± 0.04 (5)	0.22 ± 0.17 (90)	0.35 ± 0.24 (206)	0.23 ± 0.05 (2)
Mg (% DM)	0.32 ± 0.11 (89)	0.13 ± 0.02 (4)	0.13 ± 0.03 (3)	0.13 ± 0.10 (87)	0.13 ± 0.09 (199)	0.20 ± 0.05 (59)
K (% DM)	2.03 ± 0.79 (96)	2.55 ± 0.85 (160)	0.25 ± 0.07 (4)	1.39 ± 0.76 (100)	1.19 ± 1.01 (216)	1.85
Na (% DM)	0.04 ± 0.05 (47)	0.05 ± 0.04 (30)		0.12 ± 0.18 (84)	0.09 ± 0.11 (196)	
Cl (% DM)	0.48 ± 0.28 (42)	0.68 ± 0.31 (124)	1.27 ± 1.58 (2)	0.17 ± 0.15 (40)	0.17 ± 0.10 (69)	
S (% DM)	0.18 ± 0.05 (79)	0.21 ± 0.05 (137)	0.11	0.12 ± 0.10 (70)	0.16 ± 0.11 (144)	0.15
Co (mg/kg)				9.74		
Cu (mg/kg)	8.80 ± 3.94 (7)	8.87 ± 2.50 (11)		11.44 ± 14.68 (43)	9.48 ± 7.09 (90)	10.31 ± 3.29 (59)
I (mg/kg)						
Fe (mg/kg)	1,452.00 ± 1,510.00 (7)	1,346.00 ± 1,306.00 (11)		339.54 ± 488.56 (39)	1,234.00 ± 1,801.00 (95)	
Mn (mg/kg)	102.48 ± 56.58 (7)	68.98 ± 35.52 (11)		23.39 ± 30.57 (42)	31.50 ± 32.99 (88)	45.00
Mo (mg/kg)						
Se (mg/kg)					0.04	
Zn (mg/kg)	44.04 ± 25.80 (7)	32.81 ± 9.17 (11)		20.23 ± 12.55 (43)	37.86 ± 35.76 (85)	24.97 ± 7.54 (59)

TABLE 18-1 Continued

Component	Feed Name					
	Potatoes	Rice Bran	Rice Grain	Rice Hay	Rice Hulls	Rice Silage
DM (% AF)	23.54 ± 7.57 (233)	91.76 ± 1.74 (1,371)	88.81 ± 1.98 (211)	92.68 ± 1.62 (97)	91.95 ± 1.49 (37)	39.70 ± 12.38 (344)
Ash (% DM)	6.30 ± 2.96 (55)	12.17 ± 4.68 (315)	3.19 ± 3.04 (87)	14.54 ± 2.48 (60)	15.71 ± 5.27 (4)	14.43 ± 3.78 (276)
TDN (% DM)	76.7 ± 3.65 (106)	83.4 ± 12.21 (409)	82.7 ± 5.83 (142)	54.7 ± 4.14 (90)	31.5 ± 21.22 (13)	56.0 ± 5.52 (458)
DE (Mcal/kg)	3.38 ± 0.16 (97)	3.68 ± 0.57 (442)	3.65 ± 0.24 (137)	2.41 ± 0.18 (90)		2.47 ± 0.25 (461)
ME (Mcal/kg)	2.77	3.01	2.99	1.98	1.14	2.03
NEm (Mcal/kg)	1.84	2.04	2.02	1.13	0.28	1.18
NEg (Mcal/kg)	1.21	1.38	1.37	0.57	0	0.61
Sugar (% DM)						
Starch (% DM)	60.87 ± 9.55 (141)	20.17 ± 8.85 (199)	77.19 ± 9.95 (112)	8.95 ± 9.80 (57)		18.65 ± 11.79 (249)
Fat (% DM)	7.52 ± 7.45 (111)	17.64 ± 5.29 (637)	1.84 ± 0.92 (81)	1.80 ± 0.44 (61)	4.31 ± 3.95 (15)	2.44 ± 0.60 (276)
NDF (% DM)	11.19 ± 7.25 (110)	26.62 ± 11.39 (448)	8.17 ± 7.63 (136)	62.31 ± 8.32 (91)	53.84 ± 17.91 (5)	55.08 ± 10.44 (462)
ADF (% DM)	7.32 ± 5.11 (123)	15.51 ± 8.74 (468)	5.90 ± 5.47 (141)	43.34 ± 6.32 (91)	52.55 ± 14.68 (22)	40.34 ± 7.34 (461)
Lignin (% DM)	1.10 ± 1.14 (28)	5.34 ± 2.52 (116)	1.88 ± 1.78 (66)	3.80 ± 0.97 (56)		4.74 ± 1.14 (259)
CP (% DM)	10.11 ± 3.27 (142)	14.72 ± 2.99 (797)	8.37 ± 1.65 (161)	7.39 ± 2.12 (94)	5.39 ± 3.66 (36)	7.28 ± 1.74 (460)
RDP (% CP)	37.60 ± 1.96 (3)	55.30 ± 18.69 (8)	40.31 ± 14.66 (9)	61.57 ± 10.88 (6)		62.64 ± 9.53 (105)
RUP (% CP)	62.40 ± 1.96 (3)	44.70 ± 18.69 (8)	59.58 ± 14.94 (9)	38.26 ± 10.96 (6)		37.15 ± 9.51 (105)
Soluble CP (% CP)	64.71 ± 19.46 (58)	25.79 ± 12.18 (184)	19.29 ± 10.62 (83)	29.99 ± 8.34 (80)		42.48 ± 12.32 (448)
ADICP (% DM)	1.77 ± 0.73 (6)	5.04 ± 2.24 (63)	2.62 ± 1.24 (48)	3.81 ± 1.76 (55)	11.25	4.26 ± 1.58 (399)
Ca (% DM)	0.14 ± 0.14 (88)	2.04 ± 1.45 (881)	0.03 ± 0.03 (87)	0.24 ± 0.11 (82)	0.18 ± 0.14 (24)	0.24 ± 0.07 (378)
P (% DM)	0.27 ± 0.09 (91)	1.61 ± 0.71 (407)	0.25 ± 0.10 (90)	0.21 ± 0.08 (82)	0.31 ± 0.36 (25)	0.23 ± 0.07 (377)
Mg (% DM)	0.11 ± 0.04 (73)	0.73 ± 0.33 (333)	0.10 ± 0.06 (89)	0.17 ± 0.04 (82)	0.19 ± 0.18 (16)	0.16 ± 0.04 (375)
K (% DM)	1.88 ± 0.81 (93)	1.28 ± 0.49 (373)	0.27 ± 0.12 (91)	1.87 ± 0.66 (83)	0.54 ± 0.30 (20)	1.48 ± 0.48 (454)
Na (% DM)	0.04 ± 0.03 (51)	0.06 ± 0.06 (269)	0.01 ± 0.01 (46)	0.09 ± 0.19 (72)	0.04 ± 0.04 (2)	0.02 ± 0.02 (351)
Cl (% DM)	0.26 ± 0.16 (11)	0.13 ± 0.07 (64)	0.08 ± 0.04 (39)	0.62 ± 0.31 (48)		0.45 ± 0.16 (144)
S (% DM)	0.14 ± 0.05 (52)	0.16 ± 0.05 (236)	0.10 ± 0.02 (74)	0.16 ± 0.06 (77)	0.09 ± 0.04 (9)	0.13 ±0.04 (363)
Co (mg/kg)	0.57 ± 0.52 (2)	1.09 ± 1.45 (6)				
Cu (mg/kg)	5.97 ± 2.02 (36)	9.53 ± 3.88 (163)	5.46 ± 2.92 (47)	17.07 ± 7.85 (26)		13.34 ± 7.49 (273)
I (mg/kg)						
Fe (mg/kg)	134.46 ± 155.68 (33)	222.41 ± 175.78 (145)	144.14 ± 161.86 (55)	328.76 ± 218.87 (24)		489.58 ± 284.93 (270)
Mn (mg/kg)	12.13 ± 10.11 (34)	186.55 ± 67.42 (169)	66.33 ± 46.25 (58)	350.34 ± 247.40 (23)		627.05 ± 327.57 (276)
Mo (mg/kg)		1.09 (2)				
Se (mg/kg)	0.01	0.08				
Zn (mg/kg)	15.03 ± 4.81 (35)	66.73 ± 29.55 (167)	23.85 ± 7.62 (58)	38.30 ± 10.09 (26)		43.61 ± 13.65 (275)

continued

TABLE 18-1 Continued

Component	Feed Name					
	Rye Annual, Fresh	Ryegrass Hay	Ryegrass Silage	Rye Grain	Safflower Meal	Soybean Hay
DM (% AF)	31.97 ± 14.49 (2,563)	90.38 ± 2.03 (6,619)	36.82 ± 12.01 (6,182)	89.90 ± 3.05 (57)	93.50 ± 3.22 (85)	90.50 ± 3.37 (1,002)
Ash (% DM)	10.24 ± 2.40 (1,563)	9.60 ± 2.42 (3,454)	10.06 ± 2.43 (5,639)	1.77 ± 0.28 (16)	4.85 ± 1.11 (44)	8.79 ± 1.75 (347)
TDN (% DM)	65.1 ± 4.20 (2,351)	63.7 ± 5.79 (4,991)	59.6 ± 4.80 (6,656)	80.8 ± 4.43 (28)	55.8 ± 5.67 (43)	60.1 ± 5.36 (648)
DE (Mcal/kg)	2.87 ± 0.21 (2,357)	2.81 ± 0.34 (4,981)	2.63 ± 0.23 (6,632)	3.56 ± 0.19 (28)	2.46 ± 0.68 (48)	2.65 ± 0.29 (644)
ME (Mcal/kg)	2.35	2.30	2.16	2.92	2.02	2.17
NEm (Mcal/kg)	1.48	1.43	1.30	1.97	1.17	1.31
NEg (Mcal/kg)	0.89	0.85	0.72	1.32	0.61	0.74
Sugar (% DM)			1.50			
Starch (% DM)	1.87 ± 1.36 (1,275)	2.26 ± 1.53 (3,175)	1.64 ± 1.17 (5,351)	58.25 ± 5.35 (16)	1.15 ± 0.81 (12)	5.37 ± 2.47 (343)
Fat (% DM)	3.98 ± 0.94 (1,340)	3.35 ± 1.35 (3,438)	3.87 ± 1.00 (5,796)	1.44 ± 0.36 (17)	12.34 ± 11.93 (62)	3.04 ± 1.29 (351)
NDF (% DM)	51.69 ± 8.26 (2,399)	51.50 ± 10.14 (5,075)	57.69 ± 7.27 (6,783)	15.39 ± 4.50 (26)	51.48 ± 9.51 (53)	44.85 ± 10.65 (774)
ADF (% DM)	32.00 ± 5.81 (2,239)	30.89 ± 7.76 (5,010)	37.47 ± 5.19 (6,913)	7.53 ± 7.51 (32)	37.62 ± 7.92 (62)	37.05 ± 9.26 (748)
Lignin (% DM)	3.70 ± 1.23 (1,337)	4.32 ± 1.62 (3,420)	4.77 ± 1.42 (5,551)	1.57 ± 0.71 (14)	13.45 ± 2.15 (13)	7.28 ± 1.37 (336)
CP (% DM)	18.67 ± 6.05 (2,527)	18.65 ± 8.30 (6,483)	14.60 ± 4.29 (7,124)	11.32 ± 1.68 (41)	23.12 ± 5.93 (82)	16.54 ± 6.41 (1,000)
RDP (% CP)	74.27 ± 5.67 (1,286)	66.52 ± 5.85 (2,618)	76.11 ± 7.32 (5,550)	77.10 ± 5.80 (11)	76.29 ± 8.33 (7)	65.58 ± 5.63 (363)
RUP (% CP)	25.57 ± 5.66 (1,286)	33.30 ± 5.82 (2,616)	23.76 ± 7.31 (5,549)	22.53 ± 5.74 (11)	23.71 ± 8.33 (7)	34.40 ± 5.64 (363)
Soluble CP (% CP)	49.06 ± 10.37 (1,426)	35.59 ± 7.63 (3,363)	60.47 ± 11.97 (6,026)	46.49 ± 13.72 (20)	33.25 ± 12.41 (23)	35.25 ± 7.40 (348)
ADICP (% DM)	5.77 ± 2.30 (1,187)	6.13 ± 2.61 (2,956)	5.80 ± 2.04 (6,138)	1.39 ± 0.61 (9)	8.47 ± 1.50 (9)	8.70 ± 2.62 (353)
Ca (% DM)	0.52 ± 0.18 (1,611)	0.51 ± 0.02 (3,074)	0.50 ± 0.16 (5,821)	0.09 ± 0.13 (28)	0.34 ± 0.08 (37)	1.38 ± 0.31 (458)
P (% DM)	0.40 ± 0.10 (1,606)	0.32 ± 0.12 (3,286)	0.34 ± 0.08 (5,865)	0.36 ± 0.04 (26)	0.63 ± 0.18 (40)	0.25 ± 0.09 (448)
Mg (% DM)	0.21 ± 0.05 (1,585)	0.22 ± 0.07 (2,911)	0.19 ± 0.09 (5,745)	0.14 ± 0.05 (25)	0.34 ± 0.11 (28)	0.38 ± 0.11 (405)
K (% DM)	3.03 ± 0.90 (1,825)	2.38 ± 0.92 (4,855)	2.78 ± 0.80 (6,529)	0.53 ± 0.05 (24)	0.96 ± 0.26 (30)	1.79 ± 0.60 (430)
Na (% DM)	0.14 ± 0.15 (609)	0.46 ± 0.43 (1,147)	0.14 ± 0.15 (1,787)	0.03 ± 0.04 (10)	0.07 ± 0.04 (18)	0.01 ± 0.01 (133)
Cl (% DM)	1.02 ± 0.54 (1,203)	1.24 ± 0.63 (2,835)	0.99 ± 0.45 (5,071)	0.13 ± 0.05 (2)	0.27 ± 0.16 (5)	0.42 ± 0.20 (330)
S (% DM)	0.25 ± 0.08 (1,231)	0.23 ± 0.09 (2,639)	0.21 ± 0.20 (5,311)	0.13 ± 0.04 (15)	0.24 ± 0.10 (19)	0.23 ± 0.06 (375)
Co (mg/kg)						
Cu (mg/kg)	9.13 ± 3.95 (291)	9.14 ± 2.57 (1,725)	10.30 ± 4.80 (741)	6.99 ± 2.04 (6)	17.00 ± 4.03 (10)	8.51 ± 1.35 (134)
I (mg/kg)						
Fe (mg/kg)	293.53 ± 252.66 (274)	475.09 ± 546.28 (1,682)	497.17 ± 403.74 (713)	185.27 ± 157.94 (6)	161.03 ± 75.38 (8)	176.54 ± 130.81 (112)
Mn (mg/kg)	67.36 ± 39.68 (293)	132.89 ± 81.01 (1,745)	90.56 ± 52.78 (743)	41.90 ± 28.92 (6)	25.67 ±7.30 (10)	80.88 ± 32.13 (136)
Mo (mg/kg)	1.11 ± 0.48 (4)	1.53 ± 0.54 (7)	0.03 ± 0.01 (4)			
Se (mg/kg)		0.08 ± 0.05 (3)			0.12	
Zn (mg/kg)	37.06 ± 10.38 (286)	35.47 ± 21.81 (1,727)	36.14 ± 12.92 (753)	37.42 ± 3.05 (6)	60.40 ± 22.35 (10)	29.47 ± 12.66 (135)

TABLE 18-1 Continued

	Feed Name					
Component	Soybean Hulls	Soybean Meal, High CP	Soybean Meal, Low CP	Soybean Meal, Heated	Soybean Silage	Soybean Stubble
DM (% AF)	90.04 ± 2.15 (2,378)	89.24 ± 1.15 (8,795)	91.68 ± 2.94	89.76 ± 1.38 (3,941)	37.35 ± 10.64 (380)	85.48 ± 3.78 (205)
Ash (% DM)	5.05 ± 0.85 (435)	7.36 ± 0.69 (1,276)	6.43 ± 1.27	6.55 ± 1.11 (307)	9.81 ± 2.64 (320)	9.27 ± 1.64 (75)
TDN (% DM)	62.6 ± 4.16 (850)	79.5 ± 1.33 (1,983)	81.1 ± 6.73	79.3 ± 2.93 (2,607)	57.8 ± 6.07 (402)	35.0
DE (Mcal/kg)	2.76 ± 0.14 (804)	3.51 ± 0.06 (1,953)	3.58 ± 0.33 (648)	3.50 ± 0.12 (2,605)	2.55 ± 0.30 (404)	
ME (Mcal/kg)	2.26	2.88	2.93	2.87	2.09	1.27
NEm (Mcal/kg)	1.40	1.93	1.98	1.92	1.24	0.41
NEg (Mcal/kg)	0.82	1.28	1.33	1.28	0.67	0
Sugar (% DM)	2.15 ± 1.06 (2)	13.30	11.55 ± 0.64			
Starch (% DM)	1.10 ± 1.12 (222)	2.02 ± 0.80 (421)	5.05 ± 1.51	1.32 ± 0.58 (177)	4.21 ± 2.67 (307)	
Fat (% DM)	2.28 ± 1.64 (553)	1.88 ± 1.12 (2,089)	8.34 ± 4.32	8.27 ± 5.39 (780)	4.29 ± 1.74 (327)	0.70 (2)
NDF (% DM)	64.81 ± 5.68 (914)	11.33 ± 2.41 (2,047)	18.78 ± 6.57	22.67 ± 3.24 (2,631)	47.53 ± 7.87 (402)	69.75 ± 3.31 (11)
ADF (% DM)	46.40 ± 4.84 (960)	7.48 ± 1.46 (2,067)	10.93 ± 3.11	10.90 ± 1.93 (2,592)	36.86 ± 5.28 (422)	55.35 ± 5.58 (58)
Lignin (% DM)	2.47 ± 0.87 (197)	1.17 ± 0.51 (430)	1.48 ± 0.67	1.81 ± 0.75 (100)	8.01 ± 1.72 (320)	
CP (% DM)	12.37 ± 2.15 (2,279)	52.85 ± 1.32 (8,973)	46.53 ± 3.06	48.85 ± 2.63 (3,962)	17.08 ± 4.22 (445)	5.53 ± 1.76 (67)
RDP (% CP)	46.84 ± 9.08 (30)	70.42 ± 6.94 (275)	55.88 ± 10.07	50.45 ± 8.00 (271)	68.42 ± 6.91 (314)	
RUP (% CP)	53.10 ± 9.06 (30)	29.45 ± 6.67 (274)	44.08 ± 10.07	49.44 ± 7.93 (271)	31.48 ± 6.90 (314)	
Soluble CP (% CP)	30.51 ± 8.61 (337)	44.11 ± 6.83 (9)	19.49 ± 7.94	13.05 ± 3.88 (379)	49.01 ± 9.67 (341)	
ADICP (% DM)	6.37 ± 2.58 (89)	10.17 ± 6.37 (434)	8.97 ± 8.92	10.78 ± 6.31 (182)	9.90 ± 3.47 (402)	9.13 ± 3.41 (5)
Ca (% DM)	0.60 ± 0.09 (690)	0.42 ± 0.11 (2,085)	0.37 ± 0.14	0.35 ± 0.09 (301)	1.31 ± 0.40 (361)	1.25 ± 0.26 (31)
P (% DM)	0.15 ± 0.08 (689)	0.75 ± 0.06 (2,068)	0.66 ± 0.13	0.69 ± 0.09 (301)	0.30 ± 0.09 (331)	0.10 ± 0.05 (30)
Mg (% DM)	0.27 ± 0.12 (544)	0.32 ± 0.03 (1,339)	0.29 ± 0.12	0.29 ± 0.05 (208)	0.38 ± 0.16 (326)	0.44 ± 0.15 (19)
K (% DM)	1.34 ± 0.23 (674)	2.36 ± 0.22 (1,531)	2.02 ± 0.46	2.13 ± 0.28 (294)	1.81 ± 0.59 (365)	1.43 ± 0.57 (24)
Na (% DM)	0.02 ± 0.02 (284)	0.02 ± 0.01 (703)	0.07 ± 0.18	0.04 ± 0.03 (130)	0.02 ± 0.02 (77)	
Cl (% DM)	0.06 ± 0.03 (133)	0.06 ± 0.03 (234)	1.09 ± 1.97	0.07 ± 0.03 (74)	0.47 ± 0.37 (272)	
S (% DM)	0.12 ± 0.04 (347)	0.41 ± 0.03 (816)	1.05 ± 3.46	0.37 ± 0.05 (236)	0.20 ± 0.05 (303)	0.13 ± 0.05 (13)
Co (mg/kg)	0.60 ± 0.35 (3)	1.07 ± 0.76 (2)	0.75 ± 0.07			
Cu (mg/kg)	7.55 ± 1.95 (218)	15.91 ± 1.86 (661)	17.61 ± 4.84	2.40 ± 4.28 (97)	0.28 ± 3.53 (57)	4.00
I (mg/kg)						
Fe (mg/kg)	436.92 ± 100.28 (203)	201.87 ± 105.64 (538)	160.56 ± 104.06	148.60 ± 96.65 (97)	1,191.00 ± 1,635.00 (57)	
Mn (mg/kg)	20.03 ± 6.62 (211)	42.30 ± 7.18 (662)	40.66 ± 21.21	30.07 ± 9.31 (97)	85.65 ± 61.35 (55)	25.00
Mo (mg/kg)	0.57 ± 0.20 (10)	5.23 ± 2.24 (19)	3.18 ± 1.27			
Se (mg/kg)		0.53 ± 0.21 (2)				
Zn (mg/kg)	44.79 ± 9.44 (213)	51.68 ± 7.47 (657)	71.84 ± 91.53	40.40 ± 11.80 (97)	46.93 ± 24.43 (55)	9.00

continued

TABLE 18-1 Continued

	Feed Name					
Component	Soybeans, Expelled	Soybeans, Extruded	Soybeans, Roasted	Soybeans, Whole	Sudangrass, Fresh	Sudangrass Hay
DM (% AF)	88.59 ± 1.59 (176)	92.51 ± 2.77 (362)	93.32 ± 2.30 (1,208)	92.85 ± 2.93 (1,098)	30.89 ± 13.91 (994)	89.01 ± 3.89 (9,012)
Ash (% DM)		6.15 ± 0.83 (184)	5.62 ± 0.69 (327)	5.49 ± 0.74 (156)	10.86 ± 2.24 (231)	9.33 ± 1.88 (4,031)
TDN (% DM)	77.0	91.9 ± 5.58 (123)	97.4 ± 2.11 (426)	91.0 ± 2.42 (401)	54.8 ± 5.05 (278)	54.5 ± 2.80 (4,470)
DE (Mcal/kg)		4.05 ± 0.24 (144)	4.30 ± 0.18 (688)	4.01 ± 0.16 (392)	2.42 ± 0.22 (260)	2.40 ± 0.12 (4,400)
ME (Mcal/kg)	2.78	3.32	3.52	3.29	1.98	1.97
NEm (Mcal/kg)	1.85	2.29	2.45	2.27	1.13	1.12
NEg (Mcal/kg)	1.22	1.60	1.73	1.57	0.58	0.57
Sugar (% DM)						
Starch (% DM)		1.29 ± 0.63 (33)	1.30 ± 0.91 (98)	1.03 ± 0.68 (128)	2.08 ± 1.56 (222)	1.60 ± 1.39 (3,802)
Fat (% DM)	8.17 ± 1.85 (175)	13.14 ± 5.52 (283)	21.03 ± 2.00 (701)	20.61 ± 2.23 (596)	2.95 ± 0.85 (235)	1.60 ± 0.36 (4,062)
NDF (% DM)	12.60	16.58 ± 3.21 (143)	21.83 ± 5.35 (690)	17.98 ± 3.89 (395)	61.02 ± 5.64 (265)	65.83 ± 4.00 (4,840)
ADF (% DM)	8.80 ± 0.42 (2)	10.85 ± 2.22 (145)	11.53 ± 3.06 (677)	10.75 ± 2.48 (480)	37.25 ± 5.57 (292)	41.10 ± 4.11 (5,902)
Lignin (% DM)		1.83 ± 0.81 (44)	2.20 ± 0.88 (100)	1.92 ± 0.86 (34)	4.74 ± 1.23 (233)	5.09 ± 1.11 (4,045)
CP (% DM)	46.54 ± 1.95 (177)	44.40 ± 4.27 (361)	40.49 ± 3.39 (1,183)	39.97 ± 3.50 (1,004)	12.92 ± 4.63 (383)	8.33 ± 2.62 (6,948)
RDP (% CP)		57.78 ± 11.91 (82)	56.32 ± 8.88 (253)	70.99 ± 14.81 (15)	67.03 ± 5.48 (237)	57.90 ± 7.96 (3,842)
RUP (% CP)		42.17 ± 11.86 (82)	43.63 ± 8.84 (253)	29.01 ± 14.81 (15)	32.80 ± 5.46 (237)	41.99 ± 7.98 (3,842)
Soluble CP (% CP)		24.68 ± 11.47 (130)	16.72 ± 5.33 (582)	52.98 ± 20.92 (170)	42.67 ± 8.21 (243)	34.58 ± 4.92(4,097)
ADICP (% DM)		6.15 ± 3.91 (39)	9.76 ± 5.87 (104)	10.61 ± 5.09 (35)	5.68 ± 1.82 (206)	4.08 ± 1.28 (3,638)
Ca (% DM)	0.35 ± 0.01 (3)	0.29 ± 0.09 (118)	0.25 ± 0.05 (493)	0.25 ± 0.08 (253)	0.50 ± 0.19 (228)	0.44 ± 0.09 (4,378)
P (% DM)	0.63 ± 0.09 (3)	0.67 ± 0.11 (119)	0.63 ± 0.08 (508)	0.62 ± 0.11 (262)	0.31 ± 0.09 (231)	0.20 ± 0.05 (4,407)
Mg (% DM)		0.28 ± 0.02 (115)	0.26 ± 0.03 (477)	0.25 ± 0.03 (249)	0.32 ± 0.34 (217)	0.29 ± 0.07 (4,158)
K (% DM)		2.05 ± 0.21 (115)	1.83 ± 0.30 (524)	1.81 ± 0.30 (271)	2.61 ± 0.76 (264)	2.05 ± 0.50 (4,716)
Na (% DM)		0.02 ± 0.02 (46)	0.03 ± 0.04 (161)	0.02 ± 0.02 (82)	0.04 ± 0.05 (31)	0.03 ± 0.04 (873)
Cl (% DM)		0.08 ± 0.04 (26)	0.08 ± 0.04 (55)	0.07 ± 0.02 (28)	1.03 ± 0.39 (206)	1.16 ± 0.32 (3,617)
S (% DM)		0.35 ± 0.03 (91)	0.32 ± 0.03 (386)	0.31 ± 0.05 (178)	0.25 ± 0.54 (207)	0.13 ± 0.16 (3,817)
Co (mg/kg)				1.01		0.42 ± 0.30 (25)
Cu (mg/kg)		4.40 ± 2.05 (42)	13.80 ± 2.12 (154)	14.59 ± 2.80 (89)	15.80 ± 12.27 (17)	8.97 ± 5.43 (403)
I (mg/kg)						
Fe (mg/kg)		142.73 ± 46.18 (41)	144.61 ± 72.43 (152)	139.14 ± 78.45 (82)	307.20 ± 128.45 (14)	218.41 ± 122.12 (370)
Mn (mg/kg)		35.08 ± 8.20 (41)	29.18 ± 6.40 (153)	31.63 ± 18.83 (87)	70.51 ± 49.09 (17)	56.07 ± 329.30 (397)
Mo (mg/kg)				1.87 ± 1.71 (2)	1.34 ± 1.85 (2)	1.26 ± 0.75 (7)
Se (mg/kg)						0.11 ± 0.09 (3)
Zn (mg/kg)		48.25 ± 7.63 (41)	44.30 ± 7.21 (153)	47.22 ± 8.85 (90)	39.15 ± 38.98 (17)	31.33 ± 9.92 (402)

TABLE 18-1 Continued

	Feed Name					
Component	Sudangrass Silage	Sugarcane Bagasse, Silage	Sugarcane Hay	Sugarcane Silage	Sunflower Hulls	Sunflower Meal
DM (% AF)	31.33 ± 10.17 (1,251)	39.80 ± 11.59 (86)	92.72 ± 1.87 (22)	34.23 ± 12.70 (57)	91.26 ± 3.00 (27)	90.44 ± 2.46 (1,742)
Ash (% DM)	11.98 ± 3.42 (960)	6.41 ± 3.93 (125)	10.21 ± 3.93 (3)	7.90 ± 5.15 (22)	5.21 ± 1.06 (11)	6.41 ± 1.23 (337)
TDN (% DM)	53.5 ± 4.92 (1,077)	45.0 ± 10.45 (226)	53.7 ± 3.56 (18)	52.0 ± 7.42 (77)	54.6	66.4 ± 9.69 (321)
DE (Mcal/kg)	2.36 ± 0.22 (1,068)	1.99 ± 0.46 (226)	2.37 ± 0.16 (18)	2.29 ± 0.32 (76)		2.93 ± 0.42 (303)
ME (Mcal/kg)	1.93	1.63	1.94	1.88	1.97	2.40
NEm (Mcal/kg)	1.09	0.79	1.10	1.04	1.13	1.52
NEg (Mcal/kg)	0.53	0.25	0.54	0.49	0.57	0.93
Sugar (% DM)	0.80					
Starch (% DM)	1.79 ± 1.14 (900)	0.87 ± 0.69 (102)	1.50 ± 0.14 (2)	3.04 ± 6.37 (19)		1.07 ± 0.88 (90)
Fat (% DM)	3.14 ± 0.80 (973)	1.14 ± 0.61 (128)	1.84 ± 0.65 (5)	1.86 ± 1.37 (24)	16.32 ± 12.23 (18)	10.80 ± 7.75 (558)
NDF (% DM)	62.31 ± 5.47 (1,112)	75.58 ± 11.24 (241)	68.92 ± 10.62 (18)	68.76 ± 12.28 (94)	72.26 ± 6.05 (8)	41.71 ± 6.46 (345)
ADF (% DM)	40.77 ± 4.64 (1,243)	62.11 ± 11.19 (241)	47.51 ± 12.83 (18)	47.88 ± 14.08 (96)	50.11 ± 9.71 (17)	30.34 ± 6.25 (376)
Lignin (% DM)	5.76 ± 1.28 (953)	17.31 ± 5.61 (131)	6.60 ± 0.70 (3)	9.72 ± 6.20 (24)	15.60	9.04 ± 2.27 (100)
CP (% DM)	12.10 ± 3.41 (1,273)	3.93 ± 1.75 (223)	5.49 ± 3.03 (19)	5.41 ± 3.50 (78)	16.90 ± 8.75 (27)	35.01 ± 5.55 (1,721)
RDP (% CP)	70.15 ± 7.01 (933)	53.71 ± 10.90 (14)	70.46 ± 5.80 (2)	52.26 ± 10.79 (4)		71.97 ± 7.51 (27)
RUP (% CP)	29.71 ± 6.96 (932)	46.30 ± 10.90 (14)	29.54 ± 5.80 (2)	47.75 ± 10.79 (4)		27.89 ± 7.50 (27)
Soluble CP (% CP)	51.87 ± 9.35 (996)	46.74 ± 15.38 (179)	36.25 ± 12.28 (13)	51.59 ± 9.52 (53)		32.39 ± 12.33 (188)
ADICP (% DM)	6.58 ± 2.19 (1,026)	9.53 ± 4.31 (223)	4.79 ± 1.91 (3)	6.71 ± 3.87 (76)		8.83 ± 3.52 (86)
Ca (% DM)	0.52 ± 0.15 (1,000)	0.29 ± 0.20 (172)	0.39 ± 0.19 (13)	0.29 ± 0.19 (61)	0.65 ± 0.40 (10)	0.44 ± 0.12 (356)
P (% DM)	0.30 ± 0.07 (1,004)	0.05 ± 0.04 (167)	0.14 ± 0.07 (13)	0.12 ± 0.10 (62)	0.48 ± 0.34 (10)	1.04 ± 0.24 (380)
Mg (% DM)	0.27 ± 0.08 (960)	0.09 ± 0.06 (169)	0.22 ± 0.07 (13)	0.16 ± 0.11 (62)	0.31 ± 0.15 (5)	0.57 ± 0.10 (323)
K (% DM)	2.67 ± 0.76 (1,099)	0.41 ± 0.40 (175)	1.36 ± 0.62 (15)	1.32 ± 1.04 (61)	1.39 ± 0.22 (6)	1.56 ± 0.29 (347)
Na (% DM)	0.03 ± 0.03 (213)	0.03 ± 0.04 (147)	0.06 ± 0.10 (12)	0.04 ± 0.06 (52)	0.44 ± 0.61 (2)	0.04 ± 0.05 (117)
Cl (% DM)	0.92 ± 0.36 (851)	0.15 ± 0.18 (86)	0.31 ± 0.19 (3)	0.81 ± 0.76 (18)		0.15 ± 0.05 (65)
S (% DM)	0.18 ± 0.04 (923)	0.09 ± 0.07 (163)	0.16 ± 0.09 (11)	0.18 ± 0.11 (51)	0.29 ± 0.16 (2)	0.40 ± 0.11 (205)
Co (mg/kg)						1.00
Cu (mg/kg)	12.16 ± 3.46 (101)	7.15 ± 3.06 (67)	16.78 ± 13.55 (9)	9.53 ± 6.91 (16)	12.00	31.55 ± 13.05 (114)
I (mg/kg)						
Fe (mg/kg)	1,239.00 ± 1,210.00 (98)	1,211.00 ± 792.71 (66)	2,429.00 ± 2,293.00 (9)	431.83 ± 230.26 (12)		243.19 ± 130.62 (95)
Mn (mg/kg)	74.89 ± 43.63 (100)	60.27 ± 26.17 (67)	86.75 ± 72.75 (9)	61.18 ± 45.94 (16)	17.00	50.30 ± 27.39 (113)
Mo (mg/kg)	1.90 ± 0.52 (3)					
Se (mg/kg)						1.00
Zn (mg/kg)	41.24 ± 12.35 (99)	18.46 ± 10.92 (67)	27.53 ± 11.30 (9)	33.44 ± 29.92 (16)	52.00	92.79 ± 39.17 (114)

continued

TABLE 18-1 Continued

Component	Feed Name					
	Sunflower Screenings	Sunflower Silage	Sunflowers	Switchgrass Hay	Teff (Lovegrass) Hay	Timothy Hay
DM (% AF)	89.33 ± 3.01 (82)	50.43 ± 28.32 (44)	94.05 ± 2.69 (495)	94.20 ± 2.66 (162)	87.95 ± 4.17 (96)	87.83 ± 4.92 (32)
Ash (% DM)	6.12 ± 2.64 (14)	11.90 ± 2.57 (13)	3.76 ± 1.02 (140)	3.42 (2)	10.50	8.50 ± 2.62 (6)
TDN (% DM)	56.5 ± 4.88 (2)	63.3 ± 5.59 (23)	71.7 ± 17.79 (15)	46.1 ± 3.32 (26)	45.0	57.0
DE (Mcal/kg)		2.79 ± 0.22 (23)	3.16 ± 0.41 (137)			
ME (Mcal/kg)	2.04	2.29	2.59	1.67	1.63	2.06
NEm (Mcal/kg)	1.19	1.42	1.69	0.83	0.79	1.21
NEg (Mcal/kg)	0.63	0.84	1.07	0.29	0.25	0.64
Sugar (% DM)						14.15 ± 8.98 (2)
Starch (% DM)	3.00	2.78 ± 2.05 (13)	0.83 ± 0.79 (29)			
Fat (% DM)	20.99 ± 9.15 (37)	8.67 ± 3.28 (19)	33.75 ± 8.19 (382)	1.20	1.10	1.93 ± 0.65 (6)
NDF (% DM)	47.24 ± 10.94 (11)	47.24 ± 4.29 (29)	37.38 ± 9.72 (164)	81.07 ± 9.41 (18)	66.40 ± 6.34 (22)	63.81 ± 3.90 (9)
ADF (% DM)	33.98 ± 9.68 (65)	38.66 ± 4.03 (38)	30.78 ± 10.62 (93)	46.69 ± 6.16 (61)	37.98 ± 5.31 (73)	38.04 ± 2.74 (29)
Lignin (% DM)		7.85 ± 1.43 (13)	8.54 ± 6.27 (12)	6.63 (2)		
CP (% DM)	15.41 ± 2.45 (77)	11.27 ± 1.73 (43)	20.08 ± 3.64 (256)	3.39 ± 1.73 (96)	10.74 ± 4.29 (94)	9.44 ± 3.19 (33)
RDP (% CP)			77.88 ± 9.65 (6)			
RUP (% CP)			22.13 ± 9.65 (6)			
Soluble CP (% CP)		51.44 ± 9.20 (17)	42.93 ± 15.07 (20)			
ADICP (% DM)	22.50	8.45 ± 3.14 (23)	6.80 ± 2.39 (8)	8.75	16.88 (1)	
Ca (% DM)	0.73 ± 0.28 (69)	1.26 ± 0.29 (24)	0.42 ± 0.31 (92)	0.34 ± 0.16 (87)	0.47 ± 0.25 (19)	0.42 ± 0.12 (17)
P (% DM)	0.42 ± 0.10 (68)	0.36 ± 0.08 (24)	0.59 ± 0.19 (94)	0.11 ± 0.03 (87)	0.24 ± 0.11 (19)	0.21 ± 0.07 (17)
Mg (% DM)	0.44 ± 0.14 (48)	0.57 ± 0.18 (24)	0.37 ± 0.09 (69)	0.47 ± 0.41 (58)	0.19 ± 0.07 (13)	0.18 ± 0.04 (13)
K (% DM)	1.61 ± 0.42 (54)	3.27 ± 0.75 (24)	1.14 ± 0.37 (75)	0.97 ± 0.55 (59)	1.61 ± 0.75 (15)	1.92 ± 0.68 (13)
Na (% DM)	0.03 ± 0.02 (3)	0.03 ± 0.01 (19)	0.02 ± 0.01 (16)	0.07 ± 0.06 (10)		
Cl (% DM)		1.18 ± 0.28 (14)	0.10 ± 0.02 (9)			
S (% DM)	0.23 ± 0.04 (22)	0.24 ± 0.05 (22)	0.23 ± 0.04 (47)	0.08 ± 0.04 (12)	0.22 ± 0.09 (6)	0.16 ± 0.03 (7)
Co (mg/kg)						
Cu (mg/kg)	14.33 ± 1.53 (3)	12.60 ± 4.17 (15)	20.88 ± 4.13 (22)	3.60 ± 0.90 (11)	7.00 ± 2.83 (2)	6.00
I (mg/kg)						
Fe (mg/kg)		839.61 ± 1,202.00 (15)	88.97 ± 64.86 (18)			
Mn (mg/kg)	40.00 ± 4.58 (3)	39.82 ± 27.16 (15)	28.02 ± 9.62 (22)	82.77 ± 28.04 (74)	52.00	
Mo (mg/kg)						
Se (mg/kg)			1.00			1.55
Zn (mg/kg)	36.33 ± 10.02 (3)	424.21 ± 673.66 (15)	56.79 ± 13.19 (22)	11.75 ± (74)	20.00 ± 14.14 (2)	17.00

TABLE 18-1 Continued

	Feed Name					
Component	Triticale Forage, Fresh	Triticale Grain	Triticale Hay	Triticale Silage	Wheat Bran	Wheat Forage, Fresh
DM (% AF)	26.01 ± 10.36 (2,529)	88.84 ± 1.83 (450)	90.28 ± 3.06 (2,528)	32.97 ± 8.11 (4,312)	90.10 ± 1.25 (208)	34.11 ± 11.26 (1,902)
Ash (% DM)	8.91 ± 2.35 (1,875)	1.96 ± 0.50 (325)	8.39 ± 2.26 (989)	10.74 ± 3.14 (2,680)	5.48 ± 0.80 (97)	8.91 ± 2.49 (802)
TDN (% DM)	61.4 ± 4.25 (2,116)	82.7 ± 1.23 (383)	58.5 ± 4.42 (1,522)	57.8 ± 4.48 (3,018)	71.9 ± 2.76 (182)	61.7 ± 3.61 (1,090)
DE (Mcal/kg)	2.71 ± 0.20 (2,111)	3.65 ± 0.05 (384)	2.58 ± 0.21 (1,439)	2.55 ± 0.20 (2,995)	3.17 ± 0.12 (182)	2.72 ± 0.19 (1,095)
ME (Mcal/kg)	2.22	2.99	2.12	2.09	2.60	2.23
NEm (Mcal/kg)	1.36	2.02	1.26	1.24	1.69	1.37
NEg (Mcal/kg)	0.78	1.37	0.69	0.67	1.08	0.79
Sugar (% DM)	0.80 ± 0.17 (3)			1.70 ± 2.12 (3)		26.30 ± 0.28 (2)
Starch (% DM)	1.70 ± 1.85 (1,485)	61.04 ± 4.39 (357)	5.06 ± 6.38 (900)	1.94 ± 2.03 (2,467)	21.17 ± 6.84 (115)	4.11 ± 4.87 (770)
Fat (% DM)	2.89 ± 0.67 (1,773)	1.65 ± 0.37 (334)	2.09 ± 0.61 (952)	3.66 ± 0.78 (2,739)	4.32 ± 0.79 (117)	3.00 ± 0.79 (792)
NDF (% DM)	56.59 ± 6.77 (2,215)	14.10 ± 2.61 (386)	57.73 ± 8.30 (1,831)	58.57 ± 5.97 (3,490)	40.09 ± 6.26 (182)	54.16 ± 7.12 (1,217)
ADF (% DM)	34.22 ± 5.03 (2,226)	4.49 ± 1.05 (386)	36.69 ± 6.34 (1,990)	38.21 ± 4.51 (3,637)	13.72 ± 2.75 (185)	32.99 ± 5.23 (1,325)
Lignin (% DM)	3.52 ± 1.17 (1,896)	1.81 ± 0.50 (323)	4.69 ± 1.28 (951)	4.71 ± 1.27 (2,661)	4.15 ± 1.03 (81)	3.87 ± 1.10 (780)
CP (% DM)	15.33 ± 4.66 (2,378)	12.13 ± 2.20 (425)	11.58 ± 4.34 (2,349)	13.90 ± 3.68 (4,446)	17.48 ± 1.77 (202)	15.32 ± 5.24 (1,567)
RDP (% CP)	77.02 ± 5.41 (1,735)	67.30 ± 8.32 (320)	68.85 ± 6.86 (907)	81.76 ± 6.24 (2,569)	64.35 ± 3.56 (13)	75.45 ± 5.53 (767)
RUP (% CP)	22.71 ± 5.30 (1,730)	32.56 ± 8.28 (320)	31.09 ± 6.86 (907)	18.12 ± 6.25 (2,569)	35.65 ± 3.56 (13)	24.36 ± 5.46 (767)
Soluble CP (% CP)	50.10 ± 8.82 (1,781)	30.21 ± 9.18 (333)	45.87 ± 9.36 (1,008)	69.68 ± 8.46 (2,750)	39.81 ± 6.06 (118)	52.84 ± 7.71 (843)
ADICP (% DM)	4.25 ± 1.78 (1,416)	2.46 ± 1.68 (271)	3.49 ± 1.51 (866)	4.96 ± 1.74 (2,727)	3.20 ± 1.20 (73)	4.50 ± 1.84 (653)
Ca (% DM)	0.34 ± 0.12 (1,572)	0.07 ± 0.06 (328)	0.31 ± 0.12 (1,269)	0.43 ± 0.17 (3,071)	0.13 ± 0.09 (156)	0.36 ± 0.13 (931)
P (% DM)	0.31 ± 0.08 (1,609)	0.36 ± 0.06 (336)	0.25 ± 0.08 (1,449)	0.34 ± 0.08 (3,120)	1.04 ± 0.25 (167)	0.31 ± 0.08 (935)
Mg (% DM)	0.17 ± 0.06 (1,513)	0.13 ± 0.02 (333)	0.15 ± 0.05 (1,191)	0.17 ± 0.04 (3,009)	0.42 ± 0.10 (146)	0.16 ± 0.05 (927)
K (% DM)	2.58 ± 0.86 (1,922)	0.49 ± 0.08 (331)	2.03 ± 0.74 (1,447)	2.92 ± 0.95 (3,387)	1.20 ± 0.27 (158)	2.42 ± 0.97 (1,115)
Na (% DM)	0.05 ± 0.04 (144)	0.01 ± 0.01 (39)	0.04 ± 0.04 (301)	0.05 ± 0.04 (901)	0.03 ± 0.06 (123)	0.06 ± 0.07 (345)
Cl (% DM)	0.75 ± 0.38 (1,426)	0.12 ± 0.03 (41)	0.71 ± 0.46 (865)	1.02 ± 0.43 (2,587)	0.10 ± 0.02 (61)	0.76 ± 0.41 (623)
S (% DM)	0.19 ± 0.05 (1,380)	0.15 ± 0.04 (292)	0.15 ± 0.05 (1,021)	0.20 ± 0.05 (2,734)	0.19 ± 0.02 (120)	0.19 ± 0.06 (759)
Co (mg/kg)			0.40 ± 0.09 (3)	1.12 ± 0.73 (7)		0.57 (2)
Cu (mg/kg)	8.41 ± 2.62 (64)	5.97 ± 1.16 (39)	7.34 ± 2.45 (215)	8.84 ± 3.51 (391)	12.43 ± 3.11 (60)	8.52 ± 2.97 (228)
I (mg/kg)						
Fe (mg/kg)	433.96 ± 352.70 (48)	52.38 ± 14.46 (33)	230.09 ± 151.15 (153)	597.78 ± 421.77 (300)	165.14 ± 59.37 (57)	552.98 ± 488.09 (192)
Mn (mg/kg)	62.74 ± 31.19 (64)	45.70 ± 12.66 (39)	42.74 ± 17.68 (210)	54.01 ± 26.69 (386)	130.59 ± 38.99 (60)	53.23 ± 28.97 (226)
Mo (mg/kg)	5.60	0.70	1.26 ± 0.76 (33)	1.25 ± 0.50 (4)		1.43 ± 0.57 (10)
Se (mg/kg)			0.07			2.38
Zn (mg/kg)	35.47 ± 10.60 (64)	35.60 ± 10.83 (39)	25.40 ± 10.18 (214)	35.89 ± 10.18 (393)	79.45 ± 21.10 (60)	32.19 ± 12.43 (233)

continued

TABLE 18-1 Continued

Component	Feed Name					
	Wheat Grain	Wheat Grain, Steam Flaked	Wheat Hay	Wheat Middlings	Wheat Silage	Wheat Straw
DM (% AF)	88.94 ± 1.88 (4,031)	82.96 ± 3.33 (1687)	89.91 ± 3.04 (6,781)	88.85 ± 1.32 (5,382)	34.11 ± 7.71 (12,712)	91.77 ± 3.33 (6,576)
Ash (% DM)	2.27 ± 0.93 (1,684)	1.97 ± 0.49 (22)	8.15 ± 2.21 (2,111)	5.37 ± 0.78 (1,010)	10.28 ± 2.83 (7,603)	7.53 ± 2.72 (3,861)
TDN (% DM)	86.8 ± 1.83 (1,708)	86.8 ± 1.83 (1,708)	58.8 ± 4.51 (2,976)	72.9 ± 1.96 (1,396)	59.1 ± 3.73 (8,701)	50.0 ± 4.70 (4,367)
DE (Mcal/kg)	3.83 ± 0.09 (1,712)		2.59 ± 0.21 (2,854)	3.22 ± 0.09 (1,370)	2.61 ± 0.17 (8,700)	2.21 ± 0.21 (4,309)
ME (Mcal/kg)	3.14	3.14	2.13	2.64	2.14	1.81
NEm (Mcal/kg)	2.15	2.14	1.27	1.73	1.28	0.97
NEg (Mcal/kg)	1.47	1.47	0.70	1.11	0.71	0.42
Sugar (% DM)	8.55 ± 0.35 (2)		9.35 ± 11.24 (2)		1.81 ± 0.97 (229)	2.50 ± 2.25 (3)
Starch (% DM)	62.42 ± 5.10 (1,884)	64.87 ± 2.86 (1,437)	4.68 ± 4.74 (1,910)	25.56 ± 5.67 (1,144)	6.62 ± 6.82 (7,421)	1.64 ± 1.23 (3,279)
Fat (% DM)	1.89 ± 0.43 (1,993)	1.88 ± 0.56 (16)	2.00 ± 0.53 (2,011)	4.14 ± 0.63 (1,288)	3.40 ± 0.73 (7,576)	1.43 ± 0.48 (3,520)
NDF (% DM)	12.36 ± 2.92 (1,712)	13.55 ± 1.57 (56)	57.89 ± 7.90 (3,802)	38.33 ± 4.65 (1,463)	56.54 ± 6.21 (10,500)	73.65 ± 7.38 (4,870)
ADF (% DM)	4.15 ± 1.46 (1,802)	5.51 ± 1.03 (65)	35.89 ± 6.14 (4,933)	13.23 ± 2.16 (1,491)	36.59 ± 4.40 (10,870)	50.23 ± 6.60 (5,434)
Lignin (% DM)	1.52 ± 0.53 (1,390)		4.82 ± 1.56 (1,996)	3.66 ± 1.14 (135)	4.77 ± 1.05 (7,648)	7.42 ± 2.04 (3,541)
CP (% DM)	13.79 ± 2.46 (3,922)	14.42 ± 1.81 (588)	11.11 ± 3.93 (5,384)	18.56 ± 1.47 (5,083)	12.67 ± 3.13 (12,472)	5.09 ± 2.14 (110)
RDP (% CP)	64.22 ± 8.06 (1,253)		66.03 ± 7.29 (1,897)	68.19 ± 7.32 (11)	82.11 ± 6.11 (7,023)	65.35 ± 10.10 (3,177)
RUP (% CP)	35.58 ± 8.08 (1,254)		33.85 ± 7.31 (1,897)	31.62 ± 7.38 (11)	17.78 ± 6.13 (7,025)	34.54 ± 10.13 (3,177)
Soluble CP (% CP)	29.32 ± 6.90 (1,440)		40.29 ± 7.53 (2,014)	40.48 ± 6.50 (350)	67.38 ± 8.28 (7,803)	40.80 ± 10.60 (3,564)
ADICP (% DM)	2.40 ± 2.06 (1,105)	3.13	3.94 ± 1.77 (1,711)	3.43 ± 1.34 (135)	5.02 ± 1.68 (7,887)	4.96 ± 1.73 (3,065)
Ca (% DM)	0.08 ± 0.06 (1,794)	0.04 ± 0.01 (86)	0.32 ± 0.15 (3,297)	0.12 ± 0.04 (1,211)	0.38 ± 0.14 (9,109)	0.33 ± 0.17 (4,423)
P (% DM)	0.36 ± 0.07 (1,824)	0.31 ± 0.05 (96)	0.21 ± 0.07 (3,329)	1.08 ± 0.17 (1,327)	0.30 ± 0.07 (9,199)	0.11 ± 0.06 (4,554)
Mg (% DM)	0.13 ± 0.02 (1,647)	0.15 ± 0.01 (4)	0.15 ± 0.05 (2,877)	0.43 ± 0.09 (858)	0.16 ± 0.05 (8,941)	0.12 ± 0.05 (4,220)
K (% DM)	0.43 ± 0.07 (1,650)	0.42 ± 0.08 (4)	1.69 ± 0.65 (3,360)	1.15 ± 0.17 (943)	2.38 ± 0.76 (9,917)	1.34 ± 0.60 (4,737)
Na (% DM)	0.02 ± 0.01 (278)		0.05 ± 0.05 (834)	0.05 ± 0.13 (538)	0.05 ± 0.06 (2,741)	0.09 ± 0.13 (1,803)
Cl (% DM)	0.23 ± 0.03 (165)		0.69 ± 0.36 (1,692)	0.14 ± 0.16 (76)	0.82 ± 0.36 (6,340)	0.53 ± 0.41 (3,104)
S (% DM)	0.15 ± 0.03 (1417)	0.14 ± 0.01 (15)	0.16 ± 0.06 (2,206)	0.20 ± 0.02 (514)	0.18 ± 0.04 (7,781)	0.11 ± 0.05 (3,519)
Co (mg/kg)				0.56 ± 0.10 (9)	1.49 ± 0.73 (5)	0.47 ± 0.40 (8)
Cu (mg/kg)	5.44 ± 1.55 (483)	4.00	7.00 ± 2.50 (570)	12.85 ± 5.68 (463)	8.17 ± 3.01 (1,240)	5.91 ± 2.90 (1,106)
I (mg/kg)						
Fe (mg/kg)	60.28 ± 28.53 (239)		332.96 ± 277.71 (389)	144.23 ± 33.93 (330)	556.23 ± 380.73 (869)	214.40 ± 160.94 (984)
Mn (mg/kg)	42.96 ± 10.61 (483)	43.00	58.35 ± 28.02 (574)	136.62 ± 28.35 (465)	55.53 ± 27.79 (1,250)	64.14 ± 54.70 (1,091)
Mo (mg/kg)	0.65 ± 0.17 (8)		1.32 ± 0.77 (64)	1.66 ± 0.20 (26)	1.56 ± 0.83 (4)	1.29 ± 0.86 (46)
Se (mg/kg)			0.05	0.10	0.18 ± 0.05 (3)	0.21 ± 0.22 (3)
Zn (mg/kg)	29.25 ± 8.35 (483)	26.00	21.04 ± 7.80 (576)	87.85 ± 20.53 (465)	31.31 ± 10.47 (1,257)	17.08 ± 10.08 (1,116)

TABLE 18-1 Continued

Component	Feed Name	
	Whey, Wet	Whey, Dry
DM (% AF)	16.60 ± 13.33 (951)	93.84 ± 2.99 (70)
Ash (% DM)	12.72 ± 5.37 (917)	12.69 ± 1.92 (134)
TDN (% DM)	80.9 ± 6.75 (664)	82.3 ± 2.63 (50)
DE (Mcal/kg)	3.578 ± 0.43 (690)	3.63 ± 0.52 (51)
ME (Mcal/kg)	2.92	2.98
NEm (Mcal/kg)	1.97	2.01
NEg (Mcal/kg)	1.32	1.36
Sugar (% DM)	50.60 ± 16.27 (568)	56.09 ± 2.16 (108)
Starch (% DM)	3.28 ± 4.86 (43)	1.28 ± 1.39 (10)
Fat (% DM)	3.98 ± 4.18 (651)	1.96 ± 1.94 (37)
NDF (% DM)	1.66 ± 1.20 (8)	0.55 ± 0.07 (2)
ADF (% DM)	4.23 ± 0.10 (4)	0.40
Lignin (% DM)	0.60 ± 0.59 (4)	0.10
CP (% DM)	6.30 ± 4.26 (1,961)	13.91 ± 18.41 (165)
RDP (% CP)	96.52 ± 6.38 (4)	80.08 ± 33.18 (7)
RUP (% CP)	6.97 ± 8.58 (2)	23.24 ± 35.05 (6)
Soluble CP (% CP)	81.38 ± 8.04 (520)	85.59 ± 27.61 (32)
ADICP (% DM)	2.81 ± 2.42 (4)	
Ca (% DM)	0.99 ± 0.54 (1,735)	0.61 ± 0.29 (35)
P (% DM)	1.01 ± 0.51 (1,714)	0.77 ± 0.39 (35)
Mg (% DM)	0.18 ± 0.05 (1,443)	0.14 ± 0.07 (27)
K (% DM)	2.99 ± 1.23 (1,576)	1.76 ± 0.96 (36)
Na (% DM)	1.04 ± 0.66 (1,558)	1.13 ± 0.45 (45)
Cl (% DM)	3.53 ± 1.92 (203)	1.36 ± 0.73 (8)
S (% DM)	0.13 ± 0.07 (1,359)	0.21 ± 0.15 (28)
Co (mg/kg)		
Cu (mg/kg)	11.12 ± 14.06 (223)	2.17 ± 1.20 (5)
I (mg/kg)		
Fe (mg/kg)	75.43 ± 122.70 (249)	44.15 ± 76.06 (7)
Mn (mg/kg)	2.76 ± 4.56 (209)	2.31 ± 1.87 (7)
Mo (mg/kg)		
Se (mg/kg)		

[a]The SD and N values are not reported for sample size = 1. Values from ME, NEm, and NEg were calculated from the TDN value using methods described in Chapter 19.

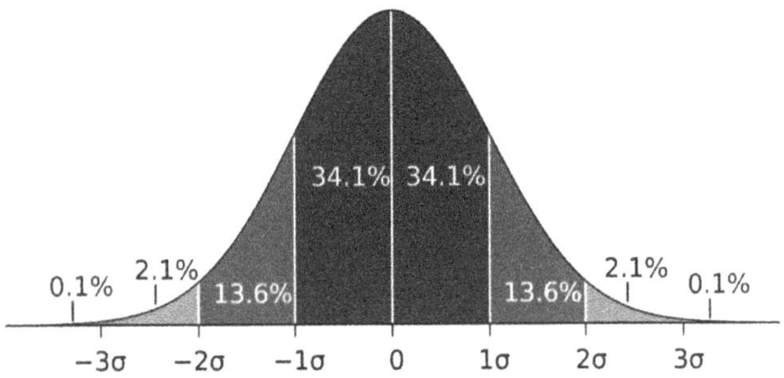

FIGURE 18-1 Normal distribution of a population based on observed values. Population standard deviation (σ) provides an indication of the range of common observations as a distance from the mean.

storage conditions, and sampling technique all influence nutrient concentrations, an average value without an estimate of the normal variation is of limited value. An estimate of the variation associated with the nutrient concentration of a given feed can also be used in stochastic programming to decrease costs of formulated diets (D'Alfonso et al., 1992). Thus, the mean, SD, and sample size (*n*) are all important criteria to evaluate when reviewing Table 18-1.

No attempt was made to differentiate near-infrared reflectance spectroscopy (NIR) data from wet chemistry methods. For some nutrients, both wet chemistry and NIR data were combined because of similarities in means and SD. In addition, using fiber analysis, energy content of common forages can be predicted (Weiss et al., 1992). As a result, energy values are provided for many common forages listed in Table 18-1; however, energy content of many feeds is based on performance studies and cattle growth.

For example, the energy content of dry-rolled corn (DRC) was assumed to be 2.17 Mcal/kg net energy required for maintenance (NEm) and 1.49 Mcal/kg net energy required for gain (NEg) based on 87.6 total digestible nutrients (TDN) for corn received at commercial labs and calculated energy value. Steam-flaking corn increases feeding value by 12% (Owens et al., 1997; Zinn et al., 2002) or more (see Chapter 17, Utilization of Byproduct Feeds by Beef Cattle; Scott et al., 2003; Macken et al., 2006) with greater increases observed when diets included wet corn gluten feed. As a result, the listed energy content of steam-flaked corn (SFC) was increased by 12% (2.43 Mcal/kg NEm and 1.67 Mcal/kg NEg), which is slightly lower than the energy values summarized by Zinn et al. (2002). The difference between results of studies for energy content of SFC relative to DRC also could relate to performance of cattle fed DRC. Likewise, the energy content of distillers grains and corn gluten feed were increased relative to DRC, with values reflecting the information provided in Chapter 17. The energy content listed for wet distillers grains plus solubles is not accurate when diets are based on SFC (Vander Pol et al., 2008; Corrigan et al., 2009). Similarly, when distillers grains are fed in forage-based diets, the energy content relative to DRC is approximately 115% (Loy et al., 2008; Nuttelman et al., 2010); however, the TDN of corn is less when supplemented in forage-based diets as a result of negative associative effects on fiber digestion (Fieser and Vanzant, 2004; Loy et al., 2007). Therefore, absolute energy values should be decreased when corn is supplemented in forage diets, and dry, wet, or modified distillers grains plus solubles energy content (TDN or NEm/NEg absolute values) should be decreased in a similar manner to corn.

The tabular TDN values of feedstuffs have two limitations. First, they are based on the physicochemical characteristics of the feedstuffs used to estimate them from either digestion trials or by equations used by feed testing laboratories that provided the data used in developing the feed composition library; thus, they may not represent the actual feedstuffs being currently fed (Tedeschi et al., 2002). Second, the values for TDN in the feed library provided in this chapter (Table 18-1) are not discounted for the level of intake above maintenance. The TDN content of the feedstuffs is usually determined or predicted by equations when animals are fed at the maintenance level of intake (Weiss et al., 1992; Tedeschi et al., 2005). For high-producing animals (i.e., feedlot) that are consuming greater amounts of feedstuffs than an animal would at maintenance, a TDN discount might be needed to reflect a decrease in digestibility, especially for fibrous feedstuffs that are not digested postruminally and concentrates that have decreased intestinal digestion because of faster transit time in the intestines. Tedeschi et al. (2005) developed equations to adjust TDN for level of intake and reported 2.3 and 5% discounts per level of intake for concentrate and forage feedstuffs, respectively. If desired, there are two options for adjusting the TDN value to specific production conditions: (1) use the mechanistic solution level in the current edition of the *Nutrient Requirements of Beef Cattle* or (2) use the equations developed by Tedeschi et al. (2005) to predict TDN values from actual feed analysis and discount them for level of intake above maintenance.

GRAZED FORAGES

Table 18-2 provides information on selected nutrients for grazed forages. Samples are provided from regions where grazed forages were sampled by either clipping or using diet collection methods with fistulated cattle. Cattle are selective grazers, which suggests that diet samples collected from fistulated cattle are generally more representative of diet quality in grazing situations and provide the best indication of energy and protein supplied by grazing. In addition, the time of year or growth stage of the grazed forages can dramatically affect energy, fiber, and protein concentrations. This list is far from exhaustive, but it provides some estimates across a wide geographic area of North America, which can be used in predicting nutrient supply of grazing cattle. It is important to note, however, that for adequate prediction of

TABLE 18-2 Energy, Fiber (NDF), and Protein (CP) Content of Forage Samples from Different Regions[a]

Forage	Location	Month, Year	Plant Stage	TDN or IVD[b]	% NDF	% CP	Sampling	Reference[c]
Blue grama	New Mexico	May 1982	Dormant	36.7	82.7	6.9	–	1
Blue grama	New Mexico	Aug 1982	Vegetative	63.1	74.9	18.8	–	1
Blue grama	New Mexico	Late Aug 1982	Vegetative	63.1	75.8	17.5	–	1
Blue grama	New Mexico	Sept 1982	Vegetative	51.6	69.9	12.5	–	1
Blue grama	New Mexico	Oct 1982	Dormant	47.9	64.9	11.9	–	1
Blue grama	New Mexico	Jan 1983	Dormant	41.6	81.2	11.3	–	1
Blue grama	New Mexico	Feb 1983	Dormant	42.0	80.2	10.0	–	1
Blue grama	New Mexico	Mar 1983	Dormant	48.0	79.2	11.9	–	1
Blue grama	New Mexico	Early Aug 1983	Vegetative	47.4	78.8	18.1	–	1
Blue grama	New Mexico	Mid Aug 1983	Vegetative	61.9	69.9	10.6	–	1
Blue grama	New Mexico	Late Aug 1983	Vegetative	60.9	80.7	12.5	–	1
Blue grama	New Mexico	Early Nov 1983	Dormant	52.2	75.8	10.0	–	1
Blue grama	New Mexico	Mid Jan 1984	Dormant	52.9	78.1	7.5	–	1
Blue grama	New Mexico	May 1984	Vegetative	73.2	83.4	15.0	–	1
Blue grama	New Mexico	Late July 1984	Vegetative	65.0	75.8	13.1	–	1
Blue grama	New Mexico	Late Aug 1984	Vegetative	69.9	89.7	16.3	–	1
Blue grama	New Mexico	Early Nov 1984	Dormant	62.3	71.2	8.8	–	1
Blue grama	New Mexico	Mid Nov 1984	Dormant	48.0	67.1	8.8	–	1
Blue grama	New Mexico	Dec 1984	Dormant	47.6	77.1	7.5	–	1
Mixed prairie	North Dakota	Late June	Vegetative	67.9	76.4	12.7	–	2
Mixed prairie	North Dakota	Late July	Vegetative	60.7	77.7	8.4	–	2
Mixed prairie	North Dakota	Late Aug	Vegetative	56.4	77.4	9.5	–	2
Mixed prairie	North Dakota	Early Oct	Dormant	57.8	80.1	8.3	–	2
Mixed prairie	North Dakota	May-June 1990	Vegetative	67.0	78.4	13.8	–	3
Mixed prairie	North Dakota	Early July 1990	Vegetative	70.1	73.4	16.3	–	3
Mixed prairie	North Dakota	Mid Aug 1990	Vegetative	54.2	82.1	11.3	–	3
Mixed prairie	North Dakota	Sept-Oct 1990	Dormant	55.6	81.2	10.6	–	3
Mixed prairie	North Dakota	May-June 1991	Vegetative	77.2	77.5	14.4	–	3
Mixed prairie	North Dakota	Early July 1991	Vegetative	62.8	84.2	13.1	–	3
Mixed prairie	North Dakota	Mid Aug 1991	Vegetative	59.2	75.4	10.0	–	3
Mixed prairie	North Dakota	Sept-Oct 1991	Dormant	49.7	77.3	11.3	–	3
Mixed prairie	North Dakota	Early June 1989	Vegetative	65.9	80.8	17.3	–	4
Mixed prairie	North Dakota	Early July 1989	Vegetative	57.7	76.7	13.4	–	4
Mixed prairie	North Dakota	Late Aug 1989	Vegetative	49.8	81.2	12.8	–	4
Mixed prairie	North Dakota	Early Oct 1989	Dormant	45.2	93.2	13.9	–	4
Mixed prairie	North Dakota	Early June 1990	Vegetative	69.8	78.7	14.8	–	4
Mixed prairie	North Dakota	Early July 1990	Vegetative	64.9	73.4	17.3	–	4
Mixed prairie	North Dakota	Late Aug 1990	Vegetative	54.7	82.1	11.3	–	4
Mixed prairie	North Dakota	Early Oct 1990	Dormant	50.7	81.2	10.8	–	4
Mixed prairie	North Dakota	Mid June 1995	Vegetative	68.1	59.5	13.6	–	5
Mixed prairie	North Dakota	Late July 1995	Vegetative	60.5	51.0	14.9	–	5

continued

TABLE 18-2 Continued

Forage	Location	Month, Year	Plant Stage	TDN or IVD[b]	% NDF	% CP	Sampling	Reference[c]
Mixed prairie	North Dakota	Early Sept 1995	Vegetative	55.6	58.7	10.2	–	5
Mixed prairie	North Dakota	Early Oct 1995	Dormant	54.3	59.6	9.7	–	5
Mixed prairie	North Dakota	Mid Nov 1995	Dormant	57.3	67.9	6.6	–	5
Mixed prairie	North Dakota	Mid Dec 1995	Dormant	53.3	72.1	6.2	–	5
Brome	North Dakota	June 1989	Vegetative	67.5	67.7	17.3	–	6
Brome	North Dakota	July 1989	Vegetative	60.1	70.4	11.4	–	6
Brome	North Dakota	Aug 1989	Vegetative	59.2	75.9	12.8	–	6
Mixed prairie	North Dakota	June 2000	Vegetative	71.1	72.4	12.2	–	7
Mixed prairie	North Dakota	July 2000	Vegetative	64.1	73.8	8.8	–	7
Mixed prairie	North Dakota	Aug 2000	Dormant	51.1	83.4	9.8	–	7
Mixed prairie	North Dakota	Sept 2000	Dormant	48.3	83.4	8.4	–	7
Mixed prairie	North Dakota	Nov 2000	Dormant	41.1	85.1	7.2	–	7
Bluestem	Oklahoma	July 2008/9	Vegetative	–	73.2	8.2	Clipped	8
Native range	Oklahoma	July 2008/9	Vegetative	–	71.5	5.2	Clipped	8
Bluestem	Oklahoma	Aug 2008/9	Vegetative	–	73.3	10.1	Clipped	8
Native range	Oklahoma	Aug 2008/9	Vegetative	–	72.2	6.3	Clipped	8
Bluestem	Oklahoma	Aug 2008/9	Vegetative	–	74.3	8.7	Clipped	8
Native range	Oklahoma	Aug 2008/9	Vegetative	–	74.9	5.4	Clipped	8
Bluestem	Oklahoma	Sept 2008/9	Vegetative	–	74.4	9.2	Clipped	8
Native range	Oklahoma	Sept 2008/9	Vegetative	–	75.2	5.5	Clipped	8
Bluestem	Oklahoma	Sept 2008/9	Vegetative	–	75.2	7.7	Clipped	8
Native range	Oklahoma	Sept 2008/9	Vegetative	–	76.5	5.0	Clipped	8
Bluestem	Oklahoma	Oct 2008/9	Vegetative	–	77.7	6.8	Clipped	8
Native range	Oklahoma	Oct 2008/9	Vegetative	–	75.2	4.5	Clipped	8
Alfalfa-brome	Iowa	May	–	64.1	–	21.7	–	9
Alfalfa-brome	Iowa	June	–	53.8	–	13.8	–	9
Alfalfa-brome	Iowa	July	–	54.3	–	15.4	–	9
Alfalfa-brome	Iowa	Aug	–	55.0	–	14.7	–	9
Alfalfa-brome	Iowa	Sept	–	53.5	–	13.8	–	9
Brome-orchardgrass-trefoil	Iowa	Apr	–	38.5	–	15.8	–	9
Brome-orchardgrass-trefoil	Iowa	May	–	57.2	–	11.8	–	9
Brome-orchardgrass-trefoil	Iowa	June	–	58.0	–	11.6	–	9
Brome-orchardgrass-trefoil	Iowa	July	–	58.7	–	14.3	–	9
Brome-orchardgrass-trefoil	Iowa	Aug	–	57.8	–	13.6	–	9
Brome-orchardgrass-trefoil	Iowa	Sept	–	56.6	–	13.0	–	9
Pasture	Iowa	Apr	–	55.8	–	17.1	–	9
Pasture	Iowa	May	–	57.1	–	14.2	–	9
Pasture	Iowa	June	–	52.0	–	10.8	–	9
Pasture	Iowa	July	–	48.7	–	10.9	–	9
Pasture	Iowa	Aug	–	47.1	–	11.4	–	9
Pasture	Iowa	Sept	–	46.3	–	11.9	–	9

TABLE 18-2 Continued

Forage	Location	Month, Year	Plant Stage	TDN or IVD[b]	% NDF	% CP	Sampling	Reference[c]
Brome	Iowa	May	–	67.1	–	22.7	–	9
Brome	Iowa	June	–	57.7	–	13.1	–	9
Brome	Iowa	July	–	57.0	–	14.0	–	9
Brome	Iowa	Aug	–	58.6	–	14.7	–	9
Brome	Iowa	Sept	–	59.4	–	16.6	–	9
Brome-red clover	Iowa	May	–	52.6	–	15.5	–	9
Brome-red clover	Iowa	June	–	57.0	–	12.5	–	9
Brome-red clover	Iowa	July	–	58.2	–	13.1	–	9
Brome-red clover	Iowa	Aug	–	56.1	–	12.7	–	9
Tall fescue-red clover	Iowa	May	–	57.4	–	15.5	–	9
Tall fescue-red clover	Iowa	June	–	58.4	–	13.2	–	9
Tall fescue-red clover	Iowa	July	–	59.3	–	14.0	–	9
Tall fescue-red clover	Iowa	Aug	–	57.1	–	14.4	–	9
Alfalfa mix	U.S.	–	–	–	–	16.4	Hay	10
Brome	U.S.	–	–	–	–	11.1	Hay	10
Bermudagrass	U.S.	–	–	–	–	9.6	Hay	10
Fescue	U.S.	–	–	–	–	10.9	Hay	10
Sudan	U.S.	–	–	–	–	7.9	Hay	10
Cereal grain forages	U.S.	–	–	–	–	10.0	Hay	10
Native grass	U.S.	–	–	–	–	9.1	–	10
Alfalfa mix	U.S.	–	–	57.8	–	15.0	–	11
Brome	U.S.	–	–	55.5	–	9.4	–	11
Bermudagrass	U.S.	–	–	57.0	–	9.2	–	11
Fescue	U.S.	–	–	54.4	–	9.8	–	11
Orchardgrass	U.S.	–	–	54.6	–	9.4	–	11
Sudan	U.S.	–	–	54.3	–	7.1	–	11
Cereal grain forages	U.S.	–	–	57.6	–	8.9	–	11
Native grass	U.S.	–	–	55.2	–	7.7	–	11
Range forage	GSL, Nebraska	Jan	–	54.2	83.3	6.9	Masticate	12
Range forage	GSL, Nebraska	Feb	–	54.6	82.5	6.2	Masticate	12
Range forage	GSL, Nebraska	Mar	–	52.6	83.0	7.4	Masticate	12
Range forage	GSL, Nebraska	Apr	–	59.5	77.1	8.0	Masticate	12
Range forage	GSL, Nebraska	May	–	65.8	68.3	12.4	Masticate	12
Range forage	GSL, Nebraska	June	–	62.8	70.1	10.8	Masticate	12
Range forage	GSL, Nebraska	July	–	55.9	65.6	11.5	Masticate	12
Range forage	GSL, Nebraska	Aug	–	55.2	64.5	8.9	Masticate	12
Range forage	GSL, Nebraska	Sept	–	51.4	69.3	8.8	Masticate	12
Range forage	GSL, Nebraska	Oct	–	53.0	74.0	7.9	Masticate	12
Range forage	GSL, Nebraska	Nov	–	51.4	74.7	7.6	Masticate	12
Range forage	GSL, Nebraska	Dec	–	53.9	77.6	7.0	Masticate	12
Meadow forage	GSL, Nebraska	–	–	63.1	64.6	13.0	Masticate	13

continued

TABLE 18-2 Continued

Forage	Location	Month, Year	Plant Stage	TDN or IVD[b]	% NDF	% CP	Sampling	Reference[c]
Meadow forage	GSL, Nebraska	–	–	58.7	65.8	11.6	Masticate	13
Brome pasture	East Nebraska	May	–	71.6	–	18.8	Masticate	14
Brome pasture	East Nebraska	June	–	63.0	–	14.4	Masticate	14
Brome pasture	East Nebraska	July	–	54.8	–	13.3	Masticate	14
Brome pasture	East Nebraska	Aug	–	54.6	–	17.1	Masticate	14
Brome pasture	East Nebraska	Sept	–	57.1	–	17.5	Masticate	14
Meadow forage	GSL, Nebraska	Summer	–	60.8	–	11.9	Masticate	15
Range forage	GSL, Nebraska	Summer	–	60.3	–	10.0	Masticate	15
Range forage	Barta, Nebraska	Early June	–	58.3	–	13.4	Masticate	15
Range forage	Barta, Nebraska	Late June	–	54.6	–	12.6	Masticate	15
Range forage	Barta, Nebraska	Early July	–	51.6	–	11.4	Masticate	15
Range forage	Barta, Nebraska	Late July	–	55.2	–	10.0	Masticate	15
Range forage	Barta, Nebraska	Early Aug	–	49.6	–	9.8	Masticate	15
Range forage	Barta, Nebraska	Late Aug	–	55.3	–	10.9	Masticate	15
Range forage	Barta, Nebraska	Early Sept	–	48.8	–	11.4	Masticate	15
Range forage	Barta, Nebraska	Late Sept	–	53.8	–	10.4	Masticate	15
Wheatgrass, awned	Canada	Jan	–	33.2	78.5	3.3	Clipped	16
Wheatgrass, awned	Canada	Feb	–	36.8	75.8	4.0	Clipped	16
Wheatgrass, awned	Canada	Mar	–	36.9	75.8	3.2	Clipped	16
Wheatgrass, awned	Canada	Apr	–	34.5	76.4	3.4	Clipped	16
Wheatgrass, awned	Canada	May	–	46.4	67.0	9.5	Clipped	16
Wheatgrass, awned	Canada	June	–	55.6	60.9	13.7	Clipped	16
Wheatgrass, awned	Canada	July	–	48.9	67.2	10.0	Clipped	16
Wheatgrass, awned	Canada	Aug	–	42.9	66.1	6.7	Clipped	16
Wheatgrass, awned	Canada	Sep	–	39.4	73.0	5.2	Clipped	16
Wheatgrass, awned	Canada	Oct	–	35.2	77.1	4.4	Clipped	16
Wheatgrass, awned	Canada	Nov	–	32.9	80.0	3.4	Clipped	16
Wheatgrass, awned	Canada	Dec	–	29.6	79.1	3.7	Clipped	16
Wheatgrass, Western	Canada	Jan	–	39.0	69.7	4.1	Clipped	16
Wheatgrass, Western	Canada	Feb	–	39.8	70.4	3.7	Clipped	16
Wheatgrass, Western	Canada	Mar	–	42.1	68.3	4.5	Clipped	16
Wheatgrass, Western	Canada	Apr	–	42.4	66.9	5.0	Clipped	16
Wheatgrass, Western	Canada	May	–	58.7	56.7	14.6	Clipped	16
Wheatgrass, Western	Canada	June	–	55.7	59.9	13.8	Clipped	16
Wheatgrass, Western	Canada	July	–	51.6	61.4	10.8	Clipped	16
Wheatgrass, Western	Canada	Aug	–	49.7	60.1	9.2	Clipped	16
Wheatgrass, Western	Canada	Sept	–	49.6	60.9	8.5	Clipped	16
Wheatgrass, Western	Canada	Oct	–	43.1	67.0	5.5	Clipped	16
Wheatgrass, Western	Canada	Nov	–	41.8	70.4	4.8	Clipped	16
Wheatgrass, Western	Canada	Dec	–	40.0	69.4	5.5	Clipped	16
Blue grama	Canada	Jan	–	39.1	74.3	4.1	Clipped	16

TABLE 18-2 Continued

Forage	Location	Month, Year	Plant Stage	TDN or IVD[b]	% NDF	% CP	Sampling	Reference[c]
Blue grama	Canada	Feb	–	38.1	72.8	4.4	Clipped	16
Blue grama	Canada	Mar	–	40.2	68.4	5.2	Clipped	16
Blue grama	Canada	Apr	–	40.1	68.2	5.1	Clipped	16
Blue grama	Canada	May	–	43.6	65.3	6.6	Clipped	16
Blue grama	Canada	June	–	48.0	65.8	9.3	Clipped	16
Blue grama	Canada	July	–	50.1	65.6	8.7	Clipped	16
Blue grama	Canada	Aug	–	48.7	67.5	7.5	Clipped	16
Blue grama	Canada	Sept	–	47.1	64.3	7.3	Clipped	16
Blue grama	Canada	Oct	–	40.3	69.9	5.7	Clipped	16
Blue grama	Canada	Nov	–	39.2	73.4	4.9	Clipped	16
Blue grama	Canada	Dec	–	41.3	70.7	5.0	Clipped	16
Green needlegrass	Canada	Jan	–	39.5	73.8	4.1	Clipped	16
Green needlegrass	Canada	Feb	–	37.5	69.0	4.8	Clipped	16
Green needlegrass	Canada	Mar	–	40.0	71.2	4.6	Clipped	16
Green needlegrass	Canada	Apr	–	40.1	70.3	4.4	Clipped	16
Green needlegrass	Canada	May	–	51.4	62.5	11.3	Clipped	16
Green needlegrass	Canada	June	–	50.4	63.4	10.7	Clipped	16
Green needlegrass	Canada	July	–	47.5	67.8	8.1	Clipped	16
Green needlegrass	Canada	Aug	–	46.2	65.3	7.5	Clipped	16
Green needlegrass	Canada	Sept	–	45.6	66.0	6.9	Clipped	16
Green needlegrass	Canada	Oct	–	40.6	71.4	5.2	Clipped	16
Green needlegrass	Canada	Nov	–	39.6	72.5	4.8	Clipped	16
Green needlegrass	Canada	Dec	–	38.2	70.7	4.8	Clipped	16
Needle and thread	Canada	Jan	–	39.4	71.8	4.1	Clipped	16
Needle and thread	Canada	Feb	–	39.2	72.4	4.1	Clipped	16
Needle and thread	Canada	Mar	–	39.4	69.5	4.6	Clipped	16
Needle and thread	Canada	Apr	–	40.6	67.8	5.1	Clipped	16
Needle and thread	Canada	May	–	46.2	61.3	8.7	Clipped	16
Needle and thread	Canada	June	–	49.9	64.9	10.5	Clipped	16
Needle and thread	Canada	July	–	46.0	68.7	7.1	Clipped	16
Needle and thread	Canada	Aug	–	45.7	69.0	7.2	Clipped	16
Needle and thread	Canada	Sept	–	44.8	69.6	7.3	Clipped	16
Needle and thread	Canada	Oct	–	39.7	69.9	5.6	Clipped	16
Needle and thread	Canada	Nov	–	40.0	70.3	5.6	Clipped	16
Needle and thread	Canada	Dec	–	39.7	72.1	5.1	Clipped	16
Sand grass	Canada	Jan	–	36.3	77.6	3.3	Clipped	16
Sand grass	Canada	Feb	–	36.8	77.5	2.8	Clipped	16
Sand grass	Canada	Mar	–	34.5	75.2	3.5	Clipped	16
Sand grass	Canada	Apr	–	40.6	73.9	3.9	Clipped	16
Sand grass	Canada	May	–	45.6	68.3	5.9	Clipped	16
Sand grass	Canada	June	–	44.4	69.0	11.0	Clipped	16

continued

TABLE 18-2 Continued

Forage	Location	Month, Year	Plant Stage	TDN or IVD[b]	% NDF	% CP	Sampling	Reference[c]
Sand grass	Canada	July	–	44.3	70.8	8.4	Clipped	16
Sand grass	Canada	Aug	–	41.5	70.1	7.2	Clipped	16
Sand grass	Canada	Sept	–	35.1	71.8	5.2	Clipped	16
Sand grass	Canada	Oct	–	35.1	72.8	4.4	Clipped	16
Sand grass	Canada	Nov	–	35.8	75.9	4.1	Clipped	16
Sand grass	Canada	Dec	–	35.8	77.0	3.8	Clipped	16
Bermudagrass	Raleigh, NC	June-Sept	–	75.3	65.7	15.4	Masticate	17
Caucasian bluestem	Raleigh, NC	June-Sept	–	74.9	70.9	9.9	Masticate	17
Gama grass	Raleigh, NC	June-Sept	–	74.6	68.7	13.6	Masticate	17
Big bluestem	Raleigh, NC	June-Sept	–	75.7	68.5	11.1	Masticate	17
Switchgrass	Raleigh, NC	June-Sept	–	79.9	68.7	12.1	Masticate	17
Bermudagrass	North Carolina	Season	–	62.8	71.7	12.3	–	18
Yellow bluestem	North Carolina	Season	–	66.1	67.4	10.8	–	18
Coastal panicgrass	North Carolina	Season	–	64.0	70.9	10.0	–	18
Coastal bermudagrass	North Carolina	–	–	49.8	71.8	8.4	Clipped	19
Tifton 44 bermuda	North Carolina	–	–	52.2	72.5	10.7	Clipped	19
Coastal bermudagrass	North Carolina	–	101 kg N/ha	72.9	61.8	14.0	Masticate	19
Coastal bermudagrass	North Carolina	–	202 kg N/ha	75.0	60.0	17.0	Masticate	19
Coastal bermudagrass	North Carolina	–	303 kg N/ha	75.0	59.5	17.5	Masticate	19
Tifton 44 bermudagrass	North Carolina	–	101 kg N/ha	75.2	61.1	15.3	Masticate	19
Tifton 44 bermudagrass	North Carolina	–	202 kg N/ha	76.0	59.7	15.9	Masticate	19
Tifton 44 bermudagrass	North Carolina	–	303 kg N/ha	78.1	58.1	17.9	Masticate	19
Flaccidgrass	North Carolina	–	Short	69.7	52.7	20.9	Masticate	20
Flaccidgrass	North Carolina	–	Medium	68.6	57.1	18.8	Masticate	20
Flaccidgrass	North Carolina	–	Tall	68.8	57.7	18.1	Masticate	20
Tall fescue	North Carolina	Spring	Short	79.7	56.4	21.5	Clipped	21
Tall fescue	North Carolina	Spring	Medium	78.5	56.9	19.9	Clipped	21
Tall fescue	North Carolina	Spring	Tall	77.1	57.7	18.2	Clipped	21
Tall fescue	North Carolina	Fall	Short	74.5	57.0	21.7	Clipped	21
Tall fescue	North Carolina	Fall	Medium	77.4	54.7	21.8	Clipped	21
Tall fescue	North Carolina	Fall	Tall	80.3	53.2	22.2	Clipped	21
Tall fescue	North Carolina	–	Continuous	87.7	46.0	23.1	Masticate	22
Tall fescue	North Carolina	–	Rotation 1	88.7	45.4	24.9	Masticate	22
Tall fescue	North Carolina	–	Rotation 2	89.1	45.6	25.4	Masticate	22
Tall fescue	North Carolina	June	–	76.3	43.1	–	Masticate	23
Tall fescue	North Carolina	May	–	79.3	45.6	–	Masticate	23
Bermudagrass	North Carolina	June	–	63.5	54.3	–	Masticate	23
Bermudagrass	North Carolina	July	–	65.0	48.7	–	Masticate	23
Switchgrass	North Carolina	June	–	71.7	53.9	–	Masticate	23
Switchgrass	North Carolina	May	–	79.9	57.4	–	Masticate	23
Switchgrass	North Carolina	July	–	70.3	49.2	–	Masticate	23

TABLE 18-2 Continued

Forage	Location	Month, Year	Plant Stage	TDN or IVD[b]	% NDF	% CP	Sampling	Reference[c]
Flaccidgrass	North Carolina	June	–	71.8	49.1	–	Masticate	23
Flaccidgrass	North Carolina	May	–	76.1	44.3	–	Masticate	23
Flaccidgrass	North Carolina	July	–	72.2	48.8	–	Masticate	23
Flaccidgrass	North Carolina	July	–	72.5	50.0	21.2	Masticate	24
Gamagrass	North Carolina	July	–	69.0	59.7	16.5	Masticate	24
Bermudagrass	North Carolina	July	–	71.8	52.6	20.9	Masticate	24
Switchgrass hay	North Carolina	Early June	–	58.8	69.3	11.3	Hay	25
Switchgrass hay	North Carolina	Late June	–	48.2	74.5	6.9	Hay	25
Switchgrass hay	North Carolina	Early July	–	42.2	76.1	5.6	Hay	25
Switchgrass hay	North Carolina	Late July	–	38.8	76.8	4.4	Hay	25
Switchgrass hay	North Carolina	Early Aug	–	33.7	78.8	4.4	Hay	25

[a]Sampling varied from harvested hays, clipped forage samples, or diet collections (masticate) using fistulated cattle.

[b]TDN or in vitro DM or OM digestibility are listed. Values should only be compared within reference for comparison purposes as no corrections were made to correct in vitro digestibility to in vivo estimates or calculated TDN.

[c]References: 1 = Krysl et al. (1987); 2 = Olson et al. (1994); 3 = Hirschfeld et al. (1996); 4 = Silcox (1991); 5 = Johnson et al. (1998); 6 = Caton et al. (1993); 7 = Cline et al. (2009); 8 = McMurphy et al. (2011); 9 = Loy (2007); 10 = Corah and Dargatz (1996); 11 = Mortimer et al. (1999); 12 = Giesert et al. (2008); 13 = Griffin et al. (2012); 14 = Greenquist et al. (2009); 15 = Buckner et al. (2013); 16 = Abouguendia (1998); 17 = Burns and Fisher (2013); 18 = Burns et al. (2012a); 19 = Burns et al. (2009); 20 = Burns et al. (2012b); 21 = Burns et al. (2011); 22 = Burns and Fisher (2011); 23 = Fisher et al. (1991); 24 = Burns et al. (1992); 25 = Burns et al. (1997).

nutrient supply, intake must be predicted accurately, which can be very challenging in grazing situations. Nonetheless, Table 18-2 provides some insights into forage composition for use when no data are available, but these values should never replace sampling on individual pastures, as many variables can dramatically change forage composition.

EFFECTS OF PROCESSING TREATMENT

Many treatments are used to improve the nutritive value of feedstuffs for beef cattle. The methods of treating or processing feeds are not reviewed in this section, but the outcomes of using some common feed processing techniques on performance of beef cattle are discussed. Further insight on the methods and details of methods are available in other reviews (e.g., Beeson and Perry, 1982; Berger et al., 1994; Zinn et al., 2002).

Forages

The nutritive value of roughages is often improved through the use of physical and, occasionally, chemical or biological treatment methods. Responses to physical processing such as steaming, chopping, wafering, and grinding (with or without pelleting) are usually in inverse proportion to the quality of the starting forage (Minson, 1963). Coarse chopping, with or without wafering, usually has only a slight influence on nutritive value, although intake might be enhanced through indirect effects such as ease of handling and presentation to the animals. Alternatively, fine grinding, with or without pelleting, can have a major influence, particularly on intake, but also on available energy. Potential benefit depends on appropriate supplementation, especially with protein (Campling and Freer, 1966; Weston, 1967). Intake of low-quality forages is also enhanced when combined with palatable, wet feeds that prevent separation in the diet and improve overall diet palatability (Peterson et al., 2009).

Pelleting also decreases particle size as well as storage and shrink loss during handling and processing. Pelleting has been shown to increase intakes, with increases of 8 to 26% compared with nonpelleted diets (Minson, 1963; Beardsley, 1964; Campling and Freer, 1966; Minson and Milford, 1968; Coleman et al., 1978). Campling and Freer (1966) evaluated the effects of grinding and pelleting roughages on intake and digestibility of grass hay and oat straw. They found that pelleting oat straw increased intake by 26% compared with nonpelleted, nonground straw, whereas dried grass intakes remained similar. In contrast, grinding decreased digestibility of both roughages compared with not grinding, presumably due to differences in rate of passage.

Increased intake is usually observed when mean particle size is decreased to 5 mm, and intake is increased in proportion to further decreases in size, with maximal intake achieved when mean particle size is 1 mm or less. Pelleting is an improvement over grinding because it produces less dust, which can result in increased intake. Pelleting increased intake by 47% and increased gain by 60% in lambs (Greenhalgh and Reid, 1974). In a summary of research with bulls, Sundstol (1991) reported that grinding by itself and grinding with pelleting enhanced intake of straw by 7 and 37%, respectively. Beardsley (1964) reviewed six studies comparing pelleted and native forms of forages and noted that pelleting increased intake by 8%, gain by as much as 100%, and feed efficiency up to 35%; this summary applies mostly to hays and straws. Silages are rarely processed as finely as dry forages, although the amount of chopping and particle size reduction that occurs during harvesting can vary significantly. Based on a summary of available literature on corn (Wilkinson, 1978) and grass silage (McDonald et al., 1991) and within the range of particle lengths commonly observed for silage (mean length, 5 to 15 mm), there is a negative relationship of length to intake; however, the intake decrease is generally less than 10%.

Digestibility of roughages is decreased by grinding, with or without pelleting, and the decrease is usually in proportion to the increased intake (Blaxter et al., 1956). Using data from 21 studies, Minson (1963) found an average 3.3% decrease in DM digestibility with grinding. Thomson and Beever (1980) reported greater decreases for ground grasses (0 to 15%) than for ground legumes (3 to 6%). Greenhalgh and Reid (1973) observed a 10-percentage-point decrease in DM digestibility for pelleted versus long-stem hay, whereas Minson and Milford (1968) observed a 6.8-percentage-point decrease. Decreased digestibility as a result of grinding or pelleting is likely a function of greater passage from the rumen leading to decreased retention time and thereby decreased digestibility, while also allowing for increased intake.

Chemical alkali is used to upgrade roughages by hydrolyzing chemical bonds between lignin and carbohydrates in grasses. Sodium hydroxide is more effective than calcium hydroxide/oxide, as well as ammonia or urea, but sodium hydroxide is more expensive and has greater environmental consequences. As a result, sodium hydroxide is less commonly used than ammoniation, followed by treatment with calcium hydroxide/oxide. Ammoniation can be beneficial because the ammonia provides a readily available ruminally degradable protein (RDP) source (Klopfenstein, 1978), and it improves intake and digestibility of forages (Morris and Mowat, 1980; Fahey et al., 1993). Berger et al. (1994) concluded, from 21 studies on crop residues and 6 on grasses, that ammoniation improved dry matter intake (DMI) by 22 and 14%, respectively. With regard to digestibility, 32 studies on ammoniated crop residues and 10 on grasses demonstrated a 15 and 16% improvement, respectively. Urea enhanced intake by 13% and digestibility by 23%. Oxidation is an alternative chemical procedure that has been used to upgrade roughages, and microbial and enzymatic methods have also been developed and tested. Steam treatment is an additional physical process that has been developed; however, none of these latter processes are widely used in North America at present. For details, the reader is referred to Berger et al. (1994).

Grains

General

Feeding grain as an energy source for growing cattle or for feedlot finishing programs is a common practice that was developed when grain was relatively inexpensive and abundant in the 1950s (Klopfenstein et al., 2013). In the United States, corn is the most widely used grain source (Vasconcelos and Galyean, 2007). Grain sorghum, barley, and wheat are common grains fed in certain geographic regions of North America, or during certain times of the year depending on their prices. The most common grain processing method in terms of number of cattle is likely steam flaking (Vasconcelos and Galyean, 2007). Dry rolling and high-moisture ensiling would be common methods of grain processing, especially for areas where grains are less expensive or in smaller operations. The reader is referred to Chapter 4 (Carbohydrates) for further details on grain processing methodologies and their effects on starch availability.

Processing can significantly improve the nutritive value of cereal grains for beef cattle. The most common physical processes used are rolling or grinding the grain, with or without additional moisture; this is done chiefly to rupture the pericarp and expose starch granules to aid digestion (Beauchemin et al., 1994). When processing is used, animal responses can be variable and unpredictable, which is likely a result of greater ruminal starch digestion and risk of subacute acidosis with increased processing (Huntington, 1997). Furthermore, processing can affect nutrient requirements by increasing the requirement for RDP because of increased energy available in the rumen, which increases the need for RDP (Cooper et al., 2002a,b).

Principles of Grain Processing

Cattle are less able than other ruminants to masticate whole grain (Theurer, 1986). Sorghum presents the greatest difficulty followed by wheat, barley, corn, and oats. Morgan and Campling (1978) found that younger cattle can digest whole grain better than older cattle; however, Campling (1991) concluded that further studies on a possible relationship between cattle age or weight and digestion of grain are needed. The ability of rumen microbes to digest grain depends on particle size (Galyean et al., 1981; Beauchemin et al., 1994)—with fine particles digested more rapidly than coarse particles. Microbial digestion proceeds from the inside to the outside of the kernel, which is why disruption of the kernel pericarp when processed or masticated is important for rate and extent of starch digestion. The protein matrix, which surrounds starch granules in the endosperm, is a barrier to digestion of starch within the kernel even after processing (McAllister et al., 1990a). For this and related reasons, there are major differences between the rates at which starch within the grains is digested. For example, barley is digested more rapidly than corn (McAllister et al., 1990b). Owens et al. (1997) and Richards and Hicks (2007) suggested that feeding whole corn leads to greater energy utilization than feeding DRC; however, their comparisons were of different studies that did not have treatments of whole corn and DRC compared within the experiment. As discussed by Richards and Hicks (2007), roughage concentration was likely less in diets containing whole corn than DRC. Other plausible explanations offered by the authors were that cattle rumination was improved by feeding whole corn, which might have aided in acidosis control, compared with feeding DRC. In more recent studies that compared whole corn and DRC in the same experiments and with diets containing byproducts, whole corn was found to have 85 to 90% the energy value of DRC (Scott et al., 2003; Vander Pol et al., 2008).

Intrinsic characteristics of grains affect the rate or extent of starch digestion and can decrease the benefits of processing. One factor is the form of starch and the other is the presence of tannins. Amylopectin is more digestible than amylose; hence, waxy grains are more digestible than other grains (Sherrod et al., 1969). Tannins present in bird-resistant grains, such as sorghums, decrease digestibility (Maxson et al., 1973). Within varieties of the same grain, TDN varied as much as 7% (Parrot et al., 1969). Grain quality for beef cattle is positively associated with grain density or fiber content, as shown for barley by Mathison et al. (1991b) and Engstrom et al. (1992).

Grain that is fermented less rapidly and extensively in the rumen can escape microbial digestion and can be digested enzymatically in the small intestine. In a review of many trials, Owens et al. (1986) estimated that cattle are 42% more efficient in utilizing starch when it is digested in the abomasum and small intestine compared with the reticulorumen. Thus, processes that cause starch to escape rumen digestion could be beneficial, provided it is effectively digested in the intestine and not passed further to the cecum, where fermentation can resume but requires degradable protein for microbial digestion, and the microbes are not digested (Owens et al., 1986). Nonetheless, starch digestion in the small intestine is limited in beef cattle (Huntington, 1997; Harmon et al., 2004; Huntington et al., 2006), which is likely a result of insufficient secretion of amylase enzyme into the small intestine (Harmon et al., 2004). Furthermore, digestion in the hindgut does not usually compensate for decreased digestion in the rumen (Goetsch et al., 1987). For these reasons, processed grain that escapes ruminal fermentation does not increase net energy (NE) of gain or improve nitrogen utilization by the animal.

There are three important points to consider that will affect digestible energy (DE) derived by the animal and could further modify the benefits of processing. First, positive effects on digestion can result by combining grains and different forms of grain, as reported between ground high-moisture corn (HMC) and dry-rolled sorghum (Stock et al., 1991); between different particle sizes of dry corn (Turgeon et al., 1983); between dry corn and HMC (Stock et al., 1987); between wheat and HMC (Bock et al., 1991); and between high-moisture sorghum grain and DRC (Streeter et al., 1989). Stock and Erickson (2006) summarized grain processing combinations and the positive associative effect observed when blending different rates of starch digestion either by processing or grain type. They suggested that with the popularity of low-starch byproducts (see Chapter 17), feeding combinations of grains with different rates of starch digestion might be less common today than in the past. In areas with limited supply or availability of byproducts, feeding different grain types or grain processed in different ways to influence rate of ruminal starch digestion can improve overall performance of finishing cattle (Stock and Erickson, 2006). The second consideration is level of feeding. Moe and Tyrrell (1979) reported that the metabolizable energy (ME) of corn grain for dairy cows was decreased from 3.58 Mcal/kg at maintenance to 2.92 Mcal/kg at 2.5 times maintenance. Bines et al. (1988) reported that intake effects on digestibility of mixed diets containing processed grain might be significant in young cattle but not in lactating cows. Zinn et al. (1995) evaluated two levels of intake (1.6 or 2.4% of body weight) in 200-kg fistulated steers fed either dry-rolled or SFC and observed numerically greater ruminal and greater total tract organic matter (OM) digestibility for steers consuming less DM. Although interest exists in restricted feeding of feedlot steers and heifers, effects on digestibility attributable to intake levels used in practice are small. The third factor is grain processing within diets containing byproducts. Extensive discussion was provided in Chapter 17 on the influence of corn processing in combination with different byproducts.

Corn. In diets containing less than 20% roughage, differences in DE and NE for corn—whole or rolled, or ground coarse or fine—are usually fairly small (Goodrich and Meiske, 1966; Vance et al., 1970, 1972; Preston, 1975). Differences in the DE and NE values of these forms of corn in low-roughage diets could be greater for the high-moisture grain (>20% water); diets containing unprocessed high-moisture grain had feeding value superior to diets containing rolled grain, and diets containing rolled grain had feeding value superior to diets containing the ground form (Mader et al., 1991).

Steam flaking corn likely started in the early 1960s in Colorado (Matsushima et al., 1964). Relative to whole dry corn, steam processing and flaking improved NE by at least 10% when roughage was included in the diet, but had no effect in an all-concentrate diet (Vance et al., 1970). From studies on diets containing 50% corn and 20% whole cottonseed, Zinn (1987) concluded that SFC contained 13.4 and 14.2% more NEm and NEg, respectively, than DRC. Zinn (1990a) reported that decreasing flake density of steam-processed corn from 0.42 to 0.30 kg/L enhanced starch digestion and improved dietary nitrogen utilization; however, the effect of flake density on corn NE was small and tended to favor flakes of intermediate density (Zinn, 1990a).

Duration of steaming before flaking was associated with increased flow of nonammonia nitrogen to the duodenum (Zinn, 1990b). Although an intermediate steaming time of 47 minutes decreased digestibility of the starch, the effect on dietary DE was very slight (<2%; Zinn, 1990a). In a review article, Zinn et al. (2002) suggested that the optimal time and amount of moisture is 30 to 45 minutes with 5 to 8% moisture added. Optimal flake density was suggested to be 0.31 kg/L to 0.36 kg/L (24 to 28 lb/bu), as lighter flaking density leads to decreased intake and ruminal acidosis (Zinn et al., 2002; Sindt et al., 2006), which could interact with moisture addition (Sindt et al., 2006).

In diets containing intermediate or higher concentrations of roughage (>25%), corn is usually ground, which adversely affects digestibility (Moe and Tyrrell, 1977, 1979). Fine-ground corn can be detrimental to utilization of the roughage (Moe et al., 1973; Ørskov, 1976, 1979). Negative associative effects of starch supplementation (i.e., corn) in silage (Joanning et al., 1981) or forage-based diets (Fieser and Vanzant, 2004; Loy et al., 2007) on forage digestibility are well established.

In many areas of North America, corn is preserved wet as a high-moisture grain. Digestible DM and energy of diets containing HMC are at least equal and could be as much as 5% greater than the same diet containing dry corn (McCaffree and Merrill, 1968; McKnight et al., 1973; Tonroy et al., 1974; Galyean et al., 1976; MacLeod et al., 1976). These results are also evident in dry corn reconstituted with moisture and stored for a short period of time before feeding (Tonroy et al., 1974). As moisture concentration increases at harvest before ensiling, starch digestibility and rate of digestion also increase (Benton et al., 2005). In addition, ME concentration also increases as moisture increases in HMC (Owens et al., 1997). A minor concern about high-moisture grain and corn in particular is that most, if not all, the vitamin E can be lost during storage (Young et al., 1975). As discussed in Chapter 17, the effect of corn processing and energy content depends on the level of corn inclusion in the diet; therefore, processing of corn is influenced by whether the corn is fed in forage-based growing diets at low inclusions or whether fibrous byproducts are included along with grain in finishing diets (Corrigan et al., 2009).

Sorghum. Grain sorghum, or milo, is the likely second most common grain source fed to feedlot cattle in the United States. Whole sorghum is not digested easily by cattle; dry grinding or steam processing and rolling significantly improve the digestibility of sorghum starch and energy. In low-roughage diets, steam processing and flaking increased starch digestibility from 3 to 5% compared to dry grinding (McNeill et al., 1971; Hinman and Johnson, 1974) and DE by 5 to 10% (Buchanan-Smith et al., 1968; Husted et al., 1968). In contrast, Garrett (1968) reported that the NE value was equal in steam-flaked sorghum and ground dry sorghum. This finding might be explained by the fact that fine grinding enhanced NE by 8% relative to the coarse-rolled sorghum (Brethour, 1980). Effectiveness of steam processing and rolling of sorghum could depend on the density of flake produced. Xiong et al. (1991) found that DMI and feed efficiency tended to be higher for diets containing sorghum grain with a density of 283 vs. 437 g/L (22 vs. 34 lb/bu). These researchers estimated the lighter grain contained 2.34 Mcal/kg NEm and 1.63 Mcal/kg NEg compared with 2.21 and 1.52 Mcal/kg, respectively, for the heavier product.

Ground, reconstituted sorghum had DE equivalent to the steam-processed and rolled product (Buchanan-Smith et al., 1968; McNeill et al., 1971; Kiesling et al., 1973); however, the latter process might enhance intake (Franks et al., 1972). Dry-heat treatments—for example, micronizing, popping, exploding, and roasting—can improve sorghum nutritive value as much as steam processing and rolling (Beeson and Perry, 1982). Starch digestibility was enhanced as much by micronizing and popping as it was by steam processing and rolling (Riggs et al., 1970; Hinman and Johnson, 1974; Croka and Wagner, 1975). Nonetheless, dry-heat treatments might not be as effective as steam processing in promoting intake. Dry-rolled sorghum is more effectively utilized in intermediate- and high-roughage diets than in low-roughage diets (Keating et al., 1965).

Barley. Barley is a common grain fed to cattle in North America and globally. Barley contains greater protein and fiber, but less starch than corn grain (Hunt, 1996). In addition, barley is more variable across varieties and growing conditions than corn (Hunt, 1996). Although cattle ate more feed when they were given diets containing whole as opposed to rolled barley, efficiency of utilization was greater for the rolled-barley diets (Mathison et al., 1991a; Mathison, 1996) compared with feeding whole barley. Dehghan-banadaky et al. (2007) provided a review of barley processing that included less traditional grain processing methods such as chemical, enzymatic, and physical processes. The most common processing method is dry rolling with or without tempering, which will be the primary focus of this section. Yaramecio et al. (1991) reported NEg values of 1.15 and 1.80 Mcal/kg for diets containing whole or rolled barley, with most of this difference attributed to improved digestibility for the rolled barely. Digestion rate and in situ disappearance were much greater for masticated or processed barley than for whole barley (Beauchemin et al., 1994). Similar improvements have been observed when rolled vs. whole barley have been fed in digestion studies (Toland, 1978). There is greater controversy about the value of steam-processed and rolled barley compared to the value of dry-rolled barley. Zinn (1993) found steam-processed barley contained 2.24 Mcal/kg NEm and 1.56 Mcal/kg NEg, respectively, vs. 2.14 and 1.47 for the dry-rolled grain. In the same experiment, the benefits of a thin flake (0.19 kg/L) as opposed to a thick flake (0.39 kg/L) were evident. In contrast, steam processing of barley failed to improve the feeding value of barley in three Canadian studies (Grimson et al., 1987; Mathison et al., 1991b; Engstrom et al., 1992). Parrot et al. (1969) reported that steam processing and rolling did not improve digestibility of barley compared to dry rolling, except when the initial DE value of the barley was low. Zinn (1993) concluded that steam-rolled barley was 2.8 to 7% greater in energy (NEm) compared to dry-rolled barley which was 92 to 96% the value of SFC. Steam processing before rolling could be useful to maximize intake of barley diets, particularly in dry areas where dry-rolled or ground barley becomes too dusty. When barley is rolled or ground, fines should be avoided to minimize digestive disturbances such as bloat (Hironaka et al., 1979). As a result, tempering or adding water before rolling is a common practice that decreases fines and likely improves uniformity of rolled barley (Mathison, 1996). Similarly, gains and feed efficiency have been equal (Mathison et al., 1997) or improved (Bradshaw et al., 1996; Wang et al., 2003) when feeding tempered vs. rolled barley. High-moisture barley has a feeding value equal to dry barley (Kennelly et al., 1988) and is superior in the rolled as opposed to whole form (Rode et al., 1986). In medium- to high-roughage diets, dry-rolled barley was equivalent to the ammoniated high-moisture whole grain (Mandell et al., 1988), and steam-rolled dry barley was superior to the whole dry grain (Morgan et al., 1991). Owens et al. (1997) concluded that barley was equivalent to corn grain when compared across studies and energy values calculated; however, in direct comparisons within the same study, barley was approximately 88 (Kennington et al., 2009) to 100% (Boss and Bowman, 1996) the energy value of corn. More recently, in diets containing 35% wet corn gluten feed, DRC had a 25% greater energy value than barley (Loe et al., 2006). In a second experiment by the same authors, cattle fed corn with 50% wet corn gluten feed gained faster and more efficiently than those fed barley, suggesting that corn had 23% more energy than barley. Barley requires processing similar to other grains, but adding water (tempering) is advised to decrease fines and dust, which has equivocal effects on performance compared to dry rolling. Greater risk of ruminal acidosis is possible for barley because of greater ruminal starch digestibility compared with corn, which likely results in variability in comparisons of barley to other grains (i.e., corn).

Oats. Starch digestibility of a high-grain whole oat diet was 61%, which contrasts to 69% when the oats were dry rolled (Ørskov et al., 1980). In mixed diets, whole oat grains seem to be well-digested by cattle, and there is little benefit in further processing (Campling, 1991).

Wheat. Wheat grain can be used for finishing beef cattle, but is much less commonly used than corn grain. In some geographic areas and during different times of the year, wheat can be substituted for corn or barley grain. Wheat grain was identified as the most common secondary grain used by feedlot consultants in the United States (Vasconcelos and Galyean, 2007). Similar to barley, wheat grain is generally greater in protein and fiber than corn, but lower in starch. Nonetheless, the rate of starch digestibility in the rumen is likely greater for wheat than corn (Stock and Erickson, 2006). Starch digestibility of a high-grain, whole-wheat diet was 83%, which increased to 99% when the wheat was rolled (Ørskov et al., 1980). In contrast to oats, digestibility of starch in mixed diets containing whole wheat was only 60%, compared to 86% for the same diet when the wheat was rolled and crushed (Toland, 1978). Finely ground wheat should be avoided in beef cattle diets to maximize intake and prevent acidosis.

MINERALS

Cattle are often supplemented with minerals to ensure that dietary supply meets the requirements. In most cases, the mineral content is estimated from the chemical structure and molecular weights of the mineral provided. Rarely is a mineral 100% pure, so other minerals are also provided in combination with most targeted minerals. Table 18-3 provides the common sources of macrominerals and the associated percentage of target mineral. For most sources of macrominerals, bioavailability is not a concern. Digestion or absorption of the minerals is associated with the relative supply compared to requirements. Cattle digest or absorb a greater proportion of a mineral as dietary supply decreases. As a result, bioavailability of macrominerals is not as great of a concern as bioavailability of microminerals. Lastly, most requirement studies accounted for source, so most dietary recommendations have an established digestion coefficient provided. Table 18-4 also provides the common sources of microminerals. With microminerals, bioavailability is dramatically impacted by its source. Therefore, bioavailability estimates are provided.

TABLE 18-3 Compositions of Common Macromineral Sources on a 100% Dry Matter Basis[a]

Mineral Element Source[b]	International Feed No.[c]	Primary Mineral Element Content[d]
Calcium Sources		Ca (%)
Bone meal, steamed, fg[e]	6-00-400	30.71
Calcium carbonate, $CaCO_3$, fg	6-01-069	39.39
Calcium chloride anhydrous, $CaCl_2$, cp[g,]*	NA[f]	36.11
Calcium chloride dihydrate, $CaCl_2{\cdot}2H_2O$, cp*	NA	27.53
Calcium hydroxide, $Ca(OH)_2$, cp	NA	54.09
Calcium oxide, CaO, cp*	NA	71.47
Calcium phosphate (monobasic), $Ca(H_2PO_4)_2$, from defluorinated phosphoric acid, fg	6-01-082	16.40
Calcium sulfate dihydrate, $CaSO_4{\cdot}2H_2O$, cp	6-01-089	23.28
Dicalcium phosphate (dibasic), $CaHPO_4$, from defluorinated phosphoric acid, fg	6-01-080	22.00
Dolomitic limestone (magnesium), fg	6-02-633	22.30
Limestone, ground, fg	6-02-632	34.00
Magnesium oxide, MgO, fg	6-02-756	3.07
Oystershell, flour (ground), fg	6-03-481	38.00

continued

TABLE 18-3 Continued

Mineral Element Source[b]	International Feed No.[c]	Primary Mineral Element Content[d]
Phosphorus Sources		P (%)
Ammonium phosphate (dibasic), $(NH_4)_2HPO_4$, fg	6-00-370	20.60
Ammonium phosphate (monobasic), $(NH_4)H_2PO_4$, fg	6-09-338	24.74
Bone meal, steamed, fg	6-00-400	12.86
Calcium phosphate (monobasic), $Ca(H_2PO_4)_2$, from defluorinated phosphoric acid, fg	6-01-082	21.60
Curacao, phosphate, fg	6-05-586	14.14
Dicalcium phosphate (dibasic), $CaHPO_4$, from defluorinated phosphoric acid, fg	6-01-080	19.30
Phosphate, defluorinated, fg	6-01-780	18.00
Phosphate rock, fg	6-03-945	13.00
Phosphate rock, low-fluorine, fg	6-03-946	14.00
Phosphoric acid, H_3PO_4, fg*	6-03-707	31.60
Sodium phosphate (monobasic) monohydrate, $NaH_2PO_4 \cdot H_2O$, fg	6-04-288	22.50
Sodium tripolyphosphate (meta- and pyrophosphate) $Na_5P_3O_{10}$, fg	6-08-076	25.00
Soft rock phosphate, colloidal clay, fg	6-03-947	9.00
Potassium Sources		K (%)
Potassium bicarbonate, $KHCO_3$, cp	6-29-493	39.05
Potassium carbonate, K_2CO_3, cp	NA	56.58
Potassium chloride, KCl, fg	6-03-755	50.00
Potassium iodide, KI, fg	6-03-759	21.00
Potassium sulfate, K_2SO_4, fg	6-06-098	41.84
Magnesium Sources		Mg (%)
Dolomitic limestone (magnesium), fg	6-02-633	9.99
Limestone, ground, fg	6-02-632	2.06
Magnesium carbonate, $MgCO_3 + Mg(OH)_2$, fg	6-02-754	30.81
Magnesium chloride hexahydrate, $MgCl_2 \cdot 6H_2O$, cp	NA	11.96
Magnesium hydroxide, $Mg(OH)_2$, cp	NA	41.69
Magnesium oxide, MgO, fg	6-02-756	56.20
Magnesium sulfate heptahydrate, $MgSO_4 \cdot 7H_2O$, fg	6-02-758	9.80
Sulfur Sources		S (%)
Ammonium phosphate (dibasic), $(NH_4)_2HPO_4$, fg	6-00-370	2.16
Ammonium phosphate (monobasic), $(NH_4)H_2PO_4$, fg	6-09-338	1.46
Ammonium sulfate, $(NH_4)_2SO_4$, fg	6-09-339	24.10
Bone meal, steamed, fg	6-00-400	2.51
Calcium phosphate (monobasic), $Ca(H_2PO_4)_2$, from defluorinated phosphoric acid, fg	6-01-082	1.22
Calcium sulfate, dihydrate $CaSO_4 \cdot 2H_2O$, fg	6-01-089	18.62
Cupric sulfate pentahydrate, $CuSO_4 \cdot 5H_2O$	6-01-720	12.84
Dicalcium phosphate (dibasic), $CaHPO_4$, from defluorinated phosphoric acid, fg	6-01-080	1.14
Ferrous sulfate heptahydrate, $FeSO_4 \cdot 7H_2O$, fg	6-20-734	12.35
Magnesium sulfate heptahydrate, $MgSO_4 \cdot 7H_2O$, fg	NA	13.31
Manganese sulfate monohydrate, $MnSO_4 \cdot H_2O$, cp	NA	18.97
Manganese sulfate pentahydrate, $MnSO_4 \cdot 5H_2O$, cp	NA	13.30
Phosphoric acid, H_3PO_4, fg*	6-03-707	1.55
Potassium sulfate, K_2SO_4, fg	6-06-098	17.35
Sodium sulfate decahydrate, $Na_2SO_4 \cdot 10H_2O$, cp	6-04-292	9.95
Zinc sulfate monohydrate, $ZnSO_4 \cdot H_2O$, fg	6-05-555	17.68

[a]NRC (2007).

[b]The compositions of hydrated mineral sources (e.g., $CaSO_4 \cdot 2H_2O$) are shown including the waters of hydration. Mineral element compositions of feed-grade sources vary by source, processing method, site of mining, and manufacturer. Sources should be analyzed or manufacturer's analyses should be used when available. Element composition of a source is listed if specific element concentration is ≥1.0% for macromineral elements, or ≥10,000 mg/kg for micromineral elements, except for fluorine concentrations which are listed because of potential toxicity.

[c]First digit denotes the class of feed: 1, dry forages and roughages; 2, pastured, range plants, and forages fed green; 3, silages; 4, energy feeds; 5, protein supplement; 6, minerals; 7, vitamins; 8, additives. The other five digits identify the individual feed.

[d]Dry matter contents have been estimated for the sources; actual analysis will be more accurate.

[e]fg = Feed-grade source.

[f]NA = Not available.

[g]cp = Chemically pure form.

*Use caution when handling and mixing, can be extremely hazardous.

TABLE 18-4 Inorganic Sources and Estimated Bioavailabilities of Trace Minerals[a]

Mineral Element and Source[b]	Chemical Formula	Mineral Content (%)	Relative Bioavailability (%)
Cobalt			
Cobalt carbonate	$CoCO_3$	43 to 47	100
Cobalt sulfate	$CoSO_4 \cdot 7H2O$	21	100
Cobalt oxide	*Co_3O_4*	*24*	*0 to 20*
Copper			
Cupric sulfate (pentahydrate)	$CuSO_4 \cdot 5H_2O$	25.2	100
Cupric chloride, tribasic	$Cu_2(OH)_3Cl$	58	100
Cupric oxide	CuO	75	0 to 10
Cupric carbonate (monohydrate)	*$CuCO_3 \cdot Cu(OH)_2 \cdot H_2O$*	*50 to 55*	*60 to 100*
Cupric sulfate (anhydrous)	*$CuSO_4$*	*39.9*	*100*
Iron			
Ferrous sulfate (monohydrate)	$FeSO_4 \cdot H_2O$	30	100
Ferrous sulfate (heptahydrate)	$FeSO_4 \cdot 7H_2O$	20	100
Ferrous carbonate	$FeCO_3$	38	15 to 80
Ferric oxide	*Fe_2O_3*	*69.9*	*0*
Ferric chloride (hexahydrate)	*$FeCl_3 \cdot 6H_2O$*	*20.7*	*40 to 100*
Ferrous oxide	*FeO*	*77.8*	—[c]

continued

TABLE 18-4 Continued

Mineral Element and Source[b]	Chemical Formula	Mineral Content (%)	Relative Bioavailability (%)
Iodine			
Ethylenediamine dihydroiodide (EDDI)	$C_2H_8N_22HI$	79.5	100
Calcium iodate	$Ca(IO_3)_2$	63.5	100
Potassium iodide	KI	68.8	100
Potassium iodate	KIO_3	*59.3*	—[c]
Cupric iodide	*CuI*	*66.6*	*100*
Manganese			
Manganous sulfate (monohydrate)	$MnSO_4{\cdot}H_2O$	29.5	100
Manganous oxide	MnO	60	70
Manganous dioxide	MnO_2	*63.1*	*35 to 95*
Manganous carbonate	$MnCO_3$	*46.4*	*30 to 100*
Manganous chloride (tetrahydrate)	$MnCl_2{\cdot}4H_2O$	*27.5*	*100*
Selenium			
Sodium selenite	Na_2SeO_3	45	100
Sodium selenate (decahydrate)	$Na_2SeO_4{\cdot}10H_2O$	*21.4*	*100*
Zinc			
Zinc sulfate (monohydrate)	$ZnSO_4{\cdot}H_2O$	35.5	100
Zinc oxide	ZnO	72	50 to 80
Zinc sulfate (heptahydrate)	$ZnSO_4{\cdot}7H_2O$	*22.3*	*100*
Zinc carbonate	$ZnCO_3$	*56*	*100*
Zinc chloride	$ZnCl_2$	*48*	*100*

[a]The mineral source listed first under each mineral element was generally the standard with which the other sources were compared to establish relative bioavailability.

[b]Less commonly used sources in italics.

[c]No data available.

REFERENCES

Abouguendia, Z. 1998. Nutrient Content and Digestibility of Saskatchewan Range Plants. Saskatchewan Agriculture Development Fund Project No. 94000114. Available online at http://www1.foragebeef.ca/$foragebeef/frgebeef.nsf/all/safrr76/$FILE/nutrient.pdf. Accessed on May 7, 2015.

Beardsley, D. W. 1964. Symposium on forage utilization: Nutritive value of forage as affected by physical form. Part II. Beef cattle and sheep studies. *Journal of Animal Science* 23:239-245.

Beauchemin, K. A., T. A. McAllister, Y. Dong, B. I. Fair, and K. J. Cheng. 1994. Effects of mastication on digestion of whole cereal grains by cattle. *Journal of Animal Science* 72:236-246.

Beeson, W. M., and T. W. Perry. 1982. Effect of processing on nutritive value of feeds: Cereal grains. Pp. 193-212 in *Handbook of Nutritive Value of Processed Food*, M. Rechcigl, ed. Boca Raton, FL: CRC Press.

Benton, J. R., G. E. Erickson, T. Klopfenstein, C. N. Macken, and K. J. Vander Pol. 2005. Effects of corn moisture and length of ensiling on dry matter digestibility and rumen degradable protein. Pp. 31-33 in *Nebraska 2005 Beef Report MP83A*. Lincoln: University of Nebraska.

Berger, L. L., G. C. Fahey, L. D. Bourquin, and E. C. Titgemeyer. 1994. Modification of forage quality after harvest. Pp. 922-966 in *Forage Quality, Evaluation, and Utilization*, G. C. Fahey, M. Collins, D. R. Mertens, and L. E. Moser, eds. Madison, WI: American Society of Agronomy, Crop Science Society, and Soil Science Society.

Bines, J. A., W. H. Broster, J. D. Stutton, V. J. Broster, D. J. Napper, T. Smith, and J. W. Siviter. 1988. Effect of amount consumed and diet composition on the apparent digestibility in cattle and sheep. *Journal of Agricultural Science* 110:249-259.

Blaxter, K. L., N. M. Graham, and F. W. Wainman. 1956. Some observations on the digestibility of food by sheep, and on related problems. *British Journal of Nutrition* 10:69-91.

Bock, B. J., R. T. Brandt, D. L. Harmon, S. J. Anderson, J. K. Elliott, and T. B. Avery. 1991. Mixtures of wheat and high-moisture corn in finishing diets: Feedlot performance and in situ rate of starch digestion in steers. *Journal of Animal Science* 69:2703-2710.

Boss, D. L., and J. G. P. Bowman. 1996. Barley varieties for finishing steers: I. Feedlot performance, in vivo diet digestion, and carcass characteristics. *Journal of Animal Science* 74:1967-1972.

Bradshaw, W. L., D. D. Hinman, R. C. Bull, D. O. Everson, and S. J. Sorenson. 1996. Effects of barley variety and processing methods on feedlot steer performance and carcass characteristics. *Journal of Animal Science* 74:18-24.

Brethour, J. R. 1980. Nutritional value of milo for cattle. Pp. 5-8 in *Report of Progress 384, Roundup 67*. Manhattan: Kansas State University.

Buchanan-Smith, J. G., R. Totusek, and A. D. Tillman. 1968. Effect of methods of processing on digestibility and utilization of grain sorghum by cattle and sheep. *Journal of Animal Science* 27:525-530.

Buckner, C. D., T. J. Klopfenstein, K. M. Rolfe, W. A. Griffin, M. J. Lamothe, A. K. Watson, J. C. MacDonald, W. H. Schacht, and P. Schroeder. 2013. Ruminally undegradable protein content and digestibility for forages using the mobile bag in situ technique. *Journal of Animal Science* 91:2812-2822.

Burns, J. C., and D. S. Fisher. 2011. Stocking strategies as related to animal and pasture productivity of endophyte-free tall fescue. *Crop Science* 51:2868-2877.

Burns, J. C., and D. S. Fisher. 2013. Steer performance and pasture productivity among five perennial warm-season grasses. *Agronomy Journal* 105:113-123.

Burns, J. C., D. S. Fisher, K. R. Pond, and D. H. Timothy. 1992. Diet characteristics, digesta kinetics, and dry matter intake of steers grazing eastern gamagrass. *Journal of Animal Science* 70:1251-1261.

Burns, J. C., K. R. Pond, D. S. Fisher, and J. M. Luginbuhl. 1997. Changes in forage quality, ingestive mastication, and digesta kinetics resulting from switchgrass maturity. *Journal of Animal Science* 75:1368-1379.

Burns, J. C., M. G. Wagger, and D. S. Fisher. 2009. Animal and pasture productivity of "Coastal" and "Tifton 44" bermudagrass at three nitrogen rates and associated soil nitrogen status. *Agronomy Journal* 101:32-40.

Burns, J. C., D. S. Fisher, and K. R. Pond. 2011. Tall fescue forage mass and canopy characteristics on steer ingestive behavior and performance. *Crop Science* 51:1850-1864.

Burns, J. C., D. S. Fisher, and K. R. Pond. 2012a. Steer performance and pasture productivity of a tall fescue-bermudagrass system compared with yellow bluestem and coastal panicgrass. *The Professional Animal Scientist* 28:272-283.

Burns, J. C., D. S. Fisher, and K. R. Pond. 2012b. Steer performance, intake, digesta kinetics, and pasture productivity of flaccidgrass at each of three forage masses. *Agronomy Journal* 104:26-35.

Campling, R. C. 1991. Processing cereal grains for cattle—A review. *Livestock Production Science* 28:223-234.

Campling, R. C., and M. Freer. 1966. Factors affecting the voluntary intake of food by cows. 8. Experiments with ground, pelleted roughages. *British Journal of Nutrition* 20:229-244.

Caton, J. S., D. O. Erickson, D. A. Carey, and D. L. Ulmer. 1993. Influence of *Aspergillus oryzae* fermentation extract on forage intake, site of digestion, in situ degradability, and duodenal amino acid flow in steers grazing cool-season pasture. *Journal of Animal Science* 71:779-787.

Cline, H. J., B. W. Neville, G. P. Lardy, and J. S. Caton. 2009. Influence of advancing season on dietary composition, intake, site of digestion, and microbial efficiency in beef steers grazing a native range in western North Dakota. *Journal of Animal Science* 87:375-383.

Coleman, S. W., O. Neri-Flores, R. J. Allen, Jr., and J. E. Moore. 1978. Effect of pelleting and of forage maturity on quality of two sub-tropical forage grasses. *Journal of Animal Science* 46:1103-1112.

Cooper, R. J., C. T. Milton, T. J. Klopfenstein, and D. J. Jordon. 2002a. Effect of corn processing on degradable intake protein requirement of finishing cattle. *Journal of Animal Science* 80:242-247.

Cooper, R. J., C. T. Milton, T. J. Klopfenstein, T. L. Scott, C. B. Wilson, and R. A. Mass. 2002b. Effect of corn processing on starch digestion and bacterial crude protein flow in finishing cattle. *Journal of Animal Science* 80:797-804.

Corah, L. R., and D. A. Dargatz. 1996. *Forage Analysis from Cow-Calf Herds in 18 States: Beef CHAPA, Cow/Calf Health and Productivity Audit.* Fort Collins, CO: U.S. Department of Agriculture, Animal and Plant Health Inspection Services, Veterinary Services, Centers for Epidemiology and Animal Health. Available online at http://www.aphis.usda.gov/animal_health/nahms/beefcowcalf/downloads/chapa/CHAPA_dr_ForageAnal.pdf. Accessed on May 8, 2015.

Corrigan, M. E., G. E. Erickson, T. J. Klopfenstein, M. K. Luebbe, K. J. Vander Pol, N. F. Meyer, C. D. Buckner, S. J. Vanness, and K. J. Hanford. 2009. Effect of corn processing method and corn wet distillers grains plus solubles inclusion level in finishing steers. *Journal of Animal Science* 87:3351-3362.

Croka, D. C., and D. G. Wagner. 1975. Micronized sorghum grain. III. Energetic efficiency for feedlot cattle. *Journal of Animal Science* 40:936-939.

D'Alfonso, T. H., W. B. Roush, and J. A. Ventura. 1992. Least-cost poultry rations with nutrient variability: A comparison of linear programming with a margin of safety and stochastic programming models. *Poultry Science* 71:255-262.

Dehghan-banadaky, M., R. Corbett, and M. Oba. 2007. Effects of barley grain processing on productivity of cattle. *Animal Feed Science and Technology* 137:1-24.

Engstrom, D. F., G. W. Mathison, and L. A. Goonewardene. 1992. Effect of β-glucan, starch, and steam vs dry rolling of barley grain on its degradability and utilization by steers. *Animal Feed Science and Technology* 37:33-46.

Fahey, G. C., L. D. Bourquin, E. C. Titgemeyer, and D. G. Atwell. 1993. Postharvest treatment of fibrous feedstuffs to improve their nutritive value. Pp. 715-766 in *Forage Cell Wall Structure and Digestibility,* H. G. Jung, D. R. Buxton, R. D. Hatfield, and J. Ralph, ed. Madison, WI: American Society of Agronomy, Crop Science Society, and Soil Society of America.

Fieser, B. G., and E. S. Vanzant. 2004. Interactions between supplement energy source and tall fescue hay maturity on forage utilization by beef steers. *Journal of Animal Science* 82:307-318.

Fisher, D. S., J. C. Burns, K. R. Pond, R. D. Mochrie, and D. H. Timothy. 1991. Effects of grass species on grazing steers: I. Diet composition and ingestive mastication. *Journal of Animal Science* 69:1188-1198.

Franks, L. G., J. R. Newsom, R. E. Renbarger, and R. Totusek. 1972. Relationship of rumen volatile fatty acids to type of grains, sorghum grain processing method and feedlot performance. *Journal of Animal Science* 35:404-409.

Galyean, M. L., D. G. Wagner, and F. N. Owens. 1976. Site and extent of starch digestion in steers fed processed corn rations. *Journal of Animal Science* 43:1088-1094.

Galyean, M. L., D. G. Wagner, and F. N. Owens. 1981. Dry matter and starch disappearance of corn and sorghum as influenced by particle size and processing. *Journal of Dairy Science* 64:1804-1812.

Garrett, W. N. 1968. Influence of the method of processing on the feeding value of milo and wheat. Pp. 36-46 in *Eighth Annual California Feeders Day Report.* Davis: University of California.

Giesert, B. G., T. J. Klopfenstein, D. C. Adams, J. A. Musgrave, and W. H. Schacht. 2008. Determination of diet protein and digestibility of native sandhills upland range. Pp. 22-24 in *Nebraska Beef Cattle Report MP91.* Lincoln: University of Nebraska.

Goetsch, A. L., F. N. Owens, M. A. Funk, and B. E. Doran. 1987. Effects of whole or ground corn with different forms of hay in 85% concentrate diets on digestion and passage rates in beef heifers. *Animal Feed Science and Technology* 18:151-164.

Goodrich, R. D., and J. C. Meiske. 1966. Whole corn grain vs. ground corn grain, long hay vs. ground hay and 10 lb. vs 15 lb. corn silage for finishing cattle. Pp. 61-67 in *Minnesota Beef Cattle Feeders Day Research Report B-76.* Minneapolis: University of Minnesota.

Greenhalgh, J. F. D., and G. W. Reid. 1973. The effects of pelleting various diets on intake and digestibility in sheep and cattle. *Animal Production* 16:223-233.

Greenhalgh, J. F. D., and G. W. Reid. 1974. Long- and short-term effects of pelleting a roughage for sheep. *Animal Production* 19:77-86.

Greenquist, M. A., T. J. Klopfenstein, W. H. Schacht, G. E. Erickson, K. J. Vander Pol, M. K. Luebbe, K. R. Brink, A. K. Schwarz, and L. B. Baleseng. 2009. Effects of nitrogen fertilization and dried distillers grains supplementation: Forage use and performance of yearling steers. *Journal of Animal Science* 87:3639-3646.

Griffin, W. A., T. J. Klopfenstein, L. A. Stalker, G. E. Erickson, J. A. Musgrave, and R. N. Funston. 2012. The effects of supplementing dried distillers grains to steers grazing cool-season meadow. *The Professional Animal Scientist* 28:56-63.

Grimson, R. E., R. D. Weisenburger, J. A. Basarab, and R. P. Stilborn. 1987. Effects of barley volume-weight and processing method on feedlot performance of finishing steers. *Canadian Journal of Animal Science* 67:43-53.

Harmon, D. L., R. M. Yamka, and N. A. Elam. 2004. Factors affecting intestinal starch digestion in ruminants: A review. *Canadian Journal of Animal Science* 84:309-318.

Hinman, D. D., and R. R. Johnson. 1974. Influence of processing methods on digestion of sorghum starch in high-concentrate beef cattle rations. *Journal of Animal Science* 39:417-422.

Hironaka, R., N. Kimura, and G. C. Kozub. 1979. Influence of feed particle size on rate and efficiency of gain, characteristics of rumen fluid and rumen epithelium, and numbers of rumen protozoa. *Canadian Journal of Animal Science* 59:395-402.

Hirschfeld, D. J., D. R. Kirby, J. S. Caton, S. S. Silcox, and K. C. Olson. 1996. Influence of grazing management on intake and composition of cattle diets. *Journal of Range Management* 49:257-263.

Hunt, C. W. 1996. Factors affecting the feeding quality of barley for ruminants. *Animal Feed Science and Technology* 62:37-48.

Huntington, G. B. 1997. Starch utilization by ruminants: From basics to the bunk. *Journal of Animal Science* 75:852-867.

Huntington, G. B., D. L. Harmon, and C. J. Richards. 2006. Sites, rates, and limits of starch digestion and glucose metabolism in growing cattle. *Journal of Animal Science* 84:E14-E24.

Husted, W. T., S. Mehen, W. H. Hale, M. Little, and B. Theurer. 1968. Digestibility of milo processed by different methods. *Journal of Animal Science* 27:531-534.

Joanning, S. W., D. E. Johnson, and B. P. Barry. 1981. Nutrient digestibility depressions in corn silage-corn mixtures fed to steers. *Journal of Animal Science* 53:1095-1103.

Johnson, J. A., J. S. Caton, W. Poland, D. R. Kirby, and D. V. Dhuyvetter. 1998. Influence of season on dietary composition, intake, and digestion by beef steers grazing mixed-grass prairie in the northern Great Plains. *Journal of Animal Science* 76:1682-1690.

Keating, E. R., W. J. Saba, W. H. Hale, and B. Taylor. 1965. Further observations on the digestion of milo and barley by steers and lambs. *Journal of Animal Science* 24:1080-1085.

Kennelly, J. J., G. W. Mathison, and G. de Boer. 1988. Influence of high-moisture barley on the performance and carcass characteristics of feedlot cattle. *Canadian Journal of Animal Science* 68:811-820.

Kennington, L. R., J. I. Szasz, C. W. Hunt, D. D. Hinman, and S. J. Sorensen. 2009. Effect of degradable intake protein level on performance of feedlot steers fed dry-rolled corn- or barley-based finishing diets. *The Professional Animal Scientist* 25:762-767.

Kiesling, H. E., J. E. McCroskey, and D. G. Wagner. 1973. A comparison of energetic efficiency of dry-rolled and reconstituted rolled sorghum grain by steers using indirect calorimetry and the comparative slaughter technique. *Journal of Animal Science* 37:790-795.

Klopfenstein, T. 1978. Chemical treatment of crop residues. *Journal of Animal Science* 46:841-848.

Klopfenstein, T. J., G. E. Erickson, and L. L. Berger. 2013. Maize is a critically important source of food, feed, energy and forage in the USA. *Field Crops Research* 153:5-11.

Krysl, L. J., M. L. Galyean, J. D. Wallace, F. T. McCollum, M. B. Judkins, M. E. Branine, and J. S. Caton. 1987. *Cattle Nutrition on Blue Grama Rangeland in New Mexico.* Bulletin 727. Las Cruces: New Mexico State University Agricultural Experiment Station.

Lesperance, A. L., E. A. Jensen, V. R. Bohman, and R. H. Madsen. 1960. Measuring selective grazing with fistulated steers. *Journal of Dairy Science* 43:1615-1622.

Loe, E. R., M. L. Bauer, and G. P. Lardy. 2006. Grain source and processing in diets containing varying concentrations of wet corn gluten feed for finishing cattle. *Journal of Animal Science* 84:986-996.

Loy, D. 2007. By-product feed utilization by grazing cattle. *Veterinary Clinics Food Animals* 23:41-52.

Loy, T. W., J. C. MacDonald, T. J. Klopfenstein, and G. E. Erickson. 2007. Effect of distillers grains or corn supplementation frequency on forage intake and digestibility. *Journal of Animal Science* 85:2625-2630.

Loy, T. W., T. J. Klopfenstein, G. E. Erickson, C. N. Macken, and J. C. MacDonald. 2008. Effect of supplemental energy source and frequency on growing calf performance. *Journal of Animal Science* 86:3504-3510.

Macken, C. N., G. E. Erickson, T. J. Klopfenstein, and R. A. Stock. 2006. Effects of corn processing method and protein concentration in finishing diets containing wet corn gluten feed on cattle performance. *The Professional Animal Scientist* 22:14-22.

MacLeod, G. K., D. N. Mowat, and R. A. Curtis. 1976. Feeding value for finishing steers and Holstein male calves of whole dried corn and of whole and rolled high moisture acid-treated corn. *Canadian Journal of Animal Science* 56:43-49.

Mader, T. L., J. M. Dahlquist, R. A. Britton, and V. E. Krause. 1991. Type and mixtures of high-moisture corn in beef cattle finishing diets. *Journal of Animal Science* 69:3480-3486.

Mandell, I. B., H. H. Nicholson, and G. I. Christison. 1988. The effects of barley processing on nutrient digestion within the gastro-intestinal tract of beef cattle fed mixed diets. *Canadian Journal of Animal Science* 68:191-198.

Mathison, G. W. 1996. Effects of processing on the utilization of grain by cattle. *Animal Feed Science and Technology* 58:113-125.

Mathison, G. W., D. F. Engstrom, and D. D. MacLeod. 1991a. Effect of feeding whole and rolled barley to steers in the morning or afternoon in diets containing differing proportions of hay and grain. *Animal Production* 53:321-330.

Mathison, G. W., R. Hironaka, B. K. Kerrigan, I. Vlach, L. P. Milligan, and R. D. Wesenburger. 1991b. Rate of starch degradation, apparent digestibility and rate and efficiency of steer gain as influenced by grain volume-weight and processing method. *Canadian Journal of Animal Science* 71:867-878.

Mathison, G. W., D. F. Engstrom, R. Soofi-Siawash, and D. Gibb. 1997. Effects of tempering and degree of processing of barley grain on the performance of bulls in the feedlot. *Canadian Journal of Animal Science* 77:421-429.

Matsushima, J. K., J. I. Sprague, and D. E. Johnson. 1964. Flaked or cracked grain—which is superior for beef cattle? *Colorado Farm and Home Research, Colorado Experiment Station* 14(4):6-8.

Maxson, W. E., R. L. Shirley, J. E. Bertrand, and A. Z. Palmer. 1973. Energy values of corn, bird-resistant and non-bird-resistant sorghum grain in rations fed to steers. *Journal of Animal Science* 37:1451-1457.

McAllister, T. M., K. J. Cheng, L. M. Rode, and J. G. Buchanan-Smith. 1990a. Use of formaldehyde to regulate digestion of barley starch. *Canadian Journal of Animal Science* 70:581-589.

McAllister, T. M., L. M. Rode, D. J. Major, K. J. Cheng, and J. G. Buchanan-Smith. 1990b. Effect of ruminal microbial colonization on cereal grain digestion. *Canadian Journal of Animal Science* 70:571-579.

McCaffree, J. D., and W. G. Merrill. 1968. High moisture corn for dairy cows in early lactation. *Journal of Dairy Science* 51:553-560.

McDonald, P., A. R. Henderson, and S. J. E. Heron. 1991. *The Biochemistry of Silage*, 2nd Ed. Marlow, UK: Chalcombe Publications.

McKnight, D. R., G. K. MacLeod, J. G. Buchanan-Smith, and D. N. Mowat. 1973. Utilization of ensiled or acid-treated high-moisture shelled corn by cattle. *Canadian Journal of Animal Science* 53:491-496.

McMurphy, C. P., E. D. Sharman, D. A. Cox, M. E. Payton, G. W. Horn, and D. L. Lalman. 2011. Effects of implant type and protein source on growth of steers grazing summer pasture. *The Professional Animal Scientist* 27:402-409.

McNeill, J. W., G. D. Potter, and J. K. Riggs. 1971. Ruminal and postruminal carbohydrate utilization in steers fed processed sorghum grain. *Journal of Animal Science* 33:1371-1374.

Minson, D. J. 1963. The effect of pelleting and wafering on the feeding value of roughage—A review. *Journal of the British Grassland Society* 18:39-44.

Minson, D. J., and R. Milford. 1968. The nutritional value of four tropical grasses when fed as chaff and pellets to sheep. *Australian Journal of Experimental Agriculture and Animal Husbandry* 8:270-276.

Moe, P. W., and H. F. Tyrrell. 1977. Effects of feed intake and physical form on energy value of corn in timothy hay diets for lactating cows. *Journal of Dairy Science* 60:752-758.

Moe, P. W., and H. F. Tyrrell. 1979. Effect of endosperm type on incremental energy value of corn grain for dairy cows. *Journal of Dairy Science* 62:447-454.

Moe, P. W., H. F. Tyrrell, and N. W. Hoover. 1973. Physical form and energy value of corn in Timothy hay diets for lactating cows. *Journal of Dairy Science* 60:752-758.

Morgan, C. A., and R. C. Campling. 1978. Digestibility of whole barley and oat grains by cattle of different ages. *Animal Production* 27:323-329.

Morgan, E. K., M. L. Gibson, M. L. Nelson, and J. R. Males. 1991. Utilization of whole or steamrolled barley fed with forages to wethers and cattle. *Animal Feed Science and Technology* 33:59-78.

Morris, P. J., and D. N. Mowat. 1980. Nutritive value of ground and/or ammoniated corn stover. *Canadian Journal of Animal Science* 60:327-336.

Mortimer, R. G., D. A. Dargatz, and L. R. Corah. 1999. *Forage Analysis from Cow-Calf Herds in 23 States.* Fort Collins, CO: U.S. Department of Agriculture, Animal and Plant Health Inspection Services, Veterinary Services, Centers for Epidemiology and Animal Health. Available online at http://www.aphis.usda.gov/animal_health/nahms/beefcowcalf/downloads/beef97/Beef97_dr_ForageAnal.pdf. Accessed on May 8, 2015.

NRC (National Research Council). 1996. *Nutrient Requirements of Beef Cattle*, 7th Rev. Ed. Washington, DC: National Academy Press.

NRC. 2000. *Nutrient Requirements of Beef Cattle: Update 2000*, 7th Rev. Ed. Washington, DC: National Academy Press.

NRC. 2007. *Nutrient Requirements of Small Ruminants.* Washington, DC: The National Academies Press.

Nuttelman, B. L., M. K. Lubbe, T. J. Klopfenstein, J. R. Benton, and G. E. Erickson. 2010. Comparing the energy value of wet distillers grains to dry rolled corn in high forage diets. Pp. 49-50 in *Nebraska Beef Cattle Report MP93.* Lincoln: University of Nebraska.

Olson, K. C., J. S. Caton, D. R. Kirby, and P. L. Norton. 1994. Influence of yeast culture supplementation and advancing season on steers grazing mixed-grass prairie in the Northern Great Plains: I. Dietary composition, intake, and in situ nutrient disappearance. *Journal of Animal Science* 72:2149-2157.

Ørskov, E. R. 1976. The effect of processing on digestion and utilization of cereals by ruminants. *Proceedings of the Nutrition Society* 35:245-252.

Ørskov, E. R. 1979. Recent information on processing of grain for ruminants. *Livestock Production Science* 6:335-347.

Ørskov, E. R., R. J. Barnes, and B. A. Lukins. 1980. A note on the effect of different amounts of NaOH application on digestibility by cattle of barley, oats, wheat and maize. *Journal of Agricultural Science* 94:271-273.

Owens, F. N., R. A. Zinn, and Y. K. Kim. 1986. Limitations to starch digestion in the ruminant small intestine. *Journal of Animal Science* 63:1634-1648.

Owens, F. N., D. S. Secrist, W. J. Hill, and D. R. Gill. 1997. The effect of grain source and grain processing on performance of feedlot cattle: A review. *Journal of Animal Science* 75:868-879.

Parrot, J. C., S. Mehen, W. H. Hale, M. Little, and B. Theurer. 1969. Digestibility of dry rolled and steam processed flaked barley. *Journal of Animal Science* 28:425-428.

Peterson, M., M. K. Luebbe, R. J. Rasby, T. J. Klopfenstein, G. E. Erickson, and L. M. Kovarik. 2009. Level of wet distillers grains plus solubles and solubles ensiled with wheat straw for growing steers. Pp. 35-36 in *Nebraska Beef Report MP92.* Lincoln: University of Nebraska.

Preston, R. L. 1975. Net energy evaluation of cattle finishing rations containing varying proportions of corn grain and corn silage. *Journal of Animal Science* 41:622-624.

Richards, C. J., and B. Hicks. 2007. Processing of corn and sorghum for feedlot cattle. *Veterinary Clinics of North America: Food Animal Practice* 23:207-221.

Riggs, J. K., J. W. Sorenson, J. L. Adame, and L. M. Schake. 1970. Popped sorghum grain for finishing beef cattle. *Journal of Animal Science* 30:634-638.

Rode, L. M., K. J. Cheng, and J. W. Costerton. 1986. Digestion by cattle of urea-treated, ammonia-treated or rolled high moisture barley. *Canadian Journal of Animal Science* 66:711-721.

Scott, T. L., C. T. Milton, G. E. Erickson, T. J. Klopfenstein, and R. A. Stock. 2003. Corn processing method in finishing diets containing wet corn gluten feed. *Journal of Animal Science* 81:3182-3190.

Sherrod, L. B., R. C. Albin, and R. D. Furr. 1969. Net energy of regular and waxy sorghum grains for finishing steers. *Journal of Animal Science* 29:997-1000.

Silcox, S. C. 1991. Cattle Diets in the Northern Great Plains: Seasonal Influences on Composition, Intake, and In Situ Degradability. M.S. Thesis, North Dakota State University, Fargo.

Sindt, J. J., J. S. Drouillard, E. C. Titgemeyer, S. P. Montgomery, E. R. Loe, B. E. Depenbusch, and P. H. Walz. 2006. Influence of steam-flaked corn moisture level and density on the site and extent of digestibility and feeding value for finishing cattle. *Journal of Animal Science* 84:424-432.

Stock, R. A., and G. E. Erickson. 2006. Associative effects and management—combinations of processed grains. Pp. 167-172 in *Proceedings of the Cattle Grain Processing Symposium, November 15-17, 2006, Tulsa, OK.* MP-177. Stillwater: University of Oklahoma. Available online at http://beefextension.com/proceedings/cattle_grains06/06-23.pdf. Accessed on May 8, 2015.

Stock, R. A., D. R. Brink, R. T. Brandt, J. K. Merrill, and K. K. Smith. 1987. Feeding combinations of high moisture corn and dry corn to finishing cattle. *Journal of Animal Science* 65:282-289.

Stock, R. A., M. H. Sindt, R. M. Cleale, and R. A. Britton. 1991. High-moisture corn utilization in finishing cattle. *Journal of Animal Science* 69:1645-1656.

Streeter, M. N., D. G. Wagner, F. N. Owens, and C. A. Hibberd. 1989. Combinations of high-moisture harvested sorghum grain and dry-rolled corn: Effects on site and extent of digestion in beef heifers. *Journal of Animal Science* 67:1623-1633.

Sundstol, F. 1991. Large scale utilization of straw for ruminant production systems. Pp. 55-60 in *Recent Advances on the Nutrition of Herbivores*, W. Ho, H. K. Wong, N. Abdullah, and Z. A. Tajuddin, eds. Kuala Lumpur: Malaysian Society of Animal Production.

Tedeschi, L. O., D. G. Fox, A. N. Pell, D. P. D. Lanna, and C. Boin. 2002. Development and evaluation of a tropical feed library for the Cornell Net Carbohydrate and Protein System model. *Scientia Agricola* 59(1):1-18.

Tedeschi, L. O., D. G. Fox, and P. H. Doane. 2005. Evaluation of the tabular feed energy and protein undegradability values of the National Research Council nutrient requirements of beef cattle. *The Professional Animal Scientist* 21:403-415.

Theurer, C. B. 1986. Grain processing effects on starch utilization by ruminants. *Journal of Animal Science* 63:1649-1662.

Thomson, D. J., and D. E. Beever. 1980. The effect of conservation on the digestion of forages by ruminants. Pp. 291-308 in *Digestive Physiology and Metabolism in Ruminants: Proceedings of the 15th International Symposium on Ruminant Physiology*, Y. Ruckebusch and P. Thivend, eds. Lancaster, UK: MTP Press.

Toland, P. C. 1978. Influence of some digestive processes on the digestion by cattle of cereal grains fed whole. *Australian Journal of Experimental Agriculture and Animal Husbandry* 18:29-33.

Tonroy, B. R., T. W. Perry, and W. M. Beeson. 1974. Dry, ensiled high-moisture, ensiled reconstituted high-moisture and volatile fatty acid treated high-moisture corn for growing-finishing beef cattle. *Journal of Animal Science* 39:931-936.

Turgeon, O. A., D. R. Brink, and R. A. Britton. 1983. Corn particle size mixtures, roughage level and starch utilization in finishing steer diets. *Journal of Animal Science* 57:739-749.

Vance, R. D., R. R. Johnson, E. W. Klosterman, B. W. Dehority, and R. L. Preston. 1970. All-concentrate rations for growing-finishing cattle. Pp. 49-60 in Beef Cattle Research—1970. Research Summary 43. Wooster: Ohio Agricultural Research and Development Center. Available online at http://kb.osu.edu/dspace/bitstream/handle/1811/59334/OARDC_research_summary_n43.pdf?sequence=1. Accessed on May 11, 2015.

Vance, R. D., R. L. Preston, E. W. Klosterman, and V. R. Cahill. 1972. Utilization of whole shelled and crimped corn grain with varying proportions of corn silage by growing-finishing steers. *Journal of Animal Science* 35:598-605.

Vander Pol, K. J., M. A. Greenquist, G. E. Erickson, T. J. Klopfenstein, and T. Robb. 2008. Effect of corn processing in finishing diets containing wet distillers grains on feedlot performance and carcass characteristics of finishing steers. *The Professional Animal Scientist* 24:439-444.

Vasconcelos, J. T., and M. L. Galyean. 2007. Nutritional recommendations of feedlot consulting nutritionists: The 2007 Texas Tech University survey. *Journal of Animal Science* 85:2772-2781.

Wang, Y., D. Greer, and T. A. McAllister. 2003. Effects of moisture, roller setting, and saponin-based surfactant on barley processing, ruminal degradation of barley, and growth performance by feedlot steers. *Journal of Animal Science* 81:2145-2154.

Weiss, W. P., H. R. Conrad, and N. R. St. Pierre. 1992. A theoretically-based model for predicting total digestible nutrient values of forages and concentrates. *Animal Feed Science Technology* 39:95-110.

Weston, R. H. 1967. Factors limiting the intake of feed by sheep. II. Studies with wheaten hay. *Australian Journal of Agricultural Research* 18:983-1002.

Wilkinson, J. M. 1978. The ensiling of forage maize: Effects on composition and nutritive value. Pp. 201-237 in *Forage Maize*, E. S. Bunting, B. F. Pain, R. H. Phipps, J. M. Wilkinson, and R. E. Gunn, eds. London: Agricultural Research Council.

Xiong, Y., S. J. Bartle, and R. L. Preston. 1991. Density of steam-flaked sorghum grain, roughage level, and feeding regimen for feedlot steers. *Journal of Animal Science* 69:1707-1718.

Yaramecio, B. J., G. W. Mathison, D. F. Engstrom, L. A. Roth, and W. R. Caine. 1991. Effect of ammoniation on the preservation and feeding value of barley grain for growing-finishing cattle. *Canadian Journal of Animal Science* 71:439-455.

Young, L. G., A. Lun, J. Pos, R. P. Forshaw, and D. Edmeades. 1975. Vitamin E stability in corn and mixed feed. *Journal of Animal Science* 40:495-499.

Zinn, R. A. 1987. Influence of lasalocid and monensin plus tylosin on comparative feeding value of steam-flaked versus dry-rolled corn in diets for feedlot cattle. *Journal of Animal Science* 65:256-266.

Zinn, R. A. 1990a. Influence of flake density on the comparative feeding value of steam-flaked corn for feedlot cattle. *Journal of Animal Science* 68:767-778.

Zinn, R. A. 1990b. Influence of steaming time on site of digestion of flaked corn in steers. *Journal of Animal Science* 68:776-781.

Zinn, R. A. 1993. Influence of processing on the comparative feeding value of barley for feedlot cattle. *Journal of Animal Science* 71:3-10.

Zinn, R. A., C. F. Adam, and M. S. Tamayo. 1995. Interaction of feed intake level on comparative ruminal and total tract digestion of dry-rolled and steam-flaked corn. *Journal of Animal Science* 73:1239-1245.

Zinn, R. A., F. N. Owens, and R. A. Ware. 2002. Flaking corn: Processing mechanics, quality standards, and impacts on energy availability and performance of feedlot cattle. *Journal Animal Science* 80:1145-1156.

19

Model Equations and Sensitivity Analyses

INTRODUCTION

One of the primary purposes of developing and applying models, such as the model presented in this edition of *Nutrient Requirements of Beef Cattle*, is to integrate and apply complex research-based knowledge to improve nutrient management through refined animal feeding and predictions of nitrogen (N) and phosphorus (P) excretion, and methane (CH_4) emission. However, the level of aggregation of the equations in the model depends on its intended use, reliability of the information available, knowledge of the user, and risk of use. Having accurate predictions of nutrient requirements for animals in a given production setting minimizes overfeeding of nutrients, increases efficiency of nutrient utilization, maximizes performance, and decreases nutrient excretion. Livestock excretion of N, P, and other minerals poses the risk of groundwater and soil contamination in areas of intensified animal production operations (73 Federal Register 70418 [2008]). Nonetheless, by using modeling techniques that predict animal requirements and by matching these predicted requirements with dietary nutrients in each unique production situation, producers have made significant strides to optimize performance while addressing environmental concerns. For example, the use of an applied nutrition model to formulate dairy cattle diets resulted in a decrease in N and P excretions of about one-third, as well as a substantial decrease in feed costs (Fox et al., 2006). Different feeding and management strategies to mitigate the impact of N and P on the environment exist for beef cattle production (Vasconcelos et al., 2007), and applied nutrition models are often needed for their implementation in practical conditions. Food-producing animals are also often identified as a source of atmospheric methane. Cattle typically lose 3 to 10% of ingested energy as eructated methane, which is equivalent to approximately 145 g methane/d for an average steer (Johnson and Johnson, 1995; IPCC, 2006). Development of management strategies, including modeling to predict nutrient requirements more accurately, can mitigate methane emissions from cattle by enhancing nutrient utilization and feed efficiency. Thus, the application of models in agricultural animal production has the potential to significantly decrease nutrient loading of the environment while providing economic benefits and tangible returns to those who implement these systems for improved animal feeding. More detailed information about beef cattle nutrition and the environment is provided in Chapter 16 (Environment).

The Beef Cattle Nutrient Requirements model in this edition provides two options (empirical and mechanistic) for predicting dietary energy- and protein-allowable beef cattle performance based on the consumption of dietary feed ingredients. These model solutions (i.e., empirical and mechanistic) use the same cattle requirement equations presented in this chapter, which can be used to compute requirements over wide variations in cattle types, body sizes, milk production levels, and environmental conditions. Similar to the previous NRC (1996, 2000) model, Level 1 solution, the current empirical level of solution (ELS) uses tabular values of total digestible nutrients (TDN) for each ingredient to compute bacterial crude protein (BCP) synthesis, ruminal bacterial requirement for ruminally degradable protein (RDP), dietary energy (i.e., TDN) supply, and tabular values of ruminally undegradable protein (RUP) to compute the supply of metabolizable protein (MP). Similar to the NRC (1996, 2000) model, Level 2 solution, the current mechanistic level of solution (MLS), uses ruminal degradation kinetics of carbohydrates (Russell et al., 1992; Sniffen et al., 1992) and protein (Ørskov and McDonald, 1979; NRC, 2001) to compute TDN, microbial crude protein (MCP), and RUP. Small intestinal digestibilities are then assigned to ruminally escaped carbohydrates and lipids, BCP, and RUP to compute dietary energy and MP supplies.

The MLS uses the same approach to calculate metabolizable energy (ME)- and MP-allowable production as the one used in the Level 2 solution of the previous NRC (1996, 2000) model, except for corrections and several modifications (e.g., three rather than five protein fractions) to meet current feeding strategies. In revising the previous model, the committee found that the application of other more mechanistic and dynamic models (Dijkstra et al., 1992; Baldwin, 1995) were restricted by the lack of field-available inputs to drive them, including feed library values. In addition, no improvement in the predictive ability of these models has been reported (Kohn et al., 1995; Pitt et al., 1996; Tylutki et al., 1994) when compared to the Level 2 solution of the Cornell Net Carbohydrate and Protein System (CNCPS; Fox et al., 2004; Tylutki et al., 2008) and the Large Ruminant Nutrition System (LRNS; TAMU, 2015), which is based on the CNCPS v.5 calculations (Fox et al., 2004). On the other hand, the major limitation of the other more highly aggregated models (INRA, 1989, 2007; CSIRO, 1990, 2007; AFRC, 1993) was the inability to use inputs available in specific production settings in North America to mechanistically predict feed net energy (NE) values and supply of amino acids. In developing the MLS option of the current Beef Cattle Nutrient Requirements model, the committee reviewed other models (AFRC, 1993; NRC, 2001; CSIRO, 2007; INRA, 2007) and adopted the one-pool, exponential kinetic to estimate RDP (Ørskov and McDonald, 1979), starch, pectin, and digestible neutral detergent fiber (NDF; Lanzas et al., 2007a; Sniffen et al., 1992). As indicated by Fox et al. (1995), the MLS can be used as

- a teaching tool to improve skills in evaluating the interactions of feed composition, feeding management, and animal requirements for varying farm conditions;
- a tool for developing tables of feed NE and MP values and adjustment factors that can extend and refine the use of conventional diet formulation programs;
- a structure to estimate feed utilization for which no values have been determined and on which to design experiments to quantify those values;
- a tool for predicting requirements and balances for nutrients for which more detailed systems of accounting are needed, such as methane and ammonia emissions and N and P excretion;
- a tool for extending research results to varying farm conditions; and
- a diagnostic tool to evaluate feeding programs and to account for more of the variation in performance in a specific production setting.

Conversely, the ELS should be used when limited information on feed composition is available and the user is not familiar with how to use, interpret, and apply the inputs and results from the MLS option.

Similar to the approach taken by the NRC (1996, 2000), the current committee placed greater emphasis on predicting the supply of nutrients because animal requirements and diet are interactive. Greater emphasis was also placed on the following: calculating feed digestibility under specific conditions, heat increment to compute lower critical temperature, calculation of efficiency of ME use for maintenance, growth, and lactation; and computing BCP from TDN. The accurate prediction of nutrient requirements and performance under specific conditions depends on the accurate description of feedstuff composition and dry matter intake (DMI). As the current committee revisited the previous NRC (1996, 2000) model, other growth models that describe some or all aspects of postabsorptive metabolism (Oltjen et al., 1986; France et al., 1987; Hoch and Agabriel, 2004) were considered. The France et al. (1987) model is mechanistic and dynamic in its approach to metabolism but it has not been extensively evaluated using field data. The NRC (1996, 2000) compared the Oltjen et al. (1986) model with their proposed model with respect to their growth predictions and found that the Oltjen et al. (1986) model did not account for as much of the variation nor was it sufficiently complete to allow prediction of requirements from common description of cattle and all conditions that must be taken into account in North America. Garcia et al. (2008) also compared the Oltjen et al. (1986) model and the Hoch and Agabriel (2004) model, and reported that although their structure and equations were valid, differences in predicting fat and protein were observed. Tedeschi et al. (2004) developed a dynamic model to predict dry matter intake required (DMIR) for a given animal's growth performance, and fat and protein deposition based on the growth equations of the NRC (1996, 2000). Their evaluation indicated that the growth model used by the NRC (1996, 2000) can predict animal performance and body composition with an acceptable degree of accuracy. Therefore, given the scope of the current edition of the *Nutrient Requirements of Beef Cattle*, the complexity of dynamic models, and the need for additional variables in predicting protein and fat deposition, the committee retained the growth equations developed by the NRC (1996, 2000).

The following sections present the calculation logistics for animal requirements and dietary supply of energy, protein, and minerals; animal intake of dry matter (DM) and water; and nutrient balances as discussed in previous chapters. A section is provided on the development of the Beef Cattle Nutrient Requirements software feed library. Subsequently, model evaluation and sensitivity analyses are provided to compare the validity of the current Beef Cattle Nutrient Requirements model for predictions under practical conditions.

REQUIREMENTS FOR ENERGY AND PROTEIN

Maintenance

Maintenance requirements are computed by adjusting the base NE for maintenance (NEm) requirement for breed, lactation, and heat loss versus heat production (HE), which is computed as ME intake minus retained energy (RE). Heat loss is affected by animal insulation factors and environmental conditions. The NEm requirement is computed based on the basal metabolism coefficient (a1) and adjustment factors for previous temperature (a2), breed (BE; Table 19-1), lactation (L; Table 19-1), gender (1.15 for bull, 1 for others), and previous plane of nutrition (COMP) as follows:

$$NEm = SBW^{0.75} \times (a1 \times BE \times L \times COMP \times SEX + a2) \quad \text{(Eq. 19-1)}$$

$$a1 = 0.077 \quad \text{(Eq. 19-2)}$$

$$a2 = 0.0007 \times (20 - Tp) \quad \text{(Eq. 19-3)}$$

$$COMP = 0.8 + (BCS - 1) \times 0.05 \quad \text{(Eq. 19-4)}$$

where

a1 is the basal metabolism coefficient, Mcal/kg$^{0.75}$/d;
a2 is the acclimatization factor, Mcal/kg$^{0.75}$/d;
BCS is body condition score (1 to 9 scale). It is used as a proxy for previous plane of nutrition;
BE is breed factor (Table 19-1);
COMP is the NEm adjustment for previous nutrition;
L is lactation factor (Table 19-1);
NEm is net energy requirement for maintenance, Mcal/d;
SBW is shrunk body weight, kg (typically 96% of full body weight);
SEX is gender effect factor (1.15 for bulls or 1 otherwise); and
Tp is previous temperature, °C.

The coefficients in Table 19-1 are used for straightbred animals and they are adjusted accordingly for two-way (the average of dam's breed and sire's breed) and three-way (the average of sire's breed and the average of maternal grandsire's breed and maternal granddam's breed) crossbred animals. The L coefficient is only used for lactating beef cows. The NRC (1996, 2000) decreased the NEm coefficient (a1) by 10% for all types of *Bos indicus* cattle breeds; however, recent evaluations of comparative slaughter experiments with Nellore cattle fed high-forage diets do not support this decrease in NEm (Tedeschi et al., 2002a; Chizzotti et al.,

TABLE 19-1 Maintenance Requirement Multipliers for Breeds and Physiological Stages, Calf Birth Weight, Peak Milk Production and Milk Composition, and Expected Body Weight at Conception as a Percentage of Mature Body Weight[a]

Code	Breed	Breed Factor (BE)	Lactation Factor (L)	Calf Birth Weight (CBW, kg)	Peak Milk Yield (PKYD, kg/d)	Milk Fat (MkFat, %)	Milk Protein (MkProt, %)	Milk Solids Not Fat (MkSNF, %)	% of Mature Weight at Conception (% MW)
1	Angus	1	1.2	31	8	4	3.8	8.3	60
2	Braford	0.95	1.2	36	7	4	3.8	8.3	55
3	Brahman	0.9	1.2	31	8	4	3.8	8.3	55
4	Brangus	0.95	1.2	33	8	4	3.8	8.3	62
5	Braunvieh	1.2	1	39	12	4	3.8	8.3	62
6	Canchim	0.9	1.2	32	6	4	3.8	8.3	62
7	Charolais	1	1.2	39	9	4	3.8	8.3	57
8	Chianina	1	1.2	41	6	4	3.8	8.3	65
9	Devon	1	1	32	8	3.5	3.3	8.3	60
10	Galloway	1	1.2	36	8	4	3.8	8.3	60
11	Gelbvieh	1	1	39	11.5	4	3.8	8.3	55
12	Gir	0.9	1.2	32	10	4	3.8	8.3	65
13	Guzerat	0.9	1.2	32	5	4	3.8	8.3	65
14	Hereford	1	1	36	7	4	3.8	8.3	60
15	Holstein	1.2	1	43	43	3.5	3.3	8.3	55
16	Jersey	1.2	1	32	34	5.2	3.9	8.3	55
17	Limousin	1	1.2	37	9	4	3.8	8.3	60
18	Longhorn	1	1.2	33	5	4	3.8	8.3	60
19	Maine Anjou	1	1.2	40	9	4	3.8	8.3	60
20	Nellore	1	1.2	32	7	4	3.8	8.3	65
21	Piedmontese	1	1.2	38	7	4	3.8	8.3	60
22	Pinzgauer	1	1.2	38	11	4	3.8	8.3	60
23	Polled Hereford	1	1.2	33	7	4	3.8	8.3	60
24	Red Poll	1	1.2	36	10	4	3.8	8.3	60
25	Sahiwal	0.9	1.2	38	8	4	3.8	8.3	65
26	Salers	1	1.2	35	9	4	3.8	8.3	60
27	Santa Gertrudis	0.95	1.2	33	8	4	3.8	8.3	62
28	Shorthorn	1	1.2	37	8.5	4	3.8	8.3	60
29	Simmental	1.2	1	39	12	4	3.8	8.3	57
30	South Devon	1	1.2	33	8	4	3.8	8.3	60
31	Tarentaise	1	1.2	33	9	4	3.8	8.3	60

[a]Adapted from NRC (1996, 2000) and Fox et al. (2004).

2008) as discussed in Chapter 11 (Maintenance Consideration: Energy and Protein). For this reason, the BE factor for Nellore cattle (Table 19-1) was reset to 1. The user may, however, override the NEm calculation through a multiplicative factor in the software. In addition, the user may override the multiplicative factor for lactating first-calf heifers in the software, as suggested in Chapter 13 (Reproduction).

The ME concentration is increased by 2.3 or 1.5% when monensin or lasalocid, respectively, are added to the diet. The net energy available for maintenance (NEma) and NE available for growth (NEga) are computed from dietary ME with and without ionophore adjustments, and, unless specified, the NE adjusted for ionophores is used throughout the calculations. The feed for maintenance (FFM) is computed by dividing the NEm by the NEma adjusted for ionophore:

$$FFM = NEm/NEma \qquad \text{(Eq. 19-5)}$$

where

FFM is feed dry matter needed to support maintenance, kg/d; and
NEma is diet net energy available for maintenance, Mcal/kg.

In addition to the acclimatization factor (a2), which is related to the previous temperature, the adjustment for cold stress is computed based on the recommendations by the NRC (1981) using heat production (HE), surface area (SA), and external (EI) and tissue insulations (TI). The total insulation (IN) and HE are then used to compute the lower critical temperature (LCT):

$$HE = \begin{cases} MEI - (DMI - FFM) \times NEga, & \text{for growing animals} \\ MEI - (DMI - FFM) \times NEma, & \text{for lactating and dry cows} \end{cases} \qquad \text{(Eq. 19-6)}$$

$$EI = (6.1816 - 0.5575 \times WS + 0.0152 \times WS^2 + 5.298 \times Hair - 0.4297 \times Hair^2 - 0.1029 \times WS \times Hair) \times Hair\ Coat \times HIDE \qquad \text{(Eq. 19-7)}$$

$$TI = \begin{cases} 2.5, & Age \leq 30 \text{ days} \\ 6.5, & 30 < Age \leq 183 \text{ days} \\ 5.1875 + 0.3125 \times BCS, & 183 < Age \leq 363 \text{ days} \\ 5.25 + 0.75 \times BCS, & Age > 363 \text{ days} \end{cases} \qquad \text{(Eq. 19-8)}$$

$$IN = EI + TI \qquad \text{(Eq. 19-9)}$$

$$SA = 0.09 \times BW^{0.67} \qquad \text{(Eq. 19-10)}$$

$$LCT = 39 - 0.85 \times IN \times HE/SA \qquad \text{(Eq. 19-11)}$$

where

BCS is body condition score;
BW is full body weight, kg;
DMI is dry matter intake, kg/d;
EI is external insulation, °C·m^2·d/Mcal (note that EI must be ≥0);
Hair Coat is 1 for no mud, 0.8 for some mud on lower body, 0.5 for mud on lower body and sides, and 0.2 for heavily covered with mud, dimensionless;
Hair is hair depth, cm;
HE is heat production, Mcal/d;
HIDE is 0.8 for thin hide thickness, 1.0 for average hide thickness, and 1.2 for thick hide thickness, dimensionless;
IN is total insulation, °C·m^2·d/Mcal;
LCT is lower critical temperature, °C;
MEI is metabolizable energy intake, Mcal/d;
NEga is dietary net energy available for growth, Mcal/kg;
SA is surface area, m^2;
TI is tissue insulation, °C·m^2·d/Mcal; and
WS is wind speed, km/h (note that WS must be ≤32 km/h).

The LCT is then compared with the current temperature (Tc) to estimate the amount of NE requirement associated with cold stress:

$$MEcs = \begin{cases} SA \times (LCT - Tc)/IN, & LCT > Tc \\ 0, & LCT \leq Tc \end{cases} \qquad \text{(Eq. 19-12)}$$

$$NEcs = MEcs \times k_m \qquad \text{(Eq. 19-13)}$$

where

k_m is the partial efficiency of use of ME to NE for maintenance;
MEcs is metabolizable energy requirement for maintenance due to cold stress, Mcal/d;
NEcs is net energy requirement for maintenance due to cold stress, Mcal/d; and
Tc is current temperature, °C.

In addition to the acclimatization factor (a2), which is related to the previous temperature, the adjustment for heat stress is computed based on the level of panting of the animal, as shown below:

$$NEhs = \begin{cases} 0.07 \times NEm, & \text{Rapid shallow panting} \\ 0.18 \times NEm, & \text{Open-mouth panting} \end{cases} \qquad \text{(Eq. 19-14)}$$

where NEhs is net energy requirement due to heat stress, Mcal/d.

The total NEm required by the animal is the sum of NE for maintenance in thermoneutral conditions plus NE required for cold stress or heat stress. The amounts of FFM and ME for maintenance are computed as shown below. The values for a2, NEcs, and NEhs are set to zero when NEm requirement is not adjusted for environmental factors:

$$NEm_{chs} = \begin{cases} NEm + NEcs, & NEcs > 0 \\ NEm + NEhs, & NEhs > 0 \end{cases} \quad \text{(Eq. 19-15)}$$

$$FFM = NEm_{chs}/NEma \quad \text{(Eq. 19-16)}$$

$$MEm_{chs} = NEm_{chs}/k_m \quad \text{(Eq. 19-17)}$$

where

NEm_{chs} is net energy requirement for maintenance plus cold stress or heat stress, Mcal/d; and
MEm_{chs} is metabolizable energy requirement for maintenance plus cold stress or heat stress, Mcal/d.

The MP requirement for maintenance is computed from the metabolic shrunk body weight (SBW) as shown below:

$$MPm = 3.8 \times SBW^{0.75} \quad \text{(Eq. 19-18)}$$

where MPm is metabolizable protein requirement for maintenance, g/d.

Lactation

Lactation requirements are calculated using age of cow, time of lactation peak, peak milk yield, day of lactation, duration of lactation, milk fat content, milk non-fat solids content, and milk protein content as described below:

$$k = 1/T \quad \text{(Eq. 19-19)}$$

$$aPKYD = (0.125 \times RMY + 0.375) \times PKYD \quad \text{(Eq. 19-20)}$$

$$a = 1/(aPKYD \times k \times e^1) \quad \text{(Eq. 19-21)}$$

$$n = DIM/7 \quad \text{(Eq. 19-22)}$$

$$\text{Age Factor} = \begin{cases} 0.74, & Age \le 2 \text{ yr} \\ 0.88, & 2 < Age \le 3 \text{ yr} \\ 1, & Age > 3 \text{ yr} \end{cases} \quad \text{(Eq. 19-23)}$$

$$E = 0.092 \times MkFat + 0.049 \times MkSNF - 0.0569 \quad \text{(Eq. 19-24)}$$

$$Yn = n/(a \times e^{k \times n}) \times \text{Age Factor} \quad \text{(Eq. 19-25)}$$

$$YEn = Yn \times E \quad \text{(Eq. 19-26)}$$

$$MEl = YEn/k_m \quad \text{(Eq. 19-27)}$$

$$YFatn = Yn \times MkFat/100 \quad \text{(Eq. 19-28)}$$

$$YProtn = Yn \times MkProt/100 \quad \text{(Eq. 19-29)}$$

$$MPl = (YProtn/0.65) \times 1{,}000 \quad \text{(Eq. 19-30)}$$

$$FFL = YEn/NEma \quad \text{(Eq. 19-31)}$$

$$TotalY = -7/(a \times k) \times (n \times e^{-k \times n} + (e^{-k \times n} - 1)/k) \times \text{Age Factor} \quad \text{(Eq. 19-32)}$$

$$TotalYE = TotalY \times E \quad \text{(Eq. 19-33)}$$

$$TotalYFat = TotalY \times MkFat/100 \quad \text{(Eq. 19-34)}$$

$$TotalYProt = TotalY \times MkProt/100 \quad \text{(Eq. 19-35)}$$

$$TotalMPl = TotalY/0.65 \quad \text{(Eq. 19-36)}$$

where

a is an intermediate variable;
Age Factor is the adjustment of milk production for cow's age, dimensionless;
aPKYD is adjusted peak milk yield, kg/d;
DIM is days in milk or days since calving;
E is energy content of milk, Mcal/kg;
e is the base of the natural (Naperian) logarithm (i.e., 2.718);
FFL is feed dry matter needed to support lactation, kg/d, and assumes partial efficiency of use of ME to net energy for lactation (k_l) is identical to k_m;
k is an intermediate variable, dimensionless;
k_m is the partial efficiency of use of ME to NE for maintenance;
MEl is metabolizable energy requirement for lactation, Mcal/d;
MkFat is milk fat content, %;
MkProt is milk protein content, %;
MkSNF is milk solids not fat content, %;
MPl is metabolizable protein requirement for lactation, g/d;
n is week in lactation;
PKYD is peak milk yield (Table 19-1), kg/d;
RMY is relative milk yield, dimensionless;
T is week of peak milk;
TotalMPl is total milk metabolizable protein for the n weeks in lactation, kg;
TotalY is total milk yield for the n weeks in lactation, kg;
TotalYE is total milk net energy requirement (milk energy) for the n weeks in lactation, Mcal;

TotalYFat is total milk fat yield for the n weeks in lactation, kg;
TotalYProt is total milk protein yield for the n weeks in lactation, kg;
YEn is net energy requirement for lactation (daily milk energy), Mcal/d;
YFatn is daily milk fat yield, kg/d;
Yn is daily milk yield, kg/d; and
YProtn is daily milk protein yield, kg/d.

Pregnancy

Calf birth weight and day of gestation are used to calculate pregnancy requirements as shown below:

$$NEy = CBW \times (0.05855 - 0.0000996 \times DP) \times e^{0.0323 \times DP - 0.0000275 \times DP^2}/1{,}000 \quad \text{(Eq. 19-37)}$$

$$MEy = NEy/0.13 \quad \text{(Eq. 19-38)}$$

$$FFP = MEy/ME \quad \text{(Eq. 19-39)}$$

$$Ypn = CBW \times (0.001669 - 0.00000211 \times DP) \times e^{0.0278 \times DP - 0.0000176 \times DP^2} \times 6.25 \quad \text{(Eq. 19-40)}$$

$$MPy = Ypn/0.65 \quad \text{(Eq. 19-41)}$$

where

CBW is calf birth weight, kg;
DP is days pregnant;
e is the base of the natural (Naperian) logarithm (i.e., 2.718);
FFP is feed dry matter needed to support pregnancy, kg/d;
MEy is metabolizable energy requirement for pregnancy, Mcal/d. It assumes a fixed partial efficiency (k_y) of 0.13;
MPy is metabolizable protein requirement for pregnancy, g/d. It assumes a fixed efficiency of 0.65;
NEy is net energy requirement for pregnancy, Mcal/d; and
Ypn is net protein retained as conceptus, g/d.

The NEm for pregnancy can be computed as shown below:

$$NEm = k_m \times MEy \quad \text{(Eq. 19-42)}$$

where k_m is the partial efficiency of use of ME to NE for maintenance; otherwise an average value of 0.6 may be used.

Growth

The growth model described below is based on the NRC (1996, 2000) model. Requirements for growth are calculated using SBW, shrunk [body] weight gain (SWG), body composition, and relative body size, which were based on a modification of the work of Tylutki et al. (1994):

$$EBW = 0.891 \times SBW \quad \text{(Eq. 19-43)}$$

$$SRW = \begin{cases} 400, & \text{22\% EBF (devoid of marbling)} \\ 435, & \text{25\% EBF (traces of marbling)} \\ 462, & \text{27\% EBF (slight marbling)} \\ 478, & \text{28\% EBF (small marbling, replacement heifers, or breeding bulls)} \end{cases} \quad \text{(Eq. 19-44)}$$

$$MSBW = 0.96 \times MW \quad \text{(Eq. 19-45)}$$

$$EQSBW = (SBW - CW) \times (SRW/MSBW) \quad \text{(Eq. 19-46)}$$

$$EQEBW = 0.891 \times EQSBW \quad \text{(Eq. 19-47)}$$

$$RE = (DMI - FFM - FFL - FFP) \times NEga \quad \text{(Eq. 19-48)}$$

$$EWG = 12.341 \times EQEBW^{-0.6837} \times RE^{0.9116} \quad \text{(Eq. 19-49)}$$

$$SWG = EWG/0.96 \quad \text{(Eq. 19-50)}$$

$$NPg = SWG \times (268 - 29.4 \times RE/SWG) \quad \text{(Eq. 19-51)}$$

$$MPg = NPg/\max(0.492, 0.834 - 0.00114 \times EQSBW) \quad \text{(Eq. 19-52)}$$

where

CW is conceptus weight, kg, and is computed in the target weight section;
EBW is empty body weight, kg;
EQEBW is equivalent EBW, kg;
EQSBW is equivalent SBW, kg;
EWG is empty [body] weight gain, kg/d;
max is the maximum between two values;
MSBW is mature SBW, kg (typically 96% of full weight), and is the actual SBW at the body fat endpoint selected for feedlot steers and heifers, at maturity for breeding heifers, or at mature weight × 0.6 for breeding bulls;
MPg is metabolizable protein requirement for gain, g/d;
MW is mature full (i.e., unshrunk) body weight, kg;
NPg is net protein requirement for gain, g/d;
RE is retained energy, Mcal/d;
SBW is shrunk body weight, kg;
SRW is standard reference weight, kg; and
SWG is shrunk [body] weight gain (i.e., ADG), kg/d.

As indicated by the NRC (1996, 2000), the incorrect

selection of mature SBW (MSBW), which is equivalent to their FSBW acronym, and quality grade (marbling score of Traces, Slight, or Small; 25, 27, and 28% empty body fat [EBF], respectively) at slaughter in the model results in inaccurate prediction of NE and net protein requirements. If the MSBW is overestimated or the degree of marbling is underestimated, the energy content of the gain will be underestimated. The opposite for each will result in an overestimation of the energy content of gain. The reason is that MSBW and quality grade are used to determine the energy and protein of the average daily gain (ADG; i.e., SWG) at any particular body weight (BW), as shown in the NE and protein requirements for growth equations. When carcass traits are available, MSBW and EBF at that MSBW (e.g., usually 28%) can be calculated using the equations devised by Garrett and Hinman (1969), Guiroy et al. (2001), and Tedeschi et al. (2004), as shown below:

$$MW = MSBW/0.96 \quad \text{(Eq. 19-53)}$$

$$MSBW = \frac{EBW_c + (28 - EBF_c) \times 14.26}{0.891} \quad \text{(Eq. 19-54)}$$

$$EBW_c = 30.26 + 1.362 \times HCW \quad \text{(Eq. 19-55)}$$

$$EBF_c = \begin{cases} 17.76207 + 4.68142 \times FT + 0.01945 \times HCW + 0.81855 \times QG - 0.06754 \times REA, & \text{REA is available} \\ 14.8796 + 4.7135 \times FT + 0.01316 \times HCW + 0.90855 \times QG, & \text{REA is not available} \end{cases} \quad \text{(Eq. 19-56)}$$

where

EBW_c is empty body weight when carcass traits were measured, kg;
EBFc is empty body fat when carcass traits were measured, %;
HCW is hot carcass weight, kg;
FT is fat thickness (i.e., backfat), cm;
MW is mature weight, kg;
QG is USDA quality grade code (4 for Select, 5 for low Choice, 6 for Choice, 7 for high Choice, and 8 for Prime); and
REA is ribeye area (i.e., longissimus dorsi muscle area), cm^2.

Target Growth of Replacement Heifers

The requirements for target growth are based on those presented in the NRC (1996, 2000) model. The coefficients for computing target breeding weights at puberty are based on the summary in Chapter 13 and can be changed by the user in the Beef Cattle Nutrient Requirements software. The coefficients for computing target breeding weights after first calving are based on U.S. Meat Animal Research Center (MARC) data summarized by Gregory et al. (1999);

$$TPW = \begin{cases} MSBW \times 0.55, & \text{Dual-purpose or dairy cattle} \\ MSBW \times 0.60, & \textit{Bos taurus} \text{ cattle} \\ MSBW \times 0.65, & \textit{Bos indicus} \text{ cattle} \end{cases} \quad \text{(Eq. 19-57)}$$

$$TPA = TCA - 280 \quad \text{(Eq. 19-58)}$$

$$ADGbp = (TPW - SBW)/(TPA - Age) \quad \text{(Eq. 19-59)}$$

$$TCW1 = MSBW \times 0.80 \quad \text{(Eq. 19-60)}$$

$$ADGap = (TCW1 - TPW)/(280 - DP) \quad \text{(Eq. 19-61)}$$

$$TCW2 = MSBW \times 0.92 \quad \text{(Eq. 19-62)}$$

$$ADGac1 = (TCW2 - TCW1)/CI \quad \text{(Eq. 19-63)}$$

$$TCW3 = MSBW \times 0.96 \quad \text{(Eq. 19-64)}$$

$$ADGac2 = (TCW3 - TCW2)/CI \quad \text{(Eq. 19-65)}$$

$$TCW4 = MSBW \quad \text{(Eq. 19-66)}$$

$$ADGac3 = (TCW4 - TCW3)/CI \quad \text{(Eq. 19-67)}$$

where

Age is the age of the animal, d;
ADGac1 is expected ADG after first calving, kg/d;
ADGac2 is expected ADG after second calving, kg/d;
ADGac3 is expected ADG after third calving, kg/d;
ADGap is expected ADG after first pregnancy, kg/d;
ADGbp is expected ADG before first pregnancy, kg/d;
CI is calving interval, d;
DP is days pregnant;
TCA is target calving age, d;
TCW1 is target first calving weight, kg;
TCW2 is target second calving weight, kg;
TCW3 is target third calving weight, kg;
TCW4 is target fourth calving weight, kg;
TPA is target pregnant age, days; and
TPW is target pregnant weight, kg.

The equations in the growth section are used to compute requirements for the target ADG. For pregnant animals, daily gain associated with gravid uterus growth (ADGpreg) should be added to predicted daily gain (i.e., SWG) and the weight of fetal and associated uterine tissue (CW; i.e., gravid uterus weight) is deducted from equivalent empty body weight (EQEBW) to compute growth requirements. The ADGpreg and CW can be calculated as follows:

$$ADGpreg = CBW \times 0.01828 \times (0.02 - 0.0000286 \times DP) \times e^{0.02 \times DP - 0.0000143 \times DP^2} \quad \text{(Eq. 19-68)}$$

$$CW = CBW \times 0.01828 \times e^{0.02 \times DP - 0.0000143 \times DP^2} \quad \text{(Eq. 19-69)}$$

where

ADGpreg is average daily gain of the gravid uterus, kg/d;
CW is gravid uterus weight, kg; and
e is the base of the natural (Naperian) logarithm (i.e., 2.718).

Energy and Protein Reserves

The body reserves model described below is different from the NRC (1996, 2000) model. It is based on the Fox et al. (1999) and Tedeschi et al. (2006) models, except that the percentage BW change per unit change of body condition score (BCS) is assumed to be 7.105% (see Chapter 13) rather than 6.85% as proposed by Fox et al. (1999). Body condition score, BW, and body composition are used to calculate energy and protein reserves. The equations were developed from data on chemical body composition and visual appraisal of BCS on 106 mature cows of diverse breed types and body sizes and were evaluated on an independent data set of 65 mature cows (C. L. Ferrell, USMARC, personal communication, 1995). The following steps are used in this calculation:

(1) Compute body fat, protein, and ash compositions for each BCS (1 to 9):

$$AFat_{BCS} = 0.037683 \times BCS \quad \text{(Eq. 19-70)}$$

$$AProt_{BCS} = 0.200886 - 0.0066762 \times BCS \quad \text{(Eq. 19-71)}$$

$$AWater_{BCS} = 0.766637 - 0.034506 \times BCS \quad \text{(Eq. 19-72)}$$

$$AAsh_{BCS} = 0.078982 - 0.00438 \times BCS \quad \text{(Eq. 19-73)}$$

where

$AAsh_{BCS}$ is the proportion of empty body ash at a given BCS;
$AFat_{BCS}$ is the proportion of empty body fat at a given BCS;
$AProt_{BCS}$ is the proportion of empty body protein at a given BCS; and
$AWater_{BCS}$ is the proportion of empty body water at a given BCS.

(2) Compute current empty body weight (EBW) from current SBW, and EBW at BCS 5, using the weight adjustment factor (WAF). The WAF can be used for SBW or EBW basis. The user may override the default coefficient of 7.105% in the software, as discussed in Chapter 13, for both mature cows and lactating primiparous females:

$$EBW = 0.851 \times SBW \quad \text{(Eq. 19-74)}$$

$$WAF_{BCS} = 1 - 0.07105 \times (5 - BCS) \quad \text{(Eq. 19-75)}$$

$$EBW_5 = EBW/WAF_{BCS} \quad \text{(Eq. 19-76)}$$

where

EBW is the current empty body weight for mature cows, kg;
EBW5 is the EBW at BCS 5, kg;
SBW is the current shrunk body weight, kg; and
WAF_{BCS} is the weight adjustment factor at a given BCS.

(3) Compute EBW for each BCS using EBW at BCS 5 (EBW_5) and WAF for each BCS:

$$EBW_{BCS} = EBW_5 \times WAF_{BCS} \quad \text{(Eq. 19-77)}$$

where EBW_{BCS} is the EBW at a given BCS, kg.

(4) Compute the fat ($TFat_{BCS}$) and protein ($TProt_{BCS}$) amounts for each BCS:

$$TFat_{BCS} = AFat_{BCS} \times EBW_{BCS} \quad \text{(Eq. 19-78)}$$

$$TProt_{BCS} = AProt_{BCS} \times EBW_{BCS} \quad \text{(Eq. 19-79)}$$

where

$TFat_{BCS}$ is the total empty body fat at a given BCS, kg; and
$TProt_{BCS}$ is the total empty body protein at a given BCS, kg.

(5) Compute the amount of total energy reserves (TE_{BCS}) for each BCS:

$$TE_{BCS} = 9.4 \times TFat_{BCS} + 5.7 \times TProt_{BCS} \quad \text{(Eq. 19-80)}$$

where TE_{BCS} is total energy of reserves at a given BCS, Mcal.

(6) Compute the amount of body energy to gain and to lose 1 BCS:

$$\text{Lose BCS: } \Delta TE_{BCS,Lose} = TE_{BCS} - TE_{BCS-1}; \quad 9 \geq BCS > 2 \quad \text{(Eq. 19-81)}$$

$$\text{Gain BCS: } \Delta TE_{BCS,Gain} = TE_{BCS+1} - TE_{BCS}; \quad 1 \le BCS < 9; \qquad \text{(Eq. 19-82)}$$

where

$\Delta TE_{BCS,Lose}$ is the loss of total energy of reserves (mobilization) at a given BCS, Mcal; and
$\Delta TE_{BCS,Gain}$ is the gain of total energy of reserves (deposition) at a given BCS, Mcal.

(7) Compute days to change 1 BCS:

$$\text{Days} = \begin{cases} \dfrac{\Delta TE_{BCS,Lose} \times 0.8}{k_m \times \Delta ME}, & \Delta ME < 0 \\ \dfrac{\Delta TE_{BCS,Gain}}{k_m \times \Delta ME}, & \Delta ME > 0 \end{cases} \qquad \text{(Eq. 19-83)}$$

where

ΔME is metabolizable energy balance (supply minus required), Mcal/d; and
k_m is the partial efficiency of use of ME to NE for maintenance, and if not available, an average value of 0.6 may be used.

During mobilization, 1 Mcal RE (i.e., TE) will substitute for 0.80 Mcal NEma, and during repletion, 1 Mcal NEma will provide 1 Mcal RE. The NE is converted to ME basis by dividing the NE by k_m (or by a fixed value of 0.6). Then, days to change 1 BCS is computed by dividing this ME equivalent RE by the ME balance (ΔME). The NE mobilized or replenished per BCS change increases as SBW or BCS increases. Table 19-2 has examples of energy mobilized or replenished per BCS change for different SBW using the submodel described above.

REQUIREMENTS FOR MINERALS AND VITAMINS

Mineral and vitamin requirements are summarized in Tables 19-3 and 19-4. Chapter 7 (Minerals) has detailed information on mineral requirements and Chapter 8 (Vitamins) has detailed information on vitamin requirements. Requirements are identified for maintenance, growth, lactation, and pregnancy.

REQUIREMENTS FOR AMINO ACIDS

Similar to that developed by the NRC (1996, 2000), the requirement of metabolizable amino acids (MAA) for maintenance, growth, and pregnancy assumes the average amino acid (AA) composition of the tissue (Table 19-5) and the requirement of MAA for lactation assumes the average composition of AA in the milk (Table 19-5). The requirements for net amino acids (NAA) for growth, lactation, and pregnancy are computed based on the net protein required for growth, lactation, and pregnancy, respectively. Then,

TABLE 19-2 Energy Reserves for Cows with Different Current Shrunk Body Weights (SBW) and Body Condition Scores (BCS)

BCS		Current SBW at BCS 5 (kg)								
Gain	Lose	400	450	500	550	600	650	700	750	800
		Mcal NE provided (lose) or required (gain) for each BCS change								
1→2	2→1	120	135	150	165	180	195	210	225	240
2→3	3→2	135	152	169	186	203	220	237	254	271
3→4	4→3	151	169	188	207	226	245	264	282	301
4→5	5→4	166	187	207	228	249	270	290	311	332
5→6	6→5	181	204	226	249	272	294	317	340	362
6→7	7→6	196	221	246	270	295	319	344	368	393
7→8	8→7	212	238	265	291	318	344	371	397	424
8→9	9→8	227	255	284	312	341	369	397	426	454

TABLE 19-3 Calcium and Phosphorus Requirements and Maximum Tolerable Concentrations (g/d)[a]

	Requirements				
Mineral	Maintenance	Growing and Finishing	Lactation	Pregnancy[b]	Maximum
Ca	0.0154 × SBW/0.5	NPg × 0.071/0.5	Yn × 1.23/0.5	CBW × (13.7/90)/0.5	0.02 × DMI
P	0.016 × SBW/0.68	NPg × 0.039/0.68	Yn × 0.95/0.68	CBW × (7.6/90)/0.68	0.007 × DMI

[a]SBW is shrunk body weight, kg; NPg is net protein requirement for gain (i.e., retained protein), g/d; Yn is milk yield, kg/d; CBW is calf birth weight, kg; and DMI is dry matter intake, g/d. The digestibility for Ca is 50% and for P it is 68%.
[b]Last 90 days of pregnancy.

TABLE 19-4 Other Mineral (DMI Basis) and Vitamin (DMI and SBW Bases) Requirements and Maximum Tolerable Concentrations

		Requirements			
Items	Unit	Growing and Finishing	Lactation	Pregnancy	Maximum Tolerable
Minerals					
Magnesium, Mg	%	0.10	0.20	0.12	0.40
Potassium, K	%	0.60	0.70	0.60	2
Sodium, Na	%	0.07	0.10	0.07	–
Sulfur, S	%	0.15	0.15	0.15	0.40
Cobalt, Co	mg/kg	0.15	0.15	0.15	25
Copper, Cu	mg/kg	10.0	10.0	10.0	40
Iodine, I	mg/kg	0.50	0.50	0.50	50
Iron, Fe	mg/kg	50.0	50.0	50.0	500
Manganese, Mn	mg/kg	20.0	40.0	40.0	1,000
Selenium, Se	mg/kg	0.10	0.10	0.10	5
Zinc, Zn	mg/kg	30.0	30.0	30.0	500
Vitamins[a]					
A	IU/kg	2,200 | 47	3,900 | 84	2,800 | 60	–
D	IU/kg	275 | 5.7	275 | 5.7	275 | 5.7	–
E[b]	IU/kg	35 | 0.73	35 | 0.73	35 | 0.73	–

[a]The reported vitamin values are the maximum of calculated values based on DMI (first value) or SBW (second value) daily recommendations. See Chapter 8 (Vitamins) for details.
[b]400 to 500 IU/d for newly received calves.

the requirements of MAA for lactation and pregnancy are computed assuming the partial efficiency of use of AA for each physiological function, as shown in Table 19-5, and the partial efficiency of use of AA for growth is computed based on the SBW.

$$\text{Maintenance: } MAAm_i = TissueAA_i \times MPm \quad \text{(Eq. 19-84)}$$

$$\text{Growth: } MAAg_i = TissueAA_i \times NPg/\max(0.492, 0.834 - 0.00114 \times EQSBW) \quad \text{(Eq. 19-85)}$$

TABLE 19-5 Amino Acid Composition of Tissue and Milk Protein (g AA/100 g Protein) and Partial Efficiency of Use of Absorbed Amino Acids for Pregnancy and Lactation

	Composition		Efficiency of Use[c]	
Amino Acid	Tissue[a]	Milk[b]	Pregnancy	Lactation
Arginine, Arg	3.3	3.40	0.66	0.85
Histidine, His	2.5	2.74	0.85	0.90
Isoleucine, Ile	2.8	5.79	0.66	0.62
Leucine, Leu	6.7	9.18	0.66	0.72
Lysine, Lys	6.4	7.62	0.85	0.88
Methionine, Met	2.0	2.71	0.85	0.98
Phenylalanine, Phe	3.5	4.75	0.85	1.00
Threonine, Thr	3.9	3.72	0.85	0.83
Tryptophan, Trp	0.6	1.51	0.85	0.85
Valine, Val	4.0	5.89	0.66	0.72

[a]Average of three studies by Ainslie et al. (1993).
[b]Based on Waghorn and Baldwin (1984).
[c]Based on Evans and Patterson (1985).

$$\text{Lactation: } MAAl_i = MilkAA_i \times YProtn \times 1{,}000/EAAl_i \quad \text{(Eq. 19-86)}$$

$$\text{Pregnancy: } MAAp_i = TissueAA_i \times Ypn/EEAp_i \quad \text{(Eq. 19-87)}$$

where

$EAAl_i$ is the partial efficiency of use of the ith metabolizable amino acid for lactation (Table 19-5), dimensionless;
$EAAp_i$ is the partial efficiency of use of the ith metabolizable amino acid for pregnancy (Table 19-5), dimensionless;
$MAAg_i$ is the requirement of the ith metabolizable amino acid for growth, g/d;
$MAAl_i$ is the requirement of the ith metabolizable amino acid for lactation, g/d;
$MAAm_i$ is the requirement of the ith metabolizable amino acid for maintenance, g/d;
$MAAp_i$ is the requirement of the ith metabolizable amino acid for pregnancy, g/d;
$MilkAA_i$ is the milk composition of the ith amino acid, g/d; and
$TissueAA_i$ is the tissue composition of the ith amino acid, g/g.

PREDICTING DRY MATTER INTAKE

The factors within angle brackets (i.e., BFAT, TEMP1, MUD1) should be used with caution as discussed in Chapter 10 (Feed Intake). These factors are calculated at the end of this section.

Growing-Finishing Calves and Yearlings

The equation recommended by the NRC (1996, 2000) is used to compute DMI for growing-finishing calves and yearlings as shown below. The NEma used to predict DMI is computed from the ME unadjusted for ionophores.

$$DMI = DMI' \times ADTV_{DMI} \times BI \times GRAZE \times \langle BFAT \times TEMP1 \times MUD1 \rangle \quad \text{(Eq. 19-88)}$$

$$DMI' = \begin{cases} SBW^{0.75} \times (0.2435 \times NEma - 0.0466 \times NEma^2 - 0.1128) / NEma' & \text{calves} \\ SBW^{0.75} \times (0.2435 \times NEma - 0.0466 \times NEma^2 - 0.0869) / NEma' & \text{yearlings} \end{cases} \quad \text{(Eq. 19-89)}$$

$$NEma' = \begin{cases} 0.95, & NEma < 1 \\ NEma, & NEma \geq 1 \end{cases} \quad \text{(Eq. 19-90)}$$

where

$ADTV_{DMI}$ is the adjustment factor for feed additive, dimensionless;
BFAT is the adjustment factor for body fat, dimensionless;
BI is the adjustment factor for breed type, dimensionless;
DMI is the dry matter intake, kg/d;
DMI′ is the auxiliary variable for dry matter intake, kg/d;
GRAZE is the DMI adjustment factor for grazing animals, dimensionless;
MUD1 is the mud depth adjustment factor, dimensionless;
NEma is the dietary net energy available for maintenance, Mcal/kg;
NEma′ is the auxiliary variable for the dietary net energy available for maintenance, Mcal/kg; and
TEMP1 is adjustment for temperature, dimensionless.

The current committee on the *Nutrient Requirements of Beef Cattle* also recommends two new equations to compute DMI, one based on NEma and another one based on initial SBW (ISBW), as shown below. The ISBW equation should be used only for feedlot animals and it is not adjusted for GRAZE. The factors within angle brackets should be used with caution, as discussed in Chapter 10.

$$DMI = \begin{cases} DMI'' \times ADTV_{DMI} \times BI \times GRAZE \times \langle BFAT \times TEMP1 \times MUD1 \rangle, & \text{grazing animals} \\ DMI'' \times ADTV_{DMI} \times BI \times \langle BFAT \times TEMP1 \times MUD1 \rangle, & \text{feedlot animals} \end{cases} \quad \text{(Eq. 19-91)}$$

$$DMI'' = SBW \times (0.012425 + 0.019218 \times NEma - 0.007259 \times NEma^2), \quad \text{(Eq. 19-92)}$$

or

$$DMI'' = \begin{cases} 3.820 + 0.0143 \times ISBW, & \text{feedlot steers or bulls} \\ 3.184 + 0.01536 \times ISBW, & \text{feedlot heifers} \end{cases} \quad \text{(Eq. 19-93)}$$

where

DMI″ is the auxiliary variable for dry matter intake, kg/d;
ISBW is the initial shrunk body weight, kg.

Beef Cows

The equations recommended by the NRC (1996, 2000) are used to compute DMI for beef cows, as shown below. The factors within angle brackets should be used with caution because DMI can be changed drastically by misusing these factors.

$$DMI = DMI' \times GRAZE \times \langle TEMP1 \times MUD1 \rangle + 0.2 \times Yn \quad \text{(Eq. 19-94)}$$

$$DMI' = \begin{cases} SBW^{0.75} \times \dfrac{0.04631 + 0.04997 \times NEma^2}{NEma'}, & DP > 93 \\ SBW^{0.75} \times \dfrac{0.0384 + 0.04997 \times NEma^2}{NEma'}, & DP \leq 93 \end{cases} \quad \text{(Eq. 19-95)}$$

$$NEma' = \begin{cases} 0.95, & NEma < 1 \\ NEma, & NEma \geq 1 \end{cases} \quad \text{(Eq. 19-96)}$$

where

DMI′ is the auxiliary variable for dry matter intake, kg/d;
DP is days pregnant;
NEma′ is the auxiliary variable for the dietary net energy available for maintenance, Mcal/kg; and
Yn is milk yield (i.e., MY), kg/d.

The dry matter intake required (DMIR) is computed using a backward calculation approach as discussed in Chapter 10. It uses the goal-seeking algorithm to iteratively find the DMI that matches the predicted animal performance with the desired animal performance.

Dry Matter Intake Adjustment Factors

Table 19-6 lists the multiplicative factors used to adjust DMI for feed additive ($ADTV_{DMI}$), dietary concentrate, and growth-promoting (i.e., anabolic) implants. The predicted DMI is decreased by 3% (default) for all diets when monensin is fed; however, when lasalocid or laidlomycin propionate

TABLE 19-6 Dry Matter Intake Adjustment Factors for Anabolic Implant and Feed Additive[a]

	Anabolic Implant	
Feed Additive	No	Yes
None or lasalocid or laidlomycin	0.94	1
Monensin	0.94 × 0.97	0.97

[a]The 0.94 and 0.97 are the default factors for anabolic implant and monensin, respectively. Users may change these adjustment factors in the computer model.

is fed, there is no adjustment for ionophore for DMI. Similarly, there is no additional adjustment of DMI for feeding a β-adrenergic agonist.

In agreement with the NRC (1996, 2000), DMI can be adjusted for breed type (BI) as follows: 1.04 for dual-purpose animals (e.g., Holstein × beef crosses), 1 for *Bos taurus* or *B. indicus*, and 1.08 for dairy animals (e.g., Holstein).

Adjustment of DMI relative to forage allowance for grazing animals (GRAZE) is performed as recommended by the NRC (1996, 2000), as described below:

$$GRAZE = \begin{cases} 1, & FORA > 4 \times GI \text{ or } FM > 1{,}150 \\ (0.17 \times FM - 0.000764 \times FM^2 + 2.4)/100, & \\ GU,\ FM, \text{ and } DOP > 0 & \end{cases} \quad \text{(Eq. 19-97)}$$

$$FORA = (1{,}000 \times GU \times FM)/(SBW \times DOP) \quad \text{(Eq. 19-98)}$$

$$GI = DMI \times 1{,}000/SBW \quad \text{(Eq. 19-99)}$$

where

DOP is days on pasture;
FORA is daily forage allowance, g DM/kg SBW;
FM is available forage mass, kg DM/ha;
GI is unadjusted grazing DMI, g DM/kg SBW; and
GU is grazing unit size, ha/animal.

The next adjustments for DMI should be used with caution as indicated in Chapter 10. The NRC (1996, 2000) recommended that DMI be adjusted for body fat (BFAT) using the multiplicative factors proposed by Fox et al. (1988). The current Beef Cattle Nutrient Requirements model uses an equation published by Fox et al. (2004) to adjust DMI for body fat as follows:

$$BFAT = \begin{cases} 1, & EQSBW < 350 \\ (0.7714 + 0.00196 \times EQSBW - 0.00000371 \times EQSBW^2), & EQSBW \geq 350 \end{cases} \quad \text{(Eq. 19-100)}$$

The adjustment for temperature (TEMP1) is based on the recommendations by the NRC (1996, 2000) with modifications proposed by Fox and Tylutki (1998) and Fox et al. (2004):

$$TEMP1 = \begin{cases} 1.16, & Tc \leq -20°C \\ 1.043 - 0.0044 \times Tc + 0.0001 \times Tc^2, & -20°C \leq Tc \leq 20°C \\ DMINC, & Tc \geq 20°C \text{ and } LNT > 20°C \\ (1 - DMINC) \times 0.75 + DMINC, & Tc \geq 20°C \text{ and } LNT < 20°C \\ 1, & \text{Storm exposure and } Tc \leq 13°C \end{cases} \quad \text{(Eq. 19-101)}$$

$$DMINC = (119.62 - 0.9708 \times CETI)/100 \quad \text{(Eq. 19-102)}$$

$$CETI = 27.88 - 0.456 \times Tc + 0.010754 \times Tc^2 - 0.4905 \times RHc + 0.00088 \times RHc^2 + 1.1507 \times WS/3.6 - 0.126447 \times (WS/3.6)^2 + 0.019867 \times Tc \times RHc - 0.046313 \times Tc \times WS/3.6 + 0.41267 \times HRS \quad \text{(Eq. 19-103)}$$

where

DMINC is the dry matter intake for night cooling adjustment factor, dimensionless;
LNT is the lowest night temperature (night cooling if LNT <20°C), °C;
CETI is the current effective temperature index, °C;
HRS is hours of sunlight (assumed to be 8 h/d); and
RHc is the current relative humidity, %.

The mud depth adjustment factor (MUD1) also affects intake by decreasing intake by 1% for each cm, as follows:

$$MUD1 = 1 - 0.01 \times \text{Mud Depth} \quad \text{(Eq. 19-104)}$$

where Mud Depth is in cm.

PREDICTING WATER INTAKE

Several factors affect water intake as discussed in Chapter 9 (Water). Empirical equations to predict water intake (WI) were developed using surface regression of the tabular data developed by Winchester and Morris (1956). The committee used the current effective temperature index (CETI; as computed above) rather than the current temperature. For growing steers, heifers, and bulls with SBW varying from 180 to 400 kg, expected ADG of 0.9 kg/d, and CETI varying from 4° to 32°C, the equation to predict WI ($R^2 = 0.997$, Akaike's Information Criterion [AIC] = 153.7) is

$$WI = 7.3 + 0.0805 \times SBW - 0.00008 \times SBW^2 - 1.225 \times CETI + 0.0411 \times CETI^2 + 0.0023268 \times SBW \times CETI \quad \text{(Eq. 19-105)}$$

For finishing steers with SBW varying from 270 to 500 kg, expected ADG of 1 kg/d, and CETI varying from 4° to 32°C, the equation to predict WI ($R^2 = 0.997$, AIC = 174.1) is

$$WI = 6.336 + 0.1057 \times SBW - 0.0000963 \times SBW^2 - 1.6 \times CETI + 0.056 \times CETI^2 + 0.00226 \times SBW \times CETI \quad \text{(Eq. 19-106)}$$

where

SBW is the shrunk body weight, kg;
CETI is the current effective temperature index, °C; and
WI is water intake, L/d.

SUPPLY OF ENERGY AND NUTRIENTS

The amounts of energy and nutrients are computed from actual DMI when available or from the predicted DMI equation (see Chapter 10) when an actual value is not available. The risk of use of predictive models increases when predicted DMI values are used in place of actual DMI.

Empirical Level of Solution (ELS)

Computing Energy Value of Feeds

Similar to the NRC (1996, 2000) model, dietary energy values are computed by summing the energy contribution of each feed to arrive at a total energy content of the diet, using tabular energy values. Tabular energy values used include TDN (% DM), ME (Mcal/kg), NEma (Mcal/kg), and NEga (Mcal/kg). By default, the ME, NEma and NEga are computed from TDN (% DM) and adjusted for ionophores, as shown below. The ME values for each feed are based on the assumption that 1 kg of TDN is equal to 4.409 Mcal digestible energy (DE; NRC, 1976) and assuming that 1 Mcal DE is equal to 0.82 Mcal ME (NRC, 1976) as adopted by the NRC (1996, 2000). The user may, however, override this DE to ME efficiency by entering the value they prefer to use. Then, the NEma and NEga are computed from the concentration of ME (NRC, 1984). The user may also override this calculation by using the user-inputted values for ME (Mcal/kg), NEma (Mcal/kg), and NEga (Mcal/kg), but in this situation, the model will not adjust ME for ionophores. Users have to correct the inputted values accordingly:

$$DE_j = (aTDN_j/100) \times DMI_j \times 4.409 \quad \text{(Eq. 19-107)}$$

$$ME_j = DE_j \times 0.82 \times ADTV_{ME} \quad \text{(Eq. 19-108)}$$

$$[ME]_j = ME_j/DMI_j \quad \text{(Eq. 19-109)}$$

$$NEma_j = 1.37 \times [ME]_j - 0.138 \times [ME]_j^2 + 0.0105 \times [ME]_j^3 - 1.12 \quad \text{(Eq. 19-110)}$$

$$NEga_j = 1.42 \times [ME]_j - 0.174 \times [ME]_j^2 + 0.0122 \times [ME]_j^3 - 1.65 \quad \text{(Eq. 19-111)}$$

where

$ADTV_{ME}$ is the effect of ionophore on ME, and if used, ME is increased by 2.3 or 1.5% for monensin or lasalocid, respectively;
$aTDN_j$ is the user-inputted apparent total digestible nutrients of the jth feedstuff, %;
DE_j is the digestible energy of the jth feedstuff, Mcal/d;
DMI_j is the dry matter intake of the jth feedstuff, kg/d;
$[ME]_j$ is the concentration of ME of the jth feedstuff, Mcal/kg, adjusted (or not) for ionophores;
ME_j is the metabolizable energy of the jth feedstuff, Mcal/d;
$NEga_j$ is the net energy available for growth of the jth feedstuff, Mcal/kg; and
$NEma_j$ is the net energy available for maintenance of the jth feedstuff, Mcal/kg.

Computing Metabolizable Protein and Amino Acid Supply

Unlike the MCP calculation proposed by the NRC (1996, 2000), the MCP is calculated, by default, using the equations developed by Galyean and Tedeschi (2014), depending on the ether extract (EE) content of the diet. The user may, however, override the default MCP yield. The supply of MP is the sum of digested RUP and digested microbial true protein (MTP, g/d). The digested RUP is assumed to be 80% for concentrates (feeds with less than 100% forage DM) or 60% for forages (feeds with 100% forage DM). The MCP is assumed to contain 80% true protein, which is assumed to have an intestinal digestibility of 80%; therefore, MP from bacteria is MCP × 0.80 × 0.80, as shown below:

$$MP_j = MPfeed_j + MPmtp_j \quad \text{(Eq. 19-112)}$$

$$MPfeed_j = \begin{cases} RUP_j \times 0.8, & \text{Forage} < 100\%\ \text{DM (concentrates)} \\ RUP_j \times 0.6, & \text{Forage} = 100\%\ \text{DM (forages)} \end{cases} \quad \text{(Eq. 19-113)}$$

$$RUP_j = (CP_j/100) \times ([RUP]_j/100) \times DMI_j \quad \text{(Eq. 19-114)}$$

$$MPmtp_j = MCP_j \times 0.8 \times 0.8 \quad \text{(Eq. 19-115)}$$

$$MCP_j = \begin{cases} (42.73 + 0.87 \times TDN_j \times DMI_j)/1{,}000, & EE < 3.9\% \\ (53.33 + 0.96 \times FFTDN_j \times DMI_j)/1{,}000, & EE \geq 3.9\% \end{cases} \quad \text{(Eq. 19-116)}$$

$$FFTDN_j = TDN_j - 2.25 \times EE_j \quad \text{(Eq. 19-117)}$$

where

CP_j is the crude protein content of the jth feedstuff, % DM;
DMI_j is the dry matter intake of the jth feedstuff, kg/d;
$FFTDN_j$ is the fat-free total digestible nutrient content of the jth feedstuff (adjusted for EE, not digestible EE), % DM;
MCP_j is the microbial crude protein supplied by the jth feedstuff, kg/d;
$MPfeed_j$ is the metabolizable protein supplied by the jth feedstuff, kg/d;
MP_j is the metabolizable protein of the jth feedstuff, kg/d;
$MPmtp_j$ is the bacterial metabolizable protein supplied by the jth feedstuff, kg/d;
MTP is microbial true protein of the diet, g/d;
RUP_j is the ruminally undegradable protein of the jth feedstuff, kg/d;
$[RUP]_j$ is the concentration of ruminally undegradable protein of the jth feedstuff, % CP; and
TDN_j is the total digestible nutrient of the jth feedstuff, % DM.

Essential AA composition of the undegradable protein of each feedstuff is used to calculate supply of AA from the feeds. Bacterial composition of essential AA is used to calculate the supply of AA from bacteria. The average bacterial AA content (Table 19-7) is used for the ELS.

$$DIGFAA_{j,i} = (FAA_{j,i}/100) \times MPfeed_j \quad \text{(Eq. 19-118)}$$

$$DIGBAA_{j,i} = (BAA_i/100) \times MPmtp_j \quad \text{(Eq. 19-119)}$$

$$DIGAA_{j,i} = DIGFAA_{j,i} + DIGBAA_{j,i} \quad \text{(Eq. 19-120)}$$

where

BAA_i is the ith bacterial amino acid content, % MP;
$DIGAA_{j,i}$ is the ith absorbed amino acid supplied by the jth feedstuff (RUP) and the bacteria of the jth feedstuff, kg/d;
$DIGBAA_{j,i}$ is the ith absorbed amino acid supplied by the bacteria of the jth feedstuff, kg/d;

TABLE 19-7 Amino Acid Composition (%) of Bacterial Cell-Wall and Non-Cell Wall Protein[a]

Amino Acid	Cell Wall	Non-Cell Wall	Ruminal Bacteria[a] Average	SD
Arginine, Arg	3.82	6.96	5.10	0.7
Histidine, His	1.74	2.69	2.00	0.4
Isoleucine, Ile	4.00	5.88	5.70	0.4
Leucine, Leu	5.90	7.51	8.10	0.8
Lysine, Lys	5.60	8.20	7.90	0.9
Methionine, Met	2.40	2.68	2.60	0.7
Phenylalanine, Phe	4.20	5.16	5.10	0.3
Tyrosine, Tyr	—	—	4.32	—
Threonine, Thr	3.30	5.59	5.80	0.5
Tryptophan, Trp[b]	1.63	1.63	1.63	—
Valine, Val	4.70	6.16	6.20	0.6

[a]Average composition and SD of 441 bacterial samples from animals fed 61 dietary treatments in 35 experiments (Clark et al., 1992). Included for comparison to the cell-wall and non-cell wall values used in this model. Average for tyrosine from Bergen et al. (1968).
[b]Data were not available; therefore, content of cell wall protein is assumed to be same as non-cell wall protein (O'Connor et al., 1993).

$DIGFAA_{j,i}$ is the ith absorbed amino acid supplied by the RUP of the jth feedstuff, kg/d; and
$FAA_{j,i}$ is the ith amino acid content of the jth feedstuff, % MP.

Computing RDP Balance and Ruminal N Balance (RNB)

The RDP requirement is computed assuming an efficiency of use of RDP by ruminal microbes of 100%. The RDP supply does not include ruminally recycled N. If dietary RDP is greater than MCP, the excess of RDP is assumed to be converted to ammonia and absorbed. When RDP balance (RDP supply minus RDP required) is negative, allowable MCP is the RDP supply, as shown below:

$$RDP_{Supply} = \sum_{j=1}^{n}\left(\left(CP_j/100\right) \times DMI_j \times \left(1-[RUP]_j/100\right)\right) \quad \text{(Eq. 19-121)}$$

$$MCP_{TDN} = \sum_{j=1}^{n} MCP_j \quad \text{(Eq. 19-122)}$$

$$RDP_{Required} = MCP_{TDN} \quad \text{(Eq. 19-123)}$$

$$RDP_{Balance} = RDP_{Supply} - RDP_{Required} \quad \text{(Eq. 19-124)}$$

$$MCP_{RDP} = \begin{cases} MCP_{TDN}, & RDP_{Balance} \geq 0 \\ RDP_{Supply}, & RDP_{Balance} < 0 \end{cases} \quad \text{(Eq. 19-125)}$$

where

CP_j is the crude protein of the jth feedstuff, % DM;
DMI_j is the dry matter intake of the jth feedstuff, kg/d;
MCP_j is the microbial crude protein of the jth feedstuff, kg/d;
MCP_{RDP} is the RDP (protein) allowable microbial crude protein, kg/d; and
MCP_{TDN} is the TDN (energy) allowable microbial crude protein, kg/d.

The ruminal N balance (RNB) is calculated from the $RDP_{Balance}$ as shown below:

$$RNB = RDP_{Balance}/6.25 \quad \text{(Eq. 19-126)}$$

where RNB is ruminal nitrogen balance, g/d.

Computing Methane Production

Enteric methane (CH_4) production (g/d) is computed using several published equations, which are assigned to three dietary forage groups: ≤20%, 20 to 40%, and ≥40% DM. Descriptive statistics (mean, median, standard deviation, minimum, and maximum) are provided for each group in the computer software.

The equations for diets with ≥40% of forage are:

Mills et al. (2003), Eq. NL2:

$$CH_4 = 45.98 \times (1 - e^{-0.003 \times MEI \times 4.184}) \times 1{,}000/55.65 \quad \text{(Eq. 19-127)}$$

Ellis et al. (2009), Eq. G:

$$CH_4 = (-1.01 + 2.76 \times NDF + 0.722 \times CB1I) \times 1{,}000/55.65 \quad \text{(Eq. 19-128)}$$

Ellis et al. (2009), Eq. N:

$$CH_4 = (2.68 - 1.14 \times (CB1I/NDFI) + 0.786 \times DMI) \times 1{,}000/55.65 \quad \text{(Eq. 19-129)}$$

P. Escobar-Bahamondes and K. Beauchemin (Agriculture and Agri-Food Canada Research Centre, Lethbridge, Alberta, Canada, personal communication, June 2014), Eq. 16.8:

$$CH_4 = 71.5 + 0.12 \times BW + 0.1 \times DMI^3 - 244.8 \times EEI^3 \quad \text{(Eq. 19-130)}$$

Moraes et al. (2014):

$$CH_4\ \text{(Animal Level)} = \begin{cases} \begin{pmatrix} -0.221 + 0.048 \\ \times GEI \times 4.184 \\ + 0.005 \times BW \end{pmatrix} \times \dfrac{1{,}000}{55.65}, & \text{Males} \\ \begin{pmatrix} -1.487 + 0.046 \\ \times GEI \times 4.184 \\ + 0.32 \times NDF \\ + 0.006 \times BW \end{pmatrix} \times \dfrac{1{,}000}{55.65}, & \text{Females} \end{cases} \quad \text{(Eq. 19-131)}$$

Moraes et al. (2014):

$$CH_4\ \text{(Diet Level)} = \begin{cases} \begin{pmatrix} 0.743 + 0.054 \\ \times GEI \times 4.184 \end{pmatrix} \times \dfrac{1{,}000}{55.65}, & \text{Males} \\ \begin{pmatrix} -0.163 + 0.051 \\ \times GEI \times 4.184 \\ +0.038 \times NDF \end{pmatrix} \times \dfrac{1{,}000}{55.65}, & \text{Females} \end{cases} \quad \text{(Eq. 19-132)}$$

The equations for diets containing between 20 and 40% of forage are:

Ellis et al. (2007), Eq. 12b:

$$CH_4 = (2.7 + 1.16 \times DMI - 15.8 \times EEI) \times 1{,}000/55.65 \quad \text{(Eq. 19-133)}$$

P. Escobar-Bahamondes and K. Beauchemin (Agriculture and Agri-Food Canada Research Centre, Lethbridge, Alberta, Canada, personal communication, June 2014), average between Eqs. 16.8 and 16.9:

$$CH_4 = 30.7 + 0.165 \times BW + 0.18 \times DMI^2 + 0.05 \times DMI^3 - 157 \times EEI^3 + 6.5 \times \frac{CPI}{NDFI} - 2.45 \times \frac{CB1I}{NDFI} \quad \text{(Eq. 19-134)}$$

And, the equations for diets with ≤20% of forage are:

Ellis et al. (2007), Eq. 9b:

$$CH_4 = (0.357 + 0.0591 \times MEI \times 4.184 + 0.05 \times Forage) \times 1{,}000/55.65 \quad \text{(Eq. 19-135)}$$

Ellis et al. (2007), Eq. 10b:

$$CH_4 = (-1.02 + 0.681 \times DMI + 0.0481 \times Forage) \times 1{,}000/55.65 \quad \text{(Eq. 19-136)}$$

Ellis et al. (2009), Eq. G:

$$CH_4 = (-1.01 + 2.76 \times NDF + 0.722 \times CB1I) \times 1{,}000/55.65 \quad \text{(Eq. 19-137)}$$

P. Escobar-Bahamondes and K. Beauchemin (Agriculture and Agri-Food Canada Research Centre, Lethbridge, Alberta, Canada, personal communication, June 2014), Eq. 16.9:

$$CH_4 = -10.1 + 0.21 \times BW + 0.36 \times DMI^2 - 69.2 \times EEI^3 + 13 \times \frac{CPI}{NDFI} - 4.9 \times \frac{CB1I}{NDFI} \quad \text{(Eq. 19-138)}$$

In addition, the calculation of CH_4 production recommended by the Intergovernmental Panel on Climate Change (IPCC, 2006) using gross energy intake (GEI) is used (Eq. 19-139). Note, however, that the coefficient of 3% (Eq. 19-142 and 19-145) is recommended for feedlot diets with less than 10% forage instead of 20% forage. The IPCC (2006) equation to compute CH_4 using DMI (Eq. 19-142) is shown below, but it is not used in the model:

$$CH_4 = \begin{cases} GEI \times 0.065 \times (4.184/55.65) \times 1{,}000, & Forage_{DMI} > 20\%\ DM \text{ (all cattle)} \\ GEI \times 0.03 \times (4.184/55.65) \times 1{,}000, & Forage_{DMI} \le 20\%\ DM \text{ (feedlot)} \end{cases} \quad \text{(Eq. 19-139)}$$

$$GEI = (4.15 \times CHO + 5.65 \times CP + 9.4 \times EE) \times DMI/100 \quad \text{(Eq. 19-140)}$$

$$CHO = 100 - (CP + EE + Ash + OA) \quad \text{(Eq. 19-141)}$$

$$CH_4 = \begin{cases} DMI \times 0.065 \times 18.45 \times 1{,}000/55.65, & Forage_{DMI} > 20\%\ DM \text{ (all cattle)} \\ DMI \times 0.03 \times 18.45 \times 1{,}000/55.65, & Forage_{DMI} \le 20\%\ DM \text{ (feedlot)} \end{cases} \quad \text{(Eq. 19-142)}$$

where

Ash is ash, % DM;
BW is body weight, kg;
CB1I is starch intake of the diet, kg/d;
CH_4 is methane production, g/d;
CHO is carbohydrate content of the diet, % DM;
CP is crude protein content of the diet, % DM;
CPI is CP intake of the diet, kg/d;
DM is dry matter, % as-fed;
DMI is dry matter intake, kg/d;
EE is ether extract (fat) content of the diet, % DM;
EEI is EE intake of the diet, kg/d;
$Forage_{DMI}$ is forage intake of the diet, kg/d;
GEI is gross energy intake, Mcal/d;
MEI is metabolizable energy intake, Mcal/d;
NDF is neutral detergent fiber, % DM;
NDFI is neutral detergent fiber intake of the diet, kg/d;
NFCI is nonfiber carbohydrate intake of the diet, kg/d; and
OA is organic acid content of the diet, % DM.

The equations above estimate daily g CH_4. The average value is converted to MJ/d, Mcal/d, and L/d using the following conversion factors: the complete combustion of CH_4 yields 890 kJ/mol or 55.65 MJ/kg or 13.3 Mcal/kg, assuming 16 g/mol. The density of CH_4 is 0.6556 g/L (kg/m^3) at 25°C and 1 atm or 0.716 g/L (kg/m^3) at 0°C and 1 atm.

Mechanistic Level of Solution (MLS)

For the MLS, predicting the energy content of the diet is accomplished by using a mechanistic rumen model to estimate apparent TDN, and RDP and RUP of each feed, then using equations and conversion factors to estimate ME, NEma, NEga, and MCP values. To calculate apparent TDN, apparent digestibilities for carbohydrates, protein, and fats are estimated. These apparent digestibilities are determined by simulating the degradation, passage, and digestion of feedstuffs in the rumen and small intestine. Excretion of N is also estimated by the model.

Carbohydrate Fractions

The carbohydrate fractionation of the NRC (1996, 2000) Level 2 is used with the addition of a fraction for pectic substances and adjustments for organic acids (OA). Equations used to calculate carbohydrate fractions of the jth feedstuff are listed below:

$$CHO_j = 100 - (CP_j + Fat_j + Ash_j + OA_j) \quad \text{(Eq. 19-143)}$$

$$CC_j = Lignin_j \times 2.4 \quad \text{(Eq. 19-144)}$$

$$CB3_j = NDF_j - CC_j \quad \text{(Eq. 19-145)}$$

$$NFC_j = CHO_j - (CB3_j + CC_j) \quad \text{(Eq. 19-146)}$$

$$CA_j = Sugars_j \quad \text{(Eq. 19-147)}$$

$$CB1_j = Starch_j \quad \text{(Eq. 19-148)}$$

$$CB2_j = NFC_j - CA_j - CB1_j \quad \text{(Eq. 19-149)}$$

where

Ash_j is the ash content of the jth feedstuff, % DM;
CA_j is the sugar content of the jth feedstuff, % DM;
$CB1_j$ is the starch content of the jth feedstuff, % DM;
$CB2_j$ is the pectin content of the jth feedstuff, % DM:
$CB3_j$ is the available (i.e., potentially fermentable) fiber content of the jth feedstuff, % DM;
CC_j is the unavailable fiber content of the jth feedstuff, % DM;
CHO_j is the carbohydrate content of the jth feedstuff, % DM;
CP_j is the crude protein content of the jth feedstuff, % DM;
Fat_j is the fat (i.e., ether extract) content of the jth feedstuff, % DM;
$Lignin_j$ is the lignin content of the jth feedstuff, % DM;
NDF_j is the neutral detergent fiber content of the jth feedstuff, % DM;
NFC_j is the nonfiber carbohydrate content of the jth feedstuff, % DM;
OA_j is the organic acid content of the jth feedstuff, % DM;
$Starch_j$ is the starch content of the jth feedstuff, % DM; and
$Sugars_j$ is the sugar content of the jth feedstuff, % DM.

Protein Fractions

The NRC (1996, 2000) committee adopted the protein fractionation used in the first version of the CNCPS model (Sniffen et al., 1992). In that model, protein was fractionated into five fractions (PA, PB1, PB2, PB3, and PC) in which PA is nonprotein N (NPN), PC is unavailable protein and determined chemically as the percentage of N bound to the acid detergent fiber (ADIN), PB1 is the protein soluble in borate-phosphate buffer, and PB2 and PB3 are the potentially degradable protein with varying levels of ruminal degradation kinetics. Similar to the NRC (2001) committee, the current committee adopted a three-pool in situ-based "ABC" system (Ørskov and McDonald, 1979; NRC, 2001) with only one fractional digestion rate (kd) for the protein B fraction, which is a composite of the protein B2 and B3 that were used in the NRC (1996, 2000) model. In the current model, "A" represents the fraction of feed CP that is instantaneously or completely degraded in the rumen, "B" represents the potentially degradable CP, and "C" represents the unavailable CP. The urea N recycled in the rumen and urea N used for anabolism (UUA) are computed as shown in the ELS section above. Based on Sniffen et al. (1992), the NRC (1996, 2000) used kd for PB1 varying from 135 to 400 %/h. Thus, at least 96% of PB1 would be degraded in the rumen if the average fractional passage rate (kp) of 6%/h and a linear and steady-state condition model for the kinetics of protein degradation in the rumen are assumed (i.e., 135/[135 + 6]). If the PA and PB1 fractions (i.e., soluble protein) were combined, on average, more than 99% of soluble protein would be degraded in the rumen. Therefore, fraction A is assumed to be the soluble protein (i.e., PA + PB1 of NRC [1996, 2000] and Sniffen et al. [1992]), fraction C is assumed to be the same as PC of NRC (1996, 2000) and Sniffen et al. (1992), and fraction B is the difference (100 – PA – PC) and its kd is calculated using the NRC (1996, 2000) kd values for PB2 and PB3, and assuming a kp of 6% for each feed. Users should modify the kd of PB accordingly.

Henceforth, the following notation is used for protein fractions in the current edition of the *Nutrient Requirements of Beef Cattle*: PA, PB, and PC as % DM.

Computing Ruminal Degradation and Escape of Carbohydrate, Protein, Fat, and Ash

Because steady-state condition and linear model premises are assumed, ruminal degradation and escape (i.e., undegraded) of carbohydrate and protein fractions can be determined by using the following generic formulas that use kd for each carbohydrate and protein B fraction, and the kp:

$$\text{Ruminally Degraded} = \frac{kd}{kd + kp} \times \text{Intake of Nutrient} \quad \text{(Eq. 19-150)}$$

$$\text{Ruminally Escaped} = \frac{kp}{kd + kp} \times \text{Intake of Nutrient} \quad \text{(Eq. 19-151)}$$

The current committee adopted the fractional passage rates (i.e., kp) estimated by the equations of Seo et al. (2006) as shown below.

$$kp_j = \begin{cases} kp_{forage,j}, & Forage_{\%DM,j} > 0 \\ kp_{concentrate,j}, & Forage_{\%DM,j} = 0 \end{cases} \quad \text{(Eq. 19-152)}$$

$$kp_{forage,j} = 2.365 + 0.0214 \times FpBW_j + 0.0734 \times CpBW_j + 0.069 \times FDMI_j \quad \text{(Eq. 19-153)}$$

$$kp_{concentrate,j} = 1.169 + 0.1375 \times FpBW_j + 0.1721 \times CpBW_j \quad \text{(Eq. 19-154)}$$

$$kp_{liquid,j} = 4.524 + 0.0223 \times FpBW + 0.2046 \times CpBW + 0.344 \times FDMI \quad \text{(Eq. 19-155)}$$

$$FpBW_j = (Forage_{\%DM,j}/100) \times DMI_j \times 1{,}000/SBW \quad \text{(Eq. 19-156)}$$

$$CpBW_j = (100 - Forage_{\%DM,j}/100) \times DMI_j \times 1{,}000/SBW \quad \text{(Eq. 19-157)}$$

$$FDMI_j = (Forage_{\%DM,j}/100) \times DMI_j \quad \text{(Eq. 19-158)}$$

where

$CpBW_j$ is the concentrate DMI as a proportion of SBW of the jth feedstuff, g/kg;
$FDMI_j$ is the forage DMI of the jth feedstuff, kg/d;
$Forage_{\%DM,j}$ is the forage content of the jth feedstuff, % DM;
$FpBW_j$ is the forage DMI as a proportion of SBW of the jth feedstuff, g/kg;
kp_j is the fractional passage rate of the jth feedstuff, %/h;
$kp_{forage,j}$ is the fractional passage rate of forage for the jth feedstuff, %/h;
$kp_{liquid,i}$ is the fractional passage rate of liquids for the jth feedstuff, %/h; and
$kp_{concentrate,j}$ is the fractional passage rate of concentrate for the jth feedstuff, %/h.

The following equations calculate the amounts of carbohydrate and protein fractions that are degraded in the rumen. Dietary fat and ash are not degraded in the rumen. Because CA is assumed to be mainly sugars and PA is assumed to be soluble protein and both are readily available in the rumen, they are completely fermented in the rumen and their kd are assumed to be extremely fast (i.e., instantaneous). The kd^*_{CB3} is adjusted for ruminal pH as shown in the next section, (*Computing Ruminal Bacterial Yield*):

$$RDCA_j = DMI_j \times CA_j \quad \text{(Eq. 19-159)}$$

$$RDCB1_j = DMI_j \times CB1_j \times \left(\frac{kd_{CB1,j}}{kd_{CB1,j} + kp_j}\right) \quad \text{(Eq. 19-160)}$$

$$RDCB2_j = DMI_j \times CB2_j \times \left(\frac{kd_{CB2,j}}{kd_{CB2,j} + kp_j}\right) \quad \text{(Eq. 19-161)}$$

$$RDCB3_j = DMI_j \times CB3_j \times \left(\frac{kd^*_{CB3,j}}{kd^*_{CB3,j} + kp_j}\right) \quad \text{(Eq. 19-162)}$$

$$RDPA_j = DMI_j \times PA_j \quad \text{(Eq. 19-163)}$$

$$RDPB_j = DMI_j \times PB_j \times \left(\frac{kd_{PB,j}}{kd_{PB,j} + kp_j}\right) \quad \text{(Eq. 19-164)}$$

where

CA_j is the sugar content of the jth feedstuff, % DM;
$CB1_j$ is the starch content of the jth feedstuff, % DM;
$CB2_j$ is the pectin content of the jth feedstuff, % DM:
$CB3_j$ is the available (i.e., potentially fermentable) fiber content of the jth feedstuff, % DM;
DMI_j is dry matter intake of the jth feedstuff, kg/d;
$kd_{CB1,j}$ is the fractional rate of degradation of CB1 of the jth feedstuff, %/h;
$kd_{CB2,j}$ is the fractional rate of degradation of CB2 of the jth feedstuff, %/h;
$kd^*_{CB3,j}$ is the adjusted fractional rate of degradation of CB3 of the jth feedstuff, %/h;
$kd_{PB,j}$ is the fractional rate of degradation of PB of the jth feedstuff, %/h;
kp_j is the fractional rate of passage of the jth feedstuff, %/h;
PA_j is the instantaneously degraded protein fraction in the rumen of the jth feedstuff, % DM;
PB_j is the potentially degradable protein fraction in the rumen of the jth feedstuff, % DM;
$RDCA_j$ is ruminally degraded carbohydrate A of the jth feedstuff, kg/d;
$RDCB1_j$ is ruminally degraded carbohydrate B1 (starch) of the jth feedstuff, kg/d;
$RDCB2_j$ is ruminally degraded carbohydrate B2 (pectin) of the jth feedstuff, kg/d;
$RDCB3_j$ is ruminally degraded carbohydrate B3 (available fiber) of the jth feedstuff, kg/d;
$RDPA_j$ is ruminally degraded protein A (soluble) of the jth feedstuff, kg/d; and
$RDPB_j$ is ruminally degraded protein B of the jth feedstuff, kg/d.

The following equations are used to calculate the amounts of carbohydrate and protein fractions, fat, and ash that escape the rumen:

$$RECB1_j = DMI_j \times CB1_j \times \left(\frac{kp_j}{kd_{CB1,j} + kp_j}\right) \quad \text{(Eq. 19-165)}$$

$$RECB2_j = DMI_j \times CB2_j \times \left(\frac{kp_j}{kd_{CB2,j} + kp_j}\right) \quad \text{(Eq. 19-166)}$$

$$RECB3_j = DMI_j \times CB3_j \times \left(\frac{kp_j}{kd^*_{CB3,j} + kp_j}\right) \quad \text{(Eq. 19-167)}$$

$$RECC_j = DMI_j \times CC_j \quad \text{(Eq. 19-168)}$$

$$REPB_j = DMI_j \times PB_j \times \left(\frac{kp_j}{kd_{PB,j} + kp_j}\right) \quad \text{(Eq. 19-169)}$$

$$REPC_j = DMI_j \times PC_j \quad \text{(Eq. 19-170)}$$

$$REFAT_j = DMI_j \times Fat_j \quad \text{(Eq. 19-171)}$$

$$REASH_j = DMI_j \times Ash_j \quad \text{(Eq. 19-172)}$$

where

Ash_j is the ash content of the jth feedstuff, % DM;
$CB1_j$ is the starch content of the jth feedstuff, % DM;
$CB2_j$ is the pectin content of the jth feedstuff, % DM:
$CB3_j$ is the available (i.e., potentially fermentable) fiber content of the jth feedstuff, % DM;
CC_j is the unavailable fiber content of the jth feedstuff, % DM;
DMI_j is dry matter intake of the jth feedstuff, kg/d;
Fat_j is the fat (i.e., ether extract) content of the jth feedstuff, % DM;
$kd_{CB1,j}$ is the fractional rate of degradation of CB1 of the jth feedstuff, %/h;
$kd_{CB2,j}$ is the fractional rate of degradation of CB2 of the jth feedstuff, %/h;
$kd^{*}_{CB3,j}$ is the adjusted fractional rate of degradation of CB3 of the jth feedstuff, %/h;
$kd_{PB,j}$ is the fractional rate of degradation of PB of the jth feedstuff, %/h;
kp_j is the fractional passage rate of the jth feedstuff, %/h.
$REASH_j$ is ruminally escaped ash of the jth feedstuff, kg/d;
$RECB1_j$ is ruminally escaped carbohydrate B1 (starch) of the jth feedstuff, kg/d;
$RECB2_j$ is ruminally escaped carbohydrate B2 (pectin) of the jth feedstuff, kg/d;
$RECB3_j$ is ruminally escaped carbohydrate B3 (available fiber) of the jth feedstuff, kg/d;
$RECC_j$ is ruminally escaped carbohydrate C of the jth feedstuff, kg/d;
$REFAT_j$ is ruminally escaped fat (i.e., ether extract) of the jth feedstuff, kg/d;
$REPB_j$ is ruminally escaped protein B of the jth feedstuff, kg/d; and
$REPC_j$ is ruminally escaped protein C of the jth feedstuff, kg/d.

Computing Ruminal Bacteria Yield

As in the CNCPS model (Russell et al., 1992), the ruminal bacterial yield is computed based on the Pirt (1965) equation. The current model uses a theoretical maximum bacterial yield of 0.4 g bacteria mass/g carbohydrate per hour for nonfiber carbohydrate (NFC)-degrading bacteria (YG1) and fiber carbohydrate (FC)-degrading bacteria (YG2), the maintenance rate of 0.15 g carbohydrate/g NFC bacteria per hour (KM1) and 0.05 g carbohydrate/g FC bacteria per hour (KM2), and kd of the carbohydrate fraction, assuming that true growth rate of ruminal bacteria is approximated by its fermentation rate.

$$Y_{CA,j} = YG1 = 0.4 \quad \text{(Eq. 19-173)}$$

$$Y_{CB1,j} = 1/(KM1/kd_{CB1,j} + 1/YG1) \quad \text{(Eq. 19-174)}$$

$$Y_{CB2,j} = 1/(KM1/kd_{CB2,j} + 1/YG1) \quad \text{(Eq.19-175)}$$

$$Y_{CB3,j} = 1/(KM2/kd_{CB3,j} + 1/YG2) \quad \text{(Eq. 19-176)}$$

where

KM1 is the maintenance rate of the nonfiber carbohydrate bacteria, 0.15 g NFC/g bacteria per hour;
KM2 is the maintenance rate of the fiber carbohydrate bacteria, 0.05 g FC/g bacteria per hour;
$Y_{CA,j}$ is the bacterial yield for sugar carbohydrate (CA) of the jth feedstuff, g bacteria/g carbohydrate;
$Y_{CB1,j}$ is the bacterial yield for starch carbohydrate (CB1) of the jth feedstuff, g bacteria/g carbohydrate;
$Y_{CB2,j}$ is the bacterial yield for pectin (CB2) of the jth feedstuff, g bacteria/g carbohydrate;
$Y_{CB3,j}$ is the bacterial yield for fiber carbohydrate (CB3) of the jth feedstuff, g bacteria/g carbohydrate;
YG1 is the theoretical maximum bacterial yield for CA, CB1, and CB2, 0.4 g bacteria/g carbohydrate; and
YG2 is the theoretical maximum bacterial yield for CB3, 0.4 g bacteria/g carbohydrate.

Because the ruminal bacteria that degrade fiber are sensitive to the ruminal pH, to maintain a viable growth rate above the fractional passage rate, the KM2 and Y_{CB3} were adjusted for ruminal pH as described by Pitt et al. (1996). Ruminal pH is computed using the forage content of the diet (Sarhan and Beauchemin, Agriculture and Agri-Food Canada, Lethbridge, Alberta, personal communication, June 2014) when physically effective NDF (peNDF) is not known (i.e., dietary peNDF = 0) or using the dietary peNDF (Pitt et al., 1996) when peNDF is known, but the user may override the ruminal pH value. Unlike the CNCPS-based models (Fox et al., 2004), the YG2 is not adjusted for ruminal pH; it is maintained at 0.4 g bacteria/g carbohydrate.

$$Y^{*}_{CB3,j} = \left(1/\left(KM2^{*}/kd_{CB3,j} + 1/YG2\right)\right) \times r_Y^{FC} \quad \text{(Eq. 19-177)}$$

$$r_Y^{FC} = \begin{cases} \left(1 - e^{-5.624 \times (pH - 5.7)^{0.909}}\right)/0.9968, & pH \geq 5.7 \\ 0, & pH < 5.7 \end{cases} \quad \text{(Eq. 19-178)}$$

$$KM2^{*} = 0.1409 - 0.0135 \times pH \quad \text{(Eq. 19-179)}$$

$$pH = \begin{cases} \begin{cases} 5.724 - 0.00963 \times Forage_P \\ BW > 200\ kg,\ Forage_P < 50\%,\ peNDF = 0 \\ 6.2,\ Forage_P \geq 50\%,\ peNDF = 0 \end{cases} \\ \begin{cases} 5.46 + 0.038 \times peNDF,\ peNDF < 26.3\%,\ peNDF > 0 \\ 6.46,\ peNDF \geq 26.3\%,\ peNDF > 0 \end{cases} \end{cases} \quad \text{(Eq. 19-180)}$$

$$BW = SBW/0.96 \quad \text{(Eq. 19-181)}$$

where

$Forage_P$ is forage proportion of the diet, % DM;
KM2* is the adjusted maintenance rate of the fiber carbohydrate (i.e., FC) bacteria, 0.05 g FC/g bacteria per hour;
peNDF is the physically effective NDF of the diet, % DM;
pH is the ruminal pH;
$Y^{*}_{CB3,j}$ is the adjusted bacterial yield for fiber carbohydrate (CB3) of the jth feedstuff, g bacteria/g carbohydrate; and
r_Y^{FC} is the relative yield of fiber carbohydrate bacteria adjusted for ruminal pH.

For comparative purposes, ruminal pH is computed with other equations (Mertens, 1997; Sauvant et al., 1999; Fox et al., 2004; Zebeli et al., 2008) using peNDF or ruminally degraded starch (RDCB1) as shown below, but as discussed in Chapter 4 (Carbohydrates), peNDF usually accounts for less than 50% of the variation in ruminal pH for feedlot beef cattle.

Mertens (1997):

$$pH = 6.67 - 14.3/peNDF \quad \text{(Eq. 19-182)}$$

Sauvant et al. (1999):

$$pH = 6.41 - 0.011 \times (RDCB1 \times 100/DMI) \quad \text{(Eq. 19-183)}$$

Fox et al. (2004):

$$pH = \begin{cases} 5.425 + 0.04229 \times peNDF,\ peNDF < 24.5\% \\ 6.46,\ peNDF \geq 24.5\% \end{cases} \quad \text{(Eq. 19-184)}$$

Zebeli et al. (2008):

$$pH = \begin{cases} 5.59 + 0.0218 \times peNDF,\ peNDF < 31.2\% \\ 6.27,\ peNDF \geq 31.2\% \end{cases} \quad \text{(Eq. 19-185)}$$

$$peNDF = \frac{\sum_{j=1}^{n}\left(DMI_j \times \left(NDF_j/100\right) \times \left(pef_j/100\right)\right)}{\sum_{j=1}^{n} DMI_j} \quad \text{(Eq. 19-186)}$$

where

pef_j is the physically effective factor of the jth feedstuff, % NDF; and
RDCB1 is the amount of ruminally degraded carbohydrate B1 (starch), kg/d.

The fractional degradation rate of fiber (kd_{CB3}) adjusted for the ruminal pH (kd^{*}_{CB3}) is then calculated using the reverse of the Pirt (1965) equation, as demonstrated by Pitt et al. (1996). The least value of kd_{CB3} or kd^{*}_{CB3} is used to calculate degraded and escaped CB3.

$$kd^{*}_{CB3,j} = Min\left(kd_{CB3,j}, \frac{KM2^{*} \times Y^{*}_{CB3,j} \times YG2}{YG2 - Y^{*}_{CB3,j}} + KM2^{*} \times YG2\right) \quad \text{(Eq. 19-187)}$$

where $kd^{*}_{CB3,j}$ is the fractional degradation rate of fiber adjusted for the ruminal pH.

Then bacterial mass (BACT) is calculated for each feedstuff as the amount of carbohydrate degraded in the rumen times its respective bacterial yield, as shown below:

$$BACT_j = RDCA_j \times Y_{CA,j} + RDCB1_j \times Y_{CB1,j} + RDCB2_j \times Y_{CB2,j} + RDCB3_j \times Y^{*}_{CB3,j} \quad \text{(Eq. 19-188)}$$

where $BACT_j$ is the amount of bacterial mass produced from the jth feedstuff, kg/d.

The bacterial crude protein (BCP), carbohydrate (BCHO), fat (BFAT), and ash (BASH) escaping the rumen are calculated using the following equations, assuming that BCP is 62.5% of bacterial mass, BCHO is 21.1% of bacterial mass, BFAT is 12% of bacterial mass, and BASH is 4.4% of bacterial mass. Of the BCP, 60% is assumed to be true protein, 25% is cell-wall protein, and 15% is nucleic acids. These values are slightly different from those used in the ELS in which the MTP would be approximately 75% (60% + 15%) of MCP (i.e., MLS) rather than 80% (i.e., ELS).

$$REBTP_j = 0.6 \times 0.625 \times BACT_j \quad \text{(Eq. 19-189)}$$

$$REBCWP_j = 0.25 \times 0.625 \times BACT_j \quad \text{(Eq. 19-190)}$$

$$REBNA_j = 0.15 \times 0.625 \times BACT_j \quad \text{(Eq. 19-191)}$$

$$REBCHO_j = 0.211 \times BACT_j \quad \text{(Eq. 19-192)}$$

$$REBFAT_j = 0.12 \times BACT_j \quad \text{(Eq. 19-193)}$$

$$REBASH_j = 0.044 \times BACT_j \quad \text{(Eq. 19-194)}$$

where

$REBASH_j$ is ruminally escaped bacterial ash of the jth feedstuff, kg/d;
$REBCHO_j$ is ruminally escaped bacterial carbohydrate of the jth feedstuff, kg/d;
$REBCWP_j$ is ruminally escaped bacterial cell-wall protein of the jth feedstuff, kg/d;
$REBFAT_j$ is ruminally escaped bacterial fat of the jth feedstuff, kg/d;
$REBNA_j$ is ruminally escaped bacterial nucleic acids of the jth feedstuff, kg/d; and
$REBTP_j$ is ruminally escaped bacterial true protein of the jth feedstuff, kg/d.

Computing Intestinal Digestibility and Absorption of Carbohydrate, Protein, Fat, and Ash

Equations for calculating digested carbohydrate, protein, fat, and ash from feed and bacterial sources are listed below. The NRC (1996, 2000) model assumed that the three pools of protein B (B1, B2, and B3) had intestinal digestibilities (ID) of 100, 100, and 80%, respectively. Because the current model assumes only one protein B fraction, an intestinal digestibility (PBID) weighted by the digested protein and RUP was computed for all feedstuffs in the library. Ninety percent of the feedstuffs were in the range of 60 to 81% of PBID, within the range assumed by the ELS and the MLS. Furthermore, if dietary RDP (RDPA + RDPB) is greater than BCP, the excess of RDP is assumed to be converted to ammonia and absorbed.

$$DIGP_j = PBID_j \times REPB_j + REBTP_j + REBNA_j \quad \text{(Eq. 19-195)}$$

Similar to the NRC (1996, 2000) model, the intestinal digestibility of fiber carbohydrate (CB3) is assumed to be 20% and the intestinal digestibility of starch (CB1) and pectic substances (CB2) are assumed to average 75%, but the CB1 can vary depending on the degree of processing as shown in Table 19-8. The bacterial carbohydrate ID is 95%:

TABLE 19-8 Postruminal Starch (CB1) Digestibilities (% of flow to the intestine)[a]

Feed and Processing Degree	Starch (CB1) ID
Corn	
Whole	30-40
Dry, rolled	65-70
Cracked	65-70
Meal	80-90
High moisture, whole	85-90
High moisture, ground	85-95
Steam flaked	92-97
Sorghum	
Dry, rolled	85-90
Dry, ground	85-90
Steam flaked	90-95
Barley	
Dry, rolled	75-85
Dry, ground	75-85
Steam flaked	85-95
Steam rolled	85-95
Wheat	
Dry, rolled	82-86
Dry, ground	82-86
Steam flaked	85-95
Steam rolled	85-95

[a]Based on Sniffen et al. (1992) and Owens and Zinn (2005).

$$DIGC_j = CB1ID_j \times RECB1_j + CB2ID_j \times RECB2_j + 0.2 \times RECB3_j + 0.95 \times REBCHO_j \quad \text{(Eq. 19-196)}$$

The ID of fat for bacterial and dietary sources is assumed to be 95%, and the ID of ash for bacterial and dietary sources is assumed to be 50%.

$$DIGF_j = 0.95 \times (REFAT_j + REBFAT_j) \quad \text{(Eq. 19-197)}$$

$$DIGA_j = 0.5 \times (REASH_j + REBASH_j) \quad \text{(Eq. 19-198)}$$

where

$DIGA_j$ is the digestible dietary and bacterial ash of the jth feedstuff, kg/d;
$DIGC_j$ is the digestible dietary and bacterial carbohydrate of the jth feedstuff, kg/d;
$DIGF_j$ is the digestible dietary and bacterial fat of the jth feedstuff, kg/d;
$DIGP_j$ is the digestible dietary and bacterial protein of the jth feedstuff, kg/d;
$PBID_j$ is the intestinal digestibility of the protein B fraction of the jth feedstuff, %;
$CB1ID_j$ is the intestinal digestibility of CB1 (i.e., starch) of the jth feedstuff (Table 19-8), %; and
$CB2ID_j$ is the intestinal digestibility of CB2 (i.e., pectin) of the jth feedstuff, %.

Computing Fecal Output and Endogenous Matter

The following equations are used to calculate undigested feed residues from dietary fiber, starch, fat, and ash fractions and indigestible bacterial cell wall, carbohydrate, fat, and ash appearing in the feces, based on the data summarized by Van Soest (1994).

$$FEP_j = (1 - PBID_j) \times REPB_j + REPC_j + REBCWP_j \quad \text{(Eq. 19-199)}$$

$$FEC_j = (1 - CB1ID_j) \times RECB1_j + (1 - CB2ID_j) \times RECB2_j + (1 - 0.2) \times RECB3_j + RECC_j + (1 - 0.95) \times REBCHO_j \quad \text{(Eq. 19-200)}$$

$$FEF_j = (1 - 0.95) \times (REFAT_j + REBFAT_j) \quad \text{(Eq. 19-201)}$$

$$FEA_j = (1 - 0.5) \times (REASH_j + REBASH_j) \quad \text{(Eq. 19-202)}$$

where

FEP_j is fecal protein of the jth feedstuff without endogenous protein, kg/d;
FEC_j is fecal carbohydrate of the jth feedstuff, kg/d;
FEF_j is fecal fat of the jth feedstuff without endogenous fat, kg/d; and
FEA_j is fecal ash of the jth feedstuff without endogenous ash, kg/d.

As discussed by Cannas et al. (2004) and Tedeschi et al. (2010), the CNCPS (Fox et al., 2004) and the NRC (1989, 1996, 2000) models may have double-accounted for microbial contribution in the endogenous matter. The amount of fecal protein is the sum of undegraded feed protein that is not digested and absorbed in the small intestine (F_U), metabolic microbial residues (F_M), and metabolic endogenous protein (F_E). These models assumed that the endogenous protein is 30 g/kg DMI, which is the intercept of the linear regression of digestible protein on intake protein (Waldo and Glenn, 1984), divided by the average indigestibility of diets (assumed 33%) to estimate the F_E component of the endogenous matter. Then, this F_E is added to F_U and F_M to estimate the endogenous matter. The problem is that the intercept of the linear regression of digestible protein on intake protein contains about 85% microbial debris (Van Soest, 1994); thus, it represents the F_{M+E} (i.e., fecal metabolic microbial residue plus metabolic endogenous protein) component of the endogenous matter, not the F_E alone. Another problem with the approach used by these models is the average indigestibility of 33%. To overcome these problems, the current model calculates the F_{M+E} using empirical equations and adds the F_U component to estimate endogenous matter for protein, fat, and ash. Fecal carbohydrate is the sum of undigested bacterial and dietary carbohydrate.

$$FEENGP_{M+E,j} = (30/1{,}000) \times DMI_j \quad \text{(Eq. 19-203)}$$

$$FEENGF_{M+E,j} = (11.9/1{,}000) \times DMI_j \quad \text{(Eq. 19-204)}$$

$$FEENGA_{M+E,j} = (17/1{,}000) \times DMI_j \quad \text{(Eq. 19-205)}$$

$$FEPROT_j = (1 - PBID_j) \times REPB_j + REPC_j + FEENGP_{M+E,j} \quad \text{(Eq. 19-206)}$$

$$FECHO_j = FEC_j \quad \text{(Eq. 19-207)}$$

$$FEFAT_j = (1 - 0.95) \times REFAT_j + FEENGF_{M+E,j} \quad \text{(Eq. 19-208)}$$

$$FEASH_j = (1 - 0.5) \times REASH_j + FEENGA_{M+E,j} \quad \text{(Eq. 19-209)}$$

where

$FEASH_j$ is the fecal ash of the jth feedstuff, kg/d;
$FECHO_j$ is the fecal carbohydrate of the jth feedstuff, kg/d;
$FEENGA_{M+E,j}$ is the fecal endogenous and microbial ash contribution of the jth feedstuff, kg/d;
$FEENGF_{M+E,j}$ is the fecal endogenous and microbial fat contribution of the jth feedstuff, kg/d;
$FEENGP_{M+E,j}$ is the fecal endogenous and microbial protein contribution of the jth feedstuff, kg/d;
$FEFAT_j$ is the fecal fat of the jth feedstuff, kg/d; and
$FEPROT_j$ is the fecal protein of the jth feedstuff, kg/d.

Computing Total Digestible Nutrients and Energy Values

The apparent TDN is potentially digestible nutrient intake minus indigestible bacterial and feed components appearing in the feces plus the OA of the feeds, assuming that OA are completely absorbed and metabolized by the animal. The ME values for each feed are based on the assumption that 1 kg of TDN is equal to 4.409 Mcal of DE (NRC, 1976) and 1 Mcal of DE is equal to 0.82 Mcal of ME (NRC, 1976) as adopted by the NRC (1996, 2000). Users may, however, override this DE to ME efficiency by entering the value they prefer to use. Then, the NEma and NEga are computed from the concentration of ME (NRC, 1984).

$$aTDN_j = (DMI_j \times CHO_j/100 - FECHO_j) + (DMI_j \times CP_j/100 - FEPROT_j) + 2.25 \times (DMI_j \times Fat_j/100 - FEFAT_j) + DMI_j \times OA_j/100 \quad \text{(Eq. 19-210)}$$

$$DE_j = aTDN_j \times 4.409 \quad \text{(Eq. 19-211)}$$

$$ME_j = DE_j \times 0.82 \times ADTV_{ME} \quad \text{(Eq. 19-212)}$$

$$[ME]_j = ME_j/DMI_j \quad \text{(Eq. 19-213)}$$

$$NEma_j = 1.37 \times [ME]_j - 0.138 \times [ME]_j^2 + 0.0105 \times [ME]_j^3 - 1.12 \quad \text{(Eq. 19-214)}$$

$$NEga_j = 1.42 \times [ME]_j - 0.174 \times [ME]_j^2 + 0.0122 \times [ME]_j^3 - 1.65 \quad \text{(Eq. 19-215)}$$

where

$ADTV_{ME}$ is the effect of ionophore on ME, and if used, ME is increased by 2.3 or 1.5% for monensin or lasalocid, respectively;
$aTDN_j$ is the apparent total digestible nutrients of the jth feedstuff, kg/d;
DE_j is the digestible energy of the jth feedstuff, Mcal/d;
$[ME]_j$ is the concentration of ME of the jth feedstuff, Mcal/kg, adjusted (or not) for ionophores;
ME_j is the metabolizable energy of the jth feedstuff, Mcal/d;
$NEga_j$ is the net energy available for growth of the jth feedstuff, Mcal/kg; and
$NEma_j$ is the net energy available for maintenance of the jth feedstuff, Mcal/kg.

Computing Metabolizable Protein and Amino Acid Supply

The metabolizable protein is the amount of digested RUP and digested bacterial true protein. It does not include the bacterial nucleic acids. Furthermore, if dietary RDP (RDPA + RDPB) is greater than BCP, the excess of RDP is assumed to be converted to ammonia and then absorbed.

$$MP_j = PBID_j \times REPB_j + REBTP_j \quad \text{(Eq. 19-216)}$$

The essential amino acid composition of the undegradable protein of each feedstuff is used to calculate the supply of amino acids from the feeds. The bacterial composition of essential amino acids is used to calculate the supply of amino acids from bacteria. The bacterial non-cell wall composition of amino acids (Table 19-7) is also used for the MLS.

$$DIGFAA_{j,i} = (FAA_{i,j}/100) \times PBID_j \times REPB_j \quad \text{(Eq. 19-217)}$$

$$DIGBAA_{j,i} = (BAA_i/100) \times REBTP_j \quad \text{(Eq. 19-218)}$$

$$DIGAA_{j,i} = DIGFAA_{j,i} + DIGBAA_{j,i} \quad \text{(Eq. 19-219)}$$

Computing RDP Balance and Ruminal N Balance (RNB)

Unlike the RDP balance calculation of the ELS, the RDP balance in the MLS can be customized by the user. By default, the MLS assumes an efficiency of use of RDP of 100% by ruminal microbes, but users may change this value to 85% (NRC, 2001). The RDP supply does not include recycled N, but to account for recycled N, users may select the N equation of the NRC (1985) or the UUA (i.e., urea N used for anabolism) equation developed by Reynolds and Kristensen (2008) or the equation in this edition of *Nutrient Requirements of Beef Cattle*. These equations are shown below. If dietary RDP is greater than MCP (positive RDP balance), the excess of RDP is assumed to be converted to ammonia and absorbed. When RDP balance (RDP supply minus RDP required) is negative, BCP is multiplied by the efficiency of use of RDP by ruminal microbes (i.e., 0.85 or 1.00).

$$RDP_{Required} = BCP_{TDN}/RDP_{Efficiency} \quad \text{(Eq. 19-220)}$$

$$BCP_{TDN} = \sum_{j=1}^{n}(0.625 \times BACT_j) \quad \text{(Eq. 19-221)}$$

$$RDP = RDP_{Supply} = \sum_{j=1}^{n}(RDPA_j + RDPB_j) + RecycledN \times 6.25/1{,}000 \quad \text{(Eq. 19-222)}$$

$$RecycledN = \begin{cases} \left(-0.1113 + 0.996 \times e^{-0.0616 \times CP}\right) \times (0.745 \times NI - 11.98), & \text{Current model} \\ [1.161 - 0.329 \times \ln(CP)] \times [0.7 \times NI - 13.74], & \text{Reynolds and Kristensen (2008)} \\ \begin{bmatrix}(121.7 - 12.01 \times CP + \\ 0.3235 \times CP^2)/100\end{bmatrix} \times NI, & \text{NRC (1985)} \end{cases} \quad \text{(Eq. 19-223)}$$

$$NI = ((CP/100)/6.25) \times DMI \times 1{,}000 \quad \text{(Eq. 19-224)}$$

$$RDP_{Balance} = RDP_{Supply} - RDP_{Required} \quad \text{(Eq. 19-225)}$$

$$BCP_{RDP} = \begin{cases} BCP_{TDN}, & RDP_{Balance} \geq 0 \\ RDP_{Supply} \times RDP_{Efficiency}, & RDP_{Balance} < 0 \end{cases} \quad \text{(Eq. 19-226)}$$

where

CP is crude protein of the diet, % DM;
NI is nitrogen intake, g/d;
$RDP_{Efficiency}$ is the efficiency of use of RDP by ruminal microbes (default =1); and
RecycledN is either urea N used for anabolism (UUA) or recycled ruminal N depending on the equation chosen (default = 0, not used), g/d.

The RNB is calculated from the $RDP_{Balance}$ as shown below:

$$RNB = RDP_{Balance}/6.25 \quad \text{(Eq. 19-227)}$$

Computing Volatile Fatty Acids and Methane Production

The calculation of the production of volatile fatty acids (VFA) and CH_4 was adapted from Pitt et al. (1996). The committee assumed that the fermentation of pectic substances is similar to starch. The end products of fermentation are essentially 1 minus the bacteria yield (Y) times the amount of ruminally degraded carbohydrate corrected for bacteria ash (4.4%) and the carbon skeletons of products used by the bacteria that did not come from carbohydrates. Because this current version of the model combines NPN (e.g., urea, ammonia) and soluble protein into one protein fraction (i.e., PA), it is assumed that feedstuffs with more than 100% CP are NPN sources. The RDP_j^* required is the amount of bacterial mass that is comprised of peptide and ammonia uptake by the NFC bacteria that did not come from degraded carbohydrates. It is adjusted for feedstuff NPN (i.e., urea) in which the amount of N, not the protein equivalent, is discounted.

$$BACT_j^{CA} = RDCA_j \times Y_{CA,j} \quad \text{(Eq. 19-228)}$$

$$BACT_j^{CB1} = RDCB1_j \times Y_{CB1,j} \quad \text{(Eq. 19-229)}$$

$$BACT_j^{CB2} = RDCB2_j \times Y_{CB2,j} \quad \text{(Eq. 19-230)}$$

$$BACT_j^{CB3} = RDCB3_j \times Y^*_{CB3,j} \quad \text{(Eq. 19-231)}$$

$$RDP_j^* = RDPB_j + \begin{cases} RDPA_j / 6.25, & CP_j \geq 100 \\ RDPA_j & CP_j < 100 \end{cases} \quad \text{(Eq. 19-232)}$$

$$EndProduct_j^{CA} = RDCA_j - (1 - 0.044) \times BACT_j^{CA} + RDP_j^* \times BACT_j^{CA} / \left(BACT_j^{CA} + BACT_j^{CB1} + BACT_j^{CB2}\right) \quad \text{(Eq. 19-233)}$$

$$EndProduct_j^{CB1} = RDCB1_j - (1 - 0.044) \times BACT_j^{CB1} + RDP_j^* \times BACT_j^{CB1} / \left(BACT_j^{CA} + BACT_j^{CB1} + BACT_j^{CB2}\right) \quad \text{(Eq. 19-234)}$$

$$EndProduct_j^{CB2} = RDCB2_j - (1 - 0.044) \times BACT_j^{CB2} + RDP_j^* \times BACT_j^{CB2} / \left(BACT_j^{CA} + BACT_j^{CB1} + BACT_j^{CB2}\right) \quad \text{(Eq. 19-235)}$$

$$EndProduct_j^{CB3} = RDCB3_j - (1 - 0.044 - 0.625/6.25) \times BACT_j^{CB3} \quad \text{(Eq. 19-236)}$$

where

$BACT_j^{CA,CB1,CB2,CB3}$ is the amount of bacteria mass produced from the jth feedstuff, kg/d;
$EndProduct_j^{CA,CB1,CB2,CB3}$ is the amount of carbohydrate or carbon skeleton of peptides not retained by the bacteria from the jth feedstuff, kg/d;
$RDCA_j$ is ruminally degraded carbohydrate A of the jth feedstuff, kg/d;
$RDCB1_j$ is ruminally degraded carbohydrate B1 (starch) of the jth feedstuff, kg/d;
$RDCB2_j$ is ruminally degraded carbohydrate B2 (pectin) of the jth feedstuff, kg/d;
$RDCB3_j$ is ruminally degraded carbohydrate B3 (available fiber) of the jth feedstuff, kg/d;
RDP_j^* is the RDP adjusted for NPN feedstuffs (i.e., urea) of the jth feedstuff, kg/d;
$RDPA_j$ is ruminally degraded protein A (soluble) of the jth feedstuff, kg/d; and
$RDPB_j$ is ruminally degraded protein B of the jth feedstuff, kg/d.

Not all end products of fermentation are converted to VFA. About 34.4% of the CA, CB1, and CB2 and 54% of the CB3 are converted to nonacid end products (Pitt et al., 1996). Thus, the VFA from the ruminally degraded carbohydrate can be computed as shown below:

$$AcidProduct_j^{CA} = EndProduct_j^{CA} - 0.344 \times RDCA_j \quad \text{(Eq. 19-237)}$$

$$AcidProduct_j^{CB1} = EndProduct_j^{CB1} - 0.344 \times RDCB1_j \quad \text{(Eq. 19-238)}$$

$$AcidProduct_j^{CB2} = EndProduct_j^{CB2} - 0.344 \times RDCB2_j \quad \text{(Eq. 19-239)}$$

TABLE 19-9 Equations to Compute Mass Fraction of Lactate, Acetate-to-Propionate Ratio (Ac:Prop), and Butyrate from Sugar, Starch, and Fiber[a]

		Substrates		
Items	Factor	A (Sugar)	B1 (Starch) and B2 (Pectin)	B3 (Fiber)
Lactate	$Factor_L$	2.444 – 0.308 × pH	3.292 – 0.473 × pH	0
Ac:Prop	$Factor_{A:P}$	0.465 × pH – 1.302	5.013 – 0.535 × pH	0.274 + 0.239 × pH
Butyrate	$Factor_B$	0.567 – 0.0575 × pH	0.737 – 0.0922 × pH	0.483 – 0.060 × pH

[a]Adapted from Pitt et al. (1996).

$$AcidProduct_j^{CB3} = EndProduct_j^{CB3} - 0.54 \times RDCB3_j \quad \text{(Eq. 19-240)}$$

The partitioning of acid products of the fermentation into acetic, propionic, and butyric depends on ruminal pH. The equations shown in Table 19-9 are used to calculate the amount of these VFA. The following equations are used to compute lactate, acetate, propionate, and butyrate production.

$$Lactate_j^{CA,CB1,CB2,CB3} = Factor_L^{CA,CB1,CB2,CB3} \times AcidProduct_j^{CA,CB1,CB2,CB3} \quad \text{(Eq. 19-241)}$$

$$VFA_j^{CA,CB1,CB2,CB3} = \left(1 - Factor_L^{CA,CB1,CB2,CB3}\right) \times AcidProduct_j^{CA,CB1,CB2,CB3} \quad \text{(Eq. 19-242)}$$

$$Butyrate_j^{CA,CB1,CB2,CB3} = Factor_B^{CA,CB1,CB2,CB3} \times VFA_j^{CA,CB1,CB2,CB3} \quad \text{(Eq. 19-243)}$$

$$Propionate_j^{CA,CB1,CB2,CB3} = \frac{\left(1 - Factor_B^{CA,CB1,CB2,CB3}\right) \times VFA_j^{CA,CB1,CB2,CB3}}{\left(1 + Factor_{A:P}^{CA,CB1,CB2,CB3}\right)} \quad \text{(Eq. 19-244)}$$

$$Acetate_j^{CA,CB1,CB2,CB3} = Factor_{A:P}^{CA,CB1,CB2,CB3} \times Propionate_j^{CA,CB1,CB2,CB3} \quad \text{(Eq. 19-245)}$$

where

$Acetate_j^{CA,CB1,CB2,CB3}$ is the amount of acetate produced from the fermentation of CA, CB1, CB2, and CB3 of the jth feedstuff, kg/d;
$AcidProduct_j^{CA,CB1,CB2,CB3}$ is the amount of acid produced from the fermentation of CA, CB1, CB2, and CB3 of the jth feedstuff, kg/d;
$Butyrate_j^{CA,CB1,CB2,CB3}$ is the amount of butyrate produced from the fermentation of CA, CB1, CB2, and CB3 of the jth feedstuff, kg/d;
$Factor_{L,B,A:P}^{CA,CB1,CB2,CB3}$ is the mass fraction coefficients to convert $AcidProduct_j^{CA,CB1,CB2,CB3}$ into lactate, acetate, propionate, and butyrate for the jth feedstuff, dimensionless;
$Lactate_j^{CA,CB1,CB2,CB3}$ is the amount of lactate produced from the fermentation of CA, CB1, CB2, and CB3 of the jth feedstuff, kg/d;
$Propionate_j^{CA,CB1,CB2,CB3}$ is the amount of propionate produced from the fermentation of CA, CB1, CB2, and CB3 of the jth feedstuff, kg/d; and
VFA_j is the amount of volatile fatty acids of the jth feedstuff, g/d.

A portion of the lactate produced in the rumen can be converted to acetate, propionate, and butyrate by *Megasphaera elsdenii* (Counotte et al., 1981). Pitt et al. (1996) developed adjustments to account for the degradation of lactate in the rumen. The following equations are used compute the yield of *M. elsdenii* and the fractional degradation rate of lactate in the rumen:

$$\mu^L = \begin{cases} -3.631 + 1.255 \times pH - 0.0925 \times pH^2, & 4.18 \le pH \le 6.07 \\ -6.906 + 2.636 \times pH - 0.2311 \times pH^2, & 6.07 \le pH \le 7.33 \end{cases} \quad \text{(Eq. 19-246)}$$

$$YG^L = \begin{cases} 0.232 \times pH - 1.176, & pH \ge 5.07 \\ 0, & pH < 5.07 \end{cases} \quad \text{(Eq. 19-247)}$$

$$m^L = 0.644 \times pH - 2.779 \quad \text{(Eq. 19-248)}$$

$$Y^L = 1/\left(\frac{m^L}{\mu^L} + \frac{1}{YG^L}\right)$$

(Eq. 19-249)

$$kd^L = (\mu^L + m^L \times YG^L) \times 100$$ (Eq. 19-250)

where

kd^L is fractional rate of degradation of lactate by *M. elsdenii*, %/h;
m^L is maintenance coefficient of *M. elsdenii* on lactate, g/g per hour;
μ^L is net growth rate of *M. elsdenii* on lactate, 1/h;
Y^L is yield of *M. elsdenii* on lactate, g/g; and
YG^L is maximum yield of *M. elsdenii* on lactate, g/d.

Then, similar to the previous equations, the amount of lactate degraded in the rumen and its end product are computed assuming 4.4% of ash and 10% of N (0.625/6.25) in the bacterial mass.

$$RDL_j = Lactate_j^{CA,CB1,CB2} \times \left(\frac{kd^L}{kd^L + kp_j}\right)$$

(Eq. 19-251)

$$EndProduct_j^L = RDL_j - (1 - 0.044 - 0.625/6.25) \times (RDL_j \times Y^L)$$

(Eq. 19-252)

where

$EndProduct_j^L$ is the end product of lactate degradation in the rumen of the jth feedstuff, kg/d; and
$RDLj$ is ruminally degraded lactate by *M. elsdenii* of the jth feedstuff, kg/d.

The end product of lactate degradation is converted to acetate, propionate, and butyrate using the following pH-dependent equations.

$$Factor_A^L = \begin{cases} 1.035 - 0.154 \times pH, & 4 \le pH < 5.65 \\ 0.222 \times pH - 1.089, & 5.65 < pH \le 8 \\ 0, & pH < 4 \end{cases}$$

(Eq. 19-253)

$$Factor_P^L = \begin{cases} 1.568 - 0.206 \times pH, & 4 \le pH < 6 \\ 0.29 \times pH - 1.408, & 6 < pH \le 8 \\ 0, & pH < 4 \end{cases}$$

(Eq. 19-254)

$$Factor_B^L = \begin{cases} 0.27 \times pH - 1.181, & 4.4 \le pH < 5.5 \\ 1.316 - 0.184 \times pH, & 5.5 < pH \le 7 \\ 0, & pH < 4.4 \end{cases}$$

(Eq. 19-255)

$$Acetate_j^L = Factor_A^L \times EndProduct_j^L$$

(Eq. 19-256)

$$Propionate_j^L = Factor_P^L \times EndProduct_j^L$$

(Eq. 19-257)

$$Butyrate_j^L = Factor_B^L \times EndProduct_j^L$$

(Eq. 19-258)

where

$Acetate_j^L$ is the amount of acetate produced from the fermentation of lactate of the jth feedstuff, kg/d;
$Butyrate_j^L$ is the amount of butyrate produced from the fermentation of lactate of the jth feedstuff, kg/d;
$Factor_{A,P,B}^L$ are the mass fraction coefficients to convert $EndProduct_j^L$ into acetate, propionate, and butyrate for the jth feedstuff, dimensionless; and
$Propionate_j^L$ is the amount of propionate produced from the fermentation of lactate of the jth feedstuff, kg/d;

The stoichiometric relationship developed by Wolin (1960) is used to calculate methane production as described below:

$$CH_{4,j}^{CA,CB1,CB2,CB3} = YCH_{4,j}^{CA,CB1,CB2,CB3} \times Propionate_j^{CA,CB1,CB2,CB3}$$

(Eq. 19-259)

$$YCH_{4,j}^{CA,CB1,CB2,CB3} = \left(\frac{1}{2 \times FracP_j^{CA,CB1,CB2,CB3}} - \frac{3}{4}\right) \times \frac{16}{64}$$

(Eq. 19-260)

$$FracP_j^{CA,CB1,CB2,CB3} = \frac{Propionate_j^{CA,CB1,CB2,CB3}}{\left(Acetate_j^{CA,CB1,CB2,CB3} + Propionate_j^{CA,CB1,CB2,CB3} + Butyrate_j^{CA,CB1,CB2,CB3}\right)}$$

(Eq. 19-261)

where

$CH_{4,j}^{CA,CB1,CB2,CB3}$ is the amount of methane, kg/d;

$FracP_j^{CA,CB1,CB2,CB3}$ is the molar fraction of propionate of VFA (acetate, propionate, and butyrate), dimensionless; and
$YCH_{4,j}^{CA,CB1,CB2,CB3}$ is the mass of methane per mass of propionate, dimensionless.

The mass of CH_4 produced daily (kg/d) is converted to Mcal/d and L/d using the following conversion factors: the complete combustion of methane yields 890 kJ/mol or 55.65 MJ/kg or 13.3 Mcal/kg, assuming 16 g/mol. The density of methane is 0.6556 g/L (kg/m^3) at 25°C and 1 atm or 0.716 g/L (kg/m^3) at 0°C and 1 atm.

NUTRIENT BALANCES

The N intake is computed as the sum of CP intake divided by 6.25, the absorbed N is computed as the sum of MP, and the retained N is computed as the sum of net protein for growth, pregnancy, and milk. The absorbed N may also be computed as N intake minus fecal N plus endogenous N.

$$N_{Intake} = \sum_{j=1}^{n}\left(\left(\frac{CP_j}{6.25 \times 100}\right) \times DMI_j \times 1{,}000\right) \quad \text{(Eq. 19-262)}$$

$$N_{Absorbed} = \sum_{j=1}^{n}(MP_j \times 1{,}000/6.25) \quad \text{(Eq. 19-263)}$$

$$N_{Retained} = (NP_g + NP_p + YProtn)/6.25 \quad \text{(Eq. 19-264)}$$

where

CP_j is the crude protein of the jth feedstuff, % DM; and
DMI_j is dry matter intake of the jth feedstuff, kg/d.

In the ELS, the urinary and fecal N are computed using the approaches derived by Waldrip et al. (2013) and Dong et al. (2014). The urinary N equation was developed using the data from Waldrip et al. (2013). The N balance is computed as the N intake minus N losses (urine and feces).

$$N_{Balance,ELS} = N_{Intake} - (N_{Urine,ELS} + N_{Feces,ELS}) \quad \text{(Eq. 19-265)}$$

$$N_{Urine,ELS} = 0.2331 \times LN(N_{Intake}) - 0.6244) \times (N_{Intake} - N_{Retained}) \quad \text{(Eq. 19-266)}$$

$$N_{Feces,ELS} = N_{Intake} - N_{Retained} - N_{Urine,ELS} \quad \text{(Eq. 19-267)}$$

In the MLS, the N excretion in feces and urine and the N balance are computed using the equations listed below:

$$N_{Balance,MLS} = N_{Intake} - N_{Urine,MLS} + N_{Feces,MLS} \quad \text{(Eq. 19-268)}$$

$$N_{Urine,MLS} = N_{Intake} - N_{Retained} - N_{Feces,MLS} \quad \text{(Eq. 19-269)}$$

$$N_{Feces,MLS} = \sum_{j=1}^{n}(FEPROT_j / 6.25) \quad \text{(Eq. 19-270)}$$

The predicted ammonia N emission from surface of a feedlot is calculated based on the recommendations by Todd et al. (2013) as discussed in Chapter 16 (Environment).

FEED LIBRARY

Development of the Feed Library

The feed library for the Beef Cattle Nutrient Requirements software was created using feed information in Chapter 18 (Composition of Selected Feeds for Beef Cattle, Table 18-1). When not found in Chapter 18, the compositional information (physical and chemical properties) and values for the MLS variables (i.e., kd_{CB1}, kd_{CB2}, kd_{CB3}, PBID, CB1ID, CB2ID, and pef), amino acids, and vitamins for a particular feed were obtained from the NRC (1996, 2000) feed library. The matching of the feed composition data from Chapter 18 and the NRC (1996, 2000) feed library was done according to a method described by Tedeschi et al. (2002b). A deviation index with the nutrients DM, CP, soluble CP, ADIN, EE, ash, starch, NDF, and lignin was calculated for each feed from Chapter 18 and the NRC (1996, 2000) feed library, using the equation below:

$$\text{Deviation Index} = \sum_{i=1}^{9}\frac{\sqrt{(N1_i - N2_i)^2}}{N2_i}; \text{ for } N2_i > 0 \quad \text{(Eq. 19-271)}$$

where N1 is the value of the ith nutrient for a feed in Chapter 18, N2 is the value of the same ith nutrient of the same feed from NRC (1996, 2000) feed library, and *i* is 1 to 9 for DM, CP, soluble CP, ADIN, EE, ash, starch, NDF, and lignin, respectively. The deviation indexes of the feeds in the NRC (1996, 2000) feed library were then ranked from least to greatest, and the top 10 feeds were selected.

Subsequently, the names of the 10 highest-ranked feeds (least deviation index values) from the NRC (1996, 2000) feed library were compared with the names of the feeds from the current software feed library. Whenever the names matched, that feed from the NRC (1996, 2000) feed library was marked; otherwise the feed from the NRC (1996, 2000) feed library with the least deviation index and of the same feed type (grass, legume, grain, etc.) was marked. Finally, each marked feed was manually evaluated and analyzed, and whenever it was judged necessary and appropriate, another feed from the NRC (1996, 2000) feed library was selected. The selected feed was used to provide the missing compositional information and the MLS variables of the feed being evaluated in the current software feed library.

For ease of use, the feed composition table in the software was organized to facilitate finding and comparing feeds of the same type and to find all values for a feed in the same column. It is arranged with feed names listed alphabetically. The international feed number (IFN) is given for each feed where appropriate for comparison with previous feed composition tables.

Measuring Feed Composition Values

Feed libraries developed for use with computer models contain feed composition values that are needed to predict the supply of nutrients available to meet animal requirements. In the current software library, feeds are described by their compositional information. The ELS uses the tabular NE and protein values. The MLS uses the feed carbohydrate and protein fractions and their digestion and passage rates to predict NE and MP values for each feed based on the interaction of these variables.

The chemical composition of feeds is described by feed carbohydrate and protein fractions that are used to predict microbial protein production, ruminal degradation and escape of carbohydrates and proteins, and ME and MP in the MLS. Feed library values for carbohydrate and protein fractions are based on Sniffen et al. (1992) and Van Soest (1994).

Feedstuffs are comprised of chemically measurable carbohydrates, protein, fat, ash, and water. The Weende system for proximate analysis has been used for more than 150 years to measure these components as crude fiber (CF), EE, DM, and total N. Nitrogen-free extract (NFE) is calculated as 100 minus the percentages of CF, EE, water, and crude protein. This system cannot be used, however, to mechanistically predict microbial growth because CF does not represent all of the fiber, NFE does not accurately represent the nonfiber carbohydrates, and protein must be described by fractions related to its ruminal degradation characteristics.

The MLS was developed to predict microbial growth and ruminal degradation and escape of carbohydrates and protein to more dynamically predict ME and MP feed values. To accomplish this objective, the detergent fiber system of feed analysis is used to compute carbohydrate (fiber carbohydrates—CHO FC, and nonfiber carbohydrates—CHO NFC) and protein fractions according to their fermentation characteristics (A = soluble and readily available, B = intermediate or slow, and C = not fermented and unavailable to the animal) based on Sniffen et al. (1992) and Lanzas et al. (2007a,b).

Partial evaluations of the system implemented in the MLS for predicting feed biological values from feed analysis of carbohydrate and protein fractions have been published (Ainslie et al., 1993; O'Connor et al., 1993; Fox et al., 1995, 2004; NRC, 2001; Schwab et al., 2003; Tedeschi et al., 2005, 2008, 2014b), but the committee recognizes that considerable research is needed to refine this structure for the MLS. The decision to implement the MLS was based on the need to have a system that will allow the application of accumulated knowledge and facilitate the accounting for more of the variation in performance by allowing the user to "manipulate" and "calibrate" critical information in ruminant nutrition. It is then assumed that further research conducted between the release of this edition and the next one will result in the refinement of sensitive coefficients to improve the accuracy of the MLS predictions under specific conditions.

The methods of CP fractionation are still under debate. Changes in the protein fractionation (Licitra et al., 1996) used by the NRC (1996, 2000) model have been proposed (Lanzas et al., 2007b) but, for the sake of simplicity, the in situ fractionation system adopted by the NRC (2001) was adopted by the current committee. The procedures used to determine each fraction are based on Sniffen et al. (1992) and they are described below.

1. Ash (AOAC, 2000).
2. Solvent-soluble fat (AOAC, 2000). This fraction is assumed to escape ruminal degradation and to have an intestinal digestibility of 95%. Only the glycerol and galactolipids are fermented and the fatty acids escape rumen digestion.
3. Carbohydrates:
   a. Residual from neutral detergent fiber (NDF) procedure is total insoluble matrix fiber (cellulose, hemicellulose, and lignin; Van Soest et al., 1991) with the use of sodium sulfite (Mertens, 2002a).
   b. Lignin, determined as acid detergent lignin, is an indicator of indigestible fiber (Van Soest et al., 1991). Then the unavailable fiber is estimated as lignin × 2.4. The factor 2.4 is not constant across feeds. It might overestimate the carbohydrate (CHO) C fraction (CC) of feeds that have low lignin content, but it seems to be sufficiently accurate for the current state of the model (Traxler et al., 1998).
   c. Available fiber (CHO fraction B2; CB2) is calculated as NDF – CC, and is used to predict ruminal fiber digestion and microbial protein production from fiber. Intestinal digestibility of the B2 fraction that escapes the rumen is assumed to be 20% (Sniffen et al., 1992); however, this quantity of NDF being fermented (i.e., disappearing) in the hindgut might not be fully used, if used at all, by the ruminant animal.
   d. Nonfiber carbohydrates (NFC: sugar, starch, pectin) are computed as 100 – CP – NDF – fat – ash.
   e. CHO fraction A (CA) is sugars. It is assumed that these nonstarch, nonpectic polysaccharides are

more rapidly degraded than most starches and pectic substances. This entire fraction is degraded in the rumen.

f. CHO fraction B1 (CB1) is starch. This fraction has variable ruminal degradability, depending on the level of intake, type of grain, degree of hydration, and type of processing. Microbial protein production is most sensitive to ruminal starch degradation in the MLS. The B1 fraction that escapes ruminal fermentation is assumed to have variable intestinal digestibility, depending on type of grain and type of processing.

g. CHO fraction B2 (CB2) is calculated as NFC – CA – CB1. This fraction represents pectic substances and it is computed by difference. Pectic substances are more rapidly degraded than starches but do not give rise to lactic acid.

4. Proteins:
   a. Total nitrogen is measured using the Kjeldahl method (AOAC, 2000).
   b. Soluble nitrogen (SP: NPN + soluble true protein) is measured to identify total N rapidly degraded in the rumen (Krishnamoorthy et al., 1983).
   c. Protein A fraction (PA) contains NPN and true rapidly degraded protein. The NPN fraction provides ammonia for both FC and NFC bacteria growth whereas the true soluble protein typically contains albumin and globulin proteins and provides peptides for meeting NFC bacteria requirements for maximum efficiency of growth. Nonetheless, these specifics were not modeled in the MLS.
   d. The detergent analysis system (Van Soest et al., 1991) was designed to analyze for carbohydrate and protein fractions in forages. It has limitations in the analysis of other feedstuffs, particularly in the case of animal byproducts, treated plant protein sources, and byproduct feeds. Nitrogen that is insoluble in neutral detergent (without sodium sulfite) and acid detergent (Van Soest et al., 1991) measures slowly degraded plus unavailable protein. Animal proteins do not contain fiber, but the connective tissues (i.e., chondroitin and hyaluronic acid) might yield unrealistic values for ADF and NDF pools. To correct for this problem, all animal proteins have been assigned ADIN values that reflect average unavailable protein resulting from heat damage and keratins. The ADIN is the protein C fraction. The residual protein fraction (B) has been assigned rates reflecting their fractional kd depending on the feedstuffs.
   e. The ADIN (Van Soest et al., 1991) was used to identify unavailable protein (protein C fraction; PC), and is assumed to have 0 ruminal and intestinal digestibility, recognizing that some studies have shown digestive disappearance of ADIN. The levels of ADIN can be adjusted where appropriate.
   f. The protein B fraction was calculated using this equation: CP – PA – PC. This fraction typically contains glutelin-, prolamin-, and extensin-type proteins with variable degrees of ruminal degradation and is assumed to have variable intestinal digestibility.
5. Feed physical characteristics are described as peNDF as reported by Sniffen et al. (1992) and Mertens (1997, 2002b); it is calculated by pef × NDF. The peNDF is the percentage of the NDF that is effective in stimulating chewing, salivation, rumination, and rumen motility. The pef is described as the percentage of the NDF remaining on a 1.18-mm screen after dry sieving (Smith and Waldo, 1969; Mertens, 1997). This value was then adjusted for density, hydration, and degree of lignification of the NDF within classes of feeds. The pef was found to be moderately correlated with ruminal pH (Pitt et al., 1996) and it is used to predict ruminal pH when available.

The ruminal pH influences microbial protein yield (Russell et al., 1992) and FC microbial growth (Pitt et al., 1996). In the ELS, no adjustment is made for MCP as a result of ruminal pH; however, in the MLS, microbial protein yield and FC fractional degradation rate are adjusted for ruminal pH. The data of Russell et al. (1992) and Pitt et al. (1996) show that ruminal pH below 6.2 results in linear decreases in microbial protein production and FC digestion. A curvilinear relationship between peNDF and ruminal pH was reported by Mertens (1997) as pH = 6.67 – 0.143/peNDF, but the committee concluded that this equation might not be suitable for beef cattle. Using data in the literature for dairy cows, beef cattle, and sheep, Pitt et al. (1996) evaluated several approaches to predict ruminal pH: dietary forage proportion, NDF content, a mechanistic submodel of rumen fermentation, or the peNDF values published by Sniffen et al. (1992). Physically effective NDF gave predictions of ruminal pH similar to the MLS, and has the advantage of simplicity and flexibility of application. Nonetheless, ruminal pH is controlled by a multitude of factors other than peNDF, as discussed in Chapter 4 (Carbohydrates), including imbalance between production and disappearance (e.g., absorption, passage, neutralization) of fermentation acids in the rumen (Allen et al., 2006). The tabular values for pef should be used as a guide, with adjustments based on field observations and experience. When peNDF content of the diet is unknown, ruminal pH can be predicted from forage proportion, as there is a strong positive relationship between these two variables, up to a point. The importance of forage particles in stimulating salivary flow in buffering the rumen is well documented (Beauchemin, 1991). Additional factors not accounted for in the peNDF system that can influence ruminal pH are total grain intake and its digestion rate, and form of grain (whole corn will stimulate rumination but processed corn might not; a higher proportion of the starch in whole corn will escape ruminal fermentation compared with processed corn and other grains). Therefore, adjustments or functional equivalents of peNDF must be assigned to feeds in these cases to make the system reflect these conditions. Ionophores will inhibit the growth of *Streptococcus bovis*, which produces lactic acid, which is 10 times stronger (i.e., more acidic) than the normal VFA produced in the rumen. Highly digestible feeds that are high in pectic substances (e.g., soybean hulls, beet pulp) will not cause the drop in ruminal pH that grains cause. Estimated peNDF requirements are provided in Table 19-10 and are based on the data of Pitt et al. (1996). For high-concentrate diets to maximize feed efficiency, the minimum peNDF is 5 to 8% DM, which will keep ruminal pH above 5.6 to 5.7. Below a ruminal pH of 5.6, the function of the rumen can be compromised and cattle might stop eating, according to Britton and Stock (1989).

Tabular Net Energy Values

The ELS uses tabular energy and protein values in traditional approaches to diet formulation. Tabular NE values are based on NRC (1984) equations and they might not be representative of feedstuffs currently being fed to beef cattle (Tedeschi et al., 2005). The NE system implemented by the 1976 Subcommittee on Beef Cattle Nutrition (NRC, 1976) for growing cattle has been successfully used since then to adjust for CH_4, urinary and heat increment losses in meeting NEm requirements, and tissue deposition. This system accounts for differences in usefulness of absorbed energy depending on source of energy and physiological function (NRC, 1984). These values are not, however, directly measurable in feeds and do not account for the variation in the ME and MP derived from feeds with varying levels of intake and extent of ruminal and intestinal digestion. The MLS allows for the prediction of NE values by accounting for these variables. Both levels of solution of the current model use the NRC (1984) equations to predict NEma and NEga values as presented in this chapter. These equations are mechanistic in predicting NE values from the standpoint of decreasing the efficiency of use of ME for maintenance and growth (with a relatively greater effect on NEga) as ME

TABLE 19-10 Estimated peNDF Requirements

Diet Type	Minimum peNDF Required, % DM
High-concentrate diet to maximize feed efficiency (gain:feed ratio)	
Fed mixed ration, good feed bunk management, and ionophores	5 to 8
Fed mixed ration, variable feed bunk management, or no ionophore	20
High-concentrate to maximize NFC use and microbial protein yield[a]	20

[a]To keep ruminal pH above 6.2 to maximize cell-wall digestion and microbial protein yield.

value of the feed decreases (NRC, 1984). Dietary NEma and NEga values determined with the body composition database described by Fox et al. (1992) were regressed against NEma and NEga predicted with the NRC (1984) equations. Dietary NEga concentrations varied from approximately 0.90 to 1.50 Mcal/kg. There was no bias in either NEma or NEga predicted values, and the r^2 was 0.89 and 0.58, respectively. The lower r^2 for NEga prediction is the result of feed for gain reflecting all cumulative errors in predicting requirements in this system because NEma requirement and feed for maintenance is computed using a fixed 0.077 Mcal/SBW$^{0.75}$. Thus, it is likely that this is a "worst-case" scenario for predicted feed NEga because maintenance requirement can be highly variable (Fox et al., 1992).

Tabular RDP/RUP Values

The system of RDP/RUP values was introduced in *Ruminant Nitrogen Usage* (NRC, 1985) and it was implemented in the *Nutrient Requirements of Dairy Cattle* (NRC, 1989) model to more accurately predict protein available to meet rumen microbial requirements and to supplement microbial protein in meeting animal requirements. The MLS allows for the determination of these values mechanistically, based on the integration of feed carbohydrate and protein fractions and microbial growth. The tabular values for use in the ELS are from various sources and represent determinations by various methods. Analytically, RDP and RUP tabular values are determined by either in vitro or in situ methods. These methods have limitations in predicting ruminal degradation and escape of protein because of inherent procedural issues and the fact that they do not account for variation in effects of fractional kd and kp values as shown by Tedeschi et al. (2005). The tabular RDP and RUP values in the previous edition and update of the *Nutrient Requirements of Beef Cattle* (NRC, 1996, 2000) were based on Van Soest (1994), NRC (1989), data in the literature, experimental data of committee members, or were generated from the Level 2 of the NRC (1996, 2000) model.

Model-Predicted Net Energy and Metabolizable Protein Values

The MLS permits the user to integrate intake, and kd and kp of carbohydrate and protein fractions, to predict ME and MP values of feeds for each unique situation; this is because MLS computes a TDN value that reflects the integration of level of intake and ruminal kd and kp. Fractional digestion rates have been assigned to each feed based on published recommendations (Sniffen et al., 1992; NRC, 2001; Lanzas et al., 2007a,b). The equations describe how these are used to predict ME and MP values. Essential AA values have been assigned to feeds to represent their concentration in the RUP based on O'Connor et al. (1993) and the findings of Tedeschi et al. (2001).

MODEL COMPARISON

The ADG predicted by the current Beef Cattle Nutrient Requirements model and the NRC (1996, 2000) model were compared; the comparison utilized steer and heifer feedlot data from 20 experiments, representing 2,539 pen-fed animals of different breeds, conducted in the Willard Sparks Beef Research Center at the Oklahoma State University, Stillwater, OK, between 2003 and 2011. Animals were fed different proportions and combinations of corn grain, alfalfa hay, molasses, tallow, cottonseed meal, cottonseed hulls, cottonseed whole, wheat grain, wet distillers grain plus solubles, or prairie hay, and supplemented with finely ground corn, wheat middlings, and minerals. Although information on initial and final BW, DMI, and proportion of feedstuff in the diet was available, compositional information on these feeds was lacking so the values from the NRC (1996, 2000) feed library were used. Similarly, the carcass information needed to predict the MW at 28% EBF was lacking; therefore, the final BW was used as MW. The proper characterization of feedstuffs and animal information is necessary to predict observed ADG correctly, but the comparison between the prediction of ADG by the current and previous NRC (1996, 2000) models provides insight on their relative predictive ability.

The current Beef Cattle Nutrient Requirements software was used to calculate the model-predicted ME and MP allowable gain for each pen (n = 379) using four solutions: the ELS and MLS of the current model using default constants and equations, and the ELS and MLS of the current model using alternative settings (13% of TDN to estimate MCP, increase of NEma by 12% when feeding ionophore, MP to NP efficiency equation, and the calculation of fecal matter) to emulate Levels 1 and 2 of the previous NRC (1996, 2000) model, respectively. The ELS of the current model and the emulated Level 1 simulations assumed no ruminal recycled N and the RDP efficiency of use for MCP was 100%. The MLS of the current model assumed an efficiency of RDP use for MCP of 85% and the ruminal recycled N was calculated using the UUA equation. The emulated Level 2 assumed an efficiency of RDP use for MCP of 100% and the NRC (1985) equation to predict ruminal recycled N.

For the ELS simulations, the dietary ME, NEma, and NEga values were calculated from TDN values of the current feed library in Chapter 18. Environmental factors were not used to adjust NEm requirement. The observed DMI was used in the simulations. The study average of observed ADG and the first-limiting ADG for these four solutions were compared using selected statistics described by Tedeschi (2006): mean square error of prediction (MSEP), coefficient of determination (r^2), concordance correlation coefficient (CCC), and the accuracy factor (Cb).

Figure 19-1A shows the predicted ME-allowable ADG for each model's solution type. The predictions of the emulated NRC (1996, 2000) Level 1 were nearly identical to those of the current model's ELS with an r^2 of 0.999, root MSEP of 0.018 kg/d, CCC of 0.998, and Cb of 0.998. In contrast, the predictions of the current model's MLS tended to predict greater ADG than the emulated NRC (1996, 2000) Level 2 by about 0.158 kg/d on average, but it had an r^2 of 0.999, root MSEP of 0.159 kg/d, CCC of 0.9, and Cb of 0.9.

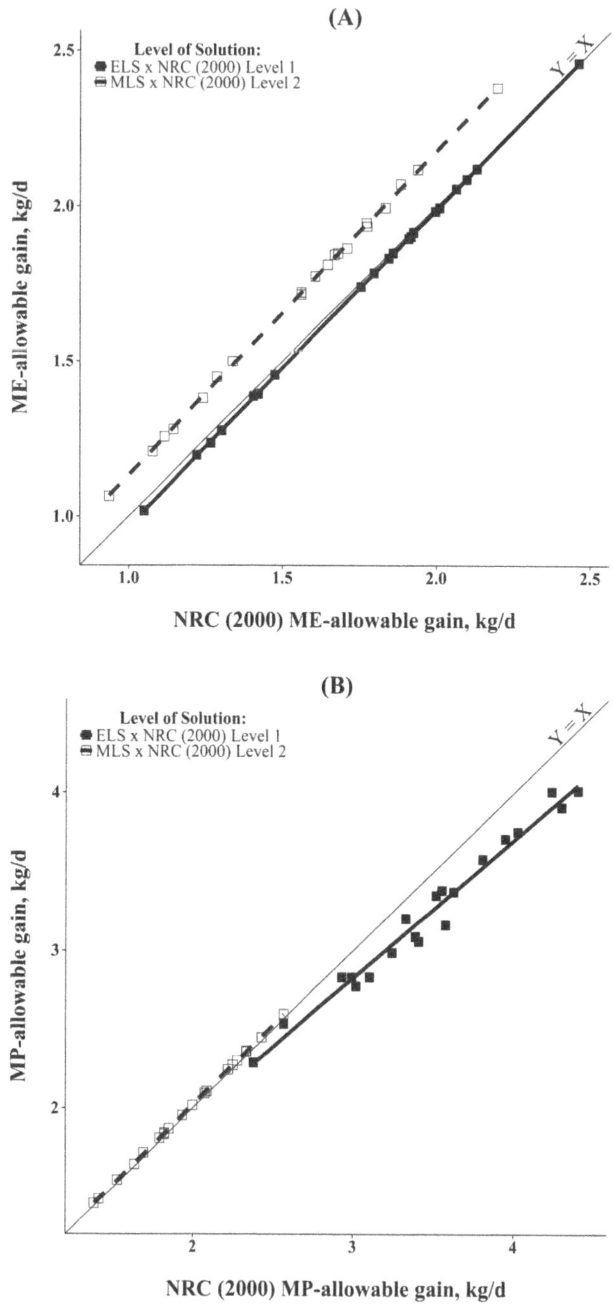

FIGURE 19-1 Comparison of the NRC (2000) and the current model's predicted (A) metabolizable energy (ME) or (B) metabolizable protein (MP) allowable gain, using data of steers and heifers from 20 feedlot studies (n = 2,539 animals). The solid line is the NRC (2000) Level 1 versus the current ELS and the dashed line is the NRC (2000) Level 2 versus the current MLS.

Figure 19-1B shows the predicted MP-allowable ADG for each model's solution type. Unlike the ME-allowable ADG predictions, the predictions of the emulated NRC (1996, 2000) Level 2 were nearly identical to those of the current MLS with an r^2 of 0.999, root MSEP of 0.023 kg/d, CCC of 0.998, and Cb of 0.998. Despite the different predictions of recycled ruminal N and RDP efficiency of use for MCP, the predictions by the current MLS and the emulated NRC (1996, 2000) Level 2 were comparable. In contrast, the predictions of the current ELS tended to predict lower MP-allowable ADG than the emulated NRC (1996, 2000) Level 1 by about 0.234 kg/d on average, but it had an r^2 of 0.975, root MSEP of 0.255 kg/d, CCC of 0.89, and Cb of 0.9. This underprediction of MP-allowable ADG by the current ELS is likely due to lower predicted MCP supply and MP to NP efficiency for growth. The predicted MP-allowable ADG by the current ELS and the emulated NRC (1996, 2000) Level 1 were greater than the current MLS and the emulated NRC (1996, 2000) Level 2 (Figure 19-1B).

Although the ME-allowable ADG difference between the emulated NRC (1996, 2000) Level 1 and the current ELS was nearly zero, the emulated NRC (1996, 2000) Level 2 and the current MLS had a systematic bias, but the precision was high. These results indicate that the predictions of ADG by the current Beef Cattle Nutrient Requirements model and the NRC (1996, 2000) model closely agree within level of solution. In situations in which adequate feed composition and animal information are not available, simple models (e.g., ELS) might yield better predictions of animal performance than complex models (e.g., MLS) (Tedeschi et al., 2014a). Mechanistic models are more suitable for understanding the mechanism underlying the animal performance, thus providing greater versatility in predicting animal performance under different production conditions. More detailed and complete data sets are necessary to fully evaluate the current model's ELS and MLS predictions, including measured physicochemical characteristics of each diet ingredient, animal performance, carcass and body compositions, and environmental factors that could alter the animal's maintenance requirement of energy.

SENSITIVITY ANALYSES

Unlike the evaluation performed on the NRC (1996, 2000) model, which relied on single point values, the sensitivity analyses performed on the current model rely on data variation. The procedure for sensitivity analyses was conducted using a stochastic approach similar to that described by Tedeschi et al. (2013). Repeated random samples of input variables were concomitantly obtained to assess their effects on the value and distribution of selected output variables. The values of the input variables were drawn from the probability density function of the normal distribution, model calculations were done with each combination of the input variables, and the output variable values were stored for statistical evaluations. This stochastic sampling was performed with @Risk v. 6.3 (Palisade Corp., Ithaca, NY) using 5,000 iterations, the Latin hypercube sampling method, and the Mersenne Twister random number generator. The initial seed for the randomization process was fixed at 123456. Input variables included BCS, initial and final SBW, mature SBW, and DMI for the animal; WS, Tp, Tc, RHp, RHc, lowest night temperature (LNT), and mud depth for the environment; and DM, CP, SP, ADIN, OA, sugars, starch, NDF, lignin, fat, ash, TDN, NEm, NEg, RUP, kd_{PB}, kd_{CB1}, kd_{CB2}, kd_{CB3}, PBID, CB1ID, and CB2ID for dietary characterization. Output variables included ME and MP required for maintenance, pregnancy, lactation, and growth; ME and MP allowable gain or milk; supplied ME and MP (RUP and MCP); ruminal nitrogen and RDP balances; and methane production (as Mcal/d, g/d, and g/kg DM). The following statistics were used to compare the simulations: distribution of the output variable, change in the output statistics, and standardized regression coefficients (SRC) as described by Kutner et al. (2005). The change in the output statistics is the mean of the output variable in the first bin (containing 10% of the iterations, in our case 500) and the tenth bin (containing 10% of the iterations, in our case 500) of the input variables ranked in ascending order (from lowest and highest). The SRC indicates that the magnitude of change (SD basis) in the output variable for each SD change in the input variable—for instance, an SRC value of 1 or −1—indicates that the output variable will change 1 or −1 SD units for each SD change in the input.

Two sensitivity analyses were performed. The first sensitivity analysis assumed fixed values for animal and diet input variables (no variation, SD = 0) and only environmental factors were allowed to change based on their mean and SD. The second sensitivity analysis assumed fixed values for animal and environmental input variables, and diet characteristics were allowed to vary according to a normal distribution.

Sensitivity Analysis of Environmental Factors

In conducting the sensitivity analysis of environmental factors, the average DMI was set to be the predicted DMI (Eq. 10-4, Chapter 10) and the SD was set to be zero (i.e., no random variation in DMI), and environmental factors were allowed to change NEm requirements and DMI predictions. Table 19-11 has the mean, SD (20% of the mean), and Spearman correlations for environmental factors. The MLS used a 300-kg growing, implanted Angus steer with BCS 5, 550 kg MW, fed a single diet ingredient, which was obtained by averaging the nutrient contents of a diet containing 15% Bermudagrass hay, 20% alfalfa hay, 55% flaked corn, 9% soybean meal, and 1% urea, formulated for an ME-allowable ADG of 1.6 kg/d. Animals were also fed monensin. Only the mean of the nutrients presented in Table 19-12 were used for this simulation, as nutrients were not allowed to vary (SD = 0). Figure 19-2 shows the results of this stochastic simulation. The variation of ME required for maintenance

TABLE 19-11 Mean, SD, and Spearman Correlations of Environmental Variables Used in the Sensitivity Analysis[a]

	WS	Tp	Tc	RHp	RHc	LNT	Mud
Mean	20	20	20	65	65	10	10
SD	4	4	4	13	13	2	2
			Spearman correlations				
WS	1						
Tp	–0.45	1					
Tc	–0.27	0.72	1				
RHp	–0.27	–0.45	–0.45	1			
RHc	–0.27	–0.45	–0.45	0	1		
LNT	0	0	0	0	0	1	
Mud	0	0	0	0	0	0	1

[a]SD is standard deviation; WS is wind speed, km/h; Tp is previous temperature, °C; Tc is current temperature, °C; RHp is previous relative humidity, %; RHc is current relative humidity; LNT is lowest night temperature, °C; Mud is mud depth, cm.

(Figure 19-2A) is expected to be narrower than that for ME available for growth (Figure 19-2B). The 90% prediction interval ($PI_{90\%}$) for ME required for maintenance was 7.5 to 8.5 Mcal/d, and the ME available for growth was between 11.1 and 12.9 Mcal/d. This corresponded to an ME-allowable gain of 1.44 ± 0.06 kg/d, with a $PI_{90\%}$ of 1.34 to 1.54 kg/d. The predicted DMI (Eq. 10-4, Chapter 10) had a mean and SD of 6.87 ± 0.17 kg/d and $PI_{90\%}$ of 6.59 to 7.14 kg/d. This variation in predicted DMI is strictly a result of the variation in environmental factors as Eq. 10-4 depends on initial SBW only and initial SBW did not vary in this sensitivity analysis.

The previous temperature (Tp) had the greatest effect on the ME required for maintenance. The change in the output statistical analysis indicated that from the least and the greatest Tp values, the mean of ME required for maintenance is expected to vary from 7.5 to 8.5 Mcal/d (Figure 19-2C), with a negative correlation ($r = -1$), indicating that as Tp increases, ME required for maintenance decreases and vice versa. For the current temperature (Tc), the mean of ME required for maintenance would be expected to vary from 7.7 to 8.4 Mcal/d (Figure 19-2C), with a negative correlation ($r = -0.69$). For the wind speed (WS), the mean ME required for maintenance would vary from 7.8 to 8.2 Mcal/d, with a positive correlation ($r = 0.42$). Dry matter intake had the least effect on ME required for maintenance under these conditions (Figure 19-2C).

Figure 19-2D shows the change in the output statistical analysis for ME available for growth. In this case, DMI had the greatest effect, with ME available for growth changing from 11.2 to 12.8 Mcal/d (Figure 19-2D), with a positive correlation ($r = 0.85$), suggesting that as DMI increases, ME available for growth increases. That is the main reason for a larger variation in ME available for growth (Figure 19-2B) compared with ME required for maintenance (Figure 19-2A). Subsequently, Tp, Tc, and WS affected ME available for growth as shown in Figure 19-2D, with the following correlations 0.50, 0.36, and –0.21, respectively. The use of Eq. 10-2 (Chapter 10), which uses NEm to predict DMI, would likely yield similar trends, but with different magnitudes.

This simulation indicated the acclimatization adjustment for Tp had the greatest effect on ME required for maintenance, and it was greater than the current temperature and wind speed variables. Mud depth, relative humidity, and lowest night temperature were not as influential as Tp, Tc, and WS under the mean and SD adopted for this sensitivity analysis (Table 19-11).

Sensitivity Analysis of Dietary Nutrients

In the sensitivity analysis of dietary nutrients, as with the previous simulation, the average DMI was again set to be the predicted DMI (Eq. 10-4, Chapter 10) and the SD was set to be zero (i.e., no random variation in predicting DMI). In this analysis, unlike in the one for environmental factors, the environmental factors were not allowed to change NEm requirements and DMI predictions. Table 19-12 has the mean, SD, and Spearman correlation for dietary nutrients. The simulation was conducted using the MLS and the same animal description as the previous sensitivity analysis for environmental factors. The mean values were obtained by averaging the nutrients of a diet containing 15% Bermudagrass hay, 20% alfalfa hay, 55% flaked corn, 9% soybean meal, and 1% urea, formulated for an ME-allowable ADG of 1.6 kg/d. Animals were also fed monensin. The Spearman correlations were obtained from the current feed library in Chapter 18. The SD were the weighted average (using the proportions above) of the SD of these individual feeds, and whenever not available (e.g., fractional degradations and intestinal digestibility), a value of 10% of the mean was used instead. The TDN, NEm, NEg, and RUP values listed in Table 19-12 were not used by the MLS as they were computed based on the nutrient composition of the diet.

Figure 19-3 shows the results of this stochastic simulation. The variation of ME required for maintenance (Figure 19-3A) is expected to be narrower than that for ME available for growth (Figure 19-3B), which agrees with the previous sensitivity analysis ($PI_{90\%}$ difference [i.e., upper $PI_{90\%}$ minus lower $PI_{90\%}$] was 0.25 and 3.16 Mcal/d, respectively). The $PI_{90\%}$ for ME required for maintenance was 7.9 to 8.15 Mcal/d and the ME available for growth was 12.8 and 16 Mcal/d. This corresponded to an ME-allowable gain of 1.7 ± 0.14 kg/d, and $PI_{90\%}$ of 1.47 to 1.93 kg/d. The predicted DMI (Eq. 10-4, Chapter 10) had a mean of 7.71 kg/d and no variation because of the reasons explained in the previous sensitivity analysis. Uncertainty in the environmental factors had a greater effect on the variation of ME required for maintenance (Figure 19-2A) than uncertainty in the nutrient composition of the diet ($PI_{90\%}$ difference was 0.96 Mcal/d in Figure 19-2A and 0.25 Mcal/d in Figure 19-3A), and their effect on ME available for growth was similar. One reason

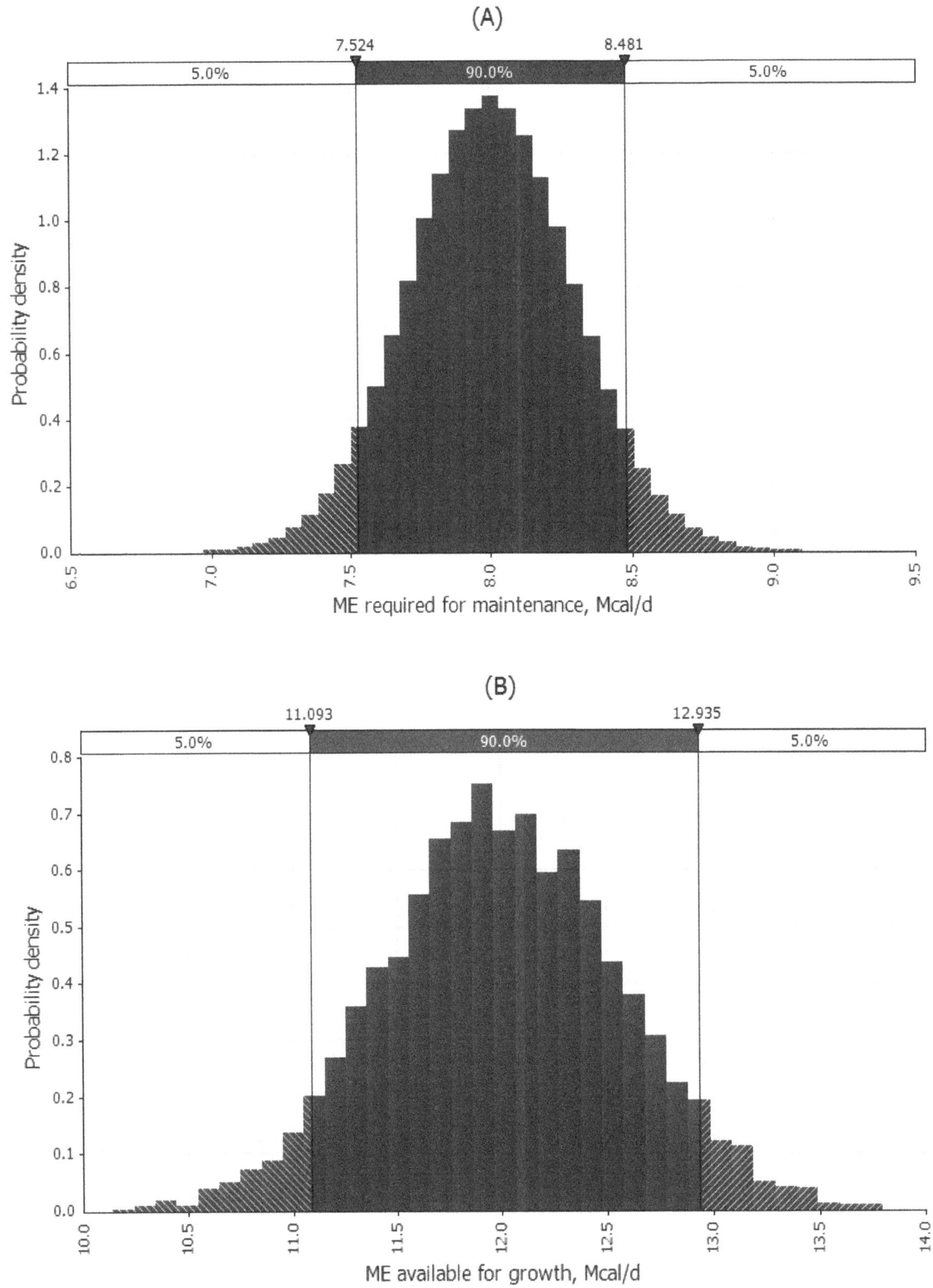

FIGURE 19-2 Effect of the variation in environmental factors on the distribution of predicted ME required for maintenance (A) and predicted ME available for growth (B), and influential variables affecting the change of predicted ME for maintenance (C) and predicted ME available for growth (D).

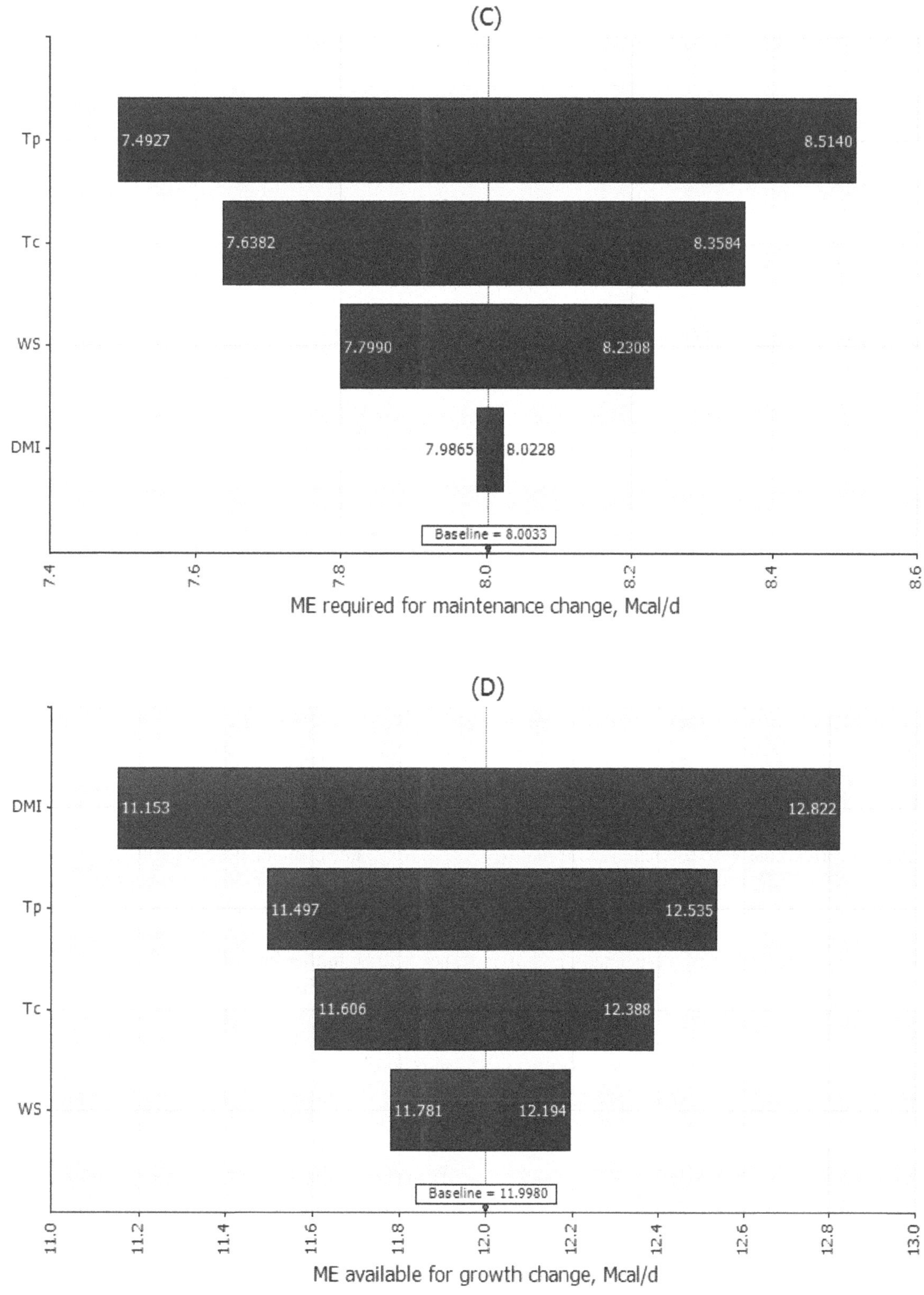

FIGURE 19-2 Continued

TABLE 19-12 Mean, SD, and Spearman Correlations of Dietary Variables Used in the Sensitivity Analysis[a]

	DM (% AF)	CP (% DM)	SP (% DM)	ADIN (% CP)	Fat (% DM)	Ash (% DM)	CB1 (% DM)	NDF (% DM)	Lignin (% DM)	TDN (% DM)	ME (Mcal/kg)	NEm (Mcal/kg)	NEg (Mcal/kg)	RUP (% CP)
Mean	79.1	19.8	32.2	4.72	3.54	4.78	42.8	23.3	2.58	77.8	3.09	2.10	1.44	23.9
SD	2.82	1.63	4.80	2.66	0.72	1.00	2.22	3.58	0.73	2.40	0.06	0.05	0.04	4.68
							Spearman Correlations							
DM	1													
CP	0.20	1												
SP	–0.32	–0.09	1											
ADIN	0.17	0.27	–0.32	1										
Fat	0.12	0.40	–0.05	0.18	1									
Ash	–0.22	–0.09	0.30	0.08	–0.24	1								
CB1	–0.17	–0.16	–0.09	–0.33	–0.09	–0.28	1							
NDF	–0.08	–0.37	0.17	0.05	–0.30	0.35	–0.34	1						
Lignin	0.12	–0.08	–0.04	0.45	0.01	0.31	–0.42	0.52	1					
TDN	0.27	0.34	–0.11	–0.17	0.32	–0.45	0.23	–0.60	–0.56	1				
ME	0.24	0.49	–0.25	0.04	0.48	–0.44	0.20	–0.78	–0.54	0.83	1			
NEm	0.15	0.37	–0.22	–0.06	0.49	–0.46	0.30	–0.78	–0.55	0.85	0.86	1		
NEg	0.15	0.37	–0.22	–0.05	0.49	–0.46	0.30	–0.78	–0.55	0.85	0.86	0.88	1	
RUP	0.16	–0.15	–0.68	0.24	0.07	–0.37	0.12	–0.17	0.00	0.21	0.17	0.18	0.18	1

[a]Mean and SD for kd_{PB}, kd_{CB1} (starch), kd_{CB2} (pectin), kd_{CB3} (NDF), PBID, CB1ID, and CB2ID were 6.7 ± 0.67 %/h, 22.7 ± 2.27 %/h, 26.2 ± 2.62 %/h, 4.59 ± 0.46 %/h, 81.7 ± 8.17%, 93.4 ± 9.34%, and 75 ± 7.5%, respectively. Their Spearman correlation was set to zero with all variables.

for this greater effect could be the larger SD of the environmental factors.

The rankings of most influential variables for both ME required for maintenance (Figure 19-3C) and ME available for growth (Figure 19-3D) were identical, but with different intensities and opposite correlations (figures not shown); this outcome was expected because ME not used for maintenance is used for growth. The dietary contents of NDF, lignin, ash (i.e., organic matter), and starch were the most influential variables, and their correlations with ME required for maintenance were 0.75, 0.73, 0.62, and –0.42, respectively, and with ME available for growth they were –0.75, –0.73, –0.62, and 0.42, respectively.

Figure 19-4A shows the distribution of predicted CH_4 production (g/d), Figures 19-4B and 19-4C show the SRC and Spearman rank of the influential variables on predicted daily CH_4 production (g/d), and Figure 19-4D shows the SRC for methane yield (g/kg of ruminally degraded carbohydrates). The predicted daily CH_4 was 143 ± 5.85 g/d with a $PI_{90\%}$ of 133 to 153 g/d. Although these values are within the observed range for typical CH_4 emissions of 36 to 145 g/d (Chapter 16), they are in the upper side of this range, likely because of the higher content of fiber of the simulated diet. Crude protein and NDF contents were the most influential variables for methane production (g/d) with CP having an SRC of 0.55 (Figure 19-4B). These values indicate that CH_4 (g/d) will increase by 55% of its SD (0.55 × 5.85 = 3.22 g/d) for each CP SD increase in the CP content of the diet (i.e., 1.63% DM units; Table 19-12), and CH_4 will decrease by 38% of its SD (0.38 × 5.85 = 2.22 g/d) for each NDF SD increase in the NDF (i.e., 3.58% DM units; Table 19-12). Figure 19-4C shows that CH_4 (g/d) was highly correlated with NDF (–0.69), lignin (–0.62), and CP (0.52). For methane yield (Figure 19-4D), protein had the greatest SRC followed by fractional rate of starch degradation. These results are bound to the correlations among nutrients (Table 19-12) in which changing one nutrient value causes other nutrients to change simultaneously. The high correlation between CH_4 and CP needs to be further investigated.

This sensitivity analysis indicated that given the average, SD, and correlations found in the feed library for Bermudagrass hay, alfalfa hay, flaked corn, and soybean meal, and the contents of NDF, lignin, and organic matter (i.e., ash) are the most influential dietary variables that can affect animal performance, and that CH_4 production is most likely affected by the variation of NDF, lignin, and CP contents in the diet. When DMI was allowed to vary (data not shown), DMI became the most influential variable in predicting methane emission (g/d).

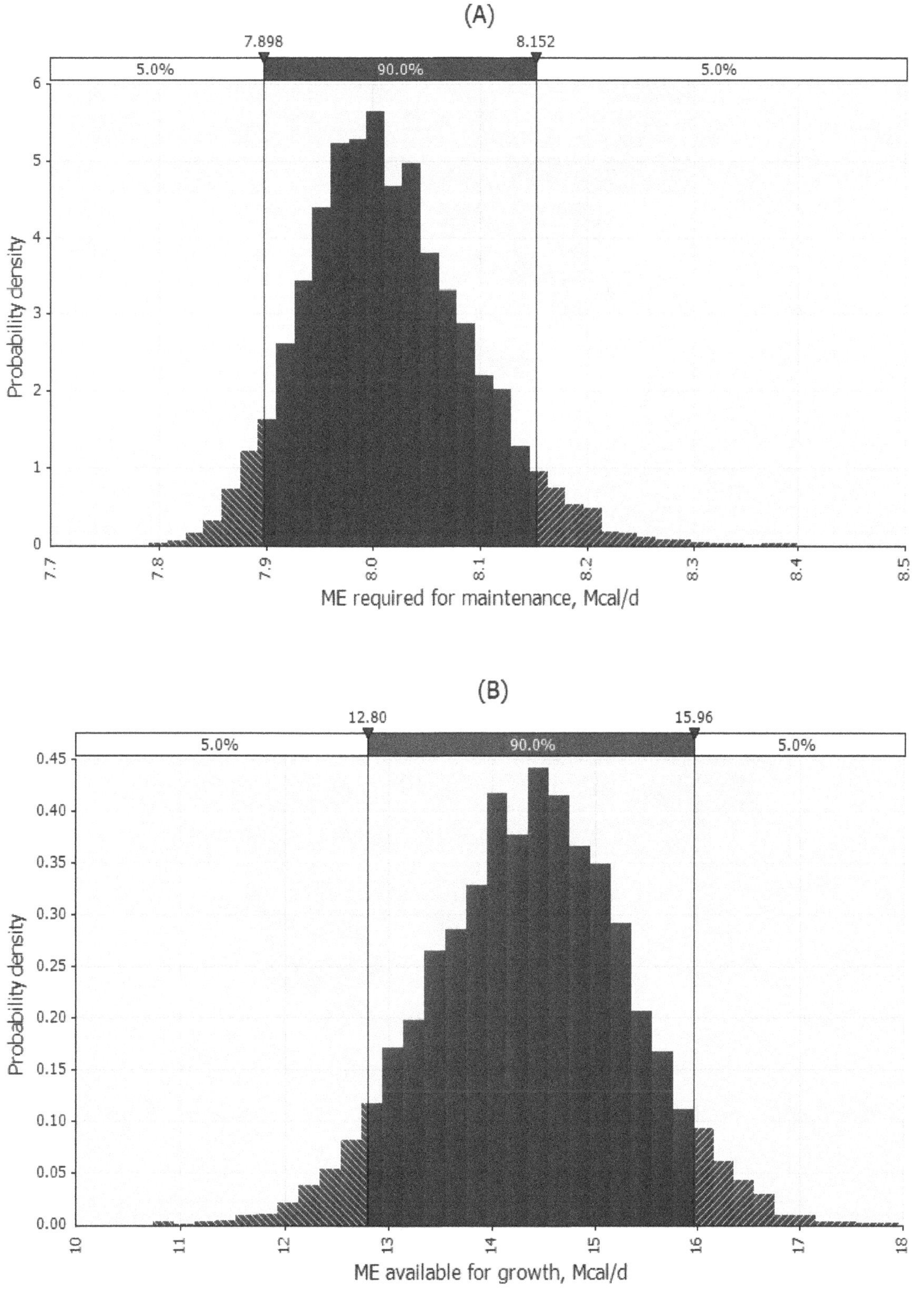

FIGURE 19-3 Effect of the variation in dietary nutrients on the distribution of predicted ME required for maintenance (A) and predicted ME available for growth (B), and influential variables affecting the change of predicted ME for maintenance (C) and predicted ME available for growth (D).

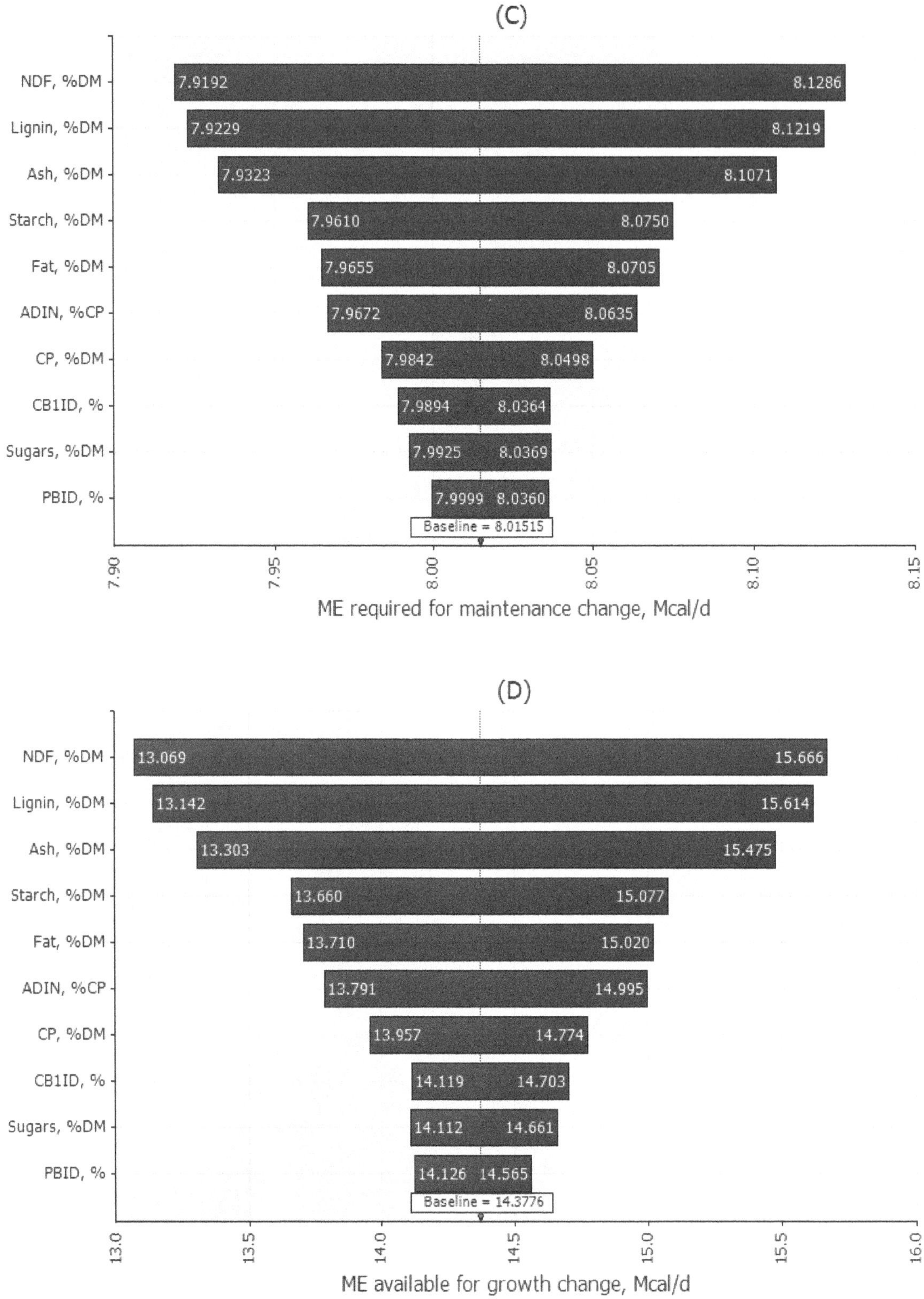

FIGURE 19-3 Continued

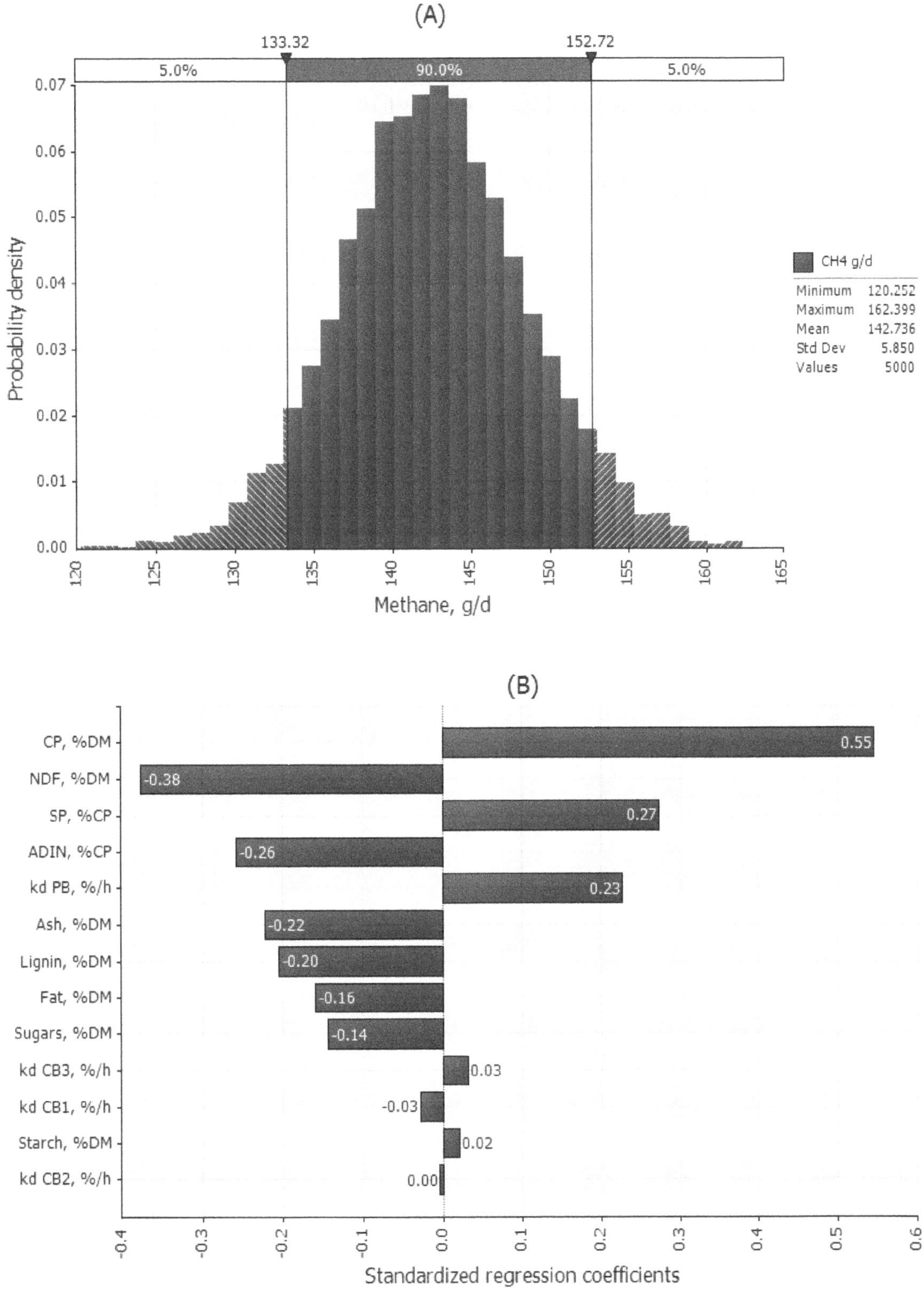

FIGURE 19-4 Effect of the variation of dietary nutrients (constant DMI) on the distribution of predicted methane production (CH_4 g/d) (A), standardized regression coefficients (B) and Spearman rank (C) of influential variables affecting methane production (CH_4 g/d) and methane yield (g/kg ruminally degraded carbohydrate) (D).

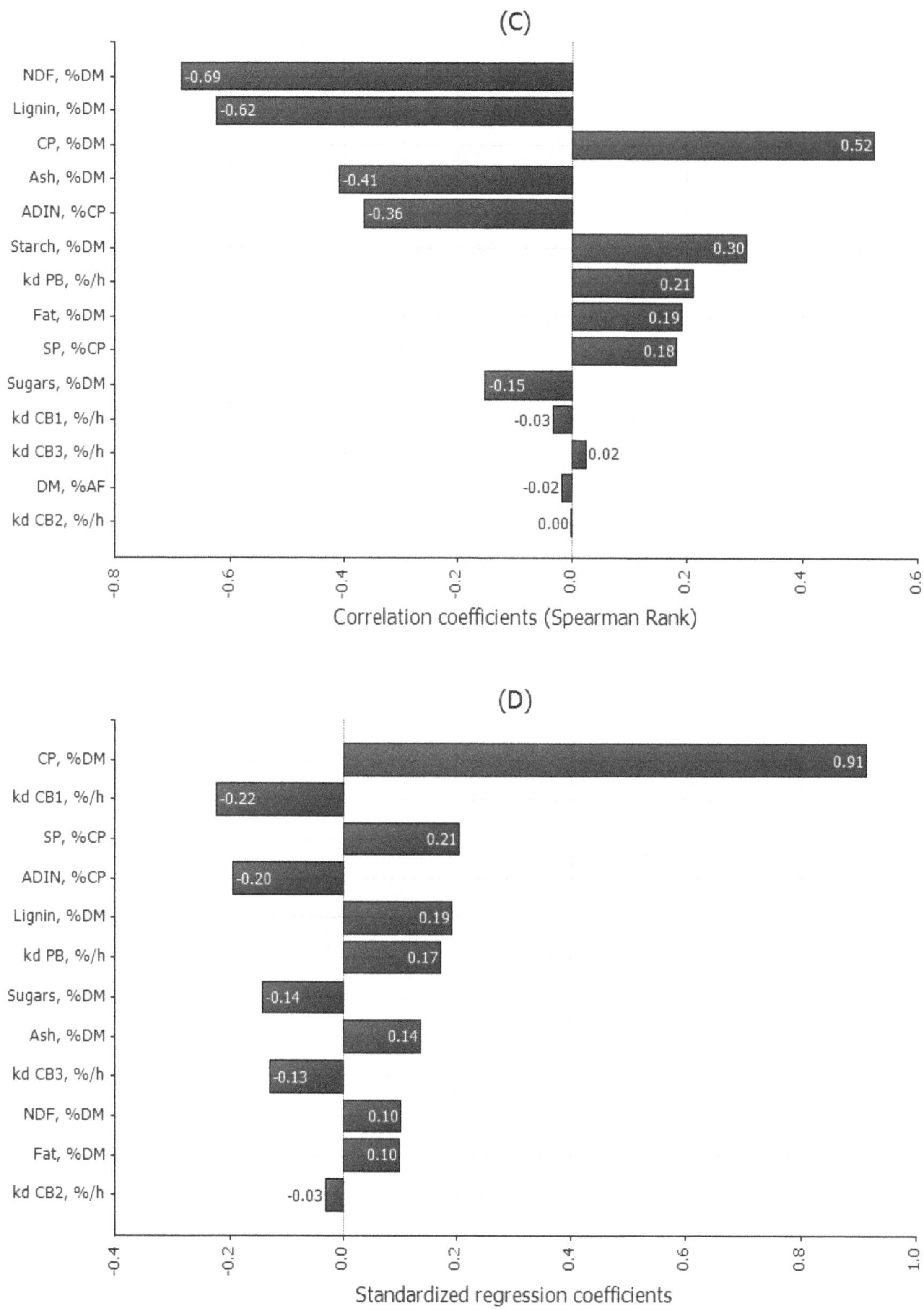

FIGURE 19-4 Continued

REFERENCES

AFRC (Agricultural and Food Research Council). 1993. *Energy and Protein Requirements of Ruminants.* Wallingford, UK: CABI International.

Ainslie, S. J., D. G. Fox, T. C. Perry, D. J. Ketchen, and M. C. Barry. 1993. Predicting amino acid adequacy of diets fed to Holstein steers. *Journal of Animal Science* 71:1312-1319.

Allen, M. S., J. A. Voelker, and M. Oba. 2006. Physically effective fiber and regulation of ruminal pH: More than just chewing. Pp. 270-278 in *Production Diseases in Farm Animals*, N. P. Joshi and T. H. Herdt, eds. Wageningen, The Netherlands: Wageningen Academic.

AOAC International. 2000. *Official Methods of Analysis of AOAC International,* 17th Ed., W. Horwitz, ed. Arlington, VA: AOAC International. Available online at http://www.aoac.org. Accessed on November 16, 2014.

Baldwin, R. L. 1995. *Modeling Ruminant Digestion and Metabolism.* New York: Chapman & Hall.

Beauchemin, K. A. 1991. Ingestion and mastication of feed by dairy cattle. *Veterinary Clinics of North America: Food Animal Practice* 7:439-463.

Bergen, W. G., D. B. Purser, and J. H. Cline. 1968. Effect of ration on the nutritive quality of rumen microbial protein. *Journal of Animal Science* 27:1497-1501.

Britton, R., and R. Stock. 1989. Acidosis: A continual problem in cattle fed high grain diets. Pp. 8-15 in *Proceedings of Cornell Nutrition Conference for Feed Manufacturers, Syracuse, NY.* Ithaca, NY: Cornell University Press.

Cannas, A., L. O. Tedeschi, D. G. Fox, A. N. Pell, and P. J. Van Soest. 2004. A mechanistic model for predicting the nutrient requirements and feed biological values for sheep. *Journal of Animal Science* 82:149-169.

Chizzotti, M. L., L. O. Tedeschi, and S. C. Valadares Filho. 2008. A meta-analysis of energy and protein requirements for maintenance and growth of Nellore cattle. *Journal of Animal Science* 86:1588-1597.

Clark, J. H., T. H. Klusmeyer, and M. R. Cameron. 1992. Microbial protein synthesis and flow of nitrogen fractions to the duodenum of dairy cows. *Journal of Dairy Science* 75:2304-2323.

Counotte, G. H. M., R. A. Prins, R. H. A. M. Janssen, and M. J. A. deBie. 1981. Role of *Megasphaera elsdenii* in the fermentation of dl-[2-13C] lactate in the rumen of dairy cattle. *Applied and Environment Microbiology* 42(4):649-655.

CSIRO (Commonwealth Scientific and Industrial Research Organisation). 1990. *Feeding Standards for Australian Livestock: Ruminants.* Melbourne, Australia: CSIRO.

CSIRO. 2007. *Nutrient Requirements of Domesticated Ruminants.* Collingwood, Australia: CSIRO.

Dijkstra, J., H. Neal, D. E. Beever, and J. France. 1992. Simulation of nutrient digestion, absorption and outflow in the rumen: Model description. *Journal of Nutrition* 122:2239-2256.

Dong, R. L., G. Y. Zhao, L. L. Chai, and K. A. Beauchemin. 2014. Prediction of urinary and fecal nitrogen excretion by beef cattle. *Journal of Animal Science* 92:4669-4681.

Ellis, J. L., E. Kebreab, N. E. Odongo, B. W. McBride, E. K. Okine, and J. France. 2007. Prediction of methane production from dairy and beef cattle. *Journal of Dairy Science* 90:3456-3466.

Ellis, J. L., E. Kebreab, N. E. Odongo, K. Beauchemin, S. McGinn, J. D. Nkrumah, S. S. Moore, R. Christopherson, G. K. Murdoch, B. W. McBride, E. K. Okine, and J. France. 2009. Modeling methane production from beef cattle using linear and nonlinear approaches. *Journal of Animal Science* 87:1334-1345.

Evans, E. H., and R. J. Patterson. 1985. Use of dynamic modelling seen as good way to formulate crude protein, amino acid requirements for cattle diets. *Feedstuffs* 57:24-27.

Fox, D. G., and T. P. Tylutki. 1998. Accounting for the effects of environment on the nutrient requirements of dairy cattle. *Journal of Dairy Science* 81:3085-3095.

Fox, D. G., C. J. Sniffen, and J. D. O'Connor. 1988. Adjusting nutrient requirements of beef cattle for animal and environmental variations. *Journal of Animal Science* 66:1475-1495.

Fox, D. G., C. J. Sniffen, J. D. O'Connor, J. B. Russell, and P. J. Van Soest. 1992. A net carbohydrate and protein system for evaluating cattle diets: III. Cattle requirements and diet adequacy. *Journal of Animal Science* 70:3578-3596.

Fox, D. G., M. C. Barry, R. E. Pitt, D. K. Roseler, and W. C. Stone. 1995. Application of the Cornell Net Carbohydrate and Protein Model for cattle consuming forage. *Journal of Animal Science* 73:267-277.

Fox, D. G., M. E. Van Amburgh, and T. P. Tylutki. 1999. Predicting requirements for growth, maturity, and body reserves in dairy cattle. *Journal of Dairy Science* 82:1968-1977.

Fox, D. G., L. O. Tedeschi, T. P. Tylutki, J. B. Russell, M. E. Van Amburgh, L. E. Chase, A. N. Pell, and T. R. Overton. 2004. The Cornell Net Carbohydrate and Protein System model for evaluating herd nutrition and nutrient excretion. *Animal Feed Science and Technology* 112:29-78.

Fox, D. G., T. P. Tylutki, L. O. Tedeschi, and P. E. Cerosaletti. 2006. Using a nutrition model to implement the NRCS feed management standard to reduce the environmental impact of a concentrated cattle feeding operation. The John Airy Symposium: Visions for Animal Agriculture and the Environment, January 2006, Kansas City, MO. Available online at http://nutritionmodels.tamu.edu/papers/FoxetalVISIONS200615.pdf. Accessed on May 11, 2015.

France, J., M. Gill, J. H. M. Thornley, and P. England. 1987. A model of nutrient utilization and body composition in beef cattle. *Animal Production* 44:371-385.

Galyean, M. L., and L. O. Tedeschi. 2014. Predicting microbial protein synthesis in beef cattle: Relationship to intakes of total digestible nutrients and crude protein. *Journal of Animal Science* 92:5099-5111.

Garcia, F., R. D. Sainz, J. Agabriel, L. G. Barioni, and J. W. Oltjen. 2008. Comparative analysis of two dynamic mechanistic models of beef cattle growth. *Animal Feed Science and Technology* 143:220-241.

Garrett, R. P., and H. Hinman. 1969. Re-evaluation of the relationship between carcass density and body composition of beef steers. *Journal of Animal Science* 28:1-5.

Gregory, K. E., L. V. Cundiff, and R. M. Koch. 1999. Composite Breeds to Use in Heterosis and Breed Differences to Improve Efficiency of Beef Production. U.S. Department of Agriculture Technical Bulletin No. 1875. Available online at http://naldc.nal.usda.gov/download/CAT10879520/PDF. Accessed on December 8, 2014.

Guiroy, P. J., D. G. Fox, L. O. Tedeschi, M. J. Baker, and M. D. Cravey. 2001. Predicting individual feed requirements of cattle fed in groups. *Journal of Animal Science* 79:1983-1995.

Hoch, T., and J. Agabriel. 2004. A mechanistic dynamic model to estimate beef cattle growth and body composition: 1. Model description. *Agricultural Systems* 81:1-15.

INRA (Institut National de la Recherche Agronomique). 1989. *Ruminant Nutrition: Recommended Allowances and Feed Tables,* R. Jarrige, ed. Montrouge, France: Institut National de la Recherche Agronomique, John Libbey Eurotext.

INRA. 2007. *Alimentation des Bovins, Ovins et Caprins: Besoins des Animaux—Valeurs des aliments.* J. Agabriel, ed. Versailles, France: Editions Quae.

IPCC (Intergovernmental Panel on Climate Change). 2006. *2006 IPCC Guidelines for National Greenhouse Gas Inventories: Agriculture, Forestry and Other Land Use,* Vol. 4, H. S. Eggleston, L. Buendia, K. Miwa, T. Ngara, and K. Tanabe, eds. Hayama, Japan: Institute for Global Environmental Strategies (IGES). Available online at: http://www.ipcc-nggip.iges.or.jp/public/2006gl/vol4.html. Accessed on November 16, 2014.

Johnson, K. A., and D. E. Johnson. 1995. Methane emissions from cattle. *Journal of Animal Science* 73:2483-2492.

Kohn, R. A., R. Boston, J. D. Ferguson, and W. Chalupa. 1995. The integration and comparison of dairy cow models. Pp. 117-128 in *Modeling Nutrient Utilization in Farm Animals: Proceedings of the 4th International*

Workshop, A. Danfaer and P. Lescoat, eds. Foulum, Denmark: National Institute of Animal Science.

Krishnamoorthy, U., C. J. Sniffen, M. D. Stern, and P. J. Van Soest. 1983. Evaluation of a mathematical model of rumen digestion and an *in vitro* simulation of rumen proteolysis to estimate the rumen-undegraded nitrogen content of feedstuffs. *British Journal of Nutrition* 50:555-568.

Kutner, M. H., C. J. Nachtsheim, J. Neter, and W. Li. 2005. *Applied Linear Statistical Models*, 5th Ed. New York: McGraw-Hill.

Lanzas, C., C. J. Sniffen, S. Seo, L. O. Tedeschi, and D. G. Fox. 2007a. A revised CNCPS feed carbohydrate fractionation scheme for formulating rations for ruminants. *Animal Feed Science and Technology* 136:167-190.

Lanzas, C., L. O. Tedeschi, S. Seo, and D. G. Fox. 2007b. Evaluation of protein fractionation systems used in formulating rations for dairy cattle. *Journal of Dairy Science* 90:507-521.

Licitra, G., T. M. Hernandez, and P. J. Van Soest. 1996. Standardization of procedures for nitrogen fractionation of ruminant feeds. *Animal Feed Science and Technology* 57:347-358.

Mertens, D. R. 1997. Creating a system for meeting the fiber requirements of dairy cows. *Journal of Dairy Science* 80:1463-1481.

Mertens, D. R. 2002a. Gravimetric determination of amylase-treated neutral detergent fiber in feeds with refluxing in beakers or crucibles: Collaborative study. *Journal of AOAC International* 85:1217-1240.

Mertens, D. R. 2002b. Measuring fiber and its effectiveness in ruminant diets. Pp. 40-66 in *Plains Nutrition Council Spring Conference, April 25-26, 2002, San Antonio, TX.* Publication AREC 02-20. Available online at http://amarillo.tamu.edu/files/2010/10/2002PNC-Proceedings.pdf. Accessed on December 8, 2014.

Mills, J. A. N., E. Kebreab, C. M. Yates, L. A. Crompton, S. B. Cammell, M. S. Dhanoa, R. E. Agnew, and J. France. 2003. Alternative approaches to predicting methane emissions from dairy cows. *Journal of Animal Science* 81:3141-3150.

Moraes, L. E., A. B. Strathe, J. G. Fadel, D. P. Casper, and E. Kebreab. 2014. Prediction of enteric methane emissions from cattle. *Global Change Biology* 20:2140-2148.

NRC (National Research Council). 1976. *Nutrient Requirements of Beef Cattle*, 5th Rev. Ed. Washington, DC: National Academy Press.

NRC. 1981. *Effect of Environment on Nutrient Requirements of Domestic Animals.* Washington, DC: National Academy Press.

NRC. 1984. *Nutrient Requirements of Beef Cattle*, 6th Rev. Ed. Washington, DC: National Academy Press.

NRC. 1985. *Ruminant Nitrogen Usage.* Washington, DC: National Academy Press.

NRC. 1989. *Nutrient Requirements of Dairy Cattle: Update 1989*, 6th Rev. Ed. Washington, DC: National Academy Press.

NRC. 1996. *Nutrient Requirements of Beef Cattle* 7th Rev. Ed. Washington, DC: National Academy Press.

NRC. 2000. *Nutrient Requirements of Beef Cattle: Update 2000*, 7th Rev. Ed. Washington, DC: National Academy Press.

NRC. 2001. *Nutrient Requirements of Dairy Cattle.* 7th Rev. Ed. Washington, DC: National Academy Press.

O'Connor, J. D., C. J. Sniffen, D. G. Fox, and W. Chalupa. 1993. A net carbohydrate and protein system for evaluating cattle diets: IV. Predicting amino acid adequacy. *Journal of Animal Science* 71:1298-1311.

Oltjen, J. W., A. C. Bywater, R. L. Baldwin, and W. N. Garrett. 1986. Development of a dynamic model of beef cattle growth and composition. *Journal of Animal Science* 62:86-97.

Ørskov, E. R., and I. McDonald. 1979. The estimation of protein degradability in the rumen from incubation measurements weighted according to rate of passage. *Journal of Agricultural Science* 92:449-503.

Owens, F., and R. A. Zinn. 2005. Corn grain for cattle: Influence of processing on site and extent of digestion. Pp. 86-112 in *Proceedings of the 20th Southwest Nutrition and Management Conference, February 24-25, 2005, Tempe, AZ,* Tucson: University of Arizona.

Pirt, S. J. 1965. The maintenance energy of bacteria in growing cultures. *Proceedings of the Royal Society of London, Series B.* 163:224-231.

Pitt, R. E., J. S. Van Kessel, D. G. Fox, A. N. Pell, M. C. Barry, and P. J. Van Soest. 1996. Prediction of ruminal volatile fatty acids and pH within the net carbohydrate and protein system. *Journal of Animal Science* 74:226-244.

Reynolds, C. K., and N. B. Kristensen. 2008. Nitrogen recycling through the gut and the nitrogen economy of ruminants: An asynchronous symbiosis. *Journal of Animal Science* 86:E293-E305.

Russell, J. B., J. D. O'Connor, D. G. Fox, P. J. Van Soest, and C. J. Sniffen. 1992. A net carbohydrate and protein system for evaluating cattle diets: I. Ruminal fermentation. *Journal of Animal Science* 70:3551-3561.

Sauvant, D., F. Meschy, and D. Mertens. 1999. Les composantes de l'acidose ruminale et les effets acidogènes des rations. *INRA Productions Animales* 12:49-60.

Schwab, C. G., T. P. Tylutki, R. S. Ordway, C. Sheaffer, and M. D. Stern. 2003. Characterization of proteins in feeds. *Journal of Dairy Science* 86:E88-E103.

Seo, S., L. O. Tedeschi, C. G. Schwab, and D. G. Fox. 2006. Development and evaluation of empirical equations to predict feed passage rate in cattle. *Animal Feed Science and Technology* 128:67-83.

Smith, L. W., and D. R. Waldo. 1969. Method for sizing forage cell wall particles. *Journal of Dairy Science* 52:2051-2053.

Sniffen, C. J., J. D. O'Connor, P. J. Van Soest, D. G. Fox, and J. B. Russell. 1992. A net carbohydrate and protein system for evaluating cattle diets: II. Carbohydrate and protein availability. *Journal of Animal Science* 70:3562-3577.

Spiess, A.-N., and N. Neumeyer. 2010. An evaluation of R^2 as an inadequate measure for nonlinear models in pharmacological and biochemical research: A Monte Carlo approach. *BMC Pharmacology* 10:6.

TAMU (Texas A&M University). 2015. Mathematical Nutrition Models: Large Ruminant Nutrition System. Available online at http://nutritionmodels.com/index.html. Accessed on April 15, 2015.

Tedeschi, L. O. 2006. Assessment of the adequacy of mathematical models. *Agricultural Systems* 89:225-247.

Tedeschi, L. O., A. N. Pell, D. G. Fox, and C. R. Llames. 2001. The amino acid profiles of the whole plant and of four residues from temperate and tropical forages. *Journal of Animal Science* 79:525-532.

Tedeschi, L. O., C. Boin, D. G. Fox, P. R. Leme, G. F. Alleoni, and D. P. D. Lanna. 2002a. Energy requirement for maintenance and growth of Nellore bulls and steers fed high-forage diets. *Journal of Animal Science* 80:1671-1682.

Tedeschi, L. O., D. G. Fox, A. N. Pell, D. P. D. Lanna, and C. Boin. 2002b. Development and evaluation of a tropical feed library for the Cornell Net Carbohydrate and Protein System model. *Scientia Agricola* 59:1-18.

Tedeschi, L. O., D. G. Fox, and P. J. Guiroy. 2004. A decision support system to improve individual cattle management. 1. A mechanistic, dynamic model for animal growth. *Agricultural Systems* 79:171-204.

Tedeschi, L. O., D. G. Fox, and P. H. Doane. 2005. Evaluation of the tabular feed energy and protein undegradability values of the National Research Council nutrient requirements of beef cattle. *The Professional Animal Scientist* 21:403-415.

Tedeschi, L. O., S. Seo, D. G. Fox, and R. Ruiz. 2006. Accounting for energy and protein reserve changes in predicting diet-allowable milk production in cattle. *Journal of Dairy Science* 89:4795-4807.

Tedeschi, L. O., W. Chalupa, E. Janczewski, D. G. Fox, C. J. Sniffen, R. Munson, P. J. Kononoff, and R. C. Boston. 2008. Evaluation and application of the CPM Dairy Nutrition model. *Journal of Agricultural Science* 146:171-182.

Tedeschi, L. O., A. Cannas, and D. G. Fox. 2010. A nutrition mathematical model to account for dietary supply and requirements of energy and nutrients for domesticated small ruminants: The development and evaluation of the Small Ruminant Nutrition System. *Small Ruminant Research* 89:174-184.

Tedeschi, L. O., D. G. Fox, and P. J. Kononoff. 2013. A dynamic model to predict fat and protein fluxes associated with body reserve changes in cattle. *Journal of Dairy Science* 96:2448-2463.

Tedeschi, L. O., L. F. L. Cavalcanti, M. A. Fonseca, M. Herrero, and P. K. Thornton. 2014a. The evolution and evaluation of dairy cattle models for predicting milk production: An agricultural model intercomparison and improvement project (AgMIP) for livestock. *Animal Production Science* 54(12):2052-2067.

Tedeschi, L. O., C. A. Ramirez-Restrepo, and J. P. Muir. 2014b. Developing a conceptual model of possible benefits of condensed tannins for ruminant production. *Animal* 8:1095-1105.

Todd, R. W., N. A. Cole, H. M. Waldrip, and R. M. Aiken. 2013. Arrhenius equation for modeling feedyard ammonia emissions using temperature and diet crude protein. *Journal of Environmental Quality* 42:666-671.

Traxler, M. J., D. G. Fox, P. J. Van Soest, A. N. Pell, C. E. Lascano, D. P. D. Lanna, J. E. Moore, R. P. Lana, M. Vélez, and A. Flores. 1998. Predicting forage indigestible NDF from lignin concentration. *Journal of Animal Science* 76:1469-1480.

Tylutki, T. P., D. G. Fox, and R. G. Anrique. 1994. Predicting net energy and protein requirements for growth of implanted and nonimplanted heifers and steers and nonimplanted bulls varying in body size. *Journal of Animal Science* 72:1806-1813.

Tylutki, T. P., D. G. Fox, V. M. Durbal, L. O. Tedeschi, J. B. Russell, M. E. Van Amburgh, T. R. Overton, L. E. Chase, and A. N. Pell. 2008. Cornell Net Carbohydrate and Protein System: A model for precision feeding of dairy cattle. *Animal Feed Science and Technology* 143:174-202.

Van Soest, P. J. 1994. *Nutritional Ecology of the Ruminant*, 2nd Ed. Ithaca, NY: Comstock Publishing Associates.

Van Soest, P. J., J. B. Robertson, and B. A. Lewis. 1991. Methods for dietary fiber, neutral detergent fiber, and nonstarch polysaccharides in relation to animal nutrition. *Journal of Dairy Science* 74:3583-3597.

Vasconcelos, J. T., L. O. Tedeschi, D. G. Fox, M. L. Galyean, and L. W. Greene. 2007. Review: Feeding nitrogen and phosphorus in beef cattle feedlot production to mitigate environmental impacts. *The Professional Animal Scientist* 23:8-17.

Waghorn, G. C., and R. L. Baldwin. 1984. Model of metabolite flux within mammary gland of the lactating cow. *Journal of Dairy Science* 67:531-544.

Waldo, D. R., and B. P. Glenn. 1984. Comparison of new protein systems for lactating dairy cows. *Journal of Dairy Science* 67:1115-1133.

Waldrip, H. M., R. W. Todd, and N. A. Cole. 2013. Prediction of nitrogen excretion by beef cattle: A meta-analysis. *Journal of Animal Science* 91:4290-4302.

Winchester, C. F., and M. J. Morris. 1956. Water intake rates of cattle. *Journal of Animal Science* 15:722-740.

Wolin, M. J. 1960. A theoretical rumen fermentation balance. *Journal of Dairy Science* 43:1452-1459.

Zebeli, Q., J. Dijkstra, M. Tafaj, H. Steingass, B. N. Ametaj, and W. Drochner. 2008. Modeling the adequacy of dietary fiber in dairy cows based on the responses of ruminal pH and milk fat production to composition of the diet. *Journal of Dairy Science* 91:2046-2066.

20

Tables of Nutrient Requirements

INTRODUCTION

The previous edition and update (NRC, 1996, 2000) and this edition of the *Nutrient Requirements of Beef Cattle* were designed to predict beef cattle requirements and performance under specific animal, environmental, and dietary conditions. Many variables (e.g., maintenance, growth, milk, ruminal microbial growth) are continuous and interact with the effects of feed composition and environmental factors. While the computer model (described in Chapter 19) is provided to allow users to calculate the effects of these variables, tables of nutrient requirements are useful and instructive for some applications. The tables of nutrient requirements provided in this edition of the *Nutrient Requirements of Beef Cattle*, however, differ from those in the 6th revised edition (NRC, 1984) because of the complexity accounted for in the current Beef Cattle Nutrient Requirements model and the previous NRC (1996, 2000) model. A restricted dedicated optimizer (RDO; i.e., *table generator*) was developed using the empirical level of solution (ELS; see Chapter 19) to compute nutrient requirements and create evaluation tables for various diets. The RDO works with the current user settings (i.e., environment, dietary compositions, management) to estimate the requirements, so users can generate tables for specific production conditions. Because the RDO uses optimization to meet the restrictions, the Microsoft® Excel® Solver Add-In is required to be installed and loaded.[1] The Beef Cattle Nutrient Requirements software will attempt to locate and load the Solver Add-In automatically. The initial values (i.e., guess values) are important for convergence (i.e., finding a suitable solution) during the optimization process. The guess values are located in cells B5 to B8 in the *Table* spreadsheet (see Figure 3 in the User's Guide, Appendix A). For the optimizations, the objective function chosen was to minimize dietary crude protein (CP) (cell D16) and other constraints, including nutrient requirements and dietary nutrients. Two of these optimization constraints are user inputs: minimum and maximum values of ruminally degradable protein (RDP) balance (cells C30 and D30) and minimum and maximum values of the ratio of the difference between metabolizable protein (MP) and metabolizable energy (ME) allowable gain (or milk) to the ME-allowable gain (or milk) (cells C31 and D31). They are initially set to be zero and 0.01 for minimum and maximum values, respectively.

The RDO allows for the determination of requirements for any body weight (BW) and level of production for growing and finishing cattle, breeding bulls, replacement/bred heifers, and lactating beef cows. This chapter includes an example for each of these classes of cattle, using the average BW of a finished steer and mature cow of 550 kg and a BW for bulls of 900 kg. For growing animals, the left side of the tables (nutrient requirement) has daily nutrient requirements for the BW and production level specified, and the right side of the tables (diet evaluator) allows the user to determine the concentration of CP, RDP, calcium (Ca), and phosphorus (P) required in a diet under specific conditions (i.e., total digestible nutrients [TDN], BW, and average daily gain [ADG]). The diet evaluator computes ME-allowable production for specified diets, balances for RDP, ruminally undegradable protein (RUP), and MP, and calculates Ca and P needed in the diet to support the dietary energy-allowable production. The optimization alters dry matter intake (DMI), CP, and RDP to meet these restrictions: RDP balance near zero and MP-allowable gain or milk that matches the ME-allowable gain or milk. The ME- and MP-allowable gains are calculated as the ME or MP balance for growth divided by their ME or MP content of the ADG, respectively. Actual DMI is set

[1]Support to load the Solver Add-In for Microsoft® Excel® can be found in the following hyperlinks:

Excel® 2007: http://office.microsoft.com/en-us/excel-help/load-the-solver-add-in-HP010021570.aspx?CTT=1,

Excel® 2010: http://office.microsoft.com/en-us/excel-help/load-the-solver-add-in-HP010342660.aspx?CTT=1, and

Excel® 2013: https://support.office.com/en-US/Article/Load-the-Solver-Add-in-ec994cd0-a396-4bf3-a5dd-feda369cef37?ui=en-US&rs=en-US&ad=US.

to match the predicted DMI, which is estimated using the equations discussed in Chapter 10 (Feed Intake).

In addition to determining nutrient density requirements, the diet evaluator allows the user to see how well a particular diet meets requirements of cattle in a feeding group with the range of weights specified for growing cattle or at each of the 12 months of the reproductive cycle for beef cows. In most beef production situations, cattle are fed in groups that vary in stage of growth or reproduction. Each group is usually fed to appetite either available forage (stocker, backgrounding, cow-calf) or high-energy, grain-based diets (growing and finishing cattle). The objective in diet formulation for high-forage diets is to determine supplemental energy, protein, and minerals needed to meet target levels of production. The objective in high-energy diets is typically to determine the protein and minerals needed to support the energy-allowable ADG. In all situations, the user attempts to develop a "best fit" diet, considering the variation of animals in a feeding group.

To use the diet evaluator, the user enters the diet TDN. Dietary CP and CP degradability are calculated by the optimization process. Dietary net energy required for maintenance (NEm), net energy required for gain (NEg), DMI, ADG, or energy balance, RDP, RUP, and MP balances (g/d) are predicted for each of the diets over a range of BW for the body size specified for growing cattle or for each month of the reproductive cycle for breeding cattle. Dietary TDN is used to predict dietary NEm and NEg. Dietary NEm and NEg can only be changed by adjusting dietary TDN because the relationship between these energy values must be kept consistent. Dietary TDN is used to predict microbial growth, which must be consistent with the energy value used to predict NEm available to meet maintenance, pregnancy, and lactation requirements and the energy value used to predict NEg-allowable ADG. To get the dietary net energy (NE) value desired, the user adjusts TDN until the desired NE value is predicted. The current Nutrient Requirements of Beef Cattle committee recognizes that the relationship between TDN, ME, NEm, and NEg might vary because of differences in the amount of intake, rates of digestion and passage, and end products of digestion in the ME and their metabolizability. Nonetheless, the relationships between them, as described in the NRC (1984), have also been used here for the reasons discussed in the previous chapters. The concentration of nutrients needed for a given level of production depends on the actual DMI of the diet being fed to support the observed level of performance in a particular production setting. The DMI predictions are from equations developed from experimental feeding period averages as reported in published feeding trials involving wide variations in cattle type and stage of growth, as discussed in Chapter 10. Thus, predicted and observed values often differ in a specific production setting. Cattle in feedlots fed finishing diets will typically consume more early in the feeding period than predicted by these equations, which is compensated for by lower DMI late in the feeding period. This behavior, however, might not be consistent for all types of cattle (e.g., Holstein steers). Furthermore, as discussed in Chapter 10, most DMI prediction equations account for only 50 to 60% of the variation, leaving 40 to 50% to be accounted for by variations in local conditions such as feeding management, cattle type, and environment. Many factors can influence the NE derived from a diet for production, including variation in maintenance requirements, rates of digestion and passage, and metabolizability. If only DMI is adjusted, predicted and observed performances might not agree. For example, unrealistically high rates and efficiencies of gain can be predicted for calves consuming high-energy diets. Conversely, when these animals approach USDA Choice grade (or an advanced maturity) at the end of the finishing period, unrealistically low ADG could be predicted if only DMI is adjusted. Given these problems of prediction early and late in growth, limits were set on the weight ranges in the diet density tables at approximately 55% of finished weight for the lightest BW and 80% of finished weight for the heaviest BW.

The primary intended use of these tables is for evaluating the interactions of BW, stage of growth, dietary energy density, and energy and protein requirements. The diet densities for CP and RDP might not be practical because the CP might need to be overfed to meet both RDP and RUP requirements. Despite their limitations as discussed in this section, simple guideline tables with requirements for dietary nutrient concentration for different classes of cattle are all that are needed in many situations.

The user is encouraged to apply the RDO with actual feed ingredients available for computing requirements for specific conditions. If cell *B18* in the *Table* spreadsheet is set to true, the composition of the first diet ingredient will be used by the RDO, except for CP, TDN, ME, NEm, and NEg, as these are needed to find an optimum solution by the RDO.

TABLE FOR GROWING AND FINISHING CATTLE

Table 20-1 shows the daily requirements (left side) and diet evaluations (right side) for growing and finishing cattle. Inputs for Table 20-1 are for an Angus steer with 550-kg finished BW at 28% fat (reference animal's empty body fat = 4), a BW range of 250 to 500 kg, and an ADG range of 0.40 to 2 kg/d. Table 20-1 (left side) shows the NEm, NEg, MP, Ca, and P daily requirement for maintenance and gain at six values of shrunk BW (SBW), which represent six different stages of growth. The predicted DMI was obtained with the equation using dietary NEm to predict DMI as a percentage of SBW described in Chapter 10. All these requirements can be used directly to formulate dietary requirements for the specified level of performance. In this edition of the *Nutrient Requirements of Beef Cattle*, the dietary NE, MP, Ca, and P needed for maintenance and growth are computed through the optimization process using the ELS calculations. The objective function is to minimize CP by changing dietary TDN subject to the following constraints:

TABLE 20-1 Nutrient Requirements (left) and Diet Evaluation (right) for Growing and Finishing Cattle

Mature SBW, kg 550

		Shrunk Body Weight (SBW), kg					
Maintenance		250	300	350	400	450	500
NEm	Mcal/d	4.8	5.6	6.2	6.9	7.5	8.1
MP	g/d	239	274	307	340	371	402
Ca	g/d	7.7	9.2	10.8	12.3	13.9	15.4
P	g/d	5.9	7.1	8.2	9.4	10.6	11.8
Growth (ADG)		*NEg Required for Gain, Mcal/d*					
0.4	kg/d	1.2	1.3	1.5	1.6	1.8	1.9
0.8	kg/d	2.5	2.8	3.2	3.5	3.8	4.1
1.2	kg/d	3.8	4.4	5	5.5	6	6.5
1.6	kg/d	5.3	6.1	6.8	7.5	8.2	8.9
2	kg/d	6.7	7.7	8.7	9.6	10.5	11.3
		MP Required for Gain, g/d					
0.4	kg/d	149	139	129	120	111	102
0.8	kg/d	288	267	246	226	207	188
1.2	kg/d	423	390	358	326	296	267
1.6	kg/d	556	510	466	423	381	341
2	kg/d	686	627	571	516	463	412
		Calcium Required for Gain, g/d					
0.4	kg/d	10.4	9.7	9	8.4	7.7	7.1
0.8	kg/d	20.1	18.6	17.2	15.8	14.4	13.1
1.2	kg/d	29.6	27.2	25	22.8	20.7	18.6
1.6	kg/d	38.9	35.6	32.5	29.5	26.6	23.8
2	kg/d	48	43.8	39.9	36.1	32.4	28.8
		Phosphorus Required for Gain, g/d					
0.4	kg/d	4.2	3.9	3.6	3.4	3.1	2.9
0.8	kg/d	8.1	7.5	6.9	6.4	5.8	5.3
1.2	kg/d	12	11	10.1	9.2	8.4	7.5
1.6	kg/d	15.7	14.4	13.1	11.9	10.8	9.6
2	kg/d	19.4	17.7	16.1	14.6	13.1	11.6

Diets	TDN (% DM)	ME (Mcal/kg)	NEm (Mcal/kg)	NEg (Mcal/kg)
A	65	2.40	1.52	0.93
B	70	2.59	1.68	1.07
C	75	2.77	1.84	1.21
D	80	2.96	2.00	1.34

SBW (kg)	Diets	DMI (kg/d)	ADG (kg/d)	CP (% DM)	RDP (% CP)	MP (g/d)	Ca (% DM)	P (% DM)
250	A	6.06	0.86	12.6	50.6	547	0.48	0.24
	B	5.93	1.04	14.2	48.1	607	0.56	0.27
	C	5.72	1.17	15.7	46.3	652	0.64	0.31
	D	5.42	1.25	17.2	45.0	680	0.71	0.34
300	A	7.27	0.94	11.3	55.4	583	0.42	0.22
	B	7.11	1.12	12.6	53.2	640	0.49	0.24
	C	6.86	1.26	13.9	51.6	682	0.55	0.27
	D	6.51	1.35	15.1	50.3	709	0.61	0.30
350	A	8.48	1.00	10.2	60.2	611	0.38	0.20
	B	8.30	1.19	11.3	58.4	664	0.43	0.22
	C	8.00	1.34	12.4	56.9	703	0.48	0.24
	D	7.59	1.43	13.5	55.8	728	0.53	0.26
400	A	9.69	1.06	9.4	65.0	632	0.34	0.18
	B	9.49	1.26	10.3	63.6	681	0.38	0.20
	C	9.15	1.41	11.2	62.4	718	0.42	0.22
	D	8.68	1.51	12.2	61.3	741	0.46	0.24
450	A	10.90	1.12	8.7	69.9	649	0.31	0.17
	B	10.67	1.32	9.4	68.9	694	0.34	0.18
	C	10.29	1.48	10.2	67.9	727	0.38	0.20
	D	9.76	1.58	11.0	67.0	748	0.41	0.22
500	A	12.12	1.17	8.0	74.8	662	0.28	0.16
	B	11.86	1.38	8.7	74.2	702	0.31	0.17
	C	11.43	1.54	9.4	73.6	731	0.34	0.18

- Actual DMI is set equal to predicted DMI using Eq. 10-2.
- ADG is set equal to the desired ADG.
- TDN, CP, and RDP can vary from 0 to 100%.
- ME (Mcal/kg), NEm (Mcal/kg), NEg (Mcal/kg), MP (g/d), Ca (g/d), and P (g/d) must be greater than 0.

Table 20-1 (right side) shows the evaluation of four diets (diets A through D) with the diet evaluator for the same animal used in the left side of Table 20-1 between 55 and 80% of final SBW. The dietary concentration of TDN was entered for each of the four diets. In this edition of the *Nutrient Requirements of Beef Cattle*, the DMI (kg/d), ADG (kg/d), CP (% DM), RDP (% CP), MP (g/d), Ca (% DM) and P (% DM) are computed through the optimization process using the ELS calculations. The objective function is to minimize CP by changing dietary CP and RDP subject to the following constraints:

- Actual DMI is set equal to predicted DMI using Eq. 10-2.
- TDN is set equal to the desired TDN.

- CP and RDP can vary from 0 to 100%.
- ADG (kg/d), ME (Mcal/kg), NEm (Mcal/kg), NEg (Mcal/kg), MP (g/d), Ca (g/d), and P (g/d) have to be greater than 0.
- RDP balance (RDP supply + 6.25 × recycled N – microbial CP) must be between 0 and 0.01, when recycled N is computed; otherwise recycled N is 0.
- The ratio of the difference between MP-allowable gain and ME-allowable gain to ME-allowable gain must be between 0 and 0.01.

Unlike in the previous edition and update of the *Nutrient Requirements of Beef Cattle* (1996, 2000), there is no ruminal microbial adjustment for dietary physically effective neutral detergent fiber (peNDF). As can be noted in Table 20-1, the simulation indicated that dietary CP (% DM) increased as ADG increased, but decreased for similar ADG as SBW increased.

TABLE 20-2 Nutrient Requirements (left) and Diet Evaluation (right) for Growing Bulls

Mature SBW, kg 900

		Shrunk Body Weight (SBW), kg					
Maintenance		300	400	500	600	700	800
NEm	Mcal/d	6.4	7.9	9.4	10.7	12.1	13.3
MP	g/d	274	340	402	461	517	572
Ca	g/d	9.2	12.3	15.4	18.5	21.6	24.6
P	g/d	7.1	9.4	11.8	14.1	16.5	18.8
Growth (ADG)		*NEg Required for Gain, Mcal/d*					
0.4	kg/d	0.9	1.1	1.3	1.5	1.7	1.9
0.8	kg/d	2	2.4	2.9	3.3	3.7	4.1
1.2	kg/d	3.1	3.8	4.5	5.1	5.8	6.4
1.6	kg/d	4.2	5.2	6.1	7	7.9	8.7
2	kg/d	5.3	6.6	7.8	9	10.1	11.1
		MP Required for Gain, g/d					
0.4	kg/d	163	150	138	126	115	104
0.8	kg/d	319	291	264	239	215	192
1.2	kg/d	471	427	386	347	310	273
1.6	kg/d	622	561	505	451	400	350
2	kg/d	770	693	621	553	487	423
		Calcium Required for Gain, g/d					
0.4	kg/d	11.4	10.5	9.6	8.8	8	7.3
0.8	kg/d	22.3	20.3	18.5	16.7	15	13.4
1.2	kg/d	32.9	29.9	27	24.2	21.6	19.1
1.6	kg/d	43.4	39.2	35.3	31.5	27.9	24.5
2	kg/d	53.8	48.4	43.4	38.6	34	29.6
		Phosphorus Required for Gain, g/d					
0.4	kg/d	4.6	4.2	3.9	3.6	3.2	2.9
0.8	kg/d	9	8.2	7.5	6.8	6.1	5.4
1.2	kg/d	13.3	12.1	10.9	9.8	8.7	7.7
1.6	kg/d	17.5	15.8	14.2	12.7	11.3	9.9
2	kg/d	21.7	19.6	17.5	15.6	13.7	11.9

Diets	TDN (% DM)	ME (Mcal/kg)	NEm (Mcal/kg)	NEg (Mcal/kg)
A	65	2.40	1.52	0.93
B	70	2.59	1.68	1.07
C	75	2.77	1.84	1.21
D	80	2.96	2.00	1.34

SBW (kg)	Diets	DMI (kg/d)	ADG (kg/d)	CP (% DM)	RDP (% CP)	MP (g/d)	Ca (% DM)	P (% DM)
300	A	7.27	1.13	13.6	45.9	718	0.55	0.27
	B	7.11	1.38	15.6	42.8	814	0.66	0.31
	C	6.86	1.57	17.6	40.7	886	0.76	0.35
	D	6.51	1.69	19.4	39.3	931	0.85	0.39
400	A	9.69	1.31	11.6	53.6	804	0.46	0.23
	B	9.49	1.58	13.1	49.9	895	0.54	0.26
	C	9.15	1.79	14.6	48.0	964	0.61	0.30
	D	8.68	1.92	16.0	46.6	1007	0.68	0.33
500	A	12.12	1.46	10.1	59.4	864	0.39	0.20
	B	11.86	1.75	11.3	57.1	949	0.45	0.23
	C	11.43	1.96	12.5	55.4	1013	0.51	0.25
	D	10.85	2.10	13.6	54.1	1053	0.56	0.28
600	A	14.54	1.58	9.0	66.2	907	0.34	0.18
	B	14.23	1.88	9.9	64.4	984	0.39	0.20
	C	13.72	2.11	10.9	62.9	1042	0.43	0.22
	D	13.02	2.26	11.8	61.7	1078	0.47	0.24
700	A	16.96	1.69	8.1	73.1	937	0.30	0.17
	B	16.60	2.01	8.8	71.8	1005	0.34	0.18
	C	16.00	2.25	9.6	70.7	1056	0.37	0.20
	D	15.19	2.40	10.4	69.6	1088	0.40	0.21
800	A	19.39	1.79	7.3	80.0	957	0.27	0.15
	B	18.97	2.12	8.0	79.4	1015	0.29	0.17
	C	18.29	2.36	8.6	78.7	1059	0.32	0.18
	D	17.36	2.52	9.3	77.8	1086	0.35	0.19

TABLE FOR BREEDING BULLS

Table 20-2 contains an example of nutrient requirement (left side) and diet evaluation (right side) for growing bulls, using a 900-kg mature SBW. Diet inputs for Table 20-2 were made as described for Table 20-1, with similar dietary TDN values. Body weight ranges were set as 55 to 80% of the 28% fat SBW of a steer of the same genotype. Optimization of CP, RDP, MP, Ca, and P were the same as described for Table 20-1.

TABLE FOR PREGNANT REPLACEMENT HEIFERS

Tables 20-3 and 20-4 contain the requirements and diet evaluations for pregnant heifers, respectively. As with the preceding tables, the animal described in the requirements table is then used in the diet evaluator. Animal descriptions entered were 550-kg mature SBW, 40-kg expected birth weight, 15 month of age at breeding, and breed code 1 (Angus). Table 20-3 shows the predicted NEm, MP, Ca, and P required daily for maintenance, growth, and pregnancy and target ADG, SBW, and expected gravid uterus weight used to compute requirements for each of the 9 months of gestation, using the equations presented in Chapter 19. As described previously, all of these tabular values can be used directly to formulate dietary requirements for the specified level of performance.

Table 20-4 shows diet evaluations for this same heifer. The diet concentration of TDN was entered for each of the four diets. Similar to the previous description, the DMI, ADG, CP (% DM), RDP (% CP), Ca (% DM) and P (% DM) are computed through the RDO process using the ELS calculations. The objective function is to minimize CP by changing dietary CP and RDP subject to the following constraints:

- Actual DMI is set equal to predicted DMI using Eq. 10-2.
- TDN is set equal to the desired TDN.

TABLE 20-3 Nutrient Requirements for Replacement Heifers

Mature SBW, kg	550								
Calf Birth BW, kg	40								
	Months Since Conception								
	1	2	3	4	5	6	7	8	9
	NEm Required, Mcal/d								
Maintenance	6.0	6.3	6.5	6.6	6.8	6.9	7.1	7.2	7.4
Growth	2.1	2.3	2.3	2.4	2.4	2.5	2.5	2.5	2.4
Pregnancy	0.0	0.1	0.2	0.4	0.7	1.4	2.4	3.9	6.2
Total	8.1	8.7	9.0	9.4	9.9	10.7	11.9	13.6	16.0
	MP Required, g/d								
Maintenance	295	310	318	326	334	342	350	357	365
Growth	130	129	127	126	124	123	123	123	125
Pregnancy	2	4	7	14	27	50	88	151	251
Total	427	443	453	466	485	515	561	632	741
	Calcium Required, g/d								
Maintenance	10.2	10.9	11.3	11.7	12.0	12.4	12.8	13.2	13.6
Growth	9.1	9.0	8.9	8.8	8.7	8.6	8.6	8.6	8.7
Pregnancy	0.0	0.0	0.0	0.0	0.0	0.0	12.2	12.2	12.2
Total	19.3	19.9	20.2	20.4	20.7	21.0	33.6	34.0	34.5
	Phosphorus Required, g/d								
Maintenance	7.8	8.3	8.6	8.9	9.2	9.5	9.8	10.1	10.4
Growth	3.7	3.6	3.6	3.5	3.5	3.5	3.5	3.5	3.5
Pregnancy	0.0	0.0	0.0	0.0	0.0	0.0	5.0	5.0	5.0
Total	11.5	12.0	12.2	12.5	12.7	13.0	18.2	18.5	18.8
	Average Daily Gain (ADG), kg/d								
Growth	0.402	0.402	0.402	0.402	0.402	0.402	0.402	0.402	0.402
Pregnancy	0.028	0.048	0.077	0.122	0.188	0.280	0.405	0.568	0.773
Total	0.430	0.450	0.479	0.524	0.590	0.682	0.807	0.970	1.175
	Weight, kg								
Shrunk BW	342	354	367	379	391	403	416	428	440
Gravid uterus	1	3	4	7	12	19	29	44	64
Total	344	357	371	386	403	423	445	472	504

TABLE 20-4 Diet Evaluation for Replacement Heifers

Diets	TDN (% DM)	ME (Mcal/kg)	NEm (Mcal/kg)	NEg (Mcal/kg)
A	55	2.03	1.18	0.62
B	60	2.22	1.36	0.78
C	65	2.40	1.52	0.93
D	70	2.59	1.68	1.07

Diets		Months Since Conception								
		1	2	3	4	5	6	7	8	9
A	DMI, kg/d	8.27	8.56	8.86	9.15	9.45	9.74	10.04	10.33	10.63
	ADG, kg/d	0.51	0.52	0.52	0.51	0.49	0.45	0.36	0.21	0.00
	CP, % DM	8.15	8.04	7.94	7.86	7.81	7.81	7.87	8.02	---
	RDP, % CP	65.1	65.8	66.4	66.8	67.1	66.9	66.2	64.8	---
	Ca, % DM	0.27	0.26	0.26	0.25	0.24	0.22	0.33	0.29	---
	P, % DM	0.15	0.15	0.15	0.15	0.14	0.14	0.18	0.16	---
B	DMI, kg/d	8.35	8.64	8.94	9.24	9.54	9.84	10.14	10.43	10.73
	ADG, kg/d	0.76	0.77	0.78	0.77	0.75	0.70	0.61	0.46	0.22
	CP, % DM	9.28	9.12	8.97	8.85	8.76	8.73	8.77	8.91	9.14
	RDP, % CP	61.8	62.7	63.5	64.2	64.7	64.8	64.3	63.2	61.4
	Ca, % DM	0.33	0.32	0.31	0.30	0.29	0.27	0.37	0.34	0.29
	P, % DM	0.18	0.17	0.17	0.17	0.16	0.16	0.20	0.18	0.16
C	DMI, kg/d	8.29	8.59	8.89	9.18	9.48	9.77	10.07	10.37	10.66
	ADG, kg/d	0.99	1.00	1.01	1.01	0.99	0.94	0.85	0.69	0.45
	CP, % DM	10.4	10.2	10.0	9.8	9.7	9.7	9.7	9.8	10.1
	RDP, % CP	59.3	60.3	61.3	62.2	62.8	63.1	62.8	61.9	60.2
	Ca, % DM	0.38	0.37	0.36	0.35	0.34	0.32	0.42	0.38	0.33
	P, % DM	0.20	0.20	0.19	0.19	0.18	0.17	0.21	0.20	0.18
D	DMI, kg/d	8.12	8.41	8.70	8.99	9.28	9.56	9.85	10.14	10.43
	ADG, kg/d	1.18	1.20	1.20	1.20	1.18	1.13	1.04	0.88	0.64
	CP, % DM	11.5	11.3	11.0	10.8	10.7	10.6	10.6	10.7	11.0
	RDP, % CP	57.4	58.5	59.6	60.5	61.3	61.7	61.6	60.8	59.2
	Ca, % DM	0.44	0.43	0.41	0.40	0.38	0.36	0.46	0.43	0.37
	P, % DM	0.22	0.22	0.21	0.21	0.20	0.19	0.23	0.22	0.20

- CP and RDP can vary from 0 to 100%.
- ADG (kg/d), ME (Mcal/kg), NEm (Mcal/kg), NEg (Mcal/kg), MP (g/d), Ca (g/d), and P (g/d) must be greater than 0.
- RDP balance (RDP supply + 6.25 × recycled N – microbial CP) must be between 0 and 0.01 when recycled N is computed; otherwise recycled N is 0.
- The ratio of the difference between MP-allowable gain and ME-allowable gain to ME-allowable gain must be between 0 and 0.01.

Because of the relationship between ME- and MP-allowable gain, the dietary CP and RDP were optimized to support ME-allowable gain, not the target ADG shown in Table 20-3.

The predicted DMI increased as pregnancy progressed because of increasing predicted SBW (shown in Table 20-3). As with growing and finishing cattle, the RDP balance was nearly constant over gestation for a given diet because microbial requirement is a relatively constant proportion of TDN; however, because the required RUP balance changes with composition of the ADG (decreased protein content of ADG with increasing SBW) and conceptus requirements, the required dietary CP changed. The CP, RDP, and RUP requirements are determined as described for growing and finishing cattle. Diet A (55% TDN) does not supply enough energy to support target heifer growth after the 5th month of pregnancy. Diet B (60% TDN) exceeds target energy-allowable ADG in all but the last 3 months of pregnancy. Diet C (65% TDN) exceeded target ADG in all months, except for the last 2 months. Diet D (70% TDN) lacked energy in the last month of pregnancy.

TABLE FOR BEEF COWS

Tables 20-5 and 20-6 contain requirements and diet evaluations for beef cows. As with the bred heifers, the animal described in the requirements is used in the diet evaluator. It computes energy and protein balances expected for each of the four diets (diets A through D) entered and CP (% DM), RDP (% CP), Ca (% DM), and P (% DM) needed in the dietary DM to meet requirements. Animal descriptions entered were 550-kg mature BW Angus, 40-kg expected birth

TABLE 20-5 Nutrient Requirements for Lactating Cows

Mature SBW, kg	550		Peak milk, kg/d	8			Milk fat, %	4.0			
Calf Birth BW, kg	40		Relative MY	5			Milk CP, %	3.4			

	Months Since Calving											
	1	2	3	4	5	6	7	8	9	10	11	12
	NEm Required, Mcal/d											
Maintenance	10.5	10.5	10.5	10.5	10.5	10.5	10.5	10.5	10.5	10.5	10.5	10.5
Pregnancy	0	0	0	0	0.1	0.2	0.4	0.7	1.4	2.4	4.0	6.2
Lactation	4.8	5.7	5.2	4.1	3.1	2.2	1.6	1.1	0.7	0.5	0.3	0.2
Total	15.3	16.2	15.7	14.7	13.7	12.9	12.4	12.3	12.6	13.4	14.8	16.9
	MP Required, g/d											
Maintenance	432	432	432	432	432	432	432	432	432	432	432	432
Pregnancy	0	0	1	2	3	7	14	27	50	88	152	251
Lactation	349	418	376	301	226	163	114	78	53	35	23	15
Total	780	850	809	734	661	601	560	536	534	555	607	697
	Calcium Required, g/d											
Maintenance	16.9	16.9	16.9	16.9	16.9	16.9	16.9	16.9	16.9	16.9	16.9	16.9
Pregnancy	0	0	0	0	0	0	0	0	0	12.2	12.2	12.2
Lactation	16.4	19.7	17.7	14.2	10.6	7.6	5.4	3.7	2.5	1.7	1.1	0.7
Total	33.3	36.6	34.6	31.1	27.6	24.6	22.3	20.6	19.4	30.8	30.2	29.8
	Phosphorus Required, g/d											
Maintenance	12.9	12.9	12.9	12.9	12.9	12.9	12.9	12.9	12.9	12.9	12.9	12.9
Pregnancy	0	0	0	0	0	0	0	0	0	5.0	5.0	5.0
Lactation	9.3	11.2	10.1	8.0	6.0	4.3	3.0	2.1	1.4	0.9	0.6	0.4
Total	22.3	24.1	23.0	21.0	19.0	17.3	16.0	15.0	14.3	18.8	18.5	18.3
	Milk Production and Gravid Uterus Gain, kg/d											
Milk	6.7	8.0	7.2	5.8	4.3	3.1	2.2	1.5	1.0	0.7	0.4	0.3
Pregnancy	0	0	0.016	0.028	0.047	0.078	0.122	0.187	0.281	0.404	0.571	0.773
	Weight, kg											
Shrunk BW	550	550	550	550	550	550	550	550	550	550	550	550
Gravid Uterus	0	0	1	2	3	5	7	12	19	29	44	64
Total	550	550	551	552	553	555	557	562	569	579	594	614

weight, 60 months age, peak milk (8 kg), milk composition (4% fat, 3.4% protein, 8.3% solids not fat), 8.5 weeks at peak milk, and 30 months' duration of lactation.

Table 20-5 shows the predicted NEm, MP, MP, Ca, and P required daily for maintenance, lactation, and pregnancy as well as SBW, daily milk production, and expected gravid uterus weight used to compute the requirements for each of the 12 months of the reproductive cycle using the equations presented in Chapter 19. As described previously, all of these values can be used directly to formulate dietary requirements for the specified level of performance.

Table 20-6 shows diet evaluations for this same cow. The diet concentration of TDN was entered for each of the four diets. The predicted DMI, ME balance (Mcal/d), CP (% DM), RDP (% CP), Ca (% DM), and P (% DM) are computed through the RDO process using the ELS calculations. The objective function is to minimize CP by changing dietary CP and RDP subject to the following constraints:

- Actual DMI is set equal to predicted DMI using Eq. 10-5.
- TDN is set equal to the desired TDN.
- CP and RDP can vary from 0 to 100%.
- ADG (kg/d), ME (Mcal/kg), NEm (Mcal/kg), NEg (Mcal/kg), MP (g/d), Ca (g/d), and P (g/d) must be greater than 0.
- RDP balance (RDP supply + 6.25 × recycled N – microbial CP) must be between 0 and 0.01 when recycled N is computed; otherwise recycled N is 0.
- The ratio of the difference between MP-allowable gain and ME-allowable gain to ME-allowable gain must be between 0 and 0.01.

Because of the relationship between ME- and MP-allowable milk, the dietary CP and RDP were optimized to support ME-allowable milk. The ME-allowable milk was computed based on the ME required for lactation. If energy supply is

TABLE 20-6 Diet Evaluation for Lactating Cows

Diets	TDN %DM	ME	NEm	NEg
		Mcal/kg		
A	55	2.03	1.18	0.62
B	60	2.22	1.36	0.78
C	65	2.40	1.52	0.93
D	70	2.59	1.68	1.07

Diets		Months Since Calving											
		1	2	3	4	5	6	7	8	9	10	11	12
A	DMI, kg/d	11.6	11.9	11.7	11.4	11.2	10.9	10.7	10.6	10.5	10.4	10.4	11.1
	ME Bal., Mcal/d	–2.6	–3.7	–3.0	–1.9	–0.8	0.1	0.6	0.6	0.0	–1.2	–3.5	–5.0
	CP, % DM	8.2	8.3	8.3	8.2	8.1	8.0	7.8	7.7	7.4	7.1	6.6	6.5
	RDP, % CP	62.6	61.9	62.3	63.1	64.0	65.0	66.2	67.7	70.1	73.6	78.9	80.1
	Ca, % DM	0.29	0.31	0.30	0.27	0.25	0.23	0.21	0.19	0.18	0.29	0.29	0.27
	P, % DM	0.19	0.20	0.20	0.18	0.17	0.16	0.15	0.14	0.14	0.18	0.18	0.16
B	DMI, kg/d	12.1	12.4	12.2	11.9	11.6	11.4	11.2	11.1	11.0	10.9	10.9	11.5
	ME Bal., Mcal/d	1.8	0.9	1.4	2.4	3.4	4.2	4.5	4.5	3.9	2.6	0.4	–1.3
	CP, % DM	10.0	10.1	10.0	10.0	9.9	9.8	9.6	9.5	9.2	8.8	8.3	8.1
	RDP, % CP	55.6	55.2	55.5	56.0	56.7	57.4	58.2	59.3	61.0	63.6	67.6	69.4
	Ca, % DM	0.28	0.30	0.28	0.26	0.24	0.22	0.20	0.19	0.18	0.28	0.28	0.26
	P, % DM	0.18	0.20	0.19	0.18	0.16	0.15	0.14	0.14	0.13	0.17	0.17	0.16
C	DMI, kg/d	12.7	12.9	12.8	12.5	12.2	11.9	11.8	11.6	11.5	11.5	11.4	12.0
	ME Bal., Mcal/d	6.3	5.4	5.9	6.8	7.7	8.3	8.7	8.6	7.9	6.6	4.3	2.6
	CP, % DM	11.8	11.8	11.8	11.7	11.6	11.5	11.4	11.2	11.0	10.6	10.1	9.7
	RDP, % CP	50.9	50.6	50.8	51.2	51.7	52.2	52.8	53.6	55.0	56.9	60.0	61.9
	Ca, % DM	0.26	0.28	0.27	0.25	0.23	0.21	0.19	0.18	0.17	0.27	0.26	0.25
	P, % DM	0.18	0.19	0.18	0.17	0.16	0.14	0.14	0.13	0.12	0.16	0.16	0.15
D	DMI, kg/d	13.3	13.5	13.4	13.1	12.8	12.6	12.4	12.2	12.1	12.1	12.0	12.5
	ME Bal., Mcal/d	10.9	10.1	10.5	11.4	12.1	12.7	12.9	12.8	12.1	10.8	8.5	6.7
	CP, % DM	13.5	13.5	13.5	13.4	13.4	13.3	13.1	13.0	12.7	12.3	11.8	11.4
	RDP, % CP	47.5	47.3	47.5	47.8	48.1	48.5	49.0	49.7	50.7	52.3	54.7	56.4
	Ca, % DM	0.25	0.27	0.26	0.24	0.22	0.20	0.18	0.17	0.16	0.25	0.25	0.24
	P, % DM	0.17	0.18	0.17	0.16	0.15	0.14	0.13	0.12	0.12	0.16	0.15	0.15

less than energy required, a negative ME balance will occur and animals will not produce the predicted milk amount unless body reserves are mobilized. Predicted DMI varies with daily milk production and forage quality. Diet A (55% TDN) met energy requirements in months 6 to 9 (cows just dry), but became deficient in energy in month 10. Diets B (60% TDN), C (65% TDN), and D (70% TDN) are adequate in energy in all months. Dietary CP required decreased with months since calving, but RDP increased. The possible reason that RDP increased was to balance RDP as both DMI and CP content decreased; thus, an increase in RDP content was needed to increase the amount of RDP.

TABLE OF ENERGY RESERVES FOR BEEF COWS

Chapter 13 (Reproduction) shows how to compute the Mcal mobilized when moving to the next lower body condition score (BCS), or when required to move from the next lower BCS to the one being considered, for cows with different mature SBW. For example, a 500-kg cow at BCS 5 that decreases to BCS 4 will mobilize 207 Mcal. If ME intake is deficient by 5 Mcal/day, this is equivalent to an NEm of 3 Mcal/d (5 × 0.6), the cow will lose 1 CS in 55 days (207 × 0.8)/3. The same cow at BCS 4 will need 207 Mcal to increase to a BCS 5. If consuming 5 Mcal ME above daily requirements, equivalent to an NEm of 3 Mcal/d (5 × 0.6), the cow will move to a BCS 5 in 69 days (207/3). To offset a negative energy balance, 1 Mcal of dietary NEm will substitute for 1 Mcal negative energy balance. To utilize energy reserves, 1 Mcal of dietary NEm can be replaced by 0.8 Mcal of tissue energy. These calculations assumed a partial efficiency of use of ME to NEm of 0.6.

REFERENCES

NRC (National Research Council). 1984. *Nutrient Requirements of Beef Cattle*, 6th Rev. Ed. Washington, DC: National Academy Press.

NRC. 1996. *Nutrient Requirements of Beef Cattle* 7th Rev. Ed. Washington, DC: National Academy Press.

NRC. 2000. *Nutrient Requirements of Beef Cattle: Update 2000*, 7th Rev. Ed. Washington, DC: National Academy Press.

21

Research Needs

INTRODUCTION

As noted by Galyean (2014), systems for describing nutrient requirements of livestock basically have two major components: (1) estimates of animal requirements for nutrients, and (2) estimates of the ability of feedstuffs to meet requirements. Consistent with this basic two-part approach, the Committee on Nutrient Requirements of Beef Cattle updated the 7th edition and update of the *Nutrient Requirements of Beef Cattle* (NRC, 1996, 2000) by conducting extensive reviews of the literature and analyzing new data from derived and experimental databases.

The Statement of Task for the Committee on Nutrient Requirements of Beef Cattle indicated that the report would also address "future areas of needed research." The process of updating nutrient requirement reports certainly lends itself to this mandate, as gaps in knowledge and areas that need additional research to solidify recommendations quickly become evident as the report is prepared. Bullet-point suggestions for areas of needed research are addressed on a topical basis in the following sections.

GENERAL

- Serial slaughter experiments at different stages of production of beef cattle (i.e., feedlot growth rate, gestational status, lactation, etc.) should be conducted across all production segments to evaluate retention of energy, protein, and minerals.

BEEF PRODUCTION SYSTEMS/BEEF QUALTY AND SAFETY

- Given the importance of forages in beef production systems, a greater focus is needed on increasing research to improve the production and nutritional value of grazed and harvested forages.
- Additional data are needed on the effects of various production systems (conventional, organic, grass-finished, etc.) on the safety and nutrient profile of beef products.
- The role of beef production systems and practices in emerging antimicrobial resistance needs to be more clearly defined.

ENERGY, CARBOHYDRATES, AND LIPIDS

- The efficiency of conversion of digestible energy (DE) to metabolizable energy (ME) is not well defined. Further evaluation of this relationship and development of an equation across the range of high-forage to high-grain diets for beef cattle is needed.
- The equations for prediction of net energy required for maintenance (NEm) and net energy required for gain (NEg) from ME should be reevaluated.
- Compositional data on carbohydrate fractions of feedstuffs are limited; additional analytical results or methods for accurately predicting these fractions are needed.
- Methods for predicting mean ruminal pH across the wide range of diets used in beef cattle production need to be improved. Predictions of the time that ruminal pH is less than a particular benchmark would be helpful and would allow beef cattle producers to more effectively manage the feeding of high-grain diets.
- Digestibility and energy values of various lipid sources are not well defined across a range of diets, and additional data are needed.
- Estimates of the effects of lipid supplementation on digestion characteristics, particularly of structural carbohydrates, would be useful for improving the accuracy of the Beef Cattle Nutrient Requirements model.

PROTEIN

- Because the standard error of prediction for equations based on total digestible nutrients (TDN) or fat-free TDN is high, equations that provide more precise predictions of microbial protein synthesis in the rumen are needed.
- The efficiency with which ruminally degradable protein (RDP) is converted to microbial crude protein (MCP) should be reevaluated.
- Equations to accurately predict recycled nitrogen (N) across a wide range of diets are needed.
- The efficiency of conversion of metabolizable protein (MP) to net protein (NP) is not well defined and deserves further study. In addition, little is known about the efficiency with which individual amino acids are used for protein deposition by beef cattle.

MINERALS, VITAMINS, AND WATER

- Research to update the calcium and phosphorus requirements would be beneficial. In particular, the role of diet type (e.g., high-roughage vs. high-grain) and phosphorus concentration on excretion route of phosphorus needs to be further evaluated.
- Research is needed to set a requirement for chromium for beef cattle in various production settings and physiological stages.
- Further evaluation of the cobalt requirement of ruminants is needed.
- The maximum tolerable concentration of selenium needs to be evaluated for practical diets provided to beef cattle under common feeding conditions.
- Newer data on the vitamin A requirements of beef cattle would be useful.
- Additional research on the vitamin requirements of beef cattle as influenced by stress (e.g., weaning, transport, disease) is needed.
- Among the water-soluble vitamins, additional data on the effects of supplementation of biotin, folic acid, pyridoxine, and thiamine at physiological and pharmacological levels would provide important information for ruminant production and for subsequent committees working to establish requirements and recommendations.
- Updated equations for predicting water intake by beef cattle in various physiological stages and production settings are needed.

FEED INTAKE

- Additional effort should be devoted to predicting feed intake by nonpregnant, pregnant, and lactating beef cows.
- The reliability of existing equations to predict feed intake by grazing ruminants is questionable, and further research is needed to develop equations that can be applied to beef cow-calf and stocker cattle operations based on grazed forages.

MAINTENANCE, GROWTH, REPRODUCTION, AND STRESS

- Additional research is needed to refine the energetic adjustments associated with changes in environmental temperature.
- Research is needed to more clearly define the effect of activity (e.g., grazing or other muscular activity) on maintenance requirements.
- Although current data suggest maintenance protein requirements might be modeled adequately, additional data to confirm this approach would be helpful.
- Additional research is needed to allow more accurate estimates of final shrunk body weight (BW), to identify transitory effects of previous plane of nutrition, gut fill, and anabolic implants on net energy and protein requirements for growth, and to provide more accurate estimates of the NEm and NEg concentrations of feeds.
- Improved equations are needed to predict NP deposited in gain.
- Research is needed to accurately define the relationship between shrunk BW and body condition score of beef cows for diverse biological types, mature sizes, and ages.
- Additional research is needed to define the maintenance energy requirements of lactating, primiparous females.
- The effects of nutrient availability and nutrient flux on hormonal regulation, follicular dynamics, embryo survival, and uterine function are not well established and merit further research, as does the role of nutrients and nutrient flow in conception and early embryo survival.
- Research to more clearly define nutrient requirements under conditions of stress associated with disease and acclimation of beef cattle to new environments would be beneficial.

COMPOUNDS THAT MODIFY DIGESTION AND METABOLISM

- Additional data on the effects of ionophores on energy utilization across a wide range of dietary conditions are needed.
- The effects of β-agonists on the estimates of final shrunk BW and energy and protein requirements of beef cattle need to be clarified.

ENVIRONMENT

- Accurate methods to predict methane production by beef cattle are needed, as are methods to predict manure composition and ammonia emissions from concentrated beef cattle feeding operations.

BYPRODUCT FEEDS AND FEED COMPOSITION

- Methods to accurately predict the energy value and estimates of the protein and carbohydrate fractions in byproduct feeds are needed.
- Estimates of the protein and carbohydrate fractions in feedstuffs used in nutrient requirement models are limited; efforts should be made to collect and summarize such data for use by future committees involved with revisions of nutrient requirement standards.
- Better information is needed on rates of digestion of protein and carbohydrate fractions for feed ingredients and methods for predicting rates of passage of ingredients.

COMPUTER MODELS FOR NUTRIENT REQUIREMENTS

- Mechanistic models would benefit from improved equations and complete data, as well as better integration of physiological and metabolic processes with modeling methods.
- A common platform for model development is needed to facilitate linking modeling efforts for large (beef and dairy) and small (sheep, goats, cervids, and camelids) ruminants.
- Development of a data repository for model development and evaluation is needed, including development of methods to ensure accuracy, error-free data, and outlier removal from the data repository.
- Development of web-based and tablet-based tools would be useful to assist stakeholders.

REFERENCE

Galyean, M. L. 2014. Invited Review: Nutrient requirements of ruminants: Derivation, validation, and application. *The Professional Animal Scientist* 30:125-128.

Appendix A

A User's Guide for Application of the Beef Cattle Nutrient Requirements Model

INTRODUCTION

The Beef Cattle Nutrient Requirements Model (BCNRM) of the *Nutrient Requirements of Beef Cattle, 8th Revised Edition*, was developed using the Microsoft® Excel® 2016 (version 16) electronic spreadsheet of the Microsoft Office 2016 to demonstrate the use of the empirical (ELS) and mechanistic (MLS) levels of solution in the calculations of the dietary supply and requirements of beef cattle for energy and nutrients, which are discussed in detail in Chapter 19 (Model Equations and Sensitivity Analyses). A basic understanding of the operations of the Microsoft® Excel® is necessary to execute the computer software. Similarly, an understanding of ruminant nutrition and knowledge of the underlying biological concepts presented in this report are essential for use and interpretation of the model's results.

Because of the many variables involved and judgments that must be made in choosing inputs and interpreting outputs, the National Academies of Sciences, Engineering, and Medicine make no claim for the accuracy of this software and the user is solely responsible for risk of use.

INSTALLATION

The BCNRM is designed to operate on a computer running Microsoft® Windows 7 or a later version, and Microsoft® Excel® 2010 or a later version. The setup file will install the software in the *[C:]\Program Files (x86)\National Academies\BCNRM2016* folder and in the *[C:]\Users\[User Name]\Documents\National Academies\BCNRM2016*. The *[C:]\Users\[User Name]\Documents\National Academies\BCNRM2016* folder contains the feed library files, the software backup, the batch template file, and user's preferences file. Computer administrator privileges are needed to install the BCNRM. The user should contact their system administrator when installation issues occur. Some files are created in the *[C:]\Users\[User Name]\Documents\National Academies\BCNRM2016* folder during the installation, but for restricted computer accounts, these files may not be created. In this case, the user will have to manually unzip them from the *[C:]\Program Files (x86)\National Academies\BCNRM2016\Default files.zip file into the [C:]\Users\[User Name]\Documents\National Academies\BCNRM2016* folder.

USING THE BEEF CATTLE NUTRIENT REQUIREMENTS MODEL

A shortcut to the BCNRM spreadsheet is created in the desktop and another shortcut is created in the *Start/Apps* screen (Windows 8 and later) or *Program Files* in the start menu (Windows 7 and earlier). After the software is installed, the user, if not previously authorized, will be requested to approve the use of active contents (i.e., programming subroutines) of the software, as shown in FIGURE 1. The user has to enable the active contents for proper functionality of the software.

Subsequently, the following will occur: (1) a splash screen, similar to that depicted in FIGURE 2, will appear, displaying information about the software and the *Terms of Use*, which the user must accept to continue using the software; (2) once the *Terms of Use* are accepted, the software will attempt to load the standard feed library (SFL; "*Beef Std 2016.FLF*" file) and the user-created feed library (UFL; "*Beef User 2016.FLF*") located in the *[C:]Users\[User Name]\Documents\National Academies\BCNRM2016* folder; (3) the software will read the user's saved preferences and make necessary modifications in the settings of the computer model; and lastly (4) the software will load the final simulation file and the message "Beef Cattle Nutrient Requirements Model 2016 is ready" will appear in the status bar at the lower left corner of the Microsoft® Excel®.

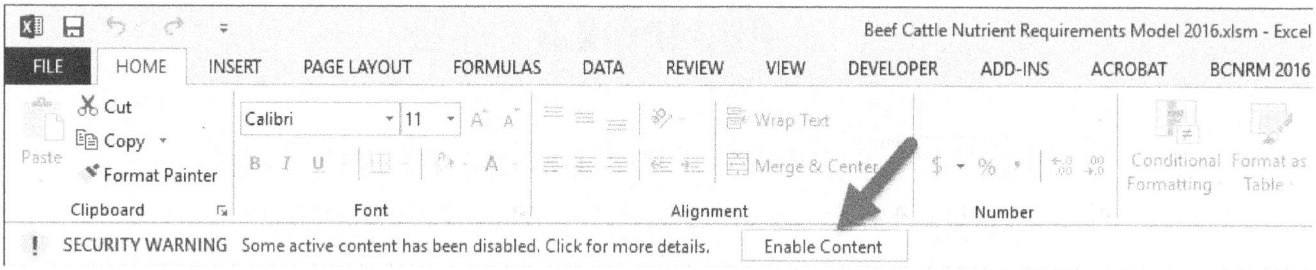

FIGURE 1 Microsoft® Excel® requesting the user to enable active content.

THE BEEF CATTLE NUTRIENT REQUIREMENTS MODEL RIBBON

The Microsoft® Excel® menu will be displayed across the top of the screen; below it is the complete ribbon as shown in FIGURE 3A. This ribbon has six distinct groups, as shown from left to right: *File* (FIGURE 3B), *Inputs* (FIGURE 3C), *Diet* (FIGURE 3D), *Feed Library* (FIGURE 3E), *Calculation* (FIGURE 3F), and *Advanced* (FIGURE 3G). The ribbon is usually located toward the right of the default ribbons of Microsoft® Excel®, as shown by the arrow in FIGURE 3A. Most of the interactive operations between the user and the software are done through the ribbon, and normal operation of the software is impossible without the ribbon. The ribbon is compatible with Microsoft® Excel® 2010, 2013, and 2016, and all icons in the ribbon are clickable.

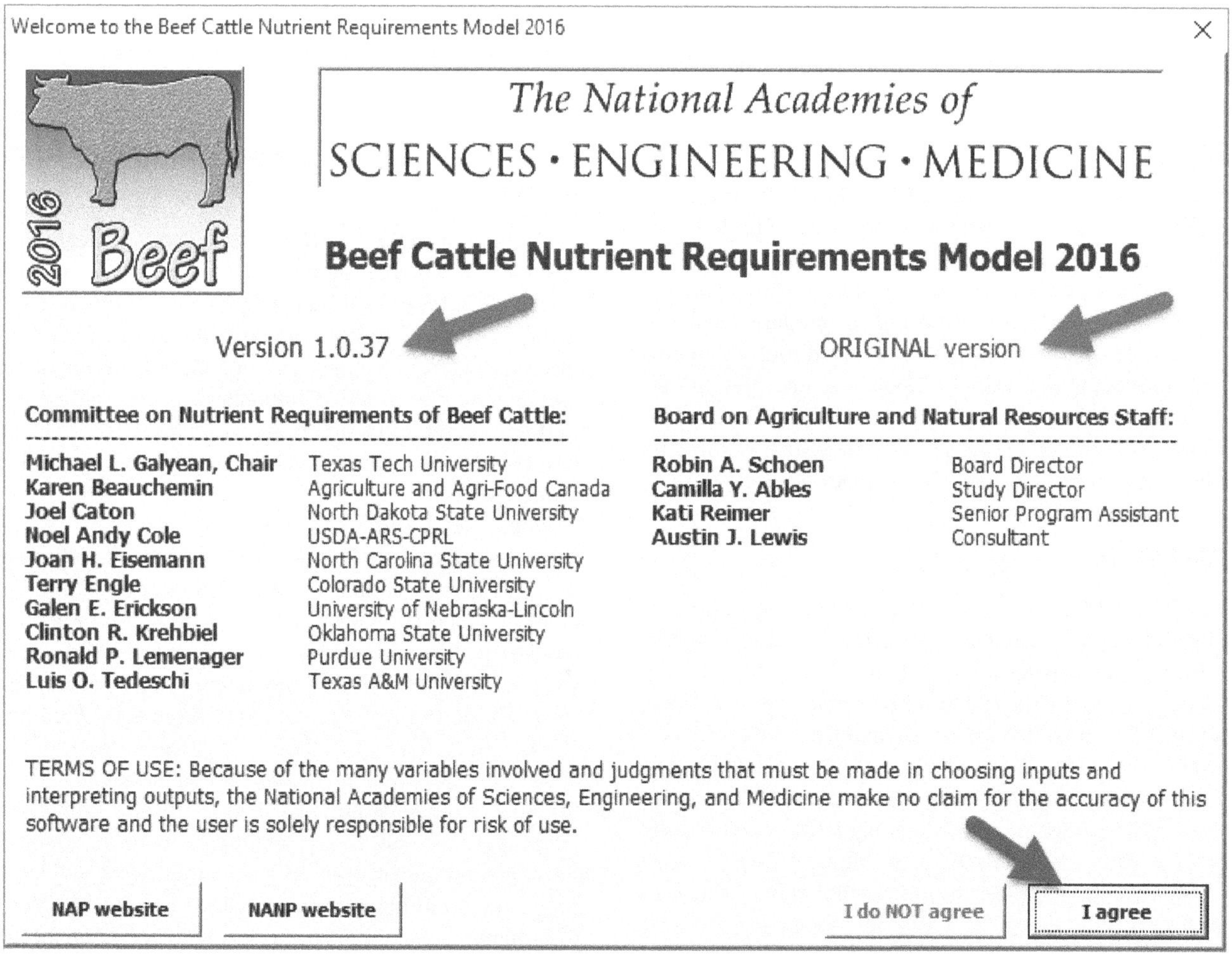

FIGURE 2 The welcome splash screen. The arrows are pointing at the major, minor, and build versions of the Beef Cattle Nutrient Requirements Model (e.g., 1.0.37), the timestamp (e.g., original or modified version), and the "*I agree*" button, which the user must click to continue loading the software. Clicking on the "*I do NOT agree*" button will terminate the session and close the software.

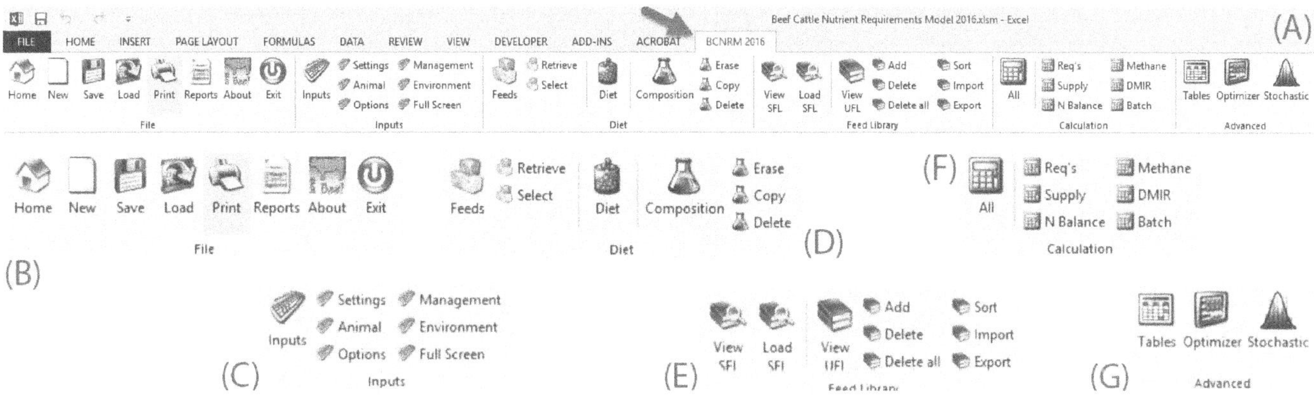

FIGURE 3 (A) The complete ribbon, (B) the *File* options, (C) the *Inputs* options, (D) the *Diet* options, (E) the *Feed Library* options, (F) the *Calculation* options, and (G) the *Advanced* options.

File Options

The *File* options are shown in FIGURE 3B and the detailed descriptions and operations of each button are discussed below. The screens shown are those for the Feedlot Case Study file 5; to retrieve that file, click on *Load* and choose the file from the list of case study files shown.

The Summary Screen (Home)

Clicking on the *Home* button will display the Home worksheet of the software from which the user can visualize important variable predictions and quickly make changes to the computer model inputs or settings. FIGURE 4 shows the Home worksheet from the Feedlot Case Study. This is the initial screen and it will appear frequently after certain operations for a quick review. *When returning to Home in the spreadsheet, be careful not to click on the Excel Home button in the top ribbon.* The top region of the Home worksheet (FIGURE 4) shows the file name (cell C2) and the creator's name, version used, and date created (cell C3). The center top region of the Home worksheet (FIGURE 4) has the calculations for metabolizable energy (ME) and metabolizable protein (MP) available, required, and the balance for five physiological stages (maintenance, pregnancy, lactation, growth, and body reserves). The right region (FIGURE 4) of the Home worksheet has the supply, requirement, and the balance of minerals, vitamins, and amino acids. The bottom portion of the right region has the nitrogen balance and the methane predictions. The center of the Home worksheet (FIGURE 4) contains the following:

(a) In the left column:
  i. The ME- and MP-allowable performance, the observed dry matter intake (DMI) and predicted DMI, water intake, and NDF intake; and

(b) The diet daily cost (cell I24), common nutrient contents (cells I25 to I36), N-to-S ratio (cell I37), Ca-to-P ratio (cell I38), physically effective neutral detergent fiber (peNDF; cell I39), and ruminal pH (cell I40). In the right column:
  i. The ruminally degradable protein (RDP) and ruminally undegradable protein (RUP);
  ii. Total digestible nutrients (TDN)- and RDP-allowable microbial crude protein (MCP);
  iii. Sources of MP (feed and ruminal microbes);
  iv. Ruminal nitrogen balance; and
  v. Diet energy values: TDN, digestible energy (DE), ME, NEma, and NEga.

The user can quickly make changes to the computer model inputs or settings in the left region of the Home worksheet (FIGURE 4). The user can do any of the following in cells D7 to D28 and cells D31, D32, D34, and D35 (shown by the arrow and box):

(a) Change the amount or percentage (cells C5 and C6) of each feed ingredient in the diet in cells D7 through D28;
(b) Set the ration intake (kg/d or lb/d) in cell D31. Type in a value and press enter;
(c) Scale the ration intake up or down in cell D32. After the intake amount of each diet ingredient (kg/d or lb/d) has been scaled up or down, the scale value changes to 100% again;
(d) Change the initial or final BW (cells D34 and D35, respectively). These are full (unshrunk) body weights;
(e) Change the solution type (1 = empirical level of solution, ELS; or 2 = mechanistic level of solution, MLS) in the dropdown box located in cell B37;
(f) Change the units (1 = metric, or 2 = imperial). As soon as the user changes the units, a query message with three options, similar to that depicted in FIGURE 5, will be shown. Choosing *YES* will change the values and units of interchangeable variables, choosing *NO* will change only the units, and values will remain the same, and choosing *CANCEL* will

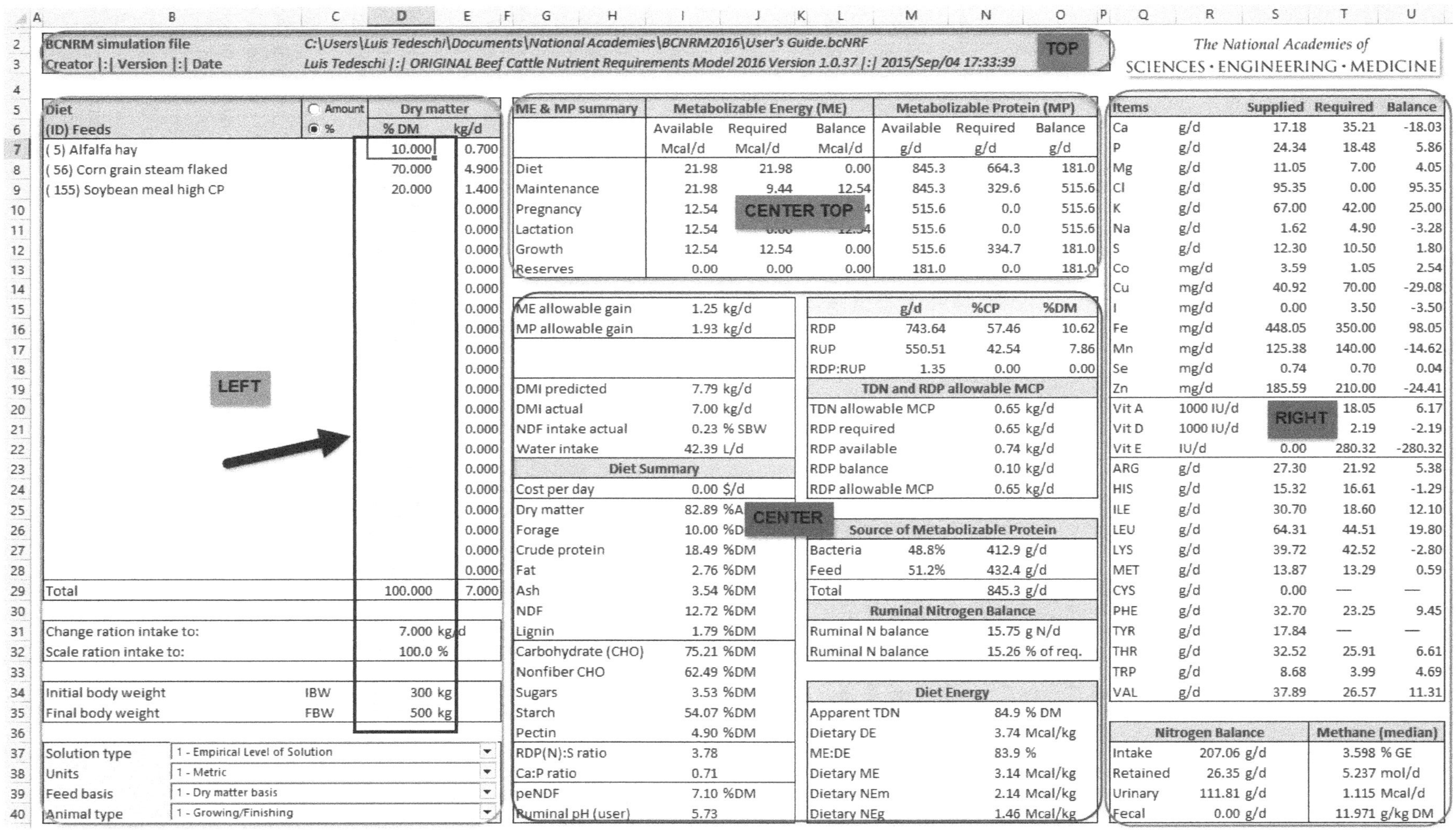

FIGURE 4 The Home worksheet.

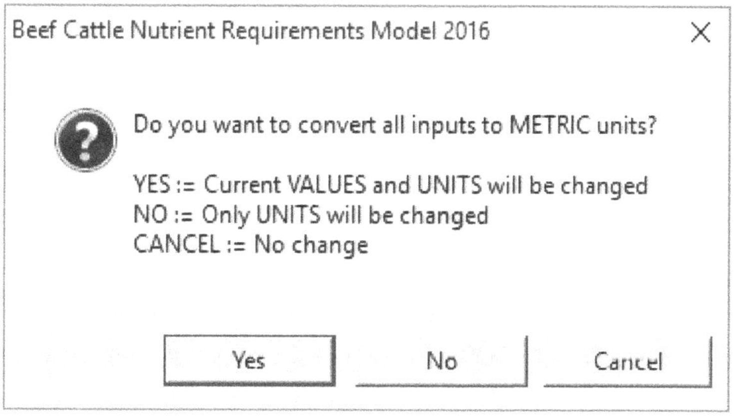

FIGURE 5 Query message for unit change.

abort the unit change process. The *YES* option is likely the one to be used most frequently. The *NO* option is generally used when the user enters values in imperial units when metric units are selected, or vice versa, and needs to match them.

(g) Change the feed basis (1 = dry matter, or 2 = as-fed basis); and
(h) Change the animal type (1 = growing/finishing, 2 = lactating cow, 3 = dry cow, or 4 = replacement heifer).

The Home summary can be printed by using the *Print* option. It is in the Reports section as shown in FIGURE 6.

Creating a New Simulation

Clicking on the *New* button will clear all user inputs and change the settings to their default values. The user's preferences are applied. This action cannot be undone, and unsaved simulations will be lost.

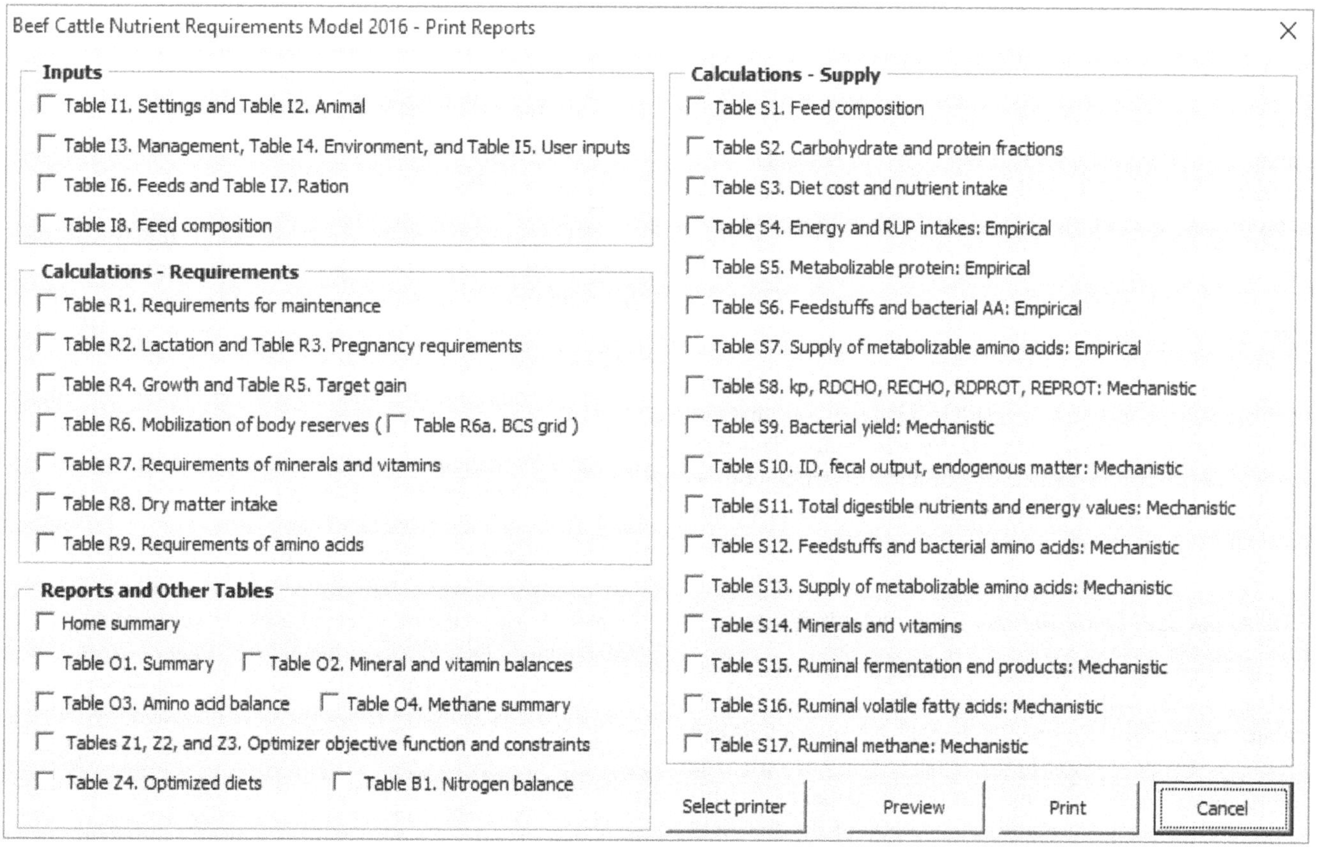

FIGURE 6 The Print Report options.

Saving and Loading Simulations

Clicking on the *Save* button will save current user inputs and settings into a BCNRM data file (*.bcNRF files). The bcNRF files can be shared, and they contain the data necessary to run the software in different computers. It saves current diet feed information but it does not save the feed library information per se (see Feed Library Options section).

Clicking on the *Load* button will retrieve the saved BCNRM data file (*.bcNRF files) into the software. Information about the creation of the saved .bcNRF simulation file will be shown to the user before effectively loading it, so the user can decide whether to continue or not with the loading process. This action cannot be undone and unsaved inputs will be lost.

Printing Tables and Reports

Clicking on the *Print* button will display the different reports that can be printed as depicted in FIGURE 6. The four buttons in this form are self-explanatory. It is recommended that the user select the printer and perform a preview before printing the reports. The printing date is added to the top, and the file name and version of the BCNRM are added to the bottom of each report. Reports are printed one by one; a combined printout will not be created.

Additional Reports

Clicking on the *Reports* button will display the Reports worksheet of the software. There are three tables in the Reports: Table O1 (summary), Table O2 (mineral and vitamin balances), and Table O3 (amino acid balance). Table O1 contains information similar to the Home worksheet. These tables can be printed using the *Print* option and selecting the appropriate tables in the Reports and Other Tables section (FIGURE 6).

Checking for Updates

Clicking on the *About* button will display a form similar to the welcome splash screen (FIGURE 2), except that the *Update* button is shown next to the *Website* button. This button will check for the availability of a newer version of the BCNRM software. If a newer version exists, the user will be prompted to download it.

Closing the Beef Cattle Nutrient Requirements Model

The *Exit* button will close the software, and unsaved inputs will be lost. This is the preferred method to close the software rather than closing Microsoft® Excel® because additional processes are performed before closing the software. The following will occur after the *Exit* button is clicked: (1) the updated UFL is saved; (2) a backup copy of the BCNRM spreadsheet is created; and (3) the user's preferences are saved. The backup copy does not have all the functionality of the BCNRM spreadsheet, and the old backup is always replaced as the user exits the program. Some of the user's preferences are saved; they include type of solution (MLS or ELS), units, diet feed basis (dry matter or as-fed basis), and feed input method (amounts or percentage).

Inputs Options

The *Inputs* options are shown in FIGURE 3C. Clicking on the *Inputs* button of the *Inputs* group will display seven tables in the Inputs worksheet. The user may view one table at a time by clicking on the respective button of the *Inputs* group (Settings, Animal, Management, Environment, and Options). Clicking on the *Full Screen* button of the *Inputs* group will hide or display the Microsoft® Excel® ribbons. This option is used to increase the visible area of the spreadsheet. The *Inputs* tables are listed below and depicted in FIGURE 7 and FIGURE 8. Information in FIGURE 7 (cells D4 to D7, D55 to D59, D64 to D75, and D80 to D91) and in FIGURE 8 (D12 to D31, D36 to D44, and L48 to S48) may be changed by the user.

(a) Table 1: Settings;
(b) Table 2: Animal;
  i. Table 2.1: Breeds, and
  ii. Table 2.2: Prediction of empty body fat (EBF) and mature full body weight (MW) from carcass traits;
(c) Table 3: Management and grazing;
(d) Table 4: Environment; and
(e) Table 5: Advanced options.

The Settings Options

The *Settings* options are in Table 1 of the Inputs worksheet (FIGURE 7). As discussed in Chapter 19, there are two levels of solution (1 = ELS and 2 = MLS). There are two unit systems (1 = metric and 2 = imperial) that can be interconverted. As soon as the user changes the units, a query message with three options similar to that depicted in FIGURE 5 will appear on the screen. Choosing *YES* will change the values and units of interchangeable variables, choosing *NO* will change the units but the values will remain the same, and choosing *CANCEL* will abort the unit change process. The *YES* option is likely the one to be used most frequently. The *NO* option is generally used when the user enters values in imperial units when metric units are selected, or vice versa, and needs to match them. Diet feeds can be entered either as dry matter or as as-fed basis. There are four animal classes (1 = growing/finishing, 2 = lactating cows, 3 = dry cows, and 4 = replacement heifers). These options can be changed in Table 1 of the *Inputs* (FIGURE 7) as well as in the *Home* screen (see the Summary Screen section).

Table 1. Settings

	Acronym	User	Unit	Metric	Unit	Description
Solution type		2	MLS			1:=Empirical, 2:=Mechanistic
Unit system		1	Metric			1:=Metric, 2:=Imperial (English)
Feed basis		1	DM			1:=Dry Matter, 2:=As-Fed
Animal class		1				1:=Grow/Finish, 2:=Lactating, 3:=Dry, 4:=Replacement

Table 3. Management and Grazing

Item	Acronym	User	Unit	Metric	Unit	Description
Feed additive		2		2		1:=None, 2:=Monensin, 3:=Lasalocid
Implant		2		2		1:=No, 2:=Yes
Grazing unit size	GU	0	ha	0.00	ha	Grazing unit size per animal
Days on pasture	DOP	0	days	0	days	Days of grazing
Available forage mass	FM	1026	kg/ha	1026.0	kg/ha	

Table 4. Environment

Item	Acronym	User	Unit	Metric	Unit	Description
Wind speed	WS	1.61	km/h	1.6	Km/h	
Temperature, previous	Tp	20.00	oC	20.0	oC	
Temperature, current	Tc	20.00	oC	20.0	oC	
RH, previous	RHp	65.00		65.0	%	
RH, current	RHc	65.00		65.0	%	
Storm exposure		1		1		1:=No, 2:=Yes
Lowest night temp.	LNT	20.00	oC	20.0	oC	with night cooling if LNT < 20 oC
Hair depth		2.54	cm	2.54	cm	
Hide		2		2		1:=Thin, 2:=Average, 3:=Thick
Hair coat		2		2		1:=No, 2:=Some, 3:=Mud lower body, 4:=Heavy
Cattle panting		1		1		1:=No, 2:=Rapid shallow, 3:=Open mouth
Mud depth		0.00	cm	0	cm	

Table 5. Advanced options

Item	Acronym	User	Unit	Metric	Unit	Description
Maintenance factor		0	%	0.00	%	Increases/decreases NE required for maintenance
Adjustments for required NEm		0		FALSE		Apply environmental adjustment factors (0:=No, 1:=Yes)
Adjustments for predicted DMI		0		FALSE		Apply environmental adjustment factors (0:=No, 1:=Yes)
Monensin adjustment for DMI		-3.00	%	0.970		Monensin effect on DMI
Implant adjustment for DMI		-6.00	%	0.940		Implant factor on DMI
Ruminal pH		0		0.00		0:=Use default equation
DE to ME efficiency		0	%	82.00	%	0:=Use default value (82%), 1:= Use Galyean et al. (2016)
Tabular ME and NE values for ELS		0		FALSE		0:=No (calculate from TDN), 1:=Yes (user input)
MCP yield (MCP/TDN, %)		0	%	0.00	%	0:=Use default equation. NRC (2000) was 13%
Desired animal ADG		1.5	kg/d	1.50	kg/d	Compare with least ME or MP allowable performance
MP to NP efficiency		0	%	49.20	%	0:=Use default value. NRC (2000) was 49.2%
Recycled ruminal N		0		0.00		0:=None, 1:=NRC (2000), 2:=R&K (2008), 3:=BCNRM (2016)

FIGURE 7 The Settings (Table 1), Management and Grazing (Table 3), Environment (Table 4), and Advanced options (Table 5) of the Inputs worksheet of the Beef Cattle Nutrient Requirements Model.

Table 2. Animal						
Item	**Acronym**	**User**	**Unit**	**Metric**	**Unit**	**Description**
Age		10		10	mo	Used to calculate tissue insulation
Age class (for DMI & WC)		1		1		1:=Calves, 2:=Yearlings
Sex		2		2		1:=Bull, 2:=Steer, 3:=Heifer, 4:=Cow
Body condition score	BCS	5		5		1:=Emaciated, 2:=Very thin, ... , 8:=Fat, 9:=Very fat
Initial body weight	IBW	300	kg	300	kg	Live or full (unshrunk) initial BW of the feeding period
Final body weight	FBW	500	kg	500	kg	Live or full (unshrunk) final BW of the feeding period
Mature body weight	MW	522	kg	522	kg	Current animal's full BW at reference animal's EBF
Reference animal's EBF		3		3		EBF: 1:=22%, 2:=25% (Std), 3:=27% (Sel), 4:=28% (Ch-)
Breed type		2		2		1:=Dual-purpose, 2:=*Bos taurus*, 3:=*Bos indicus*, 4:=Dairy
Breeding system		3		3		1:=Straightbred, 2:=2-Way X, 3:=3-Way X
Sire's breed		1		1		Angus
Maternal grandsire's breed		2		2		Braford
Maternal granddam's breed		3		3		Brahman
Days pregnant	DP	0		0	days	
Calf birth weight	CBW	30.8446	kg	30.8	kg	Recommended is 32.3 kg
Days since calving		0		0	days	
Relative milk yield	RMY	5		5		1:=Low, ..., 5:=Average, ..., 9:=High
Milk production	MY	0	kg/d	0.0	kg/d	User-inputted milk yield (MY)
Target calving age	TCA	30		30	mo	Age at first calving
Target calving interval	TCI	12		12	mo	Herd calving interval

Table 2.2. Prediction of empty body fat (EBF) and mature full (unshrunk) body weight (MW) from carcass traits						
Item	**Acronym**	**User**	**Unit**	**Metric**	**Unit**	**Description**
Hot carcass weight	HCW	0	kg	0.00	kg	
Fat thickness	FT	0	cm	0.00	cm	Backfat
Ribeye area	REA	0	cm^2	0.00	cm^2	*Longissimus dorsi* muscle area
Quality grade code	QG			0.00		4:=Select, 5:=Choice–, 6:=Choice, 7:=Choice+, 8:=Prime
% Prime	Pr			—		The sum of Pr, Ch+, Ch, Ch-, and Se has to be 100%
% high Choice	Ch+			—		The sum of Pr, Ch+, Ch, Ch-, and Se has to be 100%
% Choice	Ch			—		The sum of Pr, Ch+, Ch, Ch-, and Se has to be 100%
% low Choice	Ch-			—		The sum of Pr, Ch+, Ch, Ch-, and Se has to be 100%
% Select	Se			—		The sum of Pr, Ch+, Ch, Ch-, and Se has to be 100%
Predicted empty body fat (EBF) from carcass traits				0.00	% EBW	Guirroy et al. (2001) and Tedeschi et al. (2004)
Predicted empty body weight (EBW) from HCW				0.00		Garrett and Hinman (1969) and Tedeschi et al. (2004)
Predicted unshrunk MW at 27% EBF				0.00	kg	Guirroy et al. (2001)

Table 2.1. Breeds									
Code	**Breed**	**BE**	**L**	**CBW, kg**	**PKYD, kg**	**MkFat**	**MkProt**	**MkSNF**	**%FBW**
1	Angus	1	1.2	31.0	8.0	4	3.8	8.3	0.6
2	Braford	0.95	1.2	36.0	7.0	4	3.8	8.3	0.55
3	Brahman	0.9	1.2	31.0	8.0	4	3.8	8.3	0.55
4	Brangus	0.95	1.2	33.0	8.0	4	3.8	8.3	0.62
5	Braunvieh	1.2	1	39.0	12.0	4	3.8	8.3	0.62
6	Canchim	0.9	1.2	32.0	6.0	4	3.8	8.3	0.62
7	Charolais	1	1.2	39.0	9.0	4	3.8	8.3	0.57
8	Chianina	1	1.2	41.0	6.0	4	3.8	8.3	0.65
9	Devon	1	1	32.0	8.0	3.5	3.3	8.3	0.6
10	Galloway	1	1.2	36.0	8.0	4	3.8	8.3	0.6
11	Gelbvieh	1	1	39.0	11.5	4	3.8	8.3	0.55
12	Gir	0.9	1.2	32.0	10.0	4	3.8	8.3	0.65
13	Guzerat	0.9	1.2	32.0	5.0	4	3.8	8.3	0.65
14	Hereford	1	1	36.0	7.0	4	3.8	8.3	0.6
15	Holstein	1.2	1	43.0	43.0	3.5	3.3	8.3	0.55
16	Jersey	1.2	1	32.0	34.0	5.2	3.9	8.3	0.55
17	Limousin	1	1.2	37.0	9.0	4	3.8	8.3	0.6
18	Longhorn	1	1.2	33.0	5.0	4	3.8	8.3	0.6
19	Maine Anjou	1	1.2	40.0	9.0	4	3.8	8.3	0.6
20	Nellore	1	1.2	32.0	7.0	4	3.8	8.3	0.65
21	Piedmontese	1	1.2	38.0	7.0	4	3.8	8.3	0.6
22	Pinzgauer	1	1.2	38.0	11.0	4	3.8	8.3	0.6
23	Polled Hereford	1	1.2	33.0	7.0	4	3.8	8.3	0.6
24	Red Poll	1	1.2	36.0	10.0	4	3.8	8.3	0.6
25	Sahiwal	0.9	1.2	38.0	8.0	4	3.8	8.3	0.65
26	Salers	1	1.2	35.0	9.0	4	3.8	8.3	0.6
27	Santa Gertrudis	0.95	1.2	33.0	8.0	4	3.8	8.3	0.62
28	Shorthorn	1	1.2	37.0	8.5	4	3.8	8.3	0.6
29	Simmental	1.2	1	39.0	12.0	4	3.8	8.3	0.57
30	South Devon	1	1.2	33.0	8.0	4	3.8	8.3	0.6
31	Tarentaise	1	1.2	33.0	9.0	4	3.8	8.3	0.6
Breeding System		3	4	5	6	7	8	9	10
Straightbred		1	1.2	31.0	8.0	4	3.8	8.3	0.6
2-Way		0.975	1.2	33.5	7.5	4	3.8	8.3	0.575
3-Way		0.9625	1.2	32.3	7.8	4	3.8	8.3	0.575
User input									
Current		0.9625	1.2	32.3	7.8	4	3.8	8.3	0.575

Breed: 0.5Angus x 0.25Braford x 0.25Brahman

FIGURE 8 The Animal (Table 2) and two auxiliary tables (Breeds and EBF/MW) of the Inputs worksheet.

The Animal Inputs

The *Animal* inputs are in Table 2 of the Inputs worksheet (FIGURE 8). Each main *Inputs* table has seven columns: item's name (column B); acronyms (column C); user-entered value (column D); current unit (column E); user-entered value in metric (column F); metric unit (column G); and additional description of the item (column H). Tables 2.1 and 2.2 are auxiliary tables.

Table 2.1 (Breeds; FIGURE 8) contains the default values for breed effect on NE required for maintenance (BE), lactating effect on NE required for maintenance (L), calf birth weight (CBW), peak milk yield (PKYD), milk fat (MkFat), protein (MkProt), and solids nonfat (MkSNF) contents, and expected target pregnant weight for 31 breeds. These factors are discussed in more detail in Chapter 19. The user may also override these default values by entering preferred values in cells L48 to S48. The peak milk yield is used to predict milk yield given days in milk as discussed in Chapter 19. The user may, however, override the predicted milk yield from peak milk by directly entering milk yield in cell D29. In this case, peak milk will be ignored and the user-inputted milk yield will be used to compute energy and protein requirements for lactation.

Table 2.2 (EBF and MW; FIGURE 8) computes the mature (unshrunk) body weight (MW) at a given empty body fat (EBF) based on carcass traits (hot carcass weight, fat thickness, longissimus dorsi muscle [ribeye] area, and USDA quality grade) as discussed in Chapter 19. The user can enter the average expected USDA quality grade for a pen in cell D39 or enter the percentage of USDA Prime, high Choice, Choice, low Choice, and Select in cells D40 to D44. The calculated MW at the selected EBF (cell D19) is shown in cell F47. The user can adjust the calculated MW and enter it in cell D18.

The Management and Grazing Inputs

The *Management and Grazing* inputs are shown in Table 3 of the Inputs worksheet (FIGURE 7). These options are discussed in detail in Chapter 19. In this case, the animals are in drylots so the grazing unit size and days on pasture are zero.

The Environment Inputs

The *Environment* inputs are in Table 4 of the Inputs worksheet (FIGURE 7). These options are discussed in detail in Chapter 19.

The Advanced Options

The *Advanced* options are in Table 5 of the Inputs worksheet (FIGURE 7).

Maintenance Factor. The maintenance factor (cell D80) is used to increase or decrease NE required for maintenance (NEm). The user needs to pay careful attention to avoid double-adjusting NEm. Only choose one of the two ways to adjust NEm: cell L48 in Table 2 (see The Animal Inputs section) or cell D80 in Table 5.

Environmental Adjustments. Environmental adjustments for NE required for maintenance and DMI can be toggled on (1) or off (0) in cells D81 and D82, respectively.

Adjustments for DMI. The magnitude of DMI adjustments for monensin and implants can be changed in cells D83 and D84, respectively.

Ruminal pH. The user may enter the ruminal pH in cell D85; if zero is entered, the software will compute ruminal pH as discussed in Chapter 19.

The DE to ME Efficiency. The user may enter DE to ME efficiency in cell D86. The default is 82%.

Calculation of ME and NE for ELS. By default (cell D87 = 0), the ELS option of the software will compute ME, NEma, and NEga for each feed using the equations discussed in Chapter 19. The tabular values of ME, NEma, and NEga will be used if cell D87 is set to 1.

Microbial Crude Protein. When cell D88 is set to zero, the Equation 19-116 will be used to compute the MCP. The user may change the MCP calculation by entering the MCP/TDN coefficient in cell D88.

Desired Animal ADG. This option is used to compute the ration dry matter intake required (DMIR) for the average animal's daily gain (ADG) entered in this cell. The value of ADG entered in cell D89 is used by the *DMIR* option in the *Calculation* group (FIGURE 3F).

The MP to NP Efficiency. The calculation of MP to NP efficiency may be altered by the user in cell D90. If zero is entered, the software will compute the MP to NP efficiency as a function of shrunk body weight (SBW) as discussed in Chapter 19; if a negative number (e.g., −1) is entered, the software will compute the MP to NP efficiency based on NRC (1996, 2000). Otherwise, the software will use the value in cell D90 as the MP to NP efficiency.

Recycled Ruminal N. This option is used by the MLS calculation only; the ELS does not account for recycled ruminal N. There are four options for recycled ruminal N: option 0 uses no recycled ruminal N; option 1 uses the NRC (1985) equation adopted by NRC (1996, 2000); option 2 uses the urea N used for anabolism (UUA) that was developed by Reynolds and Kristensen (2008); and option 3 uses the UUA equation (Eq. 6-5) described in Chapter 6.

Diet Options

The *Diet* options are shown in FIGURE 3D. There are four components in the Diet group that are used to evaluate and formulate a diet: *Feeds* shows the feeds imported from the feed libraries (Table 6; FIGURE 9), *Diet* is used to enter amounts or percentage of diet feed ingredients (Table 7; FIGURE 9), *Feed Composition* is used to modify diet feed composition values (Table 8; FIGURE 10), and *Feed*

Table 6. Feed imports

#	ID	Feed library feed's name
1	2	*Alfalfa hay*
2	23	*Corn grain steam flaked*
3	57	Soybean meal high CP
4		
5		
6		
7		
8		
9		
10		
11		
12		
13		
14		
15		
16		
17		
18		
19		
20		
21		
22		
Count	3	

Table 7. Diet

Feed quantity as: ○ Amount ◉ Percentage

Feed name	% DM	kg/d	As-fed (AF) kg/d	As-fed (AF) %	Dry Matter (DM) kg/d	Dry Matter (DM) %
Alfalfa hay	10.000	0.700	0.804	9.52%	0.700	10.00%
Corn grain steam flaked	70.000	4.900	6.072	71.90%	4.900	70.00%
Soybean meal high CP	20.000	1.400	1.569	18.58%	1.400	20.00%
		0.000	0.000	0.00%	0.000	0.00%
		0.000	0.000	0.00%	0.000	0.00%
		0.000	0.000	0.00%	0.000	0.00%
		0.000	0.000	0.00%	0.000	0.00%
		0.000	0.000	0.00%	0.000	0.00%
		0.000	0.000	0.00%	0.000	0.00%
		0.000	0.000	0.00%	0.000	0.00%
		0.000	0.000	0.00%	0.000	0.00%
		0.000	0.000	0.00%	0.000	0.00%
		0.000	0.000	0.00%	0.000	0.00%
		0.000	0.000	0.00%	0.000	0.00%
		0.000	0.000	0.00%	0.000	0.00%
		0.000	0.000	0.00%	0.000	0.00%
		0.000	0.000	0.00%	0.000	0.00%
		0.000	0.000	0.00%	0.000	0.00%
		0.000	0.000	0.00%	0.000	0.00%
		0.000	0.000	0.00%	0.000	0.00%
		0.000	0.000	0.00%	0.000	0.00%
		0.000	0.000	0.00%	0.000	0.00%
Diet (sum or average)	100.000	7.000	8.445	100.00%	7.000	100.00%

Change ration intake to:	7.000 kg/d
Scale ration intake to:	100.0 %

FIGURE 9 The Feed imports (Table 6) and Diet components (Table 7) of the Inputs worksheet tables of the Beef Cattle Nutrient Requirements Model.

Table 8. Feed composition								
Locked	ID	Feed name	Cost, AF $/Tonne	Forage %DM	DM %AF	CP %DM	SP %CP	ADICP %CP
☑	5	Alfalfa hay	0	100	87.03	19.81	25.19	12.21
☑	56	Corn grain steam flaked	0	0	80.70142	8.480428	8.200536	4.110104
☐	155	Soybean meal high CP	0	0	89.23522	52.85348	44.10889	10.16993
Diet (sum or average)			0.00	10.00	82.89	18.49	30.55	8.44

FIGURE 10 The Feed composition (Table 8) of the Inputs worksheet of the Beef Cattle Nutrient Requirements Model, highlighting the row and column of the cell selected by the user. The arrow is pointing at the "locked" checkboxes used to prevent modification of feed composition by the software.

Library is used to view or select the feed library to be used and to import feeds into the user's library. Setting up a diet begins with retrieving feeds from the software feed library that have the most similar and complete chemical, biological, and physical composition values to the feeds used in the diet. Then, the user can click on *Composition* and replace the values retrieved from the feed library with their own feed analysis values.

Feed Retrieval from Feed Library

There are two methods to retrieve feeds from the feed libraries. The *first method* involves the following: (1) viewing the feed library in the feed library subgroup; (2) finding the feeds to be used in the diet and their associated feed IDs; (3) entering the feed IDs in cells W5 to W26 of Table 6 (FIGURE 9); and (4) clicking on the *Retrieve* button in the *Feeds* subgroup of the *Diet* group (FIGURE 3D). Diet feed numbers are shown in row W, and they are numbered from 1 to 22 in cells V5 to V26. If there are already feeds in the diet, the retrieval process will replace the feed composition in Table 8 (FIGURE 10) for all feeds that are not locked (cells AH5 to AH26; FIGURE 10).

During the retrieval process, the user is encouraged to lock the feeds (cells AH5 to AH26; FIGURE 10) after changing their composition values so that the feeds are protected and their composition values are not accidentally replaced by their original composition values in the feed library. For identification purposes only, locked feeds have a red background in Table 6 (FIGURE 9).

The *second method* involves using the *Select* button in the *Feeds* subgroup in the *Diet* group (FIGURE 3D). Clicking on this option will display the form in FIGURE 11.

The user may select feeds from the SFL and UFL using the dropdown boxes and lock/unlock feeds. *It is important to understand that the position of each feed is fixed, so feed #1 will always be retrieved and placed in the first feed position (row 5) and so forth.*

Diet Ingredients

After the feeds have been retrieved from the feed library, the user can click on *Diet* and enter the intake amount or percentage of each diet ingredient in Table 7 (cells AA5 to AA26; FIGURE 9). Whenever feeds are entered as percentage in Table 7 (cells AA5 to AA26; FIGURE 9), a ration intake amount must be entered in cell AA29 so that the intake amount of each diet ingredient can be calculated. Enter a value in cell AA29 to set the ration daily intake amount (kg/d or lb/d). To scale up or down the ration intake, change the value to the percentage desired in cell AA30. After the intake amount for each diet ingredient (kg/d or lb/d) has been scaled up or down, the scale value will revert to 100%. Amounts and percentages are calculated on dry-matter and as-fed bases and shown on the right side of the Table 7 (FIGURE 9).

Entering New Feeds and Changing Feed Composition

Users may enter their own feed composition values in Table 8 (FIGURE 10) or edit the composition values of a feed retrieved from a feed library as discussed in the Feed Retrieval from Feed Library section. During the retrieval process, the user is encouraged to lock the feeds (cells AH5 to AH26; FIGURE 10) after changing their composition so that feeds are protected and their composition is not replaced by their original composition values in the feed library. The feed ID must be greater than 9000 for user-entered feeds so that the software can identify them correctly.

Erasing a Feed from the Diet

The user can erase (i.e., remove) a feed from the diet by erasing its ID, name, and composition values in Table 8 (FIGURE 10) or by selecting any cell in the row of the chosen feed within Table 8 and clicking on the *Delete* button in the *Composition* subgroup in the *Diet* group (FIGURE 3D). The erased feed is unlocked automatically.

Adding or Deleting Feeds from the User-Created Feed Library

The *Copy* and *Delete* buttons in the *Composition* subgroup in the *Diet* group (FIGURE 3D) can be used to copy a feed from Table 8 (FIGURE 10) to the UFL and to delete a feed from the UFL, as explained in the next section.

Feed Library Options

The values included in the BCNRM software feed composition library are based on previously published values (NRC, 1996, 2000) and values in Table 18-1 (Means and Standard Deviations for the Composition Data of Feeds Commonly Used in Beef Diets) of the 8th Revised Edition of the *Nutrient Requirements of Beef Cattle*. A compositional value in the software feed composition library that has a value of zero does not necessarily indicate lack of the nutrient, but that reliable composition data were not available for inclusion in the library. Although these tabular values can provide useful data on nutrient composition of common feeds, the user is ultimately responsible for ensuring the accuracy of

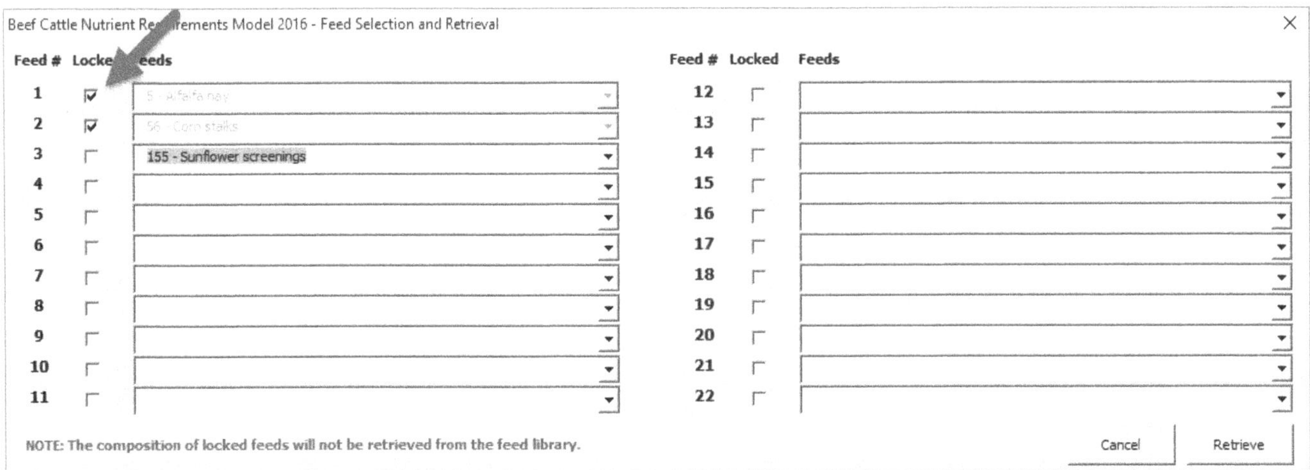

FIGURE 11 The *Select* form to retrieve feeds from feed libraries. The arrow is pointing at the "locked" checkboxes used to prevent modification of feed composition values by the software.

feed composition data used in model simulations. Thus, when available, users should input their own measured feed composition data.

The *Feed Library* options are shown in FIGURE 3E. It is through this ribbon group that the user can interact with the standard (i.e., SFL) and user-created (i.e., UFL) feed libraries. Feed library files (*.FLF) are located in the *[C:]Users\[User Name]\Documents\National Academies\BCNRM2016* folder. Feed libraries contain values that are averages obtained from commercial laboratories (i.e., DM, CP, protein solubility, NDF, ADF, lignin, fat, minerals, vitamins, amino acids) and published research (i.e., physically effective fiber, carbohydrate and protein digestion rates). The SFL (*Beef Std 2016.FLF*) is read only: it cannot be edited by the user; feeds cannot be added or deleted. Its only purpose is to allow the user to retrieve feed compositional information into the simulation file. The user may, however, load a different standard feed library by clicking on the *Load SFL* button of the *Feed Library* group. Note that the UFL cannot be opened as an SFL. A modified NRC (1996, 2000) feed library (*Beef Std 2000.FLF*) that can be used with the BCNRM is also available in the *[C:]Users\[User Name]\Documents\National Academies\BCNRM2016* folder.

After clicking on the *View SFL* button of the *Feed Library* option (FIGURE 3E), the user will be able to browse the SFL (Table F1). Only the first 10 feeds are shown in FIGURE 12. The user can scroll vertically or horizontally using the scroll bars and can annotate the feed ID (first column; FIGURE 12) to be able to retrieve feeds into the simulation file, as explained below.

Adding Feeds to the User-Created Feed Library

There are two methods to add feeds to the UFL. The first method is the direct way and requires typing the values into the UFL, whereas the second method is the indirect way and is done through Table 8 (Feed Composition).

Adding Feeds Directly. Clicking on the *View UFL* button in the *Feed Library* group (FIGURE 3E) will display Table F2 (FIGURE 13). To add one feed, click on the *Add* button in the *View UFL* subgroup (FIGURE 3E).

As an example, enter the information shown in FIGURE 13. The feed ID has to be greater than 9000, and the units of the feed components must be in metric, including the cost ($/tonne as-fed), ME, NEma, and NEga (Mcal/kg), and microminerals (mg/kg or ppm). To convert $/ton to $/tonne, multiply by 1.1023 and vice versa (e.g., $200/ton – $220.46/tonne), as 1 tonne = 1,000 kg and 1 ton = 2,000 lb. To convert Mcal/lb to Mcal/kg, multiply by 2.2046 and vice versa (e.g., 1.5 Mcal/lb = 3.3069 Mcal/kg).

Alternatively, the user may enter [sequentially] the feed ID numbers at the end of the UFL in the column BJ (e.g., cell BJ8 in FIGURE 13) to automatically add feed rows to the UFL.

Adding Feeds Indirectly. The user can add a feed and its complete compositional information to the UFL by clicking on *Composition*, then selecting a feed from Table 8 (i.e., select any cell in the row of the chosen feed within Table 8; FIGURE 10) and clicking on the *Copy* button of the *Composition* subgroup of the *Diet* group (FIGURE 3D). The new feed ID will be the current feed ID plus 9000. In case the new feed ID already exists in the UFL, the user would need to decide whether to replace it. To replace the existing feed in the UFL, the user should select *YES*. If the user does not want to replace the existing feed in the UFL, select *NO* and renumber the current feed ID (column AI) to a feed ID that does not exist in the UFL. Once the feed is copied into the UFL, the feed in the Diet (Table 8) will become locked (column AH) and the UFL will be shown. This option is very useful as the user can retrieve a feed from the SFL, replace

Table F1. Standard feed library

ID	Feed	IFN	Cost, $/Ton	Forage, %D	DM, %AF	CP, %DM
1	Alfalfa cubes		0	100	91.0	18.1
2	Alfalfa dehy		0	100	93.8	18.5
3	Alfalfa fresh		0	100	30.7	23.1
4	Alfalfa greenchop		0	100	40.5	23.1
5	Alfalfa hay		0	100	87.0	19.8
6	Alfalfa haylage		0	100	41.0	20.1
7	Almond hulls		0	100	89.2	5.5
8	Apple pomace		0	100	18.5	6.4
9	Bakery products		0	0	88.9	13.1
10	Barley grain		0	0	89.7	12.8

FIGURE 12 The first 10 feeds of the standard feed library (Table F1) of the Beef Cattle Nutrient Requirements Model.

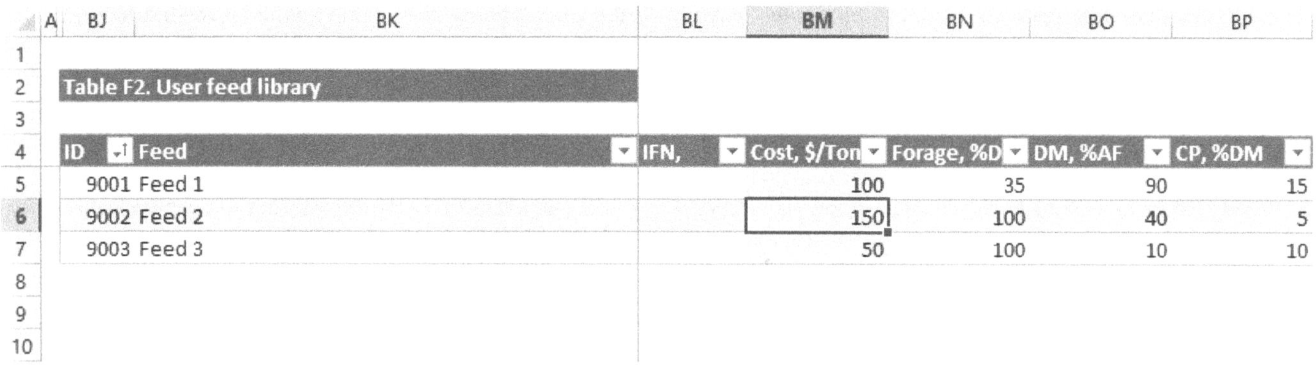

FIGURE 13 The first three feeds of the user-created feed library (Table F2) of the Beef Cattle Nutrient Requirements Model.

composition values with those available from their own analyses, and save it into the UFL for recurring use.

Deleting Feeds from the User-Created Feed Library

Similarly, there are two methods to delete feeds from the UFL. The first method is the direct way and requires selecting a feed in the UFL, and the second method is the indirect way and is done through Table 8 (Feed Composition).

Deleting Feeds Directly. Click on the *View UFL* button in the *Feed Library* option (FIGURE 3E) to display Table F2 (FIGURE 13). To delete a feed, select the feed you wish to delete permanently (remove from the UFL) by clicking on any column in the row of the feed ID. For instance, to delete Feed 2 shown in FIGURE 13, select the cell BK6 and click on the *Delete* button in the *View UFL* subgroup (FIGURE 3E). Select *YES* to delete the feed. This action cannot be undone.

Deleting Feeds Indirectly. The user can delete a feed from the UFL by clicking on *Composition* and selecting a feed in Table 8 (i.e., select any cell in the row of the chosen feed within Table 8; FIGURE 10) and then clicking on the *Delete* button of the *Composition* subgroup of the *Diet* group (FIGURE 3D). The feed will be deleted from the UFL, but it will remain in the diet (Table 8). To remove it from the diet, select a feed in Table 8 and click on the *Erase* button of the *Composition* subgroup of the *Diet* group (FIGURE 3D). These actions cannot be undone.

Deleting all Feeds from the User-Created Feed Library

Click on the *View UFL* button in the *Feed Library* option (FIGURE 3E) to display Table F2 (FIGURE 13). To delete all feeds permanently, click on the *Delete all* button in the *View UFL* subgroup (FIGURE 3E). Select *YES* to delete all feeds. This action cannot be undone.

Sorting Feeds in the User-Created Feed Library

Click on the *View UFL* button in the *Feed Library* option (FIGURE 3E) to display Table F2 (FIGURE 13). To sort feeds, click on the *Sort* button in the *View UFL* subgroup (FIGURE 3E). Feeds will be sorted ascendingly by their ID. Alternatively, the user may sort feeds from the STL or UFL by clicking on the dropdown arrow button in the feed library's column name (heading row) and selecting Sort A to Z or Sort Z to A.

Exporting the User-Created Feed Library

Click on the *View UFL* button in the *Feed Library* option (FIGURE 3E) to display Table F2 (FIGURE 13). To export the UFL, click on the *Export* button in the *View UFL* subgroup (FIGURE 3E). The complete UFL will be saved into an FLF file that can be shared among users.

Importing a User-Created Feed Library

Click on the *View UFL* button in the *Feed Library* option (FIGURE 3E) to display Table F2 (FIGURE 13). The user has two importation options: append or replace the current UFL. The append option will add the feeds from the external UFL to the end of the current UFL. The user must ensure that there are no duplicate feed ID values, as this will cause incorrect identification of feeds in the UFL. Note that SFL cannot be imported to the UFL.

Calculations Options

The *Calculations* options are shown in FIGURE 3F. Clicking on the *All* button in the *Calculation* group will display the requirement calculations (cell B2) and the supply calculations (cell AG2). All calculations are performed using the metric system.

The Requirement Calculations

Clicking on the *Req's* button in the *Calculation* group (FIGURE 3F) will only display the requirement calculations. There are nine tables in the requirement calculations: maintenance (Table R1, cell B4); lactation (Table R2, cell

B44); pregnancy (Table R3, cell B69); growth (Table R4, cell B85); target gain for replacement heifers (Table R5, cell B102); body reserves for mature cows (Table R6, cell B124); minerals and vitamins (Table R7, cell B151); dry matter intake, water intake, and dry matter required (Table R8, cell B171); and amino acids requirement (Table R9, cell B194). Auxiliary tables are also available for these tables and they are located on the right side of these tables. Although not recommended, the user can change the target pregnant weight and target calving weight (cells I104 to K106), the method to compute body reserves (cell L142), and the change in SBW/BCS (cell Q142).

The Supply Calculations

Clicking on the *Supply* button in the *Calculation* group (FIGURE 3F) will only display the supply calculations. The supply section is extensive and contains 17 tables for the ELS and MLS calculations, as listed below.

(a) Table S1: Feed composition (ELS and MLS, cell AG4);
(b) Table S2: Carbohydrate and protein fractions (ELS and MLS, cell AG31), and methane (ELS, cell BJ31);
(c) Table S3: Diet cost and nutrient intake (ELS and MLS, cell AG59);
(d) Table S4: Energy and ruminally undegradable protein (RUP) (ELS, cell AG87);
(e) Table S5: Metabolizable protein (ELS, cell 115);
(f) Table S6: Feedstuffs and bacterial amino acid (ELS, cell AG143);
(g) Table S7: Supply of metabolizable amino acids (ELS, cell AG171);
(h) Table S8: Fractional passage rate, ruminally degraded carbohydrate and protein, and ruminally escaped carbohydrate and protein (MLS, cell AG199);
(i) Table S9: Bacterial yield (MLS, cell AG227);
(j) Table S10: Intestinal digestibility, fecal output, and endogenous matter (MLS, cell AG255);
(k) Table S11: Total digestible nutrients and energy values (MLS, cell AG283);
(l) Table S12: Feedstuff and bacterial amino acids (MLS, cell AG311);
(m) Table S13: Supply of metabolizable amino acids (MLS, cell AG339);
(n) Table S14: Minerals and vitamins (ELS and MLS, cell AG367);
(o) Table S15: Ruminal fermentation end products (MLS, cell AG395);
(p) Table S16: Ruminal volatile fatty acids (MLS, cell AG423); and
(q) Table S17: Ruminal methane (MLS, cell AG451).

Although not recommended, the user may change the coefficients and factors that will modify the supply calculations such as microbial composition (cells BH89 to BH97), bacterial amino acid composition (cells AZ172 to BB183), bacterial yield factors (cells BC241 to BC247), and intestinal digestibility coefficients (cells BC256 to BC265). These coefficients and factors revert to their default values when a new simulation is initiated (see Creating a New Simulation section). In some tables, incorrect or infeasible calculations will be marked in red font.

Nitrogen Balance Calculation

Clicking on the *N Balance* button in the *Calculation* group (FIGURE 3F) will display the nitrogen balance calculations (Table B1, cell DA2), as discussed in Chapter 19, and depicted in FIGURE 14. Calculations are performed for ELS and MLS concurrently.

NUTRIENT BALANCE SECTION

Table B1. Nitrogen fluxes and balance

N Fluxes, g/d	ELS	MLS
Intake N	*207.1*	*207.1*
Metabolizable N	*135.2*	*132.4*
Net N required		
Growth	26.3	26.3
Pregnancy	0.0	0.0
Lactation (milk)	0.0	0.0
Total	26.3	26.3
Urinary N	*111.8*	*123.9*
Fecal N	*68.9*	*56.8*

N balance, g/d	ELS	MLS
Intake	207.1	207.1
Urinary	111.8	123.9
Milk	0.0	0.0
Fecal	68.9	56.8
Balance	26.3	26.3

Ammonia-N, g/d	ELS	MLS
Feedlot (1)	158.5	158.5
% of excreted N	87.7%	87.7%

(1) Only valid for feedlot animals with *ad libitum* access to feed (Todd et al., 2013).

FIGURE 14 Nitrogen balance calculations for the Empirical Level of Solution (ELS) and Mechanistic Level of Solution (MLS).

Methane Calculations

Clicking on the *Methane* button in the *Calculation* group (FIGURE 3F) will display the methane calculations (cell BD31), as discussed in Chapter 19, and depicted in FIGURE 15. Calculations for the MLS are shown in Table S17 (cell AG451), and those for the ELS are shown in cell BD31. For the ELS, the BCNRM reports the median of the prediction of methane production by different empirical equations (Chapter 19).

Although not recommended, the user can change the coefficients that will modify the calculations of methane in cells BO32 to BO35 (see top arrow on the right side of FIGURE 15). A comparative table for the MLS and ELS calculations is shown on the right side of FIGURE 15 (middle arrow). A table on the left side of FIGURE 15 (black arrow) has some descriptive statistics of the methane prediction by the ELS option. The graph on the right side of FIGURE 15 (bottom arrow) shows the relative position of methane prediction for high, medium, and low levels of forage in the diet.

Dry Matter Required Calculation

Clicking on the *DMIR* button in the *Calculation* group (FIGURE 3F) will display the dry matter required calculation (Table R8, cell H171). The user must enter the desired animal ADG in Table 5 (*Options* button in the *Inputs* group, FIGURE 3C). The DMIR uses the goal-seek tool of Microsoft® Excel® to change the actual DMI until the first-limiting ME- or MP-allowable gain matches the desired animal ADG. FIGURE 16 will be displayed at the end of the goal-seek process, and the user can then decide to keep the calculated DMIR or the actual (original) DMI. Subsequent calculations of DMIR may enhance the predicted DMIR.

Running Multiple Simulations as Batches

Clicking on the *Batch* button in the *Calculation* group (FIGURE 3F) will initiate the batch calculation mode. The batch calculation mode is used to perform multiple consecutive calculations by reading the input information from a

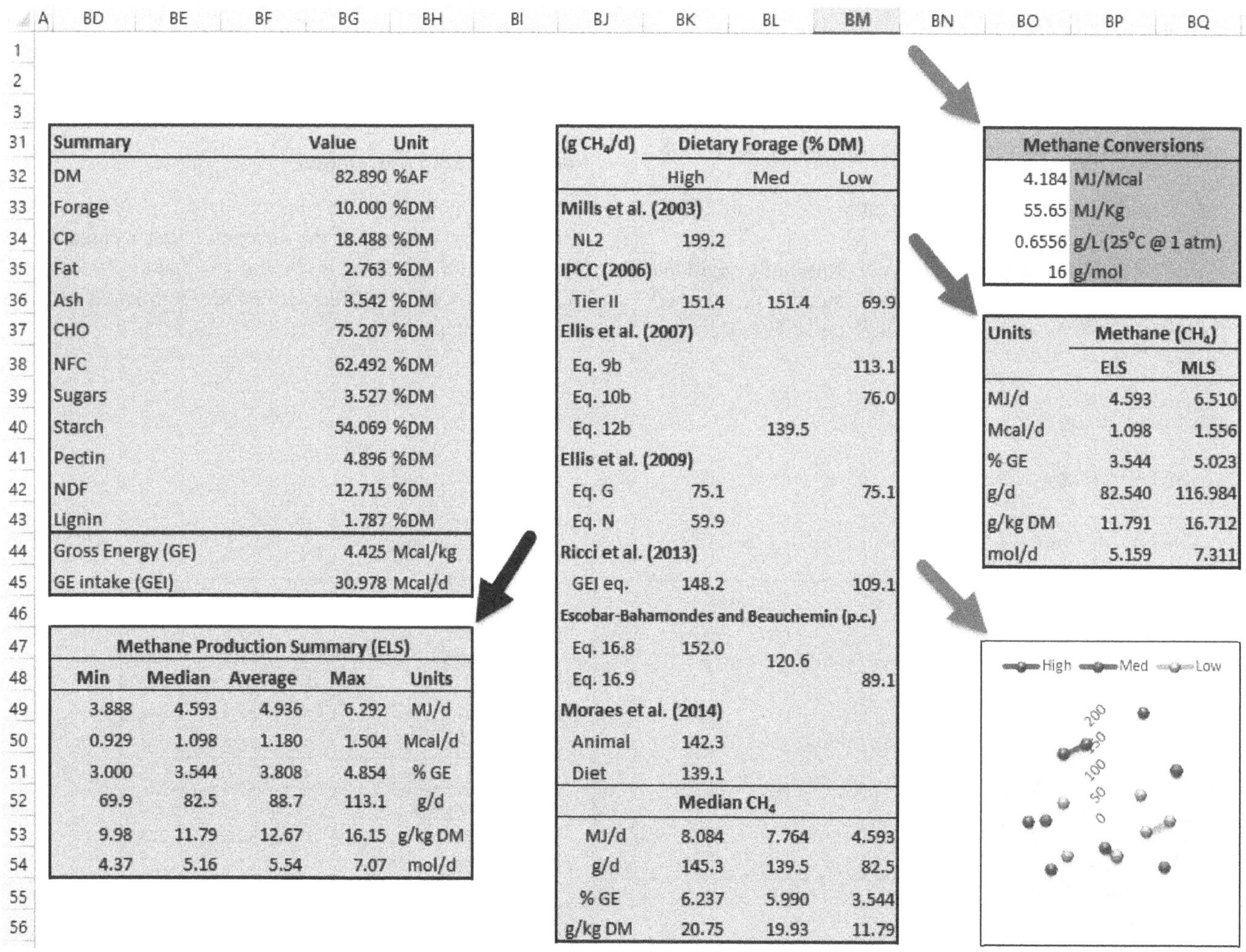

Summary	Value	Unit
DM	82.890	%AF
Forage	10.000	%DM
CP	18.488	%DM
Fat	2.763	%DM
Ash	3.542	%DM
CHO	75.207	%DM
NFC	62.492	%DM
Sugars	3.527	%DM
Starch	54.069	%DM
Pectin	4.896	%DM
NDF	12.715	%DM
Lignin	1.787	%DM
Gross Energy (GE)	4.425	Mcal/kg
GE intake (GEI)	30.978	Mcal/d

Methane Production Summary (ELS)				
Min	Median	Average	Max	Units
3.888	4.593	4.936	6.292	MJ/d
0.929	1.098	1.180	1.504	Mcal/d
3.000	3.544	3.808	4.854	% GE
69.9	82.5	88.7	113.1	g/d
9.98	11.79	12.67	16.15	g/kg DM
4.37	5.16	5.54	7.07	mol/d

($g\ CH_4/d$)	Dietary Forage (% DM)		
	High	Med	Low
Mills et al. (2003)			
NL2	199.2		
IPCC (2006)			
Tier II	151.4	151.4	69.9
Ellis et al. (2007)			
Eq. 9b			113.1
Eq. 10b			76.0
Eq. 12b		139.5	
Ellis et al. (2009)			
Eq. G	75.1		75.1
Eq. N	59.9		
Ricci et al. (2013)			
GEI eq.	148.2		109.1
Escobar-Bahamondes and Beauchemin (p.c.)			
Eq. 16.8	152.0	120.6	
Eq. 16.9			89.1
Moraes et al. (2014)			
Animal	142.3		
Diet	139.1		
Median CH_4			
MJ/d	8.084	7.764	4.593
g/d	145.3	139.5	82.5
% GE	6.237	5.990	3.544
g/kg DM	20.75	19.93	11.79

Methane Conversions	
4.184	MJ/Mcal
55.65	MJ/Kg
0.6556	g/L (25°C @ 1 atm)
16	g/mol

Units	Methane (CH_4)	
	ELS	MLS
MJ/d	4.593	6.510
Mcal/d	1.098	1.556
% GE	3.544	5.023
g/d	82.540	116.984
g/kg DM	11.791	16.712
mol/d	5.159	7.311

FIGURE 15 Methane calculations for the Empirical Level of Solution (ELS).

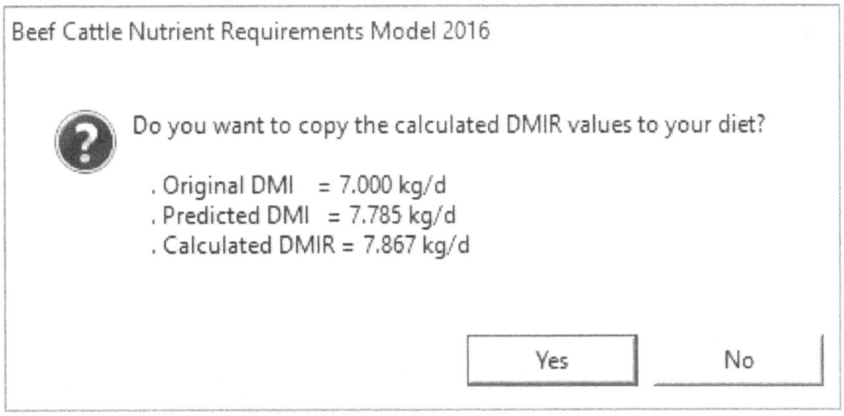

FIGURE 16 Dry matter intake required (DMIR) result.

batch file (Microsoft® Excel® file) and saving the calculations (i.e., outputs) in another Microsoft® Excel® file. A blank batch file is provided in the *[C:]\Users\[User Name]\Documents\National Academies\BCNRM2016* folder with an example. Each column (after the B column) in the batch file is a complete BCNRM simulation and the simulation ID (row 1) must be consecutive and unique.

Advanced Options

The *Advanced* options are shown in FIGURE 3G. The *Tables* and *Optimizer* options require the Microsoft® Excel® Solver Add-In to be installed and loaded.[1] The *Stochastic* option requires the Palisade @Risk 6.3 or later (Palisade Corp., Ithaca, NY) to be installed and loaded.

Table Generator

Clicking on the *Tables* button in the *Advanced* group (FIGURE 3G) will display the Table worksheet. The message "Solver was found and loaded" will appear on the lower left corner of Microsoft® Excel® if the Solver Add-in was loaded successfully. This advanced option was used to create the tables in Chapter 20 (Tables of Nutrient Requirements).

There are six main tables in the Table worksheet: requirements and diet evaluation for growing and finishing cattle (Tables T1.1 and T1.2); requirements and diet evaluation for replacement heifers and dry cows (Tables T2.1 and T2.2); and requirements and diet evaluation for lactating cows (Tables T3.1 and T3.2). At the bottom of Tables T1, T2, and T3 are three buttons in row 36 that are used to solve or to copy these tables. The tables on the left side of the Table worksheet are auxiliary tables used during the optimization process. Information in the yellow-shaded cells may be changed by the user. The user can alter the objective function in cell D16, and the appropriate equation to predict DMI should be changed in cell A20, depending on which table is used (T1, T2, or T3) so that DMI is predicted for the correct physiological stage. The table optimization uses the ELS and, during the optimization process, only the first feed's composition of the diet (Table 8) is changed. Although not recommended, the user may set the table optimization to use the current composition of feed #1 (set cell B18 to true).

Optimizing Diets

Clicking on the *Optimizer* button in the *Advanced* group (FIGURE 3G) will display the Optimizer worksheet. The message "Solver was found and loaded" will appear on the lower left corner of Microsoft® Excel® if the Solver Add-in was loaded successfully.

There are four tables in the Optimizer worksheet: objective function (Table Z1); constraints for diet ingredients (Table Z2); constraints for performance, required and supplied nutrients (Table Z3); and optimized diets (Table Z4). These tables can be printed by clicking on the *Print* button in the *File* group. Information in the yellow-shaded cells may be changed by the user. As a rule of thumb, the user should avoid restrictive constraints in the beginning, and start optimizations with relaxed constraints. Once the optimizer finds a feasible solution, and if needed, the user can set more restrictive constraints. The following steps should be followed for successful optimization:

(a) Set the objective function in cell E4 in Table Z1. The default is 1.

[1] Support to load the Solver Add-In for Microsoft® Excel® can be found in the following hyperlinks:

Excel® 2007: http://office.microsoft.com/en-us/excel-help/load-the-solver-add-in-HP010021570.aspx?CTT=1,

Excel® 2010: http://office.microsoft.com/en-us/excel-help/load-the-solver-add-in-HP010342660.aspx?CTT=1, and

Excel® 2013 and Excel® 2016: https://support.office.com/en-US/Article/Load-the-Solver-Add-in-ec994cd0-a396-4bf3-a5dd-feda369cef37?ui=en-US&rs=en-US&ad=US.

(b) Set the minimum and maximum for each diet ingredient in Table Z2 (cells J5 to K26). The default is zero for minimum and 9999 for maximum.
(c) Set the minimum and maximum for each constraint in Table Z3 (cells C16 to D34). Make sure the values match the units in column A.
(d) Use the buttons in row 34, below Table Z2, to manage the optimization:
   i. Click on the *Optimize* button to start the optimization.
   ii. Click on the *Clear diet* button to erase one saved diet or all the saved, optimized diets listed in Table Z4.
   iii. Click on the *Retrieve diet* button to copy a saved, optimized diet from Table Z4 to the solution range (cells H5 to H26) in Table Z2.
   iv. Click on the *Save solution* button to save the current solution to Table Z4. The solution will be saved in the next available slot; if no empty slot is available in Table Z4, the user will be asked what Diet ID should be replaced. The user can save up to 10 optimized solutions.
   v. Click on the *Accept solution* button to copy the current solution to the diet table (Table 7) in the Inputs worksheet.

Stochastic Simulation

Clicking on the *Stochastic* button in the *Advanced* group (FIGURE 3G) will display the Stochastic worksheet. The message "@Risk was found and loaded" will appear on the lower left corner if Palisade @Risk Add-in was successfully loaded. This advanced option was used to create the sensitivity analysis in Chapter 19. There are four tables in the Stochastic worksheet: mean, SD, and correlation matrix of animal weight and intake (Table K1); mean, SD, and correlation matrix of environment factors (Table K2); mean, SD, and correlation matrix of feed composition (Table K3); and stochastic results (Table K4). FIGURE 17 depicts Tables K1, K2, and K4. Information in the cells F and I (2 to 4), C to G (8 and 9), C to G (12 to 16), C to I (20 and 21), and C to G (24 to 30) may be changed by the user. All variables are assumed to follow the Gaussian (i.e., normal) distribution. Additional analysis can be performed with the Palisade @Risk Add-in. The Palisade @Risk Add-in is a third-party, commercial Add-in that is not included with the BCNRM.

FEEDLOT CASE STUDY

This feedlot case study is the same as the one presented in the *Nutrient Requirements of Beef Cattle*, 7th revised edition and update (NRC, 1996, 2000). The purpose of this study is to compare the predictions of the BCNRM and the NRC (1996, 2000) computer model.

Settings, Animal, Management and Grazing, Environment, and Advanced Inputs

Begin the tutorial by opening the BCNRM; agree with the *Terms of Use*; click on the *BCNRM 2016* tab (if not automatically selected); click on the *New* button of the *File* group (FIGURE 3B); and click on the *Inputs* button of the *Inputs* group (FIGURE 3C).

This case study is of a 20,000-head-capacity western Canada feedlot. Cattle are fed in open dirt lots surrounded by windbreaks. Typical pens contain 250 head. The basal diet is dry-rolled barley and barley silage. Data from closeouts will be used to adjust the model so it predicts accurately for that feedlot, and then inputs will be changed to answer the questions. The data set is from 1969 Hereford × Charolais crossbred steers fed in 8 pens in the fall, with an initial full weight of 872 lb and final weight of 1,337 lb, with an average grade of Canadian AA. The cattle received an estrogenic implant and were fed an ionophore. The average full weight during the feeding period was 1,104 lb, with an ADG of 3.48 and conversion of 6.98 lb DM/lb gain. The average daily dietary DMI was 5 lb coarsely chopped barley silage, 19 lb coarsely rolled barley grain, and 0.3 lb minerals. The available feed analysis indicated that the barley silage contains 48.7% NDF with 65% estimated to be peNDF, 10.4% CP, 3% fat, and 8% ash; and barley grain was 19% NDF with 34% estimated to be peNDF, 13% CP, 2.1% fat, and 3% ash. The average wind speed in the pens was 5 mph, the previous month's average temperature was 40°F, and the average temperature during the feeding period was 30°F. Other inputs were average hide thickness, hair depth of 0.2 inch (typical of early summer-fall; 0.5 inch is typical of winter), and average hair coat condition is clean and dry.

Settings (Table 1). Confirm that solution type is 1 (ELS), unit system is 2 (Imperial), feed basis is 1 (DM), and animal class is 1 (growing/finishing cattle). It is often practical to adjust the diet until balanced with the ELS, and then evaluate it with the MLS to get the predicted feed net energy values and amino acid balances, based on actual feed analysis for carbohydrate and protein fractions.

Animal (Table 2). Enter the following information in this table:

(a) Age (cell D12) = 14 mo; this value influences tissue insulation.
(b) Age class (cell D13) = 2; this value influences expected DMI.
(c) Sex (cell D14) = 2; this value influences maintenance requirement.
(d) Body condition score (cell D15) = 5; see Chapters 3 (Energy Terms and Concepts) and 13 (Reproduction) for a detailed discussion of the 1 to 9 condition scoring system used and its biological basis. The choices are 1 through 9 (1 = emaciated, 5 = moderate, 9 = very fat). Body condition score (BCS) is used to

Is stochastic simulation on? (FALSE=default)	TRUE	# of Iterations:	5000
Use predicted DMI?	TRUE	Initial seed:	123456
Create summary report	FALSE	pDMI (1, 2, or 3)	2

pDMI
9.29
1 = iSBW
2 = NEm
3 = pDMI

Table K1. Animal weight and intake

	BCS	IBW	FBW	MBW	DMI
Mean	5	300	300	550	5.4
SD	0.5	30.0	30.0	55.0	0.5
RiskNormal	5	300	300	550	5.4
@RISK Correlations	BCS in C10	IBW in D10	FBW in E10	MBW in F10	DMI in G10
BCS in C10	1				
IBW in D10	0	1			
FBW in E10	0	0.7660425	1		
MBW in F10	0	0	0.612834	1	
DMI in G10	-0.3830212	0.3830212	0.3830212	0.3830212	1

Table K2. Environment factors

	WS	Tp	Tc	RHp	RHc	LNT	Mud
Mean	20.00	20.00	20.00	65.00	65.00	10.00	10.00
SD	4.00	4.00	4.00	13.00	13.00	2.00	2.00
RiskNormal	20.00	20.00	20.00	64.86	64.86	10.00	10.00
@RISK Correlations	WS in C22	Tp in D22	Tc in E22	RHp in F22	RHc in G22	LNT in H22	Mud in I22
WS in C22	1						
Tp in D22	-0.4496487	1					
Tc in E22	-0.2697892	0.7194378	1				
RHp in F22	-0.2697892	-0.4496487	-0.4496487	1			
RHc in G22	-0.2697892	-0.4496487	-0.4496487	0	1		
LNT in H22	0	0	0	0	0	1	
Mud in I22	0	0	0	0	0	0	1

Table K4. Stochastic results

	Output	Mean	SD	90% Prediction Lower	90% Prediction Upper
Metabolizable energy					
Required for maintenance	11.0	7.60	0.57	6.67	8.55
Required for pregnancy	0.00	0.00	0.00	0.00	0.00
Required for lactation	0.00	0.00	0.00	0.00	0.00
Required for growth	-0.01	10.77	2.71	6.53	15.42
Allowable gain/milk	0.00	1.38	0.35	0.84	1.99
Supplied	11.00	18.37	2.72	14.13	23.11
Metabolizable Protein					
Required for maintenance	329.63	265.44	18.74	234.06	295.73
Required for pregnancy	0.00	0.00	0.00	0.00	0.00
Required for lactation	0.00	0.00	0.00	0.00	0.00
Required for growth	47.75	244.49	82.69	118.23	389.02
Allowable gain/milk	0.00	0.75	0.26	0.36	1.22
Supplied					
Undegraded feed	185.35	225.15	56.60	139.36	323.70
Microbial protein	192.03	284.78	37.56	226.07	350.73
Ruminal balance					
Nitrogen	94.01	95.82	20.82	63.93	131.67
RDP	0.53	0.52	0.12	0.33	0.73
Methane					
Mcal/d	1.43	1.58	0.20	1.27	1.93
g/d	107.70	118.91	15.00	95.17	145.10
g/kg DM	20.03	20.02	0.52	19.21	20.92
Dry matter intake		5.96	0.86	4.62	7.44

Copy diet average composition to Table K3	Set inputs before start	Set inputs after start	Start simulation	Get results

FIGURE 17 Selected tables (K1, K2, and K4) of the Stochastic worksheet of the Beef Cattle Nutrient Requirements Model.

describe tissue insulation, the potential for compensatory growth in growing cattle, and energy reserves in cows.

(e) Initial body weight (cell D16) = 872.
(f) Final body weight (cell D17) = 1,337. This is the unshrunk final body weight that best represents the group being fed. The average between initial and final weights is a major determinant of DMI, maintenance, and growth requirements.
(g) Mature body weight (cell D18) = 1337; this is the expected average unshrunk weight at the USDA grade selected in cell D19. For cows, replacement heifers, or breeding bulls, enter the expected mature weight at BCS 5. The weight that best corresponds to the cattle in question based on the user's experience for the type of growing animal, implant strategy, and ration should be entered. A general guide is that the finishing weight should be reduced by 50 to 75 lb if diets that contain more than 70% grain are fed continuously after weaning or if anabolic steroids are not used. Weight at the target grade should be increased by 50 to 75 lb if animals are grown at a slow rate for an extended period or if they are implanted with estrogen in combination with trenbolone acetate. More information in Chapters 12 (Growth) and 14 (Compounds That Modify Digestion and Metabolism).
(h) Reference animal's EBF (cell D19) at slaughter = 3; choices are 1 (22% body fat), 2 (USDA Standard or Canadian A, which are related to 25% body fat), 3 (USDA Select or Canadian AA, which are related to 27% body fat), and 4 (USDA Choice or Canadian AAA, which are related to 28% body fat). The software uses this grading system to identify the standard reference weight. The standard reference weight is divided by the finished weight, and this result is multiplied by the actual weight. These calculations provide the weight used in the equations that compute net energy and protein in the gain.
(i) Breed type (cell D20) = 2.
(j) Breeding system (cell D21) = 2; choices are 1 (straightbred), 2 (two-way crossbred), and 3 (three-way crossbred). Cell D22 is used for animal breed if straightbred, cells D22 and D23 are used if describing two-way crossbred, and cells D22, D23, and D24 are used when describing three-way crossbred. Breeding system for growing cattle influences maintenance energy requirement and predicted DMI.
(k) Breed cells (D22 and D23) = 14 (dam's breed is Hereford) and 7 (sire's breed is Charolais), respectively; valid breed codes are shown in Table 2.1. Stored breed values are used to determine maintenance energy requirements and defaults for calf birth weight and peak milk production. See Chapter 19 for the biological basis for these breed adjustments. Since this is a two-way crossbred (cell D21), the software will ignore the values in cell D24, which is used only for three-way crossbreeding system.
(l) Days pregnant (cell D25) = 0.
(m) Calf birth weight (cell D26) = 0.
(n) Days since calving (cell D27) = 0.
(o) Relative milk yield (cell D28) = 5; although not used by software as this is a growing/finishing steer (cells D7 and D17), a number between 1 and 9 must be entered in this cell.
(p) Milk production (cell D29) = 0.
(q) Target calving age (cell D30) = 0.
(r) Target calving interval (cell D31) = 0.

Save the file by clicking on the *Save* button in the *File* group (FIGURE 3B) and name it as *Feedlot case study 1.* Once the *Home* screen is displayed, click on *Inputs* button of the *Inputs* group. FIGURE 18 shows the data entered so far.

Management and Grazing (Table 3). Enter the following information in this table:

(a) Feed additive (cell D55) = 2; choices are 1 (none), 2 (monensin), or 3 (lasalocid). These affect the dietary ME values and predicted DMI.
(b) Implant (cell D56) = 2; it affects the predicted DMI.
(c) Grazing unit size (cell D57) = 0, because the cattle are fed in confinement. For animals that are on pasture, enter the number of hectares (metric) or acres (Imperial) per animal grazed in the pasture. If the distance traveled is minimal, enter zero. This input is used to adjust energy maintenance requirements for walking activity.
(d) Days on pasture (cell D58) = 0. For grazing cattle, enter the number of days on the pasture.
(e) Available forage mass (cell D59) = 0. For grazing cattle, enter the kg DM/hectare (metric) or lb DM/acre (Imperial) when the cattle are turned into the pasture. This can be estimated from hay harvesting experience, clippings, or calibrated measuring devices such as height and/or density estimates, Plexiglas weight plates, or electronic pasture probes. The available forage mass (cell D59), number of days on pasture (cell D58), and the grazing unit size (cell D57) are used to predict pasture DMI.

Environment (Table 4). Enter the following information in this table:

(a) Wind speed (cell D64) = 5; this is the average wind speed the cattle are exposed to. Wind speed influences maintenance requirements by reducing the external insulation of the animal. Increasing wind speed decreases the external insulation value of the animal and thus results in increased energy maintenance requirements. The model is very sensitive to this input after the lower critical temperature is

Table 1. Settings					
	Acronym	User	Unit	Metric	Unit
Solution type		1	ELS		
Unit system		2	Imperial		
Feed basis		1	DM		
Animal class		1			

Table 2. Animal					
Item	Acronym	User	Unit	Metric	Unit
Age		14		14	mo
Age class (for DMI & WC)		2		2	
Sex		2		2	
Body condition score	BCS	5		5	
Initial body weight	IBW	872	lb	396	kg
Final body weight	FBW	1337	lb	606	kg
Mature body weight	MW	1337	lb	606	kg
Reference animal's EBF		3		3	
Breed type		2		2	
Breeding system		2		2	
Dam's breed		14		14	
Sire's breed		7		7	
N/A		1		1	
Days pregnant	DP	0		0	days
Calf birth weight	CBW	0	lb	0.0	kg
Days since calving		0		0	days
Relative milk yield	RMY	5		5	
Milk production	MY	0	lb/d	0.0	kg/d
Target calving age	TCA	0		0	mo
Target calving interval	TCI	0		0	mo

FIGURE 18 Settings and Animal tables after data entry.

reached, so choose carefully. Sensitivity analysis was performed in Chapter 19.

(b) Temperature, previous (cell D65) = 40; this is the average temperature for the previous month. This value is used to increase NEm requirement, as it gets colder or to reduce it as it gets warmer. The model is extremely sensitive to this input after the lower critical temperature is reached, so choose carefully. Sensitivity analyses were performed in Chapter 19.

(c) Temperature, current (cell D66) = 30; this is the average temperature the cattle are exposed to. In most situations, the current average daily temperature is the most practical to use. This value is used to adjust predicted DMI for temperature effects and is used in the calculations for the effects of cold stress on energy maintenance requirements. The model is very sensitive to this input after the lower critical temperature is reached, so choose carefully. Sensitivity analysis was performed in Chapter 19.

(d) Relative humidity, previous (cell D67) = 65; this value can be used to compute the previous effective temperature index.

(e) Relative humidity, current (cell D68) = 65; this value is used to compute the current effective temperature index to predict DMI, as discussed in Chapter 19.

(f) Storm exposure (cell D69) = 1; this value is used to predict DMI, as discussed in Chapter 19.

(g) Lowest night temperature (cell D70) = 30; if greater than 68°F (20°C) there is no night cooling and DMI is reduced with hot daytime temperatures; otherwise, it is assumed that cattle can dissipate heat at night and DMI is not affected.

(h) Hair depth (cell D71) = 0.2; this is the average hair depth. This input is used to compute the external insulation of the animal. Enter the effective hair coat depth of the animal, in increments of 0.1. As hair length increases, so does the external insulation value provided by the animal's hair coat. A general guide to use is an effective coat depth of 0.25 inches (0.6 cm) during the summer and 0.5 inches (1.3 cm) during the winter. The model is very sensitive to this input after the lower critical temperature is reached, so choose carefully. Sensitivity analysis was performed in Chapter 19.

(i) Hide (cell D72) = 2; enter either 1 (thin hide—i.e., dairy or Bos indicus types); 2 (average—i.e., most European breeds); or 3 (thick—i.e., Hereford or similar breeds). This value influences the external insulation value of the animal. Increased hide thickness implies increased external insulation. The model is very sensitive to this input after the lower critical temperature is reached, so choose carefully. Sensitivity analysis was performed in Chapter 19.

(j) Hair coat (cell D73) = 1; enter either 1 (clean and dry), 2 (some mud on lower body), 3 (some mud on lower body and sides), or 4 (heavily covered with mud). This value is used to adjust external insulation. The model is very sensitive to this input after the lower critical temperature is reached, so choose carefully. Sensitivity analysis was performed in Chapter 19.

(k) Cattle panting (cell D74) = 1; enter either 1 (no panting, not heat stressed), 2 (rapid shallow panting, or 3 (open-mouth panting). This value is used to adjust maintenance energy requirements for the energy cost of dissipating heat.

(l) Mud depth (cell D75) = 0; this value affects predicted DMI. Because the animals in this study have clean and dry hair coat, there was no mud present in this situation.

Save the file by clicking on the *Save* button in the *File* group (FIGURE 3B) and name it as *Feedlot case study 2*. Once *Home* screen is displayed, click on the *Inputs* button of the *Inputs* group. FIGURE 19 depicts the management and environment data entered so far.

Advanced options (Table 5). Let us use the following values in this table to make the BCNRM calculations of environment factors more compatible with the NRC (1996, 2000):

(a) Maintenance factor (cell D80) = 0 (default).
(b) Adjustments for required NEm (cell D81) = 1.
(c) Adjustments for predicted DMI (cell D82) = 1.

Table 3. Management and Grazing

Item	Acronym	User	Unit	Metric	Unit
Feed additive		2		2	
Implant		2		2	
Grazing unit size	GU	0	acre	0.00	ha
Days on pasture	DOP	0	days	0	days
Available forage mass	FM	0	lb/acre	0.0	kg/ha

Table 4. Environment

Item	Acronym	User	Unit	Metric	Unit
Wind speed	WS	5.00	miles/h	8.0	Km/h
Temperature, previous	Tp	40.00	oF	4.4	oC
Temperature, current	Tc	30.00	oF	-1.1	oC
RH, previous	RHp	65.00		65.0	%
RH, current	RHc	65.00		65.0	%
Storm exposure		1		1	
Lowest night temp.	LNT	30.00	oF	-1.1	oC
Hair depth		0.2	in	0.51	cm
Hide		2		2	
Hair coat		1		1	
Cattle panting		1		1	
Mud depth		0.00	in	0	cm

FIGURE 19 Management and Grazing and Environment tables after data entry.

(d) Monensin adjustment for DMI (cell D83) = -3 (default).
(e) Implant adjustment for DMI (cell D84) = -6 (default).
(f) Ruminal pH (cell D85) = 0 (default).
(g) DE to ME efficiency (cell D86) = 82 (default).
(h) Tabular ME and NE values for ELS (cell D87) = 0 (default).
(i) MCP yield (MCP/TDN, %) (cell D88) = 0 (default).
(j) Desired animal ADG (cell D89) = 1.5 lb/d (default).
(k) MP to NP efficiency (cell D90) = 0 (default).
(l) Recycled ruminal N (cell D91) = 0 (default).

Save the file by clicking on the *Save* button in the *File* group (FIGURE 3B) and name it as *Feedlot case study 3*.

Dietary Inputs

We will use the NRC (2000) feed library to retrieve the same feeds used in the original (NRC, 1996, 2000) case study. Click on the *Load SFL* (SFL = standard feed library) of the *Feed Library* group (FIGURE 3E). Locate the *Beef Std 2000.FLF* file in the *[C:]\Users\[User Name]\National Academies\BCNRM2016* folder. Select and open it. Click on *YES* to load it. Scroll down and confirm that feeds 301, 402, and 800 are barley silage, barley grain–heavy, and minerals, respectively. The following steps are used to retrieve these feeds into the diet of the current simulation:

(a) Click on the *Feeds* button of the *Diet* group (FIGURE 3D).
(b) Enter 301 in cell W5, 402 in cell W6, and 800 in cell W7. Confirm that the correct feed names are shown in column X, right next to the feed ID.
(c) Click on the *Retrieve* button of the *Diet* group (FIGURE 3D). The feeds (barley silage, barley grain–heavy, and minerals) should appear in Table 8.
(d) Check the checkboxes for these feeds in cells AH5, AH6, and AH7 (Table 8) to ensure their compositions are not changed by the software.
(e) The user may click on the *Composition* button of the *Diet* group (FIGURE 3D) or simply scroll to the right to display Table 8. The first option will freeze the feed names, making it easier to browse the composition table. Click on the *Composition* button of the *Diet* group (FIGURE 3D) and modify their composition in Table 8 as per the following analytical values:

Feed	Cost, $/Ton	NDF, % DM	peNDF, % NDF	CP, % DM	Fat, % DM	Ash, % DM
Barley silage	25	48.7	65	10.4	3	8
Barley grain	120	19	34	13	2.1	3
Minerals	200	0	0	0	0	100

(f) To enter the feed amounts, click on the *Diet* button of the *Diet* group (FIGURE 3D) to display Table 7 (FIGURE 9). Make sure that the *Amount* option is selected in cell Z3 and then enter the following amounts: 5, 19, and 0.3 lb/d for barley silage, barley grain, and minerals, respectively.

Save the file by clicking on the *Save* button in the *File* group (FIGURE 3B) and name it as *Feedlot case study 4*.

Performance Summary

After saving the simulation file, the *Home* screen will be displayed; it shows the animal performance given the current information entered. The actual DMI (24.3 lb/d) is nearly identical to the predicted DMI (24.2 lb/d). The predicted DMI using the NRC (1996, 2000) equations was 24.7 lb/d as shown in Table R8 (for quick view, click on the *DMIR* button of the *Calculation* group; FIGURE 3F) as shown in FIGURE 20 (11.208 kg/d × 2.2046 lb/kg = 24.7 lb/d).

The predicted DMI would be reduced even further if the environmental factors were not taken into account (Table 5; FIGURE 7). By disabling the adjustments for predicted DMI (cells D82 = 0), the predicted DMI would be 23.5 lb/d. As a rule of thumb, if actual and predicted DMI differ by more than 5 to 10%, the user should carefully check all inputs that influence DMI (breed, body weight, mature size, temperature, mud and storm exposure, diet energy density, ionophores, implant). However, the diet is evaluated with actual DMI.

FIGURE 21 has the BCNRM predictions as shown in the *Home* screen.

The predicted MP-allowable gain is greater than the ME-allowable gain; therefore, ME is likely the first limiting factor. The predicted ADG (3.59 lb/d; FIGURE 21) was slightly greater (+3.2%) than the observed ADG of 3.48 lb/d. The NRC (1996, 2000) predicted a greater ADG of 3.67 lb/d (+5.5%). The following table lists the selected differences between the predictions of the *previous* model (NRC, 1996, 2000) and the BCNRM.

Variables	Models		
	Previous[a]	BCNRM	Δ%
Predicted DMI, lb/d	24	24.2	0.8
ME-allowable gain, lb/d	3.67	3.59	−2.2
MP-allowable gain, lb/d	4.68	3.91	−16
Dietary ME, Mcal/lb	1.28	1.31	2.3
Dietary NEm, Mcal/lb	0.96	0.88	−8.3
Dietary NEg, Mcal/lb	0.57	0.59	3.5
MP feed, g/d	323	316	−2.2
MP bacteria, g/d	583	506	−13

[a]As reported by the NRC (1996, 2000). Change the units to metric to get the g/d values for MP.

	A	B	C	D	E	F
1						
2	REQUIREMENT SECTION					
3						
171	Table R8. Dry matter intake and water intake					
172	Item		Acronym	Value	Units	Description
173	Are adjustment factors being used?			TRUE		Except for ADTV that is always applied
174	Age effect			1.000	dmnl	Effect of age on intake
175	Empty body effect		BFAT	0.979	dmnl	Effect of body fat on intake
176	Breed type effect		BI	1.000	dmnl	Effect of breed type on intake
177	Effective temperature		CETI	4.742	°C	Current effective temperature index
178	Effect of night cooling		DMINC	1.150	dmnl	Effect of night cooling on intake
179	Temperature effect		TEMP1	1.048	dmnl	Effect of temperature on intake
180	Mud depth effect		MUD1	1.000	dmnl	Effect of mud depth on intake
181	Additive effect		$ADTV_{DMI}$	0.970	dmnl	Effect of monensin or anabolic implant on intake
182	Forage daily allowance		FA	0.000	g/kg SBW/d	
183	Grazing intake		GI	23.405	g/kg SBW	Unadjusted predicted DMI for grazing animals
184	Grazing effect		GRAZE	1.000	dmnl	Effect of grazing and forage allowance on intake
185	Initial SBW		ISBW	379.715	kg	Initial shrunk body weight
186	Beef calf or yearling			11.207	kg/d	NRC (1996, 2000), with adjustments
187	Growing-finishing cattle (NEm)			10.954	kg/d	Current Beef Committee, with adjustments
188	Growing-finishing cattle (ISBW)			9.219	kg/d	Current Beef Committee, with adjustments
189	Beef cow			12.325	kg/d	Lactating, pregnant, or non-pregnant beef cows
190	Beef cow (1.1 % of BW/dietary NDF)			21.267	kg/d	Uses cow SBW minus conceptus weight
191	Predicted DMI		DMI	10.954	kg/d	For growing-finishing, it uses the NEm equation
192	Predicted WI		WI	33.724	L/d	Water intake

FIGURE 20 Predicted dry matter intake and water intake (Table R8).

Predictions with the Mechanistic Level of Solution

Change the Solution Level from ELS to MLS by clicking on the *Settings* button of the *Inputs* group (FIGURE 3C) and enter 2 in cell D4. Go back to the *Home* screen (by clicking on the *Home* button of the *File* group; FIGURE 3B). Conversely, change from ELS to MLS in cell B37 in the *Home* screen. Save the file by clicking on the *Save* button in the *File* group (FIGURE 3B) and name it as *Feedlot case study 5.* FIGURE 22 has the MLS predictions. Compare the results.

The predictions of ME, NEma, NEga, ME-allowable ADG, and RUP using the Solution Level 2 of the *previous* model (NRC, 1996, 2000) and the MLS of the BCNRM were less than the Solution Level 1 and the ELS predictions, as shown below. However, the *current* model predicted identical ME-allowable gain and lower MP-allowable gain compared to the *previous* model (NRC, 1996, 2000). Although the dietary ME and NEg values predicted by the *current* model were nearly identical to the *previous* model's predictions, the NEm values were different likely because of the ionophore adjustment of the *previous* model. The MP values from feed and bacteria predicted by the BCNRM were less than the values predicted by *previous* model.

	Models		
Variables	Previous[a]	BCNRM	Δ%
Predicted DMI, lb/d	24.4	24.5	0.8
ME-allowable gain, lb/d	3.46	3.46	0
MP-allowable gain, lb/d	6.03	5.27	–14
Dietary ME, Mcal/lb	1.25	1.29	3.1
Dietary NEm, Mcal/lb	0.93	0.86	–8.1
Dietary NEg, Mcal/lb	0.54	0.57	5.3
MP feed, g/d	241	208	–14
MP bacteria, g/d	819	767	–6.4

[a]As reported by the NRC (1996, 2000). Change the units to metric to get the g/d values for MP.

For this scenario, the ME-allowable ADG predicted by the ELS and the MLS over- and underpredicted observed ADG (3.48 lb/d) by about 0.11 and 0.02 lb/d, respectively. Several factors can affect ME-allowable gain, and the actual characterization of the diet is needed to fully understand what factors are mostly affecting the predictive performance of these models.

ME & MP summary	Metabolizable Energy (ME)			Metabolizable Protein (MP)		
	Available Mcal/d	Required Mcal/d	Balance Mcal/d	Available lb/d	Required lb/d	Balance lb/d
Diet	31.81	31.81	0.00	1.8	1.7	0.1
Maintenance	31.81	13.46	18.35	1.8	0.9	1.0
Pregnancy	18.35	0.00	18.35	1.0	0.0	1.0
Lactation	18.35	0.00	18.35	1.0	0.0	1.0
Growth	18.35	18.35	0.00	1.0	0.9	0.1
Reserves	0.00	0.00	0.00	0.1	0.0	0.1

ME allowable gain	3.59 lb/d
MP allowable gain	3.91 lb/d
DMI predicted	24.15 lb/d
DMI actual	24.30 lb/d
NDF intake actual	0.57 % SBW
Water intake	8.91 gal/d
Diet Summary	
Cost per day	1.49 $/d
Dry matter	70.00 %AF
Forage	20.58 %DM
Crude protein	12.30 %DM
Fat	2.26 %DM
Ash	5.23 %DM
NDF	24.88 %DM
Lignin	2.13 %DM
Carbohydrate (CHO)	80.21 %DM
Nonfiber CHO	55.33 %DM
Sugars	0.00 %DM
Starch	48.00 %DM
Pectin	7.33 %DM
RDP(N):S ratio	0.01
Ca:P ratio	0.44
peNDF	11.56 %DM
Ruminal pH (calc.)	5.90

	lb/d	%CP	%DM
RDP	2.10	70.25	8.64
RUP	0.89	29.75	3.66
RDP:RUP	2.36	0.00	0.00

TDN and RDP allowable MCP	
TDN allowable MCP	1.74 lb/d
RDP required	1.74 lb/d
RDP available	2.10 lb/d
RDP balance	0.36 lb/d
RDP allowable MCP	1.74 lb/d

Source of Metabolizable Protein		
Bacteria	61.6%	1.1 lb/d
Feed	38.4%	0.7 lb/d
Total		1.8 lb/d

Ruminal Nitrogen Balance	
Ruminal N balance	0.06 lb N/d
Ruminal N balance	20.45 % of req.

Diet Energy	
Apparent TDN	78.0 % DM
Dietary DE	1.56 Mcal/lb
ME:DE	83.9 %
Dietary ME	1.31 Mcal/lb
Dietary NEm	0.88 Mcal/lb
Dietary NEg	0.59 Mcal/lb

Items		Supplied	Required	Balance
Ca	g/d	16.10	42.53	-26.43
P	g/d	36.74	22.51	14.23
Mg	g/d	14.65	11.02	3.63
Cl	g/d	28.03	0.00	28.03
K	g/d	107.41	66.13	41.28
Na	g/d	3.58	7.72	-4.13
S	g/d	18.37	16.53	1.84
Co	mg/d	4.65	1.65	3.00
Cu	mg/d	63.14	110.22	-47.08
I	mg/d	0.43	5.51	-5.08
Fe	mg/d	1363.29	551.12	812.17
Mn	mg/d	259.32	220.45	38.87
Se	mg/d	1.89	1.10	0.79
Zn	mg/d	167.60	330.67	-163.07
Vit A	1000 IU/d	32.75	24.25	8.50
Vit D	1000 IU/d	0.00	3.03	-3.03
Vit E	IU/d	225.80	385.78	-159.98
ARG	g/d	34.51	25.97	8.54
HIS	g/d	14.44	19.67	-5.23
ILE	g/d	35.88	22.04	13.85
LEU	g/d	53.78	52.73	1.06
LYS	g/d	44.60	50.37	-5.77
MET	g/d	14.33	15.74	-1.41
CYS	g/d	0.00	—	—
PHE	g/d	36.49	27.54	8.95
TYR	g/d	20.69	—	—
THR	g/d	34.82	30.69	4.13
TRP	g/d	10.54	4.72	5.82
VAL	g/d	40.43	31.48	8.95

Nitrogen Balance		Methane (median)
Intake	217.00 g/d	5.918 % GE
Retained	31.23 g/d	12.973 mol/d
Urinary	116.97 g/d	2.763 Mcal/d
Fecal	0.00 g/d	8.542 g/lb DM

FIGURE 21 Summary screen showing the Empirical Level of Solution (ELS) predictions.

COW-CALF CASE STUDY

Similar to the previous case study, this cow-calf case study is the same as the one presented in the *Nutrient Requirements of Beef Cattle*, 7th revised edition and update (NRC, 1996, 2000). The purpose of this study is to compare the prediction of both computer models.

Settings, Animal, Management and Grazing, Environment, and Advanced Inputs

Begin the tutorial by opening the BCNRM; agree with the *Terms of Use*; click on the *BCNRM 2016* tab (if not automatically selected); click on the *New* button of the *File* group (FIGURE 3B); and click on the *Inputs* button of the *Inputs* group (FIGURE 3C).

The ranch used in this case study is in the Northern Plains and carries approximately 600 beef cows and 100 replacement heifers. Cows are predominantly Simmental-sired females from Angus × Hereford cows with a mature size of approximately 1,300 lb at BCS 5. Calving season for mature cows is March and April, and calf birth weight averages 80 lb. The approximate weaning date for calves was October 15. Average steer calf weaning weight at 200 days is 575 lb, and average heifer calf weaning weight at 200 days is 525 lb. Replacement heifers wean at 45% of mature weight in the middle of October, conceive at 60% of mature weight during the first week of May, and are 85% of mature weight at calving. Body condition scores average 3 to 4 at weaning. The goal is to have them back to BCS 5 by December 1 to provide insulation for winter and maintain them at BCS 5 until calving. They will lose a score by pasture turnout

ME & MP summary	Metabolizable Energy (ME)			Metabolizable Protein (MP)		
	Available Mcal/d	Required Mcal/d	Balance Mcal/d	Available lb/d	Required lb/d	Balance lb/d
Diet	31.28	31.28	0.00	2.1	1.7	0.4
Maintenance	31.28	13.51	17.77	2.1	0.9	1.3
Pregnancy	17.77	0.00	17.77	1.3	0.0	1.3
Lactation	17.77	0.00	17.77	1.3	0.0	1.3
Growth	17.77	17.77	0.00	1.3	0.8	0.4
Reserves	0.00	0.00	0.00	0.4	0.0	0.4

ME allowable gain	3.46 lb/d
MP allowable gain	5.27 lb/d
DMI predicted	24.48 lb/d
DMI actual	24.30 lb/d
NDF intake actual	0.57 % SBW
Water intake	8.91 gal/d
Diet Summary	
Cost per day	1.49 $/d
Dry matter	70.00 %AF
Forage	20.58 %DM
Crude protein	12.30 %DM
Fat	2.26 %DM
Ash	5.23 %DM
NDF	24.88 %DM
Lignin	2.13 %DM
Carbohydrate (CHO)	80.21 %DM
Nonfiber CHO	55.33 %DM
Sugars	0.00 %DM
Starch	48.00 %DM
Pectin	7.33 %DM
RDP(N):S ratio	0.01
Ca:P ratio	0.44
peNDF	11.56 %DM
Ruminal pH (user)	5.90

	lb/d	%CP	%DM
RDP	2.25	75.11	9.24
RUP	0.74	24.89	3.06
RDP:RUP	3.02	0.00	0.00

TDN and RDP allowable MCP	
TDN allowable MCP	2.82 lb/d
RDP required	2.82 lb/d
RDP available	2.25 lb/d
RDP balance	-0.57 lb/d
RDP allowable MCP	2.25 lb/d

Source of Metabolizable Protein		
Bacteria	78.6%	1.7 lb/d
Feed	21.4%	0.5 lb/d
Total		2.1 lb/d

Ruminal Nitrogen Balance	
Ruminal N balance	-0.09 lb N/d
Ruminal N balance	-20.27 % of req.

Diet Energy	
Apparent TDN	76.7 % DM
Dietary DE	1.53 Mcal/lb
ME:DE	82.0 %
Dietary ME	1.29 Mcal/lb
Dietary NEm	0.86 Mcal/lb
Dietary NEg	0.57 Mcal/lb

Items		Supplied	Required	Balance
Ca	g/d	16.10	41.60	-25.49
P	g/d	36.74	22.13	14.61
Mg	g/d	14.65	11.02	3.63
Cl	g/d	28.26	0.00	28.26
K	g/d	107.41	66.13	41.28
Na	g/d	3.58	7.72	-4.13
S	g/d	18.37	16.53	1.84
Co	mg/d	4.65	1.65	3.00
Cu	mg/d	63.14	110.22	-47.08
I	mg/d	0.43	5.51	-5.08
Fe	mg/d	1363.29	551.12	812.17
Mn	mg/d	259.32	220.45	38.87
Se	mg/d	1.88	1.10	0.78
Zn	mg/d	167.60	330.67	-163.07
Vit A	1000 IU/d	32.75	24.25	8.50
Vit D	1000 IU/d	0.00	3.03	-3.03
Vit E	IU/d	225.80	385.78	-159.98
ARG	g/d	60.10	25.53	34.58
HIS	g/d	23.89	19.34	4.54
ILE	g/d	50.69	21.66	29.03
LEU	g/d	67.36	51.83	15.53
LYS	g/d	67.31	49.51	17.80
MET	g/d	21.77	15.47	6.29
CYS	g/d	0.00	—	—
PHE	g/d	47.50	27.08	20.43
TYR	g/d	33.12	—	—
THR	g/d	47.43	30.17	17.26
TRP	g/d	14.31	4.64	9.67
VAL	g/d	54.28	30.94	23.33

Nitrogen Balance		Methane (median)
Intake	217.00 g/d	5.311 % GE
Retained	30.18 g/d	11.654 mol/d
Urinary	113.21 g/d	2.480 Mcal/d
Fecal	73.61 g/d	7.674 g/lb DM

FIGURE 22 Summary screen showing the Mechanistic Level of Solution (MLS) predictions.

(approximately May 1); but to meet the goal, they have to gain one score by the start of breeding on May 15. Over the 12-month reproductive cycle, the energy balance should average near zero. The winter-feed resources available include two qualities of hay. Corn and range cake are fed as needed to supplement the hay.

We will use this information to demonstrate how to use the ELS and MLS of the BCNRM to evaluate the feeding program for this herd, beginning with an evaluation of the winter feeding program.

Settings (Table 1). Confirm that solution type is 1 (ELS), unit system is 2 (Imperial), feed basis is 1 (DM), and animal class is 3 (dry cow). It is often practical to adjust the diet until balanced with the ELS, and then evaluate it with the MLS to get predicted feed net energy values and amino acid balances, based on actual feed analysis for carbohydrate and protein fractions.

Animal (Table 2). Enter the following information in this table:

(a) Age (cell D12) = 60 mo; this value influences tissue insulation.
(b) Age class (cell D13) = 2; this value influences expected DMI for growing/finishing cattle; it does not apply to this type of animals.
(c) Sex (cell D14) = 4; this value influences maintenance requirement. A heifer is entered as a cow after calving the first time.
(d) Body condition score (cell D15) = 5; see Chapters 3 and 13 for a detailed discussion of the 1 to 9 condition scoring system used and its biological basis. The choices are 1 through 9 (1 = emaciated, 5 = moderate, 9 = very fat). Body condition score is used to describe tissue insulation, the potential for compensatory

growth in growing cattle, and energy reserves in cows.

(e) Initial body weight (cell D16) = 1,354. For cows, the initial (D16) and final (D17) body weights are recommended to be the same.

(f) Final body weight (cell D17) = 1,354. This is the unshrunk final body weight that best represents the group being fed. The average between initial and final weights is a major determinant of DMI, maintenance, and growth requirements.

(g) Mature body weight (cell D18) = 1,354; this is the expected average unshrunk weight at the USDA grade selected in cell D19. For cows, replacement heifers, or breeding bulls, enter the expected mature weight at BCS 5. The weight that best corresponds to the cattle in question based on the user's experience for the type of growing animal, implant strategy, and ration should be entered.

(h) Breed type (cell D20) = 2.

(i) Breeding system (cell D21) = 3; choices are 1 (straightbred), 2 (two-way crossbred), and 3 (three-way crossbred). Cell D22 is used for animal breed if straightbred, cells D22 and D23 are used if describing two-way crossbred, and cells D22, D23, and D24 are used when describing three-way crossbred. The breeding system for growing cattle influences maintenance energy requirement and predicted DMI.

(j) Breed cells (D22, D23, and D24) = 29 (sire's breed is Simmental), 1 (maternal grandsire's breed is Angus), and 14 (maternal granddam's breed is Hereford), respectively; valid breed codes are shown in Table 2.1. Stored breed values are used to determine maintenance energy requirements and defaults for calf birth weight and peak milk production. See Chapter 19 for the biological basis for these breed adjustments. Since this is a three-way crossbred (cell D21), the software will use all cells (D22, D23, and D24) to calculate the defaults.

(k). Days pregnant (cell D25) = 190. This information is used along with expected birth weight (cell D26) to compute pregnancy requirements, conceptus weight, and ADG as described in Chapter 13.

(l) Calf birth weight (cell D26) = 80. This information is used along with days pregnant (cell D25) to compute pregnancy requirements, conceptus weight, and ADG as described in Chapter 13.

(m) Days since calving (cell D27) = 0 because the cows are dry. For lactating cows, this information is used along with peak milk and lactation number to predict milk production for the day entered. As this is a dry cow, there is no milk yield.

(n) Relative milk yield (cell D28) = 5; because the cows are dry, the default value of 5 (average) is left in this cell to avoid leaving it blank. For lactating cows, a number between 1 and 9 can be entered to adjust the peak milk up or down from the default value shown in the breed defaults table.

(o) Milk production (cell D29) = 0 because the cows are dry. For lactating cows in this case study, the default peak milk value of 21.5 lb/d (cell O49) for the breed type would be used along with the days in milk to compute requirements for lactation. Fox et al. (1988) calculated a male calf weaning weight of approximately 587 lb at 7 months for a 1,300-lb cow at 21.5-lb peak milk compared to the actual steer 200-d weaning weight of 575 lb. Thus, the default milk production is acceptable. The peak milk along with time of peak and duration of lactation is used to develop a lactation curve for predicting milk production for the day entered, as described in Chapter 13. If evaluating lactating cows, the default value of 21.5 would be used. However, the user may override this estimated peak milk yield by entering a value in cell O48 or D29.

(p) Target calving age (cell D30) = 24. Age at first calving (cell D30) and herd calving interval (cell D31) are used to compute growth requirements as described in Chapters 13 and 19.

(q) Target calving interval (cell D31) = 12. Herd Calving interval (cell D31) and age at first calving and (cell D30) are used to compute growth requirements as described in Chapters 13 and 19.

Save the file by clicking on the *Save* button in the *File* group (FIGURE 3B) and name it as *Cow-calf case study 1*. Once the *Home* screen is displayed, click on the *Inputs* button of the *Inputs* group. FIGURE 23 depicts the data entered so far.

Management and Grazing (Table 3). Enter the following information in this table:

(a) Feed additive (cell D55) = 1; choices are 1 (none), 2 (monensin), or 3 (lasalocid).

(b) Implant (cell D56) = 1; choices are 1 (no) or 2 (yes).

(c) Grazing unit size (cell D57) = 0 because this example is for a dry cow being wintered on harvested forage. When grazed forage is the primary feed used to meet requirements, the number of hectares (metric) or acres (Imperial) per head grazed in the pasture has to be entered. If the distance traveled is minimal, enter 0. This input is used to adjust energy maintenance requirements for walking activity.

(d). Days on pasture (cell D58) = 0 because the animals are being fed harvested feed in drylot. For grazing cattle, the number of days on the pasture would be entered.

(e) Available forage mass (cell D59) = 0. For grazing cattle, enter the kg DM/hectare (metric) or lb DM/acre (Imperial) when the cattle are turned into the pasture. This can be estimated from hay harvest-

	A	B	C	D	E	F	G
1							
2		**Table 1. Settings**					
3			**Acronym**	**User**	**Unit**	**Metric**	**Unit**
4		Solution type		1	ELS		
5		Unit system		2	Imperial		
6		Feed basis		1	DM		
7		Animal class		3			
8							
9							
10		**Table 2. Animal**					
11		**Item**	**Acronym**	**User**	**Unit**	**Metric**	**Unit**
12		Age		60		60	mo
13		Age class (for DMI & WC)		2		2	
14		Sex		4		4	
15		Body condition score	BCS	5		5	
16		Initial body weight	IBW	1354	lb	614	kg
17		Final body weight	FBW	1354	lb	614	kg
18		Mature body weight	MW	1354	lb	614	kg
19		Reference animal's EBF		3		3	
20		Breed type		2		2	
21		Breeding system		3		3	
22		Sire's breed		29		29	
23		Maternal grandsire's breed		1		1	
24		Maternal granddam's breed		14		14	
25		Days pregnant	DP	190		190	days
26		Calf birth weight	CBW	80	lb	36.3	kg
27		Days since calving		0		0	days
28		Relative milk yield	RMY	5		5	
29		Milk production	MY	0	lb/d	0.0	kg/d
30		Target calving age	TCA	24		24	mo
31		Target calving interval	TCI	12		12	mo

FIGURE 23 Settings and Animal tables after data entry.

ing experience, clippings, or calibrated measuring devices such as height and/or density estimates, Plexiglas weight plates, or electronic pasture probes. The available forage mass (cell D59), number of days on pasture (cell D58), and the grazing unit size (cell D57) are used to predict pasture DMI.

Environment (Table 4). Enter the following information in this table:

(a) Wind speed (cell D64) = 5; this is the average wind speed the cattle are exposed to. Wind speed influences maintenance requirements by reducing the external insulation of the animal. Increasing wind speed decreases the external insulation value of the animal and thus results in increased energy maintenance requirements. The model is very sensitive to this input after the lower critical temperature is reached, so choose carefully. Sensitivity analysis was performed in Chapter 19.

(b) Temperature, previous (cell D65) = 40; this is the average temperature for the previous month. This value is used to increase NEm requirement, as it gets colder or to reduce it as it gets warmer. The model is extremely sensitive to this input after the lower critical temperature is reached, so choose carefully. Sensitivity analyses were performed in Chapter 19.

(c) Temperature, current (cell D66) = 30; this is the average temperature the cattle are exposed to. In most situations, the current average daily temperature is the most practical to use. This value is used to adjust predicted DMI for temperature effects and is used in the calculations for the effects of cold stress on energy maintenance requirements. The model is very sensitive to this input after the lower critical temperature is reached, so choose carefully. Sensitivity analysis was performed in Chapter 19.

(d) Relative humidity, previous (cell D67) = 65; this value can be used to compute the previous effective temperature index.

(e) Relative humidity, current (cell D68) = 65; this value is used to compute the current effective temperature index to predict DMI, as discussed in Chapter 19.

(f) Storm exposure (cell D69) = 1; this value is used to predict DMI, as discussed in Chapter 19.

(g) Lowest night temperature (cell D70) = 30; if greater than 68°F (20°C) there is no night cooling and DMI is reduced with hot daytime temperatures; otherwise, it is assumed that cattle can dissipate heat at night and DMI is not affected.

(h) Hair depth (cell D71) = 0.5; this is the average hair depth. This input is used to compute the external insulation of the animal. Enter the effective hair coat depth of the animal, in increments of 0.1. As hair length increases, so does the external insulation value provided by the animal. A general guide to use is an effective coat depth of 0.25 inches (0.6 cm) during the summer and 0.5 inches (1.3 cm) during the winter. The model is very sensitive to this input after the lower critical temperature is reached, so choose carefully. Sensitivity analysis was performed in Chapter 19.

(i) Hide (cell D72) = 2; enter either 1 (thin hide—i.e., dairy or Bos indicus types), 2 (average—i.e., most European breeds), or 3 (thick—i.e., Hereford or similar breeds). This value influences the external insulation value of the animal. Increased hide thickness implies increased external insulation. The model is very sensitive to this input after the lower critical temperature is reached, so choose carefully. Sensitivity analysis was performed in Chapter 19.

(j) Hair coat (cell D73) = 1; enter either 1 (clean and dry), 2 (some mud on lower body), 3 (some mud on lower body and sides), or 4 (heavily covered with mud). This value is used to adjust external insulation. The model is very sensitive to this input after the lower critical temperature is reached, so choose carefully. Sensitivity analysis was performed in Chapter 19.

(k) Cattle panting (cell D74) = 1; enter either 1 (no panting, not heat stressed), 2 (rapid shallow panting), or 3 (open-mouth panting). This value is used to adjust maintenance energy requirements for the energy cost of dissipating heat.

(l) Mud depth (cell D75) = 0; this value affects predicted DMI. Because the animals in this study have clean and dry hair coat, there was no mud present in this situation.

Save the file by clicking on the *Save* button in the *File* group (FIGURE 3B) and name it as *Cow-calf case study 2*. Once the *Home* screen is displayed, click on the *Inputs* button of the *Inputs* group. FIGURE 24 shows the data entered for management and environment.

Advanced options (Table 5). Let us use the following values in this table to make the software calculations of environment factors more compatible with NRC (1996, 2000):

(a) Maintenance factor (cell D80) = 0 (default).
(b) Adjustments for required NEm (cell D81) = 1.
(c) Adjustments for predicted DMI (cell D82) = 1.
(d) Monensin adjustment for DMI (cell D83) = –3 (default).
(e) Implant adjustment for DMI (cell D84) = –6 (default).
(f) Ruminal pH (cell D85) = 0 (default).
(g) DE to ME efficiency (cell D86) = 82 (default).
(h) Tabular ME and NE values for ELS (cell D87) = 0 (default).
(i) MCP yield (MCP/TDN, %) (cell D88) = 0 (default).
(j) Desired animal ADG (cell D89) = 1.5 lb/d (default).
(k) MP to NP efficiency (cell D90) = 0 (default).
(l) Recycled ruminal N (cell D91) = 0 (default).

Save the file by clicking on the *Save* button in the *File* group (FIGURE 3B) and name it as *Cow-calf case study 3*.

Table 3. Management and Grazing

Item	Acronym	User	Unit	Metric	Unit
Feed additive		1		1	
Implant		1		1	
Grazing unit size	GU	0	acre	0.00	ha
Days on pasture	DOP	0	days	0	days
Available forage mass	FM	0	lb/acre	0.0	kg/ha

Table 4. Environment

Item	Acronym	User	Unit	Metric	Unit
Wind speed	WS	5.00	miles/h	8.0	Km/h
Temperature, previous	Tp	40.00	oF	4.4	oC
Temperature, current	Tc	30.00	oF	-1.1	oC
RH, previous	RHp	65.00		65.0	%
RH, current	RHc	65.00		65.0	%
Storm exposure		1		1	
Lowest night temp.	LNT	30.00	oF	-1.1	oC
Hair depth		0.5	in	1.27	cm
Hide		2		2	
Hair coat		1		1	
Cattle panting		1		1	
Mud depth		0.00	in	0	cm

FIGURE 24 Management and Grazing and Environment tables after data entry.

Dietary Inputs

We will use the NRC (2000) feed library to retrieve the same feeds used in the original (NRC, 1996, 2000) case study. Click on the *Load SFL* (SFL = standard feed library) of the *Feed Library* group (FIGURE 3E). Locate the *Beef Std 2000.FLF* file in the *[C:]\Users\[User Name]\Documents\National Academies\BCNRM2016* folder. Select and open it. Click on *YES* to load it. Scroll down and confirm that feeds 105, 107, and 800 are brome hay mid bloom, brome hay mature, and minerals, respectively. The following steps are used to retrieve these feeds and to enter them into the diet of the current simulation:

(a) Click on the Select button of the Diet group (FIGURE 3D).
(b) For feed #1, click on the list dropbox) and highlight feed 105; for feed #2, clock on the list dropbox and highlight feed 107; and for feed #3, click on the list dropbox, scroll down, and highlight feed 800. Click on the Retrieve button of the Feed Selection and Retrieval form (FIGURE 11). The feeds (brome hay mid bloom, brome hay mature, and minerals) should show up in Table 8.
(c) Check the checkboxes for these feeds in cells AH5, AH6, and AH7 (Table 8) to ensure their composition are not changed by the software.
(d) The user may click on the Composition button of the Diet group (FIGURE 3D) or simply scroll to the right to display Table 8. The first option will freeze the feed names, making it easier to browse the composition table. Click on the *Composition* button of the *Diet* group (FIGURE 3D) and modify their composition in Table 8 as per the following analytical values:

Feed	Cost, $/Ton	NDF, % DM	CP, % DM	RUP, % CP	TDN % DM
Brome hay, mid bloom	70	56	12	23	57
Brome hay, mature	50	65	7	25	50
Minerals	200	0	0	0	0

(e) To enter feed amounts, click on the *Diet* button of the *Diet* group (FIGURE 3D) to display Table 7 (FIGURE 9). Make sure that the *Amount* option is selected in cell Z3, and enter 26.4 lb/d for mature brome hay and 0.3 lb/d for minerals.

Save the file by clicking on the *Save* button in the *File* group (FIGURE 3B) and name it as *Cow-calf case study 4*.

Performance Summary

After saving the simulation file, the *Home* screen will be displayed; it shows the animal performance given the current information entered. The actual DMI (26.7 lb/d) is the predicted DMI by the NRC (1996, 2000) equation. The DMI predicted by the BCNRM was 26.6 lb/d (12.041 kg/d × 2.2046 lb/kg). Table R8 (for a quick view, click on the *DMIR* button of the *Calculation* group; FIGURE 3F) shows the predicted DMI assuming the 1.1% of BW/NDF was 21.6 lb/d as shown in FIGURE 25 (9.788 kg/d × 2.2046 lb/kg = 21.6 lb/d).

The predicted DMI would be reduced even further if the environmental factors were not taken into account (Table 5; FIGURE 7). By disabling the adjustments for predicted DMI (cells D82 = 0 in Table 5), the predicted DMI would be 25.32 lb/d. As a rule of thumb, if actual and predicted DMI differ by more than 5 to 10%, the user should carefully check all inputs that influence DMI (breed, body weight, mature size, temperature, mud and storm exposure, dietary energy density, ionophores, implant). However, the diet is evaluated with actual DMI. FIGURE 26 shows the BCNRM predictions as shown in the *Home* screen.

The predicted ME balance is −1.63 Mcal/d and the predicted MP balance is zero (FIGURE 26). Given the small deficit of ME, these cows are expected to lose 1 BCS (from 5 to 4) in 120 days if the conditions remain the same. The current model's predicted TDN (49.4%) is nearly identical to that predicted by the *previous* model (NRC, 1996, 2000; 49%). The RDP (+0.14 lb/d) and ruminal N (+0.02 lb/d) balances indicate that this diet met the requirements of microbial protein (FIGURE 26). The following table lists the selected differences between the predictions of the *previous* model (NRC, 1996, 2000) and the BCNRM.

Variable	Model		Δ%
	Previous[a]	BCNRM	
Predicted DMI, lb/d	26.65	26.55	–0.56
Days to change BCS	3.199 (lose)	120 (lose)	–
RDP balance, g/d	30	70	57
Dietary ME, Mcal/lb	0.81	0.81	0
Dietary NEm, Mcal/lb	0.43	0.43	0
Dietary NEg, Mcal/lb	0.19	0.19	0
MP feed, g/d	168	126	–25
MP bacteria, g/d	383	361	–5.7

[a]As reported by the NRC (1996, 2000). Change the units to metric to get the g/d values for MP and RDP.

The *previous* model (NRC, 1996, 2000) predicted a negative RDP balance (–150 g/d) when the default coefficient of 13% was used to estimate MCP. The NRC (1996, 2000) recommended reducing this coefficient to 10% for cows or calves consuming low-quality diets. When that adjustment was made, the *previous* model predicted a positive RDP balance of 30 g/d as shown above. The BCNRM ELS predicted greater RDP balance without any adjustment. Furthermore, the BCNRM ELS predicted greater ME, NEma, and NEga and lesser MP from feed and ruminal bacteria than the *previous* model's Level 1.

	A/B	C	D	E	F
1					
2	**REQUIREMENT SECTION**				
3					
171	**Table R8. Dry matter intake and water intake**				
172	**Item**	**Acronym**	**Value**	**Units**	**Description**
173	Are adjustment factors being used?		TRUE		Except for ADTV that is always applied
174	Age effect		1.000	dmnl	Effect of age on intake
175	Empty body effect	BFAT	0.905	dmnl	Effect of body fat on intake
176	Breed type effect	BI	1.000	dmnl	Effect of breed type on intake
177	Effective temperature	CETI	2.655	°C	Current effective temperature index
178	Effect of night cooling	DMINC	1.170	dmnl	Effect of night cooling on intake
179	Temperature effect	TEMP1	1.048	dmnl	Effect of temperature on intake
180	Mud depth effect	MUD1	1.000	dmnl	Effect of mud depth on intake
181	Additive effect	$ADTV_{DMI}$	0.940	dmnl	Effect of monensin or anabolic implant on intake
182	Forage daily allowance	FA	0.000	g/kg SBW/d	
183	Grazing intake	GI	24.119	g/kg SBW	Unadjusted predicted DMI for grazing animals
184	Grazing effect	GRAZE	1.000	dmnl	Effect of grazing and forage allowance on intake
185	Initial SBW	ISBW	589.604	kg	Initial shrunk body weight
186	Beef calf or yearling		11.456	kg/d	NRC (1996, 2000), with adjustments
187	Growing-finishing cattle (NEm)		12.677	kg/d	Current Beef Committee, with adjustments
188	Growing-finishing cattle (ISBW)		10.912	kg/d	Current Beef Committee, with adjustments
189	Beef cow		12.041	kg/d	Lactating, pregnant, or non-pregnant beef cows
190	Beef cow (1.1 % of BW/dietary NDF)		9.788	kg/d	Uses cow SBW minus conceptus weight
191	Predicted DMI	DMI	12.041	kg/d	For growing-finishing, it uses the NEm equation
192	Predicted WI	WI	34.865	L/d	Water intake

FIGURE 25 Predicted dry matter intake using NEma (right arrow) or BW-to-NDF ratio (left arrow).

Predictions with the Mechanistic Level of Solution

Go back to the *Home* screen (by clicking on the *Home* button of the *File* group; FIGURE 3B). Change from ELS to MLS in cell B37 in the *Home*. Save the file by clicking on the *Save* button in the *File* group (FIGURE 3B) and name it as *Cow-calf case study 5*. FIGURE 27 has the MLS predictions and the following table lists the selected differences between the predictions of the *previous* model (NRC, 1996, 2000) and the BCNRM.

Variable	Model		Δ%
	Previous[a]	BCNRM	
Predicted DMI, lb/d	26.65	25.15	–5.63
Days to change BCS	170 (lose)	64 (lose)	–
Ruminal N balance, g/d	–24	–44.7	46.3
Dietary ME, Mcal/lb	0.77	0.79	2.5
Dietary NEm, Mcal/lb	0.39	0.41	4.88
Dietary NEg, Mcal/lb	0.15	0.16	6.25
MP feed, g/d	248	209	–15.7
MP bacteria, g/d	660	467	–29.2

[a]As reported by the NRC (1996, 2000). Change the units to metric to get the g/d values for MP and RDP.

Similar to the *previous* model's Level 1 and the *current* model's ELS, the *current* model's MLS predicted nearly identical ME, NEma, and NEga, but lesser MP from feed and ruminal bacteria than the *previous* model's Level 2 solution. Given this outcome, the *previous* model estimated that cows would lose one BCS (from 5 to 4) in 170 days whereas the *current* model estimated that cows would take 64 days to lose one BCS. In addition to the more negative ME balance predicted by the *current* model (–3.05 Mcal/d; FIGURE 27), as compared to the *previous* model's prediction (–1.8 Mcal/d), the body reserves submodel differs between these models as discussed in Chapters 13 and 19.

For this scenario, both ruminal N balance (–44.7 g/d) and RDP balance (–0.280 g/d) were negative, suggesting that ruminal bacteria N requirement was not met. However, no recycled ruminal N was taken into account. In this scenario, the RDP-allowable MCP was only 1.10 lb/d.

Differences Between the ELS and MLS

The ME-allowable BCS change is computed in the MLS from energy availability based on simulations of ruminal fer-

ME & MP summary	Metabolizable Energy (ME)			Metabolizable Protein (MP)		
	Available Mcal/d	Required Mcal/d	Balance Mcal/d	Available lb/d	Required lb/d	Balance lb/d
Diet	21.65	23.27	-1.63	1.1	1.1	0.0
Maintenance	21.65	21.37	0.28	1.1	1.0	0.1
Pregnancy	0.28	1.91	-1.63	0.1	0.1	0.0
Lactation	-1.63	0.00	-1.63	0.0	0.0	0.0
Growth	-1.63	0.00	-1.63	0.0	0.0	0.0
Reserves	0.00	1.63	-1.63	0.0	0.0	0.0

Days to lose 1 BCS	120 d
DMI predicted	26.55 lb/d
DMI actual	26.70 lb/d
NDF intake actual	1.32 % SBW
Water intake	9.21 gal/d
Diet Summary	
Cost per day	0.75 $/d
Dry matter	92.07 %AF
Forage	98.88 %DM
Crude protein	6.92 %DM
Fat	1.98 %DM
Ash	8.24 %DM
NDF	64.27 %DM
Lignin	7.85 %DM
Carbohydrate (CHO)	82.86 %DM
Nonfiber CHO	18.59 %DM
Sugars	0.00 %DM
Starch	5.94 %DM
Pectin	12.65 %DM
RDP(N):S ratio	0.00
Ca:P ratio	1.18
peNDF	62.98 %DM
Ruminal pH (calc.)	6.46

	lb/d	%CP	%DM
RDP	1.39	75.00	5.19
RUP	0.46	25.00	1.73
RDP:RUP	3.00	0.00	0.00

TDN and RDP allowable MCP	
TDN allowable MCP	1.24 lb/d
RDP required	1.24 lb/d
RDP available	1.39 lb/d
RDP balance	0.14 lb/d
RDP allowable MCP	1.24 lb/d

Source of Metabolizable Protein		
Bacteria	74.2%	0.8 lb/d
Feed	25.8%	0.3 lb/d
Total		1.1 lb/d

Ruminal Nitrogen Balance	
Ruminal N balance	0.02 lb N/d
Ruminal N balance	11.54 % of req.

Diet Energy	
Apparent TDN	49.4 % DM
Dietary DE	0.99 Mcal/lb
ME:DE	82.0 %
Dietary ME	0.81 Mcal/lb
Dietary NEm	0.43 Mcal/lb
Dietary NEg	0.19 Mcal/lb

Items		Supplied	Required	Balance
Ca	g/d	31.13	18.16	12.98
P	g/d	26.34	13.87	12.47
Mg	g/d	14.37	14.53	-0.16
Cl	g/d	0.00	0.00	0.00
K	g/d	221.54	72.67	148.87
Na	g/d	1.20	8.48	-7.28
S	g/d	0.00	18.17	-18.17
Co	mg/d	0.00	1.82	-1.82
Cu	mg/d	124.54	121.11	3.43
I	mg/d	0.00	6.06	-6.06
Fe	mg/d	958.00	605.55	352.44
Mn	mg/d	874.17	484.44	389.73
Se	mg/d	0.00	1.21	-1.21
Zn	mg/d	287.40	363.33	-75.93
Vit A	1000 IU/d	179.62	35.38	144.25
Vit D	1000 IU/d	11.97	3.36	8.61
Vit E	IU/d	0.00	430.41	-430.41
ARG	g/d	21.49	16.50	4.99
HIS	g/d	8.28	12.25	-3.96
ILE	g/d	23.49	14.00	9.49
LEU	g/d	35.81	33.51	2.30
LYS	g/d	30.83	31.36	-0.53
MET	g/d	9.75	9.80	-0.05
CYS	g/d	0.00	—	—
PHE	g/d	22.57	17.15	5.42
TYR	g/d	14.40	—	—
THR	g/d	23.83	19.11	4.72
TRP	g/d	12.62	2.94	9.68
VAL	g/d	26.78	20.00	6.77

Nitrogen Balance		Methane (median)
Intake	134.12 g/d	6.361 % GE
Retained	4.80 g/d	14.522 mol/d
Urinary	66.92 g/d	3.093 Mcal/d
Fecal	0.00 g/d	8.702 g/lb DM

FIGURE 26 Summary screen showing the Empirical Level of Solution predictions.

mentation and intestinal digestion. The simulations account for the effects of (1) individual feed content of carbohydrate and protein fractions, (2) ruminal rates of digestion and passage, (3) effect of rumen pH on fiber digestibility, and (4) intestinal starch, pectin, and fiber digestibility. The *previous* model (NRC, 1996, 2000) also included the urea cost.

The MP from bacteria in the MLS is computed from bacterial growth on fiber and nonfiber carbohydrates, which are sensitive to feed amounts of fiber and nonfiber carbohydrates and their digestion rates, and rumen pH. The MP from feeds is computed from feed protein escaping digestion in the rumen, which is sensitive to feed amounts of protein fractions with medium and slow digestion rates. In this example, the MP balance is greater in the MLS (FIGURE 27) than in the ELS (FIGURE 26) evaluations because both RUP and microbial protein production were greater (0.4 versus 0 lb/d, respectively).

Rumen N balances are given as total bacterial N balance. The total N balance was greater in the ELS than the MLS (0.02 vs. –0.1 g/d, respectively) because of a greater predicted microbial yield in the MLS. The essential amino acid balances indicate that histidine was negative for the ELS simulation which is likely due to a lesser predicted microbial yield.

GENERAL RECOMMENDATIONS

The following recommendations are provided as troubleshooting guidelines in using the BCNRM in addition to the recommendations presented in the previous chapters.

Dry Matter Intake

Compare total feed dry matter entered vs. model-predicted DMI. If there is more than 5 to 10% difference, check input variables that influence predicted DMI (rations DM and quality control; accuracy of weights; body weight; current temperature; ionophore use; dietary energy density; feed processing). The actual DMI must be accurately determined,

ME & MP summary	Metabolizable Energy (ME)			Metabolizable Protein (MP)		
	Available Mcal/d	Required Mcal/d	Balance Mcal/d	Available lb/d	Required lb/d	Balance lb/d
Diet	21.01	24.05	-3.04	1.5	1.1	0.4
Maintenance	21.01	22.14	-1.14	1.5	1.0	0.5
Pregnancy	-1.14	1.91	-3.04	0.5	0.1	0.4
Lactation	-3.04	0.00	-3.04	0.4	0.0	0.4
Growth	-3.04	0.00	-3.04	0.4	0.0	0.4
Reserves	0.00	3.04	-3.04	0.4	0.0	0.4

Days to lose 1 BCS	64 d
DMI predicted	25.15 lb/d
DMI actual	26.70 lb/d
NDF intake actual	1.32 % SBW
Water intake	9.21 gal/d
Diet Summary	
Cost per day	0.75 $/d
Dry matter	92.07 %AF
Forage	98.88 %DM
Crude protein	6.92 %DM
Fat	1.98 %DM
Ash	8.24 %DM
NDF	64.27 %DM
Lignin	7.85 %DM
Carbohydrate (CHO)	82.86 %DM
Nonfiber CHO	18.59 %DM
Sugars	0.00 %DM
Starch	5.94 %DM
Pectin	12.65 %DM
RDP(N):S ratio	0.00
Ca:P ratio	1.18
peNDF	62.98 %DM
Ruminal pH (calc.)	6.46

	lb/d	%CP	%DM
RDP	1.10	59.46	4.12
RUP	0.75	40.54	2.81
RDP:RUP	1.47	0.00	0.00

TDN and RDP allowable MCP	
TDN allowable MCP	1.72 lb/d
RDP required	1.72 lb/d
RDP available	1.10 lb/d
RDP balance	-0.62 lb/d
RDP allowable MCP	1.10 lb/d

Source of Metabolizable Protein		
Bacteria	69.1%	1.0 lb/d
Feed	30.9%	0.5 lb/d
Total		1.5 lb/d

Ruminal Nitrogen Balance	
Ruminal N balance	-0.10 lb N/d
Ruminal N balance	-35.94 % of req.

Diet Energy	
Apparent TDN	48.0 % DM
Dietary DE	0.96 Mcal/lb
ME:DE	82.0 %
Dietary ME	0.79 Mcal/lb
Dietary NEm	0.41 Mcal/lb
Dietary NEg	0.16 Mcal/lb

Items		Supplied	Required	Balance
Ca	g/d	31.13	18.16	12.98
P	g/d	26.34	13.87	12.47
Mg	g/d	14.37	14.53	-0.16
Cl	g/d	12.50	0.00	12.50
K	g/d	221.54	72.67	148.87
Na	g/d	1.20	8.48	-7.28
S	g/d	0.00	18.17	-18.17
Co	mg/d	0.00	1.82	-1.82
Cu	mg/d	124.54	121.11	3.43
I	mg/d	0.00	6.06	-6.06
Fe	mg/d	958.00	605.55	352.44
Mn	mg/d	874.17	484.44	389.73
Se	mg/d	0.00	1.21	-1.21
Zn	mg/d	287.40	363.33	-75.93
Vit A	1000 IU/d	179.62	35.38	144.25
Vit D	1000 IU/d	11.97	3.36	8.61
Vit E	IU/d	0.00	430.41	-430.41
ARG	g/d	39.96	16.50	23.46
HIS	g/d	15.25	12.25	3.00
ILE	g/d	34.92	14.00	20.92
LEU	g/d	49.70	33.51	16.19
LYS	g/d	45.75	31.36	14.39
MET	g/d	14.30	9.80	4.50
CYS	g/d	0.00	—	—
PHE	g/d	33.35	17.15	16.20
TYR	g/d	20.17	—	—
THR	g/d	33.56	19.11	14.46
TRP	g/d	19.56	2.94	16.62
VAL	g/d	38.91	20.00	18.91

Nitrogen Balance		Methane (median)
Intake	134.12 g/d	2.917 % GE
Retained	4.80 g/d	6.660 mol/d
Urinary	50.28 g/d	1.419 Mcal/d
Fecal	79.04 g/d	3.991 g/lb DM

FIGURE 27 Summary screen showing the Mechanistic Level of Solution (MLS) predictions.

taking into account feed wasted, moisture content of feeds, and scale accuracy. The accuracy of any model prediction is highly dependent on the DMI used. Intake of each feed must be as uniform as possible over the day because to the committee's knowledge, all field application models assume a total mixed ration with steady-state conditions.

ME-Allowable Performance

If predicted ADG or predicted days for BCS change are not as expected for the conditions described (cattle type, diet type, environment, and management conditions), carefully check all inputs for errors as they are the greatest source of prediction errors. Mistakes or incorrect judgments about inputs such as body size, milk production and its composition, environmental conditions, or feed additives are often made.

- Adjust feed carbohydrate fractions and their digestion rates as necessary. If inputs are correct and performance is still not as expected, predicted dietary energy values are likely the cause. First, check if the predicted total dietary net energy values for each feed are near the expected values. The predicted energy values for individual feeds can be obtained by clicking on the *Supply* button in the *Calculation* subgroup (FIGURE 3F). Feed factors may be influencing energy derived from the diet as the result of feed compositional changes and, possibly, effects on digestion and passage rates. The net energy derived from forages is most sensitive to NDF amount and percentage of the NDF that is lignin, available NDF digestion rate (CHO B3), and peNDF value. For example, if the NDF of a feed is increased, the starch and sugar fractions in the feed will be decreased automatically by the software, more feed dry matter will escape digestion, and the feed will have a lower net energy value. Dry matter digestibility can be further decreased by lowering the NDF digestion rate. After making sure that the feed composition values are appropriate, the digestion rate is considered. The major factors influencing energy derived from feeds high in nonfiber carbohydrates are ruminal and intestinal starch digestion rate (CHO B1). This is mainly a concern when feeding corn grain, corn silage, sorghum grain, or sorghum silage.

- Check postruminal starch digestibility (CB1ID) to make sure that it is appropriate for the starch source being fed. Intestinal digestibilities for starch, pectin, and protein B fraction can be modified in Table 8. The BCNRM assumes an average starch digestibility (CHO B1) of 75%; however, this may not be appropriate for all starch sources. Chapters 2 (Anatomy, Digestion, and Nutrient Utilization) and 19 provide additional values for the effects of processing on corn sources. For example, if cows were supplemented with whole corn, the intestinal starch digestibility should be lowered.

Physically Effective Fiber

This is generally not a problem with high-forage-based beef cow diets. Chapters 2 and 19 provide additional information on this topic.

Ruminally Degradable Protein Balance and Rumen N Balance

If the RDP and ruminal N balances are less than zero and MP balance is negative, feeds such as soybean meal that are high in degradable true protein can be added until these balances are positive. Adjustments can also be made with nonprotein nitrogen or soluble protein until total rumen N is balanced. Because of the number of assumptions required to adequately predict total N balance, including "recycled N" (i.e., urea N used for anabolism), it may be desirable under some conditions to have supply exceed requirements by about 5% (105% of requirement) to allow for prediction errors.

Metabolizable Protein

This component represents an aggregate of nonessential amino acids and essential amino acids. The MP requirement is determined in cows by the body weight and growth requirement, conceptus growth rate, and milk amounts and composition. The adequacy of the diet to meet these requirements will depend on the microbial protein produced from fiber and nonfiber carbohydrate fermentation and feed protein escaping fermentation. If MP balance appears to be unreasonable, check the starch (CHO B1) digestion rates first. Altering the amount of degradable starch will also alter the peptide and total rumen N balance because of altered microbial growth. Often, the most economical way to increase MP supply is to increase microbial protein production by adding highly degradable sources of starch, such as processed grains. Further adjustments are made with feeds high in slowly degraded or rumen escape (bypass) protein (low protein fraction B digestion rate).

- Check total ration protein degradability and individual values for the feeds to compare with the tabular values. If considerably different, the user may have an entry error or need to adjust the protein fraction B digestion rate. First, check protein entered; then, check the protein fraction B digestion rate.

Essential Amino Acids

The amino acid with the least supply as a percentage of requirement is assumed the most limiting for the specified performance. Because of the number of assumptions required to predict amino acid adequacy, it may be desirable under some conditions to have first limiting amino acids exceed requirements by about 5% (105% of requirement) to allow for prediction errors. The adjustment for amino acids is done last because the amino acid balance is affected by the preceding steps. Essential amino acid balances can be estimated with both the ELS and MLS, but the MLS takes into account the effects of the interactions of intake, digestion, and passage rates on microbial yield and available undegraded feed protein, and estimates of their amino acid composition can be predicted along with microbial, body tissue, and milk amino acid composition. However, the development of more accurate feed composition and digestion rates, and more mechanistic approaches to predict utilization of absorbed amino acids will result in improved predictability of diet amino acid adequacy for cattle. To improve the amino acid profile of a ration, use feeds high in the first-limiting amino acids.

Dietary Crude Protein

After all of the above are correctly evaluated, the dietary CP content will be the CP requirement. The CP requirement represents the amount of RDP needed in the rumen and the amount of RUP needed to supplement the microbial protein to meet the MP requirement.

REFERENCES

Fox, D. G., C. J. Sniffen, and J. D. O'Connor. 1988. Adjusting nutrient requirements of beef cattle for animal and environmental variations. *Journal of Animal Science* 66:1475-1495.

NRC (National Research Council). 1985. *Ruminant Nitrogen Usage*. Washington, DC: National Academy Press.

NRC. 1996. *Nutrient Requirements of Beef Cattle* 7th Rev. Ed. Washington, DC: National Academy Press.

NRC. 2000. *Nutrient Requirements of Beef Cattle: Update 2000*, 7th Rev. Ed. Washington, DC: National Academy Press.

Reynolds, C. K., and N. B. Kristensen. 2008. Nitrogen recycling through the gut and the nitrogen economy of ruminants: An asynchronous symbiosis. *Journal of Animal Science* 86 (14 Suppl.):E293-E305.

Appendix B

Committee Statement of Task

A committee will prepare a report that reviews the scientific literature on the nutrition of beef cattle. All life phases and types of production will be addressed. The report will include the following elements: a comprehensive analysis of recent research on feeding and nutrition of beef cattle including research on the amounts of amino acids, lipids, minerals, vitamins, and water needed by growing and reproducing beef cattle; a summary of recent research on energy systems used in beef cattle nutrition; an update of the nutrient requirements contained in the 1996 NRC publication Nutrient Requirements of Beef Cattle; a summary of the composition of feed ingredients, mineral supplements, and feed additives routinely fed to beef cattle; a summary of information about coproducts from the biofuels industry, which will include information about the various types of products and their most effective use and information about phosphorus and sulfur contents; a review of nutritional and feeding strategies to minimize nutrient losses in manure and reduce greenhouse gas production; a discussion of the effect of feeding on the nutritional quality and food safety of beef; new information about nutrient metabolism and utilization (e.g., rumen bypass of protein and lipids, nutrient encapsulation, and starch processing); new information on feed additives that alter rumen metabolism and postabsorptive metabolism; and future areas of needed research. Depending on the extent of new information available, an update of the current computer model to calculate nutrient requirements may be developed.

Appendix C

Abbreviations and Acronyms

AA	amino acid(s)
$AAsh_{BCS}$	proportion of empty body ash at BCS, %
ACP	acyl carrier protein
ADF	acid detergent fiber
ADG	average daily gain
ADGac1	average daily gain after first calving, kg/d
ADGac2	average daily gain after second calving, kg/d
ADGac3	average daily gain after third calving, kg/d
ADGap	average daily gain after first pregnancy, kg/d
ADGbp	average daily gain before first pregnancy, kg/d
ADGpreg	daily gain associated with gravid uterus growth
ADIN	acid detergent insoluble nitrogen (also referred to as ADIP)
ADIP	acid detergent insoluble protein
ADL	acid detergent lignin
$AFat_{BCS}$	proportion of empty body fat at BCS, %
AFRC	Agricultural and Food Research Council
AFSBW	adjusted final shrunk body weight
AGA	American Grassfed Association
Age	current age, days
AI	artificial insemination
ALA	α-linoleic acid
aNDF	amylase-treated neutral detergent fiber
AOAC	Association of Official Analytical Chemists
aPKYD	adjusted peak milk yield, kg/d
$AProt_{BCS}$	proportion of empty body protein at BCS, %
AR	adrenergic receptor
ARA	arachidonic acid
ARC	Agricultural Research Council
ASABE	American Society of Agricultural and Biological Engineers
$AWater_{BCS}$	proportion of empty body water at BCS, %
BCP	bacterial crude protein
BCS	body condition score
BCNRM	Beef Cattle Nutrient Requirements Model
BH	biohydrogenation
BHV-1	bovine herpes virus-1
BIF	Beef Improvement Federation

Bo	maximum CH_4 production potential of manure under ideal conditions
BOLD	Beef in an Optimal Lean Diet
BRD	bovine respiratory disease
BVDV	bovine viral diarrhea virus
BW	body weight
CAFO	concentrated animal feeding operation(s)
CAST	Council for Agricultural Science and Technology
CBW	calf birth weight, kg
CGF	corn gluten feed
CI	calving interval, days
CLA	conjugated linoleic acid
CNCPS	Cornell Net Carbohydrate and Protein System
CO_2e	carbon dioxide equivalents
COMP	NEm adjustment for previous nutrition
CP	crude protein
CPI	crude protein intake
CPM	Cornell-Penn-Miner
CS	condition score
CSIRO	Commonwealth Scientific and Industrial Research Organisation
CV	coefficient of variation
CW	conceptus body weight, kg
D	digestibility of dry matter (as a decimal)
DASH	Dietary Approaches to Stop Hypertension
$Days_{Gain}$	days to gain 1 BCS
$Days_{Lose}$	days to lose 1 BCS
DCAB	dietary cation-anion balance
DCGF	dry corn gluten feed
DDGS	dried distillers grains with solubles
DE	digestible energy
DFM	direct-fed microbials
DGS	distillers grains with solubles
DHA	docosahexaenoic acid
dietCP	dietary crude protein as a percentage of DM
DIM	days in milk or days since calving
DM	dry matter
DMI	dry matter intake
DMIR	dry matter intake required
DP	days pregnant
DRC	dry-rolled corn
DS	dietary salt, %
DWI	daily water intake, L/d
e	base of the natural (Naperian) logarithm (i.e., 2.718)
E	energy content of milk, Mcal/kg
EAT	effective ambient temperature
EBF	empty body fat
EBG	empty body gain
EBW	empty body weight
ECF	extracellular fluid
EE	ether extract
EFA	essential fatty acids
Eh	oxidation-reduction potential

EI	external insulation
ELS	empirical level of solution
EN	endogenous nitrogen
eNDF	effective NDF
EPA	eicosapentaenoic acid
EPD	expected progeny difference(s)
EQSBW	the SBW equivalent to the NRC (1984) medium-framed size steer
FA	fatty acid(s)
FAO	Food and Agriculture Organization of the United Nations
FDA	U.S. Food and Drug Administration
FE	fecal energy
FFA	free fatty acids
FFL	feed dry matter needed to support lactation
FFP	feed dry matter needed to support pregnancy, kg/d
FFTDNI	fat-free total digestible nutrient intake
FHP	fasting heat production
FM	available forage mass
FORA	daily forage allowance
FSBW	final shrunk BW at the expected final body fat
FSSA	Food Safety Standards Agency
G:F	gain-to-feed ratio or feed efficiency, expressed as units of gain divided by units of dry matter intake
GASE	gaseous energy
GE	gross energy
GEI	gross energy intake
GER	gastrointestinal urea-N entry rate, g N/d
GF	green forage availability (1,000 kg/ha)
GHG	greenhouse gas
GIT	gastrointestinal tract
GnRH	gonadotrophin-releasing hormone
GWP	global warming potential
HACCP	Hazard Analysis and Critical Control Point
HAD	Healthy American Diet
HCW	hot carcass weight
HD	effective hair depth, cm
H_e	minimal total evaporative heat loss
HE	heat energy or heat production
H_eE	heat production at zero feed intake
H_iE	heat increment of intake energy
HIDE	adjustment for hide thickness
H_jE	heat of activity associated with obtaining feed
HMC	high-moisture corn
HVFA	undissociated VFA
ICF	intracellular fluid
IE	intake energy
IFSM	Integrated Farm System Model
IFT	Institute of Food Technologists
IGF-I	insulin-like growth factor-I
IgG	immunoglobulin G
I_m	amount of feed consumed at RE = 0

IN	total insulation
INRA	Institut National de la Recherche Agronomique
IOM	Institute of Medicine
IPCC	Intergovernmental Panel on Climate Change
ISAAA	International Service for the Acquisition of Agribiotech Applications
ISBW	initial shrunk body weight
IU	International Units
kd	digestion rate
k_g	efficiency of utilization of ME for RE and in growing animals
k_m	dietary NEm/dietary ME or efficiency of ME utilization for maintenance
kp	rate of passage
k_y	efficiency of ME utilization for pregnancy
LA	linoleic acid
LCFA	long-chain fatty acids
LCPUFA	long-chain omega-3 fatty acids
LCT	lower critical temperature
LH	luteinizing hormone
LPS	lipopolysaccharides
MAFF	Ministry of Agriculture, Fisheries, and Food
MARC	U.S. Meat Animal Research Center
MCF	methane conversion factor
MCP	microbial crude protein
MDGS	modified distillers grains with solubles
MDV	mesenteric-drained viscera
ME	metabolizable energy
MEcs	increase in maintenance energy requirement because of cold stress, Mcal/d
MEI	metabolizable energy intake
MEl	metabolizable energy for lactation, Mcal/d
MEm	ME required for maintenance
MEy	metabolizable energy requirement for pregnancy, Mcal/d
MGA	melengesterol acetate
MkFat	milk fat content, %
MkProt	milk protein content, %
MkSNF	milk solids not fat content, %
MP	metabolizable protein
MPl	metabolizable protein requirement for lactation, g/d
MPm	MP for maintenance
MPN	most probable number
MPy	metabolizable protein requirement for pregnancy, g/d
MT	maximum temperature, °C
MTY	monensin plus tylosin
MTP	microbial true protein
MUD	adjustment for mud thickness
MUFA	monounsaturated fatty acid
MW	mature body weight, kg
NAHMS	National Animal Health Monitoring System
NARMS	U.S. National Antibiotic Resistance Monitoring System
NDF	neutral detergent fiber

NE	net energy
NEcs	increase in net energy requirement associated with cold stress
NEFA	non-esterified fatty acids
NEg	net energy required for gain
NEga	net energy available for growth
NEl	net energy requirement for lactation (or net energy for lactation)
NEm	net energy required for maintenance
NEm_{pa}	net energy required for activity
NEma	net energy available for maintenance
NEmcs	net energy for maintenance under conditions of cold stress
NEr	net energy retained
NEu	energy content of the gravid uterus, kcal
NEy	net energy requirement for pregnancy
NFC	nonfiber carbohydrates, also referred to as non-NDF
NFE	nitrogen-free extract
NI	N intake(s)
NIR	near-infrared reflectance spectroscopy
NP	net protein
NPN	nonprotein nitrogen
NRCS	Natural Resources Conservation Service
OA	organic acid(s)
OM	organic matter
OMD	organic matter digestibility
OMI	organic matter intake
PA	proanthocyanidins
PAC	pharmacologically active compounds
PD	potentially digestible
PDV	portal-drained viscera
pef	physical effectiveness factor
PEM	polioencephalomalacia
peNDF	physically effective neutral detergent fiber
PG	prostaglandins
PHA	phytohaemagglutinin
PI	processing index
PKYD	peak milk yield, kg/d
PM	particulate matter
PP	daily precipitation, cm
PPI	postpartum interval
PSA	phenol-sulfuric acid
PUFA	polyunsaturated fatty acid(s)
RCS	runoff control structures
RDP	ruminally degradable protein
RDO	restricted dedicated optimizer
RE	retained energy
RFI	residual feed intake
RH	relative humidity
RMSE	root mean square error(s)
RMY	relative milk yield, kg/d
ROC	urea N recycled to the ornithine cycle
RSA	reducing sugar assay
RUP	ruminally undegradable protein

SA	surface area
SBW	shrunk body weight
SBW_5	shrunk body weight at BCS 5, kg
SD	standard deviation
SE	standard error
SeCys	selenocysteine
SeMet	selenomethionine
SFA	saturated fatty acid(s)
SFC	steam-flaked corn
SGLT	sodium-dependent glucose transporter
SI	small intestine
SNF	solids not fat
SR	solar radiation
SRC	standardized regression coefficients
SRW	standard reference weight
STEC	shiga toxin-producing *Escherichia coli*
SWG	shrunk weight gain
$\Delta TE_{BCS,Gain}$	gain of total body energy reserves (deposition) per BCS, Mcal
$\Delta TE_{BCS,Lose}$	loss of total body energy reserves (mobilization) per BCS, Mcal
Ta	mean ambient temperature
TAI	timed AI
TBA	trenbolone acetate
TBW	total body water
TCA	target calving age, days
TCW1	target first calving weight, kg
TCW2	target second calving weight, kg
TCW3	target third calving weight, kg
TCW4	target fourth calving weight, kg
TDN	total digestible nutrients
TDNI	total digestible nutrient intake
TDS	total dissolved solids
TE_{BCS}	total body energy reserves at BCS, Mcal
TF	total forage available
$TFat_{BCS}$	empty body fat at BCS, kg
THI	temperature-humidity index
TI	internal or tissue insulation
TMDL	total maximum daily loads
Tmin	daily minimum ambient temperature,°C
TotalY	total milk yield for the n weeks of lactation, kg
TotalYE	total milk net energy requirement (milk energy) for the n weeks in lactation, Mcal
TotalYFat	total milk fat yield for the n weeks in lactation, kg
TotalYProt	total milk protein yield for the n weeks in lactation, kg;
TotalMPl	total milk metabolizable protein for the n weeks in lactation, kg
T_p	previous ambient temperature, °C
TPA	target pregnancy age, days
$TProt_{BCS}$	empty body protein at BCS, kg
TPW	target first pregnancy weight, kg
TROMD	true ruminal organic matter digested
TSS	total soluble salts
UCT	upper critical temperature
UE	urinary energy
UER	urea nitrogen entry rate

UFE	nitrogen from urea excreted in the feces, g N/d
UN	United Nations
USDA	U.S. Department of Agriculture
USP	United States Pharmacopeia
UT-B	ruminal urea transporter
UUA	urea nitrogen used for anabolism, g N/d
UUE	urinary urea nitrogen excretion, g N/d
VFA	volatile fatty acid(s)
VOC	volatile organic compound(s)
WAF_{BCS}	weight adjustment factor at BCS
WCGF	wet corn gluten feed
WDGS	wet distillers grains with solubles
WHO	World Health Organization
WI	water intake
WS	wind speed, km/h
WSC	water-soluble carbohydrates
YE	tissues of the conceptus
YEn	net energy requirement for lactation (daily milk energy), Mcal/kg
YFatn	daily milk fat yield, kg/d
Y_m	methane conversion rate used to express enteric methane production as a percentage of gross energy intake
Yn	daily milk yield at week n postpartum, kg/d
Ypn	net protein retained as conceptus, g/d
YProtn	daily milk protein yield, kg/d

CHAPTER 19 ACRONYMS, ABBREVIATIONS, AND EQUATION TERMS

a	intermediate variable
a1	basal metabolism coefficient, $Mcal/kg^{0.75}$
AA	amino acid(s)
a2	acclimatization factor, $Mcal/kg^{0.75}$
$AAsh_{BCS}$	proportion of empty body ash at a given BCS, %
$Acetate_j^{CA,CB1,CB2,CB3}$	amount of acetate produced from the fermentation of CA, CB1, CB2, and CB3 of the jth feedstuff, kg/d
$Acetate_j^{L}$	amount of acetate produced from the fermentation of lactate of the jth feedstuff, kg/d
$AcidProduct_j^{CA,CB1,CB2,CB3}$	amount of acids produced from the fermentation of CA, CB1, CB2, and CB3 of the jth feedstuff, kg/d
ADGac1	average daily gain after first calving, kg/d
ADGac2	average daily gain after second calving, kg/d
ADGac3	average daily gain after third calving, kg/d
ADGap	average daily gain after first pregnancy, kg/d
ADGbp	average daily gain before first pregnancy, kg/d
ADGpreg	average daily gain of the gravid uterus, kg/d
ADIN	acid detergent fiber
ADTV	feed additive
$ADTV_{DMI}$	adjustment factor for feed additive, dimensionless
$ADTV_{ME}$	effect of ionophore on metabolizable energy
$AFat_{BCS}$	proportion of empty body fat at a given BCS, %
Age	age of the animal, d
Age Factor	adjustment of milk production for cow's age, dimensionless

AIC	Akaike's Information Criterion
aPKYD	adjusted peak milk yield, kg/d
$AProt_{BCS}$	proportion of empty body protein at a given BCS, %
Ash	ash, % DM
Ash_j	ash content of the jth feedstuff, % DM
$aTDN_j$	apparent total digestible nutrients of the jth feedstuff, kg/d
$AWater_{BCS}$	proportion of empty body water at a given BCS, %
BAA_i	i[th] bacterial amino acid content, % MP
BACT	bacterial mass
$BACT_j$	amount of bacterial mass produced from the jth feedstuff, kg/d
$BACT_j^{CA,CB1,CB2,CB3}$	amount of bacteria mass produced from the jth feedstuff, kg/d
BASH	bacterial ash
BCNRM	Beef Cattle Nutrient Requirements Model
BCP	bacterial crude protein
BCS	body condition score (1 to 9 scale); used as a proxy for previous plane of nutrition
BCHO	bacterial carbohydrate
BE	breed factor
BFAT	adjustment factor for body fat, dimensionless
BI	adjustment factor for breed type, dimensionless
$Butyrate_j^{CA,CB1,CB2,CB3}$	amount of butyrate produced from the fermentation of CA, CB1, CB2, and CB3 of the jth feedstuff, kg/d
$Butyrate_j^L$	amount of butyrate produced from the fermentation of lactate of the jth feedstuff, kg/d
BW	full (nonshrunk) body weight, kg
CA_j	sugar content of the jth feedstuff, % DM
Cb	accuracy factor
$CB2_j$	pectin content of the jth feedstuff, % DM
$CB1_j$	starch content of the jth feedstuff, % DM
CB1I	starch intake of the diet, kg/d
CB1ID	intestinal digestibility of CB1 (i.e., starch)
CB2ID	intestinal digestibility of CB2 (i.e., pectin)
$CB1ID_j$	intestinal digestibility of CB1 (i.e., starch) of the jth feedstuff, %
$CB2ID_j$	intestinal digestibility of CB2 (i.e., pectin) of the jth feedstuff, %
$CB3_j$	available (i.e., potentially fermentable) fiber content of the jth feedstuff, % DM
CBW	calf birth weight, kg
CCC	concordance correlation coefficient
CC_j	unavailable fiber content of the jth feedstuff, % DM
CETI	current effective temperature index, °C
CF	crude fiber
CH_4	methane production, g/d
$CH_{4j}^{A,CB1,CB2,CB3,L}$	amount of methane, kg/d
CHO	carbohydrate content of the diet, % DM
CHO FC	fiber carbohydrates
CHO NFC	nonfiber carbohydrates
CHO_j	carbohydrate content of the jth feedstuff, % DM
CI	calving interval, d
CNCPS	Cornell Net Carbohydrate and Protein System
COMP	NEm adjustment for previous nutrition
CP	crude protein content of the diet, % DM

$CpBW_j$ concentrate DMI as a proportion of SBW of the jth feedstuff, g/kg

CPI CP intake of the diet, kg/d

CP_j crude protein content of the jth feedstuff, % DM

CW conceptus weight, kg

DE digestible energy

DE_j digestible energy of the jth feedstuff, Mcal/d

$DIGAA_{j,i}$ i^{th} absorbed amino acid supplied by the jth feedstuff, kg/d

$DIGA_j$ digestible dietary and bacterial ash of the jth feedstuff, kg/d

$DIGBAA_{ji}$ i^{th} absorbed amino acid supplied by the bacteria of the jth feedstuff, kg/d

$DIGFAA_{j,i}$ i^{th} absorbed amino acid supplied by the RUP of the jth feedstuff, kg/d

$DIGC_j$ digestible dietary and bacterial carbohydrate of the jth feedstuff, kg/d

$DIGF_j$ digestible dietary and bacterial fat of the jth feedstuff, kg/d

$DIGP_j$ digestible dietary and bacterial protein of the jth feedstuff, kg/d

DIM days in milk or days since calving

DM dry matter, % as-fed

DMI dry matter intake, kg/d

DMI_j dry matter intake of the jth feedstuff, kg/d

DMI′ auxiliary variable for dry matter intake, kg/d

DMI″ auxiliary variable for dry matter intake, kg/d

DMINC dry matter intake for night cooling adjustment factor, dimensionless

DOP days on pasture

DP days pregnant

e base of the natural (Naperian) logarithm (i.e., 2.718)

E energy content of milk, Mcal/kg

$EAAl_i$ partial efficiency of use of the ith metabolizable amino acid for lactation (Table 19-5), dimensionless

$EAAp_i$ partial efficiency of use of the ith metabolizable amino acid for pregnancy (Table 19-5), dimensionless

EAT effective ambient temperature adjusted for thermal radiation, °C

EBF empty body fat

EBF_c empty body fat when carcass traits are measured, %

EBW empty body weight, kg

EBW_5 empty body weight at BCS 5, kg

EBW_{BCS} empty body weight at a given BCS, kg

EBW_c empty body weight when carcass traits are measured, kg

EE ether extract (fat) content of the diet, % DM

EEI ether extract intake, kg/d

EI external insulation, $°C \cdot m^2 \cdot d/Mcal$; note that EI must be ϑ0

ELS empirical level of solution

$EndProduct_j^{CA,CB1,CB2,CB3}$ amount of carbohydrate or carbon skeleton of peptides not retained by the bacteria from the jth feedstuff, kg/d

$EndProduct_j^{L}$ end product of lactate degradation in the rumen of the jth feedstuff, kg/d

EQEBW equivalent empty body weight, kg

EQSBW shrunk body weight equivalent to the NRC (1984) medium-framed size steer

EWG empty [body] weight gain, kg/d

$FAA_{j,i}$	i^{th} amino acid content of the jth feedstuff, % MP
$Factor_{L,B,A:p}^{CA,CB1,CB2,CB3}$	mass fraction coefficients to convert $AcidProduct_j^{CA,CB1,CB2,CB3}$ into lactate, acetate, propionate, and butyrate for the jth feedstuff, dimensionless
$Factor_{A,P,B}^{L}$	mass fraction coefficients to convert $EndProduct_j^{L}$ into acetate, propionate, and butyrate for the jth feedstuff, dimensionless
Fat_j	fat (i.e., ether extract) content of the jth feedstuff, % DM
FC	fiber carbohydrate
$FDMI_j$	forage DMI of the jth feedstuff, kg/d
F_E	metabolic endogenous protein
FEA_j	fecal ash of the jth feedstuff without endogenous ash, kg/d
$FEASH_j$	fecal ash of the jth feedstuff, kg/d
FEC_j	fecal carbohydrate of the jth feedstuff, kg/d
$FECHO_j$	fecal carbohydrate of the jth feedstuff, kg/d
$FEENGA_{M+E,j}$	fecal endogenous and microbial ash contribution of the jth feedstuff, kg/d
$FEENGF_{M+E,j}$	fecal endogenous and microbial fat contribution of the jth feedstuff, kg/d
$FEENGP_{M+E,j}$	fecal endogenous and microbial protein contribution of the jth feedstuff, kg/d
FEF_j	fecal fat of the jth feedstuff without endogenous fat, kg/d
$FEFAT_j$	fecal fat of the jth feedstuff, kg/d
FEP_j	fecal protein of the jth feedstuff without endogenous protein, kg/d
$FEPROT_j$	fecal protein of the jth feedstuff, kg/d
FFL	feed dry matter needed to support lactation, kg/d; assumes partial efficiency of use of metabolizable energy to net energy for lactation (k_l) is identical to k_m
FFM	feed dry matter needed to support maintenance, kg/d
FFP	feed dry matter needed to support pregnancy, kg/d
FFTDNI	fat-free total digestible nutrient intake, g/d
$FFTDN_j$	fat-free total digestible nutrient content of the jth feedstuff, % DM
F_M	metabolic microbial residues
F_{M+E}	fecal metabolic microbial residue plus metabolic endogenous protein
FM	available forage mass, kg DM/ha
FORA	daily forage allowance, g DM/kg SBW
$Forage_{\%DM}$	forage content of the jth feedstuff, % DM
$Forage_{DMI}$	forage intake of the diet, kg/d
$Forage_P$	forage of the diet, % DM
$FpBW_j$	forage DMI as a proportion of SBW of the jth feedstuff, g/kg
$FracP_j^{CA,CB1,CB2,CB3,L}$	molar fraction of propionate of VFA (acetate, propionate, and butyrate), dimensionless
FT	fat thickness (i.e., "backfat" at the 12th rib), cm
F_U	undegraded feed protein that is not digested and absorbed in the small intestine
GEI	gross energy intake, Mcal/d
GI	unadjusted grazing DMI, g DM/kg SBW
GRAZE	DMI adjustment factor for grazing animals, dimensionless
GU	grazing unit size, ha/animal

Hair	hair depth, cm
Hair Coat	1 for no mud, 0.8 for some mud on lower body, 0.5 for mud on lower body and sides, and 0.2 for heavily covered with mud, dimensionless
HCW	hot carcass weight, kg
HE	heat production, Mcal/d
HIDE	0.8 for thin hide thickness, 1.0 for average hide thickness, and 1.2 for thick hide thickness, dimensionless
HRS	hours of sunlight (assumed to be 8 h/d)
ID	intestinal digestibilities
IN	total insulation, °C·m²·d/Mcal
IPCC	Intergovernmental Panel on Climate Change
ISBW	initial shrunk body weight, kg
k	intermediate variable, dimensionless
kd	digestion rate
$kd_{CD1,j}$	fractional rate of degradation of CB1 of the jth feedstuff, %/h
$kd_{CD2,j}$	fractional rate of degradation of CB2 of the jth feedstuff, %/h
$kd^{*}_{CB3,j}$	adjusted fractional rate of degradation of CB3 of the jth feedstuff, %/h
kd^{L}	fractional rate of degradation of lactate by *Megasphaera elsdenii*, %/h
$kd_{PB,j}$	fractional rate of degradation of PB of the jth feedstuff, %/h
k_m	partial efficiency of use of ME to NE for maintenance
KM1	maintenance rate of the nonfiber carbohydrate bacteria, 0.15 g NFC/g bacteria per hour
KM2	maintenance rate of the fiber carbohydrate bacteria, 0.05 g FC/g bacteria per hour
KM2*	adjusted maintenance rate of the fiber carbohydrate (i.e., FC) bacteria, 0.05 g FC/g bacteria per hour
kp	fractional passage rate
$kp_{concentrate,j}$	fractional passage rate of concentrate for the jth feedstuff, %/h
$kp_{forage,j}$	fractional passage rate of forage for the jth feedstuff, %/h
kp_j	fractional passage rate of the jth feedstuff, %/h
$kp_{liquid,j}$	fractional passage rate of liquids for the jth feedstuff, %/h
L	lactation factor
$Lactate_j^{CA,CB1,CB2,CB3}$	amount of lactate produced from the fermentation of CA, CB1, CB2, and CB3 of the jth feedstuff, kg/d
LCT	lower critical temperature, °C
$Lignin_j$	lignin content of the jth feedstuff, % DM
LNT	lowest night temperature (night cooling if LNT < 20°C), °C
μ^{L}	net growth rate of *M. elsdenii* on lactate, 1/h
MAA	metabolizable amino acid
$MAAg_i$	requirement of the ith metabolizable amino acid for growth, g/d
$MAAl_i$	requirement of the ith metabolizable amino acid for lactation, g/d
$MAAm_i$	requirement of the ith metabolizable amino acid for maintenance, g/d
$MAAp_i$	requirement of the ith metabolizable amino acid for pregnancy, g/d
MARC	U.S. Meat Animal Research Center
MCP	microbial crude protein, g/d
MCP_j	microbial crude protein supplied by the jth feedstuff, kg/d

MCP_{RDP}	RDP (protein) allowable microbial crude protein, kg/d
MCP_{TDN}	TDN (energy) allowable microbial crude protein, kg/d
ΔME	metabolizable energy balance (supply minus required), Mcal/d
$[ME]_j$	concentration of ME of the jth feedstuff, Mcal/kg, adjusted (or not) for ionophores
MEcs	metabolizable energy requirement for maintenance due to cold stress, Mcal/d
MEI	metabolizable energy intake, Mcal/d
ME_j	metabolizable energy of the jth feedstuff, Mcal/d
MEl	metabolizable energy requirement for lactation, Mcal/d
MEm_{chs}	metabolizable energy requirement for maintenance plus cold stress or heat stress, Mcal/d
MEy	metabolizable energy requirement for pregnancy, Mcal/d
$MilkAA_j$	milk composition of the ith amino acid, g/d
MkFat	milk fat content, %;
MkProt	milk protein content, %
MkSNF	milk solids not fat content, %
m^L	maintenance coefficient of *M. elsdenii* on lactate, g/g per hour
MLS	mechanistic level of solution
MP	metabolizable protein
$MPfeed_j$	metabolizable protein supplied by the jth feedstuff, kg/d
MPg	metabolizable protein requirement for gain, g/d
MP_j	metabolizable protein of the jth feedstuff, kg/d
MPl	metabolizable protein requirement for lactation, g/d
MPm	metabolizable protein requirement for maintenance, g/d
$MPmtp_j$	bacterial metabolizable protein supplied by the jth feedstuff, kg/d
MPy	metabolizable protein requirement for pregnancy, g/d
MSBW	mature SBW (typically 96% of full weight), kg
MSEP	mean square error of prediction
MTP	microbial true protein of the diet, g/d
MUD1	mud depth adjustment factor, dimensionless
MW	mature body weight
n	week in lactation
N1	value of the ith nutrient of the feed library developed in Chapter 18
N2	value of the same i^{th} nutrient of the feed library from NRC (1996, 2000)
NAA	net AA
NDF	neutral detergent fiber, % DM
NDF_j	neutral detergent fiber content of the jth feedstuff, % DM
NDFI	neutral detergent fiber intake of the diet, kg/d
NEcs	net energy requirement for maintenance due to cold stress, Mcal/d
NEga	dietary net energy available for growth, Mcal/kg
$NEga_j$	net energy available for growth of the jth feedstuff, Mcal/kg
NEhs	net energy requirement due to heat stress, Mcal/d
NEm	net energy requirement for maintenance, Mcal/d
NEma	diet net energy available for maintenance, Mcal/kg
NEma′	auxiliary variable for the dietary net energy available for maintenance, Mcal/kg
NEm_{chs}	net energy requirement for maintenance plus cold stress or heat stress, Mcal/d
$NEma_j$	net energy available for maintenance of the jth feedstuff, Mcal/kg

NEy	net energy requirement for pregnancy, Mcal/d
NFC	nonfiber carbohydrate
NFC_j	nonfiber carbohydrate content of the jth feedstuff, % DM
NFCI	nonfiber carbohydrate intake of the diet, kg/d
NFE	nitrogen-free extract
NI	nitrogen intake, g/d
NP	net protein
NPg	net protein requirement for gain, g/d
NPN	nonprotein nitrogen
OA	organic acid content of the diet, % DM
OA_j	organic acid content of the jth feedstuff, % DM
PA	proanthocyanidins
PA_j	instantaneously degraded protein fraction in the rumen of the jth feedstuff, % DM
PB_j	potentially degradable protein fraction in the rumen of the jth feedstuff, % DM
PBID	protein B fraction intestinal digestibility
$PBID_j$	intestinal digestibility of the protein B fraction of the jth feedstuff, %
pef	physical effectiveness factor
pef_j	physically effective factor of the jth feedstuff, % NDF
peNDF	physically effective NDF of the diet, % NDF
pH	ruminal pH
PKYD	peak milk yield (Table 19-1), kg/d
$Propionate_j^{CA,CB1,CB2,CB3}$	amount of propionate produced from the fermentation of CA, CB1, CB2, and CB3 of the jth feedstuff, kg/d
$Propionate_j^{L}$	amount of propionate produced from the fermentation of lactate of the jth feedstuff, kg/d
QG	USDA quality grade code (4 for Select, 5 for low Choice, 6 for Choice, 7 for high Choice, and 8 for Prime)
$[RUP]_j$	concentration of ruminally undegradable protein of the jth feedstuff, % CP
$RDCA_j$	ruminally degraded carbohydrate A of the jth feedstuff, kg/d
RDCB1	amount of ruminally degraded carbohydrate B1 (starch), kg/d
$RDCB1_j$	ruminally degraded carbohydrate B1 (starch) of the jth feedstuff, kg/d
RDCB2	ruminally degraded carbohydrate B2 (pectin) of the jth feedstuff, kg/d
RDCB3	ruminally degraded carbohydrate B3 (available fiber) of the jth feedstuff, kg/d
RDL_j	ruminally degraded lactate by *M. elsdenii* of the jth feedstuff, kg/d
RDP	ruminally degradable protein
$RDPA_j$	ruminally degraded protein A (soluble) of the jth feedstuff, kg/d
$RDPB_j$	ruminally degraded protein B of the jth feedstuff, kg/d
$RDP_{Efficiency}$	efficiency of use of RDP by ruminal microbes (default = 1)
RDP_j^*	RDP adjusted for NPN feedstuffs (i.e., urea) of the jth feedstuff, kg/d
RE	retained energy, Mcal/d
REA	ribeye area (i.e., longissimus dorsi muscle area), cm^2
$REASH_j$	ruminally escaped ash of the jth feedstuff, kg/d

$REBASH_j$	ruminally escaped bacterial ash of the jth feedstuff, kg/d
$REBCHO_j$	ruminally escaped bacterial carbohydrate of the jth feedstuff, kg/d
$REBCWP_j$	ruminally escaped bacterial cell wall protein of the jth feedstuff, kg/d
$REBFAT_j$	ruminally escaped bacterial fat of the jth feedstuff, kg/d
$REBNA_j$	ruminally escaped bacterial nucleic acids of the jth feedstuff, kg/d
$REBTP_j$	ruminally escaped bacterial true protein of the jth feedstuff, kg/d
$RECB1_j$	ruminally escaped carbohydrate B1 (starch) of the jth feedstuff, kg/d
$RECB2_j$	ruminally escaped carbohydrate B2 (pectin) of the jth feedstuff, kg/d
$RECB3_j$	ruminally escaped carbohydrate B3 (available fiber) of the jth feedstuff, kg/d
$RECC_j$	ruminally escaped carbohydrate C of the jth feedstuff, kg/d
RecycledN	either urea N used for anabolism (UUA) or recycled ruminal N depending on the equation chosen (default = 0, not used), g/d
$REFAT_j$	ruminally escaped fat (i.e., ether extract) of the jth feedstuff, kg/d
R_Y^{FC}	relative yield of fiber carbohydrate bacteria adjusted for ruminal pH
$REPB_j$	ruminally escaped protein B of the jth feedstuff, kg/d
$REPC_j$	ruminally escaped protein C of the jth feedstuff, kg/d
RHc	current relative humidity, %
RHp	previous relative humidity
RMY	relative milk yield, dimensionless
RNB	ruminal nitrogen balance, g/d
RUP	ruminally undegradable protein
RUP_j	ruminally undegradable protein of the jth feedstuff, kg/d
SA	surface area, m^2
SBW	shrunk body weight, kg (typically 96% of full body weight)
SEX	gender effect factor (1.15 for bulls or 1 otherwise)
SRC	standardized regression coefficients
SRW	standard reference weight, kg
$Starch_j$	starch content of the jth feedstuff, % DM
$Sugars_j$	sugar content of the jth feedstuff, % DM
SWG	shrunk [body] weight gain (i.e., shrunk ADG), kg/d
$\Delta TE_{BCS,Gain}$	gain of total energy of reserves (deposition) at a given BCS, Mcal
$\Delta TE_{BCS,Lose}$	loss of total energy of reserves (mobilization) at a given BCS, Mcal
T	week of peak milk
Tc	current temperature, °C
TCA	target calving age, days
TCW1	target first calving weight, kg
TCW2	target second calving weight, kg
TCW3	target third calving weight, kg
TCW4	target fourth calving weight, kg
TDN	total digestible nutrients
TDN_j	total digestible nutrient of the jth feedstuff, % DM
TE_{BCS}	total body energy reserves at BCS, Mcal
TEMP1	adjustment for temperature, dimensionless
$TFat_{BCS}$	total empty body fat at a given BCS, kg

TI — tissue insulation, °C·m^2·d/Mcal
$TissueAA_i$ — tissue composition of the ith amino acid, g/g
TotalMP1 — total milk metabolizable protein for the n weeks in lactation, kg
TotalY — total milk yield for the n weeks in lactation, kg
TotalYE — total milk net energy requirement (milk energy) for the n weeks in lactation, Mcal
TotalYFat — total milk fat yield for the n weeks in lactation, kg
TotalYProt — total milk protein yield for the n weeks in lactation, kg
Tp — previous temperature, °C
TPA — target pregnant age, days
$TProt_{BCS}$ — total empty body protein at a given BCS, kg
TPW — target pregnant weight, kg

USDA — U.S. Department of Agriculture
UUA — urea N used for anabolism

VFA — volatile fatty acid(s)
VFA_j — amount of volatile fatty acids of the jth feedstuff, g/d
WAF — weight adjustment factor
WAF_{BCS} — weight adjustment factor at a given BCS
WI — water intake, L/d
WS — wind speed, km/h; note that WS must be ≤32 km/h

$Y_{CA,j}$ — bacterial yield for sugar carbohydrate (CA) of the jth feedstuff, g bacteria/g carbohydrate
$Y_{CD1,j}$ — bacterial yield for starch carbohydrate (CB1) of the jth feedstuff, g bacteria/g carbohydrate
$Y_{CB2,j}$ — bacterial yield for pectin (CB2) of the jth feedstuff, g bacteria/g carbohydrate
$Y_{CB3,j}$ — bacterial yield for fiber carbohydrate (CB3) of the jth feedstuff, g bacteria/g carbohydrate
$Y^*_{CB3,j}$ — adjusted bacterial yield for fiber carbohydrate (CB3) of the jth feedstuff, g bacteria/g carbohydrate
$YCH_{4k}^{A,CB1,CB2,CB3,L}$ — mass of methane per mass of propionate, dimensionless
YEn — net energy requirement for lactation (daily milk energy), Mcal/d
YFatn — daily milk fat yield, kg/d
YG1 — theoretical maximum bacterial yield for CA, CB1, and CB2, 0.4 g bacteria/g carbohydrate
YG2 — theoretical maximum bacterial yield for CB3, 0.4 g bacteria/g carbohydrate
YG^L — maximum yield of *M. elsdenii* on lactate, g/d
Y^L — yield of *M. elsdenii* on lactate, g/g
Yn — milk yield, kg/d
Ypn — net protein retained as conceptus, g/d
YProtn — daily milk protein yield, kg/d

Appendix D

Committee Member Biographies

Michael L. Galyean is the dean of the College of Agricultural Sciences and Natural Resources (April 2012-present), a Horn Professor (since 2006), and the Thornton Distinguished Chair in beef cattle nutrition at the Department of Animal and Food Sciences (1998-present) at Texas Tech University. He joined the faculty at Texas Tech in 1998 and served as the chair of the university's Animal Care and Use Committee from 2002 to 2006. Previously, he was assistant professor, associate professor, and professor in the Department of Animal and Range Sciences at the New Mexico State University (NMSU) from 1977 to 1990; professor and superintendent on the NMSU Clayton Livestock Research Center from 1990 to 1996; and professor of animal science at West Texas A&M University and the Texas Agricultural Experiment Station in Amarillo from 1996 to 1997. Dr. Galyean and his students and colleagues have authored 239 peer-reviewed journal articles, 59 invited papers and book chapters, and numerous other published proceedings, progress reports, experiment station articles, and abstracts. Twenty-nine MS students, 33 PhD students, and 9 postdoctoral research associates have worked under his guidance, and his research has been supported by more than $2.6 million in grant funds. He is a member of the American Society of Animal Science (ASAS), the American Registry of Professional Animal Scientists, the American Dairy Science Association, and the American Society for Nutritional Sciences. His awards include the College of Agricultural Sciences and Natural Resources Outstanding Researcher Award (2004, 2005), the President's Academic Achievement Award (2005) from Texas Tech, and his appointment as a Horn Professor in 2006. He also received the Animal Management Award from ASAS in 2006, and the College of Agricultural Sciences and Natural Resources Service and Outreach Award from Texas Tech in 2007. He was named a Fellow of the ASAS (2010) and was given the Morrison Award by the ASAS (2012) and the Federation of Animal Science Societies New Frontiers in Animal Nutrition Award (2013). Dr. Galyean served as a member of the National Research Council's Committee on Animal Nutrition (1997-2000) and Subcommittee on Beef Cattle Nutrition (1991-1996, 1997-1998). He was President of the Western Section of the ASAS and three times a member of the ASAS Board of Directors. In addition, he served three terms on the Editorial Board of the *Journal of Animal Science*, one 3-year term as a Section Editor, and as Editor-in-Chief from 2002 to 2005. From 2006 to 2009, he served as President-Elect, President, and Past President of the ASAS. He is currently Past President of the American Registry of Professional Animal Scientists, serving as President-Elect and President from 2012 to 2014. He holds a BS degree in agriculture from NMSU, and an MS degree in animal science and a PhD in animal nutrition, both from Oklahoma State University.

Karen A. Beauchemin is a senior research scientist at Agriculture and Agri-Food (AAFC) Canada's Lethbridge Research Centre in Alberta and an adjunct professor at the University of Alberta and the University of Saskatchewan. Before her career in research, Dr. Beauchemin spent several years in the feed industry. She obtained a PhD in ruminant nutrition at the University of Guelph (1988), an MSc in animal nutrition at Université Laval (1982), and a BSc in agriculture at McGill University (1978). Dr. Beauchemin has developed a broad-based research program to improve feed utilization of cattle and reduce environmental impact of beef and dairy production. She is recognized for her expertise in the areas of digestion kinetics, acidosis, rumen function, fiber requirements of cattle, and mitigation of enteric methane emissions. Throughout her career, Dr. Beauchemin has published more than 275 original refereed scientific papers, 20 book chapters, 23 authoritative reviews, and numerous other nonrefereed publications. She has been a speaker at numerous scientific and industry meetings. She is a member of the American Society of Animal Science (ASAS), the American Society of Dairy Science (ADSA), and the Canadian Society of Animal Science (CSAS). She is the recipient of numerous prestigious awards, including the ADSA Forage Award

(2005), the AAFC Gold Harvest Award of Excellence (2007), the CSAS Excellence in Nutrition and Meat Science Award (2009), the ADSA Applied Dairy Nutrition Award (2010), the Bertebos Prize from the Royal Swedish Academy of Agriculture and Forestry (2011), and the CSAS Fellowship Award (2014).

Joel Caton was raised in central Missouri on a diversified livestock and grain farm. He received his BS in animal science from New Mexico State University in 1981, his MS in animal science from University of Missouri in 1983, and his PhD in ruminant nutrition from New Mexico State University in 1987. In 1988, he completed a postdoctoral fellowship at University of Missouri and became an assistant professor at North Dakota State University (NDSU). While progressing through the ranks at NDSU he completed sabbaticals at the University of Reading, UK and the Rowett Research Institute in Aberdeen, Scotland. He is currently an Engberg Endowed Professor in animal sciences at NDSU and holds an 80% research, 20% teaching appointment. He has given over 60 invited presentations including national and international (11 different countries) venues. His nutrition research program has attracted over $5.1 million in grant funds and resulted in over 620 total publications (162 refereed articles and/or book chapters, 76 edited and reviewed proceedings, 306 abstracts, and 83 other nonrefeered publications). To date, he has mentored, as major or co-major advisor, 5 postdoctoral fellows, 41 graduate students, and over 30 undergraduate research experiences. He has served on study sections/peer-review panels, and as ad hoc reviewer for numerous granting agencies including USDA, NIH, and NSF. He has also served as Associate Editor, Division Editor, and Associate Editor-in-Chief for the *Journal of Animal Science* and as ad hoc reviewer for 16 other journals. He received the NDSU College of Agriculture's Research Award in 2003 and the National AFIA Ruminant Nutrition Award from the American Society of Animal Science in 2004. In 2013, he received the American Society of Animal Science Gary L. Cromwell Award in Mineral Nutrition.

Noel Andy Cole worked as a supervisory research animal scientist and laboratory director at the Conservation and Production Research Laboratory of the USDA Agricultural Research Service (ARS) in Bushland, Texas, from 1977 to 2015. For 38 years he conducted research on the effects of nutrition and management on the stress response and nutrient status of transported feeder cattle and subsequent effects on animal health and performance (1977-1996) and on the effects of nutrition and management on nutrient excretion, manure management, and ammonia and greenhouse gas emissions from beef cattle feedyards (1996-2015). Dr. Cole has received research grants from industry, and state and federal governments totaling more than $3 million. He has authored or coauthored over 350 publications including 105 refereed journal papers and has given over 120 oral presentations about his research to producer and scientific groups in 20 states, Canada, Brazil, Mexico, and Thailand. His honors include the West Texas A&M University (WTAMU) Agricultural Department Graduate of Distinction Award (2006), the Advanced Degree Graduate of Distinction Award from the Department of Animal Science at Oklahoma State University (2007), the WTAMU Distinguished Alumnus Award (2010), and the USDA-ARS Southern Plains Area Senior Scientist of the Year Award (2010). Dr. Cole is also the recipient of three awards from the American Society of Animal Science: the Animal Management Research Award (2005), the Research Fellow Award (2009), and the Ruminant Nutrition Research Award (2011). He obtained his BS degree in agriculture (animal science option) with a minor in biology from WTAMU (Canyon) in 1971, and his MS degree in animal science and PhD in animal nutrition with a minor in biochemistry from Oklahoma State University (Stillwater) in 1973 and 1975, respectively.

Joan Eisemann is a professor of nutrition at North Carolina State University (NCSU). She received her BS degree from the University of Connecticut in nutritional sciences, and her MS and PhD degrees from Cornell University in animal nutrition and nutritional biochemistry, respectively. After completing her degrees, she went to the Beltsville Agricultural Research Center as a postdoctoral research associate. In 1984, she joined the Agricultural Research Service at the U.S. Meat Animal Research Center in Clay Center, Nebraska. In 1991, she joined the Department of Animal Science at NCSU with responsibilities in research and teaching. Her research has encompassed aspects of nutrition, physiology, and endocrinology of growing ruminants and swine. Her primary research interest is regulation of nutrient partitioning in growing animals with emphasis on nitrogen metabolism. She received an Outstanding Teacher Award (2007) and the Gertrude M. Cox Award for Innovation in Teaching with Technology (2012) from NCSU. She served on numerous committees of the American Society of Animal Science (ASAS) and the American Society for Nutritional Sciences. Activities associated with the ASAS include service on the editorial board and as an Associate Editor of the *Journal of Animal Science*. She also served as a member of the ASAS Board of Directors. She served as a member of the National Research Council Board on Agriculture and Natural Resources (2007-2012).

Terry Engle is a professor of animal science at Colorado State University where he teaches vitamin and mineral metabolism, animal metabolism, and other courses. His research interests include mineral metabolism in ruminants, with primary emphasis on the role of minerals and other nutrients on immune response, disease resistance, and lipid metabolism. He is also interested in the molecular aspects of mineral absorption and transport. He has published 99 refereed manuscripts, 84 abstracts, and 33 other technical and lay publications, and has given 78 invited presentations at symposia, professional meetings, or workshops. He is a member of the American Society of Animal Science (ASAS) and was a member (2005) and chair (2006) of the ASAS Western Section Symposium Committee, and member of the American Society for Nutritional Sciences, the Society for Experimental Biology and Medicine, Gamma Sigma Delta Honor Society of Agriculture, the American Registry of Professional Animals Scientists, and Sigma Xi Scientific Research Society. In 2010, he received the Early Career Achievement Award from the ASAS. He earned his BS and MS degrees in animal science from Colorado State University in 1993 and 1996, and his PhD in nutrition from North Carolina State University in 1999.

Galen Erickson is a professor and beef feedlot extension specialist and the Nebraska Cattle Industry Professor of Animal Science, at the Animal Science Department, University of Nebraska-Lincoln. His program focuses on byproduct feeds utilized by beef cattle, nutrition-environmental interactions with a focus on nitrogen and phosphorus, and grain feeding and growth promotion for feedlot cattle. His research group has established feeding values for numerous different types of co-product feeds that are widely used by the beef industry (primarily finishing beef cattle) and is currently evaluating optimal dietary ingredients such as grain processing, forage type and amounts, as well as other nutritional and management challenges related to finishing cattle. His awards include the University of Nebraska-Lincoln Institute of Agriculture and Natural Resources Dinsdale Faculty Award (2005); Wendall Burgher Beef Industry Award (2007-2009); Midwest American Society of Animal Science Young Outstanding Researcher Award (2009); the American Society of Animal Science Early Career Achievement Award (2009); and the American Feed Industry Association Ruminant Animal Nutrition Award (2015). Dr. Erickson is a member of the American Society of Animal Science (ASAS); the American Registry of Professional Animal Scientists; Sigma Xi, Scientific Research Society; Gamma Sigma Delta Honor Society of Agriculture; and Phi Beta Delta Honor Society for International Scholars. He was Chair of the Midwest ASAS Beef Extension Committee (2007-2009) and the Odor and Nutrient Management Committee (2004-2006). He has supervised 49 graduate student degree programs, and has authored 111 journal articles, 6 book chapters, 340 abstracts, and 342 producer publications. He received his BS degree from Iowa State University (Ames) in 1995 and his MS and PhD degrees from the University of Nebraska-Lincoln in 1997 and 2001.

Clinton Krehbiel is a regents professor of animal science and the Dennis and Marta Endowed Chair (2009-present) at Oklahoma State University (OSU). Before joining OSU as an assistant professor in 2000, Dr. Krehbiel was an assistant professor at New Mexico State University from 1996 to 1999. His research interests include tissue and whole-animal energy and protein metabolism in ruminants; regulation of lipid metabolism; impact of animal health and immune function on animal growth and carcass merit; nutritional/management strategies of adapting and subsequently feeding beef cattle on high-concentrate diets, while minimizing risk of metabolic disorders; and systems research to improve efficiency of nutrient utilization by growing and finishing ruminants. He received the American Society of Animal Science Southern Section Outstanding Young Animal Scientist Award (Research) in 2005; the Oklahoma State University Department of Animal Science Tyler Award in 2005; the James A. Whatley Award for Meritorious Research in Agricultural Science in 2006; the Gamma Sigma Delta Experienced Research Scientist Award of Merit in 2008; the Sarkey's Distinguished Professor Award in 2014; and the Oklahoma State University Service Award in 2014. He is a member of the American Dairy Science Association; the American Registry of Professional Animal Scientists; the American Society of Animal Science (he served as Director-at-Large from 2009 to 2012); Gamma Sigma Delta, OSU (he served as Chapter President from 2006 to 2007); Plains Nutrition Council (he served as President from 2007 to 2008); and Sigma Xi, OSU Chapter (he served as President from 2007 to 2008). Dr. Krehbiel has published 105 journal articles, 5 book chapters, 59 proceedings papers, and over 105 experiment station publications. He obtained his BS degree in animal science and industry and his MS degree in animal science and industry (animal nutrition) from Kansas State University in 1988 and 1990, and his PhD in animal science (animal nutrition) from the University of Nebraska-Lincoln in 1994.

Ronald Lemenager is a professor and beef extension specialist in the Department of Animal Sciences at Purdue University. His research program is focused on the interaction of nutrition and reproduction in beef cow management. His extension program is aimed at developing beef production systems and marketing strategies that will not only increase animal performance and producer profitability, but also meet consumer expectations for safe, wholesome, high-quality beef in an environmentally and socially responsible manner. Dr. Lemenager is a member of the American Society of Animal Science (ASAS) and the American Registry of Professional Animal Scientists. He is a Charter Diplomat of the American College of Animal Nutrition. He has served as a director and president of the Midwest Section of the ASAS, a member of the Production Research and Producer Education Committees of the National Cattlemen's Beef Association, and as a director on the Executive Beef Board of the Indiana Beef Cattle Association. Dr. Lemenager is a recipient of the Purdue University Outstanding Teacher and Counselor Awards as well as the ASAS Midwest Section Teaching Award. He also received the Graduate of Distinction Award from Oklahoma State University (2001); two Purdue University, College of Agriculture Team Awards (Ethanol Co-products, 2009; and Moldy Corn Rapid Response, 2010);

and the Special Award from Purdue University Cooperative Extension Specialist's Association (2009). The Indiana Beef Cattle Association recognized him with the Friend of the Beef Industry Award (1992), Special Service Award (2008), and Outstanding Cattleman Award (2012). He holds a BS degree in animal science from the University of Illinois and MS and PhD degrees from Oklahoma State University in ruminant (beef) nutrition.

Luis O. Tedeschi is a professor in the Department of Animal Science at Texas A&M University and Texas A&M AgriLife Research. He is also a member of the Department of Nutrition and Food Science. He conducts research on energy and nutrient requirements of grazing and feedlot animals, growth biology and bioenergetics, chemical composition and kinetics of fermentation of feeds, modeling and simulation of decision support systems, and evaluation of mathematical models. He has also collaborated with researchers overseas to develop mathematical models for small ruminants. Dr. Tedeschi has been an active developer of submodels and a contributor to the Cornell Net Carbohydrate and Protein System and the Cattle Value Discovery System. He has published more than 165 peer-reviewed articles and book chapters, and has presented in more than 70 conferences and workshops in modeling nutrition worldwide. Dr. Tedeschi was a recipient of the prestigious McMaster Fellowship to conduct modeling research at the Commonwealth Scientific and Industrial Research Organisation Livestock Industries, Australia, in 2011. In 2012, he was conferred the Texas A&M AgriLife Vice-Chancellor's Award in Excellence. In 2013, he was awarded the distinguished J. William Fulbright Scholarship to collaborate with the Brazilian Research Company, EMBRAPA, Brazil, in the development of nutrition models to enhance the understanding of the impact of ruminant production on global warming through the emission of methane. He is a member of the American Society of Animal Science, the American Dairy Science Association, the Brazilian Society of Animal Science, the American Registry of Professional Animal Scientists, and the System Dynamics Society. He received his BS degree in agronomy engineering in 1991 and his MS degree in animal and forage sciences in 1996 from the University of São Paulo (ESALQ/USP, Piracicaba, Brazil), and his PhD degree in animal science from Cornell University in 2001.

Appendix E

Recent Publications of the Board on Agriculture and Natural Resources

POLICY AND RESOURCES

Achievements of the National Plant Genome Initiative and New Horizons in Plant Biology (2008)
Achieving Sustainable Global Capacity for Surveillance and Response to Emerging Diseases of Zoonotic Origin: Workshop Report (2008)
Agricultural Biotechnology and the Poor: Proceedings of an International Conference (2000)
Agriculture, Forestry, and Fishing Research at NIOSH (2008)
Agriculture's Role in K-12 Education (1998)
Air Emissions from Animal Feeding Operations: Current Knowledge, Future Needs (2003)
An Evaluation of the Food Safety Requirements of the Federal Purchase Ground Beef Program (2010)
Animal Biotechnology: Science-Based Concerns (2002)
Animal Care and Management at the National Zoo: Final Report (2005)
Animal Care and Management at the National Zoo: Interim Report (2004)
Animal Health at the Crossroads: Preventing, Detecting, and Diagnosing Animal Diseases (2005)
Biological Confinement of Genetically Engineered Organisms (2004)
California Agricultural Research Priorities: Pierce's Disease (2004)
Changes in the Sheep Industry in the United States: Making the Transition from Tradition (2008)
Countering Agricultural Bioterrorism (2003)
Critical Needs for Research in Veterinary Science (2005)
Critical Role of Animal Science Research in Food Security and Sustainability (2015)
Designing an Agricultural Genome Program (1998)
Diagnosis and Control of Johne's Disease (2003)
Direct and Indirect Human Contributions to Terrestrial Carbon Fluxes (2004)
Ecological Monitoring of Genetically Modified Crops (2001)
Emerging Animal Diseases: Global Markets, Global Safety: Workshop Summary (2002)
Emerging Technologies to Benefit Farmers in Sub-Saharan Africa and South Asia (2008)
Enhancing Food Safety: The Role of the Food and Drug Administration (2010)
Ensuring Safe Food: From Production to Consumption (1998)
Environmental Effects of Transgenic Plants: The Scope and Adequacy of Regulation (2002)
Evaluation of a Site-Specific Risk Assessment for the Department of Homeland Security's Planned National Bio- and Agro-Defense Facility in Manhattan, Kansas (2010)
Evaluation of the Updated Site-Specific Risk Assessment for the National Bio- and Agro-Defense Facility in Manhattan, Kansas (2012)
Exploring a Vision: Integrating Knowledge for Food and Health (2004)
Exploring Horizons for Domestic Animal Genomics (2002)
Feasibility of Using Mycoherbicides for Controlling Illicit Drug Crops (2011)
A Framework for Assessing Effects of the Food System (2015)
Frontiers in Agricultural Research: Food, Health, Environment, and Communities (2003)
Future Role of Pesticides for U.S. Agriculture (2000)
Genetically Engineered Organisms, Wildlife, and Habitat: A Workshop Summary (2008)
Genetically Modified Pest-Protected Plants: Science and Regulation (2000)
Global Challenges and Directions for Agricultural Biotechnology (2008)
The Impact of Genetically Engineered Crops on Farm Sustainability in the United States (2010)

Incorporating Science, Economics, and Sociology in Developing Sanitary and Phytosanitary Standards in International Trade (2000)
Letter Report to the Florida Department of Citrus on the Review of Research Proposals on Citrus Greening (2008)
Meeting Critical Laboratory Needs for Animal Agriculture: Examination of Three Options (2012)
National Capacity in Forestry Research (2002)
The National Plant Genome Initiative (2002)
National Research Initiative: A Vital Competitive Grants Program in Food, Fiber, and Natural-Resources Research (2000)
Predicting Invasions of Nonindigenous Plants and Plant Pests (2002)
Professional Societies and Ecologically Based Pest Management (2000)
The Potential Consequences of Public Release of Food Safety and Inspection Service Establishment-Specific Data (2011)
The Public Health Effects of Food Deserts: Workshop Summary—joint study with Institute of Medicine (2009)
Publicly Funded Agricultural Research and the Changing Structure of U.S. Agriculture (2002)
Renewable Fuel Standard: Potential Economic and Environmental Effects of U.S. Biofuel Policy (2011)
Review of the Methodology Proposed by the Food Safety and Inspection Service for Risk-Based Surveillance of In-Commerce Activities: A Letter Report (2009)
Review of the Methodology Proposed by the Food Safety and Inspection Service for Followup Surveillance of In-Commerce Businesses: A Letter Report (2009)
Review of Proposals to the Bureau of Land Management on Wild Horse and Burro Sterilization or Contraception: A Letter Report (2015)
Review of the U.S. Department of Agriculture's Animal and Plant Health Inspection Service Response to Petitions to Reclassify the Light Brown Apple Moth as a Non-Actionable Pest: A Letter Report (2009)
Safety of Genetically Engineered Foods: Approaches to Assessing Unintended Health Effects (2004)
Scientific Advances in Animal Nutrition: Promise for a New Century (2001)
The Scientific Basis for Estimating Emissions from Animal Feeding Operations: Interim Report (2002)
The Scientific Basis for Predicting the Invasive Potential of Nonindigenous Plants and Plant Pests in the United States (2002)
Scientific Criteria to Ensure Safe Food (2003)
Spurring Innovation in Food and Agriculture: A Review of the USDA Agriculture and Food Research Initiative Program (2014)
Status of Pollinators in North America (2007)
Strategic Planning for the Florida Citrus Industry: Addressing Citrus Greening (2010)
Sustainable Development of Algal Biofuels in the United States (2012)
Sustaining Global Surveillance and Response to Emerging Zoonotic Disease (2009)
Toward Sustainable Agricultural Systems in the 21st Century (2010)
Transforming Agricultural Education for a Changing World (2009)
The Use of Drugs in Food Animals: Benefits and Risks (2000)
Using Science to Improve the BLM Wild Horse and Burro Program: A Way Forward (2013)
Workforce Needs in Veterinary Medicine (2013)

ANIMAL NUTRITION PROGRAM—NUTRIENT REQUIREMENTS OF DOMESTIC ANIMALS SERIES AND RELATED TITLES

Mineral Tolerance of Animals: Second Revised Edition (2005)
Nutrient Requirements of Beef Cattle, Seventh Revised Edition, Update (2000)
Nutrient Requirements of Dairy Cattle, Seventh Revised Edition (2001)
Nutrient Requirements of Dogs and Cats (2006)
Nutrient Requirements of Fish and Shrimp (2011)
Nutrient Requirements of Horses: Sixth Revised Edition (2007)
Nutrient Requirements of Nonhuman Primates, Second Revised Edition (2002)
Nutrient Requirements of Small Ruminants: Sheep, Goats, Cervids, and New World Camelids (2007)
Nutrient Requirements of Swine, Tenth Revised Edition (1998)
Nutrient Requirements of Swine, Eleventh Revised Edition (2012)
Safety of Dietary Supplements for Horses, Dogs, and Cats (2009)
Scientific Advances in Animal Nutrition: Promise for a New Century (2001)
The First Seventy Years 1928-1998: Committee on Animal Nutrition (1998)
The Scientific Basis for Estimating Emissions from Animal Feeding Operations: Interim Report (2002)

Further information and prices are available from the National Academies Press website at http://www.nap.edu. To order any of the titles above, go to http://www.nap.edu/order.html or contact the Customer Service Department at (888) 624-6242 or (202) 334-3313. Inquiries and orders may also be sent to the National Academies Press, 500 Fifth Street, NW, Keck 360, Washington, DC 20001.

Index

A

B

D

G

M

N

O

P

Q

R

S

T

U